REFERENCE BOOK
NOT TO BE TAKEN
FROM THE LIBRARY

Environmental Engineering Dictionary
3rd Edition

Compiled and Edited by

C. C. Lee, Ph. D.

Research Program Manager
National Risk Management Research Laboratory
U.S. Environmental Protection Agency
Cincinnati, OH 45268

Chairman of Executive Steering Committee
National Technical Workgroup on
Mixed Waste Treatment

Founder
International Congress on Toxic Combustion Byproducts
Washington, DC 20036

Member of Policy Review Group
Center for Clean Technology
University of California, Los Angeles (UCLA)
Los Angeles, California 90024

Adjunct Professor
Civil and Environmental Engineering Department
University of Cincinnati
Cincinnati, OH 45221

Government Institutes
Rockville, MD

Government Institutes, Inc., 4 Research Place, Suite 200, Rockville, Maryland, 20850, USA

Copyright ©1998 by Government Institutes. All rights reserved.

02 01 00 99 98   5 4 3 2 1

No part of this work may be reproduced or transmitted in any form or by any means, electronic or mechanical, including photocopying, recording, or any information storage and retrieval system, without permission in writing from the publisher. All requests for permission to reproduce material from this work should be directed to Government Institutes, Inc., 4 Research Place, Rockville, Maryland 20850, USA.

The reader should not rely on this publication to address specific questions that apply to a particular set of facts. The author and publisher make no representation or warranty, express or implied, as to the completeness, correctness, or utility of the information in this publication. In addition, the author and publisher assume no liability of any kind whatsoever resulting from the use of or reliance upon the contents of this book.

ISBN: 0-86587-620-7

Printed in the United States of America

Table of Contents

Preface v

About the Author vii

Environmental Engineering Dictionary 1

Appendix: Environmental Acronyms 635

References 667

Preface

Because environmental problems are uncovered almost daily, both new regulations and technologies are continuously being developed to meet the challenge of solving them. As a result, many past definitions are revised and many new definitions are generated every year. The main efforts of this edition are, therefore, to upgrade the definitions contained in previous editions and to incorporate the new definitions collected from numerous sources available to the author.

This dictionary provides a comprehensive source of environmental engineering definitions along with their origins. It covers environmental subjects from many fields. Major subjects include:

(1) Statute-related engineering definitions from all major United States environmental laws such as the Clean Air Act, the Clean Water Act, the Safe Drinking Water Act, the Pollution Prevention Act, the Resource Conservation and Recovery Act, and the Superfund Act.

(2) Regulation-related engineering definitions from the entire set of United States environmental stipulations which are contained in 40 CFR (Code of Federal Regulations) from Part 1 to Part 1517.

(3) Environmental engineering definitions from countless U.S. Environmental Protection Agency reports. Major areas covered include: (A) air; (B) drinking water; (C) industrial effluents; (D) underground water; (E) landfill; (F) radioactive material contaminated waste; (G) solid waste (hazardous waste, municipal waste, medical waste, etc.); (H) Superfund waste; (I) pollution prevention/waste minimization; (J) monitoring, sampling and analysis; (K) risk analysis/health/safety; and (L) environmental assessment.

(4) Environmentally-related science and engineering terms synthesized from many handbooks and textbooks (sciences such as biology, chemistry, health, and physics; engineering fields such as chemical, civil, electrical, mechanical, and nuclear).

(5) Environmental acronyms including most abbreviations of EPA offices (*see* Appendix).

(6) References that provide the sources of information used.

A special feature of this Dictionary is the citing of references for the terms collected. This can expeditiously assist users in locating additional information, if needed. This Dictionary provides not only exact official (EPA) definitions, but also the definition's origin. It is an essential tool and will make many environmental jobs much easier for those who are involved in the environmental protection of air, water, and land resources.

C. C. Lee, Ph.D.
February 10, 1998

About the Author

Dr. C.C. Lee is a research program manager at the National Risk Management Research Laboratory of the U.S. Environmental Protection Agency in Cincinnati, Ohio. In addition, he is currently a member of the Policy Review Group to the Center for Clean Technology at the University of California, Los Angeles (UCLA). He is also the Chairman of the Sponsoring Committee to the International Congress on Toxic Combustion By-Products (ICTCB). He initiated the ICTCB and served as the Chairman of the First and Second Congresses, which were held in 1989 and 1991, respectively.

Dr. Lee has more than 22 years experience in conducting various engineering and research projects, which often involve multi-environmental issues ranging from clean air and clean water control to solid waste disposal. He has been recognized as a worldwide expert in the thermal treatment of medical and hazardous wastes—witness his leading discussions on medical waste disposal technologies at a Meeting conducted by the Congressional Office of Technology Assessment. Also, at the initiation of the U.S. State Department, he served as head of the U.S. delegation to the "Conference on National Focal Points for the Low- and Non-Waste Technology" (sponsored by the United Nations and held in Geneva, Switzerland on August 28-30, 1978). He lectured on various issues regarding solid waste disposal in numerous national and international conferences, and has published more than 100 papers and reports in various environmental areas.

He received a B.S. from the National Taiwan University in 1964, and a M.S. and Ph.D. from the North Carolina State University in 1968 and 1972, respectively. Before joining EPA in 1974, he was an assistant professor at the North Carolina State University.

NOTICE

This book was written and edited by Dr. C. C. Lee in his private capacity. No official support or endorsement by the U.S. Environmental Protection Agency is intended nor should be inferred.

Environmental Engineering Dictionary

3rd Edition

Aa

11-AA: The chemical substance 11-aminoundecanoic acid, CAS Number 2432-997 (40 CFR 704.25-1).

1987 Montreal Protocol: The Montreal Protocol, as originally adopted by the parties in 1987 (under the Clean Air Act) (40 CFR 82.3-y).

7Q10: Seven-day, consecutive low flow with a ten-year return frequency; the lowest stream flow for seven consecutive days that would be expected to occur once in ten years (EPA-94/04).

A-scale sound level: A measurement of sound approximating the sensitivity of the human ear, used to note the intensity or annoyance level of sounds (*see* sound level for more related terms) (EPA-94/04).

A-weighted decibel: A unit of weighted sound pressure level, measured by the use of a metering characteristic and the A-weighting specified in American National Standard ANSI SI.4-1971(R176) (*see* decibel for more related terms).

Abandoned mine: A mine where mining operations have occurred in the past and:
(1) The applicable reclamation bond or financial assurance has been released or forfeited; or
(2) If no reclamation bond or other financial assurance has been posted, no mining operations have occurred for five years or more (40 CFR 434.11-91).

Abandoned vehicle: The automobiles, buses, trucks and trailers that are no longer useful as such and have been left on city streets and other public places (*see* vehicle for more related terms) (EPA-83).

Abandoned well:
(1) A well whose use has been permanently discontinued or which is in a state of disrepair such that it cannot be used for its intended purpose or for observation purposes (40 CFR 146.3-91).
(2) A well whose use has been permanently discontinued or which is in a state of such disrepair that it cannot be used for its intended purpose (EPA-94/04).
(3) *See* well for more related terms.

Abatement:
(1) Reducing the degree or intensity of, or eliminating, air, water or land pollution through waste reuse, process modification or pollution control (OME-88/12).
(2) The method of reducing the degree or intensity of pollution, also the use of such a method (LBL-76/07-air).
(3) Reducing the degree or intensity of, or eliminating, pollution (EPA-94/04).

Abiotic factor: All nonliving environmental elements, factors or substances, such as climate, geology, soil, water, and atmosphere, which influence living organisms.

Abiotic: Nonliving, especially the nonliving elements in ecological systems (Course 165.6).

Abnormal occurrence: Any accidental, unplanned, or uncontrolled release of radioactivity (DOE-91/04).

Aboveground release: Any release to the surface of the land or to surface water. This includes, but is not limited to, releases from the aboveground portion of an UST system and aboveground releases associated with overfills and transfer of operations as the regulated substance moves to or from an UST system (*see* release for more related terms) (40 CFR 280.12-91).

Aboveground storage facility: A tank or other container, the bottom of which is on a plane not more than 6 inches below the surrounding surface (40 CFR 113.3-91).

Aboveground tank: A device meeting the definition of tank in 40 CFR 260.10 and that is situated in such a way that the entire surface area of the tank is completely above the plane of the adjacent surrounding surface and the entire surface area of the tank (including the tank bottom) is able to be visually inspected (*see* tank for more related terms) (40 CFR 260.10-91).

Abrasion:
(1) Wearing away of surface materials, such as refractories in an incinerator or parts of solid waste handling equipment, by the scouring action of moving solids, liquids, or gas (SW-108ts).
(2) The removal of surface material from any solid through the frictional action of another solid, a liquid, or a gas or combination thereof (cf. corrosion or erosion) (EPA-83).

Abrasive: A solid substance used in an abrasive blasting operation (29 CFR 1910.94a-91).

Abrasive blasting: The forcible application of an abrasive to a surface by pneumatic pressure, hydraulic pressure, or centrifugal force (29 CFR 1910.94a-91).

Abrasive blasting respirator: A continuous flow air-line respirator constructed so that it will cover the wearer's head, neck, and shoulders to protect him from rebounding abrasive (29 CFR 1910.94a-91).

Abrasive machining: A general machining process of using abrasive tools to finish a product, e.g., abrasive belt grinding is to rough and/or finish a workpiece by means of a power-driven belt coated with an abrasive, usually in particle form, which removes materials by scratching the surface and which makes the surface a good finish (EPA-83/06a, *see also* 29 CFR 1910.94b-91 under OSHA).

Abscission: A process by which a leaf or other part is separated from a plant (EPA-85/10).

Absent: The most sensitive analytical procedure in Standard Methods for the Examination of Water and Wastewater, 13th edition (or other approve procedure), does not show the presence of the subject constituent (LBL-76/07-water).

Absolute humidity: *See* synonym, humidity ratio (*see also* humidity for more related terms).

Absolute method: A method in which characterization is based entirely on physical (absolute) defined standards (*see* method for more related terms) (ACS-87/11).

Absolute pressure (psia): The total pressure exerted on a surface. It is a measure of pressure referred to a complete vacuum or zero pressure. Absolute pressure = atmospheric pressure + gauge pressure. Ideal gas law calculations employ absolute pressure term (*see* pressure for more related terms).

Absolute temperature: The temperature measured on the thermodynamic scale, designated as degree Kelvin (K) or Rankine (R). It is measured from absolute zero (-273.15°C or -459.76°F) and has the following relationship with degree C and F (EPA-83/06):
- Kelvin scale (K): K = °C + 273.15;
- Rankine scale (R): R = °F + 459.76; and R = 1.8 K
- $(T_2/T_1)_{Rankine} = (T_2/T_1)_{Kelvin}$

Absolute toxicity: The toxicity of the effluent without considering dilution (EPA-91/03)

Absolute zero temperature:
(1) The zero points of both absolute temperature scales, namely, Kelvin and Rankine scales (Holman-69-p5).
(2) If an externally reversible heat engine operates between two energy reservoirs, absorbing a constant heat input from the hotter reservoir, and the temperature of the colder reservoir is successively lowered, the amount of heat rejected decreases. As the amount of heat rejected approaches zero, the temperature of the colder reservoir approaches absolute zero. The thermodynamic second law alone does not lead to the conclusion that it is impossible for the temperature of any system to be absolute zero. This conclusion is in the realm of the third law of thermodynamics (Jones-60-p258).
(3) *See* temperature for more related terms.

Absorbance (A):
(1) The logarithm to the base 10 of the ratio of the initial intensity (I_o) of a beam of radiant energy to the intensity (i) of the same beam after passage through a sample at a fixed wavelength. Thus, $A = \log_{10}(I_o/I)$ (cf. Beer's Law) (40 CFR 796.3700-91).
(2) The logarithm to the base 10 of the reciprocal of the transmittance (T), $A = \log_{10}(1/T)$ (LBL-76/07-air).

Absorbate (or solute):
(1) A term used to describe the pollutant in absorber studies. It is a dissolved substance, especially the smaller component of a solution (EPA-84/09).
(2) The gaseous pollutant being absorbed, such as SO_2, H_2S, etc. (EPA-84/03b).
(3) Material that has been retained by the process of absorption (LBL-76/07-air).

Absorbed dose: The amount of chemical that enters the body of an exposed organism (*see* dose for more related terms) (EPA-94/04).

Absorbency: Property of a paper to imbibe liquids (EPA-83).

Absorbent:
(1) Any substance which takes in or absorbs other substances. In air pollution control, the liquid, usually water, into which the pollutant is absorbed (EPA-84/03b).
(2) Material in which absorption occurs (cf. adsorbent) (LBL-76/07-air).

Absorber: Also known as the wet scrubber. It is a control device for carrying on the process of absorption (EPA-84/09).

Absorptiometer: An instrument which is used to measure the concentration of the absorbing constituents in a gas or liquid.

Absorptiometric analysis: The use of absorptiometer to analyze a gas or liquid by measurement of the peak electromagnetic absorption wavelengths that are unique to a specific material or element.

Absorption:
(1) A process in which one material (the absorbent) takes up and retains another (the absorbate) with the formation of a homogeneous mixture having the attributes of a solution. Chemical reaction may accompany or follow absorption (LBL-76/07-air).
(2) The uptake of water or dissolved chemicals by a cell or an organism (as tree roots absorb dissolved nutrients in soil) (EPA-94/04).
(3) In air pollution control, a process by which a liquid material (absorbent) is used to remove one or more soluble gas (absorbate) components from a gaseous mixture, usually without chemical reaction. Typical absorbents are: water, dilute basic or acidic solution, and lean (low molecular weight) hydrocarbon oils (cf. adsorption) (EPA-84/09). Absorption is a mass transfer in which a gas (gaseous

pollutant) is dissolved in a liquid. A contaminant (pollutant) exhaust stream contacts a liquid, and the contaminant diffuses from the gas phase into the liquid phase (EPA-81/12, p1-5; 84/03b, p1-7).

(4) In chemical absorption, the absorption through chemical combination of one substance by another.
(5) In mining, the process by which a liquid is drawn into and tends to fill permeable pores in a porous solid body; also the increase in weight of a porous solid body resulting from the penetration of liquid into its permeable pores (EPA-82/05).
(6) In physical absorption, the penetration of a substance into or through another, or a process by which gas molecules are transferred to (dissolved in) a liquid phase.
(7) In radiation, the uptake of radiant energy by a substance. During this process, the radiant energy is irreversibly transformed into some other form of energy, e.g., thermal, mechanical or electrical energy.
(8) In radiation, the process by which the number and energy of particles or photons entering a body of matter are reduced by interaction with the matter (DOE-91/04).
(9) In risk assessment, the uptake of water or dissolved chemicals by a cell or an organism (Course 165.6).
(10) See absorption mechanism for more related information

Absorption factor: The fraction of a chemical making contact with an organism that is absorbed by the organism (Course 165.6).

Absorption mechanism: The process of dissolving gaseous pollutants in a liquid is referred to as absorption. Absorption is a mass transfer operation. Mass transfer can be compared to heat transfer in that both occur because a system is trying to reach equilibrium conditions. For example, in heat transfer, if a hot slab of metal is placed on top of a cold slab, heat energy will be transferred from the hot slab to the cold slab until both are at the same temperature (equilibrium). In absorption, mass instead of heat is transferred as a result of a concentration difference, rather than a heat-energy difference. Absorption continues as long as a concentration differential exists between the liquid and the gas from which the contaminant is being removed. In absorption, equilibrium depends on the solubility of the pollutant in the liquid.

To remove a gaseous pollutant by absorption, the exhaust stream must be passed through (brought in contact with) a liquid. Absorption involves three steps. In the first step, the gaseous pollutant diffuses from the bulk area of the gas phase to the gas--liquid interface. In the second step, the gas moves (transfers) across the interface to the liquid phase. This step occurs extremely rapidly once the gas molecules (pollutant) arrive at the interface area. In the third step, the gas diffuses into the bulk area of the liquid, thus making room for additional gas molecules to be absorbed. The rate of absorption (mass transfer of the pollutant from the gas phase to the liquid phase) depends on the diffusion rates of the pollutant in the gas phase (first step) and in the liquid phase (third step).

To enhance gas diffusion and, therefore, absorption:
(1) Provide a large interfacial contact area between the gas and liquid phases.
(2) Provide good mixing of the gas and liquid phases (turbulence).
(3) Allow sufficient residence, or contact, time between the phases for absorption to occur (EPA-84/03b, p1-7).

Absorption spectrophotometer: An instrument for measuring the absorption spectral lines and bands in a gas or liquid.

Absorption spectroscopy: The study of the absorption spectral lines and bands in a gas or liquids.

Absorption spectrum: The array of absorption lines and absorption bands (*see* spectrum for more related terms).

Absorption toxicokinetics: Refers to the bioavailability, i.e., the rate and extent of absorption of the test substance, and metabolism and excretion rates of the test substance after absorption (40 CFR 795.235-91).

Absorptivity: Absorbance divided by the product of the concentration of the substance and the sample path length (cf. molar absorptivity) (LBL-76/07-bio).

Abstraction (or metathesis): A bimolecular reaction where an atom (or radical) is abstracted (transferred) from the reactant to the other by a radical species, e.g., H abstraction from monochloromethane by radical OH can be expressed as $CH_3Cl + OH \longrightarrow CH_2Cl + HOH$ (cf. displacement) (EPA-88/12).

Acaricide (or miticide): An agent that destroys mites and ticks (EPA-85/10).

Accelacota: A pharmaceutical coating operation which consists of a horizontally rotating perforated drum in which tablets are placed, a coating is applied by spraying, and the coating is dried by the flow of air across the drum through the perforations (40 CFR 52.741-91).

Accelerated depreciation: In pollution abatement, an incentive arrangement to encourage an industry to install pollution abatement equipment. The industry can deduct from its taxable income the entire cost of such equipment over a shorter period of time (perhaps only one to three years) than in the case of other types of capital investment (DOI-70/04).

Accelerated erosion: The erosion of soil materials at a faster than natural rate. It occurs when vegetal cover is destroyed or is affected by some activities of man (*see* erosion for more related terms) (SW-108ts).

Accelerator: In radiation science, a device that speeds up charged particles such as electrons or protons (EPA-89/12). These fast particles can penetrate matters and are known as radiation.

Accelerator pump (diaphragm pump or plunger pump): A device used to provide a supplemental supply of fuel during increasing throttle opening as required (40 CFR 85.2122(a)(3)(ii)-91).

Acceptable daily intake: An estimate of the daily exposure dose that is likely to be without deleterious effect even if continued exposure occurs over a lifetime (EPA-92/12).

Acceptable of a batch: The number of non-complying compressors in the batch sample is less than or equal to the acceptance number as determined by the appropriate sampling plan (40 CFR 204.51-91).

Acceptable quality level (AQL): The maximum percentage of failing engines or vehicles (or samples), that for purposes of sampling inspection, can be considered satisfactory as a process average (40 CFR 86.1002.84-91).

Acceptance of a batch sequence: That the number of rejected batches in the sequences is less than or equal to the acceptance number as determined by the appropriate sampling plan (40 CFR 204.51-91).

Acceptance of a compressor: The measured noise emissions of the compressor, when emissions measured in accordance with the applicable procedure, conforms to the applicable standard (*see* compressor for more related terms) (40 CFR 204.51-91).

Acceptance of a vehicle: The measured emissions of the vehicle when measured in accordance with the applicable procedure, conforms to the applicable standard (*see* vehicle for more related terms) (40 CFR 205.51-91).

Access: The right and opportunity to examine and copy (29 CFR 1910.20-91).

Accessible environment:
(1) The atmosphere
(2) Land surfaces
(3) Surface waters
(4) Oceans
(5) All of the lithosphere that is beyond the controlled area (40 CFR 191.12-91)
(6) *See* environment for more related terms

Accessible when referring to ACM: That the material is subject to disturbance by school building occupants or custodial or maintenance personnel in the course of their normal activities (40 CFR 763.83-91).

Accident: An unexpected, undesirable event, caused by the use or presence of a pesticide, that adversely affects man or the environment (40 CFR 171.2-91).

Accident site: The location of an unexpected occurrence, failure or loss, either at a plant or along a transportation route, resulting in a release of hazardous materials (EPA-94/04).

Accidental air ingress (or ingress of secondary or other fluids): The inadvertent admission of air (or a fluid such as water, steam, or other non-primary fluid) to the primary cooling system of a modular high-temperature gas-cooled reactor (DOE-91/04).

Accidental occurrence: An accident, including continuous or repeated exposure to conditions, which results in bodily injury or property damage neither expected nor intended from the standpoint of the insured (40 CFR 264.141-91). Other accidental occurrence related terms include:
- Non-sudden accidental occurrence
- Sudden accidental occurrence

Accidental release:
(1) Any sudden or nonsudden release of petroleum from an underground storage tank that results in a need for corrective action and/or compensation for bodily injury or property damage neither expected nor intended by the tank owner or operator (40 CFR 280.92-91).
(2) The unintentional spilling, leaking, pumping, purging, emitting, emptying, discharging, escaping, dumping, or disposing of a toxic material into the environment in a manner that is not in compliance with a plant's federal, state, or local environmental permits and that creates toxic concentrations in the air that are a potential health threat to the surrounding community (EPA-8/87b).
(3) *See* release for more related terms.

Acclimated organism seed: Organisms especially grown to cope with bio-oxidation-resistant organic wastes (LBL-76/07-water).

Acclimated: Adapted to environmental change (LBL-76/07-water).

Acclimation (or acclimatization):
(1) The physiological or behavioral adaptation of organisms to one or more environmental conditions associated with the test method (e.g., temperature, hardness, pH) (40 CFR 797.1600-).
(2) A physiological change occurring within the lifetime of an organism which reduces the strain caused by stressful changes in the natural climate (e.g., seasonal or geographical) (NIOSH-84/10).
(3) The physiological and behavioral adjustments of an organism to changes in its environment (EPA-94/04).

Acclimatization: *See* synonym, acclimation.

Accredited or accreditation: When referring to a person or laboratory means that such person or laboratory is accredited in accordance with section 206 of Title II of the Act (TSCA) (40 CFR 763.83).

Accrual date: The date of the incident causing the loss or damage or the date on which the loss or damage should have been discovered by the employee through the exercise of reasonable care (40 CFR 14.2-91).

Accumulated speculatively: A material is "accumulated speculatively" if it is accumulated before being recycled. A material is not accumulated speculatively, however, if the person accumulating it can show that the material is potentially recyclable and has a feasible means of being recycled; and that--during the calendar year (commencing on January 1)--the amount of material that is recycled, or transferred to a different site for recycling, equals at least 75 percent by weight or volume of the amount of that material accumulated at the beginning of the period. In calculating the percentage of turnover, the 75 percent requirement is to be applied to each material of the same type (e.g., slags from a single smelting process) that is recycled in the same way (i.e., from which the same material is recovered or that is used in the same way). Materials accumulating in units that would be exempt from regulation under 40 CFR 261.4(c) are not be included in making the calculation. (Materials that are already defined as solid wastes also are not to be included in making the calculation.) Materials are no longer in this category once they are removed from accumulation for recycling, however (40 CFR 261.1-8-91).

Accumulation:
(1) In air pollution, the increase of concentration levels in a region, due to the combined effect of emissions and meteorological conditions for which the dispersion is poor, e.g., during a stagnation period (NATO-78/10).
(2) In a biological system, the concentration of a substance which collects in a tissue or organism and which does not disappear over time (EPA-82/11e).

Accumulator: The reservoir of a condensing unit receiving the condensate from a surface condenser (40 CFR 52.741-91).

Accuracy: *See* definition under uncertainty.

Acetate fiber: A fiber made from cellulose acetate (*see* fiber for more related terms) (EPA-74/06b).

Acetic acid (or named as ethanoic acid, vinegar acid, and methanecarboxylic acid, CH_3COOH): Glacial acetic acid is the pure compound (99.8% min.), as distinguished from the usual water solutions known as acetic acid. Vinegar is dilute acetic acid (EPA-83/06a).

Acetylation: A technique to combine an acetyl radical $(CH_3CO)^-$ onto an organic molecule.

Acetylcholine ($C_7H_{17}O_3N$): A substance in the human body having important neurotransmitter effects on various internal systems; often used as a bronchoconstrictor (EPA-89/12).

Acetylene (C_2H_2): A gas which can be prepared by the action of water on calcium carbide. The starting material for large-scale synthesis of important organic compounds (EPA-77/07).

Acetylene cylinder filler: An asbestos-containing product which is intended for use as a filler for acetylene cylinders (40 CFR 763.163-91).

acfm (actual cubic feet per minute): A flow rate unit (cubic feet per minute) measured under actual pressure and temperature. In reference to the standard conditions (i.e., pressure = 1 atm and temperature = 60° F), acfm is the gas flow rate (volumetric flow rate) obtained under the actual operating conditions (cf. scfm). Predicting performance and design calculations for control devices are based on actual flow rates (EPA-84/09).

Using the ideal gas equation, acfm and scfm (standards cubic feet per minute) can be related as $Qa = Qs \times Ta \times Ps/(Ts \times Pa)$, where: Qa and Qs = actual and standard flow rate respectively; Ta and Ts = actual and standard temperature respectively.
Example:
- Determine an actual flow rate from a given standard flow rate.

Data:
- Standard flow rate of a gas = 2000 scfm, standard conditions = 60° F and 1 atm, and actual operating conditions = 700° F and 1 atm.

Solution:
- $Qa = 2000(700 + 460) \times 1/[(60 + 460) \times 1] = 4462$ acfm.

Acid:
(1) Any compound that can react (neutralize) with a base to form a salt.
(2) A substance which dissolves in water forming hydrogen ions (cf. pH) (EPA-87/10).
Other acid related terms include:
- Bronsted acid
- Fatty acid
- Inorganic acid
- Lewis acid
- Strong acid

Acid aerosol: Acidic liquid or solid particles that are small enough to become airborne. High concentrations of acid aerosols can be irritating to the lungs and have been associated with some respiratory diseases, such as asthma (EPA-88/09b).

Acid base catalysis: The catalytic effects in increasing certain chemical reactions due to the presence of acid and base elements.

Acid base titration: A titration in which an acid of known concentration is added to a base of unknown concentration solution until the desired end point is reached, or vice verses (*see* titration for more related terms).

Acid cleaning: Using acid solutions to clean materials. Some methods of acid cleaning are pickling and oxidizing (EPA-83/06a).

Acid copper: A copper electrode deposited from an acid solution of a copper salt, usually copper sulfate (EPA-82/05).

Acid cure: In uranium extraction, sulfation of moist ore before leach (EPA-82/05).

Acid deposition (acid precipitation or acid rain): A complex chemical and atmospheric phenomenon that occurs when emissions of sulfur and nitrogen compounds and other substances are transformed by chemical processes in the atmosphere, often far from the original sources, and then deposited on earth in either wet or dry form. The wet forms, popularly called "acid rain," can fall as rain, snow, or fog. The dry forms are acidic gases or particulates (EPA-94/04).

Acid dip: An acidic solution for activating the workpiece surface prior to electroplating in an acidic solution, especially after the workpiece has been processed in an alkaline solution (EPA-83/06a).

Acid dye: A type of dye commonly used to color wool and nylon but may be used on other fibers (EPA-74/06b).

Acid fast: Describes a cell or bacterium that retains a dye that has a negatively charged molecule (LBL-76/07-bio).

Acid fog: *See* synonym, acid gas.

Acid furnace: A furnace lined with acid brick as contrasted to one lined with basic brick. In this instance, the terms acid and basic are in the same relationship as the acid anhydride and basic anhydride that are found in aqueous chemistry. The most common acid brick is silica brick or chrome brick (*see* furnace for more related terms) (EPA-74/06a).

Acid gas (acid fog or acid mist):
(1) Corrosive gases, such as:
   (A) Hydrogen chloride (HCl), formed during combustion of chlorinated or halogenated compounds or
   (B) Sulfuric acid (H_2SO_4), formed by the reaction of sulfur dioxide, in the atmosphere, with moisture and oxygen ($SO_2 + 1/2\ O_2 + H_2O \longrightarrow H_2SO_4$) (EPA-89/03b).
(2) A gas stream of hydrogen sulfide (H_2S) and carbon dioxide (CO_2) that has been separated from sour natural gas by a sweetening unit (40 CFR 60.641-91).
(3) Acidic components such as SO_2, HCl, HBr and HF of gaseous emissions from combustion or incineration processes. They can be removed by water or caustic scrubbing (CRWI-89/05).

Acid leach: A metallurgical process for extracting metals by means of acid dissolution, e.g., in copper industry, it is a technology employed to recover copper from low grade ores and mine dump materials when oxide mineralization is present, by dissolving the copper minerals with either sulfuric acid or sulfuric acid containing ferric iron (EPA-82/05).

Acid mine drainage (or ferruginous mine drainage):
(1) The mine drainage which, before any treatment, either has a pH of less than 6.0 or a total iron concentration equal to or greater than 10 mg/L (*see* mine drainage for more related terms) (40 CFR 434.11-91).
(2) Drainage of water from areas that have been mined for coal of other mineral ores. The water has a low pH because of its contact with sulfur-bearing material and is harmful to aquatic organisms (EPA-94/04).

Acid mine water: The mine water which contains free sulfuric acid, mainly due to the weathering of iron pyrites where sulfide minerals break down under the chemical influence of oxygen and water, the mine water becomes acidic and can corrode ironwork (*see* water for more related terms) (EPA-82/05).

Acid mist: *See* synonym, acid gas.

Acid neutralizing capacity: Measure of ability of water or soil to resist changes in pH (EPA-94/04).

Acid precipitation: *See* synonym, acid deposition.

Acid pulp: *See* synonym, sulfite pulp.

Acid rain: *See* synonym, acid deposition (EPA-94/04).

Acid recovery: Those sulfuric acid pickling operations that include processes for recovering the unreacted acid from spent pickling acid solutions (40 CFR 420.91-91).

Acid refractory: The refractories containing a substantial amount of silica that may react chemically with basic refractories, basic slags, or basic fluxes at high temperatures (*see* refractory for more related terms) (EPA-83).

Acid regeneration: Those hydrochloric acid pickling operations that include processes for regenerating acid from spent pickling acid solutions (40 CFR 420.91-91).

Acid solution: A solution with a pH of less than 7.00 in which the activity of the hydrogen ion is greater than the activity of the hydroxyl ion (EPA-76/03).

Acid washed activated carbon: Carbon which has been contacted with an acid solution with the purpose of dissolving ash in the activated carbon (EPA-82/11f).

Acidic:
(1) Possessing the ability to donate a proton (e.g., HCl) or accept a pair of electrons (e.g., SO_2); or having a pH less than 7 (ETI-92).
(2) The condition of water or soil that contains a sufficient amount of acid substances to lower the pH below 7.0 (EPA-94/04).

Acidic titrant: The use of a standard acid solution (a solution with known concentration) to measure the basicity of another substance by titration.

Acidification: Addition of an acid solution to a solution to increase its hydrogen ions until the solution becomes acidic (pH < 7).

Acidimeter: A volumetric analysis meter which is used to measure the amount of acid in a sample.

Acidimetry: The study of using an acidimeter to determine the amount of base present in a solution.

Acidity:
(1) The capacity of a wastewater for neutralizing a base. It is normally associated with the presence of carbon dioxide, mineral and organic acids and salts of strong acids or weak bases. It is reported as equivalent of $CaCO_3$ because many times it is not known just what acids are present (EPA-76/03).
(2) The quantitative capacity of aqueous solutions to react with hydroxyl ions. It is measured by titration with a standard solution of a base to a specified end point. Usually expressed as milligrams per liter of calcium carbonate; that is, the amount of calcium carbonate that would be required to exactly neutralize the sample (LBL-76/07-water).

Acidulate: To make acidic (EPA-76/03).

Acoustic descriptor: The numeric, symbolic, or narrative information describing a products acoustic properties as they are determined according to the test methodology that the Agency prescribes (40 CFR 211.102-91).

Acoustical assurance period (AAP): A specified period of time or miles driven after sale to the ultimate purchaser during which a newly manufactured vehicle or exhaust system, properly used and maintained, must continue in compliance with the Federal standard (40 CFR 205.151-91).

Acquired immune deficiency syndrome (AIDS): A fatal disease caused by human immunodeficiency virus (HIV).

Acre foot (of water): A quantity of water that would cover 1 acre to a depth of 1 foot, that is 43,560 cubic feet, or 325,850 gallons

(DOI-70/04).

Acrilan fiber: A trademark of Monsanto for acrylic fibers (*see* fiber for more related terms) (EPA-74/06b).

Acrylic fiber: A manufactured synthetic fiber in which the fiber-forming substance is any long-chain synthetic polymer composed of at least 85 percent by weight of acrylonitrile units (40 CFR 60.601-91).

Acrylic resin: A synthetic resin used as sand binders in core making. These resins are formed by the polymerization of acrylic acid or one of its derivatives using benzoyl peroxide or a similar catalyst. The most frequently used starting materials for acrylic resins include acrylic acid, methacrylic acid, or acrylonitrile. Exposure of these binder materials to hot metal temperatures can cause breakdown of the chemical bonds within the resin molecules and subsequent generation of cyanide. Acrylic resin is also a paint ingredient (*see* resin for more related terms) (EPA-85/10a).

Acrylic: A manufactured fiber which the fiber-forming substance is any long chain synthetic polymer composed of at least 85% by weight of acrylonitrile units. Made in both filament and staple form (EPA-74/06b).

Acrylonitrile (C_3H_3N): A colorless liquid used as a pesticide fumigant for stored grain and in the manufacture of acrylic rubbers and fibers.

Act: *See* definition under law related terms.

Act of God:
(1) Under CWA, an act occasioned by an unanticipated grave natural disaster (CWA311-33USC1321).
(2) Under CERCLA, an unanticipated grave natural disaster or other natural phenomenon of an exceptional, inevitable, and irresistible character, the effects of which could not have been prevented or avoided by the exercise of due care or foresight (SF101-42USC9601).

Act or AEA: The Atomic Energy Act of 1954 (68 Stat. 919), including any amendments thereto (other related information is provided in 10 CFR 20.3; 30.4; 40.4; 70.4).

Act or CAA: The Clean Air Act of 1970 (42USC7401-7626, Public Law 159, July 14, 1955). The Act:
(1) Was enacted in 1970 and was substantially revised in 1977 and 1991. CAA, which was required by congress to have the air safe enough to protect the public's health by May 31, 1975 and requires the setting of National Ambient Air Quality Standards (NAAQS) for major primary air pollutants.
(2) Set initial deadlines for automobile emission standards and gave the EPA administrator power to establishing the standards.
(3) Authorized EPA to list and regulated various hazardous air pollutants (CAA112; 40 CFR 52.741-91)

The main CAA regulations are the provisions of Sections 108 and 109 under which EPA has set forth for:
(1) The six most common pollutants (criteria pollutants, *see* pollutants, criteria); and
(2) The maximum level of pollution permitted in the ambient air in designated areas called air quality control regions. Such standards are called ambient standards, officially called National Ambient Air Quality Standards (cf. air quality control region, state implementation plans, or emissions trading) (40 CFR 50.4-50.12). The standards include two major items:
   (A) **National primary standards:** The levels of air quality necessary, with an adequate margin of safety, to protect the public health.
   (B) **National secondary standards:** The levels of air quality necessary to protect the public welfare (such as agricultural crops, livestock, and deterioration of materials and property) from any known or anticipated adverse effects of a pollutant.

Major subsequent amendments of CAA include:
- CAAA (Clean Air Act Amendments) of 1990.

Act or CAAA: The Clean Air Act Amendment of 1990 (Pub. L. 101-549, November 15, 1990). In 1990, Congress amended the Clean Air Act by revamping the system of hazardous air pollution regulations and by addressing new air pollution problems such as acid deposition, and added a fourth program--a comprehensive operating program to focus in one place all of the Clean Air Act requirements that apply to a given source of air pollution (Sullivan-95/04, p101).
- Under Title III of the Act, the congress establishes a list of hazardous air pollutants (HAPs).

Act or CERCLA: The Comprehensive Environmental Response, Compensation, and Liability Act of 1980 (42USC9601-9657). The Act:
- Is commonly known as Superfund.
- Provides for liability, compensation, cleanup, and emergency response for hazardous substances released into the environment and the cleanup of inactive hazardous waste disposal sites.
- Created a $1.6 billion Hazardous Substance Response Trust Fund "Superfund" from a special tax on crude oil and commercial chemicals.
- Allowed the EPA to use the money in Superfund to investigate and clean up abandoned or uncontrolled hazardous waste sites. The EPA can either pay for the site cleanup itself or take legal action to force the parties responsible for the contamination to pay for the cleanup.
- Required owners and operators of vessels or facilities handling hazardous wastes to show evidence of financial responsibility. This provision ensures that if a hazardous waste is released, the responsible person can pay the costs of removing the contaminant and restoring damaged natural resources. Persons responsible for the release are liable for all costs incurred from the cleanup and restoration of the environment. Only when a financially responsible defendant cannot be found will the Superfund absorb the costs of removing the released hazardous waste costs.

Major subsequent amendments of CERCLA include:
- SARA (Superfund Amendments and Reauthorization Act of 1986).
- EPCRA (Emergency Planning and Community Right-to-Know Act of 1986)
- RGIAQ (Radon Gas and Indoor Air Quality Research of

1986).
(Other related information is provided in 40 CFR 2.310-1; 35.6015-6; 280.12; 300.5; 300-AA; 302.3; 304.12-a; 355.20; 372.3.)

Act or CWA: The Clean Water Act (formerly referred to as the Federal Water Pollution Control Act or Federal Water Pollution Control Act Amendments of 1972), Pub. L. 92-500, 33USC1251 et seq. The Act:
- Is to restore and maintain the chemical, physical, and biological integrity of the Nation's waters.
- Requires EPA to establish a system of national effluent standards for major water pollutants.
- Requires all municipalities to use secondary sewage treatment by 1988.
- Sets interim goals of making all U.S. waters safe for fishing and swimming.
- Allows point source discharges of pollutants into waterways only with a permit from EPA.
- Requires all industries to use the best practicable technology (BPT) for control of conventional and non-conventional pollutants and to use the best available technology (BAT) that is reasonable or affordable.

CWA has five main elements:
- A permit program.
- A system of minimum national effluent standards for each industry.
- Water quality standards.
- Provisions for special problems such as toxic chemicals and oil spills.
- A construction grant program for publicly owned treatment works (POTWs).

CWA established the national goals to:
- Achieve a level of water quality which provides for the protection and propagation of fish, shellfish, and wildlife and for recreation in and on the water by July 1, 1983.
- Eliminate the discharge of pollutants into United States waters by 1985.

CWA regulated the discharge of 65 categories of priority pollutants [including at least 129 specific chemical substances (*see* Appendix 1)] by 34 industry:
1. Adhesives and sealants.
2. Aluminum forming.
3. Asbestos manufacturing.
4. Auto and other laundries.
5. Battery manufacturing.
6. Coal mining.
7. Coil coating.
8. Copper forming.
9. Electric and electronic components.
10. Electroplating.
11. Explosives manufacturing.
12. Ferroalloys.
13. Foundries.
14. Gum and wood chemicals.
15. Inorganic chemicals manufacturing.
16. Iron and steel manufacturing.
17. Leather tanning and finishing.
18. Mechanical products manufacturing.
19. Nonferrous metals manufacturing.
20. Ore mining.
21. Organic chemicals manufacturing.
22. Pesticides.
23. Petroleum refining.
24. Pharmaceutical preparations.
25. Photographic equipment and supplies.
26. Plastic and synthetic materials manufacturing.
27. Plastic processing.
28. Porcelain enamelling.
29. Printing and publishing.
30. Pulp and paperboard mills.
31. Soap and detergent manufacturing.
32. Steam electric power plants.
33. Textile mills.
34. Timber products processing (cf. existing stationary facility).

(Other related information is provided in 40 CFR 2.302-1; 35.905; 35.2005-1; 35.1605-1; 51.392; 122.2; 124.2; 125.2; 130.2-a; 131.3-a; 133.101; 233.2; 136.2-a; 230.3-a; 270.2; 406.61-c; 501.2; 503.9.)

Act or ERDDAA: The Environmental Research, Development, and Demonstration Authorization Act of 1980 (42USC1857 et seq.). The Act authorized research, development and demonstration activities in the areas of air, water, solid waste, pesticides, toxic substances and radiation.

Act or EPCRA: The Emergency Planning and Community Right-to-Know Act (EPCRA) of 1986 (42USC11001-11050). The Act established programs to provide the public with important information on the hazardous and toxic chemicals in their communities, and established emergency planning and notification requirements to protect the public in the event of a release of extremely hazardous substances (GII-96/03, p1187).

Act or FFCA: The Federal Facility Compliance Act (FFCA) of 1992 (Pub. L. 102-386, 106 Statutes 1505). FFCA amended the Resource Conservation and Recovery Act. The Act was to ensure that there was a complete and unambiguous waiver of sovereign immunity with regard to the imposition of administrative and civil fines and penalties against Federal Facilities. This allowed the state environmental agencies and the federal Environmental Protection Agency to impose civil penalties and administrative fines on Federal facilities under RCRA section 6001 for violations of federal, state and local solid and hazardous waste laws (Sullivan-95/04, p333).

Act or FFDCA: The Federal Food, Drug and Cosmetic Act (FFDCA) of 1938 as amended (21USC301-392) (other related information is provided in 40 CFR 2.308-1160.3; 163.2-a; 177.3; 178.3; 179.3; 710.2).

Act or FIFRA: The Federal Insecticide, Fungicide and Rodenticide Act of 1947, as amended, and its predecessor, 7USC135 et seq. (7USC136 et seq.)
- In 1910, Congress passed the Insecticide Act to regulate the manufacture of insecticide, Paris green, lead arsenate or fungicide. In 1947, Congress replaced the Insecticide Act with the more comprehensive FIFRA to regulate economic poisons which include not only insecticides and fungicides, but also rodenticides, herbicides, and preparations intended to control other forms or pests which were not subject to the Insecticide Act.
- Major subsequent amendments include:
  - FEPCA (Federal Environmental Pesticide Control

Act) of 1972.
(Other related information is provided in 40 CFR 2.307-1; 152.3-a; 154.3-a; 164.2-a; 165.1-a; 166.3-a; 167.3; 160.3; 152.3-a; 171.2-2; 172.1-a; 177.3; 179.3.)

Act or FWPCA: The Federal Water Pollution Control Act of 1972 (33USC1151, et seq.)
- The objective of this FWPCA was to restore and maintain the chemical, physical, and biological integrity of the Nation's waters. The Act was amended by the Clean Water Act of 1977 and is commonly known as the Clean Water Act.
- Major subsequent amendments include:
  - CWA (Clean Water Act) of 1977
  - WQA (Water Quality Act) of 1987
  - Ocean Dumping Ban Act of 1988
  - OPA (Oil Pollution Act) of 1990

(Other related information is provided in 40 CFR 21.2-e; 20.2-a; 35.3105-a; 39.105-a; 104.2-a; 108.2-a; 110.1; 113.3-b; 116.3; 121.1-f; 129.2; 220.2; 401.11-a; 403.3-b.)

Act or HSWA: The Hazardous and Solid Waste Amendment of 1984 (42USC6901 et seq.). The Act is the 1984 amendment to the Resource Conservation and Recovery Act (RCRA) of 1976. The Act:
- Established strict limits on the land disposal of hazardous waste.
- Established a strict timeline for restricting untreated hazardous waste from land disposal.
- Regulated for the first time more than 100,000 companies that produce only small quantities of hazardous wastes (less than 1,000 kilograms per month). These small quantity generators were exempted from RCRA requirements before the 1984 Amendments.
- Established the following four major policies (OSWER-87):
  - Land disposal restriction policy.
  - Deep-well injection policy.
  - Domestic sewage sludge policy.
  - Waste minimization policy.

Act or MPRSA: The Marine Protection, Research, and Sanctuaries Act of 1972, as amended, also known as the **Ocean Dumping Act** (33USC1401-1434). It outlawed dumping of waste in oceans without an EPA permit and required the EPA to designate sites to be used by permit holders. EPA claimed authority to regulate incineration at sea and had issued research permits until 1987. Because this method was considered extremely controversial, EPA stopped issuing permits since then (Winthrop-89/09, other identical or similar definitions are provided in 40 CFR 2.309-1; 220.2-a).

Act or MVICSA: The Motor Vehicle Information and Cost Savings Act, as amended, 15USC1901 et seq (other identical or similar definitions are provided in 40 CFR 2.311-1; 600.002.85-1).

Act or MWTA: The Medical Waste Tracking Act of 1988 (RCRA subtitle J). The Act:
(1) Was passed by the Congress in October and signed by the President to make it a law on November 2, 1988. MWTA is Subtitle J of RCRA.
(2) Authorized EPA to implement a two year demonstration program to track certain medical wastes and to establish requirements for the segregation, handling, and labeling of these wastes from 3 mandated States plus other opt in States. The three States are New Jersey, New York, and Connecticut (Lee-89/09).

Act or NCA: The Noise Control Act of 1972 (42USC4901-4918). The Act:
- Gave EPA the authority to set national noise emission standards for: Commercial products (NCA Sec. 6); Aircrafts (NCA Sec. 7); Railroads (NCA Sec. 17); Motor carriers (NCA Sec. 18).
- Required the EPA to assist the Federal Aviation Administration in developing noise regulations for airports and aircrafts.
- Consists of Public Law 95-574 (October 27, 1972) and the amendments by Public Law 95-609 (November 8, 1978).

(Other related information is provided in Pub. L. 92-574, 86 Stat. 1234; 40 CFR 201.1-a; 202.10-a; 203.1-1; 204.2-1; 205.2-1; 209.3-a; 211.102-b; 2.303-1.)

Act or NEPA: The National Environmental Policy Act of 1970 (Public Law 91-190, January 1, 1970; 42USC4321 et seq). The Act:
- Established the Council on Environmental Quality (CEQ) and required the development of a national policy on the environment.
- Required that Federal agencies include in their decision-making processes appropriate and careful consideration of all environmental effects of proposed actions, analyze potential environmental effects of proposed actions and their alternatives for public understanding and security, avoid or minimize adverse effects of proposed actions, and restore and enhance environmental quality as much as possible.

The purposes of the Act are:
- To declare a national policy which will encourage productive and enjoyable harmony between man and his environment.
- To promote efforts which will prevent or eliminate damage to the environment and biosphere and stimulate the health and welfare of man.
- To enrich the understanding of the ecological systems and natural resources important to the Nation; and to establish a Council on Environmental Quality.

(Other related information is provided in 40 CFR 51.852; 51.392; 93.101; 51.392; 1508.2.)

Act or OPA: The Oil Pollution Act of 1990 (Pub. L. 101-380, August 18, 1990). The Act:
(1) Imposes strict liability for a comprehensive and expansive list of damages from an oil spill into the water from vessels and facilities.
(2) Creates a $1 billion supplemental compensation fund for oil spills and details procedures for obtaining access to it (Sullivan-95/04, p169).

Act or OSHA: The Occupational Safety and Health Act of 1970 (29USC651 et seq). The purpose of this Act is to regulate commerce among the several States and with foreign nations and to provide for the general welfare, to assure so far as possible every working man and woman in the Nation safe and healthful working conditions and preserve human resources.

Act or PHSA: The Public Health Service Act (Pub. L. 93-523;

42USC300f). The Public Health Service in the Department of Health, Education, and Welfare shall be administered by the Surgeon General under the supervision and direction of the Secretary (Secretary of the Health, Education, and Welfare Department). The service shall consist of:
(1) Office of the Surgeon General
(2) National Institutes of Health
(3) Bureau of Medical Services
(4) Bureau of State Services (PHSA201)
Major subsequent amendments include:
- SDWA (Safe Drinking Water Act) of 1974 (Title XIV of PHSA) (other related information is provided in 40 CFR 141.2; 142.2; 149.101-a).

Act or PPA: The Pollution Prevention Act of 1990 (Public Law 101-508, November 5, 1990). The congress declares it to be the national policy of the United States that pollution should be prevented or reduced at the source whenever feasible; pollution that cannot be prevented should be recycled in an environmentally safe manner, whenever feasible; pollution that cannot be prevented or recycled should be treated in an environmentally safe manner whenever possible; and disposal or other release into the environmental should be employed only as a last resort and should be conducted in an environmental safe manner (Sullivan-95/04, p452).

Act or RCRA: The Resource Conservation and Recovery Act of 1976 (PL 95-609; 42USC6901 et seq). The Act:
- Was passed in 1976, as an amendment to the Solid Waste Disposal Act of 1965 to ensure that solid wastes are managed in an environmentally sound manner. The broad goals set by RCRA are to:
  - Protect human health and the environment
  - Reduce waste and conserve energy and natural resources
  - Reduce or eliminate the generation of hazardous waste as expeditiously as possible
- Establishes a regulatory system to track hazardous substances from the time of generation to disposal. It was designed to prevent new CERCLA sites from ever being created. The congress declares it to be the national policy of the United States that, whenever feasible, the generation of hazardous waste is to be reduced or eliminated as expeditiously as possible. Waste that is nevertheless generated should be treated, stored, or disposed of so as to minimize the present and future threat to human health and the environment (40USC6902);
- Promotes the protection of health and the environment and to conserve valuable material and energy resources (42USC6902); and
- Requires the use of safe and secure procedures in the treatment, transport, storage, and disposal of hazardous wastes to ensure that solid wastes are managed in an environmentally sound manner.

RCRA's major components include:
- Subtitle C: A Federal program to establish a management system that regulates hazardous waste from the time it is generated until its ultimate disposal, in effect from "cradle to grave."
- Subtitle D: A Federal program to promote and encourage the environmentally sound disposal of non-hazardous waste such as municipal garbage. It includes minimum Federal technical standards, guidelines for State solid waste plans and financial assistance to States.
- Subtitle J: A Federal program to promote and encourage the environmentally sound disposal of medical waste.
- Subtitle I: A Federal program, established by the Hazardous and Solid Waste Amendment of 1984 (HSWA), that regulates petroleum products and hazardous substances (as defined under Superfund) stored in underground tanks (EPA-86/01, *see also* 40 CFR 124.2; 144.3; 146.3; 248.4; 249.04; 250.4; 252.4; 253.4; 260.10; 270.2-91).

RCRA's major standards include:
- Maximum extraction procedure (EP) toxicity (40 CFR 261.24).
- Hazardous waste generators (RCRA Sec. 3002; 40 CFR 262).
- Hazardous waste transporters (RCRA Sec. 3003; 40 CFR 263).
- Owners and operators of TSD (RCRA Sec. 3004; 40 CFR 264):
  - Tank system (Subpart J, 40 CFR 264.190).
  - Surface impoundment (Subpart K, 40 CFR 264.220).
  - Waste pile (Subpart L, 40 CFR 264.250).
  - Land treatment (Subpart M, 40 CFR 264.270).
  - Landfill (Subpart N, 40 CFR 264.300).
  - Incinerator (Subpart O, 40 CFR 264.340).
  - Corrective action (Subpart S, 40 CFR 246.552).
  - Miscellaneous units (Subpart X, 40 CFR 264.600).

RCRA's major amendments include:
- HSWA (Hazardous and Solid Waste Amendment) of 1984.

(Other related information is provided in 40 CFR 2.305-1; 124.2; 144.3; 146.3; 248.4-a; 249.04-a; 250.4-a; 252.4-a; 253.4-a; 248.4-a; 249.04-a; 250.4-a; 252.4-a; 253.4; 256.06; 260.10; 270.2.)

Act or SARA: The Superfund Amendments and Reauthorization Act of 1986. The Act:
- Operated under the legislative authority of CERCLA and SARA that funds and carries out the EPA solid waste emergency and long-term removal remedial activities. These activities include establishing the National Priorities List, investigating sites for inclusion on the list, determining their priority level on the list, and conducting and/or supervising the ultimately determined cleanup and other remedial actions.
- In addition to certain free-standing provisions of the law, the Act includes amendments to CERCLA, the Solid Waste Disposal Act, and the Internal Revenue Code. Among the free-standing provisions of the law is Title III of SARA, also known as the "Emergency Planning and Community Right-to-Know Act of 1986" and Title IV of SARA, also known as the "Radon Gas and Indoor Air Quality Research Act of 1986." Title V of SARA amending the Internal Revenue Code is also known as the "Superfund Revenue Act of 1986."

(Other related information is provided in 40 CFR 280.12; 300.5.)

Act or SDWA: The Safe Drinking Water Act of 1974 (42USC300f-300j-9). The Act:
- Amended the Public Health Service Act in 1974 under Title XIV, Safety of Public Water Systems and was amended in 1976, 1977, 1979, 1980, 1984 and 1986. In effect, SDWA was also amended by a provision of the Hazardous and Solid

Waste Amendments of 1984 (RCRA amendment) because toxic pollutants may enter the drinking water in many ways.
- Requires EPA to regulate 9 contaminants with a year of enactment, another 40 within 2 years, and the rest within 3 years for a total of 83 new maximum contaminant levels. In addition to the above 83, at least 25 more primary standards must be required by 1991; with 25 more expected every 3 years thereafter. By 1990, U.S. EPA must specify criteria for disinfection of all water supplies.
- Requires EPA to develop the **corrosion control regulation:** Copper and lead in drinking water are from the water source and corrosion. EPA is proposing elimination of source water copper and lead through treatment and is trying to limit the copper and lead as corrosion by-products in the water distribution system (JMM-88).
- Requires EPA to establish the **drinking water priority list (DWPL)** which may have adverse effects on the health of persons and which are known or anticipated to occur in public water systems and may require regulations under the SDWA. It further requires EPA promulgate regulations for 25 new contaminants every three years. The first list was required by January 1, 1988. Because the Congress made no stipulation to the length of the DWPL, EPA may place a contaminant on the list indefinitely (JMM-88).
- Requires EPA to develop the **National Primary Drinking Water Regulation** specifying filtration criteria for surface water supplies by December 1987 and requiring disinfection of all public water supplies by 1989 (JMM-88).
- Requires EPA to develop the **total coliform regulation:** A coliform is an indicator of contaminant. If the coliform bacteria density exceeds a specific level according the method, then two check samples are required. If they exceed method levels then the system has failed the coliform MCL (JMM-88)
- Requires EPA to develop the **underground injection control (UIC):** The disposal of wastes, particularly petroleum extraction wastes, by injecting them into deep dry well, is a relatively common practice, particularly in the Southwest. UIC is to assure that underground sources of drinking water are not endangered by any underground injection (Winthrop-89/09).
- Set standards for maximum allowable levels (MCL) of certain chemicals and bacteriological pollutants in public drinking water systems; and regulated underground injection systems including deep-well injection. The standards include:
    (1) Maximum contaminant level (MCL): MCL has two control levels:
        (A) Primary maximum contaminant levels
        (B) Secondary maximum contaminant levels
    (2) Maximum contaminant level goal (MCLG)
        (A) Primary drinking water regulation
        (B) Secondary drinking water regulation

(Other related information is provided in 40 CFR 143; SDWA1401.)

Act or SWDA: The Solid Waste Disposal Act of 1965 (42USC 6901-6991i).
- The Act was to promote the protection of health and environment and to conserve valuable material and energy resources.
- The Act provided research funds and technical assistance for State and local planners to mainly manage the municipal waste.
- The Act consists of Title II of Public Law 89-272 (July 14, 1955) and subsequent amendments including:
    - RCRA (Resource Conservation and Recovery Act) of 1976.
    - HSWA (Hazardous and Solid Waste Act) of 1984.

Act or TSCA: The Toxic Substances Control Act of 1976 (15USC2601-2629). The Act:
- Created a screening mechanism for the production of new chemicals and new uses of existing chemicals.
- Prescribed EPA's range of actions to control manufacturing, use, and disposal once information gathered under the first two sections indicates a need for an action; and (4) conferred on EPA extraordinary authority to deal with imminent hazards.
- Banned manufacture and use of polychlorinated biphenyls (PCBs).
- Gives EPA power to require testing of chemical substances that present a risk of injury to health and the environment.
- Regulates the production and distribution of new chemicals and governs the manufacture, processing, distribution, and use of existing chemicals. Among the chemicals controlled by TSCA regulations are PCBs, chloroflurocarbons, and asbestos. In specific cases, there is an interface with RCRA regulations. For example, PCB disposal is generally regulated by TSCA. However, hazardous wastes mixed with PCBs are covered under RCRA.
- Regulated most heavily in the 1980s were asbestos and PCBs.
- PCBs at concentrations of 500 ppm or greater must be disposed of in an incinerator which complies with TSCA's incineration requirements. *See* PCB incineration standards in 40 CFR 761.70.
- The Act consists of Public Law 94-469 (October 11, 1976) and subsequent amendments including:
    - AHERA (Asbestos Hazard Emergency Response Act) of 1986.

(Other related information is provided in 40 CFR 2.306-1; 704.3; 707.63; 710.2-d; 712.3-p; 716.3; 720.3-b; 723.50; 723.175-1; 723.250-1; 747.115-1; 747.195-1; 747.200-1; 749.68-1; 766.3; 792.3; 763.83; 763.103-a; 763.163; 790.3; 791.3-a.)

Act or UMTRCA (40 CFR 192.00) for purposes of subparts A, B, and C of this part, means the Uranium Mill Tailings Radiation Control Act of 1978.

Act or WSOSHA (29 CFR 1910.2) means the Williams Steiger Occupational Safety and Health Act of 1970 (84 Stat. 1590).

ACTH: Adrenocorticotropic hormone, a hormone secreted by the pituitary gland that stimulates the adrenal cortex (LBL-76/07-bio).

Actinium (Ac): A radioactive metal with a.n. 89; a.w. 227; m.p. 1050° C and b.p. 3300° C. The element belongs to group IIIB of the periodic table.

Actinomycetes: A large group of moldlike microorganisms which give off an odor characteristic of rich earth and are the significant organisms involved in the stabilization of solid wastes by composting (SW-108ts).

Action level:
(1) A quantitative limit of a chemical, biological, or radiological agent at which actions need to be taken to prevent or reduce exposure or contact (Course 165.5).
(2) In the asbestos program, airborne concentration of asbestos of 0.1 fiber per cubic centimeter (f/cc) of air calculated as an 8-hour time-weighted average (40 CFR 763.121-91).
(3) (A) (In the pesticide program), regulatory levels recommended by EPA for enforcement by FDA and USDA when pesticide residues occur in food or feed commodities for reasons other than the direct application of the pesticide. As opposed to "tolerances" which are established for residues occurring as a direct result of proper usage, action levels are set for inadvertent residues resulting from previous legal use or accidental contamination (EPA-94/04).
 (B) In the Superfund program, the existence of a contaminant concentration in the environment high enough to warrant action or trigger a response under SARA and the National Oil and Hazardous Substances Contingency Plan. The term is also used in other regulatory programs (*see* tolerances) (EPA-94/04).

Action level contents: EPA recommended action levels are as follows:
(1) **Oxygen:** 19.5% (oxygen in normal ambient air is 20.5%) (measured by oxygen indicators).
(2) **Combustible gas:** 25% LEL (lower explosive level as measured by combustion gas indicators, *see also* lower explosive level).
(3) **Radiation:** 10 mR/hr (milli-roentgen per hour) (measured by radiation detectors).
(4) **Toxic gas:** The toxic gas (measured by flame ionized detectors or photo ionized detectors such as, HNU system) and the level of protection is as follows (*see also* level of protection):
- **Level A:** Level A is the highest level of respiratory, skin, and eye protection. The concentration of the toxic gas range is 500 ppm above background to 1000 ppm above background.
- **Level B:** Level B is the highest level of respiratory protection, but lesser level of skin protection. The concentration of the toxic gas is 5 ppm above background to 500 ppm above background.
- **Level C:** Level C requires air-purifying respirator. The concentration of the toxic gas range is from background concentration to 5 ppm above background.
- **Level D:** Level D requires work uniform under normal conditions.

Activate: In electronics, to treat the cathode or target of an electron tube in order to create or increase the emission of electrons (EPA-83/03).

Activated alumina: A form of aluminum oxide which adsorbs moisture readily and is used as a drying agent.

Activated carbon (or activated charcoal):
(1) Carbon which has been treated by high temperature heating with steam or carbon dioxide to produce an internal porous structure (EPA-87/10a).
(2) A highly adsorbent form of carbon used to remove odors and toxic substances from liquid or gaseous emissions. In (advanced) waste treatment it is used to remove dissolved organic matter from waste water. It is also used in motor vehicle evaporative control systems (EPA-94/04).
(3) *See* carbon for more related terms.

Activated carbon adsorption: *See* definition under adsorption process.

Activated carbon filter: A filter used to remove dissolved organic matter from water for taste and odor control. Dissolved gases, liquids and finely divided solids may also be removed (*see* filter for more related terms) (LBL-76/07-water).

Activated carbon regeneration: The regeneration of carbon after its adsorptive capacity has been reached, involving removal of organic matter from the carbon surface by oxidation or steam heating (cf. carbon regeneration) (EPA-87/10a).

Activated charcoal: *See* synonym, activated charcoal.

Activated complex: Intermediate compounds between reactants and products.

Activated sludge:
(1) Sludge that contains living organisms (DOI-70/04).
(2) A gelatinous matrix imbedded with filamentous and unicellular bacteria which serve as food for protozoa. The bacterial genera which predominate depend on the characteristics of the wastewater being treated. The activated sludge treatment of wastewater purification is one of the most common secondary waste treatment processes (LBL-76/07-water).
(3) Product that results when primary effluent is mixed with bacteria-laden sludge and then agitated and aerated to promote biological treatment, speeding the breakdown of organic matter in raw sewage undergoing secondary waste treatment (EPA-94/04).
(4) *See* sludge for more related terms.

Activated sludge process: *See* synonym, activated sludge treatment.

Activated sludge treatment (or activated sludge process): The process of using biologically active sewage sludge to hasten breakdown of organic matter in raw sewage during secondary waste treatment. The activated sludge is subsequently separated from the treated wastewater (mixed liquor) by sedimentation (EPA-74/11).

Activating agent: *See* synonym, activator.

Activation:
(1) Notification by telephone or other expeditious manner or, when required, the assembly of some of all appropriate members of the RRT (Regional Response Team) or NRT (National Response Team) (40 CFR 300.5-91).

(2) The process of treating a substance by heat, radiation or the presence of another substance so that the first mentioned substance will undergo chemical or physical change more rapidly or completely (EPA-83/06a).
(3) To make a solid surface active like the activated carbon or catalyst.

Activation energy (or heat of activation):
(1) The quantity of heat needed to destabilize molecular bonds and form reactive intermediates so that the reaction will proceed (EPA-81/09).
(2) The minimum heat needed for a chemical reaction to take place.

Activator (or activating agent):
(1) A substance which when added to a mineral pulp promotes flotation in the presence of a collecting agent. It can be used to increase the floatability of a mineral in a froth, or to reflect a depressed (sunk mineral) (EPA-82/05).
(2) Chemical substance, usually stannous chloride, that triggers the electroless deposition process on a non-conducting surface (EPA-74/03d).
(3) A chemical added to a pesticide to increase its activity (EPA-94/04).

Active alkali: A measure of the strength of alkaline pulping liquor indicating the sum of caustic soda and sodium sulfide expressed as Na_2O (EPA-87/10).

Active ingredient:
(1) An ingredient of a pesticide which is intended to prevent, destroy, repel, or mitigate any pest (40 CFR 455.10).
(2) In any pesticide product, the component that kills, or otherwise controls, target pests. Pesticides are regulated primarily on the basis of active ingredients (EPA-94/04).

Active institutional control:
(1) Controlling access to a disposal site by any means other than passive institutional controls
(2) Performing maintenance operations or remedial actions at a site
(3) Controlling or cleaning up releases from a site
(4) Monitoring parameters related to disposal system performance (40 CFR 191.12-91)

Active life of a facility: The period from the initial receipt of hazardous waste at the facility until the Regional Administrator receives certification of final closure (40 CFR 260.10-91).

Active material: An electrode material that reacts chemically to produce electrical energy when a cell discharges. Also, such material in its original composition, as applied to make an electrode (EPA-84/08).

Active mine: An underground uranium mine which is being ventilated to allow workers to enter the mine for any purpose (*see* mine for more related terms) (40 CFR 61.21-91; 61.21-91).

Active mining area: The area, on and beneath land, used or disturbed in activity related to the extraction, removal, or recovery of coal from its natural deposits. This term excludes coal preparation plants, coal preparation plant associated areas and post-mining areas (*see* mine for more related terms) (40 CFR 434.11-91, *see also* 40 CFR 440.132-91).

Active portion (facility): The portion of a facility where treatment, storage, or disposal operations are being or have been conducted after the effective date of Part 261 of this chapter and which is not a closed portion (*see* facility for more related terms, *see also* closed portion and inactive portion) (40 CFR 260.10-91).

Active service: That a drain is receiving refinery wastewater from a process unit that will continuously maintain a water seal (40 CFR 60.691-91).

Active transport: An energy-expending mechanism by which a cell moves a chemical across the cell membrane from a point of lower concentration to a point of higher concentration, against the diffusion gradient (Course 165.6).

Active use: Refers to an SO_2 constant control system installed at a smelter before August 7, 1977 and not totally removed from regular service by that date (40 CFR 57.103-91).

Active waste disposal site: Any disposal site other than an inactive site (40 CFR 61.141-91).

Activity: A set of CERCLA-funded tasks that makes up a segment of the sequence of events undertaken in determining, planning, and conducting a response to a release or potential release of a hazardous substance. These include core program, pre-remedial (i.e., preliminary assessments and site inspections), remedial investigation/feasibility studies, remedial design, remedial action, removal, enforcement, and Core Program activities (40 CFR 35.6015-91).

Activity coefficient: An auxiliary thermodynamic function to express the volatile properties of binary systems that exhibit non-ideal vapor equilibrium behavior. It may also be regarded as a correction factor that may be applied to ideal conditions to obtain real system properties under proper temperature and pressure conditions (EPA-85/10).

Activity median diameter (AMD): Refers to the median of the distribution of radioactivity, toxicological, or biological activity with respect to particle size (EPA-90/08).

Activity plans: Written procedures in a school's asbestos-management plan that detail the steps a Local Education Agency (LEA) will follow in performing the initial and additional cleaning, operation and maintenance program tasks; periodic surveillance; and reinspections required by the Asbestos Hazard Emergency Response Act (AHERA) (EPA-94/04).

Actual cubic feet per minute (acfm): *See* acronym, acfm.

Actual emissions:
(1) The actual rate of emissions of a pollutant from an emissions unit as determined in accordance with paragraphs (a)(1)(xii)(B) through (D) of this section.
(2) In general, actual emissions as of a particular date shall equal the average rate, in tons per year, at which the unit actually

emitted the pollutant during a two-year period which precedes the particular date and which is representative of normal source operation. The reviewing authority shall allow the use of a different time period upon a determination that it is more representative of normal source operation. Actual emissions shall be calculated using the unit's actual operating hours, production rates, and types of materials processed, stored, or combusted during the selected time period.

(3) The reviewing authority may presume that the source-specific allowable emissions for the unit are equivalent to the actual emissions of the unit.

(4) For any emissions unit (other than an electric utility steam generating unit specified in paragraph (a)(1)(xii)(E) of this section) which has not begun normal operations on the particular date, actual emissions shall equal the potential to emit of the unit on that date.

(5) For an electric utility steam generating unit (other than a new unit or the replacement of an existing unit) actual emissions of the unit following the physical or operational change shall equal the representative actual annual emissions of the unit, provided the source owner or operator maintains and submits to the reviewing authority, on an annual basis for a period of 5 years from the date the unit resumes regular operation, information demonstrating that the physical or operational change did not result in an emissions increase. A longer period, not to exceed 10 years, may be required by the reviewing authority if it determines such a period to be more representative of normal source post-change operations (40 CFR 51.165-xii-91).

Acute:
(1) Acute means short. In a toxicity test, it involves a stimulus severe enough to rapidly induce a response. A response observed in 96 hours or less typically is considered acute. An acute effect is not always measured in terms of lethality; it can measure a variety of effect (EPA-85/09).
(2) A stimulus severe enough to rapidly induce an effect; in aquatic toxicity tests, an effect observed in 96 hours or less typically is considered acute. When referring to aquatic toxicology or human health, an acute affect is rot always measured in terms of lethality (EPA-91/03).

Acute chronic ratio (ACR): The ratio of the acute toxicity (expressed as an LC50) of an effluent or a toxicant to its chronic toxicity (expressed as an NOEL). It is used as a factor for estimating chronic toxicity on the basis of acute toxicity data (EPA-85/09).

Acute delayed neurotoxicity: A prolonged, delayed-onset locomotor ataxia resulting from single administration of the test substance, repeated once if necessary (40 CFR 798.6540-91).

Acute dermal LD50: A statistically derived estimate of the single dermal dose of a substance that would cause 50 percent mortality to the test population under specified conditions (40 CFR 152.3-91).

Acute dermal toxicity: The adverse effects occurring within a short time of dermal application of a single dose of a substance or multiple doses given within 24 hours (40 CFR 798.1100-91).

Acute dose: *See* synonym, acute exposure.

Acute effect: An effect that can cause a disease in a short period of time. Other acute effect related terms include:
- Chronic effect
- Serious acute effect

Acute exposure (or acute dose):
(1) A one-time or short-term exposure with a duration of less than or equal to 24 hours (EPA-90/08).
(2) One dose or multiple doses occurring within a short time (24 hours or less) (EPA-92/12).
(3) A single exposure to a toxic substance which results in severe biological harm or death. Acute exposures are usually characterized as lasting no longer than a day, as compared to longer, continuing exposure over a period of time (EPA-94/04).

Acute toxicity:
(1) A deleterious response (e.g., mortality, disorientation, immobilization) to a stimulus observed in 96 hours or less (40 CFR 131.35-91, *see also* 40 CFR 300-App/A; 797.1440; 797.1800-91).
(2) The ability of a substance to cause poisonous effects resulting in severe biological harm or death soon after a single exposure or dose. Also, any severe poisonous effect resulting from a single short-term exposure to a toxic substance (*see* chronic toxicity, toxicity) (EPA-94/04).

Acute inhalation LC50: A statistically derived estimate of the concentration of a substance that would cause 50 percent mortality to the test population under specified conditions (40 CFR 152.3-91).

Acute inhalation toxicity: The adverse effects caused by a substance following a single uninterrupted exposure by inhalation over a short period of time (24 hours or less) to a substance capable of being inhaled (40 CFR 798.1150-91).

Acute lethal toxicity: The lethal effect produced on an organism within a short period of time of exposure organism within a short period of time (days) of exposure to a chemical (40 CFR 797.1350-91).

Acute oral LD50: A statistically derived estimate of the single oral dose of a substance that would cause 50 percent mortality to the test population under specified conditions (40 CFR 152.3-91).

Acute oral toxicity: The adverse effects occurring within a short time of oral administration of a single dose of a substance or multiple doses given within 24 hours (40 CFR 798.1175-1-91).

Acute toxicity endpoints (ATE): Toxicity test results, such as an LC_{50} (96 hours) and EC_{50} (48 hours), which describe a stimulus severe enough to rapidly induce an effect on aquatic organisms (EPA-91/03).

Acute toxicity test: A method used to determine the concentration of a substance that produces a toxic effect on a specified percentage of test organisms in a short period of time (e.g., 96 hours). In this guideline, death is used as the measure of toxicity (40 CFR 797.1400-91).

Acute toxicity:
(1) A deleterious response (e.g., mortality, disorientation, immobilization) to a stimulus observed in 96 hours or less (40 CFR 131.35-91, *see also* 40 CFR 300-App/A; 797.1440; 797.1800-91).
(2) The ability of a substance to cause poisonous effects resulting in severe biological harm or death soon after a single exposure or dose. Also, any severe poisonous effect resulting from a single short-term exposure to a toxic substance (EPA-12/89).
(3) *See* toxicity for more related terms.

Acute-to-chronic ratio (ACR): The ratio of the acute toxicity of an effluent or a toxicant to its chronic toxicity. It is used as a factor for estimating chronic toxicity on the basis of acute toxicity data, or for estimating acute toxicity on the basis of chronic toxicity data (EPA-91/03).

Acutely toxic chemicals: Chemicals which can cause both severe short- and long-term health effects after a single, brief exposure (short duration). These chemicals can cause damage to living tissue, impairment of the central nervous system, severe illness or in extreme cases, death when ingested, inhaled, or absorbed through the skin (EPA-85/11).

Acutely toxic conditions: Those acutely toxic to aquatic organisms following their short-term exposure within an affected area (EPA-91/03).

Acutely toxic effects: A chemical substance produces acutely toxic effects if it kills within a short time period (usually 14 days):
(1) At least 50 percent of the exposed mammalian test animals following oral administration of a single dose of the test substance at 25 milligrams or less per kilogram of body weight (LD50).
(2) At least 50 percent of the exposed mammalian test animals following dermal administration of a single dose of the test substance at 50 milligrams or less per kilogram of body weight (LD50).
(3) At least 50 percent of the exposed mammalian test animals following administration of the test substance for 8 hours or less by continuous inhalation at a steady concentration in air at 0.5 milligrams or less per liter of air (LC50) (40 CFR 721.3-91).

Adaptation:
(1) The process by which a substance induces the synthesis of any degradative enzymes necessary to catalyze the transformation of that substance (40 CFR 796.3100-91).
(2) Changes in an organism's structure or habit that help it adjust to its surroundings (EPA-94/04).

Add on control device: An air pollution control device such as carbon absorber or incinerator that reduces the pollution in an exhaust gas. The control device usually does not affect the process being controlled and thus is "add-on" technology, as opposed to a scheme to control pollution through altering the basic process itself (EPA-94/04).

Added ingredients: The prepared sauces (prepared from items such as dairy products, starches, sugar, tomato sauce and concentrate, spices, and other related pre-processed ingredients) which are added during the canning and freezing of fruits and vegetables (40 CFR 407.81-91).

Added risk: The difference between the cancer incidence under the exposure condition and the background incidence in the absence of exposure; AR = P(d) - P(O) (EPA-92/12).

Addict: Any person who habitually uses any habit-forming narcotic drugs so as to endanger the public morals, health, safety, or welfare, or who is or has been so far addicted to the use of such habit-forming narcotic drugs as to have lost the power of self-control with reference to his addiction (PHSA2-k).

Addition agent: A substance, usually an organic material, added to an electroplating solution to improve the properties of an electroplate (EPA-74/03d).

Additional polymerization: The combination of monomers by the direct addition or combination of the monomer molecules with one another to form polymers (EPA-87/10a).

Additive effect (or additivity): The combined effects of two or more chemicals equal to the sum of their individual effects (cf. synergistic effect) (Course 165.6; EPA-85/09).

Additive manufacturer: Any person who produces or manufactures an additive for use as an additive and/or sells an additive under his own name (40 CFR 79.2-91).

Additive:
(1) A chemical substance that is intentionally added to another chemical substance to improve its stability or impart some other desirable quality (40 CFR 790.3-91, *see also* 40 CFR 79.2-91).
(2) A substance added to plastic resins that imparts physical properties to meet specific applications and improve processing (OTA-89/10).
(3) In the plywood industry, any material introduced prior to the final consolidation of a board to improve some property of the final board or to achieve a desired effect in combination with another additive. It includes binders and other materials. Sometimes a specific additive may perform more than one function. Fillers and preservatives are included under this term (EPA-74/04).
(4) Materials added to ink in small amounts to alter one or more of its properties. They include driers, anti-skinning agents, dispersing agents, waxes, lubricants, surface active agents, etc. (EPA-79/12a).

Additivity: The characteristic property of a mixture of toxicants that exhibits a total toxic effect equal to the arithmetic sum of the effects of the individual toxicants (EPA-91/03).

Adenine ($C_5H_5N_5$): A needle crystal used in research on heredity, virus diseases and cancer.

Adenosine diphosphate (ADP) ($C_{10}H_{15}N_5O_{10}P_2$): A coenzyme which is an important intermediate in cellular metabolism.

Adenosine triphosphate (ATP) ($C_{10}H_{16}N_5O_{13}P_3$): A high energy biochemical intermediary in enzyme catalyzed processes (LBL-76/07-water).

Adequate evidence: The information sufficient to support the reasonable belief that a particular act or emission has occurred (40 CFR 32.105-91).

Adequate SO_2 emission limitation: A SIP emission limitation which was approved or promulgated by EPA as adequate to attain and maintain the NAAQS in the areas affected by the stack emissions without the use of any unauthorized dispersion technique (40 CFR 57.103-91).

Adequate storage: Placing of pesticides in proper containers and in safe areas as per 40 CFR 165.10 as to minimize the possibility of escape which could result in unreasonable adverse effects on the environment (40 CFR 165.1-91).

Adequately wet: Asbestos containing material that is sufficiently mixed or penetrated with liquid to prevent the release of particulates (EPA-94/04).

Adequately wetted: Sufficiently mixed or coated with water or an aqueous solution to prevent dust emissions (40 CFR 61.141-91).

Adhesion: Molecular attraction which holds the surfaces of two substances in contact, such as a gaseous pollutant and a solid adsorbent or water and rocks (EPA-84/09; 89/12).

Adhesive:
(1) Any substance or mixture of substances intended to serve as a joining compound (40 CFR 52.741-91).
(2) A chemical system used in the bonding of geomembranes. The adhesive residue results in an additional element in the seamed area. (Manufacturers and installers should be consulted for the various types of adhesives used with specific geomembranes) (EPA-91/05).

Adiabatic flame temperature: The temperature of combustion products under the conditions of no change in heat, work, kinetic energy and potential energy between the combustion system and its surroundings. The adiabatic flame temperature is the maximum temperature that a combustion system can reach, because any heat transfer from the reacting substances and any incomplete combustion would tend to lower the temperature of the products. For a given fuel and given pressure and temperature of the reactants, the maximum adiabatic flame temperature that can be achieved is with a stoichiometric mixture (*see* temperature for more related terms).

Adiabatic lapse rate: The rate of decrease of temperature with height in the atmosphere when adiabatic upward or downward motions of the air neither enhanced nor suppressed (NATO-78/10).

Adiabatic process:
(1) A process in which no heat is transferred between a system and its surroundings.
(2) A process during which no heat is extracted from or added to a system. In the atmosphere, an adiabatic upward movement of an air parcel results in cooling through expansion, an adiabatic downward movement results in warming through compression. This process explains the adiabatic lapse rate in a neutral atmosphere (NATO-78/10).
(3) *See* thermodynamic process for more related terms.

Adiabatic saturation: A process in which an air or gas stream is saturated with water vapor without adding or subtracting heat from the system (EPA-89/02).

Adiabatic saturation temperature: The temperature which results from adiabatically adding water to a gas-vapor-water mixture in a steady flow until it becomes saturated, the water being supplied at the final temperature of the mixture (*see* temperature for more related terms).

Adit:
(1) A horizontal or nearly horizontal passage driven from the surface for the working or dewatering of a mine.
(2) A passage driven into a mine from the side of a hill (EPA-82/05).

Adjacent: The bordering, contiguous, or neighboring, wetlands separated from other waters of the United States by man-made dikes or barriers, natural river berms, beach dunes, and the like are adjacent wetlands (40 CFR 230.3-91).

Adjustable capacitor: A device capable of holding an electrical charge at any one of several discrete values (EPA-83/03).

Adjusted configuration: The test configuration after adjustment of engine calibrations to the retrofit specifications, but excluding retrofit hardware installation (40 CFR 610.11-91).

Adjusted loaded vehicle weight: The numerical average of vehicle curb weight and GVWR (40 CFR 86.094.2-91).

Administrative order on consent: A legal agreement signed by EPA and an individual, business, or other entity through which the violator agrees to pay for correction of violations, take the required corrective or cleanup actions, or refrain from an activity. It describes the actions to be taken, may be subject to a comment period, applies to civil actions, and can be enforced in court (EPA-94/04).

Administrative order: A legal document signed by EPA directing an individual, business, or other entity to take corrective action or refrain from an activity. It describes the violations and actions to be taken, and can be enforced in court. Such orders may be issued, for example, as a result of an administrative complaint whereby the respondent is ordered to pay a penalty for violations of a statute (EPA-94/04).

Administrative order types: There are four types of RCRA orders: compliance orders; corrective action orders; monitoring and analysis orders and imminent hazard orders (cf. civil action) (OSWER-87):
(1) **Compliance action/order:** An order or action issued under Section 3008 (a) of RCRA, requires any person who is not complying with a requirement of RCRA to take steps to come into compliance.
(2) **Corrective action/order/policy:** In RCRA, an order EPA issues that requires corrective action under RCRA Section (h)

at a facility when there has been a release of hazardous waste or constituents into the environment. Corrective action may be required beyond the facility boundary and can be required regardless of when the waste was placed at the facility.
- (A) Corrective action involves cleaning up soils, sludges and groundwater contaminated with hazardous wastes at hundreds of RCRA facilities. Corrective actions may be required where hazardous waste releases have occurred at hazardous waste treatment, storage, and disposal facilities (TSDs) and solid waste management units (SWMUs). Early estimates suggest there are about 7 SWMUs per TSD facility. Corrective action may be required at many of these facilities.
- (B) Leaking underground storage tanks represent another potentially large universe for corrective action projects. There are approximately one million underground storage tanks, approximately 5-20 percent may be leaking.
  - (C) The volume of waste that will be generated from corrective action projects could be large. Although most waste probably will be treated on-site, some concentrated wastes may require off-site treatment. These wastes will likely compete for existing commercial treatment and land disposal capacity.
- (3) **Imminent hazard order:** Used by the responsible agency under the authority of RCRA Section 7003 to force any person contributing to an imminent and substantial endangerment to human health or the environment caused by the handling of non-hazardous or hazardous solid waste to take steps to clean up the problem.
- (4) **Monitoring and analysis order:** Used to evaluate the nature and extent of a substantial hazard to human health or the environment that exists at a TSD (treatment, storage, and disposal). It can be issued to either the current owner or to a past owner or operator if the facility is not currently in operation or the present owner could not be expected to have actual knowledge of the potential release.

Administrative procedures act (APA): A law that spells out procedures and requirements related to the promulgation of regulations. Administrative Record: All documents which EPA considered or relied on in selecting the response action at a Superfund site, culminating in the record of decision for remedial action or, an action memorandum for removal actions (EPA-94/04).

Adsorbate: Material that has been retained (adsorbed) by the process adsorption (LBL-76/07-air).

Adsorbent: A material which has the ability to cause molecules of gases, liquids, or solids to adhere to its internal surfaces without changing the adsorbent physically or chemically. Example adsorbents include activated carbon, alumina, bauxite, bone char, decolorizing carbon, Fuller's earth, magnesia, silica gel and strontium sulfate (selective adsorbent) (AP-40, p191).

Adsorber: A control device or piece of equipment for carrying on the process of adsorption. The unit is usually a vessel or a pipe containing activated carbon or another adsorbent. Generally it has a means of admitting and exhausting fluids plus whatever other piping (connections) might be needed for the operation of the unit, including desorption if involved (cf. absorber) (EPA-84/09).

Adsorption:
(1) A mass transfer process that involves removing a gaseous contaminant by adhering it to the surface of a solid. Adsorption can be classified as physical or chemical. In physical adsorption, a gas molecule adheres to the surface of the solid due to an imbalance of natural forces (electron distribution). In chemisorption, once the gas molecule adheres to the surface, it reacts chemically with it. The major distinction is that physical adsorption is readily reversible while chemisorption is not (EPA-81/12, p1-6).
(2) A process by which a solid material is used to remove one or more components from a liquid or gaseous stream, usually without chemical reaction. The removal takes place through adherence to the surface. Typical adsorbents are activated carbon, molecular sieves, silica gel, and activated alumina (EPA-84/09).
(3) The adhesion of substance to the surface of a solid (DOE-91/04).
(4) An advanced method of treating waste in which activated carbon removes organic matter from wastewater (EPA-94/04).

Other adsorption related terms include:
- Contact adsorption
- Physical adsorption

Adsorption isobar: A plot showing adsorption against various parameters, such as temperature, while holding pressure constant.

Adsorption isotherm: A plot used in evaluating the effectiveness of activated carbon treatment by showing the amount of impurity adsorbed versus the amount remaining. It is determined at a constant temperature by varying the amount of carbon used or the concentration of the impurity in contact with the carbon (EPA-87/10a).

Adsorption process: In waste treatment, there are two basic adsorption processes:
(1) **Activated carbon adsorption:** A process of using activated carbon to adsorb dissolved organic matter from wastewater. Adsorption on activated carbon occurs when a molecule is brought up to its surface and held there by physical and/or chemical forces. This process is reversible, thus allowing activated carbons to be regenerated (and reused) by the proper application of heat and steam, or solvent.
(2) **Resin adsorption:** Resin adsorption uses synthetic polymeric materials as adsorbents which are hard, insoluble spheres of high surface area, porous polymer. It involves two basic steps:
  - (A) Contacting the liquid waste stream with resins and allowing the resins to adsorb the solutes from the solution; and
  - (B) Subsequently regenerating the resins by removing the adsorbed chemicals, often effected by simply washing with the proper solvent.

The difference between carbon and resin adsorption is that resins are always chemically regenerated by the use of caustic or organic solvents to wash the adsorbed organics on the surface of the resins, while carbons, because the adsorption forces are stronger, must usually be thermally regenerated thus eliminating the possibility of material recovery. Thermal regeneration of activated carbon is to use heat to evaporate the adsorbed organics on the surface of the activated carbon and eventually burn the evaporated organics in somewhere, thus the carbon can be re-used again. The resin adsorption is mostly used for the purpose for material recovery, otherwise, it is not likely to be competitive with carbon for the treatment of high volume waste streams containing moderate to high concentrations of mixed wastes with no recovery value.

Adsorption ratio (K_d): The amount of test chemical adsorbed by a sediment or soil (i.e., the solid phase) divided by the amount of test chemical in the solution phase, which is in equilibrium with the solid phase, at a fixed solid/solution ratio (40 CFR 796.2750-91).

Adulterants: Chemical impurities or substances that by law do not belong in a food, or pesticide (EPA-94/04).

Adulterated:
(1) Applies to any pesticide if:
 (A) Its strength or purity falls below the professed standard of quality as expressed on its labeling under which it is sold;
 (B) Any substance has been substituted wholly or in part for the pesticide; or
 (C) Any valuable constituent of the pesticide has been wholly or in part abstracted (FIFRA2-7USC136-91).
(2) (A) Any pesticide whose strength or purity falls below the quality stated on its label (EPA-94/04).
 (B) A food, feed, or product that contains illegal pesticide residues (EPA-94/04).

Advanced air emission control devices: The air pollution control equipment, such as electrostatic precipitators and high energy scrubbers, that are used to treat an air discharge which has been treated initially by equipment including knockout chambers and low energy scrubbers (40 CFR 426.11-91).

Advanced electrical reactor (AER): AER is a double wall cylindrical reactor with six carbon resistance heaters located between the two cylindrical walls. An inert gas heated by the heaters flows radially inward through the inner porous reactor wall (a core) carrying high rates of heat to accomplish the heating of the porous carbon core to incandescence so that the predominant mode of heat transfer is by radiation from the core to the waste stream. During operation, the waste material to be pyrolyzed is finely ground to about a 20 mesh and introduced into the top of the reactor. As the material falls through the tubular space, it is exposed to radiation energy with power densities of over 1200 watts/in^2. The finely-divided reactants are heated through the direct impingement of electromagnetic radiation and are eventually destroyed (Lee-83/07).

Advanced treatment: A level of wastewater treatment more stringent than secondary treatment; requires an 85-percent reduction in conventional pollutant concentration or a significant reduction in nonconventional pollutants (EPA-94/04).

Advanced waste treatment:
(1) A treatment method or process employed following biological treatment:
 (A) To increase the removal of pollution load;
 (B) To remove substances which may be deleterious to receiving waters or the environment; and
 (C) To produce a high-quality effluent suitable for reuse in any specific manner or for discharge under critical conditions. The term tertiary treatment is commonly used to denote advanced waste treatment methods (EPA-82/10).
(2) Any treatment of sewage that goes beyond the secondary or biological water treatment stage and includes the removal of nutrients such as phosphorus and nitrogen and a high percentage of suspended solids (*see* primary, secondary treatment) (EPA-94/04).
(3) *See* wastewater treatment for more related terms.

Advection: The process of transport of a property (e.g., pollution or momentum) solely by the velocity field in a fluid. For a turbulent flow, advection refers only to the transport by the mean motion, where mean is defined as the time average. In meteorology, advection refers only to the horizontal or isobaric component of the atmospheric motion. In groundwater, advection refers to the natural state of motion (EPA-87/03; NATO-87/11).

Adversary adjudication: An adjudication required by statute to be held pursuant to 5USC554 in which the position of the United States is represented by counsel or otherwise, but excludes an adjudication for the purpose of granting or renewing a license (40 CFR 17.2-91).

Adverse environmental effect: Any significant and widespread adverse effect, which may reasonably be anticipated, to wildlife, aquatic life, or other natural resources, including adverse impacts on populations of endangered or threatened species or significant degradation of environmental quality over broad areas (CAA112-42USC7412-91).

Adverse impact on visibility: For purposes of section 307, visibility impairment which interferes with the management, protection, preservation, or enjoyment of the visitor's visual experience of the Federal Class I area. This determination must be made on a case-by-case basis taking into account the geographic extent, intensity, duration, frequency and time of visibility impairments, and how these factors correlate with (1) times of visitor use of the Federal Class I area, and (2) the frequency and timing of natural conditions that reduce visibility. This term does not include effects on integral vistas (40 CFR 51.301-91, *see also* 40 CFR 52.21-91).

Adverse effect level (AEL): That exposure level at which there are statistically or biologically significant increases in frequency or severity of adverse effects between the exposed population and its appropriate control (EPA-90/08).

Advertised engine displacement: The rounded volumetric engine capacity used for marketing purposes by the motorcycle manufacturer (40 CFR 205.151-91).

Advisory: A non-regulatory document that communicates risk information to those who may have to make risk management decisions (EPA-94/04).

Aerated aerobic lagoon: *See* synonym, aerated lagoon.

Aerated lagoon (aerated aerobic lagoon or aerated pond):
(1) Bacterial stabilization of wastewater in a natural or artificial

wastewater treatment pond in which mechanical or diffused air aeration is used to supplement the oxygen supply (*see* lagoon for more related terms) (EPA-87/10a).
(2) A holding and/or treatment pond that speeds up the natural process of biological decomposition of organic waste by stimulating the growth and activity of bacteria that degrade organic waste (EPA-94/04).

Aerated pond: *See* synonym, aerated lagoon.

Aeration:
(1) The bringing about of intimate contact between air and water by methods including:
 (A) Spraying water into the air over a collecting basin or,
 (B) Causing water to flow over baffles (LBL-76/07-water).
(2) The process of exposing bulk materials, such as compost, to air, (or of charging a liquid with a gas or a mixture of gases). Forced aeration refers to the use of blowers in compost piles (EPA-89/11; EPA-83).
(3) A process which promotes biological degradation of organic matter in water. The process may be passive (as when waste is exposed to air), or active (as when a mixing or bubbling device introduces the air) (EPA-94/04).

Other aeration related terms include:
- De-aeration
- Extended aeration
- Mechanical aeration
- Post aeration
- Pre-aeration
- Re-aeration
- Stage aeration
- Step aeration
- Surface aeration
- Tapered aeration

Aeration period:
(1) The theoretical time, usually expressed in hours, that the mixed liquor is subjected to aeration in an aeration tank undergoing activated-sludge treatment. It is equal to the volume of the tank divided by the volumetric rate of flow of wastes and return sludge.
(2) The theoretical time that liquids are subjected to aeration (EPA-87/10a).

Aeration tank: A chamber used to inject air into water (*see* tank for more related terms) (EPA-94/04)

Aerator: A mechanical device which provides turbulence at the air and liquid interface to increase the dissolved oxygen level in wastewater. Other aerator related terms include:
- Bed cascade aerator (*see* definition under cascade aerator)
- Cascade aerator
- Contact aerator
- De-aerator
- Free fall cascade aerator (*see* definition under cascade aerator)
- Injection aerator
- Surface aerator
- Tray aerator

Aeration tank: A chamber used to inject air into water (EPA-94/04).

Aerobe (or obligate aerobe): Organisms or bacteria which can only live in the environment with the presence of oxygen (cf. facultative aerobe or anaerobe) (EPA-89/12a).

Aerobic:
(1) Taking place in the presence of free molecular oxygen (EPA-87/10a).
(2) A biochemical process or condition occurring in the presence of oxygen (EPA-89/11).
(3) Life or processes that require, or are not destroyed by, the presence of oxygen (*see* anaerobic) (EPA-94/04).

Aerobic bacteria: Bacteria that require free elemental oxygen for their growth. Their metabolic demands can severely deplete the dissolved oxygen (LBL-76/07-water).

Aerobic biological oxidation: *See* synonym, aerobic treatment.

Aerobic digestion: *See* synonym, aerobic treatment.

Aerobic lagoon: A pond for aerobic digestion (*see* lagoon for more related terms).

Aerobic organotrophy: Organism which utilizes organic materials under aerobic conditions for its growth (LBL-76/07-water).

Aerobic oxidation: *See* synonym, aerobic treatment.

Aerobic respiration: Oxidation of organic compounds (foodstuffs) by organisms (EPA-83). The reaction can be summarized by the equation $C_6H_{12}O_6 + 6O_2 \rightarrow 6CO_2 + 6H_2O$ + energy (the release of chemical energy).

Aerobic treatment (aerobic biological oxidation, aerobic digestion or aerobic oxidation):
(1) A process in which microorganisms obtain energy by endogenous or auto-oxidation of their cellular protoplasm. The biologically degradable constituents of cellular materials are slowly oxidized to carbon dioxide, water and ammonia, with the ammonia being further converted into nitrates during the process (EPA-87/10a). The two basic systems used in the aerobic biological treatment are activated sludge and trickling filter. Activated sludge is a suspended growth, mixed culture system and trickling filter is a fixed film, attached growth system. *See* activated sludge and trickling filter for more information.
(2) Process by which microbes decompose complex organic compounds in the presence of oxygen and use the liberated energy for reproduction and growth. (Such processes include extended aeration, trickling filtration, and rotating biological contactors) (EPA-94/04).

Aerochlorination: The use of compressed air and chlorine gas to treat wastewater for removing fatty substances.

Aerodynamic (viscous) resistance diameter (D_{ar}): The Lovelace definition for aerodynamic diameter. Characteristic expression based on terms describing a particle in the Stokes' regime. Refer to

Raabe (1976) for equation (EPA-90/08).

Aerodynamic diameter:
(1) Term used to describe particles with common inertial properties to avoid the complications associated with the effects of particle size, shape and physical density (EPA-92/12).
(2) Aerodynamic diameter applies to the size of particles of aerosols. It is the diameter of a sphere of unit density which behaves aerodynamically as the particle of the test substance. It is used to compare particles of different size and densities and to predict where in the respiratory tract such particles may be deposited. This term is used in contrast to measured or geometric diameter which is representative of actual diameters which in themselves cannot be related to deposition within the respiratory tract (40 CFR 798.1150-91).

In air pollution, it is defined as the diameter of a sphere of a unit density (g/m^3) having the same falling speed in air as the particle in question (EPA-84/09). It is related to the physical diameter according to the equation below: $d_{pa} = d_p (pC)^{1/2}$, where: d_{pa} = aerodynamic diameter; d_p = physical diameter; p = particle density; and C = Cunningham correction factor.

Aerodynamic equivalent diameter (D_{ae}): Aerodynamic diameter generally used. The diameter of a unit density sphere ($d_p = 1$ g/cm^3) having the same settling velocity (due to gravity) as the particle of interest whatever shape and density. Refer to Raabe (1976) for equation (EPA-90/08).

Aerodynamic (viscous) resistance diameter (D_{ar}): The Lovelace definition for aerodynamic diameter. Characteristic expression based on terms describing a particle in the Stokes' regime. Refer to Raabe (1976) for equation (EPA-90/08).

Aerodynamics: A branch of dynamics that treats the motion of air and other gaseous fluids and deals with the forces acting on solids in motion relative to such fluids.

Aeroflocs: Synthetic water-soluble polymers used as flocculating agents (EPA-82/05).

Aeronautics: The science and art of flight (Markes-67).

Aerosol:
(1) A two phase medium consisting of a gas (usually air) and either particulates or small droplets of liquid (EPA-84/09).
(2) A particle of solid or liquid matter (e.g., dust, fog and smoke) that can remain suspended in the air because of its small size. The diameters of particles may vary from 100 micron to 0.01 micron or less. Toxic chemical aerosols (particles) can be carried into the lungs by a gas (Course 165.5; OME-88/12).
(3) A dispersion of solid or liquid particles in gaseous media and the dispersed phase is not large enough to settle out under the influence of gravity (LBL-76/07-bio).
(4) All-inclusive term. A suspension of liquid or solid particles in air (EPA-90/08).
(5) A suspension of liquid or solid particles in a gas (EPA-94/04).

Aerosol monitor: A direct-reading instrument that measures aerosols (suspended solid or liquid particles), e.g., dust, mist, fume, smoke, fog, spray (not a gas). Most of aerosol monitors use a light source and a light sensor that measures the amount of light scattered by the aerosol. Read-outs are in milligrams per cubic meter (mg/m^3) (Course 165.5).

Aerosol particles: Solid particles from 10^{-12} to 10^{-1} um in diameter dispersed in a gas (LBL-76/07-bio).

Aerosol propellant: A liquefied or compressed gas in a container where the purpose of the liquefied or compressed gas is to expel from the container liquid or solid material different from the aerosol propellant (40 CFR 762.3-91).

Affected public: The people who live and/or work near a hazardous waste site (EPA-94/04).

Affected source: A source that includes one or more affected units (*see* source for more related terms) (CAA402-42USC7651a-91).

Affected unit: A unit that is subject to emission reduction requirements or limitations under this title (CAA402-42USC7651a-91).

Affecting: Will or may have an effect on (40 CFR 1508.3-91).

Affiliated entity: A person who directly, or indirectly through one or more intermediaries, controls, is controlled by, or is under common control with the owner or operator of a source (40 CFR 66.3-91).

Affinity: The force that causes the atoms of certain elements to combine and stay combined (Webster-68).

Affluent: Flowing freely, rich or tributary (DOI-70/04).

Aflatoxin: The toxin produced by fungus which is considered as the most potent carcinogen yet discovered.

Afterburner:
(1) Afterburner is also known as the secondary burner. It is a control device in which materials in gaseous effluent are combusted (40 CFR 52.741-91). There are two types of afterburners, namely:
   (A) Direct flame afterburners which use high temperature flames to burn the combustible materials; and
   (B) Catalytic afterburners which use catalysts to oxidize residual combustible materials at an elevated temperature (without a flame) (AP-40, p171).
(2) In incinerator technology, a burner located so that the combustion gases are made to pass through its flame in order to remove smoke and odors. It may be attached to or be separated from the incinerator proper (EPA-94/04).
(3) *See* burner for more related terms.

Agar: An extract from certain red seaweeds used as a gelling agent in culture media.

Age tank: A tank used to store a chemical solution of known concentration for feed to a chemical feeder. Also called a day tank (EPA-94/04).

Aged catalytic converter: A converter that has been installed on a vehicle or engine stand and operated through a cycle specifically designed to chemically age, including exposure to representative lead concentrations, and mechanically stress the catalytic converter in a manner representative of in-use vehicle or engine conditions (cf. catalytic converter) (40 CFR 85.2122(a)(15)(ii)(F)-91).

Agent: In toxicology, *See* toxicant.

Agent orange: A toxic herbicide and defoliant used in the Vietnam conflict, containing 2,4,5-trichlorophenoxyacetic acid (2,4,5-T) and 2-4 dichlorophenoxyacetic acid (2,4-D) with trace amounts of dioxin (EPA-94/04).

Agglomeration:
(1) The process by which precipitation particles grow larger by collision or contact with cloud particles or other precipitation particles (EPA-89/12).
(2) The coalescence of dispersed suspended matter into larger flocs or particles which settle more rapidly (EPA-82/10).

Agglutination: The process of uniting solid particles coated with a thin layer of adhesive material or of arresting solid particles by impact on a surface coated with an adhesive (EPA-89/12).

Aggregate costs: The total cost of all research, surveys, studies, modeling, and other technical work completed by a Management Conference during a fiscal year to develop a Comprehensive Conservation and Management Plan for the estuary (40 CFR 35.9010-91).

Aggregate facility: An individual drain system together with ancillary downstream sewer lines and oil water separators, down to and including the secondary oil water separator, as applicable (40 CFR 60.691-91).

Aggregate: To gather particles into a mass.

Aggressive carbon dioxide: A method to measure the corrosivity and scaling properties of water. The method involves the use of excessive carbon dioxide to precipitate a specified concentration of calcium ions as calcium carbonate.

AGI impinger: An all gas impinger sampling device. Aerosol particles are drawn through an inlet tube and then through a critical orifice into a liquid menstruum (EPA-88/09c).

Aging:
(1) In electronics, storaging a permanent magnet, capacitor, meter or other device (sometimes with a voltage applied) until the characteristics of the device become essentially constant (EPA-83/03).
(2) In metal finishing, changing properties (e.g., increase in tensile strength and hardness) that occur in certain metals at atmospheric temperature after heat treatment (EPA-83/06a).

Agitation of parts: The irregular movement given to parts when they have been submerged in a plating or rinse solution (EPA-83/06a).

Agreement State: Any state with which the Atomic Energy Commission or the Nuclear Regulatory Commission has entered into an effective agreement under subsection 274b. of the Act. **Non-Agreement State** means any other State (10 CFR 30.4-91).

Agricultural pollution: Farming wastes, including runoff and leaching of pesticides and fertilizers; erosion and dust from plowing; improper disposal of animal manure and carcasses; crop residues, and debris (*see* pollution for more related terms) (EPA-94/04).

Agricultural solid waste (or agricultural waste): The solid waste that is generated by the rearing of animals, and the producing and harvesting of crops or trees (*see* waste for more related terms) (40 CFR 243.101-91, *see also* 40 CFR 246.101-91).

Agricultural waste: *See* synonym, agricultural solid waste.

Agro-ecosystem: Land used for crops, pasture, and livestock; the adjacent uncultivated land that supports other vegetation and wildlife; and the associated atmosphere, the underlying soils, groundwater, and drainage networks (EPA-94/04).

AHERA designated person (ADP): A person designated by a Local Education Agency to ensure that the AHERA requirements for asbestos management and abatement are properly implemented (EPA-94/04).

Air: A gaseous mixture appearing in the atmosphere. The compositions of air consist of:
(1) Nitrogen (78.09% by volume or mole)
(2) Oxygen (20.95%)
(3) Argon (0.93%)
(4) Others (0.03%) including: CO_2, krypton, neon, helium, H_2, xenon, and ozone (EPA-84/09).

Other air related terms include:
- Ambient air
- Cooling air
- Dry air
- Infiltration air
- Overfire air
- Open air (*see* synonym, ambient air)
- Saturated air
- Stable air
- Tempering air (*see* synonym, cooling air)
- Underfire air

Air binding: Situation where air enters the filter media and harms both the filtration and backwash processes (EPA-94/04).

Air calculation: In calculations, air is assumed that it contains 21% oxygen and 79% nitrogen. The volumetric and gravimetric analyses of air are as follows:
(1) Volumetric analysis of dry air
- **Oxygen:** 21 percent of air or 0.21 mole, or 0.21 lb-mole, when air is assumed at 1 mole.
- **Nitrogen:** 79 percent of air or 0.79 mole, or 0.79 lb-mole, when air is assumed at 1 mole. Thus,
- 0.21 moles of O_2 + 0.79 moles of N_2 = 1 mole of air; or
- 1 mole of O_2 + 3.76 moles of N_2 = 4.76 moles of air. This equation shows that for each volume (mole) of

oxygen, 3.76 volumes (moles) of nitrogen or 4.76 volumes (moles) of air are involved.

(2) Gravimetric analysis of dry air
- Air molecular weight

$0.21 O_2$ +	$0.79 N_2$ =	1 mole of air
0.21x32 +	0.79x28 =	28.84 lb/lb-mole
6.72	22.12	28.84
1	3.2917	4.2917 (A)
0.2330	0.7670	1 -- air base calculation
0.23	0.77	1
1	3.35	4.35 (B)

Both equations (A) and (B) are based on unit oxygen calculation. However, comparing equations (A) and (B), the different nitrogen values, 3.2917 and 3.35, are because of the approximation in equation (B).
- Oxygen mass fraction = 0.21x32/28.84 = 23%
- Nitrogen mass fraction = 0.79x28/28.84 = 77%

Thus, one pound of air can be expressed by:
- 0.23 lb O_2 + 0.77 lb N_2 = 1 lb air; or
- 1 lb O_2 + 3.35 lb N_2 = 4.35 lbs air. This equation shows that for each 1b of oxygen, 3.35 lbs of nitrogen or 4.35 lbs of air are involved.

(3) **Air standard volume:** Any gas at standard conditions has a volume of 359 ft^3/lb-mole (*see* standard condition).

(4) **Air density** = weight/volume = 28.84/359 = 0.0808 lb/scf, where: scf = standard cubic feet.

Air conditioner fever: *See* synonym, humidifier fever.

Air agitation: The agitation of a medium, e.g., liquid through the use of air pressure injected into the medium (EPA-83/06a).

Air analyzer (or air monitor): Typical air analyzers include:
(1) Combustible atmosphere indicator
(2) Colorimetric indicator tube (*see* synonym, toxic atmosphere monitor)
(3) Oxygen indicator
(4) Toxic atmosphere monitor

Air assisted airless spray: A spray coating method which combines compressed air with hydraulic pressure to atomize the coating material into finer droplets than is achieved with pure airless spray. Lower hydraulic pressure is used than with airless spray (40 CFR 52.741-91).

Air avid: To increase by addition of chemicals the affinity of fine particles for air bubbles (EPA-88/08a).

Air blast: Forced air circulation.

Air bleed to intake manifold retrofit: A system or device (such as a modification to the engine's carburetor or positive crankcase ventilation system) that results in engine operation at an increased air-fuel ratio so as to achieve reductions in exhaust emissions of hydrocarbons and carbon monoxide from 1967 and earlier light-duty vehicles of at least 21 percent and 58 percent, respectively (40 CFR 52.2039-91, *see also* 40 CFR 52.2490-91).

Air change per hour (ACH): The movement of a volume of air in a given period of time; if a house has one air change per hour, it means that all of the air in the house will be replaced in a one-hour period (cf. air exchange rate) (EPA-94/04).

Air classification: A process in which a stream of air is used to separate mixed materials according to the size, density, and aerodynamic drag of the pieces (EPA-89/11).

Air classifier:
(1) A system that uses a forced air stream to separate mixed materials according to the size, density, and aerodynamic drag of pieces (OME-88/12).
(2) An appliance for approximately sizing crushed minerals or ores employing currents of air (EPA-75/10c).

Air cleaner filter element: A device to remove particulates from the primary air that enters the air induction system of the engine (40 CFR 85.2122(a)(16)(ii)(A)-91).

Air compressor: An energy conversion device which converts mechanical work into a stored energy with high pressure.

Air contaminant:
(1) Any solid, liquid, or gaseous matter, any odor, or any form of energy, that is capable of being released into the atmosphere from an emission source (40 CFR 52.741-91).
(2) Any particulate matter, gas, or combination thereof, other than water vapor (*see* air pollutant) (EPA-94/04).
(3) *See* specified air contaminant for more related terms.

Air cooled furnace wall: A refractory that has a lane directly behind it through which cool air can flow (*see* furnace wall for more related terms) (SW-108ts).

Air cooled slag: In the steel industry, a slag which is cooled slowly in large pits in the ground. Light water sprays are generally used to accelerate the cooling over that which would occur in air alone. The finished slag is generally gray in color and looks like a sponge (*see* slag for more related terms) (EPA-74/06a).

Air curtain: A method of containing oil spills. Air bubbling through a perforated pipe causes an upward water flow that slows the spread of oil. It can also be used to stop fish from entering polluted water (EPA-94/04).

Air cyclone separator: A cylindrical and conical structure, without moving parts, which utilizes centrifugal force to remove solids entrained in an air stream (*see* collector for more related terms).

Air deficiency: A lack of air, in an air-fuel mixture, to supply the quantity of oxygen stoichiometrically (theoretically) required to completely oxidize the fuel (SW-108ts).

Air density: *See* definition under air.

Air dried coating: Any coatings that dry by use of air or forced air at temperatures up to 363.15 K (194° F) (40 CFR 52.741-91).

Air dry: Paper or paperboard is air dry when its moisture content is in equilibrium with atmospheric conditions to which it is exposed. According to trade custom, air dry pulps are assumed to contain 10 percent moisture, and are sold on this basis (EPA-83).

Air dry loss: That moisture gain or loss from a sample that has been partially dried to bring its moisture content close to equilibrium with the atmosphere in the room in which further reduction and division of the sample are to take place (*see* analytical parameters--laboratory for more related terms) (EPA-83).

Air drying: A process of partial drying of the sample to bring its moisture content near to equilibrium with atmosphere in the room in which further reduction, division and characterization of the sample are to take place (*see* analytical parameters--laboratory for more related terms) (EPA-83).

Air dry ton (ADT): A measurement of (pulp, paper, and paperboard) production including a moisture content of 10 percent by weight (EPA-87/10).

Air emission: The release or discharge of a toxic pollutant by an owner or operator into the ambient air either:
(1) By means of a stack; or
(2) As a fugitive dust, mist or vapor as a result inherent to the manufacturing or formulating process (40 CFR 129.2-91).
(3) *See* emission for more related terms.

Air erosion: The passage of air over friable ACBM which may result in the release of asbestos fibers (40 CFR 763.83-91).

Air exchange rate:
(1) The rate at which the house air is replaced with outdoor air. Commonly expressed in terms of air changes per hour (ACH).
(2) The movement of a volume of air in a given period of time; if a house has one air change per hour, it means that all of the air in the house is replaced in a one-hour period (EPA-88/08).
(3) cf. air change per hour.

Air fuel ratio (or air to fuel ratio): *See* synonym, combustion air fuel ratio.

Air gap: Open vertical gap or empty space that separates drinking water supply to be protected from another water system in a treatment plant or other location. The open gap protects the drinking water from contamination by backflow or backsiphonage (EPA-94/04).

Air heater (or air preheater): A heat exchanger through which air passes and is heated by a medium of a higher temperature, such as hot combustion gases in metal tubes (SW-108ts).

Air jet: A stream of high-velocity air that usually issues from a nozzle (SW-108ts).

Air knife: A jargon for a blower device intended to separate steel cans from more massive pieces of iron and steel. Experimentation is required to discover the best design and location for application (EPA-83).

Air lance: A commonly used nondestructive test method performed with a stream of air forced through a nozzle at the end of a hollow metal tube to determine seam continuity and tightness of relatively thin, flexible geomembranes (EPA-91/05).

Air liquid interface: The interface layer between the air and the liquid in which mass transfer is diffusion controlled (EPA-83/06a).

Air mass: A large volume of air with certain meteorological or polluted characteristics-e.g., a heat inversion or smogginess-while in one location. The characteristics can change as the air mass moves away (EPA-94/04).

Air modeling: The mathematical description of the movement and dispersion of airborne constituents (CRWI-89/05).

Air monitor: *See* synonym, air analyzer.

Air monitoring: *See* synonym, monitoring (EPA-94/04).

Air padding: Pumping dry air into a container to assist with the withdrawal of liquid or to force a liquified gas such as chlorine out of the container (EPA-94/04).

Air plenum: Any space used to convey air in a building, furnace, or structure. The space above a suspended ceiling is often used as an air plenum (EPA-94/04).

Air oxidation reactor: Any device or process vessel in which one or more organic reactants are combined with air, or a combination of air and oxygen, to produce one or more organic compounds. Ammoxidation and oxychlorination reactions are included in this definition (40 CFR 60.611-91).

Air oxidation reactor recovery train: An individual recovery system receiving the vent stream from at least one air oxidation reactors feeding vent streams into this system (40 CFR 60.611-91).

Air oxidation unit process: A unit process, including ammoxidation and oxychlorination unit process, that uses air, or a combination of air and oxygen, as an oxygen source in combination with one or more organic reactants to produce one or more organic compounds (40 CFR 60.611-91).

Air permeability: The property that allows passage of air through a mass (EPA-83).

Air pollutant:
(1) Any air pollutant agent or combination of such agents, including any physical, chemical, biological, radioactive (including source material, special nuclear material, and byproduct material) substance or matter which is emitted into or otherwise enters the ambient air (CAA302-42USC7602-91).
(2) Any substance in air which could, if in high enough concentration, harm man, other animals, vegetation, or material. Pollutants may include almost any natural or artificial composition of matter capable of being airborne. They may be in the form of solid particles, liquid droplets, gases, or in combinations of these forms (EPA-94/04).
(A) Those emitted directly from identifiable sources (EPA-94/04).
(B) Those produced in the air by interaction between two or more primary pollutants, or by reaction with normal atmospheric constituents, with or without photoactivation. Exclusive of pollen, fog, and dust,

which are of natural origin, about 100 contaminants have been identified and fall into the following categories: solids, sulfur compounds, volatile organic chemicals, nitrogen compounds, oxygen compounds, halogen compounds, radioactive compounds, and odors (EPA-94/04).
(3) See pollutant for more related terms.

Air pollutant types: Typical air pollutants (grouped into four classes), their sources, and their potential health effects are shown below (AP-40):
(1) Organic gases which include:
  (A) Paraffins[2,3,4] from the source of processing and transfer of petroleum products; use of solvents; motor vehicles, etc.
  (B) Olefins[1,2,3,4] from the source of processing and transfer of gasoline; motor vehicles, etc.
  (C) Aromatics[1,2,3,4]: The source is the same as for paraffins
  (D) Others:
    (a) Oxygenated hydrocarbons[2,3,4] (aldehydes, ketones, alcohols, acids) from the use of solvents & motor vehicles
    (b) Halogenated hydrocarbons[2,3,4] (carbon tetrachloride, perchlororethylene, etc.) from the use of solvents
(2) Inorganic gases which include:
  (A) Oxides of nitrogen[1,2,3,4,5] (nitric oxide, nitrogen dioxide) from the combustion of fuels; motor vehicles, etc.
  (B) Oxides of sulfur[1,2,4,5] (sulfur dioxide, sulfur trioxide) from the combustion of fuels; chemical industry, etc.
  (C) Carbon monoxide[5] from the use of motor vehicles; petroleum; metals industry; etc.
(3) Aerosols (particulate matter) which include:
  (A) Solid particles[4,5] (carbon or soot) from the source of combustion of fuels; motor vehicles, etc.
  (B) Metal oxides & salts[4] from the source of catalyst dusts from refineries; motor vehicle; combustion of fuel oil, etc.
  (C) Silicates and mineral dusts[4] from the source of minerals industry; construction
  (D) Metallic fumes[4] from the source of metals industry
(4) Liquid particles which include:
  (A) Acid drops[4] from the source of combustion of fuels; plating; battery manufacture
  (B) Oily or tarry droplets[4] from the source of motor vehicles; asphalt paving and roofing; asphalt saturators; petroleum refining
  (C) Paints and surface coatings[6] from the source of various industries

Where:
[1]: Plant damage;
[2]: Eye irritation;
[3]: Oxidant formation
[4]: Visibility reduction;
[5]: Danger to other health; and
[6]: Other

Air pollution:
(1) The presence in the atmosphere of one or more air contaminants in sufficient quantities and of such characteristics and duration as to be injurious to human, plant, or animal life, to health, or to property, or to unreasonably interfere with the enjoyment of life or property (40 CFR 52.741-91).
(2) The presence of contaminant or pollutant substances in the air that do not disperse properly and interfere with human health or welfare, or produce other harmful environmental effects (EPA-94/04).
(3) See pollution for more related terms.

Air pollution control agency: Any of the following:
(1) A single State agency designated by the Governor of that State as the official State air pollution control agency for purposes of this Act;
(2) An agency established by two or more States and having substantial powers or duties pertaining to the prevention and control of air pollution;
(3) A city, county, or other local government health authority, or, in the case of any city, county, or other local government in which there is an agency other than the health authority charged with responsibility for enforcing ordinances or laws relating to the prevention and control of air pollution, such other agency; or
(4) An agency of two or more municipalities located in the same State or in different States and having substantial powers or duties pertaining to the prevention and control of air pollution (cf. interstate air pollution control agency) (CAA302-42USC7602-91).

Air pollution control device: Mechanism or equipment that cleans emissions generated by an incinerator by removing pollutants that would otherwise be released to the atmosphere (EPA-94/04).

Air pollution control device (for cement industry): The devices used to remove particulates from the gases of the cement kiln before release to the environment (ETI-92).

Air pollution control equipment: Any equipment or facility of a type intended to eliminate, prevent, reduce or control the emission of specified air contaminants to the atmosphere (40 CFR 52.741-91). Types of air pollution control equipment (APCE) include:
(1) Particulate matter control
  (A) Inertial separator such as cyclone separator
  (B) Wet collection device
    (a) Spray chamber
    (b) Cyclone type scrubber
    (c) Orifice type scrubber
    (d) Mechanical scrubber
    (e) Mechanical, centrifugal collector with water spray
    (f) High pressure spray
    (g) Venturi scrubber
    (h) Packed tower
    (i) Wet filter
  (C) Baghouse (fabric filter)
  (D) Electrostatic precipitator
(2) Organic vapor control
  (A) Afterburner
    (a) Direct-fire afterburner
    (b) Catalytic afterburner
  (B) Boiler used as an afterburner

(C) Adsorption equipment
(D) Vapor condenser
(E) Absorption equipment
  (a) Packed tower
  (b) Plate or Tray tower
  (c) Spray tower and spray chamber
  (d) Venturi absorber (AP-40)
(3) Acid gas control
  (A) Sulfur dioxide control
  (B) Nitrogen oxide control

Air quality control region: Federally designated area that is required to meet and maintain federal ambient air quality standards. May include nearby locations in the same state or nearby states that share common air pollution problems (EPA-94/04).

Air pollution emergency: This regulation is designed to prevent the excessive buildup of air pollutants during air pollution episodes, thereby preventing the occurrence of an emergency due to the effects of these pollutants on the health of persons (40 CFR 51-App/L-91).

Air pollution emission inventory: An information collection and processing system containing data on emissions of, and sources of, air pollution from both man-made and natural causes (NATO-78/10).

Air pollution episode criteria: *See* synonym, episode criteria.

Air pollution episode:
(1) A period of abnormally high concentration of air pollutants, often due to low winds and temperature inversion, that can cause illness and death (*see* episode, pollution) (EPA-94/04).
(2) *See also* synonym, episode.

Air pollution index:
(1) In air pollution modeling, a measure of the dispersion conditions in the atmosphere based on a function of meteorological and air pollution parameters.
(2) A scheme that transforms the values of individual air pollution related parameters (e.g., concentrations of several pollutants or visibility) into a single number, or set of numbers (NATO-78/10).

Air preheater: *See* synonym, air heater (EPA-83).

Air purifying respirator (APR): A device which supplies or purifies air or a facepiece which covers the nose and mouth and seals out the contaminants (Course 165.5).

Air quality assessment: Collection, handling, evaluation, analysis and presentation of data necessary to understand the air pollution problem of a certain area and its causes. These data normally refer to geography, topography, land use, sources and emissions, ambient air quality, meteorology, climatology, atmospheric chemistry, etc. (NATO-78/10).

Air quality control region (AQCR):
(1) An area designated by the Federal Government pursuant to Section 107 of the Clean Air Act in which communities, either in the same or different states, share a common air pollution problem (EPA-94/04).
(2) An interstate area designated by the U.S. Environmental Protection Agency as necessary or appropriate or the attainment and maintenance of national ambient air quality standards (DOE-91/04).

Air quality criteria: The levels of pollution and lengths of exposure above which adverse health and welfare effects may occur (EPA-94/04).

Air quality impact statement: A document, intended for decision making, in which the impact of proposed major activities on air quality in the near and the more distant environment is described (NATO-78/10).

Air quality index: *See* synonym, pollutant standard index.

Air quality management system: A system comprising coordinated measures necessary to reach and maintain an acceptable level of ambient air quality. The system consists of:
(1) The assessment of present air quality, emissions and related factors;
(2) The comparison of projected emissions and ambient air quality with standards, criteria and guidelines; and
(3) The development, implementation, and revision of abatement strategy plans, including economic aspects and interactions with other environmental media (NATO-78/10).

Air quality restricted operation of a spray tower: An operation utilizing formulations (e.g., those with high non-ionic content) which require a very high rate of wet scrubbing to maintain desirable quality of stack gases, and thus generate much greater quantities of waste water than can be recycled to process (40 CFR d417.151-91).

Air quality simulation model: A model usually in the form of a set of mathematical equations, which relates the air quality in an area to emissions (NATO-78/10).

Air quality standards:
(1) The legally prescribed level of constituents in the outside air that cannot be exceeded during a specific time in a specified area (DOE-91/04).
(2) The level of pollutants prescribed by regulations that may not be exceeded during a given time in a defined area (EPA-94/04).

Air quality: The levels of constituents in the outside air, often in comparison to regulatory standards (DOE-91/04).

Air scrubbing: A method of removing air impurities such as dust or fume by contact with sprayed water or an aqueous chemical solution (EPA-84/08).

Air setting binder: The sand binders which harden upon exposure to air. Sodium silicate, Portland cement, and oxychloride are the primary constituents of such binders. Air setting binders that are composed primarily of oxychloride contain up to 10 percent finely divided metallic copper. The copper is added to off-set the effects of such impurities as calcium oxide, calcium hydroxide, and calcium silicate, which may be introduced during the blending of oxychloride. These impurities otherwise would decrease mold strength and durability (EPA-85/10a).

Air setting refractory mortar: A finely ground material that, when it dries, develops a strong bond between refractory materials, even when heated to working furnace temperature. Also known as **cold setting refractory mortar** (*see* mortar for more related terms) (EPA-83).

Air standard volume: *See* standard volume under standard condition.

Air stripping operation: A desorption operation employed to transfer one or more volatile components from a liquid mixture into a gas (air) either with or without the application of heat to the liquid. Packed towers, spray towers, and bubble-cap, sieve, or valve-type plate towers are among the process configurations used for contacting the air and a liquid (40 CFR 264.1031-91).

Air stripping:
(1) A treatment system that removes volatile organic compounds (VOCs) from contaminated ground water or surface water by forcing an airstream through the water and causing the compounds to evaporate (EPA-94/04).
(2) *See also* definition under stripping.

Air suspension coater/dryer: A pharmaceutical coating operation which consists of vertical chambers in which tablets or particles are placed, and a coating is applied and then dried while the tablets or particles are kept in a fluidized state by the passage of air upward through the chambers (40 CFR 52.741-91).

Air sweetening: A method of using air or oxygen to oxidize lead mercaptides to manufacture disulfide.

Air to cloth ratio: The ratio of volumetric flow rate of contaminated air to the total filtering area of the filter bags in a baghouse system. It is equivalent to filtering velocity through the bags in ft/min. The usual notation is A/C or G/C (EPA-84/09).

Air to fuel ratio: *See* synonym, combustion air fuel ratio (40 CFR 60-App/A(method 28A)-91).

Air toxics:
(1) Any air pollutant for which a national ambient air quality standard (NAAQS) does not exist (i.e., excluding ozone, carbon monoxide, PM-10, sulfur dioxide, nitrogen oxide) that may reasonably be anticipated to cause cancer, developmental effects, reproductive dysfunctions, neurological disorders, heritable gene mutations, or other serious or irreversible chronic or acute health effects in humans (EPA-94/04).
(2) *See also* synonym, toxic air pollutant.

Airborne particulates: Total suspended particulate matter found in the atmosphere as solid particles or liquid droplets. Chemical composition of particulates varies widely, depending on location and time of year. Airborne particulates include: windblown dust, emissions from industrial processes, smoke from the burning of wood and coal, and motor vehicle or non-road engine exhausts, exhaust of motor vehicles (*see* particulate for more related terms) (EPA-94/04).

Airborne radioactive material: Any radioactive material dispersed in the air in the form of dusts, fumes, mists, vapors, or gases (*see* radioactive material for more related terms) (10 CFR 20.3-91).

Airborne release: Release of any chemical into the air (EPA-94/04).

Aircraft: Any airplane for which a U.S. standard airworthiness certificate or equivalent foreign airworthiness certificate is issued (40 CFR 87.1-91).

Aircraft engine: A propulsion engine which is installed in or which is manufactured for installation in an aircraft (*see* engine for more related terms) (40 CFR 87.1-91).

Aircraft gas turbine engine: A turboprop, turbofan, or turbojet aircraft engine (40 CFR 87.1-91).

Aircraft noise: The sound noise associated with aircraft operation (*see* noise for more related terms).

Airflow bypass (or thermal bypass): Any openings through the floors between stories of a house (or through the ceiling between the living area and the attic) which facilitates the upward movement of house air under the influence of the stack effect. By facilitating the upward movement, airflow bypasses in effect facilitate exfiltration at the upper levels, which in turn increases infiltration of outdoor air and soil gas (EPA-88/08).

Airless spray: A spray coating method in which the coating is atomized by forcing it through a small opening at high pressure. The coating liquid is not mixed with air before exiting from the nozzle (40 CFR 52.741-91).

Airport: The public-use airport open to the public without prior permission and without restrictions within the physical capabilities of available facilities (40 CFR 257.3.8-91).

Aitken nuclei: Tiny particles suspended in the atmosphere which serve as condensation nuclei for rain droplet.

Alachlor: A herbicide, marketed under the trade name Lasso, used mainly to control weeds in corn and soybean fields (EPA-94/04).

Alar: Trade name for daminozide, a pesticide that makes apples redder, firmer, and less likely to drop off trees before growers are ready to pick them. It is also used to a lesser extent on peanuts, tart cherries, concord grapes, and other fruits (EPA-94/04).

Alaskan north slope: The approximately 69,000 square mile area extending from the Brooks Range to the Arctic Ocean (40 CFR 60.591; 60.631-91).

Albuminuria: Presence in urine of albumin, a protein that is a normal constituent of blood (LBL-76/07-bio).

Alcohol (ethanol, ethyl alcohol or grain alcohol): A solvent which contains the hydroxyl (OH)⁻ radical and whose name ends with (suffix) -ol. For example, methanol (CH_3OH) and ethanol

(C_2H_5OH). Other alcohol related terms include:
- Primary alcohol
- Secondary alcohol
- Tertiary alcohol

Alcohol abuse: Any misuse of alcohol which demonstrably interferes with a person's health, interpersonal relations or working ability (40 CFR 7.25-91).

Alcove: A narrow channel through which molten glass is conveyed (EPA-83).

Aldehyde group: A group of various highly reactive compounds typified by actaldehyde and characterized by the group CHO (EPA-83/06a).

Aldicarb: An insecticide sold under the trade name Temik. It is made from ethyl isocyanate (EPA-94/04).

Aldrin/dieldrin ambient water criterion: Ambient water criterion for aldrin/dieldrin in navigable waters is 0.003 µg/L (40 CFR 129.100-91).

Aldrin/dieldrin formulator: A person who produces, prepares or processes a formulated product comprising a mixture of either aldrin or dieldrin and inert materials or other diluents, into a product intended for application in any use registered under the Federal Insecticide, Fungicide and Rodenticide Act, as amended (7USC135, et seq.) (40 CFR 129.100-91).

Aldrin/dieldrin manufacturer: A manufacturer, excluding any source which is exclusively an aldrin/dieldrin formulator, who produces, prepares or processes technical aldrin or dieldrin or who uses aldrin or dieldrin as a material in the production, preparation or processing of another synthetic organic substance (40 CFR 129.100-91).

Aldrin/dieldrin:
(1) **Aldrin ($C_{12}H_8C_{16}$):** The compound aldrin as identified by the chemical name, 1,2,3,4,10,10-hexachloro-1,4,4a,5,8,8a-hexahydro-1,4-endo-5, 8-exo-dimethanon-aphthalene.
(2) **Dieldrin:** The compound dieldrin as identified by the chemical name 1,2,3,4,10,10-hexachloro-6,7-epoxy-1,4,4a,5,6,7,8,8a-octahydro-1,4-endo-5,8-exo-dimethanonaphth-alene (40 CFR 129.4-91).

Alert: Events may occur, are in progress, or have occurred that could lead to a release of radioactive material but that the release is not expected to require a response by offsite response organizations to protect persons offsite (10 CFR 30.4-91, *see also* 10 CFR 40.4; 70.4-91).

Algae:
(1) Simple plants, many microscopic, containing chlorophyll. Freshwater algae are diverse in shape, color, size, and habitat. They are the basic link in the conversion of inorganic constituents in water into organic constituents (LBL-76/07-water).
(2) Simple rootless plants that grow in sunlit waters in proportion to the amount of available nutrients. They can affect water quality adversely by lowering the dissolved oxygen in the water. They are food for fish and small aquatic animals (EPA-94/04).

Algal bloom:
(1) Algal blooms are associated with nutrient-rich runoff from composting facilities or landfills (EPA-89/11).
(2) Sudden spurts of algal growth, which can affect water quality adversely and indicate potentially hazardous changes in local water chemistry (EPA-94/04).

Algicidal: Having the property of killing algae (40 CFR 797.1050-91).

Algicide:
(1) A chemical (such as copper sulfate) used to kill or inhibit the growth of algae (phytoplankton) in a water body (DOI-70/04).
(2) A specific chemical highly toxic to algae. Algicides are often applied to water to control nuisance algal blooms (LBL-76/07-water).
(3) Substance or chemical used specifically to kill or control algae (EPA-94/04).

Algistatic: Having the property of inhibiting algal growth (40 CFR 797.1050-91).

Algorithm: A system of mathematical steps which is to be followed in prescribed order for solving a specific type of problems (EPA-75/10).

Aliphatic nitriles: Alkanes, alkenes, or alkynes which contain nitrile(s) substituent, e.g., acrylonitrile (C_3H_3N) (EPA-88/12).

Alkali metal: Chemical elements of group Ia in the periodical table, e.g., sodium and potassium.

Alkali: A water-soluble hydroxide that ionizes strongly (EPA-87/10). Any compound with strong basic qualities.

Alkali flame ionization detector: *See* definition under GC/AFID.

Alkaline:
(1) The condition of a water solution having a pH concentration greater than 7.0 and having the properties of a base (EPA-82/11f).
(2) Of, relating to, or containing an alkali, such as a carbonate or hydroxide of an alkali metal (DOE-91/04).
(3) Possessing the ability to accept a proton or donate a pair of electrons; or having a pH greater than 7 (ETI-92).
(4) The condition of water or soil which contains a sufficient amount of alkali substance to raise the pH above 7.0 (EPA-94/04).

Alkaline cleaning bath: A bath consisting of an alkaline cleaning solution through which a workpiece is processed (40 CFR 468.02-91).

Alkaline cleaning rinse for forged parts: A rinse following an alkaline cleaning bath through which a forged part is processed. A rinse consisting of a series of rinse tanks is considered as a single rinse (40 CFR 468.02-91).

Alkaline cleaning rinse: A rinse following an alkaline cleaning bath through which a workpiece is processed. A rinse consisting of a series of rinse tanks is considered as a single rinse (40 CFR 468.02-91).

Alkaline cleaning: Using a solution (bath), usually detergent, to remove lard, oil, and other such compounds from a metal surface. Alkaline cleaning is usually followed by a water rinse. The rinse may consist of single or multiple stage rinsing. For the purposes of this part, an alkaline cleaning operation is defined as a bath followed by a rinse, regardless of the number of rinse stages. Each alkaline cleaning bath and rinse combination is entitled to a discharge allowance (40 CFR 471.02-91).

Alkaline earth metal: Metal elements in group IIA of the periodic table. The elements include barium, calcium, and strontium.

Alkaline mine drainage: The mine drainage which, before any treatment, has a pH equal to or greater than 6.0 and total iron concentration of less than 10 mg/L (*see* mine drainage for more related terms) (40 CFR 434.11-91).

Alkaline polyethylene glycol (APEG): This is the chemical class on which a new technology for waste detoxification is based. In this process, potassium hydroxide reacts with polyethylene glycol to form an alkoxide. The alkoxide in turn reacts with one or more of the chlorine atoms of the organic molecule of the contaminant to be destroyed, to produce an ether and alkali (potassium or sodium) metal chloride salt. This process may proceed to complete dechlorination, although replacement of a single chlorine is sufficient to make the reaction products water soluble. In some APEG reagent formulations, dimethylsulfoxide (DMSO) is added as a co-solvent to enhance reaction rate kinetics.

Alkalinity:
(1) The measurable ability of solutions or aqueously suspended solids to neutralize an acid (SW-108ts).
(2) The capacity of water to neutralize acids, a property imparted by the water's content of carbonates, bicarbonates, hydroxides, and occasionally borates, silicates, and phosphates. Natural waters are generally neutral or slightly alkaline. The alkalinity of water may range from a few milligrams per liter to several hundred. Domestic sewage is usually slightly more alkaline than the water from which it is derived. Alkalinity is expresses in milligrams per liter of equivalent calcium carbonate (EPA-82/10).
(3) The capacity of water to neutralize acids (EPA-94/04).

Alkaloids: Basic (alkaline) nitrogenous botanical products which produce a marked physiological action when administered to animals or humans (EPA-83/09).

Alkane:
(1) One of saturated aliphatic hydrocarbons having the empirical formula C_nH_{2n+2}.
(2) A chemical compound consisting only of carbon and hydrogen in which the carbon atoms are joined to each other by single bonds (EPA-87/07a).

Alkene: One of unsaturated aliphatic hydrocarbons containing one or more than one carbon-to-carbon double bonds.

Alkyd resin: A synthetic resin made from polyhydric alcohols and polybasic acids (EPA-79/12b). Alkyd resin is a paint ingredient (*see* resin for more related terms).

Alkyd resin binders: Cold set resins used in the formation of cores. This type of binders is a three component system using alkyd-isocyanate, cobalt naphthenate, and di-penyl methane di-isocyanate. Cobalt naphthenate is the drier, and di-phenyl methane di-isocyanate is the catalyst. Exposure of these binders to hot metal temperatures can cause the breakdown of these binder materials, and the resulting degradation products might include naphthalenes, phenols, and cyanides (EPA-85/10a).

Alkyl: A monovalent radical C_nH_{2n+1} which may be formed when an alkane losses a hydrogen, usually expressed by R.

Alkyl benzene sulfonate (ABS): A chemical surface-active agent used in synthetic detergents that causes foaming; its compounds do not readily decompose biologically through bacterial actions (DOI-70/04).

Alkyl halide: A compound containing one of alkyl compounds and a halogen, e.g., ethylbromide.

Alkylation: A process wherein an alkyl group (-R) is added to a molecule (EPA-87/10).

Alkylene: A radical formed from an unsaturated aliphatic hydrocarbon, e.g., ethylene radical (C_2H_3).

Alkyne: An organic compound containing a carbon-to-carbon triple bond.

All electric melter: A glass melting furnace in which all the heat required for melting is provided by electric current from electrodes submerged in the molten glass, although some fossil fuel may be charged to the furnace as raw material only (40 CFR 60.291-91).

All sliming:
(1) Crushing all the ore in a mill to so fine a state that only a small percentage will fail to pass through a 200-mesh screen.
(2) Term used for treatment of gold ore which is ground to a size sufficiently fine for agitation as a cyanide pulp, as opposed to division into coarse sands for static leaching and fine slimes for agitation (EPA-82/05).

Allegation: A statement, made without formal proof or regard for evidence, that a chemical substances or mixture has caused significant adverse reaction to health or the environment (40 CFR 717.3-91).

Allergen:
(1) A substance capable of causing an allergic reaction because of an individual's sensitivity to that substance (EPA-88/09b).
(2) *See also* definition under toxicant and effect.

Allergic contact dermatitis: *See* synonym, skin sensitization (40 CFR 798.4100-91).

Allethrin $C_{19}H_{26}O_3$: An insecticide, toxic symptoms similar to those of the pyrethrin.

Alley collection:
(1) The collection of solid waste from containers placed adjacent to or in an alley (40 CFR 243.101-91).
(2) The picking up of solid waste from containers placed adjacent to an alley (SW-108ts).
(3) *See* waste collection for more related terms.

Allotment: An amount representing a State's share of funds requested in the President's budget or appropriated by Congress for an environmental program, as EPA determines after considering any factors indicated by this regulation. The allotment is not an entitlement but rather the objective basis for determining the range for a State's planning target (40 CFR 35.105-91).

Allowable 1985 emissions rate: A federally enforceable emissions limitation for sulfur dioxide or oxides of nitrogen, applicable to the unit in 1985 or the limitation applicable in such other subsequent year as determined by the Administrator if such a limitation for 1985 does not exist (for complete definition, *see* CAA402-42USC7651a-91).

Allowable emission: The emissions rate of a stationary source calculated using the maximum rated capacity of the source (for complete definition, *see* 40 CFR 51.165; 51-App/S; 51.166; 52.21; 52.24; 52.741-91 and *see* emission for more related terms).

Allowable 1985 emissions rate: A federally enforceable emissions limitation for sulfur dioxide or oxides of nitrogen, applicable to the unit in 1985 or the limitation applicable in such other subsequent year as determined by the Administrator if such a limitation for 1985 does not exist. Where the emissions limitation for a unit is not expressed in pounds of emissions per million Btu, or the averaging period of that emissions limitation is not expressed on an annual basis, the Administrator shall calculate the annual equivalent of that emissions limitation in pounds per million Btu to establish the allowable 1985 emissions rate (CAA402-42USC7651a).

Allowable costs: Those project costs that are eligible, reasonable, necessary, and allocable to the project; permitted by the appropriate Federal cost principles, and approved by EPA in the assistance agreement (40 CFR 30.200-91).

Allowable emissions: The emissions rate of a stationary source calculated using the maximum rated capacity of the source (unless the source is subject to federally enforceable limits which restrict the operating rate, or hours of operation, or both) and the most stringent of the following:
(1) The applicable standards set forth in 40 CFR part 60 or 61;
(2) Any applicable State Implementation Plan emissions limitation including those with a future compliance date; or
(3) The emissions rate specified as a federally enforceable permit condition, including those with a future compliance date (40 CFR 51.165-xi-91).

Allowable pressure change: The allowable amount of decrease in pressure during the static pressure test, within the time period t, as specified in the appropriate regulation, in mm H_2O (40 CFR 60-App/A(method 27)-91).

Allowable vacuum change: The allowable amount of decrease in vacuum during the static vacuum test, within the time period t, as specified in the appropriate regulation, in mm H_2O (40 CFR 60-App/A(method 27)-91).

Allowance: An authorization, allocated to an affected unit by the Administrator under this title, to emit, during or after a specified calendar year, one ton of sulfur dioxide (CAA402, *see also* 40 CFR 35.2005-91).

Alloy steel: Steels with carbon content between 0.1% to 1.1% and containing elements such as nickel, chromium, molybdenum and vanadium. (The total of all such alloying elements in these type steels is usually less than 5%) (EPA-83/06a).

Alloy: A mixture having metallic properties, composed of two or more chemical elements at least one of which is an elemental metal (EPA-85/10a).

Alloy(ing) element: An element added to a metal to affect changes in properties, and which remains within the metal. The following is a list of materials known to be used as alloying materials or additives in foundry metals: aluminum; beryllium; bismuth; boron; cadmium; calcium; carbon; cerium; chloride; chromium; cobalt; columbium; copper; hydrogen; iron; lead; lithium; magnesium; manganese; molybdenum; nickel; nitrogen; oxygen; phosphorus; potassium; selenium; silicon; sulfur; tantalum; tin; titanium; tungsten; vanadium; zinc; and zirconium (EPA-10/85a).

Alluvial: Relating to mud and/or sand deposited by flowing water (EPA-94/04).

Alluvial deposit (or placer deposit): Earth, sand, gravel or other rock or mineral materials transported by and laid down by flowing water. Alluvial deposits generally take the form of:
(1) Surface deposits;
(2) River deposits;
(3) Deep leads; and
(4) Shore deposits (EPA-82/05).

Alluvial valley floors: The unconsolidated stream laid deposits holding streams where water availability is sufficient for sub-irrigation or flood irrigation agricultural activities but does not include upland areas which are generally overlain by a thin veneer of colluvial deposits composed chiefly of debris from sheet erosion, deposits by unconcentrated runoff or slope wash, together with talus, other mass movement accumulation and windblown deposits (SMCRA701-30USC1291-90).

Allyxycarb ($C_{16}H_{22}N_2O_2$): A yellow crystal used in insecticides for fruit orchards, vegetable, rice and citrus.

Alpha cellulose: The true cellulose content of a fibrous material (EPA-87/10).

Alpha decay: A transformation process of radioactive materials in which an alpha particle is emitted by a nuclide.

Alpha particle: A positively charged particle, consisting of two protons and two neutrons, that is emitted during radioactive decay from the nucleus of certain nuclides. It is the least penetrating of the three common types of radiation (alpha, beta, and gamma) (DOE-91/04).

Alpha particle: A positively-charged subatomic particle emitted during decay of certain radioactive elements. For example, an alpha particle is released when radon-222 decays to polonium-218. An alpha particle is indistinguishable from a helium atom nucleus and consists of two protons and two neutrons (*see* particle for more related terms) (EPA-88/08a).

Alpha radiation:
(1) A positively charged particle emitted by certain radioactive materials. Alpha particle is identical with the nucleus of helium atom. It is the least penetrating of the three common types of radiation (alpha, beta and gamma) and usually not dangerous to plants, animals or man (EPA-74/11).
(2) The least penetrating type of radiation. Alpha radiation can be stopped by a sheet of paper or outer dead layer of skin (EPA-88/08a).
(3) *See* radiation for more related terms.

Alpha ray: A radioactive stream of alpha particle.

Alpha TNT: A symmetrical isomer form, 2, 4, 6-TNT which is the desired isomer for use in explosives manufacturing end products (EPA-76/03).

Alphanaphthol test: A test for sucrose concentration in condensate and condenser water. The method is based on a color change which occurs in the reaction of alphanaphthol with sucrose (EPA-75/02d).

Altered discharge: Any discharge other than a current discharge or improved discharge, as defined in this regulation (40 CFR 125.58-91).

Alternate method: Any method of sampling and analyzing for an air pollutant that is not a reference or equivalent method but that has been demonstrated in specific cases-to EPA's satisfaction-to produce results adequate for compliance monitoring (EPA-94/04).

Alternative fuels: Substitutes for traditional liquid, oil-derived motor vehicle fuels like gasoline and diesel. Includes methanol, ethanol, compressed natural gas, and others (EPA-94/04).

Alternative remedial contract strategy contractors: Government contractors who provide project management and technical services to support remedial response activities at National Priorities List sites (EPA-94/04).

Alternative technology: Approach that aims to use resources efficiently or to substitute resources in order to do minimum damage to the environment. This approach permits a large degree of personal user control over the technology (EPA-94/04).

Alternating current (ac): An electrical current that alternates its directions at a constant frequency (cf. direct current).

Alternating double filtration (ADF): Two biological filters are placed in series and are used as the primary filter and the secondary filter alternatively. This arrangement can increase the organic loading and avoid pounding. When the primary filter shows signs of pounding, or after a set time (e.g., 1 week), the two filters are reversed, with the second filter operating before the first one (*see* filtration for more related terms) (Scott-81).

Alternative courses of action: All alternatives and thus is not limited to original] project objectives and agency jurisdiction (ESA3-16USC1531-90).

Alternative effluent limitations: All effluent limitations or standards of performance for the control of the thermal component of any discharge which are established under section 316(a) and this subpart (40 CFR 125.71-91).

Alternative method: Any method of sampling and analyzing for an air pollutant which is not a reference or equivalent method but which has been demonstrated to the Administrator's satisfaction to, in specific cases, produce results adequate for his determination of compliance (*see* method for more related terms) (40 CFR 60.2-91, *see also* 40 CFR 61.02-91).

Alternative method of compliance: A method of compliance in accordance with one or more of the following authorities: (A) a substitution plan submitted and approved in accordance with subsections 404 (b) and (c); (B) a Phase I extension plan approved by the Administrator under section 404(d), using qualifying Phase I technology as determined by the Administrator in accordance with that section; or (C) repowering with a qualifying clean coal technology under section 409 (CAA402-42USC7651a-91).

Alternative technology: The proven wastewater treatment processes and techniques which provide for the reclaiming and reuse of water, productively recycle wastewater constituents or otherwise eliminate the discharge of pollutants, or recover energy. Specifically, alternative technology includes land application of effluent and sludge; aquifer recharge; aquaculture; direct reuse (non-potable); horticulture; revegetation of disturbed land; containment ponds; sludge composting and drying prior to land application; self-sustaining incineration; and methane recovery (40 CFR 35.2005-91).

Alternative water supply: Includes, but is not limited to, drinking water and household water supplies (SF101-42USC9601-91, *see also* 40 CFR 300.5-91).

Alternative: A potentially applicable remedial treatment technology or treatment train. Alternatives are developed and screened during scoping of the RI/FS (remedial investigation/feasibility study) and throughout the RI/FS process. Alternatives are investigated by performing treatability studies and selected as remedies after a detailed analysis of each alternative is conducted (EPA-89/12a).

Altitude performance adjustments: The adjustments or modifications made to vehicle, engine, or emission control functions in order to improve emission control performance at altitudes other than those for which the vehicles were designed (40 CFR 86.1602-91).

Alum: A settling agent (coagulant). It is a chemical substance (usually potassium aluminum sulfate or ammonium aluminum sulfate), gelatinous when wet, used in water-treatment plants for settling out small particles of foreign matter (cf. aluminum sulfate) (EPA-87/10; DOI-70/04).

Alum sludge: A wastewater sludge formed when alum is used as coagulant (*see* sludge for more related terms).

Alumina diaspore fireclay brick: A brick consisting mainly of diaspore or nodule clay and having an alumina content of 50, 60 or 70% (plus or minus 2.5%) (*see* brick for more related terms) (SW-108ts).

Alumina fiber: A refractory made of alumina Al_2O_3.

Alumina:
(1) Any forms of aluminum oxide, Al_2O_3, occurring naturally as corundum, in a hydrated form in bauxite, and with various impurities as ruby, sapphire, and emery (EPA-74/03b).
(2) Common name for aluminum oxide (Al_2O_3); a necessary ingredient in the manufacture of cement (ETI-92).

Aluminizing: Forming an aluminum or aluminum alloy coating on a metal by hot dipping, hot spraying or diffusion (EPA-83/06a).

Aluminothermic process: The reduction of oxides in an exothermic reaction with finely divided aluminum (EPA-82/05).

Aluminum (Al): A metal used to increase the energy of a propellant and explosive (EPA-76/03). It is a metallic element with a.n. 13; a.w. 26.98; d. 2.70 g/cc; m.p. 660° C and b.p. 2450° C. The element belongs to group IIIA of the periodic table. Major aluminum compounds include aluminum sulfate ($Al_2(SO_4)_3 \bullet 18H_2O$): A colorless salt used in papermaking, water purification, also know as alum.

Aluminum basis material: The aluminum, aluminum alloys and aluminum coated steels which are processed in coil coating (40 CFR 465.02-91).

Aluminum bronze: Copper aluminum alloys, 4 to 11% aluminum. High tensile strength, cast or cold worked, resists corrosion (EPA-83).

Aluminum casting: The remelting of aluminum or an aluminum alloy to form a cast intermediate or final product by pouring or forcing the molten metal into a mold, except for ingots, pigs, or other cast shapes related to nonferrous (primary and secondary) metals manufacturing (40 CFR 421) and aluminum forming (40 CFR 467). Processing operations following the cooling of castings not covered under aluminum forming, except for grinding scrubber operations which are covered here, are covered under the electroplating and metal finishing point source categories (40 CFR 413 and 433) (40 CFR 464.02-91).

Aluminum equivalent: An amount of aluminum which can be produced from a Mg of anodes produced by an anode bake plant as determined by 40 CFR 60.195(g) (40 CFR 60.191-91).

Aluminum foil: An aluminum sheet with thickness not exceeding 0.005 inch (EPA-83/03).

Aluminum forming: A set of manufacturing operations in which aluminum and aluminum alloys are made into semifinished products by hot or cold working (40 CFR 467.02-91).

Aluminum scrap: *See* synonym, municipal aluminum scrap.

Aluminum turnings: Oily aluminum chips produced by machining operations (EPA-83).

Alunite: A basic potassium aluminum sulfate, closely resembles kaolinite and occurs in similar locations (EPA-82/05).

Amalgamation: The process by which mercury is alloyed with some other metal to produce amalgam. It was used extensively at one time for the extraction of gold and silver from pulverized ores, now is largely superseded by the cyanide process (EPA-82/05).

Ambient air (or open air):
(1) The portion of the atmosphere, external to buildings, to which the general public has access (40 CFR 50.1-91).
(2) Any unconfined portion of the atmosphere (EPA-84/09).
(3) The surrounding atmosphere, usually the outside air, as it exists around people, plants, and structures. (It is not the air in immediate proximity to emission sources) (DOE-4/91).
(4) Any unconfined portion of the atmosphere: open air, surrounding air (EPA-94/04).
(5) *See* air for more related terms.

Ambient air quality criteria: The quantitative relationship between a pollutant dose, concentration, deposition rate or any other quality-related factors, and the related direct and/or indirect effects on receptors, e.g., humans, animals, plants, or materials. The criteria serve as the scientific basis for formulating ambient air quality standards or objectives (NATO-78/10).

Ambient air quality: Refers only to concentrations of sulfur dioxide in the ambient air, unless otherwise specified (40 CFR 57.103-91).

Ambient air quality standards:
(1) Those standards designed to protect the public health and welfare codified in 40 CFR 50 and promulgated from time to time by the USEPA pursuant to authority contained in Section 108 of the Clean Air Act, 42USC7401 et seq., as amended from time to time (40 CFR 52.741-91).
(2) *See* Criteria Pollutants and National Ambient Air Quality Standards (EPA-94/04).

Ambient aquatic life advisory concentrations (AALACS): EPA's advisory concentration limit for acute or chronic toxicity to aquatic organisms as established under section 304(a)(1) of the Clean Water Act, as amended (40 CFR 300-App/A-91).

Ambient concentration: The appropriately time-averaged concentration of a substance at a location to which the general public has access (EPA-81/09).

Ambient standard: Ambient air standards.

Ambient temperature: Temperature of the surrounding air or other medium (EPA-94/04).

Ambient temperature range: The range of ambient temperature over which the instrument will meet stated performance specifications (LBL-76/07-bio).

Ambient toxicity:
(1) The toxicity manifested by a sample collected from an

aquatic receiving system (EPA-85/09).
(2) Measured by a toxicity test on a sample collected from a waterbody (EPA-91/03).

Ambient water criterion: That concentration of a toxic pollutant in a navigable water that, based upon available data, will not result in adverse impact on important aquatic life, or on consumers of such aquatic life, after exposure of that aquatic life for periods of time exceeding 96 hours and continuing at least through one reproductive cycle; and will not result in a significant risk of adverse health effects in a large human population based on available information such as mammalian laboratory toxicity data, epidemiological studies of human occupational exposures, or human exposure data, or any other relevant data (40 CFR 129.2-91).

Ambient water criterion for aldrin/dieldrin in navigable waters: 0.003 μg/L (40 CFR 129.100-3-91).

Ambient water criterion for benzidine in navigable waters: 0.1 μg/L (40 CFR 129.104-3-91).

Ambient water criterion for DDT in navigable waters: 0.001 μg/L (40 CFR 129.101-3-91).

Ambient water criterion for endrin in navigable waters: 0.004 μg/L (40 CFR 129.102-3-91).

Ambient water criterion for PCBs in navigable waters: 0.001 μg/L (40 CFR 129.105-4-91).

Ambient water criterion for toxaphene in navigable waters: 0.005 μg/L (40 CFR 129.103-3-91).

Ambient water quality criteria (AWQC): EPA's maximum acute or chronic toxicity concentrations for protection of aquatic life and its uses as established under section 304(a)(1) of the Clean Water Act, as amended (40 CFR 300-App/A-91).

Ambient water quality: The existing stream or impoundment water quality (EPA-80/08).

Amendment review: Review of any application requiring Agency approval to amend the registration of a currently registered product, or for which an application is pending Agency decision, not entailing a major change to the use pattern of an active ingredient (40 CFR 152.403-91).

Americium (Am): A radioactive metal with a.n. 95; a.w. 243; d. 11.7 g/cc; m.p. 994° C and b.p. 2607° C. The element belongs to group IIIB of the periodic table.

Ames test (or Bruce Ames test): A bacteria bioassay used to determine the mutagenicity and potential carcinogenicity caused by a chemical or physical agent.

Amido: Indicating the presence of radicals NH_2 and CO simultaneously in one compound.

Amine: A class of organic compounds of nitrogen that may be considered as derived from ammonia (NH_3) by replacing one or more of the hydrogen atoms by organic radicals, such as CH_3 or C_6H_5, as in methylamine and aniline. The former is a gas at ordinary temperature and pressure, but other amines are liquids or solids. All amines are basic in nature and usually combine readily with hydrochloric or other strong acids to form salts (EPA-83/06a). Other amine related terms include:
- Primary amine
- Secondary amine
- Tertiary amine

Amino acid: Organic compounds containing one or more than one basic amino groups and one or more than one acidic carboxyl groups. It is an important component of proteins.

Amino- (or amin-): A compound with a property of NH_2 being attached to a radical.

Ammine: A complex compound formed by coordination of ammonia molecules with metal ions.

Ammonia (NH_3): A colorless gaseous alkaline compound with pungent odor. It is lighter than air, and is very soluble in water.

Ammonia liquor: Ammonia liquor is primarily water condensed from the coke oven gas, an aqueous solution of ammonium salts of which there are two kinds: free and fixed. The free salts are those which are decomposed on boiling to liberate ammonia. The fixed salts are those which require boiling with an alkali such as lime to liberate the ammonia (*see* liquor for more related terms) (EPA-74/06a).

Ammonia nitrogen:
(1) All nitrogen in wastewaters existing as the ammonium ion (EPA-74/01a).
(2) A gas released by the microbiological decay of plant and animal proteins. When ammonia nitrogen is found in waters, it is indicative of incomplete treatment (EPA-87/10a).

Ammonia still waste: The treated effluents from an ammonia still (EPA-74/06a).

Ammonia still: A steam stripping operation where ammonia gas is removed from ammonia liquor. The fixed still is similar except lime is added to the liquor to force the combined ammonia out of its compounds so it can be steam stripped (EPA-74/06a).

Ammonia stripping: A modification of the aeration process for removing gases in water. Ammonium ions in wastewater exist in equilibrium with ammonia and hydrogen ions. As pH increases, the equilibrium shifts to the right and above pH 9 ammonia may be liberated as a gas by agitating the wastewater in the presence of air. This is usually done in a packed tower with an air blower (cf. stripping) (EPA-87/10a).

Ammonia water (NH_4OH): A water solution of ammonia.

Ammonia-N (or ammonia-nitrogen): The value obtained by manual distillation (at pH 9.5) followed by the Nesslerization method specified in 40 CFR 136.3 (40 CFR 420.02-91).

Ammonification: The process in which ammonium is liberated from organic compounds by microorganisms (EPA-87/10a).

Ammonium molybdate ((NH_4)$_2$$MoO_4$): A white, crystalline salt which can be used as an analytical reagent and pigment.

Ammonium sulfate dryer: A unit or vessel into which ammonium sulfate is charged for the purpose of reducing the moisture content of the product using a heated gas stream. The unit includes foundations, superstructure, material charger systems, exhaust systems, and integral control systems and instrumentation (40 CFR 60.421-91).

Ammonium sulfate feed material streams: The sulfuric acid feed stream to the reactor/crystallizer for synthetic and coke oven by-product ammonium sulfate manufacturing plants; and means the total or combined feed streams (the oximation ammonium sulfate stream and the rearrangement reaction ammonium sulfate stream) to the crystallizer stage, prior to any recycle streams (40 CFR 60.421-91).

Ammonium sulfate manufacturing plant: Any plant which produces ammonium sulfate (40 CFR 60.421-91).

Ammonium: (NH_4)$^+$ radical.

Ammonolysis: The formation of an amino compound using aqueous ammonia (EPA-87/10a).

Ammoxidation: The introduction of a cyanide group into an organic compound via interaction with ammonia and oxygen to form nitriles (EPA-87/10a).

Amoeba: A microscopic, one celled animal.

Amorphous: Without apparent crystalline form (cf. crystalline) (EPA-75/01a).

Amortization: The allocation of a cost over a specified period of time by the use of regular payments. The size of the payments is based on the principal, the interest charged, and the length of time over which the cost is allocated (EPA-85/10a).

Amount of pesticide or active ingredient: The weight or volume of the pesticide or active ingredient used in producing a pesticide expressed as weight for solid or semi-solid products and as weight or volume of liquid products (40 CFR 169.1-91).

Amount of pesticide product: The quantity, expressed in weight or volume of the product, and is to be reported in pounds for solid or semi-solid pesticides and active ingredients or gallons for liquid pesticides and active ingredients, or number of individual retail units for devices (40 CFR 167.3-91).

Ampere hours: Product of amperes of electricity being used and time of that use (EPA-74/03d).

Ampere: Unit of electricity, amount of which is the current that will deposit silver at the rate of 0.0011180 gram per second (EPA-74/03d).

Amperometric titration:
(1) A titration by measuring electric current or current change during titration.
(2) A way of measuring concentrations of certain substances in water using an electric current that flows during a chemical reaction (EPA-94/04).
(3) See titration for more related terms.

Ampholytic detergent: A detergent which becomes cationic in acidic solutions and anionic in basic solutions.

Amphoteric (or amphiprotic): Showing both acidic and basic characteristics.

Amplitude: The voltage excursion recorded during the process of recording the compound nerve action potential. It is an indirect measure the number of axons firing (40 CFR 798.6850-91).

Ampoule: A glass container designed to be sealed by fusion of the glass neck (EPA-83).

Ampule: A sealed glass or plastic bulb containing solutions for hypodermic injection (EPA-83/09).

Amyotrophic lateral sclerosis: A disease characterized by a hardening of the lateral columns of the spinal cord with muscular atrophy (LBL-76/07-bio).

Anadromous:
(1) Fish that spend their adult life in the sea but swim upriver to fresh-water spawning grounds to reproduce (EPA-89/12).
(2) Fish that migrate from salt to fresh water to spawn (DOE-91/04).

Anaerobe (or obligate anaerobe): Organisms or bacteria which can only live in the environment without the presence of oxygen (cf. aerobe or facultative anaerobe) (EPA-89/12a).

Anaerobic:
(1) Able to live and grow in the absence of free oxygen (SW-108ts).
(2) A biological process in which chemically combined oxygen is used for microorganism respiration needs. It is related to biological degradation of waste matter in the absence of dissolved oxygen (EPA-82/11).
(3) Anaerobic conditions in bodies of water are often responsible for major fish kills. An anaerobic process is one of wastewater treatment technologies (EPA-74/11).
(4) A life or process that occurs in, or is not destroyed by, the absence of oxygen (EPA-94/04).

Anaerobic bacteria: Bacteria that grow only in the absence of free elemental oxygen (LBL-76/07-water).

Anaerobic decomposition: Reduction of the net energy level and change in chemical composition of organic matter caused by microorganisms in an oxygen free environment (EPA-94/04).

Anaerobic digestion process (or anaerobic contact process):
(1) A sequential, biological treatment process in which hydrocarbons are converted, in the absence of free oxygen, from complex to simple molecules and ultimately to carbon dioxide and methane. The process involves **two stage digestion**: the primary digester serves mainly to reduce

volatile suspended solids (VSS), while the secondary digester is mainly for solids-liquid separation, sludge thickening, and storage (EPA-87/10a).
(2) The breakdown of organic components by microbial components in the absence of oxygen (cf. aerobic digestion) (EPA-83).

Anaerobic fermentation: *See* synonym, anaerobic sludge digestion.

Anaerobic lagoon: A waste stabilization pond which is devoid of dissolved oxygen and has few algae. The pond is used to treat high organic load, e.g., 100 g of BOD5 per m^3 of pond per day. Some 50 % of the BOD is removed in 24 hr. (*see* lagoon for more related terms) (Scott-81).

Anaerobic organism: An organism that thrives in the absence of oxygen (LBL-76/07-water).

Anaerobic respiration: A type of respiration among some bacteria in which an inorganic oxidant (NO_3, SO_4) other than oxygen is used (EPA-83).

Anaerobic sludge digestion (or anaerobic fermentation): Anaerobic decomposition of sewage sludge (Scott-81).

Anaerobic stabilization: Decomposition of waste by an anaerobic microbe population in the continuous absence of oxygen (LBL-76/07-water).

Analysis: The ascertainment of the identity and/or concentration, or both, of the constituents or components of a sample (EPA-83). Other analysis related terms include:
- Cause analysis
- Chemical analysis
- Dimensional analysis
- Directed analysis
- Gravimetric analysis
- Elemental analysis (*see* synonym, ultimate analysis)
- Molar analysis
- Proximate analysis
- Sensitivity analysis
- Survey analysis
- Ultimate analysis
- Volumetric analysis

Analysis matrix spike: A sample created by spiking target analytes into a prepared portion of a sample just prior to analysis. It only provides information on matrix effects encountered during analysis, i.e., suppression or enhancement of instrument signal levels (*see* spike for more related terms) (EPA-84/03).

Analysis sample: Final subsample prepared from the air dried laboratory sample but reduced by passing through a mill with a 0.5 mm (0.02 in.) size or smaller final screen (*see* analytical parameters--laboratory for more related terms) (EPA-83).

Analysis using exposure or medical records: Any compilation of data or any statistical study based at least in part on information collected from individual employee exposure or medical records of information collected from health insurance claims records, provided that either the analysis has been reported to the employer or not further work is currently being done by the person responsible for preparing the analysis (29 CFR 1910.20-91).

Analyte: A chemical substance whose presence and/or concentration in a sample is determined.

Analytical batch: The basic unit for analytical quality control. It is defined as the samples which are analyzed together with the same method sequence and the same lots of reagents and with the manipulations common to each sample within the same time period or in continuous sequential time periods. Samples in each batch is of similar composition.

Analytical chemistry: A science dealing with various techniques to produce information about a chemical system.

Analytical methods for stack gases in RCRA trial burns: Sampling and analytical methods for stack gases in RCRA trial burns: (EPA-89/06, p13 & 22)

Analytical parameters--fuels: Key parameters include:
(1) As-determined basis
(2) As-received basis
(3) Ash
(4) Ash fluid temperature
(5) Ash hemispherical temperature
(6) Ash initial deformation temperature
(7) Ash softening temperature
(8) Decimal percent
(9) Dry basis
(10) Dry ash free basis
(11) Higher heating value (HHV)
(12) Proximate analysis
(13) Ultimate analysis of fuels
(14) Volatile matter

Analytical parameters--laboratory: Key parameters include:
(1) Air dry loss
(2) Air drying
(3) Analysis sample
(4) Bias
(5) Gross sample
(6) Laboratory sample:
(7) Precision
(8) Representative sample
(9) Residual moisture
(10) Residual moisture
(11) Sample division
(12) Sample preparation
(13) Sample reduction
(14) Significant loss
(15) Size consistence
(16) Total moisture

Analytical quantification level: The minimum concentration at which quantification of a specified pollutant can be reliably measured (EPA-83/03a).

Analytical sensitivity: The airborne asbestos concentration represented by each fiber counted under the electron microscope. It is determined by the air volume collected and the proportion of the filter examined. This method requires that the analytical sensitivity be no greater than 0.005 structures/cm^3 (40 CFR 763-App/A-91).

Analytical study epidemiology: The study tests a specific hypothesis regarding the etiology of the disease in question (*see* epidemiology for more related terms) (Course 165.6).

Analyzer (or monitor): A system that senses the concentration of a substance and generates an output proportional to the concentration of the substance. Typical analyzers or monitors include the following groups, *see* each group for more information:
(1) Aerosol monitor
(2) Air analyzer
(3) Continuous emission monitor (CEM)--CEMs can be grouped into (i) continuous inorganic compound analyzers and (II) continuous organic compound analyzers (EPA-81/09; 84/03a)
(4) Metal analyzer

Analyzer calibration error: The difference between the gas concentration exhibited by the gas analyzer and the known concentration of the calibration gas when the calibration gas is introduced directly to the analyzer (40 CFR 60-App/A(method 6C & 7E)-91).

Ancillary equipment: Any device including, but not limited to, such devices as piping, fittings, flanges, valves, and pumps, that is used to distribute, meter, or control the flow of hazardous waste from its point of generation to a storage or treatment tank(s) between hazardous waste storage and treatment tanks to a point of disposal onsite, or to a point of shipment for disposal off-site (40 CFR 260.10-91, *see also* 40 CFR 280.12-91).

Anderson sampler: An aerosol sampling device consisting of a series of stacked stages and collection surfaces. It determines the particle size distribution of a gas sample containing particulates (*see* particle size measurement for more related terms) (EPA-84/09).

Anecdotal data: Data based on descriptions of individual cases rather than on controlled studies (EPA-92/12).

Anemometer:
(1) An instrument for measuring wind velocity (29 CFR 1910.66-91).
(2) An instrument for measuring the wind speed.
(3) The most common type is the cup anemometer and the propeller anemometer (NATO-78/10).
(4) A gas velocity measuring meter, including the following types:
 (A) Heated thermocouple anemometer
 (B) Hot film anemometer
 (C) Hot wire anemometer
 (D) Vane anemometer
(5) *See* flow velocity meter for more related terms.

Angle of projection: The angle that contains all of the radiation projected from the lamp assembly of the analyzer at a level of greater than 2.5 percent of the peak illuminance (40 CFR 60-App/B-91).

Angle of repose:
(1) The maximum acute angle that the inclined surface of a pile of loosely divided material can make with the horizontal (SW-108ts).
(2) The angle at which matter will lie or stack in a stationary configuration (EPA-81/09).

Angle of view: The angle that contains all of the radiation detected by the photodetector assembly of the analyzer at a level greater than 2.5 percent of the peak detector response (40 CFR 60-App/B-91).

Anglesite: A mineral occurring in crystalline form or as a compact mass (EPA-83/03a).

Angstrom: A unit of length equal to 1/6438.4696 of wavelength of red line of Cd. For practical purposes, it is considered equal to 10^{-8} cm (LBL-76/07-bio).

Anhydride:
(1) A product compound after a reactant reacts with water; e.g., $SO_3 + H_2O \longrightarrow H_2SO_4$.
(2) A product compound formed after removal of its water component (cf. hydrate).

Anhydrous product: The theoretical product that would result if all water were removed from the actual product (40 CFR 417.11-91).

Anhydrous: Containing no water (EPA-83/06a).

Aniline (C_6H_7N): A brown color liquid used in perfumes, varnishes and shoe blacks.

Animal: Appropriately sensitive animals which carry out respiration by means of a lung structure permitting gaseous exchange between air and the circulatory system (40 CFR 116.3-91, *see also* FIFRA2-7USA136-91).

Animal bedding: Materials, usually organic, that are placed on the floor of livestock quarters for animal comfort and to absorb excreta (SW-108ts).

Animal bioassay: *See* synonym, bioassy.

Animal feed: Any crop grown for consumption by animals, such as pasture crops, forage, and grain (40 CFR 257.3.5-91, *see also* 40 CFR 122.23-91).

Animal hair fiber: The fibers obtained from animals for the purposes of weaving, knitting, or felting into fabric. They include goat hair, camel hair, cashmere, fur, etc. (EPA-82/09).

Animal size: The gelatinous size from animal hides used in paper-making (EPA-83).

Animal studies:
(1) Investigations using animals as surrogates for humans with the expectation that the results are pertinent to humans (EPA-94/04).
(2) An investigation using animals as surrogates for humans, on the expectation that results in animals are pertinent to humans (Course 165.6).

Anion exchange process: A reversible exchange of negative ions between functional groups of the ion exchange medium and the solution in which the solid is immersed. Used as a wastewater treatment process for removal of anions, e.g., carbonate (EPA-82/10).

Anion: *See* definition under ion.

Anionic polymer: An organic compound characterized by a large molecular weight and a net negative charge, formed by the union of two or more polymeric compounds. Certain polymers act as coagulants or coagulant aids. Added to wastewater, they enhance settlement of small suspended particles. The large molecules attract the suspended matter to form a large floc (EPA-75/10).

Anionic surfactant: An ionic type of surface-active substance that has been widely used in cleaning products. The hydrophilic group of these surfactants carries a negative charge in washing solution (EPA-82/11f).

Anneal: To treat metal, alloy, plastics, or glass by a process of heating and slow cooling in order to remove internal stresses and to make the material less brittle (EPA-83/03).

Annealing point: The temperature corresponding to a rate of elongation of 0.0136 cm/min when measured by the method of test for annealing point and strain point of glass (ASTM C 336) (EPA-83).

Annealing range: The range of glass temperature in which stresses in glass articles can be relieved at a commercially desirable rate. For purposes of comparing glasses, the annealing range is assumed to correspond with the temperatures between the annealing point and the strain point (EPA-83).

Annealing with oil: The use of oil to quench a workpiece as it passes from an annealing furnace (40 CFR 468.02-91).

Annealing with water: The use of a water spray or bath, of which water is the major constituent, to quench a workpiece as it passes from an annealing furnace (40 CFR 468.02-91).

Annual: The corporate fiscal year (40 CFR 704.3-91).

Annual average: The maximum allowable discharge of BOD5 or TSS as calculated by multiplying the total mass (kkg or 1000 lb) of each raw commodity processed for the entire processing season or calendar year by the applicable annual average limitation (cf. fruit products production or vegetable products production) (40 CFR 407.61-91, *see also* 40 CFR 407.71; 407.81-91).

Annual capacity factor: The ratio between the actual heat input to a steam generating unit from an individual fuel or combination of fuels during a period of 12 consecutive calendar months and the potential heat input to the steam generating unit from all fuels had the steam generating unit been operated for 8,760 hours during that 12-month period at the maximum design heat input capacity. In the case of steam generating units that are rented or leased, the actual heat input shall be determined based on the combined heat input from all operations of the affected facility during a period of 12 consecutive calendar months (40 CFR 60.41c-91).

Annual coke production: The coke produced in the batteries connected to the coke by-product recovery plant over a 12-month period. The first 12-month period concludes on the first December 31 that comes at least 12 months after the effective date or after the date of initial startup if initial startup is after the effective date (40 CFR 61.131-91).

Annual document log: The detailed information maintained at the facility on the PCB waste handling at the facility (40 CFR 761.3-91).

Annual flood: The maximum daily flow during 12 consecutive months, that is, the highest daily flood peak for a year of record (*see* flood for more related terms) (DOI-70/04).

Annual maximum electric demand: The greatest of all demands of the load under consideration which occurred during a prescribed demand interval in a calendar year (*see* electric demand for more related terms) (EPA-83).

Annual precipitation and annual evaporation: The mean annual precipitation and mean annual lake evaporation, respectively, as established by the U.S. Department of Commerce, Environmental Science Services Administration, Environmental Data Services, or equivalent regional rainfall and evaporation data (40 CFR 440.132.-91).

Annual report: The written document submitted each year by each disposer and commercial storer of PCB waste to the appropriate EPA Regional Administrator. The annual report is a brief summary of the information included in the annual document log (40 CFR 761.3-91).

Annual research period: The time period from August 1 of a previous calendar year to July 31 of the given calendar year, e.g., the 1981 annual research period would be the time period from August 1, 10 to July 31, 11 (40 CFR 85.402-91).

Annual system maximum electric demand: The greatest demand on an electric system during a prescribed demand interval in a calendar year (*see* electric demand for more related terms) (EPA-83).

Annual work plan: The plan, developed by the Management Conference each year, which documents projects to be undertaken during the upcoming year. The Annual Work Plan is developed within budgetary targets provided by EPA (40 CFR 35.9010-91).

Anode (plate or positive electrode):
(1) The collector of electrons in an electron tube (EPA-83/03).
(2) The positive pole of a conducting terminal or electrode in an electrochemical process that takes electrons from the anions in solution and is connected to the positive terminal of the direct current source (cf. cathode) (EPA-74/03d; 83/06a).
(3) A positive or negative electrode used in a battery, generally consisting of active materials deposited on or in a current-collecting support (EPA-84/08).

Anode bake plant: A facility which produces carbon anodes for use in a primary aluminum reduction plant (40 CFR 60.191-91).

Anodic polarization: The change in anode potential due to current flow.

Anodic stripping voltammetry: An electrochemical method of analysis (LBL-76/07-bio).

Anodizing: Producing a protective oxide film (a hard and transparent oxide up to several mils in thickness) on aluminum or other light metals by passing a high voltage electric current through a bath in which the metal is suspended (EPA-83/06a).

Anolyte: In a two-solution electrolytic cell, the plating solution at the anode that (ions) is relatively exhausted and being replaced by the incoming cell feed. It is usually acidic (cf. catholyte) (EPA-75/02a).

Anorexia: Loss of appetite (LBL-76/07-bio).

Anoxia: Relative lack of oxygen; may be due to lack of blood carrying normal amounts of oxygen or to normal perfusion of blood carrying reduced amounts of oxygen (LBL-76/07-bio).

ANSI S3.19-1974: A revision of the ANSI Z24.22-1957 measurement procedure using one-third octave band stimuli presented under diffuse (reverberant) acoustic field conditions (40 CFR 211.203-91).

ANSI Z24.22-1957: A measurement procedure published by the American National Standards Institute (ANSI) for obtaining hearing protector attenuation values at nine of the one-third octave band center frequencies by using pure tone stimuli presented to ten different test subjects under anechoic conditions (40 CFR 211.203-91).

Antagonism:
(1) An interference or inhibition of the effect of one chemical by the action of another chemical (Course 165.6).
(2) A situation where the effect of one agent cancels all or part of the effect of another agent, e.g., when a strong (corrosive) acid is mixed with a strong (corrosive) base, the result is not a stronger corrosive agent but a nearly harmless mixture. The combined effect of two or more toxic substances acting together that is less adverse than their sum would be if each were acting separately or independently (cf. synergism) (DOI-70/04).
(3) The characteristic property of a mixture of toxicants that exhibits a less-than-additive total toxic effect (EPA-91/03).
(4) Interference or inhibition of the effect of one chemical by the action of another (EPA-94/04).

Antagonistic effect: The simultaneous action of separate agents mutually opposing each other (EPA-76/03).

Antarctic (ozone hole): Refers to the seasonal depletion of ozone in a large area over Antarctica (EPA-94/04).

Anterior pituitary (syn. anterior hypophysis): Part of the gland of internal secretion at the base of the brain, producing hormones that act on adrenal cortex, thyroid gland, gonads, and skeleton (LBL-76/07-bio).

Anthracene ($C_{14}H_{10}$): A white crystal used in manufacture of dyes.

Anthracite: The coal that is classified as anthracite according to the American Society of Testing and Materials' (ASTM) Standard Specification for Classification of Coals by Rank D388-77 (incorporated by reference-see 40 CFR 60.17) (40 CFR 60.41a-91).

Anthracite coal: Coal that is classified as anthracite according to the American Society of Testing and Materials' (ASTM) Standard Specification for Classification of Coals by Rank D388-77. It is a hard natural coal of high luster which contains little volatile matter (see coal for more related terms) (EPA-82/11f).

Anti-degradation claim (or clause): A provision in air quality and water quality laws that prohibits deterioration of air or water quality in areas where the pollution levels are presently below those allowed (EPA-74/11).

Anti-degradation clause: Part of federal air quality and water quality requirements prohibiting deterioration where pollution levels are above the legal limit (EPA-94/04).

Antibiosis: An association between organisms that is injurious to one of them.

Antibiotic: A substance, such as penicillin, produced by a microorganism which has the power, in dilute solution, to inhibit or destroy other organisms (EPA-83/09).

Antibodies: Proteins produced in the body by immune system cells in response to antigens, and capable of combining with antigens (EPA-89/12). It is a protein produced by certain white blood cells in response to the entry of foreign substance into the body in order to render it harmless.

Anticatalyst: A substance, e.g., lead, that can slow down the action of a catalyst.

Antichlor: A compound used to remove excess chlorine or bleaching solution in the processes such as paper or textile manufacturing.

Anticipated operational occurrences: Those conditions of normal operation that are expected to occur one or more times during the life of a nuclear reactor (DOE-91/04).

Anticyclone: A circulation of winds around a central region of high atmospheric pressure, clockwise in the Northern Hemisphere, counterclockwise in the South Hemisphere.

Antidegradation policies: Part of each State's water quality standards. These policies are designed to protect water quality and provide a method of assessing activities that may impact the integrity of the waterbody (EPA-91/03).

Antiforming agent: A substance, e.g., silicones and alcohols, that inhibit the bubbles formation in a liquid during its agitation by reducing its surface tension.

Antifreeze: A substance, e.g., ethylene glyco, that can lower the freezing point, of a liquid if the antifreeze substance is added to the liquid.

Antigen: A substance that causes production of antibodies when introduced into animal or human tissue (EPA-89/12).

Antimonial lead: An alloy composed of lead and up to 25 percent antimony (EPA-83/03a).

Antimony (Sb): The total antimony present in the process wastewater stream exiting the wastewater treatment system (40 CFR 415.651-91). Antimony with a.n. 51; a.w. 121.75; d. 6.62 g/cc; m.p. 630.5° C and b.p. 1380° C is a brittle, hexagonal mineral and belongs to group VA of the periodic table.

Antioxidant: An organic compound added to materials such as rubber to retard oxidation or deterioration (EPA-74/12a).

Antiozonant: A compound added to rubber to retard the ozone deterioration.

Antistatic agent: An agent applied to materials such as fabric to overcome deleterious effects of static electricity. Compounds commonly used include styrene-base resins, polyalkylene glycols, polyvinyl acetate, etc. (EPA-82/09).

Antitack agent: A substance used to prevent materials such as rubber stocks from sticking together during periods of storage (EPA-74/12a).

Anvil: In hot wedge seaming of FMLs (flexible membrane liners for containing solids, liquid, and vapor matter of landfill waste), an anvil is the wedge of metal above and below which the sheets (generally, high density polyethylene materials) to be joined must pass. The temperature controllers and thermocouples of most hot wedge devices are located within the anvil (EPA-89/09; 91/05).

API separator:
(1) A primary physical wastewater treatment process capable of removing free oils and settleable solids from water (EPA-87/10a).
(2) A separator used in American Petroleum Industry (API).
(3) *See* separator for more related terms.

Apparent density: The weight per unit volume of activated carbon (EPA-82/11f).

Apparent plateau: *See* definition under steady state (40 CFR 797.1560-91).

Appliance: Any device which contains and uses a class I or class II substance as a refrigerant and which is used for household or commercial purposes, including any air conditioner, refrigerator, chiller, or freezer (CAA601-42USC7671-91).

Applicable implementation plan or applicable plan: The portion (or portions) of the implementation plan, or most recent revision thereof, which has been approved under section 110 of the Clean Air Act, 42USC7410, or promulgated under section 110(c) of the CAA, 42USC7410(c) (40 CFR 51.138-91).

Applicable legal requirements: Any of the following:

(1) In the case of any major source, any emission limitation, emission standard, or compliance schedule under any EPA-approved State implementation plan (regardless of whether the source is subject to a Federal or State consent decree);
(2) In the case of any source, an emission limitation, emission standard, standard of performance, or other requirement (including, but not limited to, work practice standards) established under section 111 or 112 of the Act;
(3) In the case of a source that is subject to a federal or federally approved state judicial consent decree or EPA approved extension, order, or suspension, any interim emission control requirement or schedule of compliance under that consent decree, extension, order or suspensions;
(4) In the case of a nonferrous smelter which has received a primary nonferrous smelter order issued or approved by EPA under section 119 of the Act, any interim emission control requirement (including a requirement relating to the use of supplemental or intermittent controls) or schedule of compliance under that order (40 CFR 66.3-c-91).

Applicable marine water quality criteria: The criteria given for marine waters in the EPA publication "Quality Criteria for Water" as published in 1976 and amended by subsequent supplements or additions (40 CFR 227.31-91).

Applicable or appropriate requirements (ARARS): Any state or federal statute that pertains to protection of human life and the environment in addressing specific conditions or use of a particular cleanup technology at a Superfund site (EPA-94/04).

Applicable or relevant and appropriate requirements (ARAR):
(1) Federal or State requirements that are legally applicable to remedial actions at CERCLA sites or, if not legally applicable, the use of which is both relevant and appropriate under the circumstances. ARARs may be chemical-, location-, or action-specific (EPA-89/12a).
(2) Applicable requirements are those clean-up standards, standards of control, and other substantive environmental protection requirements, criteria, or limitations promulgated under federal or state law that specifically address a hazardous substance, pollutant, contaminant, remedial action, location, or other circumstance at a Comprehensive Environmental Response, Compensation, and Liability Act (CERCLA) site. Relevant and appropriate requirements are those clean-up standards which, while not applicable at a CERCLA site, address problems or situations sufficiently similar to those encountered at the CERCLA site that their use is well-suited to the particular site. ARARs can be action-specific, location-specific, or chemical-specific (EPA-91/12).

Applicable plan: The plan, or most recent revision thereof, which has been approved under 40 CFR 60.27(b) or promulgated under 40 CFR 60.27(d) (40 CFR 60.21-91).

Applicable requirements: Those cleanup standards, standards of control, and other substantive requirements, criteria, or limitations promulgated under federal environmental or state environmental or facility siting laws that specifically address a hazardous substance, pollutant, contaminant, remedial action, location, or other circumstance found at a CERCLA site. Only those state standards that are identified by a state in a timely manner and that are more stringent than federal requirements may be applicable (40 CFR 300.5-91).

Applicable standard:
(1) For the purposes of listing waters under 40 CFR 130.10-(d)(2), means a numeric criterion for a priority pollutant promulgated as part of a state water quality standard. Where a state numeric criterion for a priority pollutant is not promulgated as part of a state water quality standard.
(2) For the purposes of listing waters, means the state narrative water quality criterion to control a priority pollutant (e.g., no toxics in toxic amounts) interpreted on a chemical-by-chemical basis by applying a proposed state criterion, an explicit state policy or regulation, or an EPA national water quality criterion, supplement with other relevant information (40 CFR 130.10-91).

Applicable standards and limitations: All State, interstate, and federal standards and limitations to which a "discharge" a "sewage sludge use or disposal practice," or a related activity is subject under the CWA, including "effluent limitations," water quality standards, standards of performance, toxic effluent standards or prohibitions, "best management practices," pretreatment standards and "standards for sewage sludge use or disposal" under sections 301, 302, 303, 304, 306, 307, 308, 403, and 405 of CWA (40 CFR 122.2-91).

Applicable underground injection control program with respect to a State: The program (or most recent amendment thereof):
(1) Which has been adopted by the State and which has been approved under subsection (b), or
(2) Which has been prescribed by the Administrator under subsection (c) (SDWA1422-42USC300h.1-91).

Applicable water quality standards: The State water quality standards adopted by the State pursuant to section 303 of the Act or promulgated by EPA pursuant to that section (40 CFR 110.1-91).

Application factor: In risk analysis, the ratio between the safe and lethal concentration. The factor is almost always within the range 0.1 to 0.01 (EPA-89/12).

Applicator: A device used in a coating line to apply coating (40 CFR 52.741-91).

Applied coating solids: The volume of dried or cured coating solids which is deposited and remains on the surface of the automobile or light-duty truck body (40 CFR 60.391-91).

Approach angle: The smallest angle in a plan side view of an automobile, formed by the level surface on which the automobile is standing and a line tangent to the front tire static loaded radius arc and touching the underside of the automobile forward of the front tire (40 CFR 86.084.2-91).

Approach temperature: The difference between the exit temperature of water from a cooling tower and the wet bulb temperature of the air (*see* temperature for more related terms) (EPA-82/11f).

Appropriate act and regulations: The Clean Air Act and applicable regulations promulgated under it (40 CFR 124.41-91, *see also* 40 CFR 124.2; 144.3-91).

Appropriate sensitive benthic marine organisms: At least one species each representing filter feeding, deposit feeding, and burrowing species chosen from among the most sensitive species accepted by EPA as being reliable test organisms to determine the anticipated impact on the site; provided, however, that until sufficient species are adequately tested and documented, interim guidance on appropriate organisms available for use will be provided by the Administrator, Regional Administrator, or the District Engineer, as the case may be (40 CFR 227.27-91).

Appropriate sensitive marine organisms: At least one species each representative of phytoplankton or zooplankton, crustacean or mollusk, and fish species chosen from among the most sensitive species documented in the scientific literature or accepted by EPA as being reliable test organisms to determine the anticipated impact of the wastes on the ecosystem at the disposal site. Bioassays, except on phytoplankton or zooplankton, shall be run for a minimum of 96 hours under temperature, salinity, and dissolved oxygen conditions representing the extremes of environmental stress at the disposal site. Bioassays on phytoplankton or zooplankton may be run for shorter periods of time as appropriate for the organisms tested at the discretion of EPA, or EPA and the Corps of Engineers, as the case may be (under the Marine Protection, Research, and Sanctuaries Act) (40 CFR 227.27-d-91).

Appropriate treatment of the recycle water: In Subpart J, Section 440.104 includes, but is not limited to pH adjustment settling and pH adjustment, settling, and mixed media filtration (40 CFR 440.132.-91).

Appropriate when used with respect to child-resistant packaging: That the packaging is chemically compatible with the pesticide contained therein (40 CFR 157.21-91).

Approval: Formal permission. Other approval related terms include:
- Final approval
- Interim approval

Approval authority: The Director in an NPDES State with an approved State pretreatment program and the appropriate Regional Administrator in a non NPDES State or NPDES State without an approved State pretreatment program (40 CFR 403.3-91).

Approval of the facilities plan: The approval of the facilities plan for a proposed wastewater treatment works pursuant to 40 CFR 35, Subpart E or I (40 CFR 6.501-91).

Approved measure: Refers to one contained in an NSO which is in effect (40 CFR 57.103-91).

Approved POTW pretreatment program or program or POTW pretreatment program: A program administered by a POTW that meets the criteria established in this regulation (40 CFR 403.8 and 403.9) and which has been approved by a Regional Administrator or State Director in accordance with 40 CFR 403.11 of this regulation (40 CFR 403.3-91).

Approved program: A State implementation plan providing for issuance of PSD permits which has been approved by EPA under the Clean Air Act and 40 CFR 51. An "approved State" is one administering an "approved program." "State Director" as used in 40 CFR 124.4 means the person(s) responsible for issuing PSD permits under an approved program, or that person's delegated representative (40 CFR 124.41-91).

Approved Section 120 program: A State program to assess and collect Section 120 penalties that has been approved by the Administrator (40 CFR 66.3-91).

Approved State primacy program: Consists of those program elements listed in 40 CFR 142.11(a) that were submitted with the initial State application for primary enforcement authority and approved by the EPA Administrator and all State program revisions thereafter that were approved by the EPA Administrator (40 CFR 142.2-91).

Approximate original contour: That surface configuration achieved by backfilling and grading of the mined area so that the reclaimed area, including any terracing or access roads, closely resembles the general surface configuration of the land prior to mining and blends into and complements the drainage pattern of the surrounding terrain, with all highwalls and spoil piles eliminated; water impoundments may be permitted where the regulatory authority determines that they are in compliance with section 1265(b)(8) of this title (SMCRA701-30USC1291-90).

Approximate original contour: That surface configuration achieved by backfilling and grading of the mined area so that the reclaimed area, including any terracing or access roads, closely resembles the general surface configuration of the land prior to mining and blends into and complements the drainage pattern of the surrounding terrain, with all highwalls and spoil piles eliminated; water impoundments may be permitted where the regulatory authority determines that they are in compliance with section 1265(b)(8) of this title (SMCRA701-30USC1291-90).

Apron conveyer: One or more continuous chains that are supported and moved by a system of sprockets and rollers. They carry overlapping or interlocking plates that move bulk materials on their upper surface (*see* conveyer for more related terms) (SW-108ts).

Aqua regia (chloroazotic acid, chloronitrous acid, nitrohydrochloric acid or nitromuriatic acid): A mixture of highly corrosive liquid (1 part concentrated nitric acid and 3 parts concentrated hydrochloric acid) which can dissolve all metals including silver and gold.

Aquaculture project: A defined managed water area which uses discharges of pollutants into that designated area for the maintenance or production of harvestable freshwater, estuarine, or marine plants or animals (40 CFR 122.25-91).

Aquatic animals: The appropriately sensitive wholly aquatic animals which carry out respiration by means of a gill structure permitting gaseous exchange between the water tn the circulatory system (40 CFR 116.3-91).

Aquatic biota: Animal and plant life, or fauna and flora, of a stream or other water (LBL-76/07-water).

Aquatic community: An association of interacting populations of aquatic organisms in a given waterbody or habitat (EPA-91/03).

Aquatic ecosystem: *See* definition under aquatic environment (40 CFR 230.3-91).

Aquatic environment and aquatic ecosystem: Waters of the United States, including wetlands, that serve as habitat for interrelated and interacting communities and populations of plants and animals (40 CFR 230.3-91).

Aquatic flora: The plant life associated with the aquatic ecosystem including, but not limited to, algae and higher plants (40 CFR 116.3-91).

Aquatic life: All living forms in natural waters, including plants, fish, shellfish, and lower forms of animal life (EPA-75/11).

Aquatic plant: A plant that grows in water either floating on the surface, growing up from the bottom of the body of water or growing under the surface of the water (EPA-74/11).

Aqueous: Something made up of, similar to, or containing water; watery (EPA-94/04).

Aqueous solution: A solution in which water is the solvent (EPA-87/10a).

Aquifer:
(1) A geological formation, group of formations, or part of a formation that is capable of yielding a significant amount of water to a well or spring (40 CFR 144.3-91).
(2) A subsurface confining unit which has sufficient permeability to permit water to flow through it with relative ease and therefore it provides a usable quantity of water to a well or spring.
(3) A saturated geologic unit that can transmit significant quantities of water under ordinary hydraulic gradients; the water can be pumped to the surface through a well, or it can emerge naturally as a spring or outcrop (DOE-91/04). Also *see* aquitard (EPA-87/03).
(4) An underground geological formation, or group of formations, containing usable amounts of groundwater that can supply wells and springs (EPA-94/04).

Other aquifer related terms include:
- Artesian aquifer (*see* synonym, confined aquifer)
- Confined aquifer
- Exempted aquifer
- Sole source aquifer
- Principal source aquifer (*see* synonym, sole source aquifer)
- Unconfined aquifer
- Water table aquifer (*see* synonym, unconfined aquifer)
- Uppermost aquifer

Aquifer storativity: The quantity of water an aquifer can release from or take into storage per unit surface area of the aquifer per unit change in head (EPA-87/03).

Aquifer test: A test involving the withdrawal of measured quantities of water from, or addition of water to, a well (or wells) and the measurement of resulting changes in head in the aquifer both during and after the period of discharge of addition (Course 165.7).

Aquifer transmissivity: *See* definition under transmissivity.

Aquitard:
(1) A subsurface confining unit which is characterized by low permeability that does not readily permit water to pass through it despite the fact that it stores large quantities of water (cf. aquifer) (EPA-87/03).
(2) A less-permeable geologic unit in a stratigraphic sequence. The unit is not permeable enough to transmit significant quantities of water. Aquitards separate aquifers (DOE-91/04).

Aquo ion: An ion containing one or more than one water molecules.

Arbitration: A process for the resolution of disputes. Decisions are made by an impartial arbitrator selected by the parties. These decisions are usually legally binding (cf. mediation) (EPA-94/04).

Arc chute: An asbestos-containing provided that acts as a chute or guidance device and is intended to guide electric arcs in applications such as motor starter units in electric generating plants (40 CFR 763.163-91).

Arc furnace: A furnace heated by the arc produced between electrodes (*see* furnace for more related terms) (EPA-77/07).

Archaeological sites (resources): Areas or objects modified or made by humans, either prehistorically or historically, and the data associated with these areas and objects (DOE-91/04).

Architectural coatings: Coverings such as paint and roof tar that are used on exteriors of buildings (EPA-94/04).

Architectural or engineering (A/E) services: The consultation, investigations, reports, or services for design-type projects within the scope of the practice of architecture or professional engineering as defined by the laws of the State or territory in which the recipient is located (40 CFR 33.005-91, *see also* 40 CFR 35.2005; 35.6015-91).

Area coated: The area of basis material covered by each coating of enamel (40 CFR 466.02-91).

Area contingency plan: An Area Contingency Plan prepared under subsection (j) (CWA311-33USC1321-91).

Area: The vertical projection of the pile upon the earth's surface (40 CFR 61.251-91).

Area method: A method in which the wastes are spread and compacted on the surface of the ground and cover materials are spread and compacted over them (*see* sanitary landfill for more related terms) (EPA-81/09; SW-108ts).

Area of concern: A geographic area located within the great lakes, in which beneficial uses are impaired and which has been officially designated as such under Annex 2 of the Great lakes Water Quality Agreement (CWA118-33USC1268-91).

Area of critical environmental concern: The areas within the public lands where special management attention is required (when such areas are developed or used or where no development is required) to protect and prevent irreparable damage to important historic, cultural, or scenic values, fish and wildlife resources or other natural systems or processes, or to protect life and safety from natural hazards (FLPMA103-43USC1702-90).

Area of review:
(1) In underground injection, the area surrounding an injection well that is reviewed during the permitting process to determine whether the injection operation will introduce flows between aquifers (40 CFR 147.3001-91).
(2) In the UIC program, the area surrounding an injection well that is reviewed during the permitting process to determine if flow between aquifers will be induced by the injection operation (EPA-94/04).

Area processed: The area actually exposed to process solutions. Usually this includes both sides of the metal strip (40 CFR 465.02-91).

Area source: Any small source of non-natural air pollution that is released over a relatively small area but which cannot be classified as a point source. Such sources may include vehicles and other small engines, small businesses and household activities (EPA-94/04, *see also* CAA112; 40 CFR 51.100-91).

Area under reclamation: A previously surface mined area where regrading has been completed and revegetation has commenced (EPA-82/10).

Areawide agency: An areawide management agency designated under section 208(c)(1) of the Act (40 CFR 21.2-91, *see also* 40 CFR 130.2-91).

Argon (Ar): A noble gaseous element with a.n. 18; a.w. 39.998; d. 1.4 g/cc; m.p. -189.4° C and b.p. -185.8° C. The element belongs to group VIIIA of the periodic table.

Argon oxygen decarburization vessel (AOD vessel): Any closed-bottom, refractory-lined converter vessel with submerged tuyeres through which gaseous mixtures containing argon and oxygen or nitrogen may be blown into molten steel for further refining (40 CFR 60.271a-91).

Arithmetic mean: In a sample of N units, the sum of the observed values in the sample divided by the number of the units in the sample. It can be expressed as: $A = [\text{sum } X_i]/N, I = 1.....N$, where: A = arithmetic mean; N = number of sample; X_i = discrete value of I sample; sum = summation (*see* mean for more related terms) (NATO-78/10).

Arnel fiber: A trademark of Celanese for cellulose triacetate fibers (*see* fiber for more related terms) (EPA-74/06b).

Aromatic:
(1) A type of hydrocarbon, such as benzene or toluene, added to gasoline in order to increase octane. Some aromatics are toxic (EPA-94/04).
(2) Having at least one benzene ring in a compound.

Aromatic amine: An aromatic compound containing an amine(s) substituent, e.g., aniline ($C_6H_5NH_2$) (EPA-88/12).

Aromatic compound: An organic compound similar in molecular structure to benzene (ETI-92).

Aromatic hydrocarbon: An organic compound which contains at least one 6-carbon, benzene ring structure (EPA-75/11).

Arrest: The restraint of an arrestee's liberty or the equivalent through the service of judicial process compelling such a person to respond to a criminal accusation (40 CFR 303.11-91).

Arrhenius equation: An equation of form, $k = A\exp(-E/RT)$, where k = rate constant of a reaction, A = constant for a given reaction, E = activation energy. The equation can be expressed as:
$$\ln(k) = \ln(A) - E/RT$$
A graph of $\ln(k)$ against $(1/T)$ is a straight line with a gradient $-E/R$ and an intercept on the $\ln(k)$ axis of $\ln A$.

Arrhenius parameters: Parameters which relate rate chemical reaction coefficients to changes in temperature and pressure (EPA-88/12).

Arsenic (As): The total arsenic present in the process wastewater stream exiting the wastewater treatment system (40 CFR 415.671-91). Arsenic is a metalloid element with a.n. 33; a.w. 74.92; d. 2.7 g/cc; m.p. 660° C and b.p. 2450° C. The element belongs to group VA of the periodic table. Major arsenic compounds include:
- Arsenic acid ($H_3AsO_4 \cdot 1/2H_2O$): A toxic crystal used to manufacture arsenates.
- Arsenic disulfide (As_2S_2): A crystal used in pigments, fireworks and tanning.
- Arsenic pentasulfide (As_2S_5): A yellow crystal used in pigments.
- Arsenic pentoxide (or arsenic oxide) (As_2O_5): A white powder used as a fungicide.
- Arsenic trichloride ($AsCl_3$): A colorless liquid used in the ceramic industry.
- Arsenic trioxide (As_2O_3): A toxic powder used in rodent and insect controls.
- Arsenic trisulfide (As_2S_3): A yellow powder used in glass, oil cloth, etc.
- Arsenate ((AsO_4)$^{-3}$): A negative ion derived from arsenic acid.
- Arsenide: A negative, trivalent arsenic such as H_3As.
- Arsenite ((AsO_3)$^{-3}$): A negative ion derived from As_4O_6.

Other arsenic related terms include:
- Commercial arsenic
- Inorganic arsenic
- Uncontrolled total arsenic emission

Arsenic containing compound: A chemical substance containing the arsenic atom, e.g., arsenic trioxide (As_2O_3) (EPA-88/12).

Arsenic containing glass type: Any glass that is distinguished from other glass solely by the weight percent of arsenic added as a raw material and by the weight percent of arsenic in the glass produced. Any two or more glasses that have the same weight percent of arsenic in the raw materials as well as in the glass produced shall be considered to belong to one arsenic containing glass type, without regard to the recipe used or any other characteristics of the glass or the method of production (40 CFR 61.161-91).

Arsenic kitchen: A baffled brick chamber where inorganic arsenic vapors are cooled, condensed, and removed in a solid form (40 CFR 61.181-91).

Arsenicals: Pesticides containing arsenic (EPA-94/04).

Artesian (aquifer or well): Water held under pressure in porous rock or soil confined by impermeable geologic formations (EPA-94/04).

Artesian spring (or artesian well): A spring (well) in which confined groundwater under pressure has a natural outlet (DOI-70/04).

Artesian well: *See* synonym, artesian spring.

Article: A manufactured item:
(1) Which is formed to a specific shape or design during manufacture;
(2) Which has end use function(s) dependent in whole or in part upon its shape or design during end use; and
(3) Which does not release a toxic chemical under normal conditions or processing or use of that item at the facility or establishments (40 CFR 372.3-91).

Articulation (of a water body): The area ratio of inlets and bays to the total area of a water body (DOI-70/04).

Artifact: An object produced or shaped by human workmanship of archaeological or historical interest (DOE-91/04).

As determined basis: Data representing the numerical values obtained for a particular moisture and/or ash content, in the sample at the time of measurement (*see* analytical parameters--fuels for more related terms) (EPA-83).

As low as reasonably achievable (ALARA): A concept applied to the quantity of radioactivity released in routine operation of a nuclear system or facility, including "anticipated operational occurrences." It takes into account the state of technology, economics of improvements in relation to benefits to public health and safety, and other societal and economic considerations in relation to the use of nuclear energy in the public interest (DOE-91/04).

As received basis: Analytical data calculated to the moisture condition of the sample as it arrived at the laboratory and before any processing or conditioning (*see* analytical parameters--fuels for more related terms) (EPA-83).

Asbestiform: A specific type of mineral fibrosity in which the fibers and fibrils possess high tensile strength and flexibility (40

CFR 763-App/A-91).

Asbestos: A mineral fiber that can pollute air or water and cause cancer or asbestosis when inhaled. EPA has banned or severely restricted its use in manufacturing and construction (EPA-94/04). It is the asbestiform varieties of (TSCA202-15USC2642-91).
- Actinolite $[Ca_2(Mg,Fe)_5(Si_8O_{22})(OH)_2]$;
- Amosite (cummingtonite-grunerite) $[Fe_5Mg(Si_8O_{22})(OH)_2]$;
- Anthophyllite $[(Mg,Fe)_7(Si_8O_{22})(OH)_2]$;
- Chrysotile (serpentinite) $[Mg_6(Si_4O_{10})(OH)_8]$;
- Crocidolite (riebeckite) $[Na_2(Fe_3)^{2+}(Fe_2)^{3+}(Si_8O_{22})(OH)_2]$; or
- Tremolite $[Ca_2Mg_5(Si_8O_{22})(OH)_2]$.

Other asbestos related terms include:
- Bulk asbestos
- Commercial asbestos
- Miner of asbestos
- Primary processor of asbestos
- Secondary processor of asbestos
- Use of asbestos
- Regulated asbestos containing material
- Vinyl asbestos floor tile

Asbestos abatement: Procedures to control fiber release from asbestos-containing materials in a building or to remove them entirely, including removal, encapsulation, repair, enclosure, encasement, and operations and maintenance programs (EPA-94/04).

Asbestos abatement project: Any activity involving the removal, enclosure, or encapsulation of friable asbestos material (40 CFR 763.121-91).

Asbestos cement (A/C) corrugated sheet: An asbestos-containing product made of cement and in the form of a corrugated sheet used as a non-flat-surfaced reinforcing or insulating material. Major applications of this product include: building siding or roofing; linings for waterways; and components in cooling towers (40 CFR 763.163-91).

Asbestos cement (A/C) flat sheet: An asbestos-containing product made of cement and in the form of a flat sheet used primarily as a flat-surfaced reinforcing or insulating material. Major applications of this product include: wall linings; partitions; soffit material; electrical barrier boards; bus bar run separators; reactance coil partitions; laboratory work surfaces; and components of vaults, ovens, safes, and broilers (40 CFR 763.163-91).

Asbestos cement (A/C) pipe: An asbestos-containing product made of cement and intended for use as pipe or fittings for joining pipe. Major applications of this product include: pipe used for transmitting water or sewage; conduit pipe for protection of utility or telephone cable; and pipes used for air ducts (40 CFR 763.163-91).

Asbestos cement (A/C) shingle: An asbestos- containing product made of cement and intended for use as a siding, roofing, or construction shingle serving the purpose of covering and insulating the surface of building walls and roofs (40 CFR 763.163-91).

Asbestos clothing: An asbestos-containing product designed to be worn by persons (40 CFR 763.163-91).

Asbestos containing building material (ACBM): The surfacing ACM, thermal system insulation ACM, or miscellaneous ACM that is found in or on interior structural members or other parts of a school building (40 CFR 763.83-91).

Asbestos containing material: Any material which contains more than 1 percent asbestos by weight (TSCA202).

Asbestos containing product: Any product to which asbestos is deliberately added in any concentration or which contains more than 1.0 percent asbestos by weight or area (40 CFR 763.163-91).

Asbestos containing waste material:
(1) The mill tailings or any waste that contains commercial asbestos and is generated by a source subject to the provisions of this subpart. This term includes filters from control devices, friable asbestos waste material, and bags or other similar packaging contaminated with commercial asbestos. As applied to demolition and renovation operations, this term also includes regulated asbestos-containing material waste and materials contaminated with asbestos including disposable equipment and clothing (40 CFR 61.141-91).
(2) Mill tailings or any waste that contains commercial asbestos and is generated by a source covered by the Clean Air Act Asbestos NESHAPS (EPA-94/04).

Asbestos debris: The pieces of ACBM that can be identified by color, texture, or composition, or means dust, if the dust is determined by an accredited inspector to be ACM (40 CFR 763.83-91).

Asbestos diaphragm: An asbestos-containing product that is made of paper and intended for use as a filter in the production of chlorine and other chemicals, and which acts as a mechanical barrier between the cathodic and anodic chambers of an electrolytic cell (40 CFR 763.163-91).

Asbestos Hazard Emergency Response Act (AHERA) of 1986: 15USC2641 et al. The Toxic Substances Control Act was amended by adding at the end of the AHERA in 1986. Its major contents include:
(1) Authorized EPA to promulgate regulations requiring inspection for asbestos containing material in the Nation's schools, development of asbestos management plans for such schools, response actions with respect to friable asbestos containing material in such schools, and for other purposes.
(2) The standards of ambient interior concentration of asbestos shall not exceed the ambient exterior concentration which is deemed to be (Sec. 204):
  (A) Less than 0.003 fibers per cubic centimeter, if a scanning electron microscope is used; and
  (B) Less than 0.005 fibers per cubic centimeter, if a transmission electron microscope is used.

Asbestos mill: Any facility engaged in converting, or in any intermediate step in converting, asbestos ore into commercial asbestos. Outside storage of asbestos material is not considered a part of the asbestos mill (40 CFR 61.141-91).

Asbestos minerals: Minerals, which have a fibrous structure, are heat resistant, chemically inert and possessing high electrical insulating qualities. The two main groups are serpentine and amphiboles. Chrysotile is the principal commercial variety. Other commercial varieties are amosite, crocidolite, actinolite, anthophyllite, and tremolite (EPA-82/05).

Asbestos mixture: A mixture which contains bulk asbestos or another asbestos mixture as an intentional component. An asbestos mixture may be either amorphous or a sheet, cloth fabric, or another structure. This term does not include mixtures which contain asbestos as a contaminant or impurity (40 CFR 763.63-91).

Asbestos program manager: A building owner or designated representative who supervises all aspects of the facility asbestos management and control program (EPA-94/04).

Asbestos tailings: Any solid waste that contains asbestos and is a product of asbestos mining or milling operations (40 CFR 61.141-91).

Asbestos waste from control devices: Any waste material that contains asbestos and is collected by a pollution control device (40 CFR 61.141-91).

Asbestosis: A disease associated with inhalation of asbestos fibers. The disease makes breathing progressively more difficult and can be fatal (EPA-94/04).

Ascorbic acid ($C_6H_8O_6$): A crystal used as a vitamin C.

Ash:
(1) The solid residue including both non-combustible inorganic (e.g., metals) and unburned organic (e.g., soot) residue that remains after a material is incinerated (EPA-81/09; EPA-89/11).
(2) Inorganic residue remaining after ignition of combustible substances. The analyses of ash for commonly determined major elements by prescribed methods for the oxides of silicon, aluminum, iron, titanium, phosphorus, calcium, sodium and potassium. Other elements such as the heavy metals may be included in these analyses (*see* analytical parameters--fuels for more related terms) (EPA-83).
(3) The mineral content of a product remaining after complete combustion (EPA-94/04).

Other ash related terms include:
- Bottom ash
- Combined ash
- Economizer ash
- Extraneous ash
- Fly ash
- Fly ash reinjection
- Grate sifting (*see* synonym, sifting)
- Inherent ash
- Sifting

Ash combustible:
(1) The fraction of combustible organic material remaining in the bottom ash as measured by the loss on combustion technique (EPA-89/03b).
(2) If an incinerator is operating properly, little material will remain in the ash. The extent of organics combustion is measured by the quantity of combustible materials remaining in the ash. Increases in ash combustible indicate that bed temperatures are too low, that combustion air is not being distributed properly in the bed, or that waste retention time is too short (*see* combustion indicator for more related terms) (EPA-89/03b).

Ash fluid temperature: The temperature at which the fused mass has spread out in a nearly flat layer with a maximum height of 1/16 inch (1.6 mm) (*see* analytical parameters--fuels for more related terms) (EPA-83).

Ash free basis: The method whereby the weight of ash weight in a fuel sample is subtracted from its total weight and the adjusted weight is used to calculate the percent of certain constituents present, e.g., the percent of fixed carbon on an ash-free basis is computed as follows: % ash-free fixed carbon = (fixed carbon)/(fuel sample-ash) (SW-108ts).

Ash hemispherical temperature: The temperature at which the cone has fused down to a hemispherical lump at which condition the height is one half the width of the base (*see* analytical parameters--fuels for more related terms) (EPA-83).

Ash initial deformation temperature: The temperature at which the first rounding of the apex of the triangular pyramid (cone) occurs. (the cone is prepared from the ash of a sample) (*see* analytical parameters--fuels for more related terms) (EPA-83).

Ash pit: A pit or hopper located below a furnace where residue is accumulated and from which it is removed (SW-108ts).

Ash removal door: A door through which ash is removed from the primary combustion chamber (EPA-89/03b).

Ash sluice: A trench or channel in which water transports ash from an ash pit to a disposal or collection point (SW-108ts).

Ash softening temperature: The temperature at which the cone of ash has fused down to a spherical lump in which the height is equal to the width at the base (*see* analytical parameters--fuels for more related terms) (EPA-83).

Aspect ration: A ratio of the length to the width of a particle. Minimum aspect ratio as defined by this method is equal to or greater than 5:1 (40 CFR 763-App/A-91).

Asphalt (pitch or tar): The dark-brown to black cementitious material (solid, semisolid, or liquid in consistency) of which the main constituents are bitumens which occur naturally or as a residue of petroleum refining (40 CFR 52.741-91).

Asphalt processing: The storage and blowing of asphalt (40 CFR 60.471-91).

Asphalt processing plant: A plant which blows asphalt for use in the manufacture of asphalt products (40 CFR 60.471-91).

Asphalt roofing plant: A plant which produces asphalt roofing products (shingles, roll roofing, siding, or saturated felt) (40 CFR 60.471-91).

Asphalt storage tank: Any tank used to store asphalt at asphalt roofing plants, petroleum refineries, and asphalt processing plants. Storage tanks containing cutback asphalts (asphalts diluted with solvents to reduce viscosity for low temperature applications) and emulsified asphalts (asphalts dispersed in water with an emulsifying agent) are not subject to this regulation (40 CFR 60.471-91).

Asphyxiant: *See* definition under toxicant and effect.

Aspirator: An apparatus, such as a squeeze bulb, fan, pump, or venturi, that produces a movement of a fluid by suction (EPA-83/06).

Asplund method: An (wood) attrition mill which combines the steaming and defibering in one unit in a continuous operation (EPA-74/04).

Assay:
(1) A test for a particular chemical or effect (EPA-94/04).
(2) *See also* synonym, bioassay.

Assessment:
(1) The process whereby the hazards which have been identified are evaluated in order to provide an estimate for the level of risk (EPA-87/07a).
(2) In the asbestos-in-schools program, the evaluation of the physical condition and potential for damage of all friable asbestos containing materials and thermal insulation systems (EPA-94/04).

Asset: All existing and all probable future economic benefits obtained or controlled by a particular entity (cf. current assest) (40 CFR 144.61-91, *see also* 40 CFR 264.141; 265.141; 280.92-91).

Assimilation:
(1) Removal of dissolved or suspended materials from a water mass by biological, chemical and physical processes.
(2) Conversion or incorporation of absorbed nutrients into body substances (cf. synthesis) (DOD-78/01).
(3) The ability of a body of water to purify itself of pollutants (EPA-94/04).

Assimilation capacity: The extent to which a body of water can receive wastes without significant deterioration of beneficial uses. Suitability for a given use is defined in terms of quality criteria, and these are still to some extent arbitrarily designated. The criterion most widely used is dissolved oxygen, although that parameter is by no means always relevant (LBL-76/07-water).

Assimilative capacity: The capacity of a natural body of water to receive wastewaters or toxic materials without deleterious effects and without damage to aquatic life or humans who consume the water (EPA-94/04).

Association: The organization offering arbitration services selected by EPA to conduct arbitrations pursuant to this part (40 CFR 304.12-91).

Association of boards of certification: An international organization representing boards which certify the operators of waterworks and waste water facilities (EPA-94/04).

Astatine (At): A radioactive halogen element with a.n. 85; a.w. 210; m.p. 302° C and b.p. 377° C. The element belongs to group VIIA of the periodic table.

At retail: The sale by a commercial owner of a wood heater to the ultimate purchaser (40 CFR 60.531-91).

At sea incineration: *See* synonym, ocean incineration.

At the source: At or before the commingling of delacquering scrubber liquor blowdown with other process or nonprocess wastewaters (40 CFR 421.31-91).

Ataxia: Failure of muscular coordination (LBL-76/07-bio).

Atlatl: A tool used to throw a spear (DOE-91/04).

Atmosphere (an): A standard unit of pressure representing the pressure exerted by a 29.92-inch column of mercury at sea level at 45' latitude and equal to 1000 grams per square centimeter (cf. free atmosphere) (EPA-94/04).

Atmosphere (the): The whole mass of air surrounding the earth, composed largely of oxygen and nitrogen (EPA-89/12).

Atmospheric crystallizer: An apparatus used to carry out crystallization under ambient pressure (EPA-77/07).

Atmospheric dispersion: The process of air pollutants being dispersed in the atmosphere by the wind that carries the pollutants away from their source and by turbulent air motion that results from solar heating of the earth's surface and air movement over rough terrain and surfaces (*see* dispersion for more related terms) (DOE-91/04).

Atmospheric evaporation: Evaporation at ambient pressure utilizing a tower filled with packing materials. Air is drawn in from the bottom of the tower and evaporated feed materials entering from the top. There is no recovery of the vapors (EPA-83/06a).

Atmospheric moisture: Moisture content in air.

Atmospheric pressure (or barometric pressure): The pressure of the air and the atmosphere at sea level (*see* pressure for more related terms) (EPA-81/12, p2-3; 85/09).

Atmospheric reaction: A chemical process in the atmosphere resulting in the transformation of the participating species (NATO-78/10).

Atmospheric stability: The ability of the atmosphere to disperse pollutants. The stability classes used in the Gaussian plume models include:
(1) Extremely unstable;
(2) Moderately unstable;
(3) Slightly unstable;
(3) Neutral;
(4) Slightly stable; and
(5) Moderately stable (EPA-88/09).

Atom: A basic chemical particle that can exist. It consists of a

dense **nucleus** of protons and neutrons surrounded by moving electrons. Other atom related terms include:
- Electron
- Ion
- Neutron
- Proton

Atomic absorption spectrometer: A quantitative chemical instrumentation used for the analysis of elemental constituents (cf. definitions under metal analyzer) (EPA-83/06a).

Atomic energy: All forms of energy released in the course of nuclear fission or nuclear transformation (10 CFR 70.4-91).

Atomic Energy Act (AEA) of 1954: *See* definition under Act or AEA.

Atomic pile: A nuclear reactor (EPA-74/11).

Atomic spectrometry: The term refers to the measurement of the absorption of radiation by atoms (LBL-76/07-bio).

Atomic weapon: Any device utilizing atomic energy, exclusive of the means for transporting or propelling the device (where such means is a separable and divisible part of the device), the principal purpose of which is for use as, or for development of, a weapon, a weapon prototype, or a weapon test device (10 CFR 70.4-91).

Atomic weight: A number expressing the ratio of the weight of one atom to that of another. Since the atomic weight is nothing more than a relative weight, the numerical value must be obtained with reference to some convenient standard. The modern chemical atomic weight scale uses the oxygen atom as the standard, giving it a weight scale of 16.000. The atomic weight of hydrogen is 1.0, of sulfur 32, of iron 56, etc. (cf. molecular weight) (EPA-84/09).

Atomization:
(1) The process in which a stream of water or gas impinges upon a molten metal stream, breaking it into droplets which solidify as powder particles (40 CFR 471.02-91).
(2) The reduction of liquid to a fine spray (EPA-89/03b).

Atomize: To divide a liquid into extremely minute particles, either by impact with a jet of steam or compressed air, or by passage through a mechanical device (EPA-89/12).

Attainment area: An area considered to have air quality as good as or better than the national ambient air quality standards as defined in the Clean Air Act. An area may be an attainment area for one pollutant and a non-attainment area for others (EPA-94/04).

Attenuation: The process by which a compound is reduced in concentration over time, through absorption, adsorption, degradation, dilution, and/or transformation (EPA-94/04).

Attractant: A chemical or agent that lures insects or other pests by stimulating their sense of smell (EPA-94/04).

Attributable risk: The difference between risk of exhibiting a certain adverse effect in the presence of a toxic substance and that risk in the absence of the substance (EPA-92/12).

Attrition: Wearing or grinding down of a substance by friction. Dust from such processes contributes to air pollution (EPA-94/04).

Attrition mill:
(1) In the plywood industry, a machine which produces particles by forcing coarse material, shavings, or pieces of wood between a stationary and a rotating disk, fitted with slotted or grooved segments (EPA-74/04).
(2) A ball mill in which pig lead is ground to a powder and oxidized to make the active material (a mixture of lead and lead oxide called lead oxide) in lead acid batteries (EPA-84/08).

Audiometer: An instrument for measuring hearing sensitivity (EPA-74/11).

Audit: A systematic check to determine the quality of operation of some function or activity. Audits may be of two basic types:
(1) Performance audits in which quantitative data are independently obtained for comparison with routinely obtained data in a measurement system; or
(2) System audits of a qualitative nature that consist of an on-site review of a laboratory's quality assurance system and physical facilities for sampling, calibration, and measurement (EPA-86/10a).

Other audit related terms include:
- Independently audited
- Performance audit
- System audit

Audit of data quality: An assessment of the methods used to collect, interpret, and report the information required to characterize data quality. The assessment of these data quality indicators requires a detailed review of:
(1) The recording and transfer of raw data;
(2) Data calculations;
(3) The documentation of procedures; and
(4) The selection and discussion of appropriate data quality indicators (EPA-85/08).

Auger: Any drilling device in which the cuttings are mechanically and continuously removed from the bore-hole without the use of fluids; usually used for shallow drilling or sampling (EPA-82/10).

Auger mining: Spiral boring for additional recovery of a coal seam exposed in a high-wall (EPA-82/10).

Austempering: A heat treating process to obtain greater toughness and ductility in certain high carbon steels. The process is characterized by interrupted quenching and results in the formation of bainite grain structure (EPA-83/06a).

Auto-ignition temperature: The auto-ignition temperature of an air-fuel mixture is the (lowest) temperature at which chemical reaction proceeds at a rate sufficient to result eventually in inflammation at ASTM specified test condition (*see* temperature for more related terms).

Auto-oxidation: A self-induced or internally catalyzed oxidation process (LBL-76/07-water).

Auto-transformer: A power transformer having one continuous winding that is tapped; part of the winding serves as the primary coil and all of it serves as the secondary coil, or vice versa (*see* transformer for more related terms) (EPA-83/03).

Autocatalysis: A catalysis process in which one of the products of the reaction is a catalyst for the reaction.

Autocatalytic: A chemical reaction which is catalyzed by one of the products of the reaction (EPA-87/07a).

Autoclave: A heavy vessel with thick walls for conducting chemical reactions under high pressure or for sterilizing equipment using steam under pressure (EPA-83/09).

Autogenous (autothermic) combustion: In the burning of organic materials, the heat of combustion of the organic materials is sufficient to maintain the combustion without auxiliary fuel except for startup of the burning (*see* combustion for more related terms) (OME-88/12).

Autolysis: A process of cell self destruction due to the enzyme action.

Automated method or analyzer: A method for measuring concentrations of an ambient air pollutant in which sample collection, analysis, and measurement are performed automatically (40 CFR 53.1-91).

Automated transmission component: An asbestos-containing product used as a friction material in vehicular automatic transmissions (40 CFR 763.163-91).

Automatic plating:
(1) Full plating: The workpieces are automatically conveyed through successive cleaning and plating tanks.
(2) Semi plating: The workpieces are conveyed automatically through only one plating tank (EPA-83/06a).
(3) *See* plating for more related terms.

Automatic temperature compensator: A device that continuously senses the temperature of fluid flowing through a metering device and automatically adjusts the registration of the measured volume to the corrected equivalent volume at a base temperature (40 CFR 60.431-91).

Automatic wash water control: Automatic solenoid operated shutoff devices which completely stop the flow of water into the processor when it is not being used, thereby avoiding excessive wash water flows (EPA-80/10).

Automobile and light duty truck body: The exterior surface of an automobile or light duty truck including hoods, fenders, cargo boxes, doors, and grill opening panels (40 CFR 60.391-91).

Automobile: A motor vehicle capable of carrying no more than 12 passengers (40 CFR 52.741-91, *see also* 40 CFR 60.391-91).

Automobile or light duty truck assembly plant: A facility where parts are assembled or finished for eventual inclusion into a finished automobile or light duty truck ready for sale to vehicle dealers, but not including customizers, body shops, and other repainters (40 CFR 52.741-91).

Automobile or light duty truck refinishing: The repainting of used automobiles and light duty trucks (40 CFR 52.741-91).

Autotrophic organism: An organism capable of constructing organic matter from inorganic substances (LBL-76/07-water).

Autotrophic: An organism that produces food from inorganic substances (EPA-89/12).

Auxiliary emission control device (AECD): Any element of design which senses temperature, vehicle speed, engine RPM, transmission gear, manifold vacuum, or any other parameter for the purpose of activating, modulating, delaying, or deactivating the operation of any part of the emission control system (40 CFR 86.082.2-91).

Auxiliary equipment: Accessory equipment necessary for the operation of a process train (EPA-83).

Auxiliary fuel: The fuel which is used to preheat an incinerator, to start incineration, and to maintain incineration temperature, if waste materials do not contain enough combustible organic contents (*see* fuel for more related terms) (EPA-89/03b).

Auxiliary fuel burner: A burner in either the primary or secondary chamber fueled by natural gas or fuel oil. It is used to maintain incinerator temperatures if waste has not enough heating value to maintain combustion temperatures (*see* burner for more related terms) (EPA-89/03b).

Auxiliary fuel firing equipment: Equipment used in an incinerator to supply additional heat by burning an auxiliary fuel so that the additional heat:
(1) Dry and ignite the waste material;
(2) Maintain ignition thereof; and
(3) Affect complete combustion of combustible solids, vapors, and gases (SW-108ts).

Available chlorine:
(1) A term used in rating chlorinated lime and hypochlorites as to their total oxidizing power. Also, a term formerly applied to residual chlorine; now obsolete (EPA-82/11f).
(2) The oxidizing power of a bleaching agent expressed in terms of elemental chlorine (EPA-87/10).
(3) A measure of the amount of chlorine available in chlorinated lime, hypochlorite compounds, and other materials used as a source of chlorine when compared with that of liquid or gaseous chlorine (EPA-94/04).
(4) *See* chlorine for more related terms.

Available energy: The portion of energy added to the system which could be converted to work in a series of reversible engines operating between the temperature of the system and the lowest available temperature (Holman-p176; Jones-p313; Wark-p269).

Available heat:
(1) The quantity of useful energy per unit of fuel available from complete combustion after deducting dry-flue-gas and water vapor losses (EPA-83).

(2) The available heat at any temperature is the gross quantity of heat released within a combustion chamber minus:
(A) The sensible heat carried away by the dry flue gases, and
(B) The latent heat and sensible heat carried away in water vapor contained in the flue gases (EPA-81/12, p3-10; 84/09).
(2) *See* heat for more related terms.

Available oxygen: The quantity of atmospheric oxygen dissolved in the water of a stream; the quantity of dissolved oxygen available for the oxidation of organic matter in sewage (*see* oxygen for more related terms) (EPA-87/10a).

Available system capacity: The capacity determined by subtracting the system load and the system emergency reserves from the net system capacity (40 CFR 60.41a-91).

Availability session: Informal meeting at a public location where interested citizens can talk with EPA and state officials on a one-to-one basis (EPA-94/04).

Aventurine: Glass containing colored, opaque spangles of nonglassy material (EPA-83).

Average: *See* arithmetic mean, ensemble average, geometric mean, or harmonic mean (NATO-78/10).

Average annual flood: The mean of the annual floods during a period of record (*see* flood for more related terms) (DOI-70/04).

Average concentration:
(1) As it relates to chlorine discharge means the average of analyses made over a single period of chlorine release which does not exceed two hours (40 CFR 423.11-91).
(2) The mean value of a measured concentration, e.g., in a chlorine discharge analysis, it means the average of analyses made over a single period of chlorine release (EPA-82/11a).

Average electric demand: The demand on, or the power output of, an electric system or any of its parts over any interval of time, as determined by dividing the total number of kilowatt hours by the number of units of time in the interval (*see* electric demand for more related terms) (EPA-83).

Average fuel economy: The unique fuel economy value as computed under 40 CFR 600.510 for a specific class of automobiles produced by a manufacturer that is subject to average fuel economy standard (40 CFR 600.002.85-91).

Average losses: The total difference in energy input and output or power input and output (due to losses), averaged over a time interval and expressed either in physical quantities or as a percentage of total input (*see* electric loss for more related terms) (EPA-83).

Average monthly discharge limitation: The highest allowable average of "daily discharges" over a calendar month, calculated as the sum of all "daily discharges" measured during a calendar month divided by the number of "daily discharges" measured during that month (40 CFR 122.2-91).

Average particle diameter: *See* synonym, mean particle diameter.

Average particle size: *See* synonym, cut size.

Average time: Considering a continuously varying function of time, it is the time period over which the function is given as an average. In dispersion modeling, the time interval of a specific length over which variations in pollutant concentration at a receptor are averaged. Because instantaneous concentrations at receptors exhibit wide temporal variations, time average concentrations provide a more convenient characterization of pollutant levels at a receptor. The averaging times for dispersion models are designed to be consistent with the air quality standards and commonly include the following: 1-, 3-, 8-, 24-hour, and annual average (EPA-88/09).

Average weekly discharge limitation: The highest allowable average of "daily discharges" over a calendar week, calculated as the sum of all "daily discharges" measured during a calendar week divided by the number of "daily discharges" measured during that week (40 CFR 122.2-91).

Averaging period: The period of time over which the receiving water concentration is averaged for comparison with criteria concentrations. This specification limits the duration of concentrations above the criteria (EPA-91/03).

Avicide: A lethal agent used to destroy birds but also refers to materials used for repelling birds (EPA-85/10).

Avoided cost: The cost that a utility may pay for electric power purchased from a waste to energy facility, based on how much it would have cost the utility to generate the power itself; or the cost not incurred because of diversion of waste from a landfill (*see* cost for more related terms) (OTA-89/10).

Avicron fiber: A trademark of FMC for rayon filament yarns (*see* fiber for more related terms) (EPA-74/06b).

Avril fiber: A trademark of FMC for staple and filament yarns (*see* fiber for more related terms) (EPA-74/06b).

Avulsion: A marked change in the shore of a water body or the course of a stream (which may result from wave erosion) involving extensive removal and redeposition of soil; such changes affect riparian property rights and raise legal questions concerning property lines and ownership of the transported and redeposited material (DOI-70/04).

Axenic: A culture of Lemna fronds free from other organisms (40 CFR 797.1160-91).

Axle clearance: The vertical distance from the level surface on which an automobile is standing to the lowest point on the ale differential of the automobile (40 CFR 86.084.2-91).

Axle ratio: All ratios within plus or minus 3% of the ale ratio specified in the configuration in the test order (40 CFR 86.602.84-91, *See also* 40 CFR 600.002.85-91).

Azeotrope: A liquid mixture of two or more substances which behaves like a single substance in that the vapor produced by partial evaporation of liquid has the same composition as the liquid (EPA-83/09).

Azide: Compounds containing the N_3 group.

Azimuth angle: The angle in the horizontal plane that designates where the laser beam is pointed. It is measured from an arbitrary fixed reference line in that plane (40 CFR 60-App/A(alt. method 1)-91).

Azo compound: A compound containing (-N=N-) to link two other groups.

Bb

Back drafting: A condition where the normal (up) movement of combustion products (a flue gas), resulting from the buoyant forces on the hot gases, is reversed, so that the combustion products can enter the house. Back-drafting of combustion appliances (such as fireplaces and furnaces) can occur when depressurization in the house overwhelms the buoyant force on the hot gases. Back-drafting can also be caused by high air pressures at the chimney or flue termination (EPA-88/08).

Back end materials recovery: An engineered system that provides for collection of discrete reusable materials from mixed wastes which have been burned or treated (EPA-83).

Back end system: A combination of system components that changes the chemical properties of the waste and/or converts its components into energy or compost (cf. front-end system) (EPA-83).

Back pressure: A pressure than can cause water to backflow into the water supply when a user's water system is at a higher pressure than the public system (EPA-94/04).

Backflow/back siphonage: A reverse flow condition created by a difference in water pressures that causes water to flow back into the distribution pipes of a drinking water supply from any source other than an intended one (EPA-94/04).

Backfill: The material used to refill a ditch or other excavation, or the process of doing so (EPA-83).

Backfilling: The transfer of previously moved material back into an excavation such as a mine or ditch, or against a constructed object (EPA-82/10).

Background condition: The biological, chemical, and physical conditions of a water body, upstream from the point or non-point source discharge under consideration. Background sampling location in an enforcement action will be upstream from the point of discharge, but not upstream from other inflows. If several discharges to any water body exist, and an enforcement action is being taken for possible violations to the standards, background sampling will be undertaken immediately upstream from each discharge (40 CFR 131.35-91).

Background level: In air pollution control, the concentration of air pollutants in a definite area during a fixed period of time prior to the starting up or on the stoppage of a source of emission under control. In toxic substances monitoring, the average presence in the environment, originally referring to naturally occurring phenomena (EPA-94/04).

Background radiation:
(1) The radioactivity in the environment, including cosmic rays from space and radiation that exists elsewhere - in the air, in the earth, and in man-made materials. In the U.S., most people receive 100 to 250 millirems of background radiation per year (EPA-88/08a).
(2) A normal radiation present in the lower atmosphere from cosmic rays and from earth sources (EPA-74/11).
(3) Ionizing radiation present in the environment from cosmic rays and natural sources in the earth; background radiation varies considerably with location (cf. radiation, natural) (DOE-91/04).
(4) *See* radiation for more related terms.

Background soil pH: The pH of the soil prior to the addition of substances that alter the hydrogen ion concentration (40 CFR 257.3.5-91).

Background: Ambient pollutant concentrations due to:
(1) Natural sources
(2) Nearby sources other than the one(s) currently under consideration
(3) Unidentified anthropogenic sources (EPA-88/09)

Backhoe tamping: A process step, often used in direct dump transfer systems, in which a conventional backhoe is used to compact waste contained in an open-top transfer trailer (SW-108ts).

Backing wind: The counter clockwise rotation of the wind direction (*see* wind for more related terms) (NATO-78/10).

Backscatter: The scattering of laser light in a direction opposite to that of the incident laser beam due to reflection from particulates along the beam's atmospheric path which may include a smoke plume (40 CFR 60-App/A(alt. method 1)-91).

Backscatter signal: The general term for the lidar return signal which results from laser light being backscattered by atmospheric and smoke plume particulates (40 CFR 60-App/A(alt. method 1)-91).

Backwashing:
(1) The process of cleaning a rapid sand or mechanical filter by reversing the flow of water (EPA-87/10a).
(2) Reversing the flow of water back through the filter media to remove the entrapped solids (EPA-94/04).

Bacquerel (Bq): *See* definition under radiation unit.

Bacteria (singulary bacterium):
(1) Single-cell, microscopic living organisms (single-celled microorganisms), that possess rigid cell walls. They may be aerobic, anaerobic, or facultative; they can cause disease; and some are important in pollution control, e.g., the stabilization of solid wastes, because they break down organic matter in the air, water or solid (cf. coliform) (SW-108ts).
(2) Microscopic living organisms that can aid in pollution control by metabolizing organic matter in sewage, oil spills or other pollutants. However, bacteria in soil, water or air can also cause human, animal and plant health problems (EPA-94/04).
(3) cf. sulfur bacteria.

Bacteria bed: *See* synonym, trickling filter (Scott-81).

Bacterial growth: All bacteria for wastewater treatment require food for their continued life and growth and all are affected by the conditions of their environment. Like human beings, they consume food, need moisture, require heat, and give off waste products, and they respire. Without an adequate food supply, bacteria will not grow (EPA-76/03).

Bacterial metabolism: The chemical change, constructive and destructive, occurring in bacteria (EPA-75/10).

Bacterial quantity unit (BQU): One measure of the total load of bacteria passing a given stream location and is particularly useful in comparing relative loads between wastewater treatment stations. The number of BQUs is derived as the product of flow in cfs and coliform density in MPN per 100 ml divided by 100,000 (EPA-74/01a)

Bactericide:
(1) A bacteria-killing chemical (EPA-85/10).
(2) A substance which kills bacteria (LBL-76/07-water).

Bacteriophage: A type of virus which attacks and destroys bacteria (EPA-83/09).

Bacteriostat: An agent which inhibits the growth of bacteria (EPA-75/01a).

Baffle: A flat board or plate, deflector, guide or similar device constructed or placed in flowing water or slurry systems to cause more uniform flow velocities, to absorb energy, and to divert, guide, or agitate liquids (EPA-94/04).

Baffle chamber:
(1) In incinerator design, a chamber designed to promote the settling of fly ash and coarse particulate matter by changing the direction and/or reducing the velocity of the gases produced by the combustion of the refuse or sludge (EPA-94/04).
(2) *See* synonym, settling chamber.

Baffle mark: A mark or seam on bottle resulting from a mold joint between blank mold and baffle (EPA-83).

Baffle:
(1) A construction used to close or deflect the delivery of a moving substance (EPA-83).
(2) Any deflector devices used to change the direction of a flow or the velocity of water, sewage or products of combustion such as fly ash or coarse particulate matter. Also used in deadening sound (EPA-74/11).

Bag barker: *See* synonym, debarker (EPA-74/04).

Bag blinding: The loading, or accumulation, of filter cake to the point where capacity rate is diminished in a bag house of air pollution control equipment (EPA-89/03b).

Bag failure: Filter bags in a baghouse fail with time. There are three basic failure mechanisms that can shorten the life of a bag; these are related to abrasion, thermal degradation, and chemical attack (EPA-84/09).

Bag filter (or fabric filter):
(1) A filter in which the medium is a fabric cylindrical bag (*see* collector for more related terms) (EPA-83).
(2) A device designed to remove particles form carrier gas or air by passage of the gas through a porous (fabric) medium (*see* filter for more related terms) (EPA-83).
(3) cf. baghouse.

Bagacillo: Fine bagasse particles (EPA-75/02d).

Bagasse:
(1) The fibrous residue that remains after juice is extracted from sugar cane or sugar beets. It is generally used as boiler fuel and, in some cases, in the manufacture of various by-products (EPA-75/02d).
(2) Plant residues used to bind explosives (EPA-76/03).

Bagging operation: The mechanical process by which bags are filled with nonmetallic minerals (40 CFR 60.671-91).

Baghouse (or baghouse filter): A type of air pollution control device. A baghouse is known as a bag filter or a fabric filter. A baghouse uses fabric filters to remove particulates from flue gases (ETI-92). This device operates in a way similar to the bag of an electric vacuum cleaner, passing the air and smaller particulate matter, while entrapping the larger particulates (cf. collector) (EPA-94/04, *see also* EPA-84/09).

The bag removes solid particulate matter from the flue gas stream by filtering the flue gas through fabric bags, usually made of cloth or glass fibers. Small particles are initially captured and retained on the fibers of the cloth by means of interception, impingement, diffusion, gravitational settling, and electrostatic attraction. Once a mat or cake of dust is accumulated, further collection is accomplished by sieving or other mechanisms. The cloth then serves mainly as a supporting structure for the dust mat responsible for the high collection efficiency. Periodically the accumulated dust is removed for disposal (EPA-81/09; AP-40, p106).

The major components of a fabric filter include:
(1) Dirty air inlet and a plate (diffuser) with holes in it that uniformly distributes the flue gas.
(2) A dirty air chamber or plenum which contains the fabric bags.
(3) A tube sheet, with holes for each bag, which supports the bags and separates the dirty air plenum from the clean air plenum.
(4) The tubular filter bags with supporting wire frame bag retainers.
(5) The bag cups and venturi's to which the individual bags are attached and which inject the pulse of cleaning air into the bags.
(6) The air compressor which supplies the compressed air for cleaning the bags.
(7) The ash hopper which holds the collected particulate after it is cleaned from the bag.
(8) A rotary valve air lock which discharges the ash from the hopper.

Baghouse compartment: Baghouses are usually constructed in modular forms. Each module consists of one or more compartments. Each compartment can contain a few or several thousand bags (EPA-84/09).

Baghouse filter:
(1) Large fabric bag, usually made of glass fibers, used to eliminate intermediate and large (greater than 20 microns in diameter) particles. This device operates like the bag of an electric vacuum cleaner, passing the air and smaller particles while entrapping the larger ones (EPA-94/04).
(2) See also synonym, baghouse.

Bahco sampler: A particle classifier used for measuring particle size distribution in a gas stream. Its working range is approximately 1 to 60 micrometers (or microns). It uses a combination of elutriation and centrifugation to separate particles (EPA-84/09).

Bailer: A long pipe with a valve at the lower end, used to remove slurry from the bottom or side of a well as it is being drilled (EPA-94/04).

Bait: The tool dipped into molten glass to start any drawing operation (EPA-83).

Bake oven: A device which uses heat to dry or cure coatings (40 CFR 60.311-91, see also 40 CFR 60.391-91).

Baked coatings: Any coating which is cured or dried in an oven where the oven air temperature exceeds 90° C (194° F) (40 CFR 52.741-91).

Balance experiments: Experiments on man or other animals that involve quantitative measurements of intake (via respiration and ingestion) and loss (via exhalation and excretion) of a specific element or substance. A positive balance means that more is taken in than is lost over a specified time (LBL-76/07-bio).

Balanced draft: In a balanced draft system, a forced draft fan is used to push (or blow) combustion air into the incinerator and an induced draft fan (or the natural draft stack) is used to pull the combustion gas through the incinerator and exit from the stack. The draft is balanced so that the incinerator is maintained at a slightly negative pressure. This negative pressure presents emissions from leaking from the combustion chamber (see draft for more related terms) (EPA-89/03b).

Bale: A standard bale of waste-paper is 72 in. long, 32 in. wide, and 28 in. deep, with a content of about 37 cubic feet and weighing 900 to 1,000 lbs. The size and weight may vary with the grade of paper. A bale of pulp varies in weight from 400 to 500 lbs and is approximately 30x30x13 in. size. A bale of rags varies in weight from 700 to 1,300 lbs and will vary in dimensions according to the press used. Typical dimensions are 26x30x72 in., 26x42x72 in., or 26x52x54 in. A bale of bags weighs 61 to 62 lbs (EPA-87/10).

Baler:
(1) A machine used to compress solid wastes, primary materials, or recoverable materials, with or without binding, to a density or from which will support handling and transportation as a material unit rather than requiring a disposable or reuseable container. This specifically excludes briquetters and stationary compaction equipment which is used to compact materials into disposable or reuseable containers (40 CFR 246.101-91).
(2) A machine used to compress and bind materials together (EPA-83).
(3) A long pipe with a valve at the lower end, used to remove slurry from the bottom or side of a well as it is being drilled (EPA-94/04).

Baling: Compacting solid waste into blocks to reduce volume and simplify handling (EPA-94/04).

Ball decks: A tray of rubber balls that bounce against the bottom surface of a screen, thus eliminating blinding (EPA-88/08a).

Ball mill: Pulverizing equipment for the grinding of raw materials. Grinding is done by steel balls, pebbles, or rods (EPA-83/03a).

Ball powder: Small arms powder made by emulsifying a mixture of propellant and solvent in a liquid in which they are not soluble. Evaporation of the emulsifying liquid and the solvent yields quite uniform round balls of powder (EPA-76/03).

Ballast:
(1) The flow of waters, from a ship, that is treated along with refinery wastewaters in the main treatment system (40 CFR 419.11-91).
(2) A circuit element that serves to limit an electric current or to provide a starting voltage, as in certain types of lamps, such as in fluorescent ceiling fixtures (EPA-83/03).

Balled (or nuggetized): Describes municipal ferrous scrap which has been processed by a machine so that individual particles have been formed into tight, high density balls or nuggets (EPA-83).

Ballistic separator:
(1) A device that drops mixed materials having different physical characteristics onto a high speed rotary impeller. They are

hurled off at different velocities and land in separate collecting bins (SW-108ts).
(2) *See* separator for more related terms.
(3) A machine that sorts organic from inorganic matter for composting (EPA-94/04).

Band application: The spreading of chemicals over, or next to, each row of plants in a field (EPA-94/04).

Bandwidth: The frequency range of a radio signal.

Banking: A system for recording qualified air emission reductions for later use in bubble, offset, or netting transactions (*see* emissions trading) (EPA-94/04).

BAP: Benzo(a)pyrene, a polycyclic aromatic hydrocarbon.

Bar: A CGS pressure unit, 1 bar = 10^6 dynes/(cm^2) = 10^5 pascals = 750 mmHg = 0.987 atmosphere = 14.504 psi.

Bar, billet and bloom: Those acid pickling operations that pickle bar, billet or bloom products (40 CFR 420.91-91).

Bar screen: In wastewater treatment, a device used to remove large solids (EPA-94/04).

Barge haul: The hauling of material by barge (cf. rail haul) (EPA-83).

Baridex: A quantitative method for determining the amount of BaCO$_3$ required to combine the water soluble sulfates in clays of shales (EPA-83).

Barium (Ba): An alkaline earth metal with a.n. 56; a.w. 137.34; d. 3.5 g/cc; m.p. 725° C; and b.p. 1640° C. The element belongs to group IIA of the periodic table.

Barking: An operation of removing bark (tree skin) from pulpwood prior to processing. This is carried out by means of a barker with knife, and drum, by mechanical abrasion, by hydraulic barker, or by chemical means (EPA-87/10).

Barminutor: Sec comminutor (DOI-70/04).

Baroclinity: State of the atmosphere in which density is no unique function of pressure. The surfaces of constant density (isopycnic) do not coincide with surfaces of constant pressure (isobaric). The main effect of baroclinity is that vorticity can be generated in the atmosphere (cf. barotropy) (NATO-78/10).

Barometer: A pressure gauge which measures atmospheric pressure (*see* pressure gauge for more related terms).

Barometric condenser:
(1) An apparatus used to condense vapor, in which the vapor are condensed by direct contact with water (EPA-75/02c).
(2) The cooling water and the vapors are in physical contact; the condensate is mixed in the cooling water (EPA-75/02d, AP-40, P199).
(3) *See* condenser for more related terms.

Barometric condensing operations: Those operations or processes directly associated with or related to the concentration and crystallization of sugar solutions (40 CFR 409.11-91).

Barometric damper: A hinged or pivoted plate that automatically regulates the amount of air entering a duct, breeching, flue connection, or stack. It thereby maintains a constant draft in the incinerator (*see* damper for more related terms) (SW-108ts).

Barometric leg:
(1) A pipe drawing water from a decker or similar piece of equipment discharging it below the surface of the water in a receiving tank. A syphon action is created thus drawing a vacuum on the decker (EPA-87/10).
(2) A long vertical pipe through which spent condenser water leaves the barometric condenser. It serves as a source of vacuum (EPA-75/02d).

Barometric leg water: Condenser cooling water (*see* water for more related terms) (EPA-75/02d).

Barometric pressure: *See* synonym, atmospheric pressure.

Barometric seal: A column of liquid used to hydraulically seal a scrubber, or any component thereof, from the atmosphere or any other part of the system (EPA-89/03b).

Barotropy: A state of the atmosphere in which the pressure is a unique function of the density. The surfaces of constant density (isopycnic) coincide with the surfaces of constant pressure (isobaric) (cf. baroclinity) (NATO-78/10).

Barrel finishing: The process of polishing a workpiece using a rotating or vibrating container and abrasive grains or other polishing materials to achieve the desired surface appearance (EPA-83/06a).

Barrel: Forty two (42) United States gallons at 60 degrees Fahrenheit (CWA311; SF101; 40 CFR 195.2; 113.3-91).

Barrel plating: The electroplating of workpieces in barrels (bulk) (*see* plating for more related terms) (EPA-83/06a).

Barrier coating: A layer of a material that obstructs or prevents passage of something through a surface that is to be protected, e.g. grout, caulk, or various sealing compounds; sometimes used with polyurethane membranes to prevent corrosion or oxidation of metal surfaces, chemical impacts on various materials, or, for example, to prevent radon infiltration through walls, cracks, or joints in a house (EPA-94/04).

Barrier: Any material or structure that prevents or substantially delays movement of water or radionuclides toward the accessible environment. For example, a barrier may be a geologic structure, a canister, a waste form with physical and chemical characteristics that significantly decrease the mobility of radionuclides, or a material placed over and around waste, provided that the material or structure substantially delays movement of water or radionuclides (40 CFR 191.12-91).

Barton pot: A reactor vessel, used in the Barton process, into which molten lead is fed and vigorously agitated to form fine lead droplets in the presence of air. The resulting mixture of un-oxidized lead and lead oxides (lead oxide) comprises an active material in lead acid batteries (EPA-84/08).

Barton process: A process for making lead oxide to be used in secondary lead oxide batteries. Molten lead is fed, agitated, and stirred in a pot with the resulting fine droplets oxidized. Materials are collected in a settling chamber where crystalline varieties of lead oxide are formed (EPA-83/03a).

Basal application: In pesticides, the application of a chemical on plant stems or tree trunks just above the soil line (EPA-94/04).

Basal diet: The food or diet as it is prepared or received from the supplier, without the addition of any carrier, diluent, or test substance (40 CFR 797.2050-91, *see also* 40 CFR 797.2130; 797.2150-91).

Base:
(1) A substance that in aqueous solution turns red litmus blue, furnishes hydroxyl ions, reacts with, and neutralizes an acid to form a salt and water only (EPA-83/09).
(2) It can remove hydrogen ions (protons) from an acid and combines with them in a chemical reaction.
(3) A substance which dissolves in water forming hydroxyl ions (EPA-87/10a).

Other base related terms include:
- Lewis base
- Strong base

Base board duct: A continuous system of sheet metal or plastic channel ducting that is sealed over the joint between the wall and floor around the entire perimeter of the basement. Holes drilled into hollow blocks in the wall allow suction to be drawn on the walls and joint to remove radon through the ducts to a release point away from the inside of the house (EPA-88/08).

Base date: The date set forth in paragraph (d) of this section as of which the base number of single-passenger commuter vehicles at a particular employment facility or educational institution must be determined (40 CFR 52.1161-91, *see also* 40 CFR 52.2297-91).

Base date period: The thirty day period immediately preceding the base date; "compliance date period" means the thirty day period immediately preceding the compliance date. In situations where the averaging periods are not appropriate, approval of an alternate period may be requested from the Regional Administrator (40 CFR 52.2297-91).

Base film: The substrate that is coated to produce magnetic tape (40 CFR 60.711-91).

Base flood: The flood which has a one percent chance of occurrence in any given year (also known as a 100-year flood). This term is used in the National Flood Insurance Program (NFIP) to indicate the minimum level of flooding to be used by a community in its floodplain management regulations (*see* flood for more related terms) (40 CFR 6-App/A-91).

Base floodplain: The land area covered by a 100-year flood (one percent chance floodplain). *Also see* definition of floodplain (40 CFR 6-App/A-91).

Base flow: The part of a stream flow contributed by groundwater which seeps into the surface streams (DOI-70/04).

Base gasoline: The gasoline which meets the specifications specified in CAA241 (*see* gasoline for more related terms) (CAA241-42USC7581-91).

Base level: A unique combination of basic engine inertia weight class and transmission class (40 CFR 600.002.85-91).

Base load:
(1) The load level at which a gas turbine is normally operated (40 CFR 60.331-91).
(2) The minimum load over a given period of time (EPA-83).
(3) *See* load for more related terms.

Base load station: An electric generating plant which is normally operated to take all or part of the base load of a system and, consequently operates at a constant output (EPA-83).

Base load unit: An electric generating facility operating continuously at a constant output with little hourly or daily fluctuation (EPA-82/11f).

Base pair mutagens: The agents which cause a base change in the DNA. In a reversion assay, this change may occur at the site of the original mutation or at a second site in the chromosome (40 CFR 798.5265-91, *see also* 40 CFR 798.5300-91).

Base temperature: An arbitrary reference temperature for determining liquid densities or adjusting the measured volume of a liquid quantity (*see* temperature for more related terms) (40 CFR 60.431-91).

Base vehicle: The lowest priced version of each body style that makes up a car line (*see* vehicle for more related terms) (40 CFR 600.002.85-91).

Based flood: A flood that has a 1 percent or greater chance of recurring in any year or a flood of a magnitude equaled or exceeded over in 100 years on the average over a significantly long period (*see* flood for more related terms) (40 CFR 257.3.1-91).

Baseline:
(1) The annual quantity of fossil fuel consumed by an affected unit, measured in millions of British Thermal Units (mmBtu), calculated as follows:
   (A) For each utility unit that was in commercial operation prior to January 1, 1985, the baseline shall be the annual average quantity of mmBtu's consumed in fuel during calendar years 1985, 1986, and 1987, as recorded by the Department of Energy pursuant to Form 767. For any utility unit for which such form was not filed, the baseline shall be the level specified for such unit in the 1985 National Acid Precipitation Assessment Program (NAPAP) Emissions Inventory, Version 2, National Utility Reference File (NURF) or in a corrected data base as established by the

Administrator pursuant to paragraph (3). For nonutility units, the baseline is the NAPAP Emissions Inventory, Version 2. The Administrator, in the Administrator's sole discretion, may exclude periods during which a unit is shutdown for a continuous period of four calendar months or longer, and make appropriate adjustments under this paragraph. Upon petition of the owner or operator of any unit, the Administrator may make appropriate baseline adjustments for accidents that caused prolonged outages.

(B) For any other nonutility unit that is not included in the NAPAP Emissions Inventory, Version 2, or a corrected data base as established by the Administrator pursuant to paragraph (3), the baseline shall be the annual average quantity, in mmBtu consumed in fuel by that unit, as calculated pursuant to a method which the administrator shall prescribe by regulation to be promulgated not later than eighteen months after enactment of the Clean Air Act Amendments of 1990.

(C) The Administrator shall, upon application or on his own motion, by December 31, 1991, supplement data needed in support of this title and correct any factual errors in data from which affected Phase II units' baselines or actual 1985 emission rates have been calculated. Corrected data shall be used for purposes of issuing allowances under the title. Such corrections shall not be subject to judicial review, nor shall the failure of the Administrator to correct an alleged factual error in such reports be subject to judicial review (CAA402-42USC7651a).

(2) A quantitative expression of conditions, costs, schedule, or technical progress to serve as a base or standard for measurement during the performance of an effort; the established plan against which the status of resources and the progress of a project can be measured. The environmental baseline is the site environmental conditions as they are projected to occur in 1995 (construction) and 2000 (Operation) (DOE-91/04).

Baseline area:
(1) Any intrastate area (and every part thereof) designated as attainment or unclassifiable under section 107(d)(1)(D) or (E) of the Act in which the major source or major modification establishing the minor source baseline date would construct or would have an air quality impact equal to or greater than 1 $\mu g/m^3$ (annual average) of the pollutant for which the minor source baseline date is established.
(2) Area redesignations under section 107(d)(1) (D) or (E) of the Act cannot intersect or be smaller than the area of impact of any major stationary source or major modification which:
 (A) Establishes a minor source baseline date; or
 (B) Is subject to 40 CFR 52.21 or under regulations approved pursuant to 40 CFR 51.166, and would be constructed in the same state as the state proposing the redesignation.
(3) Any baseline area established originally for the TSP increments shall remain in effect and shall apply for purposes of determining the amount of available PM-10 increments, except that such baseline area shall not remain in effect if the permit authority rescinds the corresponding minor source baseline date in accordance with paragraph (b)(14)(iv) of this section (40 CFR 51.166-15-91).

Baseline concentration: With respect to a pollutant, the ambient concentration levels which exist at the time of the first application for a permit in an area subject to this part, based on air quality data available in the Environmental Protection Agency or a State air pollution control agency and on such monitoring data as the permit applicant is required to submit. Such ambient concentration levels shall take into account all projected emissions in, or which may affect, such area from any major emitting facility on which construction commenced prior to January 6, 1975, but which has not begun operation by the date of the baseline air quality concentration determination. Emissions of sulfur oxides and particulate matter from any major emitting facility on which construction commenced after January 6, 1975, shall not be included in the baseline and shall be counted against the maximum allowable increases in pollutant concentrations established under this part (CAA169-42USC7479).

Baseline configuration: The unretrofitted test configuration, tuned in accordance with the automobile manufacturers specifications (40 CFR 610.11-91).

Baseline consumption allowance: The consumption allowances apportioned under 40 CFR 82.6 (40 CFR 82.3-91).

Baseline date: Baseline date related terms include:
- Major source baseline date
- Minor source baseline date (*see* definition under major source baseline date)

Baseline gasoline:
(1) Summertime--The term baseline gasoline means in the case of gasoline sold during the high ozone period (as defined by the Administrator) a gasoline which meets the following specifications:

BASELINE GASOLINE FUEL PROPERTIES
- API Gravity — 57.4
- Sulfur, ppm — 339
- Benzene, % — 1.53
- RVP, psi — 8.7
- Octane, R+M/2 — 87.3
- IBP, F — 91
- 10%, F — 128
- 50%, F — 218
- 90%, F — 330
- End Point, F — 415
- Aromatics, % — 32.0
- Olefins, % — 9.2
- Saturates, % — 58.8

(2) Wintertime--The administrator shall establish the specifications of baseline gasoline for gasoline sold at times other than the high ozone period (as defined by the Administrator). Such specifications shall be the specifications of 1990 industry average gasoline sold during such period (CAA211.k-42USC7545-91).
(3) *See* gasoline for more related terms.

Baseline model year: With respect to any pollutant emitted from any vehicle or engine, or class or category thereof, the model year immediately preceding the model year in which Federal standards applicable to such vehicle or engine, or class or category thereof, first applied with respect to such pollutant (CAA202).

Baseline or trend assessment survey: The planned sampling or measurement of parameters at set stations or in set areas in and near disposal sites for a period of time sufficient to provide synoptic data for determining water quality, benthic, or biological conditions as a result of ocean disposal operations. The minimum requirements of rush surveys are given in 40 CFR 228.13 (40 CFR 228.2-91).

Baseline production allowances: The production allowances apportioned under 40 CFR 82.5 (40 CFR 82.3-91).

Baseline tritium production: Tritium production in sufficient quantity to meet any future production requirements based on historical fluctuations and unspecified projections into the next century (DOE-91/04).

Baseline vehicle: The representative model year 1990 vehicles (*see* vehicle for more related terms) (CAA211.k-42USC7545-91).

Basement rock: The undifferentiated rocks, commonly igneous and metamorphic, that underlie the rocks of interest in a given area (DOE-91/04).

Basement: A type of house construction where the bottom livable level has a slab (or earthen floor) which averages 3 ft or more below grade level on one or more sides of the house and is sufficiently high to stand in (cf. crawl space) (EPA-88/08).

Basic brick: A brick made of a material which is a basic anhydride such as MgO or mixed MgO plus CaO (*see* brick for more related terms) (EPA-74/06a).

Basic component: A reactor structure, system, component, or part thereof that is necessary to ensure the:
(1) Integrity of a reactor coolant or moderator pressure boundaries
(2) Capability to shut down a reactor and maintain it in a safe shutdown, condition
(3) Capability to prevent or mitigate the consequences of accidents that could result in a large release of radiation (DOE-91/04).

Basic engine: The unique combination of manufacturer, engine displacement, number of cylinders, fuel system (as distinguished by number of carburetor barrels or use of fuel injection), catalyst usage, and other engine and emission control system characteristics specified by the Administrator (*see* engine for more related terms) (40 CFR 86.082.2-91, *see also* 40 CFR 600.002.85-91).

Basic furnace: A furnace in which the refractory material is composed of dolomite or magnesite (*see* furnace for more related terms) (EPA-74/06a).

Basic oxygen furnace steelmaking: The production of steel from molten iron, steel scrap, fluxes, and various combinations thereof, in refractory lined fuel-fired furnaces by adding oxygen (*see* furnace for more related terms) (40 CFR 420.41-91).

Basic oxygen process furnace (BOPF): Any furnace with a refractory lining in which molten steel is produced by charging scrap metal, molten iron, and flux materials or alloy additions into a vessel and by introducing a high volume of oxygen-rich gas. Open hearth, blast, and reverberatory furnaces are not included in this definition (*see* furnace for more related terms) (40 CFR 60.141; 60.141a-91).

Basic oxygen steelmaking: The basic oxygen process is carried out in a basic lined furnace which is shaped like a pear. High pressure oxygen is blown vertically downward on the surface of the molten iron through a water cooled lance (EPA-74/06a).

Basic refractory: The major constituent is lime, magnesia, or both, and which may react chemically with acid refractories, acid slags, or acid fluxes at high temperatures. Includes refractories made of chrome ore or combinations of chrome ore and dead burned magnesite (*see* refractory for more related terms) (EPA-83).

Basic vehicle frontal area: The area enclosed by the geometric projection of the basic vehicle along the longitudinal axis, which includes tires but excludes mirrors and air deflectors, onto a plane perpendicular to the longitudinal axis of the vehicle (40 CFR 86.082.2-91).

Basin: A region in which the strata or layers of rock dip in all directions toward a central point. Thus, it is any hollow or trough in the earth's crust, whether filled by water or not. A river basin is the total area drained by a river and its tributaries (DOI-70/04).

Basis material:
(1) The coiled strip which is processed (40 CFR 465.02-91).
(2) The metal part or base onto which porcelain enamel is applied (40 CFR 466.02-91).
(3) The substance of which the workpieces are made and receive the treatments in preparation for coating (EPA-82/11e).

Basis weight: The weight of a sheet of paper of a given area. It is effected by the density and thickness of the sheet (EPA-87/10).

Basophilic stippling: The characteristic appearance of some erythrocytes that contain cytoplasmic material that stains deeply with basic dyes (LBL-76/07-bio).

BAT or BAT effluent limitations: *See* definition under best available technology.

Batch distillation operation: A noncontinuous distillation operation in which a discrete quantity or batch of liquid feed is charged into a distillation unit and distilled at one time. After the initial charging of the liquid feed, no additional liquid is added during the distillation operation (40 CFR 60.661-91).

Batch fed incinerator: An incinerator that is periodically charged with waste; one charge is allowed to burn out before another is charged (*see* incinerator for more related terms) (SW-108ts).

Batch house: The place where batch materials are received, handled, weighed, and mixed, for delivery to melting units (EPA-83).

Batch loader: A type of enclosed compactor truck equipped with a loading hopper at the rear and a large mechanized panel which sweeps the solid wastes into the body of the unit (EPA-83).

Batch lot: A definite quantity of samples collected under conditions that are considered uniform (cf. lot size) (EPA-84/03).

Batch MWC: An MWC unit designed such that it cannot combust MSW continuously 24 hours per day because the design does not allow waste to be fed to the unit or ash to be removed while combustion is occurring (40 CFR 60.51a-91).

Batch, pipe and tube: Those descaling operations that remove surface scale from pip tube products in batch processes (40 CFR 420.81-91).

Batch process: A process which has an intermittent flow of raw materials into the process and, consequently, produces an intermittent flow of product and process waste from the process. This is in contrast to a continuous process in which material is continuously processed without interruptions (EPA-87/10a).

Batch, rod and wire: Those descaling operations that remove surface scale from rod and wire products in batch processes (40 CFR 420.81-91).

Batch sample: The collection of samples (compressors) that are drawn from a batch (*see* sample for more related terms) (40 CFR 204.51-91, *see also* 40 CFR 205.51-91).

Batch sample size: The number of compressors of the same category or configuration which is randomly drawn from the batch sample and which will receive emissions tests (*see* sample for more related terms) (40 CFR 204.51-91, *see also* 40 CFR 205.51-91).

Batch, sheet and plate: Those descaling operations that remove surface scale from sheet and plate products in batch processes (40 CFR 420.81-91).

Batch size: The number as designated by the Administrator in the test request, of compressors of the same category or configuration in a batch (40 CFR 204.51-91, *see also* 40 CFR 205.51-91).

Batch treatment:
(1) Any treatment in which the process is completed and the products are discharged before more raw material is charged (Scott-81).
(2) A waste treatment method where wastewater is collected over a period of time, and the collected wastewater is treated in a tank or lagoon prior to discharge. Wastewater collection may be continuous when treatment is batch (EPA-85/10a).
(3) *See* treatment for more related terms.

Batch: Any manufacturing or treatment process which accumulates a fixed volume of materials (e.g., wastewater) for processing, treatment or discharge (cf. continuous) (EPA-79/12b, *see also* 40 CFR 160.3; 169.1; 204.51; 205.51; 420.81; 420.91; 420.111; 792.3-91).

Bating: A manufacturing step following liming and preceding pickling. The purpose of this operation is to delime the hides, reduce swelling, peptize fibers, and remove protein degradation products from the hide (EPA-82/11).

Batt insulation: *See* definition under blanket insulation (40 CFR 248.4-91).

Battery:
(1) A modular electric power source where part or all of the fuel is contained within the unit and electric power is generated directly from a chemical reaction rather than indirectly through a heat cycle engine. In this regulation there is no differentiation between a single cell and a battery (40 CFR 461.2-91).
(2) A device that transforms chemical energy into electrical energy. This term usually applies to two or more cells connected in series, parallel or a combination of both. Common usage has blurred the distinction between the terms cell and battery and frequently the term battery is applied to any finished entity sold as a single unit, whether it contains one cell, as do most flashlight batteries, or several cells, as do automotive batteries (EPA-84/08).

Other battery related terms include:
- Primary battery
- Secondary battery
- Secondary cell (*see* synonym, secondary battery)
- Storage battery
- Trucked battery

Battery configuration: The electrochemical type, voltage, capacity (in Watt-hours at the c/3 rate), and physical characteristics of the battery used as the tractive energy storage device (40 CFR 600.002.85-91).

Battery furnace wall: A double or common wall between two combustion chambers both faces are exposed to heat (*see* furnace wall for more related terms) (SW-108ts).

Battery manufacturing operation: All of the specific processes used to produce a battery including the manufacture of anodes and cathode and associated ancillary operations. These manufacturing operations are excluded from regulation under any other point source category (40 CFR 461.2-91).

Battery separator: An asbestos containing product used as an insulator or separator between the negative and positive terminals in batteries and fuel cells (40 CFR 763.163-91).

Bauxite: The ore containing alumina monohydrate or alumina trihydrate which serves as the principal raw material for the production of alumina by the Bayer process or by the combination process (40 CFR 421.11-91).

Bauxite treating: A catalytic process to convert sulfur compounds in petroleum into hydrogen sulfide by passing vaporized petroleum fraction through bauxite beds.

Bayer process: A process in which impure aluminum in bauxite is dissolved in a hot, strong, alkalai solution (normally NaOH) to form sodium aluminate. Upon dilution and cooling, the solution hydrolyzes and forms a precipitate of aluminum hydroxide (EPA-82/05).

BCT or BCT effluent limitations: *See* definition under best conventional pollutant control technology.

Bead cementing operation: The system that is used to apply cement to the bead rubber before or after it is wound into its final circular form. A bead cementing operation consists of a cement application station, such as a dip tank, spray booth and nozzles, cement trough and roller or swab applicator, and all other equipment necessary to apply cement to would beads or bead rubber and to allow evaporation of solvent from cemented beads (40 CFR 60.541-91).

Bead:
(1) Rubber-covered strands of wire, wound into a circular form, which ensure a seal between a tire and the rim of the wheel onto which the tire is mounted (40 CFR 60.541-91).
(2) An enlarged, rounded edge of a tumbler or other glass article, or any raised section extending around the article.
(3) A small piece of glass tubing used around a lead wire (EPA-83).

Beamhouse: The portion of the tannery where the hides are washed, limed, fleshed, and unhaired when necessary prior to the tanning process (EPA-82/11).

Bearing capacity: The maximum load that a material can support before failing (SW-108ts).

Beater add gasket: An asbestos-containing product that is made of paper intended for use as a gasket, and designed to prevent leakage of liquids, solids, or gases and to seal the space between two sections of a component in circumstances not involving rotary, reciprocating, and helical motions. Major applications of beater-add gaskets include: gaskets for internal combustion engines; carburetors; exhaust manifolds; compressors; reactors; distillation columns; and other apparatus (*see* gasket for more related terms) (40 CFR 763.163-91).

Beater: In the pulp or paper industry, a machine consisting of a tank or tub, usually with a partition or midfeather, and containing a heavy roll revolving against a bedplate. Its function is to separate the wood materials and frees the fibers preparatory to further processing. Fillers, dyestuffs, and sizing materials may be added to the beater and thus incorporated with the paper stock. Many modifications in design have been developed without changing the basic principles (cf. refiner) (EPA-87/10).

Beccari process: A composting process developed by Dr. Giovanni Beccari in 1922. Anaerobic fermentation is followed by a final stage in which decomposition proceeds under partially aerobic conditions; the process was later modified by Verdier and Bordas (SW-108ts).

Becquerel (Bq): A rate of radioactive decay. One becquerel is the quantity of any radioactive nuclide which undergoes one disintegration per second (*see* radiation unit for more related terms) (40 CFR 302.4-App/B-91).

Bed: In mining, the smallest division of a stratified series and marked by a more or less well-defined divisional plane from the materials above and below (cf. fixed bed) (EPA-82/05).

Bed depth: In wastewater treatment, the amount of activated carbon expressed in length units which is parallel to the flow of the stream and through which the stream must pass (EPA-82/11f).

Bed load: Sediment particles resting on or near the channel bottom that are pushed or rolled along by the flow of water (EPA-94/04).

Bedding: The arrangement of rocks in layers or strata (DOI-70/04).

Bedrock: The solid rock beneath the loose material (soil and subsoil) with which most of the land surface of the earth is covered. It is sometimes several hundred feet beneath the surface, but is usually found at a much smaller depth; in places, especially on steep slopes, it has no soil cover at all (DOI-70/04).

Beehive cokemaking: Those operations in which coal is heated with the admission of air in controlled amounts for the purpose of producing coke. There are no by-product recovery operations associated with beehive cokemaking operations (40 CFR 420.11-91).

Beer Lambert law: The law states that the absorbance of a solution of a given chemical species, at a fixed wavelength, is proportional to the thickness of the solution, or the light pathlength, and the concentration of the absorbing species (*see* law for more related terms) (40 CFR 796.3700-91).

Beer's law: The absorbance of a homogeneous sample containing an absorbing substance is directly proportional to the concentration of the absorbing substance. The absorbance, A, is given by the expression, $A = \log_{10}(I_o/I)$, Where I_o is the radiant power incident on the sample and I is the radiant power transmitted through the sample (*see* law for more related terms) (LBL-76/07-air).

Beet pulp: The vegetable matter left after sugar is extracted from cosssettes. Used, wet, dehydrated, or pelleted as commercial cattle feed (*see* pulp for more related terms) (EPA-74/01a).

Begin actual construction: In general, initiation of physical on-site construction activities on an emissions unit which are of a permanent nature. Such activities include, but are not limited to, installation of building supports and foundations, laying of underground pipework, and construction of permanent storage structures. With respect to a change in method of operating this term refers to those on-site activities other than preparatory activities which mark the initiation of the change (40 CFR 51.165-91, *see also* 40 CFR 51.166; 52.21; 52.24-91).

Behavioral toxicity: Motor function (motor activity, coordination strength), sensory function (vision, audition), integrative systems (learning and memory) (*see* endpoint for more related terms) (Course 165.6; EPA-92/12).

Belowground release: Any release to the subsurface of the land and to ground water. This includes, but is not limited to, releases from the belowground portions of an underground storage tank system and belowground releases associated with overfills and transfer operations as the regulated substances move to or from an underground storage tank (*see* release for more related terms) (40 CFR 280.12-91).

Belowground storage facility: A tank or other container located other than as defined as Aboveground (40 CFR 113.3-91).

Belt conveyer: A conveying device that transports material from one location to another by means of an endless belt that is carried on a series of idlers and routed around a pulley at each end (*see* conveyer for more related terms) (40 CFR 60.671-91).

BEN: EPA's computer model for analyzing a violator's economic gain from not complying with the law (EPA-94/04).

Bench: The surface of an excavated area at some point between the material being mined and the original surface of the ground on which equipment can be set, move or operate. A working road or base below a highwall as in contour stripping for coal (EPA-82/10).

Bench scale testing: A treatability study designed to provide quantitative information for the evaluation of a technologies performance for an operable unit. A bench-scale study serves to verify that the technology can meet the anticipated ROD cleanup goals and provides information in support of remedy evaluation (EPA-89/12a).

Bench-scale tests: Laboratory testing of potential cleanup technologies (*see* treatability studies. (EPA-94/04))

Bending strength: Ability of a sheet to bend but not crease (EPA-83).

Beneath the surface of the ground: Beneath the ground surface or otherwise covered with earthen materials (40 CFR 280.12-91).

Beneficial organism: Any pollinating insect, or any pest predator, parasite, pathogen or other biological control agent which functions naturally or as part of an integrated pest management program to control another pest (40 CFR 166.3-91).

Beneficiation area: The area of land used to stockpile ore immediately before the beneficiation process, the area of land used for the beneficiation process, the area of land used to stockpile the tailings immediately after the beneficiation process, and the area of land from the stockpiled tailings to the treatment system (e.g., holding pond or settling pond, and the area of the treatment system) (40 CFR 440.141-91).

Beneficiation:
(1) The process of washing the rock to remove impurities or to separate size fractions (40 CFR 60.401-91).
(2) Concentration or other preparation of ore for smelting by drying, flotation, or magnetic separation (EPA-82/05).

Beneficiation process: The dressing or processing of gold bearing ores for the purpose of:
(1) Regulating the size of, or recovering, the ore or product;
(2) Removing unwanted constituents from the ore; and
(3) Improving the quality, purity, or assay grade of a desired product (40 CFR 440.141-91).

Benefit cost analysis: The economic analysis of a resource development project, taking into account both known and projected factors with a view to discovering the relative efficiency of the project (DOI-70/04).

Benefit: In the context of statement, anything of value, including but not limited to any advantage, preference, privilege, license, permit, favorable decision, ruling status, or loan guarantee (40 CFR 27.2-91).

Benign: Not malignant; remaining localized (EPA-92/12).

Bent glass: A flat glass that has been shaped, while hot, into cylindrical or other curved shapes (*see* glass for more related terms) (EPA-83).

Benthic macroorganism: An organism associated with the bottom material of a lake or stream, or with sludge and deposits in a trickling filter, large enough to be retained by a relatively coarse mesh screen (No.30 sieve, having openings of 0.589 mm) (cf. organism, macro) (DOI-70/04).

Benthic microorganism: A bottom dwelling organism small enough so that it will be retained only by a relatively fine mesh screen (No.100, having openings of 0.149 mm) (cf. organism, micro) (DOI-70/04).

Benthic organism (benthos): A form of aquatic plant or animal life that is found on or near the bottom of a stream, lake or ocean (EPA-89/12).

Benthic region: The bottom layer of a body of water (EPA-89/12). This region supports the benthos, a type of life that not only lives upon, but contributes to the character of the bottom.

Benthic:
(1) Relating to the bottom of a water body, e.g., mud-dwelling mollusks are benthic organisms (DOI-70/04).
(2) Plants and animals dwelling at the bottom of oceans, lakes, rivers, and other surface waters (DOE-91/04).

Benthiocarb ($C_{12}H_{16}NOCl$): A liquid used as a herbicide to control aquatic weeds in rice crops.

Benthos: Organisms such as fauna and flora that live on the bottoms of a water body (EPA-76/03). These include:
(1) Sessile animals, such as the sponges, barnacles, mussels, oysters, some of the worn, and many attached algae;
(2) Creeping forms, such as insects, snails, and certain clan; and
(3) Burrowing forms, which include most clams and worms (LBL-76/07-water).

Bentonite: Highly colloidal clay materials that are added to enamel slips to improve their susceptibility to the action of electrolytes (EPA-82/11e).

Benzalkonium chloride: A yellow white powder used as a fungicide and bactericide.

Benzene (C_6H_6): A colorless liquid used as a solvent (*See also* 40 CFR 420.02-91).

Benzene concentration: The fraction by weight of benzene in a waste as determined in accordance with the procedures specified in

40 CFR 1.355 of this subpart (40 CFR 61.341-91).

Benzene storage tank: Any tank, reservoir, or container used to collect or store refined benzene (40 CFR 61.131-91).

Benzenesulfonic acid ($C_6H_6O_3S$): A strong organic acid used in detergents.

Benzidine based dye applicator: An owner or operator who uses benzidine-based dyes in the dyeing of textiles, leather or paper (40 CFR 129.104-91).

Benzidine manufacturer: A manufacturer who produces benzidine or who produces benzidine as an intermediate product in the manufacture of dyes commonly used for textile, leather and paper dyeing (40 CFR 129.104-91).

Benzidine: The compound benzidine and its salts as identified by the chemical name 4,4 diaminobiphenyl (40 CFR 129.4-91).

Benzoic acid ($C_7H_6O_2$): An aromatic acid used in preserving foods, fats, and juices.

Berkelium (Bk): A radioactive metal with a.n. 97; a.w. 247. The element belongs to group IIIB of the periodic table.

Berm: A horizontal shelf built for the purpose of strengthening and increasing the stability of a slope or to catch or arrest slope slough material (cf. bench) (EPA-75/10c).

Bernoulli equation for a incompressible flow: The basic energy equation of a frictionless, incompressible fluid for the case of steady flow along a single streamline is expressed as: The sum of the pressure energy, the kinetic energy, and the potential energy of a given mass of the fluid is constant. The concept can be expressed as $P/w + V^2/(2g) + z$ = constant. Where static pressure head (P/w) = pressure/(specific weight); velocity head ($V^2/2g$) = velocity square/(2x(acceleration due to gravity, 32.17 ft/sec^2)); and potential head (z) = height (AP-40, p25).

Beryllium (Be):
(1) The element beryllium. Where weights or concentrations are specified, such weights or concentrations apply to beryllium only, excluding the weight or concentration of any associated elements (40 CFR 61.31-91).
(2) An alkaline earth metal with a.n. 4; a.w. 6.939; d. 1.85 g/cc; m.p. 1277° C and b.p. 2770° C. The element belongs to group IIA of the periodic table.
(3) An airborne metal hazardous to human health when inhaled. It is discharged by machine shops, ceramic and propellant plants, and foundries (EPA-94/04).

Beryllium alloy: Any metal to which beryllium has been added in order to increase its beryllium content and which contains more than 0.1 percent beryllium by weight (40 CFR 61.31-91).

Beryllium containing waste: The material contaminated with beryllium and/or beryllium compounds used or generated during any process or operation performed by a source subject to this subpart (40 CFR 61.31-91).

Beryllium copper alloy: Any copper alloy that is alloyed to contain 0.10 percent or greater beryllium (40 CFR 468.02-91).

Beryllium ore: Any naturally occurring material mined or gathered for its beryllium content (40 CFR 61.31-91).

Beryllium propellant: Any propellant incorporating beryllium (40 CFR 61.41-91).

Best available control measures (BACM):: A term used to refer to the most effective measures (according to EPA guidance) for controlling small or dispersed particulates from sources such as roadway dust, soot and ash from woodstoves and open burning of rush, timber, grasslands, or trash (EPA-94/04).

Best available control technology (BACT):
(1) An emission limitation based on the maximum degree of reduction of each pollutant subject to regulation under this Act emitted from or which results from any major emitting facility, which the permitting authority, on a case by case basis, taking into account energy, environmental, and economic impacts and other costs, determines is achievable for such facility through application of production processes and available methods, systems, and techniques, including fuel cleaning or treatment or innovative fuel combustion techniques for control of each such pollutant. In no event shall application of best available control technology result in emissions of any pollutants which will exceed the emissions allowed by any applicable standard established pursuant to section 111 or 112 of this Act (CAA169, *see also* 40 CFR 51.166,; 52.21).
(2) For any specific source, the necessary technology that would produce the greatest reduction of each pollutant regulated byu the Clean Air Act, taking into account energy, environmental, economic and other costs (EPA-94/04).

Best available controls: The degree of emissions reduction that the Administrator determines, on the basis of technological and economic feasibility, health, environmental, and energy impacts, is achievable through the application of the most effective equipment, measures, processes, methods, systems or techniques, including chemical reformulation, product or feedstock substitution, repackaging, and directions for use, consumption, storage, or disposal (CAA183.e-42USC7511b-91).

Best available demonstrated technology (BADT): Under Sec. 306 of CWA, BADT or BADCT (Best Available Demonstrated Control Technology) is a required level of treatment, technology-based limits, for new source direct discharges. New source direct discharges are governed by technology-based limits requiring zero discharge, if possible, by use of BADT. New source is defined as any facility or major modification, the construction of which is commenced after the publication of the proposed regulation (EPA-87/10a; Arbuckle-89).

Best available retrofit technology (BART): An emission limitation based on the degree of reduction achievable through the application of the best system of continuous emission reduction for each pollutant which is emitted by an existing stationary facility. The emission limitation must be established, on a case-by-case basis, taking into consideration the technology available, the costs of compliance, the energy and nonair quality environmental impacts of compliance, any pollution control equipment in use or in existence at the source, the remaining useful life of the source, and the degree of improvement in visibility which may reasonably be anticipated to result from the use of such technology (40 CFR 51.301-91).

Best available technology (BAT): The best technology, treatment techniques, or other means which the Administrator finds, after examination for efficacy under field conditions and not solely under laboratory conditions, area available (taking cost into consideration). For the purposes of setting MCLs for synthetic organic chemicals, any BAT must be at least as effective as granular activated carbon (40 CFR 141.2, *see also* 40 CFR 141.61.b-93).

Best available technology economically achievable (BAT): Under Sec. 301(b)(2)(C) of CWA, BAT is a required level of treatment, technology-based limits, for existing source direct discharges of priority, toxic and non-conventional pollutants. EPA defined BAT as the very best control and treatment measures that have been or are capable of being achieved. Although EPA is required to consider the cost of achieving the required reduction in determining whether a BAT limitation is economically achievable, it is not required to balance cost against effluent reduction benefit as required to the case of the BPT standards. The engineering factors required to be considered-age of equipment and facility, process employed, process changes, non-water quality environmental impacts and so forth-are the same for BAT as for BPT. In general, this technology level represents the best economically achievable performance in any industry category or subcategory. Moreover as a result of 1977 CWA, the emphasis has shifted from control of classical pollutants to the control of toxic substances (EPA-82/05, 10/85a; Arbuckle-89).

Best conventional pollutant control technology (BCT) is defined under section 304(b)(4) of the Act (40 CFR 430.223-91, *see also* 40 CFR 467.02-91; EPA-85/10a; Arbuckle-89).

Best demonstrated available technology (BDAT): Under the Superfund program, EPA defines BDAT in the following manner:
(1) **Best:** Refers to the highest treatment value reflecting well designed and operated treatment technologies.
(2) **Demonstrated:** Refers to a full-scale facility known to be in operation for a waste or wastes with similar treatment characteristics.
(3) **Available:** Refers to the criteria:
　(A) The technology does not present a greater total risk than land disposal;
　(B) If the technology is a proprietary or patented process, whether it can be purchased or licensed from the proprietor; and
　(C) The technology provides substantial treatment (EPA-11/86).

Best demonstrated available technology (BDAT): As identified by EPA, the most effective commercially available means of treating specific types of hazardous waste. The BDATs may change with advances in treatment technologies (EPA-94/04).

Best demonstrated technology (BDT): Under Sec. 111 of CAA, EPA must identify BDT. BDT is defined as the best system of continuous emission reduction that has been adequately demonstrated taking into account costs and other environmental and energy impacts (Johnston, 6/90).

Best management practice (BMP):
(1) The methods, measures or practices selected by an agency to meet its nonpoint source control needs. BMPs include but are not limited to structural and nonstructural controls and operation and maintenance procedures. BMPs can be applied before, during and after pollution-producing activities to reduce or eliminate the introduction of pollutants into receiving waters (40 CFR 130.2-91, *see also* 40 CFR 122.2; 232.2-91).
(2) Methods that have been determined to be the most effective, practical means of preventing or reducing pollution from nonpoint sources (EPA-94/04).

Best performance treatment technology (BPTT): Those treatment technologies selected by EPA and currently in-use in the pesticide industry. They are biological oxidation, activated carbon, hydrolysis, metals separation, chemical oxidation, resin adsorption, and steam stripping (EPA-85/10).

Best practicable control technology currently available (BPT): Is defined under section 304(b)(1) of the Act (40 CFR 430.222-91, *see also* 40 CFR 467.02-91).

BPT is a required level of treatment, technology-based limits, for existing source direct discharges required by Section 301(b) of the 1972 FWPCA. The Act provided for the establishment of nationally applicable technology-based effluent limitations on an industry-by-industry basis. For existing industrial discharges, Section 301 directs the achievement:

　…by July 1, 1977, of effluent limitations which will require application of the best practicable control technology (BPT) currently available, and by July 1, 1983, of effluent limitations which will require application of the best available technology (BAT) economically achievable.

EPA defined BPT as an average of the best existing performance by well-operated plants within each industrial category or subcategory. In establishing the 1977 effluent guidelines for BPT, EPA emphasized end-of-pipe treatment rather than in-plant control measures. In establishing the 1983 effluent limitations for BAT, EPA considered both in-plant process changes and end-of-pipe treatment measures. As the time for achievement of Best Practicable Technology (BPT) is long past, its primary relevance now is a basis for setting subsequent standards. BPT effluent limitation are established based on the degree of effluent reduction that this technology can attain. In general, this technology level represents the existing performances of well-known technologies for control of familiar pollutants associated with industry. The BPT definition was essentially unchanged by the 1977 amendments (EPA-82/05; 10/85a; Arbuckle-89).

Best practicable waste treatment technology (BPWTT): The cost-effective technology that can treat wastewater, combined sewer overflows and nonexcessive infiltration and inflow in publicly owned or individual wastewater treatment works, to meet the applicable provisions of:
(i) 40 CFR 133--secondary treatment of wastewater;
(ii) 40 CFR 125, Subpart G--marine discharge waivers;

(iii) 40 CFR 122.44(d)--more stringent water quality standards and State standards; or
(iv) 41 FR 610 (February 11, 1976)--Alternative Waste Management Techniques for Best Practicable Waste Treatment (treatment and discharge, land application techniques and utilization practices, and reuse) (40 CFR 35.2005-91).

Beta attenuation monitor: An instrument that measures the absorption of beta radiation as it traverses a small area onto which aerosol particles are collected by means of inertial impaction (EPA-81/09).

Beta particle:
(1) An elementary particle emitted by radioactive decay, that may cause skin burns. It is halted by a thin sheet of paper (EPA-89/12).
(2) A negatively-charged subatomic particle emitted during decay of certain radioactive elements. A beta particle is identical to an electron (cf. radiation, beta) (EPA-88/08a).
(3) An elementary particle emitted from a nucleus during radioactive decay; it is negatively charged, identical to an electron, and easily stopped, as by a thin sheet of metal (DOE-91/04).
(4) *See* particle for more related terms.

Beta radiation:
(1) Emitted from a nucleus during fission. Beta radiation can be stopped by an inch of wood or a thin sheet of aluminum (cf. particle, beta) (EPA-88/08a).
(2) Electrons emitted from the decay of some radioactive elements. The beta particles that may cause skin burns can be stopped by a thin metal sheet (EPA-88/09a).
(3) *See* radiation for more related terms.

Bias:
(1) A systematic error that is consistently negative or consistently positive. The mean of errors resulting from a series of observations that does not tend toward zero (*see* analytical parameters--laboratory for more related terms) (EPA-83).
(2) *See also* definition under uncertainty.

Biennial report: In RCRA, a report (EPA Form 8700-13A) submitted by generators of hazardous waste to the Regional Administrator due March 1 of each even numbered year. The report includes information on the generators activities during the previous calendar. The owner or operator of a treatment, storage, and disposal facility must also prepare and submit a biennial report using EPA Form 8700-1313 (EPA-86/01).

BIER V (Biological Effects of Ionizing Radiation V): Referring to the fifth in a series of committee reports from the National Research Council (National Research Council 1990) (DOE-91/04).

Bifurcation: A phenomenon where a stack plume divides, sometimes visibly, into separate plumes (NATO-78/10).

Billet: A long, round slender cast product used as a raw material in subsequent forming operations (EPA-83/03a).

Billing electric demand: The demand upon which billing to a customer is based, as specified in a rate schedule or contract. It may be based on the contract year, a contract minimum, or a previous maximum and, therefore, does not necessarily coincide with the actual measured demand of the billing period (*see* electric demand for more related terms) (EPA-83).

Bimetal: Beverage containers with steel bodies and aluminum tops; handled differently from pure aluminum in recycling (EPA-94/04).

Bimolecular pathway: *See* synonym, bimolecular reaction.

Bimolecular reaction (or bimolecular pathway): A chemical transformation where two species collide and change chemical identity. It is a function of temperature, pressure, reaction atmosphere, and concentration of reaction species (EPA-88/12).

Binder:
(1) Organic materials and resins which do not contain VOMs (40 CFR 52.741-91).
(2) A material used to promote cohesion between particles of carbon or graphite to produce solid carbon and graphite rods of pieces (EPA-83/03).
(3) The film forming ingredient in paint that binds the pigment particles together (EPA-79/12b).

Binding commitment: A legal obligation by the State to a local recipient that defines the terms for assistance under the SRF (40 CFR 35.3105-91).

Bioaccumulants: Substances that increase in concentration in living organisms as they take in contaminated air, water, or food because the substances are very slowly metabolized or excreted (*see* biological magnification) (EPA-94/04).

Bioaccumulation: *See* definition under biota.

Bioaccumulation factor (BAF): The ratio of a substance's concentration in tissue versus its concentration in ambient water, in situations where the organism and the food chain are exposed (EPA-91/03).

Bioaccumulative: Capable of increasing in concentration in living organisms (that are very slowly metabolized or excreted) as they breathe contaminated air, drink contaminated water, or eat contaminated food (cf. biological magnification) (EPA-89/12).

Bioaerosol: An aerosol that contains microorganisms (cf. *see* aerosol) (EPA-88/09a).

Bioassay (assay or animal bioassay):
(1) A test for evaluating the relative potency of a chemical by comparing its effect on a living organism with the effect of a standard preparation on the same type of organism. Bioassays are frequently used in the pharmaceutical industry to evaluate the potency of vitamins and drugs. Bioassay and toxicity test are not synonymous (cf. Ames test or toxicity test) (EPA-85/09).
(2) A test used to evaluate the relative potency of a chemical or a mixture of chemicals by comparing its effect on a living organism with the effect of a standard preparation on the

same type of organism. Bioassays frequently are used in the pharmaceutical industry to evaluate the potency of vitamins and drugs (EPA-91/03).
(3) The determination of the potency (bioactivity) or concentration of a test substance by noting its effects in live animals or in isolated organ preparations, as compared with the effect of a standard preparation (EPA-92/12).
(4) Study of living organisms to measure the effect of a substance, factor, or condition by comparing before-and-after exposure or other data (EPA-94/04).

Bioavailability: Refers to
(1) The rate and extent to which the administered compound is absorbed, i.e., reaches the systemic circulation (40 CFR 795.223-91, *see also* 40 CFR 795.228; 795.230; 795.231; 795.232-91).
(2) The property of a toxicant that governs its effect on exposed organisms. A reduced bioavailability would have a reduced toxic effect. (Material that is not bioavailable is not available to cause toxic effects) (EPA-85/09).
(3) A measure of the physicochemical access that a toxicant has to the biological processes of an organism. The less the bioavailability of a toxicant, the less its toxic effect on an organism (EPA-91/03).
(4) The degree to which a drug or other substance becomes available to the target tissue after administration or exposure (EPA-92/12).

Biochemical: Concerned with life including growth and other activities.

Biochemical oxidation: This usually refers to the biological treatment.

Biochemical oxygen demand (BOD) (biological oxygen demand or oxygen depleting effect):
(1) The amount of oxygen required for aerobic bacteria to completely oxidize the organic matter within a specified time and at a given temperature - an index to the degree of organic pollution in the water. When discharged to a watercourse, waste containing BOD constituents will consume dissolved oxygen in the water; the BOD indicates the rate at which the oxygen is used up. Waters that receive high BOD waste undergo reduction of oxygen and consequent damage to aquatic life (DOI-70/04).
(2) Quantity of dissolved oxygen utilized (consumed) in the biochemical oxidation of organic matter in a specified time (e.g., for 5 days quantity, BOD is then expressed as BOD5) and at a specified temperature. It is not related to the oxygen requirements in chemical combustion; it is determined entirely by the biodegradability of the material and by the amount of oxygen utilized by the microorganisms during oxidation (EPA-87/10).
(3) A measure of the amount of oxygen consumed in the biological processes that break down organic matter in water. The greater the BOD, the greater the degree of pollution (EPA-94/04).
(4) *See* oxygen for more related terms.

Biochemical specific locus mutation: A genetic change resulting from a DNA lesion causing alterations in proteins that can be detected by electrophoretic methods (40 CFR 798.5195-91)

Biochemistry: The chemistry of living beings.

Biocide:
(1) A chemical toxic to a biological life (EPA-75/07).
(2) A compound that has the ability to inactivate microorganisms. In a general term, biocide refers to compounds that can be used as an algicide (inactivate algae), bactericide (inactivate bacteria), fungicide (inactivate fungi), and viricide (inactivate viruses), or some combination of the above. It is equivalent to disinfectant (EPA-88/09a).

Bioconcentration factor (BCF):
(1) The measure of the tendency for a substance to accumulate in the tissue of an aquatic organism. BCF is determined by the extent of partitioning of a substance, at equilibrium, between the tissue of an aquatic organism and water. As the ratio of concentration of a substance in the organism divided by the concentration in water, higher BCF values reflect a tendency for substances to accumulate in the tissue of aquatic organisms (unitless) (40 CFR 300-App/A-91, *see also* 40 CFR 797.1520; 797.1560; 797.1830-91).
(2) The ratio of the concentration of a chemical in aquatic organisms (ug chemical/g organism) to the amount in water at equilibrium (ug chemical/g water) (EPA-10/85). BCF = [Concentration of chemical at equilibrium in organism (net weight)]/(mean concentration of chemical in water).
(3) The ratio of a substance's concentration in tissue versus its concentration in water, in Situations where the food chain is not exposed or contaminated. For non-metabolized substances, it represents equilibrium partitioning between water and organisms (EPA-91/03).

Bioconcentration:
(1) The accumulation of a chemical in tissues of an organism (such as a fish) to levels greater than in the surrounding medium in which the organism lives (EPA-94/04).
(2) *See also* definition under biota.

Bioconversion: The biological conversion of wastes. Also known as biological treatment.

Biocriteria: *See* synonym, biological criteria.

Biodegradability: The susceptibility of substance to decomposition by microorganisms; specifically, the rate at which compounds may be chemically broken down by bacterial and/or natural environmental factors (EPA-83/06a).

Biodegradable:
(1) Degradable (breaking down of the physical and/or chemical structure of a compound) by microbial (microorganism) actions (SW-108ts).
(2) Capable of decomposing rapidly under natural conditions (EPA-94/04).

Biodegradable detergent: One that decomposes quickly as a result of the action of organisms, eliminating form in wastewater. Biodegradable is defined as having at least 90 percent surfactant reduction, or as having surfactant concentration no higher than 0.5

mg/l (*see* detergent for more related terms) (DOI-70/04).

Biodegradable material: Waste material which is capable of being broken down by microorganisms into simple, stable compounds such as carbon dioxide and water. Most organic wastes, such as food wastes and paper, are biodegradable (EPA-89/11).

Biodegradable plastic: A plastic that can be broken down by microorganisms such as bacteria and fungi; as generally used, the term does not necessarily mean complete degradation into carbon dioxide and water (OTA-89/10).

Biodegradation:
(1) The chemical reaction of a substance induced by enzymatic activity of microorganisms (cf. photodegradation) (40 CFR 300-App/A-91).
(2) Decomposition or break-down of an organic substrate in the environment by microbial actions. It is the decay caused by the environmental factors such as light, temperature, humidity, and microorganisms. Concentration prediction of a compound over time due to biodegradation is $L_t = L_o \times e^{-kt}$. Where: L_o = Load (or concentration) at time 0; L_t = Load remaining at time t; k = biodegradation constant for a specific compound; and e = exponent (natural log system) (Course 165.6).

Biodisc: The trade name of a rotating biological contactor.

Biodiversity: Refers to the variety and variability among living organisms and the ecological complexes in which they occur. Diversity can be defined as the number of different items and their relative frequencies. For biological diversity, these items are organized at many levels, ranging from complete ecosystems to the biochemical structures that are the molecular basis of heredity. Thus, the term encompasses different ecosystem, species, and genes (EPA-94/04).

Biofilter: *See* synonym, trickling filter.

Biofiltration: The action of a fixed film reactor.

Biogas: The gas from anaerobic sludge digestion, mainly carbon dioxide and methane.

Biogasification: A resource recovery process for the extraction of methane resulting from anaerobic decomposition of organic material (EPA-83).

Biogeochemical cycle: The chemical interactions between the atmosphere, hydrosphere, lithosphere, and biosphere.

Biological (or biologicals):
(1) The preparations made from living organisms and their products, including vaccines, cultures, etc., intended for use in diagnosing, immunizing or treating humans or animals or in research pertaining thereto (40 CFR 259.10-91).
(2) Vaccines, cultures and other preparations made from living organisms and their products, intended for use in diagnosing, immunizing, or treating humans or animals, or in related research (EPA-94/04).

Biological additive: The microbiological cultures, enzymes, or nutrient additives that are deliberately introduced into an oil discharge for the specific purpose of encouraging biodegradation to mitigate the effects of the discharge (40 CFR 300.5-91).

Biological assessment: An evaluation of the biological condition of a waterbody using biological surveys and other direct measurements of resident biota in surface waters (EPA-91/03).

Biological concentration factor: *See* synonym, concentration factor.

Biological control agent: Any living organism applied to or introduced into the environment that is intended to function as a pesticide against another organism declared to be a pest by the Administrator (40 CFR 152.3-91).

Biological control: In pest control, the use of animals and organisms that eat or otherwise kill or out-compete pests (EPA-94/04).

Biological cooling tower: A cooling tower which is seeded with microorganisms and fed with nutrients in which biological degradation of organics occurs (EPA-74/04c).

Biological criteria (or biocriteria): Narrative expressions or numeric values of the biological characteristics of aquatic communities based on appropriate reference conditions. Biological criteria serve as an index of aquatic community health (EPA-91/03).

Biological factor: *See* synonym, biotic factor.

Biological film (or microbial film): A film of slime formed in biological treatment reactors such as rotating biological contactor.

Biological filter: *See* synonym, trickling filter.

Biological filtration: The process of passing a liquid through a biological filter containing media on the surfaces of which zoogleal films develop which absorb fine suspended, colloidal, and dissolved solids, and release end products of biochemical action (EPA-74/01a).

Biological indicator (index of pollution or index organism): Various classifications of organisms indicating levels of organic pollution, e.g., river classifications as follows:
(1) Clean water with BOD5 below 3 mg/liter, well oxygenated and without evidence of pollution.
(2) Slightly de-oxygenated and known to receive some pollution.
(3) Less than 50% saturated with dissolved oxygen or with substances believed to be at toxic concentrations.
(4) Grossly polluted water, offensive in smell or appearance or with a BOD5 of 12 mg/liter or more, or completely de-oxygenated in any reach, or with toxic substances, incapable of supporting fish life (cf. pollution indicator organism).

Biological inhibitor: A chemical that inhibits or disrupts biological processes (EPA-75/07).

Biological integrity: The condition of the aquatic community inhabiting unimpaired waterbodies of a specified habitat as measured by community structure and function (EPA-91/03).

Biological magnification: Refers to the process whereby certain substances such as pesticides or heavy metals move up the food chain, work their way into rivers or lakes, and are eaten by aquatic organisms such as fish, which in turn are eaten by large birds, animals or humans. The substances become concentrated in tissues or internal organs as they move up the chain (*see* bioaccumulative) (EPA-94/04).

Biological monitoring (or biomonitoring):
(1) The determination of the effects on aquatic life, including accumulation of pollutants in tissue, in receiving waters due to the discharge of pollutants:
 (A) By techniques and procedures, including sampling of organisms representative of appropriate levels of the food chain appropriate to the volume and the physical, chemical, and biological characteristics of the effluent, and
 (B) At appropriate frequencies and locations (CWA502-33USC1362-91).
(2) The living organisms in water quality surveillance used to indicate compliance with water quality standards or effluent limits and to document water quality trends. Methods of biological monitoring may include, but are not limited to, toxicity testing such as ambient toxicity testing or whole effluent toxicity testing (EPA-91/03).
(3) The use of living organisms to test the suitability of effluent for discharge into receiving waters and to test the quality of such waters downstream from the discharge.
(4) Analysis of blood, urine, tissues, etc., to measure chemical exposure in humans (EPA-89/12).

Biological oxidation:
(1) The way bacteria and microorganisms feed on and decompose complex organic materials. Used in self-purification of water bodies and in activated sludge wastewater treatment, such as activated sludge, trickling filter and aerated stabilization depends on this principle (EPA-87/10).
(2) Decomposition of complex organic materials by microorganisms. Occurs in self-purification of water bodies and in activated sludge wastewater treatment (EPA-94/04).

Biological oxygen demand (BOD):
(1) An indirect measure of the concentration of biologically degradable material present in organic wastes. It usually reflects the amount of oxygen consumed in five days by biological processes breaking down organic waste (EPA-94/04).
(2) cf. biochemical oxygen demand.

Biological product: In the pharmaceutical industry, a medicinal product derived from animals or humans (e.g., vaccines, toxoids, antisera, and human blood fractions) (EPA-83/09).

Biological slime: Biological film (Scott-81).

Biological stabilization (or cell stablizaton):
(1) Reduction in the net energy level of organic matter as a result of the metabolic activity of organisms (EPA-74/04c).
(2) Occurs when the initially available substrate has been utilized by the bacteria in a bio-system to produce relatively stable, oxidation-resistant cellular end products. At this point the BOD curve approaches zero slope (in the absence of nitrification) (LBL-76/07-water).

Biological survey (or biosurvey): The collecting, processing, and analyzing of a representative portion of the resident aquatic community to determine its structural and/or functional characteristics (EPA-91/03).

Biological treatment (or biological wastewater treatment):
(1) A treatment process that uses microorganisms to break down toxic organic waste contaminants into simple less-toxic compounds (EPA-89/12a).
(2) A waste treatment process by placing an aqueous waste stream in contact with a mixture of microorganisms which decompose the organic compounds in the waste stream. The organics may be either solvent or solid in the influent waste stream to be amenable to biodegradation. The microorganisms rely on enzymes to catalyze organic decomposition reactions, and the enzymes require water to remain active. Biological treatment processes do not alter or destroy inorganics. In fact, concentrations of soluble inorganics should be kept low so that enzymatic activity is not inhibited (cf. significant biological treatment) (EPA-83/03).
(3) A treatment technology that uses bacteria to consume organic waste (EPA-94/04).
(4) *See* treatment for more related terms.

Biological wastewater treatment: *See* synonym, biological treatment.

Biologicals: *See* synonym, biological (EPA-94/04)

Biologically available: Utilizable as a nutrient by the biota in the environment (LBL-76/07-water).

Biologically refractive: A substance which is partially or totally non-biodegradable in biological waste treatment processes (EPA-87/10).

Bioluminescence: The emission of light from living organisms.

Biomagnification: *See* definition under biota.

Biomass (or biota):
(1) The living material in a botanical ecosystem. Organic residue from the processing of agricultural and forestry products, and consumer discards (EPA-83).
(2) The plant and animal life of a region (pertaining to biota) (DOE-91/04).
(3) All of the living material in a given area; often refers to vegetation (EPA-94/04).

Biome:
(1) Entire community of living organisms in a single major ecological area (*see* biotic community) (EPA-94/04).
(2) A large naturally occurring biomass that grows under similar general type of environmental conditions.

Biomedical waste: All waste streams that require special handling including pathological waste, infectious waste, hazardous waste, and other waste generated in health care facilities and laboratories (*see* medical waste for more related terms) (OME-87/05).

Biomonitoring:
(1) The use of living organisms to test the suitability of effluents for discharge into receiving waters and to test the quality of such waters downstream from the discharge (EPA-94/04).
(2) Analysis of blood, urine, tissues, etc., to measure chemical exposure in humans (EPA-94/04).

Biopolymer: A polymer directly produced by living or once living cells or cellular components (40 CFR 723.250-91).

Bioremediation: Use of living organisms to clean up oil spills or remove other pollutants from soil, water, or wastewater; use of organisms such as non-harmful insects to remove agricultural pests or counteract diseases of trees, plants, and garden soil (EPA-94/04).

Bioretention: *See* definition under biota.

Biosorption: A contact stabilization method. The organic pollutants in the sewage are adsorbed on to the activated sludge in the contact tank and then metabolized by the bacteria during the re-aeration (stabilization) stage (Scott-81).

Biosphere:
(1) The part of the earths crust, waters, and atmosphere where living organisms can subsist (LBL-76/07-bio).
(2) The portions of the earth and its atmosphere where living things exist (DOE-91/04).
(3) The portion of Earth and its atmosphere that can support life (EPA-94/04).

Biostabilizer: A machine that converts solid waste into compost by grinding and aeration (EPA-94/04).

Biostat: A substance which inhibits the biological activity (growth) of bacteria (LBL-76/07-water).

Biostimulation: A general term used to describe the complex set of factors involved in the growth of algae (and other organisms) in a receiving waterway due to the addition of nutrients (DOD-78/01).

Biota: The animal and plant life of a given region (EPA-94/04).

Biota behavior: Behavior of chemicals in biota include (Course 165.6):
(1) **Bioaccumulation (or bioretention):**
  (A) The uptake and, at least temporary, storage of a chemical by an exposed animal. The chemical can be retained in its original form and/or as modified by enzymatic and non-enzymatic reactions in the body (40 CFR 798.7100-91).
  (B) The retention and concentration of a substance by an organism. Accumulation of a large amount of a substance in the body by ingesting small amounts of the substance over an extended time (Course 165.6).
  (C) The process by which a compound is taken up by an aquatic organism, both from water and through food (EPA-91/03).
(2) **Bioconcentration:**
  (A) The net accumulation of a substance directly from water into and onto aquatic organisms (*see also* biota) (40 CFR 797.1520-91, *see also* 40 CFR 797.1560; 797.1830-91).
  (B) The process by which a compound is absorbed from water through gills or epithelial tissues and is concentrated in the body (EPA-91/03).
  (C) The accumulation of a chemical in tissues of an organism (such as fish) to levels that are greater than the level in the medium (such as water) in which the organism residues (Course 165.6).
(3) **Biomagnification:**
  (A) Processes by which tissue levels of the material increase via bioaccumulation as the material is transported through two or more trophic levels (cf. biological magnification) (Course 165.6).
  (B) The process by which the concentration of a compound increases in species occupying successive trophic levels (EPA-91/03).
(4) **Biotransformation:** Conversion of a substance into other compounds by organisms; including biodegradation (Course 165.6).

Biotechnology:
(1) The application of engineering and technological principles to the life sciences (EPA-88/09a).
(2) Techniques that use living organisms or parts of organisms to produce a variety of products (from medicines to industrial enzymes) to improve plants or animals or to develop microorganisms to remove toxics from bodies of water, or act as pesticides (EPA-94/04).

Biotic community: A naturally occurring assemblage of plants and animals that live in the same environment and are mutually sustaining and interdependent (*see* biome) (EPA-94/04).

Biotic environment: The environment of living organisms.

Biotic factor (or biological factor): In ecology, those environmental factors, e.g., competition, predation, etc.; which are the result of living organisms and their activities (DOD-78/01).

Biotic potential: The growth rate of living organisms under ideal conditions.

Biotoxicity: Toxic to flora and fauna (EPA-89/12a).

Biotransfer: The process by which living organisms, such as bacteria, can convert a chemical compound into another (LBL-76/07-bio).

Biotransformation:
(1) Conversion of a substance into other compounds by organisms; includes biodegradation (EPA-94/04).
(2) *See also* definition under biota.

Biphenyl $C_{12}H_{10}$): A white crystal used as a heat-transfer medium.

Bird hazard: An increase in the likelihood of bird/aircraft collisions that may cause damage to the aircraft or injury to its occupants (40 CFR 257.3.8-91).

bis-: A prefix indicating double or twice.

Bisexual: *See* synonym, hermaphrodite.

Bismuth (Bi): A metallic element with a.n. 83; a.w. 208.98; d. 9.8 g/cc; m.p. 271° C and b.p. 1560° C. The element belongs to group VA of the periodic table.

Bisulfate: *See* synonym, hydrogen sulphate.

Bitterns: The saturated brine solution remaining after precipitation of sodium chloride in the solar evaporation process (40 CFR 415.161-91).

Bituminous coal: Solid fossil fuel classified as bituminous coal by ASTM Designation D38877 (incorporated by reference-*see* 40 CFR 60.17) (40 CFR 60.251-91). It is intermediate hardness containing between 50 and 92 percent fixed carbon (*see* coal for more related terms) (EPA-82/10).

Bituminous coatings: The black or brownish coating materials which are soluble in carbon disulfide, which consist mainly of hydrocarbons, and which are obtained from natural deposits or as residues from the distillation of crude oils or of low grades of coal (40 CFR 52.741-91).

Black and white film: A film consisting of a support, usually a plastic film which is coated with a light sensitive emulsion and an outer protective layer. The emulsion is adhered to the supporting base with a special layer called a sub. The emulsion contains: gelatin, silver salts of bromide, iodide, chloride, sensitizers, hardeners, and emulsion plasticizers (EPA-10/80).

Black body: A hypothetical body that absorbs all incident radiation and reflects none.

Black core (or blackheart): Most fireclays and brick-clays contain carbonaceous matter; if a brick shaped from such clays is fired too rapidly, this carbonaceous matter will not be burned out before vitrification begins. The presence of carbon and the consequent reduced state of any iron compounds in the center of the fired brick result in a black core or black heart (EPA-83).

Blackheart: *See* synonym, black core.

Black liquor: The used cooking liquor recovered from a (wood pulp processing) digester. It may also be referred to as spent cooking liquor. Strong black liquor refers to the liquor after it has been concentrated by an evaporator to a level suitable for combustion. Prior to evaporation, it is referred to as weak black liquor (*see* liquor for more related terms) (EPA-87/10).

Black liquor oxidation system: The vessels used to oxidize, with air or oxygen, the black liquor, and associated storage tank(s) (40 CFR 60.281-91).

Black liquor solids: The dry weight of the solids which enter the recovery furnace in the black liquor (40 CFR 60.281-91).

Black lung: A disease of the lungs caused by habitual inhalation of coal dust (EPA-89/12).

Black water: The water that contains animal, human, or food waste (cf. night soil and *see* water for more related terms) (EPA-94/04).

Blade: The broad flat or concave part of a machine. Other blade related terms include:
- Earth blade
- Landfill blade
- U-blade

Blank:
(1) A sample analyzed for the purpose of determining and assessing background contaminants in the field or laboratory. There are many types of blanks for selection depending on sampling applications, e.g., deionized water is one of blanks which is used to rinse automatic sampler prior to collection of samples (EPA-79/12).
(2) Optical glass formed by pressing into the rough shape and size required in the finished article. Also known as pressing (EPA-83).

Other blank related terms include:
- Calibration blank
- Equipment blank
- Field blank (*see* synonym, equipment blank)
- Trip blank (*see* synonym, equipment blank)
- Laboratory blank
- Reagent blank
- Site blank

Blank mold (or parison mold): The metal mold which first shapes the glass in the manufacture of hollow ware (EPA-83).

Blanket assembly: In a heavy-water reactor, lithium-aluminum alloy clad tubes positioned in a ring surrounding the radial reflector zone. They prevent neutron damage to the reactor vessel's metal wall by absorbing neutrons from the reflector zone, and they produce tritium (DOE-91/04).

Blanket feed: A method for charging to produce an even distribution of batch material across the width of the furnace (EPA-83).

Blanket insulation: The relatively flat and flexible insulation in coherent sheet form, furnished in units of substantial area. Batt insulation is included in this term (40 CFR 248.4-91).

Blanking: Any cutting desired shapes out of sheet metal by means of dies (EPA-6/83a).

Blast cleaning barrel: A complete enclosure which rotates on an axis, or which has an internal moving tread to tumble the parts, in order to expose various surfaces of the parts to the action of an automatic blast spray (29 CFR 1910.94a-91).

Blast cleaning barrel: A complete enclosure which rotates on an axis, or which has an internal moving tread to tumble the parts, in order to expose various surfaces of the parts to the action of an automatic blast spray (29 CFR 1910.94a-91).

Blast cleaning room: A complete enclosure in which blasting operations are performed and where the operator works inside of the room to operate the blasting nozzle and direct the flow of the abrasive material (29 CFR 1910.94a-91).

Blast cleaning room: A complete enclosure in which blasting operations are performed and where the operator works inside of the room to operate the blasting nozzle and direct the flow of the abrasive material (29 CFR 1910.94a-91).

Blast furnace: Any furnace used to recover metal from slag (*see* furnace for more related terms) (40 CFR 60.131-91, *see also* 40 CFR 60.181-91).

Blast furnace slag: The slag produced in iron blast-furnaces. It may be cooled slowly in air or more rapidly by quenching in water resulting in granulation or fritting. The structure and properties of the slag can be changed markedly by the method of cooling (EPA-83).

Blast gate: A slide metal damper in a duct, usually used to regulate the flow of forced air (SW-108ts).

Blasting cabinet: An enclosure where the operator stands outside and operates the blasting nozzle through an opening or openings in the enclosure (29 CFR 1910.94a-91).

Bleach fix or blix: A solution used in some color processing that functions both as a bleach and as a fix (EPA-80/10).

Bleach:
(1) A step in color film processing whereby the silver image is converted back to silver halides (EPA-80/10).
(2) Treating pulp so as to remove any coloring. *See also* bleeding (EPA-83).

Bleached paper: The paper made of pulp that has been treated with bleaching agents (*see* paper for more related terms) (40 CFR 250.4-91).

Bleaching agent: A chemical compound with an oxidizing or reducing characteristics, e.g., sodium hypochlorite, sulfur dioxide, sodium acid sulfite, or hydrogen peroxide.

Bleaching:
(1) Removing colored components from a textile. Common bleaches are hydrogen peroxide, sodium hypochloride, and sodium chlorite (EPA-74/06b).
(2) The brightening and delignification of pulp by the addition of oxidizing chemicals such as chlorine or reducing chemicals such as sodium hypo-chloride (EPA-87/10).

Bleed: *See* synonym, blowdown.

Bleeding: Dissolving a color from paper or pulp. *See* Brown Stock (EPA-83).

Blend fertilizer: A mixture of dry, straight and mixed fertilizer materials (cf. mixed fertilizer) (40 CFR 418.71-91).

Blinding: Plugging of the screen apertures with slightly oversized particles (EPA-88/08a).

Blister copper: Copper with 96 to 99 percent purity and appearing blistered; made by forcing air through molten copper matte (EPA-83/03a).

Blister: An imperfection in glass; a relatively large bubble or gaseous inclusion (EPA-83).

Bloach: An imperfection resulting from incompletely grinding plate glass, caused by a low place in the plate which retains part of the original rough surface (EPA-83).

Block reek: A scratch imperfection caused by cullet lodged in the felt in the polishing operation (EPA-83).

Block void: *See* synonym, top void.

Blood: Human blood, human blood components and products made from human blood (regulated under the Medical Waste Tracking Act) (FR-89/05).

Blood brain barrier: The barrier created by semipermeable cell walls and membranes to passage of some molecules from the blood to the cells of the central nervous system (LBL-76/07-bio).

Blood fractionation: The separation of human blood into its various protein fractions (EPA-83/09).

Blood products:
(1) Any product derived from human blood, including but not limited to blood plasma, platelets, red or white blood corpuscles, and other derived licensed products, such as interferon, etc (40 CFR 259.10-91).
(2) Any product derived from human blood, including but not limited to blood plasma, platelets, red or white corpuscles, and derived licensed products such as interferon (EPA-94/04).

Blood-to-air partition coefficient: Ratio of concentrations for a given chemical achieved between blood and air at equilibrium (EPA-92/12).

Bloodborne pathogens: Pathogenic microorganisms that are present in human blood and can cause disease in humans. These pathogens include but are not limited to, hepatitis B virus (HBV) and human immunodeficiency virus (HIV) (29 CFR 1910).

Bloom:
(1) An excessive growth of algae in a body of water due to an oversupply of dissolved nutrients (pollutants). It may impart a disagreeable odor to the water, cause fish to die, and impair the use of the water for drinking or recreation (cf. eutrophication) (DOI-70/04).
(2) Large, readily visible masses of microscopic and macroscopic plant life, such as green algae, occurring in bodies of water (LBL-76/07-water).
(3) A surface film on glass resulting from attack by the atmosphere or from the deposition of smoke or other vapors (EPA-83).

(4) A proliferation of algae and/or higher aquatic plants in a body of water; often related to pollution, especially when pollutants accelerate growth (EPA-94/04).

Blow and blow process: The process of forming hollow ware in which both the preliminary and final shapes are formed by air pressure (EPA-83).

Blow mold: The metal mold in which a blown glass article is finally shaped (EPA-83).

Blow pit: A large tank under a (wood processing) digester which receives the discharged chips and liquor from the digester. A constructed stainless steel plate within the blow pit acts to break up the chip structure into individual fibers of pulp upon impact (EPA-87/10).

Blowby: Leakage of liquid or gas between cylinder and piston during operation.

Blowdown (bleed or purge):
(1) The minimum discharge of recirculating water for the purpose of discharging materials contained in the water, the further buildup of which would cause concentration in amounts exceeding limits established by best engineering practice (40 CFR 401.11-91, *see also* 40 CFR 423.11-91).
(2) The amount of concentrated liquor wasted (ejected) in a recycle system in order to maintain an acceptable equilibrium of contaminants in any process liquor (EPA-82/11).

Blower (or air blower): A fan used to force air or gas under pressure (SW-108ts).
Force air blower (*see* definition under burner component)
Primary combustion chamber air blower
Secondary combustion chamber air blower

Blower door: A device consisting of an instrumented fan which can be mounted in an existing doorway of a house. By determining the air flows through this fan required to achieve different degrees of house pressurization and depressurization, the blower door permits determination of the tightness of the house shell, and an estimation of the natural in-filtration rate (EPA-88/08).

Blowing: The injection of air or oxygen-enriched air into a molten converter bath (40 CFR 61.171-91).

Blowing still: The equipment in which air is blown through asphalt flux to change the softening point and penetration rate (40 CFR 60.471-91).

Blowing tap: Any tap in which an evolution of gas forces or projects jets of flame or metal sparks beyond the ladle, runner, or collection hood (40 CFR 60.261-91).

Blowing tower: *See* synonym, spray drying tower.

Blue flame: *See* definition under flame combustion.

Blue green algae: One of major unicellular algae. It can use atmospheric nitrogen as a nutrient in developing its cells and can give water an unpleasant taste and smell.

Blue stain: A biological reaction caused by a stain producing fungi which causes a blue discoloration of sapwood, if not dried within a short time after cutting (EPA-74/04).

Blue: In the leather industry, the state or condition of hides subsequent to chromium tanning and prior to retanning. Hides in this stage of processing are characteristically blue in color (EPA-82/11).

Board insulation: The semirigid insulation preformed into rectangular units having a degree of suppleness, particularly related to their geometrical dimensions (40 CFR 248.4-91).

Bobbin: An assembly of the positive current collector and cathode material, usually molded into a cylinder (EPA-84/08).

BOD exertion: Oxygen requirement of a bio-system; biochemical oxygen demand equivalent (LBL-76/07-water).

BOD5:
(1) Five-day biochemical oxygen demand (40 CFR 401.11-91).
(2) A measure of biological decomposition of organic matter in a water sample. It is determined by measuring the oxygen required by microorganisms to oxidize the organic contaminants of a water sample under standard laboratory conditions. The standard conditions include incubation for five days at temperature 20° C (EPA-74/04).
(3) The amount of dissolved oxygen consumed in five days by biological processes breaking down organic matter (EPA-94/04).

BOD5 input: The biochemical oxygen demand of the materials entered into process. It can be calculated by multiplying the fats, proteins and carbohydrates by factors of 0.890, 1.031 and 0.691 respectively. Organic acids (e.g., lactic acids) should be included as carbohydrates. Composition of input materials may be based on either direct analyses or generally accepted published values (40 CFR 405.11-91, *see also* 40 CFR 405.21; 405.31; 405.41; 405.51; 405.61; 405.71; 405.81; 405.91; 405.101; 405.111; 405.121-91).

BOD7: The biochemical oxygen demand as determined by incubation at 20° C for a period of 7 days using an acclimated seed. Agitation employing a magnetic stirrer set at 200 to 500 rpm may be used (40 CFR 417.151-91, *see also* 40 CFR 417.161; 417.171; 417.181-91).

Bodied chemical fusion agent: A chemical fluid containing portion of the parent geomembrane that, after the application of pressure and after the passage of a certain amount of time, results in the chemical fusion of two essentially similar geomembrane sheets, leaving behind only that portion of the parent material. (Manufacturers and installers should be consulted for the various types of chemical fluids used with specific geomembranes in order to inform workers and inspectors) (EPA-91/05).

Body burden: The total amount of a specific substance (for example, lead) in an organism, including the amount stored and the amount absorbed (LBL-76/07-bio).

Body fluid: The liquid emanating or derived from humans and limited to blood: cerebrospinal, synovial, pleural, peritoneal and pericardial fluids; and semen and vaginal secretions (40 CFR 259.10-91).

Body resistance: The sum total of body mechanisms which glace barriers to the progress of invasion of pathogenic organisms (EPA-83).

Body style: The level of commonality in vehicle construction as defined by number of doors and roof treatment (e.g., sedan, convertible, fastback, hatchback) (40 CFR 86.082.2-91, *see also* 40 CFR 600.002.85-91).

Body type: The name denoting a group of vehicles that are either in the same car line or in different car lines provided the only reason the vehicles qualify to be considered in different car lines is that they are produced by a separate division of a single manufacturer (40 CFR 86.082.2-91).

Bog: A type of wetland that accumulates appreciable peat deposits. Bogs depend primarily on precipitation for their water source, and are usually acidic and rich in plant residue with a conspicuous mat of living green moss (EPA-94/04).

Boil out: A procedure, usually utilizing heat and chemicals, to clean equipment such as evaporators, heat-exchangers, and pipelines (EPA-87/10).

Boiler:
(1) A solid fuel burning appliance used primarily for heating spaces, other than the space where the appliance is located, by the distribution through pipes of a gas or fluid heated in the appliance. The appliance must be tested and listed as a boiler under accepted American or Canadian safety testing codes. A manufacturer may request an exemption in writing from the Administrator by stating why the testing and listing requirement is not practicable and by demonstrating that his appliance is otherwise a boiler (40 CFR 60.531-91, *see also* 40 CFR 60.561; 60.611; 60.661; 260.10-91).
(2) A vessel in which a liquid is vaporized, usually at constant pressure, as a result of the controlled application of heat (EPA-83).

Main components of a boiler include (EPA-81/12, p7-6):
(1) Fire wall or water wall
(2) Superheater
(3) Convection tubes and steam drum
(4) Economizer
(5) Preheater

Other boiler related terms include:
- Packaged boiler
- Wall fired boiler
- Waste heat boiler

Boiler, chemical process, furnace, and incinerator (CRWI-89/05):
(1) Boiler
  - Purpose: Energy recovery
  - Input: Fossil fuels, refuse derived fuels, ignitable wastes
  - Output: Recovered energy, wastes
(2) Chemical process
  - Purpose: Chemical production
  - Input: Raw materials, intermediates, byproducts
  - Output: Products, secondary materials, fuels, wastes
(3) Furnace
  - Purpose: Material recovery
  - Input: Undiscarded materials, spent solvents, spent catalysts, other spent materials
  - Output: Reclaimed materials, fuels, wastes
(4) Incinerator
  - Purpose: Waste treatment (burn waste)
  - Input: Discarded materials, solid wastes, hazardous wastes
  - Output: Waste disposal

Boiler and Industrial and Industrial Furnaces Rule (BIF): Regulations passed by the U.S. EPA in 1991 that govern the actual burning of hazardous waste boilers and industrial furnaces including cement kilns (ETI-92).

Boiler blowdown: Wastewater resulting from purging of solid and waste materials from the boiler system. A solids build up in concentration as a result of water evaporation (steam generation) in the boiler (cf. blowdown) (EPA-76/03).

Boiler feedwater: Water supplied to a boiler (EPA-82/11f).

Boiler operating day: A 24-hour period during which fossil fuel is combusted in a steam generating unit for the entire 24 hours (40 CFR 60.41a-91, *see also* 40 CFR 60-App/G-91).

Boiler scale: An incrustation of salts deposited on the waterside of a boiler as a result of the evaporation of water (EPA-82/11f).

Boiler tube: Tubes contained in a boiler. For a **water tube boiler**, water passes through the tubes during its conversion into steam. For a **fire tube boiler**, flames pass the tubes and transfer the heat to the water outside the tubes (EPA-82/11f).

Boiler water: The circulating water in a boiler after the generated steam has been separated and before the incoming feedwater or added chemical becomes mixed with it (*see* water for more related terms) (EPA-83).

Boiler water tube: A tube in a boiler having the water and steam on the inside and heat applied to the outside (EPA-83).

Boiling: The transition of a substance from the liquid to the gaseous phase usually at constant pressure and temperature (EPA-83).

Boiling point temperature: The temperature at which a liquid changes to a vapor. It is the temperature where the pressure of the liquid equals atmospheric pressure. The opposite change in phases is the condensation point temperature (*see* temperature for more related terms) (Course 165.5).

Boltzmann's constant (k): k = the ratio of the universal gas constant to the Avogadro number = 1.380658×10^{-23} J/K.

Bond: *See* synonym, chemical bond.

Bond fission (or bond homolysis): A unimolecular reaction cleavage of a chemical bond (typically the lowest energy bond in a molecule) between two atoms, a molecule and an atom, or two molecules produces two radical species (EPA-88/12).

Bond homolysis: *See* synonym, bond fission.

Bond paper:
(1) A generic category of paper used in a variety of end use applications such as forms (*see* form bond), offset printing, copy paper, stationery, etc. In the paper industry, the term was originally very specific but is now very general (40 CFR 250.4-91).
(2) A term originally meant paper used for printing bonds and stocks, now generally refers to high grade papers used for letters and high quality printed work. It is surface-sized for better writing and printing quality (EPA-83).
(3) *See* paper for more related terms.

Bond release: The time at which the appropriate regulatory authority returns a reclamation or performance bond based upon its determination that reclamation work (including, in the case of underground mines, mine sealing and abandonment procedures) has been satisfactorily completed (40 CFR 434.11-91).

Bond rupture: Same as bond fission (EPA-88/12).

Bonding: The process of uniting using an adhesive or fusible ingredient (EPA-83/06a).

Book paper:
(1) A generic category of papers produced in a variety of forms, weights, and finishes for use in books and other graphic arts applications, and related grades such as tablet, envelope, and converting papers (40 CFR 250.4-91).
(2) A general term for a group of papers of suitable quality for the printing trade, excluding newsprint (EPA-83).
(3) *See* paper for more related terms.

Boom:
(1) A floating device used to contain oil on a body of water.
(2) A piece of equipment used to apply pesticides from ground equipment such as a tractor or truck (*see* sonic boom) (EPA-94/04).

Booster charge: A charge that is ignited by the electric match and, in turn, initiates combustion or detonation in the propellant (EPA-76/03).

Booster cycle: The period during which additional hydraulic pressure is exerted to push the last charge of solid waste into a transfer trailer or a container attached to a stationary compactor (SW-108ts).

Booster pump: A pump which is used to increase the pressure of liquid or gas in a pipe.

Bore: Inside diameter of a cylinder.

Boreal life zone: The living organisms in the zone of northern geographic regions.

Boric acid (H_3BO_3): A white or colorless solid used for weatherproofing wood and fireproofing fabrics.

Boride: A class of boron-containing compounds, primarily calcium boride, used as a constituent in refractory materials. Metallic impurities that often accompany the use of these materials include titanium, zirconium, hafnium, vanadium, niobium, tantalum, chromium, molybdenum, tungsten, thorium, and uranium (EPA-85/10a).

Boring and turning: Scrap metals from machining of casting, rods, bars, and forgings (EPA-76/12).

Boring: Enlarging a hole by removing metals with a single or occasionally a multiple point cutting tool moving parallel to the axis of rotation of the work or tool. It includes:
(1) **Single point boring:** Cutting with a single point tool.
(2) **Precision boring:** Cutting to tolerances held within narrow limits.
(3) **Gun boring:** Cutting of deep holes.
(4) **Jig boring:** Cutting of high recision and accurate location holes.
(5) **Groove boring:** Cutting accurate recesses in hole walls (EPA-83/06a).

Boron (B): A nonmetallic element with a.n. 5; a.w. 10.81; d. 2.34 g/cc; m.p. 2030° C and b.p. 2550° C. The element belongs to group IIIA of the periodic table.

Borosilicate recipe: The glass product composition of the following approximate ranges of weight proportions: 60 to 80 percent silicon dioxide, 4 to 10 percent total R_2O (eg, Na_2O and K_2O), 5 to 35 percent boric oxides, and 0 to 13 percent other oxides (40 CFR 60.291-91).

Borosilicate vial: Glass vial used to store organic samples (EPA-88/12).

Botanical: Drugs made from a part of a plant, such as roots, bark or leaves (EPA-83/09).

Botanical pesticide:
(1) A pesticide produced from naturally occurring chemicals found in some plants. Examples include nicotine, pyrethrum, strychnine, and rotenone (EPA-85/10).
(2) A pesticide whose active ingredient is a plant-produced chemical such as nicotine or strychnine. Also called a plant-derived pesticide (EPA-94/04).
(3) *See* pesticide for more related terms.

Bottle bill: Proposed or enacted legislation which requires a returnable deposit on beer or soda containers and provides for retail store or other redemption. Such legislation is designed to discourage use of throwaway containers (EPA-94/04).

Bottom ash:
(1) In waste incineration, the solid material that remains on a hearth or falls off the grate after thermal processing is complete (40 CFR 240.101-91).
(2) The ash that drops out of the furnace gas stream in the furnace and in the economizer sections. Economizer ash is included when it is collected with bottom ash (40 CFR

423.11-91).
(3) In fossil fuel combustion, the solid residue remaining from combustion of fuel in a boiler furnace (EPA-75/02d). Economizer ash is included when it is collected with bottom ash (EPA-82/11a).
(4) The non-airborne combustion residue from burning pulverized coal in a boiler; the material which falls to the bottom of the boiler and is removed mechanically; a concentration of the non-combustible materials, which may include toxics (EPA-94/04).
(5) *See* ash for more related terms.

Bottom blown furnace: Any basic oxygen process furnace (BOPF) in which oxygen and other combustion gases are introduced to the bath of molten iron through tuyeres in the bottom of the vessel or through tuyeres in the bottom and sides of the vessel (*see* furnace for more related terms) (40 CFR 60.141a-91).

Bottom land hardwood: Forested freshwater wetlands adjacent to rivers in the southeastern United States, especially valuable for wildlife breeding, nesting and habitat (EPA-94/04).

Bottom receiver: A container or tank used to receive and collect the heavier bottoms fractions of the distillation feed stream that remain in the liquid phase (40 CFR 264.1031-91).

Botulinus organism: Those organisms that cause acute food poisoning (*see* organism for more related terms) (EPA-74/06).

Bounce: The unscheduled point contact opening(s) after initial closure and before scheduled reopening (40 CFR 85.2122(5)(ii)(B)-91)

Bound water: The water that is held by a system such as water in tissues or in soil (*see* water for more related terms).

Boundary condition: A set of mathematical conditions which must be satisfied by the solutions of a differential equation on the boundaries of the region in which the solution is sought (NATO-78/10).

Boundary layer model: A mathematical model, mostly in the form of a set of partial differential equations, which describes the structure of the boundary layer as a function of time and space (NATO-78/10).

Boundary layer: When a fluid passes a surface, the fluid tangential velocity at the immediate vicinity of the surface is zero. The velocity increases from zero to the velocity of the stream. The transition region between zero velocity and stream velocity is the boundary layer.

Bounding accident: A postulated accident that is defined to encompass the range of anticipated accidents and used to evaluate the consequences of accidents at facilities for fuel and target fabrication and processing, waste management, and hazardous materials handling (DOE-91/04).

Bourdon gage: A device for measuring the gas pressure in a system. The unit is usually denoted by psig (pound force per square inch gage) (Jones-60, p15).

Boussinesq approximation: The assumption that a fluid, in which density variations are caused by temperature variations, can be considered as incompressible. The density variations are only taken into account when producing buoyancy forces (NATO-78/10).

Box model: A simulation model of atmospheric dispersion for which it is assumed that the concentration is uniformly distributed over a specified volume or a box. The height of the box is determined by the mixing height. The horizontal dimensions are determined by the size of the area in which the emissions take place. Across the sides of the box, transport of the air pollution takes place. For steady conditions, the concentration c in the box becomes: $c=Q/F$, where Q is the total emission strength and F the ventilation factor (NATO-78/10).

Boxboard cuttings: The term consists of baled new cuttings of paperboard such as are used in the manufacture of folding paper cartons, set-up boxes and similar boxboard products (EPA-83).

Boyle's law: Boyle's law states that, when the temperature (T) is held constant, the volume (V) of a given mass of a perfect gas of a given composition varies inversely as the absolute pressure varies (*see* law for more related terms) (EPA-81/12, p2-7).

Brackish: Mixed fresh and salt water (EPA-94/04).

Brackish water: A mixture of fresh and salt water (*see* water for more related terms) (EPA-89/12).

Brackish: Saline, unpalatable (LBL-76/07-water).

Braille paper: The paper designed to emboss well so that blind can read by touch (*see* paper for more related terms) (EPA-83).

Brake block: An asbestos-containing product intended for use as a friction material in drum systems for vehicles rated at 26,001 pounds gross vehicle weight rating (GVWR) or more (40 CFR 763.163-91).

Brake horsepower: The fan power the user pays for. It is the power to operate the unit. The term, gas horsepower, is used to describe the power delivered by the fan to the polluted gas stream. The efficiency of the fan is given by the ratio of the gas horsepower to the brake horsepower (EPA-84/09).

Branched chain: *See* definition under chain.

Brass aluminum: The brass to which aluminum has been added to improve resistance to corrosion (EPA-83).

Brass (or bronze):
(1) Any metal alloy containing copper as its predominant constituent, and lesser amounts of zinc, tin, lead, or other metals (40 CFR 60.131-91).
(2) Primarily an alloy of about 88 copper and 8 to 10% tin, plus zinc. The name is now used to describe other alloys such as aluminum bronze, manganese bronze, and beryllium bronze (EPA-83).

Other brass related terms include:
- Red brass
- Yellow brass

Brazing: Joining metals by flowing a thin layer, capillary thickness, of non-ferrous filler metal into the space between them. Bonding results from the intimate contact produced by the dissolution of a small amount of base metal in the molten filler metal, without fusion of the base metal. The term brazing is used where the temperature exceeds 425-C (EPA-83/06a).

Breach of containment: The loss of containment through accidental means, deliberate means, or equipment failure causing the escape of recombinant organisms from containment barriers (EPA-88/09a).

Break point of chlorination (or chlorination break point):
(1) The process in which chlorine is added to the media (water, sewage, or industrial wastes) in amounts large enough to cause all of the nitrogen in ammonia and ammonia-derivatives present to be reduced to elemental nitrogen such that a free chlorine residual is available (*see* chlorination for more related terms) (EPA-88/09a).
(2) Addition of chlorine to water until the chlorine demand has been satisfied (EPA-94/04).

Break through: A crack or break in a filter bed that allows the passage of floc or particulate matter through a filter; will cause an increase in filter effluent turbidity (EPA-94/04).

Break through point: The point at which impurities first appear in the effluent of a granular activated carbon adsorption bed (EPA-87/10a).

Break through time: The elapsed time between initial contact of the hazardous liquid chemical with the outside surface of a protective clothing materia and the time at which the chemical can be detected at the inside surface of the material by means of the chosen analytical technique (NIOSH-84/10).

Break water: A natural or artificial barrier that serves to break the force of waves and thereby shelters craft in a harbor or protects a beach from erosion (*see* water for more related terms) (DOI-70/04).

Breakdown voltage:
(1) The voltage level at which the capacitor fails (40 CFR 85.2122(a)-(6)(ii)(C)-91).
(2) The voltage at which a discharge occurs between two electrodes (EPA-83/03).

Breaker point: The mechanical switch operated by the distributor cam to establish and interrupt the primary ignition coil current (40 CFR 85.2122(a)(5)(ii)(A)-91).

Breakout tank: A tank used to:
(1) Relieve surges in a hazardous liquid pipeline system; or
(2) Receive and store hazardous liquid transported by a pipeline for reinjection and continued transportation by pipeline (40 CFR 195.2-91).
(3) *See* tank for more related terms.

Breakover angle: The supplement of the largest angle, in the plan side view of an automobile, that can be formed by two lines tangent to the front and rear static loaded radii arcs and intersecting at a point on the underside the automobile (40 CFR 86.084.2-91).

Breakthrough capacity: The adsorption capacity of a packed bed where traces of pollutants begin to appear in the exit gas stream. The same units as equilibrium capacity (ratio of the weight of the adsorbate retained to the weight of adsorbent) are employed for breakthrough capacity (EPA-84/09).

Breathing zone:
(1) The location in the atmosphere at which persons breathe (EPA-83/06).
(2) The area of a room in which occupants breathe as they stand, sit, or lie down (EPA-88/09b).

Breeching bypass: An arrangement whereby breechings and dampers permit the intermittent use of two or more passages to direct or divert the flow of the products of combustion (SW-108ts).

Breeching: A passage that conducts the products of combustion to a stack or chimney (SW-108ts).

Breeder: A nuclear reactor that produces more fuel than it consumes (EPA-74/11).

Breeze: Breeze related terms include:
- Land breeze
- Sea breeze

Brick: A refractory brick made from fireclay. Other brick related terms include:
- Alumina diaspore fireclay brick
- Basic brick
- Firebrick
- High duty fireclay brick
- Insulating brick
- Intermediate duty fireclay brick
- Super duty fireclay brick

Brick veneer: A single layer or tier of masonry or similar materials securely attached to a wall for the purposes of providing ornamentation, protection, or insulation, but not bonded or attached to intentionally exert common action under load (*see* veneer for more related terms) (EPA-88/08).

Bridge furnace wall: A partial partition between combustion chambers over which pass the products of combustion (*see* furnace wall for more related terms) (SW-108ts).

Bridge wall cover: Refractory blocks spanning the space between the bridge walls (EPA-83).

Bright dipping: The immersion of all or part of a (metal) workpiece in a medium designed to clean or brighten the surface and leave a protective surface coating on the workpiece (EPA-83/06a).

Brightness (or whiteness): The degree of whiteness of a pulp. The standard used is magnesium oxide with a reflectance of 100%. Pulp brightness is expressed as a % of the MgO reflectance measured at a dominant wave length of 458 millimicrons (EPA-83).

Brightness unit: An increment of measurement to assess the brightness of paper (EPA-87/10).

Brine: Water saturated with a salt (cf. brackish water) (EPA-82/11f).

Brine mud: Waste material, often associated with well-drilling or mining, composed of mineral salts or other inorganic compounds (EPA-94/04).

Briquetter: A machine that compresses a material, such as metal turnings or coal dust, into small pellets (SW-108ts).

Bristols: Heavier papers, 0.006-inch or more in thickness (EPA-83).

British thermal unit (BTU): The unit for measuring quantity of heat energy, such as the heat content of fuel. It is the amount of heat energy necessary to raise the temperature of one pound of water one degree Fahrenheit. *See also* Heating Value (EPA-83).

Brittleness: Point of breaking when bent (EPA-83).

Brix: A hydrometer scale calibrated to read percent sugar by weight in pure sugar (juice) solutions. Originated by Balling, improved and corrected by Brix (EPA-74/01a; 74/03).

Broadcast application: The spreading of pesticides over an entire area (EPA-94/04).

Broker: An individual or group of individuals that act as an agent or intermediary between the sellers and buyers of recyclable materials (EPA-89/11).

Bromine (Br): A halogen element with a.n. 35; a.w. 79.909; d. 3.12 g/cc; m.p. -7.2° C; and b.p. 58° C. The element belongs to group VIIA of the periodic table.

Bromine water: A nonmetallic halogen liquid, normally deep red, corrosive and toxic, which is used as an oxidizing agent (*see* water for more related terms) (EPA-83/06a).

Bromofenoxim ($C_{13}H_7N_3O_6Br_2$): A powder used as a herbicide to control weeds.

Bronsted acid: A chemical substance which can act as a proton source (*see* acid for more related terms).

Bronze: *See* synonym, brass.

Brood stock: The animals which are cultured to produce test organisms through reproduction (40 CFR 797.1300-91, *see also* 40 CFR 797.1330-91).

Brown earth: A type of soil which is rich in organic matter derived from the annual leaf fall of trees and shrubs.

Brown mud lake: The diked reservoir (tailing pond) used to impound brown mud (EPA-74/03b).

Brown mud: The final solid waste remaining after the alumina is leached (EPA-74/03b).

Brown paper: The paper usually made from unbleached kraft pulp and used for bags, sacks, wrapping paper, and so forth (see paper for more related terms) (40 CFR 250.4-91).

Brown stock washer system: The brown stock washers and associated knotters, vacuum pumps, and filtrate tanks used to wash the pulp following the digestion system. Diffusion washers are excluded from this definition (40 CFR 60.281-91).

Brown stock: Unbleached pulp obtained from the digesters (EPA-83, see also EPA-87/10).

Brownian motion: Continuous random motion of microscopic particles in a gas or liquid medium.

Bruce Ames test: See synonym, Ames test.

Brush or wipe coating: A manual method of applying a coating using a brush, cloth, or similar object (40 CFR 52.741-91).

BTX storage tank: Any tank, reservoir, or container used to collect or store benzene-toluene-xylene or other light-oil fractions (40 CFR 61.131-91).

Bubble: A system under which existing emissions sources can propose alternate means to comply with a set of emissions limitations; under the bubble concept, sources can control more than required at one emission point where control costs are relatively low in return for a comparable relaxation of controls at a second emission point where costs are higher (EPA-94/04).

Bubble policy:
(1) An EPA policy that allows a plant complex with several facilities to decrease pollution from some facilities while increasing it from others, so long as total results are equal to or better than previous limits. Facilities where this is done are treated as if they exist in a bubble in which total emissions are averaged out. Complexes that reduce emissions substantially may bank their credits or sell them to other industries (cf. emission trading).
(2) See emissions trading (EPA-94/04).

Bubbler: An absorber in which the sample gas is introduced below the surface of a liquid absorbent. For increased efficiency, the gas stream may be broken up into small bubbles by being forced through restricted openings as in the fritted gas tip (LBL-76/07-air).

Bubbling fluidized bed combustor: A fluidized bed combustor in which the majority of the bed material remains in a fluidized state in the primary combustion zone (see fluidized bed combustion for more related terms) (40 CFR 60.51a-91).

Bucket elevator: A conveying device for metallic minerals consisting of a head and foot assembly that supports and drives an endless single or double strand chain or belt to which buckets are attached (40 CFR 60.381-91, *see also* 40 CFR 60.671-91).

Bucket: An open container affixed to the movable arms of a wheeled or tracked vehicle to spread solid waste and cover material, and to excavate soil; also a type of grapple used with an overhead crane (EPA-83).

Budget: The financial plan for the spending of all Federal and matching funds (including in kind contributions) for a technical assistance grant project as proposed by the applicant, and negotiated with and approved by the Award Official (40 CFR 35.4010-91).

Buffer: A solution or liquid whose chemical makeup neutralizes acids or bases without a great change in pH (EPA-94/04).

Buffer strips: Strips of grass or other erosion-resisting vegetation between or below cultivated strips or fields (EPA-94/04).

Buffer zone: A neutral area which acts as a protective barrier separating two conflicting forces. An area which acts to minimize the impact of pollutants on the environmental or public welfare. For example, a buffer zone is established between a composting facility and neighboring residents to minimize odor problems (EPA-89/11).

Buffer: A solution containing either a weak acid and its salt or a weak base and its salt which thereby resists changes in acidity or basicity, i.e., resists changes in pH (stabilizes the pH values, acidity or alkalinity) (EPA-87/10a).

Buffing dust: Small pieces of leather removed in the buffing operation. Buffing dust also includes small particles of abrasive used in the operation and is of a coarse powder consistency (EPA-82/11).

Buffing:
(1) In metal finishing, an operation to provide a high luster to a surface. The operation, which is not intended to remove much materials, usually follows polishing (EPA-83/06a).
(2) In leather finishing, a light sanding operation applied to the grain or underside of leather and also to splits. Buffing smooths the grain surface and improves the nap of the underside of the leather (EPA-82/11).
(3) In constructing a landfill liner, an inaccurate term often used to describe the grinding of polyethylene flexible membrane liners to remove surface oxides and waxes in preparation of extrusion seaming (EPA-89/09, *see also* EPA-91/05).

Building block technique: A method of allocating effluent limitations guidelines to multi-subcategory plants where the effluent limitations guidelines for that given plant would represent a production-weighted sum of effluent limitations guidelines which apply to each specific subcategory (EPA-76/03).

Building cooling load: The hourly amount of heat that must be removed from a building to maintain indoor comfort (measured in British Thermal Units BTUs) (EPA-94/04).

Building completion: The date when all but minor components of a project have been built, all equipment is operational and the project is capable of functioning as designed (40 CFR 35.2005-91).

Building effects on dispersion: The influence by the flow patterns and turbulence around buildings on the dispersion of emissions from or in the neighborhood of these buildings. For instance, pollution can get trapped inside the flow separation region behind a building (NATO-78/10).

Building insulation: A material, primarily designed to resist heat flow, which is installed between the conditioned volume of a building and adjacent unconditioned volumes or the outside. This term includes but is not limited to insulation products such as blanket, board, spray-in-place, and loose-fill that are used as ceiling, floor, foundation, and wall insulation (40 CFR 248.4-91).

Building: The erection, acquisition, alteration, remodeling, improvement or extension of treatment works (40 CFR 35.2005-91, *see also* 40 CFR 60.671-91).

Building related illness: A discrete, identifiable disease or illness. It can be traced to a specific pollutant or source within a building (Contrast with sick building syndrome) (EPA-88/09b).

Building, structure, facility, or installation: All of the pollutant-emitting activities which belong to the same industrial grouping, are located on one or more contiguous or adjacent properties, and are under the control of the same person (or persons under common control) except the activities of any vessel. Pollutant-emitting activities shall be considered as part of the same industrial grouping if they belong to the same Major Group (i.e., which have the same two-digit code) as described in the Standard Industrial Classification Manual, 1972, as amended by the 1977 Supplement (U.S. Government Printing Office stock numbers 4101-0065 and 003-005-00176-0, respectively) (40 CFR 51.165-ii-91).

Bulb: A glass envelope which encloses an incandescent lamp or an electronic tube (EPA-83/03).

Bulk: The term denotes thickness of a sheet of paper (EPA-83).

Bulk asbestos: Any quantity of asbestos fiber of any type or grade, or combination of types or grades, that is mined or milled with the purpose of obtaining asbestos. This term does not include asbestos that is produced or processed as a contaminant or an impurity (*see* asbestos for more related terms) (40 CFR 763.63-91).

Bulk bed washer: A type of wet dust collector consisting of a bed of lightweight spheres through which the dust laden air must pass while being sprayed by water or another scrubbing liquor (EPA-85/10a).

Bulk container: A large container that can either be pulled or lifted mechanically onto a service vehicle or emptied mechanically into a service vehicle (40 CFR 246.101-91).

Bulk density: The mass of bulky materials per unit volume (*see* density for more related terms) (EPA-84/09).

Bulk gasoline plant: A gasoline storage and distribution facility with an average throughput of 76,000 liters (20,000 gal) or less on a 30-day rolling average that distributes gasoline to gasoline dispensing facilities (40 CFR 52.741-91, *see also* 40 CFR 60.111b-91).

Bulk gasoline terminal: Any gasoline facility which receives gasoline by pipeline, ship or barge, and has a gasoline throughput greater than 75,700 liters per day. Gasoline throughput shall be the maximum calculated design throughput as may be limited by compliance with an enforceable condition under Federal, State or local law and discoverable by the Administrator and any other person (40 CFR 60.501-91).

Bulk resin: A resin which is produced by a polymerization process in which no water is used (*see* resin for more related terms) (40 CFR 61.61-91).

Bulk sample: A small portion (usually thumbnail size) of a suspect asbestoscontaining building material collected by an asbestos inspector for laboratory analysis to determine asbestos content (EPA-94/04).

Bulk terminal: Any facility which receives liquid product containing benzene by pipelines, marine vessels, tank trucks, or railcars, and loads the product for further distribution into tank trucks, railcars, or marine vessels (40 CFR 61.301-91).

Bulkhead: An air-restraining barrier constructed for a long-term control of radon-222 and radon-222 decay products.

Bulking agent: A material used to add volume to another material to make it more porous to air flow. For example, municipal solid waste may act as a bulking agent when mixed with water treatment sludge (EPA-89/11).

Bulking sludge: The sludge produced due to the inability of an activated sludge to settle. It is thus difficult to separate from the effluent and affect the quality of wastewater treatment (*see* sludge for more related terms).

Bulky waste (or oversize waste):
(1) Large items of solid waste such as household appliances, furniture, large auto parts, trees, branches, stumps, and other oversize wastes whose large size precludes or complicates their handling by normal solid wastes collection, processing, or disposal methods (*see* waste for more related terms) (40 CFR 243.101-91).
(2) Large items of waste materials, such as appliances, furniture, large auto parts, trees, stumps (EPA-94/04).

Bulldozer: A tracked vehicle equipped with an earth blade (SW-108ts).

Bundle: A structure composed of three or more fibers in a parallel arrangement with each fiber closer than one fiber diameter (40 CFR 763-App/A-91).

Bunker C oil: A general term used to indicate a heavy viscous fuel oil (OME-88/12).

Buoyancy:
(1) The ability or tendency to float or rise in liquid or air.
(2) An upward force upon a parcel of a fluid in a gravitational field due to a density difference between the parcel and surrounding fluid (NATO-78/10).
(3) The (buoyant) force exerted on a particle suspended in an air stream. The magnitude of the force is given by the weight of the fluid displaced by the particle. In air pollution calculations, this force is generally neglected (EPA-84/09).

Buret: A marked glass used to deliver variable volumes of liquid.

Burial ground (graveyard): A disposal site for radioactive waste materials that uses earth or water as a shield (EPA-94/04).

Burial operation: *See* synonym, trenching (40 CFR 257.3.6-91).

Burn:
(1) Combustion or incineration of substances.
(2) Connection of terminals, posts, or connectors in a lead acid battery by welding (EPA-84/08).
(3) In brick manufacturing, the degree of heat treatment to which refractory bricks are subjected when manufactured (SW-108ts).

Burn off: The process of severing an unwanted portion of a glass article by fusing the glass (EPA-83).

Burn rate (burning rate or combustion rate):
(1) The rate at which test fuel is consumed in a wood heater. Measured in kilogram of wood (dry basis) per hour (kg/hr) (40 CFR 60-App/A(method 28 & 28A)-91).
(2) The quantity of solid waste incinerated or the amount of heat released during incineration. The rate is usually expressed in pounds of solid waste per square foot of burning area per hour or in Btus per square foot of burning area per hour (SW-108ts, *see also* EPA-89/03b).

Burndown period: The period of time in an incinerator's operating cycle during which no additional waste is charged to the incinerator and the primary combustion chamber temperature is maintained above a minimum temperature (using auxiliary burners as necessary) to facilitate the solid phase combustion of the waste bed (cf. cooldown period) (EPA-89/03b).

Burner: A triggering mechanism used to ignite and oxidize hydrocarbon fuels (AP-40). Other burner related terms include:
- Afterburner
- Auxiliary fuel burner
- Catalytic afterburner (*see* synonym, secondary burner)
- Conical burner
- Duct burner
- Flat flame burner
- Fume incinerator (*see* synonym, secondary burner)
- Gas burner
- Manual burner
- Oil burner
- Pilot
- Primary burner
- Primary combustion chamber burner (*see* synonym, primary burner)
- Refuse burner
- Residential burner
- Secondary burner
- Secondary combustion chamber burner
- Teepee burner (*see* synonym, conical burner)
- Thermal oxidizer (*see* synonym, secondary burner)
- Vapor incinerator
- Windbox burner

Burner block: A refractory block with one or more orifices through which fuel is admitted to a furnace (EPA-83).

Burner component: The main components of a burner system include a forced air blower, a fuel train, pilot and main burners, and most importantly a flame safeguard system (EPA-89/03b).

Burning agent: Those additives that, through physical or chemical means, improve the combustibility of the materials to which they are applied (40 CFR 300.5-91).

Burning area: The horizontal projection of a grate, a hearth, or both (SW-108ts).

Burning hearth: A solid surface to support the solid fuel or solid waste in a furnace during drying, ignition, or combustion, without air openings in it. The surface upon which materials are placed for combustion (*see* hearth for more related terms) (SW-108ts).

Burning rate: *See* synonym, burn rate.

Burnishing: A surface finishing process in which minute surface irregularities are displaced rather than removed (40 CFR 471.02-91).

Burnout: A measure of ash quality; it is the percentage of the ash that is inorganic materials (EPA-89/03b).

Burnt lime: Calcined limestone (CaO), MgO (dolomitie), or a mixture of these (cf. dolomite or limestone) (EPA-83).

Bursting strength: A measure of the ability of a paper sheet to resist rupture when pressure is applied to one of its sides. Bursting strength is tested by a specified instrument, under specified conditions - the Mullen or pop test (EPA-83).

Busbar: A heavy rigid, metallic conductor, usually uninsulated, used to carry a large current or to make a common connection between several circuits (EPA-83/03).

Bushing: An insulating structure including a central conductor with provision for mounting on a barrier (conducting or otherwise), for the purpose of insulating the conductor from the barrier and conducting current from one side of the barrier to the other (EPA-83/03).

Business machine: A device that uses electronic or mechanical methods to process information, perform calculations, print or copy information, or convert sound into electrical impulses for transmission, such as:
(1) Products classified as typewriters under SIC Code 3572;
(2) Products classified as calculating and accounting machines under SIC Code 3574;
(3) Products classified as calculating and accounting machines under SIC Code 3574;
(4) Products classified as telephone and telegraph equipment under SIC Code 3661;
(5) Products classified as office machines, not elsewhere classified, under SIC Code 3579; and
(6) Photocopy machines, a subcategory of products classified as photographic equipment under SIC Code 3861 (40 CFR 60.721-91).

Butterfly damper: A plate or blade installed in a duct, breeching, flue connection, or stack that rotates on an axis to regulate the gas flow (*see* damper for more related terms) (OME-88/12).

Button cell: A tiny, circular battery made for a watch or for other microelectronic applications (EPA-84/08).

Butyl rubber: A synthetic rubber made by the solution polymerization of isobutylene and isoprene (*see* rubber for more related terms) (EPA-74/12a).

Buy back (or buy back center): A facility that pays individuals for recyclable materials and further processes them for market (EPA-89/03b; 89/11).

Buy back center: *See* synonym, buy back.

By compound: By individual stream components, not carbon equivalents (40 CFR 60.611-91, *see also* 40 CFR 60.661-91).

Bypass:
(1) Any system that prevents all or a portion of the kiln or clinker cooler exhaust gases from entering the main control device and ducts the gases through a separate control device. This does not include emergency systems designed to duct exhaust gases directly to the atmosphere in the event of a malfunction of any control device controlling kiln or clinker cooler emissions (40 CFR 60.61-91, *see also* 40 CFR 122.41; 403.17-91).
(2) An act of intentional noncompliance during which waste treatment facilities are circumvented in emergency situations (EPA-85/10).

Bypass stack: The stack that vents exhaust gases to the atmosphere from the bypass control device (*see* stack for more related terms) (40 CFR 60.61-91).

Bypass the control device: To operate the glass melting furnace without operating the control device to which that furnace's emissions are directed routinely (40 CFR 61.161-91).

Byproduct coke process: A process in which coal is carbonized in the absence of air to permit recovery of the volatile compounds and to produce coke (EPA-74/06a).

Byproduct cokemaking: Those cokemaking operations in which coal is heated in the absence of air to produce coke. In this process, byproducts may be recovered from the gases and liquids driven from the coal during cokemaking (40 CFR 420.11-91).

Byproduct material:
(1) Any radioactive material (except special nuclear material) yielded in or made radioactive by exposure to the radiation incident to the process of producing or utilizing special nuclear material (10 CFR 20.3-91, *see also* 10 CFR 30.4; 40.4-91).
(2) By-product material, which is any radioactive material that is made radioactive by exposure to the radiation incident to the process of producing or using special nuclear material (DOE-91/04).
(3) *See* nuclear material for more related terms.

Byproduct/waste: Any liquid or gaseous substance produced at chemical manufacturing plants or petroleum refineries (except natural gas, distillate oil, or residual oil) and combusted in a steam generating unit for heat recovery or for disposal. Gaseous substances with carbon dioxide levels greater than 50 percent or carbon monoxide levels greater than 10 percent are not byproduct/waste for the purposes of this subpart (40 CFR 60.41b-91).

Byproduct:
(1) A material that is not one of the primary products of a production process and is not solely or separately produced by the production process. Examples are process residues such as slags or distillation column bottoms. The term does not include a co-product that is produced for the general public's use and is ordinarily used in the form it is produced by the process (40 CFR 261.1-91, *see also* 40 CFR 704.3; 710.2; 712.3; 716.3; 720.3; 723.50; 723.175; 747.200; 761.3; 791.3-91).
(2) Material produced or generated in an industrial process in addition to the principal product (EPA-85/11).
(3) Material, other than the principal product, generated as a consequence of an industrial process (EPA-94/04).

Cc

Cab over axle or cab over engine: The cab which contains the operator-passenger compartment is directly above the engine and front ale and the entire cab can be tilted forward to permit access to the engine compartment (40 CFR 205.51-91).

Cab over engine: *See* synonym, cab over axle.

Cable pullout unloading method: A procedure in which a landfill tractor empties a transfer trailer by pulling a cable network from the front to the rear within the vehicle (EPA-83).

Cadmium (Cd):
(1) The total cadmium present in the process wastewater stream exiting the wastewater treatment system (40 CFR 415.451-91, *see also* 40 CFR 415.631-91).
(2) Cadium is a soft metallic element with a.n. 48; a.w. 112.40; d. 8.65 g/cc; m.p. 320.9° C and b.p. 765° C and belongs to group IIB of the periodic table.
(3) A heavy metal element that accumulates in the environment (EPA-94/04).

Cake: The solids discharged from a dewatering apparatus (cf. filter cake).

Calandria: The steam belt or heating element in an evaporator or vacuum pan, consisting of vertical tube sheets constituting the heating surface (EPA-75/02d).

Calandria evaporator: An evaporator using a calandria; the standard evaporator in current use in the sugar industry (EPA-75/02d).

Calandria vacuum pan: A vacuum pan using a calandria; the standard vacuum pan in current use in the sugar industry (EPA-75/02d).

Calcination:
(1) The formation of solid materials (e.g., calcium carbinate from hard water).
(2) The conversion of wastes into solid materials at elevated temperature (800-2000° C) and atmospheric pressure, without any interactions with the gaseous phase (such as air oxidation which occurs during incineration).
For an aqueous waste, the first reaction that can occur during calcination is vaporization of the water, leaving the solid materials as the granular and free flowing or a compact solid. A similar process occurs in the initial treatment of a de-watered sludge (i.e., after filtration and centrifugation). In many instances, it is possible to proceed further with the calcination to drive off volatile materials from the partially calcined solid, e.g., a salt, to form an oxide that will be more stable or reusable.

Calcine: The solid materials produced by a roaster (40 CFR 60.161-91).

Calciner: A unit in which the moisture and organic matter of phosphate rock is reduced within a combustion chamber (40 CFR 60.401-91).

Calcined pozzolan: The materials that are produced by calcination of natural siliceous or alumino-siliceous earths for the purpose of activation of pozzolanic properties (*see* pozzolan for more related terms) (EPA-83).

Calciner or nodulizing kiln: A unit in which phosphate rock is heated to high temperatures to remove organic material and/or to convert it to a nodular form. For the purpose of this subpart, calciners and nodulizing kilns are considered to be similar units (40 CFR 61.121-91).

Calcining: Heating to a high temperature without melting or fusing, e.g., to heat unformed ceramic materials in a kiln, or to heat ores, precipitates, concentrates or residues so that hydrates, carbonates or other compounds are decomposed and volatile material is expelled, e.g., to heat limestone to make lime (EPA-83/03).

Calcining zone: The thermal zone in the cement kiln in which carbon dioxide is liberated from carbonate species such as calcium carbonate (ETI-92).

Calcium (Ca): An alkaline earth metal with a.n. 20; a.w. 40.08; d. 1.55 g/cc; m.p. 838° C and b.p. 1440° C. The element belongs to group IIA of the periodic table. Major calcium compounds include:
- Calcium arsenate ($Ca_3(AsO_4)_2$): A powder used as an insecticide and molluscicide.
- Calcium arsenite ($Ca_3(AsO_3)_2$): A white granules used as an insecticide, germicide and molluscicide.
- Calcium carbonate ($CaCO_3$): A white powder used in paints, rubbers and plastics.
- Calcium cyanamide ($CaCN_2$): A colorless solid used as a herbicide, pesticide and defoliant.
- Calcium cyanide ($Ca(CN)_2$): A white powder used as an

insecticide, rodenticide and fumigant.

- Calcium cyclamate ($C_{12}H_{24}O_6N_2S_2Ca_2H_2O$): A crystal used as a low calorie sweetening agent.
- Calcium hydroxide ($Ca(OH)_2$): White crystals used in the coagulation of water.
- Calcium hypochlorite ($Ca(OCl)_2 \cdot 4H_2O$): A white powder used as bleaching agent and disinfectant for swimming pools.
- Calcium orthoarsenate ($Ca(AsO_4)_2$): A white powder used as a insecticide and herbicide.
- Calcium oxide (or quicklime, CaO): An insoluble solid used in ceramic glazes.
- Calcium phosphate ($Ca_3(PO_4)_2$): A white powder, a main component of animal bones.
- Calcium sulphate ($Ca(SO_4)$): A white solid powder, a cause of water permanent hardness. The compound includes several hydration states, e.g., anhydrite ($CaSO_4$), gypsum ($CaSO_4 \cdot 2H_2O$), and hemihydrate ($CaSO_4 \cdot 1/2H_2O$).

Calcium carbide: The material containing 70 to 85 percent calcium carbide by weight (40 CFR 60.261-91).

Calcium carbonate ($CaCO_3$): The major ingredient needed to manufacture cement (ETI-92).

Calcium hypo-chlorite: A chemical commonly used in the paper industry for bleaching pulp, and in water treatment as a germicide (EPA-87/10).

Calcium silicon: That alloy as defined by ASTM Designation A495-76 (incorporated by reference-*see* 40 CFR 60.17) (40 CFR 60.261-91).

Calcium sulfate storage pile runoff: The calcium sulfate transport water runoff from or through the calcium sulfate pile, and the precipitation which falls directly on the storage pile and which may be collected in a seepage ditch at the base of the outer slopes of the storage pile, provided such seepage ditch is protected from the incursion of surface runoff from areas outside of the outer perimeter of the seepage ditch (40 CFR 418.11-91).

Calculated level: The level of production, exports or imports of controlled substances determined for each Group of controlled substances by:
(1) Multiplying the amount (in kilograms) of production, exports or imports of each controlled substance by that substance's ozone depletion weight listed in Appendix A to this Part; and
(2) Adding together the resulting products for the controlled substances within each Group (40 CFR 82.3-91).

Calendar quarter: Not less than 12 consecutive weeks nor more than 14 consecutive weeks. The first calendar quarter of each year shall begin in January and subsequent calendar quarters shall be such that no day is included in more than one calendar quarter or omitted from inclusion within a calendar quarter. No licensee shall change the method observed by him of determining calendar quarters except at the beginning of a calendar year (10 CFR 20.3-91).

Calendering: Forming a continuous sheet by squeezing the material between two or more parallel rolls to impart the desired finish or to insure uniform thickness (EPA-83/06a).

Calgon ($NaPO_3)_6$: A corrosion inhibitor and used as water softening agent.

Calibrated accuracy: The calibrated accuracy is the difference between the indicated parameter value and its actual value. It is determined by calibrating the measuring system with an uncontaminated sensor (cf. system calibration (LBL-76/07-water).

Calibration:
(1) The set of specifications, including tolerances, unique to particular design, version, or application of a component or components assembly capable of functionally describing its operation over its working range (40 CFR 86.082.2-91, *see also* 40 CFR 600.002.85-91).
(2) The process of adjusting the instrument read-out so that it corresponds to the actual concentration. It involves checking the instrument with a known concentration of a gas or vapor to see that the instrument gives the proper response (Course 165.5). The process includes:
 (A) **Initial calibration verification standard:** A certified or independently prepared material or mixture used to verify the accuracy of the initial calibration.
 (B) **Continuing calibration verification:** Used to assure calibration accuracy during each analysis run. It must be run for each analyte at a frequency of 10% or every 2 hours during the run, whichever is more frequent. It must also be analyzed at the beginning of the run and after the last analytical sample. Its concentration must be at or near the mid-range levels of the calibration curve.

Other calibration related terms include:
- Dynamic calibration
- Static calibration

Calibration blank: A volume of deionized, distilled water acidified with HNO_3 and HCl (40 CFR 136-App/C-91). A blank of known concentration which is used to establish the adequacy of samples taken. Examples of the calibration blank include a volume of Type II water acidified with the same amounts of acids as were the standards and samples (*see* blank for more related terms).

Calibration curve: A graph or other systematic method of establishing the relationship between the analyzer response and the actual gas concentration introduced to the analyzer (40 CFR 60-App/A(method 6C)-91).

Calibration drift (deviation):
(1) The difference in the measurement system output reading from the initial calibration response at a mid-range calibration value after a stated period of operation during which no un scheduled maintenance, repair, or adjustment took place [40 CFR 60-App/A(method 6C & 7E)-91, *see also* 40 CFR 60-App/A(method 25A); App/B; App/F-91).
(2) The change in response or output of an instrument from a reference value over a period of time. Drift is measured by comparing the responses to a reference standard over a period of time with no adjustment of instrument settings (EPA-90/04).
(3) *See* calibration for more related terms.

Calibration drift test period: A monitoring system should be operated for some time before attempting drift checks, because most systems need a period of equilibration and adjustment before the performance is reasonably stable. At least one week (168 hours) of continuous operation is recommended before attempting drift tests. While a facility is operating at normal conditions, calibration drift test can be performed once each day (*see* calibration for more related terms) (EPA-90/04).

Calibration equipment: Equipment used for calibration of instruments (EPA-83/06a).

Calibration error:
(1) The difference between the gas concentration indicated by the measurement system and the known concentration of the calibration gas (40 CFR 60-App/A(method 25A)-91, *see also* 40 CFR App/B-91).
(2) A measure of the deviation of a measured value at the analyzer mid range from a reference value.
(3) The difference between the gas concentration indicated by the measurement system and the known concentration of the calibration gas (EPA-90/04).

Calibration gas:
(1) A gas of known concentration which is used to establish the response curve of an analyzer (40 CFR 86.082.2-91, *see also* 40 CFR 60-App/A(method 6C and 7E); method 21; method 25A-91).
(2) A known concentration of a gas in an appropriate diluent gas. For total hydrocarbon (THC) measurement, the concentration is expressed in terms of propane (EPA-90/04):
The calibration gas for THC measurement includes:
(1) Fuel
(2) Zero gas
(3) Low-, mid-, and high-level calibration gas

Calibration of air flow: The basic equipment required for calibrating air flow measuring instruments. The instruments include a standard meter, an air mover, and often a source of constant power. Air flow calibration methods include (EPA-83/06):
(1) Primary standard
(2) Intermediate standard
(3) Secondary standard

Calibration of equipment: The measurement of dispersal or output of application equipment and adjustment of such equipment to control the rate of dispersal, and droplet or particle size of pesticide dispersed by the equipment (40 CFR 171.2-91).

Calibration precision: The degree of agreement between measurements of the same known value, expressed as the relative percentage of the average difference between the meter readings and the known concentration to the known concentration (40 CFR 60-App/A(method 21)-91).

Calibration reference: A standard for comparison with the measured values to determine the degree of accuracy of the measurement.

Calibration standard:
(1) A series of known standard solutions used by the analyst for calibration of the instrument (i.e., preparation of the analytical curve) (40 CFR 136-App/C-91).
(2) A standard used to quantitate the relationship between the output of a sensor and a property to be measured. Calibration standards should be traceable to Standard Reference Materials or primary standard (LBL-7/76-air).
(3) *See* standard for more related terms.

Calibration valve assembly: A heated three-way assembly to direct the zero gas and calibration gases to the analyzers is recommended. Other methods, such as quick-connect lines, to route calibration gas to the analyzers are applicable (*see* total hydrocarbon concentration measurement system for more related terms) (EPA-90/04).

California list waste (52 Federal Register 25760, July 8, 1987):
(1) **Liquid hazardous wastes:** Including free liquid associated with any solid or sludge, containing free cyanides at concentrations greater than or equal to 1,000 mg/L.
(2) **Liquid hazardous wastes:** Including free liquid associated with any solid or sludge, containing the following metals (or elements) or compounds of these metals (or elements) at concentrations greater than or equal to those specified below:
- Arsenic and/or compounds (as As) 500 mg/L;
- Cadmium and/or compounds (as Cd) 100 mg/L;
- Chromium (VI and/or compounds (as Cr VI) 500 mg/L;
- Lead and/or compounds (as Pb) 500 mg/L;
- Mercury and/or compounds (as Hg) 20 mg/L;
- Nickel and/or compounds (as Ni) 134 mg/L;
- Selenium and/or compounds (as Se) 100 mg/L; and
- Thallium and/or compounds (as Th) 130 mg/L.
(3) Liquid waste having a pH less than or equal to two (2.0).
(4) Liquid hazardous wastes containing polychlorinated biphenyls at concentrations greater than or equal to 50 ppm.
(5) Hazardous wastes containing halogenated organic compounds in total concentration greater than or equal to 1,000 mg/kg.

Californium (Cf): A radioactive metal with a.n. 98; a.w. 249. The element belongs to group IIIB of the periodic table.

Calm: In meteorology, the absence of apparent motion of the air. It usually refers to wind speeds less than two knots. In the meteorological data used in air pollution modeling, the periods of calm are frequently determined by the threshold value of the anemometer (NATO-78/10).

Calomel electrode (saturated calomel electrode or standard calomel electrode): An electrode of known potential in a half cell. The electrode is mercury and the electrolyte is a solution of potassium chloride and saturated mercury chloride (calomel). It is used as a standard or reference electrode to measure pH and electromotive force (*see* electrode for more related terms).

Calorific value: *See* synonym, heat of combustion.

Calorimeter:
(1) Any of several apparatuses used for measuring quantities of absorbed or evolved heat from a specific quantity of material (EPA-83).
(2) A device for measuring heat quantities, such as combustion heat, and specific heat.

Can: Any metal container, with or without a top, cover, spout or handles, into which solid or liquid materials are packaged (40 CFR 52.741-91, *see also* 40 CFR 465.02-91).

Can coating: Any coating applied on a single walled container that is manufactured from metal sheets thinner than 29 gauge (0.0141 in.) (40 CFR 52.741-91).

Can coating facility: A facility that includes one or more can coating line(s) (40 CFR 52.741-91).

Can coating line: A coating line in which any protective, decorative, or functional coating is applied onto the surface of cans or can components (40 CFR 52.741-91).

Canal: An artificial watercourse cut to facilitate transportation, drainage, or irrigation (cf. ship canal) (DOI-70/04).

Cancellation: Refers to Section 6 (b) of the Federal Insecticide, Fungicide and Rodenticide Act (FIFRA) which authorizes cancellation of a pesticide registration if unreasonable adverse effects to the environment and public health develop when a product is used according to widespread and commonly recognized practice, or if its labeling or other material required to be submitted does not comply with FIFRA provisions (EPA-94/04).

Cancer:
(1) A disease characterized by the rapid and uncontrolled growth of aberrant cells into malignant tumors (Course 165.6).
(2) The name given to a group of diseases characterized by uncontrolled cellular growth (DOE-91/04).

Cancer potency slope factor (ql*): An indication of a chemical's human cancer-causing potential derived using animal studies or epidemiological data on human exposure. It is based on extrapolating high-dose levels over short periods of time to low-dose levels and a lifetime exposure period through the use of a linear model (EPA-91/03).

Cancer risk: Incremental probability of an individual's developing cancer over a lifetime as a result of exposure to a potential carcinogen (EPA-91/12).

Candidate method: A method of sampling and analyzing the ambient air for an air pollutant for which an application for a reference method determination or an equivalent method determination is submitted in accordance with 40 CFR 53.4, or a method tested at the initiative of the Administrator in accordance with 40 CFR 53.7 (*see* method for more related terms) (40 CFR 53.1-91).

Canmaking: The manufacturing process or processes used to manufacture a can from a basic metal (40 CFR 465.02-91).

Canned meat processor: An operation which prepares and cans meats (such as stew, sandwich spreads, or similar products) alone or in combination with other finished products at rates greater than 2730 kg (6000 lb.) per day (40 CFR 432.91-91).

Canyon technique: An area method in a depression where cover material is obtained within the depression (*see* sanitary landfill for more related terms) (EPA-83).

Cap:
(1) Another name for crown.
(2) A type of bottle closure (EPA-83).
(3) A layer of clay, or other impermeable material installed over the top of a closed landfill to prevent entry of rainwater and minimize leachate (EPA-94/04).

Capability: The maximum load which a generating unit, generating station, or other apparatus can carry under specified conditions for a given period of time, without exceeding approved limits of temperature and stress (EPA-83).

Capability margin: The difference between net system capability and system maximum load requirements (peak load). It is the margin of capability available to provide for scheduled maintenance, emergency outages, system operating requirements, and unforeseen loads. On a regional or national basis, it is the difference between aggregate net system capability of the various systems in the region or nation and the sum of system maximum (peak) loads without allowance for time diversity between the loads of the several systems. However, within a region, account is taken of diversity between peak loads of systems that are operated as a closely coordinated group (EPA-83).

Capacitance:
(1) The property of a device which permits storage of electrically-separated charges when differences in electrical potential exist between the conductors and measured as the ratio of stored charge to the difference in electrical potential between conductors (40 CFR 85.2122(a)(6)(ii)(A)-91).
(2) The ratio of the charge on one of the plates of a capacitor to the potential difference between the plates (EPA-83/03).

Capacitor: A device for accumulating and holding a charge of electricity and consisting of conducting surfaces separated by a dielectric. Other capacitor related terms include:
- Small capacitor (40 CFR 761.3)
- Large high voltage capacitor (40 CFR 761.3)
- Large low voltage capacitor (40 CFR 761.3)
- Variable capacitor
- Wet capacitor

Capacitor/condenser: A device for the storage of electrical energy consisting of No oppositely charged conducting plates separated by a dielectric and which resists the flow of direct current (40 CFR 85.2122(a)(6)(ii)(D)-91).

Capacity: The cumulative rated capacity of all initial crushers that are part of the plant (40 CFR 60.671-91). Other capacity related terms include:
- Carrying capacity
- Design capacity
- Firm capacity
- Gross capacity (*see* synonym, design capacity)
- Maximum capability (*see* synonym, firm capacity)
- Peak capability (*see* synonym, firm capacity)
- Nameplate capacity (*see* synonym, design capacity)
- Net capacity (*see* synonym, design capacity)
- Rated capacity

Capacity assurance plan: A statewide plan which supports a state's ability to manage the hazardous waste generated within its boundaries over a twenty year period (EPA-94/04).

Capacity factor:
(1) The ratio between the actual electric output from a unit and the potential electric output from that unit (CAA402, *see also* 40 CFR 40 CFR 51.100-91).
(2) The ratio of energy actually produced to that which would have been produced in the same period, if the unit had been operated continuously at rated capacity (EPA-82/11f).
(3) (Net energy produced in megawatts)/(design power in megawatts)x(calendar year) (DOE-91/04).

Capillary: A small diameter tube.

Capillary action:
(1) The tendency of liquids to penetrate, migrate, or drawn into small openings, such as cracks, pits, or fissures (EPA-83).
(2) Movement of water through very small spaces due to molecular forces called capillary forces (EPA-94/04).

Capillary flow reactor: A tubular flow reactor with internal diameter of 1.0 mm like a capillary tube. It was used by the University of Dayton Research Institute (UDRI) to test the thermal stability index (incinerability ranking) of Appendix VIII compounds in 40 CFR 261 (EPA-88/12).

Capillary fringe: The porous material just above the water table which may hold water by capillarity (a property of surface tension that draws water upwards) in the smaller void spaces (EPA-94/04).

Capillary gas chromatography: An analytical method in which sample gases pass through a capillary tube (approximately 0.2-0.5 millimeter ID and 100 meter length), adsorption occurs on a medium that is spread on the inner wall of the tube (*see* chromatography for more related terms).

Capillary pressure: The pressure difference due to capillary actions.

Capillary water: Underground water that is held above the water table by capillary attraction (cf. zone of capillarity) (SW-108ts).

Capital costs: Expenditures which result in the acquisition of, or the addition to, capital or fixed assets. Costs associated with the installation of such assets are included in capital costs (EPA-79/12b).

Capital expenditure: An expenditure for a physical or operational change to an existing facility which exceeds the product of the applicable annual asset guideline repair allowance percentage specified in the latest edition of Internal Revenue Service (IRS) Publication 534 and the existing facility's basis, as defined by section 1012 of the Internal Revenue Code. However, the total expenditure for a physical or operational change to an existing facility must not be reduced by any excluded additions as defined in IRS Publication 534, as would be done for tax purposes (40 CFR 60.2-91, *see also* 40 CFR 60.481; 60.561; 61.02-91).

Capitalization grant: The assistance agreement by which the EPA obligates and awards funds allotted to a State for purposes of capitalizing that State's revolving fund (40 CFR 35.3105-91).

Caprolactam byproduct ammonium sulfate manufacturing plant: Any plant which produces ammonium sulfate as a by-product from process streams generated during caprolactam manufacture (40 CFR 60.421-91).

Capsule: A gelatinous shell used to contain medicinal chemicals (EPA-83/09).

Captive manufacturing site (or captive operation): A plant which only manufacturers items for internal use or use by other divisions of a parent organization (EPA-79/12b).

Captive scrap (or runaround scrap): Aluminum scrap metals retained by fabricator and remelted (EPA-74/03f).

Capture: The containment or recovery of emissions from a process for direction into a duct which may be exhausted through a stack of sent to a control device. The overall abatement of emissions from a process with an add-on control device is a function both of the capture efficiency and of the control device (40 CFR 52.741-91).

Capture device: A hood, enclosed room floor sweep or other means of collecting solvent or other pollutants into a duct. The pollutant can then be directed to a pollution control device such as an afterburner or carbon adsorber. Sometimes the term is used loosely to include the control device (40 CFR 52.741-91).

Capture efficiency:
(1) The fraction of all VOM generated by a process that are directed to an abatement or recovery device (40 CFR 52.741-91).
(2) The fraction of organic vapors generated by a process that are directed to an abatement or recovery device (EPA-94/04).

Capture rate: The tonnage of recyclables collected divided by total tonnage of MSW (municipal solid waste) generated by participating households or commercial establishments (OTA-89/10).

Capture system: All equipment (including, but not limited to, hoods, ducts, fans, ovens, dryers, etc.) used to contain, collect and transport an air pollutant to a control device (40 CFR 52.741-91).

Car coupling sound: A sound which is heard and identified by the observer as that of car coupling impact, and that causes a sound level meter indicator (FAST) to register an increase of at least ten decibels above the level observed immediately before hearing the sound (40 CFR 201.1-91).

Car line: A name denoting a group of vehicles within a make or car division which has a degree of commonality in construction (e.g., body, chassis). Car line does not consider any level of decor or opulence and is not generally distinguished by characteristics as roof line, number of doors, seats, or windows, except for station wagons or light-duty trucks. Station wagons and light-duty trucks are considered to be different car lines than passenger cars (40 CFR 86.082.2).

Car sealed: For purposes of these standards, a seal that is placed on the device used to change the position of a valve (e.g., from opened to closed) such that the position of the valve cannot be changed without breaking the seal and requiring the replacement of the old seal once broken with a new seal (40 CFR 60.561-91, *see also* 40 CFR 61.301-91).

Carbamate: A group of insecticides which act on the nervous system by inhibiting the acetylcholinesterase enzyme at the nerve synapse (EPA-85/10).

Carbaryl ($C_{10}H_7OOCNHCH_3$): A colorless, crystalline insecticide.

Carbide: A general class of pressed and sintered tungsten carbide cutting tools which contain tungsten carbide plus smaller amounts of titanium and tantalum carbides along with cobalt which acts as a binder. (It is also used to describe hard compounds in steels and cast irons) (EPA-83/06a).

Carbohydrate: A compound of carbon, hydrogen, and oxygen, in which the ratio of hydrogen to oxygen is usually two to one (EPA-83/09).

Carbon (C): A nonmetallic, chiefly tetravalent element found native or as a constituent of coal, petroleum, asphalt, limestone, etc. (*See also* carbon hot forming operation) (EPA-83/03). The element with a.n. 6; a.w. 12.01; d. 2.26 g/cc; m.p. 3727° C and b.p. 4830° C belongs to group IVA of the periodic table. Other carbon related terms include:
- Activated carbon
- Fixed carbon
- Fly carbon
- Total organic carbon (TOC)

Carbon absorber: An add-on control device that uses activated carbon to absorb volatile organic compounds from a gas stream. (The VOCs are later recovered from the carbon (EPA-94/04).

Carbon adsorption (or carbon sorption):
(1) A process used to remove pollutants from wastewater by contacting the wastewater with activated carbon (EPA-87/10a). Carbon can adsorb but can not absorb.
(2) The process in which a substance (the sorbate) is brought into contact with a solid (the sorbet), usually activated carbon that can remove one or more of the gaseous contaminants (cf. activated carbon adsorption under adsorption).
(3) A treatment system that removes contaminants from ground water or surface water by forcing it through tanks containing activated carbon treated to attract the contaminants (EPA-94/04).

Carbon bed catalytic destruction: A non-electrolytic process for the catalytic oxidation of cyanide wastes using filters filled with low-temperature coke (EPA-83/06a).

Carbon black: An intensely black, finely divided pigment obtained by burning natural gas or oil with a restricted air supply (cf. soot) (EPA-79/12a).

Carbon chloroform extract (CC): An organic compound removal method in which an organic compound in water is adsorbed on activated carbon and then extracted by the solvent chloroform.

Carbon column A: A column filled with granular activated carbon whose primary function is the preferential adsorption of a particular type or types of molecules (EPA-82/11f).

Carbon cycle: The cycle of carbon (as carbon dioxide) in the biosphere system. Plants convert carbon dioxide to organic compounds by the photosynthesis. The organic compounds are then consumed by animals or other plants. By respiration of animals or decay of plants, the carbon dioxide is returned to the biosphere. Combustion of fossil fuel (e.g., oil) also releases carbon dioxide to the atmosphere.

Carbon dioxide (CO_2): A colorless, odorless, non-poisonous gas, which results from fossil fuel combustion, (thermal degradation and microbial decomposition of wastes) and is normally a part of the ambient air (EPA-89/12). Some researchers have theorized that excess CO_2 raises atmospheric temperatures (SW-108ts).

Carbon dioxide absorption tube: The Pragl combustion procedure in which an absorbent packed tube is used to capture CO_2 formed during the determination of carbon-hydrogen quantities.

Carbon dioxide recorder: An instrument that continuously monitors the volume concentration (in percent) of carbon dioxide in a flue gas (SW-108ts).

Carbon hot forming operation (or carbon): Those hot forming operations which produce a majority, on a tonnage basis, of carbon steel products (40 CFR 420.71-91).

Carbon monoxide (CO):
(1) The gas has d. 1.25 g/m³, m.p. 199° C and b.p. 192° C.
(2) A colorless, odorless, poisonous gas produced by incomplete fossil fuel combustion (EPA-94/04).

Carbon monoxide national ambient air quality standard (CO NAAQS): The standards for carbon monoxide promulgated by the Administrator under section 109, 42USC7409, of the Clean Air Act and found in 40 CFR 50.8 (40 CFR 51.138-91).

Carbon monoxide (CO) and total hydrocarbon (THC) emission limits and permit formats: Under the Omnibus Authority of HSWA, EPA has issued guidance to incinerator permit writers for the control of metals, HCl, and products of incomplete combustion (PICs) (*see* Omnibus Authority). CO and total hydrocarbon (THC) have been selected as surrogates to control PICs. The CO and THC controls are based on two-tiered approach (EPA-90/04):
(1) **Tier I limits:** The applicant can demonstrate compliance by meeting the recommended de minimis CO limit of 100 ppmv (corrected to 7 percent oxygen, dry basis) on an hourly rolling average (HRA).
Example:
- Find HRA and waste feed cut-off time during incineration, if CO is greater than 100 ppmv.

Data and calculation are shown below:

Time(hr)	CO	HRA calculation(ppmv)	value
0	30	(60x30)/60	30
3:57	30	(60x30)/60	30
3:58	1500	(59x30+1x1500)/60	56

Time(hr)	CO	HRA calculation(ppmv)	value
3:59	1500	(58x30+2x1500)/60	79
4:00	1500	(57x30+3x1500)/60	103*
4:01	30	(56x30+3x1500+1x30)/60	103*
4:02	30	(55x30+3x1500+2x30)/60	103*
4:57	30	(3x1500+57x30)/60	103*
4:58	30	(2x1500+58x30)/60	79**
4:59	30	(1x1500+59x30)/60	56
5:00	30	(60x30)/60	30

* Waste feed cut-off; and ** waste feed re-start

(2) **Tier II limits:** The minimum CO limit would be waived if the applicant demonstrates that THC emissions are not likely to pose unacceptable health risk. There are two alternative approaches to demonstrate that THC emissions are acceptable:
  (A) a health-based approach based on site-specific risk assessment; or
  (B) a technology-based approach where the applicant demonstrates that THC levels do not exceed a good operating practice-based level of 20 ppmv (hourly rolling average, corrected to 7% oxygen, dry basis and reported as propane).

The CO limits for either Tier I or Tier II must be corrected to dry stack gas and 7% oxygen in the stack gas. The correction to dry stack gas is necessary only for instruments that measure CO on a wet basis. This correction factor for humidity and oxygen would initially be determined during the trail burn and annually thereafter unless specified more frequently in the permit. The oxygen and humidity correction factors would be applied continuously.

The CO limits can be implemented in either of two formats:
(1) An hourly rolling average format; or
(2) A cumulative hourly time-above-a-level format. The cumulative hourly time-above-a-level format is designed to allow hourly average CO emissions equivalent to the hourly rolling average format. This alternative format is provided to minimize the cost of instrumentation needed to monitor, analyze, and record CO levels.

Carbon nitrogen ratio (CN) ratio: The ratio of the weight of carbon to the weight of nitrogen present in a compost or in materials that are being composted (SW-108ts).

Carbon nitrogen phosphorus ratio (CNP ratio): The average CNP ratio in a living matter is 106:10:1. In biological treatment, the ratio is usually expressed by BOD5. Nitrogen or phosphorus deficiency can be corrected by adding ammonium salts or phosphates respectively.

Carbon reduction: The process of using coke carbon as a reducing agent in a blast furnace (EPA-85/10a).

Carbon regeneration: The process of reactivating exhausted or spent carbon by thermal means (cf. activated carbon regeneration) (EPA-85/10).

Carbon regeneration unit: any enclosed thermal treatment device used to regenerate spent activated carbon (40 CFR 260.10-91).

Carbon sorption: *See* synonym, carbon adsorption.

Carbon steel:
(1) Those steel products other than specialty steel products (40 CFR 420.71-91).
(2) A steel which owes it properties chiefly to various percentage of carbon without substantial amounts of other alloying elements (EPA-83/06a).

Carbonaceous matter: Pure carbon or a compound containing or composed of carbon (SW-108ts).

Carbonaceous oxygen demand (carbonaceous OD): Oxygen demand exerted by organic carbon compounds present; oxygen required to convert organic carbon to CO_2 (LBL-76/07-water).

Carbonate: A compound containing the anion radical of carbonic acid (CO_3^- group) (EPA-83/06a).

Carbonate hardness:
(1) Hardness of water caused by the presence of carbonates and bicarbonates of calcium and magnesium (EPA-82/11f).
(2) Hardness caused by the presence of carbonates and bicarbonates of calcium and magnesium in water. Such hardness may be removed to the limit of solubility by boiling the water. When the hardness in numerically greater than the sum of the carbonate alkalinity and the bicarbonate alkalinity, that amount of hardness which is equivalent to the total alkalinity is called carbonate hardness. *See* hardness (LBL-76/07-water).

Carbonation: The process of treatment with carbon dioxide gas (EPA-74/01a).

Carbonic acid (H_2CO_3): The weak acid formed when carbon dioxide is dissolved in water.

Carbonyl: (O=C<), a radical CO consisting of one carbon atom and one oxygen atom connected by a double bond.

Carboxyhemoglobin (COHb):
(1) A compound formed by the combination of carbon monoxide and hemoglobin in the blood when carbon monoxide is inhaled. COHb reduces the ability of the blood to carry oxygen.
(2) Hemoglobin in which the iron is associated with carbon monoxide (CO). The affinity of hemoglobin for CO is about 300 times greater than for oxygen (EPA-94/04).

Carburetor: In automobile application, a device for mixing the incoming gas and air for proper combustion in the cylinders. The device is gradually replaced by the newer system, i.e., fuel injection.

Carburizing: Increasing the carbon content of a metal by heating with a carburizing medium (which may be solid, liquid or gas) usually for the purpose of producing a hardened surface by subsequent quenching (EPA-83/06a).

Carcinogen:
(1) An agent capable of inducing a cancer response (EPA-92/12).

(2) A cancer causing substance (EPA-87/07a).
(3) Any substance that can cause or aggravate cancer (EPA-94/04).
(4) *See also* definition under toxicant and effect.

Carcinogenesis: The origin or production of cancer, very likely a series of steps. The carcinogenic event so modifies the genome and/or other molecular control mechanisms in the target cells that these can give rise to a population of altered cells (EPA-92/12).

Carcinogenic: Cancer-producing (EPA-89/12).

Carcinogenic effect: Shall have the meaning provided by the Administrator under Guidelines for Carcinogenic Risk Assessment as of the date of enactment. Any revisions in the existing Guidelines shall be subject to notice and opportunity for comment (CAA112-42USC7412-91).

Carcinogenic potential: In SDWA, EPA classifies compounds for carcinogenic potential according to the weight of evidence of carcinogenicity measured against EPA's Guidelines for Carcinogen Risk Assessment. A compound may be classified into one of five Groups based on existing evidence (JMM-88).
(1) **Human carcinogen:** There is sufficient evidence from epidemiological studies.
(2) **Probable human carcinogen:**
  (A) Limited evidence of carcinogenicity in humans.
  (B) Sufficient animal evidence of carcinogenicity, but insufficient human evidence.
(3) **Possible human carcinogen:** Limited evidence of carcinogenicity in the absence of human data.
(4) **No classifiable:** Evidence inadequate to state one way or other.
(5) **No evidence of carcinogenicity for humans:** Either two animal species were tested or one animal species and humans.

Carcinogenicity:
(1) Tumor frequency in tissues, detected by gross observation or histological examination (*see* endpoint for more related terms) (Course 165.6; EPA-92/12).
(2) The potential of a substance to cause cancer (ETI-92).

Carnivore: Animals (e.g., tigers and wolves) that eat meat for living.

Carnot cycle: A hypothetical cycle which is supposed to be the most efficient heat engine to produce work. The cycle consists of four reversible processes:
(1) Constant temperature expansion (heat addition or combustion process)
(2) Isentropic (reversible adiabatic) expansion (work process)
(3) Constant temperature compression
(4) Isentropic compression
The thermal efficiency of a Carnot cycle is $E_c = 1 - T_L/T_H$, where: E_c = efficiency of a Carnot cycle; T_L = low temperature of a reservoir; T_H = high temperature of a reservoir (Holman-p170; Jones-p236; Wark-p248).

Carnot efficiency: *See* definition under thermal efficiency (Wark-p276).

Carnot engine: A heat engine which is operated between two fixed temperature limits (Wark-p276).

Carotene ($C_{40}H_{56}$): Any member of carotenoid pigments. Some pigments, for example, color carrot roots and ripe tomato fruits.

Carrier:
(1) A living organism which harbors a specific infectious agent in the absence of discernible clinical disease and serves as a potential source or reservoir of infection for man (EPA-83).
(2) For gasoline carrier, *see* 40 CFR 52.137; 80.2-91.
(3) Any material carrier, *see* 40 CFR 160.3; 792.3-91.
(4) For railroad carrier, *see* 40 CFR 201.1-91.
(5) For solvent carrier, *see* 40 CFR 797.1400; 797.1520; 797.1600-91.
(6) The inert liquid or solid material added to an active ingredient in a pesticide (EPA-94/04).

Carrier gas: The inert portion of the gas stream, usually air, from which the pollutant is to be moved (EPA-84/03).

Carrier of contaminant: The dredged or fill material that contains contaminants (40 CFR 230.3-91).

Carry cloth: A large piece of canvas or burlap used to transfer solid waste from a residential solid waste storage area to a collection vehicle. *See also* carrying container (EPA-83).

Carry over: The chemical solids and liquid entrained with the steam from a boiler.

Carrying capacity:
(1) In recreation management, the amount of use a recreation area can sustain without deterioration of its quality (EPA-94/04).
(2) In wildlife management, the maximum number of animals an area can support during a given period of the year (EPA-94/04).
(3) *See* capacity for more related terms.

Carrying container:
(1) A barrel, can or other receptacle carried by the collector in backyard carry out service. Usually of 30-55 gallon capacity and usually constructed of aluminum (EPA-83).
(2) A receptacle of 35 to 50 gallons capacity, usually constructed of plastic or aluminum, that is carried by a collector in a backyard carryout service; frequently called a tote barrel (SW-108ts).
(3) *See* container for more related terms.

Carryout collection:
(1) Carryout collection means collection of solid waste from a storage area proximate to the dwelling unit(s) or establishment (40 CFR 243.101-91).
(2) The crew collection of solid waste from an on-premise storage area using a carrying container, carry-cloth, or a mechanical method (SW-108ts).
(3) *See* waste collection for more related terms.

Cartridge filter: A discrete filter unit containing both filter paper and activated carbon that traps and removes contaminants from petroleum solvent, together with the piping and ductwork used in the installation of this device (*see* filter for more related terms) (40 CFR 60.621-91).

CAS Number: Chemical Abstracts Service Registry Number (40 CFR 704.3-91, *see also* 40 CFR 721.3-91).

CAS registration number: A number assigned by the Chemical Abstracts Service to identify a chemical (EPA-94/04).

Cascade aerator: There are two types of cascade aerators:
(1) **Bed cascade aerator:** Air is circulating from the bottom through the packing materials.
(2) **Free fall cascade aerator:** Water flows down the open aerated steps.
(3) *See* aerator for more related terms.

Cascade impactor: A device commonly used to measure particle size distributions of gas streams. It consists of a series of stacked stages and collection surfaces. Each stage consists of from one to as many as 400 precisely drilled jet orifices, identical in diameter in each stage but decreasing in diameter with each succeeding stage (*see* particle size measurement for more related terms) (EPA-84/09).

Cascade process:
(1) The basic process in turbulence by which turbulent energy generated on the largest space scale is transported successively across various space scales by the mechanism of vorticity stretching and eventually is dissipated at the smallest scales by viscous forces (NATO-78/10).
(2) A process that occurs in several steps, because a single step is inefficient to produce the desired result.

Case-control study: An epidemiologic study that looks back in time at the exposure history of individuals who have the health effect (cases) and at a group who do not (controls), to ascertain whether they differ in proportion exposed to the chemical under investigation (EPA-92/12).

Case hardening: A heat treating method by which the surface layer of alloys is made substantially harder than the interior. (Carburizing and nitriding are common ways of case hardening steels) (EPA-83/06a).

Casein: Phosphoproteins found in milk. It is the protein of milk, a white solid soluble in acids.

Casing: A pipe or tubing of appropriate material, of varying diameter and weight, lowered into a borehole during or after drilling in order to support the sides of the hole and thus prevent the walls from caving, to prevent loss of drilling mud into porous ground, or to prevent water, gas, or other fluid from entering or leaving the hole (40 CFR 146.3-91, *see also* 40 CFR 147.2902-91).

Cask (or coffin): A thick-walled container (usually lead) used to transport radioactive material. Also called a coffin (EPA-94/04).

Cast(ing):
(1) A state of a substance after solidification of the molten substance (EPA-83/06a).
(2) Pouring a molten metal into a mold to produce an object of desired shape (40 CFR 471.02-91), e.g., the process by which grids for lead acid batteries are made by pouring molten lead into molds and allowing solidification (EPA-84/08).

Cast coating: A process whereby coated paper is firmly pressed against a polished, steam heated drum (EPA-83).

Cast iron: An iron containing carbon in excess of the solubility in the austenite that exists in the alloy at the eutectic temperature. Cast iron also is defined here to include any iron-carbon alloys containing 1.2 percent or more carbon by weight (40 CFR 464.31-91, *see also* 40 CFR 471.02-91).

Castable refractory: The hydraulic-setting, suitable for casting or being pneumatically formed into heat-resistant shapes or walls (*see* refractory for more related terms) (EPA-83).

Casthouse: The facility which melts metal, holds it in furnaces for degassing (fluxing) and alloying and then casts the metal into pigs, ingots, billets, rod, etc. (EPA-83/06a).

Casting core: A very firm shape of sand used to obtain a hollow section in a casting. The core is placed in a mold cavity to give interior shape to the casting (EPA-85/10a).

Casting powder: Small particles of powder used in formulating cast propellant grains; contains nitrocellulose, stabilizer, plasticizer, and usually nitroglycerin.

Cat eye: An imperfection in glass; an elongated bubble containing a piece of foreign matter (EPA-83).

Cat scratch: An imperfection; surface irregularities on glassware resembling the marks of a cat's claws (EPA-83).

Catalysis: A process of catalyzing or promoting a reaction by a chemical agent that does not participate in the reaction. It is a surface phenomenon like adsorption (EPA-84/09). Catalytic reactions can be divided into two general types, i.e., **homogeneous catalysis** and **heterogeneous catalysis**. Other catalysis related terms include:
- Homogeneous catalysis
- Heterogeneous catalysis
- Radiation catalysis

Catalyst:
(1) A substance which will speed up or, sometimes, slow down the rate of a chemical reaction when added to a reaction system and which itself undergoes no permanent chemical change.
(2) A catalyst is an agent that alters the rate of a chemical transformation process but is chemically unchanged at the end of reaction. It is neither destroyed during the processing nor incorporated into the product.
(3) (In asphalt application), means a substance which, when added to asphalt flux in a blowing still, alters the penetrating-softening point relationship or increases the rate of oxidation of the flux (40 CFR 60.471-91).
(4) A substance that changes the speed or yield of a chemical

reaction without being consumed or chemically changed by the chemical reaction (EPA-94/04).

Catalyst carrier: A porous material used to support catalysts. Catalysts are placed upon the porous surface or into the voids.

Catalytic afterburner: *See* synonym, secondary burner.

Catalytic bath: A bath containing a substance used to accelerate the rate of chemical reaction (EPA-83/06a).

Catalytic combustion:
(1) A process in which a catalyst is used to assist in burning or oxidizing hydrocarbons or odorous contaminants; the catalyst itself remains intact (unchanged). Generally, it oxidizes hydrocarbons at a lower temperature (in the range of 600-900 °F) than required for thermal combustion. Examples of catalysts include a platinum or vanadium (EPA-84/09).
(2) One of NO_x emission reduction techniques (*see* nitrogen oxide emission control for control structure). A catalyst is used to achieve oxidation of the fuel rather than using high flame temperatures. These systems have been used in gas turbines to reduce NO_x emissions well below 10 ppm (EPA-81/12, p7-13).
(3) *See* combustion for more related terms.

Catalytic combustion device: *See* synonym, catalytic converter.

Catalytic converter (catalytic combustion device, catalytic oxidizer, or catalytic reactor):
(1) A device installed in the exhaust system of an internal combustion engine that utilizes catalytic action to oxidize hydrocarbon (HC and carbon monoxide (CO) emissions to carbon dioxide (CO_2) and water (H_2O) (40 CFR 85.2122(a)(15) (ii)(A)-91).
(2) An air pollution abatement device that removes pollutants from motor vehicle exhaust, either by oxidizing them into carbon dioxide and water or reducing them to nitrogen and oxygen (EPA-94/04).
(3) *See* aged catalytic converter for more information.

Catalytic incinerator: A control device that oxidizes volatile organic compounds (VOCs) by using a catalyst to promote the combustion process. Catalytic incinerators require lower temperatures than conventional thermal incinerators, thus saving fuel and other costs (*see* incinerator for more related terms) (EPA-94/04).

Catalytic oxidizer: *See* synonym, catalytic converter.

Catalytic reactor: *See* synonym, catalytic converter.

Catanadramous: Fish that swim downstream to spawn (EPA-89/12).

Catastrophic collapse: The sudden and utter failure of overlying strata caused by removal of underlying materials (40 CFR 146.3-91).

Catch basin: An open basin which serves as a single collection point for a stormwater runoff received directly from refinery surfaces and for refinery wastewater from process drains (40 CFR 60.691-91).

Catchment basin: *See* synonym, watershed (DOI-70/04).

Categorical exclusion:
(1) A class of actions which either individually or cumulatively would not have a significant effect on the human environment and therefore would not require preparation of an environmental assessment or environmental impact statement under the National Environmental Policy Act (NEPA) (EPA-94/04).
(2) *See* 40 CFR 1508.4-91.

Categorical pretreatment standard: A technology-based effluent limitation for an industrial facility discharging into a municipal sewer system. Analogous in stringency to Best Availability Technology (BAT) for direct dischargers (EPA-94/04).

Category:
(1) For compressor configurations, *see* 40 CFR 204.51; 205.51; 205.151; 205.165-91).
(2) For hearing protectors, *see* 40 CFR 211.203-91.

Category I nonfriable asbestos-containing material (ACM): The asbestos-containing packings, gaskets, resilient floor covering, and asphalt roofing products containing more than 1 percent asbestos as determined using the method specified in appendix A, subpart F, 40 CFR 763, section 1, Polarized Light Microscopy (40 CFR 61.141-91).

Category II nonfriable ACM: Any material, excluding Category I nonfriable ACM, containing more than 1 percent asbestos as determined using the methods specified in appendix A, subpart F, 40 CFR 763, section 1, Polarized Light Microscopy that, when dry, cannot be crumbled, pulverized, or reduced to powder by hand pressure (40 CFR 61.141-91).

Category of chemical substances: A group of chemical substances the members of which are similar in molecular structure, in physical, chemical, or biological properties, in use, or in mode of entrance in to the human body or into the environment, or the members of which are in some other way suitable for classification as such for purposes of this Act, except that such term does not mean a group of chemical substances which are grouped together solely on the basis of their being new chemical substances (TSCA26, *see also* 40 CFR 723.50; 723.175; 723.250-91).

Category of mixtures: A group of mixtures the members of which are similar in molecular structure, in physical, chemical, or biological properties, in use, or in the mode of entrance into the human body or into the environment, or the members of which are in some other way suitable for classification as such for purposes of this Act (TSCA26-15USC2625-91).

Category of water: The applicability of the instrumentation to the analysis of a pollutant or parameter in fresh, waste, and/or saline water samples (*see* water for more related terms) (LBL-76/07-water).

Cathode (or negative electrode):
(1) The primary source of electrons in an electron tube; in directly heated tubes the filament is the cathode, and in indirectly heated tubes a coated metal cathode surrounds a heater (EPA-83/03).
(2) The negative pole of an electrode or conducting terminal (cf. anode) (EPA-75/10).

Cathode ray tube: The electronic devices in which electrons focus through a vacuum to generate a controlled image on a luminescent surface. This definition does not include receiving and transmitting tubes (40 CFR 469.31-91, *see also* 40 CFR 469.32-91).

Cathodic inhibitor: An inhibitor of metal corrosion. The inhibitor such as calcium bicarbonate or sodium phosphate is deposited on the metal surface to reduce the rate of the cathodic reaction in a conducting medium.

Cathodic polarization: An electrical connection of a nickel electrode plaque to promote deposition of active nickel materials (EPA-84/08).

Cathodic protection:
(1) A technique to prevent corrosion of a metal surface by making that surface the cathode of an electrochemical cell. For example, a tank system can be cathodically protected through the application of either galvanic anodes or impressed current (40 CFR 280.12-91).
(2) A technique to prevent corrosion of a metal surface by making that surface the cathode of an electrochemical cell (EPA-94/04).

Cathodic protection tester: A person who can demonstrate an understanding of the principles and measurements of all common types of cathodic protection systems as applied to buried or submerged metal piping and tank systems. At a minimum, such persons must have education and experience in soil resistivity, stray current, structure-to-soil potential, and component electrical isolation measurements of buried metal piping and tank systems (40 CFR 280.12-91).

Catholyte: In a two-solution electrolytic cell, the incoming cell feed containing a relatively high concentration of the metal to be plated on the cathode (cf. anolyte) (EPA-75/02a).

Cation: *See* definition under ion.

Cation analysis: An analysis of cations in a solution.

Cation exchange capacity: The sum of exchangeable cations a soil can absorb expressed in milli-equivalents per 100 grams of soil as determined by sampling the soil to the depth of cultivation or solid waste placement, whichever is greater, and analyzing by the summation method for distinctly acid soils or the sodium acetate method for neutral, calcareous or saline soils (Methods of Soil Analysis, Agronomy Monograph No.9. C.A. Black, ed., American Society of Agronomy, Madison, Wisconsin. pp 891-901, 1965) (40 CFR 257.3.5-91, *see also* 40 CFR 796.2700; 796.2750-91).

Cation exchange material: A material capable of the reversible exchange of positively charged ions.

Cation exchange process: The reversible exchange of positive ions between functional groups of the ion exchange medium and the solution in which the solid is immersed. Used as a wastewater treatment process for removal of cations, e.g., calcium (EPA-82/11f).

Cationic collectors: In flotation, amines and related organic compounds capable of producing positively charged hydrocarbon-bearing ions for the purpose of floating miscellaneous minerals, especially silicates (EPA-82/05).

Cationic dye: The colored component of dye which bears a positive charge (EPA-74/06b).

Cationic flocculant: In flocculation, surface active substances which have the active constituent in the positive ion. Used to flocculate and neutralize the negative charge residing on colloidal particles (EPA-82/10).

Cationic polymer: A polymer that contains one or more covalently linked subunits that bear a net positive charge (40 CFR 723.250-91).

Cationic reagents: In flotation, surface active substances which have the active constituents in the positive ion. Used to flocculate and to collect minerals that are not flocculated by the reagents, such as oleic acid or soaps, in which the surface-active ingredient is the negative ion (EPA-82/05).

Cationic surfactant: A surfactant in which the hydrophilic groups are positively charged; usually a quaternary ammonium salt such as cetyl trimethyl ammonium bromide (CeTAB), $C_{16}CH_{33}N + (CH_3)_3Br$. Cationic surfactants, as a class, are poor cleaners but exhibit remarkable disinfectant properties (EPA-11/82f).

Caulking compound: A soft plastic material, consisting of pigment and vehicle, used for sealing joints in buildings and other structures where normal structural movement may occur (EPA-79/12b).

Cause analysis: An analysis by which a relation is obtained between found concentration levels and emission sources (*see* analysis for more related terms) (NATO-78/10).

Causeway: A raised way or road across wet or marshy ground, across the surface of a water body, or from a shore to an island (DOI-70/04).

Caustic: Capable of destroying or eating away by a chemical action. Applies to strong bases and characterized by the presence of hydroxyl ions in solution (EPA-83/06a).

Caustic alkalinity: Alkalinity caused by strongly alkaline compounds such as NaOH or $Ca(OH)_2$.

Caustic rinse: The cleaning of residue from a workpiece such as an ink or paint tub with a caustic solution (EPA-79/12a; 79/12b).

Caustic scrubbing: Use of a caustic solution such as sodium hydroxide to wash or to remove residues or HCl.

Caustic soda: Sodium hydroxide, a strong alkaline substance used as the cleaning agent in some detergents (EPA-89/12).

Caustic system: Caustic system related terms include:
- Closed loop caustic system
- Open caustic system
- Partial recycle caustic system

Causticizing: In the paper industry, a process of making white liquor from green liquor by addition of slaked lime. Most Na_2CO_3 is thereby converted to NaOH (EPA-87/10).

Cavitation:
(1) Emulsification produced by rapid formation and collapse of vapor or gas bubbles.
(2) The formation and collapse of vapor bubbles in a flowing liquid. Specifically, the formation and collapse of vapor cavities in a pump or compressor when there is sufficient resistance to flow at the inlet side (EPA-8/87b).
(3) The formation and collapse of gas pockets or bubbles on the blade of an impeller or the gate of a valve; collapse of these pockets or bubbles drives water with such force that it can cause pitting of the gate or valve surface (EPA-94/04).

CBOD5:
(1) The five day measure of the pollutant parameter carbonaceous biochemical oxygen demand (CBOD5) (cf. BOD5 or oxygen demand, chemical) (40 CFR 133.101-91).
(2) The amount of dissolved oxygen consumed in 5 days from the carbonaceous portion of biological processes breaking down in an effluent. The test methodology is the same as for BOD5, except that nitrogen demand is suppressed (EPA-89/12).

CCA type (wood) preservative: Any one of several inorganic salt formulations based on salts of copper, chromium, and arsenic used for the purpose of protecting wood quality (EPA-74/04).

Ceiling insulation: A material, primarily designed to resist heat flow, which is installed between the conditioned area of a building and an unconditioned attic as well as common ceiling floor assemblies between separately conditioned units in multi-unit structures. Where the conditioned area of a building extends to the roof, ceiling insulation includes such a material used between the underside and upperside of the roof (40 CFR 248.4-91).

Cell:
(1) Compacted solid wastes that are enclosed by natural soil or cover material in a land disposal site (40 CFR 241.101-91).
(2) The basic building block of a battery. It is an electrochemical device consisting of an anode and a cathode in a common electrolyte kept apart with a separator. This assembly may be used in its own container as a single cell battery or be combined and interconnected with other cells in a container to form a multicelled battery (EPA-84/08).
(3) (A) In solid waste disposal, holes where waste is dumped, compacted, and covered with layers of dirt on a daily basis (EPA-94/04).
(B) The smallest structural part of living matter capable of functioning as an independent unit (EPA-94/04).

Cell height: The vertical distance between the top and bottom of the compacted solid waste enclosed by natural soil or cover material in a sanitary landfill (cf. cell thickness) (SW-108ts).

Cell landfill: Compacted solid wastes that are enclosed by natural soil or cover material in a sanitary landfill (EPA-83).

Cell membrane: *See* synonym, cell wall.

Cell room: A structure(s) housing one or more mercury electrolytic chlor-alkali cells (40 CFR 61.51-91).

Cell stabilization: *See* synonym, biological stabilization (LBL-76/07-water).

Cell synthesis: The formation of new cells by bacteria (EPA-75/10).

Cell thickness: The perpendicular distance between the cover materials placed over the last working faces of two successive cells in a sanitary landfill (cf. cell height) (SW-108ts).

Cell type incinerator: An incinerator whose grate areas are divided into cells, each of which has its own ash drop, underfire air control, and ash grate (*see* incinerator for more related terms) (SW-108ts).

Cell wall (or cell membrane): A semipermeable structure that forms the outer limit of a cell. The structure is made of cellulose, lignin, etc.

Cellular polyisocyanurate insulation: The insulation produced principally by the polymerization of polymeric polyisocyanates, usually in the presence of polyhydroxl compounds with the addition of catalysts, cell stabilizers, and blowing agents (40 CFR 248.4-91).

Cellular polystyrene insulation: An organic foam composed principally of polymerized styrene resin processed to form a homogenous rigid mass of cells (40 CFR 248.4-91).

Cellular polyurethane insulation: The insulation composed principally of the catalyzed reaction product of polyisocyanurates and polyhydroxl compounds, processed usually with a blowing agent to form a rigid foam having a predominantly closed cell structure (40 CFR 248.4-91).

Cellulose (($C_6H_{10}O_5)_n$):
(1) Vegetable fiber such as paper, wood, and cane (40 CFR 248.4-91).
(2) The major polysaccharide component of the cell walls of all woods, straws, bast fibers and seed hairs. It is the main solid constituent of wood plants and is the principal raw material of pulp, paper and paperboard (EPA-87/10).

Cellulose fiber fiberboard: The insulation composed principally of cellulose fibers usually derived from paper, paperboard stock, cane, or wood, with or without binders (40 CFR 248.4-91).

Cellulose fiber loose-fill: A basic material of recycled wood-based cellulosic fiber made from selected paper, paperboard stock, or ground wood stock, excluding contaminated materials which may reasonably be expected to be retained in the finished product, with suitable chemicals introduced to provide properties such as flame resistance, processing and handling characteristics. The basic cellulosic material may be processed into a form suitable for installation by pneumatic or pouring methods (40 CFR 248.4-k-91).

Celsius (C): A temperature scale in which water boiling point is 100° C and water freezing (or ice melting) point is 0° C. Celsius was a Swedish astronomer in 17th century (cf. temperature).

Cement: Mixture of ground clinker and a small amount of gypsum (cf. Portland cement) (ETI-92).

Cement(ing): The operation whereby a cement slurry is pumped into a drilled hole and/or forced behind the casing (40 CFR 146.3-91, *see also* 40 CFR 147.2902-91).

Cement based process: Incorporation of common (Portland) cement as a stabilization agent for isolating hazardous wastes. Portland cement is produced by firing a charge of limestone and clay or other silicate mixtures at high temperatures. The resulting clinker is then ground to a fine powder to yield a cement consisting of about 50% tricalcium and 25% dicalcium silicates, 10% tricalcium aluminate, and 10% calcium aluminoferrite. The cementation process, which is initiated by the addition of water to the anhydrous cement powder, first produces a colloidal calcium silicate hydrate gel of indefinite composition and structure. Hardening of the gel occurs slowly and gradually through the interlacing of thin, densely packed, silicate fibrils growing from the individual cement particles. This fibrillar matrix incorporates the added aggregates and/or waste into a monolithic, rocklike mass (*see* solidification and stabilization for more related terms).

Cement copper: Copper precipitated by iron from copper sulfate solutions (EPA-82/05).

Cement kiln dust (CKD): The collection of particulate matter removed from the kiln gases by the air pollution control devices (ETI-92).

Cement kiln for waste destruction: In cement manufacture, the kiln operates at higher temperatures and for longer residence times than those used in waste incinerators. It is also a common practice to add chlorides to the kiln to reduce the alkali concentration of the cement final product. Use of chlorinated hydrogen wastes in a cement kiln would provide useful recovery of chlorine and energy.

A typical cement manufacturing process include that materials containing calcium, silicon, aluminum and iron are ground to a fine powder called raw meal. The raw meals required for wet and dry cement processes are similar except that the raw meal for the wet process is in the form of a slurry containing approximately 35% water, while for the dry process, it contains less than 0.5% water. The raw meal is fed into the kiln and is burned in the kiln to produce an intermediate product called clinker.

The kiln slopes towards the burning zone and rotates slowly, causing the raw material to gradually move into the burning zone. Reactions which occur during gradual heating include the combination of lime with silica, alumina and iron to form the desired compounds in the clinker at a final temperature near 2650°F. Major clinker compounds are tricalcium silicate ($3CaO \bullet SiO_2$), dicalcium silicate ($2CaO \bullet SiO_2$), tricalcium aluminate ($3CaO \bullet Al_2O_3$), and tetracalcium aluminoferrite ($4CaO \bullet Al_2O_3 \bullet Fe_2O_3$) (EPS-3/77).

Cementation:
(1) A process in which a metal is added to a solution to initiate the precipitation of another metals, e.g., iron may be added to a copper sulfate solution to precipitation copper (Cu). It can be expressed as $Fe + CuSO_4 ---> Cu + FeSO_4$ (EPA-83/03a).
(2) The electrochemical reduction of metal ions by contacting with a metal of higher oxidation potential. It is usually used for the simultaneous recovery of copper and reduction of hexavalent chromium with the aid of scrap iron (EPA-83/06a).

Cementitious: Densely packed and nonfibrous friable materials (EPA-94/04).

Center drilling: Drilling a conical hole in the end of a workpiece (*see* drilling for more related terms) (EPA-83/06a).

Central collection point:
(1) A location where a generator consolidates regulated medical waste brought together from original generation points prior to its transport off-site or its treatment-site (e.g., incineration) (40 CFR 259.10-91).
(2) Location were a generator of regulated medical waste consolidates wastes originally generated at various locations in his facility. The wastes are gathered together for treatment on-site or for transportation elsewhere for treatment and/or disposal. This term could also apply to community hazardous waste collections, industrial and other waste management systems (EPA-94/04).

Central incinerator: A conveniently located facility that burns solid waste collected from many different sources (*see* incinerator for more related terms) (SW-108ts).

Central limit theorem: A statistical result which states that for a sufficiently large sample size, the distribution of means of random samples from a population with a finite variance will be approximately normal in forms, regardless of the form of the underlying population distribution (EPA-87/10a).

Central nervous system (CNS): A portion of the nervous system which consists of the brain and spinal cord (Course 165.6).

Central garbage grinder: A conveniently located facility that mechanically (with flushing water) pulverizes food wastes collected from many sources (EPA-83).

Central treatment facility: A treatment plant which co-treats process wastewater from more than one manufacturing operation or co-treats process wastewater with non-contact cooling water or with non-process wastewater, e.g., utility blow-down, miscellaneous run-off, etc. (*see* wastewater treatment for more related terms) (EPA-83/03).

Centralized yard waste composting: A system utilizing a central facility within a politically defined area with the purpose of composting yard wastes (EPA-89/11).

Centrate: The liquid fraction that is separated from the solids fraction of a slurry through centrifugation (EPA-87/10a).

Centri-cleaners: Cone-shaped devices into which dilute pulp slurry is pumped under high pressure. By centrifugal action heavy foreign material is forced to the outside of the tube and is rejected out the bottom. Accepted stock rises up through the center position of the tube and passes out the top (EPA-83).

Centrifugal collector:
(1) A mechanical system using centrifugal force to remove aerosols from a gas stream or to de-water sludge (EPA-94/04).
(2) *See also* synonym, mechanical separator.

Centrifugation: *See* synonym, mechanical separator.

Centrifuge:
(1) The machine used to separate solids by centrifugal force (EPA-87/10a).
(2) The treatment process whereby solids such as sludge can be separated from a liquid by the use of centrifugal force.

CEQ Regulations: The regulations issued by the Council on Environmental Quality (CEQ) on November 29, 1978 (*see* 43 FR 55978), which implement Executive Order 11991. The CEQ Regulations will often be referred to throughout this regulation by reference to 40 CFR 1500 et al (40 CFR 6.101-91).

Cermet: An alloy of a heat-resistant ceramic compound and a metal (DOE-91/04).

Ceramic: A product made by the baking or firing of a nonmetallic mineral such as tile, cement, refractories, and brick (EPA-83/03).

Ceramic plant: A manufacturing plant producing ceramic items (40 CFR 61.31-91).

CERCLA hazardous substance: A substance on the list defined in Section 101(14) of CERCLA. Note: Listed CERCLA hazardous substances appear in Table 302.4 of 40 CFR 302 (*see* hazardous substance for more related terms) (40 CFR 355.20-91).

CERCLIS: The abbreviation of the CERCLA Information System, EPA's comprehensive data base and management system that inventories and tracks releases addressed or needing to be addressed by the Superfund program. CERCLIS contains the official inventory of CERCLA sites and supports EPA's site planning and tracking functions. Sites that EPA decides do not warrant moving further in the site evaluation process are given a "No Further Response Action Planned" (NFRAP) designation in CERCLIS. This means that no additional federal steps under CERCLA will be taken at the site unless future information so warrants. Sites are not removed from the data base after completion of evaluations in order to document that these evaluations took place and to preclude the possibility that they be needlessly repeated. Inclusion of a specific site or area in the CERCLIS data base does not represent a determination of any party's liability, nor does it represent a finding that any response action is necessary. Sites that are deleted from the NPL (national priority list) are not designated NFRAP sites. Deleted sites are listed in a separate category in the CERCLIS data base (cf. record of decision) (40 CFR 300.5-91).

Cerebellum: A large dorsally projecting part of the brain having the special function of muscle coordination and maintenance of equilibrium (LBL-76/07-bio).

Cerebral anoxia: Relative lack of oxygen in the brain (LBL-76/07-bio).

Ceremonial and religious water use: The activities involving traditional Native American spiritual practices which involve, among other things, primary (direct) contact with water (40 CFR 131.35-91).

Cerium (Ce): A rare earth metal with a.n. 58; a.w. 140.12; d. 6.67 g/cc; m.p. 795° C and b.p. 3468° C. The element belongs to group IIIB of the periodic table.

Cerium metal: Any of a group of rare-earth metals separable as a group from other metals occurring with them and in addition to cerium, it includes lanthanum, praseodymium, neodymium, promethium, samarium and sometimes europium (EPA-82/05).

Cerium mineral: Rare earths, the important one is monazite (EPA-82/05).

Certificate holder: The entity in whose name the certificate of conformity for a class of motor vehicles or motor vehicle engines has been issued (40 CFR 85.1502-91).

Certificate of conformity: The document issued by the Administrator under section 206(a) of the Act (40 CFR 85.1502-91).

Certification: A statement of professional opinion based upon knowledge and belief (40 CFR 26.102-91, *see also* 40 CFR 163.2; 171.2; 260.10; 761.3-91).

Certification or audit test: A series of at least four test runs conducted for certification or audit purposes that meets the burn rate specifications in Section 5 (40 CFR 60-App/A(method 28)-91).

Certification vehicle: A vehicle which is selected under 40 CFR 86.084.24; (b)(1) and used to determine compliance under 40 CFR 86.084.30 for issuance of an original certificate of conformity (40 CFR 600.002.85; 91).

Certification vehicle emission margin for a certified engine family: The difference between the EPA emission standards and the average FTP emission test results of that engine family's emission-data vehicles at the projected applicable useful life mileage point (i.e., useful life mileage for light-duty vehicles is 50,000 miles and for light-duty trucks is 120,000 miles for 1985 and later model years or 50,000 miles for 1984 and earlier model years (40 CFR 85.2113-91).

Certified aftermarket part: Any aftermarket part which has been certified pursuant to this subpart (40 CFR 85.2113-91).

Certified applicator, etc. (or FIFRA applicator):
(1) **Certified applicator** means any individual who is certified

under section 4 as authorized to use or supervise the use of any pesticide which is classified for restricted use. Any applicator who holds or applies registered pesticides, or uses dilutions of registered pesticides consistent with section 2(ee) of this Act, only to provide a service of controlling pests without delivering any unapplied pesticide to any person so served is not deemed to be a seller or distributor of pesticides under this Act.

(2) **Private applicator** means a certified applicator who uses or supervises the use of any pesticide which is classified for restricted use for purposes of producing any agricultural commodity on property owned or rented by him or his employer or (if applied without compensation other than trading of personal services between producers of agricultural commodities) on the property of another person.

(3) **Commercial applicator** means an applicator (whether or not the applicator is a private applicator with respect to some uses) who uses or supervises the use of any pesticide which is classified for restricted use for any purpose or on any property other than as provided by paragraph (2).

(4) "Under The Direct Supervision of A Certified Applicator"-- Unless otherwise prescribed by its labeling, a pesticide shall be considered to be applied under the direct supervision of a certified applicator if it is applied by a competent person acting under the instructions and control of a certified applicator who is available if and when needed, even though such certified applicator is not physically present at the time and place the pesticide is applied (FIFRA2-7USC136).

Certified part: A part certified in accordance with the aftermarket part certification regulations contained in this subpart (40 CFR 85.2102-91).

Certified reference material: *See* definition under reference material.

Cesium (Cs): A alkali metal that is the most electropositive element known. Used especially in photoelectric cells (EPA-89/12). The element with a.n. 55; a.w. 132.90; d. 1.90 g/cc; m.p. 28.7° C and b.p. 690° C belongs to group IA of the periodic table.

Cesspool: An underground structure designed to hold sewage from a residence. The waste is permitted to percolate from the cesspool into the surrounding soil (DOI-70/04).

Cetane number: A number to rate the ignition characteristics of a diesel fuel combusted in a diesel engine (cf. octane number).

Chain: A chemical structure in which similar atoms are linked by bonds. Types of chemical chains include:
(1) **Straight chain:** A chain to link single atoms, not groups, in a chemical structure.
(2) **Branched chain (or side chain):** A chain to link side groups of atoms.
(3) **Closed chain:** A ring of atoms in a molecule.
(4) **Open chain:** If a molecule is not a closed chain, the molecule is an open chain (cf. chemical bond).

Chain grate: A stoker which has a massive moving chain as a grate surface; the grate consisting of links mounted on rods to form a continuous belt like surface that is generally pulled by sprockets on the front shaft (*see* grate for more related terms) (EPA-83).

Chain grate stoker: A stoker with a moving chain as a grate surface. The grate consists of links mounted on rods to form a continuous surface that is generally driven by a shaft with sprockets (*see* stoker for more related terms) (SW-108ts).

Chain of custody: In QA/QC, chain of custody includes:
(1) Sample safeguards;
(2) Sample subdivision;
(3) Laboratory records; and
(4) Sample custodian (ACS-87/11).

Chain of infection: *See* synonym, infection process.

Chain reaction: Sequential chemical reactions in which one initial chemical reaction results in (or activate) reactions of other chemicals.

Chalcocite: Copper sulfite, Cu_2S (EPA-82/05).

Chalcopyrite: A sulfide of copper and iron, $CuFeS_2$ (EPA-82/05).

Challenge exposure: An experimental exposure of a previously treated subject to a test substance following an induction period, to determine whether the subject will react in a hypersensitive manner (40 CFR 798.4100-91).

Chamber: An enclosed space inside an incinerator (OME-88/12).

Change order: A written order issued by a recipient, or its designated agent, to its contractor authorizing an addition to, deletion from, or revision of, a contract, usually initiated at the contractors request (40 CFR 35.6015-91).

Channel:
(1) The water-filled groove through which run-off water flows. In a narrow valley the channel may include the entire valley floor, but ordinarily it occupies only a small fraction of the valley (DOI-70/04).
(2) That part of a forehearth which carries the glass from the tank to the flow spout and in which temperature adjustments are made (EPA-83).

Channel drain: *See* synonym, French drain.

Channelization: Straightening and deepening streams so water will move faster, a marsh-drainage tactic that can interfere with waste assimilation capacity, disturb fish and wildlife habitats, and aggravate flooding (EPA-94/04).

Char: Carbonaceous material resulting from incomplete combustion (EPA-83).

Character displacement: The change of morphology or characteristics of two species living in the same area to reduce the competition for food resources.

Characteristics:
(1) In RCRA, EPA has identified four characteristics of a hazardous waste: ignitability, corrosivity, reactivity, and EP

toxicity (EP: extraction procedure). Any solid waste that exhibits one or more of these characteristics is classified as a hazardous waste (cf. definition under hazardous waste) (EPA-86/01).
(2) Any one of the four categories used in defining hazardous waste: ignitability, corrosivity, reactivity, and toxicity (EPA-94/04).

Charcoal: A product of the destructive distillation of wood. Used as a fuel and as a source of carbon in the foundry industry. Because of the nature of the destructive distillation process, charcoal may contain residuals of toxic pollutants such as phenol, benzene, toluene, naphthalene, and nitrosamines (EPA-85/10a).

Charge:
(1) (In iron industry), the addition of iron and steel scrap or other materials into the top of an electric arc furnace (EAF) (40 CFR 60.271-91, see also 40 CFR 60.271a-91).
(2) In incineration, the quantity of solid waste introduced into an incinerator or a furnace at one time (SW-108ts).
(3) In battery industry, the conversion of electrical energy into chemical energy within a cell-battery. This restoration of active electronic materials is done by forcing a current through the cell-battery in the opposite direction to that during discharge (EPA-84/08).

Charge chrome: That alloy containing 52 to 70 percent by weight chromium, 5 to 8 percent by weight carbon, and 3 to 6 percent by weight silicon (40 CFR 60.261-91).

Charge door: A door or an opening through which waste is charged to the incinerator (EPA-89/03b).

Charge rate: The quantity of waste materials loaded into an incinerator over a unit of time but which is not necessarily burned. Usually expressed in pounds of waste per hour (EPA-89/03b).

Charging: The addition of a molten or solid material to a copper converter (40 CFR 61.171-91).

Charging chute: An overhead passage through which waste materials drop into an incinerator (SW-108ts).

Charging cutoff gate: A modified charging gate used in continuous feed furnaces that do not have high temperatures near the charging hopper. A sliding steel plate at the bottom of the charging hopper closes on a machined seat at the top of the charging chute (SW-108ts).

Charging gate: A horizontal, movable cover on a top charging furnace. The cover opens for waste feed and closes during incineration. It may also apply to similar devices on side charging furnaces (OME-88/12).

Charging hopper: An enlarged opening at the top of a charging chute (SW-108ts).

Charging period: The time period commencing at the moment an EAF starts to open and ending either three minutes after the EAF roof is returned to its closed position or six minutes after commencement of opening of the roof, whichever is longer (40 CFR 60.271-91).

Charles' law: The Charles' law states that, when the volume is held constant, the absolute pressure of a given mass of a perfect gas of a given composition varies directly as the absolute temperature varies (see law for more related terms) (EPA-81/12, p8-6) (EPA-81/12, p8).

Check: An imperfection; a surface crack (EPA-83).

Check sample: A blank which has been spiked with the analyte(s) from an independent source in order to monitor the execution of the analytical method. The level of the spike is at the regulatory action level when applicable. Otherwise, the spike should be at 5 times the estimate of the quantification limit. The matrix used should be phase matched with the samples and well characterized, e.g., reagent grade water is appropriate for an aqueous sample (see sample for more related terms).

Check valve: A valve to limit a flow in a pipe to a single direction, i.e., no reverse flow direction may takes place.

Checker work: A pattern of multiple openings in a refractory structure through which the products of combustion pass to accelerate the turbulent mixing of gases (SW-108ts).

Checkers: The firebrick, alternating with openings, in the chambers of regenerative furnace (EPA-83).

Chelated compound: A compound containing a metal as an integral part of a ring structure and is not readily ionized (EPA-83/06a).

Chelating: Forming a compound containing a metal ion in a ring-like molecular configuration (EPA-87/10a).

Chelating agent:
(1) A coordinate compound in which a central atom (usually a metal) is joined by covalent bonds to two or more other molecules or ions (called ligands) so that heterocyclic rings are formed with the central (metal) atom as part of each ring. Thus, the compound is suspending the metal in solution (EPA-83/06a).
(2) A chemical agent whose atoms form more than one coordinate bonds with metals in a solution.

Chelating resin: In an ion-exchange solution, resins with high selectivity for certain specific cations.

Chelation: The formation of coordinate covalent bonds between a central metal ion and a liquid that contains two or more sites for combination with the metal ion (EPA-83/03a).

Chelatometry: See synonym, complexometric analysis (see also tritation for more related terms).

Chemical: A chemical substance or mixture (40 CFR 790.3-91).

Chemical adhesive fusion agent: A chemical fluid that may or may not contain a portion of the parent geomembrane and an adhesive that, after the application of pressure and after passage of a certain amount of time, results in the chemical fusion of two geomembrane sheets, leaving behind an adhesive layer that is dissimilar from the parent liner material. (Manufacturers and installers should be consulted for the various types of chemical fluids used with specific geomembrane to inform workers and inspectors) (EPA-91/05).

Chemical agent: Those elements, compounds, or mixtures that coagulate, disperse, dissolve, emulsify, foam, neutralize, precipitate, reduce, solubilize, oxidize, concentrate, congeal, entrap, fix, make the pollutant mass more rigid or viscous, or otherwise facilitate the mitigation of deleterious effects or the removal of the pollutant from the water (40 CFR 300.5-91).

Chemical analysis: The use of a standard chemical analytical procedure to determine the concentration of a specific pollutant in a wastewater sample (*see* analysis for more related terms) (EPA-82/11f).

Chemical bond (or bond): The attraction force holding atoms together in a molecule or crystal (cf. chain). Other chemical bond related terms include:
- Coordinate bond
- Coordinate valence (*see* synonym, coordinate bond)
- Covalent bond
- Double bond
- Electron pair bond (*see* synonym, covalent bond)
- Electrovalent bond (*see* synonym, ionic bond)
- Ionic bond
- Metallic bond
- Multiple bond
- Peptide bond
- Single bond
- Triple bond

Chemical brightening: A process utilizing an addition agent that leads to the formation of a bright plate or that improves the brightness of the deposit (EPA-83/06a).

Chemical coagulation: The destabilization and initial aggregation of colloidal and finely divided suspended matter by the addition of a floc-forming chemical (EPA-83/03).

Chemical coal cleaning: One of SO_2 emission reduction techniques (*see* sulfur oxide emission control for control structure). Chemical coal cleaning methods that reduce the organic bound sulfur include two technologies:
(1) Microwave desulfurization
(2) Hydrothermal desulfurization (EPA-81/12, p8-4).

Chemical composition: The name and percentage by weight of each compound in an additive and the name and percentage by weight of each element in an additive (40 CFR 79.2-91).

Chemical deposition: A process used to deposit a metal oxide on a substrate. The film is formed by hydrolysis of a mixture of chlorides at the hot surface of the substrate. Careful control of the water mixture insures that the oxide is formed on the substrate surface (EPA-83/06a).

Chemical durability (or durability): The lasting quality (both physical and chemical) of a glass surface. Frequently evaluated, after prolonged weathering or storing, in terms of chemical and physical changes in the glass surface, or in the contents of a vessel (EPA-83).

Chemical emergency preparedness program (CEPP): CEPP is developed by EPA to address accidental releases of acutely toxic chemicals (NRT-87/03).

Chemical energy: Energy stored in a chemical compound.

Chemical equilibrium: A chemical reaction condition in which the rate of the forward reaction equals the rate of the backward reaction. Under this condition, the concentration of reactants and products reaches a steady state (cf. reversible reaction).

Chemical etching: Dissolving a part of the surface of a metal or all of the metal laminated to a base (EPA-83/06a).

Chemical fixation: *See* synonym, solidification and stabilization.

Chemical formula: A formula containing symbols and numbers to represent a chemical composition (or structure) (e.g., C_6H_6 for benzene).

Chemical fusion: The chemically-induced reorganization in the polymeric structure of the surface of a polymer geomembrane that, after the application of pressure and the passage of a certain amount of time, results in the chemical fusion of two essentially similar geomembrane sheets being permanently joined together (EPA-91/05).

Chemical fusion agent: A chemical fluid that, after the application of the passage of a certain amount of time, results in the chemical fusion of two essentially similar geomembrane sheets without any other polymeric or adhesive additives. (Manufacturers and installers should be consulted for the various types of chemical fusion agents used with specific geomembranes to inform workers and inspectors) (EPA-91/05).

Chemical glass: A chemically durable glass suitable for use in making laboratory apparatus (*see* glass for more related terms) (EPA-83).

Chemical half-life: The time required to reduce 50 percent of the original concentration of a chemical in a specific medium (Course 165.6).

Chemical hazards response information system/hazard assessment computer system (CHRIS/HACS): CHRIS/HACS is developed by the U.S. Coast Guard. HACS is a computerized model of the four CHRIS manuals that contain chemical-specific data. Federal OSCs use HACS to find answers to specific questions during a chemical spill/response. State and local officials and industry representatives may ask an OSC to request a HACS run for contingency planning purposes (NRT-87/03).

Chemical inhibitor: A compound which is capable of reducing (or stopping) the rate of a chemical reaction.

Chemical kinetics: A science dealing with the mechanisms and rates of chemical reactions.

Chemical machining: A production of derived shapes and dimensions through selective or overall removal of metal by controlled chemical attack or etching (EPA-83/06a).

Chemical manufacturing plant: Any facility engaged in the production of chemicals by chemical, thermal, physical, or biological processes for use as a product, co-product, byproduct, or intermediate including but not limited to industrial organic chemicals, organic pesticide products, pharmaceutical preparations, paint and allied products, fertilizers, and agricultural chemicals. Examples of chemical manufacturing plants include facilities at which process units are operated to produce one or more of the following chemicals: benzenesulfonic acid, benzene, chlorobenzene, cumene, cyclohexane, ethylene, ethylbenzene, hydroquinone, linear alkylbenzene, nitrobenzene, resorcinol, sulfolane, or styrene (40 CFR 61.341-91).

Chemical metal cleaning waste: Any wastewater resulting from the cleaning of any metal process equipment with chemical compounds, including, but not limited to, boiler tube cleaning (*see* waste for more related terms) (40 CFR 423.11-91).

Chemical metal coloring: The production of desired colors on metal surfaces by appropriate chemical or electrochemical actions (EPA-83/06a).

Chemical milling: Removing large amounts of stock by etching selected areas of complex workpieces. This process entails cleaning, masking, etching, and demasking (EPA-83/06a).

Chemical name: The scientific designation of a chemical substance in accordance with the nomenclature system developed by the International Union of Pure and Applied chemistry or the Chemical Abstracts Service's rules of nomenclature, or a name which will clearly identify a chemical substance for the purpose of conducting a hazard evaluation (40 CFR 721.3-91).

Chemical oxidation: *See* synonym, oxidation.

Chemical oxidation demand (COD): *See* synonym, chemical oxygen demand.

Chemical oxygen demand (COD) (or chemical oxidation demand):
(1) A measure of the amount of oxygen required to oxidize organic and oxidizable inorganic compounds in water. The COD test, like the BOD test, is used to determine the degree of pollution in an effluent (EPA-74/11).
(2) A measure of the oxygen required to oxidize all compounds, both organic and inorganic, in water (EPA-94/04).
(3) *See* oxygen for more related terms.

Chemical potential: In a thermodynamic system, the change of Gibbs free energy with respect to the change in the amount of a component. The rate of change can be expressed as (dG/dn), where dG is the differential change of the Gibbs free energy and dn is the differential change of a component in the system.

Chemical precipitation:
(1) Formation of insoluble materials generated by addition of chemicals to a solution.
(2) The process of softening water by addition of lime and soda ash as the precipitants (EPA-3/83).

Chemical prewash: A salt bath between the fix and final wash which chemically removes the fix from the emulsion at a faster rate than can be done by washing, thereby reducing the after fix wash water, time and volume (EPA-10/80).

Chemical process: A particular method of manufacturing or making a chemical, usually involving a number of steps or operations (EPA-85/11).

Chemical protective clothing: The items of clothing that provide a protective barrier to prevent dermal contact with chemical substances of concern. Examples can include, but are not limited to, full body protective clothing, boots, coveralls, gloves, jackets, and pants (40 CFR 721.3-91).

Chemical pulping: A chemical method to dissolve the wood lignin to separate wood fibers for producing paper pulp. The lignin holds fibers together.

Chemical reaction: A chemical change that a mixture of chemical compounds (called reactants) forms a new mixture of chemical compounds (called products). Other chemical reaction related terms include:
- First order reaction
- Second order reaction
- Third order reaction

Chemical reactor: A container such as a tank or a vessel where chemical reactions take place.

Chemical recovery system: A chemical treatment system to remove metals or other materials from wastewater (EPA-83/06a).

Chemical reduction: *See* synonym, reduction.

Chemical resistance: The ability to resist chemical attack. (Note-- The attack is dependent on the method of test and its severity is measured by determining the changes in physical properties. Time, temperature, stress, and reagent, may all be factors that affect the chemical resistance of a material) (NIOSH-84/10).

Chemical sensitivity: A health problem characterized by effects such as dizziness, eye and throat irritation, chest tightness and nasal congestion that appear whenever a person is exposed to certain chemicals. People may react to even trace amounts of chemicals to which they become sensitized (EPA-88/09b).

Chemical sludge: The sludge produced due to the addition of chemical coagulation to help the sedimentation of suspended matter in a wastewater treatment process (*see* sludge for more related terms).

Chemical stratification: The formation of layers of water in a lake that are of different densities. This density difference is caused by changes in the concentrations of dissolved substances in the water at different depths (DOD-78/01).

Chemical structure: The molecular structure of a compound in an additive (40 CFR 79.2-91).

Chemical substance:
(1) Except as provided in subparagraph (2), the term "chemical substance" means any organic or inorganic substance of a particular molecular identity, including:
- (A) Any combination of such substances occurring in whole or in part as a result of a chemical reaction or occurring in nature, and
- (B) Any element or uncombined radical.

(2) Such term does not include:
- (A) Any mixture,
- (B) Any pesticide (as defined in the Federal Insecticide, Fungicide, and Rodenticide Act) when manufactured, processed, or distributed in commerce for use as a pesticide,
- (C) Tobacco or any tobacco product,
- (D) Any source material, special nuclear material, or byproduct material (as such terms are defined in the Atomic Energy Act of 1954 and regulations issued under such Act),
- (E) Any article the sale of which is subject to the tax imposed by section 4181 of the Internal Revenue Code of 1954 (determined without regard to any exemptions from such tax provided by section 4182 or 4221 or any other provision of such Code), and
- (F) Any food, food additive, drug, cosmetic, or device (as such terms are defined in section 201 of the Federal Food, Drug, and Cosmetic Act) when manufactured, processed, or distributed in commerce for use as a food, food additive, drug, cosmetic, or device (TSCA3-15USC2602).

(3) cf. new chemical substance.

Chemical synthesis: The process of chemically combining two or more constituent substances into a single substance (EPA-83/09).

Chemical thermodynamics: A branch of thermodynamics which applies thermodynamic principles to the chemical problems of concern.

Chemical transportation emergency center (CHEMTREC): CHEMTREC is operated by the Chemical Manufacturers Association. Provides information and/or assistance to emergency responders. CHEMTREC contacts the shipper or producer of the material for more detailed information, including on-scene assistance when feasible. Can be reached 24 hours a day by calling 800-424-9300 (in the United States only). (*See also* HIT) (NRT-87/03).

Chemical treatment:
(1) A treatment process that alters the chemical structure of a toxic waste contaminant to reduce the waste's toxicity, mobility, or volume (EPA-89/12a).
(2) Any one of a variety of technologies that use chemicals or a variety of chemical processes to treat waste (EPA-94/04).
(3) *See* treatment for more related terms.

Chemical waste landfill: A landfill at which protection against risk of injury to health or the environment from migration of PCBs to land, water, or the atmosphere is provided from PCBs and PCB Items deposited therein by locating, engineering, and operating the landfill as specified in 40 CFR 761.75 (40 CFR 761.3-91).

Chemical wood pulp: The wood pulp obtained by digestion of wood with solutions of various chemicals. The principal chemical processes are the sulfite, sulfate (kraft), and soda processes (*see* pulp for more related terms) (EPA-87/10).

Chemiluminescence analyzer (CA): One of continuous emission monitors (*see* continuous emission monitor or luminescence analyzer for various types). Chemiluminescence is the emission of light energy that results from a chemical reaction. It has been proven reliable and is the most widely used method for source NO/NO_x analysis (EPA-84/03a).

Chemi-mechanical pulp: The pulp produced mechanically by grinding or refining after presoaking of wood with caustic soda/sodium sulfite solution (*see* pulp for more related terms) (EPA-74/05a).

Chemisorption:
(1) An adsorption where the forces holding the adsorbate to the adsorbent are chemical (valance) instead of physical (van der Waals) (EPA-82/11f).
(2) Adsorption, especially when irreversible, by means of chemical forces in contrast with physical forces (LBL-76/07-air).

Chemnet:
(1) A mutual aid network of chemical shippers and contractors. CHEMNET has more than fifty participating companies with emergency teams, twenty-three subscribers (who receive services in an incident from a participant and then reimburse response and cleanup costs), and several emergency response contractors. CHEMNET is activated when a member shipper cannot respond promptly to an incident involving that company's product(s) and requiring the presence of a chemical expert. If a member company cannot go the scene of the incident, the shipper will authorize a CHEMNET-contracted emergency response company to go. Communications for the network are provided by CHEMTREC, with the shipper receiving notification and details about the incident from the CHEMTREC communicator (NRT-87/03).
(2) Mutual aid network of chemical shippers and contractors that assigns a contracted emergency response company to provide technical support if a representative of the firm whose chemicals are involved in an incident is not readily available (EPA-94/04).

Chemoautotroph: Any autotroph bacteria and protozoans which do not carry out photosynthesis.

Chemosterilant: A chemical that controls pests by preventing reproduction (EPA-94/04).

Chemosynthesis: Synthesis of organic compounds for energy by some microorganisms.

Chemterc: The industry-sponsored Chemical Transportation Emergency Center; provides information and/or emergency assistance to emergency responders (EPA-94/04).

Child resistant packaging: The packaging that is designed and constructed to be significantly difficult for children under 5 years of age to open or obtain a toxic or harmful amount of the substance contained therein within a reasonable time, and that is not difficult for normal adults to use properly (40 CFR 157.21-91).

Chill mark (or settle mark): A wrinkled surface condition on glassware resulting from uneven cooling in the forming process (EPA-83).

Chilled water loop: Any closed cooling water system that transfers heat from air handling units or refrigeration equipment to a refrigeration machine, or chiller (40 CFR 749.68-91).

Chilling effect: The lowering of the Earth's temperature because of increased particles in the air blocking the sun's rays (cf. greenhouse effect) (EPA-94/04).

Chimney: *See* synonym, stack.

Chimney effect: A phenomenon consisting of a vertical movement of a localized mass of air or other gases due to temperature differences (EPA-83/06).

China clay: Top quality, high-grade clay used for paper coating and filling (EPA-83).

Chip:
(1) An imperfection due to breakage of a small fragment out of an otherwise regular surface.
(2) A small piece of wood produced by a chipper in a form suitable for processing into pulp or particle board.
(3) A term used to designate a paperboard (chipboard) made from waste paper--usually mixed papers (EPA-83).

Chipped glass: A glass article with chipped surface produced intentionally (*see* glass for more related terms) (EPA-83).

Chipper: A size-reduction device having sharp blades attached to a rotating shaft (mandrel) that shave or chip off pieces of certain objects, such as tree branches or brush (*see* size reduction machine for more related terms) (SW-108ts).

Chipping: The process of removing thin extra glass prior to grinding (EPA-83).

Chisel plowing: Preparing croplands by using a special implement that avoids complete inversion of the soil as in with conventional plowing. Chisel plowing can leave a protective cover or crop residues on the soil surface to help prevent erosion and improve filtration (EPA-94/04).

Chloramines ($CH_3C_6H_4SO_2NClNa \bullet 3H_2O$): Compounds obtained by chlorine disinfection from the action of hypochlorite solutions (weak acidic easily decomposed) on compounds containing NH and NH_2 groups (EPA-75/10).

Chlordane ($C_{10}H_6Cl_8$): A volatile liquid used as a insecticide.

Chlorendic acid ($C_9H_4Cl_6O_4$): A white powder used in fungicides, insecticides, fire-resistant polyester resins.

Chloride: *See* definition under halide.

Chlorinalysis: A process of converting chlorinated hydrocarbons into carbon tetrachloride at the temperatures around 500° C and pressure above 50 atm. It is not really a waste treatment process, but a carbon tetrachloride manufacturing process which can accept a large quantity of toxic waste.

Chlorinated aromatic: An aromatic compound containing a chlorine (Cl) atom(s) substituent; e.g., chlorobenzene (C_6H_5Cl) (EPA-88/12).

Chlorinated ethylene: An alkene containing a chlorine (Cl) atom(s) substituent; e.g., chloroethylene (C_2H_3Cl) (EPA-88/12).

Chlorinated hydrocarbons: These include a class of persistent, broad-spectrum insecticides that linger in the environment and accumulate in the food chain. Among them are DDT, aldrin, dieldrin, heptachlor, chlordane, lindane, endrin, mirex, hexachloride, and toxaphene. Other examples include TCE, used as an industrial solvent (EPA-94/04).

Chlorinated polyethylene (CPE): Family of polymers produced by the chemical reaction of chlorine with polyethylene. The resultant polymers presently contain 25-45% chlorine by weight and 0-25% crystallinity (EPA-91/05).

Chlorinated polyethylene-reinforced (CPE-R): Sheets of CPE with an encapsulated fabric reinforcement layer, called a scrim (EPA-91/05).

Chlorinated solvent: An organic solvent containing chlorine atoms, e.g., methylene chloride and 1,1,1-trichloromethane, used in aerosol spray containers and in highway paint (EPA-94/04).

Chlorinated terphenyl: A chemical substance, CAS No. 61788-33-6, comprised of chlorinated ortho-, meta-, and paraterphenyl (40 CFR 704.45-91).

Chlorinated waste: The wastes that are composed primarily of organic compounds containing chlorine (ETI-92).

Chlorination: The application of chlorine to drinking water, sewage, or industrial waste to disinfect or to oxidize undesirable compounds (EPA-94/04). Other chlorination related terms include:
- Break point chlorination
- Free residual chlorination

Chlorination break point: *See* synonym, break point of chlorination.

Chlorinator: A device that adds chlorine, in gas or liquid form, to water or sewage to kill infectious bacteria (EPA-94/04).

Chlorine (Cl):
(1) A halogen element with a.n. 17; a.w. 35.45; d. 1.56 g/cc; m.p. -101.0° C and b.p. -34.7° C. The element belongs to group VIIA of the periodic table.

(2) The total residual chlorine present in the process wastewater stream exiting the wastewater treatment system (40 CFR 415.651-91).

Other chlorine related terms include:
- Available chlorine
- Combined chlorine
- Combined available residual chlorine
- Free chlorine
- Free available chlorine
- Free residual chlorine
- Residual chlorine
- Total residual chlorine
- Total residual oxidants for intake water with bromide (*see* synonym, total residual chlorine)

Chlorine contact chamber: That part of a water treatment plant where effluent is disinfected by chlorine (EPA-94/04).

Chlorine contact tank: In the leather industry, a detention basin designed to allow sufficient time for the diffusion and reaction of chlorine in a liquid for disinfection purposes (EPA-82/11).

Chlorine demand:
(1) The quantity of chlorine absorbed by wastewater (or water) in a given length of time (EPA-82/11f).
(2) The difference between the amount of chlorine applied to a treated supply and the amount of free, combined, or total available chlorine remaining at the end of the contact period. The chlorine demand is determined by the amount of oxidizable material present in the water (DOI-70/04).
(3) The difference between the amount of chlorine added to water or wastewater and the amount of residual chlorine remaining at the end of a specified contact period (LBL-76/07-water).

Chlorine dioxide (ClO_2): A chemical used in pulp bleaching as a water solution, usually in one or more of the latter stages of a multistage sequence. It is prepared by a variety of processes at the plant site usually from sodium chlorate, acid, and a reducing agent (EPA-87/10). Chlorine dioxide with d. 3.09 g/cc; m.p. -59.5° C and b.p. 9.9° C is a poisonous green gas that is even a stronger oxidizing agent than chlorine.

Chlorine emergency plan (CHLOREP): CHLOREP is operated by the Chlorine Institute. A 24-hour mutual aid program. Response is activated by a CHEMTREC call to the designated CHLOREP contact, who notifies the appropriate team leader, based upon CHLOREP's geographical sector assignments for teams. The team leader in turn calls the emergency caller at the incident scene and determines what advice and assistance are needed. The team leader then decides whether or not to dispatch his team to the scene (NRT-87/03).

Chlorinity: A measurement of the content of chlorine and other halogen elements in a solution.

Chlorofluorocarbons (CFCs):
(1) Organic compounds containing chlorine and/or fluorine atoms within the molecule (EPA-8/87b).
(2) A family of inert, nontoxic, and easily liquified chemicals used in refrigeration, air conditioning, packaging, insulation, or as solvents and aerosol propellants. Because CFCs are not destroyed in the lower atmosphere they drift into the upper atmosphere where their chlorine components destroy ozone (EPA-94/04).

Chlorophenoxy: A class of herbicides that may be found in domestic water supplies and cause adverse health effects (EPA-94/04).

CLO: A unit to express the relative thermal insulation values of various clothing assemblies. 1 clo = 0.180 C m² h C^{-1} (NIOSH-84/10).

Chloroform (or trichloromathane, $CHCl_3$): A colorless solvent used in fire extinguishers.

Chlorosis: Discoloration of normally green plant parts caused by disease, lack of nutrients, or various air pollutants (EPA-94/04).

Chlorosulfonated polyethylene (CSPE): Family of polymers produced by the reaction of polyethylene with chlorine and sulphur dioxide. Present polymers contain 23.5 to 43% chlorine and 1.0 to 1.4% sulphur. A low water absorption grade is identified as significantly different from standard grades (EPA-91/05).

Chlorophenol: Any compounds resulted from the reaction of phenol and chlorine.

Chlorophyll: One of two pigments (chlorophyll a ($C_{55}H_{72}O_5N_4Mg$) and chlorophyll b ($C_{55}H_{70}O_6N_4Mg$)) responsible for the green color of plants and for the process of photosynthesis.

Chlorotrifluoromethane (monochlorotrifluoromethane, or trifluorochloromethane, $CClF_3$): A colorless gas used in refrigerants and aerosol propellants.

Chlorosulfonated polyethylene-reinforced (CSPE-R): Sheets of CSPE with an encapsulated fabric reinforcement layer, called a scrim (EPA-91/05).

Choke pull-off: *See* vacuum break (40 CFR 85.2122(a)(1)(ii)(F)-91).

Cholinesterase: An enzyme found in animals that regulates nerve impulses. Cholinesterase inhibition is associated with a variety of acute symptoms such as nausea, vomiting, blurred vision, stomach cramps, and rapid heart rate (EPA-94/04).

Chromate: *See* definition under chromium.

Chromate conversion coating: A protective coating formed by an immersing metal in an aqueous acidified solution consisting substantially of chromic acid or water soluble salts of chromic acid together with various catalysts or activators (EPA-83/06a).

Chromatid type aberration: The damage expressed as breakage of single chromatids or breakage and/or reunion between chromatids (40 CFR 798.5300-91, *see also* 40 CFR 798.5385-91).

Chromatizing: To treat or impregnate with a chromate (salt of ester of chromic acid) or dichromate, especially with potassium dichromate (EPA-83/06a).

Chromatography (or column chromatography): A method of separation (of gases, liquids, or dissolved substances), based on adsorbing the gases, vapors, or substances of a mixture and then driving each off, one by one. This permits the separation of the mixture into individual compounds. A detector is usually employed to identify/quantify the separated components (cf. mass spectrometer) (EPA-84/09). The method involves that a tube is packed with stationary phase materials (adsorbing materials; e.g., alumina). The sample (moving phase) is injected into the column and is continuously washed through with a solvent. Different components of the sample are adsorbed to different extends and move down the column at different rates. Other chromatography related terms include:
- Capillary gas chromatography
- Column chromatography (*see* synonym, chromatography)
- Ion exchange chromatography
- Liquid chromatography
- Thin layer chromatography

Chrome pickle process: A process of forming a corrosion-resistant oxide film on the surface of magnesium base metals by immersion in a bath of an alkaline bichromate (EPA-83/06a).

Chrome pigments: The chrome yellow, chrome orange, molybdate chrome orange, anhydrous and hydrous chromium oxide, chrome green, and zinc yellow (40 CFR 415.341-91).

Chrome sand (or chrome-iron ore): A dark material containing dark brown streaks with sub-metallic to metallic luster. Usually found as grains disseminated in perioditite rocks. Used in the preparation of molds (EPA-85/10a).

Chrome tan: The process of converting hide into leather using a form of chromium (40 CFR 425.02-91).

Chromic acid (H_2CrO_4): A hydrate of CrO_3.

Chromite: Chrome iron ore, $FeCr_2O_4$ (EPA-82/05).

Chromite flour: Chrome sand ground to 200 mesh or finer which can be used as a filler material for mold coatings for steel castings (EPA-85/10a).

Chromium (Cr):
(1) Chromium is a hard transition metal with a.n. 24; a.w. 51.99; d. 7.19 g/cc; m.p. 1875° C and b.p. 2665° C and belongs to group VIB of the periodic table.
(2) The total chromium present in the process wastewater stream exiting the wastewater treatment system (cf. heavy metals) (40 CFR 415.661-91, *see also* 40 CFR 420.02; 428.101-91).
(3) *See* heavy metals (EPA-94/04).
Major chromium compounds include:
- Chromate: A salt containing ion $(CrO_4)^{-2}$.
- Dichromate (VI): A salt containing ion $(Cr_2O_7)^-$.
- cf. total chromium.

Chromium catalyst: Plating bath constituent that in small amounts makes possible the continuing capability to electro-deposit chromium. Usually fluoride, fluorosilicate and/or sulfate (EPA-74/03d).

Chromium VI (or hexavalent chromium): The value obtained by the method specified in 40 CFR 136.3 (40 CFR 420.02-91).

Chromogen: A reagent which produces a colored product (LBL-76/07-water).

Chromophore: An element that provides color in a dye.

Chromosome:
(1) Cellular elements regarded as the carriers of hereditary characteristics (LBL-76/07-bio).
(2) A rodlike structure in the nucleus of a cell that forms during mitosis; composed of DNA and protein; chromosomes contain the genes responsible for heredity. It is a gene carrier (Course 165.6).

Chromosome mutations: The chromosomal changes resulting from breakage and reunion of chromosomes. Chromosomal mutations are also produced through nondisjunction of chromosomes during cell division (40 CFR 798.5955-91).

Chromosome type aberrations: The changes which result from damage expressed in both sister chromatids at the same time (40 CFR 798.5300-91, *see also* 40 CFR 798.5385-91).

Chronic:
(1) Chronic means long. Occurring over a long period of time, either continuously or intermittently; used to describe ongoing exposures and effects that develop only after a long exposure (Course 165.6).
(2) Characterized by a slow progressive course of indefinite duration, usually especially of degenerative invasive diseases, some infections, psychoses, inflammations and the carrier state (LBL-76/07-bio).
(3) A stimulus that lingers or continues for a relatively long period of time, often one-tenth of the life span or more. Chronic should be considered a relative term depending on the life span of an organism. The measurement of a chronic effect can be reduced growth, reduced reproduction, etc., in addition to lethality (EPA-91/03).

Chronic dose: *See* synonym, chronic exposure.

Chronic effect:
(1) An effect that is manifest after some time has elapsed from initial exposure (EPA-92/12).
(2) An adverse effect on a human or animal in which symptoms recur frequently or develop slowly over a long period of time (EPA-94/04).
Other chronic effect related terms include:
- Acute effect
- Serious chronic effect

Chronic exposure (or chronic dose):
(1) Doses repeatedly received by a body over a long period of time (Course 165.5).
(2) Multiple exposures occurring over an extended period of time, or a significant fraction of the animal's or the individual's lifetime (EPA-90/08; 92/12).

Chronic nephritis: Chronic inflammation of the kidneys (LBL-76/07-bio).

Chronic study: A toxicity study designed to measure the (toxic) effects of chronic exposure to a chemical (EPA-92/12).

Chronic toxicity:
(1) The lowest concentration of a constituent causing observable effects (i.e., considering lethality, growth, reduced reproduction, etc.) over a relatively long period of time, usually a 28-day test period for small fish test species (40 CFR 131.35-91, *see also* 40 CFR 300-App/A-91).
(2) The capacity of a substance to cause long term poisonous human health effects (cf. toxicity, acute) (EPA-94/04).
(3) *See* toxicity for more related terms.

Chronic toxicity endpoints (CTE): Results, such as a no observed effect concentration, lowest observed effect concentration, effect concentration, and inhibition concentration based on observations of reduced reproduction, growth, and/or survival from life cycle, partial life cycle, and early life stage tests with aquatic animal species (EPA-91/03).

Chronic toxicity test: A method used to determine the concentration of a substance in water that produces an adverse effect on a test organism over an extended period of time. In this test guideline, mortality and reproduction (and optionally, growth) are the criteria of toxicity (40 CFR 797.1330-91, *see also* 40 CFR 797.1950-91).

Chrysocolla: Hydrated copper silicate, $CuSiO_3 \cdot 2H_2O$ (EPA-82/05).

Chrysotile: A metamorphic mineral, an asbestos, the fibrous variety of serpentine. A silicate of magnesium, with silica tetrahedra arranged in sheets (EPA-82/05).

Chute fed incinerator: An incinerator that is charged through a chute that extends two or more floors above it (*see* incinerator for more related terms) (SW-108ts).

Cinchona: An alkaloid-containing bark of trees.

Cinder: Another name for slag (cf. slag) (EPA-74/06a).

Cinnabar: Mercury sulfide, HgS (EPA-82/05).

Circle of influence: The circular outer edge of a depression produced in the water table by the pumping of water from a well (*see* cone of influence, cone of Depression (EPA-94/04).

Circuit breaker: A device capable of making, carrying, and breaking currents under normal or abnormal circuit conditions (EPA-83/03).

Circulating bed combustor: The circulating bed combustor (CBC) uses high velocity air to entrain circulating solids in a highly turbulent combustion loop. The combustion chamber is typically 30 ft high and has a 12 in. thick ceramic liner. Solid feed is introduced into the combustor loop at the loop seal where it immediately contacts the hot recirculating solids stream exiting the hot cyclone. Liquid feeds are typically injected directly into the combustion zone of the CBC. Upon entering the CBC, hazardous materials are rapidly heated and continue to be exposed to high temperatures (1450 to 1600°F) throughout their stay in the CBC. Residence times in the combustor range from 2 seconds for gases to 30 minutes for larger feed materials (<1.0 in.).

The high combustion air velocity for circulating solids create a uniform temperature (+/- 50°F) around the combustion loop (combustion chamber, hot cyclone, return leg), reportedly resulting in extremely efficient combustion and eliminating the need for an afterburner. During operation, ash is periodically removed from the CBC by means of a water-cooled ash removal system. The hot gas leaving the cyclone is cooled in a flue gas cooler and particulates escaping the cyclone are collected in fabric filter baghouses (*see* incinerator for more related terms) (Lee-88/08).

Circulating fluidized bed combustor: A fluidized bed combustor in which the majority of the fluidized bed material is carried out of the primary combustion zone and is transported back to the primary zone through a recirculation loop (*see* fluidized bed combustion for more related terms) (40 CFR 60.51a-91).

Circulating water pump: A pump which delivers cooling water to the condensers of a powerplant (EPA-82/11f).

Circulating water system: A system which conveys cooling water from its source to the main condensers and then to the point of discharge (*see* cooling water for more related terms) (EPA-82/11f).

cis: A form of isomerism indicating that atoms are on the same side of an asymmetric molecule.

Cistern: Small tank or storage facility used to store water for a home or farm; often use to store rain water (EPA-94/04).

Citraturia: The presence in the urine of citric acid. The term usually implies the presence of increased quantities of citric acid in the urine (LBL-76/07-bio).

Citric acid ($C_6H_8O_7 \cdot H_2O$): A white crystalline acid mainly present in citrus fruits.

Citrus pulp (dried) citrus peel (dried): Chopped peel, seeds and other non-juice parts of the fruit that have been limed and dried for cattle feed (EPA-74/03).

City fuel economy: The fuel economy determined by operating a vehicle (or vehicles) over the driving schedule in the Federal emission test procedure (40 CFR 600.002.85-91).

City fuel economy test: *See* synonym, federal test procedure (40 CFR 610.11-91).

Civil action: In RCRA, a law suit filed in court against a person who has either failed to comply with statutory or regulatory requirements or an administrative order or has contributed to a release of hazardous wastes or constituents. There are four types of civil actions: compliance, corrective, monitoring and analysis, and imminent hazard (cf. administrative order) (EPA-86/01).

Cladding (or metal cladding):
(1) The art of producing a composite metal containing two or

more layers that have been metallurgically bonded together by roll bonding (co-rolling), solder application (or brazing), or explosion bonding (40 CFR 471.02-91).
(2) In a reactor, the material that covers each tubular fuel and target assembly (DOE-91/04).

Claim:
(1) A request, made in writing for a sum certain, for compensation for damages or removal costs resulting from an incident (OPA1001-91, *see also* SF101; 40 CFR 14.2; 27.2; 35.6015; 211.203; 300.5; 304.12-91).
(2) *See* business confidentiality claim (40 CFR 2.201-91).

Claimant: Any person or government who presents a claim for compensation under this title (OPA1001, *see also* SF101; 40 CFR 350.1-91).

Clamshell bucket: A vessel used to hoist and convey materials. It has two jaws that clamp together when the vessel is lifted by specially attached cables (SW-108ts).

Clarification: Clearing action that occurs during wastewater treatment when solids settle out. This is often aided by centrifugal action and chemically induced coagulation in wastewater (EPA-94/04).

Clarifier: A tank in which solids are settled to the bottom and are subsequently removed as sludge (EPA-94/04). Other clarifier related terms include:
- Green liquor clarifier
- Mechanical clarifier
- Primary clarifier
- Primary sedimentation tank (*see* synonym, primary clarifier)
- Primary settling tank (*see* synonym, primary clarifier)
- Secondary clarifier
- Secondary settling tank (*see* synonym, secondary clarifier)
- Secondary sedimentation tank (*see* synonym, secondary clarifier)

Clarity: Degree of opaque due to the suspended particles in a solution (liquid).

Clark process: One of methods for softening water in which calcium hydroxide is used to convert (neutralize) acid carbonates into normal carbonates.

Class: A group of vehicles which are identical in all material aspects with respect to the parameters listed in 40 CFR 205.155 of this subpart (40 CFR 86.402.78-91, *see also* 40 CFR 205.151-91).

Class 1E: The safety classification of the electric equipment and systems that are essential to emergency reactor shutdown, containment isolation, reactor core cooling and containment, and reactor heat removal or that are otherwise essential in preventing a significant release of radioactive material to the environment (DOE-91/04).

Class A facility: A facility engaged in administrative activities considered essential to the overall direction and continuity of the protection program; engaged in research and development, manufacture, production, assembly, or storage of nuclear weapons, weapon assemblies, or military reactors; engaged--in major research and development in uranium enrichment or operation of major uranium enrichment facilities; involved in research and development, manufacture, production, or assembly of non-nuclear weapon components, assemblies, and parts essential to the weapons or military reactor programs; or receiving, handling, or storing Top Secret documents (exclusive of keying material for secure communications) over an extended period. Or, a facility that possesses specified quantities of special nuclear material (DOE-91/04).

Class I area: Under the Clean Air Act, a Class I area is one in which visibility is protected more stringently than under the national ambient air quality standards; includes national parks, wilderness area, monuments and other areas of special national and cultural significance (EPA-94/04).

Class I landfill disposal site: That complete protection is provided for all time for the quality of ground and surface waters from all wastes deposited therein and against hazard to public health and wildlife resources (*see* landfill for more related terms) (EPA-83).

Class I sludge management facility: Any POTW (public owned treatment works) identified under 40 CFR 403.8(a) as being required to have an approved pretreatment program (including such POTWs located in a State that has elected to assume local program responsibilities pursuant to 40 CFR 403.10(e)) and any other treatment works treating domestic sewage classified as a Class I sludge management facility by the Regional Administrator, or, in the case of approved State programs, the Regional Administrator in conjunction with the State Director, because of the potential for its sludge use or disposal practices to adversely affect public health and the environment (40 CFR 122.2-91).

Class I substance: Each of the substances listed as provided in section 602(a) (CAA601-42USC7671-91).

Class II landfill disposal site: That protection is provided to water quality from Group 2 and Group 3 wastes. The types of physical features and the extent of protection of groundwater quality can divide class ti sites into several categories. Mixed municipal refuse is usually in Class II (*see* landfill for more related terms) (EPA-83).

Class II substance: Each of the substances listed as provided in section 602(b) (CAA601-42USC7671-91).

Class II well: The wells which inject fluids:
(1) Which are brought to the surface in connection with conventional oil or natural gas production and may be commingled with waste waters from gas plants which are an integral part of production operations, unless those waters would be classified as a hazardous waste at the time of injection;
(2) For enhanced recovery of oil or natural gas; and (c) For storage of hydrocarbons which are liquid at standard temperature and pressure (*see* well for more related terms) (40 CFR 147.2902-91).

Class III landfill disposal site: That protection is provided to water quality from Group 3 wastes (essentially the inert wastes) by location, construction, and operation which prevent erosion of deposited material (*see* landfill for more related terms) (EPA-83).

Class T3: All aircraft gas turbine engines of the JT3D model family (40 CFR 87.1-91).

Class T8: All aircraft gas turbine engines of the JT8D model family (40 CFR 87.1-91).

Class TF: All turbofan or turbojet aircraft engines except engines of Class T3, T8, and TSS (40 CFR 87.1-91).

Class TP: All aircraft turboprop erines (40 CFR 87.1-91).

Class TSS: All aircraft gas turbine engines employed for propulsion of aircraft designed to operate at supersonic flight speeds (40 CFR 87.1-91).

Classification: The separation and rearrangement of waste materials according to composition (e.g., organic or inorganic), size, weight, color, shape, the like, etc. using specialized equipment (OME-88/12).

Classification of railroads: The division of railroad industry operating companies by the Interstate Commerce Commission into three categories. As of 1, Class I railroads must have annual revenues of $50 million or greater, Class II railroads must have annual revenues of between $10 and $50 million and Class III railroads must have less than $10 million in annual revenues (40 CFR 201.1-91).

Classified information: The official information which has been assigned a security classification category in the interest of the national defense or foreign relations of the United States (40 CFR 11.4-91).

Classified material: Any document, apparatus, model, film, recording, or any other physical object from which classified information can be derived by study, analysis, observation, or use of the material involved (40 CFR 11.4-91).

Classified waste: Waste material that has been given security classification in accordance with 50USC401 and Executive Order 11652 (*see* waste for more related terms) (40 CFR 246.101-91).

Classifier (or grit washer):
(1) A machine or device for separating the constituents of a material according to relative sizes and densities thus facilitating concentration and treatment. Classifiers are also used to separate sand from slime, water from sand, and water from slime. It is used in particular where an upward current of water is used to remove fine particles form coarser materials.
(2) In mineral dressing, a device that takes the ball-mill discharge and separates it into two portions-the finished product which is ground as fine as desired and oversize material (EPA-82/05).

Other classifier related terms include:
- Air classifier
- Rake classifier
- Spiral classifier

Claus sulfur recovery plant: A process unit which recovers sulfur from hydrogen sulfide by a vapor-phase catalytic reaction of sulfur dioxide and hydrogen sulfide (40 CFR 60.101-91).

Clay:
(1) Aluminum silicates less than 0.002 mm (2.0 micro meter) in size. Because of their size, most clay types can go into colloidal suspension (EPA-87/10a).
(2) A natural, earthy, fine-grained material which develops plasticity (putty-like properties) when wetted, but is hard when baked or fired. Used as filler and for coating paper sheets (EPA-87/10, *see also* EPA-83; SW-108ts).

Clay mineral analysis: The estimation or determination of the kinds of clay-size minerals and the amount present in a sediment or soil (40 CFR 796.2750-91).

Clay soil: Soil material containing more than 40 percent clay, less than 45 percent sand, and less than 40 percent silt (EPA-94/04).

Clean air: The air of such purity that it will not cause harm or discomfort to an individual if it is inhaled for extended periods of time (29 CFR 1910.94a-91).

Clean Air Act (CAA) of 1970: *See* the term of Act or CAA.

Clean Air Act Amendment (CAAA) of 1990: *See* the term of Act or CAAA.

Clean air standards: Any enforceable rules, regulations, guidelines, standards, limitations, orders, controls, prohibitions, or other requirements which are contained in, issued under, or otherwise adopted pursuant to the CAA or Executive Order 11738, an applicable implementation plan as described in section 110(d) of the CAA, and approved implementation procedure or plan under section 111(c) or section 111(d), respectively, of the CAA or an approved implementation procedure under section 112(d) of the CAA (40 CFR 15.4-91).

Clean alternative fuel: Any fuel (including methanol, ethanol, or other alcohols (including any mixture thereof containing 85 percent or more by volume of such alcohol with gasoline or other fuels), reformulated gasoline, diesel, natural gas, liquefied petroleum gas, and hydrogen) or power source (including electricity) used in a clean-fuel vehicle that complies with the standards and requirements applicable to such vehicle under this title when using such fuel or power source. In the case of any flexible fuel vehicle or dual fuel vehicle, the term "clean alternative fuel" means only a fuel with respect to which such vehicle was certified as a clean-fuel vehicle meeting the standards applicable to clean-fuel vehicles under section 243(d)(2) when operating on clean alternative fuel (or any CARB standards which replaces such standards pursuant to section 243(e)) (CAA241-42USC7581).

Clean area: A controlled environment which is maintained and monitored to assure a low probability of asbestos contamination to materials in that space. Clean areas used in this method have HEPA filtered air under positive pressure and are capable of sustained operation with an open laboratory blank which on subsequent analysis has an average of less than 18 structures/mm^2 in an area of 0.057 mm^2 (nominally 10 200-mesh grid openings) and a maximum of 53 structures/mm^2 for any single preparation for that same area (40 CFR 763-AA-5-91).

Clean coal technology: Any technology not in widespread use prior to the Clean Air Act amendments of 1990. This Act will achieve significant reductions in pollutants associated with the burning of coal (EPA-94/04).

Clean fuel: Blends or substitutes for gasoline fuels, including compressed natural gas, methanol, ethanol, liquified petroleum gas, and others (EPA-94/04).

Clean fuel vehicle: A vehicle in a class or category of vehicles which has been certified to meet for any model year the clean-fuel vehicle standards applicable under this part for that model year to clean-fuel vehicles in that class or category (*see* vehicle for more related terms) (CAA241-42USC7581-91).

Clean Water Act (CWA) of 1987: *See* the term of Act or CWA.

Clean water standards: Any enforceable limitation, control, condition, prohibition, standard, or other requirement which is established pursuant to the CWA or contained in a permit issued to a discharger by the United States Environmental Protection Agency, or by a State under an approved program, as authorized by section 402 of the CWA, or by a local government to ensure compliance with pretreatment regulations as required by section 307 of the Clean Water Act (40 CFR 15.4-91).

Cleaner: A device which creates a cyclone effect to remove dirt and other rejects from pulp using the differences in density to aid in separation (EPA-87/10).

Cleaned steel cans: Tin coated or tin free, hydraulically or mechanically compressed to the size, weight and shape requirements of the customer. May include aluminum tops of beverage cans, but must be free of aluminum cans, loose tin or terne plate, dirt, garbage, non-ferrous metals (except those used in can construction), and non-metallics of any kind (EPA-83).

Cleaning (or etching):
(1) A chemical solution bath and a rinse or series of rinses designed to produce a desired surface finish on the workpiece. This term includes air pollution control scrubbers which are sometimes used to control fumes from chemical solution baths. Conversion coating and anodizing when performed as an integral part of the aluminum forming operations are considered cleaning or etching operations. When conversion coating or anodizing are covered here they are not subject to regulation under the provisions of 40 CFR 433, Metal Finishing (40 CFR 467.02-91).
(2) A process where material is removed by chemical action (EPA-83/06a).

Cleaning agent and degreaser (or cleaning solvent): Solvents used to clean oil and grease or dirt from the surface of a metal. Common cleaning and degreasing agents include ethylene dichloride, perchloroethylene, and trichloroethylene (EPA-85/10a).

Cleaning solvent: *See* synonym, cleaning agent and degreaser.

Cleaning water: The process water used to clean the surface of an intermediate or final plastic product or to clean the surfaces of equipment used in plastics molding and forming that contact an intermediate or final plastic product. It includes water used in both the detergent wash and rinse cycles of a cleaning process (*see* water for more related terms) (40 CFR 463.2-91).

Cleanout door: Openings in the primary and secondary chamber that are used to remove ash.

Cleanout/inspection door: A door in the primary or secondary combustion chamber which can be opened when an incinerator is shut down for either removing ash or inspecting refractory conditions (EPA-89/03b).

Cleanup: Actions taken to deal with a release or threat of release of a hazardous substance that could affect humans and/or the environment. The term cleanup is sometimes used interchangeably with the terms remedial action, removal action, response action, or corrective action (EPA-94/04).

Clear coating: The coatings that lack color and opacity or are transparent using the undercoat as a reflectant base or undertone color (40 CFR 52.741-91).

Clear cut: Harvesting all the trees in one area at one time, a practice that can encourage fast rainfall or snowmelt runoff, erosion, sedimentation of streams and lakes, flooding, and destroys vital habitat (EPA-94/04).

Clear topcoat: The final coating which contains binders, but not opaque pigments, and is specifically formulated to form a transparent or translucent solid protective film (40 CFR 52.741-91).

Clear well: A reservoir for storing filtered water of sufficient quantity to prevent the need to vary the filtration rate with variations in demand. Also used to provide chlorine contact time for disinfection (EPA-94/04).

Clearing bath: A processing solution that removes most residual fixer from a processed film or paper prior to washing, minimizing the water requirement (EPA-80/10).

Clearinghouse: A location to serve as an information dissemination point (EPA-80/08).

Climatological frequency distribution: A frequency distribution of climatological parameters such as wind direction sectors, wind speed classes and diffusion categories, which is used in air quality simulation models for long term average (NATO-78/10).

Climax community: A relatively stable ecological community in an area. The community will remain about the same under the current climate.

Cline: The gradual change of characteristics exhibited by members of a series of adjacent populations of the same species.

Clinical laboratory: A workplace where diagnostic or other screening procedures are performed on blood or other potentially infectious materials (29 CFR 1910).

Clinical study: A study of humans suffering from symptoms induced by chemical exposure (Course 165.6).

Clinker:
(1) Hard, sintered, or fused pieces of residue formed in a furnace by the agglomeration of ash, metals, glass and ceramics (SW-108ts, *see also* EPA-83).
(2) Lumps composed of tricalciumsilicate, dicalciumsilicate, tricalciumaluminate, and tetracalcium aluminoferrite that form in the sintering zone of the cement kiln (ETI-92).

Clipper: A machine which cuts veneer sheets to various sizes and also may remove defects (EPA-74/04).

Clipping and forging: Metal scraps from industrial manufacturing plants such as aircraft and metal fabricators (EPA-76/12).

Climograph: A climatic diagram to show climatic data such as annual rain and temperature distribution.

Cloning: In biotechnology, obtaining a group of genetically identical cells from a single cell; making identical copies of a gene (EPA-94/04).

Closed chain: *See* definition under chain.

Closed circuit apparatus: An apparatus of the type in which the exhalation is rebreathed by the wearer after the carbon dioxide has been effectively removed and a suitable oxygen concentration restored from sources composed of compressed oxygen, or chemical oxygen, or liquid-oxygen (NIOSH-84/10).

Closed circulating water system:
(1) Any configuration of equipment in which any heat is transferred by circulating water that is contained within the equipment and not discharged to the air; chilled water loops are included (*see* cooling water for more related terms) (40 CFR 749.68-91).
(2) A system which passes water through the condensers then through an artificial cooling device and keeps recycling it (EPA-82/11f).
(3) *See* cooling water for more related terms.

Closed cooling water system: Any configuration of equipment in which any heat is transferred by circulating water that is contained within the equipment and not discharged to the air; chilled water loops are included (*see* cooling water for more related terms) (40 CFR 749.68-91).

Closed course competition event: Any organized competition event covering an enclosed, repeated or confined route intended for easy viewing of the entire route by all spectators. Such events include short track, dirt track, drag race, speedway, hillclimb, ice race, and the Bonneville Speed Trials (40 CFR 205.151-91).

Closed formation: Formation of lead battery plates done with the plates already in the battery case (EPA-84/08).

Closed loop caustic system:
(1) A tank cleaning system which recycles and uses all of a secondary water rinse as make-up water for the caustic (EPA-79/12a).
(2) A tank cleaning system which recycles a primary caustic rinse and uses all of a secondary water rinse as make-up water for the caustic (EPA-79/12b).
(3) *See* caustic system for more related terms.

Closed loop evaporation system: A system used for the recovery of chemicals and water from a chemical finishing process. An evaporator concentrates flow from the rinse water holding tank. The concentrates rinse solution is returned to the bath, and distilled water is returned to the final rinse tank. The system is designed for recovering 100 percent of chemicals normally lost in dragout for reuse in the process (EPA-83/06a).

Closed loop recycling: Reclaiming or reusing wastewater for non-potable purposes in an enclosed process (EPA-94/04).

Closed loop rinsing: The recirculation of rinse water without the introduction of additional make-up water (EPA-83/06a).

Closed portion (facility): The portion of a facility which an owner or operator has closed in accordance with the approved facility closure plan and all applicable closure requirements (*see* facility for more related terms, *see also* active portion and inactive portion) (40 CFR 260.10-91).

Closed steaming: A method of steaming in which the steam required is generated in the retort by passing steam through heating coils that are covered with water. The water used for this purpose is recycled (EPA-74/04).

Closed vent system: A system that is not open to the atmosphere and is composed of piping, connections, and, if necessary, flow inducing devices that transport gas or vapor from an emission source to a control device (40 CFR 52.741-91, *see also* 40 CFR 60.481; 60.561; 60.691; 61.241; 61.341; 264.1031-91).

Closed system: A thermodynamic system that no mass crosses its boundaries as the system changes from one state to another. In a closed system, energy is allowed to cross its boundaries in various forms. When a closed system executes a cycle, the net amount of heat added to the system during the cycle is equal to the net amount of work done by the system (*see* thermodynamic system for more related terms) (Holman-69; Jones-p97).

Closeout: The final EPA or recipient actions taken to assure satisfactory completion of project work and to fulfill administrative requirements, including financial settlement submission of acceptable required final reports, and resolution of any outstanding issues under the cooperative agreement and/or Superfund State Contract (40 CFR 35.6015-91).

Closing rpm: The engine speed in Figure 2 of Appendix I (40 CFR 205.151-91).

Closure:
(1) The act of securing a Hazardous Waste Management facility pursuant to the requirements of 40 CFR 264 (40 CFR 270.2-91).
(2) The procedure a landfill operator must follow when a landfill reaches its legal capacity for solid waste: ceasing acceptance

of solid waste and placing a cap on the landfill site (EPA-94/04).
Other closure related terms include:
- Closure period
- Closure plan
- Final closure
- Partial closure
- Post-closure
- Post-closure plan
- Current closure cost estimate
- Current post-closure cost estimate

Closure period: The period of time beginning with the cessation, with respect to a waste impoundment, of uranium ore processing operations and ending with completion of requirements specified under a closure plan (*see* closure for more related terms) (40 CFR 192.31-91).

Closure plan: The plan required under 40 CFR 264.112 of this chapter (*see* closure for more related terms) (40 CFR 192.31-91, *see also* 40 CFR 264.141; 265.141-91).

Closure problem: In the Eulerian description for mean turbulent quantities, more unknown terms are present than the number of equations. These additional terms originate from cross correlations between turbulent quantities (e.g., Reynolds stresses). The term indicates that these extra terms must be expressed as function of the other terms in the equation in order to make a solution of the equation possible (NATO-78/10).

Cloud: A visible dispersion occupying a discrete portion of space, with apparent boundaries (EPA-83/06).

Cloud seeding: A technique to add certain substances to an atmosphere to convert cloud into rain.

Cluster: A structure with fibers in a random arrangement such that all fibers are intermixed and no single fiber is isolated from the group. Groupings must have more than two intersections (40 CFR 763-App/A-91).

Cluster analysis: A statistical procedure which solves the problem of separating objects into groups so that each object is more like objects in the same group than like objects in other groups (EPA-79/12c).

Clutch facing: An asbestos-containing product intended for use as a friction material or lining in the clutch mechanisms or manual transmission vehicles (40 CFR 763.163-91).

Co-composting: Simultaneous composting of two or more diverse waste streams (EPA-89/11).

Co-disposal:
(1) The disposal of two or more than two types of waste in one area, e.g., disposal of unprocessed MSW (municipal solid waste) and incinerator ash in the same landfill (OTA-89/10).
(2) The technique in which sludge is combined with other combustible materials (e.g., refuse, or refuse-derived fuel) to form a furnace feed with a higher heating value than the original sludge (*see* land disposal for more related terms) (OME-88/12).

Co-fire: Burning of two fuels in the same combustion unit, e.g., coal and natural gas, or oil and coal (EPA-94/04).

Co-incineration: The joint incineration of hazardous waste, in any form, with refuse and/or sludge (*see* incineration for more related terms) (EPA-81/09).

Co-permittee: A permittee to a NPDES permit that is only responsible for permit conditions relating to the discharge for which it is operator (40 CFR 122.26-91).

Co-precipitation of metal: The precipitation of a metal with another metal (*see* metal for more related terms) (EPA-83/06a).

Coagulant: A substance which forms a precipitate or floc when added to water. Suspended solids adhere to the large surface area of the floc, thus increasing their weight and expediting sedimentation (EPA-82/11).

Coagulation:
(1) A process using coagulant chemicals and mixing by which colloidal and suspended materials are destabilized and agglomerated into flocs (40 CFR 141.2-91).
(2) Clumping of particles in wastewater to settle out impurities, often induced by chemicals such as lime, alum, and iron salts (EPA-94/04).

Coagulation chemicals: Hydrolyzable divalent and trivalent metallic ions of aluminum, magnesium, and iron salts. They include alum (aluminum sulfate), quicklime (calcium oxide), hydrated lime (calcium hydroxide), sulfuric acid, anhydrous ferric chloride. Lime and acid affect only the solution pH which in turn causes coagulant precipitation, such as that of magnesium (EPA-76/03).

Coagulator: A soluble substance, such as lime, which when added to a suspension of very fine solid particles in water causes these particles to adhere in clusters which will settle easily. Used to assist in reclaiming water used in flotation (EPA-82/05).

Coal: All solid fuels classified as anthracite, bituminous, subbituminous, or lignite by the American Society and Testing and Materials, Designation D388-77 (incorporated by reference-*see* 40 CFR 60.17) (40 CFR 60.41-91, *see also* 40 CFR 60.41b; 60.41c; 60.251-91). Other coal related terms include:
- Anthracite coal
- Bituminous coal
- Subbituminous coal
- Lignite coal

Coal bunker: A single or group of coal trailers, hoppers, silos or other containers that:
(1) Are physically attached to the affected facility; and
(2) Provide coal to the coal pulverizers (40 CFR 60-App/G-91).

Coal cleaning technology: Precombustion process by which coal is physically or chemically treated to remove some of its sulfur so as to reduce sulfur dioxide emissions (EPA-94/04).

Coal equivalent of fuels: The quantity of coal of stated kind and heating value which would be required to supply the Btu equivalent to the comparative fuel(s). The Btu content of fuels is generally divided by the representative heating value of coal (EPA-83).

Coal gasification:
(1) One of SO_2 emission reduction techniques (*see* sulfur oxide emission control for control structure). Coal gasification means any processes that are used to convert solid coals to gas fuels for cleaner combustion (EPA-81/12, p8-6).
(2) Conversion of coal to a gaseous product by one of several available technologies (EPA-94/04).

Three basic steps are common to all coal gasification processes:
(1) Coal pretreatment
(2) Gasification
(3) Gas cleaning

Coal laboratory: As used in subchapter VIII of this chapter, a university coal research laboratory established and operated pursuant to a designation made under section 1311 of this title (SMCRA701-30USC1291-90).

Coal liquefaction: One of SO_2 emission reduction techniques (*see* sulfur oxide emission control for control structure). Coal liquefaction is a process for changing coal into synthetic oil. It is similar to coal gasification. Two basic approaches for liquefaction are used. One involves using a gasifier to convert coal to carbon monoxide, hydrogen and methane; followed by condensation to convert the gases to oils. The second approach uses a solvent or slurry to liquefy pulverized coal and then processes this liquid into a heavy fuel oil. Some processes produce both a synthetic gas and synthetic oil.

Hydrogen is used to convert sulfur in the coal to hydrogen sulfide gas. Hydrogen sulfide is partially oxidized to form elemental sulfur and water. More than 85% of the sulfur is removed from coal by liquefaction (EPA-81/12, p8-6).

Coal mine: An area of land with all property placed upon, under or above the surface of such land, used in or resulting from the work of extracting coal from its natural deposits by any means or method including secondary recovery of coal from refuse or other storage piles derived from mining, cleaning, or preparation of coal (EPA-82/10).

Coal mine drainage: Any water drained, pumped or siphoned from a coal mine (EPA-82/10).

Coal only heater: An enclosed, coal-burning appliance capable of space heating, or domestic water heating, which has all of the following characteristics:
(1) An opening for emptying ash that is located near the bottom or the side of the appliance,
(2) A system that admits air primarily up and through the fuel bed,
(3) A grate or other similar device for shaking or disturbing the fuel bed or power-driven mechanical stoker,
(4) Installation instructions that state that the use of wood in the stove, except for coal ignition purposes, is prohibited by law, and
(5) The model is listed by a nationally recognized safety-testing laboratory for use of coal only, except for coal ignition purposes (40 CFR 60.531-91).

Coal pile drainage: *See* synonym, coal pile runoff.

Coal pile runoff (or coal pile drainage): The rainfall runoff from or through any coal storage pile (40 CFR 423.11-91).

Coal preparation plant: Any facility (excluding underground mining operations) which prepares coal by one or more of the following processes: breaking, crushing, screening, wet or dry cleaning, and thermal drying (40 CFR 60.251-91, *see also* 40 CFR 434.11-91).

Coal preparation plant associated areas: The coal preparation plant yards, immediate access roads, coal refuse piles and coal storage piles and facilities (40 CFR 434.11-91).

Coal preparation plant water circuit: All pipes, channels, basins, tanks, and all other structures and equipment that convey, contain, treat, or process any water that is used is coal preparation processes within a coal preparation plant (40 CFR 434.11-91).

Coal pretreatment: Coal pretreatment involves coal pulverizing and washing (*see* coal gasification for more related terms) (EPA-81/12, p8-6).

Coal processing and conveying equipment: Any machinery used to reduce the size of coal or to separate coal from refuse, and the equipment used to convey coal to or remove coal and refuse from the machinery. This includes, but is not limited to, breakers, crushers, screens, and conveyor belts (40 CFR 60.251-91).

Coal rate: The weight in pounds of coal (including the coal equivalent of other fuels) burned for steam or electric generation divided by the resulting net generation (cf. heat rate) (EPA-83).

Coal RDF mixed fuel fired combustor: A combustor that fires coal and RDF simultaneously (40 CFR 60.51a-91).

Coal refuse: The waste products of coal mining, cleaning, and coal preparation operations (e.g. culm, gob, etc.) containing coal, matrix material, clay, and other organic and inorganic material (40 CFR 60.41-91, *see also* 40 CFR 60.41a; 60.41b; 60.41c-91).

Coal refuse disposal pile: Any coal refuse deposited on the earth and intended as permanent disposal or long-term storage (greater than 180 days) of such material, but does not include coal refuse deposited within the active mining area or coal refuse never removed from the active mining area (40 CFR 434.11-91).

Coal remining operation: A coal mining operation which begins after the date of the enactment of this subsection at a site on which coal mining was conducted before the effective date of the Surface Mining Control and Reclamation Act of 1977 (CWA301.p).

Coal storage system: Any facility used to store coal except for open storage piles (40 CFR 60.251-91).

Coalescence: The merging of two droplets to form a larger droplet (NATO-78/10).

Coalition for responsible waste incineration (CRWI): A coalition of waste generators with on-site facilities and academic institutions, dedicated to the advancement of high-temperature incineration for the destruction of wastes (CRWI-89/05).

Coarse paper: The paper used for industrial purposes, as distinguished from those used for cultural or sanitary purposes (*see* paper for more related terms) (40 CFR 250.4-91).

Coast guard district response group: A Coast Guard District Response Group established under subsection (j) (CWA311-33USC1321-91).

Coastal:
(1) Any body of water landward of the territorial seas as defined in 40 CFR 125.1(gg),1 or
(2) Any wetlands adjacent to such waters (40 CFR 435.41-91).

Coastal energy activity: Any of the following activities if, and to the extent that:
(1) The conduct, support, or facilitation of such activity requires and involves the siting, construction, expansion, or operation of any equipment or facility; and
(2) Any technical requirement exists which, in the determination of the Secretary, necessitates that the siting, construction, expansion, or operation of such equipment or facility be carried out in, or in close proximity to, the coastal zone of any coastal state:
(A) Any outer Continental Shelf energy activity.
(B) Any transportation, conversion, treatment, transfer, or storage of liquefied natural gas.
(C) Any transportation, transfer, or storage of oil, natural gas, or coal (including, but not limited to, by means of any deepwater port, as defined in section 1502(10) of title 33). For purposes of this paragraph, the siting, construction, expansion, or operation of any equipment or facility shall be "in close proximity to" the coastal zone of any coastal state if such siting, construction, expansion, or operation has, or is likely to have, a significant effect on such coastal zone (CZMA304-16USC1453).

Coastal resource of national significance: Any coastal wetland, beach, dune, barrier island, reef, estuary, or fish and wildlife habitat, if any such area is determined by a coastal state to be of substantial biological or natural storm protective value (CZMA304-16USC1453-90).

Coastal state: A state of the United States in, or bordering on, the Atlantic, Pacific, or Arctic Ocean, the Gulf of Mexico, Long Island Sound, or one or more of the Great Lakes. For the purposes of this chapter, the term also includes Puerto Rico, the Virgin Islands, Guam, the Commonwealth of the Northern Mariana Islands, and the Trust Territories of the Pacific Islands, and American Samoa (CZMA304-16USC1453-90).

Coastal water: For the purposes of classifying the size of discharges, the waters of the coastal zone except for the Great Lakes and specified ports and harbors on inland rivers (*see* water for more related terms) (40 CFR 300.5-91, *see also* CZMA304).

Coastal zone: Lands and waters adjacent to the coast that exert an influence on the uses of the sea and its ecology, or whose uses and ecology are affected by the sea (EPA-94/04).

Coated: A term generally applied to a workpeice such as metal, wood, paper and paperboard, whose surface has been treated with chemicals such as paint, varnish, clay or some other pigment and adhesive mixture or other suitable materials, to improve the surface properties of the workpiece.

Coated paper: The paper surface that has been treated with clay or other materials (*see* paper for more related terms) (EPA-83).

Coating:
(1) A material applied onto or impregnated into a substrate for protective, decorative, or functional purposes. Such materials include, but are not limited to, paints, varnishes, sealers, adhesives, thinners, diluents, and inks (40 CFR 52.741-91, *see also* 40 CFR 60.461-91).
(2) Dipping, enrobing, glazing, icing, panning, and so forth (AP-40, p790).
Other coating related terms include:
• Fog coating
• Mist coating (*see* synonym, fog coating)
• Uniforming coating (*see* synonym, fog coating)

Coating application station: That portion of the large appliance surface coating operation where a prime coat or a top coat is applied to large appliance parts or products (e.g., dip tank, spray booth, or flow coating unit) (40 CFR 60.451-91, *see also* 40 CFR 60.461-91).

Coating applicator: The equipment used to apply a coating (40 CFR 52.741-91, *see also* 40 CFR 60.441; 60.711; 60.741-91).

Coating blow: The process in which air is blown through hot asphalt flux to produce coating asphalt. The coating blow starts when the air is turned on and stops when the air is turned off (40 CFR 60.471-91).

Coating color: Coating mixture consisting mainly of pigments and adhesives (EPA-83).

Coating line: An operation consisting of a series of one or more coating applicators and any associated flash-off areas, drying areas, and ovens wherein a surface coating is applied, dried, or cured. (It is not necessary for an operation to have an oven, or flash-off area, or drying area to be included in this definition) (40 CFR 52.741-91, *see also* 40 CFR 60.441-91).

Coating mix preparation equipment: All mixing vessels in which solvent and other materials are blended to prepare polymeric coatings (40 CFR 60.741-91, *see also* 40 CFR 60.711-91).

Coating operation: Any coating applicator, flashoff area, and drying oven located between a base film unwind station and a base film rewind station that coat a continuous base film to produce magnetic tape (40 CFR 60.711-91, *see also* 40 CFR 60.721; 60.741; 466.02-91).

Coating plant: Any plant that contains one or more coating line(s) (40 CFR 52.741-91).

Coating solids applied: The solids content of the coated adhesive, release, or precoat as measured by Reference Method 24 (40 CFR 60.441-91, *see also* 40 CFR 60.721-91).

Cobalt (Co): The total cobalt present in the process wastewater stream exiting the wastewater treatment system (40 CFR 415.651-91). Cobalt hard transition metal with a.n. 27; a.w. 58.93; d. 8.9 g/cc; m.p. 1495° C and b.p. 2900° C and belongs to group VIII of the periodic table. Major compounds include:
- Anhydrous cobalt chloride ($CoCl_2$): Blue crystals.
- Hydrated cobalt chloride ($CoCl_2 \bullet 6H_2O$): Red crystals.

Coburning: The burning of waste and a fuel at same time (EPA-81/09).

Cocurrent or concurrent: Flow of scrubbing liquid in the same direction as the gas stream (EPA-89/03b).

Code of Federal Regulation (CFR): A document containing all finalized regulations (EPA-86/01).

Coefficient of haze (COH): A measurement of visibility interference in the atmosphere (EPA-94/04).

Coefficient of octanol/water partition (Kow): An indicator for (1) bioaccumulation potential and (2) adsorption potential. Kow can be expressed as (Course 165.6): kow = [x] octanol/[x] aqueous = non-polar/polar.

Coefficient of organic carbon partition (Koc): Tendency indication of an organic chemical to be adsorbed. It ranges from 1 to 10^7; higher values reflect greater sorption potential. Koc = (mg compound adsorbed/Kg organic carbon)/(mg compound dissolved/liter solution) (Course 165.6).

Coefficient of performance (COP): A measurement of a refrigeration cycle. A refrigeration cycle is the opposite of a Carnot cycle (Jones-p100). COP can be expressed as:
- COP = (refrigeration effect)/(work input) = $Q_{in}/W = Q_{in}/(Q_{out} - Q_{in})$
- For a reversed Carnot cycle, COP = $1/(T_H/T_L - 1)$

Where:
Q_{in} = heat absorbed from a low temperature region
Q_{out} = heat rejected to a high temperature region
W = net work input of cycle
T_L = low temperature of a reservoir
T_H = high temperature of a reservoir

Coefficient of variation:
(1) A measure of relative dispersion. It is equal to the standard deviation divided by the mean and multiplied by 100 to give a percentage value (EPA-84/03).
(2) A standard statistical measure of the relative variation of a distribution or set of data, defined as the standard deviation divided by the mean (EPA-91/03).

Coenzyme: The non-protein portion of an enzyme system (LBL-76/07-water).

Coenzyme A ($C_{21}H_{36}O_{16}N_7P_3S$): A complex organic compound that acts with enzymes involved in various biochemical reactions.

Coevolution: An evolutionary phenomena among major groups or organisms in the ecological system, e.g., plants and plant eaters.

Coexistence: The act or state of living peacefully among different species.

Coffin: *See* synonym, cask.

Cofired combustor: A unit combusting MSW or RDF with a non-MSW fuel and subject to a Federally enforceable permit limiting the unit to combusting a fuel feed stream, 30 percent or less of the weight of which is comprised, in aggregate, of MSW or RDF as measured on a 24 hour daily basis. A unit combusting a fuel feed stream, more than 30 percent of the weight of which is comprised, in aggregate, of MSW or RDF shall be considered an MWC unit and not a cofired combustor. Cofired combustors which fire less than 30 percent segregated medical waste and no other municipal solid waste are not covered by this subpart (40 CFR 60.51a-91).

Cogeneration: The production of both electricity and steam at one facility from the same primary fuel source (OTA-89/10).

Cogeneration system: A power system which simultaneously produces both electrical (or mechanical) and thermal energy from the same energy source.

Cogeneration steam generating unit: A steam generating unit that simultaneously produces both electrical (or mechanical) and thermal energy from the same primary energy source (40 CFR 60.41c-91).

Coherence: The act or state of natural or logical connection.

Cohort study: An epidemiologic study that observes subjects in differently exposed groups and compares the incidence of symptoms. Although ordinarily prospective in nature, such a study is sometimes carried out retrospectively, using historical data (EPA-92/12).

Coil:
(1) Any flat metal sheet or strip that is rolled or wound in concentric rings (40 CFR 52.741-91, *see also* 40 CFR 85.2122(a)(ii)(A); 465.02-91).
(2) A number of turns of a wire used to introduce inductance into an electric circuit, to produce magnetic flux, or to react mechanically to a changing magnetic flux (EPA-83/03).

Coil coating: Any coating applied on any flat metal sheet or strip that comes in rolls or coils (40 CFR 52.741-91, *see also* 40 CFR 465.02-91).

Coil coating facility: A facility that includes one or more coil coating line(s) (40 CFR 52.741-91).

Coil coating line: A coating line in which any protective, decorative or functional coating is applied onto the surface of flat metal sheets, strips, rolls, or coils for industrial or commercial use (40 CFR 52.741-91).

Coil condensate: The condensate formed in steam lines and heating coils (EPA-74/04).

Coil core assembly: A unit made of the coil windings of a transformer placed over the magnetic core (EPA-83/03).

Coincident electric demand: The sum of two or more demands which occur in the same demand interval (*see* electric demand for more related terms) (EPA-83).

Coke: The carbon residue left when the volatile matter is driven off of coal by high temperature distillation (cf. coking) (EPA-74/06a). Other coke related terms include:
- Foundry coke
- Furnace coke
- Petroleum coke
- Pitch coke

Coke breeze: Small particles of coke; these are usually used in the coke plants as boiler feed or screened for domestic trade (EPA-74/06a).

Coke burn-off: The coke removed from the surface of the fluid catalytic cracking unit catalyst by combustion in the catalyst regenerator. The rate of coke burn-off is calculated by the formula specified in 40 CFR 60.106 (40 CFR 60.101-91).

Coke byproduct recovery plant: Any plant designed and operated for the separation and recovery of coal tar derivatives (by-products) evolved from coal during the coking process of a coke oven battery (40 CFR 61.131-91, *see also* 40 CFR 61.341-91).

Coke oven: An industrial process which converts coal into coke, one of the basic materials used in blast furnaces for the conversion of iron ore into iron (EPA-94/04).

Coke oven byproduct ammonium sulfate manufacturing plant: Any plant which produces ammonium sulfate by reacting sulfuric acid with ammonia recovered as a by-product from the manufacture of coke (40 CFR 60.421-91).

Coke wharf: The place where coke is discharged from quench cars prior to screening (EPA-74/06a).

Coking:
(1) The conversion by heating in the absence or near absence of air, of a carbonaceous fuel, particularly certain bituminous coals, to a coherent, firm, cellular product known as coke.
(2) A process for thermally converting the heavy residual bottoms of crude oil entirely to lower-boiling petroleum products and by-product petroleum coke (cf. coke) (EPA-83/03).

Colburn chart: A graph providing a means for calculating overall gas transfer unit for absorbers (EPA-84/09).

Cold air return: The registers and ducting which withdraw house air from various parts of the house and direct it to a central forced air furnace or heat pump. The return ducting is at low pressure relative to the house because the central furnace fan draws air out of the house through this ducting (EPA-88/08).

Cold blooded animals (or poikilothermic animals): Animals that lack a temperature regulating mechanism that offsets external temperature changes. Their temperature fluctuates to a large degree with that of their environment. Examples are fish, shellfish, and aquatic insects (LBL-76/07-water).

Cold cleaning: The process of cleaning and removing soils from surfaces by spraying, brushing, flushing, or immersion while maintaining the organic solvent below its boiling point. Wipe cleaning is not included in this definition (40 CFR 52.741-91).

Cold crucible arc melting: Melting and purification of metals in a cold refractory vessel or pot (EPA-83/03a).

Cold drawing: A process of forcing materials through dies or other mandrels to produce wire, rod, tubular and some bars (EPA-83/06a).

Cold drying hearth: A surface upon which unheated waste materials are placed to dry or burn; hot combustion gases are then passed over the materials (*see* hearth for more related terms) (SW-108ts).

Cold joint: The contact joint between two adjacent concrete slabs or parts of a slab that were poured at different times (EPA-88/08).

Cold lime soda process: One of methods for softening water in which hydrated lime (sometimes in combination with soda ash) is used to dissolve calcium or magnesium in hard water to form settleable sludge for removal.

Cold metal furnace: A furnace that is usually charged with two batches of (cold) solid material (*see* furnace for more related terms) (EPA-74/06a).

Cold rolling: The process of rolling a workpiece below the recrystallization temperature of the copper or copper alloy (40 CFR 468.02-91).

Cold set resin: A resin that sets or hardens without the application of heat. Used in foundry operations as sand binders (*see* resin for more related terms) (EPA-85/10a).

Cold setting refractory mortar: *See* definition under air setting refractory mortar.

Cold standby: Maintenance of protected reactor condition in which the fuel is removed, the moderator is stored in tanks, and equipment and system layup is performed to prevent deterioration, such that future refueling and restart are possible (DOE-91/04).

Cold temperature CO: A standard for automobile carbon monoxide (CO) emissions to be met at a low temperature (i.e. 20 degrees Fahrenheit). Conventional automobile catalytic convertors are less efficient upon start-up at low temperatures (EPA-94/04).

Cold vapor atomic absorption spectrometry (CVAAS): CVAAS uses a chemical reduction to reduce mercury selectively. The procedure is extremely sensitive but is subject to interferences from some volatile organics, chlorine, and sulfur compounds (*see* metal analyzer for more related terms) (SW-846).

Cold worked pipe and tube: Those cold forming operations that process unheated tube products using either water or oil solutions for cooling and lubrication (40 CFR 420.101-91).

Colic: A paroxysmal pain in the abdomen, due to spasm, distention, or obstruction of any one of the hollow viscera (LBL-76/07-bio).

Coliform (coliform bacterium, coliform organism or colon bacillus):
(1) Any organisms common to the intestinal tract of animals whose presence in wastewater is an indicator of pollution and of potentially dangerous bacterial contamination (EPA-74/11).
(2) Microorganisms found in the intestinal tract of humans and animals. Their presence in water indicates fecal pollution and potentially dangerous bacterial contamination by disease-causing microorganisms (EPA-89/12).
(3) cf. total coliform.

Coliform bacterium: *See* synonym, coliform.

Coliform group bacteria: A group of bacteria predominantly inhabiting the intestines of man or animal, but also occasionally found elsewhere. It includes all aerobic and facultative anaerobic, Gram-negative, non-spore-forming bacilli that ferment lactose with production of gas. Also included are all bacteria that produce a dark, purplish-green colony with metallic sheen by the membrane-filter techniques used for coliform identification. The two groups are not always identical, but they are generally of equal sanitary significance (LBL-76/07-water).

Coliform index: A rating of the purity of water based on a count of fecal bacteria (EPA-94/04).

Coliform organism:
(1) Microorganisms found in the intestinal tract of humans and animals. Their presence in water indicates fecal pollution and potentially adverse contamination by pathogens (EPA-94/04).
(2) *See also* synonym, coliform.

Collecting sewer: A sewer that collects wastewater from lateral sewers and connects to a trunk sewer (*see* sewer for more related terms) (DOI-70/04).

Collection: *See* synonym, waste collection.

Collection frequency: The number of times collection is provided in a given period of time (40 CFR 243.101-91).

Collection method: *See* synonym, waste collection method.

Collection stop: A stop made by a vehicle and crew to collect solid waste from one or more service sites (SW-108ts).

Collection theory: The principal mechanisms by which liquids may be used to remove aerosols from gas streams are as follows:
(1) Wetting of the particles by contact with a liquid droplet; and
(2) Impingement of wetted or unwetted particles on collecting surfaces followed by their removal from the surfaces by a flush with a liquid.
Mechanisms for wetting the particle include:
(1) Impingement by spray droplets;
(2) Diffusion;
(3) Condensation; and
(4) Humidification and electrostatic precipitation (AP-40, p100).

Collective CEDE: The committed effective dose equivalent (CEDE) of radiation for a population (DOE-91/04).

Collector:
(1) In the area of ore mining and dressing, a collector is a heteropolar compound containing a hydrogen carbon group and an ionizing group, chosen for the ability to adsorb selectively in froth flotation processes and render the adsorbing surface relatively hydrophobic. A promoter (EPA-82/05).
(2) In the control of gaseous and particulate emissions, a collector is a device for removing and retaining contaminants from air or other gases. Usually this term is applied to cleaning devices in exhaust systems (cf. air pollution control equipment) (EPA-84/09).
Types of collectors include (EPA-83):
(1) Bag filter (EPA-83)
(2) Cyclone (EPA-83)
(3) Dust (EPA-83)
(4) Fly ash (EPA-83)
(5) Mechanical collector (EPA-83)
(6) Multicyclone collector (EPA-83)

Collector sewer: The common lateral sewers, within a Publicly owned treatment system, which are primarily installed to receive wastewaters directly from facilities which convey wastewater from individual systems, or from private property, and which include service Y connections designed for connection with those facilities including:
(1) Crossover sewers connecting more than one property on one side of a major street, road, or highway to a lateral sewer on the other side when more cost effective than parallel sewers; and
(2) Except as provided in paragraph (b)(10)(iii) of this section, pumping units and pressurized lines serving individual structures or groups of structures when such units are cost effective and are owned and maintained by the grantee.
(3) This definition excludes other facilities which convey wastewater from individual structures, from private property to the public lateral sewer, or its equivalent and also excludes facilities associated with alternatives to conventional treatment works in small communities (40 CFR 35.2005-10).

Collector sewers: Pipes used to collect and carry wastewater from individual sources to an interceptor sewer that will carry it to a treatment facility (EPA-94/04).

Collision theory: In a chemical reaction, the product formation rate is proportional to the number of reactant molecule collisions.

Colloids:
(1) Microscopic suspended particles (approximately 1 to 1000 nanometers) which do not settle in a standing liquid and can only be removed by coagulation or biological action (EPA-82/11).
(2) Very small, finely divided solids (that do not dissolve) that remain dispersed in a liquid for a long time due to their small

size and electrical charge (EPA-94/04).

Colloid chemistry: A scientific study of colloid.

Colloidal dispersion: A mixture resembling a true solution but containing one or more substances that are finely divided but large enough to prevent passage through a semipermeable membrane. It consists of particles which are larger than molecules, which settle out very slowly with time, which scatter a beam of light, and which are too small for resolution with an ordinary light microscope (40 CFR 796.1840-91).

Colloidal matter: In wastewater, fine suspended particles that do not settle out except very slowly, and hence require special treatment such as sedimentation with coagulants or dialysis (DOI-70/04).

Colon bacillus: *See* synonym, coliform (DOI-70/04).

Colony: An aggregate of mother and daughter fronds attached to each other (40 CFR 797.1160-91).

Colony forming unit (CCU): A unit of enumeration used in microbiology, each of which represents the existence of one or more viable cells (EPA-88/09a).

Colophon rosin: *See* synonym, natural rosin.

Color:
(1) A measure of the light-absorbing capacity of a wastewater after turbidity has been removed. One unit of color is that produced by one mg/L of platinum as K_2PtCl_6 (EPA-82/11).
(2) Refers to standard Platinum Cobalt Test, using standards for color intensity of water samples. Commonly, standards are prepared at various concentrations which later may be referenced as units of color, derived from flow and concentration standard (EPA-87/10).

Color coat: The coat applied to a part that affects the color and gloss of the part, not including the prime coat or texture coat. This definition includes fog coating, but does not include conductive sensitizers or electromagnetic interference/radio frequency interference shielding coatings (40 CFR 60.721-91).

Color coupler: A group of organic chemicals which reacts with the oxidized components of the developers to form color dyes. They are either incorporated in the film emulsion at the time of manufacture or they are included in the color developing solution (EPA-80/10).

Color film: Generally, a color film has three separate light sensitive emulsion layers, which after inclusion of the appropriate sensitizing dyes, record an image of the blue light components on one layer, the green light components on another, and the red light components on the third layer (EPA-80/10).

Color plant: The portion of a fine paper-mill where pulp is dyed or colored prior to being made into paper (EPA-87/10).

Color reversal (DC) process: A color reversal film process in which the color couplers are added during development (EPA-80/10).

Color reversal (IC) process: A color reversal film and paper process in which the color couplers which form the color dye image are incorporated into the emulsion layers at the time of manufacture (EPA-80/10).

Color unit: A measure of color concentration in water (EPA-87/10).

Colorant: A concentrated coloring agent which is added to a base paint to produce the desired final color. Colorants are usually added to the paint by the retailer for the customer (EPA-79/12b).

Coloring: In the leather industry, a process step in the tannery whereby the color of the tanned hide is changed to that of the desired marketable product by dyeing or painting (EPA-82/11).

Colorimeter: An instrument used for color measurement based on optical comparison with standard colors (LBL-76/07-air).

Colorimetric (method): A procedure for establishing the concentration of impurities in water by comparing its color to a set of known color impurity standards (EPA-83/06a).

Colorimetric indicator tube: *See* synonym, toxic atmosphere monitor.

Colorimetry: *See* synonym, molecular absorption spectrophotometry (LBL-76/07-water).

Column dryer: Any equipment used to reduce the moisture content of grain in which the grain flows from the top to the bottom in one or more continuous packed columns between two perforated metal sheets (40 CFR 60.301-91).

Column chromatography: *See* synonym, chromatography.

Combination: Those cold rolling operations which include recirculation of rolling solutions at one or more mill stands, and once through use of rolling solutions at the remaining stand or stands (40 CFR 420.101-91).

Combination acid pickling: Those operations in which steel products are immersed in solutions of more than one acid to chemically remove scale and oxides, and those rinsing steps associated with such immersions (40 CFR 420.91-91).

Combination tanned: Leathers tanned with more than one tanning agent, e.g., initially chromium tanned followed by a second tannage (called a RETAN) with vegetable materials (EPA-82/11).

Combined ash: The mixture of bottom ash and fly ash (*see* ash for more related terms) (OTA-89/10).

Combined available residual chlorine: The total residual chlorine remaining in water, sewage, or industrial wastes at the end of specified contact period which reacts chemically and biologically as chloramines or organic chloramines (*see* chlorine for more related terms) (EPA-82/11f).

Combined chlorine: The chlorine present in solution in forms other than those free chlorine. These forms include the chloramines, organo-chloramines, and chlorinated organics as well as other compounds containing chlorine (*see* chlorine for more related terms) (EPA-88/09a).

Combined cycle gas turbine: A stationary turbine combustion system where heat from the turbine exhaust gases is recovered by a steam generating unit (*see* turbine for more related terms) (40 CFR 60.41a-91, *see also* 40 CFR 60.331-91).

Combined cycle system: A system in which a separate source such as a gas turbine, internal combustion engine, kiln, etc., provides exhaust gas to a heat recovery steam generating unit (40 CFR 60.41b-91, *see also* 40 CFR 60.41c-91).

Combined fuel economy:
(1) The fuel economy value determined for a vehicle (or vehicles) by harmonically averaging the city and highway fuel economy values, weighted 0.55 and 0.45 respectively, for gasoline-fueled and diesel vehicles.
(2) For electric vehicles, the term means the equivalent petroleum-based fuel economy value as determined by the calculation procedure promulgated by the Secretary of Energy (40 CFR 600.002.85-91).

Combined metal: The total of gold, platinum and palladium (*see* metal for more related terms) (40 CFR 421.261-91).

Combined sewer: A sewer system that carries both sewage and storm-water runoff. Normally, its entire flow goes to a waste treatment plant, but during a heavy storm, the volume of water may be so great as to cause overflows of untreated mixtures of storm water and sewage into receiving waters. Storm-water runoff may also carry toxic chemicals from industrial areas or streets into the sewer system (*see* sewer for more related terms) (EPA-94/04).

Combined sewer overflows: Discharge of a mixture of storm water and domestic waste when the flow capacity of a sewer system is exceeded during rainstorms (EPA-94/04).

Combined wastewater: The wastewater from all sources in a facility, e.g., combined wastewater from pesticide industry may include pesticide, pesticide intermediate, and non-pesticide (*see* wastewater for more related terms) (EPA-85/10).

Combustibility: The ability of a material to act as a fuel (Course 165.5).

Combustible: The materials that can be ignited at a specific temperature in the presence of air to release heat energy (cf. total combustible) (40 CFR 240.101-91).

Combustible atmosphere indicator: If air is contaminated with combustible gases, the indicator is to indicate if the combustible gases have proper concentration for combustion, if ignited. The gas will not combust, if it is either too lean or too rich. The indicator uses a combustion chamber containing a filament that combusts the flammable gas. To facilitate combustion the filament is heated or is coated with a catalyst (like platinum or palladium), or both. The filament is part of a balanced resistor circuit called a Wheatstone Bridge. The hot filament combusts the gas on the immediate surface of the element, thus raising the temperature of the filament. As the temperature of the filament increases so does its resistance. This change in resistance causes an imbalance in the Wheatstone Bridge which in turn measures the ratio of combustible gas present compared to the total required to reach the combustible level (*see* air analyzer for more related terms).

Combustible rubbish: The miscellaneous burnable materials. In general, the organic component of rubbish. Also referred to as Trash (*see* rubbish for more related terms) (EPA-83).

Combustible waste: Discarded material capable of combustion includes paper, cardboard, cartons, wood, boxes, excelsior, plastic, rags, bedding, leather, trimmings, household waste (*see* waste for more related terms) (EPA-83).

Combustion:
(1) Burning, or rapid oxidation, accompanied by release of energy in the form of heat and light. A basic cause of air pollution (EPA-94/04).
(2) Refers to controlled burning of waste, in which heat chemically alters organic compounds, converting into stable inorganics such as carbon dioxide and water (EPA-94/04).

Combustion (or incineration): The phenomena of burning, or rapid oxidation, accompanied by release of energy in the form of heat and light. It is accompanied by the release of energy in the form of heat. Scientifically, the terms, combustion and incineration, have the same definition and have been used interchangeably in waste incineration documents. Combustion, however, is generally used more often in the area of fossil-fuel burning for steam or power generation and incineration is used more often when referring to waste destruction.

Combustion involves heat and light. It is a process of burning resulting from the rapid oxidation of organic (fuel) compounds which can be expressed in terms of fundamental chemical reaction equations. Consider the oxidation of carbon: $C + O_2 \longrightarrow CO_2$, where C and O_2 are the reactants and CO_2 is the products.

This equation states that one mole of carbon reacts with one mole of oxygen to form one mole of carbon dioxide. This also means that 12 lbs of carbon react with 32 lbs of oxygen to form 44 lbs of carbon dioxide. All feed substances that undergo the combustion process are called the reactants, and the substances that result from the combustion process are called the products.

Obviously, combustion has to follow the law of mass conservation and the law of energy conservation. Therefore, during combustion, chemical elements can react with each other but the mass and the energy level of the entire combustion system must remain the same. Mass and energy balance calculations are two key ways to define a combustion system. The mass balance determines the levels of products formed by the reactants and the energy balance determines the amount of energy transfer within of a combustion system. Other combustion or incineration related terms include:
- Autogenous (or autothermic) combustion
- Catalytic combustion
- Complete combustion
- Flame combustion
- Incomplete combustion
- Lean combustion
- Open combustion
- Rich combustion

- Stoichiometric combustion
- Suppressed combustion
- Theoretical combustion (*see* synonym, stoichiometric combustion)

Combustion air: The ambient air associated with oxygen needed for burning a fuel or waste materials. Air is a typical source of oxygen for combustion. It is considered as an ideal gas in many incineration calculations. Air is supplied through air ports by a forced draft fan, by an induced draft fan or by a natural draft. Generally, about 1 standard cubic foot of combustion air is required per 100 Btu's heat input to an incinerator (SW-108ts; EPA-89/03b). Other combustion air related terms include:
- Deficient air (*see* synonym, substoichiometric combustion air)
- Excess air (*see* synonym, excess combustion air)
- Excess combustion air
- Primary air (*see* synonym, primary combustion air)
- Primary combustion air
- Secondary air (*see* synonym, secondary combustion air)
- Secondary combustion air
- Stoichiometric air: (*see* synonym, stoichiometric combustion air)
- Stoichiometric combustion air
- Starved air (*see* synonym, substoichiometric combustion air)
- Starved air combustion (*see* synonym, substoichiometric combustion air)
- Starved air incineration (*see* synonym, substoichiometric combustion air)
- Theoretical air (*see* synonym, stoichiometric combustion air)
- Theoretical combustion air (*see* synonym, stoichiometric combustion air)
- Substoichiometric combustion air

Combustion air calculation: For the combustion of methane, CH_4, determine combustion air requirements.

$$CH_4 + 2(1 + a)(O_2 + 3.76N_2) ---> CO_2 + 2H_2O + 2aO_2 + 2(1 + a)(3.76)N_2$$

- For this example: aa = 2(1 + a) and ta = 2;
  TA = 2(1 + a)/2(%) = (1 + a)(%) and EA = [2(1 + a) - 2]/2(%) = a(%)
- For more examples, the combustion of propane is as follows:
  - 100% TA: $C_3H_8 + 5 \times 1.0(O_2 + 3.76N_2) --> 3CO_2 + 4H_2O + 18.8N_2$
  - 80% TA: $C_3H_8 + 5 \times 0.8(O_2 + 3.76N_2) --> 2CO + CO_2 + 4H_2O + 15.04N_2$
  - 150% TA: $C_3H_8 + 5 \times 1.5(O_2 + 3.76N_2) --> 3CO_2 + 4H_2O + 2.5O_2 + 28.2N_2$
    The 150% TA is equivalent to 50% excess air.

Combustion air fuel ratio (air fuel ratio, air to fuel ratio or combustion air to fuel ration): The ratio of the mass of dry combustion air introduced into the firebox, to the mass of dry fuel consumed (grams of dry air per gram of dry wood burned) (40 CFR 60-App/A(method 28A)-91). The air/fuel ratio can be expressed by the following two options:
(1) AFm = (Air/Fuel) = n(1 + 3.76)/nf, by mole
(2) AFw = (Air/Fuel) = n(1 + 3.76) x 29/(nf x Mf), by weight, where n = moles of oxygen; nf = moles of fuel; Mf = molecular weight of fuel; 29 = molecular weight of air (actually 28.84)

Example:
- Calculate the theoretical air/fuel ratio for the combustion of octane (C_8H_{18}).

Solution:
- The combustion equation is:
  $C_8H_{18} + 12.5O_2 + 12.5(3.76)N_2 ---> 8CO_2 + 9H_2O + 47.0N_2$
  - The theoretical air/fuel ratio on a mole basis is: AFm = (12.5)(1+3.76)/1 = 59.5 moles air/mole fuel
  - The theoretical air/fuel ratio on a weight basis is found by introducing the molecular weights of the air and fuel. That is: AFw = 59.5(28.84)/114.2 = 15 lb air/lb fuel

Combustion air to fuel ratio: *See* synonym, combustion air fuel ratio.

Combustion analysis (or incineration analysis): Combustion or incineration is a very complicated subject. It involves (Lee-88/11):
- Mass balance (it determines the amounts of the products formed the reactants).
- Energy balance (it determines the energy transferred within a combustion system or how much auxiliary fuel is needed for an incinerator to reach a certain temperature).
- Thermodynamic analysis (it reveals information about the changes of the chemical components of a combustion system; however, it does not reveal how rapidly these changes will occur).
- Kinetic analysis (it provides information on how quickly changes can occur but does not predict the extent of change that is ultimately possible).
- Heat transfer (it determines the temperature distribution within a combustion system).
- Turbulent mixing (it determines whether the waste compounds are effectively put in contact with oxygen for reaction).
- Residence time (it determines the volumetric size of a combustor).

Combustion byproduct: Any other compounds, other than the combustion products under the complete combustion process emitted from combustors, e.g., waste incinerators or fossil fuel boilers. In general, combustion byproducts are the product species produced under the incomplete combustion process (cf. byproduct) (Lee-11/90).

Combustion chamber:
(1) The space in which combustion takes place.
(2) The actual compartment where waste is burned in an incinerator (EPA-94/04).

Other combustion chamber related terms include:
- Primary chamber (*see* synonym, primary combustion chamber)
- Primary combustion chamber
- Primary combustion chamber water spray
- Primary combustion chamber underfire steam injection
- Secondary chamber (*see* synonym, secondary combustion chamber)
- Secondary combustion chamber
- Secondary combustion chamber air port

Combustion component and incineration component: Fundamentally, there are three major components [namely (1) fuel; (2) oxidizer; and (3) diluent] in a combustion system and four major components [namely (1) fuel; (2) oxidizer; (3) diluent; and (4) waste] in an incineration system.

(1) **Fuel:** A fuel is a mixture of hydrocarbons containing energy-rich bonds such as the carbon and carbon-hydrogen bonds. These hydro-carbons are a common source of chemical potential energy (*see* hydrocarbon for more information).

(2) **Oxidizer:** An oxidizer is the chemical species that reacts with the fuel or waste compounds during incineration. Its function is to transform the chemical potential energy stored in the fuel into thermal energy or to convert heavy-molecule waste compounds into light, simple compounds such as CO_2, H_2O and HCl. Most commonly, the oxidizer is molecular oxygen, a constituent of air.

(3) **Diluent:** A diluent is a substance that does not participate chemically in the combustion reaction either as a fuel substance or as an oxidizer. It is physically present and often does influence the combustion process. For example, diluents have heat capacity and while they do not make a positive contribution to the total energy released, they do act as a thermal sink and limit the temperature rise achieved by combustion. A diluent can be thought of as a substance that participates principally in the physical aspects of the combustion process. There are several possible diluents in an incineration system, some are:

   (A) **Nitrogen:** Nitrogen which comprises almost 79% of air, is the most common diluent.

   (B) **Excess amount of oxygen:** Incineration normally takes place at about 150 to 200% of the amount of theoretical air needed for combustion. The excess oxygen (the amount over 100% of theoretical air) will act as a diluent. That is, its participation in the combustion process will be physical but not chemical.

   (C) Water vapor contained either in combustion air or in the waste feed (or, for that matter, the amount formed during the incineration process).

   (D) Inorganic ash compounds such as trace heavy metals present in the waste or in the fuel.

(4) **Waste:** The waste streams that are incinerable include hazardous waste, municipal waste, toxic substances (PCB), medical waste, spent pesticides, sludges from both municipal and industrial wastewater treatment processes, and other unclassified waste such as non-hazardous military waste.

Combustion correction factor (or correction factor): For comparison purposes, it is sometimes necessary to correct a measured value of a compound's concentration at the stack to a certain desired concentration. The way to make such a correction follows. The correction factor (CF) for oxygen is defined as:

$$CF = (21 - \text{desired } O_2)/(21 - \text{measured } O_2)$$

Example 1: CF calculations

- $CH_4 + 2(O_2 + 3.76N_2) \longrightarrow CO_2 + 2H_2O + 7.52N_2$ (A)
- $CH_4 + 2(1.5)(O_2 + 3.76N_2) \longrightarrow CO_2 + 2H_2O + O_2 + 11.28N_2$ (B)
- Equation (A) depicts combustion at 0% excess air and Equation B shows 150% air combustion.
- The following calculation is to show how CO_2 measured at 150% of theoretical air combustion (50% excess air) is converted to 0% excess air combustion.
- O_2 measured at the wet condition of Equation B = $1/(1 + 2 + 1 + 11.28) = 6.5\%$.
- O_2 measured at the dry condition of Equation B = $1/(1 + 1 + 11.28) = 7.5\%$.
- To correct the measured (dry) O_2 value to a desired (0% O_2 in products): $CF = (21 - 0)/(21 - 7.5) = 1.55$
- CO_2 measured at a dry condition (Eq. B) = $1/(1 + 1 + 11.28) = 7.5\%$
- To correct the measured CO_2 to a desired O_2 level of 0% excess air (i.e., at the theoretical air condition): $C = CF \times C_m$; where: C = corrected parameter value at the desired O_2 level; C_m = measured parameters such as CO or CO_2 which are required to be corrected to a desired O_2 level.
- For this example: CO_2 (at 0% O_2 level) = $CF \times (CO_2)_m$ = $1.55 \times 7.5\% = 11.625\%$
- This figure can be checked with the dry CO_2 level from the theoretical air combustion of CH_4 from Equation A. That is: $CO_2 = 1/(1 + 7.52) = 11.73\%$. The discrepancy between the 11.625% and 11.73% values is probably due to rounding off certain values; in any case, the values are very close.

Example 2: Conversion between two correction factors

Given: CO at 7% O_2 level

Determine: CO at 10% O_2 level

- CO at 7% O_2 is:
  $$(CO)_7 = [(21-7)/(21-O_m)](CO)_m \quad (C)$$
  where: subscript m refers to the measured value.
- CO at 10% O_2 is:
  $$(CO)_{10} = [(21-10)/(21-O_m)](CO)_m \quad (D)$$
- From the ratio (D)/(C),
  $$(CO)_{10} = (11/14) \times (CO)_7 = 0.7857 \times (CO)_7$$

Combustion efficiency (CE): Combustion efficiency is defined as: $CE = CO_2/(CO_2 + CO)$, where: CO_2 = carbon dioxide; and CO = carbon monoxide.

Combustion equivalence ratio (or equivalence ratio): Often it is desirable to compare the richness or leanness of combustion for different fuels. The equivalence ratio (ER) is convenient for this type of comparison, and it may be defined as:

(1) The actual fuel-air ratio divided by the stoichiometric fuel-air ratio (equation); or

(2) Alternatively, it is sometimes defined in terms of air-fuel ratios (equation B):

$$ER = \frac{(F/A)_{actual}}{(F/A)_{theoretical}} \quad (A)$$

$$ER = \frac{(A/F)_{actual}}{(A/F)_{theoretical}} \quad (B)$$

By either definition, for a stoichiometric mixture, ER = 1.0. However, it should be recognized that even though the equivalence ratios defined above have identical values at stoichiometric conditions, they are not identical for "off-stoichiometric" (rich or lean) mixtures because (F/A) and (A/F) ratios are reciprocals. This is illustrated in the following Table:

- Per Equation (A)
  - Rich .... ER > 1
  - Stoichiometric .. ER = 1
  - Lean .... ER < 1
- Per Equation (B)
  - Rich .... ER < 1

- Stoichiometric .. ER=1
- Lean .... ER>1

Combustion excursion: The volatilized combustible compounds, particulate and partially oxidized combustible compounds in the exhaust gas emissions created when combustible material volatilizes too rapidly and creates an organic cloud which can not thoroughly mixed with oxygen in the combustion chamber and under extreme conditions creates an oxygen demand which exceeds the oxygen supply (Session 7 of the 1st ICTCB-89).

Combustion for air pollution control: Four basic combustion systems that can be used to control combustible gaseous emissions. They are:
(1) Flare;
(2) Thermal oxidizer;
(3) Catalytic oxidizer; and
(4) Process boiler (EPA-81/12, p3-1).

Combustion gas (flue gas or stack gas):
(1) The exhaust gas from a combustion process (EPA-83). It may contain nitrogen oxides, carbon oxides, water vapor, sulfur oxides, particles and many chemical pollutants (EPA-89/12).
(2) The gas discharged from a stack after combustion. It can include nitrogen oxides, carbon oxides, water vapor, sulfur oxides, particles, and many other chemical pollutants (EPA-84/09).

Combustion indicator: An instrument to monitor the quality of combustion. It includes:
(1) Opacity
(2) Stack gas oxygen concentration
(3) Stack gas CO concentration
(4) Combustion temperature
(5) Ash combustible

Combustion limit: Combustion limit related terms include:
- Lower explosive limit (LEL)
- Upper explosive limit (UEL)

Combustion mode: In general, combustion or incineration involves three simultaneous chemical reaction modes: (1) Strong oxidation; (2) weak pyrolysis; and (3) weak radical attack.
(1) **Oxidation:** The oxidation of waste is shown in the following example for which dichloromethane is oxidized to produce harmless products: $CH_2Cl_2 + O_2 + 3.76N_2 ---> CO_2 + 2HCl + 3.76N_2$

A generalized formula for the complete incineration of a typical waste, $C_xH_yCl_z$, (ie, "theoretical air combustion") can be expressed as follows for y > z: $C_xH_yCl_z + [x + (y - z)/4](O_2 + 3.76N_2) ---> xCO_2 + zHCl + (y-z)/2H_2O + 3.76[x + (y - z)/4]N_2$, where: x, y and z represent the relative number of atoms of carbon, hydrogen, and chlorine respectively.

(2) **Pyrolysis:** Pyrolysis is a thermal degradation process wherein carbonaceous materials are destroyed or chemically rearranged in the absence or near absence of oxygen or air. It uses heat to break the bonds of the elements contained in a compound.

Although incineration requires about 50 to 150% excess air to ensure enough oxygen in the combustion chamber to effectively contact with the waste, some small fraction of the waste still may not have a chance to contact the oxygen. These small waste fractions that remain in the high-temperature environment may undergo pyrolysis, e.g., the pyrolysis of cellulose and PCB may follow:
- Cellulose: $C_6H_{10}O_5 ---> 2CO + CH_4 + 3H_2O + 3C$
- PCB: $C_{12}H_7Cl_3 ---> 12C + 3HCl + 2H_2$

The degraded compounds generally produce simpler compounds such as CO, CH_4, and H_2O which will be in the gaseous phase, and carbon (C), char, which will be in the solid or liquid phase.

(3) **Radical attack:** During incineration, flames are characterized by temperatures usually in the neighborhood of 1000 C and a radical-rich gas flow. This gas flow consists primarily of atomic hydrogen (H), atomic oxygen (O), atomic chlorine (Cl), hydroxyl radicals (OH.), possibly methyl radicals (CH_3.) in carbon-hydrogen-oxygen systems, and chloroxy radicals (ClO.) in chlorine-containing systems. The radical attack on waste compounds facilitates the decomposition of the waste.

Combustion modification: One of NO_x emission reduction techniques (*see* nitrogen oxide emission control for control structure). Nitrogen oxide emissions result from operating conditions in the furnace, the amount of nitrogen in the combustion air, and the amount of nitrogen in the fuel. Combustion conditions in the furnace can be modified to reduce NO_x emissions (EPA-81/12, p7-5).

Combustion of coal and limestone mixtures: One of SO_2 emission reduction techniques (*see* sulfur oxide emission control for control structure). Sulfur oxides can be removed by burning coal and limestone mixtures in a boiler. There are two potential burning technologies:
(1) Fluidized bed combustion
(2) Limestone coal pellets as fuel (EPA-81/12, p8-5).

Combustion product:
(1) The product species produced under the complete combustion process. In fossil fuel combustion, the combustion products are CO_2, H_2O and N_2. In waste incineration, the combustion products are CO_2, H_2O, N_2, HCl and Cl_2 (cf. combustion byproduct) (Lee-85/11).
(2) Material produced or generated during the burning or oxidation of a material (EPA-85/11).
(3) Substance produced during the burning or oxidation of a material (EPA-94/04).

Combustion rate: *See* synonym, burn rate.

Combustion residual oxygen in flue gas: The residual oxygen (% O_2) in flue gas can be calculated by the following equation: Excess air (%) = % $O_2/(21\% - \% O_2)$.

In this equation, excess air is a given condition which is the ratio of excess air to the theoretical air amount. Thus, the residual O_2 after combustion can be determined. Example, a 50% excess air combustion of propane is: $C_3H_8 + 5 \times 1.5(O_2 + 3.76N_2) ---> 3CO_2 + 4H_2O + 2.5O_2 + 28.2N_2$.

The residual O_2 in flue gas is: $50\% = \%O_2/(21\% - \% O_2)$. Thus, $O_2 = 7\%$. This value can be checked by actual calculation of O_2 fraction in the flue gas, i.e., $O_2 = 2.5/(3 + 4 + 2.5 + 28.2) = 6.6\%$. The discrepancy between the 7% and 6.6% values is probably due to rounding off certain values; in any case, the values are very close.

Combustion temperature: Rapid increases or decreases in combustion gas temperature indicate potential combustion problems. Rising temperatures indicate that the heat input is increasing and/or airflow is decreasing which can lead to insufficient air for complete combustion. Falling temperatures indicate problems in sustaining combustion (*see* combustion indicator for more related terms) (EPA-89/03b).

Comfort cooling towers: The cooling towers that are dedicated exclusively to and are an integral part of heating, ventilation, and air conditioning or refrigeration systems (40 CFR 749.68-91).

Command post: Facility located at a safe distance upwind from an accident site, where the on-scene coordinator, responders, and technical representatives make response decisions, deploy manpower and equipment, maintain liaison with news media, and handle communications (EPA-94/04).

Commence: As applied to construction of a major stationary source or major modification means that the owner or operator has all necessary preconstruction approvals or permits and either has:
(1) Begun, or caused to begin, a continuous program of actual on-site construction of the source, to be completed within a reasonable time; or
(2) Entered into binding agreements or contractual obligations, which cannot be canceled or modified without substantial loss to the owner or operator, to undertake a program of actual construction of the source to be completed within a reasonable time (40 CFR 51.165-91, *see also* 40 CFR 51-App/S; 51.166; 52.01; 52.21; 52.24; 52.2486; 60.2; 60.666; 61.02-91).

Commence construction: To engage in a continuous program of on-site construction including site clearance, grading, dredging, or land filling specifically designed for a parking facility in preparation for the fabrication, erection, or installation of the building components of the facility. For the purpose of this paragraph, interruptions resulting from acts of God, strikes, litigation, or other matters beyond the control of the owner shall be disregarded in determining whether a construction or modification program is continuous (40 CFR 52.1135-91, *see also* 40 CFR 52.1135-91).

Commenced commercial operation: To have begun to generate electricity for sale (CAA402-42USC7651a-91).

Commencement of construction: Any clearing of land, excavation, or other substantial action that would adversely affect the natural environment of a site but does not include changes desirable for the temporary use of the land for public recreational uses, necessary borings to determine site characteristics or other preconstruction monitoring to establish background information related to the suitability of a site or to the protection of environmental values (10 CFR 30.4-91, *see also* 10 CFR 40.4; 70.4-91).

Commensalism: An interaction (e.g., habitation together) of two species (animals or plants) in which one species benefits from the association while the other species is not affected.

Comment period: Time provided for the public to review and comment on a proposed EPA action or rulemaking after publication in the Federal Register (EPA-94/04).

Commerce:
(1) Commerce between any place in any State and any place outside thereof; and
(2) Commerce wholly within the District of Columbia (CAA216, *see also* 40 CFR 205.2; 710.2; 720.3; 747.115; 747.195; 747.200; 761.3; 763.163-91).

Commerce clause: A constitutional clause granting Congress the power to regulate all commerce; the dormant commerce clause makes it explicit that State lines cannot be made barriers to the free flow of commerce (OTA-89/10).

Commercial activity: All activities of industry and trade, including but not limited to, the buying or selling of commodities and activities conducted for the purpose of facilitating such buying and selling: Provided, however, That it does not include exhibition of commodities by museum or similar cultural or historical organizations (ESA3-16USC1531-90).

Commercial aircraft engine: Any aircraft engine used or intended for use by an air carrier (including those engaged in intrastate air transportation) or a commercial operator (including those engaged in intrastate air transportation) as these terms are defined in the Federal Aviation Act and the Federal Aviation Regulations (*see* engine for more related terms) (40 CFR 87.1-91).

Commercial aircraft gas turbine engine: A turboprop, turbofan, or turbojet commercial aircraft engine (40 CFR 87.1-91).

Commercial and industrial friction product: An asbestos-containing product, which is either molded or woven, intended for use as a friction material in braking and gear changing components in industrial and commercial machinery and consumer appliances. Major applications of this product include: hand brakes; segments; blocks; and other components used as brake linings, rings and clutches in industrial and commercial machinery and consumer appliances (40 CFR 763.163-91).

Commercial applicator: A certified applicator (whether or not he is a private applicator with respect to some uses) who uses or supervises the use of any pesticide which is classified for restricted use for any purpose or on any property other than as provided by the definition of private applicator (40 CFR 171.2-91).

Commercial arsenic: Any form of arsenic that is produced by extraction from any arsenic containing substance and is intended for sale or for intentional use in a manufacturing process. Arsenic that is a naturally occurring trace constituent of another substance is not considered commercial arsenic (*see* arsenic for more related terms) (40 CFR 61.161-91).

Commercial asbestos: Any material containing asbestos that is extracted from ore and has value because of its asbestos content (*see* asbestos for more related terms) (40 CFR 61.141-91).

Commercial establishment: Stores, offices, restaurants, warehouses and other non-manufacturing activities (40 CFR 246.101-91).

Commercial hexane: For purposes of this section, is a product obtained from crude oil, natural gas liquids, or petroleum refinery processing in accordance with the American Society for Testing and Materials Designation D 1836-83 (ASTM D 1836), consists primarily of six-carbon alkanes or cycloalkanes, and contains at least 40 liquid volume percent n-hexane (CAS No. 110-543) and at least 5 liquid volume percent methylcyclopentane (MCP; CAS No. 96-377). ASTM D 1836, formally entitled "Standard Specification for Commercial Hexanes," is published in 1986 Annual Book of ASTM Standards: Petroleum Products and Lubricants, ASTM D 1836-83, pp. 966-967, 1986, is incorporated by reference, and is available for public inspection at the Office of the Federal Register, Room 8301, 1100 L Street NW. Washington DC (40 CFR 799.2155-1-91).

Commercial hexane test substance: For purposes of this section, is a product which conforms to the specifications of ASTM D 1836 and contains no more than 40 liquid volume percent n-hexane and no less than 10 liquid volume percent MCP (40 CFR 799.2155-91).

Commercial incinerator: A privately owned incinerator used to burn wastes from all sources for profit purposes (*see* incinerator for more related terms).

Commercial item descriptions: A series of simplified item descriptions under the Federal specifications and standards program used in the acquisition of commercial off-the-shelf and commercial type products (40 CFR 248.4-91).

Commercial operator (Ohio's definition): All persons, firms or corporations who own or operate stores, restaurants, industries, institutions and other similar places, public or private, charitable or non-charitable, and includes all responsible persons other than householders, upon whose premises putrescible wastes, other refuse or both is, or are, created (EPA-83).

Commercial paper: An asbestos-containing product which is made of paper intended for use as general insulation paper or muffler paper. Major applications of commercial papers are insulation against fire, heat transfer, and corrosion in circumstances that require a thin, but durable, barrier (*see* paper for more related terms) (40 CFR 763.163-91).

Commercial refuse: All solid wastes originating in businesses and multiple unit rental structures, such as office buildings, apartment houses, stores, markets, theaters and privately owned hospitals and other institutional units (*see* refuse for more related terms) (EPA-83).

Commercial solid waste (or commercial waste): All types of solid wastes generated by stores, offices, restaurants, warehouses, and other non-manufacturing activities, excluding residential and industrial wastes (*see* waste for more related terms) (40 CFR 243.101-91, *see also* 40 CFR 245.101; 246.101-91).

Commercial waste:
(1) All solid waste emanating from business establishments such as stores, markets, office buildings, restaurants, shopping centers, and theaters (EPA-94/04).
(2) *See also* synonym, commercial solid waste.

Commercial waste management facility: A treatment, storage, disposal, or transfer facility which accepts waste from a variety of sources, as compared to a private facility which normally manages a limited waste stream generated by its own operations (EPA-94/04).

Commercial storer of PCB waste: The owner or operator of each facility which is subject to the PCB storage facility standards of 40 CFR 761.65, and who engages in storage activities involving PCB waste generated by others, or PCB waste that was removed while servicing the equipment owned by others and brokered for disposal. The receipt of a fee or any other form of compensation for storage services is not necessary to qualify as a commercial storer of PCB waste. It is sufficient under this definition that the facility stores PCB waste generated by others or the facility removed the PCB waste while servicing equipment owned by others. A generator who stores only the generator's own waste is subject to the storage requirements of 40 CFR 761.65, but is not required to seek approval as a commercial storer. If a facility's storage of PCB waste at no time exceeds 500 liquid gallons of PCBs, the owner or operator is not required to seek approval as a commercial storer of PCB waste (40 CFR 761.3-91).

Commercial use: The use of a chemical substance or any mixture containing the chemical substance in a commercial enterprise providing saleable goods or a service to consumers (e.g., a commercial dry cleaning establishment or painting contractor) (40 CFR 721.3-91).

Commercial use request: Refers to a request from or on behalf of one who seeks information for a use or purpose that furthers the commercial, trade or profit interests of the requestor or the person on whose behalf the request is made. In determining whether a requestor properly belongs in this category, EPA must determine the use to which a requestor will put the documents requested. Moreover, where EPA has reasonable cause to doubt the use to which a requestor will put the records sought, or where that use is not clear from the request itself, EPA may seek additional clarification before assigning the request to a specific category (40 CFR 2.100-e-91).

Commercial vessel: Those vessels used in the business of transporting property for compensation or hire, or in transporting property in the business of the owner, lessee, or operator of the vessel (*see* vessel for more related terms) (CWA312-33USC1322-91).

Commingled recyclable:
(1) A recyclable material separated from mixed MSW (municipal solid waste) at the point of generation. Further separation into individual components occurs at the collection vehicle or centralized processing facility (OTA-89/10).
(2) A mixture of several recyclable materials contained in one container (EPA-89/11).
(3) Mixed recyclables that are collected together (EPA-94/04).

Comminuter: A machine that shreds or pulverizes solids to make waste treatment easier (EPA-94/04).

Comminution: Mechanical shredding or pulverizing of waste. Used in both solid waste management and wastewater treatment (EPA-94/04).

Commission finishing: The finishing of textile materials, 50 percent or more of which are owned by others, in mills that are 51 percent or more independent (i.e., only a minority ownership by company(ies) with greige or integrated operations); the mills must process 20 percent or more of their commissioned production through batch, noncontinuous processing operations with 50 percent or more of their commissioned orders processed in 5000 yard or smaller lots (40 CFR 410.01-91).

Commission scouring: The scouring of wool, 50 percent or more of which is owned by others, in mills that are 51 percent or more independent (i.e., only a minority ownership by company(ies) with greige or integrated operations); the mills must process 20 percent or more of their commissioned production through batch, noncontinuous processing operations (40 CFR 410.11-91).

Committee or local emergency planning committee: The local emergency planning committee appointed by the emergency response commission (40 CFR 355.20; 370.2-91).

Common carrier by motor vehicle: Any person who holds himself out to the general public to engage in the transportation by motor vehicle in interstate or foreign commerce of passengers or property or any class or classes thereof for compensation, whether over regular or irregular routes (40 CFR 202.10-91).

Common emission control device: A control device controlling emissions from the coating operation as well as from another emission source within the plant (40 CFR 60.711-91, *see also* 40 CFR 60.741-91).

Common exposure route: A likely way (oral, dermal, respiratory) by which a pesticide may reach and/or enter an organism (40 CFR 171.2-91).

Common law: Common law is derived from the application of natural reason, an innate sense of justice and the dictates of conscience. It is not the result of legislative enactment. Rather, its authority is derived solely from usages and customs which have been recognized, affirmed and enforced by the courts through judicial decisions (*see* environmental law system for more related terms) (Sullivan-95/04, p6). Other common law related terms include:
- Tort: Three types of torts most commonly encountered in the environmental field are:
    - Negligence
    - Nuisance
    - Trespass

Common metal: Copper, nickel, chromium, zinc, tin, lead, cadmium, iron, aluminum, or any combination thereof (*see* metal for more related terms) (EPA-83/06a).

Common name: Any designation or identification such as code name, code number, trade name, brand name, or generic chemical name used to identify a chemical substance other than by its chemical name (40 CFR 721.3-91).

Common pesticide name: A common chemical name given to a pesticide by a recognized committee on pesticide nomenclature. Many pesticides are known by a number of trade or brand names but have only one recognized common name, e.g., the common name for Sevin insecticide is carbaryl (EPA-85/10).

Common rosin: *See* synonym, natural rosin.

Communicable disease: All illness due to an infectious agent or its toxic products which is transmitted directly or indirectly to a healthy person through the agency of an infected entity, intermediate host, vector or inanimate environment (EPA-83).

Communicable period: The time during which an infection may be transferred (EPA-83).

Community:
(1) An aggregation of species (plants or animals) living in the same area (habitat).
(2) In ecology, a group of interacting populations in time and space. Sometimes, a particular subgrouping may be specified, such as the fish community in a lake or the soil arthropod community in a forest (EPA-94/04).

Community (biotic): All plant and animal populations occupying a specific area under relatively similar conditions (DOE-91/04).

Community awareness and emergency response (CAER): Program developed by the Chemical Manufacturers Association. Guidance for chemical plant managers to assist them in taking the initiative in cooperating with local communities to develop integrated (community/industry) hazardous materials response plans (NRT-87/03).

Community component: A general term that may pertain to the biotic guild (fish, invertebrates, algae), the taxonomic category (order, family, genus, species), the feeding strategy (herbivore, omnivore, predator), or the organizational level (individual, population, assemblage) of a biological entity within the aquatic community (EPA-91/03).

Community relation (or public participation): EPA's program to inform and encourage public participation in the Superfund process and to respond to community concerns (for complete definition, *See* 40 CFR 300.5-91).

Community relations:
(1) EPA's program to inform and encourage public participation in the Superfund process and to respond to community concerns. The term "public" includes citizens directly affected by the site, other interested citizens or parties, organized groups, elected officials, and potentially responsible parties (40 CFR 300.5-91).
(2) The EPA effort to establish two-way communication with the public to create understanding of EPA programs and related actions, to assure public input into decision-making processes related to affected communities, and to make certain that the

Agency is aware of and responsive to public concerns. Specific community relations activities are required in relation to Superfund remedial actions (EPA-94/04).

Community Right-To-Know Act of 1986: *See* the term of Act or EPCRA.

Community water system:
(1) A public water system which serves at least 15 service connections used by year-round residents or regularly serves at least 25 year-round residents (40 CFR 141.2-91, *see also* 40 CFR 191.12-91).
(2) A public water system which serves at least 15 service connections used by year-round residents or regularly serves at least 25 year-round residents (EPA-94/04).

Commuter: An employee who travels regularly to a place of employment (40 CFR 52.1161-91, *see also* 40 CFR 52.2294; 52.2297-91).

Compact:
(1) To reduce size or dimensions or increase density without adding or subtracting matter.
(2) To treat glass in a manner, such as by heat-treatment, to approach maximum density (EPA-83).

Compaction: Reduction of the bulk of solid waste by rolling and tamping (EPA-94/04).

Compaction pit transfer system: A transfer system in which solid waste is compacted in a storage pit by a crawler tractor before being pushed into an open-top transfer trailer (SW-108ts).

Compactor: A power-driven device used to compress materials to a smaller volume (EPA-89/11; SW-108ts). Other compactor related terms include:
- Compactor collection vehicle (mobile)
- Sanitary landfill compactor
- Stationary compactor

Compactor collection truck: An enclosed vehicle provided with special mechanical devices for loading the refuse into the main compartment of the body and for compressing and distributing the refuse, within the body (EPA-83).

Compactor collection vehicle (mobile): A vehicle with an enclosed body containing mechanical devices that convey solid waste into the main compartment of the body and compress it into a smaller volume of greater density (*see* compactor for more related terms) (40 CFR 243.101-91).

Comparability: *See* definition under quality indicator.

Comparative method: A method in which characterization is based on chemical standards (i.e., comparison with such standards) (*see* method for more related terms) (ACS-87/11).

Compartment: A liquid-tight division of a delivery tank (40 CFR 60-App/A(method 27)-91).

Compartmentalized vehicle: A collection vehicle which has two or more compartments for placement of solid wastes or recyclable materials. The compartments may be within the main truck body or on the outside of that body as in the form of metal racks (40 CFR 246.101-91).

Compatibility:
(1) That property of a pesticide which permits its use with other chemicals without undesirable results being caused by the combination (40 CFR 171.2-91).
(2) If two or more hazardous materials remain in contact indefinitely without reaction, they are compatible (cf. waste compatibility) (Course 165.5).

Compatible: The ability of two or more substances to maintain their respective physical and chemical properties upon contact with one another for the design life of the tank system under conditions like to be encountered in the UST (40 CFR 280.12-91).

Compatible pollutant: The pollutants which can be adequately treated in publicly-owned treatment works without upsetting the treatment process (*see* pollutant for more related terms) (EPA-83/06a).

Compensation point: The oxygen released in photosynthesis equals the amount used in respiration and the carbon dioxide released in respiration equals the amount used in photosynthesis.

Competent: Properly qualified to perform functions associated with pesticide application, the degree of capability required being directly related to the nature of the activity and the associated responsibility (40 CFR 171.2-91).

Complementary cumulative distribution function (CCDF): The CCDF is used to provide information on the frequency of consequence magnitudes exceeding specific value (DOE-91/04).

Complete: In reference to an application for a permit, that the application contains all the information necessary for processing the application. Designating an application complete for purposes of permit processing does not preclude the reviewing authority from requesting or accepting any additional information (40 CFR 51.166-91, *see also* 40 CFR 52.21-91).

Complete combustion: A combustion process that all carbon and hydrogen elements in reactants are converted into only carbon dioxide and water in products under a stoichiometric condition (or 100% combustion air). The conversion of carbon and hydrogen can be expressed as follows: $C + O_2 \longrightarrow CO_2$; and $H_2 + 1/2\, O_2 \longrightarrow H_2O$. An example of complete combustion of a propane gas is: $C_3H_8 + 5 \times 1.0\,(O_2 + 3.76N_2) \longrightarrow 3CO_2 + 4H_2O + 18.8\,N_2$

A generalized formula for complete combustion of a conventional fuel, C_xH_y (theoretical air combustion), can be expressed as follows: $C_xH_y + (x + y/4)O_2 + 3.76(x + y/4)N_2 \longrightarrow xCO_2 + y/2 H_2O + 3.76(x + y/4)N_2$, where x and y represent the relative number of atoms of carbon, and hydrogen, respectively.

To obtain the optimum temperature during combustion, it is desirable to convert all the chemical energy stored in the reactants into thermal energy. To reach this goal, all carbon and hydrogen elements in a combustion system must be fully oxidized and become only carbon dioxide and water (*see* combustion for more related terms).

Complete destruction of pesticides: An alteration (of pesticides) by physical or chemical processes to inorganic forms (40 CFR 165.1-91).

Complete incineration: The ultimate goal of the incineration of waste is to convert the waste materials into harmless combustion products so that they can be safely emitted to the environment. When a waste is completely incinerated, the elements in the waste are generally assumed to follow the following reaction patterns:
- Hydrogen, H ------> H_2O
- Carbon, C ------> CO_2
- Chloride, Cl ------> HCl or Cl_2
- Fluoride, F ------> HF or F_2
- Sulfur, S ------> SO_2
- Nitrogen, N ------> N_2
- Alkali metals------> Carbonate
- Sodium, Na ------> Na_2CO_3
- Potassium, K ------> KOH
- Non-alkali metals--> Oxides
- Copper, Cu ------> CuO
- Iron, Fe ------> Fe_2O

However, complete incineration is solely a theoretical concept. In actual practice, partially oxidized products of incomplete combustion (PICs) are formed. These PICs may include carbon monoxide (CO), soot, and a whole myriad of other organics. It is always possible to over design an incinerator or to use extra fuel for higher flame temperatures to ensure sufficiently complete combustion. However, either of these corrective measures increase the cost of incineration (*see* incineration for more related terms).

Complete recycle: The complete re-use of a stream, with make-up water added for evaporation losses. There is no blowdown stream from a totally recycled flow and the process water is not periodically or continuously discharged (EPA-85/10a).

Complete treatment: A method of treating water that consists of the addition of coagulant chemicals, flash mixing, coagulation-flocculation, sedimentation, and filtration. Also called conventional filtration (EPA-94/04).

Complete wastewater treatment system: The system consists of all the treatment works necessary to meet the requirements of title III of the Act (for complete definition, *see* 40 CFR 35.905; 35.2005-91).

Complete waste treatment system: Consists of all the treatment works necessary to meet the requirements of title III of the Act, involving:
(1) The transport of wastewater from individual homes or buildings to a plant or facility where treatment of the wastewater is accomplished;
(2) The treatment of the wastewater to remove pollutants; and
(3) The ultimate disposal, including recycling or reuse, of the treated wastewater and residues which result from the treatment Process (40 CFR 35.2005-12-91).

Completely closed drain system: An individual drain system that is not open to the atmosphere and is equipped and operated with a closed vent system and control device complying with the requirements of 40 CFR 60.692-5 (40 CFR 60.691-91).

Completely mixed batch reactor: A closed reactor system where the mixing is provided to ensure that no concentration gradient exist (EPA-88/09a).

Completely mixed condition: No measurable difference in the concentration of a pollutant exists across a transect of the waterbody (e.g., does not vary by 5 percent) (EPA-91/03).

Completely mixed flow through reactor: A reactor system which allows the reaction fluids to flow in and out of the system, while inside the reactor, the fluids are completely mixed.

Completeness: *See* definition under quality indicator.

Complex: A collection of facilities dedicated to the same set of activities (DOE-91/04).

Complex compound: A compound capable of forming chemical bonds with metal atoms or ions, e.g., $(Cu(NH_3))^{+2}$.

Complex cyanide: Complex ions containing cyanide ions and a cation such as iron, e.g., ferrocyanide $[Fe(CN)_6]^{-4}$ and/or ferricyanide $(Fe(CN)_6)^{-3}$.

Complex manufacturing operation: The simple unit processes (desizing, fiber preparation and dyeing) plus any additional manufacturing operations such as printing, water proofing, or applying stain resistance or other functional fabric finishes (cf. simple manufacturing operation) (40 CFR 410.41-91, *see also* 40 CFR 410.51; 410.61-91).

Complex salt: A metal salt in which there are no detectable metal ions in a solution, because the metal ions are strongly bound in the complex ion.

Complex slaughterhouse: A slaughterhouse that accomplishes extensive byproduct processing, usually at least three of such operations as rendering, paunch and viscera handling, blood processing, hide processing, or hair processing (40 CFR 432.21-91).

Complex terrain: A terrain whose rise within 5 kilometers of a stack is greater than the physical stack height (*see* terrain for more related terms) (EPA-90/04).

Complexation: *See* synonym, complexing.

Complexing (or complexation):
(1) Forming a compound containing a number of parts, often used to describe a metal atom associated with a set of organic ligands (EPA-87/10a).
(2) In chemistry, the process of incorporation into other compounds, such as through hydration, oxygenation, halogenation, and chelation (LBL-76/07-bio).

Complexing agent: A compound that joins with a metal to form an ion which has a molecular structure consisting of a central atom (the metal) bonded to other atoms by coordinate covalent bonds (EPA-83/06a).

Complexometric analysis (complexometry titration or chelatometry): A method of volumetric analysis in which the formation of colored complex is used to indicate the result of a titration (cf. titration).

Complexometric titration: *See* synonym, complexometric analysis (*see also* tritation for more related terms).

Compliance: The compliance with clean air standards or clean water standards. For the purpose of these regulations, compliance also shall mean compliance with a schedule or plan ordered or approved by a court of competent jurisdiction, the United States Environmental Protection Agency, or an air or water pollution control agency, in accordance with the requirements of the CAA or the CWA and regulations issued pursuant thereto (40 CFR 15.4-91).

Compliance action/order: *See* definition under administrative order.

Compliance coal: Any coal that emits less than 1.2 pounds of sulfur dioxide per million Btu when burned. Also known as low sulfur coal (EPA-94/04).

Compliance coating: A coating whose volatile organic compound content does not exceed that allowed by regulation (EPA-94/04).

Compliance cycle:
(1) The nine-year calendar year cycle during which public water systems must monitor. Each compliance cycle consists of three three-year compliance periods. The first calendar year cycle begins January 1, 1993 and ends December 31, 2001; the second begins January 1, 2002 and ends December 31, 2010; the third begins January 1, 2011 and ends December 31, 2019 (40 CFR 141.2-91).
(2) The nine-year calendar year cycle, beginning January 1, 1993, during which public water systems must monitor. Each cycle consists of three three-year compliance periods (EPA-94/04).

Compliance level: An emission level determined during a Production Compliance Audit Pursuant to Subpart L of this part (40 CFR 86.1002.84-91, *see also* 40 CFR 86.1102.87-91).

Compliance monitoring: Collection and evaluation of data, including self-monitoring reports, and verification to show whether pollutant concentrations and loads contained in permitted discharges are in compliance with the limits and conditions specified in the permit (EPA-94/04).

Compliance period: A three-year calendar year period within a compliance cycle. Each compliance cycle has three three-year compliance periods. Within the first compliance cycle, the first compliance period runs from January 1, 1993 to December 31, 1995; the second from January 1, 1996 to December 31, 1998; the third from January 1, 1999 to December 31, 2001 (40 CFR 141.2-91).

Compliance plan: For purposes of the requirements of this title, either:
(1) A statement that the source will comply with all applicable requirements under this title, or
(2) Where applicable, a schedule and description of the method or methods for compliance and certification by the owner or operator that the source is in compliance with the requirements of this title (CAA402-42USC7651a-91).

Compliance schedule: A negotiated agreement between a pollution source and a government agency that specifies dates and procedures by which a source will reduce emissions and, thereby, comply with a regulation (EPA-94/04).

Composite board: Any combination of different types of board, either with another type board or with another sheet material. The composite board may be laminated in a separate operation or at the same time as the board is pressed. Examples of composite boards include veneer-faced particle board, hardboard-faced insulation board and particle board, and metal-faced hardboard (EPA-74/04).

Composite liner: A liner system composed of an engineered soil layer overlain by a synthetic flexible membrane liner (OTA-89/10).

Composite NO_x standard: For a manufacturer which elects to average light-duty trucks subject to the NO_x standard of 40 CFR 86.088.9(a)(iii)(A) together with those subject to the NO_x standard of 40 CFR 86.088.9(a)(iii)(B) in the light-duty truck NO_x averaging program, means that standard calculated according to the equation (set forth in 40 CFR 86.088.2) and rounded to the nearest one-tenth gram per mile (40 CFR 86.088.2-91).

Composite particulate standard: For a manufacturer which elects to average diesel light-duty vehicles and diesel light-duty trucks with a loaded vehicle weight equal to or less than 3,750 lbs (LDDT1s) together in the particulate averaging program, means that standard calculated according to the equation (set forth in 40 CFR 86.087.2 under this definition) and rounded to the nearest one-hundredth (0.01) gram per mile (40 CFR 86.087.2-91, *see also* 40 CFR 86.085.2; 86.090.2-91).

Composite sample (or mixed sample):
(1) A series of small samples taken over a given time period and combined as one sample in order to provide a representative analysis of the average constituent levels during the sampling period (EPA-82/11).
(2) A mixture of grab samples collected at the same sampling point at different times (EPA-87/10).
(3) A sample composed of no less than eight grab samples taken over the compositing period (40 CFR 471.02-91).
(4) *See* sample for more related terms.

Composite wastewater sample: A combination of individual samples of water or wastewater taken at selected intervals and mixed in proportion to flow or time to minimize the effect of the variability of an individual samples (*see* sample for more related terms) (EPA-83/03).

Compost:
(1) The relatively stable decomposed organic materials resulting from the composting process. Also referred to as humus (cf. humus) (EPA-89/11).
(2) The relatively stable humus material that is produced from a composting process in which bacteria in soil mixed with

garbage and degradable trash break down the mixture into organic fertilizer (EPA-94/04).

Compost toilet: A tank for holding domestic waste where anaerobic and aerobic digestion take place.

Composting: The controlled biological decomposition of organic material in the presence of air to form a humus-like material. Controlled methods of composting include mechanical mixing and aerating, ventilating the materials by dropping them through a vertical series of aerated chambers, or placing the compost in piles out in the open air and mixing it or turning it periodically (EPA-94/04). Types of composting include (SW-108ts):
(1) Dano Biostabilizer system composting
(2) Indoor process composting
(3) Mechanical composting method
(4) Mechanical process composting
(5) Ventilated cell composting
(6) Windrow composting
(7) Municipal solid waste composting

Compound imbibition: The most common type of imbibition involves the addition and recirculation of water and juices to the bagasse at different points in a four mill network in order to dissolve sucrose (EPA-75/02d).

Comprehensive Environmental Response, Compensation and Liability Act (CERCLA) of 1980: *See* the term of Act or CERCLA.

Comprehensive development: The basin wide development of water and land resources for optimum beneficial uses of a river system and its watershed (DOI-70/04).

Comprehensive pharmaceutical date base: Combined data base containing the first 308 survey of PMA-member companies and the second, or supplemental 308 survey (EPA-83/09).

Compressed liquid or subcooled liquid: The liquid at a temperature lower than the saturation temperature at a given pressure (Jones-p143).

Compressed natural gas (CNG): An alternative fuel for motor vehicles; considered one of cleanest because of low hydrocarbon emissions and its vapors are relatively non-ozone producing. However, it does emit a significant quantity of nitrogen oxides (EPA-94/04).

Compressible fluid: A fluid for which the density varies. The factors causing the density change include the pressure and the temperature (*see* truck for more related terms) (NATO-78/10).

Compressor: A mechanical device:
(1) To provide the desired pressure for chemical and physical reactions;
(2) To control boiling points of fluids, as in gas separation, refrigeration, and evaporation;
(3) To evacuate enclosed volumes;
(4) To transport gases or vapors;
(5) To store compressible fluids as gases or liquids under pressure and assist in recovering them from storage or tank cars; and
(6) To convert mechanical energy to fluid energy for operating instruments, air agitation, fluidization, solid transport, blowcases, air tools, and motors (AP-40, p67).

Other compressor related terms include:
- Acceptance of a compressor
- Failing compressor
- Reciprocating compressor
- Test compressor

Compressor configuration: The basic classification unit of a manufacturers product line and is comprised of compressor lines, models or series which are identical in all material respects with regard to the parameters listed in 40 CFR 204.55.3 (40 CFR 204.51-91).

Compressor type: Types of compressors include:
(1) Positive displacement compressors which produce pressure by reducing the gas volume. This type including reciprocating and rotary machines can discharge up to 35,000 psia pressure.
(2) Dynamic compressors which produce pressure by accelerating the gas and converting the velocity into pressure in a receiving chamber. This type including centrifugal- or axial-flow machines can discharge up to 4200 psia pressure (cf. portable air compressor) (AP-40).

Computer paper: A type of paper used in manifold business forms produced in rolls and/or fan folded. It is used with computers and word processors to print out data, information, letters, advertising, etc. It is commonly called computer printout (*see* paper for more related terms) (40 CFR 250.4-91).

Computer printout paper: The paper consists of white sulphite or sulphate papers in forms manufactured for use in data processing machines. This grade may contain colored stripes and/or computer printing, and may contain not more than 5% of ground-wood in the packing. All stock must be untreated and uncoated (*see* paper for more related terms) (EPA-83).

Concentrate:
(1) In mining, the product of concentration.
  (A) To separate ore or metal from its containing rock or earth.
  (B) The enriched ore after removal of waste in a beneficiation mill, the clean product recovered in froth flotation (EPA-82/05).
(2) In sampling, the enriched concentration of a substance which has been investigated.

Concentrated animal feeding operation: An animal feeding operation which meets the criteria in Appendix B of this part, or which the Director designates under paragraph (c) of this section (40 CFR 122.23-91).

Concentrated aquatic animal production facility: A hatchery, fish farm, or other facility which meets the criteria in Appendix C of this part, or which the Director designates under paragraph (c) of this section (40 CFR 122.24-91).

Concentration:
(1) To an exposure level. Exposure is expressed as weight or volume of test substance per volume of air (mg/L), or as parts per million (ppm) (40 CFR 798.4350-91).
(2) The amount of a substance, expressed as mass or volume, in a unit volume of subjects such as air. The units of concentration include ppm or ppb.
(3) Quantity of the substance contained in a unit quantity of sample. (In absorption spectrometry, it is usually expressed in grams per liter, however for environmental samples g/mL is preferred (LBL-7/76-bio).

Other concentration related terms include:
- Molal solution
- Molality
- Molar solution
- Molarity

Concentration cycle: The multiplicative factor by which the dissolved solids in cooling water are allowed to concentrate due to evaporation of the water (DOE-91/04).

Concentration factor (or biological concentration factor): A factor which is the ratio of the concentration within the tissue or organism to the concentration outside the tissue or organism (EPA-82/11e).

Concentration of a solution: The amount of solute in a given amount of solvent and can be expressed as a weight/weight or weight/volume relationship. The conversion from a weight relationship to one of volume incorporates density as a factor. For dilute aqueous solutions, the density of the solvent is approximately equal to the density of the solutions; thus, concentrations in mg/L are approximately equal to $10^{-3}g/10^3g$ or parts per million (ppm); ones in $\mu g/L$ are approximately equal to $10^{-6}g/10^3g$ or parts Per billion (ppb). In addition, concentration can be expressed in terms of molarity, normality, molality, and mole fraction. For example, to convert from weight/volume to molarity one incorporates molecular mass as a factor (40 CFR 796.1840-91, *see also* 40 CFR 796.1860-91).

Concentration time product: The product of the residual biocide concentration and contact time for a given level of microbial inactivation (e.g., 99 percent inactivation) (EPA-88/09a).

Concentration vs. time study: Results in a graph which plots the measured concentration of a given compound in a solution as a function of elapsed time. Usually, it provides a more reliable determination of equilibrium water solubility of hydrophobic compounds than can be obtained by single measurements of separate samples (40 CFR 796.1840-91).

Concentrator:
(1) A plant where ore is separated into values (concentrates) and rejects (tails). An appliance in such a plant, e.g., flotation cell, jig, electromagnet, shaking table. Also called mill.
(2) In chemical analysis, an apparatus used to increase the concentration of substances under study, e.g., distillation, stripping, etc.

Concentric winding: Transformer windings in which the low voltage winding is in the form of a cylinder next to the core, and the high voltage winding, also cylindrical, surrounds the low-voltage winding (EPA-83/03).

Conceptual design: Efforts to develop project scope that will satisfy program needs; ensure project feasibility and attainable performance levels of the project for congressional consideration and develop project criteria and design par meters for all engineering disciplines; and identify applicable codes and standards, quality assurance requirements, environmental studies, construction materials, space allowances, energy conservation features, health, safety, safeguards, and security requirements and any other features or requirements necessary to describe the project (DOE-91/04).

Conceptual site model: A model of a site developed at scoping using readily available information. Used to identify all potential or suspected sources of contamination, types and concentrations of contaminants detected at the site, potentially contaminated media, and potential exposure pathways, including receptors. This model is also known as conceptual evaluation model (EPA-91/12).

Concerted molecular elimination: A unimolecular reaction involving the elimination of a stable molecule from the initial reactant via a three-, four-, or six-center transition state (EPA-88/12).

Concrete: A mixture of cement, sand, and gravel (ETI-92).

Concrete curing compounds: Any coating applied to freshly poured concrete to retard the evaporation of water (40 CFR 52.741-91).

Concurrent: The construction of a control device is commenced or completed within the period beginning 6 months prior to the date construction of affected coating mix preparation equipment commences and ending 2 years after the date construction of affected coating mix preparation equipment is completed (40 CFR 60.711-91, *see also* 40 CFR 60.741-91).

Condensate:
(1) Volatile organic liquid separated from its associated gases, which condenses due to changes in the temperature or pressure and remains liquid at standard conditions (40 CFR 52.741-91, *see also* 40 CFR 60.111; 60.111a; 60.111b-91).
(2) Liquid or solid matter formed by condensation of a vapor phase. In sampling, the term is applied to the components of an atmosphere that have been isolated by simple cooling (EPA-83/06).
(3) In steam heating, water condensed from steam. In air conditioning, water extracted from air, as condensation on the cooling coil of a refrigeration machine.

Condensate polisher: An ion exchanger used to adsorb minute quantities of cations and anions present in condensate as a result of corrosion and erosion of metallic surfaces (EPA-82/11f).

Condensate stripper system: A column, and associated condensers, used to strip, with air or steam, TRS compounds from condensate streams from various processes within a kraft pulp mill (40 CFR 60.281-91).

Condensation:
(1) The process of converting a material in its gaseous phase to a liquid or solid state by the removal of heat or by the application pressure, or by both. Usually in air sampling, only cooling is used (EPA-83/06).
(2) The change of a pure substance from a vapor phase to its liquid phase.
(3) *See* latent heat for more related terms.

Condensation point temperature: The temperature at which a vapor changes to a liquid. It is the temperature where the pressure of the vapor equals atmospheric pressure. The opposite change in phases is the boiling point temperature (*see* temperature for more related terms) (Course 165.5).

Condensation sampling: A process consisting of the collection of one or several components of a gaseous mixture by simple cooling of the gas stream in a device that retains the condensate (*see* sampling for more related terms) (EPA-83/06).

Condenser:
(1) A heat transfer device that reduces a thermodynamic fluid from its vapor phase to its liquid phase (40 CFR 264.1031-91).
(2) A heat exchange device used for condensation.
(3) *See* evaporator for comparison
Types of condensers include:
- Barometric condenser
- Surface condenser
- Jet condenser
- Wet condenser (*see* synonym, jet condenser)

Condenser stack gases: The gaseous effluent evolved from the stack of processes utilizing heat to extract mercury metal from mercury ore (40 CFR 61.51-91).

Condenser water: The water used for cooling in a condenser (*see* water for more related terms) (EPA-75/02d).

Condensoid: The particles of a dispersion formed by condensation (EPA-83/06).

Conditional registration: Under special circumstances, the Federal Insecticide, Fungicide, and Rodenticide Act (FIFRA) permits registration of pesticide products that is conditional upon the submission of additional data. These special circumstances include a finding by the EPA Administrator that a new product or use of an existing pesticide will not significantly increase the risk of unreasonable adverse effects. A product containing a new (previously unregistered) active ingredient may be conditionally registered only if the Administrator finds that such conditional registration is in the public interest, that a reasonable time for conducting the additional studies has not elapsed, and the use of the pesticide for the period of conditional registration will not present an unreasonable risk (EPA-94/04).

Conditionally exempt generators (CEG): Persons or enterprises which produce less than 220 pounds of hazardous waste per month. Exempt from most regulation, they are required merely to determine whether their waste is hazardous, notify appropriate state or local agencies, and ship it by permitted facility for proper disposal (*see* an authorized transporter to a small quantity generator (EPA-94/04).

Conditioned: Heated and/or mechanically cooled (40 CFR 248.4-91).

Conditioning:
(1) The exposure of construction materials, test chambers, and testing apparatus to dilution water or to test solutions prior to the start of a test in order to minimize the sorption of the test substance onto the test facilities or the leaching of substances from the test facilities into the dilution water or test solution (40 CFR 797.1400-91, *see also* 40 CFR 797.1600-91).
(2) In leather industry, introducing controlled amounts of moisture to the dried leather giving it varying degrees of softness (EPA-82/11).
(3) In pulp industry, the practice of heating logs prior to cutting in order to improve the cutting properties of the wood and in some cases to facilitate debarking (EPA-74/04).
(4) In mining industry, a stage of froth-flotation processes in which the surfaces of the mineral species present in a pulp are treated with appropriate chemicals to influence their reaction when the pulp is aerated (EPA-82/05).

Conductance:
(1) A rapid method of estimating the dissolved-solids content of a water supply by determining the capacity of a water sample to carry an electrical current (EPA-94/04).
(2) *See also* synonym, electrical conductivity (EPA-83/06a).

Conduction: The transfer of heat from one part of a body to another part or to another body by short-range interaction of molecules and/or electrons (*see* heat transfer for more related terms) (Markes-67).

Conduction velocity: The speed at which the compound nerve action potential traverses a nerve (40 CFR 798.6850-91).

Conductive sensitizer: A coating applied to a plastic substrate to render it conductive for purposes of electrostatic application of subsequent prime, color, texture, or touch-up coats (40 CFR 60.721-91).

Conductivity:
(1) A measurement of electrolyte concentration by determining electrical conductance in a water sample (EPA-87/10a).
(2) A measure of the ability of water in conducting an electrical current. In practical terms, it is used for approximating the salinity or total dissolved solids content of water (EPA-74/01a).
(3) A measure of the ability of a solution to carry an electrical current: (EPA-94/04)t

Conductivity meter: An instrument which displays a quantitative indication of conductance (EPA-83/06a).

Conductivity surface: A surface that can transfer heat or electricity (EPA-83/06a).

Conductometer: A device for measuring thermal conductivity.

Conductometric titration: A titration technique in which the electrical conductivity of the reactant mixture is continuously measured as the reactant is added (*see* titration for more related terms).

Conductor: A wire, cable, or other body or medium suitable for carrying electric current (EPA-83/03).

Conduit: Tubing of flexible metals or other materials through which insulated electric wires are run (EPA-83/03).

Cone of influence: That area around the well within which increased injection zone pressures caused by injection into the hazardous waste injection well would be sufficient to drive fluids into a underground source of drinking water (USDW) (40 CFR 146.61-91).

Cone of depression:
(1) When a well is pumped, the water level in its vicinity declines to provide a gradient to drive water toward the discharged point. The gradient becomes steeper as the well is approached empty, because the flow is converging as a cone shape from all directions and the area through which the flow is occurring gets smaller. This results in a cone of depression around the well (EPA-87/03).
(2) A depression in the water table that develops around a pumped well (EPA-94/04).

Cone of influence: The depression, roughly conical in shape, produced in the water table by the pumping of water from a well (EPA-94/04).

Confidence interval: The range of values which is believed, with a preassigned probability called the confidence level, to include the particular value of some parameter being estimated (NATO-78/10).

Confidence limit: The limits within which, at some specified level of probability, the true value of a result lies (40 CFR 797.1350-91, *see also* 40 CFR 797.1440-91).

Confidential: *See* definition under security classification category.

Configuration:
(1) A subclassification of an engine-system combination on the basis of engine code, inertia weight class, transmission type and gear ratios, final drive ratio, and other parameters which may be designated by the Administrator (40 CFR 86.082.2-91, *see also* 40 CFR 86.602.84; 86.1002.84; 86.1102.87; 205.51; 205.151-91).
(2) The mechanical arrangement, calibration and condition of a test automobile, with particular respect to carburetion, ignition timing, and emission control systems (40 CFR 610.11-91).
(3) The functional or physical characteristics of hardware or software, as set forth in technical documentation and achieved in a project (DOE-91/04).

Configuration management: The systematic evaluation, coordination, approval (or disapproval), documentation, implementation, and audit of all approved changes in the configuration of a product after formal identification of its configuration (DOE-91/04).

Confined aquifer:
(1) An aquifer bounded above and below by impermeable beds or by beds of distinctly lower permeability than that of the aquifer itself; an aquifer containing confined ground water (40 CFR 260.10-91).
(2) An aquifer that carries water under pressure (DOI-70/04; DOE-91/04).
(3) An aquifer in which ground water is confined under pressure which is significantly greater than atmospheric pressure (EPA-94/04).
(4) *See* aquifer for more related terms

Confined flow: A fluid flowing through a bounded boundary such as a pipe (*see* flow for more related terms).

Confined groundwater: The groundwater that is bounded by impermeable soil or rock (*see* groundwater for more related terms).

Confining bed: A body of impermeable or distinctly less permeable material stratigraphically adjacent to one or more aquifers (40 CFR 146.3-91, *see also* 40 CFR 147.2902-91).

Confining zone: A geological formation, group of formations, or part of a formation that is capable of limiting fluid movement above an injection zone (40 CFR 146.3-91, *see also* 40 CFR 147.2902-91).

Confluence: The point at which one stream flows into another or where two streams converge and unite (DOI-70/04).

Confluent: A tributary, a stream that joins another (DOI-70/04).

Confluent growth:
(1) A continuous bacterial growth covering the entire filtration area of a membrane filter, or a portion thereof, in which bacterial colonies are not discrete (40 CFR 141.2-91).
(2) A continuous bacterial growth covering the entire filtration area of a membrane filter, or a portion thereof, in which bacterial colonies are not discrete (EPA-94/04).

Confounder: A condition or variable that may be a factor in producing the same response as the agent under study. The effects of such factors may be discerned through careful design and analysis (EPA-92/12).

Confounding factors: Variables other than the chemical exposure level which can affect the incidence or degree of a parameter being measured, e.g., smoking confounds studies of occupational exposure to other agents (Course 165.6).

Congener: Any one particular member of a class of chemical substances. A specific congener is denoted by unique chemical structure, for example 2,3,7,8- tetrachlorodibenzofuran (40 CFR 766.3-91).

Congenital: Resulting from or developing during one's prenatal environment. The condition is acquired during development in the womb and is not inherited from the parents (EPA-165.5).

Conical burner (or teepee burner): A hollow, cone-shape combustion chamber that has an exhaust vent at its point and a door at its base through which waste materials are charged and air is delivered to the burning solid waste inside the cone (*see* burner for more related terms) (SW-108ts).

Coning: A type of dispersion of a stack plume under nearly neutral atmospheric conditions, with average to high wind speeds. The horizontal and vertical dispersions of the plume are comparable (NATO-78/10).

Conjugate acid and base pair: By definition, an acid is a proton donor and a base is a proton acceptor. A conjugate acid and base pair refers to two compounds if they are related by loss or gain of a proton.

Connate water: The water imprisoned in sedimentary rocks at the time of their formation and held there; sometimes called fossil water (*see* water for more related terms) (DOI-70/04).

Connected load: The sum of the capacities or ratings of the electric power consuming apparatus connected to a supplying system, or any part of the system under consideration (*see* load for more related terms) (EPA-83).

Connected piping: All underground piping including valves, elbows, joints, flanges, and flexible connectors attached to a tank system through which regulated substances flow. For the purpose of determining how much piping is connected to any individual UST system, the piping that joins two UST systems should be allocated equally between them (40 CFR 280.12-91).

Connector: The flanged, screwed, welded, or other joined fittings used to connect two pipe lines or a pipe line and a piece of process equipment (40 CFR 60.481-91, *see also* 40 CFR 61.241; 264.1031-91).

Consent Agreement: Any written document, signed by the parties, containing stipulations or conclusions of fact or law and a proposed penalty or proposed revocation or suspension acceptable to both complainant and respondent (40 CFR 22.03-91).

Consent decree:
(1) In Superfunds, a legal document, approved by a judge, that formalizes an agreement reached between EPA and potentially responsible parties (PRPs) through which PRPs will conduct all or part of a cleanup action at a Superfund site; cease or correct actions or processes that are polluting the environment; or otherwise comply with regulations where the PRP's failure to comply caused EPA to initiate regulatory enforcement actions. The consent decree describes the actions PRP's will take and may be subject to a public comment period (EPA-89/12).
(2) In CWA, the Settlement Agreement entered into by EPA with the Natural Resources Defense Council and approved by the U.S. District Court for the District of Columbia on June 7, 1976 (8 ERC 2120, D.D.C. 1976), modified on March 9, 1979 and again by Order of the Court dated October 26, 1982, August 2, 1983, January 6, 1984, July 5, 1984, January 7, 1985, April 24, 1986 and January 8, 1987. One of the principal provisions of the Settlement Agreement was to direct EPA to consider an extended list of 65 classes of toxic pollutants in 21 industrial categories in the development of effluent limitations guidelines and new source performance standards. This list has since limited to 126 specific toxic pollutants and expanded to 34 industrial categories (EPA-87/10a).
(3) A legal document, approved by a judge, that formalizes an agreement reached between EPA and potentially responsible parties (PRPs) through which PRPs will conduct all or part of a cleanup action at a Superfund site; cease or correct actions or processes that are polluting the environment; or otherwise comply with EPA initiated regulatory enforcement actions to resolve the contamination at the Superfund site involved. The consent decree describes the actions PRPs will take and may be subject to a public comment period (EPA-94/04).

Conservation: Preserving and renewing, when possible, human and natural resources. The use, protection, and improvement of natural resources according to principles that will assure their highest economic or social benefits (EPA-94/04).

Conservation and management: The collection and application of biological information for the purposes of increasing and maintaining the number of animals within species and populations of marine mammals at their optimum sustainable population. Such terms include the entire scope of activities that constitute a modern scientific resource program, including. but not limited to, research, census, law enforcement, and habitat acquisition and improvement. Also included within these terms, when and where appropriate, is the periodic or total protection of species or populations as well as regulated taking (MMPA3-16USC1362-90).

Conservation of energy: The total energy of a system remains constant.

Conservation of mass: The total mass of a system remains constant.

Conservative pollutant: The pollutants that do not decay, are persistent and are not biodegradable. Examples include heavy metals and many pesticides (*see* pollutant for more related terms) (EPA-85/09).

Conserve, conserving, and conservation: To use and the use of all methods and procedures which are necessary to bring any endangered species or threatened species to the point at which the measures provided pursuant to this chapter are no longer necessary. Such methods and procedures include, but are not limited to, al] activities associated with scientific resources management such as research, census, law enforcement, habitat acquisition and maintenance, propagation, live trapping, and transplantation, and, in the extraordinary case where population pressures within a given ecosystem cannot be otherwise relieved, may include regulated taking (ESA3-16USC1531-90).

Consignee: The ultimate treatment, storage or disposal facility in a receiving country to which the hazardous waste will be sent (40 CFR 262.51-91).

Consistency: A percentage of pulp present in any combination of water and pulp (EPA-83).

Consistent removal: The average of the lowest 50 percent of the removal measured according to paragraph (b)(2) of this section. All sample data obtained for the measured pollutant during the time period prescribed in paragraph (b)(2) of this section must be reported and used in computing Consistent Removal. If a substance is measurable in the influent but not in the effluent, the effluent level may be assumed to be the limit of measurement, and those data may be used by POTW at its discretion and subject to approval by the Approval Authority. If the substance is not measurable in the influent, the date may not be used. Where the number of samples with concentrations equal to or above the limit of measurement is between 8 and 12, the average of the lowest 6 removals shall be used. If there are less than 8 samples with concentrations equal to or above the limit of measurement, the Approval Authority may approve alternate means for demonstrating consistent removal (40 CFR 403.7(b)(1)).

Consolidated assistance: An assistance agreement awarded under more than one EPA program authority or funded together with one or more other Federal agencies. Applicants for consolidated assistance submit only one application (40 CFR 30.200-91).

Consolidated PMN: *See* synonym, consolidated premanufacture notice (40 CFR 700.43-91).

Consolidated premanufacture notice or Consolidated PMN: Any PMN submitted to EPA that covers more than ore chemical substance (each being assigned a separate PMN number by EPA) as a result of a prenotice agreement with EPA. (*See* 4 FR 2134) (40 CFR 700.43-91).

Consortium: An association of manufacturers and/or Processors who have made an agreement to jointly sponsor testing (40 CFR 790.3-91).

Conspecific: Referring to the population of the same species.

Constant control, control technology, and continuous emission reduction technology: The systems which limit the quantity, rate, or concentration, excluding the use of dilution, and emissions of air pollutants on a continuous basis (40 CFR 57.103-91).

Constituent: Organic or inorganic material, dissolved gas, debris or organisms present in fresh, saline or wastewater (LBL-76/07-water).

Construction: The erection, building, acquisition, alteration, remodeling, modification, improvement, or extension of any facility; Provided, that it does not mean preparation or undertaking of: Plans to determine feasibility; engineering, architectural, legal, fiscal, or economic investigations or studies; surveys, designs, plans, writings, drawings, specifications or procedures (40 CFR 21.2-91).

Construction and demolition waste:
(1) The waste building materials, packaging, and rubble resulting from construction, remodeling, repair, and demolition operations on pavements, houses, commercial buildings, and other structures (40 CFR 243.101-91, *see also* 40 CFR 246.101-91).
(2) Waste building materials, dredging materials, tree stumps, and rubble resulting from construction, remodeling, repair, and demolition of homes, commercial buildings and other structures and pavements. May contain lead, asbestos, or other hazardous substances (EPA-94/04).
(3) *See* waste for more related terms.

Construction ban: If, under the Clean Air Act, EPA disapproves an area's planning requirements for correcting nonattainment, the Agency can ban the construction or modification of any major stationery source of the pollutant for which the area is in non-attainment (EPA-94/04).

Construction material: Any article, material, or supply brought to the construction site for incorporation in the building or work (40 CFR 35.936.1-91).

Construction quality assurance (CQA): In constructing a landfill liner, an action that provides a means of controlling and measuring the characteristics of the manufactured and installed product (*see* quality assurance for more related terms) (EPA-89/09, *see also* EPA-91/05).

Construction quality control (CQC): In constructing a landfill liner, a planned system of activities whose purpose is to provide a continuing evaluation of the quality control program, initiating corrective action where necessary (*see* quality control for more related terms) (EPA-89/09, *see also* EPA-91/05).

Construction work: The construction, rehabilitation, alteration, conversion, extension, demolition or repair of buildings, highways, or other changes or improvements to real property, including facilities providing utility services. The term also includes the supervision, inspection, and other on-site functions incidental to the actual construction (40 CFR 8.2-91).

Consumer: Any person who purchases a beverage in a beverage container for final use or consumption (40 CFR 244.101-91, *see also* 40 CFR 721.3; 721.1750-91).

Consumer product: A chemical substance that is directly, or as part of a mixture, sold or made available to consumers for their use in or around a permanent or temporary household or residence, in or around a school, or in recreation (CAA183.e, *see also* 40 CFR 302.3; 721.3; 721.1750-91).

Consumer waste: Materials which have been used and discarded by the buyer, or consumer, as opposed to house waste created in the manufacturing process (*see* waste for more related terms) (EPA-83).

Consumptive use: Water removed from available supplies without return to a water resource system (uses such as manufacturing, agriculture, and food preparation) (EPA-94/04).

Consumptive use of water: The water use resulting in a large proportion of loss to the atmosphere by evapotranspiration (as in irrigation), or by combination with a manufactured product (*see* water for more related terms) (DOI-70/04).

Consumptive water use: The difference in the volume of water withdrawn from body of water and the amount released back into the body of water. Water that is not returned to the body of water can enter the atmosphere through evaporation (DOE-91/04).

Consumptive use: With respect to heating oil, means consumed on the premises (40 CFR 280.12-91).

Contact adsorption: The removal of pollutants in fluids by direct contact with adsorbents (*see* adsorption for more related terms).

Contact aerator: An aerator in which compressed air is used to provide air for a variety of submerged bed aeration to assist biological digestion of waste (*see* aerator for more related terms).

Contact angle: The angle between a liquid surface and a solid surface.

Contact bed: *See* synonym, sand filter.

Contact condenser: A heat exchanger unit where the coolant and vapor streams (high temperature medium) are physically mixed (EPA-84/09).

Contact cooling and heating water: The process water that contacts the raw materials or plastic product for the purpose of heat transfer during the plastics molding and forming (*see* water for more related terms) (40 CFR 463.2-91).

Contact cooling water: Any wastewater which contacts the aluminum workpiece or the raw materials used in forming aluminum (*see* cooling water for more related terms) (40 CFR 467.02-91, *see also* 40 CFR 471.02-91).

Contact filter: *See* synonym, sand filter.

Contact insecticide: *See* synonym, contact pesticide.

Contact material: Any substance formulated to remove metals, sulfur, nitrogen, or any other contaminant from petroleum derivatives (40 CFR 60.101-91).

Contact period: *See* synonym, contact time.

Contact pesticide (or contact insecticide): A chemical that kills pests when it touches them, instead of by ingestion. Also, soil that contains the minute skeletons of certain algae that scratch and dehydrate waxy-coated insects (*see* pesticide for more related terms) (EPA-94/04).

Contact power theory: A theory predicting the collection efficiency of a scrubber system. The contact power theory assumes that the collection efficiency of a scrubber is solely a function of the total pressure loss for the unit (EPA-84/09).

Contact process wastewater: The process-generated wastewater which has come in direct or indirect contact with the reactants used in the process. It includes such streams as contact cooling water, filtrates, concentrates, wash waters, etc. (*see* wastewater for more related terms) (EPA-76/03).

Contact resistance: The opposition to the flow of current between the mounting bracket and the insulated terminal (40 CFR 85.2122(a)(5)(ii)(D)-91).

Contact stabilization: Aerobic digestion (EPA-87/10a).

Contact time (or contact period): The length of time for which a given disinfectant concentration must be maintained to assure a given level of inactivation (EPA-88/09a).

Contact water: The water or oil that comes into direct contact with the metal being cast, or with a mold that has been in direct contact with the metal. The metal contacted may be raw material, intermediate product, waste product, or finished product (*see* water for more related terms) (EPA-85/10a).

Contagion: The communication of a disease by direct or indirect contact.

Container:
(1) Any portable waste management unit in which a material is stored, transported, treated, or otherwise handled. Examples of containers are drums, barrels, tank trucks, barges, dumpsters, tank cars, dump trucks, and ships (40 CFR 61.341-91, *see also* 40 CFR 259.10; 260.10; 266.111-91).
(2) Any package, can, bottle, bag, barrel, drum, tank, or other containing-device (excluding sprag applicator tanks) used to enclose a pesticide or pesticide-related waste (40 CFR 165.1-91, *see also* 40 CFR 749.68-91).

Other container related terms include:
- Carrying container
- Disposable container
- Lift and carry container
- Roll-on/roll-off container
- Waste storage container

Container deposit legislation: Laws that require monetary deposits to be levied on beverage containers. The money is returned to the consumer when the containers are returned to the retailer. Also called Bottle Bills (EPA-89/11).

Container glass:
(1) A glass made of soda-lime recipe, clear or colored, which is pressed and/or blown into bottles, jars, ampoules, and other products listed in Standard Industrial Classification 3221 (SIC 3221) (40 CFR 60.291-91).
(2) Container glass includes beverage containers such as food, liquor, wine, beer, soft drinks, medical, toiletries and chemicals (EPA-83).
(3) *See* glass for more related terms.

Container train (or refuse train): Small trailers, hitched in series that are pulled by a motor vehicle. They are utilized to collect and transport solid waste (SW-108ts).

Containerboard: A general term designating:
(1) The component materials used in the fabrication of corrugated paperboard and solid fiber paperboard: linerboard, corrugating medium, chipboard; and
(2) Solid fiber or corrugated combined paperboard used in the manufacture of shipping containers and related products

(EPA-83).

Containment: Response actions that involve construction of a barrier to prevent the migration of contaminated wastes (EPA-89/12a). Other containment related terms include:
- Primary containment
- Secondary containment

Containment/control: A system to which toxic emissions from safety relief discharges are routed to be controlled. A caustic scrubber and/or flare can be containment/control devices. These systems may serve the dual function of destructing continuous process exhaust gas emissions (EPA-87/07a; 86/12).

Containment design basis: For a nuclear reactor, those bounding conditions for the design of the containment, including temperature, pressure, and leakage rate. Because the containment is provided as an additional barrier to mitigate the consequences of accidents involving the release of radio active materials, the containment design basis may include an additional specified margin above those conditions expected to result from the plant design-basis accidents to ensure the containment design can mitigate unlikely or unforseen events (DOE-91/04).

Containment integrity: The state in which barriers intended to prevent the escape of recombinant organisms are in sound unimpaired, or perfect conditions (EPA-88/09a).

Contaminant:
(1) Any physical, chemical, biological, or radiological substance or matter in water (SDWA1401, *see also* 40 CFR 141.2; 142.2; 143.2; 144.3; 146.3; 147.2902; 149.101; 230.3-91).
(2) Any physical, chemical, biological, or radiological substance or matter that has an adverse affect on air, water, or soil (EPA-94/04).

Contaminate: To introduce a substance that would cause:
(1) The concentration of that substance in the ground water to exceed the maximum contaminant level specified in Appendix I, or
(2) An increase in the concentration of that substance in the ground water where the existing concentration of that substance exceeds the maximum contaminant level specified in Appendix I (40 CFR 257.3.4-91).

Contamination:
(1) Presence of an alien organism on a body surface (EPA-83).
(2) Intrusion of undesirable elements to air, water or land (cf. pollution) (EPA-83/06a).
(3) Contamination is any fouling or sensor which causes its calibrated output to shift by a discernible amount (LBL-76/07-water).
(4) Introduction into water, air and soil of microorganisms, chemicals, toxic substances, wastes, or wastewater in a concentration that makes the medium unfit for its next intended use. Also applies to surfaces of objects and buildings, and various household and agricultural use products (EPA-94/04).

Contaminated nonprocess wastewater: Any water which, during manufacturing or processing, comes into incidental contact with any raw material, intermediate product, finished product, by-product or waste product by means of:
(1) Rainfall runoff;
(2) Accidental spills;
(3) Accidental leaks caused by the failure of process equipment, which are repaired within the shortest reasonable time not to exceed 24 hours after discovery; and
(4) Discharges from safety showers and related personal safety equipment: Provided, that all reasonable measures have been taken:
  (A) To prevent, reduce and control such contact to the maximum extent feasible; and
  (B) To mitigate the effects of such contact once it has occurred (40 CFR 415.91-91).
(5) *See* wastewater for more related terms.

Contaminated runoff: The runoff which comes into contact with any raw material, intermediate product, finished product, byproduct or waste product located on petroleum refinery property (40 CFR 419.11-91).

Contiguous zone: The entire zone established or to be established by the United States under article 24 of the Convention on the Territorial Sea and the Contiguous Zone (CWA311, *see also* CWA502; 40 CFR 110.1; 116.3; 117.1; 122.2; 300.5-91).

Contingency plan:
(1) A document setting out an organized, planned, and coordinated course of action to be followed in case of a fire, explosion, or release of hazardous waste or hazardous waste constituents which could threaten human health or the environment (40 CFR 260.10-91).
(2) A plan which describes the actions that facility personnel will take to minimize the hazards to human health or the environment from fires, explosions or accidental releases of hazardous materials (EPA-7/87a).
(3) A document developed to identify and catalog all the elements required to respond to an emergency, to define responsibilities and specific tasks, and to serve as a response guide (EPA-85/11).
(4) A document setting out an organized, planned, and coordinated course of action to be followed in case of a fire, explosion, or other accident that releases toxic chemicals, hazardous waste, or radioactive materials that threaten human health or the environment (*see* National Oil and Hazardous Substances Contingency Plan) (EPA-94/04).

Continuing calibration verification: *See* definition under calibration.

Continuity equation (or equation of continuity): According to the principle of mass conservation, mass can be neither created nor destroyed between sections A_1 and A_2. Thus the continuity equation is: $d_1 A_1 V_1 = d_2 A_2 V_2$, where: d = density; A = cross-sectional area; and V = velocity (AP-40, p27).

Continuous: Those descaling operations that remove surface scale from the sheet or wire products in continuous processes (40 CFR 420.81-91, *see also* 40 CFR 420.91; 420.111-91).

Continuous casting: The production of sheet, rod, or other long shapes by solidifying the metal while it is being poured through an open ended mold using little or no contact cooling water. Continuous casting of rod and sheet generates spent lubricants and rod casting also generates contact cooling water (40 CFR 467.02-91, *see also* 40 CFR 471.02-91).

Continuous discharge:
(1) A discharge which occurs without interruption throughout the operating hours of the facility, except for infrequent shutdowns for maintenance, process changes, or other similar activities (*see* discharge for more related terms) (40 CFR 122.2-91).
(2) A routine release to the environment that occurs without interruption, except for infrequent shutdowns for maintenance, process changes, etc (EPA-94/04).

Continuous disposal: A method of tailings management and disposal in which tailings are dewatered by mechanical methods immediately after generation. The dried tailings are then placed in trenches or other disposal areas and immediately covered to limit emissions consistent with applicable Federal standards (*see* disposal for more related terms) (*see* discharge for more related terms) (40 CFR 61.251-91).

Continuous emission: Any gas stream containing VOC that is generated essentially continuously when the process line or any piece of equipment in the process line is operating (*see* emission for more related terms) (40 CFR 60.561-91).

Continuous emission monitor (CEM): Types of CEM of include:
(1) Continuous inorganic emission monitor
  (A) Electroanalytical method
    (a) Electrocatalytic oxygen analyzer (EPA-81/09)
    (b) Polarographic analyzer (EPA-81/09; 84/03a)
  (B) Fourier transform infrared spectroscopy analyzer (FTS-IR) (EPA-84/03a)
  (C) Gas filter correlation analyzer (EPA-84/03a)
  (D) Infrared spectrophotometer (Course 165.5)
  (E) Luminescence analyzer
    (a) Chemiluminescence analyzer (EPA-84/03a)
    (b) Fluorescence analyzer (EPA-84/03a)
    (c) Photoluminescence (flame photometric) (EPA-84/03a)
  (F) Nondispersion infrared (NDIR) (EPA-81/09)
  (G) Nondispersion ultraviolet (NDUV) (EPA-81/09)
  (H) Paramagnetic oxygen analyzer (EPA-81/09)
    (a) Magnetic wind instrument (EPA-84/03a)
    (b) Magneto-dynamic instrument (EPA-84/03a)
    (c) Magnetopneumatic instrument (EPA-84/03a)
  (i) Second derivative absorption analyzer (EPA-84/03a)
(2) Continuous organic emission monitor:
  (A) Gas chromatography (GC)--compound separator (EPA-84/03a)
  (B) GC/compound detector or compound separator/compound detector
    (a) GC/AFID (gas chromatography/alkali flame ionization detector) (EPA-84/03a)
    (b) GC/DD (gas chromatography/dual detector) (EPA-84/03a)
    (c) GC/ECD (gas chromatography/electrolytic conductivity detector) (EPA-84/03a)
    (d) GC/ECD (gas chromatography/electron capture detector) (EPA-84/03a)
    (e) GC/FID (gas chromatography/flame ionization detector) (Course 165.5)
    (f) GC/HECD (gas chromatography/hall electrolytic conductivity detector) (EPA-84/03a)
    (g) GC/PID (gas chromatography/photoionization detector) (Course 165.5)
    (h) GC/TCD (gas chromatography/thermal conductivity detector) (EPA-84/03a)
    (i) GC/TSD (gas chromatography/thermionic specific detector) (EPA-84/03a)
  (C) Hybrid chromatograph monitor (EPA-84/03a)
    (a) GC/IR (gas chromatography/infrared absorption spectrometer) (EPA-84/03a)
    (b) GC/MS (gas chromatography/mass spectrometry) (EPA-84/03a)
  (D) Radiation emission-absorption instrumentation
    (a) Infrared absorption (EPA-84/03a)
    (b) Ultraviolet absorption (EPA-84/03a)

Continuous emission monitoring system (CEMS):
(1) A CEMS is comprised of all the equipment used to generate data and includes the sampling extraction and transport hardware, the analyzer(s), and the data recording/processing hardware and software (EPA-90/04)
(2) The equipment as required by section 412, used to sample, analyze, measure, and provide on a continuous basis a permanent record of emissions and flow (expressed in pounds per million British thermal units (lbs/mmBtu), pounds per hour (lbs/hr) or such other form as the Administrator may prescribe by regulations under section 412) (CAA402, *see also* 40 CFR 60.51a-91).
(3) The total equipment required for the determination of opacity. The system consists of the following major subsystems:
  (A) **Sample interface:** That portion of CEMS that protects the analyzer from the effects of the stack effluent and aids in keeping the optical surfaces clean.
  (B) **Analyzer:** That portion of the CEMS that senses the pollutant and generates an output that is a function of the opacity.
  (C) **Data recorder:** That portion of the CEMS that provides a permanent record of the analyzer output in terms of opacity. The data recorder may include automatic data-reduction capabilities (40 CFR 60-App/B-91).
(4) The total equipment required for the determination of a gas concentration or emission rate (40 CFR 60-App/F-91).

Continuous feed incinerator: An incinerator into which waste is almost charged continuously to maintain a steady rate of burning (*see* incinerator for more related terms) (SW-108ts).

Continuous filter: A type of filters used in wastewater treatment for filtration of suspended solids in a continuous basis (*see* filter for more related terms).

Continuous flow stirred tank: A tank with a well mixed flow (*see* tank for more related terms).

Continuous length processor: An automatic processing machine whereby long rolls of film or paper are fed into successive photoprocessing tanks via a series of appropriate crossover connections between racks. The starting end of the material to be processed is attached to a leader which guides the material through the machine (EPA-80/10).

Continuous monitoring: Monitoring without interruption throughout a given period. The sample to be analyzed passes the measurement section of the analyzer without interruption and which evaluates the detector response to the sample at least once each 15 seconds and records the average of these observations each and every minute (cf. rolling average) (*see* monitoring for more related terms) (EPA-90/04).

Continuous monitoring system: The total equipment, required under the emission monitoring sections in applicable subparts, used to sample and condition (if applicable), to analyze, and to provide a permanent record of emissions or process parameters (40 CFR 60.2-91).

Continuous operations: That the industrial user introduces regulated wastewaters to the POTW throughout the operating hours of the facility, except for infrequent shutdowns for maintenance, process changes, or other similar activities (40 CFR 471.02-91).

Continuous process:
(1) A process which has a constant flow of raw materials into the process and consequently a constant flow of product from the process (EPA-87/10a).
(2) With respect to polystyrene resin, a method of manufacture in which the styrene raw material is delivered on a continuous basis to the reactor in which the styrene is polymerized to polystyrene (40 CFR 52.741-91, *see also* 40 CFR 60.561-91).

Continuous recorder: A data recording device recording an instantaneous data value at least once every 15 minutes (40 CFR 60.611-91, *see also* 40 CFR 60.661; 264.1031-91).

Continuous release: A release that occurs without interruption or abatement or that is routine, anticipated, and intermittent and incidental to normal operations or treatment processes (40 CFR 302.3-91).

Continuous sample (or sampling):
(1) Withdrawal of a portion of a sample over a period of time with continuous analysis or with separation of the desired material continuously in a linear form. Examples include continuous withdrawal of the atmosphere accompanied by absorption of a component in a flowing stream of absorbent or by filtration on a moving strip or paper. Such a sample may be obtained with a considerable concentration of the contaminant, but it still indicates fluctuations in concentration that occur during the period of sampling (EPA-83/06).
(2) Sampling without interruptions throughout an operation or for a predetermined time (EPA-83/06).
(3) A flow of water from a particular place in a plant to the location where samples are collected for testing; may be used to obtain grab or composite samples (EPA-94/04).
(4) *See* sample for more related terms.

Continuous simulation model: A fate and transport model that uses time series input data to predict receiving water quality concentrations in the same chronological order as that of the input variables (EPA-91/03).

Continuous source: A source which emits pollution continuously over a time period much larger than the travel time to a point where concentration is considered. Usually it is assumed that during this time period, the emission is constant (*see* source for more related terms).

Continuous stimulation model: A fate and transport model that uses time-series input data to predict receiving water quality concentrations in the same chronological order as that of the input variables (EPA-85/09).

Continuous treatment: The treatment of waste streams operating without interruption (as opposed to batch treatment). Sometimes referred to as flow through treatment (*see* treatment for more related terms) (EPA-85/10a).

Continuous vapor processing system: A vapor processing system that treats total organic compounds vapors collected from gasoline tank trucks on a demand basis without intermediate accumulation in a vapor holder (40 CFR 60.501-91).

Contour line: An ideal line connecting all points at which the elevation is equal (DOI-70/04).

Contour plowing: Soil tilling method that follows the shape of the land to discourage erosion (EPA-94/04).

Contour strip farming: A kind of contour farming in which row crops are planted in strips, between alternating strips of close-growing, erosion resistant forage crops (EPA-94/04).

Contouring: Plowing and planting land across a slope, rather than up and down hill, in order to control erosion (DOI-70/04).

Contract: Any Government contract or any federally assisted construction contract (40 CFR 8.2-91, *see also* 40 CFR 15.4; 31.3; 35.6015-91).

Contract carrier by motor vehicle: Any person who engages in transportation by motor vehicle of passengers or property in interstate or foreign commerce for compensation (other than transportation referred to in Paragraph (b) of this section) under continuing contracts with one person or a limited number of persons either:
(1) For the furnishing of transportation services through the assignment of motor vehicles for a continuing period of time to the exclusive use of each person served or
(2) Or the furnishing of transportation services designed to meet the distinct need of each individual customer (40 CFR 202.10-91).
(3) *See* vehicle for more related terms.

Contract collection: The collection of solid waste carried out in accordance with a written agreement in which the rights and duties of the contractual parties are set forth (*see* waste collection for more related terms) (SW-108ts).

Contract disposal: The disposal of waste products through an outside party for a fee (EPA-87/10a).

Contract hauling: The collection of wastewater or sludge by a private disposal service, scavenger, or purveyor in tank trucks or by other means for transportation from the site (EPA-79/12b).

Contract laboratory: A laboratory under contract to EPA, which analyzes samples taken from wastes, soil, air, and water or carry out research projects (EPA-94/04).

Contract laboratory program (CLP): An analytical program developed for CERCLA waste site samples to fill the need for legally defensible analytical results supported by a high level of quality assurance and documentation (40 CFR 300-App/A-91).

Contract required detection limit (CRDL): The term equivalent to contract-required quantitation limit, but used primarily for inorganic substances (40 CFR 300-App/A-91).

Contract required quantitation limit (CRQL): The substance-specific level that a CLP laboratory must be able to routinely and reliably detect in specific sample matrices. It is not the lowest detectable level achievable, but rather the level that a CLP laboratory should reasonably quantify. The CRQL may or may not be equal to the quantitation limit of a given substance in a given sample. For HRS purposes, the term CRQL refers to both the contract-required quantitation limit and the contract-required detection limit (40 CFR 300-App/A-91).

Contract specification: The set of specifications prepared for an individual construction project, which contains design, performance, and material requirements for that project (40 CFR 249.04-91).

Contractor: Unless otherwise indicated, a prime contractor or subcontractor (cf. prime contractor) (40 CFR 8.2-91, *see also* 40 CFR 15.4; 30.200; 33.005; 35.936.1; 35.4010; 35.6015-91).

Contractor removal: Disposal of oils, spent solutions, or sludge by a scavenger service (EPA-83/06a).

Contrail: Long, narrow clouds caused when high-flying jet aircraft disturb the atmosphere (EPA-89/12).

Contribute: Resulting in measurably higher average 8-hour ambient CO concentrations over the NAAQS or an increased number of violations of the NAAQS in an area which currently experiences CO levels above the standard (40 CFR 51.138-91).

Control including the terms controlling, controlled by, and under common control with:
(1) The power to direct or cause the direction of the management and policies of a person or organization, whether by the ownership of stock, voting rights, by contract, or otherwise (40 CFR 66.3-91).
(2) Any remedial action intended to stabilize, inhibit future misuse of, or reduce emissions of effluents from residual radioactive materials (40 CFR 192.01-91, *see also* 40 CFR 192.31-91).
(3) *See* possession (40 CFR 704.3; 720.3; 723.50; 723.250-91).

(4) An exposure of test organisms to dilution water only or dilution water containing the test solvent or carrier (no toxic agent is intentionally or inadvertently added) (40 CFR 797.1600-91).
(5) Controlling air humidity, temperature, pressure, and velocity; inspecting, measuring, tempering, weighing, and so forth (AP-40, p790).

Control device: The equipment (such as an afterburner or adsorber) used to remove or prevent the emission of air pollutants from a contaminated exhaust stream (40 CFR 52.741-91, *see also* 40 CFR 60.261; 60.271; 60.271a; 60.381; 60.481; 60.561; 60.671; 60.691; 60.711; 60.741; 61.171; 61.181; 61.241; 61.301; 61.341; 264.1031-91).

Control device efficiency: The ratio of pollution prevented by a control device and the pollution introduced to the control device, expressed as a percentage (40 CFR 52.741-91).

Control device shutdown: The cessation of operation of a control device for any purpose (40 CFR 264.1031-91).

Control group: A group of subjects observed in the absence of agent exposure or, in the instance of a case/control study, in the absence of an adverse response (EPA-92/12).

Control mass: The quantity of the mass within a thermodynamic system (*see* thermodynamic system for more related terms).

Control method: A method of controlling and/or eliminating air pollutants. It may involve process modification but generally refers to a device specifically designed to reduce the substance from the exhaust (or stream) (EPA-84/09).

Control rod:
(1) The element of a nuclear reactor that absorb slow neutrons and are used to increase, decrease, or maintain the neutron density in the reactor (DOE-91/04).
(2) A rod (e.g., boron or cadmium which are capable of absorbing neutrons) used to control the chain reactions of a nuclear reactor. The rod can be moved into or out of a reactor core to control the rate of reaction.

Control strategy: A combination of measures designated to achieve the aggregate reduction of emissions necessary for attainment and maintenance of national standards including, but not limited to, measures such as:
(1) Emission limitations.
(2) Federal or State emission charges or taxes or other economic incentives or disincentives.
(3) Closing or relocation of residential, commercial, or industrial facilities.
(4) Changes in schedules or methods of operation of commercial or industrial facilities or transportation systems, including, but not limited to, short-term changes made in accordance with standby plans.
(5) Periodic inspection and testing of motor vehicle emission control systems, at such time as the Administrator determines that such programs are feasible and practicable.
(6) Emission control measures applicable to in-use motor vehicles, including, but not limited to, measures such as

mandatory maintenance, installation of emission control devices, and conversion to gaseous fuels.
(7) Any transportation control measure including those transportation measures listed in section 108(f) of the Clean Air Act as amended.
(8) Any variation of, or alternative to any measure delineated herein.
(9) Control or prohibition of a fuel or fuel additive used in motor vehicles, if such control or prohibition is necessary to achieve a national primary or secondary air quality standard and is approved by the Administrator under section 211(c)(4)(C) of the Act (40 CFR 51.100-n-91).

Control substance: Any chemical substance or mixture or any other material other than a test substance that is administered to the test system in the course of a study for the purpose of establishing a basis for comparison with the test substance for known chemical or biological measurements (40 CFR 160.3-91, *see also* 40 CFR 792.3-91).

Control system: A system designed to automatically maintain all controlled process variables within a prescribed range (EPA-87/07a).

Control technique guidelines (CTG): EPA documents issued to assist state and local pollution control authorities to achieve and maintain air quality standards for certain sources through reasonably available control technologies. CTGs can be quite specific, e.g., one was written to control organic emissions from solvent metal cleaning, known as degreasing (EPA-94/04).

Control volume: A fixed, well defined region of a space in which uniform conditions are assumed (EPA-88/09a).

Controlled reaction: A chemical reaction under temperature and pressure conditions maintained within safe limits to produce a desired product or process (EPA-94/04).

Controlled air: Controlling an air flow to attain desired rate of combustion which is used in starved air combustion processes.

Controlled air incinerator: An incinerator with two or more combustion areas in which the amounts and distribution of air are controlled. Partial combustion takes place in the first zone, and hydrocarbon gases are burned in a subsequent zone or zones (cf. starved air incinerator; *see* incinerator for more related terms) (EPA-83).

Controlled area:
(1) A surface location, to be identified by passive institutional controls, that encompasses no more than 100 square kilometers and extends horizontally no more than five kilometers in any direction from the outer boundary of the original location of the radioactive wastes in a disposal system; and
(2) The subsurface underlying such a surface location (40 CFR 191.12-91).

Controlled burning dump: Refuse trucks are unloaded onto a prepared dirt bank, usually about 12 feet high, with a slope of approximately 40 degrees. The dump operator uses a hook to distribute any piles of refuse evenly and then lights each load on the downwind edge. Now considered illegal (*see* land disposal for more related terms) (EPA-83).

Controlled substance: A controlled substance in schedules I through V of the Controlled Substances Act (21USC812), and as further defined by regulation at 21 CFR 1308.11 through 1308.15 (40 CFR 32.605-91, *see also* 40 CFR 82.3-91).

Controlled surface mine drainage: Any surface mine drainage that is pumped or siphoned from the active mining area (*see* mine drainage for more related terms) (40 CFR 434.11-91).

Controlled vehicle: The light duty vehicles sold nationally (except in California) in the 1968 model year and later and light duty vehicles sold in California in the 1966 model-year and later (*see* vehicle for more related terms) (40 CFR 51-App/N-91).

Controlling interest: The direct ownership of at least 50 percent of the voting stock of another entity (40 CFR 280.92-91).

Convection:
(1) In meteorology, it is the vertical transport and mixing of atmospheric properties (NATO-78/10).
(2) In heat transfer, the transfer of heat by the combined mechanisms of fluid mixing and conduction. It can be natural or forced convection (*see* heat transfer for more related terms) (Markes-67).

Other convection related terms include:
- Free convection
- Forced convection
- Local free convection
- Natural convection

Convention: The Convention on International Trade in Endangered Species of Wild Fauna and Flora, signed on March 3. 1973, and the appendices thereto (ESA3-16USC1531-90).

Conventional filtration: *See* complete treatment (EPA-94/04).

Conventional filtration treatment: A series of processes including coagulation, flocculation, sedimentation, and filtration resulting in substantial particulate removal (*see* filtration for more related terms) (40 CFR 141.2-91).

Conventional fuel: The fossil fuels including: coal, oil, or gas (*see* fuel for more related terms) (EPA-83).

Conventional gasoline: Any gasoline which does not meet specifications set by a certification under this subsection (*see* gasoline for more related terms) (CAA211.k-42USC7545-91).

Conventional mine: An open pit or underground excavation for the production of minerals (*see* mine for more related terms) (40 CFR 146.3-91).

Conventional pollutant: The pollutants designated pursuant to section 304.a.4 of the CWA:
(1) Biochemical oxygen demand (BOD);
(2) Total suspended solids (nonfilterable) (TSS);
(3) pH;

(4) Fecal coliform bacteria; and
(5) Oil and grease (40 CFR 401.16-92).
(6) *See* pollutant for more related terms.

Conventional pollutants: Statutorily listed pollutants understood well by scientists. These may be in the form of organic waste, sediment, acid, bacteria, viruses, nutrients, oil and grease, or heat (EPA-94/04).

Conventional systems: Systems that have been traditionally used to collect municipal wastewater in gravity sewers and convey it to a central primary or secondary treatment plant prior to discharge to surface waters (EPA-94/04).

Conventional technology:
(1) A proven technology which is not alternative or innovative.
(2) Wastewater treatment processes and techniques involving the treatment of wastewater at a centralized treatment plant by means of biological or physical/chemical unit processes followed by direct point source discharge to surface waters (40 CFR 35.2005-91).
(3) Wet flue gas desulfurization (FGD) technology, dry FGD technology, atmospheric fluidized bed combustion technology, and oil hydrodesulfurization technology (40 CFR 60.41b-91, *see also* 40 CFR 60.41c-91).

Conventional tilling: Tillage operations considered standard for a specific location and crop and that tend to bury the crop residues; usually considered as a base for determining the cost effectiveness of control practices (EPA-94/04).

Conventional wastewater treatment: Systems that have been traditionally used to collect municipal wastewater in gravity sewers and convey it to a central primary or secondary treatment plant prior to discharge to surface waters (EPA-89/12). The systems of wastewater treatment include screening, sedimentation, coagulation, rapid sand filtration, and disinfection with chlorine (*see* wastewater treatment for more related terms) (DOI-70/04).

Convergence: The gain per unit of time of a property in a volume due to the transport of this property by fluid motions through the surface into the volume. The loss of a property is called **divergence** (NATO-78/10).

Convergence distance: The distance from the lidar to the point of overlap of the lidar receiver's field-of-view and the laser beam (40 CFR 60-App/A(alt. method 1)-91).

Convergent evolution: The development of similar characteristics in different species, because the species live in the similar environmental conditions.

Conversion (or converting): Any process or operation applied to paper or paperboard after the normal papermaking operations. Printing, box making, waxing, envelope making, and the like, are all converting operations. A paper converter is an organization that manufacture products from paper (EPA-83).

Conversion coating: A coating produced by chemical or electrochemical treatment of a metallic surface that gives a superficial layer containing a compound of the metal, e.g., chromate coating on zinc and cadmium, oxide coatings on steel (EPA-83/06a).

Conversion efficiency: The measure of the catalytic converters ability to oxidize HC/CO to CO_2/H_2O under fully warmed-up conditions stated as a percentage calculated by the formula set forth in Section 5.2122(a)(15)(ii)(B) (40 CFR 85.2122(a)(15)(ii)(B)-91).

Conversion process: A process of chemical change in materials so that the identity of the original material is lost. Examples are the recovery of energy from organics by combustion and biological conversion of cellulose to sugars (cf. processing) (EPA-83).

Converter: Any vessel to which lead concentrate or bullion is charged and refined (40 CFR 60.181-91).

Converter arsenic charging rate: The hourly rate at which arsenic is charged to the copper converters in the copper converter department based on the arsenic content of the copper matte and of any lead matte that is charged to the copper converters (40 CFR 61.171-91).

Converting:
(1) The process of blowing air through molten metal to oxidize impurities (EPA-83/03a).
(2) Any operation in which paper is made into a product, not necessarily the final product to be made (EPA-87/10).

Convective movement: The bulk flow of radon-containing soil gas into the house as the result of pressure differences between the house and the soil. Distinguished from diffusive movement (EPA-88/08).

Conveyance loss: Water loss in pipes, channels, conduits, ditches by leakage or evaporation (EPA-94/04).

Conveyer: Types of conveyers include:
(1) Apron conveyer
(2) Belt conveyer
(3) Drag conveyer
(4) Flight conveyer
(5) Inclined plate conveyer
(6) Residue conveyer
(7) Screw conveyer

Conveyer weighing system: One of solid or sludge flow rate meters. This method includes belt weighers, weigh belts/augers, and loss-in-weight feeders. All conveyor weighing systems are fairly similar in operation, mainly differing because of placement locations of the weighing device. In general, the accuracy of these systems is around +/- 2% but tends to decrease as particles become larger and less uniform in size. Sludges can be monitored with the systems, provided that wet material does not drain off the conveyor belt. Screw augers, however, can often be used in such cases to replace the conventional conveyor belt (*see* flow rate meter for more related terms) (EPA-89/06).

Conveying system: A device for transporting materials from one piece of equipment or location to another location within a plant. Conveying systems include but are not limited to the following: Feeders, belt conveyors, bucket elevators and pneumatic systems

(40 CFR 60.671-91).

Conveyor belt transfer point: A point in the conveying operation where the metallic mineral or metallic mineral concentrate is transferred to or from a conveyor belt except where the metallic mineral is being transferred to a stockpile (40 CFR 60.381-91).

Conveyorized degreasing: The continuous process of cleaning and removing soils from surfaces utilizing either cold or vaporized solvents (40 CFR 52.741-91).

Conviction: A finding of guilt (including a plea of nolo contendere) or imposition of sentence, or both, by any judicial body charged with the responsibility to determine violations of the Federal or State criminal drug statutes (40 CFR 32.605-91, *see also* 40 CFR 32.105; 303.11-91).

Cooking: Heating of wood, water, and chemicals in a closed vessel under pressure to a temperature sufficient to separate the fibrous portion of wood by dissolving lignin and other non-fibrous constituents (EPA-87/10).

Cooking liquor: A mixture of chemicals and water used to dissolve lignin in wood chips (*see* liquor for more related terms) (EPA-87/10).

Cookstove: A wood-fired appliance that is designed primarily for cooking food and that has the following characteristics:
(1) An oven, with a volume of 0.028 cubic meters (1 cubic foot) or greater, and an oven rack;
(2) A device for measuring oven temperatures;
(3) A flame path that is routed around the oven;
(4) A shaker grate;
(5) An ash pan;
(6) An ash clean-out door below the oven; and
(7) The absence of a fan or heat channels to dissipate heat from the appliance (40 CFR 60.531-91).

Coolant:
(1) A liquid or gas used to reduce the heat generated by power production in nuclear reactors, electric generators, various industrial and mechanical processes, and automobile engines (EPA-89/12).
(2) A substance, either gas or water, circulated through nuclear reactor or processing plant to remove heat (DOE-91/04).

Cooldown period: The period of time at the end of an incinerator's operating cycle during which the incinerator is allowed to cool down. The cooldown period follows the burndown period (cf. burndown period) (EPA-89/03b).

Cooling air (or tempering air): The ambient air added for cooling by dilution (*see* air for more related terms) (EPA-83).

Cooling canal: A canal in which warm water enters at one end, is cooled by contact with air, and is discharged at the other end (EPA-82/11f).

Cooling electricity use: Amount of electricity used to meet the building cooling load (*see* building cooling load) (EPA-94/04).

Cooling pond: A water reservoir equipped with spray aeration equipment from which cooling water is drawn and to which it is returned (*see* pond for more related terms) (EPA-74/04).

Cooling spray: A water spray which is directed into flue gases to cool them and, in most cases, to remove some fly ash (SW-108ts).

Cooling tower: A structure that helps remove heat from water used as a coolant; e.g., in electric power generating plants (EPA-94/04). Other cooling tower related terms include:
- Biological cooling tower
- Comfort cooling tower
- Dry cooling tower
- Dry tower (*see* synonym, dry cooling tower)
- Natural draft cooling tower
- Wet cooling tower

Cooling tower approach: The difference in temperature between the water off the tower and the design air inlet wet bulb (w.b.); for 75°F water outlet and 65° F w.b. air inlet, the approach is 10° F (Gurney-66).

Cooling tower basin: A basin located at the bottom of a cooling tower for collecting the falling water (EPA-82/11f).

Cooling tower packing: A media providing large surface areas for the purpose of enhancing mass and heat transfer, usually between a gas vapor and a liquid (EPA-82/11f).

Cooling tower range: The difference in temperature between the water onto and water off the tower; for 90° F inlet and 75° F outlet water temperature, the range is 15° F (Gurney-66).

Cooling water: Water used for cooling in an industrial or manufacturing process, since its temperature after use is normally higher than that of the lake or stream into which it is discharged, it may constitute a source of thermal pollution (DOI-70/04). Other cooling water related terms include:
- Circulating water system
- Closed circulating water system
- Closed cooling water system
- Contact cooling water
- Non-contact cooling water
- Non-contact cooling water system
- Non-process wastewater cooling water
- Once through cooling water
- Process wastewater cooling water
- Recirculated cooling water
- *See* water for more related terms

Cooling water load: The energy in the form of heat dissipated by cooling water (DOI-70/04).

Cooling zone: In cement manufacturing, the last 10 to 20 feet of the cement kiln in which the clinker begins to cool (ETI-92).

Cooperative agreement: An assistance agreement whereby EPA transfers money, property, services or anything of value to a state for the accomplishment of CERCLA-authorized activities or tasks (EPA-94/04).

Coordinate bond (or coordinate valence): A chemical bond between two atoms in which only one atom provides a shared pair of electrons to form the chemical bond (*see* chemical bond for more related terms).

Coordinate valence: *See* synonym, coordinate bond.

Coordination chemistry: Chemistry that deals with the interaction of metal ions with other molecules by means of coordinate bonds.

Coordination site: Chemical configuration of a molecule where interaction between it and another molecule occurs (LBL-76/07-bio).

Copolymer:
(1) A polymer that has two different repeat units in its chain (40 CFR 60.561-91).
(2) The polymer obtained when two or more monomers are involved in the polymerization reaction (EPA-75/01a).

Copper (Cu):
(1) A common, reddish, chiefly univalent and bivalent metallic element that is ductile and malleable and one of the best conductors of heat and electricity (EPA-83/03). Copper is a transition metal with a.n. 29; a.w. 63.54; d. 8.96 g/cc; m.p. 1083° C and b.p. 2595° C and belongs to group IB of the periodic table.
(2) The total copper present in the process wastewater stream exiting the wastewater treatment system (40 CFR 415.361-91, *see also* 40 CFR 415.471; 415.651-91).
(3) Total copper and is determined by the method specified in 40 CFR 136.3 (40 CFR 420.02-91).
(4) Major copper compounds include copper sulfate ($CuSO_4$): A compound used to eliminate blooms of algae. However, too much quantity of $CuSO_4$ is harmful to water life (e.g., fish).

Copper casting: The remelting of copper or a copper alloy to form a cast intermediate or final product by pouring or forcing the molten metal into a mold, except for ingots, pigs, or other cast shapes related to nonferrous (primary and secondary) metals manufacturing (40 CFR 421). Also excluded are casting of beryllium alloys in which beryllium is present at 0.1 or greater percent by weight and precious metals alloys In which the precious metal is present at 30 or greater percent by weight. Except for grinding scrubber operations which are covered here, processing operations following the cooling of castings are covered under the electroplating and metal finishing point source categories (40 CFR 413 and 433) (40 CFR 464.02-91).

Copper converter: Any vessel to which copper matte is charged and oxidized to copper (40 CFR 60.161-91, *see also* 40 CFR 61.171-91).

Copper converter department: All copper converters at a primary copper smelter (40 CFR 61.171-91).

Copper flash: Quick preliminary deposition of copper for making surface acceptable for subsequent plating (EPA-83/06a).

Copper matte: An impure sulfide mixture formed by smelting the sulfide ores in copper (EPA-83/03a).

Copper mineral: Those of the oxidized zone of copper deposits (zone of oxidized enrichment) which include azurite, chrysocolla, copper metal, cuprite, and malachite. Those of the underlying zone (that of secondary sulfide enrichment) include bornite, chalcocite, chalcopyrite, covellite. The zone of primary sulfides (relatively low in grade) includes the unaltered minerals bornite and chalcopyrite (EPA-82/05).

Copper matte: Any molten solution of copper and iron sulfides produced by smelting copper sulfide ore concentrates or calcines (40 CFR 61.171-91).

Coppera: *See* synonym, ferrous sulfate.

Coppering: The treatment of water with a copper compound to prevent algal growths that cause noxious taste and odor (DOI-70/04).

Coprecipitation: Precipitation of more than one compounds simultaneously.

Coproduct: A chemical substance produced for a commercial purpose during the manufacture, processing, use or disposal of another chemical substance or mixture (40 CFR 704.3-91, *see also* 40 CFR 716.3-91).

Copy of study: The written presentation of the purpose and methodology of a study and its results (40 CFR 716.3-91).

Copy paper: *See* definition under xerographic paper (40 CFR 250.4-91).

Cord: An attenuated glassy inclusion possessing optical and other properties differing from those of the surrounding glass (cf. stria) (EPA-83).

Core: The uranium-containing heart of a nuclear reactor, where energy is released (EPA-94/04).

Core and mold washes: A mixture of various materials, primarily graphite, used to obtain a better finish on castings, including smoother surfaces, less scabbing and buckling, and less metal penetration. The filler material for washes should be refractory type composed of silica flour, zircon flour or chromite flour (EPA-85/10a).

Core binder: Bonding and holding materials used in the formation of sand cores. The three general types consist of those that harden at room temperature, those that require baking, and the natural clays. Binders that harden at room temperature include sodium silicate, Portland cement, and chemical cements such as oxychloride. Binders that require baking include the resins, resin oils, pitch, molasses, cereals, sulfide liquor, and proteins. Fire-clay and bentonite are the natural clay binders (EPA-85/10a).

Core binder accelerator: Used in conjunction with furan resins to cause hardening of the resin sand mixture at room temperature. The most commonly used accelerator is phosphoric acid (EPA-85/10a).

Core damage: Damage to a reactor core that would result in exceeding core safety limits (DOE-91/04).

Core drilling: Enlarging a hole with a chamfer-edged, multiple-flute drill (*see* drilling for more related terms) (EPA-83/06a).

Core furnace wall: The center courses of brick in a battery wall, which are not exposed directly to furnace heat (*see* furnace wall for more related terms) (SW-108ts).

Core grade: The quality ratings, based on standard evaluation criteria established by the Office of Pesticide Programs, given to toxicological studies after submission by registrants (EPA-92/12).

Core of the drawing with neat oils subcategory: Include drawing using neat oils, stationary casting, artificial aging, annealing, degreasing, sawing, and swaging (40 CFR 467.51-91, *see also* 40 CFR 467.61-91).

Core of the drawing with emulsions or soaps subcategory: Include drawing using emulsions or soaps, stationary casting, artificial aging, annealing, degreasing, sawing, and swaging (40 CFR 467.61-91).

Core of the extrusion subcategory: Include extrusion die cleaning, dummy block cooling, stationary casting, artificial aging, annealing, degreasing, and sawing (40 CFR 467.31-91).

Core of the forging subcategory: Include forging, artificial aging, annealing, degreasing, and sawing (40 CFR 467.41-91).

Core of the rolling with neat oils subcategory: Include rolling using neat oils, roll grinding, sawing, annealing, stationary casting, homogenizing artificial aging, degreasing, and stamping (40 CFR 467.11-91).

Core of the rolling with emulsions subcategory: Include rolling using emulsions, roll grinding, stationary casting, homogenizing, artificial aging, annealing, and sawing (40 CFR 467.21-91).

Core oil: Core oil is used in oil-sand cores as a parting agent to prevent the core material from sticking to the cast metal. Core oils are generally classified as mineral oils (refined petroleum oils) and are available as proprietary mixtures or can be ordered to specification. Typical core oils have specific gravities of 0.93 to 0.965 and contain a minimum of 70 percent non-volatiles at 177° C (350° F) (*see* oil for more related terms) (EPA-85/10a).

Core program cooperative agreement: An assistance agreement whereby EPA supports states or tribal governments with funds to help defray the cost of nonitem-specific administrative and training activities (EPA-94/04).

Coriolis acceleration: The apparent acceleration of a body in a relative coordinate system, which rotates with respect to an inertial coordinate system (NATO-78/10).

Coriolis flowmeter: *See* synonym, mass flowmeter.

Corium: The layer of hide between the epidermis and the flesh. Also called the dermis (EPA-82/11).

Corona: A phenomenon of gaseous (electric) discharge in which there occurs ionization of gas molecules by electron collision in regions of a high electric field. The electric field is aided by the use of an irregular shaped discharge electrode which possesses pointed protrusions. These projections develop a high intensity point field and initiate corona. When electrons enter this area of high field strength, they are accelerated to high velocities and energy levels. These electrons, upon impact with gas molecules in the area, cause orbital electrons to be released from the gas molecules. These released electrons are also accelerated and continue the ionization process. This process continues until the electric field decreases to the point where there is insufficient energy to perpetuate ionization (Cheremisinoff-77).

Corona discharge: A discharge of electricity appearing as a bluish-purple glow on the surface of and adjacent to a conductor when the voltage gradient exceeds a certain critical value; caused by ionization of the surrounding air by the high voltage (EPA-83/03).

Corpus striatum: A subcortical mass of gray and white substance in each cerebral hemisphere, containing the caudate nucleus and the lentiform nucleus (LBL-76/07-bio).

Correction factor (CF): *See* synonym, combustion correction factor.

Corrective action/order/policy: *See* definition under administrative order.

Correlation coefficient:
(1) In quality control, a number between -1 and 1 that indicates the degree of linear relationship between two sets of numbers (EPA-84/03).
(2) In statistics, with respect to linear correlation between two variables, it indicates the dependence between the two variables. Complete dependence is given by a correlation coefficient of one and no dependence is given by a correlation coefficient of zero. The value of the correlation coefficient follows from the covariance divided by the product of the standard deviations of both variables (NATO-78/10).

Correlation function: In turbulence, the average relation between two quantities as a function of time and/or space. It can be expressed both in a Eulerian frame of reference and a Lagrangian frame of reference (NATO-78/10).

Corresponding onshore area: With respect to any OCS source, the onshore attainment or nonattainment area that is closest to the source, unless the Administrator determines that another area with more stringent requirements with respect to the control and abatement of air pollution may reasonably be expected to be affected by such emissions. Such determination shall be based on the potential for air pollutants from the OCS source to reach the other onshore area and the potential of such air pollutants to affect the efforts of the other onshore area to attain or maintain any Federal or State ambient air quality standard or to comply with the provisions of part C of title I (CAA328-42USC7627-91).

Corrosion:
(1) Chemical or electrochemical oxidation of the surface of metal which can result in loss of material or accumulation of deposits (cf. abrasion or erosion) (EPA-83).

(2) The dissolution and wearing away of metal caused by a chemical reaction such as between water and the pipes, chemicals touching a metal surface, or contact between two metals (EPA-94/04).

Corrosion expert: A person who, by reason of his knowledge of the physical sciences and the principles of engineering and mathematics, acquired by a professional education and related practical experience, is qualified to engage in the practice of corrosion control on buried or submerged metal piping systems and metal tanks. Such a person must be certified as being qualified by the National Association of Corrosion Engineers (NACE) or be a registered professional engineer who has certification or licensing that includes education and experience in corrosion control on buried or submerged metal piping systems and metal tanks (40 CFR 260.10-91, *see also* 40 CFR 280.12-91).

Corrosion inhibitor: A chemical agent which slows down or prohibits a corrosion reaction (EPA-82/11f).

Corrosion resistant steels: The stainless steels with high nickel and chromium alloy content (EPA-83/06a).

Corrosive: A chemical agent that reacts with the surface of a material causing it to deteriorate or wear away (EPA-94/04).

Corrosive to the tank/valve: A lading meets the criteria for corrosivity specified in 40 CFR 173.240 of this subchapter, for the material of construction of the tank or valve; or the lading has been shown through experience to be corrosive to the tank or valve (40 CFR 180.403-91).

Corrosivity:
(1) One of the four U.S. EPA hazardous waste characteristics (ETI-92).
(2) Aqueous and has a pH of less than or equal to 2.0 or greater than or equal to 12.5; or
(3) A liquid and corrodes steel at a rate greater than 6.35 millimeters (.250 inches) per year under specified testing procedures.
(4) *See* hazardous waste characteristics for more related terms.

Corrugated box: A container for goods which is composed of an inner fluting of material (corrugating medium) and one or two outer liners of material (linerboard) (40 CFR 246.101-91, *see also* 40 CFR 250.4-91).

Corrugated container waste: Discarded corrugated boxes (*see* waste for more related terms) (40 CFR 246.101-91).

Corrugated container wastepaper: *See* definition under waste paper.

Corrugated paper:
(1) Paper or cardboard manufactured in a series of wrinkles or folds, or into alternating ridges and grooves (EPA-89/11).
(2) An asbestos-containing product made of corrugated paper, which is often cemented to a flat backing, may be laminated with foils or other materials, and has a corrugated surface. Major applications of asbestos corrugated paper include: thermal insulation for pipe coverings; block insulation; panel insulation in elevators; insulation in appliances; and insulation in low-pressure steam, hot water, and process lines (40 CFR 763.163-91).
(3) *See* paper for more related terms.

Corrugating medium: A paperboard used at corrugating plants to form the corrugated or fluted (wave-like) member in making such products as corrugated combined board and corrugated wrapping materials (EPA-87/10, *see also* EPA-83).

Corrugating medium furnish subdivision mills: The mills where only recycled corrugating medium is used in the production of paperboard (cf. noncorrugating medium furnish subdivision mills) (40 CFR 430.51-91).

Cortical atrophy: Wasting away of the outer layer(s), e.g., of the brain or kidney (LBL-76/07-bio).

Cosmetic: Shall have the meanings contained in the Federal Food, Drug, and Cosmetic Act, 21USC321 et seq., and the regulations issued under it (40 CFR 710.2-91, *see also* 40 CFR 720.3-91).

Cosmic radiation (or cosmic ray): High energy particles (e.g., electrons and atom nuclei) which impinge upon the earth from space with nearly the speed of light.

Cosmic ray: *See* synonym, cosmic radiation.

Cost: The amount or equivalent amount paid or charged for something. Other cost related terms include:
- Avoided cost
- Direct cost (*see* synonym, direct labor cost)
- Direct labor cost
- Fixed cost
- Fixed capital cost
- Obligation
- Opportunity cost
- Response cost

Cost analysis: The review and evaluation of each element of subagreement cost to determine reasonableness, allocability and allowability (40 CFR 33.005-91, *see also* 40 CFR 35.6015-91).

Cost and profit center:
(1) A cost center is a business whose objective is to accomplish its mission within cost or expenses parameters. A cost center realizes no income.
(2) A profit center is a business whose objective is to contribute income over and above its expenditures and allocated charges (EPA-79/12b).

Cost benefit analysis:
(1) A quantitative evaluation of the costs which would be incurred versus the overall benefits to society of a proposed action such as the establishment of an acceptance dose of a toxic chemical (Course 165.6).
(2) An evaluation of the costs and the benefits of a proposed action, serving as a tool in decision making (NATO-78/10).
(3) A quantitative evaluation of the costs which would be incurred versus the overall benefits to society of a proposed action such as the establishment of an acceptable dose of a

toxic chemical (EPA-94/04).

Cost benefit ratio: The ratio of the costs invested to the benefits gained.

Cost effective alternative: An alternative control or corrective method identified after analysis as being the best available in terms of reliability, performance, and cost. Although costs are one important consideration, regulatory and compliance analysis does not require EPA to choose the least expensive alternative. For example, when selecting or approving a method for cleaning up a Superfund site the Agency balances costs with the long-term effectiveness of the methods proposed and the potential danger posed by the site (EPA-94/04).

Cost of capital: Capital recovery costs minus the depreciation (EPA-83/06a).

Cost per ton per minute: A unit that is often used in cost comparisons between transfer and direct haul operations (SW-108ts).

Cost recovery: A legal process by which potentially responsible parties who contributed to contamination at a Superfund site can be required to reimburse the Trust Fund for money spent during any cleanup actions by the federal government (EPA-94/04).

Cost share: The portion of allowable project costs that a recipient contributes toward completing its project (i.e., non-Federal share, matching share) (40 CFR 30.200-91, *see also* 40 CFR 35.6015-91).

Cost sharing: A publicly financed program through which society, as a beneficiary of environmental protection, shares part of the cost of pollution control with those who must actually install the controls. In Superfund, the government may pay part of the cost of a cleanup action with those responsible for the pollution paying the major share (EPA-94/04).

Cost sharing or matching: The value of the third party in-kind contributions and the portion of the costs of federally assisted project or program not borne by the Federal Government (40 CFR 31.3-91).

Cost type contract: A contract or subcontract under a grant in which the contractor or subcontractor is paid on the basis of the costs it incurs, with or without a fee (40 CFR 31.3-91).

Cotton content: Paper that contains fibers from raw cotton, as compared to rag content (EPA-83).

Cotton fiber content paper: The paper that contains a minimum of 25 percent and up to 100 percent cellulose fibers derived from lint cotton, cotton linters, and cotton or linen cloth cuttings. It is also known as rag content paper or rag paper. It is used for stationery, currency, ledgers, wedding invitations, maps, and other specialty papers (*see* paper for more related terms) (40 CFR 250.4-91).

Cotton fiber furnish subdivision mills: Those mills where significant quantities of cotton fibers (equal to or greater than 4 percent of the total product) are used in the production of fine papers (cf. wood fiber furnish subdivision mills) (40 CFR 430.181-91).

Cotton kier liquor: Waste liquor from cotton processing (*see* liquor for more related terms) (LBL-76/07-water).

Cotton linters: Short fibers surrounding the cotton seed (EPA-87/10).

Couch roll: A roll primarily involved in dewatering and picking off, or couching, of the newly formed paper web from the wire on which it was formed and partially dewatered. The couch roll is involved in the transfer of the web to the wet press felt for further dewatering (EPA-87/10).

Coulomb's law (or law of electrostatic attraction): The force (F, attraction or repulsion) between two charged particles is proportional to the charge of the particles (Q_1 and Q_2) and inversely proportional to the square of the distance (d) between them. The relationship can be expressed as:
- $F = (Q_1 Q_2)/(4(pi)c_1 d^2)$, if the two charged particles are in a vacuum space. where pi is 3.1416; c_1 is the absolute permitivity or electric constant.
- $F = (Q_1 Q_2)/(4(pi)c_2 d^2)$, if the two charged particles are in a non-vacuum space. Where c_2 is absolute permitivity.

The relative permitivity $c_r = c_2/c_1$ is also called the **dielectric constant** (*see* law for more related terms).

Coulometer (or voltameter): A type of an electrolytic cell for measuring the electric charge by determining the deposition of metals on the cathode from a salt solution.

Count median size: A measurement of particle size for samples of particulate matter, consisting of that diameter of particle such that one half of the number of particles is larger and half is smaller (EPA-83/06).

Counter current:
(1) Opposite directions.
(2) Two fluids flow in opposite directions) (cf. counter flow).

Counter current cascade rinsing: A method of rinsing or washing using a segmented tank system in which water flows from one tank segment to the next, counter to the direction of movement of the material being washed (EPA-84/08).

Counter current extraction: A liquid-liquid extraction technique in which liquid 1 containing solutes to be moved and liquid 2 (e.g., solvent) to extract the solutes from liquid 1 flow in opposite directions (cf. liquid-liquid extraction).

Counter current washing:
(1) A method of washing used on the bleach plant or brownstock washers where fresh water is applied on the last stage showers, and the effluent from each stage is used on the washer showers of the preceding stage (EPA-87/10).
(2) A method of washing film or paper using a segmented tank system in which water is cascaded progressively from one tank segment to the next, counter to the movement of the film paper (EPA-80/10).

Counter flow: A process in which two media flow through a system in opposite directions (EPA-82/11f).

Covalent bond (or electron pair bond): A chemical bond between two atoms in which each atom contributes an electron to form the chemical bond (*see* chemical bond for more related terms).

Cover:
(1) With respect to coating mix preparation equipment, a device that lies over the equipment opening to prevent VOC from escaping and that meets the requirements found in 40 CFR 60.712(c)(1)-(5) (40 CFR 60.711-91, *see also* 40 CFR 60.741-91).
(2) A device or system which is placed on or over a waste placed in a waste management unit so that the entire waste surface area is enclosed and sealed to minimize air emissions. A cover may have openings necessary for operation, inspection, and maintenance of the waste management unit such as access hatches, sampling ports, and gauge wells provided that each opening is closed and sealed when not in use. Example of covers include a fixed roof installed on a tank, a lid installed on a container, and an air-supported enclosure installed over a waste management unit (40 CFR 61.341-91).
(3) Vegetation or other material providing protection as ground cover (EPA-89/12).

Other cover related terms include:
- Daily cover
- Final cover
- Intermediate cover

Cover coat: The final coat of porcelain enamel (EPA-82/11e)

Cover crop: A crop that provides temporary protection for delicate seedlings and/or provides a cover canopy for seasonal soil protection and improvement between normal crop production periods (EPA-94/04).

Cover material:
(1) Soil or other suitable material that is used to cover compacted solid wastes in a land disposal site (40 CFR 241.101-91).
(2) Soil used to cover compacted solid waste in a sanitary landfill (EPA-94/04).

Cover stock or cover paper: A heavyweight paper commonly used for covers, books, brochures, pamphlets, and the like (40 CFR 250.4-91).

Covered Federal action: Any of the following Federal actions:
(1) The awarding of any Federal contract;
(2) The making of any Federal grant;
(3) The making of any Federal loan;
(4) The entering into of any cooperative agreement; and
(5) The extension, continuation, renewal, amendment, or modification of any Federal contract, grant, loan, or cooperative agreement.

Covered Federal action does not include receiving from an agency a commitment providing for the United States to insure or guarantee a loan. Loan guarantees and loan insurance are addressed independently within this part (40 CFR 34.105-91).

Covered fleet: 10 or more motor vehicles which are owned or operated by a single person. In determining the number of vehicles owned or operated by a single person for purposes of this paragraph, all motor vehicles owned or operated, leased or otherwise controlled by such person, by any person who controls such person, by any person controlled by such person, and by any person under common control with such person shall be treated as owned by such person. The term "covered fleet" shall not include motor vehicles held for lease or rental to the general public, motor vehicles held for sale by motor vehicle dealers (including demonstration vehicles), motor vehicles used for motor vehicle manufacturer product evaluations or tests, law enforcement and other emergency vehicles, or nonroad vehicles (including farm and construction vehicles) (CAA241-42USC7581).

Covered fleet operator: A person who operates a fleet of at least ten covered fleet vehicles (as defined in section 241(6) of the Act) and that fleet is operated in a single covered area (even if the covered fleet vehicles are garaged outside of it). For purposes of this definition, the vehicle types described in the definition of covered fleet (section 241(5) of the Act) as exempt from the program will not be counted toward the ten-vehicle criterion (40 CFR 88.302.94-91).

Covered fleet vehicle: Only a motor vehicle which is:
(1) In a vehicle class for which standards are applicable under this part; and
(2) In a covered fleet which is centrally fueled (or capable of being centrally fueled). No vehicle which under normal operations is garaged at a personal residence at night shall be considered to be a vehicle which is capable of being centrally fueled within the meaning of this paragraph (CAA241-42USC7581-91).
(3) *See* vehicle for more related terms.

Covered furnace: An electric furnace with a water-cooled cover over the top to limit the introduction of air which would burn the gases from the reduction process. The furnace may have sleeves at the electrodes (fixed seals or sealed furnaces) with the charge introduced through ports in the furnace cover, or the charge may be introduced through annular spaces surrounding the electrodes (mixed seals or semi-closed furnace) (*see* furnace for more related terms) (EPA-75/02).

Coverage period: A time span which is 1 day less than 2 years, as identified in Subpart D, and is the time span which a person uses to determine his/her reporting year. Subject manufacturing or processing activities may or may not have occurred during the coverage period (40 CFR 704.203-91).

Covered area: The 9 ozone nonattainment areas having a 1980 population in excess of 250,000 and having the highest ozone design value during the period 1987 through 1989 shall be 'covered areas' for purposes of this subsection. Effective one year after the reclassification of any ozone nonattainment area as a Severe ozone nonattainment area under section 181(b), such Severe area shall also be a "covered area" for purposes of this subsection (CAA211.k-42USC7545-91).

Covered States: Those States that are participating in the demonstration medical waste tracking program and includes: Connecticut, New Jersey, New York, Rhode Island, and Puerto

Rico. Any other State is a Non-Covered State (40 CFR 259.10-91).

CPG 1-3: Federal Assistance Handbook: Emergency Management, Direction and Control Programs, prepared by FEMA. Provides States with guidance on administrative and programmatic requirements associated with FEMA funds (NRT-87/03).

CPG 1-35: Hazard Identification, Capability Assessment, and Multi-Year Development Plan for Local Governments, prepared by FEMA. As a planning tool, it can guide local jurisdictions through a logical sequence for identifying hazards, assessing capabilities, setting priorities, and scheduling activities to improve capability over time (NRT-87/03).

CPG 1-5: Objectives for Local Emergency Management, prepared by FEMA. Describes and explains functional objectives that represent a comprehensive and integrated emergency management program. Includes recommended activities for each objective (NRT-87/03).

CPG 1-8: Guide for Development of State and Local Emergency Operations Plans, prepared by FEMA (*see* EOP below) (NRT-87/03).

CPG 1-8A: Guide for the Review of State and Local Emergency Operations Plans, prepared by FEMA. Provides FEMA staff with a standard instrument for assessing EOPs that are developed to satisfy the eligibility requirement to receive Emergency Management Assistance funding (NRT-87/03).

Cr(+6): Hexavalent chromium (40 CFR 415.171-91).

Cr(T): Total chromium (40 CFR 415.171-91).

Cr,VI: Hexavalent chromium (40 CFR 413.02-91).

Cracking:
(1) The thermal decomposition of complex hydrocarbons into simpler compounds or elements.
(2) A process wherein heat and pressure are used for the rearrangement of the molecular structure of hydrocarbons or low-octane petroleum fractions (EPA-87/10a).

Crackled: Glassware, the surface of which has been intentionally cracked by water immersion and partially healed by reheating before final shaping (EPA-83).

Cradle-to-grave or manifest system: A procedure in which hazardous materials are identified and followed as they are produced, treated, transported, and disposed of by a series of permanent, linkable, descriptive documents (e.g., manifests). Commonly referred to as the cradle-to-grave system (EPA-94/04).

Crane:
(1) **Bridge crane:** A lifting unit that can maneuver horizontally in two directions (EPA-83).
(2) **Monorail crane:** A lifting unit, suspended from a single rail, that can only move in one horizontal direction (EPA-83).

Crane lift: Maximum safe vertical distance through which a crane bucket can move (EPA-83).

Crankcase: The housing for an automobile crankshaft.

Crankcase emission: Airborne substances emitted to the atmosphere from any portion of the engine crankcase ventilation or lubrication (*see* emission for more related terms) (40 CFR 86.082.2-91, *see also* 40 CFR 86.402.78-91).

Crawl space: In some types of houses which are constructed so that the floor is raised slightly above the ground, an area beneath the floor which allows access to utilities and other services. This is in contrast to slab-on-grade or basement construction houses (cf. basement) (EPA-89/12).

Creasability: An ability of paper to be folded without cracking along the folded edge (EPA-83).

Crease resistant: Fabrics that have been treated to make them resistant to wrinkling. One of the most common methods is to incorporate a resin (EPA-74/06b).

Creative checklist: A list of major hazards and nuisances designed so that when an individual item from the list is associated with a particular material or a significant part of a unit, an image of a specific hazard or nuisance is generated as a stimulus to the imagination of members of a multi-disciplinary team (EPA-87/07a).

Creative checklist hazard and operability study: A hazard and operability study which uses a creative checklist to stimulate a systematic, yet creative search for hazards (EPA-87/07a).

Credible accident: An accident that has a probability of occurrence greater than or equal to 10^{-6}/yr (DOE-91/04).

Creditor agency: The Federal agency to which the debt is owed (40 CFR 13.2-91).

Creep failure: Failure of a piece of metal as a result of creep. Creep is time dependent deformation as a result of stress. Metals will deform when exposed to stress. High levels of stress can result in rapid deformation and rapid failure. Lower levels of stress can result in slow deformation and protracted failure (EPA-8/87b).

Crenothrix: *See* synonym, iron bacteria.

Creslan fiber: A trademark of American Cyanamid for acrylic fibers (*see* fiber for more related terms) (EPA-74/06b).

Cresol (methylphenol, $CH_3C_6H_4OH$): A compound used as a germicide and antiseptic. There three isomers, differing in the relative positions of the methyl and hydroxyl.

Creosote:
(1) A complex mixture of organic materials obtained as a by-product from coking and petroleum refining operations that is used as a wood preservative (EPA-74/04).
(2) Distillate from tar (EPA-74/06a).

Creped: A light crinkled characteristic imparted to paper by a creping device to increase surface area, absorption, and elasticity. This is a customary procedure in tissue papers and fine decorative papers (EPA-87/10).

Criminal action: A prosecutorial action taken by the United States Government or a State towards any person(s) who has knowingly and willfully not complied with the law. Such an action can result in the imposition of fines or imprisonment (EPA-86/01).

Criminal drug statute: A Federal or non-Federal criminal statute involving the manufacture, distribution, dispensing, use, or possession of any controlled substance (40 CFR 32.605-91).

Criteria:
(1) Elements of State water quality standards, expressed as constituent concentrations, levels, or narrative statements, representing a quality of water that supports a particular use. When criteria are met, water quality will generally protect the designated use (40 CFR 131.3-91, *see also* 40 CFR 220.2; 256.06-91).
(2) Descriptive factors taken into account by EPA in setting standards for various pollutants. These factors are used to determine limits on allowable concentration levels, and to limit the number of violations per year. When issued by EPA, the criteria provide guidance to the states on how to establish their standards (EPA-94/04).

Criteria continuous concentration (CCC): The EPA national water quality criteria recommendation for the highest in-stream concentration of a toxicant or an effluent to which organisms can be exposed indefinitely without causing unacceptable effect (EPA-85/09; 91/03).

Criteria maximum concentration (CMC): The EPA national water quality criteria recommendation for the highest in-stream concentration of a toxicant or an effluent to which organisms can be exposed for a brief period of time without causing unacceptable effect (EPA-85/09; 91/03).

Criteria pollutant: The 1970 amendments to the Clean Air Act required EPA to set National Ambient Air Quality Standards for certain pollutants known to be hazardous to human health. EPA has identified and set standards to protect human health and welfare for six pollutants: ozone, carbon monoxide, total suspended particulates, sulfur dioxide, lead, and nitrogen oxide. The term, criteria pollutants derives from the requirement that EPA must describe the characteristics and potential health and welfare effects of these pollutants. It is on the basis of these criteria that standards are set or revised (*see* pollutant for more related terms) (EPA-94/04).

Critical aquatic organism: The aquatic organisms that are commercially or recreationally valuable, rare or endangered, of specific scientific interest, or necessary to the well-being of some significant species or to the balance of the ecological system (*see* organism for more related terms) (EPA-76/04).

Critical aquifer protection area: Either of the following:
(1) All or part of an area located within an area for which an application or designation as a sole or principal source aquifer pursuant to section 1424(e), has been submitted and approved by the Administrator not later than 24 months after the enactment of the Safe Drinking Water Act Amendments of 1986 and which satisfies the criteria established by the Administrator under subsection (d).
(2) All or part of an area which is within an aquifer designated as a sole source aquifer as of the enactment of the Safe Drinking Water Act Amendments of 1986 and for which an areawide ground water quality protection plan has been approved under section 208 of the Clean Water Act prior to such enactment (SDWA1427-42USC300h.6-91).

Critical diameter (or critical size): Particle diameters equal to or greater than this size are collected with 100% efficiency (EPA-83/06).

Critical effect: The first adverse effect, or its known precursor, that occurs as the dose rate increases (EPA-90/08; 92/12).

Critical emission related component: Those components which are designed primarily for emission control, or whose failure may result in a significant increase in emissions accompanied by no significant impairment (or perhaps even an improvement) in performance, driveability, and/or fuel economy as determined by the Administrator (40 CFR 86.088.2-91).

Critical emission related maintenance: That maintenance to be performed on critical emission related components (40 CFR 86.088.2-91).

Critical habitat:
(A) For a threatened or endangered species means:
  (i) the specific areas within the geographical area occupied by the species at the time it is listed in accordance with the provisions of section 1533 of this title, on which are found those physical or biological features (i) essential to the conservation of the species and (II) which may require special management considerations or protection and
  (ii) specific areas outside the geographical area occupied by the species at the time it listed in accordance with the provisions of section 1533 of this title, upon a determination by the Secretary that such areas are essential for the conservation of the species.
(B) Critical habitat may be established for those species now listed as threatened or endangered species for which no critical habitat has heretofore been established as set forth in subparagraph (A) of this paragraph.
(C) Except in those circumstances determined by the Secretary, critical habitat shall not include the entire geographical area which can be occupied by the threatened or endangered species (ESA3-16USC1531).

Critical humidity: A humidity condition above which a water soluble salt absorbs water vapor from the atmosphere and becomes damp and below which the salt releases water vapor and stays dry (*see* humidity for more related terms).

Critical life stage: The period of time in an organism's life-span in which it is the most susceptible to adverse effect caused by exposure to toxicants, usually during early development (egg, embryo, larvae). Chronic toxicity tests are often run on critical life stages to replace long duration, life cycle tests since the toxic effect occurs during the critical life stage (EPA-85/09; 91/03).

Critical mass:
(1) The smallest mass of fissionable material that will support a self-sustaining nuclear chain reaction under specified conditions (DOE-91/04).
(2) The minimum material needed to sustain the chain reaction of a nuclear reactor.

Critical measurements: Those measurement, data gathering, or data generation activities that directly impact the technical objectives of a project (EPA-85/08).

Critical organ: The most exposed human organ or tissue exclusive of the integumentary system (skin) and the cornea (*see* organ for more related terms) (40 CFR 191.02-91).

Critical point (or critical state): The conditions where the states (temperature, pressure and specific volume) of saturated-liquid and saturated-vapor are identical.

Critical point (or critical state): The conditions where the states (temperature, pressure and specific volume) of saturated-liquid and saturated-vapor are identical. At the critical point, the difference of the following properties between the saturated liquid and saturated vapor is zero (cf. quality) (EPA-82/11f):
- Enthalpy: $h_{fg} = 0$
- Entropy: $s_{fg} = 0$
- Internal energy: $u_{fg} = 0$
- Specific volume: $v_{fg} = 0$

Critical pollutant: The pollutant with the highest subindex during the reporting period (*see* pollutant for more related terms) (40 CFR 58-App/G-91).

Critical pressure: The pressure of the critical point where the states (temperature, pressure and specific volume) of saturated-liquid and saturated-vapor are identical (*see* pressure for more related terms).

Critical streamflow: The amount of water available for the generation of water power during the most adverse streamflow period (DOI-70/04).

Critical temperature: The temperature of the critical point where the states (temperature, pressure and specific volume) of saturated-liquid and saturated-vapor are identical (*see* temperature for more related terms).

Critical velocity: The velocity above which fluid is turbulent.

Critical volume: The volume at the critical point.

Critical size: *See* synonym, critical diameter.

Critical state: *See* synonym, critical point.

Criticality: The condition in which a nuclear reactor sustains a chain reaction.

Crop for direct human consumption: The crops that are consumed by humans without processing to minimize pathogens prior to distribution to the consumer (40 CFR 257.3.6-91).

Crop consumptive use: The amount of water transpired during plant growth plus what evaporated from the soil surface and foliage in the crop area (EPA-94/04).

Crop rotation: Planting a succession of different crops on the same land area as opposed to planting the same crop time after time (EPA-94/04).

Cross check editing: The determination of a value's validity based on the occurrence of a value of some other variables (cf. magnitude of entry editing) (EPA-79/12c).

Cross connection: Any actual ore potential connection between a drinking water system and an unapproved water supply or other source of contamination (EPA-94/04).

Cross link: A comparatively short connecting unit (such as a chemical bond or a chemically bonded atom or group) between neighboring polymer chains (EPA-75/12a).

Cross flow: The flow of scrubbing liquid normal (perpendicular) to the gas stream (*see* flow for more related terms) (EPA-89/03b).

Cross recovery furnace: A furnace used to recover chemicals consisting primarily of sodium and sulfur compounds by burning black liquor which on a quarterly basis contains more than 7 weight percent of the total pulp solids from the neutral sulfite semichemical process and has a green liquor sulfidity of more than 28 percent (*see* furnace for more related terms) (40 CFR 60.281-91).

Cross wind: A direction perpendicular to the wind vector. Usually applied to the horizontal direction (*see* wind for more related terms) (NATO-78/10).

Crossband:
(1) Used as a noun, the layers of veneer whose grain direction is at right angles to that of the face piles, applied particularly to five-ply plywood and lumber core panels, and more generally to all layers between the core and the faces (EPA-74/04).
(2) Used as a verb, to place the grain of the layers of veneer at right angles in order to minimize swelling and shrinking (EPA-74/04).

Crossbedding: The disposition of rocks in layers with minor strata lying oblique to the main strata (DOI-70/04).

Crown flint glass: An optical crown glass bordering on optical flint glass because of the addition of a substantial content of lead oxide and with somewhat higher dispersion than optical crown glass (EPA-83).

Crucible: A highly refractory vessel used to melt metals (EPA-85/10a).

Crucible furnace: A furnace fired with coal, coke, oil or gas, in which metal contained in crucibles is melted (EPA-83). Types of crucible furnaces include:
(1) Tilting furnace;
(2) Pit crucible furance;
(3) Stationary crucible furance; and
(4) Pot furnace (AP-40, p238).

(5) *See* furnace for more related terms.

Crude intermediate plastic material: The plastic material formulated in an on-site polymerization process (40 CFR 463.2-91).

Crude oil:
(1) A naturally occurring mixture which consists of hydrocarbons and sulfur, nitrogen, or oxygen derivatives of hydrocarbons and which is a liquid at standard conditions (40 CFR 52.741-91).
(2) Unrefined petroleum.
(3) *See* oil for more related terms.

Crude oil gathering: The transportation of crude oil or condensate after custody transfer between a production facility and a reception point (40 CFR 52.741-91).

Crude tall oil: A dark brown mixture of fatty acids, rosins, and neutral materials liberated by the acidification of soap skimmings (*see* oil for more related terms) (EPA-79/12).

Crusher: A machine used to crush any metallic mineral and includes feeders or conveyors located immediately below the crushing surfaces. Crushers include, but are not limited to, the following types: jaw, gyratory, cone, and hammermill (40 CFR 60.381-91, *see also* 40 CFR 60.671-91). Other crusher related terms include:
- Jaw crusher
- Roll crusher

Crustacean (pl. crustacea): The small animals ranging in size form 0.2 to 0.3 millimeters long which move very rapidly through the water in search of food. They have recognizable head and posterior sections and are a principal source of food for small fish (EPA-83/09).

Cryogenic focusing: The process whereby gas phase solutes, which comprise a small fraction of a flowing gas stream, are deposited on a low (sub-ambient) temperature surface. In thermal decomposition experiments, thermal reaction products are produced in a gas stream (air or nitrogen) and are then deposited on the head of a capillary GC column held at cryogenic temperatures (EPA-88/12).

Crystal: A clear, transparent natural mineral or man-made, glass-like material (cf. electronic crystal).

Crystal glass: A colorless glass, highly transparent, frequently used for art or tableware (*see* glass for more related terms) (EPA-83).

Crystal structure: The geometrical arrangement of the molecules that occupy the space lattice of the crystalline portion of a polymer (EPA-91/05).

Crystalline: A substance with regular arrangement of atoms in a space lattice, an ordered structure, such as a crystal--as opposed to amorphous (cf. amorphous) (EPA-83/06a; 75/01a).

Crystallization (or suspension freezing): A freezing (dewatering) process used to separate substances because of their different freezing temperature. When an aqueous solution containing dissolved salts or suspended solids is frozen, fresh water ice crystals form and the salts are concentrated in the remaining brine solution. The ice crystals can be separated from the brine by mechanical means, washed, and melted to yield fresh water. The remaining brine may contain potentially hazardous substances which can be treated further or disposed of.

The freezing process resemble the distillation process in that heat transfer is a very important element of the process, and a change of phase is involved. With a suitable means of separating the ice crystals from the mother liquor, freezing can be made into a process which yields a concentrated waste product stream and high-quality water (99+% pure). Theoretically, at least, these processes can be applied to non-aqueous streams as well.

CT or CTcalc: The product of residual disinfectant concentration (C) in mg/L determined before or at the first customer, and the corresponding disinfectant contact time (T) in minutes, i.e., C x T (for complete definition, *see* 40 CFR 141.2-91).

CT or CTcalc: The product of "residual disinfectant concentration." (C) in mg/l determined before or at the first customer, and the corresponding "disinfectant contact time" (T) in minutes, i.e., "C" x "T." If a public water system applies disinfectants at more than one point prior to the first customer, it must determine the CT of each disinfectant sequence before or at the first customer to determine the total percent inactivation or total inactivation ratio. In determining the total inactivation ratio, the public water system must determine the residual disinfectant concentration of each disinfection sequence and corresponding contact time before any subsequent disinfection application point(s). "CT99.9" is the CT value required for 99.9 percent (3-log) inactivation of Giardia lamblia cysts. CT99.9 for a variety of disinfectants and conditions appear in Tables 1.1-1.6, 2.1, and 3.1 of 40 CFR 141.74(b)(3).
- (CTcalc)/(CT99.9) is the inactivation ratio.
- The sum of the inactivation ratios, or total inactivation ratio shown as Σ = (CTcalc)/(CT99.9) is calculated by adding together the inactivation ratio for each disinfection sequence.
- A total inactivation ratio equal to or greater than 1.0 is assumed to provide a 3-log inactivation of Giardia lamblia cysts (40 CFR 141.2-91).

Cubic feet per minute (CFM): A measure of the volume of a substance flowing through air within a fixed period of time. With regard to indoor air, refers to the amount of air, in cubic feet, that is exchanged with indoor air in a minute's time, i.e., the air exchange rate (EPA-94/04).

Cubic feet or cubic meters of production in Subpart A: The cubic feet or cubic meters of logs from which bark is removed (40 CFR 429.11-91).

Curie (Ci):
(1) A rate of radioactive decay. One curie is the quantity of any radioactive nuclide which undergoes 3.7×10^{10} disintegrations per second. One curie is equal to 3.7×10^{10} becquerel (*see* radiation unit for more related terms) (40 CFR 302.4-App/B-91).
(2) The amount of radioactive material which disintegrates at the rate of 37 billion atoms per second, i.e., 1 Ci = 3.7×10^{10} (37 billion) transformations per second. This is a useful quantity, e.g., when comparing the relative rates of equal masses of

different nuclides. Thus, one gram of Cesium-134 (half-life = 2.06 years) will decay fifteen times as fast as one gram of Cesium-137 (half-life = 30.2 years). Cesium-134's activity is said to be 15 times as great, for equal weights (*see* radiation unit for more related terms) (10 CFR 30.4; 192.01-91, *see also* 40 CFR 300-App/A-91; LBL-76/07-rad).
Other curie related terms include:
- Millicurie (mCi)
- Microcurie (μCi)
- Picocurie (pCi)
- Picocurie per liter (pCi/L)

Curium (Cm): A radioactive metal with a.n. 96; a.w. 247; m.p. 1340° C. The element belongs to group IIIB of the periodic table.

Cullet: Crushed glass (EPA-94/04).

Cullet (or glass cullet):
(1) Any broken glass generated in the manufacturing process (40 CFR 426.21-91, *see also* 40 CFR 426.31; 426.41; 426.101; 61.161-91).
(2) Waste or broken glass, usually suitable as an addition to raw glass melt materials. Types of cullets include:
  (A) Foreign cullet--cullet from an outside source.
  (B) domestic cullet (factory cullet)--cullet from within the plant.
  (C) The portion of a glass article which will later be cut off and discarded or remelted (EPA-83).
(3) cf. raw cullet.

Cullet water: The water which is exclusively and directly applied to molten glass in order to solidify the glass (*see* water for more related terms) (40 CFR 426.11-91).

Cultural eutrophication: Increasing rate at which water bodies "die" by pollution from human activities (EPA-94/04).

Cultural resource: Areas or objects that are of cultural significance to Native Americans and other defined ethnic groups (DOE-91/04).

Culture: A mass of microorganisms growing in a medium (EPA-83/09).

Cultured product: Fermentation-type dairy products manufactured by inoculating different forms of milk with a bacterial culture. This designation includes yogurt, cultured buttermilk, sour cream, and cultured cream cheese, among other products (EPA-74/05).

Cultures and stocks: Infectious agents and associated biologicals including: cultures from medical and pathological laboratories; cultures and stocks of infectious agents from research and industrial laboratories; waste from the production of biologicals; discarded live and attenuated vaccines; and culture dishes and devices used to transfer, inoculate, and mix cultures (*see* regulated medical waste (EPA-94/04).

Cumulative distribution: A type of (particle) size distribution, generated by plotting particle diameter versus cumulative percent. For log-normal distributions, the particle diameter is plotted on logarithmic coordinates and percent less than stated size (% LTSS) or percent greater than stated size (% GTSS) is plotted on probability (normal) coordinates. Thus, if 10% of the particles have a size equal to or less than 3.2 micrometers (microns, u), the point (3.2 u, 10%) would fall on a cumulative distribution curve (cf. log-normal distribution) (EPA-84/09).

Cumulative exposure:
(1) The summation of exposure of an organism to a chemical over a period of time (Course 165.6).
(2) The summation of exposures of an organism to a chemical over a period of time (EPA-94/04).

Cumulative impact: The impact on the environment which results from the incremental impact of the action when added to other past, present, and reasonably foreseeable future actions regardless of what agency (Federal or non-Federal) or person undertakes such other actions. Cumulative impacts can result from individually minor but collectively significant actions taking place over a period of time (40 CFR 1508.5-91; DOE-91/04).

Cumulative toxicity: The adverse effects of repeated doses occurring as a result of prolonged action on, or increased concentration of, the administered substance or its metabolites in susceptible tissue (*see* toxicity for more related terms) (40 CFR 795.260-91, *see also* 40 CFR 798.2250; 798.2450; 798.2650; 798.2675-91).

Cumulative working level months (CWLM): The sum of lifetime exposure to radon working levels expressed in total working level months (EPA-94/04).

Cunningham correction factor: A correction factor applied to Stokes law for particle less than 2.0 microns (or micrometers) in diameter. Particles below this size are affected by collisions by air molecules which alter the settling velocity of the particle. (*See also* Stokes law) (EPA-84/09).

Cupelled: Refined by means of a small shallow porous bone cup that is used in assaying precious metals (EPA-83/03a).

Cupola or cupola furnace:
(1) A vertical shaft furnace consisting of a cylindrical steel shell lined with refractories and equipped with air inlets at the base and an opening near the top for charging fuel and melting stock (EPA-85/10a).
(2) A melting furnace for producing gray iron. It is also used to melt or reduce copper, brasses, bronzes, and lead (AP-40, p234).
(3) *See* furnace for more related terms.

Cuprite: A secondary copper mineral, Cu_2O (EPA-82/05).

Curb collection (or curbside collection): The collection of solid waste placed adjacent to a street (*see* waste collection for more related terms) (40 CFR 243.101-91).

Curbside collection: *See* synonym, curb collection.

Curb idle: For manual transmission code heavy-duty engines, the manufacturers recommended engine speed with the transmission in neutral or with the clutch disengaged. For automatic transmission code heavy-duty engines, curb-idle means the manufacturers recommended engine speed with the automatic transmission in gear and the output shaft stalled (40 CFR 86.082.2-91, see also 40 CFR 86.084.2-91).

Curb mass: The actual or manufacturers estimated mass of the vehicle with fluids at nominal capacity and with all equipment specified by the Administrator (40 CFR 86.402.78-91).

Curb stop: A water service shutoff valve located in a water service pipe near the curb and between the water main and the building (EPA-94/04).

Curbside collection:
(1) Method of collecting recyclable materials at homes, community districts or businesses (EPA-94/04).
(2) See also definition under collection.

Curcumine or carmine method: A standard method of measuring the concentration of boron (B) within a solution (EPA-83/06a).

Curie (Ci): See definition under radiation unit.

Curing:
(1) A heating/drying process carried out in an elevated temperature enclosure (EPA-83/03).
(2) The strength gain over time of a chemically fused, bodied chemically fused, or chemical adhesive geomembrane seam due primarily to evaporation of solvents or crosslinking of the organic phase of the mixture (EPA-91/05).

Curing oven: A device that uses heat to dry or cure the coating(s) applied to large appliance parts or products (40 CFR 60.451-91, see also 40 CFR 60.461-91).

Curing time: The time required for full curing as indicated by no further increase in strength over time (EPA-91/05).

Currency paper: The paper used for printing currency, bonds, and other government securities (see paper for more related terms) (EPA-83).

Current: Movements or flows of water, set in motion by winds and waves or by differences in temperature (DOI-70/04).

Current asset: The cash or other assets or resources commonly identified as those which are reasonably expected to be realized in cash or sold or consumed during the normal operating cycle of the business (see asset for more related terms) (40 CFR 144.61-91).

Current carrying capacity: The maximum current that can be continuously carried without causing permanent deterioration of electrical or mechanical properties of a device or conductor (EPA-83/03).

Current closure cost estimate: The most recent of the estimates prepared in accordance with 40 CFR 264.142 (a), (b), and (c) (see closure for more related terms) (40 CFR 264.141-91, see also 40 CFR 265.141-91).

Current discharge: The volume, composition, and location of an applicant's discharge as of anytime between December 27, 1977, and December 29, 1982, as designated by the applicant (40 CFR 125.58-91).

Current collector: The grid portion of the electrode which conducts the current to the terminal (EPA-84/08).

Current liability: The obligations whose liquidation is reasonably expected to require the use of existing resources properly classifiable as current assets or the creation of other current liabilities (see liability for more related terms) (40 CFR 144.61-91, see also 40 CFR 264.141; 265.141-91).

Current meter: A liquid velocity measuring meter in an open channel. Types of the current meter include:
(1) **Pitot tube:** For measuring clean gas and liquid velocity confined in pipes by measuring the difference between impact pressure and static pressure.
(2) **Turbine flowmeter:** Use magnetic detection to measure flow rotation speed.
(3) See flow velocity meter for more related terms.

Current plugging and abandonment cost estimate: The most recent of the estimates prepared in accordance with 40 CFR 144.62(a), (b), and (c) of this title (40 CFR 264.141-91, see also 40 CFR 265.141-91).

Current plugging cost estimate: The most recent of the estimates prepared in accordance with 40 CFR 144.62(a), (b) and (c) (40 CFR 144.61-91).

Current post closure cost estimate: The most recent of the estimates prepared in accordance with 40 CFR 264.144 (a), (b), and (c) (see closure for more related terms) (40 CFR 264.141-91, see also 40 CFR 265.141-91).

Current production: The amount of planned production in the calendar year ir which the pesticides report is submitted, including new products not previously sold or distributed (40 CFR 167.3-91).

Currying: Incorporating oils into leathers in a low water system, classically used to lubricate the vegetable tanned leather (EPA-82/11).

Curtail: To cease operations to the extent technically feasible to reduce emissions (40 CFR 61.181-91).

Curtain damper: A damper, composed of flexible materials, moving in a vertical plane as it is rolled (see damper for more related terms).

Curtain furnace wall: A hanging or arched refractory construction or baffle that deflects combustion gases downward (see furnace wall for more related terms) (SW-108ts).

Custody transfer: The transfer of produced petroleum and/or condensate after processing and/or treating in the producing operations, from storage tanks or automatic transfer facilities to pipelines or any other forms of transportation (40 CFR 52.741-91, see also 40 CFR 60.111; 60.111a; 60.111b-91).

Custom blender: Any establishment which provides the service of mixing pesticides to customer's specifications, usually a pesticide(s)-fertilizer(s), pesticide-pesticide, or pesticide-animal feed mixture, when:

(1) The blend is prepared to the order of the customer and is not held in inventory by the blender;
(2) The blend is to be used on the customer's property (including leased or rented property);
(3) The pesticide(s) used in the blend bears end-use labeling directions which do not prohibit use of the product in such a blend;
(4) The blend is prepared from registered pesticides;
(5) The blend is delivered to the end-user along with a copy of the end-use labeling of each pesticide used in the blend and a statement specifying the composition of mixture; and
(6) No other pesticide production activity is performed at the establishment (40 CFR 167.3-91).

Custom molded device: A hearing protective device that is made to conform to a specific ear canal. This is usually accomplished by using moldable compound to obtain impression of the ear and ear canal. The compound is subsequently permanently hardened to retain this shape (40 CFR 211.203-91).

Custom territory of the United States: The 50 States, Puerto Rice, and the District of Columbia (40 CFR 720.3-91, *see also* 40 CFR 372.3; 763.63-91).

Cut: The portion of a land surface or an area from which earth or rock has been or will be excavated. The distance between an original ground surface and an excavated surface (SW-108ts).

Cut and fill: A method of disposing of refuse on land by utilizing the principles of engineering to confine the refuse to the smallest practical area, reducing it to the smallest practical volume and covering it with a layer of earth at the conclusion of each days operation, or at more frequent intervals. *See* Sanitary Landfill (EPA-83).

Cut diameter: *See* synonym, cut size.

Cut glass: The glassware decorated by grinding figures or patterns on its surface by abrasive means, followed by polishing (*see* glass for more related terms) (EPA-83).

Cut off trench: A trench that is filled with barrier materials. The barrier is used to prevent the movement of gas or water (from landfill) or to intercept them and to direct them to another location (cf. gas barrier) (SW-108ts).

Cut power rule: A generalized method of calculating overall efficiency of a wet scrubber. The method assumes the predominant collection mechanism is inertial impaction (EPA-84/09).

Cut size (or cut diameter) and average particle size: The cut size is the diameter of those particles collected with 50% efficiency, e.g., cut diameter 0.2 micron meter (μm) means half of the 0.2 um particles are captured and half pass through the scrubber. Collection efficiency for particles larger than the cut size will be greater than 50% while that for smaller particles will be less. Average particle size is the average of the particle size range, e.g., if the size range is 10 to 15 microns, the average size is 12.5 microns (AP-40; Hesketh-79).

Cutie-pie: An instrument used to measure radiation levels (EPA-94/04).

Cutout or by-pass or similar devices: The devices which vary the exhaust system gas flow so as to discharge the exhaust gas and acoustic energy to the atmosphere without passing through the entire length of the exhaust system, including all exhaust system sound attenuation components (40 CFR 202.10-91).

Cutting: To penetrate with a sharp-edged instrument and includes sawing, but does not include shearing, slicing, or punching (40 CFR 61.141-91).

Cutting fluids: Lubricants employed to ease metal and machining operations, produce surface smoothness and extend tool life by providing lubricity and cooling. Fluids can be emulsified oils in water, straight mineral oils when better smoothness and accuracy are required, or blends of both (EPA-83/06a).

Cyanic acid (HCNO): An unstable, poisonous and explosive acid liquid.

Cyanidation: A process of extracting gold and silver as cyanide slimes from their ores by treatment with dilute solutions of potassium cyanide and sodium cyanide (EPA-82/05).

Cyanidation vat: A large tank, with a filter bottom, in which sands are treated with sodium cyanide solution to dissolve out gold (EPA-82/05).

Cyanide: The total cyanide and is determined by the method specified in 40 CFR 136.3 (40 CFR 420.02-91). It is a compound containing the CN^- ion. Other cyanide related terms include:
- Free cyanide
- Oxidizable cyanide
- Total cyanide

Cyanide A: Those cyanides amenable to chlorination and is determined by the methods specified in 40 CFR 136.3 (40 CFR 415.91-91, *see also* 40 CFR 415.421-91).

Cyanide destruction unit: A treatment system designed specifically to remove cyanide (40 CFR 439.1-91).

Cyaniding: A process of case hardening an iron-base alloy by the simultaneous absorption of carbon and nitrogen by heating in a cyanide salt. Cyaniding is usually followed by quenching to produce a hard case (EPA-83/06a).

Cyanogen $(CN)_2$: A very toxic, colorless gas with a pungent odor.

Cyanogen chloride (ClCN): A toxic, colorless gas or liquid used in military poison gas.

Cycle (or cyclic process): In thermodynamics, a process or a series of processes which returns the system to the state of the original conditions (*see* thermodynamic process for more related terms).

Cycle of concentration: The average concentration of dissolved solids in cooling water due to the evaporative process in a cooling tower (DOE-91/04).

Cyclethrin ($C_{21}H_{28}O_3$): A viscous, brown liquid used as an insecticide for flies, roaches and grain pests.

Cycling plant: An electrical generating facility which operates between peak load and base load conditions (EPA-82/11f).

cyclo-: A prefix indicating a compound structure containing a ring made of carbon atoms.

Cyclone:
(1) In weather, a region of low pressure, winds blow counter-clockwise about the low pressure center in the northern hemisphere and blow clockwise about the low pressure center in the southern hemisphere.
(2) In particle separation, an inlet gas stream is made to move vortically; centrifugal forces tend to drive suspended particles to the wall of the cyclone (*see* collector for more related terms) (EPA-83).
(3) A cyclone is known as a **cyclone collector** or a **cyclone separator**. It is a typical **mechanical separator**. A cyclone obtains its name by similarity with natural atmospheric storms. It is one of the oldest type of inertial separator. A cyclone which is a conical shaped apparatus is an inertial separator without moving parts. It separates particulate matter from a carrier gas by transforming the velocity of an inlet streams into a double vortex confined within the cyclone. In the double vortex, the gas entering the tangential inlet near the top of the cylindrical body creates a vortex or spiral flow downward between the walls of the gas discharge outlet and the body of the cyclone. This votex, called the main votex continues downward even below the walls of the gas outlet, and at some region near the bottom of the cone, the vortex reverses its direction of axial flow but maintains its direction of rotation, so that a secondary or inner vortex core is formed traveling upward to the gas outlet. The particulates, because of their inertia, tend to move toward the outside wall, from which they are let to a receiver (*see* mechanical separator for more related terms) (AP-40).

Cyclone collector:
(1) A device that uses centrifugal force to pull large particles from polluted air (EPA-94/04).
(2) *See also* synonym, cyclone separator.

Cyclone for pulp application: A classifying (or concentrating) separator into which pulp is fed, so as to take a circular path. Coarser and heavier fractions of solids report at the apex of long cone while finer particles overflow from central vortex (EPA-82/05).

Cyclone furnace: A water-cooled horizontal cylinder in which fuel is fired, heat is released at extremely high rates, and combustion is completed. The hot gases are then ejected into the main furnace. The fuel and combustion air enter tangential imparting a whirling motion to the burning fuel, hence the name Cyclone Furnace. Molten slag forms on the cylinder walls and flows off for removal (*see* furnace for more related terms) (EPA-82/11f).

Cyclone mist eliminator: A mist eliminator used to collect heavy liquid loadings (*see* mist eliminator for more related terms) (EPA-81/09).

Cyclone separator (or cyclone collector): A cyclone is a typical mechanical separator. It obtains its name by similarity with natural atmospheric storms. It is one of the oldest type of inertial separator. A cyclone which is a conical shaped apparatus is an inertial separator without moving parts. It separates particulate matter from a carrier gas by transforming the velocity of an inlet streams into a double vortex confined within the cyclone. In the double vortex, the gas entering the tangential inlet near the top of the cylindrical body creates a vortex or spiral flow downward between the walls of the gas discharge outlet and the body of the cyclone. This votex, called the main votex continues downward even below the walls of the gas outlet, and at some region near the bottom of the cone, the vortex reverses its direction of axial flow but maintains its direction of rotation, so that a secondary or inner vortex core is formed traveling upward to the gas outlet. The particulates, because of their inertia, tend to move toward the outside wall, from which they are let to a receiver (*see* separator for more related terms) (AP-40, p91).

Cyclone type scrubber: One of air pollution control devices. The scrubber ranges from a simple dry cyclone with a spray nozzle to specially constructed multistage devices. All feature a tangential inlet to a cylindrical body, and many feature additional vanes that accentuate the cyclonic action and also act as impingement and collection surfaces. Since centrifugal force is the principal collecting mechanism, efficiency is promoted by comparatively high gas velocities. Pressure drops varies from 2 to 8 inches water gage, and water rates vary from 4 to 10 gpm per 1000 cfm gas handled (*see* scrubber for more related terms) (AP-40).

Cyclonic flow: A spiraling movement of exhaust gases within a duct or stack (*see* flow for more related terms) (40 CFR 60.251-91).

Cylinder condensate: The condensation that forms on the walls of the retort during steaming operations (EPA-74/04).

Cylinder process: A process for manufacture of window glass wherein molten glass is blown and drawn into the form of a cylinder, which is subsequently split longitudinally, reheated in a flattening kiln, and flattened (EPA-83).

Cystine ($C_{16}H_{12}N_2S_2$): An amino acid, important in stabilizing the structure of protein molecules.

Cytology: Branch of biology concerned with the study of cells as vital units with reference to their structure, function, multiplication, pathology, and life history (LBL-76/07-bio).

Cytotoxic compounds: Substances that are generally used in chemotherapy (cancer treatment). The compounds are highly toxic to cells. They can impair, injure, or kill cells and can cause cell mutations, cancer, and birth defects. Seven cytotoxic compounds which are regulated under the "U Group" of the RCRA (Resource Conservation and Recovery Act) are:
(1) Chlorambucil
(2) Cyclophosphamide

(3) Duanamycin
(4) Mmelphalan
(5) Mitomycin
(6) Streptozotocin
(7) Uracil mustard (EPA-90/06).

Dd

Dacron fiber: A trademark of Du Pont for polyester filaments and staple fibers (*see* fiber for more related terms) (EPA-74/06b).

Dag (or aquadaq): A conductive graphite coating on the inner and outer side walls of some cathode-ray tubes (EPA-83/03).

Daily cover: The cover material that is spread and compacted on the top and side slopes of compacted solid waste at least at the end of each operating day in order to control vectors, fire, moisture, and erosion and to assure an aesthetic appearance (*see* cover for more related terms) (40 CFR 241.101-91).

Daily data: The flow and pollutant measurements (BOD, COD, TOC, pH, etc.) taken by certain plants on a daily basis for extended periods of time (EPA-87/10a).

Daily discharge: The discharge of a pollutant measured during a calendar day or any 24-hour period that reasonably represents the calendar day for purposes of sampling. For pollutants with limitations expressed in units of mass, the daily discharge is calculated as the total mass of the pollutant discharged over the day. For pollutants with limitations expressed in other units of measurement, the daily discharge is calculated as the average measurement of the pollutant over the day (*see* discharge for more related terms) (40 CFR 122.2-91).

Daily flood peak: The maximum flow on any one day during a flood event (*see* flood for more related terms) (DOI-70/04).

Daily maximum limitation: A value that should not be exceeded by any one effluent measurement (40 CFR 429.11-11).

Daily route method: A method in which each collection crew is assigned a weekly route that is divided into daily routes (*see* waste collection for more related terms) (SW-108ts).

Daily weighted average VOM content: The average VOM (volatile organic matter) content of two or more coatings as applied on a coating line during any day, taking into account the fraction of total coating volume that each coating represents, as calculated with the equation (provided in 40 CFR 52.741) (40 CFR 52.741-91).

Dalton's atomic theory: Matter is made of atom particles. All atoms are alike in the same element and are different in different elements. Chemical reactions take place between atoms of elements.

Dalton's law of partial pressure: When gases or vapors (having no chemical interaction) are present as a mixture in a given space, pressure exerted by a component of the gas mixture at a given temperature is the same as it would exert if it filled the whole space alone (Dalton model in mixture property) (*see* law for more related terms) (EPA-81/12, p2-9).

Dam:
(1) An artificial barrier for impounding water or sediment.
(2) A natural barrier created by the lodgment of driftwood across a stream channel, by alluvial deposition, by a landslide, or by the work of beavers (DOI-70/04).
(3) cf. dry dam.

Damage: The damages specified in section 1002(b) of this Act, and includes the cost of assessing these damages (OPA1001, *see also* SF101).

Damaged friable miscellaneous ACM: The friable miscellaneous ACM (asbestos containing material) which has deteriorated or sustained physical injury such that the internal structure (cohesion) of the material is inadequate or, if applicable, which has delaminated such that its bond to the substrate (adhesion) is inadequate or which for any other reason lacks fiber cohesion or adhesion qualities. Such damage or deterioration may be illustrated by the separation of ACM into layers; separation of ACM from the substrate; flaking, blistering, or crumbling of the ACM surface; water damage; significant or repeated water stains, scrapes, gouges, mars or other signs of physical injury on the ACM. Asbestos debris originating from the ACBM (asbestos containing building material) in question may also indicate damage (40 CFR 763.83-91).

Damaged friable surfacing ACM: The friable surfacing ACM which has deteriorated or sustained physical injury such that the internal structure (cohesion) of the material is inadequate or which has delaminated such that its bond to the substrate (adhesion) is inadequate, or which, for any other reason, lacks fiber cohesion or adhesion qualities. Such damage or deterioration may be illustrated by the separation of ACM into layers; separation of ACM from the substrate; flaking, blistering, or crumbling of the ACM surface; water damage; significant or repeated water stains, scrapes, gouges, mars or other signs of physical injury on the ACM. Asbestos debris originating from the ACBM in question may also indicate damage (40 CFR 763.83-91).

Damaged or significantly damaged thermal system insulation ACM: The thermal system insulation ACM on pipes, boilers, tanks, ducts, and other thermal system insulation equipment where the insulation has lost its structural integrity, or its covering, in whole or in part, is crushed, water-stained, gouged, punctured, missing, or not intact such that it is not able to contain fibers. Damage may be further illustrated by occasional punctures, gouges or other signs of physical injury to ACM; occasional water damage on the protective coverings/jackets; or exposed ACM ends or joints. Asbestos debris originating from the ACBM in question may also indicate damage (40 CFR 763.83).

Damper: A manually or automatically adjustable valve or plate installed in a breeching, duct, or stack to regulate the flow of air or other fluids (EPA-89/03b). Other damper related terms include:
- Barometric damper
- Butterfly damper
- Curtain damper
- Flap damper
- Guillotine damper
- Louvre damper
- Sliding damper

Dano Biostabilizer system composting: An aerobic, thermophilic composting process in which optimum conditions of moisture, air, and temperature are maintained in a single, slowly revolving cylinder that retains the compostable solid waste for one to five days. The material is later windowed (*see* composting for more related terms) (SW-108ts).

Darcy's law: An equation used to predict pressure drop across a filter, pipe or fitting. The equation is "pressure drop = fabric resistance factor x filtration velocity" (*see* law for more related terms) (EPA-84/09).

Data auditing: The process of examining data after they have been stored or archived (NATO-78/10).

Data bank: A collection of data files (cf. data file or data set) (NATO-78/10).

Data call in: A part of the Office of Pesticide Programs (OPP) process of developing key required test data, especially on the long-term, chronic effects of existing pesticides, in advance of scheduled Registration Standard reviews. Data Call-In from manufacturers is an adjunct of the Registration Standards program intended to expedite re-registration (EPA-94/04).

Data collection portfolio (DCP): The written questionnaire used to survey the metal molding and casting industry (EPA-85/10a).

Data correlation: The process of the conversion of reduced data into a functional relationship and the development of the significance of both the data and the relationship for the purpose of process evaluation (EPA-82/11f).

Data file: A collection of data sets (cf. data bank or data set) (NATO-78/10).

Data fleet: A fleet of automobiles tested at zero device-miles in baseline configuration, the retrofitted configuration and in some cases the adjusted configuration, in order to determine the changes in fuel economy and exhaust emissions due to the retrofitted configuration, and where applicable the changes due to the adjusted configuration, as compared to the fuel economy and exhaust emissions of the baseline configuration (40 CFR 610.11-91).

Data gap: The absence of any valid study or studies in the Agency files which would satisfy a specific data requirement for a particular pesticide product (40 CFR 152.83-91).

Data precision (or relative precision): A measure of how exactly the result is determined, without reference to what the result means; measure of reproducibility of result; i.e., uncertainty in terms of a fraction of the value of the result (EPA-88/12).

Date of completion: The date when all field work has been completed and all deliverables (e.g., lab results, technical expert reports) have been received by the local government (40 CFR 310.11-91).

Data processing: A systematic conversion of data (manual or automatic) from one state or condition to another (NATO-78/10).

Data quality: The total features and characteristics of data that bear on the ability to satisfy a given purpose. The characteristics of major importance are accuracy, precision, comparability, completeness, and representativeness and method detection limit (*see* quality for more related terms) (EPA-86/10a).

Data quality objective (DQO): The sum of characteristics of a data set that describe its utility for satisfying a given purpose. Characteristics may be precision, accuracy, completeness, representativeness, and comparability, but they may also include experimental design and statistical confidence issues. The objectives for data quality, DQOs are established before the study is conducted (*see* quality for more related terms) (EPA-89/12a).

Data recorder:
(1) A strip chart recorder, analog computer, or digital recorder for recording measurement data from the analyzer output (*see* gas concentration measurement system for more related terms) (40 CFR 60-App/A(method 6C), *see also* 40 CFR 60-App/A(method 7E); 60-App/A(method 25A)-91).
(2) The system that computes the hourly rolling averages, displays and reports a permanent record of the measurement values. The minimum data recording requirement is one measurement value per minute. A recorder can be a strip-chart, analog computer, or digital recorder (*see* total hydrocarbon concentration measurement system for more related terms) (EPA-90/04).

Data reduction: The process from the conversion of raw field data into a systematic flow which assists in recognizing errors, omissions, and the overall data quality (EPA-82/11f).

Data retrieval: The process of recovering data previously stored (NATO-78/10).

Data set: A collection of individual items (cf. data bank or data file) (NATO-78/10).

Data significance: The result of the statistical analysis of a data group or bank wherein the value or significance of the data receives a thorough appraisal (cf. statistical significance) (EPA-82/11f).

Data storage: The process of placing data into an organized file or any other repository from which the data can be retrieved on demand (NATO-78/10).

Data submitters list: The current Agency list, entitled Pesticide Data Submitters by Chemical, of persons who have submitted data to the Agency (40 CFR 152.83-91).

Data validation: A systematic process for reviewing a body of data against a set of criteria to provide assurance that the data are adequate for their intended use. Data validation consists of data editing, screening, checking, auditing, verification, certification, and review (EPA-86/10a).

Datum: A base for reference (e.g., calculation or measurement).

Daughter: *See* synonym, decay product.

Daughter product: *See* synonym, decay product.

Day: 24 hours (40 CFR 60.51-91).

Day night sound level: The 24 hour time of day weighted equivalent sound level, in decibels, for any continuous 4 hour period, obtained after addition of ten decibels to sound levels produced in the hours from 10 p.m. to 7 a.m. (2200-000). It is abbreviated as L_{dn} (*see* sound level for more related terms) (40 CFR 201.1-91).

Day tank: *See* age tank (EPA-94/04).

Day to day uncertainty: Precision of result, taken on different days (EPA-88/12).

dB (or dB(A)): An abbreviation meaning A-weighted sound level in decibels, reference: 20 micropascals (cf. decibel) (40 CFR 201.1-91, *see also* 40 CFR 202.10; 204.2; 205.2-91)

DDD (2,2-bis(para-chlorophenyl)-1,1-dichloroethane, $C_{14}H_{10}Cl_4$): An insecticide, also known as DDE.

DDT (($ClC_6H_4)_2CH(CCl_3$)):
(1) The compounds DDT, DDD, and DDE as identified by the chemical names: (DDT)-1,1,1-trichloro-2,2-bis(p-chlorophenyl) ethane and some o,p-isomers; (DDD) or (TDE)-1,1-dichloro-2,2-bis(p-chlorophenyl) ethane and some o,p-isomers; (DDE)-1,1-dichloro-2,2-bis(p-chlorophenyl) ethylene (40 CFR 129.4-91).
(2) The first chlorinated hydrocarbon insecticide (chemical name: Dichloro-Diphenyl-Trichloromethane). It has a half-life of 15 years (non-biodegradable) and can collect in fatty tissues of certain animals. EPA banned registration and interstate sale of DDT for virtually all but emergency uses in the United States in 1972 because of its persistence in the environment and accumulation in the food chain (EPA-94/04).

DDT ambient water criterion: The ambient water criterion for DDT in navigable waters is 0.001 µg/L (40 CFR 129.101-91).

DDT formulator: A person who produces, prepares or processes a formulated product comprising a mixture of DDT and inert materials or other diluents into a product intended for application in any use registered under the Federal Insecticide, Fungicide and Rodenticide Act, as amended (7USC135, et seq.) (40 CFR 129.101-91).

DDT manufacturer: A manufacturer, excluding any source which is exclusively a DDT formulator, who produces, prepares or processes technical DDT, or who uses DDT as a material in the production, preparation or processing of another synthetic organic substance (40 CFR 129.101-91).

De-aeration (or deaeration):
(1) A process by which dissolved air and oxygen are stripped from water either by physical or chemical methods (EPA-82/11f).
(2) Removal of oxygen from commodities such as juices or fruit slices to prevent adverse effects on properties of the final products by aerobic decomposition (EPA-75/10).
(3) *See* aeration for more related terms.

De-aerator (or deaerator): A device for the removal of oxygen, carbon dioxide, and other gases from water (*see* aerator for more related terms) (EPA-82/11f).

De-sludge: A centrifuge designed to remove the coarse particles from a peel oil emulsion (*see* sludge for more related terms) (EPA-74/03).

Deacidification: Reducing the acidity by adding base components.

Deactivation: Reducing the reactivity (e.g., radioactivity) of a substance.

Dead end: The end of a water main which is not connected to other parts of the distribution system (EPA-94/04).

Dead plate grate: A stationary grate through which no air passes (*see* grate for more related terms) (EPA-83).

Dead reservoir storage: The volume of water in a reservoir below the lowest outlet or operating level (DOI-70/04).

Dead rinse: A rinse step in which water is not replenished or discharged (EPA-83/06a).

Deadheading: Closing or nearly closing or blocking the discharge outlet or piping of an operating pump or compressor (EPA-8/87b).

Deaeration: *See* synonym, de-aeration.

Deaerator: *See* synonym, de-aerator.

Dealkylation: The removal of an alkyl group (-R) from a molecule (EPA-87/10a).

Death: The lack of reaction of a test organism to gentle prodding

(40 CFR 795.120-91, *see also* 40 CFR 797.1400; 797.1930; 797.1950; 797.1970-91).

Debarker (or bag barker): A machine which removes bark from logs. A debarker may be wet or dry, depending on whether or not water is used in the operation. There are several types of debarkers including drum barkers, ring barkers, bag barkers, hydraulic barkers, and cutterhead barkers. With the exception of the hydraulic barker, all use abrasion or scraping actions to remove bark. Hydraulic barkers utilize high pressure streams of water. All types may utilize water, and all wet debarking operations may use large amounts of water and produce effluents with high solids concentrations (EPA-74/04).

Debarment: An action taken by a debarring official in accordance with these regulations to exclude a person from participating in covered transactions. A person so excluded is debarred (40 CFR 32.105-91).

Debris: The woody material such as bark, twigs, branches, heartwood or sapwood that will not pass through a 2.54 cm (1.0 in) diameter round opening and is present in the discharge from a wet storage facility (40 CFR 429.11-91, *see also* EPA-89/12a).

Decant: To draw off the upper layer of liquid after the heavier material (a solid or another liquid) has settled (EPA-94/04).

Decant structure: An apparatus for removing clarified water from the surface layers of tailings or setting points. Commonly used structure include towers in which surface waters flow over a gate (adjustable in height) and down the tower to a conduit generally buried beneath the tailings, decant weirs over which water flows to a channel external to the tailings pond, and floating decant barges which surface water out of the pond (EPA-82/05).

Decantation: A method for mechanical dewatering of a wet solid by pouring off the liquid without disturbing the underlying sediment or precipitate (EPA-84/08).

Decanting: Separating liquids from solids by drawing off the upper layer after the heavier material has settled (EPA-75/02d).

Decarbonate: To remove carbon by a chemical means.

Decay:
(1) The decrease of chemical strength or physical quality with time, e.g., the decrease of an air pollutant concentration with time due to chemical and physical processes involving this pollutant (NATO-78/10).
(2) Disintegration of wood substance due to action of wood-destroying fungi. It is also known as dote and rot (29 CFR 1910.25-91).
(3) The transformation of a radioactive substance into a decay product by the relationship: $I = I_0 \exp(-ct)$. Where I is the radioactive intensity at time t, I_0 is the initial radioactive intensity; c is the **decay constant**.

Decay constant: *See* definition under decay.

Decay product (daughter or daughter product): An isotope formed by the radioactive decay of some other isotope. This newly formed isotope possesses physical and chemical properties that are different from those of its parent isotope, and may also be radioactive (cf. fission) (40 CFR 300-App/A-91).

Decay products: Degraded radioactive materials, often referred to as "daughters" or "progeny"; radon decay products of most concern from a public health standpoint are polonium-214 and polonium-218 (EPA-94/04).

Dechlorination: Removal of chlorine from a substance by chemically replacing it with hydrogen or hydroxide ions in order to detoxify a substances (EPA-94/04).

Dechlorination process: A process by which excess chlorine is removed from water to a desired level, e.g., 0.1 mg/l maximum limit. Usually accomplished by passage through carbon beds or by aeration at a suitable pH value (EPA-82/11f).

Decibel (dB):
(1) The unit measure of sound level, abbreviated as dB (40 CFR 201.1-91).
(2) A unit of sound measurement. In general, a sound doubles in loudness for every increase of ten decibels (EPA-89/12).
(3) A unit for describing the ratio of two powers or intensities, or the ratio of a power to a reference power. In the measurement of sound intensity (noise), the pressure of the reference sound is usually taken as 2×10^{-4} dynes/cm, equal to 0.1 bel or 1 dB (DOE-91/04).
(4) cf. A-weighted decibel.

Decker: A piece of equipment commonly used to thicken pulp. It consists of a wire-covered drum in a pulp vat. A vacuum is applied to the center of the drum, commonly by a barometric leg, to pull water out of the stock slurry (EPA-87/10).

Decimal percent: A percentage expressed in decimal form, i.e., 8.12% = 0.0812 decimal percent (*see* analytical parameters--fuels for more related terms) (EPA-83).

Decisional body: Any Agency employee who is or may reasonably be expected to be involved in the decisional process of the proceeding including the Administrator, Judicial Officer, Presiding Officer, the Regional Administrator (if he does not designate himself as a member of the Agency trial staff), and any of their staff participating in the decisional process. In the case of a non-adversary panel hearing, the decisional body shall also include the panel members whether or not permanently employed by the Agency (40 CFR 57.809-2-91).

Decolorizer: A compound used for removing color from a material.

Decommission: To remove (as a facility) safely from service and reduce residual radioactivity to a level that permits release of the property for unrestricted use and termination of license (10 CFR 30.4-91, *see also* 10 CFR 40.4; 70.4-91).

Decommissioning: Removing facilities contaminated with radiation (such as processing plants, waste tanks, and burial grounds) from service and reducing or stabilizing radioactive contamination. Decommissioning includes the following concepts:
(1) Decontamination, dismantling, and return of an area to its

original condition without restrictions on use or occupancy; and

(2) Partial decontamination, isolation of remaining residues, and continued surveillance and restrictions on use or occupancy (DOE-91/04).

Decomposition:
(1) For the purposes of these standards, an event in a polymerization reactor that advances to the point where the polymerization reaction becomes uncontrollable, the polymer begins to break down (decompose), and it becomes necessary to relieve the reactor instantaneously in order to avoid catastrophic equipment damage or serious adverse personnel safety consequences (40 CFR 60.561-91).
(2) The chemical breakdown of organic waste materials by bacteria. Aerobic process refers to one using oxygen-breathing bacteria, while anaerobic refers to a process using bacteria which breathe in an inorganic oxidant. Total decomposition of organic waste produces carbon dioxide, water, and inorganic solids (EPA-83).
(3) The breakdown of matter by bacteria and fungi, changing the chemical makeup and physical appearance of materials (EPA-94/04).

Decomposition emissions: Those emissions released from a polymer production process as the result of a decomposition or during attempts to prevent a decomposition (*see* emission for more related terms) (40 CFR 60.561-91).

Decomposition product: Material produced or generated by the physical or chemical degradation of a parent material (EPA-85/11).

DECON: Equipment, structures, and portions of a facility and site containing radioactive contaminates are removed or decontaminated to a level that permits the facility to be released for unrestricted use shortly after cessation of operations (DOE-91/04).

Decontamination:
(1) The process of reducing or eliminating the presence of harmful substances, such as infectious agents, so as to reduce the likelihood of disease transmission from those substances (40 CFR 259.10-91).
(2) The removal of radioactive contamination from facilities, equipment, or soils by washing, heating, chemical or electrochemical action, mechanical cleaning, or other techniques (DOE-91/04).
(3) Those procedures taken to minimize contamination of personnel and equipment, minimize translocation of hazardous materials by external contamination, and reduce external contamination by removal, neutralization or chelation (NIOSH-84/10).
(4) Removal of harmful substances such as noxious chemicals, harmful bacteria or other organisms, or radioactive material from exposed individuals, rooms and furnishings in buildings, or the exterior environment (EPA-94/04).

Decontamination/detoxification:
(1) Processes which will convert pesticides into nontoxic compounds (40 CFR 165.1-91).
(2) A process which removes or destroys chemical, biological, or radiological contamination from, or neutralizing of it on, a person, object, or area (EPA-88/09a).

Decontamination area: An enclosed are adjacent and connected to the regulated area and consisting of an equipment room, shower area, and clean room, which is used for the decontamination of workers, materials, and equipment contaminated with asbestos (40 CFR 763.121-91).

Decorating: Embossing, imprinting, sugaring, topping, and so forth (AP-40, p790).

Decorative chrome: The multi-layer electroplate of copper, nickel and chromium in that order on the basis material to provide the bright decorative appearance (EPA-74/03d).

Dedicated incinerator: A privately owned incinerator used to burn only the owner's wastes (*see* incinerator for more related terms) (EPA-81/09).

Deep bed filtration: The common removal of suspended solids from wastewater streams by filtering through a relatively deep (0.3-0.9m) granular bed. The porous bed formed by the granular media can be designed to remove practically all suspended particles by physical- chemical effects (*see* filtration for more related terms) (EPA-83/06a).

Deep mine: An underground mine (*see* mine for more related terms) (EPA-82/10).

Deep shaft system: An activated sludge treatment system. Air is injected into a deep shaft 60-300 meter deep well. Because every 10 meter deep water column, the pressure increases approximately 1 atm. The high pressure will results in high oxygen solubility and thus increasing treatment efficiency.

Deep water port: A facility licensed under the Deepwater Port of Act of 1974 (33USC1501-1524) (OPA1001-91, *see also* 40 CFR 110.1-91).

Deep well injection:
(1) The disposal of raw or treated hazardous wastes by pumping them into deep wells for filtration through porous or permeable substance rock and then containment within surrounded layers of impermeable rock or clay for permanent storage (*see* well injection for more related terms) (EPA-87/10a).
(2) Deposition of raw or treated, filtered hazardous waste by pumping it into deep wells, where it is contained in the pores of permeable subsurface rock (EPA-94/04).

Deeper saturated zone: Groundwater occurs in this zone where all pores are filled with fluid that is under pressure greater than atmosphere (*see* water table for more related terms) (EPA-87/03).

Defeat device: An AECD that reduces the effectiveness of the emission control system under conditions which may reasonably be expected to be encountered in normal urban vehicle operation and use, unless:
(1) Such conditions are substantially included in the Federal emission test procedure;
(2) The need for the AECD is justified in terms of protecting the

vehicle against damage or accident; or
(3) The AECD does not go beyond the requirements of engine starting (40 CFR 86.082.2-91).

Defendant: Any person alleged in a complaint under 40 CFR 27.2 to be liable for a civil penalty or assessment under 40 CFR 27.2 (40 CFR 27.2-91).

Deferrization: A removal of iron elements from a compound.

Deficient air : *See* synonym, substoichiometric combustion air.

Defiberization: The reduction of wood materials to fibers (EPA-74/04).

Defiberize: Separating fibrous woody bundles into individual fibers (EPA-83).

Definite working day collection method: A variation of the large route method. Definite routes are laid out and a crew assigned to each. Collection proceeds along a route for the length of time adopted for a working day. The next day, collection begins where the crew stopped the day before. This continues until the route is completely collected, whereupon the crew starts collection again at the beginning of the route, without interruption (*see* waste collection for more related terms) (EPA-83).

Definite working day method: A variation of the large-route method in which definite routes are laid out and a crew assigned to each. Collection proceeds along a route for the length of time adopted for a working day. The next day, collection begins where the crew stopped the day before. This procedure continues until the whole route is covered, whereupon the crew returns to the beginning of the route (*see* waste collection method for more related terms) (SW-108ts).

Deflagration:
(1) The act of burning very suddenly and violently, but without a resultant shock wave (detonation) (EPA-81/09).
(2) An explosion in which the fame propagates through the unburned gas-air mixture at subsonic velocities (cf. detonation or propagation rate) (EPA-83).
(3) Rapid oxidation and burning with intense heat (DOE-91/04).

Deflaker: A high-speed mixing and agitating machine through which a fibrous stock suspension in water is pumped to obtain complete separation and dispersion of each individual fiber, and break up of any fiber lumps, knots, or bits of undefibered paper (EPA-87/10).

Defoliant:
(1) Any substance or mixture of substances intended for causing the leaves or foliage to drop from a plant, with or without causing abscission (FIFRA2-7USC136-91).
(2) A herbicide that removes leaves from trees and growing plants (EPA-94/04).

Deflocculating agent: A material added to a suspension to prevent settling (EPA-94/04).

Defluoridation: The removal of excess fluoride in drinking water to prevent the staining of teeth (EPA-94/04).

Deformation: Changes in form produced by external forces or loads on non-rigid bodies (Markes-67).

Deformer: An agent used to reduce the formation of forms.

Degasification:
(1) The removal of a gas from a liquid (EPA-82/11f).
(2) A water treatment that removes dissolved gases from he water (EPA-94/04).

Degassing: The removal of dissolved hydrogen from the molten aluminum prior to casting. Chemicals are added and gases are bubbled through the molten aluminum. Sometimes a wet scrubber is used to remove excess chlorine gas (40 CFR 467.02-91).

Deglasser: A separator used to remove small particles of glass, metal and other products from compost. In addition, it utilizes a pulsed, rising column of air to separate heavy items contained in compost (EPA-83).

Degradable: Capable of being reduced, broken down or chemically separated (EPA-83/06a).

Degradation:
(1) The process by which a chemical is reduced to a less complex form (EPA-89/12).
(2) A deleterious change in the chemical structure of a plastic (NIOSH-84/10).

Degradation products: Those chemicals resulting from partial decomposition or chemical breakdown of pesticides (40 CFR 165.1-91).

Degreaser: Any equipment or system used in solvent cleaning (40 CFR 52.741-91).

Degreasing:
(1) The removal of oils and greases from the surface of the metal workpiece. This process can be accomplished with detergents as in alkaline cleaning or by the use of solvents (40 CFR 471.02-91).
(2) The process of removing greases and oils from sewage, waste and sludge (EPA-87/10a).
(3) In leather industry, a solvent of detergent is added to the drum containing washed hides. Grease is removed from the hides and recovered as a by-product (EPA-82/11).

Degree day: The difference between the mean temperature of a certain day and a reference temperature, expressed in degrees (NATO-78/10).

Degree of hazard: A relative measure of how much harm a substance can do (Course 165.5).

Dehalogenation: A removal of halogen elements from a compound.

Dehydration: The removal of water from a material (EPA-87/10a).

Dehydrogenation: The removal of one or more hydrogen atoms from an organic molecule (EPA-87/10a).

Deinking:
(1) The operation of reclaiming fiber from waste paper by removing ink, coloring materials, and fillers (EPA-87/10).
(2) Removing ink, filler and other extraneous material from reusable paper by mechanical, hydraulic and chemical treatment (EPA-83).

Deionized water (type II water or demineralized water):
(1) Reagent, analyte-free, or laboratory pure water means distilled or deionized water or Type II reagent water which is free of contaminants that may interfere with the analytical test in question.
(2) Water treated to remove most of the cations (metal ions) and anions (EPA-75/02c).
(3) *See* water for more related terms.

Deionizer (or demineralizer): A process for treating water by removal of cations and anions (EPA-82/11f).

Delaney amendment: The amendment under the Food, Drug and Cosmetics Act prohibits adding a known carcinogen to food.

Delayed compliance order: An order issued by the State or by the Administrator to an existing stationary source, postponing the date required under an applicable implementation plan for compliance by such source with any requirement of such plan (CAA302, *see also* 40 CFR 65.01-91).

Delegated state: A state (or other governmental entity such as a tribal government) that has received authority to administer an environmental regulatory program in lieu of a federal counterpart. As used in connection with NPDES, UIC, and PWS programs, the term does not connote any transfer of federal authority to a state (EPA-94/04).

Deliming: The manufacturing step in the tan-yard that is intended to remove the lime from hides coming from the beam-house (EPA-82/11).

Delinquent debt: Any debt which has not been paid by the date specified by tie Government for payment or which has not been satisfied in accordance with a repayment agreement (40 CFR 13.2-91).

Deliquescence: The process of changing a crystal from a solid state to a saturated solution by absorbing atmospheric moisture.

Delist: Use of the petition process to have a facility's toxic designation rescinded (EPA-94/04).

Delisting petition: A petition to exclude a waste generated at a particular facility from listing as a hazardous waste (RCRA Sec. 3001).

Delivery tank: Any container, including associated pipes and fittings, that is attached to or forms a part of any truck, trailer, or railcar used for the transport of gasoline (40 CFR 60-App/A(method 27)-91).

Delivery tank vapor collection equipment: Any piping, hoses, and devices on the delivery tank used to collect and route gasoline vapors either from the tank to a bulk terminal vapor control system or from a bulk plant or service station into the tank (40 CFR 60-App/A(method 27)-91).

Delivery vessel: The tank trucks and tank trailers used for the delivery of gasoline (40 CFR 52.2285-91, *see also* 40 CFR 52.2286-91).

Delusterant: A compound (usually an inorganic mineral) added to reduce gloss or surface reflectivity of plastic resins or fibers (EPA-75/01a).

Demagging: Removing magnesium from aluminum alloys by chemical reaction (EPA-76/12).

Demand charge: The specified charge to be billed on the basis of the billing demand, under an applicable rate schedule or contract (EPA-83).

Demand interval: The period of time during which the electric energy flow is averaged in determining demand, such as 60-minute, 30-minute, 15-minute or instantaneous (EPA-83).

Demand limited material: A secondary material for which buyers are relatively scarce even though supplies may be available (OTA-89/10).

Demand-side waste management: Prices whereby consumers use purchasing decisions to communicate to product manufacturers that they prefer environmentally sound products packaged with the least amount of waste, made from recycled or recyclable materials, and containing no hazardous substances (EPA-94/04).

Demand type apparatus: An apparatus in which the pressure inside the facepiece in relation to the immediate environment is positive during exhalation and negative during inhalation (NIOSH-84/10).

Demanganization: The removal of manganese from water or a solution.

Demineralization:
(1) The process of removing dissolved minerals from water by ion exchange, reverse osmosis, electrodialysis, or other processes (EPA-82/11).
(2) Removal of mineral impurities from a substance such as sugar (EPA-75/02d).
(3) A treatment process that removes dissolved minerals from water (EPA-94/04).

Demineralized water: *See* synonym, deionized water.

Demineralizer: *See* synonym, deionizer (EPA-82/11f).

Demister: A mechanical device used to remove entrained water droplets from a scrubbed gas stream (EPA-89/03b).

Demography: The study of the characteristics of human populations such as size, growth, density, distribution and vital statistics

(Course 165.6).

Demolition: The wrecking or taking out of any load-supporting structural member of a facility together with any related handling operations or the intentional burning of any facility (40 CFR 61.141-91, *see also* 40 CFR 763.121-91).

Demolition waste: *See* construction and demolition waste.

Demonstration: The initial exhibition of a new technology process or practice or a significantly new combination or use of technologies, processes or practices, subsequent to the development stage, for the purpose of proving technological feasibility and cost effectiveness (RCRA1004-42USC6903-91).

Demyelination: Destruction of the myelin, a fatlike substance forming a sheath around some nerve fibers (LBL-76/07-bio).

Denitrification:
(1) Bacterial mediated reduction of nitrate to nitrite. Other bacteria may further reduce the nitrite to ammonia and finally nitrogen gas. This reduction of nitrate occurs under anaerobic conditions. The nitrate replaces oxygen as an electron acceptor during the metabolism of carbon compounds under anaerobic conditions. The heterotrophic microorganisms which participate in this process include pseudomonades, achromobacters and bacilli (EPA-87/10a).
(2) The anaerobic biological reduction of nitrate to nitrogen gas (EPA-94/04).

Dense media separation (heavy media separation or sink float): Separation of solid wastes into heavy and light fractions in a fluid medium whose density lies between theirs (*see* separation for more separation methods) (SW-108ts).

Densified refuse derived fuel (d-RDF): A refuse-derived fuel that has been processed to produce briquettes, pellets, or cubes (*see* fuel for more related terms) (EPA-89/11).

Density:
(1) The mass of a substance in the system divided by its volume or the mass per unit volume. It can be expressed as: d = (mass)/(volume) = m/V. For water at temperature 60° F, the density is 62.4 lb/ft^3 in British units and 1.0 g/cm^3 in cgs units (EPA-84/09, p23).
(2) The mass of a unit volume of liquid, expressed as grams per cubic centimeter, kilograms per liter, or pounds per gallon, at a specified temperature (40 CFR 60.431-91, *see also* 40 CFR 796.1840; 796.1860-91).
(3) In sanitary landfill, the ratio of the combined weight of solid waste and the soil cover to the combined volume of the solid waste and the soil cover ($W_{sw} + W_{soil}$)/($V_{sw} + V_{soil}$) (SW-108ts).
(4) A measure of how heavy a solid, liquid, or gas is for its size (EPA-94/04).
(5) cf. Bulk density

Denuder: A horizontal or vertical container which is part of a mercury chlor-alkali cell and in which water and alkali metal amalgam are converted to alkali metal hydroxide, mercury, and hydrogen gas in a short-circuited, electrolytic reaction (40 CFR 61.51-91).

Deodorant: A chemical used to remove odor.

Deoxidizing:
(1) Removing an oxide film from an alloy such as aluminum oxide (EPA-83/06a).
(2) Removing oxygen by any means; or reducing from the state of an oxide.

Deoxygenation: Removal of oxygen from a solution or a substance. Deoxygenation is desirable in the feed water of a high pressure boiler but not undesirable in most biological treatment of wastewater (cf. reoxygenation).

Deoxyribonucleic acid: *See* DNA (EPA-89/12).

Departure angle: The smallest angle, in a plan side view of an automobile, formed by the level surface on which the automobile is standing and a line tangent to the rear tire static loaded radius arc and touching the underside of the automobile rearward of the rear tire. This definition applies beginning with the 14 model year (40 CFR 86.084.2-91).

Dependent variable: A variable whose value is a function of one or more than one variables (*see* variable for more related terms) (EPA-79/12c).

Dephenolizer: A facility in which phenol is removed from the ammonia liquor and recovers it as sodium phenolate; this is usually accomplished by liquid extraction and vapor recirculation (EPA-74/06a).

Depleted uranium:
(1) The source material uranium in which the isotope uranium-235 is less than 0.711 weight percent of the total uranium present. Depleted uranium does not include special nuclear material (10 CFR 40.4-91).
(2) Uranium whose content of the isotope U-235 is less than 0.7%, which is the U-235 content of natural uranium (DOE-91/04).
(3) *See* uranium for more related terms.

Depletion curve: In hydraulics, a graphical representation of water depletion from storage-stream channels, surface soil, and groundwater. A depletion curve can be drawn for base flow, direct runoff, or total flow (EPA-94/04).

Depletion of oxygen: Deficiency of dissolved oxygen in water because of oxygen consumption by organic substances.

Depletion or depleted: Any case in which:
(1) The Secretary, after consultation with the Marine Mammal Commission and the Committee of Scientific Advisors on Marine Mammals established under subchapter III of this chapter, determines that a species or population stock is below its optimum sustainable population;
(2) A State to which authority for the conservation and management of a species or population stock is transferred under section 1379 of this title, determines that such species or stock is below its optimum sustainable population; or

(3) A species or population stock is listed as an endangered species or a threatened species under the Endangered Species Act of 1973 [16USC1531 et seq.] (MMPA3-16USC1362-90).

Depletion or loss: The volume of water which is evaporated, embodied in product, or otherwise disposed of in such a way that it is no longer available for reuse in the plant or available for reuse by another outside the plant (EPA-74/01a).

Depolarizer: A term often used to denote the cathode active material (EPA-84/08).

Deposit:
(1) The sum paid to the dealer by the consumer when beverages are purchased in returnable beverage containers, and which is refunded when the beverage container is returned (40 CFR 244.101-91).
(2) In mining, mineral or ore deposit is used to designate a natural occurrence of a useful mineral or an ore in sufficient extent and degree of concentration to invite exploitation (EPA-82/05).
(3) In electroplating, the material formed on the electrode or workpiece, i.e., a metal in electroplating (EPA-74/03d).

Deposition: The adsorption or absorption of an air pollutant at a ground, vegetation or water surface (NATO-78/10).

Deposition area: The area of the surface available for aerosol deposition in a control volume (EPA-88/09a).

Depository site: A disposal site (other than a processing site) selected under section 104(b) or 105(b) of the Act (40 CFR 192.01-91).

Depreciation: The reduction in value of an item caused by the elapse of time between the date of acquisition and the date of loss or damage (40 CFR 14.2-91).

Depressing agent (depressor or depressant): In the froth flotation process, a substance which reacts with particle surface to render it less prone to stay in the froth, thus causing it to wet down as a tailing product (contrary to activator) (EPA-82/05).

Depression storage: The water contained in minor natural depressions in the land surface such as puddles (DOI-70/04).

Depressurization: A condition that occurs when the air pressure inside a structure is lower that the air pressure outside. Depressurization can occur when household appliances such as fireplaces or furnaces, that consume or exhaust house air, are not supplied with enough makeup air. Radon may be drawn into a house more rapidly under depressurized conditions (EPA-94/04).

Depuration: The elimination of a test substance from test organism (40 CFR 797.1520-91, *see also* 40 CFR 797.1830-91).

Depuration or clearance or elimination: The process of losing test material from the test organisms (40 CFR 797.1560-91).

Depuration phase: The portion of a bioconcentration test after the uptake phase during which the organisms are in flowing water to which no test substance is added (40 CFR 797.1520-91, *see also* 40 CFR 797.1830-91).

Depuration rate constant (K_2): The mathematically determined value that is used to define the depuration of test material from previously exposed test animals when placed in untreated dilution water, usually reported in units per hour (40 CFR 797.1560-91).

Derivative: A substance extracted from another body or substance (EPA-76/03).

Dermal: Of the skin; through or by the skin (Course 165.6).

Dermal corrosion: The production of irreversible tissue damage in the skin following the application of the test substance (40 CFR 798.4470-91).

Dermal exposure:
(1) The contact between a chemical and the skin (Course 165.6).
(2) Contact between a chemical and the skin (EPA-94/04).

Dermal irritation: The production of reversible inflammatory changes in the skin following the application of a test substance (40 CFR 798.4470-91).

Dermal toxicity: The ability of a pesticide or toxic chemical to poison people or animals by contact with the skin (*see* toxicity for more related terms) (EPA-94/04).

DES: A synthetic estrogen, diethylstilbestrol is used as a growth stimulant in food animals. Residues in meat are thought to be carcinogenic (EPA-94/04).

Desalination (or desalinization):
(1) Removing salts from ocean or brackish water by using various technologies (EPA-94/04).
(2) Removal of salts from soil by artificial means, usually leaching (EPA-94/04).

Desalinization:
(1) Any mechanical procedure or process where some or all of the salt is removed from lake water and the freshwater portion is returned to the lake (40 CFR 35.1605.7-91).
(2) *See also* synonym, desalination.

Descaling: The removal of scale and metallic oxides from the surface of a metal by mechanical or chemical means. The former includes the use of steam, scale-breakers and chipping tools, the latter method includes pickling in acid solutions (EPA-83/06a).

Descriptive study epidemiology: The study describes the amount and distribution of disease in a population (*see* epidemiology for more related terms) (Course 165.6).

Desert: A large region where only a few forms of life can exist because of lack of water.

Desiccant:
(1) Any substance or mixture of substances intended for artificially accelerating the drying of plant tissue (FIFRA2-7USC136-91).

(2) A chemical agent that absorbs moisture; some desiccants are capable of drying out plants or insects, causing death (EPA-94/04).

Desiccator: An apparatus for drying substances.

Design basis: For nuclear facilities, information that identifies the specific functions to be performed by a structure, system, or component and the specific values (or ranges of values) chosen for controlling parameters for reference bounds for design. These values may be:
(1) Restraints derived from generally accepted state-of-the-art practices for achieving functional goals;
(2) Requirements derived from analysis (based on calculation and/or experiments) of the effects of a postulated accident for which a structure, system, or component must meet its functional goals; or
(3) Requirements derived from federal safety objectives, principles, goals, or requirements (DOE-91/04).

Design basis accident (DBA): For nuclear facilities, a postulated abnormal event that is used to establish the performance requirements of structures, systems, and components that are necessary to:
(1) Maintain them in a safe shutdown condition indefinitely; or
(2) Prevent or mitigate the consequences of the DBA so that the general public and operating staff are not exposed to radiation in excess of appropriate guideline values (DOE-91/04).

Design basis depressurization accidents: Postulated accidents in a modular high-temperature gas-cooled reactor in which a rapid reduction in primary coolant pressure occurs as a result of egress of a portion of the primary coolant system inventory from a breach of the primary coolant system boundary up to the maximum credible flow area (DOE-91/04).

Design capacity (gross capacity, nameplate capacity or net capacity):
(1) (In an electric generating station), the maximum continuous electrical output (gross) for which a station is designed under specified conditions. Manufacturer's name plate rating usually verified by performance tests before commercial operation. Net output is less than gross because of electrical use by generating station auxiliaries. The station is usually operated below design capacity to avoid exceeding equipment warranties which may occur because of normal fluctuations in steam pressure and temperature (EPA-83).
(2) (In a waste incineration facility), the weight of solid waste of a specified gross calorific value that a thermal processing facility is designed to process in 24 hours of continuous operation; usually expressed in tons per day (40 CFR 240.101-91).
(3) (In a general system), the quantity of material that a designer anticipates his system will be able to process in a specified time period under specified conditions (EPA-83).
(4) The average daily flow that a treatment plant or other facility is designed to accommodate (EPA-94/04).
(5) *See* capacity for more related terms.

Design effluent level: The long-term average effluent level demonstrated or judged achievable for recommended treatment technologies (EPA-85/10).

Design flow: The flow used for steady-state waste load allocation modeling (EPA-91/03; 85/09).

Design life: The period for which equipment or an entire facility can be expected to perform adequately (EPA-80/08).

Design period: The number of years for which the planned facility is expected to provide service, generally twenty years under the construction grants program (EPA-80/08).

Design power: The thermal power rating of a tritium-production reactor that, when combined with the anticipated capacity factor, will produce the goal tritium requirement (DOE-91/04).

Design value: The monitored reading used by EPA to determine an area's air quality status, e.g., for ozone, the fourth highest reading measured over the most recent three years is the design value, for carbon monoxide, the second highest non-overlapping 8-hour concentration for one year is the design value (EPA-94/04).

Designated facility:
(1) Any existing facility (*see* 40 CFR 60.2(aa)) which emits a designated pollutant and which would be subject to a standard of performance for that pollutant if the existing facility were an affected facility (*see* 40 CFR 60.2) (40 CFR 60.21-91).
(2) A hazardous waste treatment, storage, or disposal facility which:
 (A) Has received a permit (or interim status) in accordance with the requirements of Parts 270 and 124 of this chapter;
 (B) Has received a permit (or interim status) from a State authorized in accordance with Part 271 of this chapter; or
 (C) Is regulated under 40 CFR 261.6(c)(2) or Subpart F of Part 266 of this chapter; and
(3) That has been designated on the manifest by the generator pursuant to 40 CFR 262.20. If a waste is destined to a facility in an authorized State which has not yet obtained authorization to regulate that particular waste as hazardous, then the designated facility must be a facility allowed by the receiving State to accept such waste (40 CFR 260.10-91).
(4) The offsite disposer or commercial storer of PCB waste designated on the manifest as the facility that will receive a manifested shipment of PCB waste (40 CFR 761.3-91).
(5) *See* facility for more related terms.

Designated management agency (DMA): An agency identified by a WQM plan and designated by the Governor to implement specific control recommendations (40 CFR 130.2-91).

Designated pollutant:
(1) Any air pollutant, emissions of which are subject to a standard of performance for new stationary sources but for which air quality criteria have not been issued, and which is not included on a list published under section 108(a) or section 112(b)(1)(A) of the Act (40 CFR 60.21-91).
(2) An air pollutant which is neither a criteria nor hazardous pollutant, as described in the Clean Air Act, but for which

new source performance standards exist. The Clean Air Act does require states to control these pollutants, which include acid mist, total reduced sulfur (TRS), and fluorides (EPA-94/04).

(3) *See* pollutant for more related terms.

Designated project area: The portions of the waters of the United States within which the permittee or permit applicant plans to confine the cultivated species, using a method or plan or operation (including, but not limited to, physical confinement) which, on the basis of reliable scientific evidence, is expected to ensure that specific individual organisms comprising an aquaculture crop will enjoy increased growth attributable to the discharge of pollutants, and be harvested within a defined geographic area (40 CFR 122.25-91).

Designated representative: A responsible person or official authorized by the owner or operator of a unit to represent the owner or operator in matters pertaining to the holding, transfer, or disposition of allowances allocated to a unit, and the submission of and compliance with permits, permit applications, and compliance plans for the unit (CAA402, *see also* 29 CFR 1910.20-91).

Designated state agency: The State agency designated by State law or other authority to be responsible for registering pesticides to meet special local needs (40 CFR 172.21-91).

Designated uses:
(1) Those uses specified in water quality standards for each water body or segment whether or not they are being attained (cf. existing uses) (40 CFR 131.3-91; EPA-91/03).
(2) Those water uses identified in state water quality standards which must be achieved and maintained as required under the Clean Water Act. Uses can include cold water fisheries, public water supply, agriculture, etc. (EPA-94/04).

Designer bugs: Popular term for microbes developed through biotechnology that can degrade specific toxic chemicals at their source in toxic waste dumps or in ground water (EPA-94/04).

Desizing facilities: For NSPS (40 CFR 410.45), those facilities that desize more than 50 percent of their total production. These facilities may also perform other processing such as fiber preparation, scouring, mercerizing, functional finishing, bleaching, dyeing and printing (40 CFR 410.41-91).

Desmutting: The removal of smut (matter that soils or blackens) generally by chemical actions (EPA-83/06a).

Desorption (or regeneration):
(1) Any process that accomplishes a partial or complete separation of either an adsorbed substance from an adsorbent or an absorbed substance from an absorbent (EPA-84/09).
(2) The reverse of adsorption. A phenomenon where an adsorbed molecule leaves the surface of the adsorbent (EPA-87/10a).
(3) The renewing or reuse of materials such as activated carbon, single ion exchange resins, and filter beds by appropriate means to remove organics, metals, solids, etc. (EPA-87/10a).
(4) The process of freeing from a sorbed state (LBL-76/07-air).
(5) The manipulation of individual cells or masses of cells to cause them to develop into whole plants (EPA-89/12).

Desorption efficiency of a particular compound applied to a sorbet and subsequently extracted with a solvent: The weight of the compound which can be recovered from the sorbet divided by the weight of the compound originally sorbed (40 CFR 796.1950-91).

Dessication: The loss of water by direct evaporation, by drainage or dredging, by escape of water through subterranean outlets, by a drop in the groundwater level, or by the removal or destruction of a dam (DOI-70/04)

Destination facility:
(1) The disposal facility, the incineration facility, or the facility that both treats and destroys regulated medical waste, to which a consignment of such is intended to be shipped, specified in Box 8 of the Medical Waste Tracking Form (40 CFR 259.10-91).
(2) The facility to which regulated medical waste is shipped for treatment and destruction, incineration, and/or disposal (EPA-94/04).

Destratification: Vertical mixing within a lake or reservoir to totally or partially eliminate separate layers of temperature, plant, or animal life (EPA-94/04).

Destroyed medical waste: Regulated medical waste that has been ruined, torn apart, or mutilated through thermal treatment, melting, shredding, grinding, tearing, or breaking, so that it is no longer generally recognized as medical waste, but has not yet been treated (excludes compacted regulated medical waste (EPA-94/04).

Destroyed regulated medical waste: The regulated medical waste that is no longer generally recognizable as medical waste because the waste has been ruined, torn apart, or mutilated (it does not mean compaction) through:
(1) Processes such as thermal treatment or melting, during which treatment and destruction could occur; or
(2) Processes such as shredding, grinding, tearing, or breaking, during which only destruction would take place (40 CFR 259.10-91).
(3) *See* medical waste or waste for more related terms.

Destruction and removal efficiency (DRE):
(1) A parameter developed by the U.S. EPA to determine the ability of a combustion device to destroy and/or remove the organic components of the hazardous waste feed (ETI-92).
(2) This term refers to a required level of performance for permitted hazardous waste incinerators. Such incinerators must destroy and remove 99.99 percent of the principal organic hazardous constituents (POHCs) to ensure the safe operation of the incinerators. The DRE required for incinerators burning dioxins, furans and PCB (polychlorinated biphenyls) is 99.9999 percent. DRE is defined by the following equation: DRE = $((W_i - W_o)/W_i)100\%$, where: W_i = Mass feed rate of POHC; and W_o = Mass emission rate of POHC in flue gas (downstream of all air pollution control equipment).
(3) A percentage that represents the number of molecules of a compound removed or destroyed in an incinerator relative to

the number of molecules entered the system (e.g., a DRE of 99.99 percent means that 9,999 molecules are destroyed for every 10,000 that enter; 99.99 percent is known as "four nines." For some pollutants, the RCRA removal requirement may be a stringent as "six nines") (EPA-94/04).

Destruction and removal efficiency example: Calculation of sample volume required to show 99.99% DRE (cf. calculation in VOST) (EPA-81/09, p5-123).

(1) **Computation of maximum W_o to satisfy 99.99% DRE**
    Given conditions:
    (A) POHC designated by the permit writer: hexachlorobenzene;
    (B) Concentration of POHC in waste feed: 1.0%;
    (C) Waste feed rate: 1000 lbs/hr.
    Solution:
    - W_{in} = (concentration of POHC in waste)(waste feed rate) = (0.01)(1000 lb/hr) = 10 lb/hr
    - $W_o = W_{in}(1-DRE)$ = (10 lb/hr)(1-0.9999) = 0.001 lb/hr

    Note: The expression of the DRE to 5 or 6 decimal places is justified because an error by as much as 25% in the W_o would affect only the fifth decimal place.

(2) **Computation of minimum weight of POHC sample that can be collected.** Given conditions:
    (A) Detection limit of hexachlorobenzene in analytical sample extract as injected in the GC/MS: 1 ng/uL or 1 ug/mL;
    (B) Average extraction efficiency: 60%.
    Solution:
    - Because the extracted sample is concentrated via evaporation before injection into the GC/MS, then the minimum weight of collected hexachlorobenzene is independent of extract liquid volume. The minimum detectable total weight of POHC collected as obtained from laboratory analysis:
    - W_{sample} = (detection limit)/(extraction efficiency) = (1 ug/mL)/(0.60) = 1.667 ug

(3) **Computation of the POHC stack gas loading**
    Given conditions: Stack gas volume flow rate at standard conditions, Q = 85382 scf/min.
    Solution:
    - POHC concentration (c_g) = (total weight of POHC in sample)/(volume of sample at standard conditions)
    - $c_g = (W_o/Q)$(1 hr/60 min) = [(0.001 lb/hr)/(85382 scf/min)][1 hr/ 60 min] = 1.95 E-10 lb/scf = 8.85 E-07 gram/scf

    Note: This computation assumes 100% collection of the POHC on the filter, resin module, and impingers.

(4) **Computation of minimum stack gas sample volume (V_{std})**
    - $V_{std} = W_{sample}/c_g$ = (1.667 E-06 grams)/(8.85 E-07 grams/scf) = 1.884 scf

Destruction efficiency (DE): Same as destruction and removal efficiency except Wo = mass emission rate of POHC leaving combustion zone of incinerator (upstream of all air pollution control equipment) (EPA-81/09).

Destruction facility:
(1) A facility that destroys regulated medical waste by ruining or mutilating it, or tearing it apart (40 CFR 259.10-91).
(2) A facility that destroys regulated medical waste by mashing or mutilating it (EPA-94/04).

Destruction or adverse modification: A direct or indirect alteration of critical habitat which appreciably diminishes the likelihood of the survival and recovery of threatened or endangered species using that habitat (40 CFR 257.3.2-91).

Destructive distillation: The airless heating of organic matter that results in the evolution of volatile substances and produces a solid char consisting of fixed carbon and ash (cf. Lantz process) (SW-108ts).

Destructive test: In seaming a landfill liner, a test performed on FML (flexible membrane liner) samples cut out of a field installation to verify specification performance requirements, e.g., shear and peel tests of FML seams during which the specimens are destroyed (*see* test for more related terms) (EPA-89/09, *see also* EPA-91/05).

Desulfurization: Removal of sulfur from fossil fuels (or removal of sulfur dioxide from combustion flue gases) to reduce pollution (EPA-94/04).

Detachable container rear loader: A detachable container system in which roll-out containers, typically 1 to 3 yard capacity are hoisted at the rear of the collection vehicle and mechanically emptied. Container is left with the customer (EPA-83).

Detachable container side loader: System similar to rear loader except loaded at side of collection vehicle (EPA-83).

Detachable container system: A partially mechanized self-service refuse removal procedure with specially constructed containers and vehicles. It is mechanized in that special equipment is used to empty the containers and haul refuse to the disposal site. It is self-service when the customer deposits the refuse in the container (cf. rear loader, detachable container) (EPA-83).

Detached house: Single family dwellings as opposed to apartments, duplexes, town-houses, or condominiums. Those dwellings which are typically occupied by one family unit and which do not share foundations and/or walls with other family dwellings (EPA-88/08).

Detailed analysis of alternative: A comparative analysis of all remedial alternatives that have successfully completed the technology screening phase. Each alternative is assessed against EPA's nine evaluation criteria before final remedy selections are made (EPA-89/12a). The nine evaluation criteria are as follows (EPA-89/12a):
(1) Overall protection of human health and the environment
(2) Compliance with ARARs
(3) Long term effectiveness and permanence
(4) Reduction of toxicity, mobility, or volume
(5) Short term effectiveness
(6) Implementability
(7) Cost
(8) State acceptance
(9) Community acceptance

Detectable leak rate: The smallest leak (from a storage tank), expressed in terms of gallons-or liters-per-hour, that a test can reliably discern with a certain probability of detection or false alarm (EPA-94/04).

Detection criterion: A predetermined rule to ascertain whether a tank is leaking or not. Most volumetric tests use a threshold value as the detection criterion (*see* volumetric tank tests) (EPA-94/04).

Detection level: The minimum concentration of a substance that can be measured with a 99% confidence that the analytical concentration is greater than zero (EPA-88/08a).

Detection limit (or minimum detectable sensitivity): The minimum concentration of an analyte (substance) that can be measured and reported with 99% confidence that the analyte concentration is greater than zero as determined by the procedure set forth at Appendix B of this part (cf. method detection limit) (40 CFR 136.2-91, *see also* 40 CFR 300-App/A-91). Other detection limit related terms include:
- Instrument detection limit (IDL)
- Limit of detection (LOD)
- Method detection limit (MDL)

Detention: The dwelling time of wastewater in a treatment unit (EPA-82/11).

Detention period: *See* synonym, detention time.

Detention tank: A tank used for temporarily holding the processing liquid (cf. tank or retention pond).

Detention time (detention period or retention period):
(1) The time allowed for solids to collect in a setting tank. Theoretically, detention time is equal to the volume of the tank divided by the flow rate. The actual detention time is determined by the purpose of the tank. Also, the design resident time in a tank or reaction vessel which allows a chemical reaction to go to completion, such as the reduction of chromium +6 or the destruction of cyanide (*see* time for more related terms) (EPA-82/11f).
(2) (A) The theoretical calculated time required for a small amount of water to pass through a tank at a given rate of flow (EPA-94/04).
(B) The actual time that a small amount of water is in a settling basin, flocculating basin, or rapid-mix chamber (EPA-94/04).
(C) In storage reservoirs, the length of time water will be held before being used (EPA-94/04).

Detergent: Synthetic washing agent that helps to remove dirt and oil. Some contain compounds which kill useful bacteria and encourage algae growth when they are in wastewater that reaches receiving waters (EPA-94/04). Other detergent related terms include:
- Biodegradable detergent
- Synthetic detergent

Deterministic model: A model which relates emission data directly to air quality data based on the simulation of a physical process (NATO-78/10).

Deterministic variable: A variable whose values are purely determined to be a function of physical facts (*see* variable for more related terms) (EPA-79/12c).

Deterrent: A propellant additive that reduces the burning rate (EPA-76/03).

Detinning: Recovering tin from tin cans by a chemical process which makes the remaining steel more easily recycled (EPA-89/11).

Detonation:
(1) The very rapid decomposition of an explosive. The reaction is propagated by a shock wave rather than by heating the area near to the flame (EPA-76/03).
(2) Spontaneous combustion of such rapidity that the reaction pressure rise is virtually instantaneous and the flame rate exceeds the speed of sound, e.g. Advancing reaction zone is preceded by a shock wave (cf. deflagration, explosion, flame front) (EPA-83).

Detoxification: The reduction or removal of toxic or hazardous compounds before release of the contaminated waste to a receiving body (e.g., sewer, river, or further treatment) (cf. decontamination).

Detrivorous: Organisms that use dead animals or decomposed organic matter as foods.

Detritus:
(1) The coarse cellular debris (e.g., dead or dying vegetation or animals).
(2) The heavier mineral debris moved by natural watercourses, usually in bed-load form (LBL-76/07-water).

Deuterium (D): *See* synonym, heavy hydrogen.

Developed country and developing country: In general,
(1) The developed country means the country with high living standards and good environmental protection.
(2) The developing country means the country with low living standards and poor environmental protection.

Developer:
(1) A person, government unit, or company that proposes to build a hazardous waste treatment, storage, or disposal facility (EPA-89/12).
(2) A chemical processing solution containing a developing agent. This solution converts the exposed portions of the photographic emulsion to silver, creating images of metallic silver (EPA-80/10).

Developing paper: *See* synonym, sensitized paper.

Development and screening of alternatives: The identification and screening of potentially applicable treatment technologies for remedy selection (EPA-89/12a).

Development effects: Adverse effects such as altered growth, structural abnormality, functional deficiency, or death observed in a developing organism (EPA-94/04).

Development work: In mining, a work undertaken to open up ore bodies as distinguished from the work of actual ore extraction or exploration (EPA-82/05).

Developmental toxicity:
(1) The property of a chemical that causes in utero death, structural or functional abnormalities or growth retardation during the period of development (40 CFR 798.4350-91, *see also* 40 CFR 798.4500; 798.4420-91).
(2) The study of adverse effects on the developing organism (including death, structural abnormality, altered growth, or functional deficiency) resulting from exposure prior to conception (in either parent), during prenatal development, or postnatally up to the time of sexual maturation (EPA-92/12).
(3) *See* toxicity for more related terms.

Device: For pest control, any instrument or contrivance (other than a firearm) which is intended for trapping, destroying, repelling, or mitigating any pest or any other form of plant or animal life (other than man and other than bacteria, virus, or other microorganism on or in living man or other living animals); but not including equipment used for the application of pesticides when sold separately therefrom (FIFRA2-7USC136-91, *see also* 40 CFR 167.3; 169.1; 610.11; 710.2; 720.3-91).

Device integrity: The durability of a device and effect of its malfunction on vehicle safety or other parts of the vehicle system (40 CFR 610.11-91).

Devitrification: Crystallization in glass (EPA-83).

Devulcanization: The softening of a vulcanizate by heat and chemical additives during reclaiming (EPA-74/12a).

Dew point temperature: The saturation temperature of the water corresponding to its partial pressure in the mixture. It is thus the temperature at which condensation begins if the mixture is cooled at constant pressure (*see* temperature for more related terms).
Example:
- Trichloroethylene (C_2HCl_3) is incinerated with methane gas at theoretical air conditions. Calculate the mole analysis of the combustion products and determine the dew point of the products for a total pressure of 14.696 psia.

Solution:
- $C_2HCl_3 + CH_4 + 3.5(O_2 + 3.76N_2) \rightarrow 3CO_2 + 3HCl + H_2O + 13.16N_2$.
- The total number of moles in the products is, $n = 3 + 3 + 1 + 13.16 = 20.16$ Moles.
- The mole fraction of the products is: $y_{CO2} = 3/20.16 = 14.88\%$; $y_{HCl} = 3/20.16 = 14.88\%$; $y_{H2O} = 1/20.16 = 4.96\%$; $y_{N2} = 13.16/20.16 = 65.28\%$.
- The dew point of the products is the temperature at which the vapor is saturated with water. It corresponds to the partial pressure of the water vapor, or: $p_{H2O} = y_{H2O} \times p = 0.0496 \times 14.696 = 0.7298$ psia.
- The saturation temperature corresponding to 0.7298 psia is about 94° F. Therefore: dew point = 94° F.
- This example illustrates two facts; namely:
  (A) Nitrogen is about 65% of the combustion product gas. This means that most of the fuel is used to heat the nitrogen in the air to the incineration temperature.
  (B) The dew point is 94° F. The moisture in the combustion gas would condense if the stack gas temperature is cooled below 94 f. Because the condensed moisture may contain HCl, the HCl will probably be corrosive to the stack or other parts in the incinerator that are downstream of the condensation point. Therefore, in designing this incinerator stack, the combustion gas should be maintained above 94° F.

Dew point pressure: The pressure corresponding to its dew point temperature (*see* pressure for more related terms).

Dewater:
(1) Remove or separate a portion of the water in a sludge or slurry to dry the sludge so it can be handled and disposed (EPA-94/04).
(2) Remove or drain the water from a tank or trench (EPA-94/04).

Dewater(ing):
(1) Removing water from any products by mechanical or evaporative methods. In mining, to remove water from a mine usually by pumping, drainage or evaporation (EPA-82/05).
(2) The removal of water by filtration, centrifugation, pressing, open air drying, or other,methods. Dewatering sludge facilitates disposal by burning or landfilling. The term is also applied when removing water from pulp (EPA-83).
(3) Pumping water from the soil to ensure proper soil characteristics for construction of facilities. May also be required during operation if the water table impinges foundations (DOE-91/04).

Dewatered: To remove the water from recently produced tailings by mechanical or evaporative methods such that the water content of the tailings does not exceed 30 percent by weight (40 CFR 61.251-91).

Dewatered sludge: The dry sludge or sludge whose water has been removed (*see* sludge for more related terms).

Dextrose (glucose or grape sugar): A monosaccharide sugar with the formula $C_6H_{12}O_6$. Dextrose is a minor component of raw sugar (cf. fructose) (EPA-75/02d).

Di-n-octyl phthalate: A liquid dielectric that is presently being substituted for a PCB dielectric fluid (EPA-83/03).

Diagnostic feasibility study: A two-part study to determine a lake's current condition and to develop possible methods for lake restoration and protection:
(1) The diagnostic portion of the study includes gathering information and data to determine the limnological, morphological, demographic, socio-economic, and other pertinent characteristics of the lake and its watershed. This information will provide recipients an understanding of the quality of the lake, specifying the location and loading characteristics of significant sources polluting the lake.
(2) The feasibility portion of the study includes:
    (A) Analyzing the diagnostic information to define

methods and procedures for controlling the sources of pollution;
(B) Determining the most energy and cost efficient procedures to improve the quality of the lake for maximum public benefit;
(C) Developing a technical plan and milestone schedule for implementing pollution control measures and in-lake restoration procedures; and
(D) If necessary, conducting pilot scale evaluations (40 CFR 35.1605-8-91).

Diakinesis and metaphase I: The stages of meiotic prophase scored cytologically for the presence of multivalent chromosome association characteristic of translocation carriers (40 CFR 798.5460-91).

Dialysis: One of several membrane separation procedures. The procedures usually consist of a barrier which will preferentially pass certain components of a fluid mixture or solution, and a driving force to cause such transfer to take place. It uses a semi-permeable membrane capable of passing small solute molecules (such as salts and small organic species) while retaining colloids and solutes of higher molecular weight. The driving force for this transfer is the concentration gradient and the difference in chemical activity of the constituents on either side of the membrane. The transfer through the membrane is by diffusion, that is, the progress of individual molecules, rather than by the hydrodynamic flow that would occur through a porous medium. Various membrane separation processes, the function of the membrane and the type of driving force in each are provided below:
(1) Reverse osmosis
 - Function of Membrane: Selective transport of water
 - Driving Force: Pressure
(2) Ultra-filtration
 - Function of Membrane: Discriminates on the basis of molecular size, shape and flexibility
 - Driving Force: Pressure
(3) Electro-dialysis
 - Function of Membrane: Selective ion transport
 - Driving Force: Electrical potential gradient
(4) Dialysis
 - Function of Membrane: Selective solute transport
 - Driving Force: Concentration
(5) Gel Permeation
 - Function of Membrane: Retard high molecular
 - Driving Force: Concentration
(6) Chromatography
 - Function of Membrane: Weight solute penetration
 - Driving Force: Concentration
(7) Liqui permeation
 - Function of Membrane: Selective transport of liquids
 - Driving Force: Concentration

Diaminobenzidene: A chemical used in the standard method of measuring the concentrations of selenium in a solution (EPA-83/06a).

Diapause: The period of suspended development or growth in some insects and mites (cf. dormancy).

Diaphragm pump: *See* synonym, accelerator pump.

Diaphragm displacement: The distance through which the center of the diaphragm moves when activated. In the case of a non-modulated stem, diaphragm displacement corresponds to stem displacement (40 CFR 85.2122(a)(1)(ii)(A)-91).

Diatom: A general name for algae or any class (bacillariophyceae) of planktonic unicellular algae.

Diatomaceous earth (or diatomite):
(1) A filter medium used for filtration of effluents from secondary and tertiary treatments, particularly when a very high grade of water for reuse in certain industrial purposes is required. Also used as an adsorbent form oils and oily emulsions in some wastewater treatment designs (EPA-82/11f).
(2) A chalk-like material (fossilized diatoms) used to filter out solid waste in wastewater treatment plants, also used as an active ingredient in some powdered pesticides (EPA-94/04).

Diatomaceous earth filtration: A process resulting in substantial particulate removal in which:
(1) A precoat cake of diatomaceous earth filter media is deposited on a support membrane (septum); and
(2) While the water is filtered by passing through the cake on the septum, additional filter media known as body feed is continuously added to the feed water to maintain the permeability of the filter cake (40 CFR 141.2-91).
(3) *See* filtration for more related terms.

Diatomite: *See* synonym, diatomaceous earth.

Diazinon: An insecticide. In 1986, EPA banned its use on open areas such as sod farms and golf courses because it posed a danger to migratory birds who gathered on them in large numbers. The ban did not apply to its use in agriculture, or on lawns of homes and commercial establishments (EPA-94/04).

Diazotization: The conversion of an amine ($-NH_2$) to a diazonium salt by reaction with nitrous acid (EPA-87/10a).

Dibasic acid: An acid capable of donating two protons (hydrogen ions) (EPA-83/06a).

Dibenzofuran:
(1) Any of a family of compounds which has as a nucleus a triple-ring structure consisting of two benzene rings connected through a pair of bridges between the benzene rings. The bridges are a carbon-carbon bridge and a carbon-oxygen-carbon bridge at both substitution positions (40 CFR 766.3-91).
(2) A group of highly toxic organic compounds (EPA-94/04).

Dibenzo-p-dioxin (or dioxin):
(1) Any of a family of compounds which has as a nucleus a triple-ring structure consisting of two benzene rings connected through a pair of oxygen atoms (40 CFR 766.3-91).
(2) Any of a family of compounds known chemically as dibenzo-p-dioxins. Concern about them arises from their potential toxicity and contaminants in commercial products. Tests on laboratory animals indicate that it is one of the more toxic

man-made chemicals known (EPA-89/12).

Dice: The more or less cubical fracture of tempered glass (EPA-83).

Dichlorodifluoromethane (CCl_2F_2): A refrigerant.

Dichlorophen ($C_{13}H_{10}Cl_2O_2$): An insecticide.

Dichromate: *See* definition under chromium.

Dichromate bleach: A bleach used in some black and white reversal and color film processing (EPA-80/10).

Dichromate reflux: A standard method of measuring the chemical oxygen demand of a solution (EPA-83/06a).

Dicofol: A pesticide used on citrus fruits (EPA-94/04).

Die: A tool or mold used to cut shapes to or form impressions on materials such as metals and ceramics (EPA-83/03).

Die casting: Casting is produced by forcing a molten metal under pressure into a metal mold (called die).
(1) In hot chamber machines, the pressure cylinder is submerged in the molten metal resulting in a minimum of time and metal cooling during casting.
(2) Vacuum feed machines use a vacuum to draw a measured amount of melt from the molten bath into the feed chamber.
(3) Pressure feed systems use a hydraulic or pneumatic cylinder to feed molten metal to the die (EPA-83/06a).

Die coatings: Oil containing lubricants or parting compounds such as carbon tetrachloride, cyclohexane, methylene chloride, xylene and hexamethylenetetramine. The coatings are used to prevent castings from adhering to the die and to provide a casting with a better finish. A correctly chosen lubricant will allow metal to flow into cavities that otherwise cannot be filled (EPA-85/10a).

Die cutting (or blanking): Cutting of plastic or metal sheets into shapes by striking with a punch (EPA-83/03).

Dieldrin ($C_{12}H_8Cl_6O$): An insecticide.

Dielectric: A material that is highly resistant to the conductance of electricity; an insulator (EPA-83/03).

Dielectric constant: *See* definition under Coulomb's law.

Dielectric material: A material that does not conduct direct electrical current. Dielectric coatings are used to electrically isolate UST systems from the surround soils. Dielectric bushings are used to electrically isolate portions of the UST system (e.g., tank from piping) (40 CFR 280.12-91).

Dielectric strength: The ability of the material of the cap and/or rotor to resist the flow of electric current (40 CFR 85.2122(a)(7)(ii)(B)-91, *see also* 40 CFR 85.2122(a)(8)(ii)(E); 85.2122(a)(9)(ii)(C)-91).

Diesel: The type of engine with operating characteristics significantly similar to the theoretical Diesel combustion cycle. The non-use of a throttle during normal operation is indicative of a diesel engine. This definition applies beginning with the 10 model year (40 CFR 86.090.2-91).

Diesel engine vehicle: A vehicle which is powered by a diesel engine (*see* vehicle for more related terms).

Dietary LC50: A statistically derived estimate of the concentration of a test substance in the diet that would cause 50 percent mortality to the test population under specified conditions (40 CFR 152.161-91).

Diethylstilbestrol (DES): A synthetic estrogen, It is used as a growth stimulant in food animals. Residues in meat are thought to be carcinogenic (EPA-94/04).

Differential equation: An equation expressing a relationship between functions and their derivatives. An ordinary differential equation is one with only one independent variable, and a partial differential equation is one with more than one independent variables (NATO-78/10).

Differential flotation: The separation of a complex ore into two or more valuable minerals and gangue by flotation, also called selectively flotation. This type of flotation is made possible by the use of suitable depressors and activators (EPA-82/05).

Differential medium (or selective medium): A selected medium which is particularly favors the growth of certain organisms.

Differential settlement: The nonuniform subsidence of material from a fixed horizontal reference plane (*see* settlement for more related terms) (EPA-83).

Differentiation: The process by which single cells grow into particular forms of specialized tissue, e.g., root, stem, leaf (EPA-89/12).

Diffused air (or aeration): A type of aeration that forces oxygen into sewage by pumping air through perforated pipes inside a holding tank (EPA-94/04).

Diffuser: An apparatus into which water and cossettes are fed, the water extracting sugar from the sugar beet cells (EPA-74/01a).

Diffusion: The movement of suspended or dissolved particles from a more concentrated to a less concentrated region as a result of the random movement of individual particles; the process tends to distribute them uniformly throughout the available volume. Its applications include:
(1) Very small particle (less than 0.1 μm in diameter) experience random movement in an exhaust stream. These particles are so tiny that they are bumped by gas molecules as they move in the exhaust stream. This bumping, or bombardment, causes them to first move one way and then another in a random manner, or diffuse, through the gas. This irregular motion can cause the particles to collide with a droplet and be collected. Because of this, in certain scrubbers, the removal efficiency of particles smaller than 0.1 μm can actually increase.

The rate of diffusion depends on relative velocity, particle diameter, and liquid-droplet diameter. As for impaction, collection due to diffusion increases with an increase in relative velocity (liquid- or gas-pressure input) and a decrease in liquid-droplet size. However, collection by diffusion increases as particle size decreases. This mechanism enables certain scrubbers to effectively remove the very tiny particles. In the particle size range of approximately 0.1 to 1.0 μm, neither of these two dominates. Particles in this size range are not collected as efficiently as are either larger particles collected by impaction or smaller particles collected by diffusion (see also particle collection mechanisms for wet scrubbing systems) (EPA-84/03b, p1-5).

(2) In atmospheric diffusion, it is primarily caused by turbulent air motions. The concept of diffusion is also frequently denoted by the word dispersion (cf. dispersion) (NATO-78/10).

(3) The movement of suspended or dissolved particles from a more concentrated to a less concentrated area. The process tends to distribute the particles more uniformly (EPA-94/04).

Diffusion category: This is to characterize different turbulence intensities and therefore different dispersions in the atmospheric boundary layer (NATO-78/10).

Diffusion current: Current which is controlled by the rate of diffusion of the species in the solution.

Diffusion diameter: Diameter of a sphere having the same diffusion mobility as the particle in question. $D_p < 0.5$ μm (EPA-90/08).

Diffusion equation: A parabolic partial differential equation which describes the process of diffusion as a function of time and space (NATO-78/10).

Diffusion flame: A flame produced by a combustion process in which the fuel and the air are not mixed prior to entering the combustion zone. Diffusion flame combustion, which primarily occurs in large-scale devices such as incinerators, boilers, and furnaces, involves very complicated processes such as precombustion, mixing, volatilization, vaporization, combustion, and post-flame reactions (see flame for more related terms).

Diffusion layer: The layer with ion concentration gradient nearby the electrode.

Diffusion model: A mathematical model describing the diffusion process (NATO-78/10).

Diffusion washing: Washing (wood) pulps with an open ended vessel by diffusing or passing the wash media through the pulp mass (EPA-87/10).

Diffusive movement: The random movement of individual atoms or molecules, such as radon atoms, in the absence of (or independent of) bulk (convective) gas flow. Atoms of radon can diffuse through tiny openings, or even through unbroken concrete slabs. Distinguished from convective movement (EPA-88/08).

Diffusivity: The coefficient of proportionality between the gradient of a property and its flux caused by molecular processes (NATO-78/10).

Digested sludge:
(1) The sludge digested under anaerobic conditions until the volatile content has been reduced, usually by approximately 50 percent or more (EPA-83/09).
(2) Sludge which has been treated by either anaerobic or aerobic digestion.
(3) See sludge for more related terms.

Digester (or sludge digestion tank): In wastewater treatment, a closed tank; in solid waste conversion, a unit in which bacterial action is induced and accelerated in order to break down organic matter and establish the proper carbon to nitrogen ratio (cf. stage digestion) (EPA-94/04).

Digester system: Each continuous digester or each batch digester used for the cooking of wood in white liquor, and associated flash tank(s), below tank(s), chip steamer(s), and condenser(s) (40 CFR 60.281-91).

Digestion: The biochemical decomposition of organic matter, resulting in partial gasification, liquefaction, and mineralization of pollutants (EPA-94/04).

Digestion chamber: A chamber or a tank used for the biochemical digestion of wastewater.

Dike:
(1) An embankment or ridge of either natural or man-made materials used to prevent the movement of liquids, sludges, solids, or other materials (40 CFR 260.10-91).
(2) A low wall that can act as a barrier to prevent a spill from spreading (EPA-94/04).

Diluent:
(1) The material added to a pesticide by the user or manufacturer to reduce the concentration of active ingredient in the mixture (40 CFR 165.1-91).
(2) Any liquid or solid material used to dilute or carry an active ingredient (EPA-94/04).

Diluent gas: A major gaseous constituent in a gaseous pollutant mixture. For combustion sources, CO_2 and O_2 are the major gaseous constituents of interest (40 CFR 60-App/F-91).

Dilution: The diminishing of the concentration of a pollutant by mixing it into an increasing volume (NATO-78/10).

Dilution factor: A parameter following from model calculations or measurements. When multiplied with an emission strength, it leads to a concentration value (NATO-78/10).

Dilution ratio: The relationship between the volume of water in a stream and the volume of incoming water. It affects the ability of the stream to assimilate waste (EPA-94/04).

Dilution water: The water to which the test substance is added and in which the organisms undergo exposure (see water for more related terms) (40 CFR 797.1520-91, see also 40 CFR 797.1600-

91).

Dilution weight: A parameter in the HRS surface water migration pathway that reduces the point value assigned to targets as the flow or depth of the relevant surface water body increases. [unitless] (40 CFR 300-App/A-91).

Dilution zone: *See* synonym, mixing zone

Dimension and unit: In engineering practice,
- Dimensions include: force, length, mass, time, temperature, electric charge, etc.
- Units include: pound, dyne, inch, cm, gram, second, etc. (Jones-p682)

Dimensional analysis: An analysis of a physical problem using only the dimensions of the dependent and independent variables and those of the basic parameters in the problem. Dimensional analysis is particularly useful for the derivation of similarity relations, where variables and parameters are grouped together in such a way that the problem can be described in terms of a few dimensionless numbers (*see* analysis for more related terms) (NATO-78/10).

Dimensional stability: Refers to ability of a paper to hold its lengthwise and crosswise dimensions under varying conditions (EPA-83).

Dimitic: Lakes and reservoirs that freeze over and normally go through two stratification and two mixing cycles a year (EPA-94/04).

Dimorphism: The existence of different forms from a same species.

Dinitrotoluene (DNT): *See* mononitrotoluene.

Dinocap: A fungicide used primarily by apple growers to control summer diseases. EPA proposed restrictions on its use in 1986 when laboratory tests found it caused birth defects in rabbits (EPA-94/04).

Dinoseb: A herbicide that is also used as a fungicide and insecticide. It was banned by EPA in 1986 because it posed the risk of birth defects and sterility (EPA-94/04).

Diode (crystal diode or crystal rectifier): In semiconductor, a two-electrode semiconductor device that utilizes the rectifying properties of a p-n junction or point contact (EPA-83/03).

Dioxin:
(1) The common name for polychlorinated dibenzo-p-dioxins (PCDDs). Production of these compounds has been associated with low-temperature combustion processes and high exposure to these compounds has been associated with adverse health effects, particularly in laboratory animals (ETI-92).
(2) Any of a family of compounds known chemically as dibenzo-p-dioxins. Concern about them arises from their potential toxicity and contaminants in commercial products. Tests on laboratory animals indicate that it is one of the more toxic man-made compounds (EPA-94/04).
(3) *See* dibenzo-p-dioxin (40 CFR 766.3-91).

Dioxin/furan:
(1) Dioxin is a generic term for a group of 75 polychlorinated dibenzo-para-dioxins (PCDDs), and furan is a generic term for a group of 135 polychlorinated dibenzofurans (PCDFs). Any member of the PCDD or PCDF family has one to eight chlorine substituents. One of the most toxic man-made substances, 2,3,7,8-TCDD (tetrachlorinated-dibenzo-p-dioxin), is an unwanted by-product resulting chiefly from the manufacture of the pesticide 2,4,5-TCP (trichlorophenol) (EPA-8/87a).
(2) total tetra-through octachlorinated dibenzo-p-dioxins and dibenzofurans (40 CFR 60.51a-91).

Dioxin waste: Includes those RCRA wastes listed as EPA hazardous waste numbers F021, F022, F023, F026 and F027. It encompasses the wastes from the production and manufacturing use of tri-, tetra-, and pentachlorophenols, wastes from the manufacturing use of tetra-, penta, and hexachlorobenzene under alkaline conditions, and also discarded, unused formulations containing tri-, tetra-, and pentachlorophenols (*see* waste for more related terms) (EPA-86/10b, p1-2).

Dip coating: A method of applying coatings in which the part is submerged in a tank filled with the coating (40 CFR 52.741-91, *see also* 40 CFR 60.311-91).

Diphenol ((C_6H_4OH)$_2$): A compound containing two phenol groups.

Diploid: A cell with two sets of chromosomes, designated as 2n (cf. haploid).

Dipolar ion: An ion with both positive and negative charges.

Direct application: Those cold rolling operations which include once-through use of rolling solutions at all mill stands (40 CFR 420.101-91).

Direct arc furnace: An electric furnace whose electrodes are spaced just below the surface of the slag cover. The current passes from ne electrode through the slag, the metal charge, the slag, and back to the other electrode. Heat is generated by radiation from the arc as well as from the resistance heat effect within the bath. The slag serves as a protective function by shielding the metal charge from vaporized carbon and the extremely high temperature (*see* furnace for more related terms) (AP-40, p236).

Direct chill casting: The pouring of molten aluminum into a water-cooled mold. Contact cooling water is sprayed onto the aluminum as it is dropped into the mold, and the aluminum ingot falls into a water bath at the end of the casting process (40 CFR 467.02-91, *see also* 40 CFR 471.02-91).

Direct cost: *See* synonym, direct labor cost.

Direct current (dc): An electric current that its charge flows in one direction only (cf. alternating current).

Direct discharge:
(1) The discharge of a pollutant (*see* discharge for more related terms) (40 CFR 122.2-91).
(2) A municipal or industrial facility which introduces pollution through a defined conveyance or system such as outlet pipes; a point source (EPA-94/04).

Direct dump transfer system: The unloading of solid waste directly from a collection vehicle into an open-top transfer trailer or container (SW-108ts).

Direct economic effect: The initial increases in output from different sectors of the economy resulting from some new activity within a predefined geographic region (DOE-91/04).

Direct fed incinerator: An incinerator that accepts solid waste directly into its combustion chambers (*see* incinerator for more related terms) (SW-108ts).

Direct filtration:
(1) A series of processes including coagulation and filtration but excluding sedimentation resulting in substantial particulate removal (*see* filtration for more related terms) (40 CFR 141.2-91).
(2) A method of treating water which consists of the addition of coagulant chemicals, flash mixing, coagulation, minimal flocculation, and filtration. Sedimentation is not used (EPA-94/04).

Direct fired heater: A heater in which heat is supplied by combustion, as distinguished from a heat exchanger where heat is supplied by a hot liquid or gas (EPA-74/04b).

Direct labor cost (or direct cost): Salaries, wages and other direct compensations earned by the employee (*see* cost for more related terms) (EPA-83/06a).

Direct photolysis: The direct absorption of light by a chemical followed by a reaction which transforms the parent chemical into one or more products (*see* photolysis for more related terms) (40 CFR 796.3700-91).

Direct runoff: Water that flows over the ground surface or through the ground directly into streams, rivers, and lakes (EPA-94/04).

Direct shell evacuation control system (DEC system): A system that maintains a negative pressure within the electric arc furnace above the slag or metal and ducts emissions to the control device (40 CFR 60.271a-91).

Direct shell evacuation system: Any system that maintains a negative pressure within the EAF above the slag or metal and ducts these emissions to the control device (40 CFR 60.271-91).

Direct toxicity: The toxicity that has an effect on organisms themselves instead of effecting an alteration of their habitat or interference with their food supply (*see* toxicity for more related terms) (DOD-78/01).

Direct transfer equipment: Any device (including, but not limited to, such devices as piping, fittings, flanges, valves, and pumps) that is used to distribute, meter, or control the flow of hazardous waste between a container (i.e., transport vehicle) and a boiler of industrial furnace (40 CFR 266.111-91).

Directed analysis: The directed analysis is the qualitative confirmation of compound presence and identity. Also, quantitative data of known quality for a set of constituents that might reasonably be expected to be present in the waste based on professional judgment and/or the results of proximate and survey analyses (*see* analysis for more related terms) (EPA-82/02).

Disc brake pad for heavy-weight vehicles: An asbestos-containing product intended for use as a friction material in disc brake systems for vehicles rated at 26,001 pounds gross vehicle weight rating (GVWR) or more (40 CFR 763.163-91).

Disc brake pad for light- and medium-weight vehicles: An asbestos-containing product intended for use as a friction material in disc brake systems for vehicles rated at less than 26,001 pounds gross vehicle weight rating (GVWR) (40 CFR 763.163-91).

Disc pulper: A machine which produces pulp or fiber through the shredding action of rotating and stationary discs (EPA-74/04).

Disc wheels: All power-driven rotatable discs faced with abrasive materials, artificial or natural, and used for grinding or polishing on the side of the assembled disc (29 CFR 1910.94b-91).

Discarded material: A discarded material is any material which is:
(1) Abandoned, as explained in paragraph (b) of this section; or
(2) Recycled, as explained in paragraph (c) of this section; or
(3) Considered inherently waste-like, as explained in paragraph (d) of this section (40 CFR 261.2-91).

Discharge:
(1) As defined by section 311(a)(2) the of CWA, includes, but is not limited to, any spilling, leaking, pumping, pouring, emitting, emptying, or dumping of oil, but excludes discharges in compliance with a permit under section 402 of the CWA, discharges resulting from circumstances identified and reviewed and made a part of the public record with respect to a permit issued or modified under section 402 of the CWA, and subject to a condition in such permit, or continuous or anticipated intermittent discharges from a point source, identified in a permit or permit application under section 402 of the CWA, that are caused by events occurring within the scope of relevant operating or treatment systems. For purposes of the NCP, discharge also means **threat of discharge** (40 CFR 300.5-91).
(2) In mining, any outflow from a pump, drill hole, piping system, channel, weir or other discernible, confined or discrete conveyance (EPA-82/05).
(3) In flow measurement, the volume of water that passes through a given cross-section of a channel during a unit of time. This flow, measured in cubic feet per second, is the amount of water fed to the stream from surface and groundwater run-off. Discharge varies according to velocity of flow, which in turn depends upon gradient (down-stream slope, usually expressed in feet per mile), volume of water, load of rock particles being carried, shape of the channel, and cross-sectional area of the channel (DOI-70/04).

(4) In battery, release of electric power from a battery (EPA-84/08).
(5) Flow of surface water in a stream or canal or the outflow of ground water from a flowing artesian well, ditch, or spring. Can also apply to discharge of liquid effluent from a facility or of chemical emissions into the air through designated venting mechanisms (EPA-94/04).

Other discharge related terms include:
- Continuous discharge
- Current discharge
- Daily discharge
- Direct discharge
- Improved discharge
- Indirect discharge
- Modified discharge

Discharge allowance: The amount of pollutant (mg per kg of production unit) that a plant will be permitted to discharge. For this category the allowances are specific to battery manufacturing operations (40 CFR 461.2-91).

Discharge area: An area in which subsurface water, including both ground water and water in the unsaturated zone, is discharged to the land surface, to surface water, or to the atmosphere (Course 165.7).

Discharge length scale: The square root of the cross-sectional area of any discharge outlet (EPA-91/03).

Discharge monitoring report (DMR): The EPA uniform national form, including any subsequent additions, revisions, or modifications for the reporting of self-monitoring results by permittees. DMRs must be used by approved States as well as by EPA. EPA will supply DMRs to any approved State upon request. The EPA national forms may be modified to substitute the State Agency name, address, logo, and other similar information, as appropriate, in place of EPA's (40 CFR 122.2-91).

Discharge of a pollutant and the term discharge of pollutants: Each means:
(1) Any addition of any pollutant to navigable waters from any point source,
(2) Any addition of any pollutant to the waters of the contiguous zone or the ocean from any point source other than a vessel or other floating craft (cf. discharge of pollutants) (CWA502, see also 40 CFR 122.2; 401.11-91).

Discharge of dredged material:
(1) Except as provided below in paragraph (2), the term discharge of dredged material means any addition of dredged material into, including any redeposit of dredged material within, the waters of the United States. The term includes, but is not limited to, the following:
 (i) The addition of dredged material to a specified discharge site located in waters of the Untied States;
 (ii) The runoff or overflow, associated with a dredging operation, from a contained land or water disposal area; and
 (iii) Any addition, including any redeposit, of dredged material, including excavated material, into waters of the United States which is incidental to any activity, including mechanized landclearing, ditching, channelization, or other excavation.
(2) The term discharge of dredged material does not include the following:
 (i) Discharges of pollutants into waters of the United States resulting from the onshore subsequent processing of dredged material that is extracted for any commercial use (other than fill). These discharges are subject to section 402 of the Clean Water Act even though the extraction and deposit of such material may require a permit from the Corps or applicable state.
 (ii) Activities that involve only the cutting or removing of vegetation above the ground (e.g., mowing, rotary cutting, and chainsawing) where the activity neither substantially disturbs the root system nor involves mechanized pushing, dragging, or other similar activities that redeposit excavated soil material.
(3) Section 404 authorization is not required for the following:
 (i) Any incidental addition, including redeposit, of dredged material associated with any activity that does not have or would not have the effect of destroying or degrading an area of waters of the U.S. as defined in paragraphs (4) and (5) of this definition; however, this exception does not apply to any person preparing to undertake mechanized landclearing, ditching, channelization and other excavation activity in a water of the United States, which would result in a redeposit of dredged material, unless the person demonstrates to the satisfaction of the Corps, or EPA as appropriate, prior to commencing the activity involving the discharge, that the activity would not have the effect of destroying or degrading any area of waters of the United States, as defined in paragraphs (4) and (5) of this definition. The person proposing to undertake mechanized landclearing, ditching, channelization or other excavation activity bears the burden of demonstrating that such activity would not destroy or degrade any area of waters of the United States.
 (ii) Incidental movement of dredged material occurring during normal dredging operations, defined as dredging for navigation in navigable waters of the United States, as that term is defined in 33 CFR part 329, with proper authorization from the Congress or the Corps pursuant to 33 CFR part 322; however, this exception is not applicable to dredging activities in wetlands, as that term is defined at 40 CFR 232.2(r) of this Chapter.
 (iii) Those discharges of dredged material associated with ditching, channelization or other excavation activities in waters of the United States, including wetlands, for which Section 404 authorization was not previously required, as determined by the Corps district in which the activity occurs or would occur, provided that prior to August 25, 1993, the excavation activity commenced or was under contract to commence work and that the activity will be completed no later that August 25, 1994. This provision does not apply to discharges associated with mechanized landclearing. For those excavation activities that occur on an ongoing basis (either continuously or periodically), e.g., mining operations, the Corps retains the authority to grant, on a case-by-case basis, an

extension of this 12-month grandfather provision provided that the discharger has submitted to the Corps within the 12-month period an individual permit application seeking Section 404 authorization for such excavation activity. In no event can the grandfather period under this paragraph extend beyond August 25, 1996.

(iv) Certain discharges, such as those associated with normal farming, silviculture, and ranching activities, are not prohibited by or otherwise subject to regulation under Section 404. *See* 40 CFR 232.3 for discharges that do not require permits.

(4) For purposes of this section, an activity associated with a discharge of dredged material destroys an area of waters of the United States if it alters the area in such a way that it would no longer be a water of the United States. Note: Unauthorized discharges into waters of the United States do not eliminate Clean Water Act jurisdiction, even where such unauthorized discharges have the effect of destroying waters of the United States.

(5) For purposes of this section, an activity associated with a discharge of dredged material degrades an area of waters of the United States if it has more than a de minimis (i.e., inconsequential) effect on the area by causing an identifiable individual or cumulative adverse effect on any aquatic function (40 CFR 232.2-91).

Discharge of fill material:

(1) The term discharge of fill material means the addition of fill material into waters of the United States. The term generally includes, without limitation, the following activities: Placement of fill that is necessary for the construction of any structure in a water of the United States; the building of any structure or impoundment requiring rock, sand, dirt, or other material for its construction; site-development fills for recreational, industrial, commercial, residential, and other uses; causeways or road fills; dams and dikes; artificial islands; property protection and/or reclamation devices such as riprap, groins, seawalls, breakwaters, and revetments; beach nourishment; levees; fill for structures such as sewage treatment facilities, intake and outfall pipes associated with power plants and subaqueous utility lines; and artificial reefs.

(2) In addition, placement of pilings in waters of the United States constitutes a discharge of fill material and requires a Section 404 permit when such placement has or would have the effect of a discharge of fill material. Examples of such activities that have the effect of a discharge of fill material include, but are not limited to, the following: Projects where the pilings are so closely spaced that sedimentation rates would be increased; projects in which the pilings themselves effectively would replace the bottom of a waterbody; projects involving the placement of pilings that would reduce the reach or impair the flow or circulation of waters of the United States; and projects involving the placement of pilings which would result in the adverse alteration or elimination of aquatic functions:

(i) Placement of pilings in waters of the United States that does not have or would not have the effect of a discharge of fill material shall not require a Section 404 permit. Placement of pilings for linear projects, such as bridges, elevated walkways, and powerline structures, generally does not have the effect of a discharge of fill material. Furthermore, placement of pilings in waters of the United States for piers, wharves, and an individual house on stilts generally does not have the effect of a discharge of fill material. All pilings, however, placed in the navigable waters of the United States, as that term is defined in 33 CFR part 329, require authorization under section 10 of the Rivers and Harbors Act of 1899 (*see* 33 CFR part 322).

(ii) [Reserved] (40 CFR 232.2-91)

Discharge or hazardous waste discharge: The accidental or intentional spilling, leaking, pumping, pouring, emitting, emptying, or dumping of hazardous waste into or on any land or water (40 CFR 260.10-91).

Discharge point: The point within the disposal site at which the dredged or fill material is released (40 CFR 230.3-91).

Discounting: The process of analyzing future costs and revenues to consider the effects of interest and inflation, and make all dollar amounts comparable on an equivalent basis (EPA-80/08).

Discrete wet scrubbing device: A distinct, stand-alone device that removes particulates and fumes from a contaminated gas stream by bringing the gas stream into contact with a scrubber liquor, usually water, and from which there is a wastewater discharge. Examples of discrete wet scrubbing devices are: spray towers and chambers, venturi scrubbers (fixed and variable), wet caps, packed bed scrubbers, quenchers, and orifice scrubbers. Semi-wet scrubbing devices where water is added and totally evaporates prior to dry air pollution control are not considered to be discrete wet scrubbing devices. Ancillary scrubber operations such as fan washes and backwashes are not considered to be discrete wet scrubber devices. These ancillary operations are covered by the mass limitations of the associated scrubber. Aftercoolers are not considered to be discrete wet scrubbing devices, and water dischargers from aftercooling are not regulated as a process wastewater in this category (40 CFR 464.31-i-91).

Disease: (In the true sense of the term) an interruption, cessation, or disorder of body functions, systems, or organs. (A disease, e.g., a genetic disorder may manifest itself without the involvement of a microorganism (EPA-5/90).

Disease vector:

(1) A carrier, usually an arthropod, that is capable of transmitting a pathogen from one organism to another (cf. vermin) (SW-108ts).

(2) Rodents, flies, and mosquitoes capable of transmitting disease to humans (40 CFR 257.3.6-91).

(3) *See* vector for more related terms.

Disinfect:

(1) To inactivate virtually all recognized pathogenic microorganisms but not necessarily all microbial forms (e.g., bacterial endospores) on inanimate objects (29 CFR 1910).

(2) To destroy pathogens but not necessarily all microbial life (cf. pasteurize or sterilize) (EPA-83).

Disinfectant:
(1) Any oxidant, including but not limited to chlorine, chlorine dioxide, chloramines, and ozone added to water in any part of the treatment or distribution process, that is intended to kill or inactivate pathogenic microorganisms (40 CFR 141.2-91).
(2) A chemical or physical process that kills pathogenic organisms in water. Chlorine is often used to disinfect sewage treatment effluent, water supplies, wells, and swimming pools (EPA-94/04).

Disinfectant time: The time it takes water to move from the point of disinfectant application (or the previous point of residual disinfectant measurement) to a point before or at the point where the residual disinfectant is measured. In pipelines, the time is calculated by dividing the internal volume of the pipe by the maximum hourly flow rate; within mixing basins and storage reservoirs it is determined by tracer studies of an equivalent demonstration (EPA-94/04).

Disinfectant contact time (T in CT calculations): The time in minutes that it takes for water to move from the point of disinfectant application or the previous point of disinfectant residual measurement to a point before or at the point where residual disinfectant concentration ("C") is measured. Where only one "C" is measured, "T" is the time in minutes that it takes for water to move from the point of disinfectant application to a point before or at where residual disinfectant concentration ("C") is measured. Where more than one "C" is measured, "T" is:
(1) For the first measurement of "C," the time in minutes that it takes for water to move from the first or only point of disinfectant application to a point before or at the point where the first "C" is measured and
(2) For subsequent measurements of "C," the time in minutes that it takes for water to move from the previous "C" measurement point to the "C" measurement point for which the particular "T" is being calculated. Disinfectant contact time in pipelines must be calculated based on "plug flow" by dividing the internal volume of the pipe by the maximum hourly flow rate through that pipe. Disinfectant contact time within mixing basins and storage reservoirs must be determined by tracer studies or an equivalent demonstration (40 CFR 141.2-1).

Disinfection: A process which inactivates pathogenic organisms in water by chemical oxidants or equivalent agents (40 CFR 141.2-91).

Disinfection by-product: A compound formed by the reaction of a disinfectant such as chlorine with organic material in the water supply (EPA-94/04).

Disintegrating: Breaking, chipping, chopping, crushing, cutting, grinding, milling, maturating, pulverizing, refining, shredding, slicing, and spraying (AP-40, p790).

Disk colorimeter: A rotating color disk used for comparing the standard color against the color to be measured.

Disk refiner: A motor-driven refiner whose working elements consist of one or more matched pairs of disks having a pattern of ribs machined into their faces and arranged so that one disk of the pair is rotated. The other disk is usually stationary, but may be driven in the opposite direction of rotation (EPA-87/10).

Dispersant:
(1) Those chemical agents that emulsify, disperse, or solubilize oil into the water column or promote the surface spreading of oil slicks to facilitate dispersal of the oil into the water column (40 CFR 300.5-91).
(2) A chemical agent used to break up concentrations of organic material such as spilled oil (EPA-94/04).

Dispenser: The permanent (intended to be refilled) or disposable (discarded when empty) container designed to hold more than one complete set of hearing protector(s) for the express purpose of display to promote sale or display to promote use or both (40 CFR 211.203-91).

Dispersal: The spreading of organisms or species from one place to more places.

Dispersed air flotation: The separation of low density contaminants from water using minute air bubbles attached to individual particles to provide or increase the buoyancy of the particle. The bubbles are generated by introducing air through a revolving impeller or porous media (EPA-83/06a).

Disperser: A mixing machine that acts to distribute the components of paint or ink (EPA-79/12b).

Dispersing agent: A reagent added to flotation circuits to prevent flocculation, especially of objectionable colloidal slimes. Sodium silicate is frequently added for this purpose (EPA-82/05).

Dispersion:
(1) In wave phenomena, variation of refractive index with wave length of light (EPA-83).
(2) In passive fluid contaminants, the growth of the dimensions of a cloud of passive particles in the fluid, which scatter due to turbulent or molecular fluid motions. The concept of dispersion in this sense is also frequently denoted by the word diffusion.
(3) In air pollution meteorology, the process of distributing air pollutant emissions by the combined action of advection and diffusion (NATO-78/10).
(4) In air modeling or **transport phenomena**, the air mass exchange between regions in space, in the lower atmosphere, dominated by eddy exchange due to turbulent air movements, the magnitude of which is generally relatable to atmospheric stability (EPA-88/09).
(5) In ground water, the spreading and mixing of chemical constituents in ground water caused by diffusion and by mixing due to microscopic variations in velocities within and between pores (Course 165.7).
(6) The dilution or removal of a substance by diffusion, turbulence, etc. Technically, a two phase system involving two substances, the first of which is uniformly distributed in a finely divided state through the second (the dispersion medium) (EPA-83).

Dispersion model: A dispersion model is a mathematical

representation of the transport and turbulent diffusion processes that occur in the atmosphere. Generally, such a model relates pollutant concentrations for specific receptors and averaging times to emissions from pollutant sources. This relationship is a function of meteorological conditions and the spatial relationship between sources and receptors. Thus the input data requirements for a dispersion model include meteorological data, source data, and receptor informations.

Based on the EPA "Guideline on Air Quality Models, EPA450-2-78-027, 86/07", the following are EPA recommended dispersion models (EPA-89/08).

terrain	urban-rural	average period	model selected
flat/rolling	urban or rural	annual or hourly	ISCLT ISCST
complex	urban	annual	LONGZ
complex	urban	hourly	SHORTZ
complex	rural	annual or hourly	COMPLEX 1

The industrial source complex model for long-term (ISCLT) and short-term (ISCST) modes were selected for flat and rolling terrain, because they can address building down-wash and elevated releases and can account for terrain differences between sources and receptors. The long-term mode is used for estimating annual average, while the short-term mode for estimating maximum hourly concentrations.

The two computer models most frequently maintained by EPA's OAQPS (Office of Air Quality, Planning and Standards) to produce exposure and risk estimates are:
(1) The human exposure model (HEM); and
(2) The atmospheric modeling subsystem (referred to as GAMS) of EPA's Graphical Exposure Modeling System (GEMS), maintained by the Office of Toxic Substances. The GEMS model has an interactive, user-friendly, format and can be easily accessed with a microcomputer and a modem to implement models such as ISCLT or ISCST.

Dispersion parameter: A parameter which describes the growth of the dimensions of a Gaussian plume or a Gaussian puff as a function of travel distance or travel time. The dispersion parameters are classified according to diffusion categories, which describe the influence of different turbulence conditions in the atmospheric boundary layer on the dispersion (NATO-78/10).

Dispersion resin: A resin manufactured in such a way as to form fluid dispersions when dispersed in a plasticizer or plasticizer/diluent mixtures (*see* resin for more related terms) (40 CFR 61.61-91).

Dispersion technique:
(1) Any technique which attempts to affect the concentration of a pollutant in the ambient air by:
 (i) Using that portion of a stack which exceeds good engineering practice stack height;
 (ii) Varying the rate of emission of a pollutant according to atmospheric conditions or ambient concentrations of that pollutant; or
 (iii) Increasing final exhaust gas plume rise by manipulating source process parameters, exhaust gas parameters, stack parameters, or combining exhaust gases from several existing stacks into one stack; or other selective handling of exhaust gas streams so as to increase the exhaust gas plume rise.
(2) The preceding sentence does not include:
 (i) The reheating of a gas stream, following use of a pollution control system, for the purpose of returning the gas to the temperature at which it was originally discharged from the facility generating the gas stream;
 (ii) The merging of exhaust gas streams where:
  (A) The source owner or operator demonstrates that the facility was originally designed and constructed with such merged gas streams;
  (B) After July 8, 1985 such merging is part of a change in operation at the facility that includes the installation of pollution controls and is accompanied by a net reduction in the allowable emissions of a pollutant. This exclusion from the definition of dispersion techniques shall apply only to the emission limitation for the pollutant affected by such change in operation; or
  (C) Before July 8, 1985, such merging was part of a change in operation at the facility that included the installation of emissions control equipment or was carried out for sound economic or engineering reasons. Where there was an increase in the emission limitation or, in the event that no emission limitation was in existence prior to the merging, an increase in the quantity of pollutants actually emitted prior to the merging, the reviewing agency shall presume that merging was significantly motivated by an intent to gain emissions credit for greater dispersion. Absent a demonstration by the source owner or operator that merging was not significantly motivated by such intent, the reviewing agency shall deny credit for the effects of such merging in calculating the allowable emissions for the source;
 (iii) Smoke management in agricultural or silvicultural prescribed burning programs;
 (iv) Episodic restrictions on residential woodburning and open burning; or
 (v) Techniques under 40 CFR 51.100(hh)(1)(iii) which increase final exhaust gas plume rise where the resulting allowable emissions of sulfur dioxide from the facility do not exceed 5,000 tons per year (40 CFR 51.100-hh-91).

Dispersoid: The particles of a dispersion system (EPA-83/06).

Displacement:
(1) The relative movement of any two sides of a fault measured in any direction (40 CFR 264.18-91).
(2) A bimolecular chemical reaction where an atom or radical is exchanged from one reactant to the other during the process (cf. abstraction) (EPA-88/12).

Displacement work: *See* synonym, flow work.

Disposable: Consumer products, other items, and packaging used once or a few times and discarded (EPA-94/04).

Disposable container: Plastic or paper sacks designed for storing solid waste (*see* container for more related terms) (SW-108ts).

Disposable device: A hearing protective device that is intended to be discarded after one period of use (40 CFR 211.203-91).

Disposable pay: That part of current basic pay, special pay, incentive pay, retired pay, retainer pay, or in the case of an employee not entitled to basic pay, other authorized pay remaining after the deduction of any amount described in 5 CFR 581.105 (b) through (f). These deductions include but are not limited to: Social security withholdings; Federal, State and local tax withholdings; health insurance premiums; retirement contributions; and life insurance premiums (40 CFR 13.2-h-91).

Disposal:
(1) In AEA, disposal means permanent isolation of spent nuclear fuel or radioactive waste from the accessible environment with no intent of recovery, wether or not such isolation permits the recovery of such fuel or waste. For example, disposal of waste in a mined geologic repository occurs when all of the shafts to the repository are backfilled and sealed (40 CFR 191.02-91).
(2) In CWA, *see* sludge disposal.
(3) In RCRA and TSCA, disposal means the discharge, deposit, injection, dumping, spilling, leaking, or placing of any solid waste or hazardous waste into or on any land or water so that such solid waste or hazardous waste or any constituent thereof may enter the environment or be emitted into the air or discharged into any waters, including ground waters (40 CFR 260.10-93). Methods of disposal include: underground injection (D80); landfill (D81); land treatment (D82); ocean disposal (D83); surface impoundment (D84); and other (D85) (40 CFR 264-App/I) (*see also* RCRA1004, 40 CFR 245.101; 257.2; 270.2; 373.4; 761.3-91).
(4) In the hazardous and Solid Waste Amendments of 1984, the Act expanded the scope of RCRA and imposed mandatory deadlines on EPA for banning certain solvents and dioxin-containing waste from landfills (Winthrop-89/09).
(5) In the Hazardous Materials Transportation Act (HMTA) (49USC1801-1813), EPA extended the scope of the DOT requirements under HMTA to apply to all shipments of hazardous wastes whether they are being transported within or between several states. If a spill or accident occurs during movement, the transporter is responsible for making sure the wastes do not spread from the scene, notifying the proper authorities, and arranging to clean-up the spilled wastes as quickly as possible. All transporters are required to obtain an EPA identification number; comply with manifest signature requirements; deliver the waste in accordance with directions on the manifest; and maintain records of waste shipments (Winthrop-89/09).
(6) In the Superfund, the transporters are required to report any spill or other release of a reportable quantity (RQ) of hazardous substance. Reportable quantities are based on toxicity, ignitability, reactivity, chronic toxicity, and susceptibility of the substance to degradation (Winthrop-89/09).
(7) Final placement or destruction of toxic, radioactive, or other wastes; surplus or banned pesticides or other chemicals; polluted soils; and drums containing hazardous materials from removal actions or accidental releases. Disposal may be accomplished through use of approved secure landfills, surface impoundments, land farming, deep-well injection, ocean dumping, or incineration (EPA-94/04).

Other disposal related terms include:
- Continuous disposal
- Phased disposal
- Safe disposal
- Solid waste disposal
- Waste disposal

Disposal area:
(1) The region within the perimeter of an impoundment or pile containing uranium by product materials to which the post-closure requirements of 40 CFR 192.32(b)(1) of this subpart apply (40 CFR 192.31-91).
(2) A site, location, tract of land, area, building, structure or premises used or intended to be used for partial or total refuse disposal (EPA-83).

Disposal facility: A facility or part of a facility at which hazardous waste is intentionally placed into or on any land or water, and at which waste will remain after closure (40 CFR 260.10-91, *see also* 40 CFR 270.2-91).

Disposal of incinerator ash: The principal rule governing the disposal of incinerator ash from hazardous waste is the "derived from" rule. Ash is a solid waste containing inorganic components of the waste feed that were not destroyed by burning. Some toxic metals may be concentrated in the ash although lead and cadmium may volatilize and emit through the stack. Basically, if a listed hazardous waste is incinerated, the ash from the waste is considered hazardous and must be disposed of in a licensed hazardous waste landfill.

The "mixture" rule may also apply to disposal of ash. Under RCRA:
(1) A mixture of listed waste is treated as hazardous waste unless the mixture qualifies for exemption or is delisted.
(2) A mixture including a characteristic waste and a solid waste is deemed hazardous only if the entire mixture continues to exhibit the hazardous characteristic (Winthrop-89/09).

Disposal site:
(1) (For dredged or fill material disposal under the Section 404 program), means that portion of the waters of the United States where specific disposal activities are permitted and consist of a bottom surface area and any overlying volume of water. In the case of wetlands on which surface water is not present, the disposal site consists of the wetland surface area (40 CFR 230.3-91).
(2) (For ocean dumping), means an interim for finally approved and precise geographical area within which ocean dumping of wastes is permitted under conditions specified in permits issued under sections 102 and 103 of the Act. Such sites are identified by boundaries established by (1) coordinates of latitude and longitude for each corner, or by (2) coordinates of latitude and longitude for the center point and a radius in nautical miles from that point. Boundary coordinates shall be identified as precisely as is warranted by the accuracy with which the site can be located with existing navigational aids or by the implantation of transponders, buoys or other means

of marking the site (40 CFR 228.2-91).

(3) For land disposal of solid waste, *see* open dump in 40 CFR 241.101.

(4) For non-hazardous solid waste disposal, *see* sanitary landfill in 40 CFR 257.2.

(5) For hazardous solid waste disposal, *see* disposal facility, (secure) landfill, land treatment facility and underground injection (well injection) in 40 CFR 260.10.

(6) (For nuclear waste disposal), means the region within the smallest perimeter of residual radioactive material (excluding cover materials) following completion of control activities (40 CFR 192.01-91, *see also* 40 CFR 192.31-91).

Disposal site designation study: The collection, analysis and interpretation of all available pertinent data and information on a proposed disposal site prior to use, including but not limited to, that from baseline surveys, special purpose surveys of other Federal agencies, public data archives, and social and economic studies and records of areas which would be affected by use of the proposed site (40 CFR 228.2-91).

Disposal site evaluation study: The collection, analysis, and interpretation of all pertinent information available concerning an existing disposal site, including but not limited to, data and information from trend assessment surveys, monitoring surveys, special purpose surveys of other Federal agencies, public data archives, and social and economic studies and records of affected areas (40 CFR 228.2-91).

Disposal system: Any combination of engineered and natural barriers that isolate spent nuclear fuel or radioactive waste after disposal (40 CFR 191.12-91).

Disposal well: A well used for the disposal of waste into a subsurface (*see* well for more related terms) (40 CFR 146.3-91, *see also* 40 CFR 147.2902-91).

Disposer of PCB waste: As the term is used in subparts J and K of this part, it means any person who owns or operates a facility approved by EPA for the disposal of PCB waste which is regulated for disposal under the requirements of subpart D of this part (*see* PCB for more related terms) (40 CFR 761.3-91).

Dispute: Refers to a present controversy between parties subject to a test rule over the amount or method of reimbursement for the cost of developing health and environmental data on the test chemical (40 CFR 791.3-91).

Dissipation: The conversion of kinetic energy of a fluid into heat by molecular internal fluid friction (NATO-78/10).

Dissociation: The reversible splitting into two or more chemical species which may be ionic. The process is indicated generally by the equation (provided in this subpart) (40 CFR 796.1370-91). The breakdown of a chemical species into smaller components.

Dissolved element: Those elements which will pass through a 0.45 μm membrane filter (cf. dissolved metal) (40 CFR 136-App/C-91).

Dissolved air flotation: A flotation process that adds air to wastewater in the form of fine bubbles which become attached to suspended sludge particles, increasing the buoyancy of the particles and producing more positive flotation (EPA-87/10a).

Dissolved metal: The concentration of metals determined in sample after the sample is filtered through a 0.45-μm filter (*see* metal for more related terms) (Method 3005, SW 846).

Dissolved oxygen (DO): The oxygen freely available in water. Dissolved oxygen is vital to fish and other aquatic life and for the prevention of odors. Traditionally, the level of dissolved oxygen has been accepted as the single most important indicator of a water body's ability to support desirable aquatic life. Secondary and advanced waste treatment are generally designed to protect DO in waste-receiving waters (*see* oxygen for more related terms) (EPA-94/04).

Dissolved oxygen unit: The units of measurement used are milligrams per liter (mg/L) and parts per million (ppm), where mg/L is defined as the actual weight of oxygen per liter of water and ppm is defined as the parts actual weight of oxygen dissolved in a million parts weight of water, i.e., a pound of oxygen in a million pounds of water is 1 ppm. For practical purposes in pollution control work, these two are used interchangeably; the density of water is so close to 1 g/cm^3 that the error is negligible. Similarly, the changes in volume of oxygen with changes in temperature are insignificant. This, however, is not true if sensors are calibrated in percent saturation rather than in mg/L or ppm. In that case, both temperature and barometric pressure must be taken into consideration (*see* oxygen for more related terms) (EPA-76/03).

Dissolved solids (DS):
(1) The total amount of dissolved materials, organic and inorganic, contained in water or wastes. Excessive dissolved solids can make water unsuitable for industrial uses, unpalatable for drinking, and even cathartic. Potable water supplies may have dissolved solid content from 20 to 1000 mg/L, but sources which have more than 500 mg/L are not recommended by the U.S. Public Health Service (DOI-70/04).
(2) Disintegrated organic and inorganic material in water. Excessive amounts make water unfit to drink or use in industrial processes (EPA-94/04).

Dissolving pulp: A special grade of chemical pulp made from wood or cotton linters for use in the manufacture of regenerated cellulose (viscose rayon and cellophane) or cellulose derivatives such as acetate and nitrate (*see* pulp for more related terms) (EPA-87/10).

Distance piece: An open or enclosed casing through which the piston rod travels, separating the compressor cylinder from the crankcase (40 CFR 60.481-91).

Distance weight: A parameter in the HRS air migration, ground water migration, and soil exposure pathways that reduces the point value assigned to targets as their distance increases from the site. [unitless] (40 CFR 300-App/A-91).

Distillate: The products of condensed vapors.

Distillate oil: The fuel oils that contain 0.05 weight percent nitrogen or less and comply with the specifications for fuel oils numbers 1 and 2, as defined by the American Society of Testing and Materials in ASTM D396-78, Standard Specifications for Fuel Oils (incorporated by reference--*see* 40 CFR 60.17) (*see* oil for more related terms) (40 CFR 60.41b-91, *see also* 40 CFR 60.41c-91).

Distillate receiver: A container or tank used to receive and collect liquid material (condensed) from the overhead condenser of a distillation unit and from which the condensed liquid is pumped to larger storage tanks or other process units (40 CFR 264.1031-91).

Distillation:
(1) Distillation is the boiling of a liquid solution and condensation of the vapor for the purpose of separating the components. It involves two basic phases, the liquid phase and the vapor phase. The components which are to be separated by distillation are present in both phases but in different concentrations. If there are only two components in the liquid, one concentrates in the condensed vapor (condensate) and the other in the residual liquid. If there are more than two components, the less volatile components concentrate in the residual liquid and the more volatile in the vapor condensate (cf. steam distillation).
(2) The act of purifying liquids through boiling, so that the steam condenses to a pure liquid and the pollutants remain in a concentrated residue (EPA-94/04).

Distillation operation: An operation separating one or more feed stream(s) into two or more exit stream(s), each exit stream having component concentrations different from those in the feed stream(s). The separation is achieved by the redistribution of the components between the liquid and vapor-phase as they approach equilibrium within the distillation unit (40 CFR 60.661-91, *see also* 40 CFR 264.1031-91).

Distillation refining: A metal with an impurity having a higher vapor pressure than the base metal can be refined by heating the metal to the point where the impurity vaporizes (EPA-83/06a).

Distillation silver nitrate titration: A standard method of measuring the concentration of cyanides in a solution (*see* titration for more related terms) (EPA-83/06a).

Distillation SPADNS: A standard method of measuring the concentration of fluoride in a solution (EPA-83/06a).

Distillation unit: A device or vessel In which distillation operations occur, including all associated internals (such as trays or packing) and accessories (such as reboiler, condenser. vacuum pump, steam jet, etc.). plus any associated recovery system (40 CFR 60.661-91).

Distilled water: The water that has been distilled for removing impurity substances (*see* water for more related terms).

Distribution coefficient (Kd): A measure of the extent of partitioning of a substance between geologic materials (for example, soil, sediment, rock) and water (or called partition coefficient). The distribution coefficient is used in the HRS in evaluating the mobility of a substance for the ground water migration pathway (40 CFR 300-App/A-91).

Distribution line: A pipeline other than a gathering or transmission line (40 CFR 192.3-91).

Distribution reservoir (or service reservoir): A water reservoir to support the fluctuation of water demand and for emergency needs.

Distribution system: The term may be used in both water supply system or electric distribution system.

Distribution transformer: An element of an electric distribution system located near consumers which changes primary distribution voltage to a lower consumer voltage (*see* transformer for more related terms) (EPA-83/03).

Distributor:
(1) A device for directing the secondary current from the induction coil to the park plugs at the proper intervals and in the proper firing order (40 CFR 85.2122(a)(11)(ii) (A)-91).
(2) *See* 40 CFR 52.137; 80.2; 244.101-91).

Distributor firing angle: The angular relationship of breaker point opening from one opening to the next in the firing sequence (40 CFR 85.2122(a)(11)(ii)(B)-91).

District heating: A central heat generating source which supplies heat by either steam or hot water to buildings in a district.

Disturbed area: An area which has had its natural condition altered in the process of mining coal, preparing coal, or other mine related activities. This includes but is not limited to all areas affected by topsoil removal; road construction; construction of mine facilities; coal mining, reclamation and preparation activities; deposition of topsoil, overburden, coal or waste materials, etc. These areas are classified as "disturbed" until said areas have been returned to approximate original contour (or post-mining land use) and topsoil (where appropriate) has been replaced (EPA-82/10).

Diurnal:
(1) Recurring daily. Applied to variations in concentration of air contaminants, diurnal indicates variations following a distinctive pattern and recurring from day to day (EPA-83/06).
(2) Pertaining to animals active during the day (DOE-91/04).

Diurnal breathing losses: The evaporative emissions as a result of the daily range in temperature (40 CFR 86.082.2-91).

Diurnal cycle: Daily variation in the photosynthesis-respiration pattern of aquatic flora resulting in a dissolved oxygen maximum during the day and a minimum at night--usually just before dawn (LBL-76/07-water).

Diurnal flow curve: A curve which depicts flow distribution over the 24-hour day (EPA-82/11f).

Diurnal peak: The highest value of air pollution in daily variations.

Diurnal variation: The variation throughout the day of a quantity in the earth's atmosphere. Usually applied to daily recurring processes (NATO-78/10).

Divalent: Describing a atom (e.g., metal atom) which is capable of combining with two hydrogen atoms.

Divergence: *See* synonym, convergence.

Diversion:
(1) Use of part of a stream flow as a water supply (EPA-94/04).
(2) A channel with a supporting ridge on the lower side constructed across a slope to divert water at a non-erosive velocity to sites where it can be used or disposed of (EPA-94/04).

Diversion (of water): The taking of water from a water body by way of a canal, pipe, or other conduit (DOI-70/04).

Diversion rate (recovery rate or recycling rate):
(1) A measure of the amount of waste material being diverted for recycling compared with the total amount that was previously thrown away (EPA-89/11).
(2) The tonnage of recyclables collected and processed into new products divided by total tonnage of municipal waste. The term is also known as the (OTA-89/10).
(3) The percentage of waste materials diverted from traditional disposal such as landfilling or incineration to be recycled, composted, or re-used (EPA-94/04).

Diversity: The number and abundance of species in a specified location (EPA-85/09; 91/03).

Diversity electric: The characteristic or variety of electric loads whereby individual maximum demands usually occur at different times. Diversity among customers' loads results in diversity among the loads (*see* electric for more related terms) (EPA-83).

Divide (or drainage divide): The boundary between one drainage basin and another; the line separating two watersheds (DOI-70/04).

DNA: Deoxyribonucleic acid, the molecule in which the genetic information for most living cells is encoded. Viruses, too, can contain RNA (EPA-89/12).

DNA hybridization: Use of a segment of DNA, called a DNA probe, to identify its complementary DNA; used to detect specific genes (EPA-94/04).

DNT (dinitrotoluene): Added as a deterrent to propellant grains; reduces burning rate (EPA-76/03).

Doctrine of appropriation (or priority of rights): The doctrine that whoever puts water to a beneficial use may continue to take it so long as the use does not conflict with use by someone with an earlier claim to the same source: First in time, first in right." In the seventeen western states this doctrine applies either exclusively or as a hybrid appropriation-riparian right doctrine (DOI-70/04).

Document glass: An ultraviolet absorbing glass used for protecting documents (*see* glass for more related terms) (EPA-83).

DOE Office of New Production Reactors: An organization within the Department of Energy, reporting directly to the Secretary, that implements the Department's strategy to provide new production reactor capacity on an urgent schedule for an ensured supply of nuclear materials, primarily tritium, to maintain the nation's nuclear deterrent capability (DOE-91/04).

Dog hairs: Defect of protruding fibers (EPA-83).

Doghouse: A small boxlike vestibule on a glass furnace into which the batch is fed or which facilitates the introduction and removal of floaters (EPA-83).

Doilies: The paper place mats used on food service trays in hospitals and other institutions (40 CFR 250.4-91).

Dollar base: A period in time in which all costs are related. Investment costs are related by the Sewage Treatment Plant Construction Cost Index. Supply costs are related by the Industrial Commodities Wholesale Price Index (EPA-83/06a).

Dolomite: A mineral having the empirical, composition, 1 mole of calcium carbonate and 1 mole of magnesium carbonate ($CaCO_3 \bullet MgCO_3$) (cf. burnt lime or limestone) (EPA-83).

Domestic construction material: An unmanufactured construction material which has been mined or produced in the United States, or a manufactured construction material which has been manufactured in the United States if the cost of its components which are mined, produced, or manufactured in the United States exceeds 50 percent of the cost of all its components (40 CFR 35.936.13-91).

Domestic garbage: *See* synonym, domestic refuse.

Domestic municipal or household waste: *See* synonym, residential waste.

Domestic or other non-distribution system plumbing problem: A coliform contamination problem in a public water system with more than one service connection that is limited to the specific service connection from which the coliform-positive sample was taken (40 CFR 141.2-91).

Domestic refuse (or domestic garbage): The solid wastes which normally originate in residential household or multifamily units (*see* refuse for more related terms) (EPA-83).

Domestic sewage: Untreated sanitary wastes that pass through a sewer system (*see* sewage for more related terms) (40 CFR 261.4-91).

Domestic sewage sludge policy: In HSWA, this policy involves:
(1) Setting concentration limits for a variety of metals and organics found in municipal sludge; and
(2) Applying the Toxicity Concentration Leaching Procedure (TCLP) test to domestic sludges. If the sludges fail the TCLP test, they would be managed as a RCRA hazardous waste (OSWER-87)

Domestic use of water: The water used in homes and on lawns, including use for laundry, washing cars, cooling, and swimming pools (*see* water for more related terms) (DOI-70/04).

Dominance hierarchy: The relative positions of forcefulness.

Dominant lethal mutation: One occurring in a germ cell which does not cause dysfunction of the gamete, but which is lethal to the fertilized egg or developing embryo (40 CFR 798.5450-91).

Dopant: An impurity element added to semiconductor materials used in crystal diodes and transistors (EPA-83/03).

Dope: Slang for mold lubricant (EPA-83).

Dore: Gold and silver bullion remaining in a cupelling furnace after oxidized lead is removed (EPA-83/03a).

Dormancy: The inactive period during the development of many animals and plants (cf. diapause, estivation or hibernation).

Dosage: The time integral of the concentration of a pollutant over a sampling time (NATO-78/10).

Dosage/dose: The actual quantity of a chemical administered to an organism or to which it is exposed (EPA-94/04).

Dose:
(1) The amount of test substance administered. Dose is expressed as weight of test substance (g, mg) per unit weight of test animal (e.g., mg/kg), or as weight of test substance per unit weight of food or drinking water (40 CFR 795.260-91, *see also* 40 CFR 798.1100; 798.1175; 798.4900; 798.2250; 798.2450; 798.2650; 798.2675-91).
(2) In radiology, the quantity of energy or radiation absorbed (EPA-89/12).
(3) The energy imparted to matter by ionizing radiation. The unit of absorbed dose is the rad, which is equal to 100 ergs per gram of irradiated material in any medium (cf. reference dose) (DOE-91/04).

Other dose related terms include:
- Absorbed dose
- Maximum tolerated dose (MTD)
- Reference dose
- Risk specific dose
- Threshold dose

Dose equivalent:
(1) The product of the absorbed dose from ionizing radiation and such factors as account for differences in biological effectiveness due to the type of radiation and its distribution in the body as specified by the International Commission on Radiological Units and Measurements (ICRU) (40 CFR 141.2-91, *see also* 40 CFR 190.02-91; EPA-4/91).
(2) The product of the absorbed does from ionizing radiation and such factors as account for biological differences due to the type of radiation and its distribution in the body as specified by the International Commission on Radiological Units and Measurements (EPA-94/04).

Dose estimate (mg/Kg-day) (Course 165.6):
(1) For inhalation, Dose = (concentration in air x contact rate x exposure duration x absorbed fraction)/body weight;
(2) For ingestion from drinking water, Dose = (concentration in water x contact rate x exposure duration x absorbed fraction)/body weight; and
(3) For dermal absorption, Dose = (concentration in soil or water x contact rate x exposure duration x exposed area x absorbed fraction)/body weight

Example:
- Determine an individual's intake daily dose.

Data:
- The concentration of intake amounts is usually expressed as mg/liter-water or ppm (1 liter-water weighs 1000 g).

Assumption:
(1) Water consumed by an individual each day: 2 liters
(2) Concentration of a substance present in the water: 10 ppm
(3) Average weight of an individual: 70 Kg (154 pounds)

Solution:
- The daily dose of the hypothetical case: 10 x 2/70 = 0.29 g/Kg-day.

Dose response:
(1) The relationship between the dose and the proportion of a population sample showing a defined effect (40 CFR 798.1100-91, *see also* 40 CFR 798.1175; 798.1150-91).
(2) Dose response is a quantitative relationship between the dose of a chemical and an effect caused by the chemical. In general, a given amount of a toxic agent will elicit a given type and intensity of response. The dose response relationship is a fundamental concept in toxicology and the basis for measurement of the relative harmfulness of a chemical. The National Institute for Occupational Safety and Health (NIOSH) defines a number of general dose-response terms in the "Restry of Toxic Substances, 1983, p. xxxii" (Course 165.5).
(3) How a biological organism's response to a toxic substance quantitatively shifts as its overall exposure to the substance changes (e.g., a small dose of carbon monoxide may cause drowsiness; a large dose can be fatal) (EPA-94/04).

Other dose response related terms include:
- Infectious dose 50
- Infective dose
- Lethal concentration
- Lethal concentration fifty (LC50)
- Lethal concentration low (LC LO)
- Lethal dose (LD)
- Lethal dose fifty (LD50)
- Lethal dose low (LD LO)
- Lethal dose zero (LD 0)
- Toxic concentration low
- Toxic dose low

Dose response assessment: Estimating the potency of a chemical (EPA-94/04).

Dose response curve: A graphical presentation of the relationship between degree of exposure to a chemical (dose) and observed biological effect or response (Course 165.6).

Dose response relationship:
(1) A relationship between the amount of an agent (either administered, absorbed, or believed to be effective) and changes in certain aspects of the biological system (usually toxic effects), apparently in response to that agent (EPA-92/12).
(2) The quantitative relationship between the amount of exposure to a substance and the extent of toxic injury or disease

produced (EPA-94/04).

Dosemeter: *See* synonym, dosimeter.

Dosimeter (or dosemeter): An instrument that measures exposure to radiation (EPA-89/12).

Dosimetry processor: An individual or an organization that processes and evaluates personnel monitoring equipment in order to determine the radiation dose delivered to the equipment (10 CFR 20.3-91).

Dosing tank: A sewage tank in which effluent is collected until there is enough in it to operate the next treatment process. Dosing tanks may have **multiple dosing tanks** for alternate applications (*see* tank for more related terms).

DOT reportable quantity: The quantity of a substance specified in U.S. Department of Transportation regulation that triggers labelling, packaging and other requirements related to shipping such substances (EPA-94/04).

Double base: A propellant which is made from two explosive substances, e.g., nitroceullulose, gelatinized with nitroglycerin (EPA-76/03).

Double base propellant: A propellant containing two energy-giving ingredients; nitrocellulose and nitroglycerin (EPA-76/03).

Double block and bleed system: Two block valves connected in series with a bleed valve or line that can vent the line between the two block valves (40 CFR 60.481; 61.241; 264.1031-91).

Double bond: Two atoms which share two pairs of electrons (*see* chemical bond for more related terms).

Double column rectifier: A system of fractional distillation employed in the air-separation process to effectively separate compressed, purified air into liquefied components of nitrogen and oxygen (EPA-77/07).

Double effect evaporators: Double effect evaporators are two evaporators in series where the vapors from one are used to boil liquid in the other (EPA-87/10a).

Double wash/rinse: A minimum requirement to cleanse solid surfaces (both impervious and nonimpervious) two times with an appropriate solvent or other material in which PCB's are at least 5 percent soluble (by weight). A volume of PCB-free fluid sufficient to cover the contaminated surface completely must be used in each wash/rinse. The wash/rinse requirement does not mean the mere spreading of solvent or other fluid over the surface, nor does the requirement mean a once-over wipe with a soaked cloth. Precautions must be taken to contain any runoff to dispose properly of wastes generated during the cleaning (40 CFR 761.123-91).

Downdraught: The capturing of a plume by the downward air motions behind buildings or other structures. The plume reaches ground level prematurely, which results in high concentrations there (NATO-78/10).

Downgradient: The direction that groundwater flows; similar to "downstream" for surface water (EPA-94/04).

Downpass: A chamber or gas passage placed between two combustion chambers to carry the products of combustion downward (SW-108ts).

Downwash: The downward from a point of reference toward which the wind is blowing (NATO-78/10).

DPD (n,n-diethyl-paraphenylene diamine): An agent for testing the residual chlorine in drinking water.

DPN [(diphosphopyridine nucleotide), or nicotinamide adenine dinucleotide (NAD)]: A coenzyme necessary for the alcoholic fermentation of glucose (LBL-76/07-water).

Draft:
(1) A gas flow resulting from pressure difference, e.g., the pressure difference between an incinerator and the atmosphere moves the products of combustion from the incinerator to the atmosphere (EPA-89/03b).
(2) The act of drawing or removing water from a tank or reservoir (EPA-94/04).
(3) The water which is drawn or removed (EPA-94/04).
Other draft related terms include:
- Forced draft
- Induced draft
- Natural draft
- Balanced draft

Draft equipment: Equipment for generating draft gas flow in a combustion system.

Draft controller: An automatic device that maintains a uniform furnace draft by regulating a damper (SW-108ts).

Draft permit:
(1) A document prepared under 40 CFR 124.6 indicating the Director's tentative decision to issue or deny, modify, revoke and reissue, terminate, or reissue a permit. A notice of intent to terminate a permit, and a notice of intent to deny a permit, as discussed in 40 CFR 124.5, are types of draft permits. A denial of a request for modification, revocation and reissuance, or termination, as discussed in 40 CFR 124.5, is not a draft permit. A proposed permit is not a draft permit (*see* permit for more related terms) (40 CFR 122.2-91, *see also* 40 CFR 124.2; 124.41; 144.3; 270.2-91).
(2) A preliminary permit drafted and published by EPA; subject to public review and comment before final action on the application (EPA-94/04).

Drag:
(1) A normalized value for pressure drop which is obtained by dividing pressure drop by the gas velocity. This parameter allows comparison of one dust/filter medium to another on a common basis and at various times during the filtration cycle.
(2) The lower half of a two piece sand mold (EPA-85/10a).

Drag coefficient:
(1) A coefficient used to calculate the drag force on a particle (EPA-84/09).
(2) For a atmospheric boundary layer, a dimensionless number equal to the ratio of the square of the friction velocity to the square of the wind speed at a given height (NATO-78/10).

Drag conveyer: A conveyer that uses vertical steel plates fastened between two continuous chains to drag material across a smooth surface (*see* conveyer for more related terms) (SW-108ts).

Drag force: The force a particle experiences whenever there is relative motion between the particle and gas stream. It is a resistive force (EPA-84/09).

Drag in: Water or solution carried into another solution by a workpiece such as a film and the associated handling equipment (EPA-83/06a).

Dragline:
(1) A piece of excavating equipment which employs a cable-hung bucket to remove overburden (EPA-82/10).
(2) A revolving shovel that carries a bucket attached only by cables and digs by pulling the bucket toward itself (SW-108ts).

Drag plate: A plate beneath a traveling or chain-grate stoker used to support the returning grates (SW-108ts).

Drag out:
(1) The solution that adheres to the objects removed from a bath, more precisely defined as that solution which is carried past the edge of the tank (EPA-83/06a).
(2) Loss of process chemicals and solution onto products during processing which are made up by periodic fresh addition of chemicals and solution (EPA-82/11).

Drag out reduction: Minimization of the amount of materials (bath or solution) removed from a process tank by adhering to the part or its transfer device (EPA-83/06a).

Dragline: A mobile crane that carries a bucket attached only by cables and digs and loads material by pulling the bucket toward itself (EPA-83).

Drain: To carry away or to empty liquid from a liquid source.

Drainage:
(1) A device, e.g., a channel or a pipe, to carry away or to empty liquid from a liquid source.
(2) Improving the productivity of some agricultural land by removing excess water from the soil by such means as ditches or subsurface drainage tiles (EPA-94/04).

Drainage basin: The area of land that drains water, sediment, and dissolved materials to a common outlet at some point along a stream channel (EPA-94/04).

Drainage divide: *See* synonym, divide.

Drainage phase: A period in which the excess plating solution adhering to the part or workpiece is allowed to drain off (EPA-83/06a).

Drainage water: The incidental surface waters from diverse sources such as rainfall, snow melt or permafrost melt (*see* water for more related terms) (40 CFR 440.141-91).

Drainage well: A well drilled to carry excess water off agricultural fields. Because they act as a funnel from the surface to the ground water below, drainage wells can contribute to ground water contamination (EPA-94/04).

Draught: Draught related terms include:
- Forced draught
- Induced draught

Draw: A tributary valley or stream that usually discharges water only after a rainstorm (DOI-70/04).

Drawdown:
(1) The lowering of the water level in a well and in the adjacent water table as a result of withdrawal by pumping; a drop in the water level of a reservoir (DOI-70/04).
(2) (A) The drop in the water table or level of water in the ground when water is being pumped from a well (EPA-94/04).
(B) The amount of water used from a tank or reservoir (EPA-94/04).
(C) The drop in the water level of a tank or reservoir (EPA-94/04).

Drawing: The process of pulling a metal through a die or succession of dies to reduce the metal's diameter or alter its cross-sectional shape (40 CFR 471.02-91, *see also* 40 CFR 467.02; 468.02-91).

Drawing compound: Oils, waxes, or greases added to facilitate stamping and forming of metal (EPA-82/11e).

Dredge:
(1) A self-contained combination of an elevating excavator (e.g., bucket line dredge), the beneficiation or gold-concentrating plant, and a tailings disposal plant, all mounted on a floating barge (40 CFR 440.141-91).
(2) In mining, a large floating device for underwater excavation of materials using either a chain of buckets, suction pumps, or other devices to elevate and wash alluvial deposits and gravel for gold, tin, platinum, heavy minerals, etc. (EPA-82/05).
(3) cf. suction dredge.

Dredged material: The material that is excavated or dredged from waters of the United States (40 CFR 232.2-91).

Dredged material permit: A permit issued by the Corps of Engineers under section 103 of the Act (*see* 33 CFR 209.120) and any Federal projects reviewed under section 103(e) of the Act. (*See* 33 CFR 209.145) (40 CFR 220.2-91).

Dredging: Removal of mud from the bottom of water bodies. This can disturb the ecosystem and causes silting that kills aquatic life.

Dredging of contaminated muds can expose biota to heavy metals and other toxics. Dredging activities may be subject to regulation under Section 404 of the Clean Water Act (EPA-94/04).

Dregs: The inert rejects from the green liquor clarifier of a pulp mill (EPA-87/10).

Dregs washer: A piece of equipment used to wash the green liquor (Na_2CO_3) off the dregs prior to their disposal (EPA-87/10).

Dressing: In mining, originally referred to the pickling, sorting, and washing of ores preparatory to reduction. The term now includes more elaborate processes of milling and concentration of ores (EPA-82/05).

Drier: A composition which accelerates the drying of oil, paint, printing ink, or varnish. Driers are available in both solid and liquid forms (EPA-79/12b).

Drift:
(1) In sampling, it means deviation.
(2) In cooling tower, entrained water carried from a cooling device by the exhaust air. Also known as carry-over (EPA-82/11f; Gurney-66).
(3) In mining, a deep mine entry driven directly into a horizontal or near horizontal mineral seam or vein when it outcrops or is exposed at the ground surface (EPA-82/10).

Other drift related terms include:
- Calibration drift
- Calibration drift test period
- Span drift
- Zero drift

Drift loss: Water lost from a cooling water system due to the carry out of water drops or vapors by the rising cooling air stream.

Drift mining: A term applied to working alluvial deposits by under ground methods of mining. The paystreak is reached through an adit or a shallow shaft. Wheelbarrows or small cars may be used for transporting the gravel to a sluice on the surface (EPA-82/05).

Drift organism: The benthic organisms temporarily suspended in the water and carried downstream by the current (DOD-78/01).

Drift velocity (or particle migration velocity): Once a particle is charged in an electrostatic precipitator, the particle will migrate toward the collection electrode with a certain drift velocity. It represents the collectability of the particles within the confines of an electrostatic precipitator (EPA-84/09).

Drilling: Making a hole with a rotary, end-cutting tool having one or more cutting lips and one or more helical or straight flutes or tubes for the ejection of chips and the passage of a cutting fluid (EPA-83/06a). Other drilling related terms include:
- Center drilling
- Core drilling
- Spade drilling
- Step drilling
- Gun drilling
- Oil hole or pressurized coolant drilling

Drilling and production facility: All drilling and servicing equipment, wells, flow lines, separators, equipment, gathering lines, and auxiliary nontransportation-related equipment used in the production of petroleum but does not include natural gasoline plants (40 CFR 60.111-91).

Drilling mud: A heavy suspension used in drilling an injection well, introduced down the drill pipe and through the drill bit (40 CFR 144.3-91).

Drinking water: *See* synonym, potable water.

Drinking water equivalent level: Protective level of exposure related to potentially non-carcinogenic effects of chemicals that are also known to cause cancer (EPA-94/04).

Drinking water standards: Standards are in terms of suspended matter, excess salts, unpleasant taste and all harmful microbes. Specific standards are enforced by State regulations.

Drinking water supply: Any raw or finished water source that is or may be used by a public water system (as defined in the Safe Drinking Water Act) or as drinking water by one or more individuals (SF101, *see also* 40 CFR 300.5-91).

Drip pad: An engineered structure consisting of a curbed, free-draining base, constructed of non-earthen materials and designed to convey preservative kick-back or drippage from treated wood, precipitation, and surface water run-on to an associated collection system at wood preserving plants (40 CFR 260.10-91).

Drip time: The period during which a part is suspended over baths in order to allow the excessive dragout to drain off (EPA-83/06a).

Drive roller: In constructing a landfill liner, a knurled or rubber roller which grips the FML (flexible membrane liner) sheets via applied pressure and propel the seaming device at a controlled rate of travel (EPA-89/09, *see also* EPA-91/05).

Drive system: Is determined by the number and location of drive axles (e.g., front wheel drive, rear wheel drive, four wheel drive) and any other feature of the drive system if the Administrator determines that such other features may result in a fuel economy difference (40 CFR 600.002.85-91).

Drive train configuration: The unique combination of engine code, transmission configuration, and axile ratio (40 CFR 86.082.2-91).

Driving cycle (or driving mode): A standardized simulation of a real driving situation in order to establish vehicle emission factors (NATO-78/10).

Driving mode: *See* synonym, driving cycle.

Drop arch: Any vertical refractory wall supported by arch construction (EPA-83).

Drop manhole: *See* synonym, manhole.

Drop off:
(1) The transport of individual MSW (municipal solid waste)

(e.g., newspaper, cans, and bottles) by individuals to a specified area for subsequent processing (OTA-89/10).
(2) Recyclable materials collection method in which individuals bring them to a designated collection site (EPA-94/04).

Drop off center: A method of collecting recyclable or compostible materials in which the materials are taken by individuals to collection sites and deposited into designated containers (EPA-89/11).

Droplet: A small liquid particle of such size and density as to fall under still conditions, but which may remain suspended under turbulent conditions (EPA-83/06).

Dross:
(1) Residues generated during the processing of molten metals or metal alloys by oxidation in air (EPA-76/12).
(2) Oxidized impurities occurring on the surface of molten metal (EPA-83/03a).

Dross reverberatory furnace: A furnace used for the removal or refining of impurities from lead bullion (*see* furnace for more related terms) (40 CFR 60.181-91).

Drug: Shall have the meanings contained in the Federal Food, Drug, and Cosmetic Act, 21USC321 et seq., and the regulations issued under it (40 CFR 710.2; 720.3-91).

Drug abuse:
(1) The use of any drug or substance listed by the Department of Justice in 21 CFR 1308.11, under authority of the Controlled Substances Act, 21USC801, as a controlled substance unavailable for prescription because:
  (A) The drug or substance has a high potential for abuse,
  (B) The drug or other substance has no currently accepted medical use in treatment in the United States,
  (C) There is a lack of accepted safety for use of the drug or other substance under medical supervision.
(2) The misuse of any drug or substance listed by the Department of Justice in 21 CFR 1308.12-1308.15 under authority of the Controlled Substances Act as a controlled substance available for prescription (40 CFR 7.25-91).

Drug dependent person: A person who is using a controlled substance (as defined in section 102 of the Controlled substances Act) and who is in a state of psychic or physical dependence, or both, arising from the use of that substance on a continuous basis. Drug dependence is characterized by behavioral and other responses which include a strong compulsion to take the substance on a continuous basis in order to experience its psychic effects or to avoid the discomfort caused by its absence (PHSA2-q).

Drug free workplace: A site for the performance of work done in connection with a specific grant at which employees of the grantee are prohibited from engaging in the unlawful manufacture, distribution, dispensing, possession, or use of a controlled substance (40 CFR 32.605-91).

Drum: A 55-gallon drum is a standard size industrial barrel used for storing and transporting raw materials, products and wastes. One (1) 55-gallon drum = 7.3524 cubic feet = 0.20820 cubic meters.

Drum brake lining: Any asbestos-containing product intended for use as a friction material in drum brake systems for vehicles rated at less than 26,001 pounds gross vehicle weight rating (GVWR) (40 CFR 763.163-91).

Drum mill: A long, inclined steel drum that rotates and grinds solid wastes in its rough interior. Smaller ground materials fall through holes near the end of the drum and larger materials drop out of the end. The drum mill is used in some composting operations (SW-108ts).

Dry air: Air containing no water vapor (*see* air for more related terms) (EPA-89/03b).

Dry ash free basis: Analytical data calculated to a theoretical base of no moisture or ash associated with the sample. Numerical values (air dry loss, residual moisture, and ash content) are used for converting the as determined data to a moisture and ash-free basis (*see* analytical parameters--fuels for more related terms) (EPA-83).

Dry basis: Analytical data calculated to a theoretical base of no moisture associated with the sample. The numerical value (residual moisture value) is used for converting the as determined data to a dry basis (*see* analytical parameters--fuels for more related terms) (EPA-83).

Dry bottom furnace: A furnace in which the ash leaves the boiler bottom as a solid (as opposed to a molten slag) (*see* furnace for more related terms) (EPA-82/11f).

Dry bulb temperature: The temperature measured by a thermometer whose bulb (mercury holder at the bottom of a thermometer) is dry (exposed to air). It is the temperature indicated by an ordinary thermometer placed in the mixture (*see* temperature for more related terms) (EPA-89/03b).

Dry charge process: A process for the manufacture of lead acid storage batteries in which the plates are charged by electrolysis in sulfuric acid, rinsed, and drained or dried prior to shipment of the battery. Charging of the plates usually occurs in separate containers before assembly of the battery but may be accomplished in the battery case. Batteries produced by the dry charge process are shipped without acid electrolyte. Also referred to as dehydrated plate or dehydrated batteries (EPA-84/08).

Dry cleaning operation: That process by which an organic solvent is used in the commercial cleaning of garments and other fabric materials (40 CFR 52.1088-91, *see also* 40 CFR 52.1107; 52.2440-91).

Dry cooling tower (or dry tower):
(1) A cooling tower in which the fluid to be cooled flows within a closed system which transfers heat to the environment using finned or extended surfaces (EPA-82/11f).
(2) Similar to a regular cooling tower, except that when water is circulating through cooling tubes, it does not evaporate.

Dry criticality: Nuclear reactor's criticality is reached without applying coolant.

Dry dam: A retarding structure in the headwater area of a stream, designed for flash flood control, no permanent storage of water is involved, and the area can be farmed or grazed between flood periods (*see* dam for more related terms) (DOI-70/04).

Dry electrolytic capacitor: An electrolytic capacitor with a paste rather than liquid electrolyte (EPA-83/03).

Dry electrostatic precipitator (ESP): A process by which particles suspended in a gas flow are electrically charged and separated (attracted to a collecting electrode of an opposite charge) from the gas stream under the influence of an electrostatic field in the order of 40,000-100,000 volts (*see* electrostatic precipitator for more related terms) (EPA-84/09).

Dry farming: A method of farming without irrigation in an area of limited rainfall, the land being treated so as to conserve the moisture it contains (DOI-70/04).

Dry flue gas desulfurization technology: A sulfur dioxide control system that is located downstream of the steam generating unit and removes sulfur oxides from the combustion gases of the steam generating unit by contacting the combustion gases with an alkaline slurry or solution and forming a dry powder material. This definition includes devices where the dry powder material is subsequently converted to another form. Alkaline slurries or solutions used in dry flue gas desulfurization technology include but are not limited to lime and sodium (40 CFR 60.41b-91, *see also* 40 CFR 60.41c-91).

Dry gas: Gas containing no water vapor

Dry impingement: The process or impingement carried out so that gases or particulate matter carried in the gas stream is retained upon the surface against which the stream is directed. The collecting surface may be treated with a film of adhesive (*see* impingement for more related terms) (LBL-76/07-air).

Dry injection: The injection of a dry reagent such as lime powder into an incinerator or the original waste to aid in the control of acid gas emission during incineration (OTA-89/10).

Dry injection adsorption system: One of air pollution control devices. This type of scrubber uses finely divided calcium hydroxide for the adsorption of acid gases. The reagent feed has particle sizes with 90% by weight solids through 325 mesh screens. The calcium hydroxide is injected countercurrently in the gas stream. The gas stream containing the entrained calcium hydroxide particles and fly ash is then vented to fabric filter. Adsorption of acid gases and organic compounds (if present) occurs primarily while the gas stream passes the dust cake on the surface of the filter bags. The calcium hydroxide feed rate for dry injection is three to four times the stoichiometric quantities needed, thus making the system unattractive for very large systems (*see* scrubber for more related terms) (EPA-89/02).

The major components of a dry injection system include:
(1) Dry sorbent storage tank
(2) Blower and pneumatic line for transfer of the sorbent
(3) Injector
(4) Particulate control device (fabric filter or ESP) for collection of the dry sorbent
(5) Expansion/reaction chamber (optional). In some cases, the expansion/reaction chamber is not included.

Dry lake: The site of a former lake, which needs not be literally dry, but may support marsh or even aquatic vegetation (*see* lake for more related terms) (DOI-70/04).

Dry limestone process: A method of controlling air pollution caused by sulfur oxides. The polluted gases are exposed to limestone which combines with oxides of sulfur to form manageable residues (EPA-74/11).

Dry lot: A confinement facility for growing ducks in confinement with a dry litter floor cover and no access to swimming areas (40 CFR 412.21-91).

Dry nitrogen: The nitrogen which has been purified by removal of water vapor (*see* nitrogen for more related terms) (EPA-88/12).

Dry oxygen: The oxygen measured at the dry condition (without water vapor) (cf. combustion correction factor) (*see* oxygen for more related terms).

Dry process:
(1) Processes which handle or process solid waste directly as received without the addition of water (EPA-83).
(2) A cement manufacturing process in which the feed material enters the kiln system in dry powdered form (*see* process for more related terms) (ETI-92).

Dry saturated steam: *See* synonym, dry steam.

Dry scrubber: One of air pollution control devices. A dry scrubber utilizes absorption and adsorption for the removal of acid gases, primarily hydrogen chloride, sulfur dioxide, hydrogen fluoride, and other acid gases. The scrubbers can be grouped into three major categories. The main differences between the various systems are the physical form of the alkaline reagent and the design of the vessel used for contacting the acid gas laden stream (*see* scrubber for more related terms) (EPA-89/02; 89/03b).

Dry sorption process: A process that involves contacting the gas stream with a solid phase that can remove one or more of the gaseous contaminants (EPA-81/09).

Dry steam (or dry saturated steam):
(1) Steam, a mixture of vapor and liquid, at the saturation temperature corresponding to the pressure and containing no water in suspension, i.e., the liquid component is zero and the quality is 100%.
(2) Steam containing no condensed moisture.
(3) *See* steam for more related terms

Dry tower: *See* synonym, dry cooling tower.

Dry transformer: A transformer having the core and coils neither impregnated with an insulating fluid nor immersed in an insulating oil (*see* transformer for more related terms) (EPA-83/03).

Dry weather flow: The rate at which wastewater flows through sewage treatment plants during periods when no storm runoff enters the sewers (*see* flow for more related terms) (DOI-70/04).

Dry well: A dry compartment of a pump structure at or below pumping level where pumps are located (*see* well for more related terms) (EPA-82/11f).

Dryer:
(1) Any facility in which a copper sulfide ore concentrate charge is heated in the presence of air to eliminate a portion of the moisture from the charge, provided less than 5 percent of the sulfur contained in the charge is eliminated in the facility (40 CFR 60.161-91).
(2) A unit in which the moisture content of phosphate rock is reduced by contact with a heated gas stream (40 CFR 60.401-91, *see also* 40 CFR 60.621-91).

Dryer felt: Continuous blanket, generally made from cotton and asbestos, at holds the wet sheet of paper against the dryer rolls of the paper machine (EPA-83).

Dryer scum: A deposit of soluble salts present in the clay and carried to the surface by the water as it escapes during drying (EPA-83).

Drying agent: An agent which is capable of absorbing water vapors from other substances.

Drying area: The area where VOC from applied cement or green tire sprays is allowed to evaporate (40 CFR 60.541-91).

Drying bed: In wastewater treatment, an area for dewatering of sludge by evaporation and seepage (EPA-83/03).

Drying hearth: A surface within the primary combustion chamber upon which wet waste materials are deposited for drying, prior to burning (*see* hearth for more related terms) (EPA-83).

Drying oil: An oil which readily takes oxygen from the air and changes it to a relatively hard, tough, elastic substance, when exposed, to form a thin, dry film. Drying oils also act as binders for pigments used in coatings (*see* oil for more related terms) (EPA-79/12b).

Drying oven: A chamber in which heat is used to bake, cure, polymerize, or dry a surface coating (40 CFR 60.711-91, *see also* 40 CFR 60.741-91).

Dual detector: *See* definition under GC/DD.

Dual media: A deep-bed filtration system utilizing two separate and discrete layers of dissimilar media (e.g., anthracite and sand) placed one on top of the other to perform the filtration function (EPA-76/03).

Dual media filtration: A deep-bed filtration system utilizing two separate and discrete layers of dissimilar media (e.g., anthracite and sand) placed one on top of the other to perform the filtration function (*see* filtration for more related terms) (EPA-87/10a).

Dual nickel plate: Two layers of nickel electroplate with different properties to enhance corrosion resistance and appearance under chromium electroplate. Requires two different nickel plating baths (EPA-74/03d).

Dual significance: The classification of priority pollutants which are:
(1) Manufactured pesticide products (primary significance) and are controlled by monitoring other pollutant of primary significance (secondary significance); or
(2) Manufactured pesticide products with zero wastewater discharge (primary significance) and lack adequate monitoring data to recommend regulation in other pesticide processes (secondary significance) (EPA-85/10).

Duct: A conduit, usually metal or fiberglass, round or rectangular in cross section, used for conveyance of air (OME-88/12).

Duct burner: A device that combusts fuel and that is placed in the exhaust duct from another source (such as a stationary gas turbine, internal combustion engine, kiln, etc.) to allow the firing of additional fuel to heat the exhaust gases before the exhaust gases enter a steam generating unit (*see* burner for more related terms) (40 CFR 60.41c-91, *see also* 40 CFR 60.41b-91).

Duct work: Any enclosed channel(s) which direct the movement of air or other gas (EPA-88/08).

Ductile iron: A cast iron that has been treated while molten with a master alloy containing an element such as magnesium or cerium to induce the formation of free graphite as nodules or spherules, which imparts a measurable degree of ductility to the cast metal (40 CFR 464.31-91).

Dulong's formula: A formula for calculating the approximate heating value of a solid fuel based on its ultimate analysis (SW-108ts). The formula is: $Q = 14544(C) + 62028(H) - 7753.3(O) + 4050(S)$, Q is the heating value of the mixture; C, H, O, and S are ultimate analysis (% by weight) of carbon, hydrogen, oxygen, and sulfur respectively.

Dummy variable: A symbol representing the statistical encoding of a qualitative value (EPA-79/12c). Encoding means to transfer words into numbers, e.g., in a "0" and "1" system, 0 can be assigned as no, negative or failure events; while "1" can be assigned as yes, positive or successful events (*see* variable for more related terms).

Dump (open dump, open dumping or surface dump):
(1) A land disposal site at which solid wastes are disposed of in a manner that does not protect the environment, are susceptible to open burning, and are exposed to the elements, vectors, and scavengers (*see* land disposal for more related terms) (40 CFR 240.101-91, *see also* RCRA1004; 40 CFR 165.1; 241.101; 257.2-91).
(2) A site used to dispose of solid waste without environmental controls (EPA-94/04).

Dump energy: Energy generated by water power that cannot be stored or conserved when such energy is beyond the immediate needs of the producing system (*see* electric energy for more related terms) (EPA-83).

Dump leaching: A term applied to dissolving and recovering minerals from subore-grade materials from a mine dump. The dump is irrigated with water, sometimes acidified, which percolates into and through the dump, and run-off from the bottom of the dump is collected, and a mineral in solution is recovered by chemical reaction. Often used to extract copper from low grade, waste material of mixed oxide and sulfide mineralization produced in open pit mining (EPA-82/05).

Dump plate: a hinged plate in an incinerator that supports residue and from which residue may be discharged by rotating the plate (SW-108ts).

Dump stack (or emergency safety stack): A quick opening, counter-balanced safety device to allow release of rapidly expanding gases (explosion) and pressure relief thereby preventing serious equipment damage (*see* stack for more related terms) (CRWI-5/89).

Dump site: A site used to dispose of solid wastes, or other containerized wastes, without environmental controls (*see* dump for more related terms).

Dumping: A disposition of material: Provided, that it does not mean a disposition of any effluent from any outfall structure to the extent that such disposition is regulated under the provisions of the FWPCA, under the provisions of section 13 of the River and Harbor Act of 1899, as amended (33USC407), or under the provisions of the Atomic Energy Act of 1954, as amended (42USC2011), nor does it mean a routine discharge of effluent incidental to the propulsion of, or operation of motor-driven equipment on, vessels: Provided further, that it does not mean the construction of any fixed structure or artificial island nor the intentional placement of any device in ocean waters or on or in the submerged land beneath such waters, for a purpose other than disposal, when such construction or such placement is otherwise regulated by Federal or State law or occurs pursuant to an authorized Federal or State program; and provided further, that it does not include the deposit of oyster shells, or other materials when such deposit is made for the purpose of developing, maintaining, or harvesting fisheries resources and is otherwise regulated by Federal or State law or occurs pursuant to an authorized Federal or State program (*see* land disposal for more related terms) (40 CFR 220.2-e-91).

Duplexing: An operation in which a lower grade of steels is produced in the basic oxygen furnace or open hearth and is then alloyed in the electric furnace (EPA-74/06a).

Duplicate sample: *See* synonym, replicate sample.

Duplicator paper: The writing papers used for masters or copy sheets in the aniline ink or hectograph process of reproduction (commonly called spirit machines) (*see* paper for more related terms) (40 CFR 250.4-91).

Durability: *See* synonym, chemical durability (EPA-83).

Durability fleet: A fleet of automobiles operated for mileage accumulation used to assess deterioration effects associated with the retrofit device (40 CFR 610.11-91).

Dust: Particles light enough to be suspended in air (EPA-89/12, *see* also EPA-83).

Dust box: A device to remove sugar dust from air, usually employing water sprays; a dust collector (EPA-74/01a).

Dust catcher: A part of the blast furnace through which the major portion of the dust is removed by mechanical separation (EPA-74/06a).

Dust collector:
(1) A device or combination of devices for separating dust from the air handled by an exhaust ventilation system (29 CFR 1910.94a-91).
(2) An air pollution control device for removing dust from air streams. Filtration, electrostatic precipitation, or cyclone principles may be utilized, but the term usually infers a dry system, not involving a water stream (cf. air pollution control equipment) (EPA-75/02c).
(3) Any device used to remove particulate from a gas stream (*see* collector for more related terms) (EPA-83).

Dust handling equipment: Any equipment used to handle particulate matter collected by the air pollution control device (and located at or near such device) serving any electric submerged arc furnace subject to this subpart (40 CFR 60.261-91, *see also* 40 CFR 60.271-91).

Dust handling system: The equipment used to handle particulate matter collected by the control device for an electric arc furnace or AOD vessel subject to this subpart. For the purposes of this subpart, the dust-handling system shall consist of the control device dust hoppers, the dust-conveying equipment, any central dust storage equipment, the dust-treating equipment (e.g., pug mill, pelletizer), dust transfer equipment (from storage to truck), and any secondary control devices used with the dust transfer equipment (40 CFR 60.271a-91).

Dust loading: The total amount of dust in a gas stream. Dust concentration usually applied to the contents of collection ducts and the emission from stacks. Its typical units include grains of dust per cubic foot of air or grams of dust per cubic meter (*see* loading for more related terms) (EPA-83/06).

Dustfall jar: An open container used to collect large particles from the air for measurement and analysis (EPA-94/04).

Dwell angle: The number of degrees of distributor mechanical rotation during which the breaker points are conducting current (40 CFR 85.2122(a)(5)(ii)(C); 85.2122(a)(11)(ii)(C)-91).

Dwell time: The time required for a chemical fusion, bodied chemical fusion or adhesive seam to take its initial tack, enabling the two opposing geomembranes to be joined together (*see* time for more related terms) (EPA-91/05).

Dye: A coloring material.

Dye penetrant testing: A nondestructive method for finding discontinuities that are open to the surface of the metal. A dye is applied to the surface of metal and the excess is rinsed off. Dye that penetrates surface discontinuities will not be rinsed away thus marking these discontinuities (40 CFR 471.02-91).

Dye image: A color image formed when the oxidized developer combines with the color couplers (EPA-80/10).

Dynamic calibration: The calibration of a measurement system by use of calibration material having characteristics similar to the unknown material to be measured. For example, the use of a gas containing sulfur dioxide of known concentrations in an air mixture could be used to calibrate a sulfur dioxide bubbler system (*see* calibration for more related terms) (LBL-76/07-air).

Dynamometer: *See* definition under shaft work.

Dynamometer idle for automatic transmission code heavy-duty engines: The manufacturer's recommended engine speed without a transmission that simulates the recommended engine speed with a transmission and with the transmission in neutral (40 CFR 86.082.2-91).

Dyne: A force unit. 1 dyne = 1 gram mass x 1 cm/sec^2 acceleration. 1 dyne = 10^{-5} newton.

Dysprosium (Dy): A rare earth metal with a.n. 66; a.w. 162.50; d. 8.54 g/cc; m.p. 1407° C and b.p. 2600° C. The element belongs to group IIIB of the periodic table.

Dystrophic lake: Acidic, shallow bodies of water that contain much humus and/or other organic matter; contain many plants but few fish (*see* lake for more related terms) (EPA-94/04).

Ee

e: The base for the natural or Naperian logarithms which equals 2.71828.. (EPA-85/10).

Ear insert device: A hearing protective device that is designed to be inserted into the ear canal, and to be held in place principally by virtue of its fit inside the ear canal (40 CFR 211.203-j-91).

Ear muff device: A hearing protective device that consists of two acoustic enclosures which fit over the ears and which are held in place by a spring-like headband to which the enclosures are attached (40 CFR 211.203-k-91).

Early life stage toxicity test: A test to determine the minimum concentration of a substance which produces a statistically significant observable effect on hatching, survival, development and/or growth of a fish species continuously exposed during the period of their early development (40 CFR 797.1600-91).

Earth blade: A heavy, broad plate that is connected to the front of a tractor and is used to push and spread soil or other materials (*see* blade for more related terms) (SW-108ts).

Earthen pond: A pond constructed with or without filtration control measures for the purpose of detention, long-term storage, or land disposal of influent waste waters (EPA-74/01a).

EC-X, ECX, or Ecx: The experimentally derived chemical concentration that is calculated to effect X percent of the test criterion (40 CFR 797.1050-91, *see also* 40 CFR 797.1060; 797.1160; 797.2750; 797.2800; 797.2850-91).

EC50: The experimentally derived concentration of test substance in dilution water that is calculated to affect 50 percent of a test population during continuous exposure over a specified period of time. In this guideline, the effect measured is immobilization (40 CFR 797.1300-91, *see also* 40 CFR 797.1330; 797.1800; 797.1830-91).

ECD: Electron capture detector, *see* definition under continuous emission monitor.

Ecological impact: The effect that a man-made or natural activity has on living organisms and their non-living (abiotic) environment (EPA-94/04).

Ecology:
(1) The science of the interrelations between living organisms and their environment (cf. ecosystem) (EPA-76/03).
(2) The relationship of living things to one another and their environment, or the study of such relationships (EPA-94/04).

Ecological indicator: A characteristic of the environment that, when measured, quantifies magnitude of stress, habitat characteristics, degree of exposure to a stressor, or ecological response to exposure. The term is a collective term for response, exposure. The term is a collective term for response, exposure, habitat, and stressor indicators (EPA-94/04).

Ecological risk assessment: The application of a formal framework, analytical process, or model to estimate the effects of human actions(s) on a natural resource and to interpret the significance of those effects in light of the uncertainties identified in each component of the assessment process. Such analysis includes initial hazard identification, exposure and doseresponse assessments, and risk characterization (EPA-94/04).

Economic poison:
(1) The same meaning as it has under the Federal Insecticide, Fungicide, and Rodenticide Act (7USC135-135k) and the regulations issued thereunder (40 CFR 163.2-91).
(2) The chemical used to control insects, rodents, plant diseases, weeds and other pests, and also to defoliate economic crops such as cotton (EPA-74/11).
(3) Chemicals used to control pests and to defoliate cash crops such as cotton (EPA-94/04).
(4) *See* poison for more related terms.

Economies of scale: Increases in production capacity that reduce the average cost per ton of output (OTA-89/10).

Economizer: A heat exchanger which uses the heat of combustion gases to raise the boiler feedwater temperature before the feedwater enters the boiler (EPA-82/11f).

Economizer ash: The carry-over ash from the boiler which due to its size and weight, settles in a hopper below the economizer (*see* ash for more related terms) (EPA-82/11f).

Economy energy: Energy produced and supplied from a more economical in one system, substituted for that being produced by a less economical source in another system (*see* electric energy for more related terms) (EPA-83).

Ecosphere: The bio-bubble that contains life on earth, in surface waters, and in the air (cf. biosphere) (EPA-94/04).

Ecosystem:
(1) That set of ecological relationships resulting from a primary focus on an organism, group of organisms, or a portion of the earth's surface (LBL-76/07-water).
(2) A complex of the community of living things and the environment forming a functioning whole in nature (DOE-91/04).
(1) The interacting system of a biological community and its non-living environmental surroundings (cf. ecology) (EPA-94/04).

Ecosystem structure: Attributes related to instantaneous physical state of an ecosystem; examples include species population density, species richness or evenness, and standing crop biomass (EPA-94/04).

Ecotone: A habitat created by the juxtaposition of distinctly different habitats; an edge habitat; or an ecological zone or boundary where two or more ecosystems meet (EPA-94/04).

Ectohumus: An accumulation of organic matter on the soil surface. It is also known as mor (cf. humus).

Ecotoxicological study: Measurement of effects of environmental toxicants on indigenous populations of organisms (Course 165.6).

ED10 (10 percent effective dose): An estimated dose associated with a 10 percent increase in response over control groups. For HRS purposes, the response considered is cancer. [milligrams toxicant per kilogram body weight per day (mg/kg-day)] (40 CFR 300-App/A-91).

Eddy: In a turbulent flow, a portion of fluid with an organized structure of its own. It can only exist for a certain time before being destroyed by breaking up into eddies of smaller size (NATO-78/10).

Eddy current separator: A device which passes a varying magnetic field through feed material, thereby inducing eddy currents in the nonferrous metals (e.g., aluminum) present in the feed. The eddy currents counteract the magnetic field and exert a repelling force on the metals, separating them from the field and the remainder of the feed (*see* separator for more related terms) (EPA-83).

Eddy diffusivity: The eddy diffusivity is the coefficient which describes the diffusive property of a turbulent flow in a form analogous to the molecular diffusion coefficient. However, molecular diffusion is a property of the fluid, while turbulent diffusion is a property of the flow and can therefore be a different function of space and time for each flow (NATO-78/10).

Eddy flow: A turbulent flow.

Eddy viscosity: The internal friction in a fluid caused by turbulent motions (NATO-78/10).

Edema: An abnormal accumulation of tissue fluid in the connective tissue of various cavities (LBL-76/07-bio).

Effect:
(1) **Direct effects**, which are caused by the action and occur at the same time and place.
(2) **Indirect effects**, which are caused by the action and are later in time or farther removed in distance, but are still reasonably foreseeable. Indirect effects may include growth inducing effects and other effects related to induced changes in the pattern of land use, population density or growth rate, and related effects on air and water and other natural systems, including ecosystems. Effects and impacts as used in these regulations are synonymous. Effects includes ecological (such as the effects on natural resources and on the components, structures, and functioning of affected ecosystems), aesthetic, historic, cultural, economic, social, or health, whether direct, indirect, or cumulative. Effects may also include those resulting from actions which may have both beneficial and detrimental effects, even if on balance the agency believes that the effect will be beneficial (40 CFR 1508.8-91).

Effect concentration (EC): A point estimate of the toxicant concentration that would cause an observable adverse effect (such as death, immobilization, or serious incapacitation) in a given percentage of the test organisms (EPA-91/03).

Effective area: Available surface areas of absorbents contacted by gas or liquid to be treated.

Effective date: The date of promulgation in the federal register of an applicable standard or other regulation under this part (40 CFR 61.02-91).

Effective date of a UIC program: The date that a State UIC program is approved or established by the Administrator (40 CFR 146.3-91).

Effective date of an NSO: The effective date listed in the Federal Register publication of EPA's issuance or approval of an NSO (40 CFR 57.103-91).

Effective dose equivalent:
(1) The sum of the products of the dose equivalent to the organ or tissue and the weighting factors applicable to each of the body organs or tissues that are irradiated. Weighting actors are: 0.25 or gonads, 0.15 for breast, 0.12 for red bone marrow, 0.12 for lungs, 0.03 for thyroid, 0.03 for bone surface, and 0.06 for each of the other five organs receiving the highest dose equivalent (10 CFR 30.4; 70.4; 40 CFR 61.21; 61.91; 61.101-91).
(2) The radiation dose to the whole body that would have the same biological effect as a given dose equivalent to a particular organ or tissue (DOE-91/04).

Effective head: *See* definition under pressure head.

Effective kilogram:
(1) For the source material uranium in which the uranium isotope uranium-235 is greater than 0.005 (0.5 weight

percent) of the total uranium present: 10,000 kilograms; and
(2) For any other source material: 20.000 kilograms (10 CFR 40.4-91).

Effective kilograms of special nuclear material:
(1) For plutonium and uranium-233 their weight in kilograms;
(2) For uranium with an enrichment in the isotope U-235 of 0.01 (1%) and above, its element weight in kilograms multiplied by the square of its enrichment expressed as a decimal weight fraction; and
(3) For uranium with an enrichment in the isotope U-235 below 0.01 (1%), by its element weight in kilograms multiplied by 0.0001 (10 CFR 70.4-91).

Effective leakage area: A parameter determined from blower door testing, giving a measure of the tightness of the house shell. Conceptually, this leakage area reflects the square inches of open area through the house shell, through which air can infiltrate or exfiltrate (EPA-88/08).

Effective porosity: Interconnected pore volumes that permit the flow of water in a porous medium (DOE-91/04).

Effective source height: The sum of the physical source height, which is the height of the source above ground level, and the plume rise (NATO-78/10).

Efficiency:
(1) The gas turbine manufacturer's rated heat rate at peak load in terms of heat input per unit of power output based on the lower heating value of the fuel (40 CFR 60.331-91).
(2) (In automobile emission control), the ability of the air cleaner or the unit under test to remove contaminant (40 CFR 85.2122(a)(16)(ii)(C)-91).
(3) In air pollution control, the ratio of the weight of pollutant collected to the total weight of pollutant entering the collector. It can be expressed by the equation: Eff(%) = 100x(inlet loading - outlet loading)/(inlet loading) (EPA-89/03b).

Example:
- Determine the collection efficiency.

Data:
- Inlet loading = 2 grains/ft^3 and outlet loading = 0.1 grains/ft^3.

Solution:
- Eff = 100(2 - 0.1)/2 = 95%

Efflorescence:
(1) A deposit of soluble salts that appear on the surface of building bricks. Salts derived from the bricks themselves can consist chiefly of $CaSO_4$, $MgSO_4$, K_2SO_4, and Na_2SO_4; soluble sulphates present in the raw clay can be rendered insoluble by the addition of $BaCO_3$ to the clay while it is being mixed; this precipitates the sulphate as insoluble $BaSO_4$. Efflorescence may arise from soluble salts in the mortar or, if a wall has no damp course or is backed by soil, from the soil itself (EPA-83).
(2) A phenomena that a solid with hydrate components, loses its hydrate content and forms a powdery deposit on its surface, when the solid is exposed to the air.

Effluence: *See* synonym, effluent.

Effluent (effluence or efflux):
(1) Any solid, liquid or gas which enters the environment as a by-product of a man-oriented process. The substances that flow out of a designated source (EPA-83). Effluent, effluence and efflux have the same meaning.
(2) Dredged material or fill material, including return flow from confined sites (40 CFR 232.2-91).
(3) A gas or fluid discharged into the environment (DOE-91/04).
(4) Wastewater-treated or untreated-that flows out of a treatment plant, sewer, or industrial outfall. Generally refers to wastes discharged into surface waters (EPA-94/04).

Other effluent related terms include:
- Primary effluent
- Secondary effluent

Effluent biomonitoring: The measurement of the biological effects of effluents (such as toxicity, biostimulation, and bioaccumulation: (EPA-85/09))

Effluent concentrations consistently achievable through proper operation and maintenance: For a given pollutant parameter,
(1) The 95th percentile value for the 30 day average effluent quality achieved by a treatment works in a period of at least two years excluding values attributable to upsets, bypasses, operational errors, or other unusual conditions; and
(2) A 7 day average value equal to 1.5 times the value derived under paragraph (f)(1) of this section (40 CFR 133.101-91).

Effluent data:
(1) With reference to any source of discharge of any pollutant (as that term is defined in section 502(6) of the Act, 33USC1362(6)):
   (A) Information necessary to determine the identity, amount, frequency, concentration, temperature, or other characteristics (to the extent related to water quality) of any pollutant which has been discharged by the source (or of any pollutant resulting from any discharge from the source), or any combination of the foregoing;
   (B) Information necessary to determine the identity, amount, frequency, concentration, temperature, or other characteristics (to the extent related to water quality) of the pollutants which, under an applicable standard or limitation, the source was authorized to discharge (including, to the extent necessary for such purpose, a description of the manner or rate of operation of the source); and
   (C) A general description of the location and/or nature of the source to the extent necessary to identify the source and to distinguish it from other sources (including, to the extent necessary or such purposes, a description of the device, installation, or operation constituting the source).
(2) Notwithstanding paragraph (a)(2)(i) of this section, the following information shall be considered to be effluent data only to the extent necessary to allow EPA to disclose publicly that a source is (or is not) in compliance with an applicable standard or limitation, or to allow EPA to demonstrate the feasibility, practicability, or attainability (or lack thereof) of

an existing or proposed standard or limitation:
(A) Information concerning research, or the results of research, on an product, method, device, or installation (or any component thereof) which was produced, developed, installed, and used only for research purposes; and
(B) Information concerning any product, method, device, or installation (or any component thereof) designed and intended to be marketed or used commercially but not yet so marketed or used (cf. emission data) (40 CFR 2.302-2).

Effluent guidelines: Technical EPA documents which set effluent limitations for given industries and pollutants (EPA-94/04).

Effluent limitation:
(1) Any restriction established by a State or the Administrator on quantities, rates, and concentrations of chemical, physical, biological, and other constituents which are discharged from point sources into navigable waters, the waters of the contiguous zone, or the ocean, including schedules of compliance (CWA502, *see also* 40 CFR 108.2; 122.2; 401.11-91).
(2) Restrictions established by a State or EPA on quantities, rates, and concentrations in wastewater discharges (Effluent limitation is also known as effluent standard) (EPA-94/04).

Effluent limitations guidelines: A regulation published by the Administrator under section 304(b) of CWA to adopt or revise effluent limitations (40 CFR 122.2; 401.11-91).

Effluent limitations for industrial dischargers: CWA has a two-part approach to establishing effluent limitations for industrial dischargers:
(1) Nationwide base-level treatment to be established through an assessment of what is technologically and economically achievable for a particular industry; and
(2) More stringent treatment requirements for specific plants where necessary to achieve water quality objectives for the particular body of water into which that plant dischargers.
Prior to 1977, EPA focused almost entirely on high-volume conventional pollutants such as biochemical oxygen demand (BOD), suspended solids (SS), and acidity and alkalinity (pH) when it developed effluent limits required by the Act. In the 1977 amendments to the Act, EPA shifted to priority pollutants which are 129 specific toxic chemicals in 34 industry categories (Winthrop-89/09).

Effluent seepage: Diffuse Discharge onto the ground of liquids that have percolated through solid waste or another medium. They contain dissolved or suspended materials (SW-108ts).

Effluent standard: Any effluent standard or limitation, which may include a prohibition of any discharge, established or proposed to be established for any toxic pollutant under section 307(a) of the Act (40 CFR 104.2-91, *see also* 40 CFR 129.2-91).

Effluent standard: *See* effluent limitation (EPA-94/04).

Efflux: *See* definition under effluent.

Eggs set: All eggs placed under incubation, i.e., total eggs minus cracked eggs and those selected for analysis of eggshell thickness. The number of eggs set, itself, is an artificial number, but it is essential for the statistical analysis of other development parameters (40 CFR 797.2130-iii-91).

Eight (8) hour time weighted average: The cumulative exposure for an 8-hour work shift computed as set forth in paragraph (a)(5) of this section (40 CFR 704.102-91).

Einsteinium (Es): A radioactive metal with a.n. 99 and a.w. 254. It belongs to group IIIB of the periodic table.

Ejector:
(1) A device that uses a fluid under pressure, such as steam, air, or, water, to move another fluid by developing suction. Suction is developed by discharging the fluid under pressure through a Venturi (EPA-83/06).
(2) A device used to disperse a chemical solution into water being treated (EPA-94/04).

Ekman layer: The atmospheric transition layer between the surface layer and the free atmosphere, in which the air motion is primarily determined by pressure forces, Coriolis forces and frictional forces due to the presence of the earth's surface (NATO-78/10).

Ekman spiral: In the Ekman layer, a balance of pressure, Coriolis and frictional forces results in a change of wind direction and speed with height. In the northern hemisphere, the wind direction change is clockwise with height. It depends primarily on the latitude and the atmospheric stability of the Ekman layer (NATO-78/10).

Elasticity: Ability of a material to return to original dimensions after deformation (EPA-83).

Electret passive environmental radon monitor (E-PERM): A device that uses an electrostatically charged plastic disk called an electret to sense radon in air. When radon decays, it produces ions, which are collected by the electred, resulting in a measurable decrease in the charge on the disk (EPA-88/08).

Electric: Electricity related phenomena. Other electric related terms include:
- Diversity electric
- Thermal electric

Electric arc furnace (EAF): A furnace that produces molten steel and heats the charge materials with electric arcs from carbon electrodes. Furnaces that continuously feed direct-reduced iron ore pellets as the primary source of iron are not affected facilities within the scope of this definition (*see* furnace for more related terms) (40 CFR 60.271-91, *see also* 40 CFR 60.271a-91).

Electric arc furnace steelmaking: The production of steel principally from steel scrap and fluxes in refractory lined furnaces by passing an electric current through the scrap or steel bath (*see* furnace for more related terms) (40 CFR 420.41-91).

Electric demand: The rate at which energy is delivered to or by a system, part of a system, or a piece of equipment. The primary source of demand is the power-consuming equipment of the

customers (cf. load) (EPA-83). Types of electric demand include:
(1) Annual maximum electric demand
(2) Annual system maximum electric demand
(3) Average electric demand
(4) Billing electric demand
(5) Coincident electric demand
(6) Instantaneous electric peak demand
(7) Integrated electric demand
(8) Maximum electric demand (peak demand or peak load)
(9) Non-coincident electric demand

Electric dispatching: The operating control of an integrated electric generating system involving operations such as:
(1) The assignment of load to specific generating stations and other sources of supply to effect the most economical supply as the total or the significant area loads rise or fall.
(2) The control of operations and maintenance of high-volume lines, substations and equipment, including administration of safety procedures.
(3) The operation of principal tie lines and switching.
(4) Tthe scheduling of energy transactions with connecting electric utilities (EPA-83).

Electric distribution: The act or process of distributing electric energy from convenient points on the transmission or bulk power system to the consumers (EPA-83).

Electric diversity factor: The ratio of the sum of the non-coincident maximum demands of two or more loads to their coincident maximum demand for the same period (EPA-83).

Electric energy: Electric energy related terms include:
- Dump energy
- Economy energy
- Interchange energy
- Net for distribution energy
- Off-peak energy
- On-peak energy
- Surplus energy

Electric furnace: Any furnace which uses electricity to produce over 50 percent of the heat required in the production of refined brass or bronze (40 CFR 60.131-91, AP-40, p236). Electric furnace has three functions:
(1) Synthesis of compounds not available in the natural state by fusing selected raw materials;
(2) Purification of ores; and
(3) Alteration of crystalline structure of ores having a satisfactory chemical purity but an undesirable crystal structure.
Types of electric arc furnaces include:
(1) Direct arc furnace
(2) Indirect arc furnace
(3) Induction furnace
(4) Resistance furnace

Electric generating plant (or generating station): A station at which are located electric generators, their drivers, and auxiliary equipment for converting mechanical, thermal or gravitational energy into electric power (cf. steam electric generating station) (EPA-83).

Electric generating station use: The kilowatt hours used at an electric generating station for such purposes as excitation and operation of auxiliary and other facilities essential to the operation of the station. Energy used is the difference between the gross generation plus any supply from outside the station and the net output of the station (EPA-83).

Electric generating unit: An electric generator together with its mechanical driver (EPA-83).

Electric generation: The act or process of transforming other forms of energy into electric energy; or to the amount to electric energy so produced, expressed in kilowatt hors. It includes:
(1) **Gross generation:** The total amount of electric energy produced by the generating units in a generating station.
(2) **Net generation:** Gross generation less kilowatt hours consumed out of gross generation for plant use (EPA-83).

Electric generation total fuel expense: After residual credit, total cost (including freight and handling) of fuel used in the production of electric energy, less fuel portion of steam transfer credit, and residual credits, such as net credits from the disposal of ashes, cinders, and nuclear byproducts (EPA-83).

Electric generation utility factor: The ratio of average load (kilowatts) over a designated period of time to the net capability of the system in the same period of time. This ratio represents the average use made of the net capability of equipment over a specific period of time. *see also* Load Factor and Capacity Factor (EPA-83).

Electric heat rate: A measure of generating station thermal efficiency, generally expressed in Btu per net kilowatt hour. It is computed by dividing the total Btu content of fuel burned for electric generation by the resulting net kilowatt hour generation (cf. coal rate) (EPA-83).

Electric load: The electric power delivered or required at any specified point or points on a system. Load originates primarily at the power consuming equipment of the customers (*see* load for more related terms) (EPA-83).

Electric losses: The general term applied to energy (kilowatt hours) and power (kilowatts) lost in the operation of an electric system. Losses occur principally as energy transformations from kilowatt hours to waste heat in electrical conductors and apparatus (EPA-83). Other electric loss related terms include:
- Average losses
- Energy losses
- Line losses
- Peak per cent losses
- System losses

Electric power: The time rate of generating, transferring or using electric energy, usually expressed in kilowatts (EPA-83).

Electric production: The act or process of generating electricity (EPA-83).

Electric smelting furnace: Any furnace in which the heat necessary for smelting of the lead sulfide ore concentrate charge is generated by passing an electric current through a portion of the molten mass in the furnace (*see* furnace for more related terms) (40 CFR 60.181-91).

Electric submerged arc furnace: Any furnace wherein electrical energy is converted to heat energy by transmission of current between electrodes partially submerged in the furnace charge (*see* furnace for more related terms) (40 CFR 60.261-91).

Electric summer peak: The greatest load on an electric system during a prescribed demand interval in the summer (or cooling) season, usually between June 1 and September 30 (EPA-83).

Electric system: The physically connected electric generation, transmission, distribution and other facilities operated as an integral unit under one control, management or operating supervision (EPA-83).

Electric system net input: Net available energy that is put into a utility's system for sale within its own service area. The net energy generated in a system's own plants, plus energy received from other systems, less energy delivered to other systems (EPA-83).

Electric system output: The net generation by the system's own plants plus purchased energy, plus or minus net interchange energy (EPA-83).

Electric traction motor: An electrically powered motor which provides tractive energy to the wheels of a vehicle (40 CFR 600.002.85-91).

Electric transmission: The act or process of transporting electric energy in bulk from a source or sources of supply to other principal parts of the system or to other utility systems (EPA-83).

Electric utility combined cycle gas turbine: any combined cycle gas turbine used for electric generation that is constructed for the purpose of supplying more than one-third of its potential electric output capacity and more than 25 MW electrical output to any utility power distribution system for sale. Any steam distribution system that is constructed for the purpose of providing steam to a steam electric generator that would produce electrical power for sale is also considered in determining the electrical energy output capacity of the affected facility (*see* turbine for more related terms) (40 CFR 60.41a-91).

Electric utility company: The largest interconnected organization, business, or governmental entity that generates electric power for sale (e.g., a holding company with operating subsidiary companies) (40 CFR 60.41a-91).

Electric utility stationary gas turbine: Any stationary gas turbine constructed for the purpose of supplying more than one third of its potential electric output capacity to any utility power distribution system for sale (40 CFR 60.331-91).

Electric utility steam generating unit: Any fossil fuel fired combustion unit of more than 25 megawatts that serves a generator that produces electricity for sale. A unit that cogenerates steam and electricity and supplies more than one-third of its potential electric output capacity and more than 25 megawatts electrical output to any utility power distribution system for sale shall be considered an electric utility steam generating unit (CAA112, *see also* 40 CFR 60.41a-91).

Electric wheeling service: The use of the transmission facilities of one system to transmit power to or from another system (EPA-83).

Electric winter peak: The greatest load on an electric system during any prescribed demand interval in the winter or heating season (EPA-83).

Electrical capacitor manufacturer: A manufacturer who produces or assembles electrical capacitors in which PCB or PCB-containing compounds are part of the dielectric (40 CFR 129.105-91).

Electrical charging system: A device to convert 60Hz alternating electric current, as commonly available in residential electric service in the United States, to a proper form for recharging the energy storage device (40 CFR 600.002.85-91).

Electrical conductivity (or conductance): The property which allows an electric current to flow when a potential difference is applied. It is the reciprocal of the resistance in ohms measured between opposite faces of a centimeter cube of an aqueous solution at a specified temperature. It is expressed as micromhos per centimeter at temperature degrees Celsius (EPA-83/06a).

Electrical equipment: The underground equipment that contains dielectric fluid that is necessary for the operation of equipment such as transformers and buried electrical cables (40 CFR 280.12-91).

Electrical resistivity (or specific resistance): An indicator of a material to oppose an electric current flow. It can be expressed as RA/L, where R = resistance of the material, A = cross section area of the material and L = length of the material.

Electrical transformer manufacturer: A manufacturer who produces or assembles electrical transformers in which PCB or PCB-containing compounds are part of the dielectric (40 CFR 129.105-91).

Electrically heated choke: A device which contains a means for applying heat to the thermostatic coil by electrical current (40 CFR 85.2122(a)(2)(iii)(C)-91).

Electrocatalytic oxygen analyzer (EOA): One of electroanalytical methods for continuous emission monitoring (*see* continuous emission monitor for various types). EOA is a method for the determination of oxygen which has developed as an outgrowth of fuel-cell technology. These so-called fuel-cell oxygen analyzers are not actually fuel cells, but simple electrolytic concentration cells that use a special solid catalytic electrolyte to aid the flow of electrons. These analyzers are available in both extractive and in-situ (in-stack) configurations.

The instruments designed to continuously monitor oxygen concentrations utilize different concentrations of oxygens expressed in terms of partial pressures. A special porous material, zirconium oxide, serves both as an electrolyte and as a high temperature catalyst to produce oxygen ions (EPA-81/09).

Electrobrightening: A process of reversed electro deposition which results in an anodic metal with a high polish surface (EPA-83/06a).

Electrochemical equivalent: The weight of metal electrodeposited (or other substance changed chemically by reduction or oxidation) per unit of time and unit of current; i.e., pound per ampere hour, grams per ampere-second (EPA-74/03d).

Electrochemical machining: Shaping of an anode by the following process--The anode and cathode are placed close together and electrolyte is pumped into the space between them. An electrical potential is applied to the electrodes causing anode metal to be dissolved selectively, producing a shaped anode that complements the shape of the cathode (EPA-83/03).

Electrochemical potential: The potential differences when two electrodes are placed in a conducting solution and also connected through an external circuit to activate the electrochemical reactions.

Electrochemical process: The process of chemical change in a conducting solution when electric current passes the solution. It involves:
(1) Discharge of the stored chemical energy to produce electricity (e.g., battery).
(2) Charge of chemical energy by electricity (e.g., to charge a battery).

Electrochemical reduction cell: A cell in which chemical reduction takes place at the cathode and chemical oxidation at the anode.

Electrochemical transducer: *See* synonym, polarographic analyzer.

Electrochemistry: A branch of chemistry that deals with the relation of electricity to chemical changes and with the interconversion of chemical and electrical energy.

Electrocleaning: The process of anodic removal of surface oxides and scale from a workpiece (EPA-83/06a).

Electrocoating: *See* synonym, electroplating.

Electrode:
(1) A conducting material for passing an electric current into or out of a solution by adding electrons to or taking electrons from ions in the solution (EPA-83/06a).
(2) Long cylindrical rods made of carbon or graphite used in electric arc furnaces to conduct electricity into the metal charge (EPA-85/10a).

Other electrode related terms include:
- Anode
- Calomel electrode
- Cathode
- Saturated calomel electrode (*see* synonym, calomel electrode)
- Standard calomel electrode (*see* synonym, calomel electrode)
- Hydrogen electrode
- Hydrogen half cell (*see* synonym, hydrogen electrode)
- Glass electrode
- Negative electrode (*see* synonym, cathode)
- pH electrode
- Plate (*see* synonym, anode)
- Platinum electrode
- Positive electrode (*see* synonym, anode)
- Quinhydrone electrode
- Reference electrode (*see* synonym, standard electrode)
- Sintered plate electrode
- Standard electrode
- Working electrode
- Standard electrode potential

Electrode potential: The potential between an electrode and electrolyte in a half cell.

Electrodeposition (EDP): A method of applying coatings in which the part is submerged in a tank filled with the coatings and in which an electrical potential is used to enhance deposition of the coatings on the part (40 CFR 60.311-91, *see also* 40 CFR 60.391; 60.451-91).

Electrodialysis:
(1) The separation of a substance from a solution through a membrane accomplished by the application of an electric potential across to the membrane (EPA-87/10a).
(2) A process that uses electrical current applied to permeable membranes to remove minerals from water. Often used to desalinize salty or brackish water (EPA-94/04).

Electrodynamic separator: Utilizes a rotating drum or other moving pole in place of one or more fixed charged plates (poles). Can be used to separate electrical conductive material (e.g., nonferrous metal) from nonconductive material (e.g., organics) (EPA-83).

Electroforming: The production or reproduction of articles by electrodeposition upon a mandrel or mold that is subsequently separated from the deposit (EPA-74/03d).

Electroless plating: The deposition of conductive material from an autocatalytic plating solution without application of electrical current (40 CFR 413.71-91).

Electrolysis: Refers to the reactions of oxidation or reduction that take place at the surface of conductive electrodes immersed in an electrolyte, under the influence of an applied potential. Electroplating and anodizing are examples of electrolysis.

Electrolyte:
(1) A nonmetallic electrical conductor in which current is carried by the movement of ions (EPA-83/03).
(2) A liquid, most often a solution, that will conduct an electric current (EPA-83/06a).
(3) The liquid or material that permits conduction of ions between cell electrodes (EPA-84/08).

Electrolytic: Relating to a chemical change produced by passage of a current through a conducting substance (such as water) (EPA-87/10a).

Electrolytic analysis: Analysis of reducible and oxidizable conducting solutions. Measurement is based on the weight of material plated onto the electrode.

Electrolytic cell: A unit apparatus in which electrochemical reactions are produced by applying electrical energy or which supplies electrical energy as a result of chemical reactions and which includes two or more electrodes and one or more electrolytes contained in a suitable vessel (EPA-83/06a).

Electrolytic chlorine: Chlorine produced as a result of electrochemical reactions.

Electrolytic conductivity detector: *See* definition under GC/ECD.

Electrolytic copper: Copper refined by the electrolytic method. This gives metal of high purity, over 99.94%, and enables precious metals, such as gold and silver, to be recovered. Used in refining most of the copper produced (EPA-83).

Electrolytic decomposition: An electrochemical treatment used for the oxidation of cyanides. The method is practical and economical when applied to concentrated solutions such as contaminated baths, cyanide dips, stripping solutions, and concentrated rinses. Electrolysis is carried out at a current density of 35 amp/sq. ft. at the anode and 70 amp/sq. ft. at the cathode. Metal is deposited at the cathode and can be reclaimed (EPA-83/06a).

Electrolytic oxidation: A reaction by an electrolyte in which there is an increase in valence resulting from a loss of electrons (cf. electrolytic reduction) (EPA-83/06a).

Electrolytic precipitation: Making a powdered active material by electrodeposition, e.g., making silver powders from silver bars (EPA-84/08).

Electrolytic process: A process that a low voltage direct current passes through an electrolyte containing metallic ions causing the metallic ions to plate on the cathode as free metal atoms. The process is used to produce chromium and manganese metal, which are included with the ferroalloys. Chromium metal produced by this process is 99+% pure (EPA-75/02a).

Electrolytic reduction: A reaction in which there is a decrease in valence resulting from a gain in electrons (cf. electrolytic oxidation) (EPA-83/06a).

Electrolytic refining: The method of producing pure metals by making the impure metal the anode in an electrolytic cell and by making the depositing a pure cathode. The impurities either remain un-dissolved at the anode or pass into solutions in the electrolyte (EPA-83/06a).

Electrolytic silver recovery: The removal of silver from silver-bearing solutions by application of a direct current to electrodes in the solution causing metallic silver to deposit on the cathode (EPA-80/10).

Electrolytic slime: Insoluble impurities removed from the bottom of an electrolytic cell during electrolytic refining (EPA-83/03a).

Electromagnetic interference/radio frequency interference (EMI/RFI) shielding coating: A conductive coating that is applied to a plastic substrate to attenuate EMI/RFI signals (40 CFR 60.721-91).

Electrometallurgical process: The application of electric current to a metallurgical process either for electrolytic deposition or as a source of heat (EPA-83/06a).

Electrometric titration: A standard method of measuring the alkalinity of a solution (EPA-83/06a).

Electromotive series of metals: A list of metals and alloys arranged according to their standard electrode potentials; which also reflects their relative corrosion potential (EPA-86/12).

Electron: A negatively charged particle and carries a negative charge of 1.602×10^{-19} coulomb. The number of electrons circulating around a nucleus is equal to the number of positive charges on the nucleus (*see* atom for more related terms).

Electron acceptor: During the process of reduction and oxidation, an atom or part of a molecule joined by a covalent bond to an electron donor.

Electron beam lithography: Similar to photo-lithograph, a fine beam of electrons which is used to scan a pattern and to expose an electron-sensitive resistance in an unmasked area of an object surface (EPA-83/03).

Electron beam machining: The process of removing materials from a workpiece by a high velocity focused stream of electrons which melt and vaporize the workpiece at the point of impingerent (EPA-83/06a).

Electron beam melting: A melting process in which an electron beam is used as a heating source (EPA-83/03a).

Electron capture detector (ECD): *See* synonym, GC/ECD.

Electron discharge lamp: An electron lamp in which light is produced by passage of an electric current through a metallic vapor or gas (EPA-83/03).

Electron gun: An electrode structure that produces and may control, focus, deflect and converge one or more electron beams in an electron tube (EPA-83/03).

Electron pair bond: *See* synonym, covalent bond.

Electron tube: An electron device in which conduction of electricity is accomplished by electrons moving through a vacuum or gaseous medium within a gas-tight envelope (EPA-83/03).

Electronic crystals: Crystals or crystalline materials which because of their unique structural and electronic properties are used in electronic devices. Examples of these crystals are crystals comprised of quartz, ceramic, silicon, gallium arsenide, and idium arsenide (*see* crystal for more related terms) (40 CFR 469.22-91).

Electronic optical sorter: Separates glass from stones and pieces of ceramics; sorts the glass according to color. Photoelectric detector determines the color or opacity of the material and blasts of air deflect the pieces into the proper containers (EPA-83).

Electrophoresis: Electrophoresis is the transport of electrically charged particles under the influence of a DC electric field. The charged particles migrate to collecting membranes, which are located between the two electrodes, for subsequent removal. Because the particles are not allowed to reach the electrode, reaction at the electrode with materials do not take place. Particles without charges, isoelectric particles are not collected. The membranes which stop the migration of particles are dialyzing membranes (e.g., cellulose) which allow only water and small ions to pass through (cf. gel electrophoresis).

Electroplating (or electrocoating):
(1) The electrodeposition of a metallic or nonmetallic coating onto the surface of a workpiece (40 CFR 471.02-91).
(2) A process used to deposit, or plate, a coating of metal upon the surface of another metal by electrochemical reactions (AP-40, p829).

Electroplating process wastewater: The process wastewater generated in operations which are subject to regulation under any of Subparts A through H of this part (*see* wastewater for more related terms) (40 CFR 413.02-91).

Electrostatic bell or disc spray: An electrostatic spray coating method in which a rapidly spinning bell- or disc-shaped applicator is used to create a fine mist and apply the coating with high transfer efficiency (40 CFR 52.741-91).

Electrostatic precipitation mechanism: The process of electrostatic precipitation consists of:
(1) Gas ions are formed by means of high-voltage corona discharge.
(2) The solid or liquid particles are charged by bombardment by the gaseous ions or electrons.
(3) The electrostatic field causes the discharged particles to migrate to a collecting electrode of opposite polarity.
(4) The charge on a particle must be neutralized by the collecting electrode.
(5) Reentrainment of the collected particles must be prevented.
(6) The collected particles must be transferred from the collecting electrode to storage for subsequent disposal (AP-40, p140).

Electrostatic precipitator (ESP):
(1) An air pollution control device in which solid or liquid particulates in a gas stream are charged as they pass through an electric field and precipitated on a collection surface (40 CFR 60.471-91).
(2) A unit used for separating dust particles and/or mist from a polluted air stream by an electrostatic field. The polluted air passes over a high voltage negative electrode, whose voltage is of the order of 40,000-100,000 volts. An electrostatic field then imparts a charge on the particles, which are then attracted to a collecting electrode of an opposite charge (EPA-84/09).
(3) A device that removes particles from a gas stream (smoke) after combustion occurs. The ESP imparts an electrical charge to the particles, causing them to adhere to metal plates inside the precipitator. Rapping on the plates causes the particles to fall into a hopper for disposal (EPA-94/04).

Electrostatic precipitator (ESP) process: Particulate matter is first charged with electricity before it can be collected in an ESP. Once the particles or liquid aerosols that makeup the particulate matter are charged, they move towards an oppositely charged surface because of electrostatic attraction (opposite charges attract each other, the same charges repel each other). The collected particles are removed by rapping or washing the collecting surface. This charging, collecting, and removal process is commonly referred to as precipitation.

ESP can be classified according to a number of design features. These features include the method of charging (single-stage or two-stage), the method of particle removal from collection surfaces (wet or dry), the temperature of operation (cold-side or hot side), and the structural design and operation of the discharge electrodes (tubular or plate) (EPA-89/03b).

Most ESP's used to reduce particulate matter emissions from boilers and other industrial processes are single-stage ESP's. These units use very high voltage to charge particles. The particles, once charged, move in a direction perpendicular to the gas flow and are collected on the oppositely charged collection surface. Because particle charging and collection occurs in the same stage, these ESP's are called single stage ESP's.

Collected particles are removed from the unit by rapping the collection electrodes or by spraying water on the electrodes to wash the particles away.

ESP's use either flat plates or cylindrical tubes to collect particulate matter. Most ESP's use plates as collection electrodes. In this arrangement, dirty gas flows into a chamber consisting of a series of small diameter discharge electrodes (wires) especially spaced between rows of plates. Discharge electrodes are approximately 0.05 to 0.15 inches (0.02 to 0.06 centimeters) in diameter. Collection plates are usually between 20 and 40 feet (6 and 12 meters) high and spaced from 4 to 12 inches (1.6 to 4.7 centimeters) apart.

Electrostatic precipitator (ESP) type:
(1) Dry electrostatic precipitator (ESP)
(2) Wet electrostatic precipitator
(3) Electrostatically augmented scrubber

Electrostatic separator: A device utilizing the principle that electrical conductors lose an induced static charge faster than insulators. In this way, an electrostatic sorter can separate conducting materials (e.g., Aluminum) from non-conducting ones (e.g., glass) after the particles are charged in a high voltage direct current electrical field (*see* separator for more related terms) (EPA-83).

Electrostatic spray: A spray coating method in which opposite electrical charges are applied to the substrate and the coating. The coating is attracted to the object due to the electrostatic potential between them (40 CFR 52.741-91).

Electrostatic spray application: A spray application method that uses an electrical potential to increase the transfer efficiency of the coatings (40 CFR 60.311-91, *see also* 40 CFR 60.391-91).

Electrostatically augmented scrubber: A control device that couples the mechanisms of electrostatic attraction and inertial separation by charging particles prior to entry into a wet collector (*see* electrostatic precipitator for more related terms) (EPA-81/09).

Electrovalent bond: *See* synonym, ionic bond.

Electrowinning: The recovery of a metal from an ore by means of electrochemical processes, i.e., deposition of metal on an electrode by passing electric current through an electrolyte (EPA-82/05).

Element: In a battery, a combination of negative and positive plates and separators to make a cell in a lead-acid storage battery (EPA-84/08). In chemistry, the particle which is theoretically the irreducible constituents of the material world. It includes electron, proton and neutron.

Elemental analysis: *See* synonym, ultimate analysis.

Elemental phosphorus plant or plant: Any facility that processes phosphate rock to produce elemental phosphorus. A plant includes all buildings, structures, operations, calciners and nodulizing kilns on one contiguous site (40 CFR 61.121-91).

Elementary neutralization unit: A device which:
(1) Is used for neutralizing wastes that are hazardous wastes only because they exhibit the corrosivity characteristic defined in 261.22 of this chapter, or they are listed in Subpart D of Part 261 of the chapter only for this reason; and
(2) Meets the definition of tank, tank system, container, transport vehicle, or vessel in 40 CFR 260.10 of this chapter (40 CFR 260.10-91, *see also* 40 CFR 270.2-91).

Elevation angle: The angle of inclination of the laser beam referenced to the horizontal plane (40 CFR 60-App/A(alt. method 1)-91).

Eleven AA (11 AA): The chemical substance 11-aminoundecanoic acid, CAS Number 2432-99-7 (40 CFR 704.25-91).

Eleven contiguous Western States: The States of Arizona, California, Colorado, Idaho, Montana, Nevada, New Mexico, Oregon, Utah, Washington, and Wyoming (FLPMA103-43USC1702-90).

Eligible coastal state: A coastal state that for any fiscal year for which a grant is applied for under this section:
(1) Has a management program approved under section 1455 of this title; and
(2) In the judgment of the Secretary, is making satisfactory progress in activities designed to result in significant improvement in achieving the coastal management objectives specified in section 1452(2)(A) through (i) of this title (CZMA306a-16USC1455a).

Eligible costs: The construction costs for waste-water treatment works upon which EPA grants are based (EPA-94/04).

Eluate: A solution used to extract collected ions from an ion exchange resin or solvent and return the its active state (EPA-82/05).

Eluent: A solution used to extract collected ions from an ion exchange resin or solvent and return the resin to its active state (EPA-82/10).

Elution:
(1) The process of washing out or removing a substance through the use of a solvent.
(2) In an ion exchange process, the stripping of adsorbed ions from an ion exchange resin by passing solutions containing other ions in relatively high concentrations through the resin (EPA-87/10a).

Elutriation:
(1) In air pollution control, a particle separation technique that is accomplished by particles settling out of a gas stream that is moving in an upward direction (EPA-84/09).
(2) In solid waste treatment, separation of solid waste into heavy and light fractions by washing (SW-108ts).
(3) In sludge conditioning, a process whereby the sludge is washed, either with fresh water or plant effluent, to reduce the sludge alkalinity and fine particles, thus decreasing the amount of required coagulant in further treatment steps or in sludge dewatering (EPA-87/10a).
(4) *See* separator for more related terms.

EMAP data: Environmental monitoring data collected under the auspices of the Environmental Monitoring and Assessment Program. All EMAP data share the common attribute of being of known quality, having been collected in the context of explicit data quality objectives (DQOs) and a consistent quality assurance program (EPA-94/04).

Embankment (or impoundment): A storage basin made to contain wastes from mines or preparation plants (EPA-82/10).

Embryo: The young spor ophytic plant before the start of germination (40 CFR 797.2750-91).

Embryo cup: A small glass jar or similar container with a screened bottom in which the embryos of some species (i.e., minnow) are placed during the incubation period and which is normally oscillated to ensure a flow of water through the cup (40 CFR 797.1600-91).

Emergency: A situation created by an accidental release or spill of hazardous chemicals which poses a threat to the safety of workers, residents, the environment, or property (EPA-85/11).

Emergency (chemical): A situation created by an accidental release or spill of hazardous chemicals which poses a threat to the safety of workers, residents, the environment, or property (EPA-94/04).

Emergency action plan: A plan for a workplace, or parts thereof, describing what procedures the employer and employees must take to ensure employee safety from fire or other emergencies (29 CFR 1910.35-91).

Emergency broadcasting system (EBS): EBS is to be used to inform the public about the nature of a hazardous materials incident and what safety steps they should take (NRT-87/03).

Emergency condition:
(1) For electric generation, *see* 40 CFR 60.41a-91.
(2) For pesticide application, *see* 40 CFR 166.3-91.
(3) For a nuclear facility, occurrences or accidents that might occur infrequently during start-up testing or operation of the

facility. The equipment, components, and structures might be deformed by these conditions to the extent that repair is required prior to reuse (DOE-91/04).

Emergency episode: *See* synonym, episode (EPA-94/04).

Emergency escape route: The route that employees are directed to follow in the event they are required to evacuate the workplace or seek a designated refuge area (29 CFR 1910.35-91).

Emergency fuel: A fuel fired by a gas turbine only during circumstances, such as natural gas supply curtailment or breakdown of delivery system, that make it impossible to fire natural gas in the gas turbine (*see* fuel for more related terms) (40 CFR 60.331-91).

Emergency gas turbine: Any stationary gas turbine which operates as a mechanical or electrical power source only when the primary power source for a facility has been rendered inoperable by an emergency situation (*see* turbine for more related terms) (40 CFR 60.331-91).

Emergency management institute (EMI): EMI is a component of FEMA's National Emergency Training Center located in Emmitsburg, Maryland. It conducts resident and nonresident training activities for Federal, State, and local government officials, managers in the private economic sector, and members of professional and volunteer organizations on subjects that range from civil nuclear preparedness systems to domestic emergencies caused by natural and technological hazards. Nonresident training activities are also conducted by State Emergency Management Training Offices under cooperative agreements that offer financial and technical assistance to establish annual training programs that fulfill emergency management training requirements in communities throughout the nation (NRT-87/03).

Emergency operations plan (EOP): Developed in accord with the guidance in CPG 1-8. EOPs are multi-hazard, functional plans that treat emergency management activities generically. EOPs provide for as much generally applicable capability as possible without reference to any particular hazard; then they address the unique aspects of individual disasters in hazard-specific appendices (NRT-87/03).

Emergency permit: A UIC permit issued in accordance with 40 CFR 144.34 (*see* permit for more related terms) (40 CFR 144.3-91, *see also* 40 CFR 270.2-91).

Emergency Planning and Community Right-to-know Act (EPCRA) of 1986: *See* the term of Act or EPCRA.

Emergency procedures: The various special procedures necessary to protect the environment from wastewater treatment plant failures caused by power outages, chemical spills, equipment failures, major storms, floods, etc. (EPA-82/11e).

Emergency project: A project involving the removal, enclosure, or encapsulation of friable asbestos-containing material that was not planned but results from a sudden unexpected event (40 CFR 763.121-91).

Emergency renovation operation: A renovation operation that was not planned but results from a sudden, unexpected event that, if not immediately attended to, presents a safety or public hazard, is necessary to protect equipment from damage, or is necessary to avoid imposing an unreasonable financial burden. This term includes operations necessitated by nonroutine failures of equipment (40 CFR 61.141-91).

Emergency response plan: A plan of action to be followed by source operators after a toxic substance has been accidentally released to the atmosphere. The plan includes notification of authorities and impacted population zones, minimizing the quantity of the discharge, etc (EPA-87/07a).

Emergency response values: Concentrations of chemicals, published by various groups, defining acceptable levels for short-term exposures in emergencies (EPA-94/04).

Emergency safety stack: *See* synonym, dump stack.

Emergency situation: For continuing use of a PCB Transformer exists when:
(1) Neither a non-PCB Transformer nor a PCB-Contaminated transformer is currently in storage for reuse or readily available (i.e., available within 24 hours) for installation.
(2) Immediate replacement is necessary to continue service to power users (40 CFR 761.3-91).

Emergency vent stream: For the purposes of these standards, an intermittent emission that results from a decomposition, attempts to prevent decompositions, power failure, equipment failure, or other unexpected cause that requires immediate venting of gases from process equipment in order to avoid safety hazards or equipment damage. This includes intermittent vents that occur from process equipment where normal operating parameters (e.g., pressure or temperature) are exceeded such that the process equipment cannot be returned to normal operating conditions using the design features of the system and venting must occur to avoid equipment failure or adverse safety personnel consequences and to minimize adverse effects of the runaway reaction. This does not include intermittent vents that are designed into the process to maintain normal operating conditions of process vessels including those vents that regulate normal process vessel pressure (40 CFR 60.561-91).

Emergent (emersed) aquatic plants: Rooted plants, such as the bulrush and cattail, that grow in shallow water with a portion of their stems and leaves rising above the water surface (DOI-70/04).

Emerging technology: Any sulfur dioxide control system that is not defined as a conventional technology under this section, and for which the owner or operator of the facility has applied to the Administrator and received approval to operate as an emerging technology under 40 CFR 60.49b(a)(4) (40 CFR 60.41b-91, *see also* 40 CFR 60.41c-91).

Emesis: Vomiting (LBL-76/07-bio).

Emetic: An agent that induces vomiting (LBL-76/07-bio).

Eminent domain: Government taking--or forced acquisition--of private land for public use, with compensation paid to the landowner (EPA-89/12).

Emission:
(1) Gas-borne pollutants released to the atmosphere (40 CFR 240.101-91).
(2) Material released into the air either by a discrete source (primary emission) or as the result of a photochemical reaction or chain of reactions (secondary emission).
(3) The total of substances exhausted to the atmosphere (EPA-83).
(4) Pollution discharged into the atmosphere from smokestacks, other vents, and surface areas of commercial or industrial facilities; from residential chimneys; and from motor vehicle, locomotive, or aircraft exhausts (EPA-94/04).

Other emission related terms include:
- Actual emission
- Air emission
- Allowable emission
- Continuous emission
- Crankcase emission
- Decomposition emission
- Evaporative emission
- Excess emission
- Exhaust emission
- Fugitive emission
- Intermittent emission
- Natural emission
- Primary emission
- Process emission
- Process fugitive emission
- Secondary emission
- Smoke emission
- Stack emission

Emission cap: A limit designed to prevent projected growth in emissions from existing and future stationary sources from eroding any mandated reduction. Generally, such provisions require any emission growth from facilities under the restrictions be offset by equivalent reductions at other facilities under the same cap (*see* emissions trading) (EPA-94/04).

Emission control device: Any solvent recovery or solvent destruction device used to control volatile organic compounds (VOC) emissions from flexible vinyl and urethane rotogravure printing lines (40 CFR 60.581-91).

Emission control system: The combination of an emission control device and a vapor capture system for the purpose of reducing VOC emissions from flexible vinyl and urethane rotogravure printing lines (40 CFR 60.581-91).

Emission critical parameters: Those critical parameters and tolerances which, if equivalent from one part to another, will not cause the vehicle to exceed applicable emissior stardards with such parts instlled (40 CFR 85.2113-91).

Emission data:
(1) With reference to any source of emission of any substance into the air:
  (A) Information necessary to determine the identity, amount, frequency, concentration, or other characteristics (to the extent related to air quality) of any emission which has been emitted by the source (or of any pollutant resulting from any emission by the source), or any combination of the foregoing;
  (B) Information necessary to determine the identity, amount, frequency, concentration, or other characteristics (to the extent related to air quality) of the emissions which, under an applicable standard or limitation, the source was authorized to emit (including, to the extent necessary for such purposes, a description of the manner or rate of operation of the source); and
  (C) A general description of the location and/or nature of the source to the extent necessary to identify the source and to distinguish it from other sources (including, to the extent necessary for such purposes, a description of the device, installation, or operation constituting the source).
(2) Notwithstanding paragraph (a)(2)(i) of this section, the following information shall be considered to be emission data only to the extent necessary to allow EPA to disclose publicly that a source is (or is not) in compliance with an applicable standard or imitation, or to allow EPA to demonstrate the feasibility, practicability, or attainability (or lack thereof) of an existing or proposed standard or limitation:
  (A) Information concerning research, or the results of research, on any project, method, device or installation (or any component thereof which was produced, developed, installed, and used only for research purposes; and
  (B) Information concerning any product, method, device, or installation (or any component thereof) designed and intended to be marketed or used commercially but not yet so marketed or used (cf. effluent data) (40 CFR 2.301-2-91).

Emission factor: The relationship between the amount of pollution produced and the amount of raw material processed. For example, an emission factor for a blast furnace making iron would be the number of pounds of particulates per ton of raw materials (EPA-94/04).

Emission flame photometry: A type of flame photometry. *See* photoluminescence under continuous inorganic compound analyzer, luminescence analyzer.

Emission frequency: The percentage of time that emissions are visible during the observation period (40 CFR 60-App/A(method 22)-91).

Emission guideline: A guideline set forth in Subpart C of this part, or in a final guideline document published under 40 CFR 60.22(a), which reflects the degree of emission reduction achievable through the application of the best system of emission reduction which (taking into account the cost of such reduction) the Administrator has determined has been adequately demonstrated for designated facilities (40 CFR 60.21-91).

Emission inventory: A listing, by source, of the amount of air pollutants discharged into the atmosphere of a community; used to establish emission standards (EPA-94/04).

Emission limitation and emission standard: A requirement established by the State or the Administrator which limits the quantity, rate, or concentration of emissions of air pollutants on a continuous basis, including any requirement relating to the operation or maintenance of a source of assure continuous emission reduction (CAA302, *see also* 40 CFR 51.100-91).

Emission measurement system: All of the equipment necessary to transport and measure the level of emissions. This includes the sample system and the instrumentation system (40 CFR 87.1-91).

Emission performance warranty: The warranty given pursuant to this subpart and section 20(b) of the Act (40 CFR 85.2102-91).

Emission rate: The amount of pollutants emitted per unit of time (OME-88/12).

Emission related defect: A defect in design, materials, or workmanship in a device, system, or assembly described in the approved Application for Certification (required by 40 CFR 86.077.22 and like provisions of Part 85 and Part 86 of Title 40 of the Code of Federal Regulations) which affects any parameter or specification enumerated in Appendix VIII (40 CFR 85.1902-91).

Emission related maintenance: The maintenance which does substantially affect emissions or which is likely to affect the emissions deterioration of the vehicle or engine during normal in-use operation, even if the maintenance is performed at some time other than that which is recommended (40 CFR 86.088.2-91, *see also* 40 CFR 86.084.2-91).

Emission related parts: Those parts installed for the specific purpose of controlling emissions or those components, systems, or elements of design which must function properly to assure continued vehicle emission compliance (40 CFR 85.2102-91).

Emission screen limit: In RCRA, emission screen limit is developed by back-calculating from the risk specific doses and the reference air concentrations, using dispersion coefficients for reasonable, worst case facilities (cf. feed rate screen limit) (EPA-90/04).

Emission short test: *See* synonym, Office Director--approved emission test.

Emission source and source: Any facility from which VOM is emitted or capable of being emitted into the atmosphere (40 CFR 52.741-91).

Emission standard:
(1) A legally enforceable regulation setting forth an allowable rate of emissions into the atmosphere, or prescribing equipment specifications for control of air pollution emissions (40 CFR 60.21-91).
(2) Legally enforceable limits on the quantities and/or kinds of air contaminants that can be emitted into the atmosphere (DOE-91/04).
(3) The maximum amount of air polluting discharge legally allowed from a single source, mobile or stationary (EPA-94/04).

Emission standard or limitation under this Act (CAA):
(1) A schedule or timetable of compliance, emission limitation, standard of performance or emission standard,
(2) A control or prohibition respecting a motor vehicle fuel or fuel additive, which is in effect under this Act (including a requirement applicable by reason of section 118) or under an applicable implementation plan, or
(3) Any condition or requirement of a permit under part C of title I (relating to significant deterioration of air quality) or part D of title I (relating to nonattainment), any condition or requirement of section 113(d) (relating to certain enforcement orders), section 119 (relating to primary nonferrous smelter orders), any condition or requirement under an applicable implementation plan relating to transportation control measures, air quality maintenance plans, vehicle inspection and maintenance programs or vapor recovery requirements, section 221 (e) and (f) (relating to fuels and fuel additives), section 169A (relating to visibility protection), any condition or requirement under part B of title I (relating to ozone protection), or any requirement under section 111 or 112 (without regard to whether such requirement is expressed as an emission standard or otherwise). [2] which is in effect under this Act (including a requirement applicable by reason of section 118) or under an applicable implementation plan. [1] So in original public law. Period probably should be a comma. [2] So in original. Subsection (a) enacted without paragraph (2) (CAA304-42USC7604).

Emission time: The accumulated amount of time that emissions are visible during the observation period (40 CFR 60-App/A(method 22)-91).

Emissions trading: The creation of surplus emission reductions at certain stacks, vents, or similar emissions sources and the use of this surplus to meet or redefine pollution requirements applicable to other emission sources. This allows one source to increase emissions when another sources reduces them, maintaining an overall constant emission level. Facilities that reduce emissions substantially may "bank" their "credits" or sell them to other industries (cf. bubble policy) (EPA-94/04).

Emission unit: Any part of a stationary source which emits or would have the potential to emit any pollutant subject to regulation under the Act (40 CFR 51.165-91, *see also* 40 CFR 51.166; 51-App/S; 52.21; 52.24-91).

Emission warranty: Those warranties given by vehicle manufacturers pursuant to section 207 of the Act (40 CFR 85.2113-91).

Emissive coating: An oxide coating applied to an electrode to enhance the emission of electrons (EPA-83/03).

Emitter: Anything or anyone causing emissions (NATO-78/10).

Employee: An employee of an employer (OSHA3).

Employee exposure: That exposure to airborne asbestos would occur if the employee were not using respiratory protective equipment (40 CFR 763.121-91).

Employee salary offset: The administrative collection of a debt by deductions at one or more officially established pay intervals from the current pay account of an employee without the employee's consent (40 CFR 13.2-91).

Emulsified oil and grease: An oil or grease dispersed in an immiscible liquid usually in droplets of larger than colloidal size. In general suspension of oil or grease within another liquid (usually water) (EPA-83/06a).

Emulsifier: An agent which promotes formation and stabilization of an emulsion, usually a surface active agent (EPA-75/01a).

Emulsifying agent: A material that increases the stability of a dispersion of one liquid in another (EPA-83/06a).

Emulsion: Stable dispersions of two immiscible liquids. In the aluminum forming category this is usually an oil and water mixture (40 CFR 467.02-91, *see also* 40 CFR 471.02-91).

Emulsion breaking: Decreasing the stability of dispersion of one liquid in another (EPA-83/06a).

Emulsion cleaning: A cleaning process using organic solvents dispersed in an aqueous medium with the aid of an emulsifying agent (EPA-83/06a).

Enamel:
(1) A coating that cures by chemical cross-linking of its base resin. Enamels can be distinguished from lacquers because enamels are not readily resoluble in their original solvent (40 CFR 52.741-91).
(2) A pigmented coating which is characterized by an ability to form an especially smooth film which is free from brush or other tool marks. Although most enamels are glossy, flat enamels are also available. Enamel is usually considered to be relatively hard coatings (EPA-79/12b).

Enameling iron: A type of steel made especially for application of porcelain enamel coatings (EPA-82/11e).

Encapsulate: To seal a pesticide, and its container if appropriate, in an impervious container made of plastic, glass, or other suitable material which will not be chemically degraded by the contents. This container then should be sealed within a durable container made from steel, plastic, concrete, or other suitable material of sufficient thickness and strength to resist physical damage during and subsequent to burial or storage (40 CFR 165.1-91).

Encapsulation:
(1) A complete enclosure of a waste in another material in such a way as to isolate it from external effects such as those of water or of air; e.g., casting toxic waste into something like concrete which is insoluble and thus inaccessible to water and other substances that might bring out the poison.
(2) In 40 CFR, the treatment of ACBM (asbestos containing building material) with a material that surrounds or embeds asbestos fibers in an adhesive matrix to prevent the release of fibers, as the encapsulant creates a membrane over the surface (bridging encapsulant) or penetrates the material and binds its components together (penetrating encapsulant) (40 CFR 763.83-91).
(3) The treatment of asbestos-containing material with a liquid that covers the surface with a protective coating or embeds fibers in an adhesive matrix to prevent their release into the air (EPA-94/04).

Encephalitis: Inflammation of the brain (LBL-76/07-bio).

Enclose: To cover any VOL (volatile organic liquids) surface that is exposed to the atmosphere (40 CFR 52.741-91).

Enclosed process: A manufacturing or processing operation that is designed and operated so that there is no intentional release into the environment of any substance present in the operation. An operation with fugitive, inadvertent, or emergency pressure relief releases remains an enclosed process so long as measures are taken to prevent worker exposure to and environmental contamination from the releases (40 CFR 704.3-91, *see also* 40 CFR 704.25; 704.104; 721.350; 721.1175-91).

Enclosed storage area: Any area covered by a roof under which metallic minerals are stored prior to further processing or loading (40 CFR 60.381-91).

Enclosed truck or railcar loading station: The portion of a nonmetallic mineral processing plant were nonmetallic minerals are loaded by an enclosed conveying system into enclosed trucks or railcars (40 CFR 60.671-91).

Enclosure:
(1) A structure that surrounds a VOC (Cement, solvent, or spray) application area and drying area, and that captures and contains evaporated VOC and vents it to a control device. Enclosures may have permanent and temporary openings (40 CFR 60.541-91, *see also* 40 CFR 60.441-91).
(2) An airtight, impermeable, permanent barrier around ACBM to prevent the release of asbestos fibers into the air (40 CFR 763.83-91).
(3) Putting an airtight, impermeable, permanent barrier around asbestos-containing materials to prevent the release of asbestos fibers into the air (EPA-94/04).

End box: A container(s) located on one or both ends of a mercury chlor-alkali electrolyzer which serves as a connection between the electrolyzer and denuder for rich and stripped amalgam (40 CFR 61.51-91).

End box ventilation system: A ventilation system which collects mercury emissions from the end-boxes, the mercury pump sumps, and their water collection systems (40 CFR 61.51-91).

End finisher: A polymerization reaction vessel operated under very low pressures, typically at pressures of 2 torr or less, in order to produce high viscosity poly(ethylene terephthalate). An end finisher is preceded in a high viscosity poly(ethylene terephthalate) process line by one or more polymerization vessels operated under less severe vacuums, typically between 5 and 10 torr. A high viscosity poly(ethylene terephthalate) process line may have one or more end finishers (40 CFR 60.561-91).

End-of-pipe (EOP) technology: The final treatment process used to remove or alter selected constituents of the wastewater from manufacturing operations (EPA-87/10a).

End-of-pipe treatment: The reduction and/or removal of pollutants by a treatment process such as chemical treatment just prior to actual discharge (EPA-83/03).

End sealing compound coat: A compound applied to can ends which functions as a gasket when the end is assembled onto the can (40 CFR 52.741-91).

End use product: A pesticide product whose labeling:
(1) Includes directions for use of the product (as distributed or sold, or after combination by the user with other substances) for controlling pests or defoliating, desiccating, or regulating the growth of plants; and
(2) Does not state that the product may be used to manufacture or formulate other pesticide products (40 CFR 152.3-91, *see also* 40 CFR 158.153-91).

Endangered species (rare species or threatened species):
(1) Any species which is in danger of extinction throughout all or a significant portion of its range other than a species of the Class Insecta determined by the Secretary to constitute a pest whose protection under the provisions of this chapter would present an overwhelming and overriding risk to man (ESA3-16USC1531-90).
(2) Any species listed as such pursuant to section 4 of the Endangered Species Act (ESA3; 40 CFR 257.3.2-91, *see also* 40 CFR 723.50-91).
(3) Species of plants and animals that are threatened with either extinction or serious depletion in an area and that are formally listed by the U.S. Fish and Wildlife Service (DOE-91/04).
(4) Any species that is likely to become an endangered species within the foreseeable future throughout all or a significant portion of its range (DOE-91/04).
(5) Animals, birds, fish, plants, or other living organisms threatened with extinction by man-made or natural changes in their environment. Requirements for declaring a species endangered are contained in the Endangered Species Act (EPA-94/04).

Endangerment assessment: A study to determine the nature and extent of contamination at a site on the National Priorities List and the risks posed to public health or the environment. EPA or the state conduct the study when a legal action is to be taken to direct potentially responsible parties to clean up a site or pay for it. An endangerment assessment supplements a remedial investigation (EPA-94/04).

Endemic: The regular occurrence of a fairly constant number of cases within given area (EPA-83).

Endogenous: endo = within; genesis = growth. The term decsribes growing due to internal causes; e.g., bacteria growth within an organism (cf. exogenous). The growth of bacterial cultures eventually becomes endogenous as the result of environmental constraints imposed on the system (e.g., disappearance of nutrient substrate, enzymes, etc.) (LBL-76/07-water).

Endogenous lead: The lead that has already entered the body (*see* lead for more related terms) (LBL-76/07-bio).

Endogenous respiration: Respiration of cells by using the nutrients from dead cells.

Endogenous state: The state of continuous physiological activity in a stationary phase, which may eventually enter a declining phase (LBL-76/07-water).

Endogenous urinary lead: Refers to the urinary excretion of lead that occurs normally in the absence of chelating agents (*see* lead for more related terms) (LBL-76/07-bio).

Endogenous viruses: The viral genetic information that is a permanent part of the host cell DNA and is stably passed on to progeny (EPA-88/09a).

Endonuclease: An enzyme that degrades nucleic acid intramolecularly (EPA-88/09a).

Endothermic reaction: The reaction system that absorbs heat from the surroundings as the reaction takes place (cf. exothermic reaction).

Endotoxin: Poison released by the degeneration (or death) of bacterial cells.

Endpoint: A biological effect used as an index of the effect of a chemical on an organism. Common toxicological endpoints and their measurement parameters are as follows (Course 165.6; EPA-92/12). Other endpoint related terms include:
- Behavioral toxicity
- Carcinogenicity
- Hematologic toxicity
- Hepatoxicity
- Inhalation toxicity
- Mutagenicity
- Neurotoxicity
- Renal toxicity
- Reproductive toxicity
- Teratogenicity

Endpoint filtrate COD: The point at which the filtrate COD has become constant (in the Mass Culture Aeration BOD method) (LBL-76/07-water).

Endpoint titration: The point indicating that a desired effect (or reaction, color change, etc.) in the titration has been reached (*see* titration for more related terms).

Endrin ($C_{12}H_8OCl_6$):
(1) A pesticide insoluble in water.
(2) In 40 CFR , it means the compound endrin as identified by the chemical name 1,2,3,4,10, 10-hexachloro-6,7-epoxy-1,4,4a,5,6,7,8,8a-octahydro-1,4-endo-5,8-endodimethanonaphthalene (40 CFR 129.4-91, *see also* 40 CFR 704.102; 721.1150-91).
(3) A pesticide toxic to freshwater and marine aquatic life that produces adverse health effects in domestic water supplies (EPA-94/04).

Endrin formulator: A person who produces, prepares or processes a formulated product comprising a mixture of endrin and inert materials or other diluents into a product intended for application in any use registered under the Federal Insecticide, Fungicide and Rodenticide Act, as amended (7USC135, et seq.) (40 CFR 129.101-91).

Endrin manufacturer: A manufacturer, excluding any source which is exclusively an endrin formulator, who produces, prepares or processes technical endrin or who uses endrin as a material in the production, preparation or processing of another synthetic organic substance (40 CFR 129.101-91).

(Endrin ambient water criterion): The ambient water criterion for endrin in navigable waters is 0.004 µg/L (40 CFR 129.101-91).

Energy: The capacity of a body to do work by its motion or configuration. Other energy related terms include:
- Kinetic energy
- Potential energy

Energy average level: A quantity calculated by taking ten times the common logarithm of the arithmetic average of the antilogs of one-tenth of each of the levels being averaged. The levels may be of any consistent type, e.g., maximum sound levels, sound exposure levels, and day-night sound levels (40 CFR 201.1-91).

Energy balance: A concept based on a fundamental law of physical science (conservation of energy) known as the first law of thermodynamics which says that although energy may be transformed it may not be destroyed. Heat balance is a loose term applied to a special form of energy balance frequently used in processes which are associated with gaseous pollutant control by combustion (cf. heat balance or material balance) (EPA-84/09).

Energy band: An energy range that an electron has.

Energy conversion area: A facility that produces electricity from steam generated by reactor heat (DOE-91/04).

Energy facilities: Any equipment or facility which is or will be used primarily:
(1) In the exploration for, or the development, production, conversion, storage, transfer, processing, or transportation of, any energy resource; or
(2) For the manufacture, production, or assembly of equipment, machinery, products, or devices which are involved in any activity described in subparagraph (A). The term includes, but is not limited to:
(A) Electric generating plants;
(B) Petroleum refineries and associated facilities;
(C) Gasification plants;
(D) Facilities used for the transportation, conversion, treatment, transfer, or storage of liquefied natural gas;
(E) Uranium enrichment or nuclear fuel processing facilities;
(F) Oil and gas facilities, including platforms, assembly plants, storage depots, tank farms, crew and supply bases, and refining complexes;
(G) Facilities including deepwater ports, for the transfer of petroleum;
(H) Pipelines and transmission facilities; and
(I) Terminals which are associated with any of the foregoing (CZMA304-16USC1453).

Energy recovery: Obtaining energy from waste through a variety of processes (e.g., combustion) (EPA-94/04).

Energy grade line: The line representing the total head (or total energy) of a flow from section to section. The total energy of a flow in any section with reference to some datum is the sum of the elevation head z, the pressure head y, and the velocity head $V^2/2g$ (M&EI-72).

Energy head loss: The difference of total energy at any two flow sections (M&EI-72).

Energy losses: The kilowatt hours lost in the operation of an electric system (*see* electric loss for more related terms) (EPA-83).

Energy metabolism: Energy transformation due to the metabolism in a cell.

Energy oxygen: Oxygen utilized by biota during their exponential growth phase for the synthesis of new cellular materials or biologically stable organic substances (LBL-76/07-water).

Energy recovery:
(1) The retrieval of energy from waste by converting heat from incineration or methane gas from landfills (OTA-89/10).
(2) Conversion of waste energy, generally through the combustion of processed or raw refuse to produce steam (EPA-89/11).

Energy recovery processes: Processes that recover the energy content of combustible wastes directly by burning, or indirectly by being converted to another fuel form such as gas or oil (EPA-83).

Energy storage device: A rechargeable means of storing tractive energy on board a vehicle such as storage batteries or a flywheel (40 CFR 600.002.85-91).

Energy summation of levels: A quantity calculated by taking ten times the common logarithm of the sum of the antilogs of one-tenth of each of the levels being summed. The levels may be of any consistent type, e.g., day-night sound level or equivalent sound level (40 CFR 201.1-91).

Enforceable requirements: Conditions or limitations in permits issued under the Clean Water Act, section 402 or 404 that, if violated, could result in the issuance of a compliance order or initiation of a civil or criminal action under federal or applicable state laws. If a permit has not been issued, the term includes any requirement which, in the Regional Administrator's judgement, would be included in the permit when issued. Where no permit applies, the term includes any requirement which the RA determines is necessary for the best practical waste treatment technology to meet applicable criteria (EPA-94/04).

Enforceable requirements of the Act (CWA): Those conditions or limitations of section 402 or 404 permits which, if violated, could result in the issuance of a compliance order or initiation of a civil or criminal action under section 30 of the Act or applicable State laws. If a permit has not been issued, the term shall include any requirement which, in the Regional Administrator's judgment, would be included in the permit when issued. Where no permit applies, the term shall include any requirement which the Regional Administrator determines is necessary for the best practicable waste treatment technology to meet applicable criteria (40 CFR 35.2005-15-91).

Enforcement: EPA, state, or local legal actions to obtain compliance with environmental laws, rules, regulations, or agreements and/or obtain penalties or criminal sanctions for violations. Enforcement procedures may vary, depending on the specific requirements of different environmental laws and related implementing regulatory requirements. Under CERCLA, for example, EPA will seek to require potentially responsible parties to clean up a Superfund site, or pay for the cleanup, whereas under the Clean Air Act the agency may invoke sanctions against cities failing to meet ambient air quality standards that could prevent certain types of construction or federal funding. In other situations, if investigations by EPA and state agencies uncover willful violations, criminal trials and penalties are sought (EPA-94/04).

Enforcement decision document (EDD): A document that provides an explanation to the public of EPA's selection of the cleanup alternative at enforcement sites on the National Priorities List. Similar to a Record of Decision (EPA-94/04).

Engine: A machine for converting thermal energy into mechanical power to produce force and motion. Other engine related terms include:
- Aircraft engine
- Basic engine
- Commercial aircraft engine
- External combustion engine
- Flexible fuel engine: (*see* synonym, flexible fuel vehicle)
- Heavy duty engine
- Internal combustion engine
- Light duty vehicle and engine
- Military engine
- New aircraft turbine engine
- Non-road engine
- Test engine

Engine code: The unique combination, within an engine system combination, of displacement, carburetor (or fuel injection) calibration, choke calibration, distributor calibration, auxiliary emission control devices, and other engine and mission control system components specified by the Administrator (40 CFR 86.082.2-91, *see also* 40 CFR 600.002.85-91).

Engine configuration: *See* synonym, vehicle or engine configuration.

Engine displacement: The volumetric engine capacity as defined in 40 CFR 205.153 (40 CFR 205.151-91).

Engine family: The basic classification unit of a vehicles product line for a single model year used for the purpose of emission-data vehicle or engine selection and as determined in accordance with 40 CFR 86.078-24 (40 CFR 85.2113-91, *see also* 40 CFR 86.082.2; 86.402.78-91).

Engine family group: A combination of engine families for the purpose of determining a minimum deterioration factor under the Alternative Durability program (40 CFR 86.082.2-91).

Engine lubricating oils: The petroleum-based oils used for reducing friction in engine parts (40 CFR 252.4-91).

Engine model: All commercial aircraft turbine engines which are of the same general series, displacement, and design characteristics and are usually approved under the same type certificate (40 CFR 87.1-91).

Engine displacement system combination: An engine family-displacement-emission control system combination (40 CFR 86.402.78-91).

Engine sidescreen: A rugged screen that fits on the engine housing of a vehicle used at a sanitary landfill to keep paper and other objects from accumulating and damaging the engine (SW-108ts).

Engine system combination: An engine family-exhaust emission control system combination (40 CFR 86.082.2-91).

Engineered safety features: For a nuclear facility, features that prevent, limit, or mitigate the release of radioactive material from its primary containment (DOE-91/04).

Engineering control: Controls that isolate or remove the hazard from the workplace (29 CFR 1910).

Enhanced inspection and maintenance: An improved automobile inspection and maintenance program-aimed at reducing automobile emissions-that contains, at a minimum, more vehicle types and model years, tighter inspection, and better management practices. It may also include annual computerized or centralized inspections, under-the-hood inspection- for signs of tampering with pollution control equipment, and increased repair waiver cost (EPA-94/04).

Enrichment: The addition of nutrients (e.g., nitrogen, phosphorus, carbon compounds) from sewage effluent or agricultural runoff to surface water, greatly increases the growth potential for algae and other aquatic plants (EPA-94/04).

Ensemble average: Considering a quantity which is a random function of time and/or space, the ensemble average is defined as the arithmetic average over N independent realizations of the process (N ---> infinite). The average is calculated at same values of the time and space coordinates with respect to each realization (NATO-78/10).

Enteric virus: Any virus known to be excreted in quantity in feces; infectious hepatitis virus is such a virus (DOI-70/04).

Enterobacter: Bacteria usually found in the intestine of human or animal bodies.

Enterprise fund: A fund for a specific purpose that is self-supporting from the revenue it generates (EPA-89/11).

Enthalpy (H): Because the combination of U + PV occurs frequently in many thermodynamic applications, the combination has been given a name as enthalpy. Mathematically, enthalpy is defined as, H = U + PV. Where H is enthalpy, U is internal energy, P is pressure, and V is volume. Enthalpy is an arbitrary combination of other properties and therefore it is a property. It should be noticed that enthalpy is not a form of energy. Enthalpy has a unit of B/lbm (Jones-p68).

Mathematically, enthalpy has the following related equations:
- By definition: H = U + PV
- Applying the First thermodynamic law: $Q = H_2 - H_1$
  Where: Q = heat; H_1 = enthalpy value at state 1; and H_2 = enthalpy value at state 2.
- Applying the ideal gas law: $Q = H_2 - H_1 = mCp(T_2 - T_1)$
  Where: m = mass; Cp = constant pressure specific heat; T_1 = temperature at state 1; and T_2 = temperature at state 2.

Enthalpy of combustion: *See* synonym, heat of combustion.

Enthalpy of formation: *See* synonym, heat of formation.

Enthalpy of hydration: *See* synonym, heat of formation.

Enthalpy of reaction: *See* synonym, heat of formation.

ENTOMB: Radioactive contaminates are encased (entombed) in a structurally sound and long-lived material (such as concrete), which is maintained and monitored until the radioactivity decays to levels permitting release of the facility for unrestricted use (DOE-91/04).

Entomology: A science dealing with the study of insects.

Entomophilous: Describing the pollination coursed by insects.

Entrain: To trap bubbles in water either mechanically through turbulence or chemically through a reaction (EPA-94/04).

Entrainment:
(1) Carry-over, e.g., the carry-over of liquid mist in air after scrubbering. Mist eliminators are used to reduce entrainment (EPA-84/09).
(2) The involuntary capture and inclusion of organisms in streams of flowing water. Term often applied to the cooling water systems of power plants/reactors. The organisms involved depend on the intake screen mesh size; they may include phyto- and zooplankton, fish eggs and larvae (ichthyoplankton), shellfish larvae, and other forms of aquatic life (DOE-91/04).

Entrainment separator: *See* synonym, mist eliminator.

Entry loss: The loss in static pressure caused by air flowing into a duct or hood. It is usually expressed in inches of water gauge (29 CFR 1910.94b-91).

Entry routes: Pathways by which soil gas can flow into a house. Openings through the flooring and walls where the house contacts the soil (EPA-88/08).

Entric: Organisms that inhabit the intestinal tract of vertebrates (EPA-88/09a).

Entropy: In thermodynamics, the ratio of the heat added to a system to the absolute temperature at which it was added. For a reversible process, entropy is defined as:
dS = dQ/T
where:
- S = entropy in differential amount
- Q = heat in differential amount
- T = temperature

Entropy is a measure of unavailable energy in a system. It shows that an increase of entropy is accompanied by a decrease in available energy.

Entropy of activation: The entropy difference between the intermediate compounds and the reactants.

Environment:
(1) Includes water, air, and land and all plants and man and other animals living therein, and interrelationship which exists among these (FIFRA2, *see also* SF101; SF329; TSCA3; 40 CFR 6.1003; 171.2; 300.5; 302.3; 355.20; 370.2; 723.50; 723.175; 723.250; 762.3-91).
(2) The sum of all external conditions affecting the life, development and survival of an organism (EPA-94/04).

Other environment related terms include:
- Accessible environment
- General environment
- Human environment
- Marine environment

Environmental accumulation: The accumulation of pollutants in various sectors of the environment (cf. bioaccumulation).

Environmental assessment:
(1) A written environmental analysis which is prepared pursuant to the National Environmental Policy Act to determine whether a federal action would significantly affect the environment and thus require preparation of a more detailed environmental impact statement (40 CFR 1508.9-91).
(2) An environmental analysis prepared pursuant to the National Environmental Policy Act to determine whether a federal action would significantly affect the environment and thus require a more detailed environmental impact statement (EPA-94/04).

Environmental audit:
(1) An independent assessment of the current status of a party's compliance with applicable environmental requirements.
(2) A documented assessment of a facility to monitor the progress of necessary corrective actions, to ensure compliance with environmental laws and regulations and to evaluate field organization practices and procedures (DOE-91/04).
(3) An independent assessment of the current status of a party's compliance with applicable environmental requirements or of a party's environmental compliance policies, practices, and controls (EPA-94/04).

Environmental document: Includes the documents specified in 40 CFR 1508.9 (environmental assessment), 40 CFR 1508.11 (environmental impact statement), 40 CFR 1508.13 (finding of no significant impact), and 40 CFR 1508.22 (notice of intent) (40 CFR

1508.10-91).

Environmental emergencies: Incidents involving the release (or potential release) of hazardous materials into the environment which require immediate corrective actions (Course 165.5).

Environmental equity: Equal protection from environmental hazards of individuals, groups or communities regardless of race, ethnicity, or economic status (EPA-94/04).

Environmental exposure: Human exposure to pollutants originating from facility emissions. Threshold levels are not necessarily surpassed, but low level chronic pollutant exposure is one of the most common forms of environmental exposure (*see* threshold level) (EPA-94/04).

Environmental fate: The destiny of a chemical after release to the environment; involves considerations such as transport through air, soil and water, bioconcentration, degradation, etc. (Course 165.6).

Environmental hazard: A condition capable of posing an unreasonable risk to air, water, or soil quality, and to plants or wildlife (Course 165.5).

Environmental impact statement (EIS): A document required of federal agencies by the National Environmental Policy Act (NEPA) for major projects or legislative proposals significantly affecting the environment. A tool for decision making, it describes the positive and negative effects of the undertaking and lists alternative actions (EPA-94/04, *see also* 40 CFR 6.201; 1508.11-91).

Environmental indicator: A measurement, statistic or value that provides a proximate gauge or evidence of the effects of environmental management programs or of the state or condition of the environment (EPA-94/04).

Environmental information document: Any written analysis prepared by an applicant, grantee or contractor describing the environmental impacts of a proposed action. This document will be of sufficient scope to enable the responsible official to prepare an environmental assessment as described in the remaining subparts of this regulation (40 CFR 6.101-91).

Environmental justice: The fair treatment of all races, cultures, incomes, and educational levels with respect to the development, implementation, and enforcement of environmental laws, regulations, and policies. Fair treatment implies that no population of people should be forced to shoulder a disproportionate share of the negative environmental impacts of pollution or environmental hazards due to a lack of political or economic strength.levels (EPA-94/04)

Environmental law: The U.S. Congress created the Environmental Protection Agency on December 2, 1970 and has passed various environmental laws since then. The purpose of the environmental laws is to protect human health and the environment. *see* the highlights of various environmental laws under the term "Act or law acronym, e.g., Act or CAA means the highlight of the Clean Air Act" (*see* law for more related terms).

Environmental law system: An organized way of using all of the laws in our legal system to minimize, prevent, punish or remedy the consequences of actions which damage or threaten the environment, public health and safety. Environmental law encompasses:
(1) Laws: federal and state statutes and local ordinances
(2) Regulations promulgated by federal, state and local agencies
(3) Court decisions interpreting these laws and regulations
(4) The common law
(5) United States constitution and state constitutions
(6) Treaties (Sullivan-95/04, p1).

Environmental noise: The intensity, duration, and the character of sounds from all sources (*see* noise for more related terms) (NCA3-42USC4902-87-91, *see also* 40 CFR 40 CFR 205.2-91).

Environmental persistence: The ability to resist biodegradation and to last, or survive, in the environment, e.g., DDT, whose half-life is 15 years, is an environmentally persistent substance.

Environmental protection: Protection of human health and the environment.

Environmental Research, Development, and Demonstration Authorization Act (ERDDAA) of 1980: *See* the term of Act or ERDDAA.

Environmental resistance: All constraint factors in the environment.

Environmental response team (ERT):
(1) A group of highly specialized experts available through EPA 24 hours a day (NRT-87/03).
(2) EPA experts located in Edison, N.J., and Cincinnati, OH, who can provide around-the-clock technical assistance to EPA regional offices and states during all types of hazardous waste site emergencies and spills of hazardous substances (EPA-94/04).

Environmental review: The process whereby an evaluation is undertaken by EPA to determine whether a proposed Agency action may have a significant impact on the preparation on the EIS (40 CFR 6.101-91).

Environmental regulations: The detailed procedures developed by U.S. EPA to implement the environmental laws enacted by U.S. Congress.

Environmental sample (or field sample): A representative sample of any material (aqueous, non-aqueous, or multimedia) collected from any source for which determination of composition or contamination is requested or required (*see* sample for more related terms).

Environmental stress crack (ESC): External or internal stress propagation in a plastic caused by environmental conditions which are usually chemical or thermal in nature (cf. stress crack) (EPA-91/05).

Environmental survey: A documented, multi-disciplined assessment (with sampling and analysis) of a facility to determine environmental conditions and to identify environmental problems requiring corrective action (DOE-91/04).

Environmental system: The interaction of an organism or group of organisms with its natural and manmade surroundings (SW-108ts).

Environmental transformation product: Any chemical substance resulting from the action of environmental processes on a parent compound that changes the molecular identity of the parent compound (40 CFR 723.50-91).

Environmentally related measurements: Any data collection activity or investigation involving the assessment of chemical, physical, or biological factors in the environment which affect human health or the quality of life. The following are examples of environmentally related measurements:
(1) A determination of pollutant concentrations from sources or in the ambient environment, including studies of pollutant transport and fate;
(2) A determination of the effects of pollutants on human health and on the environment;
(3) A determination of the risk/benefit of pollutants in the environment;
(4) A determination of the quality of environmental data used in economic studies; and
(5) A determination of the environmental impact of cultural and natural processes (40 CFR 30.200-91).

Environmentally transformed: A chemical substance is environmentally transformed when its chemical structure changes as a result of the action of environmental processes on it (40 CFR 721.3-91).

Enzyme:
(1) A protein that acts as a catalyst in a chemical reaction. During the reaction, the enzyme becomes a part of the chemical reaction; but after the reaction is complete, the enzyme is usually split off un-changed, and ready to catalyze another reaction. Nearly all biochemical reactions are catalyzed by enzymes which are specific to the substrate being acted upon. There are nearly 2,000 known enzymes. No single species or cell has all of them.
(2) A class of complex proteinaceous substances produced by living cells and essential to life processes. They act similarly to catalysts in that they promote a variety of usually reversible cellular reactions (e.g., oxidation, hydrolysis) at cell temperature without themselves undergoing permanent change. Frequently the presence of activators such as metal ions or coenzymes is required for reaction to occur (LBL-76/07-water).

EP toxicity: Extraction procedure toxicity, *see* definition under hazardous waste.

EPA:
(1) The United States Environmental Protection Agency (40 CFR 2.100-91).
(2) The U.S. Environmental Protection Agency; established in 1970 by Presidential Executive Order, bringing together parts of various government agencies involved with the control of pollution (EPA-89/12).

EPA acknowledgement of consent: The cable sent to EPA from the U.S. Embassy in a receiving country that acknowledges the written consent of the receiving country to accept the hazardous waste and describes the terms and conditions of the receiving country's consent to the shipment (40 CFR 262.51-91).

EPA hazardous waste number: The number assigned by EPA to each hazardous waste listed in Part 261, Subpart D, of this chapter and to each characteristic identified in Part 261, Subpart C, of this chapter (40 CFR 260.10-91).

EPA identification number:
(1) The number assigned by EPA to each generator, transporter, and treatment, storage, or disposal facility (40 CFR 260.10-91).
(2) The 12-digit number assigned to a facility by EPA upon notification of PCB waste activity under 40 CFR 761.205 (40 CFR 761.3-91).

EPA record or record: Any document, writing, photograph, sound or magnetic recording, drawing, or other similar thing by which information has been preserved, from which the information can be retrieved and copied, and over which EPA has possession or control. It may include copies of the records of other Federal agencies (*see* 40 CFR 2.111(d)). The term includes informal writings (such as drafts and the like), and also includes information preserved in a form which must be translated or deciphered by machine in order to be intelligible to humans. The term includes documents and the like which were created or acquired by EPA, its predecessors, its officers, and its employees by use of Government funds or in the course of transacting official business. However, the term does not include materials which are the personal records of an EPA officer or employee. Nor does the term include materials published by non-Federal organizations which are readily available to the public, such as books, journals, and periodicals available through reference libraries, even if such materials are in EPA's possession (40 CFR 2.100-b-91).

Ephemeral stream: *See* synonym, intermittent stream.

Ephippium: A resting egg which develops under the carapace in response to stress conditions in daphnids (40 CFR 797.1300; 797.1330-91).

Epidemic: Widespread outbreak of a disease, or a large number of cases of a disease in a single community or relatively small area (EPA-89/12).

Epidemiologic study: An investigation of elements contributing to disease or toxic effects in human populations. Epidemiology is an observational science not experimental. The study purpose is to (1) Discover and interpret the patterns and occurrences of diseases and related health conditions. (2) Develop quantitative assessments of these conditions. (3) Relate assessments of risk to personal and environmental characteristics which determine the degree of risk (Course 165.6).

Epidemiology:
(1) The study of the sources and distribution of disease in a population (ETI-92).
(2) Study of the distribution of disease, or other health-related states and events in human populations, as related to age, sex, occupation, ethnic, and economic status in order to

identify and alleviate health problems and promote better health (EPA-94/04).

Other epidemiology related terms include:
- Analytic study epidemiology
- Descriptive study epidemiology

Epidermis: The top layer of skin; animal hair is an epidermal outgrowth (EPA-82/11).

Epilimnion:
(1) The warmer layer of a water body due to mixing by wind actions and turbulence.
(2) That region of a body of water that extends from the surface to the thermocline and does not have a permanent temperature stratification (LBL-76/07-water).
(3) The upper layer of water in a thermally stratified lake or reservoir. This layer consists of the warmest water (EPA-94/04).

Epiphyte: A plant that grows upon another objects such as a plant or a building.

Epiphyton: A non-parasitic secondary plant growth (LBL-76/07-water).

Episode (air pollution episode or emergency episode):
(1) Although episode is most commonly used in relation to air pollution, the term may also be used in connection with other kinds of environmental events such as a massive water pollution situation.
(2) An air pollution incident in a given area caused by a concentration of atmospheric pollutants under meteorological conditions that may result in a significant increase in illnesses or deaths. May also describe water pollution events or hazardous material spills (EPA-94/04).

Episode criteria: The conditions justifying the proclamation of an air pollution alert, air pollution warning, or air pollution emergency shall be deemed to exist whenever the Director determines that the accumulation of air pollutants in any place is attaining or has attained levels which could, if such levels are sustained or exceeded, lead to a substantial threat to the health of persons. In making this determination, the Director will be guided by the following criteria:
(a) Air pollution forecast;
(b) Alert;
(c) Warning;
(d) Emergency; and
(e) Termination (40 CFR 51-AL-1.1-91).

Epoxy resin: A plastic or resinous material used for strong, fast-setting adhesives, as heat resistant coatings and binders, etc. (*see* resin for more related terms) (EPA-79/12a).

Equal opportunity clause: The contract provisions set forth in section 4 (a) or (b), as appropriate (40 CFR 8.2-91).

Equalization:
(1) The process whereby waste streams from different sources varying in pH, chemical constituents, and flow rates are collected in a common container. The effluent stream from this equalization container will have a fairly constant flow and pH level and will help prevent unnecessary shock to the waste treatment system (EPA-83/03).
(2) Any process for averaging variations in a flow and/or composition of wastewater so as to affect a more uniform discharge (EPA-79/12b).

Equalization basin: A holding basin in which variations in flow and composition of a liquid are averaged (EPA-87/10a).

Equalizing reservoir: A water reservoir located between the main supply and consumers to maintain the fluctuation of water use.

Equation of continuity: *See* synonym, continuity equation.

Equation of state: An equation to express the thermodynamic state of a substance. It relates pressure p, volume V, absolute temperature T and number of moles of a substance in an equation. For example, $PV = mRT$ is an equation of state for an ideal gas. Where, P = pressure; V = volume; m = mass; R = gas constant; and T = absolute temperature (Jones-p151).

Equilibrium: A thermodynamic state is in equilibrium when all its thermodynamic properties such as pressure, temperature and density are uniform throughout its system (cf. thermodynamic equilibrium).
(1) In air pollution control, a state in which there is no tendency toward spontaneous change. Equilibrium exists when the net interchange of material between phases is zero. Thus, in the absorption process, equilibrium exists when gas molecules leave a liquid (desorption) at the same rate which they enter (absorption) (EPA-84/09).
(2) In chemical reaction, the forward and reverse reactions are producing at the same rate; defined by the equilibrium constant, K (EPA-2/88).
(3) In relation to radiation, the state at which the radioactivity of consecutive elements within a radioactive series is neither increasing nor decreasing (EPA-94/04).

Equilibrium capacity: A term used in adsorption studies to describe the maximum amount of pollutant retained by an adsorbent at equilibrium and at a given operating condition. Common units are lb/lb adsorbent or lb/100 lb adsorbent (EPA-84/09).

Equilibrium concentration: A state at which the concentration of chemicals in a solution remain in a constant proportion to one another (EPA-83/06a).

Equilibrium constant (K): A constant at a given temperature for which a reversible chemical reaction reaches its chemical equilibrium. For a reaction $aA + bB = cC + dD$, $K = C^c D^d / A^a B^b$.

Equilibrium partitioning (EP): A method for generating sediment criteria that focuses on the chemical interaction between sediments and contaminants (EPA-91/03).

Equilibrium line: A plot describing the mole fraction of the solute (pollutant) in the gas phase in equilibrium with the mole fraction of solute in the liquid phase. This is often required in absorber calculations (EPA-84/09).

Equipment: The tangible, nonexpendable, personal property having a useful life of more than one year and an acquisition cost of $5,000 or more per unit. A grantee may use its own definition of equipment provided that such definition would at least include all equipment defined above (40 CFR 31.3-91, *see also* 40 CFR 35.6015; 60.481; 60.591; 60.631; 61.131; 61.241; 264.1031-91).

Equipment blank (field blank or trip blank): Equipment blanks, field blanks and trip blanks are used to check on sample contamination originating from sample transport, shipping and from site conditions. Equipment blanks are opened in the field and the contents are poured appropriately over or through the sample collection device, collected in a sample container, and returned to the laboratory as a sample. Equipment blanks are a check on sampling device cleanliness. Field blanks are also opened in the field and nearby the sampling platform. The purpose is to check surrounding conditions. Trip blanks are not opened in the field. The purpose is to check sample shipping conditions (*see* blank for more related terms).

Equipment room (change room): A contaminated room located within the decontamination area that is supplied with impermeable base or containers for the disposal of contaminated protective clothing and equipment (40 CFR 763.121-91).

Equivalence point: The point in a chemical titration at which the quantities of titrants and substances being titrated are equivalent.

Equivalence ratio: *See* synonym, combustion equivalence ratio.

Equivalent: That a chemical substance or mixture is able to represent or substitute for another in a test or series of tests, and that the data from one substance can be used to make scientific and regulatory decisions concerning the other substance (40 CFR 790.3-91).

Equivalent diameter: Four times the area of an opening divided by its perimeter (40 CFR 60.711-91, *see also* 40 CFR 60.741-91).

Equivalent hide: A statistical term used to relate the production of tanneries using various types of raw materials. An equivalent hide is represented by 3.7 sq m of surface area and is the average size for a cattle-hide (EPA-1/82).

Equivalent live weight killed (ELWK): The total weight of the total number of animals slaughtered at locations other than the slaughterhouse or packinghouse, which animals provide hides, blood, viscera or renderable materials for processing at that slaughterhouse, in addition to those derived from animals slaughtered on site (40 CFR 432.11-91, *see also* 40 CFR 432.21; 432.31; 432.41-91).

Equivalent method:
(1) Any method of sampling and analyzing for an air pollutant which has been demonstrated to the Administrator's satisfaction to have a consistent and quantitatively known relationship to the reference method, under specified conditions (40 CFR -91, *see also* 40 CFR 50.1; 53.12; 60.2; 61.02; 260.10-91).
(2) Any method of sampling and analyzing for air pollution which has been demonstrated to the EPA Administrator's satisfaction to be, under specific conditions, an acceptable alternative to the normally used reference methods (EPA-94/04).
(3) *See* method for more related terms.

Equivalent P_2O_5 feed: The quantity of phosphorus, expressed as phosphorous pentoxide, fed to the process (40 CFR 60.201-91, *see also* 40 CFR 60.211; 60.221; 60.231; 60.241-91).

Equivalent petroleum-based fuel economy value: A number which represents the average number of miles traveled by an electric vehicle per gallon of gasoline (40 CFR 600.502.81-91).

Equivalent sound level:
(1) The level, in decibels, of the mean-square A-weighted sound pressure during a stated time period, with reference to the square of the standard reference sound pressure of 20 micropascals. It is the level of the sound exposure divided by the time period and is abbreviated as L_{eq} (40 CFR 201.1-91).
(2) The equivalent steady sound level that, if continuous during a specified time period, would contain the same total energy as the actual time-varying sound. For example, L_{eq} (1-h) and L_{eq} (24-h) are the 1-hour and 24-hour equivalent sound level, respectively (DOE-91/04).
(3) *See* sound level for more related terms.

Equivalent test weight: The weight, within an inertia weight class, which is used in the dynamometer testing of a vehicle and which is based on its loaded vehicle weight or adjusted loaded vehicle weight in accordance with the provisions of subparts A and B of this part (40 CFR 86.094.2-91).

Equivalent weight: The weight of a compound (or element) that can reacts with or displaces 1.008 grams of hydrogen (or 8 grams of oxygen or 35.5 grams of chlorine) (cf. gram equivalent weight).

Erbium (Er): A rare earth metal with a.n. 68; a.w. 167.26; d. 9.05 g/cc; m.p. 1497° C and b.p. 2900° C. The element belongs to group IIIB of the periodic table.

Erosion:
(1) Physical wearing away of a surface (cf. abrasion or corrosion) (EPA-83).
(2) The wearing away of land surface by wind or water, intensified by land-clearing practices related to farming, residential or industrial development, road building, or logging (EPA-94/04).
Other erosion related terms include:
- Accelerated erosion
- Refractory erosion

Error: *See* definition under uncertainty.

Erythroid hypoplasia: Decreased formation of erythroid elements of the blood (i.e., red blood cells) (LBL-76/07-bio).

Erythropoiesis: Formation of red blood cells (LBL-76/07-bio).

Esparto: A grass whose bast fibers are used to produce high-class book and printing papers and medium class writing papers (EPA-87/10).

Essential element: Elements such as carbon, hydrogen, oxygen, nitrogen, etc. needed by living organisms to maintain their normal growth.

Essential oil:
(1) Oils composed mainly of terpene hydrocarbons (turpentine), which are obtained by steam distillation of wood chips, bark, or leaves of selected trees (EPA-79/12).
(2) The oil in citurs peel, peel oil (EPA-74/03).
(3) *See* oil for more related terms.

Established Federal standard: Any operative occupational safety and health standard established by an agency of the United States and presently in effect, or contained in any Act of Congress in force on the date of enactment of this (OSHA3, *see* also 29 CFR 1910.2-91).

Establishment: Any place where a pesticide or device or active ingredient used in producing a pesticide is produced, or held, for distribution or sale (FIFRA2; 40 CFR 167.3; 372.3-91).

Ester: An organic compound corresponding in structure to a salt in inorganic chemistry. Esters are considered as derived from the acids by the exchange of the replaceable hydrogen of the latter for an organic alkyl radical. Esters are not ionic compounds, but salts usually are (EPA-83/06a).

Ester gum: A resin made from rosin or rosin acids and a polyhydric alcoho, such as glycerin or pentaerythritol (EPA-79/12).

Esterification:
(1) The production of esters from carboxylic acids by the replacement of the hydrogen of the hydroxyl group with a hydrocarbon group (EPA-87/10a).
(2) The combination of an alcohol and an organic acid to produce an ester and water. The reaction is carried out in the liquid phase, with aqueous sulfuric acid as a catalyst (EPA-83/09).

Estimated exposure dose (EED): The measured or calculated dose to which humans are likely to be exposed considering exposure by all sources and routes (EPA-92/12).

Estivation: Dormancy of animals during summer time (cf. dormancy or hibernation).

Estuarine sanctuary: A research area which may include any part or all of an estuary and any island, transitional area, and upland in, adjoining, or adjacent to such estuary, and which constitutes to the extent feasible a natural unit, set aside to provide scientists and students the opportunity to examine over a period of time the ecological relationships within the area (CZMA304-16USC1453-90).

Estuary:
(1) That part of a river or stream or other body of water having unimpaired connection with the open sea, where the sea water is measurably diluted with fresh water derived from land drainage. The term includes estuary-type areas of the Great Lakes (CZMA304-16USC1453-90).
(2) A confined body of water where fresh water flows mix with seawater due to tidal action (DOE-91/04).
(3) Regions of interaction between rivers and near-shore ocean waters, where tidal action and river flow mix fresh and salt water. Such areas include bays, mouths of rivers, salt marshes, and lagoons. These brackish water ecosystems shelter and feed marine life, birds, and wildlife (*see* wetlands) (EPA-94/04).

Etchant: The material used in the chemical process of removing glass fibers and epoxy between neighboring conductor layers of a PC board for a given distance (EPA-83/06a).

Etching: *See* synonym, cleaning.

Ethanol:
(1) An alternative automotive fuel derived from grain and corn; usually blended with gasoline to form gasohol (EPA-94/04).
(2) *See also* synonym, alcohol.

Ethanol blender: Any person who owns, leases, operates, controls, or supervises an ethanol blending plant (40 CFR 52.137; 80.2-91).

Ethanol blending plant: Any refinery at which gasoline is produced solely through the addition of ethanol to gasoline, and at which the quality and quantity of gasoline is not altered in any other manner (40 CFR 52.137-91, *see also* 40 CFR 80.2-91).

Ether ($(C_2H_5)_2O$): A colorless solvent.

Ethene: *See* synonym, ethylene.

Ethical products: Pharmaceuticals promoted by advertising to the medical, dental, and veterinary professions (EPA-83/09).

Ethiofencarb ($C_{11}H_{15}O_2NS$): A yellow liquid used as an insecticide.

Ethnic foods: The production of canned and frozen Chinese and Mexican specialties utilizing fresh and pre-processed bean sprouts, bamboo shoots, water chestnuts, celery, cactus, tomatoes, and other similar vegetables necessary for the production of the various characteristic product styles (40 CFR 407.81-91).

Ethnography: A branch of anthropology dealing with the study of human culture (DOE-91/04).

Ethyl alcohol: *See* synonym, alcohol.

Ethylene (ethene or olefiant gas, C_2H_4): A colorless, flammable gas.

Ethylene dibromide (EDB): A chemical used as an agricultural fumigant and in certain industrial processes. Extremely toxic and found to be a carcinogen in laboratory animals, EDB has been banned for most agricultural uses in the United States (EPA-94/04).

Ethylene dichloride plant: Icludes any plant which produces ethylene dichloride by reaction of oxygen and hydrogen chloride with ethylene (40 CFR 61.61-91).

Ethylene dichloride purification: Icludes any part of the process of ethylene dichloride purification following ethylene dichloride formation, but excludes crude, intermediate, and final ethylene dichloride storage tanks (40 CFR 61.61-91).

Ethylene interpolymer alloy (EIA): A blend of ethylene vinyl acetate and polyvinyl chloride resulting in a thermoplastic elastomer (EPA-91/05).

Ethylene interpolymer alloy-reinforced (EIA-R): Sheets of EIA with an encapsulated fabric reinforcement layer (EPA-91/05).

Ethylene vinyl acetate (EVA): A chemical commonly used in hot melt adhesives (EPA-83).

Ethylenediamine tetraacetic acid ($HOOCCH_2)_2NCH_2CH_2N-(CH_2COOH)_2$): A chelating compound used in some bleach and bleach-fix solutions. It is also used in other solutions to sequester some ions present in hard water (EPA-80/10).

Ethylenediamine tetra acetic acid (EDTA) titration: A standard method of measuring the hardness of a solution (EPA-83/06a). It is a chelating agent which forms complex compounds with many cations, is used to determine the concentration of calcium or magnesium ions, and is thus used to measure the hardness of water (*see* titration for more related terms).

Eukaryote: A cell with distinct nucleus.

Eukaryotic cell: A complex cell type, characterized by having a nuclear membrane, mitochondria, and numerous chromosomes. All living cells except bacteria and blue green algae are eucaryotic (EPA-88/09a).

Etiological agent: The pathogenic organism or its toxin causing a specific disease in a living body (EPA-83).

Eulerian frame of reference: A coordinate system in which the properties of a flow are described as a function of time and space where the coordinate system remains fixed (NATO-78/10).

Eulittoral zone: The shore zone of a body of water between the limits of water-level fluctuation (LBL-76/07-water).

Euphotic zone: The lighted region that extends vertically from the water surface to the level at which photosynthesis fails to occur because of ineffective light penetration (LBL-76/07-water).

Europium (Eu): A rare earth metal with a.n. 63; a.w. 151.96; d. 5.26 g/cc; m.p. 826° C and b.p. 1439° C. The element belongs to group IIIB of the periodic table.

Eurytopic organisms: Organisms with a wide range of tolerance to a particular environmental factor. Examples are sludgeworms and bloodworms (LBL-76/07-water).

Eustacy: The fluctuation of the sea level.

Eutectic: An alloy or solution of materials that has the lowest possible constant melting temperature (EPA-83).

Eutrophic lake:
(1) A lake that exhibits any of the following characteristics:
 (A) Excessive biomass accumulations of primary producers;
 (B) Rapid organic and/or inorganic sedimentation and shallowing; or
 (C) Seasonal and/or diurnal dissolved oxygen deficiencies that may cause obnoxious odors, fish kills, or a shift in the composition of aquatic fauna to less desirable forms (40 CFR 35.1605.5-91).
(2) Shallow, murky bodies of water that have excessive concentrations of plant nutrients causing excessive algal production (cf. dystrophic lakes) (EPA-94/04).
(3) *See* lake for more related terms.

Eutrophication: The slow aging process during which a lake, estuary, or bay evolves into a bog or marsh and eventually disappears. During the later stages of eutrophication the water body is choked by abundant plant life due to higher levels of nutritive compounds such as nitrogen and phosphorus. Human activities can accelerate the process (EPA-94/04).

Evacuation: Removal of residents from an area or danger (EPA-85/11).

Event tree: A logic diagram which depicts all pathways (success and failure) originating from an initiating event (EPA-87/07a).

Evaluation program or program: The sequence of analyses and tests prescribed by the Administrator as described in 40 CFR 610.13 in order to evaluate the performance of a retrofit device (40 CFR 610.11-91).

Evaporation: Evaporation is the process by which a substance changes from the liquid to the gas or vapor state. The opposite process is condensation. In phase separation, it is the vaporization of a liquid from a solution or a slurry for separation of liquid from a dissolved or suspended solid or liquid. The term evaporation can only be applied to a unit operation in which heat energy is transferred to a solution or suspension of a solid or liquid and where one of the components of the substance to which heat is applied is not appreciably volatile. The process and the equipment are similar to that of the stills or reboilers of distillation except, in evaporation, no attempt is made to separate components of the vapor. The objective of evaporation is to concentrate a solution consisting of a non-volatile solute and a volatile solvent.

Evaporation differs from drying in that the residue is usually a highly viscous liquid, rather than a solid; it differs from distillation in that the vapor is usually a single component, and even when the vapor is a mixture, no attempt is made in the evaporation step to separate the vapor into fractions; it differs from crystallization in that the emphasis is placed on concentrating a solution rather than forming and building crystals.

Evaporation chamber: In the steel industry, a method used for cooling gases to the precipitators in which an exact heat balance is maintained between water required and gas cooled; no effluent is discharged as all of the water is evaporated (EPA-74/06a).

Evaporation pond:
(1) An open holding facility which depends primarily on climatic

conditions such as evaporation, precipitation, temperature, humidity, and wind velocity to effect dissipation (evaporation) of wastewater. External means such as spray recirculation or heating can be used to increase the rate of evaporation (EPA-87/10a).
(2) Areas where sewage sludge is dumped and dried (EPA-94/04).
(3) *See* pond for more related terms.

Evaporative emission: Hydrocarbons emitted into the atmosphere from a motor vehicle, other than exhaust and crankcase emissions (*see* emission for more related terms) (40 CFR 86.082.2-91).

Evaporative emission code: A unique combination, in an evaporative emission family evaporative emission control system combination, of purge system calibrations, fuel tank and carburetor bowl vent calibrations and other fuel system and evaporative emission control system components and calibrations specified by the Administrator (40 CFR 86.082.2-91).

Evaporative vehicle configuration: A unique combination of basic engine, engine code, body type, and evaporative emission code (40 CFR 86.082.2-91).

Evaporator: A device which is used to convert a liquid into a gaseous state or vapor. Sizing of evaporator heating surfaces can in many cases make use of individual film coefficients.
Most evaporators deal with concentrated solutions, frequently containing high concentrations of crystallizing solids in suspension, that have properties far different from those of pure liquids. Care must be exercised in defining the temperature differences used in computing the heat transfer characteristics (cf. condenser).

Evapotranspiration:
(1) The combined processes by which water is transferred from the earth's surface to the atmosphere; evaporation of liquid or solid water plus transpiration from plants (DOE-91/04).
(2) The loss of water from the soil both by evaporation and by transpiration from the plants growing in the soil (EPA-94/04).

Evase duct: An expanding duct connection on the outlet of a fan or in an air/gas flow passage; its purpose is to convert kinetic energy into static pressure (EPA-83).

Evase stack: An expanding connection on the outlet of a fan or in an air flow passage. Its purpose is to convert kinetic energy into static pressure (*see* stack for more related terms) (SW-108ts).

Evidence: Any matter in the rule-making record (TSCA Sec. 19).

Evoked response technique: A technique widely used in electrophysiology whereby a stimulus (e.g., electric shock, light flash, click) is applied peripherally to the electrode used to detect the response (LBL-76/07-bio).

Evolution: The gradual progress of living organisms.

Ex parte communication: Any communication, written or oral, relating to the merits of the proceeding between the decisional body and an interested person outside the Agency or the Agency trial staff which was not originally filed or stated in the administrative record or in the hearing. Ex parte communications do not include:
(1) Communications between Agency employees other than between the Agency trial staff and the members of the decisional body;
(2) Discussions between the decisional body and either:
 (A) Interested persons outside the Agency, or
 (B) The Agency trial staff, if all parties have received prior written notice of the proposed communications and have been given the opportunity to be present and participate therein (40 CFR 124.78-3).

Excavation zone: The volume containing the tank system and backfill material bounded by the ground surface, walls, and floor of the pit and trenches into which the UST system is placed at the time of installation (40 CFR 280.12-91).

Exceedance: Violation of the pollutant levels permitted by environmental protection standards (EPA-94/04).

Exception report: In RCRA, a report that generators who transport waste off-site must submit to the Regional Administrator if they do not receive a copy of the manifest signed and dated by owner or operator of the designated facility to which their waste was shipped within 45 days from the date on which the initial transporter accepted the waste (EPA-86/01).

Excess activated sludge (excess sludge or surplus sludge): Activated sludge which is in excess of the return sludge and needs to be disposed of (*see* sludge for more related terms).

Excess air: *See* synonym, excess combustion air.

Excess air incinerator: An excess-air incinerator is also known as the multiple chamber incinerator because it consists of two or more combustion chambers; the primary chamber for solid phase combustion and the secondary chamber for gas phase combustion. Because its waste feed is always operated in a batch mode, an excess-air incinerator is also commonly referred as a batch or retort incinerator. As the name implies, an excess-air incinerator is operated with excess air levels well above stoichiometric (typically 60-200% excess air) in both the primary and secondary combustion chambers (cf. incinerator, starved-air) (OME-86/10). The traditional designs of excess-air incinerators include:
(1) **In-line incinerator:** The flow of combustion gases travels through different combustion chambers in a straight manner. To enter from one chamber to another, it turns in the vertical (upward and downward) direction only. This type of incinerator may have three chambers, i.e., primary chamber, mixing chamber and secondary chamber. The combustion gases flow upward from primary chamber, pass the chamber wall, flow downward in mixing chamber, pass the chamber wall and flow upward in secondary chamber for further burning (EPA-89/03b).
(2) **Retort incinerator:** The flow of combustion gases turns in the vertical direction as in the in-line incinerator, but also turns sideways as it travels through the incinerator. Because the secondary chamber is adjacent to the primary chamber (they share a wall) and the gases turn in a "U" shape, the incinerator is more compact. In-line incinerators perform better in the capacity range greater than 750 lb/hr. The retort

design performs more efficiently than the in-line design in the capacity range of less than 750 lb/hr (cf. retort) (EPA-89/03b).
(3) See incinerator for more related terms.

Excess ammonia liquor storage tank: Any tank, reservoir, or container used to collect or store a flushing liquor solution prior to ammonia or phenol recovery (40 CFR 61.131-91).

Excess capacity factor: A multiplier on a process size to account for shutdown for cleaning and maintenance (EPA-83/06a).

Excess combustion air (or excess air): The combustion air supplied in addition to that theoretically required (in excess of that necessary to burn compounds completely or greater than stoichiometric air requirements). The excessive amounts appear in the products of combustion. The amount of excess air is normally expressed as a percentage of the theoretical (stoichiometric) air required for complete combustion of the compound. Typically, a waste incinerator operates with an overall 140 to 200% excess air level. Calculations of theoretical and excess air are as follows:
- TA (Percent of theoretical air) = aa/ta(%)
- EA (Percent of excess air) = (aa - ta)/ta(%)
- where: aa = moles of actual air used in combustion; ta = moles of theoretical (stoichiometric) air used in combustion; and aa - ta = excess air (see combustion air for more related terms).

Excess emission: Emissions of an air pollutant in excess of an emission standard (see emission for more related terms) (40 CFR 51.100-91).

Excess emissions and monitoring systems performance report: A report that must be submitted periodically by a source in order to provide data on its compliance with stated emission limits and operating parameters, and on the performance of its monitoring systems (40 CFR 60.2-91).

Excess lifetime risk: The additional or extra risk incurred over the lifetime of an individual by exposure to a toxic substance (EPA-92/12).

Excess lime softening: Excessive lime is added to remove various hard water salts and to maintain the required alkalinity.

Excess pesticide: See synonym, pesticide (40 CFR 165.1-91).

Excess property: Any property under the control of a Federal agency that is not required for immediate or foreseeable needs and thus is a candidate for disposal (40 CFR 35.6015-91).

Excess sludge: See synonym, excess activated sludge.

Excessive concentration: Is defined for the purpose of determining good engineering practice stack height under 40 CFR 51.100(ii)(3) and means:
(1) For sources seeking credit for stack height exceeding that established under 40 CFR 51.100(ii)(2) a maximum ground-level concentration due to emissions from a stack due in whole or part to downwash, wakes, and eddy effects produced by nearby structures or nearby terrain features which individually is at least 40 percent in excess of the maximum concentration experienced in the absence of such downwash, wakes, or eddy effects and which contributes to a total concentration due to emissions from all sources that is greater than an ambient air quality standard. For sources subject to the prevention of significant deterioration program (40 CFR 51.166 and 52.21), an excessive concentration alternatively means a maximum ground-level concentration due to emissions from a stack due in whole or part to downwash, wakes, or eddy effects produced by nearby structures or nearby terrain features which individually is at least 40 percent in excess of the maximum concentration experienced in the absence of such downwash, wakes, or eddy effects and greater than a prevention of significant deterioration increment. The allowable emission rate to be used in making demonstrations under this part shall be prescribed by the new source performance standard that is applicable to the source category unless the owner or operator demonstrates that this emission rate is infeasible. Where such demonstrations are approved by the authority administering the State implementation plan, an alternative emission rate shall be established in consultation with the source owner or operator.
(2) For sources seeking credit after October 11, 1983, for increases in existing stack heights up to the heights established under 40 CFR 51.100(ii)(2), either:
  (A) A maximum ground-level concentration due in whole or part to downwash, wakes or eddy effects as provided in paragraph (kk)(1) of this section, except that the emission rate specified by any applicable State implementation plan (or, in the absence of such a limit, the actual emission rate) shall be used, or
  (B) The actual presence of a local nuisance caused by the existing stack, as determined by the authority administering the State implementation plan; and
(3) For sources seeking credit after January 12, 1979 for a stack height determined under 40 CFR 51.100(ii)(2) where the authority administering the State implementation plan requires the use of a field study or fluid model to verify GEP stack height, for sources seeking stack height credit after November 9, 1984 based on the aerodynamic influence of cooling towers, and for sources seeking stack height credit after December 31, 1970 based on the aerodynamic influence of structures not adequately represented by the equations in 40 CFR 51.100(ii)(2), a maximum ground-level concentration due in whole or part to downwash, wakes or eddy effects that is at least 40 percent in excess of the maximum concentration experienced in the absence of such downwash, wakes, or eddy effects (40 CFR 51.100-kk-91).

Excessive infiltration/inflow: The quantities of infiltration-inflow which can be economically eliminated from a sewerage system by rehabilitation, as determined in a cost effectiveness analysis that compares the costs for correcting the infiltration/inflow conditions to the total costs for transportation and treatment of the infiltration/inflow, subject to the provisions in Section 35.2 (see infiltration/inflow for more related terms) (40 CFR 35.905-91, see also 40 CFR 35.2005-91).

Excessive release: A discharge of more than 295 g (0.65 lbs) of mercaptans and/or hydrogen sulfide into the atmosphere in any 5-

minute period (40 CFR 52.741-91).

Exchange coefficient: *See* definition under exchange coefficient hypothesis.

Exchange coefficient hypothesis: The assumption that the turbulent flux of a property is proportional to its mean gradient. The coefficient of proportionality is called the **exchange coefficient** (NATO-78/10).

Excited state: The state that an ion has a higher energy level than that of its normal level.

Exclusion: In the asbestos program, one of several situations that permit a Local Education Agency (LEA) to delete one or more of the items required by the Asbestos Hazard Emergency Response Act (AHERA), e.g., records of previous asbestos sample collection and analysis may be used by the accredited inspector in lieu of AHERA bulk sampling (EPA-94/04).

Exclusionary ordinance: Zoning that excludes classes of persons or businesses from a particular neighborhood or area (EPA-94/04).

Exclusion zone: Area of site possessing the highest concentration of contaminants, also called the "hot" zone (EPA-89/12a).

Exclusionary: Any form of zoning ordinance that tends to exclude specific classes of persons or businesses from a particular district or area (EPA-89/12).

Exclusive use study: A study that meets each of the following requirements:
(1) The study pertains to a new active ingredient (new chemical) or new combination of active ingredients (new combination) first registered after September 30, 1978;
(2) The study was submitted in support of, or as a condition of approval of, the application resulting in the first registration of product containing such new chemical or new combination (first registration), or an application to amend such registration to add a new use; and
(3) The study was not submitted to satisfy a data requirement imposed under FIFRA section 3(c)(2)(B); provided that, a study is an exclusive use study only during the 10-year period following the date of the first registration (40 CFR 152.83-c-91).

Excretion: The discharge of excessive or unusable substances from a living organisms.

Excursion: An unintentional noncompliance occurring for reasons beyond the reasonable control of the permittee (cf. upset) (EPA-85/10).

Executive Order 12856 of August 3, 1993:
(1) Federal compliance with Right-to-Know laws: The Federal Government should be a good neighbor to local communities by becoming a leader in providing information to the public concerning toxic and hazardous chemicals and extremely hazardous substances at Federal facilities, and in planning for and preventing harm to the public through the planned or unplanned releases of chemicals.
(2) Federal compliance with pollution prevention requirements: The Federal Government should be a leader in the field of pollution prevention through the management of its facilities, its acquisition practices, and in supporting the development of innovative pollution prevention programs and technologies (GII-96/03, p1187).

Executive Order 12873 of October 20, 1993: The Federal Acquisition, Recycle and Waste Prevention. Consistent with the demands of efficiency and cost effectiveness, the head of each Executive agency shall incorporate waste prevention and recycling in the agency's daily operations and work to increase and expand markets for recovered materials through greater Federal Government preference and demand for such products (GII-96/03, p1194).

Exempt solvent: The specific organic compounds that are not subject to requirements of regulation because they have been deemed by EPA to be of negligible photochemical reactivity (*see* solvent for more related terms) (EPA-94/04).

Exempted aquifer:
(1) An aquifer or its portion that meets the criteria in the definition of underground source of drinking water but which has been exempted according to the procedures in 40 CFR 144.7 (40 CFR 144.3-91).
(2) Underground bodies of water defined in the Underground Injection Control program as aquifers that are sources of drinking water (although they are not being used as such) and that are exempted from regulations barring underground injection activities (EPA-94/04).
(3) *See* aquifer for more related terms

Exemption: A state with primacy may relieve a public water system from a requirement respecting an MCL, treatment technique, or both by granting an exemption if the system cannot comply due to compelling economic or other factors; the system was in operation on the effective date of the requirement or MCL; and the exemption will not create an unreasonable public health risk (*see* variance) (EPA-94/04).

Exemption application: Any application submitted to EPA under section 5(h)(2) of the Act (40 CFR 700.43-91).

Exemption category: A category of chemical substances for which a person(s) has applied for or been granted an exemption under section 5(h)(4) of the Act (15USC2604) (40 CFR 723.175-91).

Exemption holder: Refers to a manufacturer or processor, subject to a test rule, that has received an exemption under sections 4(c)(1) or 4(c)(2) of TSCA from the requirement to conduct a test and submit data (40 CFR 791.3-91).

Exemption notice: Any notice submitted to EPA under Section 40 CFR 723.15 of this chapter (40 CFR 700.43-91).

Exergonic: Describing a biochemical reaction in which the products possess less free energy than that of reactants.

Exfiltration:
(1) The opposite of infiltration. The exhaust of gases from a

building or structure, due to wind velocity and thermal effects, through defects in the structure and normal leakage around openings (EPA-84/09).
(2) The movement of indoor air out of the house (EPA-88/08a).

Exhaust emission: The substances emitted to the atmosphere from the exhaust discharge nozzle of an aircraft or aircraft engine (*see* emission for more related terms) (40 CFR 87.1-91).

Exhaust fan: A fan oriented so that it blows indoor air out of the house. Exhaust fans cause outdoor air (and soil gas) to infiltrate at other locations in the house, to compensate for the exhausted air (EPA-88/08).

Exhaust gas: Any offgas (the constituents of which may consist of any fluids, either as a liquid and/or gas) discharged directly or ultimately to the atmosphere that was initially contained in or was in direct contact with the equipment for which exhaust gas limits are prescribed in 40 CFR 61.62(a) and (b); 40 CFR 61.63(a); 40 CFR 61.64(a)(1), (b), (c), and (d); 40 CFR 61.65(b)(1)(ii), (b)(2), (b)(3), (b)(5), (b)(6)(ii), (b)(7), and (b)(9)(ii); and 61.65(d). A leak as defined in paragraph (w) of this section is not an exhaust gas. Equipment which contains exhaust gas is subject to 40 CFR 61.65(b)(8), whether or not that equipment contains 10 percent by volume vinyl chloride (40 CFR 61.61-x-91).

Exhaust gas recirculation (EGR)-air bleed: A system or device (such as modification of the engine's carburetor or positive crankcase ventilation system) that results in engine operation at an increased airfuel ratio so as to achieve reductions in exhaust emissions of hydrocarbons and carbon monoxide of 25 percent and 40 percent, respectively, from light-duty vehicles of model years 1968-1970 (40 CFR 52.2491-91).

Exhaust header pipe: Any tube of constant diameter which conducts exhaust gas from an engine exhaust port to other exhaust system components which provide noise attenuation. Tubes with cross connections or internal baffling are not considered to be exhaust header pipes (40 CFR 205.165-91).

Exhaust system:
(1) A system consisting of branch pipes connected to hoods or enclosures, one or more header pipes, an exhaust fan, means for separating solid contaminants from the air flowing in the system, and a discharge stack to outside (29 CFR 1910.94b-91).
(2) The system comprised of a combination of components which provides for enclosed flow of exhaust gas from engine parts to the atmosphere (40 CFR 202.10-91, *see also* 40 CFR 205.51; 205.151-91).

Exhaust ventilation system: A system for removing contaminated air from a space, comprising two or more of the following elements:
(1) Enclosure or hood
(2) Duct work
(3) Dust collecting equipment
(4) Exhauster
(5) Discharge stack (29 CFR 1910.94a-91)

Exhaust wash: Water used to trap droplets and solubles from air passed to remove spray, vapor, and gasses from electroplating and process tanks (EPA-74/03d).

Exhauster: A fan located between the inlet gas flange and outlet gas flange of the coke oven gas line that provides motive power for coke oven gases (40 CFR 61.131-91).

Existing class II well: Wells that were authorized by BIA (Bureau of Indian Affairs) and constructed and completed before the effective date of this program (*see* well for more related terms) (40 CFR 147.2902-91).

Existing control device: For the purposes of these standards, an air pollution control device that has been in operation on or before September 30, 1987, or that has been in operation between September 30, 1987, and January 10, 1989, on those continuous or intermittent emissions from a process section that is marked by an -- in Table 1 of this subpart (40 CFR 60.561-91).

Existing control device is reconstructed: For the purposes of these standards, the capital expenditure of at least 50 percent of the replacement cost of the existing control device (40 CFR 60.561-91).

Existing control device is replaced: For the purposes of these standards, the replacement of an existing control device with another control device (40 CFR 60.561-91).

Existing facility: With reference to a stationary source, any apparatus of the type for which a standard is promulgated in this part, and the construction or modification of which was commenced before the date of proposal of that standard; or any apparatus which could be altered in such a way as to be of that type (40 CFR 60.2; 260.10-91).

Existing hazardous waste management (HWM) facility or existing facility: A facility which was in operation or for which construction commenced on or before November 19, 1980. A facility has commenced construction if:
(1) The owner or operator has obtained the Federal, State and local approvals or permits necessary to begin physical construction; and either (2)(i) A continuous on-site, physical construction program has begun; or
(2) The owner or operator has entered into contractual obligations--which cannot be cancelled or modified without substantial loss--for physical construction of the facility to be completed within a reasonable time (40 CFR 260.10-91).

Existing impoundment: Any uranium mill tailings impoundment which is licensed to accept additional tailings and is in existence as of December 15, 1989 (40 CFR 61.251-91).

Existing indirect dischargers: Only those two iron blast furnace operations with discharges to publicly owned treatment works prior to May 27, 1982 (40 CFR 420.31-91).

Existing injection well: An injection well other than a new injection well (*see* well for more related terms) (40 CFR 144.3-91, *see also* 40 CFR 146.3-91).

Existing OCS source: Any OCS (outer continental shelf) source other than a new OCS source (*see* source for more related terms)

(CAA328-42USC7627-91).

Existing portion:
(1) For uranium waste, the land surface area of an existing surface impoundment on which significant quantities of uranium by-product materials have been placed prior to promulgation of this standard (40 CFR 192.11-91).
(2) For hazardous waste, the land surface area of an existing waste management unit, included in the original Part A permit application, on which wastes have been placed prior to the issuance of a permit (40 CFR 260.10-91).

Existing solid waste incineration unit: A solid waste unit which is not a new or modified solid waste incineration unit (*see* solid waste incineration unit for more related terms) (CAA129.g-42USC7429-91).

Existing source:
(1) In CAA, Any stationary source other than a new source (CAA111, *see also* CAA112; 40 CFR 61.02; 122.29; 129.2-91).
(2) In CWA, an existing source is any facility from which there is or may be a discharge of pollutants, the construction of which is commenced before the publication of proposed regulations prescribing a standard of performance under Section 306 of CWA (EPA-87/10a).
(3) *See* source for more related terms.

Existing stationary facility: Any of the following stationary sources of air pollutants, including any reconstructed source, which was not in operation prior to August 7, 1962, and was in existence on August 7, 1977, and has the potential to emit 250 tons per year or more of any air pollutant. In determining potential to emit, fugitive emissions, to the extent quantifiable, must be counted:
1. Fossil-fuel fired steam electric plants of more than 250 million British thermal units per hour heat input,
2. Coal cleaning plants (thermal dryers),
3. Kraft pulp mills,
4. Portland cement plants,
5. Primary zinc smelters,
6. Iron and steel mill plants,
7. Primary aluminum ore reduction plants,
8. Primary copper smelters,
9. Municipal incinerators capable of charging more than 250 tons of refuse per day,
10. Hydrofluoric, sulfuric, and nitric acid plants,
11. Petroleum refineries,
12. Lime plants,
13. Phosphate rock processing plants,
14. Coke oven batteries,
15. Sulfur recovery plants,
16. Carbon black plants (furnace process.,
17. Primary lead smelters,
18. Fuel conversion plants,
19. Sintering plants,
20. Secondary metal production facilities,
21. Chemical process plants,
22. Fossil-fuel boilers of more than 250 million British thermal units per hour heat input,
23. Petroleum storage and transfer facilities with a capacity exceeding 300,000 barrels,
24. Taconite ore processing facilities,
25. Glass fiber processing plants, and
26. Charcoal production facilities (40 CFR 51.301-e-91).

Existing tank system or existing component: A tank system or component that is used for the storage or treatment of hazardous waste and that is in operation or for which installation has commenced on or prior to July 14, 1986. Installation will be considered to have commenced if the owner or operator has obtained all Federal, State, and local approvals or permits necessary to begin physical construction of the site or installation of the tank system and if either
(1) A continuous on-site physical construction or installation program has begun, or
(2) The owner or operator has entered into contractual obligations--which cannot be canceled or modified without substantial loss--for physical construction of the site or installation of the tank system to be completed within a reasonable time (40 CFR 260.10-91).

Existing unit: A unit (including units subject to section 111) that commenced commercial operation before the date of enactment of the Clean Air Act Amendments of 1990. Any unit that commenced commercial operation before the date of enactment of the Clean Air Act Amendments of 1990 which is modified, reconstructed, or repowered after the date of enactment of the Clean Air Act Amendments of 1990 shall continue to be an existing unit for the purposes of this title. For the purposes of this title, existing units shall not include simple combustion turbines, or units which serve a generator with a nameplate capacity of 25 MWe or less (CAA402-42USC7651a).

Existing uses: Those uses actually attained in the water body on or after November 28, 1975, whether or not they are included in the water quality standards (cf. designated uses) (40 CFR 131.3-91).

Existing vapor processing system: A vapor processing system [capable of achieving emissions to the atmosphere no greater than 80 milligrams of total organic compounds per liter of gasoline loaded], the construction or refurbishment of which was commenced before December 7, 1980, and which was not constructed or refurbished after that date (40 CFR 60.501-91).

Existing vessel: Existing vessel includes every description of watercraft or other artificial contrivance used, or capable of being used, as a means of transportation on the navigable waters, the construction of which is initiated before promulgation of standards and regulations under this section (*see* vessel for more related terms) (CWA312).

Existing well: A Class I well which was authorized prior to August 25, 1988 by an approved State program, or an EPA-administered program or a well which has become a Class I well as a result of a change in the definition of the injected waste which would render the waste hazardous under 40 CFR 261.3 of this Part (*see* well for more related terms) (40 CFR 146.61-91).

Exit velocity: The velocity of the effluent at the exit of a source (NATO-78/10).

Exo: A prefix showing outside or outer.

Exogenous: Growing due to external causes (cf. endogenous).

Exotherm: The curve (e.g., temperature vs. time) of an exothermic reaction.

Exothermic: A term used to characterize the evolution of heat. Specifically refers to chemical reactions from which heat is evolved (EPA-87/07a).

Exothermic reaction: The reaction system that liberates heat to the surroundings as the reaction takes place (cf. endothermic reaction).

Exotic species: A species that is not indigenous to a region (EPA-94/04).

Exotic stream: A perennial stream that flows through a desert region but has its source in an area outside the desert. Typically it loses volume through evaporation and seepage, and has few or no tributaries as it crosses the desert. Examples include the Nile and the Colorado (DOI-70/04).

Expandable polystyrene: A polystyrene bead to which a blowing agent has been added using either an in-situ suspension process or a post-impregnation suspension process (40 CFR 60.561-91).

Expansion chamber: Any chamber designed to reduce the velocity of the products of combustion and promote the settling of fly ash from the gas stream. *See* settling chamber (EPA-83).

Experimental animal: The individual animals or groups of animals, regardless of species, intended for use and used solely for research purposes and does not include animals intended to be used for any food purposes (40 CFR 172.1-91).

Experimental flame failure mode: Generally based on an experimental determination of destruction efficiency (or flame failure) in bench-scale flame systems; compound ranking dependent on failure mode of flame. Example of failure modes includes temperature distribution, oxygen availability in the combustion chamber (EPA-88/12).

Experimental furnace: A glass melting furnace with the sole purpose of operating to evaluate glass melting processes, technologies, or glass products. An experimental furnace does not produce glass that is sold (except for further research and development purposes) or that is used as a raw material for nonexperimental furnaces (*see* furnace for more related terms) (40 CFR 60.291-91).

Experimental process line: A polymer or copolymer manufacturing process line with the sole purpose of operating to evaluate polymer manufacturing processes, technologies, or products. An experimental process line does not produce a polymer or resin that is sold or that is used as a raw material for nonexperimental process lines (40 CFR 60.561-91).

Experimental start date: The first date the test substance is applied to the test system (40 CFR 160.3; 792.3-91).

Experimental technology: A technology which has not been proven feasible under the conditions in which it is being tested (40 CFR 146.3-91).

Experimental termination date: The last date on which data are collected directly from the study (40 CFR 792.3-91).

Experimental use permit: Obtained by manufacturers for testing new pesticides or uses of thereof whenever they conduct experimental field studies to support registration on 10 acres or more on land or one acre or more of water (EPA-94/04).

Experimental use permit review: The review of an application for a permit pursuant to section 5 of FIFRA to apply a limited quantity of a pesticide in order to accumulate information necessary to register the pesticide. The application may be for a new chemical or for a new use of an old chemical. The fee applies to such experimental uses of a single unregistered active ingredient (no limit on the number of other active ingredients, in a tank mix, already registered for the crops involved) and no more than three crops. This fee does not apply to experimental use permits required for small-scale field testing of microbial pest control agents (40 CFR 172.3) (40 CFR 152.403-f-91).

Exploitation: The extraction of useful substances following the exploration.

Exploration: The location of the presence of economic deposits and establishing their nature, shape, and grade and the investigation may be divided into:
(1) Preliminary; and
(2) Final (EPA-82/05).

Explosion: A chemical reaction or change of state occurring virtually instantly with the release of high temperature and generally a large quantity of gas. *See also* Deflagration, Detonation and Flame Front (EPA-83).

Explosion suppression: A technique by which burning in a confined space is detected and arrested during incipient stages, preventing development of pressure which could result in an explosion (EPA-83).

Explosion venting: The provision of an opening for the release of pressure and heated (explosion) gases. Thus preventing the development of destructive pressure (EPA-83).

Explosive: A substance (mixture) capable of rapid conversion into more stable products, with the liberation of heat and usually the formation of gases (EPA-76/03).

Explosive gas:
(1) Gases that can react and produce a large amount of heat in a very short period of time causing a sudden increase in pressure.
(2) In 40 CFR , it means methane (CH_4) (40 CFR 257.3.8-91).

Explosive limits: The amounts of vapor in the air that form explosive mixtures; limits are expressed as lower and upper limits and give the range of vapor concentrations in air that will explode if an ignition source is present (EPA-94/04).

Explosive waste: The waste which is unstable and may readily undergo violent chemical change or explode. Such waste may react violently on contact with air or water, or form explosive mixtures, or generate toxic fumes when mixed with water (OME-12/88).

Exponential decay: The decay of physical properties in accordance with an exponential function (e^{-t}), where t denotes time. Exponential means a variable which varies as the power of another variable. For an equation $y = c^t$, y varies exponentially with t.

Exponential growth: The growth of a population (e.g., bacteria) in accordance with the power of a variable (c^t) or an exponential function (e^t), where c is a constant and t time).

Exposure: The amount of radiation or pollutant present in a given environment that represents a potential health threat to living organisms (cf. occupational exposure) (EPA-94/04).

Exposure assessment:
(1) An assessment to determine the extent of exposure of, or potential for exposure of, individuals to chemical substances based on such factors as the nature and extent of contamination and the existence of or potential for pathways of human exposure (including ground or surface water contamination, air emissions, and food chain contamination), the size of community within the likely pathways of exposure, and the comparison of expected human exposure levels to the short-term and long-term health effects associated with identified contaminants and any available recommended exposure or tolerance limits for such contaminants. Such assessment shall not delay corrective action to abate immediate hazards or reduce exposure (cf. risk assessment) (RCRA9003).
(2) Identifying the pathways by which toxicants may reach individuals, estimating how much of a chemical an individual is likely to be exposed to, and estimating the number likely to be exposed (EPA-94/04).

Exposure coefficient: A term which combines information on the frequency, mode, and magnitude of contact with contaminated medium to yield a quantitative value of the amount of contaminated medium contacted per day (Course 165.6).

Exposure dosage: *See* synonym, exposure level.

Exposure incident: A specific eye, mouth, other mucous membrane, non-intact skin, or parenteral contact with blood or other potentially infectious materials that results from the performance of an employee's duties (29 CFR 1910).

Exposure indicator: A characteristic of the environment measured to provide evidence of the occurrence or magnitude of a response indicator's exposure to a chemical or biological stress (EPA-94/04).

Exposure level: The amount (concentration) of a chemical at the absorptive surfaces of an organism (EPA-94/04).

Exposure level (or exposure dosage): The concentration of the contaminant to which the population in question is exposed (LBL-76/07-bio).

Exposure level, chemical: The amount (concentration) of a chemical at the absorptive surface of an organism (Course 165.6).

Exposure or exposed: An employee is subjected to a toxic substance or harmful physical agent in the course of employment through any route of entry (inhalation, ingestion, skin contact or absorption, etc.), and includes past exposure and potential (e.g., accidental or possible) exposure, but does not include situations where the employer can demonstrate that the toxic substance or harmful physical agent is not used, handled, stored, generated, or present in the workplace in any manner different from typical non-occupational situations (29 CFR 1910.20-91).

Exposure parameters: Variables used in the calculation of intake (e.g., exposure duration, inhalation rate, average body weight) (EPA-91/12).

Exposure pathway: The course a chemical or physical agent takes from a source to an exposed organism. An exposure pathway describes a unique mechanism by which an individual or population is exposed to chemicals or physical agents at or originating rom a site. Each exposure pathway includes a source or release from a source, an exposure point, and an exposure route. If the exposure point differs from the source, a transport/exposure medium (e.g., air) or media (in cases of intermedia transfer) also would the indicated (EPA-91/12).

Exposure period (or exposure time):
(1) The time that a person is exposed to a toxic substance.
(2) In 40 CFR, it is the 5-day period during which test birds are offered a diet containing the test substance (40 CFR 797.2050-91).

Exposure point: A location of potential contact between an organism and a chemical or physical agent (EPA-91/12).

Exposure route:
(1) Any route of exposure entry such as inhalation, ingestion, skin contact or absorption, etc.
(2) The way a chemical or physical agent comes in contact with an organism (i.e., by ingestion, inhalation, dermal contact) (EPA-91/12).

Exposure scenario: A set of conditions or assumptions about sources, exposure pathways, concentrations of toxic chemicals and populations (numbers, characteristics and habits) which aid the investigator in evaluating and quantifying exposure in a given situation (Course 165.6).

Exposure time: *See* synonym, exposure period.

Extended aeration: A modification of the activated sludge process that employs long retention time of wastewater in the presence of activated sludge and air, usually for greater than 24 hours (*see* aeration for more related terms) (EPA-75/10).

Extender:
(1) A pigment which is usually inexpensive and inert in nature, used to give opacity and extend or increase the bulk of a paint, thus reducing its unit cost, and modifying its consistency (EPA-79/12b).
(2) A low specific gravity substance used in rubber formulations

chiefly to reduce costs (cf. filler) (EPA-74/12a).

Extensive property:
(1) A thermodynamic property which is derived from an inherent property, e.g., the total pressure of a multi-component system equals the sum of the pressure of each component in the system. The total pressure is the extensive property and the pressure of each component is the inherent property.
(2) Also known as the **extrinsic property**. It is a property whose value for a system equals to the sum of its values for the various parts of the system. Examples of the extensive property include mass, volume, weight, energy, enthalpy, entropy, etc. (Jones-p12; Wark-p5).
(3) See property for more related terms.

Extensor muscles: The muscles under voluntary control that, when contracted, extend the limbs (LBL-76/07-bio).

Extent of chlorination: The percent by weight of chlorine (40 CFR 704.43; 704.45-91).

Exterior base coat: A coating applied to the exterior of a can body, or flat sheet to provide protection to the metal or to provide background for any lithographic or printing operation (40 CFR 52.741-91).

Exterior base coating operation: The system on each beverage can surface coating line used to apply a coating to the exterior of a two-piece beverage can body. The exterior base coat provides corrosion resistance and a background for lithography or printing operations. The exterior base coat operation consists of the coating application station, flashoff area, and curing oven. The exterior base coat may be pigmented or clear (unpigmented) (40 CFR 60.491-91).

Exterior end coat: A coating applied to the exterior end of a can to provide protection to the metal (40 CFR 52.741-91).

Exterior paint: A coating for the outside surfaces of a structure (EPA-79/12b).

External floating roof: A cover over an open top storage tank consisting of a double deck or pontoon single deck which rests upon and is supported by the volatile organic liquid being contained and is equipped with a closure seal or seals to close the space between the roof edge and tank shell (40 CFR 52.741-91, see also 40 CFR 61.341-91).

External combustion engine: An engine in which heat is supplied from external sources such as a furnace or a reactor (see engine for more related terms).

External radiation: The radiation originating from a source outside the body, such as cosmic radiation. The source of external radiation can be either natural or man-made (see radiation for more related terms) (EPA-88/08a).

External reference material: See definition under reference material.

Extinction lake (or senescent lake): The gradual permanent loss of the water or water surface of a lake by the encroachment of vegetation, transforming the lake into a bog, marsh, or swamp (see lake for more related terms) (DOI-70/04).

Extra risk: The added risk to that portion of the population that is not included in measurement of background tumor rate; $ER(d) = [P(d) - P(O)]/[1-P(O)]$ (EPA-92/12).

Extractable metals: The concentration of metals in an unfiltered sample following digestion with hot dilute mineral acid (see metal for more related terms) (Ref. 19, LBL-76).

Extraction:
(1) The separation of specific constituents from a matrix of solids or a solution, employing mechanical and/or chemical methods (cf. waste minimization) (EPA-83).
(2) In waste treatment, extraction is to extract hazardous constituents from contaminated soil, thus circumventing the need to incinerate or otherwise treat the soil itself. In principal, extraction can be done by physical or chemical means. Preferably, this can be done in-situ to eliminate the costs and risks associated with excavating the soils. Several new extraction applications are under development for actual applications. A conceptual in-situ vacuum extraction of volatile contaminants from soils and groundwater uses wells which are placed in the unsaturated soil zone above the water table. A vacuum source is applied to the well and volatile organic compounds are drawn off for further treatment.
(3) In mining:
  (A) The process of mining and removal of ore from a mine.
  (B) The separation of metal or valuable mineral from an ore or concentrate.
  (C) Used in relation to all processes that are used in obtaining metals from their ores. Broadly, these processes involve the breaking down of the ore both mechanically (crushing) and chemically (decomposition), and the separation of the metal from the associated gangue (EPA-82/05).

Extraction column: A column such as a vertical tank where the desired products are extracted from raw materials.

Extraction plant: A facility chemically processing berylium ore to beryllium metal, alloy, or oxide, or performing any of the intermediate steps in these processes (40 CFR 61.31-91).

Extraction procedure (EP toxic): Determining toxicity by a procedure which simulates leaching; if a certain concentration of a toxic substance can be leached from a waste, that waste is considered hazardous, i.e., "E P Toxic" (EPA-94/04).

Extraction procedure toxicity (EP toxicity):
(1) One of the four U.S. EPA hazardous waste characteristics (ETI-92).
(2) A test that is designed to identify wastes likely to leach hazardous concentrations of particular toxic constituents into the ground water as a result of improper management (EPA-86/01).
(3) See hazardous waste characteristics for more related terms.

Extraction site: The place from which the dredged or fill material proposed for discharge is to be removed (40 CFR 230.3-91).

Extraction water: The water removed during a wood pulp manufacturing process (*see* water for more related terms) (EPA-87/10).

Extractor column: The column is used to extract the solute from the aqueous solution produced by the generator column (40 CFR 796.1720-91, *see also* 40 CFR 796.1860-91).

Extraneous ash: After combustion, that portion of the residue (ash non-combustible) which is derived from entrained materials which were mixed with the combustible material (*see* ash for more related terms) (EPA-83).

Extrapolation:
(1) An estimation of a numerical value of an empirical (measured) function at a point outside the range of data which were used to calibrate the function. The quantitative risk estimates for carcinogens are generally low-dose extrapolations based on observations made at higher doses. Generally one has a measured dose and measured effect (EPA-92/12).
(2) Estimation of unknown values by extending or projecting from known values.

Extreme environmental conditions: The exposure to any or all of the following: ambient weather conditions; temperatures consistently above 95° C (203° F); detergents; abrasive and scouring agents; solvents; or corrosive atmospheres (40 CFR 52.741-91).

Extreme performance coating: Any coating which during intended use is exposed to extreme environmental conditions (40 CFR 52.741-91).

Extremely hazardous substance: Any of 406 chemicals identified by EPA on the basis of toxicity, and listed under SARA Title III. The list is subject to revision (*see* hazardous substance for more related terms) (EPA-94/04, *see also* SF329; 40 CFR 355.20; 370.2-91).

Extrudate: In constructing a landfill liner, the molten polymer which is emitted from an extruder during seaming using either extrusion fillet or extrusion flat methods. The extrudate is initially in the form of a ribbon, rod, bead or pellets (EPA-89/09, *see* EPA-91/05).

Extruded propellant: Any propellant made by pressing solvent-softened or gelatinized nitrocellulose through a dye to form grains (EPA-76/03).

Extruder:
(1) A machine producing a densified product by forcing the material under pressure, through a die, into the desired shapes or form (EPA-83).
(2) (In factory), a stationary machine with a driver screw for continuous forming of polymeric compounds by forcing through a die. It is used to manufacture films and sheeting (EPA-89/09).
(3) (In lanffill liner), a portable device with a driver screw for continuous forming of a ribbon, rod or bead of extrudate for making FML (flexible membrane liner) seams (EPA-89/09).

Extrusion: The application of pressure to a billet of aluminum, forcing the aluminum to flow through a die orifice. The extrusion subcategory is based on the extrusion process (40 CFR 467.02-91, *see also* also 40 CFR 468.02; 471.02-91).

Extrusion die cleaning: The process by which the steel dies used in extrusion of aluminum are cleaned. The term includes a dip into a concentrated caustic bath to dissolve the aluminum followed by a water rinse. It also includes the use of a wet scrubber with the die cleaning operation (40 CFR 467.31-91).

Extrusion heat treatment: The spray application of water to a workpiece immediately following extrusions for the purpose of heat treatment (40 CFR 468.02-91).

Extrusion seam: A seam between two geomembrane sheets achieved by heat-extruding a polymer material between or over the overlap areas followed by the application of pressure (*see* seam for more related terms) (EPA-91/05).

Exxon thermal DeNO$_x$ process: One of NO$_x$ emission reduction techniques (*see* nitrogen oxide emission control for control structure). Ammonia is injected into the post combustion zone of the boiler. The ammonia reacts with NO$_x$ (which is 95% NO) to reduce the oxides to molecular nitrogen and water: $4NH_3 + 4NO + O_2 ---> 4N_2 + 6H_2O$. This reaction is extremely temperature dependent. In a boiler, this reaction successfully takes place at approximately 950° C (1740° F). At temperatures (above 1090° C) the ammonia is oxidized, forming additional NO$_x$. At lower temperatures (below 850° C) the ammonia passes through the boiler unreacted (EPA-81/12, p7-13).

Eye corrosion: The production of irreversible tissue damage in the eye following application of a test substance to the anterior surface of the eye (40 CFR 798.4500-91).

Eye irritation: The production of reversible changes in the eye following the applcation of a test substance to the anterior surface of the eye (40 CFR 798.4500-91).

Ff

Fabric: A collective term applied to cloth (cf. fiber).

Fabric coating: Any coating applied on textile fabric. Fabric coating includes the application of coatings by impregnation (40 CFR 52.741-91).

Fabric coating facility: A facility that includes one or more fabric coating lines (40 CFR 52.741-91).

Fabric coating line: A coating line in which any protective, decorative, or functional coating or reinforcing material is applied on or impregnated into a textile fabric (40 CFR 52.741-91).

Fabric filter: *See* synonym, bag filter.

Fabric filter:
(1) A cloth device that catches dust particles from industrial emissions (EPA-94/04).
(2) *See also* synonym, bag filter.

Fabricating: Any processing (e.g., cutting, sawing, drilling) of a manufactured product that contains commercial asbestos, with the exception of processing at temporary sites (field fabricating) for the construction or restoration of facilities. In the case of friction products, fabricating includes bonding, debonding, grinding, sawing, drilling, or other similar operations performed as part of fabricating (40 CFR 61.141-91).

Face: The working side of a landfill or mining operation (EPA-83).

Facial tissue: A class of soft absorbent papers in the sanitary tissue group (40 CFR 250.4-91).

Facility:
(1) Any structure, group of structures, equipment, or device (other than a vessel) which is used for one or more of the following purposes: exploring for, drilling for, producing, storing, handling, transferring, processing, or transporting oil. This term includes any motor vehicle, rolling stock, or pipeline used for one or more of these purposes (OPA1001, *see also* RCRA9003).
(2) Equipment, systems, buildings, utilities, services, and related activities whose use is directed to a common purpose at a single location (DOE-91/04).
(3) A location at which a process or set of processes are used to produce, refine or repackage chemicals, or a location where a large enough inventory of chemicals are stored so that a significant accidental release of a toxic chemical is possible (EPA-87/07a).

Other facility related terms include:
- Active portion (facility)
- Closed portion (facility)
- Designated facility
- Existing stationary facility
- Inactive facility
- Inactive portion (facility)
- Off-shore facility
- Off-site facility
- On-shore facility
- On-site facility

Facility component: Any part of a facility including equipment (40 CFR 61.141-91).

Facility emergency coordinator: Representative of a facility covered by environmental law (e.g., a chemical plant) who participates in the emergency reporting process with the Local Emergency Planning Committee (LEPC) (EPA-94/04).

Facility mailing list: The mailing list for a facility maintained by EPA in accordance with 40 CFR 124.10(c)(viii) (40 CFR 270.2-91).

Facility or activity: Any NPDES (national pollutant discharge elimination system) point source or any other facility or activity (including land or appurtenances thereto) that is subject to regulation under the NPDES program (40 CFR 122.2-91, *see also* 40 CFR 124.2; 124.41; 144.3; 144.3; 144.70; 146.3; 270.2-91).

Facility or equipment: The buildings, structures, process or production equipment or machinery which form a permanent part of the new source and which will be used in its operation, if these facilities or equipment are of such value as to represent a substantial commitment to construct. It excludes facilities or equipment used in connection with feasibility, engineering, and design studies regarding the source or water pollution treatment for the source (40 CFR 122.29-91).

Facility plans: Plans and studies related to the construction of treatment works necessary to comply with the Clean Water Act or RCRA. A facilities plan investigates needs and provides information on the cost effectiveness of alternatives, a recommended plan, an environmental assessment of the recommendations, and descriptions of the treatment works, costs, and a completion schedule (EPA-94/04).

Facility structure: Any buildings and sheds or utility or drainage lines on the facility (40 CFR 257.3.8-91).

Factory seam: In constructing a landfill liner, the seaming of FML (flexible membrane liner) rolls together to make large panels for transportation and field installation: Note that this is rarely done for polyethylene which is made in relatively wide sheets (EPA-89/09, *see also* EPA-91/05).

Facultative:
(1) Having the ability to live under both aerobic (with free oxygen) and anaerobic (without free oxygen) conditions (EPA-87/10a).
(2) Able to live and grow with or without free oxygen (SW-108ts).

Facultative aerobe: An organism that although fundamentally an aerobe can grow in the presence of free oxygen (cf. obligate aerobe) (EPA-74/06).

Facultative anaerobe:
(1) An organism that although fundamentally an aerobe can grow in the absence of free oxygen (cf. obligate anaerobe) (LBL-76/07-water).
(2) Bacteria that grow under aerobic conditions by utilizing oxygen as the terminal electron acceptor (oxygen respiration), but under anaerobic conditions utilize an organic compound as a terminal electron acceptor (fermentation) (EPA-88/09a).

Facultative anaerobic bacteria: *See* synonym, facultative bacteria.

Facultative bacteria (or facultative anaerobic bacteria):
(1) Bacteria which can grow (exist and reproduce) under either aerobic or anaerobic conditions (under the presence or absence of dissolved oxygen) (EPA-75/01). It is either anaerobic or aerobic according to the conditions around it.
(2) Bacteria that can live under aerobic or anaerobic conditions (EPA-94/04).

Facultative decomposition: The decomposition of organic matter by facultative microorganisms (EPA-75/01).

Facultative lagoon (or facultative pond): A combination of the aerobic and anaerobic logoons. It is divided by loading and thermal stratifications into an aerobic surface and an anaerobic bottom, therfore the principles of both the aerobic and anaerobic processes apply (EPA-76/03).

Fahrenheit (F): A temperature scale in which water boiling point is 212° F and water freezing (or ice melting) point is 32° F (cf. temperature).

Failing compressor: The measured noise emissions of the compressor, when measured in accordance with the applicable procedure, exceeds the applicable standard (*see* compressor for more related terms) (40 CFR 204.51-91).

Failing exhaust system: When installed on any Federally regulated motorcycle for which it is designed and marketed, that motorcycle and exhaust system exceed the applicable standards (40 CFR 205.165-91).

Failing vehicle: The measured emissions of the vehicle, when measured in accordance with the applicable procedure, exceeds the applicable standard (*see* vehicle for more related terms) (40 CFR 205.51-91, *see also* 40 CFR 205.151-91).

Failure mode: An excursion from normal operating conditions which controls incineration efficiency (EPA-88/12).

Fair market value: The amount at which property would change hands between a willing buyer and a willing seller, neither being under any compulsion to buy or sell and both having reasonable knowledge of the relevant facts. Fair market value is the price in cash, or its equivalent, for which the property would have been sold on the open market (40 CFR 35.6015-91).

Fall time:
(1) The time interval between initial response and 95 percent of final response after a step decrease in input concentration (40 CFR 53.23-91).
(2) The time interval between the initial response and a 90% response (unless otherwise specified) after a step decrease in the inlet concentration. This measurement is usually, but not necessarily, the same as the rise time (LBL-76/07-water).
(3) *See* time for more related terms.

Fallout: The sedimentation of dust or fine particle in the atmosphere. Especially used with reference to radioactive debris (NATO-78/10).

Family NO_x emission limit: The NO_x emission level to which an engine family is certified in the light-duty truck NO_x averaging program, expressed to one-tenth of a gram per mile accuracy (*see* nitrogen oxide for more related terms) (40 CFR 86.088.2-91).

Fan: A device to move air from one point to another. It has two major types:
(1) Radial flow or centrifugal type, in which the airflow is at right angles to the axis of rotation of the rotor; and
(2) Axial flow or propeller type, in which the airflow is parallel to the axis of rotation of the rotor (AP-40, p60).
Other fan related terms include:
- Forced draft fan
- Induced draft fan
- Overfire air fan
- Overfire fan (*see* synonym, overfire air fan)
- Primary air fan
- Secondary air fan

Fan efficiency: The efficiency of a fan is given by the ratio of the gas (air) horsepower to the brake horsepower. Gas horsepower is the power delivered to the moving gas. Brake horsepower is the power delivered to the fan. The fan efficiency term accounts for irreversibilities in fan operation; it must be taken into account in fan calculation since no fan is 100% efficient (EPA-84/09).

Fan law: A law used to predict how a change in one fan variable can affect other fan variables. Fan laws are based on the premise that if two fans are geometrically similar (homologous) their performance curves are similar and the ratio of fan variables can then be related through the fan laws (*see* law for more related terms) (EPA-84/09).

Fanconi syndrome: There are several Fanconi syndromes. As used in this document, the term refers to the triad of glycosuria, hyperaminoaciduria, and hypophosphatemia in the presence of hyperphosphaturia. This triad is associated with injury to proximal renal tubular cells (LBL-76/07-bio).

Fanning: The fanning of a stack plume is characterized by very slow vertical diffusion during stable conditions (NATO-78/10).

Far region: The region of the atmosphere's path along the lidar line-of-sight beyond or behind the plume being measured (40 CFR 60-App/A(alt. method 1)-91).

Faraday: The number of coulombs (96,490) required for an electrochemical reaction involving one chemical equivalent (EPA-74/03d).

Faraday's law of electrolysis: The amount of chemical change in an electrolysis process is proportional to the electrical charge passed (*see* law for more related terms).

Farm pond: A shallow structure for the impoundment of water to meet agricultural needs, such as irrigation, stock watering, spraying, and fire protection; the pond site may be a natural depression deepened to store surface run-off or utilize groundwater, or it may be created by building a dam on a small stream; it may be temporary or permanent, and may provide collateral recreational benefits such as fishing, boating, bathing, wildlife habitat, picnicking, and aesthetic values (DOI-70/04).

Farm tank: A tank located on a tract of land devoted to the production of crops or raising animals, including fish, and associated residences and improvements. A farm tank must be located on the farm property (40 CFR 280.12-91).

Fast meter response: That the fast response of the sound level meter shall be used. The fast dynamic response shall comply with the meter dynamic characteristics in paragraph 5.3 of the American National Standard Specification for Sound Level Meters, ANSI S1.4-1971. This publication is available from the American National Standards Institute, Inc., 1430 Broadway, New York, New York 1001 (cf. slow meter response) (40 CFR 201.1-91, *see also* 40 CFR 202.10; 205.2-91).

Fast turnaround operation of a spray drying tower: The operation involving more than 6 changes of formulation in a 30 consecutive day period that are of such degree and type (e.g., high phosphate to no phosphate) as to require cleaning of the tower to maintain minimal product quality (40 CFR 417.151-91).

Fast turnaround operation of automated fill lines: An operation involving more than 8 changes of formulation in a 30 consecutive day period that are of such degree and type as to require thorough purging and washing of the fill line to maintain minimal product quality (40 CFR 417.161-91).

Fastness: Resistance to change in color (EPA-83).

Fat: The glycerol esters of long chain fatty acids of animal or vegetable (EPA-74/04c).

Fat refining: Purifying fats by treatment with clay, caustic, etc. (EPA-74/04c).

Fat splitting: Splitting fatty triglycerides to fatty acids and glycerine by hydrolysis (EPA-74/04c).

Fatliquoring: A process by which oils and related fatty substances replace natural oils lost in the beam-house and chromium tanning processes. It regulates the softness and pliability of the leather (EPA-82/11).

Fatty acid:
(1) An organic acid obtained by the hydrolysis (saponification) of natural fats and oils, e.g., stearic and palmitic acids. These acids are monobasic and may not contain some double bonds. They usually contain sixteen or more carbon atoms (EPA-75/12a).
(2) A naturally occurring organic compound of wood (EPA-87/10).
(3) *See* acid for more related terms.

Fatty oil: The triglycerides which are liquid at room temperature (*see* oil for more related terms) (EPA-74/04c).

Fault: A surface or zone of rock fracture along which there has been displacement (40 CFR 146.3-91, *see also* 40 CFR 147.2902; 264.18; 147.2902-91).

Fault plane: A surface along which a fault taken place (DOI-70/04).

Fault tree: A logic diagram which depicts the inter-relationships of various primary events and sub-events to an undesired top event (EPA-87/07a).

Fault tree analysis: A means of analyzing hazards. Hazardous events are first identified by other techniques such as HAZOP. Then all combinations of individual failures that can lead to that hazardous event are shown in the logical format of the fault tree. By estimating the individual failure probabilities, and then using the appropriate arithmetical expressions, the top-event frequency can be calculated (NRT-87/03).

Faulted condition: Occurrences or accidents that are not expected to occur during the start-up testing or operation of nuclear facilities, but that are postulated because of their consequences and because they represent upper bounds on failures or accidents with a probability of occurrence sufficiently high to require a consideration in design. The equipment, components, and structures might be grossly deformed by these conditions without a loss of their nuclear safety function. It is likely that damage would be extensive enough that repair would be, required prior to reuse (DOE-91/04).

Fauna:
(1) The animal life adapted for living in a specified environment (EPA-83/09).

(2) The entire animal life of a region (LBL-76/07-water).

Feasibility study (FS):
(1) A study undertaken by the lead agency to develop and evaluate options for remedial action. The FS emphasizes data analysis and is generally performed concurrently and in an interactive fashion with the remedial investigation (RI), using data gathered during the RI. The RI data are used to define the objectives of the response action, to develop remedial action alternatives, and to undertake an initial screening and detailed analysis of the alternatives. The term also refers to a report that describes the results of the study (40 CFR 300.5-91).
(2) Analysis of the practicability of a proposal; e.g., a description and analysis of potential cleanup alternatives for a site such as one on the National Priorities List. The feasibility study usually recommends selection of a cost-effective alternative. It usually starts as soon as the remedial investigation is underway; together, they are commonly referred to as the "RI/FS" (EPA-94/04).
(3) A small-scale investigation of a problem to ascertain whether a proposed research approach is likely to provide useful data (EPA-94/04).

Fecal coliform bacteria:
(1) Those organisms associated with the intestines of warm-blooded animals that are commonly used to indicate the presence of fecal material and the potential presence of organisms capable of causing human disease (40 CFR 140.1-91).
(2) Bacteria found in the intestinal tracts of mammals. Their presence in water or sludge is an indicator of pollution and possible contamination by pathogens (EPA-94/04).

Fecundity: The potential reproductivity of organisms.

Federal act: The Federal Water Pollution Control Act, as amended, 33USC1151, et seq (40 CFR 109.2-91).

Federal agency: Any department, agency, or instrumentality of the United States (ESA3, *see also* NCA3; RCRA1004; SDWA1401).

Federal class I area: Any Federal land that is classified or reclassified Class I (40 CFR 51.301-91).

Federal cooperative agreement: A cooperative agreement entered into by an agency (40 CFR 34.105-91).

Federal delayed compliance order: A delayed compliance order issued by the Administrator under section 113(d) (1), (3), (4) or (5) of the Act (40 CFR 65.01-91).

Federal emission test procedure: Refers to the dynamometer driving schedule, dynamometer procedure, and sampling and analytical procedures described in Part 6 for the respective model year, which are used to derive city fuel economy data for gasoline-fueled or diesel vehicles (40 CFR 600.002.85-91).

Federal facility: A facility that is owned or operated by a department. agency, or instrumentality of the United States (40 CFR 35.4010-91, *see also* 40 CFR 61.101; 243.101; 244.101; 246.101-91).

Federal Facility Compliance Act (FFCA) of 1992: *See* the term of Act or FFCA.

Federal financial assistance: Any financial benefits provided directly as aid to a project by a department, agency, or instrumentality of the Federal government in any form including contracts, grants, and loan guarantees. Actions or programs carried out by the Federal government itself such as dredging performed by the Army Corp of Engineers do not involve Federal financial assistance. Actions performed for the Federal government by contractors, such as construction of roads on Federal lands by a contractor under the supervision of the Bureau of Land Management, should be distinguished from contracts entered into specifically for the purpose of providing financial assistance, and will not be considered programs or actions receiving Federal financial assistance. Federal financial assistance is limited to benefits earmarked for a specific program or action and directly award to the program or action. Indirect assistance, e.g., in the form of a loan to a developer by a lending institution which in turn receives Federal assistance not specifically related to the project in question is not Federal financial assistance under section 1424(e) (40 CFR 149.101-g-91).

Federal Food, Drug and Cosmetic Act (FFDCA): *See* the term of Act or FFDCA.

Federal government: Includes the legislative, executive, and judicial branches of the Government of the United States, and the government of the District of Columbia (NCA15, *see also* 40 CFR 40 CFR 203.1-91).

Federal grant: An award of financial assistance in the form of money, or property in lieu of money, by the Federal Government or a direct appropriation made by law to any person. The term does not include technical assistance which provides service instead of money, or other assistance in the form of revenue sharing, loans, loan guarantees, loan insurance, interest subsidies, insurance, or direct United States cash assistance to an individual (40 CFR 34.105-e-91).

Federal highway fuel economy test procedure: Refers to the dynamometer driving schedule, dynamometer procedure, and sampling and analytical procedures described in Subpart B of this part and which are used to derive highway fuel economy data for gasoline-fueled or diesel vehicles (40 CFR 600.002.85-91).

Federal implementation plan:
(1) A plan (or portion thereof) promulgated by the Administrator to fill all or a portion of a gap or otherwise correct all or a portion of an inadequacy in a State implementation plan, and which includes enforceable emission limitations or other control measures, means or techniques (including economic incentives, such as marketable permits or auctions of emissions allowances), and provides for attainment of the relevant national ambient air quality standard (CAA302-42USC7602-91).
(2) Under current law, a federally implemented plan to achieve attainment of air quality standards, used when a state is unable to develop an adequate plan (EPA-94/04).

Federal Insecticide, Fungicide, and Rodenticide Act of 1947: *See* the term of Act or FIFRA.

Federal motor vehicle control program: All federal actions aimed at controlling pollution from motor vehicles by such efforts as establishing and enforcing tailpipe and evaporative emission standards for new vehicles, testing methods development, and guidance to states operating inspection and maintenance programs (EPA-94/04).

Federal on-scene coordinator: A Federal On-Scene Coordinator designated in the National Contingency Plan (CWA311-33USC1321-91).

Federal register document: A document intended for publication in the Federal Register and bearing in its heading an identification code including the letters FRL (40 CFR 23.1-91).

Federal standards: For the purpose of this subpart, the standards specified in 40 CFR 205.152(a)(1), (2) and (3) (40 CFR 205.165-91).

Federal, State and local approvals or permits necessary to begin physical construction: The permits and approvals required under Federal, State or local hazardous waste control statutes, regulations or ordinances (40 CFR 260.10; 270.2-91).

Federal test procedure (FTP) or city fuel economy test: The test procedures specified in 40 CFR 86, except as those procedures are modified in these protocols (40 CFR 610.11-6).

Federal Water Pollution Control Act of 1972: *See* the term of Act or FWPCA.

Federally enforceable: All limitations and conditions which are enforceable by the Administrator, including those requirements developed pursuant to 40 CFR 60 and 61, requirements within any applicable State implementation plan, and any permit requirements established pursuant to 40 CFR 52.21 or under regulations approved pursuant to this section, 40 CFR 51 subpart I, including operating permits issued under an EPA-approved program that is incorporated into the State implementation plan and expressly requires adherence to any permit issued under such program (40 CFR 51-AS-12-91).

Federally permitted release:
(1) Discharges in compliance with a permit under section 402 of the Federal Water Pollution Control Act,
(2) Discharges resulting from circumstances identified and reviewed and made part of the public record with respect to a permit issued or modified under section 402 of the Federal Water Pollution Control Act and subject to a condition of such permit,
(3) Continuous or anticipated intermittent discharges from a point source, identified in a permit or permit application under section 402 of the Federal Water Pollution Control Act, which are caused by events occurring within the scope of relevant operating or treatment systems,
(4) Discharges in compliance with a legally enforceable permit under section 404 of the Federal Water Pollution Control Act,
(5) Releases in compliance with a legallyfinal permit issued pursuant to section 3005 (a) through (d) of the Solid Waste Disposal Act from a hazardous waste treatment, storage, or disposal facility when such permit specifically identifies the hazardous substances and makes such substances subject to a standard of practice, control procedure or bioassay limitation or condition, or other control on the hazardous substances in such release,
(6) Any release in compliance with a legally enforceable permit issued under section 102 or section 103 of the Marine Protection, Research, and Sanctuaries Act of 1972, (G) any injection of fluids authorized under Federal underground injection control programs or State programs submitted for Federal approval (and not disapproved by the Administrator of the Environmental Protection Agency) pursuant to Part C of the Safe Drinking Water Act,
(7) Any emission into the air subject to a permit or control regulation under section 111, section 112, title I part C, title I part D, or State implementation plans submitted in accordance with section 110 of the Clean Air Act (and not disapproved by the Administrator of the Environmental Protection Agency), including any schedule or waiver granted, promulgated, or approved under these sections,
(8) Any injection of fluids or other materials authorized under applicable State law:
   (A) For the purpose of stimulating or treating wells for the production of crude oil, natural gas, or water,
   (B) For the purpose of secondary, tertiary, or other enhanced recovery of crude oil or natural gas, or
   (C) Which are brought to the surface in conjunction with the production of crude oil or natural gas and which are reinjected,
(9) The introduction of any pollutant into a publicly owned treatment works when such pollutant is specified in and in compliance with applicable pretreatment standards of section 307 (b) or (c) of the Clean Water Act and enforceable requirements in a pretreatment program submitted by a State or municipality for Federal approval under section 402 of such Act, and
(10) Any release of source, special nuclear, or byproduct material, as those terms are defined in the Atomic Energy Act of 1954, in compliance with a legally enforceable license, permit, regulation, or order issued pursuant to the Atomic Energy Act of 1954 (SF101-42USC9601).

Federally recognized Indian tribal government: The governing body or a governmental agency of any Indian tribe, band, nation, or other organized group or community (including any Native village as defined in section 3 of the Alaska Native Claims Settlement Act, 85 Stat 688) certified by the Secretary of the Interior as eligible for the special programs and services provided by him through the Bureau of Indian Affairs (40 CFR 31.3-91).

Federally registered: Currently registered under sec. 3 of the Act, after having been initially registered under the Federal Insecticide, Fungicide, and Rodenticide Act of 1947 (Pub. L. 86-139; 73 Stat. 286; June 25, 1947) by the Secretary of Agriculture or under FIFRA by the Administrator (40 CFR 162.151-91).

Federally regulated motorcycle: For the purpose of this subpart, any motorcycle subject to the noise standards of Subpart D of this part (*see* motocycle for more related terms) (40 CFR 205.165-91).

Fee: Fee related terms include:
- Product fee
- Tipping fee

Feed gas: The chemical composition of the exhaust gas measured at the converter inlet (40 CFR 85.2122(a)(15)(ii)(E)-91).

Feed rate screen limit: In RCRA, feed rate screen limit is developed by back calculating from the emission screen limits, assuming that all metals and chlorine fed to the device are emitted (i.e., no partitioning or removal by air pollution control equipment) (cf. emission screen limit) (EPA-90/04).

Feedlot:
(1) A new concentrated confined animal or poultry growing operation for meat, milk or egg production, or stabling, in pens or houses wherein the animals or poultry are fed at the place of confinement and crop or forage growth or production is not sustained in the area of confinement (40 CFR 412.11-91, *see also* 40 CFR 412.21-91).
(2) A confined area for the controlled feeding of animals. Tends to concentrate large amounts of animal waste that cannot be absorbed by the soil and, hence, may be carried to nearby streams or lakes by rainfall runoff (EPA-94/04).

Feedstock: The crude oil and natural gas liquids fed to the topping units (40 CFR 419.11-91).

Feedwater heater: Heat exchangers in which boiler feedwater is preheated by steam extracted from the turbine (EPA-82/11f).

Felt: Endless woven belt, usually of wool, used as a conveyor during paper formation and provides a cushion between drying press rolls of the papermaking machine (EPA-83).

FEMA-REP-5: Guidance for Developing State and Local Radiological Emergency Response Plans and Preparedness for Transportation Accidents, prepared by FEMA. Provides a basis for State and local governments to develop emergency plans and improve emergency preparedness for transportation accidents involving radioactive materials (NRT-87/03).

Fen: A type of wetland that accumulates peat deposits. Fens are less acidic than bogs, deriving most of their water from groundwater rich in calcium and magnesium (*see* wetlands) (EPA-94/04).

Fermentation:
(1) Chemical reactions accompanied by living microbes that are supplied with nutrients and other critical conditions such as heat, pressure, and light that are specific to the reaction at hand (EPA-89/12).
(2) In anaerobic biodegradation, a large molecule is transformed to smaller ones with the extraction of energy (e.g., sugar is converted into alcohol).

Fermentor broth: A slurry of microorganisms in water containing nutrients (carbohydrates, nitrogen) necessary for the microorganisms growth (EPA-83/09).

Fermium (Fm): A radioactive metal with a.n. 100 and a.w. 253. The element belongs to group IIIB of the periodic table.

Ferric arsenate ($FeAsO_4 \cdot 2H_2O$): A green powder used as an insecticide.

Ferric chloride ($FeCl_3$): A brown crystal used in coagulants, photography, pigments and ink.

Ferric sulfate ($Fe_2(SO_4)_3 \cdot 9H_2O$): A yellow crystal used in textile dyeing, disinfectant and soil conditioners.

Ferricyanide: The ion, usually in the form of potassium ferricyanide, is used as a bleach for oxidizing metallic silver to ionic silver in some color processes. Ferri-cyanide is reduced to ferrocyanide as it oxidizes silver in the film emulsions (EPA-80/10).

Ferricyanide bleach: A processing solution containing the ferricyanide ion. This is used to convert metallic silver to ionic silver, which is removed in the fixing step (EPA-80/10).

Ferrite:
(1) A chemical compound containing iron (EPA-87/10a).
(2) A solid solution in which alpha iron is present (EPA-83/06a).

Ferroalloy: An intermediate material, used as an addition agent or charge material in the production of steel and other metals. Historically, these materials were ferrous alloys, hence the name. In modern usage, however, the term has been broadened to cover such materials as silicon metal, which are produced in a manner similar to that used in the production of ferroalloys (EPA-75/02).

Ferrochrome silicon: The alloy as defined by ASTM Designation A482-76 (incorporated by reference-*see* 40 CFR 60.17) (40 CFR 60.261-91).

Ferrocyanide: The ion that results when ferricyanide oxidizes silver or reacts with various reducing agents (EPA-80/10).

Ferrofluids: A stable suspension of magnetic particles in a liquid medium which, when placed in a magnetic field of sufficient magnitude, causes the apparent density of the fluid to increase. Controlling the magnetic field enables the selective separation of materials according to their density (EPA-83).

Ferromanganese blast furnace: Those blast furnaces which produce molten iron containing more than fifty percent manganese (*see* furnace for more related terms) (40 CFR 420.31-91).

Ferromanganese silicon: The alloy containing 63 to 66 percent by weight manganese, 28 to 32 percent by weight silicon, and a maximum of 0.08 percent by weight carbon (40 CFR 60.261-91).

Ferrosilicon: The alloy as defined by ASTM Designation A100-69 (Reapproved 1974) (incorporated by reference-*see* 40 CFR 60.17) grades A, B, C, D, and E, which contains 50 or more percent by weight silicon (40 CFR 60.261-91).

Ferrous: Relating to or containing iron (EPA-83/06a).

Ferrous arsenate ($Fe_3(AsO_4)_2 \cdot 6H_2O$): A green powder used in insecticides.

Ferrous casting: The remelting of ferrous metals to form a cast intermediate or finished product by pouring the molten metal into a mold. Except for grinding scrubber operations which are covered here, processing operations following the cooling of castings are covered under the electroplating and metal finishing point source categories (40 CFR 413 and 433) (40 CFR 464.02-91).

Ferrous chloride ($FeCl_2 \bullet 4H_2O$): A green crystal used as a mordant in dyeing.

Ferrous metal: The metals that are derived from iron. They can be removed using large magnets at separation facilities (EPA-89/11). Iron and steel are ferrous metals (*see* metal for more related terms).

Ferrous sulfate (coppera or green vitriol, $FeSO_4 \bullet 7H_2O$): A blue-green crystal used in wood preservative, weed killers, etc.

Ferruginous: Containing iron.

Ferruginous chert: A sedimentary deposit consisting of chalcedony or of fine-grained quartz and variable amounts of hematite, magnetite, or limonite (EPA-82/05).

Ferruginous deposit: A sedimentary rock containing enough iron to justify exploitation as iron ore. The iron is present, in different cases, in silicate, carbonate, or oxide form, occurring as the minerals chamosite, thuringite, siderite, hematitie, limonite, etc. (EPA-82/05).

Ferruginous mine drainage: *See* synonym, acid mine drainage.

Fertile water body: A water body that has a prolific growth of aquatic plants and an abundance of aquatic fauna; extreme acidity, alkalinity, salinity, or the presence of toxic matter may interfere with the population of a water body (DOI-70/04).

Fertility: The capacity of reproductivity.

Fertilizer: Materials such as nitrogen and phosphorus that provide nutrients for plants. Commercially sold fertilizers may contain other chemicals or may be in the form of processed sewage sludge (EPA-89/12).

Fescue: Grasses cultivated for meadows or lawns (EPA-75/10).

FGD citrate process: One of SO_2 emission reduction techniques (*see* sulfur oxide emission control for control structure). The citrate precess is a regenerable FGD process. It was developed by the U.S. Bureau of Mines. The citrate process uses sodium citrate and citric acid as buffering agents to attain a higher solubility of the SO_2 in an aqueous absorbent solution. The chemistry of this process is very complex. The absorption of SO_2 is pH dependent, increasing with higher pH. SO_2 forms H_2SO_3 when absorbed by water, resulting in decreasing pH values. This creates a more acidic solution that inhibits additional absorption of SO_2 gas. By using a buffering agent to prevent a pH drop, a substantially higher amount of SO_2 can be absorbed (EPA-81/12, p8-22).

FGD double-alkali process: One of SO_2 emission reduction techniques (*see* sulfur oxide emission control for control structure). Dual-, or double-, alkali scrubbing is a nonregenerable or throwaway FGD process that uses a sodium based alkali solution to remove SO_2 from combustion exhaust gas. The sodium alkali solution absorbs SO_2, and the spent absorbing liquor is regenerated with lime or limestone. Calcium sulfites and sulfates are precipitated and discarded as sludge. The regenerated sodium scrubbing solution is returned to the absorber loop. The dual-alkali process has reduced plugging and scaling problems in the absorber because sodium scrubbing compounds are very soluble. Dual-alkali systems are capable of 95% SO_2 reduction. Particulate matter is removed prior to SO_2 scrubbing by an electrostatic precipitator or a venturi scrubber. This is done to prevent fly ash erosion of the absorber internals and to prevent any appreciable oxidation of the sodium solution in the absorber due to catalytic elements in the fly ash (EPA-84/03b, p8-15).

Although the double-alkali process regenerates the scrubbing reagent, it is classified as throwaway since it does not produce a salable product and generates solids that must be disposed of in a landfill (EPA-84/03b, p8-6).

(1) Process chemistry: The sodium alkali solution is usually a mixture of three components:
- (A) Sodium carbonate (Na_2CO_3), also called soda ash;
- (B) Sodium sulfite (Na_2SO_3); and
- (C) Sodium hydroxide (NaOH), also called caustic.

The SO_2 reacts with the alkaline components to primarily form two salts: (a) sodium sulfite (Na_2SO_3); and (b) sodium bisulfite ($NaHSO_3$) as indicated in the following main absorption reactions:

- $2NaOH + SO_2 ---> Na_2SO_3 + H_2O$
- $NaOH + SO_2 ---> NaHSO_3$
- $Na_2CO_3 + 2SO_2 + H_2O <---> 2NaHSO_3 + CO_2$
  where <---> means reversible reactions
- $Na_2CO_3 + SO_2 ---> 2NaSO_3 + CO_2$
- $Na_2SO_3 + SO_2 + H_2O ---> 2NaHSO_3$

In addition to the above reactions, some of the SO_3 present may react with alkaline components to produce sodium sulfate. For example:

- $2NaOH + SO_3 ---> Na_2SO_4 + H_2O$

Throughout the system, some sodium sulfite is oxidized to sulfate by:

- $2Na_2SO_3 + O_2 ---> 2Na_2SO_4$

After reaction in the absorber, spent scrubbing liquor is bled to a reactor tank for regeneration. Sodium bisulfite and sodium sulfate are inactive salts and do not absorb any SO_2. Actually, it is the hydroxide ion (OH^-), sulfite ion (SO_3)$^-$ and carbonate ion (CO_3)$^-$ that absorb SO_2 gas. Sodium bisulfite and sodium sulfate are reacted with lime or limestone to produce a calcium sludge and a regenerated sodium solution (EPA-84/03b, p8-16).

- $2NaHSO_3 + Ca(OH)_2 ----> Na_2SO_3 + CaSO_3 \bullet 1/2H_2O + 3/2H_2O$ (lime) (sludge)
- $Na_2SO_3 + Ca(OH)_2 + 1/2H_2O ---> CaSO_3 \bullet 1/2H_2O + 2NaOH$ (lime) (sludge)
- $Na_2SO_4 + Ca(OH)_2 ---> 2NaOH + CaSO_4$ (lime) (sludge)

(2) Process description: The dual-alkali process uses two loops: (A) absorption; and (B) regeneration.
- (A) Absorption loop: In the absorption loop, the sodium solution contacts the flue gas in the absorber to remove

SO_2.

(B) Regeneration loop: The products (Na_2SO_3 and Na_2SO_4) from the absorption loop are mixed with lime slurry to produce a calcium sludge and a regenerated sodium solution (NaOH). In this example, the regenerated sodium solution is the mixture of (a) sodium sulfite (Na_2SO_3); and (b) sodium hydroxide (NaOH). The third component of the sodium solution, sodium carbonate (Na_2CO_3) (also called soda ash), is not regenerated and therefore new soda ash is needed for the system (EPA-84/03b, p8-16).

FGD dry injection: One of SO_2 emission reduction techniques (*see* sulfur oxide emission control for control structure). In dry injection systems, a dry alkaline material is injected into a flue gas stream. This is accomplished by pneumatically injecting the dry sorbent into a flue gas duct, or by precoating or continuously feeding sorbent onto a fabric filter surface. Most dry injection systems use pneumatic injection of dry alkaline material in the boiler furnace area or in the duct that precedes the ESP or baghouse. Sodium based sorbents are used more frequently than lime. Many dry injection systems have used nahcolite, a naturally occurring mineral which is 80% sodium bicarbonate found in large reserves in Colorado. Sodium carbonate (soda ash) is also used but is not as reactive as sodium bicarbonate. The major problem of using nahcolite is that it is not presently being mined on a commercial scale. Large investments must be made before it will be mined commercially. Other natural minerals such as raw trona have been tested; trona contains sodium bicarbonate and sodium carbonate (EPA-81/12, p8-27).

FGD dry scrubbing process: One of SO_2 emission reduction techniques (*see* sulfur oxide emission control for control structure). In dry FGD, the flue gas containing SO_2 is contacted with an alkaline material to produce a dry waste product for disposal. This technology includes:

(1) **FGD spray dryer with a baghouse or ESP:** Injection of an alkaline slurry in a spray dryer with collection of dry particles in a baghouse or electrostatic precipitator (ESP);
(2) **FGD dry injection:** Dry injection of alkaline material into the flue gas stream with collection of dry particles in a baghouse or ESP; and
(3) **FGD other dry SO_2 processes:** Addition of alkaline material to the fuel prior to combustion.

These technologies are capable of SO_2 emission reduction ranging from 60 to 90% depending on which system is used (EPA-81/12, p8-23).

FGD lime scrubbing: One of SO_2 emission reduction techniques (*see* sulfur oxide emission control for control structure). Lime scrubbing uses an alkaline slurry made by adding lime (CaO), usually 90% pure, to water. The alkaline slurry is sprayed in the absorber and reacts with the SO_2 in the flue gas. Insoluble calcium sulfite ($CaSO_3$) and calcium sulfate ($CaSO_4$) salts are formed in the chemical reaction that occurs in the scrubber and are removed as sludge. The sludge produced can be stablized to produce an inert landfill material or can be stored in sludge ponds.

(1) Process chemistry: A number of reactions take place in the absorber. Before the calcium can react with the SO_2, both must be broken down into their respective ions.
   (A) SO_2 dissociation: SO_2 is absorbed in the water and form sulfite $(SO_3)^-$ and sulfate $(SO_4)^-$ ions.
   - $SO_{2(gaseous)} \longrightarrow SO_{2(aqueous)}$
   - $SO_2 + H_2O \longrightarrow H_2SO_3$ or
   - $SO_2 + H_2O + 1/2\ O_2 \longrightarrow H_2SO_4$, if there is excess oxygen.
   - $H_2SO_3 \longrightarrow 2H^+ + (SO_3)^-$ or
   - $H_2SO_4 \longrightarrow 2H^+ + (SO_4)^-$
   (B) Lime (CaO) dissolution: Lime is slaked with water to produce a calcium slurry of CaO and H_2O or calcium hydroxide $[Ca(OH)_2]$. The calcium hydroxide/water mixture is a solution containing calcium ions $(Ca)^{++}$ and hydroxide ions $(OH)^-$.
   - $CaO_{(solid)} + H_2O \longrightarrow Ca(OH)_{2(aqueous)}$
   - $Ca(OH)_2 \longrightarrow Ca^{++} + 2(OH)^-$
   (C) Overall reactions: Now that the SO_2 and lime are broken into their ions, namely, $(SO_3)^-$ and Ca^{++}. Calcium ions combine with sulfate and sulfite ions to produce a calcium sulfate and calcium sulfate sludge. The basic reactions occurring are:
   - $Ca^{++} + (SO_3)^- + 2H^+ + 2(OH)^- \longrightarrow CaSO_{3(solid)} + 2H_2O$ or
   - $Ca(OH)_2 + H_2SO_3 \longrightarrow Ca(SO_3) + 2H_2O$
   - $Ca^{++} + (SO_4)^- + 2H^+ + 2(OH)^- \longrightarrow CaSO_{4(solid)} + 2H_2O$ or
   - $Ca(OH)_2 + H_2SO_4 \longrightarrow Ca(SO_4) + 2H_2O$

   From the above relationships and assuming that the lime is 90% pure, it will take 1.1 moles of lime to remove 1 mole of SO_2 gas (EPA-84/03b, p8-6).

(2) Process equipment: The equipment necessary for SO_2 emission reduction comes under four operations:
   (A) Scrubbing or absorption: Accomplished with scrubbers, holding tanks, liquid-spray nozzles, and circulation pumps.
   (B) Lime handling and slurry preparation: Accomplished with lime unloading and storage equipment, lime processing and slurry preparation equipment.
   (C) Sludge processing: Accomplished with sludge clarifiers for dewatering, sludge pumps and handling equipment, and sludge solidifying equipment.
   (D) Flue gas handling: Accomplished with inlet and outlet ductwork, dampers, fans, and stack gas reheaters (EPA-84/03b, p8-7).

(3) Process description: A typical schematic lime FGD system is as follows:
   (A) Flue gas from the boiler first passes through a particulate emission removal device [such as a variable-throat venturi scrubber (EPA-81/12, p8-8)] then into the absorber [such as a fixed-throat venturi scrubber (EPA-81/12, p8-8)] where the SO_2 is removed. The gas then passes through the entrainment separator to a reheater and is finally exhausted out of the stack. Individual FGD systems vary considerably, depending on the FGD vendor and the plant layout. ESPs or scrubbers can be used for particle removal with the various absorbers used for SO_2 removal.
   (B) A slurry of spent scrubbing liquid and sludge from the absorber then goes to a recirculation tank. From this tank, a fixed amount of the slurry is bled off to process the sludge, and, at the same time, an equal amount of fresh lime is added to the recirculation tank. Sludge is sent to a clarifier, where a large portion of

water is removed from the sludge and sent to a holding tank. Makeup water is added to the process-water holding tank, and this liquid is returned to the recirculation tank. The partially dewatered sludge from the clarifier is sent to a vacuum filter, where most of the water is removed (and sent to the process-water holding tank) and the sludge is sent to a settling pond (EPA-84/03b, p8-7).

FGD limestone scrubbing: One of SO_2 emission reduction techniques (*see* sulfur oxide emission control for control structure). Limestone scrubbers are very similar to lime scrubbers. Limestone scrubbing uses an alkaline slurry from limestone ($CaCO_3$) in an absorber to react with SO_2 in the flue gas. Calcium sulfite ($CaSO_3$) and calcium sulfate ($CaSO_4$) salts are formed in the reaction and are removed as sludge. Two major equipment differences between lime and limestone scrubbing are:
- Their uses of feed preparation equipment; and
- Their higher liquid-to-gas ratios (since limestone is less reactive than lime).

Even with these differences, the processes are so similar that an FGD system can be set up to use either lime or limestone in the scrubbing liquid (EPA-81/12, p8-11).

(1) Process chemistry: Limestone scrubbing's process chemistry is also very similar to lime scrubbing's. Limestone ($CaCO_3$) is slaked with water to form aqueous ($CaCO_3$) and is sprayed in the absorber. Sulfite and sulfate ions are produced as SO_2 gas contacts the water. These ions combine with calcium ions to produce calcium sulfite and calcium sulfate sludge. The basic reactions are (EPA-81/12, p8-11).:
- $SO_2 + CaCO_3 + H_2O + O_2 ---> CaSO_3 + H_2O + CO_2 + O_2$
- $SO_2 + CaCO_3 + H_2O + 1/2\ O_2 ---> CaSO_4 + H_2O + CO_2$

The only difference is in the dissolution reaction that generates the calcium ion. When limestone is mixed with water, the following reaction occurs (EPA-84/03b, p8-11).:
- $CaCO_{3(solid)} + H_2O ---> Ca^{++} + (HCO_3)^- + OH^-$.

(2) Process description: The equipment necessary for SO_2 absorption is the same as that for lime scrubbing, except in the slurry preparation. The limestone feed (rock) is reduced in size by crushing it in a ball mill. Limestone is sent to a size classifier. Pieces larger than 200 mesh are sent back to the ball mill for recrushing. Limestone is mixed with water in a slurry supply tank. Limestone is generally a little cheaper than lime, making it more popular for use in large FGD systems (EPA-81/12, p8-11).

FGD magnesium oxide: One of SO_2 emission reduction techniques (*see* sulfur oxide emission control for control structure). Magnesium oxide scrubbing is a regenerable FGD process used to remove SO_2 from combustion exhaust gas. Magnesium oxide (MgO) slurry absorbs SO_2 and forms magnesium sulfite. Magnesium sulfite solids are separated by centrifugation and dried to remove moisture. The mixture is calcined to regenerate magnesium oxide and produce concentrated SO_2 gas for production of sulfuric acid or elemental sulfur (EPA-81/12, p8-21).

Particulate matter is removed from boiler exhaust by a precipitator or wet scrubber prior to entering the absorber. Magnesium oxide slurry is sprayed and absorbs SO_2 according to the following simplified reactions:

- $Mg(OH)_2 + 5H_2O + SO_2 ---> MgSO_3 \bullet 6H_2O$
- $MgSO_3 \bullet 6H_2O + SO_2 ---> Mg(HSO_3)_2 + 5H_2O$
- $Mg(HSO_3)_2 + MgO ---> 2MgSO_3 + H_2O$
- $2MgSO_3 + O_2 + 7H_2O ---> 2MgSO_4 \bullet 7H_2O$

The aqueous slurry used for scrubbing contains the hydrated crystals of MgO, $MgSO_3$ and $MgSO_4$. A continuous side stream of this recycled slurry is sent to a centrifuge where partial dewatering produces a moist cake. The liquor removed from the crystals is returned to the main slurry stream. The moist cake is dried at 350 to 450° F in a direct contact or rotary bed dryer. The dried cake is then sent to a calciner where coke is burned at very high temperatures (1250 to 1340° F) to regenerate magnesium oxide crystals according to the following reactions:

(1) Cake dryer
- $MgSO_3 \bullet 6H_2O ---> MgSO_3 + 6H_2O$ (vapor)
- $MgSO_4 \bullet 7H_2O ---> MgSO_4 + 7H_2O$ (vapor)

(2) MgO regeneration in calciner
- $MgSO_3 ---> MgO + SO_2$ (concentrated gas)
- $C + 1/2 O_2 ---> CO$ (gas)
- $CO + MgSO_4 ---> CO_2 + MgO + SO_2$ (concentrated gas)

FGD nonregenerable (throwaway) process: One of SO_2 emission reduction techniques (*see* sulfur oxide emission control for control structure). Nonregenerable FGD processes, also known as throwaway FGD processes, are those which generate a sludge or waste product as a result of SO_2 emission reduction. The sludge is not re-usable and must be disposed of properly in a pond or landfill (EPA-81/12, p8-7). Types of FGD nongenerable (throwaway) processes include:
(1) FGD lime scrubbing (EPA-84/03b, p8-6)
(2) FGD limestone scrubbing (EPA-81/12, p8-11)
(3) FGD double alkali scrubbing (EPA-84/03b, p8-15)
(4) FGD sodium-based once-through scrubbing (EPA-84/03b, p8-20)

FGD other dry SO_2 process: One of SO_2 emission reduction techniques (*see* sulfur oxide emission control for control structure). Coal and limestone fuel mixture have been tested as a method for reducing SO_2 emissions. This technology uses limestone injection with a low NO_x burner. Another promising dry process, the Shell UOP process, uses a copper oxide catalyst for SO_2 emission reduction. Shell UOP is a dry process that simultaneously removes both NO_x and SO_x emissions. This process can also be designed to remove either compound separately (EPA-81/12, p8-28).

FGD regenerable process: One of SO_2 emission reduction techniques (*see* sulfur oxide emission control for control structure). Regenerable FGD processes are wet scrubbing processes that remove SO_2 from the flue gas and generate salable (usable) products. Regenerable products include elemental sulfur, sulfuric acid, or in the case of lime or limestone scrubbing, gypsum (used for wallboard). Regenerable processes do not produce a sludge, thereby eliminating the sludge disposal problem (EPA-81/12, p8-17). Types of FGD regenerable processes inlcude:
(1) FGD Wellman-Lord (EPA-84/03b, p8-27)
(2) FGD magnesium oxide (EPA-81/12, p8-21)
(2) FGD citrate (EPA-81/12, p8-22)

Most regenerable processes also:
(1) Have the potential for consistently obtaining a high SO_2 removal efficiency, usually exceeding 90%.

(2) Utilize the scrubbing reagent more efficiently than nonregenerable processes.
(3) Use scrubbing liquors that do not cause scaling and plugging problems in the scrubber.

The major drawback of using these processes is that these systems are usually more complicated in design and are more expensive to install and operate (EPA-84/03b, p8-26).

FGD sodium-based once-through scrubbing: One of SO_2 emission reduction techniques (*see* sulfur oxide emission control for control structure). Sodium-based throwaway (once-through) scrubbing systems are the overwhelming choice for FGD systems installed on industrial boilers. These systems use a clear liquid absorbent of either sodium carbonate (Na_2CO_3), sodium hydroxide (NaOH), or sodium bicarbonate ($NaHSO_3$). Sodium-based systems are favored for treating flue gas from industrial boilers because:

- Sodium alkali is the most efficient of the commercial reagents in removing SO_2 and the chemistry is relatively simple.
- They are soluble systems--as opposed to slurry systems - making for scale-free operation and fewer components.
- Such systems can handle the wider variations in flue-gas composition resulting from the burning of many different fuels by industry.
- The systems are often smaller, and operating costs are a small percentage of total plant costs.
- In some cases, these plants have a waste caustic stream or soda ash available for use as the absorbent.

These systems have been applied to only a few large utility boilers because:

- The process consumes a premium chemical (NaOH or Na_2CO_3) that is much more costly per pound than calcium-based reagents.
- The liquid wastes contain highly soluble sodium salt compounds. Therefore, the huge quantities of liquid wastes generated by large utilities would have to be sent to ponds to allow the water to evaporate.

(1) Process chemistry: The process chemistry is very similar to that of the double-alkali process, except the absorbent is not regenerated.
(2) Process description: Exhaust gas from the boiler may first pass through an ESP or baghouse to remove particulate matter. Sodium chemicals are mixed with water and sprayed into the absorber. The solution reacts with the SO_2 in the flue gas to form sodium sulfite, sodium bisulfite, and a very small amount of sodium sulfate. A bleed stream is taken from the scrubbing liquor recirculation stream at a rate equal to the amount of SO_2 that is being absorbed. The bleed stream is sent to a neutralization tank and aeration tower before being sent to a lined disposal pond.

Some coal-fired units use ESPs or baghouses to remove fly ash before the gas enters the scrubber. In these cases, the absorber can be a plate tower or spray tower that provides good scrubbing efficiency at low pressure drops. For simultaneous SO_2 and fly ash removal, venturi scrubbers can be used. In fact, many of the industrial sodium-based throwaway systems are venturi scrubbers originally designed to remove particulate matter. These units were slightly modified to inject a sodium-based scrubbing liquor. Although removal of both particles and SO_2 in one vessel can be economically attractive, the problems of high pressure drops and using a scrubbing medium to remove fly ash must be considered. However, in cases where the particle concentration is low, such as from oil-fired units, simultaneous particulate and SO_2 emission reduction can be effective (EPA-84/03b, p8-20).

FGD spray dryer with a baghouse or ESP: One of SO_2 emission reduction techniques (*see* sulfur oxide emission control for control structure). In a FGD spray process, alkaline is injected into a spray dryer with dry particle collection in a baghouse or ESP. Spray dryers have been used in the chemical, food processing, and mineral preparation industries over the past 40 years. **Spray dryers** are vessels where hot flue gases are contacted with a finely atomized wet alkaline spray. The high temperatures of the flue gas, 250 to 400° F, evaporate the moisture from the wet alkaline sprays, leaving a dry powdered product. The dry product is collected in a baghouse or ESP.

Flue gas enters the top of the spray dryer and is swirled by a fixed vane ring to cause intimate contact with the slurry spray. The slurry is atomized into extremely fine droplets by rotary atomizers. The turbulent mixing of the flue gas with the fine droplets results in rapid SO_2 absorption and evaporation of the moisture. A small portion of the hot flue gas is added to the spray-dryer-discharge duct to maintain the temperature of the gas above the dew point. Reheat prevents condensation and corrosion in the duct. Reheat also prevents bags in the baghouse from becoming plugged or caked with moist particles.

Sodium carbonate solutions and lime slurries are the most common absorbents used. A sodium carbonate solution will generally achieve a higher level of SO_2 removal than lime slurries. When sodium carbonate is used, SO_2 removal efficiencies are approximately 75 to 90%, lime removal efficiencies are 70 to 85%. However, vendors of dry scrubbing systems claim that their units are capable of achieving 90% SO_2 reduction using a lime slurry in a spray dryer. Lime is very popular for two reasons:

(1) Lime is less expensive than sodium carbonate.
(2) Sodium carbonate and SO_2 form sodium sulfite and sodium sulfate which are very soluble causing leaching problems when landfilled.

Some of the evaporated alkaline spray will fall into the bottom of the spray dryer and be recycled. The majority of the spray reacts with SO_2 in the flue gas to form powdered sulfates and sulfites. The particles, along with fly ash in the flue gas, are then collected in a baghouse or electrostatic precipitator. Baghouses have an advantage because unreacted alkaline material collected on the bags can react with any remaining SO_2 in the flue gas. Some process developers have reported SO_2 removal on bag surfaces on the order of 10%. However, since bags are sensitive to wetting, a 35 to 50° F margin above the saturation temperature of the flue gas must be maintained. ESPs have the advantage of not being as sensitive to moisture as baghouses. However, SO_2 removal is not quite as efficient using ESPs.

The major differences between wet absorption SO_2 scrubbers and spray dryer systems is in the scrubbing method and the amount of moisture during the scrubbing action. In a wet scrubber, flue gas is saturated with liquid sprays (usually alkaline). SO_2 is absorbed by the water and also reacts with the chemical. Absorption increases as temperature decreases. Flue gas is cooled and saturated with the scrubbing liquid to remove SO_2. At optimum operating temperatures efficiency is usually > 90%.

In a spray dryer, finely atomized alkaline droplets are

contacted with flue gas which is at air preheater outlet temperatures (250 to 400° F). The flue gas is humidified to within 50° F of its saturation temperature by the moisture evaporating from the alkaline slurry. Reaction of the SO_2 with the alkaline material proceeds both during and following the drying process, although to what degree is not completely understood. Since the flue gas temperature and humidity are set by air preheater outlet conditions, the amount of moisture that can be evaporated into the flue gas is also set. This means that the amount of alkaline slurry that can be evaporated in the dryer is limited by flue gas conditions. Alkaline slurry sprayed into the dryer must be carefully controlled to avoid moisture in the flue gas from condensing in the ducting, particulate emission control equipment, or the stack. SO_2 removal efficiencies are generally < 85%.

The major problems with dry injection systems are the low sodium utilization in the process and the disposal of leachable sodium sulfur compounds. U.S. EPA reports that only 40 to 60% of the dry alkaline injected material is used at high SO_2 removal conditions (EPA-81/12, p8-24).

FGD system comparison: Comparisons between dry and wet FGD systems can be made in four major areas:
(1) Waste disposal: Lime, limestone, double-alkali, and dry FGD systems produce a sludge that must be disposed of properly. New installations must meet all solid waste regulations including stringent RCRA regulations. The regenerable FGD processes, including Wellman-Lord, magnesium oxide, and citrate, generate a usable product (sulfur or sulfuric acid) that has a commercial value. The dry FGD systems produce a dry waste product that can be discarded using conventional fly ash removal systems instead of sludge removal systems. However, sodium-based dry FGD systems are undesirable because they produce waste products that easily leach from conventional landfills.
(2) Chemical reagent requirements: Lime and limestone are cheaper than sodium based absorbents and are readily available. In dry scrubbing systems, a higher stoichiometric ratio of lime is necessary to achieve SO_2 reduction efficiencies similar to wet lime and limestone scrubbing efficiencies. Consequently, costs for lime or limestone materials will be somewhat higher in dry systems.
(3) SO_2 removal efficiencies: Regarding SO_2 removal efficiencies, all of the wet scrubbing processes are capable of at least 90% removal efficiency. Most units are capable of 95% removal efficiency. These systems have been installed and operating for a number of years with good reliability. Dry scrubbing is capable of removing at least 75 to 85% SO_2. Vendors of dry FGD equipment claim 90% capabilities. This technology is very attractive for utility boilers burning low sulfur western coal. With improvements, dry scrubbing may be used on boilers burning medium sulfur coal (2 to 3%) in the near future.
(4) Economics: Cost estimates for FGD systems are not easy to obtain. For a reference purpose, the following cost information is provided (EPA-81/12, p8-29).

FGD system	(A)*	(B)*
Limestone	98	4.02
Lime	90	4.25
Double alkali	101	4.19
Wellman-Lord	131-141	5.11-6.03
Magnesium oxide	132	5.08
Citrate	143	6.44

(A) = Mid-1979 capital investment ($/kw)
(B) = Mid-1980 first year operating cost (mills/kWh)
*Cost basis: 500 MW power plant burning 3.5% sulfur coal achieving 90% SO_2 removal efficiency.

FGD Wellman-Lord: One of SO_2 emission reduction techniques (*see* sulfur oxide emission control for control structure). The Wellman-Lord process is a regenerable FGD process used to reduce SO_2 emissions from utility and industrial boilers and produces a usable or salable product. This process is sometimes referred to as the Wellman-Lord/Allied Chemical process, Allied Chemical referring to the regeneration step.
(1) Process chemistry: In the Wellman-Lord process, the SO_2 is absorbed by an aqueous sodium sulfite solution which forms a sodium bisulfite solution according to the following equation:
- $SO_2 + Na_2SO_3 + H_2O ---> 2Na_2HSO_3$

Some oxidation occurs in the absorber to form sodium sulfate, which is unreactive with SO_2 gas.
- $Na_2SO_3 + 1/2 O_2 ---> Na_2SO_4$

The formation of sodium sulfate depletes the supply of sodium sulfite available for scrubbing. This can be made up by adding sodium carbonate to the scrubbing slurry to combine with sodium bisulfite according to the following chemical reaction:
- $Na_2CO_3 + 2NaHSO_3 ---> 2Na_2SO_3 + CO_2 + H_2O$.

The absorbent is then regenerated by evaporating the water from the bisulfite solution.
- $2NaHSO_3 ---> Na_2SO_3 + H_2O + SO_2$ (concentrated gas)

The concentrated SO_2 produced in the regeneration step is then sent to the Allied process for conversion to elemental sulfur or sulfuric acid (EPA-84/03b, p8-26).
(2) Process description: The process equipment includes an electrostatic precipitator for removing particulate matter; a venturi scrubber for cooling flue gas and removing SO_3 and chlorides; an SO_2 absorber; an evaporator-crystallizer for regenerating the absorbent; and the Allied Chemical process for reducing concentrated SO_2 gas into elemental sulfur or sulfuric acid. The absorber is a plate tower. SO_2 gas is scrubbed with a sodium sulfite solution at each plate. A mist eliminator removes entrained liquid droplets from gas exiting the absorber. There is a direct-fired natural gas reheating system in the absorber stack to reheat cleaned gas for good dispersion of the steam plume.

The solution (sodium bisulfite), collected at the bottom of the absorber, overflows into an absorber surge tank. This solution is pumped through a filter to remove any collected particulate matter. A small side-stream is sent to a purge treatment system where sodium sulfate is removed. The solution is then pumped to the evaporator for regeneration of the sodium sulfite solution.

The evaporator is a forced-circulation vacuum evaporator. Solution is recirculated in the evaporator, where low-pressure steam evaporates water from the sodium bisulfite solution. When sufficient water is removed, sodium sulfite crystals form and precipitate. Concentrated so, gas (95% by volume) is removed by the steam. The sodium sulfite crystals form a slurry that is withdrawn continuously and sent to a dissolving tank, where condensate from the evaporator is used to dissolve the sodium sulfite crystals into a solution. This solution is pumped back into the top stage of the absorber. The water vapor is removed from the evaporator s overhead SO_2/H_2O vapors by water-cooled condensers. The SO_2 is compressed by a liquid-ring compressor and sent to the Allied Chemical SO_2 reduction plant (EPA-84/03b, p8-27).

FGD wet scrubbing process: One of SO_2 emission reduction techniques (*see* sulfur oxide emission control for control structure). Wet scrubbing processes use a liquid absorbent to absorb the SO_2 gases. Wet scrubbing can be further categorized into:
(1) FGD nongenerable (throwaway) process (EPA-81/12, p8-7)
    (A) FGD lime scrubbing (EPA-84/03b, p8-6)
    (B) FGD limestone scrubbing (EPA-81/12, p8-11)
    (C) FGD double alkali scrubbing (EPA-84/03b, p8-15)
    (D) FGD sodium-based once-through scrubbing (EPA-84/03b, p8-20)
(2) FGD regenerable process (EPA-81/12, p8-17)
    (A) FGD Wellman-Lord (EPA-84/03b, p8-27)
    (B) FGD magnesium oxide (EPA-81/12, p8-21)
    (C) FGD citrate (EPA-81/12, p8-22)

Fiber:
(1) The dry wool and other fibers as received at the wool finishing mill for processing into wool and blended products (40 CFR 410.21-91, *see also* 40 CFR 763-App/A; 763.121-91).
(2) The slender thread-like elements of wood or similar cellulosic materials, which, when separated by chemical and/or mechanical means, as in pulping, can be formed into fiberboard (cf. fabric) (EPA-74/04).
(3) An individual filament made by attenuating molten glass. A continuous filament is a glass fiber of great or indefinite length. A staple fiber is a glass fiber of relatively short length (generally less than 17 in.) (432 mm) (EPA-83).

Other fiber related terms include:
- Acetate fiber
- Acrilan fiber
- Arnel fiber
- Avicron fiber
- Avril fiber
- Creslan fiber
- Dacron fiber
- Fortrel fiber
- Herculon fiber
- Kodel fiber
- Leaf fiber
- Lycra fiber
- Mitin fiber
- Nylon fiber
- Orlon fiber
- Polyester fiber
- Rayon fiber
- Synthetic fiber

Fiber bed mist eliminator: A mist eliminator used to collect fine acid mists (*see* mist eliminator for more related terms) (EPA-81/09).

Fiber or fiberboard boxes: The boxes made from containerboard, either solid fiber or corrugated paperboard (general term) or boxes made from solid paperboard of the same material throughout (specific term) (40 CFR 250.4-91).

Fiber preparation: The reduction of wood to a fibrous condition for hardboard manufacture, utilizing mechanical, thermal, or explosive methods (EPA-74/04).

Fiber release episode: Any uncontrolled or unintentional disturbance of ACBM resulting in visible emission (40 CFR 763.83-91).

Fiber resource: Sources of fiber available to the paper industry for the manufacture of paper and paperboard (EPA-83).

Fiberboard: A sheet material manufactured from fibers of wood or other ligno-cellulosic materials with the primary bond deriving from the arrangement of the fibers and their inherent adhesive properties. Bonding agents or other materials may be added during manufacture to increase strength, resistance to moisture, fire, insects, or decay, or to improve some other property of the product. Alternative spelling: fibreboard (EPA-74/04).

Fiberglass insulation: The insulation which is composed principally of glass fibers, with or without binders (40 CFR 248.4-91).

Fibre fuel: A process where the combustible fraction of solid waste is extracted, shredded and used as a fuel (*see* fuel for more related terms) (EPA-83).

Fibrosis: A condition marked by the formation of thread-like tissue.

Fibrosis producers: *See* definition under toxicant and effect.

Fick's law: The diffusion rate across a plane is proportional to the negative rate of the concentration change.

Fickian diffusion: A diffusion described by the diffusion equation in which the diffusivities or exchange coefficients can be considered as constant with respect to the time and space coordinates (NATO-78/10).

FID: Flame ionization detection, *see* definition under continuous emission monitor.

FID analyzer: A flame ionized detector (FID) analyzer senses the organic concentration and generates an output proportional to the gas concentration (*see* total hydrocarbon concentration measurement system for more related terms) (EPA-90/04).

Field: Any area treated land area, or part thereof, upon which one or more pesticides are used for agricultural purposes, all as specified by this part (40 CFR 170.2-91)

Field blank: *See* synonym, equipment blank.

Field capacity of solid waste (or moisture holding capacity): The amount of water retained in solid waste after it has been saturated and has drained freely. (beyond which the application of additional water will cause it to drain rapidly to underlying material). Also known as moisture-holding capacity (SW-108ts, *see also* EPA-83).

Field effect transistor: A transistor made by the metal-oxide-semiconductor (MOS) technique, differing from bipolar ones in that only one kind of charge carrier is active in a single device. Those that employ electrons are called n-MOS transistors; those that employ holes are p-MOS transistors (EPA-83/03).

Field gas: The feedstock gas entering the natural gas processing plant (40 CFR 60.631-91).

Field matrix spike: A sample created by spiking target analytes into a portion of a sample in the field at the point of sample acquisition. This data quality assessment provides information on the target analyte stability after collection, during transport, and storage, as well as on losses, etc., during sample preparation and on error of analysis (*see* spike for more related terms) (EPA-84/03).

Field method: A method applicable to non-laboratory situations (*see* method for more related terms) (ACS-87/11).

Field operation: Usually denotes the use of portable equipment capable of analysis under a wide range of environmental conditions (*see* instrument use for more related terms) (LBL-76/07-water).

Field sample: *See* synonym, environmental sample.

Field seam: In constructing a landfill liner, the seaming of FML rolls or panels together in the field making a continuous liner system (EPA-89/09, *see also* EPA-91/05).

Field testing: The practical and generally small-scale testing of innovative or alternative technologies directed to verifying performance and/or refining design parameters not sufficiently tested to resolve technical uncertainties which prevent the funding of a promising improvement in innovative or alternative treatment technology (40 CFR 35.2005-91).

FIFRA applicator: *See* synonym, certified applicator.

FIFRA pesticide ingredient: An ingredient of a pesticide that must be registered with EPA under the Federal Insecticide, fungicide, and Rodenticide Act. Products making pesticide claims mst register under FIFRA and may be subject to labeling and use requirements (EPA-94/04).

Filament:
(1) A metallic wire which is heated in an incandescent lamp to produce light by passing an electron current through it.
(2) A cathode in a fluorescent lamp that emits electrons when electric current is passed through it (EPA-83/03).

Filamentous organism: Lean or thread organisms such as algae and fungi which may result in bulking sludge.

Filamentous sludge: Sludge with filamentous organisms (*see* sludge for more related terms).

Fill:
(1) The introduction of VOL into a storage vessel but not necessarily to complete capacity (cf. packing) (40 CFR 60.111b-91).
(2) Deposits made by man of natural soils and/or waste materials (cf. sanitary landfilling) (EPA-83).

Fill material: Any pollutant which replaces portions of the waters of the United States with dry land or which changes the bottom elevation of a water body for any purpose (40 CFR 232.2-91).

Filler:
(1) An inert substance in a composition to increase the bulk, strength, and/or lower the cost, etc (EPA-79/12a).
(2) A high specific gravity (2 - 4.5) compound used in rubber mixtures to provide a certain degree of stiffness and hardness and used to decrease costs. Fillers have neither reinforcing nor coloring properties and are similar to extenders in their cost-reducing function (cf. extender) (EPA-74/12a).
(3) A material, such as clay, added to paper (EPA-83).

Filling: Depositing dirt and mud or other materials into aquatic areas to create more dry land, usually for agricultural or commercial development purposes. Such activities often damage the ecology of the area (EPA-94/04).

Film badge: A piece of masked photographic film worn like a badge by nuclear workers to monitor an exposure to radiation. Nuclear radiation darkens the film (EPA-74/11).

Film stripping: A separation of silver-bearing materials from scrap photographic films (EPA-83/03a).

Filter:
(1) A porous device through which a gas or liquid is passed to remove suspended particles or dust.
(2) Wet solids generated by the filtration of solids from a liquid. This filter cake may be a pure material (product) or a waste material containing additional fine solids (i.e., diatomaceous earth) that has been added to aid in the filtration (cf. baghouse) (EPA-87/10a).

Other filter related terms include:
- Activated carbon filter
- Bag filter
- Bacteria bed (*see* synonym, trickling filter)
- Biofilter (*see* synonym, trickling filter)
- Biological filter (*see* synonym, trickling filter)
- Fabric filter (*see* synonym, bag filter)
- Cartridge filter
- Contact bed (*see* synonym, sand filter)
- Contact filter (*see* synonym, sand filter)
- Continuous filter
- Granular filter
- High rate filter
- Intermittent filter
- Liquid filter
- Low rate filter
- Mud filter
- Multi-media filter
- Negative pressure fabric filter
- Percolating filter
- Positive pressure fabric filter
- Precoat filter
- Preliminary filter
- Pressure filter

- Rapid sand filter
- Rotary vacuum filter
- Sand filter
- Solvent filter
- Sprinkling filter (*see* synonym, trickling filter)
- Stage trickling filter
- Trickling filter
- Vacuum filter

Filter aid (or filtration aid): A coagulant or flocculant which is added to a sludge to enhance its filterability.

Filter background level: The concentration of structures per square millimeter of filter that is considered indistinguishable from the concentration measured on a blank (filters through which no air has been drawn). For this method the filter background level is defined as 70 structures/mm2 (40 CFR 763-App/A-91).

Filter bed: A device for removing suspended solids from water, consisting of granular material placed in a layer(s) and capable of being cleaned hydraulically by reversing the direction of the flow (EPA-82/11f).

Filter cake:
(1) The accumulation of dust on the surface of a baghouse.
(2) Wet solids generated by the filtration of solids from a liquid (EPA-87/10a).

Filter cake washing: Washing of a filter cake to remove unwanted residuals.

Filter cloth: A fabric cloth used for filtration.

Filter dam: A pervious barrier of loose stones, or stones and brush, placed in the outlet of a water body to prevent fish from moving out or in the inlet of a water body to prevent fish from entering (DOI-70/04)

Filter loading:
(1) The flow of wastewater per unit area per day applied to a sand filter.
(2) The amount (or rate) of organics (e.g., kg of BOD) in wastewater applied to a trickling filter (cf. loading, hydraulic).
(3) *see* loading for more related terms.

Filter media: The permeable barrier employed in the filtration process to separate the particles from the fluid stream and also to act as the substrate for dust cake development.

Filter press: In the past, the most common type of filter used to depart solids from sludge. It consists of a simple and efficient plate and frame filter which allows filtered juice to mix with clarified juice and be sent to the evaporators (EPA-75/02d).

Filter press cake: A residual waste product (e.g., from the process of grease recovery following the wool scouring process); filter-press cake may contain organic matter, dirt, grit, or other residue (DOI-70/04).

Filter rate: The volume (or weight) of wastewater that passes through a filter per unit time.

Filter run: The duration between two backwashings in a filter.

Filter strip: Strip or area of vegetation used for removing sediment, organic matter, and other pollutants from runoff and waste water (EPA-94/04).

Filterability:
(1) The ability for a substance to pass through a filter.
(2) The ability of removing suspended solid from an effluent by a filter.

Filterable (dissolved) metals: Those metals which will pass through a 0.45 um membrane filter (LBL-76/07-water).

Filtering capacitor: A capacitor used in a power-supply filter system to provide a low-reactant path for alternating currents and thereby suppress ripple currents, without affecting direct currents (EPA-83/03).

Filtrate: Liquid after passing through a filter (EPA-83/06a).

Filtration:
(1) A process for removing particulate matter from water by passage through porous media (40 CFR 141.2-91).
(2) A treatment process, under the control of qualified operators, for removing solid (particulate) matter from water by means of porous media such as sand or a man-made filter; often used to remove particles that containing pathogens (EPA-94/04).
(3) Filtration is a physical process whereby particles suspended in a fluid are separated by forcing the fluid through a porous medium. As the fluid passes through the medium, the suspended particles are trapped on the surface of the medium and/or within the body of the medium. Filter media can be a thick barrier of a granular material, such as sand, coke, coal, or porous ceramic. The pressure differential to move the fluid through the medium can be induced by gravity, positive pressure, or vacuum. The intended application has a great influence on both the type of filter and its physical features.

Other filtration related terms include:
- Alternating double filtration
- Conventional filtration treatment
- Deep bed filtration
- Diatomaceous earth filtration
- Direct filtration
- Dual media filtration
- Gravity filtration
- Membrane filtration
- Micro filtration (*see* synonym, ultra filtration)
- Pressure filtration
- Sand filtration
- Slow sand filtration
- Ultra filtration
- Vacuum filtration

Filtration aid: *See* synonym, filter aid.

Final acute value (FAV):
(1) An acute toxicity limit in which the measured acute toxicity

expressed as an LC50 of an effluent or a toxicant is adjusted by a factor (0.3 is recommended) to eliminate mortality (EPA-85/09).

(2) An estimate of the concentration of the toxicant corresponding to a cumulative probability of 0.05 in the acute toxicity values for all genera for which acceptable acute tests have been conducted on the toxicant (EPA-91/03).

Final approval: The approval received by a state program that meets the requirements in 40 CFR 281.11(b) (*see* approval for more related terms) (40 CFR 281.12-91).

Final authorization: The approval by EPA of a State program which has met the requirements of section 3006(b) of RCRA and the applicable requirements of Part 271, Subpart A (40 CFR 270.2-91).

Final closure: The closure of all hazardous waste management units at the facility in accordance with all applicable closure requirements so that hazardous waste management activities under Parts 264 and 265 of this Chapter are no longer conducted at the facility unless subject to the provisions in 40 CFR 262.34 (*see* closure for more related terms) (40 CFR 260.10-91).

Final contour: The surface shape or contour of a surface mine (or section thereof) after all mining and earth moving (regrading) operations have been completed (EPA-82/10).

Final cooler: In the steel industry, a hurdle packed tower that cools the coke oven gas by direct contact. The gas must be cooled to 30 C for recovery of light oil (EPA-74/06a).

Final cover: The cover material that serves the same functions as daily cover but, in addition, may be permanently exposed on the surface (*see* cover for more related terms) (40 CFR 241.101-91).

Final order:
(A) An order issued by the Administrator after an appeal of an initial decision, accelerated decision, decision to dismiss, or default order, disposing of a matter in controversy between the parties; or
(B) An initial decision which becomes a final order under 40 CFR 22.2(c) (40 CFR 22.03-91).

Final printed labeling: The label or labeling of the product when distributed or sold. Final printed labeling does not include the Package of the product, unless the labeling is an integral part of the package (40 CFR 152.3-91).

Final product: A new chemical substance (as new chemical substance is defined in 40 CFR 720.3 of this chapter) that is manufactured by a person for distribution in commerce, or for use by the person other than as an intermediate (40 CFR 700.43-91).

Final repair coat: The repainting of any topcoat which is damaged during vehicle assembly (40 CFR 52.741-91).

Final remediation levels: Chemical-specific clean-up levels that are documented in the Record of Decision (ROD). They may differ from preliminary remediation goals (PRGs) because of modifications resulting from consideration of various uncertainties, technical 1 nd exposure factors, as well as all nine selection-of-remedy criteria outlined in the National Oil and Hazardous Substances Pollution Contingency Plan (NCP) (EPA-91/12).

Final sedimentation: The settling of partly settled, flocculated or oxidized sewage in a final tank. (The term settling is preferred) (EPA-76/03).

Financial assurance for closure: Documentation or proof that an owner or operator of a facility such as a landfill or other waste repository is capable of paying the projected costs of closing the facility and monitoring it afterwards as provided in RCRA regulations (EPA-94/04).

Finding of no significant impact (FNSI): A document prepared by a federal agency that presents the reasons why a proposed action would not have a significant impact on the environment and thus would not require preparation of an Environmental Impact Statement. An FNSI is based on the results of an environmental assessment (EPA-94/04, *see also* 40 CFR 1508.13-91).

Fine: Very short pulp fibers or fiber fragments and ray cells. They are sometimes referred to as flour or wood flour (cf. flour) (EPA-87/10).

Fine paper: Fine paper includes printing, writing and cover papers (*see* paper for more related terms) (EPA-83).

Fine rack: Fine screen with approximately 25 mm square openings.

Fine sand: Sand particles with approximately 0.25 millimeter in diameter.

Finger lake: A long narrow glacial lake (DOI-70/04).

Finish coat operation: The coating application station, curing oven, and quench station used to apply and dry or cure the final coating(s) on the surface of the metal coil. Where only a single coating is applied to the metal coil, that coating is considered a finish coat (40 CFR 60.461-91).

Finished product: The final manufactured product as fresh meat cuts, hams, bacon or other smoked meats, sausage, luncheon meats, stew, canned meats or related products (40 CFR 432.51-91, *see also* 40 CFR 432.61; 432.71; 432.81; 432.91-91).

Finished water: Water that has passed through a water treatment plant; all the treatment processes are completed or "finished". The water is ready to be delivered to consumers (EPA-94/04).

Finishing water: The processed water used to remove waste plastic material generated during a finishing process or to lubricate a plastic product during a finishing process. It includes water used to machine or assemble intermediate or final plastic products (*see* water for more related terms) (40 CFR 463.2-91).

Finite difference method: A numerical method of solving differential equations by expressing the derivatives of the function as differences between values of the function at discrete points, usually called grid points (NATO-78/10).

Fire: The phenomenon of combustion manifested in light, flame and heat. It may involve:
(1) Constructive burning of a raw material for purposes of synthesizing a final product;
(2) Constructive burning of a fuel for purposes of releasing usable energy; or
(3) Destructive burning of a waste for purposes of destroying specific constituent(s) (CRWI-89/05).

Fire engine: A truck which uses water to extinguish fires.

Fire extinguishers: Portable fire extinguishers are classified by the National Fire Protection Association according to the types of burning materials they are designed to extinguish. Classes of the fire extinguisher are as follows:
- Class A: Ordinary combustibles, such as wood, paper, textiles;
- Class B: Flammable liquids, such as oil, grease, paint;
- Class C: Fires involving electrical wiring and equipment where safety requires the use of electrically non-conductive extinguishing media; and
- Class D: Combustible metals, such as magnesium, sodium, zinc, powdered aluminum (Course 165.5)

Fire fighting turbine: Any stationary gas turbine that is used solely to pump water for extinguishing fires (40 CFR 60.331-91).

Fire hydrant: A water supply device to provide water for a fire engine.

Fire monitor: A mechanical device holding a rotating nozzle, which emits a stream of water for use in fire-fighting. Fire monitors may be fixed in place or may be portable. A fire monitor allows one person to direct water on a fire whereas a hose of the same flow-rate would require more than one person (EPA-87/07a).

Fire point temperature: The lowest temperature at which an oil vaporizes readily enough to burn at least five seconds after ignition (*see* temperature for more related terms) (OME-88/12).

Fire polish: To make glass smooth, rounded, or glossy by heating in a fire (EPA-83).

Fire retardant: A formulation of inorganic salts that imparts fire resistance when injected into wood in high concentrations (EPA-74/04).

Fire tube boiler: *See* definition under boiler tubes.

Firebox: The chamber or compartment of a boiler or furnace in which materials are burned, but not the combustion chamber of afterburner of an incinerator (40 CFR 52.741-91, *see also* 40 CFR 60-App/A(method 28 & 28A)-91).

Firebrick: A refractory brick made from fireclay (*see* brick for more related terms) (OME-88/12).

Fireclay: A sedimentary clay containing only small amounts of fluxing impurities, high aluminum silicate and capable of withstanding high temperature (EPA-83).

Fireclay mortar: A mortar made of high-fusion-point fireclay and water. It is often used to fill joints in refractory walls to stop air or gas leaks without forming a strong bond (*see* mortar for more related terms) (EPA-83).

Fireside cleaning: Cleaning of the outside surface of boiler tubes and combustion chamber refractories to remove deposits formed during the combustions (EPA-82/11f).

Firm capacity (maximum capability or peak capability):
(1) (In an electric generating station), the maximum continuous electrical output of a generating station as determined by demonstration tests and assured by the station owner/operator. Firm capacity in older stations is less than the design capacity because of equipment degeneration, change in fuel character or type. Derating of newer stations may occur because of deficiencies in the original design and/or installed equipment (EPA-83).
(2) (In a general system), An assumed facility processing capacity accounting for equipment vulnerability (EPA-83).
(3) *See* capacity for more related terms.

First attempt at repair: To take rapid action for the purpose of stopping or reducing leakage of organic material to atmosphere using best practices (40 CFR 60.481, *see also* 40 CFR 61.241; 264.1031-91).

First draw: The water that immediately comes out when a tap is first opened. This water is likely to have the highest level of lead contamination from plumbing materials (EPA-94/04).

First federal official: The first federal representative of a participating agency of the National Response Team to arrive at the scene of a discharge or a release. This official coordinates activities under the NCP and may initiate, in consultation with the OSC, any necessary actions until the arrival of the predesignated OSC. A state with primary jurisdiction over a site covered by a cooperative agreement will act in the stead of the first federal official for any incident at the site (40 CFR 300.5-91).

First food use: Refers to the use of a pesticide on a food or in a manner which otherwise would be expected to result in residues in a food, if no permanent tolerance, exemption from the requirement of a tolerance, or food additive regulation for residues of the pesticide on any food has been established for the pesticide under section 408 (d) or (e) or 409 of the Federal Food, Drug, and Cosmetic Act (40 CFR 166.3-91).

First order inactivation kinetics: The rate of microbial inactivation ($-dN/dt$) is directly proportional to the concentration of surviving organisms (N) at a given time (t) (EPA-88/09a).

First order reaction:
(1) A reaction in which the rate of disappearance of a chemical is directly proportional to the concentration of the chemical and is not a function of the concentration of any other chemical present in the reaction mixture (cf. chemical reaction) (40 CFR 796.3700-91).
(2) The rate of concentration C decrease with time for a compound is proportional to concentration C, ie, $dC/dt = -kC$, where d is the differential symbol, t is the time and k is

a constant (*see* chemical reaction for more related terms).

First responder: The first trained personnel to arrive on the scene of a hazardous material incident. Usually officials from local emergency services, firefighters, and police (Course 165.5)

Fish: Fish related terms include:
- Game fish
- Rough fish

Fish eyes: Translucent spots in the finished paper. These imperfections are caused by under-fibered portions of stock (EPA-83).

Fish kill: The destruction of fish in a water body -- in winter, due to prolonged ice and snow cover or freezing of the water; in summer, due to oxygen deficiency resulting from excessive organic matter; in any season, due to toxic pollutants or disease (DOI-70/04).

Fish ladder: A device to facilitate the movement of migrating fish over a dam; it may consist of a stairlike series of small ponds connected by flowing water (DOI-70/04).

Fish meal: A ground, dried product made from fish or shellfish or parts thereof, generally produced by cooking raw fish or shellfish with steam and pressing the material to obtain the solids which are then dried (EPA-74/06).

Fish solubles: A product extracted from the residual press liquor (stick water) after the solids are removed for drying (fish meal) and the oil extracted by centrifuging. This residue is generally condensed to 50% solids and marketed as condensed fish solubles (EPA-74/06).

Fish or wildlife: Any member of the animal kingdom, including without limitation any mammal, fish, bird (including any migratory, nonmigratory, or endangered bird for which protection is also afforded by treaty or other international agreement), amphibian, reptile, mollusk, crustacean, arthropod or other invertebrate, and includes any part, product, egg, or offspring thereof, or the dead body or parts thereof (ESA3-16USC1531).

Fishmouth: The uneven mating of two geomembranes to be joined wherein the upper sheet has excessive length that prevents it from being bonded flat to the lower sheet. The resultant opening is often referred to as a fishmouth (EPA-91/05).

Fission (or nuclear fission):
(1) The splitting of a heavy atomic nucleus into two nuclei of lighter elements, accompanied by the release of energy and generally one or more neutrons. Fission can occur spontaneously or be induced by neutron bombardment (DOE-91/04).
(2) The division of a heavy radioactive atomic nucleus into smaller parts. The smaller parts are the so-called **fission products** (cf. fussion).

Fission product: *See* definition under fission (cf. decay product).

Five day BOD (BOD5): *See* definition under biochemical oxygen demand.

Five (5)-year, 6-hour precipitation event: The maximum 6-hour precipitation event with a probable recurrence interval of once in 5 years as established by the U.S. Department of Commerce, National Oceanic and Atmospheric Administration, National Weather Service, or equivalent regional or rainfall probability information (40 CFR 440.141-91).

Fix: A step in photoprocessing whereby the silver halides are removed from the emulsion using a solvent such as sodium thiosulfate (EPA-80/10).

Fix, sample: A sample is "fixed" in the field by adding chemicals that prevent water quality indicators of interest in the sample from changing before laboratory measurements are made (EPA-94/04).

Fixed bed: A filter or adsorption bed where the entire media is exhausted before any of the media is cleaned (EPA-75/02d).

Fixed capacitor: A capacitor having a definite capacitance value that can not be adjusted (cf. capacitor) (EPA-83/03).

Fixed capital cost: The capital needed to provide all of the depreciable components (*see* cost for more related terms) (40 CFR 51.301-91).

Fixed capital cost of the new components: As used in 40 CFR 60.15, includes the fixed capital cost of all depreciable components which are or will be replaced pursuant to all continuous programs of component replacement which are commenced within any 2-year period following December 30, 1983 (40 CFR 60.666-91).

Fixed carbon:
(1) The ash free carbonaceous material that remains after volatile matter is driven off a dry solid sample (OME-88/12).
(2) The non-volatile organic portion of waste. For fixed carbon, the combustion reaction is a solid-phase reaction that occurs primarily in the waste bed (although some materials may burn in suspension). Key parameters are bed parameters, solid retention time, and mechanical turbulence in the bed. The solid retention time is the time that the waste bed remains in the primarily chamber. Mechanical turbulence of the bed is needed to expose all the solid waste to oxygen for complete burnout. Without mechanical turbulence, the ash formed during combustion can cover the unburned waste and prevent the oxygen necessary for combustion from contacting the waste (cf. volatile matter) (EPA-89/03b).
(3) *see* carbon for more related terms.

Fixed cost: The cost that does not vary with level of output of a production facility (e.g., administrative costs, building rent, and mortgage payments) (*see* cost for more related terms) (OTA-89/10).

Fixed film biological process (or fixed film reactor): A layer of a biological film which is adhered to a solid medium in a wastewater treatment reactor.

Fixed hearth incinerator: *See* synonym, starved air incinerator.

Fixed grate (or stationary grate): A grate without moving parts, also called a stationary grate (OME-88/12). A fixed grate through which no air passes is called a dead plate (*see* grate for more related terms).

Fixed onshore and offshore oil well drilling facilities: Including all equipment and appurtenances related thereto used in drilling operations for exploratory or development wells, but excluding any terminal facility, unit or process integrally associated with the handling or transferring of oil in bulk to or from a vessel (cf. mobile onshore...) (40 CFR 112, App.).

Fixed packer: An adjunct of a refuse container system which compacts refuse at the site of generation into a detachable container (EPA-83).

Fixed plant: Any nonmetallic mineral processing plant at which the processing equipment specified in 40 CFR 60.670(a) is attached by a cable, chain, turnbuckle, bolt or other means (except electrical connections) to any anchor, slab, or structure including bedrock (40 CFR 60.671-91).

Fixed roof: A cover that is mounted to a tank or chamber in a stationary manner and which does not move with fluctuations in wastewater levels (40 CFR 60.691-91, *see also* 40 CFR 61.341-91).

Fixed roof tank: A cylindrical shell with a permanently affixed roof (40 CFR 52.741-91).

Fixed source: For the purpose of these guidelines, a stationary facility that converts fossil fuel into energy, such as steam, hot water, electricity, etc. (*see* source for more related terms) (40 CFR 247.101-91).

Flame:
(1) A visible radiation emitted from the rapid oxidation of materials. When oxidation is rapid, the temperature of the material rises rapidly due to the inability of transferring heat to the surroundings as rapidly as it is produced by the oxidation reaction. The material thus emits visible radiation (light) which is referred as a flame.
(2) A chemically reacting, radical-rich gas flow which propagates through space at temperatures generally above 1800° F. This gas flow may consist of atomic hydrogen (H), atomic oxygen (O), atomic chlorine (Cl), hydroxyl radicals (OH.), and methyl radicals (CH_3.) in carbon-hydrogen-oxygen systems, and chloroxy radicals (ClO.) in chlorine-containing systems. There are always several sequential reaction pathways with the reaction rate controlled by many chemical and physical factors such as chemical kinetics, mixing, air-fuel ratio, etc.

Other flame related terms include:
- Diffusion flame
- Laminar flame
- Premixed flame
- Turbulent flame

Flame and non-flame destruction: An incineration process involves two fundamental modes, namely the flame mode and the non-flame mode of thermal decomposition. Flame destruction of waste involves reaction and destruction within the flame itself, while for the non-flame mode, the flame serves as the heat source for the hot combustion gas to continue decomposition beyond the flame. The high temperatures required to promote flame or non-flame mode reactions may be generated from actual combustion of the organic constituents in the waste or of co-fired auxiliary fuel.

The non-flame zone of an incinerator surrounds and extends beyond the flame. This non-flame zone is characterized by temperatures generally between 650° F and the flame temperature. It is also a reacting gas flow consisting of a mixture of nitrogen, oxygen, carbon dioxide, water, hydrogen chloride, chlorine, organic compounds, intermediates, and radicals. Although the reaction conditions are much less severe compared to flame conditions, the time spent by the reactants under the non-flame thermal decomposition conditions is much greater than the time spent in the flame (a few seconds versus milliseconds).

Flame atomic absorption spectrometry (FAAS): FAAS or direct aspiration determinations, as opposed to ICP, are normally completed as single element analyses and are relatively free of inter element spectral interferences. Either a nitrous oxide acetylene or air-acetylene flame is used as an energy source for dissociating the aspirated sample into the free atomic state making analyte atoms available for absorption of light. In the analysis of some elements the temperature or type of flame used is critical. If the proper flame and analytical conditions are not used, chemical and ionization interferences can occur (*see* metal analyzer for more related terms) (SW-846).

Flame combustion: Flame combustion includes yellow flame and blue flame:
(1) **Yellow flame:** When mixing fuel and air, two different mechanisms of combustion can occur. A **luminous (yellow) flame** results when air and fuel flowing through separate ports are ignited at the burner nozzle. The yellow flame results from thermal cracking of the fuel. Cracking occurs when hydrocarbons are intensely heated before they have a chance to combine with oxygen. The cracking releases both hydrogen and carbon which diffuse to the flame to form CO_2 and H_2O. The carbon particles give the flame the yellow appearance. If incomplete combustion occurs from temperature cooling or if there is insufficient oxygen, soot and black smoke will form.
(2) **Blue flame:** Blue flame combustion occurs when the fuel and air are premixed in front of the burner nozzle. This produces a short, intense, blue flame. The reason for the different flame is that the fuel-air mixture is gradually heated. The hydrocarbon molecules are slowly oxidized, going from aldehydes and ketones to CO_2 and H_2O. No cracking occurs and no carbon particles are formed. Incomplete combustion results in the release of the intermediate, partially oxidized compounds. Blue haze and odors are emitted from the stack (EPA-81/12, p3-7).
(3) *See* combustion for more related terms.

Flame front: The hypothetical moving plane across which rapid combustion is occurring (cf. deflagration or detonation) (EPA-83).

Flame hardened: A surface hardened by controlled torch heating followed by quenching with water or air (EPA-83/06a).

Flame incineration: Burning of combustible pollutants whose concentration is in the flammability limit. Operating temperatures are in the 2000-2500° F range (cf. flare) (EPA-84/09).

Flame ionization detector (FID): *See* definition under GC/FID.

Flame photometric analyzer: *See* synonym, photoluminescence

Flame port: An opening between the primary chamber and mixing chamber of a multiple chamber incinerator through which combustion gases pass (EPA-89/03b).

Flame safeguard: The device which controls the burner ignition process. When the burner is first started, the burner blower starts and when it reaches full speed, a purge timer starts. When the purge timer times out, the flame safeguard energizes the pilot relay that opens the pilot fuel supply and ignitor. When the pilot lights, a flame detector (either an ultraviolet scanner [gas or oil] or flame rod circuit [gas only]) detects the pilot flame and causes the main flame relay to activate the fuel supply to the main burner. The pilot then ignites the main burner. The flame detector continues to operate and shuts the burner down if the main burner fails. Additionally, if the air supply is lost, both pilot and flame relays shut off the fuel supply. The pilot usually is ignited for no more than 15 seconds (interrupted pilot). If the main burner does not ignite during the pilot ignition period, the flame safeguard system shuts the entire system down by closing the solenoid shut-off valve and turning off the burner blower (*see* burner component for more related terms) (EPA-89/03b).

Flame spraying: The process of applying a metallic coating to a workpiece whereby finely powdered fragments or wire, together with suitable fluxes, are projected through a cone of flame onto the workpiece (EPA-83/06a).

Flame zone:
(1) That portion of the combustion chamber in a boiler occupied by the flame envelope (40 CFR 60.561-91, *see also* 40 CFR 60.611; 60.661; 264.1031-91).
(2) A three-dimensional space in an incinerator where emission of visible light is accompanied by large exothermic chemical reaction, radical concentrations, temperature gradient and rapid reaction rate (EPA-88/12).

Flap damper: A damper consisting of one or more blades each pivoted about one edge (*see* damper for more related terms).

Flare:
(1) A (tall) stack for burning excess quantities of waste combustible gases typically present at oil refineries (EPA-84/09). Flares are simply burners that have been designed to handle varying rates of fuel while burning smokelessly. In general, flares can be classified as either elevated or ground level (EPA-81/12, p3-16). Flares are widely used in the petroleum and chemical industries to dispose of waste gases. Increased emphasis on potentially hazardous (volatile organic compound) emission will require high flare efficiencies.
(2) A control device that burns hazardous materials to prevent their release into the environment; may operate continuously or intermittently, usually on top a stack (EPA-94/04).

Flareback: A burst of flame from a furnace in a direction opposed to the normal gas flow. It usually occurs when accumulated combustible gases ignite (SW-108ts).

Flaring: The burning of methane emitted from collection pipes at a landfill (OTA-89/10).

Flash dry: The process of drying a wet organic material by passing through a high temperature zone at such a rate that the water is rapidly evaporated but the organic material, protected by the boiling point of water, is not overheated (OME-88/12).

Flash evaporation: An evaporation using steam heated tubes with feed material under high vacuum. The feed material flashes off when it enters the evaporation chamber (EPA-83/06a).

Flash flood: A sudden and violent flood after a heavy rain (*see* flood for more related terms) (DOI-70/04).

Flash point temperature:
(1) The lowest temperature at which a material (a liquid or solid) gives off sufficient vapor to form a ignitable vapor-air mixture near the surface of the liquid or solid (to flash into a momentary flame when ignited) (OME-88/12).
(2) The minimum temperature at which the vapor-air mixture above a volatile substance ignites when exposed to a flame (ETI-92).
(3) *See* temperature for more related terms.

Flashing:
(1) In die casting, the fin of metal that results from leakage between the mating die surfaces (EPA-85/10a).
(2) In constructing a landfill liner, the molten extrudate or sheet material which is extruded beyond the die edge or molten edge, also called squeeze-out (EPA-89/09).
(3) In molding, the molten extrudate or sheet material which is extruded beyond the die edge or molten edge, also called **squeeze-out** (EPA-91/05).

Flashoff area: The portion of a surface coating operation between the coating application area and bake oven (40 CFR 60.311-91, *see also* 40 CFR 60.391; 60.441; 60.451; 60.711; 60.741-91).

Flashover: The discharge of ignition voltage across the surface of the distributor cap and/or rotor rather than at the spark plug gap (40 CFR 85.2122 (a)(7)(ii)(A)-91, *see also* 40 CFR 85.2122(a)(8)(ii)(F); 85.2122(a)(9)(ii)(B)-91).

Flask:
(1) A rectangular frame open at top and bottom used to retain molding sand a pattern (EPA-85/10a).
(2) A unit of measurement for mercury, 76 pounds (EPA-82/05).

Flat flame burner: A burner terminating in a substantially rectangular nozzle, from which fuel and air discharged in a flat stream (*see* burner for more related terms).

Flat glass:
(1) A glass made of soda-lime recipe and produced into continuous flat sheets and other products listed in SIC 3211 (40 CFR 60.291-91).
(2) Included in this category are sheet or window glass, plate glass, and laminated glass (EPA-83).

(3) *See* glass for more related terms.

Flat mill: Those steel hot forming operations that reduce heated slabs to plates, strip and sheet, or skelp (40 CFR 420.71-91).

Flat terrain: A terrain whose rise within 5 kilometers of a stack is less than or equal to 10 percent of the physical stack height (*see* terrain for more related terms) (EPA-90/04).

Fleshing: The mechanical removal of flesh and fatty substances from the underside of a hide prior to tanning. In the case of sheepskin tanning, fleshing is often accomplished after the tanning process (EPA-82/11).

Flexible fuel engine: *See* synonym, flexible fuel vehicle.

Flexible fuel vehicle (or flexible fuel engine): Any motor vehicle (or motor vehicle engine) engineered and designed to be operated on a petroleum fuel, a methanol fuel, or any mixtures of the two (*see* vehicle for more related terms) (40 CFR 86.090.2-91).

Flexible membrane liner (FML): A synonymous term for geomembrane, generally, a high density polyethylene sheet used in the containment of solid, liquid and vapor materials of landfill waste (*see* liner for more related terms) (EPA-89/09).

Flexible vinyl and urethane product: Those products, except for resilient floor coverings (977 Standard Industry Code 3996) and flexible packaging, that are more than 50 micrometers (0.002 inches) thick, and that consist of or contain a vinyl or urethane sheet or a vinyl or urethane coated web (40 CFR 60.581-91).

Flexographic ink: A quick drying, low viscosity ink based on volatile solvents that are used in the flexographic printing process. Flexographic inks can be water based (*see* ink for more related terms) (EPA-79/12a).

Flexographic printing: The application of words, designs, and pictures to a substrate by means of a roll printing technique in which the pattern to be applied is raised above the printing roll and the image carrier is made of elastomeric materials (40 CFR 52.741-91).

Flexographic printing line: A printing line in which each roll printer uses a roll with raised areas for applying an image such as words, designs, or pictures to a substrate. The image carrier on the roll is made of rubber or other elastomeric material (40 CFR 52.741-91).

Flight conveyer: A drag conveyer that has rollers interspersed in its pull chains to reduce friction (*see* conveyer for more related terms) (SW-108ts).

Flint glass:
(1) Clear glass (OTA-89/10).
(2) A lead-containing colorless glass (EPA-83).
(3) *See* glass for more related terms.

Flint glass cutlet: A particulate glass material that contains no more than 0.1 weight % FeO_3, or 0.0015 weight % Cr_2O_3 as determined by chemical analysis (EPA-83).

Float: The proper level or volume of chemicals, and water that is maintained in any wet process unit (vats, drums, or processors) within the tannery (EPA-82/11).

Floating aquatic plants: Rooted plants that wholly or partly float on the surface of the water, e.g., water lilies, water hyacinth and duckweed (DOD-78/01; LBL-76/07-water).

Floating roof: A roof on a stationary tank, reservoir, or other container which moves vertically upon change in volume of the stored material (40 CFR 52.741-91, *see also* 40 CFR 60.111; 61.341; 60.691-91).

Floating slab: *See* synonym, French drain.

Float gauge: A device for measuring the elevation of the surface of a liquid, the actuating element of which is a buoyant float that rests on the surface of the liquid and rises or falls with it. The elevation of the surface is measured by a chain or tape attached to the float (EPA-83/03).

Floating matter: In wastewater, it includes froth, oil, and floating solids. Froth results from detergent cleaning, certain mineral flotation processes, pulp and paper manufacture, and municipal sewage. Oil results from chemical processes, refining, machinery lubrication, and metal-working. Floating solids can be pulp or textile fibers, fine coke, food pulps, bark, or sawdust (DOE-70/04).

Floating slab: *See* synonym, French drain.

Floc: A clump of solids formed in sewage by biological or chemical action (EPA-94/04).

Flocculant: An agent that induces or promotes flocculation or produces floccules or other aggregate formation, especially in clays and soils (flocculant flocculating agent) (EPA-82/05).

Flocculant flocculating agent: A substance which produces flocculation (EPA-82/05).

Flocculate: To cause to aggregate or to coalesce into small lumps or loose clusters, e.g., the calcium ion tends to flocculate clays (EPA-82/05).

Flocculation (or coagulation):
(1) A process to enhance agglomeration or collection of smaller floc particles into larger, more easily settleable particles through gentle stirring by hydraulic or mechanical means (40 CFR 141.2-91).
(2) Process by which clumps of solids in water or sewage aggregate through biological or chemical action so they can be separated from water or sewage (EPA-94/04).

Flocculator: An apparatus designed for the formation of floc in water or sewage (EPA-83/03).

Flock: A natural or synthetic fiber added to lead-acid battery paste as a stiffening agent (EPA-84/08).

Flood (or flooding):
(1) A general and temporary condition of partial or complete

inundation of normally dry land areas from the overflow of inland and/or tidal waters, and/or runoff of surface waters from any source, or flooding from any other source (40 CFR 6-App/A-91).
(2) A stream flow that greatly exceeds the average stream flow, whether or not it overtops the channel banks (DOI-70/04).
(3) The excessive entrainment of liquid in a packed column or tray tower. This leads directly to a condition of high pressure drop (EPA-84/09).

Other flood related terms include:
- Annual flood
- Average annual
- Base flood
- Base floodplain
- Based flood
- Daily flood peak
- Flash flood
- Momentary flood peak

Flood coating: The generous application of a bodied chemical compound, or chemical adhesive compound to protect exposed yarns in scrim reinforced geomembranes (EPA-91/05).

Flood crest: The highest elevation reached by flood waters in a flood event. It is commonly measured in feet above an accepted datum, such as flood stage (DOI-70/04).

Flood damage: The economic loss caused by flood, including inundation, erosion, and sediment deposition; the loss may be evaluated in terms of cost of replacement, repair, or rehabilitation; decrease in market or sales value; or resulting decrease in income or production (DOI-70/04).

Flood event: A series of flows constituting a distinct progressive rise culminating in a crest, together with the recession that follows the crest (DOI-70/04).

Flood stage: The elevation of the water surface (selected by local usage or by an investigator) above which the stream is considered to be in flood. Commonly it is the stage at which damage begins (DOI-70/04).

Flooding: *See* synonym, flood.

Flooding velocity: The gas velocity or narrow range of gas velocities in a packed bed or plate tower scrubber at which (for a given packing or plate design and liquid flow rate) the liquid flow down the column is impeded, and a liquid layer is formed at tip of the column. Eventually, liquid is blown out of top of the column (EPA-81/09).

Floodplain:
(1) The lowland and relatively flat areas adjoining inland and coastal waters and other floodprone areas such as offshore islands, including at a minimum, that area subject to a one percent or greater chance of flooding in any given year. The base floodplain shall be used to designate the 100-year floodplain (one percent chance floodplain). The critical action floodplain is defined as the 500-year floodplain (0.2 percent chance floodplain) (40 CFR 6-App/A-91, *see also* 40 CFR 257.3.1-91).
(2) The portion of a river valley that becomes covered with water when the river overflows its banks at flood stage (DOE-91/04).
(3) The flat or nearly flat land along a river or stream or in a tidal area that is covered by water during a flood (EPA-94/04).

Floor insulation: A material, primarily designed to resist heat flow, which is installed between the first level conditioned area of a building and an unconditioned basement, a crawl space, or the outside beneath it. Where the first level conditioned area of a building is on a ground level concrete slab, floor insulation includes such a material installed around the perimeter of or on the slab. In the case of mobile homes, floor insulation also means skirting to enclose the space between the building and the ground (40 CFR 248.4-p-91).

Floor sweep: Capture of heavier-than-air gases that collect at floor level (EPA-94/04).

Floodproofing: The modification of individual structures and facilities, their sites, and their contents to protect against structural failure, to keep water out or to reduce effects of water entry (40 CFR 6-App/A-91).

Flooring felt: An asbestos-containing product which is made of paper felt intended for use as an underlayer for floor coverings, or to be bonded to the underside of vinyl sheet flooring (40 CFR 763.163-91).

Flora:
(1) The plant life characteristic of a region (EPA-76/03).
(2) The entire plant life of a region (LBL-76/07-water).

Flotation:
(1) In wastewater treatment, an EPA listed treatment technology, list code, T42. Flotation is the collection of substances immersed in wastewater by taking advantage of differences in specific gravities, or else by entrapment of solid particles with air causing them to rise to the surface for subsequent disposal (DOI-70/04).
(2) In mining, the process is a physical-chemical method (carried out in a wet environment) for concentrating finely ground ores. A necessary and important step to prepare the ore for flotation is crushing and grinding, which is designed for each particular application to separate the individual mineral particles that make up the ore by reducing them to their natural grain size. Also, in some cases, the removal of slimes is necessary before flotation.
(3) cf. gravity flotation.

Flotation agent: A substance or chemical which alters the surface tension of water or which makes it froth easily. The reagents used in the flotation process include pH regulators, slime dispersants, re-surfacing agents, wetting agents, conditioning agents, collectors, and frothers (EPA-82/10).

Flour: A term applied to the fine fibers or fiber fragments of a pulp. They are also known as fines (cf. fine) (EPA-87/10).

Flow: Moving along in a stream as water or different liquid. Other

flow related terms include:
- Confined flow
- Open channel flow
- Cross flow
- Cyclonic flow
- Dry weather flow
- Pipe flow
- Saturated flow
- Steady and unsteady flow
- Stoke flow
- Uniform and non-uniform flow

Flow backwater: The surface curve of a stream of water when backed up by a dam or other obstruction (M&EI-72).

Flow channel: The appendages used for conditioning and distributing molten glass to forming apparatuses and are a permanently separate source of emissions such that no mixing of emissions occurs with emissions from the melter cooling system prior to their being vented to the atmosphere (40 CFR 60.291-91).

Flow chute: A chute is a channel with so deep in a grade that a uniform flow can take place at a depth less than the flow critical depth (M&EI-72).

Flow coating: A method of applying coatings in which the part is carried through a chamber containing numerous nozzles which direct unatomized streams of coatings from many different angles onto the surface of the part (40 CFR 60.311-91).

Flow control: A legal or economic means by which waste is directed to particular destinations. For example, an ordinance requiring that certain wastes be sent to a combustion facility is waste flow control (EPA-89/11).

Flow control ordinance: The ordinance that requires delivery of collected MSW (municipal solid waste) to a specific management facility (OTA-89/10).

Flow critical depth: The depth of a flow at the discharge outlet either from a free discharge or a channel. This depth has the minimum specific energy (M&EI-72).

Flow drawdown: The transition from a condition of a uniform flow in the conduit to the discharge at a free outlet (M&EI-72).

Flow head meter: The meter uses Bernoulli's theorem to calculate the pipe flow pressure head. It's types include critical flow nozzle, elbow meter, flow nozzle, orifice meter, venturi meter.

Flow (hydraulic) jump: When water moving at a high velocity in a comparatively shallow stream strikes water having a substantial depth, there is likely to be a rise in the water surface of the stream, forming the so called a hydraulic jump (M&EI-72).

Flow indicator: A device which indicates whether gas flow is present in a vent stream (40 CFR 60.611-91, *see also* 40 CFR 60.661; 264.1031-91).

Flow line (or streamline): The path that a particle of water follows in its movement (Course 165.7).

Flow measurement: The direct-discharge and velocity-area are two major methods used for the measurements of flowing fluids. For the direct-discharge measurements, the following methods and apparatuses have been used, the choice depending on the flow conditions encountered:
(1) Weighing the discharge
(2) Volumetric discharge measurement
(3) Orifice
(4) Standard weirs

Flow meter: An instrument for measuring the rate of flow of a fluid moving through a pipe or duct system. The instrument is calibrated to give volume or mass rate of flow (cf. flow rate meter) (LBL-76/07-air).

Flow model: A mathematical model of the effluent wastewater flow, developed through the use of multiple linear regression techniques (EPA-82/10).

Flow nozzle meter: A water meter of the differential medium type in which the flow through the primary element or nozzle produces a pressure difference or differential head, which the secondary element or float tube then uses as an indication of the rate of flow (EPA-82/11f).

Flow proportional composite sample is composed of grab samples collected continuously or discretely in proportion to the total flow at time of collection or to the total flow since collection of the previous grab sample. The grab volume or frequency of grab collection may be varied in proportion to flow (40 CFR 471.02-91).

Flow rate:
(1) The volume per time unit given to the flow of gases or other fluid substance which emerges from an orifice, pump, turbine or passes along a conduit or channel (40 CFR 146.3-91).
(2) The rate, expressed in gallons-or liters-per-hour, at which a fluid escapes from a hole or fissure in a tank. Such measurements are also made of liquid waste, effluent, and surface water movement (EPA-94/04).

Flow rate meter (or flowmeter): An instrument for measuring the rate of flow of a fluid moving through a pipe or duct system. The instrument is calibrated to give volume or mass rate of a flow (EPA-83/06). A flow can be expressed in volumetric or weight units per unit time. Thus, gases are generally measured in cubic feet per minute, steam in pounds per minutes, liquid in gallons per minute. The flow rate may be measured directly by attaching a rate device such as tachometer to volumetric meter (cf. liquid rate meter or gas rate meter). Types of the flow rate meter include:
(1) Liquid flow rate meter
  (A) Coriolis flowmeter (*see* synonym, mass flowmeter)
  (B) Mass flowmeter
  (C) Orifice meter
  (D) Positive displacement meter
  (E) Rotameter
  (F) Vortex shedding meter
(2) Gaseous flow rate meter
(3) Solid (sludge) rate meter
  (A) Conveyer weighing system
  (B) Level indicator

(C) Momentum flowmeter
(D) Nuclear absorption
(E) Stationary weight indicator
(F) Volumetric method

Flow regime: A term applied to describe the characteristics of particle motion in a fluid flow. It includes (cf. Reynolds Number) (EPA-84/09):
(1) Laminar (Stokes) flow
(2) Transition (intermediate) flow
(3) Turbulent (Newton) flow

Flow through: A continuous or an intermittent passage of test solution or dilution water through a test chamber or a holding or acclimation tank, with no recycling (40 CFR 795.120-91, *see also* 40 CFR 797.1300; 797.1330; 797.1400; 797.1930; 797.1950; 797.1600; 797.1970-91).

Flow through process tank: A tank that forms an integral part of a production process through which there is a steady, variable, recurring, or intermittent flow of materials during the operation of the process. Flow-through process tanks do not include tanks used for the storage of materials prior to their introduction into the production process or for the storage of finished products or by-products from the production process (40 CFR 280.12-91).

Flow through test: A toxicity test in which water is renewed continuously in the test chambers, the test chemical being transported with the water used to renew the test medium (40 CFR 797.1350-91, *see also* 40 CFR 797.1350-91).

Flow velocity meter: Flow velocity meter related terms include:
- Anemometer
- Current meter
- Lysimeter

Flow work (or displacement work): The work which is required to push the fluid across the system boundary or to push the fluid in or out of a control volume (Jones-p98; Wark-p124).

Flowmeter:
(1) A gauge that shows the speed of wastewater moving through a treatment plant. Also used to measure the speed of liquids moving through various industrial processes (EPA-94/04).
(2) *See alos* synonym, flow rate meter.

Flue: Any passage designed to carry combustion gases and entrained particulates (cf. stack) (SW-108ts).

Flue dust: Solid particles (smaller than 100 microns) carried in the products of combustion (SW-108ts).

Flue fed incinerator: An incinerator that is charged through a shaft that functions as a chute for charging waste and has a flue to carry the products of combustion (*see* incinerator for more related terms) (SW-108ts).

Flue gas:
(1) The air coming out of a chimney after combustion in the burner it is venting. It can include nitrogen oxides, carbon oxides, water vapor, sulfur oxides, particles and many chemical pollutants (EPA-94/04).
(2) *See also* synonym, combustion gas.

Flue gas desulfurization (FGD):
(1) A technology that employs a sorbent, usually lime or limestone, to remove sulfur dioxide from the gases produced by burning fossil fuels. Flue gas desulfurization is current state-of-the art technology for major SO_2 emitters, like power plants (EPA-94/04).
(2) One of SO_2 emission reduction techniques (*see* sulfur oxide emission control for control structure). FGD refers to the removal of SO_2 from the process exhaust stream. It is the most popular technology used for controlling sulfur oxide emissions from combustion sources. The majority of FGD systems have been applied to combustion sources such as utility and some industrial coal-fired boilers. FGD systems are also used to reduce SO_2 emissions from some industrial plants such as smelters, acid plants, refineries, and pulp and paper mills.

FGD scrubbing processes can either be wet or dry and can be further grouped into:
(1) FGD wet scrubbing process
   (A) FGD throwaway (nonregenerable) process
   (B) FGD regenerable process
(2) FGD dry scrubbing process

Flue gas recirculation (FGR): One of NO_x emission reduction techniques (*see* nitrogen oxide emission control for control structure). A portion (10 to 30%) of the flue gas exhaust is recycled back into the main combustion chamber by removing it from the stack breeching and mixing it with the secondary air windbox. The gas enters directly into the combustion zone. This recirculated gas lowers the flame temperature and dilutes the oxygen content of the combustion air, thus lowering thermal NO_x emissions (EPA-81/12, p7-10).

Flue gas washer or scrubber (or gas washer): Equipment for removing objectionable constituents from the products of combustion by means of spray, wet baffles, etc (EPA-83).

Flue gas desulfurization: A technology which uses a sorbent, usually lime or limestone, to remove sulfur dioxide from the gases produced by burning fossil fuels. Flue gas desulfurization is current the state-of-the art technology in use by major SO2 emitters, e.g., power plants (EPA-94/04).

Flue gas scrubber or washer: A type of equipment that removes fly ash and other objectionable materials from flue gases by the use of sprays, wet baffles, or other means that require water as the primary separation mechanism (SW-108ts).

Fluid: Any material or substance which flows or moves whether in a semisolid, liquid, sludge, gas, or any other form or state (40 CFR 144.3-91, *see also* 40 CFR 144.3; 146.3; 147.2902-91). Other fluid related terms include:
- Compressible fluid
- Incompressible fluid

Fluid bed roaster: A type of roasters in which the material is suspended in air during roasting (EPA-83/03a).

Fluid catalytic cracking unit: A refinery process unit in which petroleum derivatives are continuously charged; hydrocarbon molecules in the presence of a catalyst suspended in a fluidized bed are fractured into smaller molecules, or react with a contact material suspended in a fluidized bed to improve feedstock quality for additional processing; and the catalyst or contact material is continuously regenerated by burning off coke and other deposits. The unit includes the riser, reactor, regenerator, air blowers, spent catalyst or contact material stripper, catalyst or contact material recovery equipment, and regenerator equipment for controlling air pollutant emissions and for heat recovery (40 CFR 60.101-91).

Fluid catalytic cracking unit catalyst regenerator: One or more regenerators (multiple regenerators) which comprise that portion of the fluid catalytic cracking unit in which coke burn-off and catalyst or contact material regeneration occurs, and includes the regenerator combustion air blower(s) (40 CFR 60.101-91).

Fluid potential: The mechanical energy per unit mass of water or other fluid at any given point in space and time, with respect to an arbitrary state and datum (Course 165.7).

Fluid statics: Fluid statics is concerned with the static properties and behavior of fluids. In the case of liquids, this subject is known as hydrostatics; in the case of gases, it is called pneumatics (Perry-73).

Fluid temperature: The temperature at which a standard ash cone fuses down into a flat layer on the test base, when heated in accordance with a prescribed procedure (*see* temperature for more related terms).

Fluidized: A mass of solid particles that is made to flow like a liquid by injection of water or gas is said to have been fluidized. In water treatment, a bed of filter media is fluidized by backwashing water through the filter (EPA-94/04).

Fluidized bed combustion (FBC):
(1) Oxidation of combustible materials within a bed of solid, inert (non-combustible) particles which under the action of vertical hot air flow will act as a fluid.
(2) A combustion process in which heat is transferred from finely divided particles, such as sand, to combustible materials in a combustion chamber. The materials are supported and fluidized by a column of moving air (SW-108ts).
(3) For the control of SO_2 emissions (*see* sulfur oxide emission control for control structure). A grid supports a bed of coal and limestone (or dolomite) in the firebox of the boiler. Combustion air is forced upward through the grid suspending the coal and limestone bed in a fluid-like motion. Natural gas is used to ignite the pulverized coal. Once the coal is ignited, the gas is turned off. The sulfur in the coal is oxidized to SO_2 and consequently combined with the limestone to form calcium sulfate ($CaSO_4$). The $CaSO_4$ and flyash particulate matter are usually collected in a baghouse or electrostatic precipitator (EPA-81/12, p8-5).

Other fluidized bed combustion related terms include:
- Bubbling fluidized bed combustor
- Circulating fluidized bed combustor

Fluidized bed combustion technology: The combustion of fuel in a bed or series of beds (including but not limited to bubbling bed units and circulating bed units) of limestone aggregate (or other sorbent materials) in which these materials are forced upward by the flow of combustion air and the gaseous products of combustion (40 CFR 60.41b-91, *see also* 40 CFR 60.41c-91).

Fluidized bed incinerator:
(1) Fluidized bed incinerators may be either circulating or bubbling bed designs. Both types consist of a single refractory lined combustion vessel partially filled with inert granular materials such as sand, alumina, calcium carbonate, or other materials. Combustion air is supplied through a distributor plate at the base of the combustor at a rate sufficient to fluidize (bubbling bed) or entrain the bed material (circulating bed). In the circulating bed design, air velocities are higher and the solids are blown overhead, separated in a cyclone, and returned to the combustion chamber. Operating temperatures are normally maintained in the 760 to 870 C (1400 to 1600° F) range and excess air requirements range from 20 to 40 percent (40 CFR 60.51a-91).
(2) An incinerator that uses a bed of hot sand or other granular material to transfer heat directly to waste. Used mainly for destroying municipal sludge (EPA-94/04).
(3) *See* incinerator for more related terms.

Fluidized bed reactor: A process in which heat is transferred from a churning suspension of finely divided particles, such as sand, to the suspended particles of introduced materials for calcining, heat treatment, regeneration or combustion (cf. incinerator, fluidized bed) (EPA-83).

Flume: A natural or man-made channel that diverts water (EPA-94/04).

Flume wastewater: The discharge of flume water which is employed to convey beets into the beet sugar processing plant (*see* wastewater for more related terms) (EPA-74/01a).

Fluorescence analyzer: One of continuous emission monitors (*see* continuous emission monitor or luminescence analyzer for various types). Fluorescence is a photoluminescent process in which light energy of a given wavelength is absorbed and light energy of a different wavelength is emitted. In this process, the molecule that is excited by the light energy typically remains excited for about 10^{-8} to 10^{-4} second. This period of time is sufficient for the molecule to dissipate some of this energy in the form of vibrational and rotational motions. When the remaining energy is re-emitted as light, the energy of the light will be lower and the light of a longer wavelength (shorter frequency) will be observed (EPA-84/03a).

Fluorescence analysis: *See* synonym, fluorometric analysis.

Fluorescent lamp: An electric discharge lamp in which phosphor materials transform ultraviolet radiation from mercury vapor ionization to a visible light (EPA-83/03).

Fluorescent light ballast: A device that electrically controls fluorescent light fixtures and that includes a capacitor containing 0.1 kg or less of dielectric (40 CFR 761.3-91).

Fluoridation:
(1) The addition of a chemical to increase the concentration of fluoride ions drinking water to reduce the incidence of tooth decay in children (EPA-94/04).
(2) The process of adding fluorine salts (e.g., NaF) to drinking water supplies to prevent dental decay.

Fluoride: Gaseous, solid, or dissolved compounds containing fluorine that result from industrial processes. Excessive amounts in food can lead to fluorosis (cf. total fluoride) (EPA-94/04).

Fluoride formation: A chemical that in concentration of approximately 1.0 mg/L is a preventive of tooth decay. Fluoride may occur naturally in water, or may be added in controlled amounts. Waters that contain excessive fluoride require defluoridation to reduce the fluoride content to an acceptable level (DOI-70/04).

Fluorination: A chemical reaction of a fluorine with other compounds.

Fluorine (F): A halogen element with a.n. 9; a.w. 18.9984; d. 1.505 g/cc; m.p. -219.6° C and b.p. -188.2° C. The element belongs to group VIIA of the periodic table.

Fluorocarbon (FC): Any of a number of organic compounds analogous to hydrocarbons in which one or more hydrogen atoms are replaced by fluorine. Once used in the United States as a propellant for domestic aerosols, they are now found mainly in coolants and some industrial processes. FCs containing chlorine are called chlorofluorocarbons (CFCs). They are believed to be modifying the ozone layer in the stratosphere, thereby allowing more harmful solar radiation to reach the Earth's surface (EPA-94/04).

Fluorometric analysis (fluorescence analysis or fluorometry): A method of chemical analysis by comparing the radiation intensity of a sample with a given radiation intensity of a fluorescing material.

Fluorometry: *See* synonym, fluorometric analysis.

Fluorosis: An abnormal condition caused by excessive intake of fluorine, characterized chiefly by mottling of the teeth (EPA-89/12).

Flush:
(1) To open a cold-water tap to clear out all the water which may have been sitting for a long time in the pipes. In new homes, to flush a system means to send large volumes of water gushing through the unused pipes to remove loose particles of solder and flux (EPA-94/04).
(2) To force large amounts of water through liquid to clean out piping or tubing, storage or process tanks (EPA-94/04).

Flushing: A method of transferring pigments from dispersions in water to dispersions in oil by displacement of the water by oil. The resulting dispersions are known as flushed colors (EPA-79/12a).

Flushing liquor: In the steel industry, the water recycled in the collecting main for the purpose of cooling the gas as it leaves the ovens (*see* liquor for more related terms) (EPA-74/06a).

Flushing liquor circulation tank: Any vessel that functions to store or contain flushing liquor that is separated from the tar in the tar decanter and is recirculated as the cooled liquor to the gas collection system (*see* liquor for more related terms) (40 CFR 61.131-91).

Fluvial: Found or produced in a river (DOE-91/04).

Flux:
(1) The rate of transport of a (air) quantity (NATO-78/10).
(2) Materials added to a fusion process (molten metal) (in a furnace to promote melting) for the purpose of removing impurities from the hot metal (EPA-74/06a).
(3) A flowing or flow (EPA-94/04).

Fluxing: Dissolving or melting of a substance by chemical actions (cf. degassing) (SW-108ts).

Fluxing salt (or covering flux): Sodium chloride or a mixture of equal parts of sodium and potassium chlorides containing varying amounts of cryolite. Used to remove and gather contaminants at the surface of molten scrap (EPA-76/12).

Fly ash:
(1) Suspended particles, charred paper, dust, soot, and other partially oxidized matter carried in the products of combustion (40 CFR 240.101-91).
(2) The component of coal which results from the combustion of coal, and is the finely divided mineral residue which is typically collected from boiler stack gases by electrostatic precipitator or mechanical collection devices (40 CFR 249.04-91).
(3) The ash that is carried out of the furnace by the gas stream and collected by mechanical precipitators, electrostatic precipitators, and/or fabric filters. Economizer ash is included when it is collected with fly ash (40 CFR 423.11-91).
(4) The ash that is noncombustible particles formed primarily during the combustion of coal and is carried out of the furnace by the stream and collected by air pollution control equipment such as mechanical precipitators, electrostatic precipitators, and/or fabric filters. Economizer ash is included when it is collected with fly ash (EPA-84/09; EPA-82/11a).
(5) Non-combustible residual particles expelled by flue gas (EPA-94/04).
(6) *See* ash for more related terms.

Fly ash collector:
(1) Equipment for removing fly ash (particulates) from the combustion gases prior to their discharge to the atmosphere (EPA-83).
(2) A device for removing fly ash from combustion gases (*see* collector for more related terms) (EPA-83).

Fly ash reinjection: Recycling of fly ash (usually containing significant carbon) to the furnace complete burnout of remaining combustibles (*see* ash for more related terms) (EPA-83).

Fly carbon: Fine particles of carbon carried by gases leaving a combustion system; usually also contain ash (*see* carbon for more related terms) (EPA-83).

F:M ratio (food to microorganism ratio): The ratio of organic material (food) to mixed liquid (microorganisms) in an aerated sludge aeration basin (EPA-85/10). Food may be expressed in pounds of suspended solids, COD, or BOD5 added per day to the aeration tank, and microorganisms may be expressed as mixed liquor suspended solids or mixed liquor volatile suspended solids in the aeration tank (EPA-74/05).

Foam: A layer of bubbles on the surface of molten glass (EPA-83).

Foam fractionation: *See* synonym, froth flotation.

Foam-in-place insulation foam: The rigid cellular foam produced by catalyzed chemical reactions that hardens at the site of the work. The term includes spray-applied and injected applications such as spray-in-place foam and pour-in-place (40 CFR 248.4-91).

Foam separation: *See* synonym, froth flotation.

Fog:
(1) Water droplets suspended in the atmosphere near the earth surface, which reduce the horizontal visibility (NATO-78/10).
(2) A loose term applied to visible aerosols in which the dispersed phase is liquid. Formation by condensation is usually implied. In meteorology, a dispersion of water or ice (LBL-76/07-air).

Fog coating (mist coating or uniforming coating): A thin coating applied to plastic parts that have molded-in color or texture or both to improve color uniformity (*see* coating for more related terms) (40 CFR 60.721-91).

Fogging: Applying a pesticide by rapidly heating the liquid chemical so that it forms very fine droplets that resemble smoke or fog. Used to destroy mosquitoes, black flies, and similar pests (EPA-94/04).

Folding boxboard: A paperboard suitable for the manufacture of folding cartons (40 CFR 250.4-91).

Folding stock: Paper made from long pulp fibers (EPA-83).

Folding strength: A test which indicates the physical resistance of the paper to creasing and recreasing (EPA-83).

Fomite: An inanimate object that can harbor or transmit pathogenic organisms (SW-108ts).

Food additive: Any substance the intended use of which results or may reasonably be expected to result, directly or indirectly, in its becoming a component of or otherwise affecting the characteristics of any food (including any such substance intended for use in producing, manufacturing, packing, processing, preparing, treating, packaging, transporting, or holding food), except that such term does not include:
(1) A pesticide chemical in or on a raw agricultural commodity.
(2) A pesticide chemical to the extent that it is intended for use or is used in the production, storage, or transportation of any raw agricultural commodity.
(3) A color additive.
(4) Any substance used in accordance with a sanction or approval granted prior to September 6, 1958, pursuant to the FFDCA, the Poultry Products Inspection Act, or the Federal Meat Inspection Act.
(5) A new animal drug.
(6) A substance that is generally recognized, among experts qualified by scientific training and experience to evaluate its safety, as having been adequately shown through scientific procedures (or, in the case of a substance used in food prior to January, 1, 1958, through either scientific procedures or experience based on common use in food) to be safe under the conditions of its intended use (40 CFR 177.3-91).

Food additive regulation: A regulation issued pursuant to FFDCA section 409 that states the conditions under which a food additive may be safely used. A food additive regulation under this part ordinarily establishes a tolerance for pesticide residues in or on a particular processed food or a group of such foods. It may also specify:
(1) The particular food or classes of food in or on which a food additive may be used.
(2) The maximum quantity of the food additive which may be used in or on such food.
(3) The manner in which the food additive may be added to or used in or on such food.
(4) Directions or other labeling or packaging requirements for the food additive (40 CFR 177.3-91).

Food and Drug Administration Action Level (FDAAL): Under section 408 of the Federal Food, Drug and Cosmetic Act, as amended, concentration of a poisonous or deleterious substance in human food or animal feed at or above which FDA will take legal action to remove adulterated products from the market. Only FDAALs established for fish and shellfish apply in the HRS (40 CFR 300-App/A-91).

Food chain:
(1) A sequence of organisms, each of which uses the next, lower member of the sequence as a food source (EPA-94/04).
(2) The pathways by which any material entering the environment passes from the first absorbing organism through plants and animals to humans (DOE-91/04).

Food chain crops: The tobacco, crops grown for human consumption, and animal feed for animals whose products are consumed by humans (40 CFR 257.3.5-91, *see also* 40 CFR 260.10-91).

Food to microorganism ratio: *See* F:M ratio.

Food processing waste: The waste resulting from operations that alter the form or composition of agriculture products for marketing purposes (*see* waste for more related terms) (EPA-83).

Food waste: The organic residues generated by the handling, storage, sale, preparation, cooking, and serving of foods; commonly called garbage (*see* waste for more related terms) (40 CFR 246.101-91).

Food waste disposer: *See* garbage grinding (EPA-83).

Food web: A relationship of food chains.

Foot: The residues of refining fats or oils which contain color bodies, insolubles, suspended matter, etc. (EPA-74/04c).

Footing: A concrete or stone base which supports a foundation wall and which is used to distribute the weight of the house over the soil or subgrade underlying the house (EPA-88/08).

Force: The action of one body on another which causes acceleration of the second body unless acted on by an equal and opposite action counteracting the effect. Force is a vector quantity (Markes-67). Force can be expressed:
- $F = mg/g_c$

Where: m = mass; g = local acceleration of gravity; and g_c = gravitational acceleration. The example values and units of g_c are as follows:

g_c = 32.17 (lbm-ft)/(lbf-sec^2)
= 1 (slug-ft)/(lbf-sec^2)
= 1 (lbm-ft)/(poundal-sec^2)
= 1 (kg mass-m)/(newton-sec^2)
= 1 (g mass-cm)/(dyne-sec^2) (Holman-p18)

The above values can be interpreted as:
1 lbf = 32.174 (lbm-ft)/sec^2
1 lbf = 1 (slug-ft)/sec^2
1 poundal (pdl) = 1 (lbm-ft)/sec^2
1 dyne = (g-cm)/sec^2
1 newton = (kg-m)/sec^2

At the sea level, g = 32.17 (ft)/(sec^2) = 9.8066 (m)/(sec^2), therefore, 1 lbm weighs 1 lbf (cf. weight).

Force account work: The use of the recipients own employees or equipment for construction, construction-related activities (including A and E services), or for repair or improvement to a facility (40 CFR 30.200-91).

Force main: A pipe in which wastewater is carried under pressure (DOI-70/04).

Forced air blower: The forced air blower provides the combustion air needed to burn the oil or gas fuel and, if oil is used, the atomizing air (*see* blower or burner component for more related terms) (EPA-89/03b).

Forced air furnace: A central unit (air conditioner or heat pump) that functions by recirculating the house air through a heat exchanger. A forced-air furnace is distinguished from a central hot-water space heating system, or electric resistance heating (*see* furnace for more related terms) (EPA-8/88).

Forced convection:
(1) In liquid, a fluid flow caused by using a pump or another similar device (EPA-88/09a).
(2) In atmosphere:
  (A) The vertical transport of atmospheric properties, e.g., caused by orographic lifting or horizontal wind convergence which results in an ascending air motion.
  (B) The vertical mixing of atmospheric properties by mechanical turbulence (NATO-78/10).
(3) *See* convection for more related terms.

Forced draft: The positive pressure created by the action of a fan or a blower which supplies the primary or secondary combustion air in an incinerator (*see* draft for more related terms) (OME-88/12).

Forced draft fan: A fan to push air or gases (*see* fan for more related terms) (EPA-83).

Forced draft tower: *See* definition under mechanical draft tower.

Forced draught: The system used in mechanical-draught cooling towers whereby the air is forced into the tower by a fan (*see* draught for more related terms) (Gurney-66).

Forced expiratory volume (FEV$_1$) at one second: The volume of air which can be forcibly exhaled during the first second of expiration following a maximal inspiration (EPA-90/08).

Forced vital capacity (FVC): The maximal volume of air which can be exhaled as forcibly and rapidly as possible after a maximal inspiration (EPA-90/08).

Foreign offshore unit: A facility which is located, in whole or in part, in the territorial sea or on the continental shelf of a foreign country and which is or was used for one or more of the following purposes: exploring for, drilling for, producing, storing, handling, transferring, processing, or transporting oil produced from the seabed beneath the foreign country's territorial sea or form the foreign country's continental shelf (OPA1001-91).

Forest: A concentration of trees and related vegetation in non-urban areas sparsely inhabited by and infrequently used by humans; characterized by natural terrain and drainage patterns (40 CFR 171.2-91).

Forest residues: Refers to all residues left in the forest after harvesting. Forest residues include logging residues, cull trees and partially rotten trees left standing after logging, and branches left on the forest floor. *See* Solid waste (EPA-83).

Forging: The exertion of pressure on dies or rolls surrounding heated aluminum stock, forcing the stock to change shape and in the case where dies are used to take the shape of the die. The forging subcategory is based on the forging process (40 CFR 467.02-91, *see also* 40 CFR 471.02-91).

Form bond: A lightweight commodity paper designed primarily for business forms including computer printout and carbonless paper forms. (*see* manifold business forms) (40 CFR 250.4-91).

Formal amendment: A written modification of an assistance agreement signed by both the authorized representative of the recipient and the award official (40 CFR 30.200-91).

Formal hearing: Any evidentiary hearing under subpart E or any panel hearing under subpart F but does not mean a public hearing conducted under 40 CFR 124.12 (40 CFR 124.2-c-91).

Formaldehyde: A colorless, pungent, irritating gas, CH_2O, used chiefly as a disinfectant and preservative and in synthesizing other compounds and resins (EPA-94/04).

Formalin: A solution of formaldehyde in water (EPA-75/01a).

Formation (forming):
(1) A body of consolidated or unconsolidated rock characterized by a degree of lithologic homogeneity which is prevailingly, but not necessarily, tabular and is mappable on the earth's surface or traceable in the subsurface (40 CFR 144.3-91, *see also* 40 CFR 146.3; 147.2902-91).
(2) An electrochemical process which converts the battery electrode material into the desired chemical condition, e.g., in a silver-zinc battery the silver applied to the cathode is converted to silver oxide and the zinc oxide applied to the anode is converted to elemental zinc. Formation is generally used interchangeably with charging, although it may involve a repeated charge-discharge cycle (EPA-84/08).
(3) Application of voltage to an electrolytic capacitor, electrolytic rectifier or semiconductor device to produce a desired permanent change in electrical characteristics as part of the manufacturing process (EPA-83/03).
(4) Arrangement of the fibers in a sheet of paper. An irregular arrangement is termed wild and a uniform arrangement is called close (EPA-83).

Formation fluid: The fluid present in a formation under natural conditions as opposed to introduced fluids, such as drilling mud (40 CFR 144.3; 146.3-91).

Formazan (triphenyltetrazolium chloride): A color indicator used in the quantitative estimation of oxidation-reduction equilibria involving hydrogen transfer (LBL-76/07-water).

Former: A machine which transforms pulp into a sheet of paper or paperboard (EPA-83).

Formic acid (or mathanoic acid, HCOOH): A toxic, corrosive liquid used in tanning, electroplating and coagulating rubber latex.

Forming:
(1) A set of manufacturing operations in which metals and alloys are made into semifinished products by hot or cold working (40 CFR 471.02-91).
(2) Casting, extruding, flaking, molding, pelletizing, rolling, shaping, stamping, and die casing (AP-40, p790).

Formula quantity: The strategic special nuclear material in any combination in a quantity of 5000 grams or more computed by the formula, grams=(grams contained U^{235}) + 2.5(grams U^{235} + grams plutonium) (10 CFR 70.4-91).

Formulating and packaging: The physical mixing of technical grade pesticide ingredients into liquids, dusts and powders, or granules and their subsequent packaging into marketable containers (40 CFR 455.41-91).

Formulation:
(1) (A) The process of mixing, blending, or dilution of one or more active ingredients with one or more other active or inert ingredients, without an intended chemical reaction, to obtain manufacturing use product or an end use product; or
(B) The repackaging of any registered product (40 CFR 158.153-91).
(2) The substance or mixture of substances which is comprised of all active and inert ingredients in a pesticide (EPA-94/04).

Fortrel fiber: The trademark of Fiber Industries for polyester fiber (*see* fiber for more related terms) (EPA-74/06b).

Forward stepping multiple linear regression: A multiple linear regression procedure in which the order that the independent variables enter into the regression model is determined by the respective contribution of each variable to explain the variability of the dependent variables (*see* regression for more related terms) (EPA-79/12c).

Fossil fuel:
(1) Natural gas, petroleum, coal, and any form of solid, liquid, or gaseous fuel derived from such materials for the purpose of creating useful heat (40 CFR 60.41; 60.41a; 60.161-91).
(2) Fuels such as coal and oil that are generated by the decay of plant and/or animal matter under conditions of heat and pressure (ETI-92).
(3) Fuel derived from ancient organic remains, e.g., peat, coal, crude oil, and natural gas (EPA-94/04).

Fossil fuel and wood residue fired steam generating unit: A furnace or boiler used in the process of burning fossil fuel and wood residue for the purpose of producing steam by heat transfer (40 CFR 60.41-91).

Fossil fuel fired steam generator: A furnace or boiler used in the process of burning fossil fuel for the primary purpose of producing steam by heat transfer (40 CFR 51.100-91).

Fossil fuel fired steam generating unit: A furnace or boiler used in the process of burning fossil fuel for the purpose of producing steam by heat transfer (40 CFR 60.41-91).

Fouling: Material accumulates in flow passages or on heat absorbing surfaces in an incinerator or other combustion chambers. (Fouling causes impedance to the flow of fluids or heat) (cf. slag) (SW-108ts).

Foundation insulation: A material, primarily designed to resist heat flow, which is installed in foundation walls between conditioned volumes and unconditioned volumes and the outside or surrounding earth, at the perimeters of concrete slab-on-grade foundations, and at common foundation wall assemblies between conditioned basement volumes (40 CFR 248.4-91).

Foundry: A facility engaged in the melting or casting of beryllium metal or alloy (40 CFR 61.31-91).

Foundry coke:
(1) Coke that is produced from raw materials with less than 26 percent volatile material by weight and that is subject to a coking period of 24 hours or more. Percent volatile material of the raw materials (by weight) is the weighted average percent volatile material of all raw materials (by weight) charged to the coke oven per coking cycle (40 CFR 61.131-91).
(2) The residue from the destructive distillation of coal. A

primary ingredient in the making of cast iron in the cupola. Because of the nature of the destructive distillation process and impurities in the coal, the coke may contain residuals of toxic pollutants such as phenol, benzene, toluene, naphthalene, and nitrosamines (EPA-85/10a).
(3) See coke for more related terms.

Foundry coke by-product recovery plant: A coke by-product recovery plant connected to coke batteries whose annual coke production is at least 75 percent foundry coke (40 CFR 61.131-91).

Fountain solution: The solution which is applied to the image plate to maintain hydrophilic properties of the non-image areas (40 CFR 52.741-91).

Four (4)-AAP colorimeteric method: An analytical method used to detect and quantify total phenols and total phenolic compounds. The method involves the reaction of phenols with a color developing agent, 4-aminoantipyrine (4-AAP) (cf. phenols 4AAP) (EPA-85/10a).

Four center transition state: A transition state where four partial (one electron or three electron) bonds are formed; primarily results in the elimination of a stable molecule from the reactant (EPA-88/12).

Four hour block average or 4-hour block average: The average of all hourly emission rates when the affected facility is operating and combusting MSW measured over 4-hour periods of time from 12 midnight to 4 a.m., 4 a.m. to 8 a.m., 8 a.m. to 12 noon, 12 noon to 4 p.m., 4 p.m. to 8 p.m., and 8 p.m. to 12 midnight (40 CFR 60.51a-91).

Four (4) 9s: 99.99 % destruction and removal efficiency (DRE) for hazardous waste incineration standards required by the Resource Conservation and Recovery Act (cf. six 9s).

Four (4) Rs: Four major activities for waste minimization, namely, reduction; reuse; recycle; and regeneration.

Four wheel drive general utility vehicle: A four-wheel-drive, general purpose automobile capable of off-highway operation that has a wheelbase not more than 110 inches and that has a body shape similar to a 1977 Jeep CJ-5 or CJ-7, or the 1977 Toyota Land Cruiser, as defined by the Secretary of Transportation at 49 CFR 553.4 (40 CFR 600.002.85-42-91).

Fourcault process: The method of making sheet glass by drawing vertically upward from a slotted debiteuse block (EPA-83).

Fourdrinier: A machine for the continuous manufacture of paper. It consists essentially of a wide endless wire mesh belt on which the pulp is dewatered and a drying section to form the paper web (EPA-83).

Fourier analysis: The representation of physical or mathematical data by Fourier series or by a Fourier integral (NATO-78/10).

Fourier transform infrared spectroscopy analyzer (FTS-IR): One of continuous emission monitors (see continuous emission monitor for various types). The analyzer uses a rapid scanning interferometer consisting of two mirrors, one movable and one fixed, and a beam splitter. The beam splitter reflects back 50% and transmits 50% of the incoming infrared radiation. These two beams are then reflected back towards the beam splitter by the two mirrors. Depending on the position of the moving mirror, these two beams recombine at the beam splitter with a specific path difference between them. This produces the interferogram. The interferogram is generated by the interferometer modulating the infrared beam as the moving mirror is translated. The modulation frequencies depend on the wavelength of the incident radiation and the velocity of the moving mirror. The interferogram is produced after absorption by a sample, the modulated radiation reaches the detector where intensity is recorded as a function of path difference. The FTS-IR is capable of measurement ranges from the near visible to less than 10 cm^{-1} and is compatible with gas chromatograph systems. Its analysis can be performed in close to real time (EPA-84/03a).

Fourteen day (14-day) old survivors: The birds that survive for weeks following hatch. Values are expressed both as a percentage of hatched eggs and as the number per pen per season (test) (40 CFR 797.2130-91, see also 40 CFR 797.2150-91).

Forward mutation: A gene mutation from the wild (parent) type to the mutant condition (cf. reverse mutation) (40 CFR 798.5140-91, see also 40 CFR 798.5250-91).

Forward mutation assay: To detects a gene mutation from the parental type to the mutant form which gives rise to a change in an enzymatic or functional protein (40 CFR 798.5300-91).

Franchise collection: The solid waste collection made by a private firm that is given exclusive right to collect for a fee paid by customers in a specific territory or from specific types of customers (see waste collection for more related terms) (SW-108ts).

Francium (Fr): A alkali, radioactive metal with a.n. 87; a.w. 223; d. 2.4 g/cc; m.p. 27° C and b.p. 677° C. The element belongs to group IIIB of the periodic table.

Fractionation (or fractional distillation):
(1) The separation of constituents, or group of constituents, of a liquid mixture of miscible and volatile substances by vaporization and recondensing at spcific boiling point ranges (EPA-87/10a).
(2) The process of separating a mixture into components having different properties (as by distillation, precipitation, or screening) (LBL-76/07-air).

Fractionation operation: A distillation operation or method used to separate a mixture of several volatile components of different boiling points in successive stages, each stage removing from the mixture some proportion of one of the components (40 CFR 264.1031-91).

Frameshift mutagens: The agents which cause the addition or deletion of single or multiple base pairs in the DNA molecule (40 CFR 798.5265; 798.5300-91).

Franchising: The exclusive right granted a contractor to collect and/or dispose of solid wastes from a district or community, generally conferred by a governing political body (EPA-83).

Frank-effect level (FEL): Exposure level which produces unmistakable adverse effects, such as irreversible functional impairment or mortality, at a statistically or biologically significant increase in frequency or severity between an exposed population and its appropriate control (EPA-90/08; 92/12).

Free atmosphere: The portion of the atmosphere above the planetary boundary layer. It is no longer influenced by the presence of the earth's surface (*see* atmosphere for more related terms) (NATO-87/10).

Free available chlorine: The value obtained using the amperometric titration method for free available chlorine described in Standard Methods for the Examination of Water and Wastewater, page 112 (13th edition) (*see* chlorine for more related terms) (40 CFR 423.11-91).

Free chlorine: The chlorine present in solutions in the aqueous (Cl_2), hypochlorous acid (HOCl), or hypochlorite ion (OCl^-) forms (*see* chlorine for more related terms) (EPA-88/09a).

Free convection: In atmosphere,
(1) The vertical transport of atmospheric properties caused by the vertical air motions induced by buoyancy forces.
(2) The vertical mixing of atmospheric properties by thermal turbulence (NATO-78/10).
(3) *See* convection for more related terms.

Free cyanide:
(1) True--the actual concentration of cyanide radical or equivalent alkali cyanide not combined in complex ions with metals in solutions.
(2) Calculated--the concentration of cyanide or alkali cyanide present in solution in excess of that calculated as necessary to form a specified complex ion with a metal or metals present in solution.
(3) Analytical--the free cyanide content of solution as determined by a specified analytical method (EPA-83/06a).
(4) *See* cyanide for more related terms.

Free leg: In the steel industry, a portion of the ammonia still from which ammonia, hydrogen sulfide, carbon dioxide, and hydrogen cyanide are steam distilled and returned to the gas stream (cf. lime leg) (EPA-74/06a).

Free groundwater: The groundwater in aquifers that are not bounded by or confined in impervious strata (*see* groundwater for more related terms) (SW-108ts).

Free liquid: The liquids which readily separate from the solid portion of a waste under ambient temperature and pressure (40 CFR 260.10-91).

Free moisture: The liquid that will drain freely by gravity from solid materials (40 CFR 240.101; 241.101-91).

Free product: Refers to a regulated substance that is present as a non-aqueous phase liquid (e.g., liquid not dissolved in water) (40FR280.12-91).

Free radical (or radical):
(1) A molecule or atom with at least one unpaired valence electron (EPA-88/12).
(2) An atom or a group of atoms, such as triphenyl methyl ($(C_6H_5)_3C^-$), characterized by the presence of at least one unpaired electron. Free radicals are effective in initiating many polymerizations (EPA-75/01a).

Free residual chlorination: The application of chlorine to water, sewage, or industrial wastes to produce directly or through the destruction of ammonia, or of certain organic nitrogenous compounds, free chlorine residual (*see* chlorination for more related terms) (EPA-82/11f).

Free residual chlorine: The unreacted hypochlorous acid (HOCl), or hypochlorite ion (OCl^-) forms (*see* chlorine for more related terms).

Free surface: The boundary between atmosphere and water.

Free stall barn shaly: The specialized facilities wherein producing cows are permitted free movement between resting and feeding areas (40 CFR 412.11-91).

Free water: The suspended water on a surface that is free to move as distinguished from absorbed or inherent water (*see* water for more related terms) (EPA-83).

Freeboard:
(1) The vertical distance between the top of a tank or surface impoundment dike, and the surface of the waste contained therein (40 CFR 260.10-91).
(2) Vertical distance from the normal water surface to the top of the confining wall (EPA-94/04).
(3) The vertical distance from the sand surface to the underside of a trough in a sand filter (EPA-94/04).

Freeness: A measure of the rate with which water drains from a stock suspension through a wire mesh screen or a perforated plate. It is also known as slowness or wetness (EPA-87/10).

Freeness tester: Instrument use for the measurement of the rate at which water drains from a suspension of water and pulp on a wire mesh (EPA-83).

Freeze: *See* 40 CFR 52.1135-91.

Freeze area: That portion of the Boston Intrastate Region enclosed within the following boundaries. The City of Cambridge; that portion of the City of Boston from the Charles River and the Boston Inner Harbor on north and northeast of pier 4 on NOrthern Avenue; by the east side of pier 4 to B Street, B Street extension of B Street to B Street, B Street, Dorchester Avenue, and the Preble Street to Old Colony Avenue, then east to the water, then by the water's edge around Columbia Point on various courses generally easterly, southerly, and westerly to the center of the bridge on Morrissey Boulevard, on the east and southeast; then due west to Freeport Street, Freeport Street, Dorchester Avenue, Southeast Expressway, Southampton Street, Reading Street, Island Street, Chadwick Street, Carlow Street, Albany Street, Hunneman Street, Madison Street, Windsor Street, Cabot Street, Ruggles Street, Parker Street, Ward

Street, Huntington Avenue, Brookline-Boston municipal boundary, Mountford Street to the Boston University Bridge on the southwest and west; and the Logan International Airport. Where a street or roadway forms a boundary the entire right-of-way of the street is within the freeze area as defined (40 CFR 52.1128-4-91).

Freeze drying (or lyophilization): A concentration process which relies on a refrigeration cycle with vacuum. At temperatures below 0° C, water is removed by sublimation as long as the partial pressure of moisture in the gas phase over the sample is less than the vapor pressure of water in the sample. The drying operation usually proceeds in two stages. In the first, most of the water as ice is sublimed from the frozen sample in a high-vacuum system. The temperature is kept well below 0° C. In the second stage, the nearly dry product is dried at a higher temperature, depending on stability, in order to attain a minimum moisture level in the shortest time. Removal of water vapor under high vacuum must be efficient to maintain low vapor pressure in the system.

Freezing out: A process consisting of the collection of one or several components of a gaseous mixture by simple cooling of the gas stream in a device which retains the condensate (LBL-76/07-air).

Freezing point temperature: The temperature at which a given liquid substance will solidify or freeze on removal of heat. Freezing point of water is 0° C (*see* temperature for more related terms).

Freezing thawing: Method ASTM C-67 for determining brick durability (EPA-83).

French drain (perimeter drain, channel drain, or floating slab): A water drainage technique installed in basements of some houses during initial construction. If present, typically consists of a 1- or 2-in. gap between the basement wall and the concrete floor slab around the entire perimeter inside the basement to allow water to drain to aggregate under the slab and then soak away (EPA-88/08).

Freon: A trade name for a group of chemical compounds, typically chloroflurocarbons, used as refrigerants (EPA-88/12).

Frequency:
(1) How often criteria can be exceeded without unacceptably affecting the community (EPA-85/09; 91/03).
(2) Number of cycles per unit time (LBL-76/07-bio).

Fresh feed: Any petroleum derivative feedstock stream charged directly into the riser or reactor of a fluid catalytic cracking unit except for petroleum derivatives recycled within the fluid catalytic cracking unit, fractionator, or gas recovery unit (40 CFR 60.101-91).

Fresh granular triple superphosphate: The granular triple superphosphate produced no more than 10 days prior to the date of the performance test (40 CFR 60.241-91).

Fresh sludge: *See* synonym, green sludge.

Fresh water:
(1) Underground source of drinking water (40 CFR 147.2902-91).
(2) Water that generally contains less than 1,000 milligrams-per-liter of dissolved solids (EPA-89/12).
(3) *See* water for more related terms.

Freshwater lake: Any inland pond, reservoir, impoundment, or other similar body of water that has recreational value, that exhibits no oceanic and tidal influences, and that has a total dissolved solids concentration of less than 1 percent (40 CFR 35.1605.2-91).

Freundlich equation: Valid for monomolecular physical and chemical adsorption. It is an empirical relationship. When it is valid, it gives a concise analytical equation for experimental facts, rather than an accurate description of the mechanism of adsorption. It is the oldest isotherm equation and is widely employed in industrial design. The equation is (Calvert-84): $\ln v = \ln k + (1/n)\ln P$, where: v = the volume adsorbed; P = pressure; $\ln$ = natural logarithm; and k = constant.

Friable:
(1) When referring to material in a school building means that the material, when dry, may be crumbled, pulverized, or reduced to powder by hand pressure, and includes previously nonfriable material after such previously nonfriable material becomes damages to the extent that when dry it may be crumbled, pulverized, or reduced to powder by hand pressure (40 CFR 763.83-91).
(2) Easy to break or crumbling naturally (EPA-82/05).
(3) Capable of being crumbled, pulverized, or reduced to powder by hand pressure (EPA-94/04).

Friable asbestos: Any material containing more than one percent asbestos, and that can be crumbled or reduced to powder by hand pressure (may include previously non-friable material which becomes broken or damaged by mechanical force) (EPA-94/04).

Friable asbestos-containing material: Any asbestos-containing material applied on ceiling, walls, structures members, piping, duct work, or any other part of a building which when dry may be crumbled, pulverized, or reduced to powder by hand pressure. The term includes non-friable asbestos-containing material after such previously non-friable material becomes damaged to the extent that when dry it may be crumbled, pulverized, or reduced to powder by hand pressure (TSCA202-15USC2642-91).

Friable asbestos material: Any material containing more than 1 percent asbestos by weight which, when dry, may be crumbled, pulverized or reduced to powder by hand pressure (40 CFR 763.121-91, *see also* 40 CFR 61.141-91).

Friable material: Any material applied onto ceilings, walls, structural members, piping, ductwork, or any other part of the building structure which, when dry, may be crumbled, pulverized, or reduced to powder by hand pressure (40 CFR 763.103-91).

Frit: Specially formulated glass in a granular or flake form (EPA-82/11e).

Frit seal: A seal made by fusing together metallic powders with a glass binder for such applications as hermatically sealing ceramic packages for integrated circuits (EPA-83/03).

Frond: A single Lemna leaf-like structure (40 CFR 797.1160-91).

Frond mortality: The dead fronds which may be identified by a total discoloration (yellow, white, black or clear) of the entire frond (40 CFR 797.1160-91).

Front end loader: A collection vehicle with arms that engage a detachable container, move it up over the cab, empty it into the vehicle's body, and return it to the ground (SW-108ts).

Front end recovery: Mechanical processing of as discarded solid wastes into separate constituents (cf. back end system) (EPA-83).

Front end system or process: Size reduction, separation and/or physical modification of solid wastes to afford practical use or reuse (cf. material recovery) (EPA-83).

Froth: In the flotation process, a collection of bubbles resulting from agitation, the bubbles being the agency for raising (floating) the particles of ore to the surface of the cell (EPA-82/05).

Froth flotation (foam fractionation or foam separation): A process for separating, in aqueous suspension, finely divided particles that have different surface characteristics. Reagents are selected which, when added to the mixture, will coat only the desired material and make their surfaces water-repellent (hydrophobic). When air is bubbled through the solution, the coated particles become affixed to the air bubbles and are buoyed to the surface where they can be removed as froth (EPA-83).

Frother: A substance used in flotation processes to make air bubbles sufficiently permanent principally by reducing surface tension. Common frothers are pine oil, creyslic acid, and amyl alcohol (EPA-82/05).

Frothing clarifier: A flotation device that separates tricalcium phosphate precipitate from the liquor (EPA-75/02d).

Fructose ($C_6H_{12}O_6$): A fruit sugar (cf. dextose).

Fructosuria: The presence in the urine of fructose, a monosaccharide formed from the breakdown of more complex sugars and normally converted ultimately (during metabolism) to carbon dioxide and water (LBL-76/07-bio).

Frugivore: Animals that eat fruits.

Fuel:
(1) Any material which is capable of releasing energy or power by combustion or other chemical or physical reaction (40 CFR 79.2-91).
(2) (A) Gasoline and diesel fuel for gasoline- or diesel-powered automobiles; or
    (B) Electrical energy for electrically powered automobiles (40 CFR 600.002.85-91).
(3) For calibration gas application, a 40% H_2 and 60% He or 40% H_2 and 60% N_2 gas mixture is recommended to avoid an oxygen synergism effect that reportedly occurs when oxygen concentration varies significantly from a mean value (*see* calibration gas for more related terms) (EPA-90/04).

Other fuel related terms include:
- Auxiliary fuel
- Clean alternative fuel
- Conventional fuel
- Densified refuse derived fuel
- Emergency fuel
- Fibre fuel
- Motor fuel
- Refuse derived fuel
- Solid derived fuel (*see* synonym, solid waste derived fuel)
- Solid waste derived fuel

Fuel bed: The layer of solid fuel or solid waste on a furnace grate or hearth (*see* fuel for more related terms) (SW-108ts).

Fuel combustion emission source: Any furnace, boiler, or similar equipment used for the primary purpose of producing heat or power by indirect heat transfer (40 CFR 52.741-91).

Fuel denitrogenation: One of NO_x emission reduction techniques (*see* nitrogen oxide emission control for control structure). It is to remove the nitrogen contained in the fuel. Amounts of nitrogen vary in a fossil fuel. Coal, shale, and residual fuel oil contain a larger amount of nitrogen than either distillate oil or natural gas. The nitrogen in the fuel can be emitted as NO_x when the fossil fuel is burned in the furnace.

Nitrogen is removed from coal, shale, or heavy fuel oil by liquefying the fuels and mixing with hydrogen gas. The mixture is heated and a catalyst is used to cause the nitrogen in the fuel and the hydrogen to unite. This reaction produces two products; ammonia and a cleaner fuel. Researchers are developing better catalysts and finding ways to reduce the deposition of carbon on the catalyst surface. Carbon deposits reduce the effectiveness of the catalyst life. This technology can reduce the nitrogen content in both natural fuels and synthetic liquid or gaseous fuel (made from shale and coal). This could become an increasingly important technology with the development and use of synthetic fuels in the future (EPA-81/12, p7-4).

Fuel desulfurization: One of SO_2 emission reduction techniques (*see* sulfur oxide emission control for control structure). Fuel desulfurization means to remove or reduce the sulfur content of coal before it is burned. Coal contains sulfur in two forms:
(1) Mineral sulfur in the form of inorganic pyrite: The amount of inorganic (and organic) sulfur in coal varies from 10 to 90%. Mineral sulfur can be removed by physical coal cleaning.
(2) Organic sulfur which is chemically bound to the coal: Organic sulfur requires chemical cleaning.

The techniques for fuel desulfurization include:
(1) Physical coal cleaning
(2) Chemical coal cleaning
    (A) Microwave desulfurization
    (B) Hydrothermal desulfurization (EPA-81/12, p8-4).

Fuel economy:
(1) The average number of miles traveled by an automobile or group of automobiles per gallon of gasoline or diesel fuel consumed as computed in 40 CFR 600.113 or 40 CFR 600.207; or
(2) The equivalent petroleum-based fuel economy for an electrically powered automobile as determined by the

Secretary of Energy (40 CFR 600.002.85-91, *see also* 40 CFR 2.311; 610.11-91).
Other fuel economy related terms include:
- Highway fuel economy
- Highway fuel economy test

Fuel economy data: Any measurement or calculation of fuel economy for any model type and average fuel economy of a manufacturer under section 503(d) of the Act, 15USC2003(d) (40 CFR 2.311-91).

Fuel economy data vehicle: A vehicle used for the purpose of determining fuel economy which is not a certification vehicle (40 CFR 600.002.85-91).

Fuel economy standard: The Corporate Average Fuel Economy Standard (CAFE) effective in 1978. It enhanced the national fuel conservation effort imposing a miles-per-gallon floor for motor vehicles (EPA-94/04).

Fuel efficiency: The proportion of the energy released on combustion of a fuel that is converted into useful energy (EPA-94/04).

Fuel element: Elements of nuclear fuel materials.

Fuel evaporative emissions: The vaporized fuel emitted into the atmosphere from the fuel system of a motor vehicle (40 CFR 86.082.2-91).

Fuel fabrication facility: The land, buildings, equipment, and processes used to make reactor fuel from raw material. This facility may also be used to assemble the fuel with other materials into components for use in reactors (DOE-91/04)

Fuel processing facility: The land, buildings, equipment, and processes used to extract uranium and other materials from used (spent) reactor fuel. This facility may also be used to treat materials that can not be recycled for transfer to waste treatment, storage, or disposal facilities (DOE-91/04)

Fuel gas: Any gas which is generated at a petroleum refinery and which is combusted. Fuel gas also includes natural gas when the natural gas is combined and combusted in any proportion with a gas generated at a refinery. Fuel gas does not include gases generated by catalytic cracking unit catalyst regenerators and fluid coking burners (40 CFR 60.101-91).

Fuel gas combustion device: Any equipment, such as process heaters, boilers and flares used to combust fuel gas, except facilities in which gases are combusted to produce sulfur or sulfuric acid (40 CFR 60.101-91).

Fuel gas system: A system for collection of refinery fuel gas including, but not limited to, piping for collecting tail gas from various process units, mixing drums and controls, and distribution piping (40 CFR 52.741-91).

Fuel manufacturer: Any person who, for sale or introduction into commerce, produces or manufactures a fuel or causes or directs the alteration of the chemical composition of, or the mixture of chemical compounds in, a bulk fuel by adding to it an additive (40 CFR 79.2-91).

Fuel/oxygen stoichiometry: Quotient of [ratio of moles (or mass) of fuel divided by moles (or mass) of oxygen] for stoichiometric oxidation (EPA-2/88).

Fuel pretreatment: A process that removes a portion of the sulfur in a fuel before combustion of the fuel in a steam generating unit (40 CFR 60.41b; 60.41c-91).

Fuel rod: A nuclear fuel rod.

Fuel system: The combination of fuel tank(s), fuel pump, fuel lines, and carburetor or fuel injection components, and includes all fuel system vents and fuel evaporative emission control system components (40 CFR 86.082.2-91, *see also* 40 CFR 86.402.78-91).

Fuel switching:
(1) A precombustion process whereby a low-sulfur coal is used in place of a higher sulfur coal in a power plant to reduce sulfur dioxide emissions (EPA-94/04).
(2) Illegally using leaded gasoline in a car designed to use only unleaded (EPA-94/04).

Fuel train: The series of components that controls the flow of fuel to the burner. The fuel train set up for gas and oil burners is basically the same. Each fuel train has a pressure gauge, a manual shutoff valve, and a solenoid shut-off valve. The only difference between the gas and oil fuel trains is the device used to control fuel flow; the oil fuel train utilizes a needle flow valve while the gas fuel train utilizes a gas orifice union (*see* burner component for more related terms) (EPA-89/03b).

Fuel venting emissions: The raw fuel, exclusive of hydrocarbons in the exhaust emissions, discharged from aircraft gas turbine engines during all normal ground and tight operations (40 CFR 87.1-91).

Fugitive dust, mist or vapor: The dust, mist or vapor containing a toxic pollutant regulated under this part which is emitted from any source other than through a stack (40 CFR 129.2-91).

Fugitive emission:
(1) Those emissions which could not reasonably pass through a stack, chimney, vent, or other functionally equivalent opening (40 CFR 51.165-91, *see also* 40 CFR 51.166; 51.301; 51-App/S; 52.21; 52.24; 57.103; 60.301; 60.671; 60-App/A(method 22)-91).
(2) Emissions not caught by a capture system (EPA-94/04).
(3) Emissions to the atmosphere from pumps, valves, flanges, seals, and other process points not vented through a stack. Also includes emissions from area sources such as ponds, lagoons, landfills, and piles of stored material (DOE-91/04).
(4) Nonstack emissions usually associated with normal plant operations or leaks at the facility (ETI-92).
(5) *See* emission for more related terms.

Fugitive emissions equipment: Each pump, compressor, pressure relief device, sampling connection system, open-ended valve or line, valve, and flange or other connector in VOC service and any devices or systems required by subpart VV of this part (40 CFR 60.561-91).

Fugitive source: Any source of emissions not controlled by an air pollution control device (*see* source for more related terms) (40 CFR 61.141-91).

Fugitive volatile organic compounds: Any volatile organic compounds which are emitted from the coating applicator and flashoff areas and are not emitted in the oven (40 CFR 60.441-91).

Full boil process: A soap making process where the neat soap is completed in the kettle and the by-product glycerine drawn off (EPA-74/04c).

Full capacity: The operation of the steam generating unit at 90 percent or more of the maximum steady-state design heat input capacity (40 CFR 60.41b-91).

Full load torque: The torque necessary for a motor to produce its rated horsepower at full load speed (*see* torque for more related terms) (EPA-83).

Fume:
(1) Tiny particles trapped in vapor in a gas stream (EPA-94/04).
(2) Suspended particles in a gas, one micron or less in diameter (SW-108ts).

Fume incinerator: *See* synonym, secondary burner.

Fume scrubber:
(1) One of air pollution control devices. A fume scrubber is a device that removes pollutant constituents from an air stream by dissolving them in a liquid solvent, specifically water (EPA-74/12a).
(2) Those pollution control devices used to remove and clean fumes originating in pickling operations (40 CFR 420.91-91, *see also* 40 CFR 420.121-91).
(3) *See* scrubber for more related terms.

Fume suppression system: The equipment comprising any system used to inhibit the generation of emissions from steelmaking facilities with an inert gas, flame, or steam blanket applied to the surface of molten iron or steel (40 CFR 60.141a-91).

Fumigant: A pesticide vaporized to kill pests. Used in buildings and greenhouses (EPA-94/04).

Fumigation: The rapid downward mixing of the plume as it flows from a layer of stable air into an area of moderate to strong turbulence. It typically occurs when a source is located near a shoreline of a relatively cool lake or ocean. It also occurs when the morning temperature inversion is rapidly dissipated by solar heating of the ground surface. The effect of fumigation is to produce a short period (30-60 minutes) of very high concentrations within several kilometers or two of a source (EPA-88/09).

Functional equivalent: Term used to describe EPA's decision-making process and its relationship to the environmental review conducted under the National Environmental Policy Act (NEPA). A review is considered functionally equivalent when it addresses the substantive components of a NEPA review (EPA-94/04).

Functional finish chemical: A substance applied to a fabric to provide desirable properties such as wrinkle-resistance, water-repellency, flame-resistance, etc. (EPA-82/09).

Functional space: A room, group of rooms, or homogeneous area (including crawl spaces or the space between a dropped ceiling and the floor or roof deck above), such as classroom(s), a cafeteria, gymnasium, hallway(s), designated by a person accredited to prepare management plans, design abatement projects, or conduct response actions (40 CFR 763.83-91).

Functionally equivalent component: A component which performs the same function or measurement and which meets or exceeds the performance specifications of another component (40 CFR 270.2-91).

Fund:
(1) The Coastal Energy Impact Fund established by section 1456a(h) of this title (CZMA304-16USC1453-90).
(2) The Oil Spill Liability Trust Fund, established by section 9509 of the Internal Revenue code of 1986 (26USC9509) (OPA1001-91).
(3) The Abandoned Mine Reclamation Fund established pursuant to section 1231 of this title (SMCRA701-30USC1291-90).

Fund (or trust fund): A fund set up under the Comprehensive Environmental Response, Compensation and Liability Act (CERCLA) to help pay for cleanup of hazardous waste sites and for legal action to force those responsible for the sites to clean them up (cf. oil pollution fund) (EPA-89/12; 40 CFR 300.5-91).

Fungi (singular, fungus):
(1) Any non-chlorophyll-bearing thallophyte (that is, any non-chlorophyll-bearing plant of a lower order than mosses and liverworts), as for example, rust, smut, mildew, mold, yeast, and bacteria, except those on or in living man or other animals and those on or in processed food, beverage, or pharmaceuticals (FIFRA2-7USC136-91).
(2) Simple plants that lack a photosynthetic pigment (SW-108ts).
(3) Molds, mildews, yeasts, mushrooms, and puffballs, a group organisms lacking in chlorophyll (i.e., are not photosynthetic) and which are usually non-mobile, filamentous, and multicellular. Some grow in soil, others attach themselves to decaying trees and other plants whence they obtain nutrients. Some are pathogens, others stabilize sewage and digest composted waste (EPA-94/04).

Fungicide:
(1) Substances or a mixture of substances intended to prevent, destroy, or mitigate any fungi (LBL-76/07-water).
(2) Pesticides which are used to control, deter, or destroy fungi (EPA-94/04).

Fungistat: A chemical that keeps fungi from growing (EPA-94/04).

Fungus: *See* fungi.

Funnel: The rear, funnel-shaped portion of the glass enclosure of a cathode ray tube (EPA-83/03).

Funnel access: A small parcel of riparian land deeded collectively to a group of land owners who have no frontage bordering the water, so as to give them legal access to the water (DOI-70/04).

Furan:
(1) The common name for polychlorinated dibenzofurans (PCDFs). Generation of these compounds has been associated with low-temperature combustion processes and high exposure to these compounds has been associated with adverse health effects, particularly in laboratory animals (ETI-92).
(2) *See* dioxin/furan.

Furan resin: A heaterocyclic ring compound formed from diene and cyclic vinyl ether. Its main use is as a cold set resin in conjunction with acid accelerators such as phosphoric or toluene sulfonic acid for making core sand mixtures that harden at room temperature. Toluene could be formed during thermal degradation of furan resins during metal pouring (*see* resin for more related terms) (EPA-85/10a).

Furfuryl alcohol: A synthetic resin used to formulate core binders.

Furnace: A chamber where drying, ignition and combustion of fuels occurs (EPA-83). In CAA, it is a solid fuel burning appliance that is designed to be located outside the ordinary living areas and that warms spaces other than the space where the appliance is located, by the distribution of air heated in the appliance through ducts. The appliance must be tested and listed as a furnace under accepted American or Canadian safety testing codes unless exempted from this provision by the Administrator. A manufacturer may request an exemption in writing from the Administrator by stating why the testing and listing requirement is not practicable and by demonstrating that his appliance is otherwise a furnace (40 CFR 60.531-91, *see also* 40 CFR 240.101-91). Other furnace related terms include:
- Acid furnace
- Arc furnace
- Basic furnace
- Basic oxygen process furnace (BOPF)
- Basic oxygen furnace steelmaking
- Blast furnace
- Bottom blown furnace
- Cold metal furnace
- Covered furnace
- Cross recovery furnace
- Crucible furnace
- Cupola furnace
- Cyclone furnace
- Dross reverberatory furnace
- Dry bottom furnace
- Electric furnace
    - Direct arc furnace
    - Indirect arc furnace
    - Induction furnace
    - Resistance furnace
- Electric arc furnace
- Electric arc furnace steelmaking
- Electric submerged arc furnace
- Experimental furnace
- Ferromanganese blast furnace
- Forced-air furnace
- Halogen acid furnace (HAF)
- Hand glass melting furnace
- Hot metal furnace
- Industrial furnace
- Iron blast furnace
- Muffle furnace
- Open arc furnace
- Open furnace
- Open hearth furnace
- Open hearth furnace steelmaking
- Outokumpu furnace
- Pit crucible furnace
- Pot furnace
- Reverberatory furnace: Reverberatory furnace includes the following types: stationary, rotating, rocking, and tilting
- Reverberatory smelting furnace
- Slag tap furnace
- Smelting electric furnace
- Smelting furnace
- Stationary crucible furnace
- Straight kraft recovery furnace
- Submerged arc furnace
- Tilting furnace
- Top blown furnace
- Traveling grate furnace

Furnace arch: A nearly horizontal structure that extends into a furnace and serves to deflect gases (SW-108ts).

Furnace charge: Any material introduced into the electric submerged arc furnace, and may consist of, but is not limited to, ores, slag, carbonaceous material, and limestone (40 CFR 60.261-91).

Furnace coke: The coke produced in by-product ovens that is not foundry coke (*see* coke for more related terms) (40 CFR 61.131-91).

Furnace coke by-product recovery plant: A coke by-product recovery plant that is not a foundry coke by-product recovery plant (40 CFR 61.131-91).

Furnace cycle: The time period from completion of a furnace product tap to the completion of the next consecutive product tap (40 CFR 60.261-91).

Furnace power input: The resistive electrical power consumption of an electric submerged arc furnace as measured in kilowatts (40 CFR 60.261-91).

Furnace pull: That amount of glass drawn from the glass furnace or furnaces (40 CFR 426.81-91).

Furnace volume: The total internal volume of combustion chambers (SW-108ts).

Furnace wall: Furnace wall related terms include:
- Air cooled furnace wall

- Battery furnace wall
- Bridge furnace wall
- Core furnace wall
- Curtain furnace wall
- Gravity furnace wall
- Insulated furnace wall
- Refractory furnace wall
- Sectionally supported furnace wall
- Unit suspended furnace wall
- Water cooled furnace wall

Furnish: The pulp used as raw materials in a paper mill (OTA-89/10).

Furrow irrigation: Irrigation method in which water travels through the field by means of small channels between each row or groups of rows (EPA-94/04).

Fuse: An over-current protective device with a circuit-opening fusible part that would be heated and severed by over-current passage (EPA-83/03).

Fused silica capillary column: A very small internal diameter (< 1 mm) fused silica tube coated with a stationary phase; in the presence of flowing inert gas, this tube acts to separate or resolve organic compounds (order of elution dependent of specific physical properties) when the tube is uniformly heated to elevated temperatures (250-300° C) (EPA-88/12).

Fusion:
(1) The union of two chemical species by melting (EPA-77/07).
(2) The heating of an enamel coated item to a continuous, uniform glass film (EPA-82/11e).
(3) In thermodynamics, the change of a pure substance from a solid phase to its liquid phase. Fusion is an endothermic process, that is energy must be added to a solid substance (solid phase) to convert it to a liquid (liquid phase). This energy is commonly referred to as the latent heat of fusion
(4) *see* fission and latent heat for more related terms.
Other fusion related terms include:
- Nuclear fusion
- Thermal fusion

Fusion point temperature: The temperature at which a particular complex mixture of minerals can flow under the weight of its own mass. Because most refractory materials have no definite fusion points but soften gradually over a range of temperature, the conditions of measurement have been standardized by the ASTM (cf. pyrometric cone equivalent and *see* temperature for more related terms) (SW-108ts).

Future liability: Refers to potentially responsible parties' obligations to pay for additional response activities beyond those specified in the Record of Decision or Consent Decree (EPA-94/04).

Gg

Gadolinium (Gd): A rare earth metal with a.n. 64; a.w. 157.25; d. 7.89 g/cc; m.p. 1312° C and b.p. 3000° C. The element belongs to group IIIB of the periodic table.

Gage pressure: *See* synonym, gauge pressure.

Gall: A layer of molten sulfates floating upon glass in a tank (EPA-83).

Gallium (Ga): A soft metallic element with a.n. 31; a.w. 69.72; d. 5.91 g/cc; m.p. 29.8° C and b.p. 2237° C. The element belongs to group IIIA of the periodic table.

Galvanic series (or electro chemical series): A sequential ranking of metal properties changing from cathodic (less corroded) to anodic (more corroded) end in a medium. The ranking is: silver (cathodic), copper, brass, cast-iron, mild steel, aluminum (anodic).

Galvanized basis material: The zinc coated steel, galvalum, brass and other copper base strip which is processed in coil coating (40 CFR 465.02-91).

Galvanizing:
(1) Coating steel products with zinc by the hot dip process including the immersion of the steel product in a molten bath of zinc metal, and the related operations preceding and subsequent to the immersion phase (40 CFR 420.121-91).
(2) The deposition of zinc on the surface of steel for corrosion protection (EPA-83/06a).

Game fish: Species like trout, salmon, or bass, caught for sport. Many of them show more sensitivity to environmental change than "rough" fish (EPA-94/04).

Gamma multi-hit model: A dose-response model of the form: $P(d)$ = Integral from 0 to d of $([a^{**}k][s^{**}(k-l)][exp(-as)]/G(u))ds$; where: $G(u)$ = integral from 0 to infinity of $[s^{**}(u-l)][exp(-s)]ds$; $P(d)$ = the probability of cancer from a dose rate d; k = the number of hits necessary to induce the tumor; a = a constant; when k = 1, *see* the one-hit model (EPA-92/12).

Gamma radiation (or gamma ray):
(1) are true rays of energy in contrast to alpha and beta radiation. The properties are similar to x-rays and other electromagnetic waves. They are the most penetrating waves of radiant nuclear energy but can be blocked by dense materials such as lead (EPA-89/12).
(2) Electromagnetic radiation of extremely short wavelength (0.001 to 0.1 micrometer) emitted from the nucleus in many radioactive decay processes, e.g., from the decay of I-125. The gamma ray is the most penetrating waves of radiant nuclear energy. It does not contain particles but may be emitted along with both alpha and beta particles and it can be stopped by dense materials like lead (EPA-88/09a).
(3) A form of electromagnetic, high-energy radiation emitted from a nucleus. Gamma rays are essentially the same as x-rays and require heavy shieldings, such as concrete or steel to be stopped (EPA-88/08a).
(4) High-energy, short-wavelength, electromagnetic radiation accompanying fission and emitted from the nucleus of an atom. Gamma rays are very penetrating and can be stopped only by dense materials (such as lead) or a thick layer of shielding materials (DOE-91/04).
(5) *See* radiation for more related terms.

Gamma ray: *See* synonym, gamma radiation.

Gap location: The position of the electrode gap in the combustion chamber (40 CFR 85.2122 (a)(8)(ii)(D)-91).

Gap spacing: The distance between the center electrode and the ground electrode where the high voltage ignition arc is discharged (40 CFR 85.2122 (a)(8)(ii)(C)-91).

Garbage:
(1) Spoiled or waste food that is thrown away, generally defined as wet food waste. It is used as a general term for all products discarded (EPA-89/11).
(2) Animal and vegetable waste resulting from the handling, storage, sale, preparation, cooking, and serving of foods (EPA-94/04).

Garbage grinding (or food waste disposer): A method of grinding food waste by a household disposal and washing it into the sewer system. Ground garbage then must be disposed of as sewage sludge (*see* size reduction machine for more related terms) (EPA-74/11, *see also* EPA-83).

Garrison facility: Any permanent military installation (40 CFR 60.331-91).

Gas:
(1) One of three states of aggregation of matter, having neither independent shape nor volume and tending to expand indefinitely (EPA-83/06).
(2) Natural gas, flammable gas, or gas which is toxic or corrosive (40 CFR 192.3-91).
(3) In air pollution control, pollutant gases include organic gases and inorganic gases (*see* pollutant, air).

Other gas related terms include:
- Heavy gas
- Saturated gas

Gas absorption: The taking in of a gas or vapor by a liquid. It includes both physical and chemical absorption processes. One of key factors in designing an absorber is the determination of the maximum concentration that a gas can be dissolved in a liquid.
- **Raoult's law** is used to calculate the maximum concentration of a dissolved gas in a concentrated solution.
- **Henry's law** is used to calculate the maximum concentration of a dissolved gas in a dilute solution (Hesketh-79, p143).

Gas analyzer: The portion of the gas concentration measurement system that senses the gas to be measured and generates an output proportional to its concentration (*see* gas concentration measurement system for more related terms).

Gas burner: A burner which is used to burn gas (*see* burner for more related terms) (AP-40).

Gas burst agitation: This is the most common method of automatic agitation found in automatic processing machines. Gas is released at controlled intervals through tiny holes in a distributor plate in the bottom of the solution tank. The gas bubbles formed during release provide the random agitation pattern necessary for uniform results (EPA-80/10).

Gas barrier: Any device or material used to divert the flow of gases produced in a sanitary landfill or by other disposal techniques (cf. cut-off trench) (SW-108ts).

Gas carburizing: The introduction of carbon into the surface layers of mill steel by heating in a current of gas high in carbon (EPA-83/06a).

Gas chromatography (GC)--compound separator: One of continuous emission monitors (*see* continuous emission monitor for various types). GC is a common technique used for separating and analyzing mixtures of gases and vapors.

GC is an instrument which uses the gas chromatography separation technique to segregate the components of organic substances in a mixture (cf. chromatograph). A gas mixture is percolated though a column of porous solids or liquid coated solids which selectively retard sample components. A carrier gas is used to bring the discreet bands to a detector and through analysis of the detector response and the component retention time, the sample can be identified and quantified. Gas chromatography has been in use in the laboratory since 1905; however, it has only recently been used in continuous monitoring applications.

The most common type of GC column is a liquid coated on an inert solid support contained in a small bore (1/8 in.) stainless steel or glass tube. The choice of the proper liquid coating is crucial and is usually similar in chemical structure to the sample components of interest. Organic compounds can be divided into five classes of solute polarity ranging from most polar (Class I) to non-polar (Class V). Compounds having similar boiling points can be separated by choosing column materials of the appropriate polarity. With the right choice, the order of elution of the sample components can be manipulated. Through selection of the best solid support/liquid phase combination, and through optimizing sample size, flow rates, column temperatures and length, and the sensitivity of the detector, a gas chromatograph can be adapted for almost any sample analysis situation (EPA-84/03a).

Once the compounds of interest have been appropriately separated by a gas chromatography, they are swept into a detector for compound identification. GC and its compound detectors include:
(1) GC/AFID (gas chromatography/alkali flame ionization detector) (EPA-84/03a)
(2) GC/DD (gas chromatography/dual detector) (EPA-84/03a)
(3) GC/ECD (gas chromatography/electrolytic conductivity detector) (EPA-84/03a)
(4) GC/ECD (gas chromatography/electron capture detector) (EPA-84/03a)
(5) GC/FID (gas chromatography/flame ionization detector) (Course 165.5)
(6) GC/HECD (gas chromatography/hall electrolytic conductivity detector) (EPA-84/03a)
(7) GC/IR (gas chromatography/infrared absorption spectrometer) (EPA-84/03a)
(8) GC/MS (gas chromatography/mass spectrometry) (EPA-84/03a)
(9) GC/PID (gas chromatography/photoionization detector) (Course 165.5)
(10) GC/TCD (gas chromatography/thermal conductivity detector) (EPA-84/03a)
(11) GC/TSD (gas chromatography/thermionic specific detector) (EPA-84/03a)

Gas chromatograph/mass spectrometer: Highly sophisticated instrument that identifies the molecular composition and concentrations of various chemicals in water and soil samples (EPA-94/04).

Gas cleaning: In coal gasification, H_2S is removed during the gas cleaning step, generally by a scrubbing process. H_2S is converted to elemental sulfur by partial oxidation and catalytic conversion. The synthetic gas produced is sulfur free and can be burned without releasing harmful pollutants (*see* coal gasification for more related terms) (EPA-81/12, p8-6).

Gas collection: The process of dissolving gaseous pollutants in a liquid is referred to as **absorption**. Absorption is a mass transfer operation. Maw transfer can be compared to heat transfer in that both occur because a system is trying to reach equilibrium conditions. For example, in heat transfer, if a hot slab of metal is placed on top of a cold slab, heat energy will be transferred from the hot slab to the cold slab until both are at the same temperature (equilibrium). In absorption, mass instead of heat is transferred as a result of a concentration difference, rather than a heat-energy difference. Absorption continues as long as a concentration differential exists between the liquid and the gas from which the contaminant is being removed. In absorption, equilibrium depends on the solubility of the pollutant in the liquid.

To remove a gaseous pollutant by absorption, the exhaust stream must be passed through (brought in contact with) a liquid. Three steps involve in absorption. In the first step, the gaseous pollutant diffuses from the bulk area of the gas phase to the gas--liquid interface. In the second step, the gas moves (transfers) across the interface to the liquid phase. This step occurs extremely rapidly once the gas molecules (pollutant) arrive at the interface area. In the third step, the gas diffuses into the bulk area of the liquid, thus making room for additional gas molecules to be absorbed. The rate of absorption (mass transfer of the pollutant from the gas phase to the liquid phase) depends on the diffusion rates of the pollutant in the gas phase (first step) and in the liquid phase (third step).

To enhance gas diffusion and, therefore, absorption:
(1) Provide a large interfacial contact area between the gas and liquid phases.
(2) Provide good mixing of the gas and liquid phases (turbulence).
(3) Allow sufficient residence, or contact, time between the phases for absorption to occur.

Two of these three gas-collection mechanisms, large contact area and good mixing, are also important for particle collection. The third factor, sufficient residence time. works in direct opposition to efficient particle collection. To increase residence time, the relative velocity of the gas and liquid streams must be reduced. Therefore, achieving a high removal efficiency for both gaseous and particulate pollutants is extremely difficult unless the gaseous pollutant is very soluble in the liquid.

As previously mentioned, a very important factor affecting the amount of a pollutant that can be absorbed is its solubility. Solubility governs the amount of liquid (liquid-to-gas ratio) required and the necessary contact time. More soluble gases require less liquid. Also, more soluble gases will be absorbed faster. Solubility is a function of both the temperature and, to a lesser extent, the pressure of the system. As temperature increases, the amount of gas that can be absorbed by a liquid decreases. From the ideal gas law: as temperature increases, the volume of a gas also increases; therefore, at a higher temperature, gas volume increases and less gas is absorbed. For this reason, some absorption systems use inlet quench sprays to cool the incoming exhaust stream, thereby increasing absorption efficiency. Pressure affects the solubility of a gas in the opposite manner. When the pressure of a system is increased, the amount of gas absorbed generally increases (EPA-84/03b, p1-7).

Gas concentration measurement system: The total equipment required for the determination of gas concentration. The measurement system consists of the following major subsystems:
(1) Sample interface
(2) Gas analyzer
(3) Data recorder
(4) *See* measurement for more related terms.

Gas/gas method: Either of two methods for determining capture which rely only on gas phase measurements. The first method requires construction of a temporary total enclosure (TTE) to ensure that all would be fugitive emissions are measured. The second method uses the building or room which houses the facility as an enclosure. The second method requires that all other VOM sources within the room be shut down while the test is performed, but all fans and blowers within the room must be operated according to normal procedures (40 CFR 52.741-91).

Gas filter correlation analyzer (GFC): One of continuous emission monitors (*see* continuous emission monitor for various types). GFC offers improved specificity over NDIR. GFC spectroscopy is based upon comparison of the detailed structure of the infrared absorption spectrum of the measured gas to that of other gases also present in the sample being analyzed. The technique is implemented by using the measured gas itself, in high concentration, as a filter for the infrared radiation transmitted through the analyzer (EPA-84/03a).

Gaseous flow rate meter: The best types of flowmeters for gases are the orifice meter and the Vortex shedding meter (*see* flow rate meter for more related terms) (EPA-89/06).

Gas meter: An instrument for measuring the quantity of a gas passing through the meter (LBL-76/07-air).

Gas nitriding: A process of hardening metals by heating and diffusing nitrogen gas into the surface (EPA-83/06a).

Gas phase process: A polymerization process in which the polymerization reaction is carried out in the gas phase; i.e., the monomer(s) are gases in a fluidized bed of catalyst particles and granular polymer (40 CFR 60.561-91).

Gas phase separation: The process of separating volatile constituents from water by the application of selective gas permeable membranes (EPA-83/06a).

Gas phase thermal decomposition kinetics: A generic term applied to high-temperature thermal decomposition (oxygen concentrations less than the stoichiometric value) in absence of significant exothermic chemical reactions radical concentrations, and temperature gradients (EPA-88/12).

Gas phase thermal oxidation kinetics: A generic term applied to high-temperature thermal decomposition in the presence of excess oxygen (greater than the stoichiometric value) without large exothermic chemical reaction, radical concentrations, and temperature gradients (EPA-88/12).

Gas service: That the component contains process fluid that is in the gaseous state at operating conditions (40 CFR 52.741-91).

Gas side pressure drop: *See* synonym, pressure drop.

Gas solubility coefficient: The amount of gas that a unit water can absorb under 1 atmospheric pressure and a given temperature.

Gas tight: Operated with no detectable emissions (40 CFR 60.691-91).

Gas tight syringe:
(1) Commercially available syringe which isolates and delivers a gases sample (cf. hypodermic syringe) (EPA-88/12).
(2) *See also* syringe, gas-tight.

Gas turbine: A system to convert heat contained in combustion gases into work, e.g., to run a shaft or an electrical generator (*see* turbine for more related terms).

Gas turbine model: A group of gas turbines having the same nominal air flow, combuster inlet pressure, combuster inlet temperature, firing temperature, turbine inlet temperature and turbine inlet pressure (*see* turbine for more related terms) (40 CFR 60.331-91).

Gas viscosity: In a gas, the molecules are too far apart for intermolecular cohesion to be effective. Thus, shear stress is predominantly the result of an exchange of momentum between flowing strata caused by molecular activity. Because molecular activity increases with temperature increases, the shear stress increases with a rise in the temperature. Therefore, gas viscosity is increased when the temperature increases (*see* viscosity for more related terms) (EPA-81/12, p2-6).

Gas washer: An apparatus used to remove entrained solids and other substances from carbon dioxide gas from a lime kiln (EPA-74/01a).

Gas washer or scrubber: *See* synonym, flue gas washer or scrubber (EPA-83).

Gas well: Any well which produces natural gas in a ratio to the petroleum liquids produced greater than 15,000 cubic feet of gas per 1 barrel (42 gallons) of petroleum liquids (*see* well for more related terms) (40 CFR 435.61-91).

Gaseous flow rate meter: The best types of flowmeters for gases are the orifice meter and the Vortex shedding meter, discussed in the liquid flow rate meter (EPA-6/89).

Gasification:
(1) The pulverized coal is gasified in a reactor with limited oxygen. Gasification produced either a low, medium, or high Btu gas by applying heat and pressure or by using a catalyst to break down the components of coal (EPA-80/08).
(2) The gas produced contains carbon monoxide (CO), hydrogen (H_2), carbon dioxide (CO_2), water (H_2O), methane (CH_4), and contaminants such as hydrogen sulfide and char. Low and medium Btu gas contains more CO and H, than high Btu gas which contains a higher coal content. Methane gas produces more heat when burned. The sulfur in the coal is converted to H_2S during gasification (EPA-81/12, p8-6).
(3) Conversion of solid material such as coal into a gas for use as a fuel (EPA-94/04).
(4) *See* coal gasification for more related terms.

Gasket: A rubber, metal or other material used to place around a joint to make the joint gas or liquid tight (cf. beater add gasket).

Gasohol:
(1) Mixture of gasoline and ethanol derived from fermented agricultural products containing at least nine percent ethanol. Gasohol emissions contain less carbon monoxide than those from gasoline (EPA-94/04).
(2) A blending fuel of gasoline (90 %) and alcohol (10 %).

Gasoline: Any petroleum distillate or petroleum distillate/alcohol blend having a Reid vapor pressure of 27.6 kPa or greater which is used as a fuel for internal combustion engines (40 CFR 52.741-91). Other gasoline related terms include:
- Base gasoline
- Baseline gasoline
- Conventional gasoline
- Leaded gasoline
- Reformulated gasoline

Gasoline blending stock or component: Any liquid compound which is blended with other liquid compounds or with lead additives to produce gasoline (40 CFR 80.2-91).

Gasoline dispensing facility: Any site where gasoline is transferred from a stationary storage tank to a motor vehicle gasoline tank used to provide fuel to the engine of that motor vehicle (40 CFR 52.741-91).

Gasoline service station: Any site where gasoline is dispensed to motor vehicle fuel tanks from stationary storage tanks (40 CFR 60.111b-91).

Gasoline tank truck: A delivery tank truck used at bulk gasoline terminals which is loading gasoline or which has loaded gasoline on the immediately previous load (40 CFR 60.501-91).

Gasoline volatility: The property of gasoline whereby it evaporates into a vapor. Gasoline vapor is a volatile organic compound (EPA-94/04).

Gasometer: An apparatus employing a calibrated volume which is used to calibrate gas measuring devices (LBL-76/07-air).

Gasometric: Pertaining to measurement of a gas parameter (LBL-76/07-water).

Gathering line: A pipeline that transports gas from a current production facility to a transmission line or main (40 CFR 192.3-91, *see also* 40 CFR 195.2; 280.12-91).

Gate:
(1) One of the electrodes in a field effect transistor (EPA-83/03).
(2) An entry passage for molten metal into a mold (EPA-85/10a).

Gauge pressure (or gage pressure) (psig):
(1) A measure of pressure expressed as a quantity above atmospheric pressure or some other reference pressure (EPA-84/09).
(2) The difference between absolute and atmospheric pressure in a particular system and is normally measured by an instrument which has atmospheric pressure as a reference, i.e., zero gauge pressure is equal to atmospheric pressure (EPA-83/06).
(3) *See* pressure for more related terms.

Gauging station: A location on a stream or conduit where measurements of discharge are customarily made. The location includes a stretch of channel through which the flow is uniform and a control downstream from this stretch. The station usually has a recording or other gauge for measuring the elevation of the water surface in the channel or conduit (EPA-82/11f).

Gaussian plume model: An approximation of the dispersion of a plume from a continuous point source. The concentration distribution perpendicular to the plume axis is assumed to be Gaussian. The plume travels with a uniform wind velocity downwind from the source. Its dimensions perpendicular to the wind direction are described by dispersion parameters as a function of distance or travel time from the source. The dispersion coefficients depend on diffusion categories and sometimes also on the source height and the surface roughness. The basic assumption underlying the Gaussian plume model is that the dispersion takes place in a stationary and homogeneous atmosphere, with a sufficient wind speed (greater than or equal to 1 meter/second) (NATO-78/10).

Gaussian puff model: An approximation of the dispersion of a puff from an instantaneous point source. The concentration distribution inside the puff is assumed to be Gaussian. The dimensions of the puff are described by dispersion parameters as a function of travel time of the puff. These dispersion coefficients depend on diffusion categories and sometimes also on the source height and the surface roughness (NATO-78/10).

GC/AFID (gas chromatography/alkali flame ionization detector): One of continuous emission monitors (*see* continuous emission monitor for various types). AFID is used to analyze pesticides. For AFID, a small pellet of Cs, Br, or Rb_2SO_4 is placed at the tip of a typical FID burner which is operated in a starved oxygen mode. The sensitivity of the detector to phosphorous containing compounds is enhanced 5000 to 1 over normal hydrocarbons. Different salt pellets have been used to enhance sensitivity to different elements, the next most common application being nitrogen compounds (EPA-84/03a).

GC/DD (gas chromatography/dual detector): One of continuous emission monitors (*see* continuous emission monitor for various types). Combining the high sensitivity of the FID detector to easily pyrolyzed, low boiling organic compounds with the high sensitivity of ECD detectors to higher boiling aromatics, etc., is one method for more complete quantification of organic emissions from an incinerator. One instrument could be chosen with two columns, each specified to enhance the separation of an response time associated with the appropriate organic compounds. One or both detector(s) could be used at a given time, depending on the waste being burned. The non-destructive nature of the PID detector allows its use in series with other GC detectors. PID/FID, PID/ECD and PID/NPD combinations have all been used for specific applications. A typical PID/FID application is the identification of hydrocarbon classes. The PID response increases with increasingly degrees of un-saturation and the FID response is mostly unaffected by double bonds. By comparison of the relative response of these two detectors, alkanes, olefins and aromatic compound classes can be identified in complex sample matrices (EPA-84/03a).

GC/ECD (gas chromatography/electrolytic conductivity detector): One of continuous emission monitors (*see* continuous emission monitor for various types). A device which measures the conductivity of an electrolyte, e.g, it can be used to measure mineral ionic concentration in a solution (cf. thermal conductivity detector) (EPA-84/03a).

GC/ECD (gas chromatography/electron capture detector): One of continuous emission monitors (*see* continuous emission monitor for various types). ECD measures the loss of electrical signal rather than the produced electrical current. Carrier gas molecules are excited by a radioactive source to produce a steady background current under a fixed applied voltage. When a sample is introduced that absorbs electrons, the current is reduced and this reduction is indicated by an amplifying electrometer. The source of the electron current in an ECD is usually the radioactive decay of nickel 63 or tritium. The high energy electrons emitted interact with the carrier gas which in turn interact with sample constituents.

Usually, a small amount of methane is added to the carrier gas (argon) to deactivate further reactions by excited molecules. However, this dilution of sample gas results in reduced sensitivity. Pure nitrogen carrier gases have been used in laboratory studies with mixed success. The ECD is well adapted for use in the analysis of 10^{-10} to 10^{-12} levels of pollutants incinerator effluents. The ECD may be operated above 350° C if the Ni63 source is used. However, it is highly sensitive to temperature changes. Baseline drifts of 50 percent have been reported for a 2° C room temperature change. This detector is extremely sensitive to sub-picogram quantities of pesticides, alkyl halides, conjugated carbonyls, nitrides and organometallics, and is quite insensitive to hydrocarbons, alcohols and ketones. In summary, the ECD is highly sensitive, not very linear, and is very sensitive to surrounding conditions. Therefore probably not well suited to the rigorous demands of continuous monitoring (EPA-84/03a).

GC/FID (gas chromatography/flame ionization detector): One of continuous emission monitors (*see* continuous emission monitor for various types). FID uses a hydrogen flame as the means to ionize organic (toxic) vapors. It responds to virtually all organic compounds, that is, compounds that contain carbon-hydrogen or carbon-carbon bonds. The flame detector analyzes the mechanisms of breaking-bonds as the following reaction indicates: $Rh + O ---> RHO^+ + e^- ---> CO_2 + H_2O$.

Inside the detector chamber, the sample is exposed to a hydrogen flame which ionizes the organic vapors. When most organic vapors burn, positively charged carbon-containing ions are produced which are collected by a negatively charged collecting electrode in the chamber. An electric field exists between the conductors surrounding the flame and a collecting electrode. As the positive ions are collected, a current proportional to the hydrocarbon concentration is generated on the input electrode. This current is measured with a preamplifier which has an output signal proportional to the ionization current. A signal conducting amplifier is used to amplify the signal from the preamp and to condition it for subsequent meter or external recorder display.

When this is compared to a photoionization device (PID), a major difference should be noted between the detectors. PID detection is dependent upon the ionization potential (eV) and the ease in which an electron can be ionized (displaced) from a molecule. This mechanism is variable, highly dependent on the individual characteristics of a particular substance. This results in a more variable response factor for the vast majority of organics that are ionizable. Therefore, in general, one does not *see* large sensitivity shifts between different substances when using an FID as compared to a PID. Flame ionization detectors are the most sensitive for saturated hydrocarbons, alkanes and unsaturated hydrocarbons alkenes. Substances that contain substituted functional groups such as hydroxide (OH^-), and chloride (Cl^-), tend to reduce the detector's sensitivity; however, overall, the detectabilities remain good (Course 165.5).

GC/HECD (gas chromatography/hall electrolytic conductivity detector): One of continuous emission monitors (*see* continuous emission monitor for various types). HECD is particularly sensitive to nitrogen, sulfur and halogen containing compounds and is more discriminating than the ECD (EPA-84/03a).

GC/IR (gas chromatography/infrared absorption spectrometer): One of continuous emission monitors (*see* continuous emission monitor or hybrid chromatograph monitor for various types). Combining the separation and detection power of a GC with the identification capabilities of an IR absorption spectrometer produces a more specific instrumental technique for identification of organic compounds. To date, the major problem with combining these techniques has been the development of suitable equipment to interface the two instruments. The volume of sample from the GC is traditionally only a tiny fraction of the volume of sample usually required by an IR instrument. Reducing the cell volume of the IR decreases sensitivity unless additional passes through the cell are provided. IR cells (light pipes) have been developed that are compatible with capillary GC instruments and through the development of computer software, GC/IR and GC/FTS-IR are gaining popularity as the "poor mans" GC/MS. Within the next few years, this technology will be developed to the point where it will be applicable to continuous air pollutant monitoring. At present, only 55 toxic substances have been detected in the laboratory with the GC/FTS-IR technique with minimum detectable quantities on the order of 1 to 10 micrograms (EPA-84/03a).

GC/MS (gas chromatography/mass spectrometry): One of continuous emission monitors (*see* continuous emission monitor or hybrid chromatograph monitor for various types). A compound in a gas matrix can be more fully identified through analysis of retention time in a GC and the mass spectrum. Identity can be established by comparing the total ion current profile of an eluted compound to a published standard spectrum. GC/MS techniques are particularly suited for analysis of organics in water through a concentration step. GC/MS has been used to identify organic ambient air contaminants. The concentration step involves passing the air sample through an absorber column that traps the organic material followed by thermal or solvent desorption of that material into the GC. This technique is semi-continuous and overall response times of a GC/MS are typically greater than three minutes. This powerful research tool could be adapted to identify and quantitate organic compounds in incinerator effluents in close to real time. At present, no GC/MS instrumentation is in routine use as a continuous monitor. Double mass spectrometry (MS/MS) and laser multi-photon ionization mass spectrometry have been identified as potential on-line or real time instruments for the identification of polycyclic aromatic hydrocarbons. These instruments do not use the GC for separation of components and therefore do not involve the same delays in response time.

A GC/MS or MS/MS technique is a rather complicated instrument, i.e., the mass spectra produced is a complex and close to real time results can only be provided through a computer with extensive library searching capabilities. The MS can scan for certain compounds within seconds; however, full spectrum scans usually take greater than three minutes. These disadvantages should be weighed against the high sensitivity and resolution capabilities of the GC/MS system (EPA-84/03a).

GC/PID (gas chromatography/photoionization detector): One of continuous emission monitors (*see* continuous emission monitor for various types). All atoms and molecules are composed of particles: electrons, protons, and neutrons. Electrons, negatively charged particles, rotate in orbit around the nucleus, the dense inner core. The nucleus consists of an equal number of protons (positively charged particles) as electrons found in the orbital cloud. The interaction of the oppositely charged particles and the laws of quantum mechanics keep the electrons in orbits outside the nucleus. The energy required to remove the outermost electron from the molecule is called the ionization potential (IP) and is specific for any compound or atomic species. Ionization potentials are measured in electron volts (eV). High frequency radiation (ultraviolet and above) is capable of causing ionization and is hence called ionizing radiation.

When a photon of ultraviolet radiation strikes a chemical compound, it ionizes the molecule if the energy of the radiation is equal to or greater than the IP of the compound. Since ions are charged particles, they may be collected on a charged plate and produce a current. The measured current will be directly proportional to the number of ionized molecules. The photoionization process can be illustrated as: $R + h\nu \longrightarrow R^+ + e^-$, where R is an organic or inorganic molecule and $h\nu$ represents a photon of UV light with energy equal to or greater than the ionization potential of that particular chemical species. R^+ is the ionized molecule (Course 165.5).

GC/TCD (gas chromatography/thermal conductivity detector): One of continuous emission monitors (*see* continuous emission monitor for various types). A detector which measures thermal conductivity. It is a common detector that is sensitive to all compounds. Since pollutants at trace levels are of interest in hazardous waste incinerators, the other components of the sample would pose an insurmountable selectivity problem with this analyzer. Highly selective detectors have been developed where sensitivity to certain compounds or elements in compounds has been enhanced (cf. electrolytic contivity detector) (EPA-84/03a).

GC/TSD (gas chromatography/thermionic specific detector): One of continuous emission monitors (*see* continuous emission monitor for various types). TSD is also a highly sensitive detector that is specific for nitrogen containing compounds. These detectors usually achieve selectivity and sensitivity at the expense of durability and reliability (EPA-84/03a).

GC: Gas chromatograph, *see* definition under continuous emission monitor.

Gear forming: A process for making small gear by rolling the gear material as it is pressed between hardened gear shaped dies (EPA-83/06a).

Gear oils: The petroleum-based oils used for lubricating machinery gears (40 CFR 252.4-91).

Geiger counter: An electrical device that detects the presence of certain types of radioactivity (EPA-89/12).

Gel: A colloidal, jelly-like solid.

Gel chromatography: A type of chromatography column in which the stationary phase is a gel and the moving phase is a liquid.

Gel electrophoresis: Electrophoresis takes place in a gel medium (*see* electrophoresis for more related terms).

Gelation: The process of forming a gel.

Gelled electrolyte: An electrolyte which may or may not be mixed with electrode material, that has been gelled with a chemical agent to immobilize it (EPA-84/08).

Gelling: A process for producing dried pulp for cattle feed from the peelings, cores and trimmings of cannery wastes. The sediment is pressed and dried and used as cattle feed (EPA-83).

Gene:
(1) A length of DNA that directs the synthesis of a protein (EPA-89/12).
(2) A sequence of DNA that codes for a specific protein such as an inherited trail (EPA-88/09a).

Gene flow: The passage of genes from one population to an another population.

Gene library: A collection of DNA fragments from cells or organisms. So far, no simple way for sorting the contents of gene libraries has been devised. However, DNA pieces can be moved into bacterial cells where sorting according to gene function becomes feasible (EPA-89/12).

General counsel: The General Counsel of the U.S. Environmental Protection Agency, or his or her designee (40 CFR 15.4-91, *see also* 40 CFR 23.1-91).

General environment: The total terrestrial, atmospheric and aquatic environments outside sites upon which any operation which is part of a nuclear fuel cycle is conducted (*see* environment for more related terms) (40 CFR 190.02-91, *see also* 40 CFR 191.02-91).

General permit: A permit applicable to a class or category of dischargers (*see* permit for more related terms) (EPA-94/04, *see also* 40 CFR 122.2; 124.2; 232.2-91).

General processing: The internal subdivision of the low water use processing subcategory for facilities described in 40 CFR 410.30 that do not qualify under the water jet weaving subdivision (40 CFR 410.31-91).

General purpose incinerator: An incinerator that burns miscellaneous types of wastes, usually from numerous sources and customers (*see* incinerator for more related terms) (EPA-81/09).

General reporting facility: A facility having one or more hazardous chemicals above the 10,000 pound threshold for planning quantities. Such facilities must file MSDS and emergency inventory information with the SERC and LEPC and local fire departments (EPA-94/04).

Generation:
(1) The act or process of producing solid waste (40 CFR 243.101-91, *see also* 40 CFR 246.101-91).
(2) The conversion of chemical or mechanical energy into electrical energy (EPA-82/11f).
(3) The branching pattern of the airways. Each division into a major daughter (larger in diameter) and minor daughter airway is termed a generation. Numbering begins with the trachea (EPA-90/08).

Generation of wastewater: The process whereby wastewater results from the manufacturing process. Wastewater may be generated but not discharged (*see* wastewater for more related terms) (EPA-79/12a).

Generator:
(1) A device that produces electricity and which is reported as a generating unit pursuant to Department of Energy Form 860 (CAA402-42USC7651a-91).
(2) Any person, by site location, whose act or process produces hazardous waste identified or listed in 40 CFR 261 (40 CFR 144.3-91, *see also* 40 CFR 146.3; 260.10; 270.10-91).
(3) Any person, by site, whose act or process produces regulated medical waste as defined in Subpart D of this part, or whose act first cause a regulated medical waste to become subject to regulation. In the case where more than one person (e.g., doctors with separate medical practices) are located in the same building, each individual business entity is a separate generator for the purposes of this part (40 CFR 259.10-91).
(4) A facility or mobile source that emits pollutants into the air or releases hazardous waste into water or soil (EPA-94/04).
(5) Any person, by site, whose act or process produces regulated medical waste or whose act first causes such waste to become subject to regulation. In a case where more than one person (e.g., doctors with separate medical practices) is located in the same building, each business entity is a separate generator (EPA-94/04).

Generator column: Is used to partition the test substance between the octanol and water phases (40 CFR 796.1720-91, *see also* 40 CFR 796.1860-91).

Generator of PCB waste: Any person whose act or process produces PCBs that are regulated for disposal under subpart D of this part, or whose act first causes PCBs or PCB Items to become subject to the disposal requirements of subpart D of this part, or who has physical control over the PCBs when a decision is made that the use of the PCBs has been terminated and therefore is subject to the disposal requirements of subpart D of this part. Unless another provision of this part specifically requires a site-specific meaning, "generator of PCB waste" includes all of the sites of PCB waste generation owned or operated by the person who generates PCB waste (40 CFR 761.3-91).

Generic process chemistry: A class of chemical reactions which share a common mechanism or yield related products (e.g., chlorination, oxidation, ammoxidation cracking and reforming, and hydrolysis). Forty-one major generic processes have been identified in the organic chemical and plastics/synthetic fibers industries (EPA-87/10a).

Generic technology: A group of technologies which use different methods to perform the same task (EPA-80/08).

Genetic engineering:
(1) A process of inserting new genetic information into existing cells in order to modify any organism for the purpose of changing one of its characteristics (EPA-94/04).
(2) The use of laboratory methods to alter a cell's genetic code so that it will produce desired chemicals or perform desired functions (EPA-88/09a).

Genetically engineered microorganism (GEM): A microorganism that has undergone external processes by which its basic set of genes has been altered (EPA-88/09a).

Genetics: A science (a branch of biology) dealing with the study of biological heredity and variation.

Genome: The total genetic information needed to reproduce a cell or a virus. In bacteria, it is equivalent to one chromosome (EPA-88/09a).

Genotoxic chemicals: Chemicals that are toxic to genetic materials, including carcinogens, teratogens, and mutagens (EPA-3/80).

Genotoxic: The ability of a substance to damage an organism's genetic material (DNA) (EPA-91/03).

Genotype: The composition of a gene in an organism.

Geographic information system (GIS): A computer system designed for storing, manipulating, analyzing, and displaying data in a geographic context (EPA-94/04).

Geological log: A detailed description of all underground features(depth, thickness, type of formations) discovered during the drilling of a well (EPA-94/04).

Geology: The science that deals with (the study of) the earth: the materials, processes, environments, and history of the planet, especially the lithosphere, including the rocks and their formation and structure (DOE-91/04).

Geomembrane: An essentially impermeable membrane used as a solid, liquid or vapor barrier with foundation, soil, rock, earth, or any other geotechnical engineering-related material as an integral part of a human-made project, structure, or system (ASTM definition) (*see* liner for more related terms) (EPA-89/09, *see also* EPA-91/05).

Geometric mean: The nth root of a production of n factors (40 CFR 131.35-91). In a sample of N discrete values, the geometric mean (G) is defined as: $\ln(G) = [\text{sum } \ln (X_i)]/N$, $i = 1....N$, where: G = geometric mean; ln = natural logarithm; N = number of samples; X_i = discrete values at i th sample (*see* mean for more related terms) (NATO-87/10).

Geometric mean diameter or median diameter: The calculated aerodynamic diameter which divides the particles of an aerosol in half based on the weight of the particles. Fifty percent of the particles by weight will be larger than the median diameter and 50 percent of the particles will be smaller than the median diameter. The median diameter describes the particle size distribution of any aerosol based on the weight and size of the particles (40 CFR 798.2450-91).

Geometric standard deviation: A measure of the variation or dispersion of data. The degree to which numerical data tend to spread about an average value is called the variation or dispersion of the data. Various measures of dispersion or variation are available, the most common being the (geometric) standard deviation (EPA-84/09).

Geophysical log: A record of the structure and composition of the earth encountered when drilling a well or similar type of test hold or boring (EPA-94/04).

Geostrophic wind: The horizontal wind velocity, for which the horizontal component of the Coriolis acceleration exactly balances the horizontal pressure force (*see* wind for more related terms) (NATO-78/10).

Geotextile: Any permeable textile used with foundation, soil, rock, earth, or any other geotechnical engineering-related material as an integral part of a human-made project, structure, or system (ASTM definition) (*see* liner for more related terms) (EPA-89/09, *see also* EPA-91/05).

Geothermal energy: Energy (or heat) within the interior of the earth.

Germ cells: Reproductive cells including sperm (spermatozoon) and egg (ovum) cells.

Germ line: Is comprised of the cells in the gonads of higher eukaryotes, which are the carriers of the genetic information for the species (40 CFR 798.5195-91, *see also* 40 CFR 798.5200-91).

Germanium (Ge): A metalloid element with a.n. 32; a.w. 72.59; d. 5.32 g/cc; m.p. 937.4° C and b.p. 2830° C. The element belongs to group IVA of the periodic table.

Germicide: Any compound that kills disease-causing microorganisms (EPA-94/04).

Germination: The resumption of active growth by an embryo (40 CFR 797.2800-91, *see also* 40 CFR 797.2750-91).

Getter:
(1) A metal coating inside a lamp which is activated by an electric current to absorb residual water vapor and oxygen (EPA-83/03).
(2) A material capable of capturing gases. In an NPR light-water reactor, a material incorporated into the tritium target to prevent buildup of pressure in the target and prevent loss of tritium (DOE-91/04).

Geyser: A thermal spring that erupts intermittently (DOI-70/04).

Giardia lamblia: Protozoan in the feces of man and animals that can cause severe gastrointestinal diseases when it contaminates drinking water (EPA-94/04).

Gibbs-Dalton law: Also known as the **Dalton law**. The law states that in a mixture of ideal gases, the properties of each component behaves as if it existed alone in the system at the volume and the temperature of the mixture (Holman-p312; Wark-p323).

Gigawatt year: Refers to the quantity of electrical energy produced at the busbar of a generating station. A gigawatt is equal to one billion watts. A gigawatt-year is equivalent to the amount of energy output represented by an average electric power level of one gigawatt sustained for one year (40 CFR 190.02-91).

Gilsonite: A material used primarily for sand binders. It is one of the purest natural bitumens (99.9 percent) and is found in lead mines. Lead may be present as an impurity in Gilsonite (EPA-85/10a).

Glacial acetic acid (CH_3COOH): A caustic liquid used as a solvent.

Gland: A device of soft wear-resistant material used to minimize leakage between a rotating shaft and the stationary portion of a vessel such as a pump (EPA-83/09).

Gland water: The water used to lubricate a gland. Sometimes called packing water (*see* water for more related terms) (EPA-87/10).

Glass: A hard, amorphous, inorganic, usually transparent, brittle substance made by fusing silicates, and sometimes borates and phosphates, with certain basic oxides and then rapidly cooling to prevent crystallization (EPA-83/03). Other glass related terms include:
- Bent glass
- Chemical glass
- Chipped glass
- Container glass
- Crystal glass
- Cut glass
- Document glass
- Flat glass
- Flint glass
- Hard glass
- Heat absorbing glass
- Heat resisting glass
- High transmission glass
- Lead glass
- Molded glass
- Opal glass
- Ophthalmic glass
- Optical crown glass
- Optical flint glass
- Oven glass
- Phosphate glass
- Plate glass
- Polished plate glass (*see* synonym, plate glass)
- Polished wire glass
- Pressed glass
- Pressed and blown glass
- Quartz glass
- Rolled glass
- Safety glass
- Sheet glass
- Silica glass
- Slab glass
- Square cut
- Structural glass
- Tempered glass
- Water glass
- Window glass

Glass blowing: The shaping of hot glass by air pressure (EPA-83).

Glass container: General term applied to glass bottles and jars (EPA-83).

Glass cullet: *see* synonym, cullet.

Glass fiber filtration: A standard method of measuring total suspended solids (EPA-83/06a).

Glass electrode: A glass tube (or bulb) containing an acidic solution which is used as an electrode or as a half cell. The tube wall is so thin that hydrogen ions can diffuse through (*see* electrode for more related terms).

Glass fiber reinforced polyisocyanurate/polyurethane foam: The cellular polyisocyanurate or cellular polyurethane insulation made with glass fibers within the foam core (40 CFR 248.4-91).

Glass melting furnace: A unit comprising a refractory vessel in which raw materials are charged, melted at high temperature, refined, and conditioned to produce molten glass. The unit includes foundations, superstructure and retaining walls, raw material charger systems, heat exchangers, melter cooling system, exhaust system, refractory brick work, fuel supply and electrical boosting equipment, integral control systems and instrumentation, and appendaees for conditioning and distributing molten glass to forming apparatuses. The forming apparatuses, including the float bath used in flat glass manufacturing and flow channels in wool fiberglass and textile fiberglass manufacturing, are not considered part of the glass melting furnace (40 CFR 60.291-91, *see also* 40 CFR 61.161-91).

Glass produced: The weight of the glass pulled from the glass melting furnace (40 CFR 60.291-91, *see also* 40 CFR 61.161-91).

Glass pull rate: The mass of molten glass utilized in the manufacture of wool fiberglass insulation at a single manufacturing line in a specified time period (40 CFR 60.681-91).

Glassification: *See* synonym, vitrification.

Glassine: Highly hydrated, transparent, calendered paper used for envelope windows (EPA-83).

Glassine paper: The paper used as protective wrapping of foodstuffs and products including tobacco products, chemicals, metal parts, as well as for purposes where its transparent features are useful (i.e., window envelopes). This paper is grease resistant and has high resistance to the passage of air and any essential oil vapors (*see* paper for more related terms) (EPA-87/10).

Glaze (glaze ice, glazed frost, verglas): A coating of ice, generally clear and smooth but usually containing some air pockets, formed on exposed objects by the freezing of a film of supercooled water deposited by rain, drizzle, or fog or possibly condensed from supercooled water vapor (DOE-91/04).

Glazed paper: The paper with a high gloss or polished finish (*see* paper for more related terms) (EPA-83).

Glassphalt: An asphalt product that uses crushed glass as a partial substitute for aggregate in the mix (OTA-89/10).

Global commons: That area (land, air, water) outside the jurisdiction of any nation (40 CFR 6.1003-91).

Global radiation: The sum of the direct solar radiation and the diffuse sky radiation received on a unit horizontal surface (*see* radiation for more related terms) (NATO-78/10).

Global scale: A space scale which is of the order of magnitude of the earth's circumference (roughly 10,000 km) (NATO-78/10).

Gloss: Surface quality that reflects light (EPA-83).

Glove bag: A sealed compartment with attached inner gloves used for the handling of asbestos-containing materials. Properly installed and used, glove bags provide a small work area enclosure typically used for small-scale asbestos stripping operations. Information on glove-bag installation, equipment and supplies, and work practices is contained in the Occupational Safety and Health Administration's (OSHA's) final rule on occupational exposure to asbestos (appendix G to 29 CFR 1926.58) (40 CFR 61.141-91).

Glovebag: A polyethylene or polyvinyl chloride bag-like enclosure affixed around an asbestos-containing source (most often thermal system insulation) permitting the material to be removed while minimizing release of airborne fibers in the surrounding atmosphere (EPA-94/04).

Gloveboxes: Chambers designed for operators to work with radioactive materials through ports, wearing special gloves that protect them from radioactivity (DOE-91/04).

Glucose: *See* synonym, dextose.

Glycosuria: The presence in the urine of glucose, a monosaccharide formed from more complex sugars and normally retained in the body as a source of energy (LBL-76/07-bio).

Gold (Au): A transition metal with a.n. 79; a.w. 196.96; d. 19.3 g/cc; m.p. 1063° C and b.p. 2970° C. The element belongs to group IB of the periodic table.

Gooch crucible: A porcelain dish used for filtration.

Good combustion practices (GCP): The set of conditions that minimize emission of organic compounds (from municipal waste combustors) (CST-90, Vol. 74, No. 1-6, p230).

Good engineering practice (GEP): A generally acceptable procedure based upon engineering experience and judgment (CRWI-89/05).

Gooseneck: A portion of a service connection between the distribution system water main and a meter. Sometimes called a pigtail (EPA-94/04).

Gouty diathesis: Predisposition to gout (LBL-76/07-bio).

Grab sample:
(1) A single sample which is collected at a time and place most representative of total discharge (40 CFR 471.02-91).
(2) Instantaneous sampling, or a sample taken at a random location and at a random time (EPA-87/10a).
(3) A single sample collected at a particular time and place that represents the composition of the water only at that time and place (EPA-94/04).
(4) *See* sample for more related terms.

Grade:
(1) A term applied to a paper or pulp which is ranked (or distinguished from other papers or pulps) on the basis of its use, appearance, quality, manufacturing history, raw materials, performance, or a combination of these factors (EPA-83).
(2) The level of the ground surrounding a house. In construction, it typically refers to the surface of the ground. Things can be located at grade, below grade, or above grade relative to the surface of the ground (EPA-88/08).

Grade of resin: The subdivision of resin classification which describes it as a unique resin, i.e., the most exact description of a resin with no further subdivision (*see* resin for more related terms) (40 CFR 61.61-91).

Grader: An earth moving equipment. It is a gas- or diesel-powered, pneumatic-wheeled machine equipped with a centrally located blade that can be to level the surface of the ground (SW-108ts).

Gradient: The degree of slope or a rate of change (SW-108ts).

Gradient wind: The horizontal wind velocity tangent to the isobars on a geopotential surface for which the centripital acceleration and the Coriclis acceleration balance the horizontal pressure force (*see* wind for more related terms) (NATO-78/10).

Grain:
(1) In measurement, a unit of mass weight equivalent to 65 milligrams, 2/1,000 of an ounce, or 1/7000 of a pound (EPA-84/09-91; 74/11).
(2) (In agriculture), corn, wheat, sorghum, rice rye, oats, barley and soybeans (40 CFR 60.301-91).
(3) In explosive, a single piece of formed propellant, regardless of size (EPA-76/03).
(4) In plywood manufacture, the direction, size, arrangement, and appearance of the fibers in wood or veneer (EPA-74/04).
(5) In leather tanning, the epidermal side of the tanned hide. The grain side is the smooth side of the hide where the hair is located prior to removal (EPA-82/11).

Grain alcohol: *See* synonym, alcohol.

Grain elevator: Any plant or installation at which grain is unloaded, handled, cleaned, dried, stored, or loaded (40 CFR 60.301-91).

Grain handling operations: Include bucket elevators or legs (excluding legs used to unload barges or ships), scale hoppers and surge bins (garners), turn heads, scalpers, cleaners, trippers, and the headhouse and other such structures (40 CFR 60.301-91).

Grain loading: The rate at which particles are emitted from a pollution source. Measurement is made by the number of grains per cubic foot of gas emitted (*see* loading for more related terms) (EPA-94/04).

Grain loading station: The portion of a grain elevator where the grain is transferred from the elevator to a truck, railcar, barge, or ship (40 CFR 60.301-91).

Grain paper: The paper with machine direction (*see* paper for more related terms) (EPA-83).

Grain storage elevator: Any grain elevator located at any wheat flour mill, wet corn mill, dry corn mill (human consumption), rice mill, or soybean oil extraction plant which has a permanent grain storage capacity of 35,200 m^3 (ca. 1 million bushels) (ca. = capacity) (40 CFR 60.301-91).

Grain terminal elevator: Any grain elevator which has a permanent storage capacity of more than 88,100 m^3 (ca. 2.5 million U.S. bushels), except those located at animal food manufacturers, pet food manufacturers, cereal manufacturers, breweries, and livestock feedlots (ca. = capacity) (40 CFR 60.301-91).

Grain unloading station: The portion of a grain elevator where the grain is transferred from a truck, railcar, barge, or ship to a receiving hopper (40 CFR 60.301-91).

Gram atomic weight: The atomic weight of an element expressed in grams. Similarly, it can be expressed in pounds (cf. pound atomic weight).

Gram equivalent weight: The equivalent weight of an element or a compound expressed in grams. Similarly, it can be expressed in pounds (cf. equivalent weight).

Gram molecular weight: The molecular weight of a compound expressed in grams. Similarly, it can be expressed in pounds.

Granular activated carbon (GAC) treatment: A filtering system often used in small water systems and individual homes to remove organics. GAC can be highly effective in removing elevated levels of radon from water (EPA-94/04).

Granular diammonium phosphate plant: Any plant manufacturing granular diammonium phosphate by reacting phosphoric acid with ammonia (40 CFR 60.221-91).

Granular filter: A device for removing suspended solids from water, consisting of granular material placed in a layer(s) and capable of being cleaned by reversing the direction of the flow (*see* filter for more related terms) (EPA-82/10).

Granular triple superphosphate storage facility: Any facility curing or storing granular triple superphosphate (40 CFR 60.241-91).

Granulated blast furnace slag: The solid product of a predominantly glassy nature formed by the water granulation of blast-furnace slag. Some granulated blast-furnace slags are used as lightweight aggregates for concrete or are finely ground to make a cementitious ingredient of concrete (*see* slag for more related terms) (EPA-83).

Granulated slag: In the steel industry, a product made by dumping liquid blast furnace slags past a high pressure water jet and allowing it to fall into a pit of water. The material looks like light tan sand (*see* slag for more related terms) (EPA-74/06a).

Granulation:
(1) The process which removes remaining moisture from sugar, and thus also separates the crystals from one another (EPA-75/02d).
(2) The process of forming small masses.

Granulator: A rotary dryer used in sugar refineries to remove free moisture from sugar crystals prior to packaging or storing (EPA-75/02d).

Grape sugar: *See* synonym, dextose.

Graphical exposure modeling system (GEMS): *See* definition under dispersion model.

Graphite: A soft black lustrous carbon that conducts electricity and is a constituent of coal, petroleum, asphalt, limestone, etc.(EPA-83/03).

Graphite furnace atomic absorption spectrometry (GFAAS): GFAAS replaces the flame with an electrically heated graphite furnace. The furnace allows for gradual heating of the sample aliquot in several stages. Thus, the processes of desolvation, drying, decomposition of organic and inorganic molecules and salts, and formation of atoms which must occur in a flame or ICP in a few milliseconds may be allowed to occur over a much longer time period and at controlled temperatures in the furnace. This allows an experienced analyst to remove unwanted matrix components by using temperature programming and/or matrix modifiers. The major advantage of this technique is that it affords extremely low detection limits. It is the easiest to perform on relatively clean samples. Because this technique is so sensitive, interferences can be a real problem; finding the optimum combination of digestion, heating times and temperatures, and matrix modifiers can be a challenge for complex matrices (*see* metal analyzer for more related terms) (SW-846).

Grapple: A clamshell-type bucket having three or more jaws, also called a star or orange peel bucket (SW-108ts).

Grass waterway: An area of grass over which run-off water can move in a thin sheet across the land surface and thus proceed more slowly than it does when it moves across cultivated crops, hence causing less erosion (DOI-70/04).

Grassed waterway: Natural or constructed watercourse or outlet that is shaped or graded and established in suitable vegetation for the disposal of runoff water without erosion (EPA-94/04).

Grassland: An area where the vegetation is dominated by grasses.

Grate: A device used to support solid waste or solid fuel in a furnace during the drying, igniting and burning process. Its openings (tuyeres) permit air to pass through it (cf. stoker) (SW-108ts). Other grate related terms include:
- Chain grate
- Dead plate grate
- Fixed grate
- Movable grate
- Oscillating grate
- Reciprocating grate
- Rocking grate
- Traveling grate
- Stationary grate (*see* synonym, fixed grate)

Grate sifting: *See* synonym, sifting.

Gravel: Rock fragments from 2 mm to 64 mm (.08 to 2.5 inches) in diameter; gravel mixed with sand, cobbles, boulders, and containing no more than 15 percent of fines (SW-108ts).

Gravimetric 103-105C: A standard method of measuring total solids in aqueous solutions (EPA-83/06a).

Gravimetric 550C: A standard method of measuring total volatile solids in aqueous solutions (EPA-83/06a).

Gravimetric analysis:
(1) A quantitative (weighing) analysis of a chemical component.
(2) The analysis of a gas mixture is based on the mass or weight of each component in the system.
(3) *See* property for more related terms.

Gravitational fall: The downward settling of particles in the atmosphere due to the effects of gravity. The rate of descent of a particle depends on the balance between the aerodynamic drag and the gravitational acceleration (Stokes law). For particles with approximately the density of water and a diameter of less than 20 micro-meter the fall velocity is too small compared with the vertical velocities in the atmosphere, so that these particles can remain aloft (NATO-78/10).

Gravitational water: The water that flows downward due to the gravitational force (*see* water for more related terms).

Gravity: The gravitational force by which terrestrial bodies tend to fall toward the center of the earth. In accordance with Newton's second law of motion, $F = ma$, F is the force producing an acceleration "a" on a body of mass "m" (cf. specific gravity).

Gravity filtration: The settling of heavier and rising of lighter constituents within a solution (*see* filtration for more related terms) (EPA-83/06a).

Gravity flotation: The separation of water and low density contaminants such as oil or grease by reduction of the wastewater flow velocity and turbulence for a sufficient time to permit separation due to difference in specific gravity. The floated material is removed by some skimming technique (EPA-83/06a).

Gravity furnace wall: A furnace wall supported directly by the foundation of floor of a structure (*see* furnace wall for more related terms) (EPA-83).

Gravity separation: A separation method which uses density differences and gravitational pull to separate mixed materials into various components (*see* separation for more separation methods) (EPA-82/05; 87/10a).

Gravity separation method: The treatment of mineral particles which exploits differences between their specific gravities. The separation is usually performed by means of sluices, jigs, classifiers, spirals, hydrocyclones, or shaking tables (40 CFR 440.141-91).

Gravity settler: *See* synonym, settling chamber.

Gravity settling tank: *See* synonym, settling chamber.

Gravity sewer: A collection system where gravity is used to transport wastewater from the homes to a centralized treatment or disposal facility. Periodically the wastewater may be pumped to a higher elevation, but energy costs are generally low since water flows downhill. Most gravity sewers are conventional technologies (*see* sewer for more related terms) (EPA-80/08).

Gravity thickening: *see* definition under sludge thickening.

Gravure cylinder: A printing cylinder with an intaglio image consisting of minute cells or indentations specially engraved or etched into the cylinder's surface to hold ink when continuously revolved through a fountain of ink (40 CFR 60.431-91, *see also* 40 CFR 60.581-91).

Gravure ink: A quick drying, low viscosity ink based on volatile solvents (*see* ink for more related terms) (EPA-79/12a).

Gray cast irons: Alloys primarily of iron, carbon and silicon along with other alloying elements in which the graphite is in flake form. (These irons are characterized by low ductility but have many other properties such as good castability and good damping capacity) (EPA-83/06a).

Gray iron: A cast iron that gives a gray fracture due to the presence of flake graphite (40 CFR 464.31-91)..

Gray water:
(1) All non-toilet household wastewater (EPA-80/08).
(2) Galley, bath, and shower water (CWA312).
(3) Domestic wastewater composed of wash water from kitchen, bathroom, and laundry sinks, tubs, and washers (EPA-94/04).
(4) *See* water for more related terms.

Grazing permit and lease: Any document authorizing use of public lands or lands in National Forests in the eleven contiguous western States for the purpose of grazing domestic livestock (FLPMA103-

43USC1702-90).

Grease:
(1) A solid or semi-solid composition made up of animal fats, alkali, water, oil and various additives (EPA-74/04b).
(2) In wastewater, a group of substances including fats, waxes, free fatty acids, calcium and magnesium soaps, mineral oil and certain other non-fatty materials. The type of solvent and method used for extraction should be stated for quantification. The grease analysis will measure both free and emulsified oils and greases. Generally expressed in mg/L (EPA-83/03).

Grease interceptor (or grease trap): A device in a drain to remove grease before it enters into a wastewater treatment system (cf. grease skimmer).

Grease proof paper: Paper which is resistant to oil and grease penetration (EPA-87/10).

Grease skimmer: A device for removing grease or scum from the surface of wastewater in a tank (EPA-83/03).

Grease trap: *See* synonym, grease interceptor.

Great Lake: Lake Ontario, Lake Erie, Lake Huron (including Lake St. Clair), Lake Michigan, and Lake Superior, and the connecting channels (Saint Mary's River, Saint Clair River, Detroit River, Niagara River, and Saint Lawrence River to the Canadian Boarder) (CWA118-33USC1268-91).

Great Lakes States: The States of Illinois, Indiana, Michigan, Minnesota, New York, Ohio, Pennsylvania, and Wisconsin (CWA118-33USC1268-91).

Great Lakes Water Quality Agreement: The bilateral agreement, between the United States and Canada which was signed in 1978 and amended by the Protocol of 1987 (CWA118-33USC1268-91).

Green algae: Algae that have pigments similar in color to those of higher green plants. Common forms produce algal mats or floating "moss" in lakes (LBL-76/07-water).

Green belts: Certain areas restricted from being used for buildings and houses; they often serve as separating buffers between pollution sources and concentrations of population (EPA-74/11).

Green body: An unbaked carbon rod or piece that is usually soft and quite easily broken (EPA-83/03).

Green clipper: A clipper which clips veneer prior to being dried (EPA-74/04).

Green function: A function associated with a given boundary value problem, which is a differential equation defined in a certain region and required to meet given boundary conditions. The green function appears as the kernel in the integral representation of the solution of the equation (NATO-78/10).

Green hide: A hide which may be cured but has not been tanned (*see* hide for more related terms) (EPA-82/11).

Green liquor:
(1) Liquor made by dissolving the smelt from the kraft process water and weak liquor preparatory to causticizing (EPA-87/10).
(2) The aluminum-bearing solution from the bauxite digesters before further processing (EPA-74/03b).
(3) *See* liquor for more related terms.

Green liquor clarifier: A piece of equipment used to separate the dregs from the green liquor, allowing recovery of the green liquor for processing into white "cooking" liquor (*see* clarifier for more related terms) (EPA-87/10).

Green liquor sulfidity: The sulfidity of the liquor which leaves the smelt dissolving tank (40 CFR 60.281-91).

Green paper: The incompletely dried or seasoned paper (*see* paper for more related terms) (EPA-83).

Green sludge (or fresh sludge): The raw sludge before treatment (*see* sludge for more related terms).

Green stock: Unseasoned wood (EPA-74/04).

Green tire: An assembled, uncured tire (40 CFR 60.541-91).

Green tire spraying operation: The system used to apply a mold release agent and lubricant to the inside and/or outside of green tires to facilitate the curing process and to prevent rubber from sticking to the curing press. A green tire spraying operation consists of a booth where spraying is performed, the spray application station, and related equipment, such as the lubricant supply system (40 CFR 60.541-91).

Green vitriol: *See* synonym, ferrous sulfate.

Green water: The water that shows green color due to the growth of green algae which flourish, particularly, fast under the sunlight in summer (*see* water for more related terms).

Greenhouse effect:
(1) The heating effect of the atmosphere upon the earth. Light waves from the sun pass through the air and are absorbed by the earth. The earth then re-radiates this energy as heat waves that are absorbed by the air, specifically by carbon dioxide. The air thus behaves like glass in a greenhouse, allowing the passage of light but not of heat. Thus many scientists theorize that an increase in the atmospheric concentration of CO_2 can eventually cause an increase in the earth's surface temperature (cf. chilling effect) (EPA-74/11).
(2) The warming of the Earth's atmosphere attributed to a build-up of carbon dioxide or other gases; some scientists think that this build-up allows the sun's rays to heat the Earth, while infra-red radiation makes the atmosphere opaque to a counterbalancing loss of heat (EPA-94/04).

Grid:
(1) An open structure for mounting on the sample to aid in its examination in the TEM. The term is used here to denote a 200-mesh copper lattice approximately 3 mm in diameter (40 CFR 763-App/A-91).

(2) A stationary support or retainer for a bed of packing in a packed bed scrubber (EPA-89/03b).
(3) An electrode located between the cathode and anode of an electron tube which has one or more openings through which electrons or ions can pass, and which controls the flow of electrons from cathode to anode (EPA-83/03).
(4) The support for the active materials and a means to conduct current from the active materials to the cell terminals; usually a metal screen, expanded metal mesh, or a perforated metal plate (EPA-84/08).

Grid casting facility: The facility which includes all lead melting pots and machines used for casting the grid used in battery manufacturing (40 CFR 60.371-91).

Grid model: The numerical solution, usually by finite difference methods, of a problem in a Eulerian frame of a reference at discrete points in space, which are called grid points. The solution at each grid point is calculated as a function of time (NATO-78/10).

Gridding of water: The distribution of water by pipeline. On a regional basis this may be done so as to achieve a more efficient adjustment between supply and demand than is found in nature (*see* water for more related terms) (DOI-70/04).

Grinder:
(1) Equipment for the size reduction of a material (cf. shredder) (EPA-83).
(2) A unit which is used to pulverize dry phosphate rock to the final product size used in the manufacture of phosphate fertilizer and does not include crushing devices used in mining (40 CFR 60.401-91).

Grinder pump: A mechanical device that shreds solids and raises sewage to a higher elevation through pressure sewers (EPA-94/04).

Grinding:
(1) To reduce to powder or small fragments and includes mechanical chipping or drilling (40 CFR 61.141-91).
(2) The process of removing stock from a workpiece by the use of a tool consisting of abrasive grains held by a rigid or semi-rigid grinder. Grinding includes surface finishing, sanding, and slicing (40 CFR 471.02-91).
(3) In mining, size reduction into relatively fine particles. It is arbitrarily divided into dry grinding performed on mineral containing only moisture as mined, and wet grinding, usually done in rod, ball or pebble mills with added water (EPA-82/05).
(4) In constructing a landfill liner, the removal of oxide layers and waxes from the surface of a polyethylene sheet in preparation of extrusion fillet or extrusion flat seaming (EPA-89/09, *see also* EPA-91/05).

Grinding fluids: Water based, straight oil, or synthetic based lubricants containing mineral oil, soaps, or fatty materials lubricants serve to cool the part and maintain the abrasiveness of the grinding wheel face (EPA-83/06a).

Grinding mill: A machine used for the wet or dry fine crushing of any nonmetallic mineral. Grinding mills include, but are not limited to, the following types: hammer, roller, rod, pebble and ball, and fluid energy. The grinding mill includes the air conveying system, air separator, or air classifier, where such systems are used (40 CFR 60.671-91).

Grinding wheels: All power-driven rotatable grinding or abrasive wheels, except disc wheels as defined in this standards, consisting of abrasive particles held together by artificial or natural bonds and used for peripheral grinding (29 CFR 1910.94b-91).

Grindstone: A natural or artificial stone which is channeled or grooved and used for the manufacture of mechanical, chemi-mechanical, and groundwood pulp (EPA-87/10).

Grit: Large size materials.

Grit chamber: *See* synonym, grit removal chamber.

Grit channel: Similar to the grit removal chamber, grit channel is a channel used for extracting grits from sewage.

Grit removal chamber (or grit chamber):
(1) A small detention chamber or an enlargement of a sewer designed to reduce the velocity of a liquid flow and permit the separation of mineral from organic solids by differential sedimentation (EPA-87/10a).
(2) A chamber used for settling out stones, cinders, and sands (DOI-70/04).

Grit washer: *See* synonym, classifier.

Grizzly screen: Screen made of heavy fixed bars, used to remove oversized stones, tree stumps, etc. (EPA-88/08a).

Grog: Calcined fireclay or clean broken fireclay brick, ground to suitable fineness. It is added to a refractory batch to reduce shrinkage in drying and firing (EPA-83).

Gross alpha/beta particle activity: The total radioactivity due to alpha or beta particle emission as inferred from measurements on a dry sample (EPA-94/04).

Gross alpha particle activity:
(1) The total radioactivity due to alpha particle emission as inferred from measurements on a dry sample (40 CFR 141.2-91).
(2) Total activity due to emission of alpha particles. Used as the screening measurement for radioactivity generally due to naturally-occurring radionuclides. Activity is commonly measured in picocuries (EPA-89/12).

Gross available head: The pressure difference between two points in a water channel.

Gross beta particle activity: Total activity due to emission of beta particles. Used as the screening measurement for radioactivity from man-made radionuclides since the decay products of fission are beta particle and gamma ray emitters. Activity is commonly measured in picocuries (EPA-89/12).

Gross boat particle activity: The total radioactivity due to beta particle emission as inferred from measurements on a dry sample

(40 CFR 141.2-91).

Gross calorific value: *See* synonym, high heating value.

Gross cane: That amount of crop material as harvested, including field trash and other extraneous material (40 CFR 409.41; 409.61; 409.81-91).

Gross combination weight rating (GCWR): The value specified by the manufacturer as the loaded weight of a combination vehicle (40 CFR 202.10; 205.51-91)

Gross capacity: *See* synonym, design capacity.

Gross heating value: *See* synonym, high heating value.

Gross national product (GNP): The total productivity of a nation.

Gross production of fiberboard products: The air dry weight of hardboard or insulation board following formation of the mat and prior to trimming and finishing operations (40 CFR 429.11-91).

Gross sample: A sample representing one lot and composed of a number of increments on which neither reduction nor division has been performed (*see* analytical parameters--laboratory for more related terms) (EPA-83).

Gross sampling: A sample representing one lot and composed of a number of increments on which neither reduction nor division has been performed (EPA-83).

Gross ton: *See* definition under ton.

Gross vehicle weight: The manufacturer's gross weight rating for the individual vehicle (40 CFR 52.741; 86.082.2-91).

Gross vehicle weight rating: The value specified by the manufacturer as the maximum design loaded weight of a single vehicle (40 CFR 52.741-91).

Grotthus Draper law: The first law of photochemistry, states that only light which is absorbed can be effective in reducing a chemical transformation (*see* law for more related terms) (40 CFR 796.3700-91).

Ground based inversion: A temperature inversion based at the ground surface. This type of inversion often forms during night time, when, due to radiation losses, the temperature of the surface drops, inducing the cooling of the air close to the surface (NATO-78/10).

Ground coat: The first coat of porcelain enamel (EPA-82/11e).

Ground cover: Plants grown to keep soil from eroding (EPA-94/04).

Ground level concentration: Applied to the concentration, calculated or observed, in the neighborhood of the ground surface (NATO-78/10).

Ground phosphate rock handling and storage system: A system which is used for the conveyance and storage of ground phosphate rock from grinders at phosphate rock plants (40 CFR 60.401-91).

Ground pressure: The weight of a machine divided by the area in square inches of the ground directly supporting it (EPA-83).

Groundwater:
(1) Water in a saturated zone or stratum beneath the surface of land or water (SF101, *see also* 40 CFR 144.3; 146.3; 147.2902; 191.12; 241.101; 257.3.4; 260.10; 270.2; 300.5-91).
(2) The supply of fresh water found beneath the Earth's surface, usually in aquifers, which supply wells and springs. Because ground water is a major source of drinking water, there is growing concern over contamination from leaching agricultural or industrial pollutants or leaking underground storage tanks (EPA-94/04).
(3) The supply of water under the earth's surface in aquifer (DOE-91/04).

Other groundwater related terms include:
- Confined groundwater
- Free groundwater
- Significant source of groundwater
- Special source of groundwater
- Unconfined groundwater

Groundwater contamination: The degradation of the natural quality of ground water as a result of man's activities (EPA-87/03).

Groundwater contamination mechanism: The contamination mechanisms include (EPA-87/03):
(1) **Infiltration:** A portion of the water which has fallen to the earth slowly infiltrates the soil through pore spaces in the soil matrix. As the water moves downward under the influence of gravity, it dissolves materials such as contaminants, organics and inorganics with which it comes into contact and forms leachate.
(2) **Direct migration:** Contaminants can migrate directly into ground water from below ground sources (e.g., storage tanks, pipelines) which lie within the saturated zone.
(3) **Interaquifer exchange:** Contaminated ground water cam mix with uncontaminated ground water through a process known as interaguifer exchange in which one water-bearing unit communicates hydraulically with another.
(4) **Recharge from surface water:** Ground water may be drawn for many applications, the contaminated surface water can then enter and contaminate the ground water system.

Groundwater contamination source: Major sources include (EPA-87/03):
(1) Septic tanks and injection wells that are intentionally designed to discharge substances.
(2) Landfills and open dumps that are not designed to release contaminants to the subsurface.
(3) Pipelines and transportation that are designed to retain substances during transmission or transport. Any pipeline breakage or traffic accident may result in contamination.

Groundwater contaminant movement mechanism: Including:
(1) **Advection:** Movement caused by the flow (advection) of ground water.

(2) **Dispersion:** Movement caused by the irregular mixing (dispersion) of waters during advection.
(3) **Retardation:** Chemical or physical mechanisms which can retard, delay or slow the movement of constituents in ground water (EPA-87/03).

Groundwater discharge: Ground water entering near coastal waters which has been contaminated by landfill leachate, deep well injection of hazardous wastes, septic tanks, etc (EPA-94/04).

Groundwater divide: A ridge in the water table or other potentiometric surface from which ground water moves away in both directions normal to the ridge line (Course 165.7).

Groundwater flow: Flow of water in an aquifer or soil. That portion of the discharge of a stream which is derived from groundwater (EPA-83).

Groundwater infiltration: In 40 CFR 440.131 means that water which enters the treatment facility as a result (if the interception of natural springs, aquifers, or run-off which percolates into the round and seeps into the treatment facility's tailings pond or wastewater holding facility and that cannot be diverted by ditching or grouting the tailings pond or wastewater holding facility (40 CFR 440.132.-91).

Groundwater model: Simulated representation of a ground water system to aid definition of behavior and decision-making (Course 165.7).

Groundwater pollution boundary: The lines at which the concentration of all pollutants have fallen below the maximum permissible concentration for potable water, or where all water properties have taken on the normal values of the environment concerned (EPA-87/03).

Groundwater reservoir: All rocks in the zone of saturation (cf. aquifer) (Course 165.7).

Groundwater run-off: Groundwater that is discharged into a stream channel as spring or seepage water (EPA-74/11).

Groundwater system: A ground-water reservoir and its contained water; includes hydraulic and geochemical features (Course 165.7).

Groundwater under the direct influence (UDI) of surface water:
(1) Any water beneath the surface of the ground with:
  (A) Significant occurrence of insects or other macroorganisms, algae, or large-diameter pathogens such as Giardia lamblia, or
  (B) Significant and relatively rapid shifts in water characteristics such as turbidity, temperature, conductivity, or pH which closely correlate to climatological or surface water conditions. Direct influence must be determined for individual sources in accordance with criteria established by the State. The State determination of direct influence may be based on site-specific measurements of water quality and/or documentation of well construction characteristics and geology with field evaluation (40 CFR 141.2-91).
(2) Any water beneath the surface of the ground with:
  (A) Significant occurrence of inserts or other microorganisms, algae, or large-diameter pathogens (EPA-94/04);
  (B) Significant and relatively rapid shifts in water characteristics such as turbidity, temperature, conductivity, or pH which closely correlate to climatological or surface water conditions. Direct influence must be determined for individual sources in accordance with criteria established by the state (EPA-94/04).

Groundwood paper: A general term for a variety of papers, mainly printing, containing a large proportion of mechanical pulp along with chemical pulps (*see* paper for more related terms) (EPA-83).

Groundwood pulp: A wood pulp produced mechanically by a grinding action that separates wood fibers from resinous binders. It is used principally for newsprint and printing papers (*see* pulp for more related terms) (EPA-83).

Group task method: A method in which the responsibility for collecting on assigned routes is shared by more than one crew. Any crew that finishes a particular route works on another until all are completed (*see* waste collection method for more related terms) (SW-108ts).

Grouser: A ridge or cleat that extends across a crawler tractor track to improve its traction (SW-108ts).

Grout: A cementing or sealing mixture of cement and water to which sand, sawdust, or other fillers (additives) may be added (SW-108ts).

Growth: A relative measure of the viability of an algal population based on the number nd/or weight of algal cells per volume of nutrient medium or test solution in a pacified period of time (40 CFR 797.1050-91).

Growth rate: An increase in biomass or cell numbers of algae per unit time (40 CFR 797.1060-91).

Guaranteed loan program: The program established pursuant to Pub. L. 4-55 which amended the Act by adding section 213 (40 CFR 39.105-91).

Guarantor: Any person, other than the responsible party, who provides evidence of financial responsibility for a responsible party under this Act (OPA1001, *see also* SF101-91).

Guidance (how to): *See* definition under law related terms.

Guidance for controlling asbestos-containing material in buildings: The Environmental Protection Agency document with such title as in effect on March 31, 1986 (TSCA202-15USC2642-91).

Guidance manual: A supplemental document to EPA regulations. It provides a detailed procedure for EPA permit writers on how to implement regulations and for industry on how to comply with regulations(cf. hazardous waste incineration guidance series).

Guide coat operation: The guide coat spray booth, flash-off area and bake oven(s) which are used to apply and dry or cure a surface coating between the prime coat and topcoat operation on the components of automobile and light-duty truck bodies (40 CFR 60.391-91).

Guide specification: A general specification--often referred to as a design standard or design guideline--which is a model standard and is suggested or required for use in the design of all of the construction projects of an agency (*see* specification for more related terms) (40 CFR 249.04-91).

Guide word hazard and operability study: A hazard and operability study which uses guide words to stimulate a systematic yet creative search for hazards (EPA-87/07a).

Guidelines for carcinogen risk assessment: U.S. EPA guidelines intended to guide Agency evaluation of suspect carcinogens in line with statutory policies and procedures. *See* FR 33992-34003, September 24, 1986 (EPA-92/12).

Guidelines for exposure assessment: U.S. EPA guidelines intended to guide Agency analysis of exposure assessment data in line with statutory policies and procedures. *See* 51 FR 34042-34054, September 24, 1986 (EPA-92/12).

Guidelines for health assessment of suspect developmental toxicants: U.S. EPA guidelines intended to guide Agency analysis of developmental toxicity data in line with statutory policies and procedures. *See* 51 FR 34028-34040, September 24, 1986 (EPA-92/12).

Guidelines for mutagenicity risk assessment: U.S. EPA guidelines intended to guide Agency analysis of mutagenicity data as related to heritable mutagenic risks, in line with statutory policies and procedures. *See* 51 FR 34006-34012, September 24, 1986 (EPA-92/12).

Guidelines for the health risk assessment of chemical mixtures: U.S. EPA guidelines intended to guide Agency analysis of information relating to health effects data on chemical mixtures in line with statutory policies and procedures. *See* 51 FR 34014-34025, September 24, 1986 (EPA-92/12).

Guillotine damper: An adjustable plate, utilized to regulate the flow of gases, installed vertically in a breeching (*see* damper for more related terms) (SW-108ts).

Gully erosion:
(1) Severe erosion in which trenches are cut to a depth greater than 30 centimeters (a foot). Generally, ditches deep enough to cross with farm equipment are considered gullies (EPA-94/04).
(2) The widening, deepening, and cutting back of small channels and waterways due to erosion (DOI-70/04).

Gum rosin: *See* synonym, natural rosin.

Gun: In constructing a landfill liner, a synonymous term for hand held extrusion fillet device (EPA-89/09, *see also* EPA-91/05).

Gun drilling: Using special straight flute drills with a single lip and cutting fluid at high pressures for deep hole drilling (*see* drilling for more related terms) (EPA-83/06a).

Gustiness: The fluctuation of wind speed over a specified time interval (NATO-78/10).

Gyp pond: In fertilizer phosphate plants, a pond receiving wastewater and acting as a recirculation, cooling and water reuse pond (EPA-6/76).

Gypsum: The common name for calcium sulfate dihydrate ($CaSO_4 \bullet 2H_2O$); an essential ingredient of cement that controls the rate of setting (ETI-92).

Gypsum cement: A group of cements consisting primarily of calcium sulfate and produced by the complete dehydration of gypsum. It usually contains additives such as aluminum sulfate or potassium carbonate. It is used in sand binder formulation (EPA-85/10a).

Hh

Habit-forming narcotic drug or narcotic: Opium and coca leaves and the several alkaloids derived therefrom, the best known of these alkaloids being morphia, heroin, and codeine, obtained from opium, and cocaine derived from the coca plant; all compounds, salts, preparations, or other derivatives obtained either from the raw material or from the various alkaloids; Indian hemp and its various forms; isonipecaine and its derivatives, compound, salts and preparations; opiates (as defined in section 3228(f) of the Internal Revenue Code) (PHSA2-j).

Habitat:
(1) The place where a population (e.g., human, animal, plant, microorganism) lives and its surroundings, both living and non-living. Habitat Indicator: A physical attribute of the environment measured to characterize conditions necessary to support an organism, population, or community in the absence of pollutants, e.g., salinity of esturine waters or substrate type in streams or lakes (EPA-94/04).
(2) The place or type of site where a plant or animal normally lives or grows (DOE-91/04).

Hafnium (Hf): A transition metal with a.n. 72; a.w. 178.49; d. 13.1 g/cc; m.p. 2222° C and b.p. 5400° C. The element belongs to group IVB of the periodic table.

Hahnium (Ha): A transactinide element (atomic number greater than 103). The element with a.n. 105 is unstable, has very short half live, and belongs to group VB of the periodic table.

Hair pulp: The removal of hair by chemical dissolution (40 CFR 425.02-91).

Hair save: The physical or mechanical removal of hair which has not been chemically dissolved, and either selling the hair as a byproduct or disposing of it as a solid waste (40 CFR 425.02-91).

Half cell: A cell with one electrode which is immersed in an electrolyte solution. For a cathode and anode electrode cell, the cell can be considered as two half cell.

Half cell potential: The relative potential between one electrode and its electrolyte. The total potential of a cell is the sum of two half cell potentials, i.e., cathodic and anodic potential.

Half life:
(1) Length of time required for an initial concentration of a substance to be halved as a result of loss through decay. The HRS considers five decay processes: biodegradation, hydrolysis, photolysis, radioactive decay, and volatilization (40 CFR 300-App/A-91).
(2) (A) The time required for a pollutant to lose half its affect on the environment. For example, the half-life of DDT in the environment is 15 years, of radium, 1,580 years.
(B) The time required for half of the atoms of a radioactive element to undergo decay.
(C) The time required for the elimination of one half a total dose from the body (EPA-94/04).

Half life of a chemical: The time required for the concentration of the chemical being tested to be reduced to one-half its initial value (40 CFR 796.3700-91).

Halide: The compound resulted from the reaction of a halogen compound (e.g., chlorine) with another element or group, e.g., reaction of chlorine with other compounds will result in chloride.

Hall electrolytic conductivity detector: *See* definition under GC/HECD.

Halocarbon:
(1) The chemical compounds $CFCl_3$ and CF_2Cl_2 and such other halogenated compounds as the Administrator determines may reasonably be anticipated to contribute to reductions in the concentration of ozone in the stratosphere (CAA111-42USC7411-91).
(2) A chemical compound such as $CFCl_3$ and CF_2Cl_2 containing both carbon and one or more than one halogen elements. Halocarbons may cause the reduction of ozone concentration in the stratosphere.

Haloethanes: Halogenated ethanes; e.g., trichloro-trifluoroethane (EPA-88/12).

Haloform: A chemical compound with a CHX_3 format, where X is a halogen element; e.g., $CHCl_3$.

Halogen:
(1) One of the chemical elements chlorine, bromine or iodine (40 CFR 141.2-91).
(2) Any of a group of 5 chemically-related nonmetallic elements that includes bromine, fluorine, chlorine, iodine, and astatine

(EPA-89/12).

Halogen acid furnace (HAF): An industrial furnace specific to the production of halogen acids, e.g., HCl from halogenated secondary materials (*see* furnace for more related terms) (CRWI-89/05).

Halogenated organic compounds or HOCs: Those compounds having a carbon-halogen bond which are listed under appendix III to this part (40 CFR 268.2-91).

Halogenated vent stream: Any vent stream determined to have a total concentration (by volume) of compounds containing halogens of 20 ppmv (by compound) or greater (40 CFR 60.611; 60.661-91).

Halogenation: The incorporation of one of the halogen elements (bromine, chlorine, or fluorine) into a chemical compound (EPA-87/10a).

Halomethane: A halogenated methane; e.g., trichlorofluoromethane (EPA-88/12).

Halon: Bromine-containing compounds with long atmospheric lifetimes whose breakdown in the stratosphere causes depletion of ozone. Halons are used in fire-fighting (EPA-94/04).

Hammer forging: Heating and pounding a metal to shape it into a desired form (EPA-83/06a).

Hammer provision: A statutory requirement that goes into effect automatically if EPA fails to issue regulations by certain dates specified in the statute (EPA-86/01).

Hammermill: A high-speed machine that hammers and cuts to crush, grind chip, or shred solid wastes (*see* size reduction machine for more related terms) (EPA-94/04).

Hand glass melting furnace: A glass melting furnace where the molten glass is removed from the furnace by a glassworker using a blowpipe or a pontil (*see* furnace for more related terms) (40 CFR 60.291-91).

Hand sheet: Test sample sheet of paper made from stock on a sheet mold (EPA-83).

Handicapped person:
(1) Handicapped person means any person who:
  (A) Has a physical or mental impairment which substantially limits one or more major life activities,
  (B) Has a record of such an impairment, or
  (C) Is regarded as having such an impairment. For purposes of employment, the term handicapped person does not include any person who is an alcoholic or drug abuser whose current use of alcohol or drugs prevents such individual from performing the duties of the job in question or whose employment, by reason of such current drug or alcohol abuse, would constitute a direct threat to property or the safety of others.
(2) As used in this paragraph, the phrase:
  (A) **Physical or mental impairment** means:
    (a) Any physiological disorder or condition, cosmetic disfigurement, or anatomical loss affecting one or more of the following body systems: Neurological; musculoskeletal; special sense organs; respiratory, including speech organs; cardiovascular; reproductive; digestive; genito-urinary; hemic and lymphatic; skin; and endocrine; and
    (b) Any mental or psychological disorder, such as mental retardation, organic brain syndrome, emotional or mental illness, and specific learning disabilities.
  (B) Major life activities means functions such as caring for one's self, performing manual tasks, walking, seeing, hearing, speaking, breathing, learning, and working.
  (C) Has a record of such an impairment means has a history of, or has been misclassified as having, a mental or physical impairment that substantially limits one or more major life activities.
  (D) Is regarded as having an impairment means:
    (a) Has a physical or mental impairment that does not substantially limit major life activities but that is treated by a recipient as constituting such a limitation;
    (b) Has a physical or mental impairment that substantially limits major life activities only as a result of the attitudes of others toward such impairment; or
    (c) Has none of the impairments defined above but is treated by a recipient as having such an impairment (40 CFR 7.25-91).

Hang up: Refers to the process of hydrocarbon molecules being adsorbed, condensed, or by any other method removed from the sample flow prior to reaching the instrument detector. It also refers to any subsequent desorption of the molecules into the sample flow when they are assumed to be absent (40 CFR 86.082.2-91).

Haploid: A cell with a single set of unpaired chromosomes (cf. diploid).

Hard chrome: A chromium electroplate applied for non-decorative use such as wear resistance in engineering applications (EPA-74/03d).

Hard glass: A glass with:
(1) Exceptionally high viscosity at elevated temperatures;
(2) High softening point;
(3) Difficult to melt; and
(4) Hard to scratch (EPA-83).
(5) *See* glass for more related terms.

Hard lead: The antimonial lead, containing 6 to 28% antimony used for roofing, gutters, chemical tank linings (*see* lead for more related terms) (EPA-83).

Hard water:
(1) Water containing dissolved minerals such as calcium, iron and magnesium. The most notable characteristic of hard water is its inability to lather soap. Some pesticide chemicals will curdle or settle out when added to hard water (EPA-74/11).
(2) Alkaline water containing dissolved salts that interfere with

some industrial processes and prevent soap from sudsing (EPA-94/04).
(3) *See* water for more related terms.

Hardboard: A panel manufactured from interfelted ligno-cellulosic fibers consolidated under heat and pressure to a density of 0.5 g/cu cm (31 lb/cu ft) or greater (40 CFR 429.11-91).

Hardboard press: A machine which completes the reassembly of wood particles and welds them into a tough, durable, grainless board (EPA-74/04).

Hardener: A chemical present in some photographic solutions that reacts with gelatin in the emulsion to protect the film from damage during or after processing. Common hardeners are potassium aluminum sulfate, potassium chromium sulfate, and formaldehyde solution (EPA-80/10).

Hardness: The total concentration of the calcium and magnesium ions in water expressed as calcium carbonate (mg $CaCO_3$/liter) (cf. hard water) (40 CFR 797.1600-91). Other hardness related terms include:
- Temporary gardness
- Permanent gardness

Hardness (water): Characteristic of water caused by presence of various salts. Hard water may interfere with some industrial processes and prevent soap from lathering (EPA-94/04).

Hardness of water:
(1) A measure of the calcium and magnesium salts present in water. Soft water is that with less than 60 ppm (parts of salts per million parts of water); temporary water is that with 60 to 120 ppm of salts; permanent water is that with salts in excess of 120 ppm. (Other salts that may occur in water include those of iron, aluminum, manganese, strontium, zinc.) (DOI-70/04).
(2) A characteristic of water, imparted by calcium, magnesium, and ion salts such as bicarbonates, carbonates, sulfates, chlorides, and nitrates. These cause curdling of soap, deposition of scale in boilers, damage in some industrial processes and sometimes objectionable taste. Hardness may be determined by standard laboratory procedure or computed from the amounts of calcium and magnesium as well as iron, aluminum, manganese, barium, strontium, and zinc, and is expressed as equivalent calcium carbonate (EPA-83/03).
(3) *See* water for more related terms.

Hardness paper: The paper with resistance to indentation (*see* paper for more related terms) (EPA-83).

Hardpan: A hardened, compacted, or cemented soil layer (SW-108ts).

Hardwood: The wood obtained from trees of the angiosperm class, such as birch, gum, maple, oak, and poplar. Hardwoods are also known as porous woods (cf. softwood) (EPA-87/10).

Hardwood veneer: Hardwood veneer can be categorized according to use, the three most important ones are (EPA-74/04):
(1) Face veneer: the highest quality used to make panels employed in furniture and interior decoration;
(2) Commercial veneer: used for crossbands, cores, backs of plywood panels and concealed parts of furniture; and
(3) Container veneer: inexpensive veneers used in the making of crates, hampers, baskets, kits, etc.
(4) *See* veneer for more related terms.

Harmonic mean: Is defined as: $(1/H) = (1/N)(sum(1/X_i))$, i = 1....N, where: H = harmonic mean; N = number of sample; X_i = Discrete value at i^{th} sample (*see* mean for more related terms).

Harmonic mean flow: The number of daily flow measurements divided by the sum of the reciprocals of the flows. Tar is, it is the reciprocal of the mean of reciprocals (EPA-91/03).

Harris process: A process in which sodium hydroxide and sodium nitrate are added to molten lead to soften or refine it. These two compounds react with impurities in the molten metal forming a slag that floats to the top of the molten metal (EPA-83/03a).

Hatch: The eggs or young birds that are the same age and that are derived from the same adult breeding population, where the adults are of the same strain and stock (40 CFR 797.2050-91, *see also* 40 CFR 797.2175-91).

Hatchability: The embryos that mature, pip the shell, and liberate themselves from the eggs on day 23 or 24 of incubation. Values are expressed as percentage of viable embryos (fertile eggs) (40 CFR 797.2130-91, *see also* 40 CFR 797.2150-91).

Hatchback: A passenger automobile where the conventional luggage compartment, i.e., trunk, is replaced by a cargo area which is open to the passenger compartment and accessed vertically by a rear door which encompasses the rear window (40 CFR 600.002.85-91).

Haul distance:
(1) The distance a collection vehicle travels from its last pickup stop to the solid waste transfer station, processing facility, or sanitary landfill.
(2) The distance a vehicle travels from a solid waste transfer station or processing facility to a point of final disposal.
(3) The distance that cover materials must be transported from an excavation or stockpile to the working face of a sanitary landfill (SW-108ts).

Haul time: The elapsed or cumulative time spent transporting solid waste between two specific locations (SW-108ts).

Hauler: Garbage collection company that offers complete refuse removal service; many also will also collect recyclables (EPA-94/04).

Hazard:
(1) A probability that a given pesticide will have an adverse effect on man or the environment in a given situation, the relative likelihood of danger or ill effect being dependent on a number of interrelated factors present at any given time (40 CFR 171.2-91).
(2) A source of danger. The potential for death, injury or other forms of damage to life and property (EPA-87/07a).

(3) Any situation that has the potential for doing damage to life, property, and/or the environment (EPA-85/11).
Other hazard related terms include:
- High-hazard content
- Low hazard content
- Ordinary hazard content

Hazard analysis: The procedures involved in:
(1) Identifying potential sources of release of hazardous materials from fixed facilities or transportation accidents;
(2) Determining the vulnerability of a geographical area to a release of hazardous materials; and
(3) Comparing hazards to determine which present greater or lesser risks to a community (EPA-89/12).

Hazard and operability study (HAZOP):
(1) A systematic technique for identifying hazards or operability problems throughout an entire facility. One examines each segment of a process and lists all possible deviations for normal operating conditions and how they might occur. The consequences on the process are assessed, and the means available to detect and correct the deviations are examined (NRT-87/03).
(2) The application of a formal systematic critical examination to the process and engineering intentions of the new facilities to assess the hazard potential of mal-operation of individual items of equipment and the consequential effects on the facility as a whole (EPA-87/07a).

Hazard category: Any of the following:
(1) "Immediate (acute) health hazard," including "highly toxic," "toxic," "irritant," "sensitizer," "corrosive," (as defined under 40 CFR 910.1200 of Title 29 of the Code of Federal Regulations) and other hazardous chemicals that cause an adverse effect to a target organ and which effect usually occurs rapidly as a result of short term exposure and is of short duration;
(2) "Delayed (chronic) health hazard," including "carcinogens" (as defined under 40 CFR 910.1200 of Title 29 of the Code of Federal Regulations) and other hazardous chemicals that cause an adverse effect to a target organ and which effect generally occurs as a result of long term exposure and is of long duration;
(3) "Fire hazard," including "flammable," combustible liquid," "pyrophoric," and "oxidizer" (as defined under 40 CFR 910.1200 of Title 29 of the Code of Federal Regulations);
(4) "Sudden release of pressure," including "explosive" and "compressed gas" (as defined under 40 CFR 910.1200 of Title 29 of the Code of Federal Regulations); and
(5) "Reactive," including "unstable reactive," "organic peroxide," and "water reactive" (as defined under 40 CFR 910.1200 of Title 29 of the Code of Federal Regulations) (40 CFR 370.2-91).

Hazard classe: A series of nine descriptive terms that have been established by the United Nations Committee of Experts to categorize the hazardous nature of chemical, physical, and biological materials. These categories are: flammable liquids, explosive, gases, oxidizers, radioactive materials, corrosives, flammable solids, poisonous and infectious substances, and dangerous substances (Course 165.5).

Hazard communication standard: An OSHA regulation that requires chemical manufacturers, suppliers, and importers to assess the hazards of the chemicals that they make, supply, or import, and to inform employers, customers, and workers of these hazards through MSDS sheets (EPA-94/04).

Hazard evaluation:
(1) A component of risk evaluation that involves gathering and evaluating data on the types of health injury or disease that may be produced by a chemical and on the conditions of exposure under which such health effects are produced (EPA-94/04).
(2) The impact or risk evaluation that a hazardous substance poses to public health and the environment (Course 165.5).

Hazard identification: Determining if a chemical can cause adverse health effects in humans and what those affects might be (EPA-94/04).

Hazard index (HI): The sum of two or more hazard quotients for multiple substances and/or multiple exposure pathways (EPA-91/12).

Hazard information transmission (HIT): HIT program provides a digital transmission of the CHEMTREC emergency chemical report to first responders at the scene of a hazardous materials incident. The report advises the responder on the hazards of the materials, the level of protective clothing required, mitigating action to take in the event of a spill, leak or fire, and first aid for victims. HIT is a free public service provided by the Chemical Manufacturers Association. Reports are sent in emergency situations only to organizations that have pre-registered with HIT. Brochures and registration forms may be obtained by writing: Manager, CHEMTREC/CHEMNET, 2501 M Street, N.W., Washington, DC, 20037 (NRT-87/03).

Hazard quotient (HQ): The ratio of a single substance exposure level over a specified time period to a reference dose for that substance derived from a similar exposure period (EPA-91/12).

Hazard ranking system (HRS): The method used by EPA to evaluate the relative potential of hazardous substance releases to cause health or safety problems, or ecological or environmental damage (40 CFR 300.5-91).

Hazard ranking system (HRS) scoring and National Priority List (NPL): An assessment of the relative degree of risk to human health and the environment posed by sites and facilities subject to review. The scoring process includes:
(1) If preliminary assessment/site investigation (PA/SI) indicates potential or actual contamination problem, site is assigned an HRS score.
(2) HRS score calculation is based on:
   (A) Observed releases of contaminants.
   (B) Quantity and toxicity of contaminants.
   (C) Potential for human exposure.
   (D) Size of population affected.
(3) Sites with HRS scores above threshold value and proposed for NPL listing become eligible for Superfund funding (SF105).

Hazard source: Hazard sources are classified by the type of energy transferred. The sources include (Course 165.5):
(1) **Kinetic/mechanical:** Striking or struck by injuries.
(2) **Thermal:** Fires, explosions, hot environment.
(3) **Electrical:** Faulty wiring, downed power lines.
(4) **Chemical:** Effects from corrosives, or toxic chemicals, etc.
(5) **Acoustic:** Explosions, loud machinery.
(6) **Biological:** Poisonous or, disease producing organisms, plants or animals.
(7) **Radioactive:** Ionizing and non-ionizing radiation.

Hazardous agent: Including:
(1) **Biological agent:** Living organisms that can cause sickness or death to exposed individuals.
(2) **Radiological material:** Radiation intensity that may be harmful to response personnel.
(3) **Chemical compound:** Chemicals that are toxic, corrosive, and active.

Hazardous air pollutant:
(1) Air pollutants which are not covered by ambient air quality standards but which, as defined in the Clean Air Act, may reasonably be expected to cause or contribute to irreversible illness or death. Such pollutants include asbestos, beryllium, mercury, benzene, coke oven emissions, radionuclides, and vinyl chloride (EPA-94/04).
(2) A list of hazardous air pollutants was provided in CAA112.b (EPA-89/12, *see also* CAA112-91).
(3) *see* pollutant for more related terms.

Hazardous and Solid Waste Act (HSWA) of 1984: *see* the term of Act or HSWA.

Hazardous chemical: Any hazardous chemical as defined under 29 CFR 1910.1200(c) of Title 29 of the Code of Federal Regulations, except that such term does not include the following substances:
(1) Any food, food additive, color additive, drug, or cosmetic regulated by the Food and Drug Administration.
(2) Any substance present as a solid in any manufactured item to the extent exposure to the substance does not occur under normal conditions of use.
(3) Any substance to the extent it is used for personal, family, or household purposes, or is present in the same form and concentration as a product packaged for distribution and use by the general public.
(4) Any substance to the extent it is used in a research laboratory or a hospital or other medical facility under the direct supervision of a technically qualified individual.
(5) Any substance to the extent it is used in routine agricultural operations or is a fertilizer held for sale by a retailer to the ultimate customer.
(Other identical or similar definitions are provided in 40 CFR 370.2. (40 CFR 355.20-91))

Hazardous chemical:
(1) An EPA designation for any hazardous material requiring an MSDS under OSHA's Hazard Communication Standard. Such substances are capable of producing fires and explosions or adverse health effects like cancer and dermatitis. Hazardous chemicals are distinct from hazardous waste.(*see* Hazardous Waste (EPA-94/04))
(2) A chemical which is explosive, flammable, poisonous, corrosive, reactive, or radioactive and requires special care in handling because of the hazards it poses to public health and the environment (EPA-85/11).

Hazardous constituent or constituents: Those constituents listed in appendix VIII to part 261 of this chapter (40 CFR 268.2-91).

Hazardous liquid: The petroleum, petroleum products, or anhydrous ammonia (40 CFR 195.2-91).

Hazardous material:
(1) A substance or material which has been determined by the Secretary of Transportation to be cable of posing an unreasonable risk to health, safety, and property when transported in commerce, and which has been so designated (Course 165.5).
(2) Any materials capable of causing an acute or chronic human illness or injury as result of an acute exposure (NIOSH-84/10).
(3) Refers generally to hazardous substances, petroleum, natural gas, synthetic gas, acutely toxic chemicals, and other toxic chemicals (NRT-87/03).

Hazardous materials incidents: A release of a hazardous material to the extent that human exposure could result in acute or chronic injury or illness (NIOSH-84/10).

Hazardous ranking system (HRS): The principle screening tool used by EPA to evaluate risks to public health and the environment associated with abandoned or uncontrolled hazardous waste sites. The HRS calculates a score based on the potential of hazardous substances spreading from the site through the air, surface water, or ground water, and on other factors such as density and proximity of human population. This score is the primary factor in deciding if the site should be on the National Priorities List and, if so, what ranking it should have compared to other sites on the list (EPA-94/04).

Hazardous sample: A sample that is considered to contain high concentrations of contaminants (Course 165.5).

Hazardous substance:
(1) Any material that poses a threat to human health and-/or the environment. Typical hazardous substances are toxic, corrosive, ignitable, explosive, or chemically reactive (EPA-94/04).
(2) Any substance designated by EPA to be reported if a designated quantity of the substance is spilled in the waters of the United States or if otherwise released into the environment (EPA-94/04).
Other hazardous substance related terms include:
- CERCLA hazardous substance
- Extremely hazardous substance
- Unregulated hazardous substance

Hazardous substance UST system: An underground storage tank system that contains a hazardous substance defined in section 101(14) of the Comprehensive Environmental Response, Compensation and Liability Act of 1980 (but not including any substance regulated as a hazardous waste under subtitle C) or any mixture of such substances and petroleum, and which is not a petroleum UST system (40 CFR 280.12-91).

Hazardous waste:
(1) A solid waste, or combination of solid wastes, which because of its quantity, concentration, or physical, chemical, or infectious characteristics may:
- (A) Cause, or significantly contribute to an increase in mortality or an increase in serious irreversible, or incapacitating reversible, illness; or
- (B) Pose a substantial present or potential hazard to human health or the environment when improperly treated, stored, transported, or disposed of, or otherwise managed (RCRA1004, *see also* 40 CFR 2.305; 144.3; 146.3; 240.101; 241.101; 243.101; 260.10; 261.3; 270.2; 302.3-91).

(2) By-products of society that can pose a substantial or potential hazard to human health or the environment when improperly managed. Possesses at least one of four characteristics (ignitability, corrosivity, reactivity, or toxicity), or appears on special EPA lists (EPA-94/04).
(3) Any solid, semisolid, liquid, or gaseous waste that is ignitable, corrosive, toxic, or reactive as defined by the Resource Conservation and Recovery Act and identified or listed in 40 CFR 261 (DOE-91/04).
(4) Material that no longer has commercial value and requires disposal and is either specifically listed by the U.S. EPA or meets one of the four hazardous waste characteristics defined by the U.S. EPA (ETI-92).
(5) *See* waste for more related terms.

Hazardous waste characteristics: *see* Subpart C in 40 CFR 261.20-91. Types of hazardous waste characteristics related terms include:
- Corrosivity
- Ignitability
- Reactivity
- Extraction Procedure (EP) Toxicity

Hazardous waste classification:
(1) Listed hazardous waste (40 CFR 261.31-261.33)
- Nonspecific waste source, F waste
- Specific waste source, K waste
- Toxic waste (T), U waste
- Acute hazardous waste (H), P waste

(2) Characteristic hazardous waste (40 CFR 261.20-261.24)
- Ignitable waste (i), D waste
- Corrosive waste (C), D waste
- Reactive waste (R), D waste
- Extraction Procedure (EP) toxicity waste (E), D waste

Hazardous waste constituent: A constituent that caused the Administrator to list the hazardous waste in Part 261, Subpart D, of this chapter, or a constituent listed in Table 1 of 40 CFR 261.24 of this chapter (40 CFR 260.10-91).

Hazardous waste discharge: *See* definition under discharge or hazardous waste discharge.

Hazardous waste exclusion:
(1) Materials which are not solid wastes.
(2) Solid wastes which are not hazardous wastes.
(3) Hazardous wastes which are exempted from certain regulations.
(4) Samples, a sample of solid waste or a sample of water, soil, or air, which is collected for the sole purpose of testing to determine its characteristics or composition, is not subject to any requirements of this parts.
(5) Treatability study samples.
(6) Samples undergoing treatability studies of laboratories and testing facilities (40 CFR 261.20-91).

Hazardous waste generation: The act or process of producing hazardous waste (RCRA1004-42USC6903-91).

Hazardous waste incineration guidance series: Documents to assist both the applicant and the permit writer in the RCRA process leading to a final operating permit for hazardous waste incinerators (cf. guidance manual) (EPA-89/01):
(1) **Volume 1:** (EPA-83/07), "Guidance Manual of Hazardous Waste Incinerator Permits," SW-966, NTIS PB84-100-577, July 1983, Volume I of the Hazardous Waste Incineration Guidance Series.
This document, which was prepared by Mitre Corporation, describes the overall incinerator permitting process, highlights the specific guidance provided by other manuals, and addresses permitting issues not covered in the other manuals such as treatment of data in lieu of a trial burn. Thus it can be viewed as a road map and a good summary of all permitting issues (EPA-89/06).

(2) **Volume 2:** (EPA-89/01), "Guidance on Setting Permit Conditions and Reporting Trial Burn Results," EPA625-6-89-019, January 1989, Volume II of the Hazardous Waste Incineration Guidance Series.
This document, which was prepared by Acurex Corporation, provides guidance on selecting key operating parameters, translating trial burn data into permit operating conditions, and reporting trial burn results. It also discusses planning a trial burn to achieve workable permit limits (EPA-89/06).

(3) **Volume 3:** (EPA-89/06), "Hazardous Waste Incineraiton Measurement Guidance Manual," EPA625-6-89-021, June 1989, Volume III of the Hazardous Waste Incineration Guidance Series.
This document, which was prepared by MRI, provides general guidance to permit writers in reviewing the measurement aspects of incineration permit applications and trial burn plans. It is oriented to how measurements are made, not what measurements to make. The guidance deals specially with commonly required measurement parameters and measurement methods for process monitoring, sampling and analysis aspects of trial burns and subsequent operation of the incinerator, and QA/QC associated with these activities. As a guidance tool, this document introduces the major elements of incineration measurements via sample checklists, general discussion, and technical references (EPA-89/06).

(4) **Volume 4:** (EPA-89/08), "Guidance on Metals and Hydrogen Chloride Controls for Hazardous Waste Incinerators," No EPA number, August 1989, Volume IV of the Hazardous Waste Incineration Guidance Series.
This document, which was prepared by Versar, provides specific guidance on limiting metals emissions from

incinerators. In particular, a risk assessment approach to setting limits on metal components in the waste is employed. It also provide guidance on HCl emissions. (Note: Earlier title was Guidance for Permit Writers for Limiting Metal and Hcl Emissions from hazardous Waste Incinerators) (EPA-89/06).

(5) **Volume 5:** (EPA-90/04), "Guidance on PIC Controls for Hazardous Waste Incinerators," EPA 530-SW-90-040, April 1990, Volume V of the Hazardous Waste Incineration Guidance Series.

This document, which was prepared by MRI Corporation, details the specific permit requirements for CO and total hydrocarbon (THC) emissions from hazardous waste incinerators in the RCRA system. Emission limits for CO and THC and the rationale for their selection are discussed. (Note: Earlier title was: Guidance on Carbon Monoxide Controls for Hazardous Waste Incineration).

(6) **Volume 6:** (EPA-89/11a), "Proposed Methods for Stack Emissions Measurement of CO, O_2 THC, HCl, and Metals at hazardous Waste Incinerators," No EPA number, November 1989, Volume VI of the Hazardous Waste Incineration Guidance Series.

This document, which was prepared by MRI Corporation, discusses proposed measurement methods that will be required to implement the proposed amendments to the incinerator rules specified in 40 CFR 264 subpart O. The amendments are to limit the emission of toxic metals, hydrogen chloride (HCl) and carbon monoxide as a surrogate for the products of incomplete combustion (EPA-89/11).

Hazardous waste landfill: An excavated or engineered site where hazardous waste is deposited and covered (EPA-94/04).

Hazardous waste leachate: The liquid that has percolated through or drained from hazardous waste disposed of in or on the ground (*see* leachate for more related terms) (OME-88/12).

Hazardous waste listing criteria: *See* Subpart B in 40 CFR 261.11-91.

Hazardous waste management: The systematic control of the collection, source separation, storage, transportation, processing, treatment, recovery, and disposal of hazardous wastes (cf. management) (RCRA1004; 40 CFR 260.10-91).

Hazardous waste management facility (HWM facility): All contiguous land, and structures, other appurtenances, and improvements on the land used for treating, storing, or disposing of hazardous waste. A facility may consist of several treatment, storage, or disposal operational units (for example, one or more landfills, surface impoundments, or combination of them) (40 CFR 144.3; 146.3; 270.2-91).

Hazardous waste management unit: A contiguous area of land on or in which hazardous waste is placed, or the largest area in which there is significant likelihood of mixing hazardous waste constituents in the same area. Examples of hazardous waste management units include a surface impoundment, a waste pile, a land treatment area, a landfill cell, an incinerator, a tank and its associated piping and underlying containment system and a container storage area. A container alone does not constitute a unit; the unit includes containers and the land or pad upon which they are placed (40 CFR 260.10-91).

Hazardous waste management unit shutdown: A work practice or operational procedure that stops operation of a hazardous waste management unit or part of a hazardous waste management unit. An unscheduled work practice or operational procedure that stops operation of a hazardous waste management unit or part of a hazardous waste management unit for less than 24 hours is not a hazardous waste management unit shutdown. The use of spare equipment and technically feasible bypassing of equipment without stopping operation are not hazardous waste management unit shutdowns (40 CFR 264.1031-91).

Hazardous wastestream: The material containing CERCLA hazardous substances (as defined in CERCLA section 101(14) that was deposited, stored, disposed, or place in, or that otherwise migrated to, a source (40 CFR 300-App/A-91).

Hazards analysis: Procedures used to:
(1) Identify potential sources of release of hazardous materials from fixed facilities or transportation accidents;
(2) Determine the vulnerability of a geographical area to a release of hazardous materials; and
(3) Compare hazards to determine which present greater or lesser risks to a community (EPA-94/04).

Hazards identification: Providing information on which facilities have extremely hazardous substances, what those chemicals are, how much there is at each facility, how the chemicals are stored, and whether they are used at high temperatures (EPA-94/04).

Haze: Fine particles suspended in the atmosphere.

Hazen number (hazen unit): Unit of water color. One hazen unit is the color exhibited by a solution with 1 mg/L of chloroplatinic acid and 2 mg/L of cobaltous chloride.

HDD or 2,3,7,8-HDD: Any of the dibenzo-p-dioxins totally chlorinated or totally brominated at the following positions on the molecular structure: 2,3,7,8; 1,2,3,7,8; 1,2,3,4,7,8; 1,2,3,6,7.8; 1,2,3,7,8,9; and 1,2,3,4,7,8,9 (40 CFR 766.3-91)

HDF or 2,3,7,8-HDF: Any of the dibenzofurans totally chlorinated or totally brominated at the following positions on the molecular structure: 2,3,7,8; 1,2,3,7,8; 2,3,4,7,8; 1,2,3,4,7,8; 1,2,3,6,7,8; 1,2,3,7,8,9; 2,3,4,6,7,8; 1,2,3,4,6,7,8; and 1,2,3,4,7,8,9 (40 CFR 766.3-91).

Head:
(1) In hydraulics, the height above a standard datum of the surface of a column of water (or other liquid) that can be supported by the static pressure at a given point (Course 165.7).
(2) In leather industry, the part of the hide which is cut off at the flare into the shoulder; i.e., the hide formerly covering the head of the animal (EPA-82/11).

Head race: The pipe or chute by which water falls into the turbine of a power plant (DOI-70/04).

Head of navigation: The farthest point up a river to be reached by vessels for the purpose of trade (DOI-70/04).

Headband: The component of hearing protective device which applies force to, and holds in place on the head, the component which is intended to acoustically seal the ear canal (40 CFR 211.203-91).

Headend fuel-processing facility: The part of a reactor fuel-processing facility used for the first stage of fuel processing in which fuel is mechanically and/or chemically treated prior to uranium extraction for recovery purposes. The headend facility may be separated from, but adjacent to, the rest of the fuel processing facility (DOE-91/04).

Headend plutonium-target processing facility: The part of a plutonium-target processing facility used for the first stage of target processing, in which targets are mechanically and/or chemically treated prior to plutonium extraction for recovery purposes. The headend facility may be separated from, but adjacent to, the rest of the plutonium-target processing facility (DOE-91/04).

Headend: That portion of a reactor-fuel or plutonium-target processing system in which the initial preparation, up to and including dissolution, is conducted (DOE-91/04).

Header: A pipe used to supply and distribute liquid to downstream outlets (EPA-89/03b).

Headset: A class of web-offset lithography which requires a heated dryer to solidify the printing inks (40 CFR 52.741-91).

Headset web offset lithographic printing line: A lithographic printing line in which a blanket cylinder is used to transfer ink from a plate cylinder to a substrate continuously feed from a roll or an extension process and an oven is used to solidify the printing inks (40 CFR 52.741-91).

Headspace: The unfilled space in a glass container fitted with a closure device (EPA-83).

Health advisory: An estimate of acceptable drinking water levels for a chemical substance based on health effects information; a Health Advisory is not a legally enforceable Federal standard, but serves as technical guidance to assist Federal, state, and local officials (EPA-92/12).

Health and safety plan: A plan that specifies the procedures that are sufficient to protect on-site personnel and surrounding communities from the physical, chemical, and/or biological hazards of the site. The health and safety plan outlines:
(1) Site hazards;
(2) Work areas and site control Procedures;
(3) Air surveillance procedures;
(4) Levels of protection;
(5) Decontamination and site emergency plans;
(6) Arrangements for weather-related problems; and
(7) Responsibilities for implementing the health and safety plan (40 CFR 35.6015-91).

Health and safety study: Any study of any effect of a chemical substance or mixture on health or the environment or on both, including underlying data and epidemiological studies, studies of occupational exposure to a chemical substance or mixture, toxicological, clinical, and ecological studies of a chemical substance or mixture, and any test performed pursuant to this Act (TSCA3, *see also* 40 CFR 716.3; 720.3; 723.50-91).

Health hazard type:
(1) **Acute toxicity:** The older term used to describe immediate toxicity. Its former use was associated with toxic effects that were severe (e.g., mortality) in contrast to the term subacute toxicity that was associated with toxic effects that were less severe. The term acute toxicity is often confused with that of acute exposure.
(2) **Allergic reaction:** Adverse reaction to a chemical resulting from previous sensitization to that chemical or to a structurally similar one.
(3) **Chronic toxicity:** The older term used to describe delayed toxicity. However, the term chronic toxicity also refers to effects that persist over a long period of time whether or not they occur immediately or are delayed. The term chronic toxicity is often confused with that of chronic exposure.
(4) **Idiosyncratic reaction:** A genetically determined abnormal reactivity to a chemical.
(5) **Immediate versus delayed toxicity:** Immediate effects occur or develop rapidly after a single administration of a substance, while delayed effects are those that occur after the lapse of some time. These effects have also been referred to as acute and chronic, respectively.
(6) **Reversible versus irreversible toxicity:** Reversible toxic effects are those that can be repaired, usually by a specific tissue's ability to regenerate or mend itself after chemical exposure, while irreversible toxic effects are those that cannot be repaired.
(7) **Local versus systemic toxicity:** Local effects refer to those that occur at the site of first contact between the biological system and the toxicant; systemic effects are those that are elicited after absorption and distribution of the toxicant from its entry point to a distant site (EPA-92/12).

Health advisory level: A non-regulatory health-based reference level of chemical traces (usually in ppm) in drinking water at which there are no adverse health risks when ingested over various periods of time. Such levels are established for one day, 10 days, long term and life-time exposure periods. They contain a large margin of safety (EPA-94/04).

Health assessment: An evaluation of available data on existing or potential risks to human health posed by a Superfund site. The Agency for Toxic Substances and Disease Registry (ATSDR) of the Department of Health and Human Services (DHHS) is required to perform such an assessment at every site on the National Priorities List (EPA-94/04).

Heap leaching: A process used in the recovery of copper from weathered ore and material from mine dumps. The liquor seeping through the beds is led to tanks, where it is treated with scrap iron to precipitate the copper from solution. This process can also be applied to the sodium slufide leaching of mercury ores (EPA-82/05).

Hearing: A hearing conducted as specified in this subpart to enable the Agency to decide whether to impose sanctions on a respondent for violations of the Executive Order and rules, regulations, and orders thereunder (40 CFR 8.33-91, *see also* 40 CFR 22.03; 123.64; 164.2-91).

Hearing protective device: Any device or material, capable of being worn on the head or in the ear canal, that is sold wholly or in part on the basis of its ability to reduce the level of sound entering the ear. This includes devices of which hearing protection may not be the primary function, but which are nonetheless sold partially as providing hearing protection to the user. This term is used interchangeably with the terms, hearing protector nd device (40 CFR 211.203-91).

Hearth: The bottom of a furnace on which waste materials are exposed to the flame (EPA-89/03b). Other hearth related terms include:
- Burning hearth
- Cold drying hearth
- Drying hearth
- Hot drying hearth

Heat (or heat load): Heat is also known as heat load. It is an interactive flow of energy between a system and its surroundings which is caused by a temperature difference between the system and the surroundings. Typical heat units include: Btu and Calorie. The conventional signs of heat are as follows:
- Heat added to the system is positive
- Heat liberated from the system is negative (cf. work)

An adiabatic process is a process in which no heat is transferred between a system and its surroundings. Applying the first thermodynamic law under the steady state, heat can be obtained:

$Q = mC_p(T_2 - T_1)$

Where:
- m = mass flow
- C_p = constant pressure specific heat
- T_1 = temperature at state 1
- T_2 = temperature at state 2

Example:
- Determine heat rate.

Data:
- Mass flow rate = 1200 lb/min, C_p = 0.26 Btu/lb-F, initial and final temperatures = 200° F and 1200° F respectively.

Solution:
- $Q = mC_p(T_2 - T_1) = 1200 \times 0.26 \times (1200 - 200) = 3.12 \times 10^5$ Btu/min.

Other heat related terms include:
- Available heat
- Latent heat
- Sensible heat
- Specific heat

Heat absorbing glass: A glass having the property of absorbing a substantial percentage of radiant energy (*see* glass for more related terms) (EPA-83).

Heat and work conversion factor
- 1 Btu = 778 ft-lbf
- 1 Btu = 1/3413 kwhr
- 1 Btu = 1/2545 hp-hr
- 1 hp = 33000 ft-lb/min
- 1 hp = 0.746 kw
- 1 calorie = 4.184 joules
- 1 Btu/((lb-mole)(F)) = 1 cal/((g-mole)(C)) (Wark-p75)
- 1 Btu/((lb)(F)) = 1 cal/((g)(C)) (Wark-p75)

Heat balance: An accounting of the distribution of the heat input and output of an incinerator, usually on an hourly basis (cf. energy balance) (SW-108ts).

Heat capacity: Heat is the amount of heat required to raise the temperature of one mole of substance by one degree (molar heat capacity) or the amount of heat required to raise the temperature of one unit mass of substance by one degree (EPA-84/09). Heat capacity and specific heat are interchangeable. It includes:

(1) **Constant pressure specific heat:** (C_p) = (enthalpy)/temperature, under the constant pressure process.
(2) **Constant volume specific heat:** (C_v) = (internal energy)/temperature, under the constant volume process.
(3) **Typical units of specific heat include:** Btu/lb-F (Btu/lb-R), Btu/(lb-mole-F), cal/g-C, or cal/(g-mole-C). Some frequently used C_p and C_v values are shown below:

Gas	C_p(Btu/lb-R)	C_v(Btu/lb-R)
Air	0.240	0.171
CO	0.249	0.178
CO_2	0.203	0.158
Steam, H_2O	0.445	0.335
Nitrogen, N	0.248	0.177
Oxygen, O_2	0.219	0.157

Heat cycle: The period beginning when scrap is charged to an empty EAF and ending when the EAF tap is completed or beginning when molten steel is charged to an empty AOD vessel and ending when the AOD vessel tap is completed (40 CFR 60.271a-91).

Heat engine: A device which receives heat and produces work while executing a cycle (Jones-p91; Wark-p276).

Heat exchanger: A device used to transfer heat from one stream to another without allowing them to mix. In an air-to-air heat exchanger for residential use, heat from exhausted indoor air is transferred to incoming outdoor air, without mixing the two streams (EPA-88/08).

Heat exchanging: Chilling, freezing, and refrigerating; heating, cooking, broiling, roasting, baking, and so forth (AP-40, p790).

Heat exchanging cyclone: A unit used in the cement production process that utilizes the kiln's hot exit gases to heat the raw materials prior to entering the kiln (ETI-92).

Heat flux:
(1) In meteorology, the heat transfer between the surface and the atmosphere.
(2) In air pollution modeling, the heat discharged by sources causing plume rise (NATO-78/10).

Heat input:
(1) The total gross calorific value (where gross calorific value is measured by ASTM Method D2015-66, D240-64, or D1826-64) of all fuels burned (40 CFR 52.01-91).
(2) Heat derived from combustion of fuel in a steam generating unit and does not include the heat input from preheated combustion air, recirculated flue gases, or exhaust gases from other sources, such as gas turbines, internal combustion engines, kilns, etc (40 CFR 60.41b-91, *see also* 40 CFR 60.41c-91).

Heat island: An area, generally an urban or industrial complex, where the temperature is higher than that of its normally rural surroundings. It is generally an air circulation problem peculiar to cities. Tall buildings, heat from pavements and concentrations of pollutants create a haze dome that prevents rising hot air from being cooled at its normal rate. The heat island effect may lead to a local circulation pattern in which air pollution can get trapped. The general effect of a heat island on the dispersion conditions is that the atmosphere is less stable over the urban or industrial complex than over the rural surroundings. It can also trap high concentrations of pollutants and present a serious health problem (EPA-74/11; NATO-78/10).

Heat island effect:
(1) A dome of elevated temperatures over an urban area caused by structural and pavement heat fluxes, and pollutant emissions from the area below the dome (EPA-89/12).
(2) A "dome" of elevated temperatures over an urban area caused by structural and pavement heat fluxes, and pollutant emissions (EPA-94/04).

Heat load: *See* synonym, heat.

Heat of activation: *See* synonym, activation energy.

Heat of adsorption: The heat given off when molecules are adsorbed (EPA-82/11f).

Heat of combustion (calorific value, enthalpy of combustion, heat of reaction or heating value): The energy liberated when a compound experiences complete combustion with oxygen with both the reactants starting, and the products ending at the same condition, usually 25° C or 60° F and 1 atm. Since all combustion processes result in a decrease of enthalpy of the system, energy in the form of heat is released during the reaction. The adopted convention is for the values of heat of combustion to be negative in sign (EPA-84/09).

The heating value can be determined experimentally by calorimeters in which the products of combustion are cooled to the initial temperature and the heat adsorbed by the cooling medium is measured. The heating value of a waste is a measure of the energy released when the waste is burned. The heat transfer of the heating value during a combustion reaction can be computed by using the thermodynamic first law (Markes-67)

Heating value is measured in units of Btu/lb (J/kg). A heating value of about 5,000 Btu/lb (11.6×10^6 J/kg) or greater is needed to sustain combustion. Wastes with lower heating values can be burned, but they will not maintain adequate temperature without the addition of auxiliary fuel. The heating value of the waste also is needed to calculate total heat input to the incinerator where: Heat Input (Btu/h) = Feed Rate (lb/h) x Heating Value (Btu/lb)

Moisture is evaporated from the waste as the temperature of the waste is raised in the combustion chamber; it passes through the incinerator, unchanged, as water vapor. This evaporation of moisture uses energy and reduces the temperature in the combustion chamber. The water vapor also increases the combustion gas flow rate, which reduces combustion gas residence time.

Heat of combustion has been proposed as one of the criteria to determine the ranking of hazardous waste incinerability. The rationale is that if a compound has more heat of combustion or can release more heat than other compounds during combustion, the compound would be easier to be incinerated

For example, dichloromethane would release 1.7 kcal/gram during combustion as shown in the following equation: $CH_2Cl_2 + O_2 + 3.76N_2 ---> CO_2 + 2HCl + 3.76N_2 + 1.70$ kcal/gram

The heat of combustion is normally written with a negative sign. This is a thermodynamic convention and indicates that energy flows out of the system. The heat of combustion of any compound at standard conditions can be calculated from the standard heats of formation of the compound and of the oxidation products.

If a fuel is only partially oxidized, the entire heat of combustion is not released; some of this heat remains bound up as potential chemical energy in the bonds of the partial oxidation species. Therefore, the final or peak temperature produced will be lower than if the fuel were completely oxidized.

Heat of formation (enthalpy of formation, enthalpy of hydration or enthalpy of reaction): The quantity of heat transferred during the formation of a compound from its elements at standard conditions (temperature = 25° C (77° F) and pressure = 1 atm) where the energy level of all elements (reactants, in this case) is assigned to be zero. Heat of formation can be determined from the change in enthalpy resulting when a compound is formed from its elements at constant temperature and pressure conditions.

Example:
- At standard conditions, methane formation may be expressed as: $C + 2H_2 ---> CH_4$. If the heat transfer were accurately measured, the heat of formation, Hf, would be -17.9 Kcal/g-mole.

Solution:
- Applying the Thermodynamic First Law to this process, and using Table 12.3 in (Wylen-73, p502), heat of formation can be obtained by:
- $Hf = h_2 - h_1 = h_{CH4} - h_C - h_{2H2} = -17.9$ Kcal/g-mole - 0 - 0 = -17.9 Kcal/g-mole
- Thus, the measurement of the heat transferred actually provides the enthalpy difference between the products and the reactants. If zero value is assigned to the enthalpy of all elements at standard conditions (25° C and 1 atm pressure), then the enthalpy of the reactants in this case is zero. The enthalpy of CH_4 at 25° C and 1 atm. pressure (relative to this base in which the enthalpy of its elements is assigned to be zero) is called its enthalpy of formation.
- The enthalpy of CH_4 at any other state would be found by adding the change of enthalpy between 25° C and 1 atm. and the given state to the enthalpy of formation. That is, the enthalpy at any temperature and pressure is:
- H(t,p) = Hf + enthalpy change from the conditions of 25° C and 1 atm. and the given conditions.

Heat of fusion: Latent heat involved in changing between the solid and the liquid states.

Heat of reaction: *See* synonym, heat of combustion.

Heat pipe: A circular-cross-section tube which is used to continuously transfer latent heat without the help of external work. The structural elements of the heat pipe include a closed outer vessel, a porous capillary wick, and a working fluid. The concept is that heating one region of the heat pipe evaporates working fluid from the wick and drives the vapor to other regions where it condenses, giving up the latent heat of the working fluid. Within the wick, capillary forces return the condensate to the evaporator region. Typical heat pipe operation is characterized by nearly isothermal conditions along its length.

Heat pump: *See* synonym, refrigerator.

Heat rating: The measurement of engine indicated mean effective pressure (IMEP) value obtained on the engine at a point when the supercharge pressure is 25.4 mm (one inch) Hg below the preignition point of the spark plug, as rated according to SAE J549A Recommended Practice (40 CFR 85.2122(a)(8)(ii)(B)-91).

Heat recovery ventilators (HRVs): Also known as air-to-air heat exchangers (EPA-88/08).

Heat release rate:
(1) The amount of heat liberated in a chamber during complete combustion; it is usually expressed in Btu per hour per cubic foot of the internal volume of the furnace where the combustion takes place (EPA-83).
(2) The steam generating unit design heat input capacity (in MW or Btu/hour) divided by the furnace volume (in cubic meters or cubic feet); the furnace volume is that volume bounded by the front furnace wall where the burner is located, the furnace side waterwall, and extending to the level just below or in front of the first row of convection pass tubes (40 CFR 60.41b-91).

Heat resistant steels: Steels with high resistance to oxidation and moderate strength at high temperatures above 500° C (EPA-83/06a).

Heat resisting glass: A glass able to withstand high thermal shock, generally because of low expansion coefficient (*see* glass for more related terms) (EPA-83).

Heat set ink: A letterpress and lithographic ink which dries under the action of heat by evaporation of their high boiling solvent (*see* ink for more related terms) (EPA-79/12a).

Heat setting refractory mortar: A mortar in which the bond is developed by the application of relatively high temperatures, which vitrify part of its constituents. Also known as **hot setting refractory mortar** (*see* mortar for more related terms) (EPA-83).

Heat stroke: Breakdown of the body's heat regulating mechanism (Course 165.5)

Heat time: The period commencing when scrap is charged to an empty EAF and terminating when the EAF tap is completed (40 CFR 60.271-91).

Heat transfer: The transfer of heat from a higher temperature point to a lower temperature point. Other heat transfer related terms include:
- Conduction
- Convection also known as mass transfer
- Radiation

Heat transfer medium: Any material which is used to transfer heat from one point to another point (40 CFR 60.41b-91, *see also* 40 CFR 60.41c-91).

Heat treatment: The application of heat of specified temperature and duration to change the physical properties of the metal (40 CFR 467.02-91, *see also* 40 CFR 471.02; 468.02-91).

Heated airless spray: An airless spray coating method in which the coating is heated just prior to application (40 CFR 52.741-91).

Heating oil: The petroleum that is No. 1, No. 2, No. 4--light, No. 4--heavy, No. 5--light, No. 5--heavy, and No. 6 technical grades of fuel oil; other residual fuel oils (including Navy Special Fuel Oil and Bunker C); and other fuels when used as substitutes for one of these fuel oils. Heating oil is typically used in the operation of heating equipment, boilers, or furnaces (*see* oil for more related terms) (40 CFR 280.12-91).

Heating season: The coldest months of the year when pollution emissions are higher in some areas because of increased fossil fuel consumption (EPA-74/11).

Heating value: Also known as the heat of combustion. It is the energy liberated when a compound experiences complete combustion with oxygen with both the reactants starting, and the products ending at the same condition, usually 25° C or 60° F and 1 atm. Other heating value related terms include:
- Gross calorific value (*see* synonym, high heating value)
- Gross heating value (*see* synonym, high heating value)
- High heating value
- Higher heating value (*see* synonym, high heating value)
- Low heating value
- Lower heating value (*see* synonym, low heating value)
- Net heating value (*see* synonym, low heating value)

Heavy duty engine: Any engine which the engine manufacturer could reasonably expect to be used for motive power in a heavy-duty vehicle (*see* engine for more related terms) (40 CFR 86.082.2-91).

Heavy duty vehicle: A truck, bus, or other vehicle manufactured primarily for use on the public streets, roads, and highways (not including any vehicle operated exclusively on a rail or rails) which has a gross vehicle weight (as determined under regulations promulgated by the Administrator) in excess of six thousand pounds. Such term includes any such vehicle which has special features enabling off street or off highway operation and use (*see* vehicle for more related terms) (CAA202, *see also* 40 CFR 86.082.2-91).

Heavy hydrogen (or deuterium):
(1) Heavy hydrogen is a hydrogen isotope with a.w. 2.0144; d. 2.0 g/cc; m.p. -254° C and b.p. -249° C.
(2) A nonradioactive isotope of the element hydrogen with one neutron and one proton in the atomic nucleus (DOE-91/04).
(3) *See* hydrogen for more related terms.

Heavy gas: A mixture of a gaseous air pollutant and air, where the density of this mixture is considerably greater than that of pure air. The dispersion of a heavy gas is influenced by gravity effects (*see* gas for more related terms) (NATO-78/10).

Heavy light duty truck: Any light duty truck rated greater than 6000 lbs GVWR (*see* truck for more related terms) (40 CFR 86.094.2-91).

Heavy liquid: The liquid with a true vapor pressure of less than 0.3 kPa (0.04 psi) at 294.3 K (70° F) established in a standard reference text or as determined by ASTM method D2879-86 (incorporated by reference as specified in 40 CFR 52.742); or which has 0.1 Reid Vapor Pressure as determined by ASTM method D323-82 (incorporated by reference as specified in 40 CFR 52.742); or which when distilled requires a temperature of 421.95 K (300° F) or greater to recover 10 percent of the liquid as determined by ASTM method D86-82 (incorporated by reference as specified in 40 CFR 52.742) (*see* liquid for more related terms) (40 CFR 52.741-91).

Heavy media separation: *See* synonym, dense media separation.

Heavy media separator: A unit process used to separate materials of differing density by float/sink in a colloidal suspension of a finely ground dense mineral. This suspension, or media, usually consists of a water-suspension of magnetite, galena or ferrosilicon (*see* separator for more related terms) (EPA-83).

Heavy metals (or toxic metals):
(1) Metallic elements of higher atomic weights, including but not limited to arsenic, cadmium, copper, lead, mercury, manganese, zinc, chromium, tin, thallium, and selenium (40 CFR 165.1-91).
(2) All uranium, plutonium, or thorium placed into a nuclear reactor (40 CFR 191.12-91).
(3) Metallic elements with high atomic weights, e.g., mercury, chromium, cadmium, arsenic, and lead; can damage living things at low concentrations and tend to accumulate in the food chain (EPA-94/04).
(4) *See* metal for more related terms.

Heavy nickel deposition: Deposition of 0.07 grams per square foot or more of total nickel on the basis metal (EPA-82/11e).

Heavy off-highway vehicle products: For the purpose of paragraph (e) of this section, heavy construction, mining, farming, or material handling equipment; heavy industrial engines; diesel-electric locomotives and associated power generation equipment; and the components of such equipment or engines (40 CFR 52.741-91).

Heavy off-highway vehicle products coating facility: A facility that includes one or more heavy off-highway vehicle products coating line(s) (40 CFR 52.741-91).

Heavy off-highway vehicle products coating line: A coating line in which any protective, decorative, or functional coating is applied onto the surface of heavy off-highway vehicle products (40 CFR 52.741-91).

Heavy passenger cars: For the 1984 model year only, a passenger car or passenger car derivative capable of seating 12 passengers or less, rated at 6,000 pounds GVW or more and having an equivalent test weight of 5,000 pounds or more (40 CFR 86.084.2-91).

Heavy water (D_2O):
(1) A form of water (a molecule with two hydrogen atoms and one oxygen atom) in which the hydrogen atoms consist largely or completely of the deuterium isotope. Heavy water has almost identical chemical properties, but quite different nuclear properties, as light water (common water) (DOE-91/04).
(2) Deuterium oxide (D_2O) with m.p. 4° C and b.p. 101° C is the water (H_2O) whose hydrogen H_2 is replaced by a heavier hydrogen isotope deuterium D_2.
(3) *See* water for more related terms

Heavy water reactor (HWR): A nuclear reactor in which circulating heavy water is used to cool the reactor core and to moderate (reduce the energy of) the neutrons created in the core by the fission reactions (DOE-91/04).

Heel:
(1) In air pollution control, the amount of pollutant retained by the adsorbent after regeneration (or desorbing) (EPA-84/09).
(2) In metallurgy, the part of the molten metal alloy remaining in the furnace to facilitate melting of scrap being charged (EPA-76/12).

Heel tap: An imperfection in which the base or bottom of a bottle is very thick in one area and very thin in another (EPA-83).

Helical spray nozzle: For this type of nozzle, it has a descending spiral impingement surface that breaks up the sprayed liquid into a cone of tiny droplets. The cones can be full or hollow with spray angles from 50° to 180°. There are no internal parts, which helps reduce nozzle plugging. These nozzles can be made of stainless steel, brass, alloys, and plastic materials (*see* spray nozzle for more types) (EPA-84/03b, p2-2).

Helium (He): A noble gaseous element with a.n. 2; a.w. 4.0026; d. 0.126 g/cc; m.p. -269.7° C and b.p. -268.9° C. The element belongs to group VIIIA of the periodic table.

Hematite: One of the most common ores of iron, Fe_2O_3, which contains about 70% metallic iron and 30% oxygen. Most of the iron produced in North America comes from the iron ranges of the Lake Superior district, especially the Mesabi Range Minnesota. The hydrated variety of this ore is called limonite (EPA-82/05).

Hematopoietic system: The system of cells in the bone marrow, spleen, and lymph nodes concerned with formation of the cellular element of the blood (LBL-76/07-bio).

Hemoglobin: A molecules of red blood cells. It contains iron and carries oxygen.

Hemogram: A clinical term used to encompass several hematologic indices, including hematocrit, hemoglobin, and red blood cell count (LBL-76/07-bio).

Hematologic toxicity: Hematocrit, hemoglobin levels, changes in cellular components (erythrocytes, leucocytes, platelets), plasma components, and foreign substances (*see* endpoint for more related terms) (Course 165.6; EPA-92/12).

Hemp: A tall plant with a strong base fiber used to make cord or rope. Also the cord or rope made from hemp fiber (EPA-83).

Henry's law:
(1) The measure of the volatility of a substance in a dilute solution of water at equilibrium. It is the ratio of the vapor pressure exerted by a substance in the gas phase over a dilute aqueous solution of that substance to its concentration in the solution at a given temperature. For HRS purposes, use the value reported at or near 25° C. [atmosphere-cubic meters per mole (atm-m^3/mole)] (40 CFR 300-App/A-91).
(2) An expression which relates the concentration of a chemical dissolved in the aqueous phase to the concentration (or pressure) of the chemical in the gaseous phase when the two are at equilibrium with each other (Course 165.6). H = Ca/Cw, where: Ca = Concentration of compound in air and Cw = concentration of compound in water.
(3) The law can be expressed in several equivalent forms, a convenient form being: $C_g = HC_l$ where C_g and C_l are the gas-(g) and liquid-(l) phase concentrations. The constant (H) is the ratio at equilibrium of the gas phase concentration to the liquid-phase concentration of the gas (i.e., moles per liter in air/moles per liter in solution) (EPA-90/08).
(4) For dilute solutions, where the components do not interact, the resulting partial pressure (p) of a component "A" in equilibrium with other components can be expressed as: $p = c_A H$; where:
- p = equilibrium partial pressure of component "A" over solution
- c = concentration of "A" in liquid phase, g-mole/cm^3
- H = Henry's law constant, (atm-cm^3)/(g-mole) of pure A at the same temperature and pressure as the solution
 In comparing with Henry's law, Raoult's law is for concentrated solutions (Hesketh-79, p145).
(5) *See* gas absorption for more related terms.

Hepatic porphyria: An inborn error of metabolism characterized by increased formation and accumulation of pyrroles in the liver (LBL-76/07-bio).

Hepatocellular carcinomas: Malignant tumors of the cells comprising the outer layer of the liver (EPA-85/10).

Hepatocellular injury: Injury to the cells of the liver (LBL-76/07-bio).

Hepatoma: A malignant tumor occurring in the liver (Course 165.6).

Hepatoxicity: Gross and microscopic examination, organ weight, liver function (bile formation, lipid metabolism, protein metabolism, carbohydrate metabolism, metabolism of foreign compounds, serum enzyme activities) (*see* endpoint for more related terms) (Course 165.6; EPA-92/12).

Heptachlor: An insecticide that was banned on some food products in 1975 and all of them 1978. It was allowed for use in seed treatment until 1983. More recently it was found in milk and other dairy products in Arkansas and Missouri where dairy cattle were illegally fed treated seed (EPA-94/04).

Herbicide: A chemical pesticide designed to control or destroy plants, weeds, or grasses (cf. selective herbicide) (EPA-89/12).

Herbivore:
(1) An animal that feeds on plants (EPA-89/12).
(2) An organism that feeds on vegetation (LBL-76/07-water).

Herculon fiber: A trademark of Hercules for polypropylene fibers (*see* fiber for more related terms) (EPA-74/06b).

Heredity: The translocation of genetics from one generation to the next generation.

Heritable translocation: Reciprocal translocations transmitted from parent to the succeeding progeny (40 CFR 798.5955-91, *see also* 40 CFR 798.5460-91).

Hermaphrodite (or bisexual): An animal or a plant with both male and female reproductive organs present in the same individual.

Herpetofauna: Reptiles and amphibians (DOE-91/04).

Hess's law: The total energy change in converting reactants to end products is independent of intermediate steps taken.

Heteroallelic diploids: The diploid strains of yeast carrying two different, inactive alleles of the same gene locus causing aa nutritional requirement (40 CFR 798.5575-91).

Heterogeneity: The conditions of being different in different populations.

Heterogeneous catalysis: This process consists of transformations of multi-phase systems containing at least one solid phase which is generally the catalyst, or at least a kinetically active component of the catalyzed system. The most important common examples of heterogeneous catalysis are those in which gaseous or liquid reaction systems are placed in the presence of a solid catalyst (*see* catalysis for more related terms).

Heterogeneous reaction: A reaction is heterogeneous if it requires the presence of at least two phases of substances to proceed (cf. homogeneous reaction).

Heterotrophic organisms:
(1) Organisms that are dependent on organic matter for food (LBL-76/07-water).
(2) Species that are dependent on organic matter for food (EPA-94/04).

HEX-BCH: The chemical substance 1,2,3,4,7,7-

hexachloronorbornadiene, CAS Number 3389-71-7 (40 CFR 704.102-91).

Hexane (C_6H_{14}): A colorless liquid used in filling for thermometers instead mercury.

Hexane solubles (or extractables): Fats, oils, and greases in wastewaters that are quantified by an analytical technique involving hexane extraction (EPA-75/04).

1,2,3,4,5,6-hexachloro-cyclohexane ($C_6H_6Cl_6$): A white powder used in insecticide for flies, cockroaches, aphids and boll weevils.

Hexavalent chromium: *See* synonym, chromium VI (40 CFR 420.02-91).

HFPO: The chemical substance hexafluoropropylene oxide, CAS Number 428-59-1. [Listed in TSCA Inventory as oxirane, trifluoro(trifluoromethyl)-] (40 CFR 704.104-91).

Hibernation: Dormancy of animals during winter time (cf. dormancy or estivation).

Hide: Any animal pelt or skin as received by a tannery as raw material to be processed (cf. green hide) (40 CFR 425.02-91).

Hierarchy: A body of data organized according to a rank (EPA-88/12).

High alumina refractory: A product containing 4.5 percent more alumina than typical refractories (*see* refractory for more related terms) (EPA-83).

High altitude: Any elevation over 1,219 meters (4,000 feet) (40 CFR 86.082.2-91).

High altitude conditions: A test altitude of 1,620 meters (5,315 feet), plus or minus 100 meters (328 feet), or equivalent observed barometric test conditions of 83.3 plus or minus 1 kilopascals (40 CFR 86.082.2-91).

High altitude reference point: An elevation of 1,620 meters (5,315 feet) plus or minus 100 meters (32 feet), or equivalent observed barometric test conditions of 83.3 kPa (24.2 inches Hg), plus or minus 1 kPa (0.30 Hg) (40 CFR 86.082.2-91).

High carbon ferrochrome: The alloy as defined by ASTM Designation A101-73 (incorporated by reference-*see* 40 CFR 60.17) grades HC1 through HC6 (40 CFR 60.261-91).

High concentration PCBs: The PCBs that contain 500 ppm or greater PCBs, or those materials which EPA requires to be assumed to contain 500 ppm or greater PCBs in the absence of testing (40 CFR 761.123-91). Low-concentration PCBs means PCBs that are tested and found to contain less than 500 ppm PCBs, or those PCB-containing materials which EPA requires to be assumed to be at concentrations below 500 ppm (i.e., untested mineral oil dielectric fluid) (*see* PCB for more related terms) (40 CFR 761.123-91).

High contact industrial surface: A surface in an industrial setting which is repeatedly touched, often for relatively long periods of time. Manned machinery and control panels are examples of high-contact industrial surfaces. High-contact industrial surfaces are generally of impervious solid material. Examples of low-contact industrial surfaces include ceilings, walls, floors, roofs, roadways and side walks in the industrial area, utility poles, unmanned machinery, concrete pads beneath electrical equipment, curbing, exterior structural building components, indoor vaults, and pipes (under TSCA) (40 CFR 761.123-91).

High contact residential commercial surface: A surface in a residential-commercial area which is repeatedly touched, often for relatively long periods of time. Doors, wall areas below 6 feet in height, uncovered flooring, windowsills, fencing, banisters, stairs, automobiles, and children's play areas such as outdoor patios and sidewalks are examples of high-contact residential-commercial surfaces. Examples of low-contact residential-commercial surfaces include interior ceilings, interior wall areas above 6 feet in height, roofs, asphalt roadways, concrete roadways, wooden utility poles, unmanned machinery, concrete pads beneath electrical equipment, curbing, exterior structural building components, (e.g., aluminum/vinyl siding, cinder block, asphalt tiles), and pipes (40 CFR 761.123-91).

High density polyethylene (HDPE):
(1) A thermoplastic polymer or copolymer comprised of at least 50 percent ethylene by weight and having a density of greater than 0.940 g/cm^3 (cf. polyethylene) (40 CFR 60.561-91).
(2) A polymer prepared by low-pressure polymerization of ethylene as the principal monomer and having the characteristics of ASTM D134 Type III and IV polyethylene. Such polymer resins have density greater than or equal to 0.41 g/cc as noted in ASTM D124 (EPA-91/05).
(3) A material used to make plastic bottles and other products that produces toxic fumes when burned (EPA-94/04).
(4) *See* polyethylene for more related terms.

High dose: Not exceed the lower explosive limit (LEL) and ideally should induce minimal toxicity (cf. low dose) (40 CFR 795.232-91).

High duty fireclay brick: A fireclay brick that has a pyrometric cone equivalent (PCE) not lower than Cone 31-23, or does not deform more than 1.5% at 2460° F in the standard local test (*see* brick for more related terms) (SW-108ts).

High efficiency particulate air (HEPA): Refers to a filtering system capable of trapping and retaining at least 99.97 percent of all monodispersed particles 0.3 um in diameter or larger (40 CFR 763.83-91).

High efficiency particulate air (HEPA) filter: A filter capable of trapping and retaining at least 99.97 percent of all monodispersed particles of 0.3 micrometer in diameter or larger (40 CFR 763.121-91).

High energy forming: A process where parts are formed at a rapid rate by using extremely high pressures, e.g., explosive forming, electrohydraulic forming (EPA-83/06a).

High energy rate forging (HERF): A closed die process where hot or cold deforming is accomplished by a high velocity ram (EPA-

83/06a).

High grade electrical paper: An asbestos-containing product that is made of paper and consisting of asbestos fibers and high-temperature resistant organic binders and used in or with electrical devices for purposes of insulation or protection. Major applications of this product include insulation for high-temperature, low voltage applications such as in motors, generators, transformers, switch gears, and other heavy electrical apparatus (*see* paper for more related terms) (40 CFR 763.163-91).

High grade paper: The letterhead, dry copy papers, miscellaneous business forms, stationery, typing paper, tablet sheets, and computer printout paper and cards, commonly sold as white ledger, computer printout and tab card grade by the wastepaper industry (*see* paper for more related terms) (40 CFR 246.101-91).

High hazard content: High hazard contents shall be classified as those which are liable to burn with extreme rapidity or from which poisonous fumes or explosions are to be feared in the event of fire (*see* hazard for more related terms) (29 CFR 1910.35-91).

High heat release rate: A heat release rate greater than 730,000 J/sec-m^3 (70,000 Btu/hour-ft^3) (40 CFR 60.41b-91).

High heating value (HHV) (gross calorific value, gross heating value or higher heating value):
(1) Heat liberated when waste is burned completely and the products of combustion are cooled to the initial temperature of the waste. Usually expressed in British thermal units per pound (40 CFR 240.101-91).
(2) The enthalpy change or heat released when a gas is stoichiometrically combusted at 60° F, with the final (flue) products at 60° F and any water present in the liquid state. Usually expressed as Btu/lb or Btu/scf fuel (EPA-84/09).
(3) The heating value produced from the combustion of a waste under the condition that the water vapor in combustion gas has condensed into the liquid state. Any fuel containing hydrogen yields water as one product of combustion. At atmospheric pressure, the partial pressure of the water vapor in the resulting combustion gas mixture is usually sufficient high to cause water to condense out if the temperature is allowed to fall below 120 to 140° F. This causes liberation of the heat of vaporization of any water condensed. Thus, HHV is the heat transfer with liquid water in the products.
(4) *See* heating value for more related terms.

High level nuclear waste facility: Plant designed to handle disposal of used nuclear fuel, high-level radioactive waste, and plutonium waste (EPA-94/04).

High level of volatile impurity: A total smelter charge containing more than 0.2 weight percent arsenic, 0.1 weight percent antimony, 4.5 weight percent lead or 5.5 weight percent zinc, on a dry basis (40 CFR 60.161-91).

High level radioactive waste (HLRW):
(1) High level radioactive waste as defined in the Nuclear Waste Policy Act of 1982 (Pub. L. 97-425) (40 CFR 191.02-91, *see also* 40 CFR 227.30-91).
(2) Waste generated in the fuel of a nuclear reactor, found at nuclear reactors or nuclear fuel reprocessing plants. It is a serious threat to anyone who comes near the wastes without shielding (cf. low level radioactive waste) (EPA-89/12).
(3) *See* radioactive waste or waste for more related terms.

High level waste: The highly radioactive waste material that results from the reprocessing of spent nuclear fuel, including liquid waste produced directly in reprocessing and any solid waste derived from the liquid. High-level waste contains a combination of transuranic waste and fission products in concentrations requiring permanent isolation (cf. radioactive waste, high level and *see* waste for more related terms) (DOE-91/04).

High-line jumpers: Pipes or hoses connected to fire hydrants and laid on top of the ground to provide emergency water service for an isolated portion of a distribution system (EPA-94/04).

High pressure distribution system: A distribution system in which the gas pressure in the main is higher than the pressure provided to the customer (40 CFR 192.3-91).

High pressure process: The conventional production process for the manufacture of low density polyethylene in which a reaction pressure of about 15,000 psig or greater is used (40 CFR 60.561-91).

High processing packinghouse: A packinghouse which processes both animals slaughtered at the site and additional carcasses from outside sources (cf. low processing packinghouse) (40 CFR 432.41-91).

High rate filter: A trickling filter operated at high average daily dosing rate. All between 10 and 30 mgd/acre, sometimes including recirculation of effluent (*see* filter for more related terms) (EPA-82/11f).

High risk community: A community located within the vicinity of numerous sites or facilities or other potential sources of environmental exposure/health hazards which may result in high levels of exposure to contaminants or pollutants. In determining risk or potential risk, factors such as total weight of toxic contaminants, toxicity, routes of exposure, and other factors may be used (EPA-94/04).

High temperature aluminum coating: A coating that is certified to withstand a temperature of 537.8° C (1000° F) for 24 hours (40 CFR 52.741-91).

High terrain: Any area having an elevation 900 feet or more above the base of the stack of a source (*see* terrain for more related terms) (40 CFR 51.166-91, *see also* 40 CFR 52.21-91).

High-to-low dose extrapolation: The process of prediction of low exposure risk to humans from the measured high exposure-high risk data involving rodents (EPA-94/04).

High transmission glass: A glass which transmits an exceptionally high percentage of the visible light (*see* glass for more related terms) (EPA-83).

High velocity air filter (HVAF): An air pollution control filtration device for the removal of sticky, oily, or liquid aerosol particulate matter from exhaust gas streams (40 CFR 60.471-91).

High viscosity poly(ethylene terephthalate): The poly(ethylene terephthalate) that has an intrinsic viscosity of 0.9 or higher and is used in such applications as tire cord and seat belts (40 CFR 60.561-91).

High volume air sampling: *See* synonym, high volume sampler.

High volume sampler (or high volume air sampling): A device used in the measurement and analysis of ambient suspended particulate pollution (EPA-74/11). Approximately 3,000 ft^3 of air is filtered through a filter for a period of time. The filter is then weighed for determining the amount of particles collected.

Higher aquatic plant: Flowering aquatic plants. (These are separately categorized as Emergent, Floating, and Submerged Aquatic Plants) (LBL-76/07-water).

Higher heating value:
(1) The heat produced by combustion of a unit quantity at constant volume, in an oxygen bomb calorimeter under specified conditions. Also known as gross calorific value (*see* analytical parameters--fuels for more related terms) (EPA-83).
(2) *See* synonym, high heating value.

Highly volatile liquid or HVL: A hazardous liquid which will form a vapor cloud when released to the atmosphere and which has a vapor pressure exceeding 276 kPa (40 psia) at 37.8° C (100° F) (40 CFR 195.2-91).

Highwall: The un-excavated face of exposed overburden and coal in a surface mine or the face or bank on the uphill side of a contour strip mine excavation (EPA-82/10).

Highway: The streets, roads, and public ways in any State (40 CFR 202.10; 205.2-91).

Highway fuel economy: The fuel economy determined by operating a vehicle (or vehicles) over the driving schedule in the Federal highway fuel economy test procedure (*see* fuel economy for more related terms) (40 CFR 600.002.85-91).

Highway fuel economy test: The test procedure described in 40 CFR 600.111(b) (*see* fuel economy for more related terms) (40 CFR 610.11-91).

Hindered settling: The settling speed of particles in a highly polluted liquid is hindered by other particles due to the density of suspended solids.

Histology: The study of the structure of cells and tissues; usually involves microscopic examination of tissue slices (Course 165.6).

Historic resource: The sites, districts, structures, and objects considered limited and nonrenewable because of their association with historic events, persons, or social or historic movements (DOE-91/04).

Hog feeding: The utilization of heat-treated food wastes as a livestock feed (SW-108ts).

Holding of a copper converter: The suspending blowing operations while maintaining in a heated state the molten bath in the copper converter (40 CFR 61.171-91).

Holding pond: A pond or reservoir, usually made of earth, built to store polluted runoff (*see* pond for more related terms) (EPA-89/12).

Holding tank: A reservoir to contain preparation materials so as to be ready for immediate service (*see* tank for more related terms) (EPA-83/03).

Hollow block wall (or block wall): A wall constructed using hollow rectangular masonry blocks. The blocks might be fabricated using a concrete base (concrete block), using ash from combustion of solid fuels (cinder block), or expanded clays. Walls constructed using hollow blocks form an interconnected network with their interior hollow cavities unless the cavities are filled with concrete (EPA-88/08).

Holocene: The most recent epoch of the quaternary period, extending from the end of the Pleistocene to the present (40 CFR 264.18-91).

Home range: The general activity range (or area) of organisms.

Home scrap (or revert scrap): The scrap that never leaves the manufacturing plant and is reprocessed there. Also known as revert scrap (*see* scrap for more related terms) (SW-108ts).

Homeostasis: Maintenance of constant internal conditions in an organism (Course 165.6).

Homeostatic plateau: The point or place at which homeostasis occurs (LBL-76/07-water).

Homeostatis: (homeo = same; stasis = condition) Condition of equilibrium or stability (LBL-76/07-water).

Homeowner water system: Any water system which supplies piped water to a single residence (EPA-94/04).

Hometherm: Maintenance of constant body temperatures in an organism.

Holmium (Ho): A rare earth metal with a.n. 67; a.w. 164.93; d. 8.80 g/cc; m.p. 1461° C and b.p. 2600° C. The element belongs to group IIIB of the periodic table.

Homogeneity: A condition or process of which the properties are independent of a space coordinate (NATO-78/10).

Homogeneous area:
(1) An area of surfacing material, thermal system insulation material, or miscellaneous material that is uniform in color and texture (40 CFR 763.83-91).
(2) In accordance with Asbestos Hazard and Emergency Response Act (AHERA) definitions, an area of surfacing

materials, thermal surface insulation, or miscellaneous material that is uniform in color and texture (EPA-94/04).

Homogeneous catalysis: A chemical reaction process under the condition that all the reactants are contained in a single phase. The primary groupings in this category are reactions between gases and vapors that are catalyzed by other gases and vapors, and reactions in a liquid medium containing a liquid catalyst or a catalyst in solution (*see* catalysis for more related terms).

Homogeneous reaction: A reaction is homogeneous if it takes place in a single phase (cf. heterogeneous reaction).

Homogeneous turbulence: The turbulence for which the statistical properties are independent of the location in the flow field (*see* turbulence for more related terms) (NATO-78/10).

Homolog: A group of isomers that have the same degree of halogenation. For example, the homologous class of tetrachlorodibenzo-p-dioxins consists of all dibenzo-p-dioxins containing four chlorine atoms. When the homologous classes discussed in this part are referred to, the following abbreviations for the prefix denoting the number of halogens are used: tetra-, T (4 atoms); penta-, Pe (5 atoms); hexa-, Hx (6 atoms); hepta-, Hp (7 atoms) (40 CFR 766.3-91).

Homologous: Pertaining to a structure relation between parts of different organisms (e.g., bird wing vs. fish fin).

Homologous chromosomes: The chromosomes which have the same structural features.

Homologous group: A group of chemicals that have the same degree of halogenation, e.g., the homologous class of tetrachlorodibenzo-p-dioxins (TCDD) consists of those PCDDs which have four chlorine atoms. Isomer is a particular member of a homologous group, e.g., 2,3,7,8- TCDD is the tetra- isomer which has chlorine atoms at the 2-, 3-, 7-, and 8- positions (EPA-8/87a).

Homology: A group of chemicals or organisms which have some degree of similarity (e.g., chemicals in the same group of the periodic table).

Homopolymer: A polymer containing only units of one single monomer (EPA-75/01a).

Hood: A partial enclosure or canopy for capturing and exhausting, by means of a draft, the organic vapors or other fumes rising from a coating process or other source (40 CFR 52.741-91).

Hood and enclosures: The partial or complete enclosure around the wheel or disc through which air enters an exhaust system during operation (29 CFR 1910.94b-91).

Hook blade: A shielded knife blade confined in such a way that the blade cuts upward or is drawn toward the person doing the cutting to avoid damage to underlying sheets (EPA-91/05).

Hood capture efficiency:
(1) The emissions from a process which are captured by the hood and directed into a control device, expressed as a percentage of all emissions (40 CFR 52.741-91).
(2) Ratio of the emissions captured by a hood and directed into a control or disposal device, expressed as a percent of all emissions (EPA-94/04).

Hood or enclosure: Any device used to capture fugitive volatile organic compounds (40 CFR 60.441-91).

Hook blade: A shielded knife blade confined in such a way that the blade cuts upward or is drawn toward the person doing the cutting (EPA-89/09)

Horizontal double-spindle disc grinder: A grinding machine carrying two power-driven, rotatable, coaxial, horizontal spindles upon the inside ends of which are mounted abrasive disc wheels used for grinding two surfaces simultaneously (29 CFR 1910.94b-91).

Horizontal single-spindle disc grinder: A grinding machine carrying an abrasive disc wheel upon one or both ends of a power-driven, rotatable single horizontal spindle (29 CFR 1910.94b-91).

Hormone: A chemical substance secreted in one part of an organism and transported to another part of that organism where it has a specific effect (Course 165.6).

Horn: In seaming a landfill liner, the vibration device used with ultrasonic seaming which vibrates at high frequency causing friction and a subsequent melting of the surfaces that it contacts (EPA-89/09, *see also* EPA-91/05).

Hosiery product: For NSPS (40 CFR 410.55), the internal subdivision of the knit fabric finishing subcategory for facilities that are engaged primarily in dyeing or finishing hosiery of any type (40 CFR 410.51-91).

Hospital waste: Waste generated from hospitals and clinics (*see* medical waste for more related terms).

Host:
(1) Any plant or animal on or in which another lives for nourishment, development, or protection (40 CFR 171.2-91).
(2) (A) In genetics, the organism, typically a bacterium, into which a gene from another organism is transplanted.
    (B) In medicine, an animal infected by or parasitized by another organism (EPA-89/12).

Host vector system: In the context of rDNA technology, the particular organism (host) into which the gene is cloned and the vehicle (vector) that carries the gene into the host (EPA-88/09a).

Hot: A colloquial term meaning highly radioactive (EPA-74/11).

Hot compression molding: In plastic processing, a technique of thermoset molding in which preheated molding compound is closed and heat and pressure (in the form of a downward moving ram) are applied until the material has cured (EPA-83/06a).

Hot dip coating: The process of coating a metallic workpiece with another metal by immersion in a molten bath to provide a protective film (EPA-83/06a).

Hot drying hearth: A surface upon which waste materials are placed to dry or burn. Hot combustion gases first pass over the wastes and then under the hearth (*see* hearth for more related terms) (SW-108ts).

Hot forming: The hot form whose steel operations in which solidified, heated steel is shaped by rolls (40 CFR 420.71-91).

Hot metal furnace: A furnace that is initially charged with (cold) solid materials followed by a second charge of melted liquid (*see* furnace for more related terms) (EPA-74/06a).

Hot metal transfer station: The facility where molten iron is emptied from the railroad torpedo car or hot metal car to the shop ladle. This includes the transfer of molten iron from the torpedo car or hot metal car to a mixer (or other intermediate vessel) and from a mixer (or other intermediate vessel) to the ladle. This facility is also known as the reladling station or ladle transfer station (40 CFR 60.141a-91).

Hot mix asphalt facility: Any facility, as described in 40 CFR 60.90, used to manufacture hot mix asphalt by heating and drying aggregate and mixing with asphalt cements (40 CFR 60.91-91).

Hot pressing: Forming a powder metallurgy compact at a temperature high enough to effect concurrent sintering (40 CFR 471.02-91).

Hot rolled: A term used to describe alloys which are rolled at temperatures above the recrystallization temperature. (Many alloys are hot rolled, and machinability of such alloys may vary because of differences in rolling conditions from lot to lot (EPA-83/06a).

Hot setting refractory mortar: *See* definition under heat setting refractory mortar.

Hot soak losses: The evaporative emissions after termination of engine operation (40 CFR 86.082.2-91).

Hot stamping: Engraving operation for marking plastics in which roll leaf is stamped with heated metal dies onto the face of the plastics. Ink compounds can also be used (EPA-83/06a).

Hot strip and sheet mill: Those steel hot forming operations that produce flat hot-rolled products other than plates (40 CFR 420.71-91).

Hot water seal: A heated water bath (heated to approximately 180° F) used to seal the surface coating on formed aluminum which has been anodized and coated. In establishing an effluent allowance for this operation, the hot water seal shall be classified as a cleaning or etching rinse (40 CFR 467.02-91).

Hot well: The reservoir of a condensing unit receiving the condensate from a barometric condenser (*see* well for more related terms) (40 CFR 52.741-91, *see also* 40 CFR 264.1031-91).

Hourly rolling average: *See* definition under rolling average.

House air: *See* synonym, indoor air.

House sewer connection: The sewer that connects a house to the sewer in the adjacent street (DOI-70/04).

Housed lot: Totally roofed buildings which may be open or completely enclosed on the sides wherein animals or poultry are housed over solid concrete or dirt floors, slotted (partially open) floors over pits or manure collection areas in pens, stalls or cages, with or without bedding materials and mechanical ventilation. For the purposes hereof, the term housed lot is synonymous with the terms slotted floor buildings (swine, beef), barn (dairy cattle) or stable (horses), houses (turkeys, chickens), which are terms widely used in the industry (40 CFR 412.11-91).

Household hazardous waste: The product used at residences that is discarded in MSW (municipal solid waste) and that contains substances already regulated under RCRA as hazardous waste (*see* waste for more related terms) (OTA-89/10).

Household solid waste: *See* synonym, household waste.

Household waste: Any material (including garbage, trash and sanitary wastes in septic tanks) derived from households (including single and multiple residences, hotels and motels, bunkhouses, ranger stations, crew quarters, campgrounds, picnic grounds and day-use recreation areas) (*see* waste for more related terms) (40 CFR 261.4-91).

Household waste (domestic waste: Solid waste, composed of garbage and rubbish, which normally originated in a private home or apartment house. Domestic waste may contain a significant amount of toxic or hazardous waste (EPA-94/04).

HRS factor: The primary rating elements internal to the HRS (hazard ranking system) (40 CFR 300-App/A-91).

HRS factor category: A set of HRS (hazard ranking system) factors (that is, likelihood of release [or exposure], waste characteristics, targets) (40 CFR 300-App/A-91).

HRS migration pathways: HRS (hazard ranking system) ground water, surface water, and air migration pathways (40 CFR 300-App/A-91).

HRS pathway: A set of HRS (hazard ranking system) factor categories combined to produce a score to measure relative risks posed by a site in one of four environmental pathways (that is, ground water, surface water, soil, and air) (40 CFR 300-App/A-91).

HRS site score: The composite of the four HRS (hazard ranking system) pathway scores (40 CFR 300-App/A-91).

Hue: Colored light due to the variation of its wavelength.

Human ecology: The study of the relationship between human beings and their environment.

Human environment: Shall be interpreted comprehensively to include the natural and physical environment and the relationship of people with that environment. (*see* the definition of effects (40 CFR 1508.8). This means that economic or social effects are not intended by themselves to require preparation of an environmental impact statement. When an environmental impact statement is prepared and economic or social and natural or physical environmental effects are interrelated, then the environmental impact statement will discuss all of these effects on the human environment (*see* environment for more related terms) (40 CFR 1508.14-91).

Human equivalent concentration: Exposure concentration for humans that has been adjusted for dosimetric differences between experimental animal species and humans to be equivalent to the exposure concentration associated with observed effects in the experimental animal species. If occupational human exposures are used for extrapolation, the human equivalent concentration represents the equivalent human exposure concentration adjusted to a - continuous basis (EPA-92/12).

Human equivalent dose: A dose which, when administered to humans, produces an effect equal to that produced by a dose in animals (EPA-94/04).

Human exposure evaluation: Describing the nature and size of the population exposed to a substance and the magnitude and duration of their exposure. The evaluation could concern past, current, or anticipated exposures (EPA-94/04).

Human health risk: The likelihood that a given exposure or series of exposures may have or will damage the health of individuals (EPA-94/04).

Human immunodeficiency virus (HIV): The virus that can cause AIDS.

Human subject: A living individual about whom an investigator (whether professional or student) conducting research obtains (1) Data through intervention or interaction with the individual, or (2) Identifiable private information (for complete definition, *see* 40 CFR 26.102-91).

Human subject: A living individual about whom an investigator (whether professional or student) conducting research obtains:
(1) Data through intervention or interaction with the individual, or
(2) Identifiable private information.
Intervention includes both physical procedures by which data are gathered (for example, venipuncture) and manipulations of the subject or the subject's environment that are performed for research purposes. Interaction includes communication or interpersonal contact between investigator and subject. "Private information" includes information about behavior that occurs in a context in which an individual can reasonably expect that no observation or recording is taking place, and information which has been provided for specific purposes by an individual and which the individual can reasonably expect will not be made public (for example, a medical record). Private information must be individually identifiable (i.e., the identity of the subject is or may readily be ascertained by the investigator or associated with the information) in order for obtaining the information to constitute research involving human subjects (40 CFR 26.102-f-91).

Humane: In the context of the taking of a marine mammal means that method of taking which involves the least possible degree of pain and suffering practicable to the mammal involved (MMPA3-16USC1362).

Human exposure model (HEM): *See* definition under dispersion model.

Humane: In the context of the taking of a marine mammal means that method of taking which involves the least possible degree of pain and suffering practicable to the mamma involved (MMPA3-16USC1362-90).

Humectant: An agent which absorbs water. It is often added to resin formulations in order to increase water absorption and thereby minimize problems associated with electrostatic charge (EPA-75/01a).

Humic acid: Organic acids derived from humus.

Humidification: The seasoning operation to which newly pressed hardboard are subjected to prevent warpage due to excessive dryness (EPA-74/04).

Humidification chamber: A chamber in which the water vapor content of a gas is increased (EPA-83/03a).

Humidifier: A device to add moisture to air.

Humidifier fever (air conditioner fever or ventilation fever): A respiratory illness that may be caused by exposure to toxins from microorganisms found in wet or moist areas in humidifiers or air conditioners (EPA-88/09b).

Humidity: Water vapor within a space. Other humidity related terms include:
- Absolute humidity: *See* synonym, humidity ratio.
- Critical humidity
- Psychrometric chart
- Relative humidity
- Specific humidity: *See* synonym, humidity ratio.

Humidity indicator: A device which shows different colors as the surrounding humidity changes.

Humidity ratio (HR) (absolute humidity or specific humidity):
HR = [water vapor mass (m_v) in air-vapor mixture] / [air mass (m_a) in air-vapor mixture]
   = [vapor molecular weight (M_v) x vapor partial pressure (p_v)] / [air molecular weight (M_a) x air partial pressure (p_a)]
   = $0.622 p_v/p_a = 0.622 \times RH \times p_g/p_a$

where:
   RH = relative humidity
   p_g = saturation vapor pressure

Example:
- Compute humidity ratio, air mass and water vapor mass.

Data:
- 2000 ft^3 air-water mixture at 14.696 psia, 90$_0$F, has a relative humidity of 70%.

Solution:
- From the Steam Tables (Wylen-73, p649), the saturation pressure for water vapor at 90° F is 0.6982 psia. The partial pressure of the water vapor in the mixture is then $p_v = RH \times p_g = 0.7(0.6982) = 0.489$ psia.
- Since the total atmospheric pressure is equal to the sum of the partial pressures, the air partial pressure equals $p_a = P - p_v = 14.696 - 0.489 = 14.207$ psia.
- $HR = 0.622 p_v/p_a = 0.622 \times 0.489/14.207 = 0.02135$ lbs of water vapor/lb of dry air = 150.5 grains of water vapor/lb of dry air.
- The result is often expressed as grains of water vapor per pound of dry air (where 1 lb = 7000 grains).
- The value of 0.0215 lbs of water vapor/lb of dry air can also be obtained from a Psychrometric Chart (Wylen-73-p708) based on the parameters of dry-bulb temperature (90° F) and 70% relative humidity. Because the water vapor content of combustion air behaves like a heat sink during the combustion process, it can significantly affect the actual combustion temperature achieved. This example also shows that a Psychrometric Chart can be a very convenient tool to use in determining the water vapor content of ambient air.
- Air mass $(m_a) = (p_a \times V)/(R_a \times T) = (14.21 \times 144 \times 2000)/(53.33 \times 550) = 139.6$ lbs. Where: R_a = air gas constant; and T absolute temperature.
- Water vapor mass $(m_v) = HR \times m_a = 0.02135 \times 139.6 = 2.98$ lbs.

Humus: Decomposed organic materials (cf. ectohumus) (EPA-89/12).

HWM facility: Hazardous Waste Management facility (40 CFR 144.3; 146.3, *see also* 40 CFR 270.2-91).

HVAC system: The heating, ventilating, and air conditioning system for a house. Generally refers to a central furnace and air conditioner (EPA-88/08).

Hybrid: A cell or organism resulting from a cross between two unlike plant or animal cells or organisms (EPA-89/12).

Hybrid chromatograph monitor: One of continuous emission monitors (*see* continuous emission monitor for various types). The combination of a GC and other instrumentation. A basic gas chromatograph consists of a series of sample separation and detection equipment. Many combinations of these basic components are possible which would allow virtually any organic compound to be identified. Numerous columns can be used in parallel or in series to tailor the sample separation and retention time to a particular need. Various detectors could be used in parallel or in series to enhance sensitivity and selectivity to compounds of interest. Since gas chromatography involves both a separation and detection step, in some cases only one of these steps may be needed. A variety of other instrumental methods; GC/MS, GC/IR and GC/FTS-IR combines a chromatographic separation with an instrumental identification of the components of the eluted materials. The combination of a GC and other instrumentation results in more of a research oriented monitor and the capital costs are orders of magnitude higher than the simple GC (EPA-84/03a). Types of hybrid chromatograph monitors include:
(1) GC/IR (gas chromatography/infrared absorption spectrometer) (EPA-84/03a)
(2) GC/MS (gas chromatography/mass spectrometry) (EPA-84/03a)

Hybrid integrated circuit: A circuit that is part integrated and part discrete (EPA-83/03).

Hybridization: The information of a duplex between two complementary DNA molecules or between RNA molecule (EPA-88/09a).

Hybridoma: A hybrid cell that produces monoclonal antibodies in large quantities (EPA-89/12).

Hydrapulper: *See* definition under pulper (EPA-83).

Hydrate: A solid or crystalline compound formed by combination of the compound with water (e.g., copper sulfate, $CuSO_4 \bullet 5H_2O$) (cf. anhydride).

Hydration: Property imparted to pulp fibers by mechanicay means which slows the degree of water drainage from the pulp mass (EPA-83).

Hydraulic barking: A wood processing operation that removes bark from wood by the use of water under a pressure of 6.8 atm (100 psia) or greater (40 CFR 429.11-91).

Hydraulic conductivity: The constant of proportionality in Darcy's Law of fluid flow that describes the ease with which a porous medium permits fluids to flow and the ease with which the fluid flows given its physical properties (cf. permeability) (DOE-91/04).

Hydraulic cyclone: A fluid classifying device that separates heavier particles from a slurry (EPA-85/10a).

Hydraulic fluids: The petroleum-based hydraulic fluids (40 CFR 252.4-91).

Hydraulic grade line: The line connecting the points to which the liquid would rise at various places along any pipe or conduit, if piezometer tubes were inserted in the liquid. It is a measure of the pressure head available at these points. In the case of water flowing in a canal or open channel, as opposed to flow in a pipe under pressure, the hydraulic grade line corresponds with the profile of the water surface (M&EI-72).

Hydraulic gradient: In general, the direction of groundwater flow due to changes in the depth of the water table (EPA-94/04).

Hydraulic gradient: The slope of the water table or the change in water level (static head) per unit of distance along the direction of the flow (EPA-87/03; Course 165.7).

Hydraulic lift tank: A tank holding hydraulic fluid for a closed-loop mechanical system that uses compressed air or hydraulic fluid to operate lifts, elevators, and other similar devices (40 CFR 280.12-91).

Hydraulic loading: In the activated sludge process, the food to microorganisms (F/M) ratio defined as the amount of biodegradable material available to a given amount of microorganisms per unit of time (*see* loading for more related terms) (EPA-87/10a).

Hydraulic mining:
(1) Mining by washing sand and soil away with water which leaves the desired mineral.
(2) The process by which a bank of gold-bearing earth and rock is excavated by a jet of water, discharged through the converging nozzle of a pipe under great pressure. The debris is carried away with the same water and discharged on lower levels into water courses below (EPA-82/05).

Hydraulic overload: A condition when the quantity of wastewater flowing into a facility exceeds its design capacity (EPA-80/08).

Hydraulic scooper: A self-propelled crawler vehicle equipped with hydraulically operated arms that lift, empty, and replace containers carried on a transfer trailer bed (SW-108ts).

Hydraulic setting mortar: A mortar that hardens or sets as a result of hydration, a chemical reaction with water. In an incinerator, the water in the mortar evaporates and a ceramic bond develops when the working furnace temperature is applied (*see* mortar for more related terms) (EPA-83).

Hydraulic tipper: A device that unloads a transfer trailer by raising its front end to a 70 degree angle (SW-108ts).

Hydraulics of sewer: The principle of sewer hydraulics. Its main factors that affect the flow of wastewater and sewage in sewers are:
(1) Slope;
(2) Cross-sectional area;
(3) Roughness of interior pipe surface;
(4) Conditions of flow, i.e., full, partly full, steady or varied flow;
(5) Presence of absence of obstructions, bends, etc.; and
(6) Character, specific gravity, and viscosity of the liquid (M&EI-72).
(7) *See* sewer for more related terms

Hydrazine treatment: An application of a reducing agent to form a conductive metal film on a silver oxide cathode (EPA-84/08).

Hydrite atomic absorption spectrometry (HAAS): HAAS utilizes a chemical reduction to reduce and separate arsenic or selenium selectively from a sample digestate. The technique therefore has the advantage of being able to isolate these two elements from complex samples which may cause interferences for other analytical procedures. Significant interferences have been reported when any of the following is present:
(1) Easily reduced metals (Cu, Ag, Hg);
(2) High concentrations of transition metals (>200 mg/L); and
(3) Oxidizing agents (oxides of nitrogen) remaining following sample digestion (SW-846).
(4) *See* metal analyzer for more related terms.

Hydrocarbon:
(1) Any organic compound consisting predominantly of carbon and hydrogen (40 CFR 60.111-91).
(2) Chemical compounds that consist entirely of carbon and hydrogen (EPA-94/04).

Hydrocarbon family: Hydrocarbon is a vast family of compounds containing carbon and hydrogen in various combinations, found especially in fossil fuels. It contains energy-rich bonds such as the carbon-carbon and carbon-hydrogen bonds. While some hydrocarbons are a common source of chemical potential energy and is a major component of a fuel, some hydrocarbons are major air pollutants, some may be carcinogenic and others contribute to photochemical smog. Several important families of hydrocarbons are summarized as follows:
- Paraffin ($C_nH_{(2n+2)}$): Saturated chain
- Olefin (C_nH_{2n}): Not saturated chain
- Diolefin ($C_nH_{(2n-2)}$): Not saturated chain
- Naphthene (C_nH_{2n}): Saturated ring
- Aromatic
  - Benzene ($C_nH_{(2n-6)}$): Not saturated ring
  - Naphthalene ($C_nH_{(2n-12)}$): Not saturated ring

Other hydrocarbon family related terms include:
- Unsaturated hydrocarbon
- Saturated hydrocarbon
- Isomer

Hydrocephalus (exvacuo): Increased volume of cerebral spinal fluid within the cranial vault, associated with decreased volume of cortical tissue, as in severe cortical atrophy (LBL-76/07-bio).

Hydrochloric acid (or hydrogen chloride, HCl): A by-product from the combustion of chlorinated compounds. It is a corrosive compound.

Hydrochloric acid pickling: Those operations in which steel products are immersed in hydrochloric acid solutions to chemically remove oxides and scale, and those rinsing operations associated with such immersions (40 CFR 420.91-91).

Hydrocyclone: A cyclone separator in which a spray of water is used (EPA-82/10).

Hydroclassifier: A machine which uses an upward current of water to remove fine particles from coarser materials (EPA-82/10).

Hydrofluoric acid: Hydrogen fluoride in an aqueous solution (EPA-83/06a).

Hydroformylation: An addition of a formyl molecule (U-CHO) across a double bond to form an aldehyde (EPA-87/10a).

Hydrogen (H): A colorless gaseous element with a.n. 1; a.w. 1.0079; d. 0.071 g/cc; m.p. -259.2° C and b.p. -252.7° C. The element belongs to group IA of the periodic table (cf. heavy hydrogen).

Hydrogen chloride: *See* synonym, hydrochloric acid.

Hydrogen electrode (or hydrogen half cell): A half cell with a noble metal (e.g., platinum) foil as the electrode and a solution of hydrogen gas as the electrolyte. The standard hydrogen electrode, which uses a platinum foil with a 1.0 M solution of hydrogen ions under 1 atmospheric pressure and 25° C temperature, is used to measure standard electrode potentials (*see* electrode for more related terms).

Hydrogen gas stream: A hydrogen stream formed in the chlor-alkali cell denuder (40 CFR 61.51-91).

Hydrogen half cell: *See* synonym, hydrogen electrode.

Hydrogen ion concentration: The weight of hydrogen ions in grams per liter of solution. Commonly expressed as the pH value that represents the logarithm of the reciprocal of the hydrogen ion concentration (EPA-84/08).

Hydrogen peroxide (H_2O_2): Unstable liquid used as an antiseptic.

Hydrogen sulfide (H_2S):
(1) A poisonous gas with the odor of rotten eggs that is produced from the putrefaction of sulfur-containing organic material. Odorous in concentrations as small as parts per billion (EPA-83).
(2) Gas emitted during organic decomposition. Also a byproduct of oil refining and burning. Smells like rotten eggs and, in heavy concentration, can kill or cause illness (EPA-94/04).

Hydrogen sulphate (or bisulfate): A compound (or salt) that contains the ion $(HSO_4)^-$ or an ester of the type $RHSO_4$, where R is an organic group.

Hydrogenation: A reaction of hydrogen with an organic compound (EPA-87/10a).

Hydrogeologic cycle: The natural process recycling water from the atmosphere down to (and through) the earthy and back to the atmosphere again (EPA-94/04).

Hydrogeology:
(1) The study of surface and subsurface water (EPA-89/11).
(2) The geology of ground water, with particular emphasis on the chemistry and movement of water (EPA-94/04).

Hydrolization: The addition of H2O to a molecule. In sugar production hydrolization of sucrose results in an inversion into glucose and fructose and represents lost production (EPA-75/02d).

Hydrologic cycle:
(1) Movement or exchange of water between the atmosphere and the earth (EPA-94/04).
(2) The continual exchange of moisture between the earth and the atmosphere, consisting of evaporation, condensation, precipitation (rain or snow), stream run-off, absorption into the soil, etc. (DOI-70/04).

Hydrology:
(1) The science dealing with the properties, distribution, and circulation of water (EPA-94/04).
(2) The science dealing with the properties, distribution, and circulation of natural water systems (DOE-91/04).

Hydrolysate (or hydrolyzate): In mining, a sediment consisting partly of chemically un-decomposed, finely ground rock powder and partly of insoluble matter derived from hydrolytic decomposition during weathering (EPA-82/05).

Hydrolysis: The chemical reaction of a substance with water (40 CFR 300-App/A-91).

Hydrolysis: A chemical reaction of water with another substance in which hydrogen (H) and hydroxyl (OH) are added to the other substance, forming usually two or more new compounds. The general formula is: $XY + H_2O \longrightarrow HY + XOH$. Examples of water reacting with an organic and inorganic substances are:
- Reacting with organic substance: $CH_{11}Cl + H_2O \longrightarrow HCl + C_5H_{11}OH$
- Racting with inorganic substance: $KCN + H_2O \longrightarrow HCN + KOH$.

Hydrolyzing: The reaction involving the decomposition of organic materials by interaction with water. Poultry feathers for example, are hydrolyzed to a proteinaceous product by heating under pressure (EPA-75/04).

Hydrometallurgical: The use of wet processes to treat metals (EPA-83/03a).

Hydrometer: An instrument responsive to humidity conditions (usually relative humidity) of the atmosphere.

Hydro-metallurgical process: The treatment of ores by wet processes such as leaching (EPA-83/06a).

Hydro-metallurgy: The treatment of ores, concentrates, and other metal-bearing materials by wet processes, usually involving the solution of some component, and its subsequent recovery from the solution (EPA-82/05).

Hydrophilic:
(1) A surface or a functional group having a strong affinity for water or being readily wettable (EPA-83/06a).
(2) Having a strong affinity for water (EPA-94/04).

Hydrophobic:
(1) A surface which is non-wettable or not readily wettable (EPA-83/06a).
(2) Having a strong aversion for water (EPA-94/04).

Hydrophytes: Plants that grow only in water or very wet earth (DOI-70/04).

Hydropneumatic: A water system, usually small, in which a water pump is automatically controlled by the air pressure in a compressed air tank (EPA-94/04).

Hydroquinone: A developing agent used to form a conductive metal film on a silver oxide cathode (EPA-84/08).

Hydroscopic: Water adsorbing (EPA-76/03).

Hydrostatic equilibrium: A condition which is realized when complete balance exists between vertical pressure forces and gravity forces (NATO-78/10).

Hydrostatic pressure: The force per unit area measured in terms of the height of a column of water under the influence of gravity (EPA-83/06a).

Hydrotesting: The testing of piping or tubing by filling with water and pressurizing to test for integrity (40 CFR 471.02-91).

Hydrothermal desulfurization: One of SO_2 emission reduction techniques (*see* sulfur oxide emission control for control structure). In hydrothermal desulfurization, coal is crushed and mixed with a solution of sodium and calcium hydroxides [NaOH and $Ca(OH)_2$]. When this mixture is heated to 275° C in a pressurized vessel, most of the pyritic sulfur and 20 to 50% of the organic sulfur is converted to sodium and calcium sulfites (Na_2SO_3 and $CaSO_3$). The coal is rinsed to remove the sulfites and the water is processed to recycle the sodium and calcium hydroxides. This process is an expensive but effective method for removing sulfur from coal (EPA-81/12, p8-5).

Hydroxide: A chemical compound containing the radical group OH^- (EPA-87/10a).

Hygroscopic:
(1) Readily absorbing and retaining moisture, usually in reference to readily absorbing moisture from the air (EPA-86/12).
(2) Readily taking up and retaining moisture (water) (EPA-87/07a).

Hygroscopicity: An act of attracting moisture from the air (EPA-81/09).

Hypergolic: Two substances which will self-ignite on contact (EPA-76/03).

Hyperaminoaciuria: Presence in the urine of above-normal amounts of amino acids (LBL-76/07-bio).

Hypereutrophy: Excessively enriched with nutrients. A hypereutrophic body of water may be inundated with algae, and is generally oxygen-deficient (LBL-76/07-water).

Hyperphosphaturia: Above-normal amounts of phosphate compounds in the urine (LBL-76/07-bio).

Hypersensitivity pneumonitis: A group of respiratory diseases, including humidifier fever, that involve inflammation of the lungs. Most forms of hypersensitivity pneumonitis are thought to be caused by an allergic reaction triggered by repeated exposure to biological contaminants. Humidifier fever may be an exception, distinguished from other forms of pneumonitis by the fact it does not appear to involve an allergic reaction (EPA-88/09b).

Hypertrophic water: The water with high nutrient contents (*see* water for more related terms) (DOI-70/04).

Hyperuricemia: Abnormal amounts of uric acid in the blood (LBL-76/07-bio).

Hypochromic anemia: A condition characterized by a disproportionate reduction of red cell hemoglobin, compared with the volume of packed red cells (LBL-76/07-bio).

Hypocotyl: The portion of the axis of an embryo or seedling situated between the cotyledons (seed leaves) and the radicle (40 CFR 797.2750-91).

Hypodermic syringe: Commercially available syringe which isolates and delivers a liquid sample (cf. gas tight syringe) (EPA-88/12).

Hypolimnion:
(1) The region of a body of water that extends from the thermocline to the bottom of the lake and is removed from surface influence (LBL-76/07-water).
(2) Bottom waters of a thermally stratified lake., The hypolimnion of a eutrophic lake is usually low or lacking in oxygen (EPA-94/04).

Hypophosphatemia: Abnormally low amount of phosphate compounds in the blood (LBL-76/07-bio).

Hypothalamus: The posterior portion of the forebrain that includes the nuclei of nerve cells that exert control over visceral activities, water balance, temperature, sleep, etc (LBL-76/07-bio).

Hypoxia: A deficiency of oxygen (Course 165.6).

Ice fog: An atmospheric suspension of highly reflective ice crystals (40 CFR 60.331-91).

Ideal gas: Imaginary (or hypothetical) gases or vapors which obey the ideal gas law at the pressure approaching to zero (0) (very low density). No real gas obeys the ideal gas exactly over all ranges of temperature and pressure. Although the lighter gases (hydrogen, oxygen, air, etc.) at ambient conditions approach ideal gas law behavior, the heavier gases such as sulfur dioxide and hydrocarbons, particularly at high pressures and low temperatures, deviate considerably from the ideal gas law. Despite these deviations, the ideal gas law is routinely used in all air pollution calculations (EPA-84/09).

Ideal gas law: A law or an equation describing the relationship among pressure, volume, and temperature of an ideal gas. A general form for any number of moles of an ideal gas is (cf. equation of state):

$$PV = nR_uT = mRT, \quad n = m/M, \quad R = R_u/M$$

Where: m = mass, lb or g; M = molecular weight, lb/lb-mole or g/g-mole; n = number of moles (lb-moles, g-moles or kg-moles); R = gas constant;

R_u = universal gas constant
 = 82.0575 (atm-cm^3)/(g-mole-K)
 = 0.0821 (atm-m^3)/(kg-mole-K)
 = 0.7302 (atm-ft^3)/(lb-mole-R)
 = 0.0821 (atm-liter)/(g-mole-K)
 = 83.144 x 10^6 (g-cm^2)/(sec^2-g-mole-K)
 = 8.3144 x 10^4 (kg-m^2)/(sec^2-kg-mole-K)
 = 1545 (lb$_f$-ft)/(lb-mole-R)
 = 4.9686 x 10^4 (lb$_m$-ft^2)/(sec^2-lb-mole-R)
 = 8.3144 x 10^6 (Pa-cm)/(sec-g-mole-K)
 = 8.3144 x 10^3 (kPa-cm)/(sec-g-mole-K)
 = 8.3144 x 10^5 (kPa-m)/(sec-g-mole-K)
 = 0.1724 (psi-ft^3)/(lb-mole-R)
 = 10.73 (psia-ft^3)/lb-mole-R
 = 21.83 (in-Hg)(ft^3)/(lb-mole-R)
 = 62.4 (mm-Hg-liter)/(g-mole-K)
 = 555 (mm-Hg)(ft^3)/(lb-mole-R)
 = 1.9872 Btu/(lb-mole-R)
 = 1.9872 cal/(g-mole-K)
 = 8.3144 joule/(g-mole-K)
 = 8.3144 x (N-m)/(g-mole-K)

P = absolute pressure
T = absolute Temperature, R or K
R = 460 + F
K = 273 + C
V = volume, ft^3 (EPA-81/12, p2-8; 84/09).

Ideal solution: A solution which obeys Raoult's law. The law means that the vapor partial pressure of a solvent in a mixture solution is proportional to its mole fraction.

Identification: The recognition of a situation, its causes and consequences relating to a defined potential, e.g., hazard identification (EPA-87/07a).

Identification code or EPA I.D. Number: The unique code assigned to each generator, transporter, and treatment, storage, or disposal facility by regulating agencies to facilitate identification and tracking of chemicals or hazardous waste (EPA-94/04).

Identifying characteristics: A description of asbestos or asbestos containing material, including:
(1) The mineral or chemical constituents (or both) of the asbestos or material by weight or volume (or both).
(2) The types or classes of the product in which the asbestos or material is contained.
(3) The designs, patterns, or textures of the product in which the asbestos or material is contained, and
(4) The means by which the product in which the asbestos or containing asbestos or asbestos-containing material (TSCA-AIA1-91).

Identity: Any chemical or common name used to identify a chemical substance or a mixture containing that substance (40 CFR 721.3-91).

Idle: The condition where all engines capable of providing motive power to the locomotive are set at the lowest operating throttle position; and where all auxiliary non-motive power engines are rot operating (40 CFR 201.1-91).

Idle adjustments: A series of adjustments which include idle revolutions per minute, idle air/fuel ratio, and basic timing (40 CFR 51-App/N-91).

Idle emission test: A sampling procedure for exhaust emissions which requires operation of the engine in the idle mode only. At a minimum, the idle test must consist of the following procedures carried out on a fully warmed up engine: A verification that the idle revolutions per minute is within manufacturer's specified limits and a measurement of the exhaust carbon monoxide and/or hydrocarbon concentrations during the period of time from 15 to 25 seconds after the engine either was used to move the car or was run at 2,000 to 2,500 r/min with no load for 2 or 3 seconds (40 CFR 51-App/N-91).

Ignitability:
(1) One of the four U.S. EPA hazardous waste characteristics (ETI-92).
(2) The waste which is capable, during routine handling, of causing a fire or exacerbating a fire once started.
(3) The ability of a material to generate enough combustion vapors for iginition.
(4) *See* hazardous waste characteristics for more related terms.

Ignitable: Capable of burning or causing a fire (EPA-94/04).

Igniter: Any device used to ignite a propellant (EPA-76/03).

Ignition: The initiation of combustion.

Ignition arch: A refractory furnace arch or surface located over a fuel bed to radiate heat and to accelerate ignition (SW-108ts).

Ignition delay time: The interval between an initial exposure to a step function change in temperature and the principal exothermicity of the reaction as indicated by a rapid increase in the temperature and pressure of the mixture (*see* time for more related terms) (EPA-88/12).

Ignition temperature: The lowest temperature of a fuel at which combustion becomes self-sustaining (*see* temperature for more related terms) (SW-108ts).

Illicit discharge: Any discharge to a municipal separate storm sewer that is not composed entirely of storm water except discharges pursuant to a NPDES permit (other than the NPDES permit for discharges from the municipal separate storm sewer) and discharges resulting from fire fighting activities (40 CFR 122.26-91).

Imbibition (or maceration): The use of water in the milling process to dissolve sucrose. Identical, in this connotation, to maceration and saturation (EPA-75/02d).

Imhoff cone:
(1) A clear cone-shaped container used to measure the volume of settleable solids in a specific volume of water (EPA-94/04).
(2) A cone shaped glass vessel of 1 liter volume. It is used to measure the settleable solids.

Imhoff tank: A combination wastewater treatment tank which allows sedimentation to take place in its upper compartment and digestion to take place in its lower compartment (EPA-87/10a).

Imhoff tank sludge: Sludge produced from the Imhoff tank.

Immediate dissolved oxygen demand (IDOD): The oxygen demand of reduced forms of N, P, S, some metallic species and some easily oxidizable organic compounds (e.g., formaldehyde: CH_2O) (*see* oxygen for more related terms) (LBL-76/07-water).

Immediate removal: An action undertaken to prevent or mitigate immediate and significant risk of harm to human life or health or to the environment. As set forth in the National Contingency Plan, the action shall be terminated after $1 million has been obligated or six months have elapsed from the date of initial response (Course 165.5).

Immediately dangerous to life and health (IDLH):
(1) Conditions that pose an immediate threat to life or health or conditions that pose an immediate threat of severe exposure to contaminants, such as radioactive materials, which are likely to have adverse cumulative or delayed effects on health (NIOSH-84/10).
(2) The maximum level to which a healthy individual can be exposed to a chemical for 30 minutes and escape without suffering irreversible health effects or impairing symptoms. Used as a "level of concern" (*see* level of concern) (EPA-94/04).

Immersed area: The total area wetted by the solution or plated area plus masked area (EPA-83/06a).

Immersion dose: Dose resulting from being surrounded by a medium (air or water) that contains radionuclides (DOE-91/04).

Immersion plate: A metallic deposit produced by a displacement reaction in which one metal displaces another from a solution, e.g., $Fe + Cu^{+2} = Cu + Fe^{+2}$.

Immersion sampling: The collection of a liquid sample by immersing a container in the liquid and by filling the container with the desired quantity (*see* sampling for more related terms).

Imminent danger to the health and safety of the public: The existence of any condition or practice, or any violation of a permit or other requirement of this chapter in a surface coal mining and reclamation operation, which condition, practice, or violation could reasonably be expected to cause substantial physical harm to persons outside the permit area before such condition, practice, or violation can be abated. A reasonable expectation of death or serious injury before abatement exists if a rational person, subjected to the same conditions or practices giving rise to the peril, would not expose himself or herself to the danger during the time necessary for abatement (SMCRA701-30USC1291-90).

Imminent hazard: A situation which exists when the continued use of a pesticide during the time required for cancellation proceedings would be likely to result in unreasonable adverse hazard to the survival of a species declared endangered by the Secretary of the Interior under Pub.L. 91-135 (FIFRA2, *see also* 40 CFR 165.1-91).

Imminent hazard order: *See* definition under administrative order.

Imminently hazardous chemical substance or mixture: A chemical substance or mixture which presents an imminent and unreasonable risk of serious or widespread injury to health or the environment. Such risk to health or the environment shall be considered imminent if it is shown that the manufacture, processing, distribution in commerce, use, or disposal of the chemical substance or mixture, or that any combination of such activities, is likely to result in such injury to health or the environment before a final rule under TSCA Sec. 6 can protect against such risk (TSCA Sec. 7).

Immission: The transfer of contaminants from the atmosphere into an acceptor such as into the lungs. It does not mean ground level concentration (NATO-78/10).

Immobilization: The lack of movement by the test organisms except for minor activity of the appendages (40 CFR 797.1300-91, *see also* 40 CFR 797.1330-91).

Immune response: Not susceptible to some disease because of the presence of the specific antibody protection.

Immunocompetence: The ability to resist infection (immune response).

Impact extrusion: A cold extrusion process for producing tubular components by striking a slug of the metal, which has been placed in the cavity of the die, with a punch moving at high velocity (EPA-83/03).

Impact deformation: The process of applying an impact force to a workpiece such that the workpiece is permanently deformed or shaped. Impact deformation operations include shot peening, peening, forging, high energy forming, heading, or stamping (EPA-83/06a).

Impact mill: A machine that grinds materials by throwing them against heavy metal projections rigidly attached to a rapidly rotating shaft (*see* size reduction machine for more related terms) (SW-108ts).

Impaction: The act of bringing matter forcibly in contact. The term is often used synonymously with impingement.
(1) In air sampling, impingement refers to a process for the collection of particulate matter in which the gas being sampled is directed forcibly against a surface (EPA-83/06).
(2) In a wet scrubbing system, dust particles will tend to follow the streamlines of the exhaust stream. However, when liquid droplets are introduced into the exhaust stream, particles cannot always follow these streamlines as they diverge around the droplet. The particle's mass causes it to break away from the streamlines and impact on the droplet. Impaction is the predominant collection mechanism for scrubbers having gas stream velocities greater than 0.3 m/s (1 ft/sec). Most scrubbers do operate with gas stream velocities well above 0.3 m/s. Therefore, at these velocities, particles having diameters greater than 1.0 μm are collected by this mechanism.

As the velocity of the particles in the exhaust stream increases relative to the liquid droplets' velocity, impaction increases. Impaction also increases as the size of the liquid droplet decreases. This is because there will be more droplets (for the same amount of liquid) within the vessel, consequently increasing the likelihood that the particles will impact on the droplets (*see also* particle collection mechanisms for wet scrubbing systems) (EPA-84/03b, p1-4).

Impactor: A sampling device that employs the principle of impaction (impingement). The cascade impactor impactions in series to collect successively smaller sizes of particles (EPA-83/06).

Impedance: The rate at which a substance can absorb and transmit sound (EPA-74/11).

Impermeable:
(1) Not easily penetrated, The property of a material or soil that does not allow, or allows only with great difficulty, the movement or passage of water (EPA-94/04).
(2) Resistant to the flow of water or other fluid (EPA-83).

Impermeable rock (or impervious rock): A rock which, being non-porous (e.g., un-fissured granite) or practically so (e.g., clay), does not allow water to soak into it or pass through it freely. A non-porous rock may be pervious, however, owing to joints and fissures (DOI-70/04).

Impervious: Resistant to penetration by fluids or gases (SW-108ts).

Impervious chemical protective clothing: Impervious to a chemical substance if the substance causes no chemical or mechanical degradation, permeation, or penetration of the chemical protective clothing under the conditions of, and the duration of, exposure (40 CFR 721.3-91).

Impervious solid surfaces: The solid surfaces which are nonporous and thus unlikely to absorb spilled PCBs within the short period of time required for cleanup of spills under this policy. Impervious solid surfaces include, but are not limited to, metals, glass, aluminum siding, and enameled or laminated surfaces (40 CFR 761.123-91).

Impingement:
(1) The act of bringing matter forcibly in contact. As used in air sampling, impingement refers to a process for the collection of gases or particulate matter in which the gas stream being sampled is directed forcibly against a surface (LBL-76/07-air).
(2) The process by which aquatic organisms too large to pass through the screens of water intake system become caught on the screens and are unable to escape (DOE-91/04).
Other impingement related terms include:
- Dry impingement
- Wet impingement

Impingement nozzle: For this type of nozzle, highly pressurized liquid passes through a hollow tube in the nozzle and strikes a pin or plate at the nozzle tip. A very fine fog of tiny, uniform-sized droplets approximately 25 to 40 μm in diameter is produced. Because there are no internal parts in the nozzle, it will not plug as long as particles larger than the opening are filtered out by a strainer. These nozzles are usually made of stainless steel or brass (*see* spray nozzle for more types) (EPA-84/03b, p2-2).

Impingement separator: A device used to remove particles from the air; a typical one is made of closely-spaced louvers (baffles), where the air stream is caused to change directions, and the particles hit and adhere to the plates (*see* separator for more related terms) (EPA-84/09).

Impinger: Broadly, a sampling instrument employing impingement for the collection of particulate matter. Commonly, this term is applied to specific instruments, the midget and standard impinger (EPA-83/06). Other impinger related terms include:
- Midget impinger
- Standard impinger

Implantation: A process of introducing impurities into the near surface regions of solids by directing a beam of ions at the solid (EPA-83/03).

Implementation: For purposes of Federal financial assistance (other than rural communities assistance), the term implementation does not include the acquisition, leasing, construction, or modification of facilities or equipment or the acquisition, leasing, or improvement of land (RCRA1004, see also 40 CFR 249.04; 256.06-91).

Implementation plan: A document of the steps to be taken to ensure attainment of environmental quality standards within a specified time period. Implementationplans are required by various laws (EPA-74/11).

Impounding lake: *See* synonym, impounding reservoir.

Impounding reservoir (or impounding lake): A reservoir for water storage. The purpose is to store water for future use and for flood control.

Impoundment (or surface impoundment):
(1) A waste management unit which is a natural topographic depression, man-made excavation, or diked area formed primarily of earthen materials (although it may be lined with man-made materials), which is designed to hold an accumulation of liquid wastes or waste containing free liquids, and which is not an injection well. Examples of surface impoundments are holding, storage, settling, and aeration pits, ponds, and lagoons (40 CFR 61.341-91, *see also* 40 CFR 260.10; 280.12-91).
(2) A body of water or sludge confined by a dam, dike, floodgate, or other barrier (EPA-94/04).

Impregnate: To force a liquid substance into the space of a porous solid in order to change its properties (EPA-83/03).

Impregnating compound: Materials of low viscosity and surface tension, used primarily for the sealing of castings. Polyester resins and sodium silicate are the two types of materials used. Phthalic anhydride and diallyl phthalate are used in the formulation of the polyester resins (EPA-85/10a).

Impregnation:
(1) The process of filling pores of a formed powder part, usually with a liquid such as a lubricant, or mixing particles of a nonmetallic substance in a matrix of metal powder (40 CFR 471.02-91).
(2) Method of making an electrode by precipitating active material on a sintered nickel plaque (EPA-84/08).
(3) The process of treating a sheet or web of paper or paperboard with a liquid such as hot asphalt or wax, a solution of some material in a volatile solvent, or a liquid such as an oil. It is also used as a term to describe a treatment in which fibrous raw materials are infused with a chemical solution prior to a digesting or fiberizing process. Sometimes called pre-impregnation (EPA-87/10).

Improved discharge: The volume, composition and location of an applicant's discharge following:
(1) Construction of planned outfall improvements, including, without limitation, outfall relocation, outfall repair, or diffuser modification; or
(2) Construction of planned treatment system improvements to treatment levels or discharge characteristics; or
(3) Implementation of a planned program to improve operation and maintenance of an existing treatment system or to eliminate or control the introduction of pollutants into the applicant's treatment works (40 CFR 125.58-91).
(4) *See* discharge for more related terms.

Impulsive noise: An acoustic vent characterized by very short rise time nd duration (40 CFR 211.203-91).

Impurity: Any chemical element present in an additive that is not included in the chemical formula or identified in the breakdown by element in the chemical composition of such additive (40 CFR 79.2-91, *see also* 40 CFR 704.3; 158.153; 710.2; 712.3; 716.3; 720.3; 723.50; 723.175; 723.250; 747.115; 747.195; 747.200; 761.3; 790.3; 791.3-91).

Impurity associated with an active ingredient:
(1) Any impurity present in the technical grade of active ingredient; and
(2) Any impurity which forms in the pesticide product through reactions between the active ingredient and any other component of the product or packaging of the product (40 CFR 158.153-91).

In-benzene service: A piece of equipment either contains or contacts a fluid (liquid or gas) that is at least 10 percent benzene by weight as determined according to the provisions of 40 CFR 61.245(d). The provisions of 40 CFR 61.245(d) also specify how to determine that a piece of equipment is not in benzene service (40 CFR 61.111-91, *see also* 40 CFR 61.131-91).

In-existence: That the owner or operator has obtained all necessary preconstruction approvals or permits required by Federal, State, or local air pollution emissions and air quality laws or regulations and either has:
(1) Begun, or caused to begin, a continuous program of physical on-site construction of the facility; or
(2) Entered into binding agreements or contractual obligations, which cannot be cancelled or modified without substantial loss to the owner or operator, to undertake a program of construction of the facility to be completed in a reasonable time (40 CFR 51.301-91).

In-gas/vapor service: That the piece of equipment contains process fluid that is in the gaseous state at operating conditions (cf. in liquid service) (40 CFR 60.481; 61.241; 264.1031-91).

In-heavy liquid service: That the piece of equipment is not in gas/vapor service or in light liquid service (cf. in light liquid service) (40 CFR 60.481-91, *see also* 40 CFR 264.1031-91).

In-hydrogen service: That a compressor contains a process fluid that meets the conditions specified in 40 CFR 60.593(b) (40 CFR 60.591-91).

In-kind contribution: The value of a non-cash contribution to meet a recipients cost sharing requirements. An in-kind contribution may consist of charges for real property and equipment or the value of goods and services directly benefiting the EPA funded project (40 CFR 30.200-91, *see also* 40 CFR 35.4010; 35.6015-91).

In-light liquid service: That the piece of equipment contains a liquid that meets the conditions specified in 60.485(e) (cf. in heavy liquid service) (40 CFR 60.481-91, *see also* 40 CFR 60.631; 60.591; 264.1031-91).

In-line filtration: Pre-treatment method in which chemical coagulants directly to the filter inlet pipe. The chemicals are mixed by the flowing water> Commonly used in pressure filtration installations. Eliminates need for flocculation and sedimentation (EPA-94/04).

In-line incinerator: *See* definition under excess air incinerator.

In-liquid service: That a piece of equipment is not in gas/vapor service (cf. in gas/vapor service) (40 CFR 61.241-91).

In-operation: Engaged in activity related to the primary design function of the source (40 CFR 51.301-91, *see also* 40 CFR 260.10; 270.2-91).

In-plant measure (in plant source control, internal control or internal measure):
(1) A technology applied within the manufacturing process to reduce or eliminate pollutants in the raw waste water (EPA-76/03).
(2) Controls or measures applied at the source of a waste to reduce or eliminate the necessity for further treatment (EPA-87/10a).

In-plant source control: *See* synonym, in plant measure.

In-process control technology:
(1) The conservation of chemicals and water throughout the production operations to reduce the amount of wastewater to be discharged (40 CFR 467.02; 471.02-91).
(2) The regulation and the conservation of chemicals and the reduction of water usage throughout the operations as opposed to end-of-pipe treatment (EPA-83/06a).

In-process tank: A container used for mixing, blending, heating, reacting, holding, crystallizing, evaporating or cleaning operations in the manufacture of pharmaceuticals (*see* tank for more related terms) (40 CFR 52.741-91).

In-situ leach: A leaching of broken ore in the subsurface as it occurs, usually in abandoned underground mines which previously employed block-caving mining methods (EPA-82/05).

In-situ leach methods: The processes involving the purposeful introduction of suitable leaching solutions into a uranium ore body to dissolve the valuable minerals in place and the purposeful leaching of uranium ore in static or semistatic condition either by gravity through an open pile, or by flooding confined ore pile. It does not include the natural dissolution of uranium by ground waters, the incidental leaching of uranium by mine drainage, nor the rehabilitation of aquifers and the monitoring of these aquifers (40 CFR 440.132.-91).

In-situ sampling systems: The nonextractive samplers or in-line samplers (40 CFR 60.481-91, *see also* 40 CFR 61.241; 264.1031-91).

In-situ stripping: Treatment system that remove or "strips" volatile organic compounds from contaminated ground or surface water by forcing an airstream through the water and causing the compounds to evaporate (EPA-94/04).

In-situ suspension process: A manufacturing process in which styrene, blowing agent, and other raw materials are added together within a reactor for the production of expandable polystyrene (40 CFR 60.561-91).

In-situ treatment: The process of treating a contaminated matrix (soil, sludge, or ground water) in place. In situ processes may use physical, chemical, thermal, or biological technologies to treat the site (*see* treatment for more related terms) (EPA-89/12a).

In-vacuum service: For the purpose of paragraph (i) of this section, equipment which is operating at an internal pressure that is at least 5 kPa (0.73 psia) below ambient pressure (40 CFR 52.741-91, *see also* 40 CFR 60.481; 61.241; 264.1031-91).

In-vessel composting: A composting method in which the compost is continuously and mechanically mixed and aerated in a large, contained area (EPA-89/11).

In-VHAP service: That a piece of equipment either contains or contacts a fluid (liquid or gas) that is at least 10 percent by weight a volatile hazardous air pollutant (VHAP) as determined according to the provisions of 40 CFR 61.245(d). The provisions of 40 CFR 61.245(d) also specify how to determine that a piece of equipment is not in VHAP service (40 CFR 61.241-91).

In-vinyl chloride service: That a piece of equipment either contains or contacts a liquid that is a least 10 percent vinyl chloride by weight or a gas that is at least 10 percent by volume vinyl chloride as determined according to the provisions of 40 CFR 61.67(h). The provisions of 40 CFR 61.67(h) also specify how to determine that a piece of equipment is not in vinyl chloride service. For the purposes of this subpart, this definition must be used in place of the definition of in VHAP service in Subpart V of this part (40 CFR 61.61-91).

In-vitro:
(1) In glass; a test-tube culture.
(2) Any laboratory test using living cells taken from an organism (EPA-89/12).

In-vivo: In the living body of a plant or animal. In vivo tests are those laboratory experiments carried out on whole animals or human volunteers (EPA-89/12).

In-VOC service: That the piece of equipment contains or contacts a process fluid that is at least 10 percent VOC by weight. (The provisions of 40 CFR 60.485(d) specify how to determine that a piece of equipment is not in VOC service) (40 CFR 60.481-91, *see also* 40 CFR 61.241-91).

in. w.c.: Inches water column, a unit to express liquid pressure.

In-wet gas service: That a piece of equipment contains or contacts the field gas before the extraction step in the process (40 CFR 60.631-91).

Inactivation: The irreversible loss of ability of an organism to propagate under any conditions or the irreversible loss of ability for biological material (organism, enzymes, viruses) to conduct its natural biological function (EPA-88/09a).

Inactive facility: A facility which no longer receives solid waste (*see* facility for more related terms) (40 CFR 256.06-91).

Inactive portion (facility): The portion of a facility which is not operated after the effective date of Part 261 of this chapter (*see* facility for more related terms, *see also* active portion and closed portion) (40 CFR 260.10-91).

Inactive stack: a stack to which no further routine additions of phosphogypsum will be made and which is no longer used for water management associated with the production of phosphogypsum. If a stack has not been used for either purpose for two years it is presumed to be inactive (*see* stack for more related terms) (40 CFR 61.201-91).

Inactive waste disposal site: Any disposal site or portion of it where additional asbestos-containing waste material will not be deposited within the past year (40 CFR 61.141-91).

Incidence: The number of new cases of a disease within a specified period of time (EPA-92/12).

Incidence rate: The ratio of the number of new cases over a period of time to the population at risk (EPA-92/12).

Incident: Any occurrence or series of occurrences having the same origin, involving one or more vessels, facilities, or any combustion thereof, resulting in the discharge or substantial threat of discharge of oil (OPA1001-91).

Incident characterization: The process of identifying the substance(s) involved in an incident, determining exposure pathways and projecting the effect it will have on people, property, wildlife and plants, and the disruption of services (Course 165.5).

Incident command post: A facility located at a safe distance from an emergency site, where the incident commander, key staff, and technical representatives can make decisions and deploy emergency manpower and equipment (EPA-94/04).

Incident command system (ICS):
(1) The combination of facilities, equipment, personnel, procedures, and communications operating within a common organizational structure with responsibility for management of assigned resources to effectively accomplish stated objectives at the scene of an incident (NRT-87/03).
(2) The organizational arrangement wherein one person, normally the Fire Chief of the impacted district, is in charge of an integrated, comprehensive emergency response organization and the emergency incident site, backed by an Emergency Operations Center staff with resources, information, and advice (EPA-94/04).

Incident evaluation: The process of assessing the impact of released or potentially released substances on public health and the environment (Course 165.5).

Incident response: All activities that are required when responding to incidents. The activities can be divided into five broad, interacting elements (Course 165.5):
(1) **Recognition:** Identification of the substance involved and the characteristics which determine its degree of hazard.
(2) **Evaluation:** Impact or risk the substance poses to the public health and the environment.
(3) **Control:** Methods to eliminate or reduce the impact of the incident.
(4) **Information:** Knowledge acquired concerning the conditions or circumstances particular to an incident.
(5) **Safety:** Protection of responders from harm.

Incinerability: *See* definition under thermal stability index.

Incineration (or thermal incineration):
(1) The controlled process which combustible solid, liquid, or gaseous wastes are burned and changed into noncombustible gases (40 CFR 240.101-91).
(2) A treatment technology involving destruction of waste by controlled burning at high temperatures, e.g., burning sludge to remove the water and reduce the remaining residues to a safe, non-burnable ash that can be disposed of safely on land, in some waters, or in underground locations (EPA-94/04).
(3) cf. combustion

Other incineration related terms include:
- Co-incineration
- At sea incineration (*see* synonym, ocean incineration)
- Complete incineration
- Ocean incineration
- Oxygen enriched incineration

Incineration at sea: Disposal of waste by burning at sea on specially-designed incinerator ships (EPA-94/04).

Incineration vessel: Any vessel which carries hazardous substances for the purpose of incineration of such substances, so long as such substances or residues of such substances are on board (SF101-42USC9601-91).

Incinerator:
(1) A combustion apparatus in which refuse (or waste) is burned (40 CFR 52.741-91)
(2) A furnace for burning waste under controlled conditions (EPA-94/04).
(3) An engineered apparatus used to burn waste substances and

in which all the factors of combustion - temperature, retention time, turbulence, and combustion air - can be controlled (EPA-83).

Other incinerator related terms include:
- Batch fed incinerator
- Catalytic incinerator
- Cell type incinerator
- Central incinerator
- Chute fed incinerator
- Circulating bed incinerator
- Circulating fluidized bed combustor: (*see* synonym, circulating bed combustor)
- Commercial incinerator
- Continuous feed incinerator
- Controlled air incinerator (*see* synonym, starved air incinerator)
- Dedicated incinerator
- Direct fed incinerator
- Excess-air incinerator
- In-line incinerator (*see* definition under excess-air incinerator)
- Retort incinerator (*see* definition under excess-air incinerator)
- Fixed hearth incinerator (*see* synonym, starved air incinerator)
- Flue fed incinerator
- Fluidized bed incinerator
- General purpose incinerator
- Industrial incinerator
- Infrared incinerator
- Liquid injection incinerator
- Mass burn incinerator
- Medical waste incinerator
- Mobile incinerator system
- Modular combustion unit (*see* synonym, modular incinerator)
- Modular incinerator
- Molten salt combustion (*see* synonym, molten salt incinerator)
- Molten salt incinerator
- Molten salt reactor
- Multiple chamber incinerator
- Multiple hearth incinerator
- Municipal incinerator (*see* synonym, municipal waste incinerator)
- Municipal waste incinerator
- On-site incinerator
- Open pit incinerator
- Pesticide incinerator
- Plasma arc incinerator
- Pyrolytic incinerator (*see* synonym, starved air incinerator)
- Qualified incinerator
- Refuse derived fuel (RDF) incinerator
- Residential incinerator
- A rotary kiln incinerator
- Similar incinerator
- Single chamber incinerator
- Spreader stoker incinerator
- Starved air incinerator
- Waterwall incinerator

Incinerator analysis: Analysis of parameters which affect the performance of incinerators. Key parameters include:
(1) Furnace temperature
(2) Supplemental fuel requirement
(3) Combustion air requirement
(4) Flue gas discharge quantity
(5) Residence time
(6) Turbulence/mixing

From these parameters, sizes of equipment associated with the incinerator system can be calculated. The analysis requires two major asssumptions:
- The total heat leaving (exiting) a system is equal to the heat entering that system. Heat entering a system includes the potential heat release of a waste, and/or fuel, that is fired with that system.
- Flue gas from an incinerator is composed of moisture plus dry air. Actual incinerator off-gas contains more carbon dioxide and less oxygen than is found in air. The assumption that dry flue gas has the properties of air, however, greatly simplifies calculations while introducing a relatively small error (less than 3%) in calculated temperatures and heat requirements.

Incinerator design capacity: The number of tons of solid waste that a designer anticipates his incinerator will be able to process in a 24-hour period if specified criteria are met (SW-108ts).

Incinerator gases: Combustion gases which may contain water vapor and excess or dilution air added after the combustion chamber (EPA-83).

Incinerator operation mode: Including:
(1) **Continuous duty:** The incinerator can be continuously operated 24 hours per day. The system is designed so that waste can be continuously fed and ash can be continuously removed without shut down the operation of the incinerator (EPA-89/03b).
(2) **Intermittent duty:** The incinerator is intermittently loaded with batches of waste, one after another, over a period of time, usually one to two work shifts. The batches might be fed at uneven intervals, when waste is available. In any event, the incinerator must be shut down to remove ash from the system, thus its operation is intermittent (EPA-89/03b).
(3) **Single batch duty:** The incinerator is loaded with a batch of waste, sealed, and turned on. After combustion is completed, the incinerator is allowed to cooled and the ash is removed. Usually, ash is not removed until the next day (EPA-89/03b).

Incinerator process: The following information summarizes the typical ranges for temperature and residence time in six most popular incinerators (EPA-81/09):
- Rotary kiln   1500-2900 (F)
- Liquid injection   1500-2900
- Fluidized bed   840-1800
- Co-incineration   300-2900
- Starved air   900-1500
- Multiple hearth
  - drying zone   600-1000
  - incinerating zone   1400-1800

In general, for gas or liquid waste, the residence time is less than 2 seconds. For solid waste, it is in the range of hours, depending on solid sizes and its nature. This general rule applies to all types of incinerators.

Incinerator residue: All of the solid material collected after an incineration process is completed (EPA-83).

Incinerator stoker: A mechanically operable moving grate arrangement for supporting, burning, and transporting solid waste in a furnace and discharging the residue (*see* stoker for more related terms) (SW-108ts).

Incipient LC50: That test substance concentration, calculated from experimentally derived mortality data, that is lethal to 50 percent of a test population when exposure to the test substance is continued until the mean increase in mortality does not exceed 10 percent in any concentration over a 24-hour period (40 CFR 797.1400-91).

Incipient lethal level: The concentration or level of an abiotic factor beyond which an organism could not survive (DOD-78/01).

Inclined plate conveyer: A separating device that operates by feeding material onto an inclined steel plate belt conveyer so that heavy and resilient materials, such as glass, bounce down the conveyer, and light and inelastic materials are carried upward by the motion of the belt (*see* conveyer for more related terms) (SW-108ts).

Inclusive bodies: Insoluble dense clusters of (recombinant) cell products (e.g., proteins) found in the cell's cytoplasm that are protected somewhat from degradation and are easily collected and purified upon cell lysis (EPA-88/09a).

Incombustible waste: The materials such as metals and stones that do not burn under the normal incinerator operating conditions (*see* waste for more related terms).

Incompatible pollutant: The pollutants which would cause harm to, would adversely affect the performance of, or would be inadequately treated in publicly owned sewage treatment works (*see* pollutant for more related terms) (EPA-76/12).

Incompatible waste:
(1) A hazardous waste which is unsuitable for:
   (A) Placement in a particular device or facility because it may cause corrosion or decay of containment materials (e.g., container inner liners or tank walls); or
   (B) Commingling with another waste or material under uncontrolled conditions because the commingling might produce heat or pressure, fire or explosion, violent reaction, toxic dusts, mists, fumes, or gases, or flammable fumes or gases (40 CFR 260.10-91).
(2) A waste unsuitable for mixing with another waste or material because it may react to form a hazard (EPA-94/04).
(3) *See* waste for more related terms.

Incomplete combustion: The combustion which is not a complete combustion (*see* combustion for more related terms).

Incomplete gasoline-fueled heavy-duty vehicle: Any gasoline-fueled heavy-duty vehicle which does not have the primary load-carrying device, or passenger compartment, or engine compartment or fuel system attached (40 CFR 86.085.2-91).

Incomplete truck: Any truck which does not have the primary load carrying device or container attached (*see* truck for more related terms) (40 CFR 86.082.2-91).

Incompressible fluid: A fluid for which the density is independent of the pressure and the temperature (NATO-78/10).

Incorporated into the soil: The injection of solid waste beneath the surface of the soil or the mixing of solid waste with the surface soil (40 CFR 257.3.5-91, *see also* 40 CFR 257.3.6-91).

Incorporated place: The District of Columbia, or a city, town, township, or village that is incorporated under the laws of the State in which it is located (under the Federal Water Pollution Control Act) (40 CFR 122.26-91).

Increments of progress: The steps toward compliance which will be taken by a specific source, including:
(1) Date of submittal of the source's final control plan to the appropriate air pollution control agency;
(2) Date by which contracts for emission control systems or process modifications will be awarded; or date by which orders will be issued for the purchase of component parts to accomplish emission control or process modification;
(3) Date of initiation of on-site construction or installation of emission control equipment or process change; (4) Date by which on-site construction or installation of emission control equipment or process modification is to be completed; and
(5) Date by which final compliance is to be achieved (40 CFR 51.100-91, *see also* 40 CFR 60.21-91).

Incubate: To maintain cultures, bacteria, or other microorganisms at the most favorable temperature for development (EPA-76/03).

Incubation period: The time interval between the infection of a susceptible person or animal and the appearance of signs or symptoms of the disease (EPA-83).

Independent commercial importer (ICI): An importer who is not an original equipment manufacturer (OEM) or does not have a contractual agreement with an OEM to act as its authorized representative for the distribution of motor vehicles or motor vehicle engines in the U.S. market (40 CFR 85.1502-91).

Independent laboratory: A test facility operated independently of any motor vehicle, motor vehicle engine, or retrofit device manufacturer capable of performing retrofit device evaluation tests. Additionally, the laboratory shall have no financial interests in the outcome of these tests other than a fee charged for each test performed (40 CFR 610.11-91).

Independent printed circuit board manufacturer: A facility which manufacturers printed circuit boards principally for sale to other companies (40 CFR 433.11-91).

Independent variable: A variable whose value is not dependent on the value of any other variables and can be arbitrarily specified (*see* variable for more related terms) (NATO-78/10).

Independently audited: Refers to an audit performed by an independent certified public accountant in accordance with generally accepted auditing standards (*see* audit for more related terms) (40 CFR 144.61-91).

Index (or index mark): A mark on a choke thermostat housing, located in a fixed relationship to the thermostatic coil tang position to aid in assembly and service adjustment of the choke (40 CFR 85.2122(a)(2)(iii)(H)-91).

Index mark: *See* synonym, index.

Index of mobility: A number that is proportional to the contaminant probability of escaping its point of origin and migrating through air and water. It can be expressed as: Mobility index = log[(water solubility x vapor pressure)/Koc], where Koc = coefficient of organic carbon partition (Course 165.6).

Index of pollution: *See* synonym, biological indicator.

Index organism: *See* synonym, biological indicator.

Indicator: In biology, an organism, species, or community whose characteristics show the presence of specific environmental conditions (EPA-89/12.

Indicator:
(1) In biology, an organism, species, or community whose characteristics show the presence of specific environmental conditions.
(2) In chemistry, a substance that shows a visible change, usually of color, at a desired point in a chemical reaction.
(3) A device that indicates the result of a measurement, e.g., a pressure gauge or a moveable scale (EPA-94/04).

Iodine (i): A halogen element with a.n. 53; a.w. 126.904; d. 4.94 g/cc; m.p. 113.7 C° and b.p. 183 C°. The element belongs to group VIIA of the periodic table.

Indirect ammonia recovery system: Those systems which recover ammonium hydroxide as a by-product from coke oven gases and waste ammonia liquors (40 CFR 420.11-91).

Indirect arc furnace: An electric furnace whose metal charge is placed below the electrodes, and the arc is formed between the electrodes and above the charge (*see* furnace for more related terms) (AP-40, p236).

Indirect discharge or discharge:
(1) The introduction of pollutants into a POTW from any non-domestic source regulated under section 307(b), (c) or (d) of the Act (40 CFR 403.3-91).
(2) Introduction of pollutants from a non-domestic source into a publicly owned waste-treatment system. Indirect dischargers can be commercial or industrial facilities whose wastes enter local sewers (EPA-94/04).
(3) *See* discharge for more related terms.

Indirect discharger: A nondomestic discharger introducing pollutants to a publicly owned treatment works (cf. new discharger) (40 CFR 122.2-91).

Indirect economic effects: Indirect effects result from the need to supply industries experiencing direct economic effects with additional outputs to allow them to increase their production. The additional output from each directly affected industry requires inputs from other industries within a region (that is, purchases of goods and services). This results in a multiplier effect to show the change in total economic activity resulting from a new activity in a region (DOE-91/04).

Indirect heat transfer: The transfer of heat in such a way that the source of heat does not come into direct contact with process materials (40 CFR 52.741-91).

Indirect impact: *See* synonym, secondary impact.

Indirect labor costs: Labor-related costs paid by the employer other than salaries, wages and other direct compensation such as social security and insurance (EPA-83/06a).

Indirect source: Any facility or building, property, road or parking facility that attracts motor vehicle traffic and, indirectly, causes pollution (EPA-94/04).

Indirect toxicity: The toxicity that affects organisms by interfering with their food supply or modifying their habitat instead of directly acting on the organisms themselves (cf. direct toxicity) (DOD-78/01).

Indium (In): A soft metallic element with a.n. 49; a.w. 114.82; d. 7.31 g/cc; m.p. 156.2° C and b.p. 2000° C. The element belongs to group IIIA of the periodic table.

Individual drain system: All process drains connected to the first common downstream junction box. The term includes all such drains and common junction box, together with their associated sewer lines and other junction boxes, down to the receiving oil water separator (40 CFR 60.691-91, *see also* 40 CFR 61.341-91).

Individual generation site: The contiguous site at or on which one or more hazardous wastes are generated. An individual generation site, such as a large manufacturing plant, may have one or more sources of hazardous waste but is considered a single or individual generation site if the site or property is contiguous (40 CFR 260.10-91).

Individual risk: The probability that an individual person will experience an adverse effect. This is identical to population risk unless specific population subgroups can be identified that have different (higher or lower) risks (EPA-92/12).

Individual systems: The privately owned alternative wastewater treatment works (including dual waterless/gray water systems) serving one or more principal residences, or small commercial establishments. Normally these are onsite systems with localized treatment and disposal of wastewater, but may be systems utilizing small diameter gravity, pressure or vacuum sewers conveying treated or partially treated wastewater. These system can also include small diameter gravity sewers carrying raw wastewater to cluster systems (40 CFR 35.2005-91).

Indo: A prefix showing within or inside a chemical formula.

Indoor air (or house air):
(1) The air that occupies the space within the interior of a house or other buildings (cf. indoor climate) (EPA-88/08).
(2) The breathing air inside a habitable structure or conveyance (EPA-94/04).

Indoor air pollution: Chemical, physical, or biological contaminants in indoor air (*see* pollution for more related terms) (EPA-94/04).

Indoor climate: Temperature, humidity, lighting, and noise levels in a habitable structure or conveyance. Indoor climate can affect indoor air pollution (EPA-94/04).

Indore process composting: An anaerobic composting method that originated in India. Organic wastes are placed in alternate layers with human or animal excreta in a pit or pile. The piles are turned twice in six months and drainage is used to keep the compost moist (*see* composting for more related terms) (EPA-83).

Induced draft: The negative pressure created by the action of a fan, blower or other gas moving device located between an incinerator and a stack (*see* draft for more related terms) (OME-88/12).

Induced draft fan:
(1) A fan that exhausts hot gases from heat-absorbing equipment such as dust collectors or scrubbers (SW-108ts).
(2) A device designed to pull combustion air and combustion products through an incinerator and associated air pollution control devices. The facility is operated under negative pressure (less than atmospheric) to reduce intermittent and fugitive releases (CRWI-89/05).
(3) *See* fan for more related terms.

Induced draft tower: *See* definition under mechanical draft tower.

Induced draught: The system used in mechanical-draught cooling towers whereby the air is drawn through the tower by a fan (*see* draught for more related terms) (Gurney-66).

Induced economic effects: The spending of households resulting from direct and indirect economic effects. Increases in output from a new economic activity lead to an increase in household spending throughout the economy as firms increase their labor inputs (DOE-91/04).

Induction exposure: An experimental exposure of a subject to a test substance with the intention of inducing a hypersensitive state (40 CFR 798.4100-91).

Induction furnace:
(1) An electric furnace consisting of a crucible within a water-cooled copper coil. An alternating current in the coil around the crucible induces eddy currents in the metal charge and thus develops heat within the mass of the charge (AP-40, p236).
(2) A furnace that induction heating is obtained by inducing an electric current in the charge and may be considered as operating on the transformer principle. Induction furnaces, which may be low frequency or high frequency, are used to produce small tonnages of specialty alloys through remelting of the required constituents (EPA-74/02a).
(3) *See* furnace for more related terms.

Induction period: A period of at least 1 week following a sensitization exposure during which a hypersensitive state is developed (40 CFR 798.4100-91).

Inductively coupled argon plasma emission spectrometry (ICP): An atomic emission spectrometric method. The basis of the method is the measurement of atomic emission by an optical spectroscopic technique. Samples are nebulized and the aerosol that is produced is transported to the plasma torch where excitation occurs. Characteristic atomic-line emission spectra are produced by radio-frequency inductively coupled plasma. The spectra are dispersed by a grating spectrometer and the intensities of the lines are monitored by photomultiplier tubes. The photocurrents from the photomultiplier tubes are processed and controlled by a computer system. A background correction technique is required to compensate for variable background contribution to the determination of trace elements (40 CFR 136-App/C-91).

ICP's primary advantage is that it allows simultaneous or rapid sequential determination of many elements in a short time. The primary disadvantage of ICP is background radiation from other elements and the plasma gases. Although all ICP instruments utilize high-resolution optics and background correction to minimize these interferences, analysis for traces of metals in the presence of a large excess of a single metal is difficult. Examples would be traces of metals in an alloy or traces of metals in a limes (high calcium) waste. ICP and Flame AA have comparable detection limits (within a factor of 4) except that ICP exhibits greater sensitivity for refractories (Al, Ba, etc.) Furnace AA, in general, will exhibit lower detection limits than either ICP or FLAA (*see* metal analyzer for more related terms) (SW-846).

Industrial alcohol: Alcohol that has been treated with acetates, ketones or gasoline to make it unfit for human drinking.

Industrial building: A building directly used in manufacturing or technically productive enterprises. Industrial buildings are not generally or typically accessible to other than workers. Industrial buildings include buildings used directly in the production of power, the manufacture of products, the mining of raw materials, and the storage of textiles, petroleum products, wood and paper products, chemicals, plastics, and metals (40 CFR 761.3-91).

Industrial cost recovery:
(1) The grantees recovery from the industrial users of a treatment works of the grant amount allocable to the treatment of waste from such users under section 204(b) of the Act and this subpart.
(2) The grantees recovery from the commercial users of an individual system of the grant amount allocable to the treatment of waste from such users under section 201(h) of the Act and this subpart (40 CFR 35.905-91).

Industrial cost recovery period: That period during which the grant amount allocable to the treatment of wastes from industrial users is recovered from the industrial users of such works (40 CFR 35.905-91).

Industrial effluent: *See* synonym, industrial wastewater.

Industrial furnace: Any of the following enclosed devices that are integral components of manufacturing processes and that use thermal treatment to accomplish recovery of materials or energy:
(1) Cement kilns
(2) Lime kilns
(3) Aggregate kilns
(4) Phosphate kilns
(5) Coke ovens
(6) Blast furnaces
(7) Smelting, melting and refining furnaces (including pyrometallurgical devices such as cupolas, reverberator furnaces, sintering machine, roasters, and foundry furnaces)
(8) Titanium dioxide chloride process oxidation reactors
(9) Methane reforming furnaces
10) Pulping liquor recovery furnaces
(11) Combustion devices used in the recovery of sulfur values from spent sulfuric acid
(12) Halogen acid furnaces (HAFs)
(13) Such other devices as the Administrator may, after notice and comment, add to this list on the basis of one or more of the following factors:
  (A) The design and use of the device primarily to accomplish recovery of material products;
  (B) The use of the device to burn or reduce raw materials to make a material product;
  (C) The use of the device to burn or reduce secondary materials as effective substitutes for raw materials, in processes using raw materials as principal feedstocks;
  (D) The use of the device to burn or reduce secondary materials as ingredients in an industrial process to make a material product;
  (E) The use of the device in common industrial practice to produce a material product; and
  (F) Other factors, as appropriate (40 CFR 260.10-91).
(14) *See* furnace for more related terms.

Industrial furnace: Any of the following enclosed devices that are integral components of manufacturing processes and that use thermal treatment to accomplish recovery of materials or energy (40 CFR 260.10-91):
1. Cement kilns
2. Lime kilns
3. Aggregate kilns
4. Phosphate kilns
5. Coke ovens
6. Blast furnaces
7. Smelting, melting and refining furnaces (including pyrometallurgical devices such as cupolas, reverberator furnaces, sintering machine, roasters, and foundry furnaces)
8. Titanium dioxide chloride process oxidation reactors
9. Methane reforming furnaces
10. Pulping liquor recovery furnaces
11. Combustion devices used in the recovery of sulfur values from spent sulfuric acid
12. Halogen acid furnaces (HAFs) for the production of acid from halogenated hazardous waste generated by chemical production facilities where the furnace is located on the site of a chemical production facility, the acid product has a halogen acid content of at least 3%, the acid product is used in a manufacturing process, and, except for hazardous waste burned as fuel, hazardous waste fed to the furnace has a minimum halogen content of 20% as-generated.

Industrial hygiene: Science of industrial safety.

Industrial incinerator:
(1) An incinerator designed to burn industrial waste (SW-108ts).
(2) An incinerator owned and operated for private use by an individual waste generator (CRWI-5/89).
(3) *See* incinerator for more related terms.

Industrial pollution prevention:
(1) The terms "industrial pollution prevention" and "pollution prevention" refer to the combination of industrial source reduction and toxic chemical use substitution. It does not include any recycling or treatment of pollutants, nor does it include substituting a nontoxic product made with nontoxic chemicals for a nontoxic product made with toxic chemicals (cf. pollution prevention) (EPA-91/10, p7).
(2) Combination of industrial source reduction and toxic chemical use substitution (EPA-94/04).

Industrial refuse: *See* synonym, industrial solid waste.

Industrial sales paint: The paint which is primarily sold to other manufacturers for factory application to such products as aircraft, appliances, furniture, machinery, etc. (*see* paint for more related terms) (EPA-79/12b).

Industrial solid waste (industrial refuse or industrial waste): The solid waste generated by industrial processes and manufacturing (*see* waste for more related terms) (40 CFR 243.101-91, *see also* 40 CFR 246.101-91).

Industrial source: A unit that does not serve a generator that produces electricity, a 'nonutility unit' as defined in this section, or a process source as defined in section 410(e) (*see* source for more related terms) (CAA402).

Industrial source complex model for long term (ISCLT): *see* definition under dispersion model.

Industrial source complex model for short term (ISCST): *see* definition under dispersion model.

Industrial source reduction:
(1) Industrial source reduction is defined in the recently enacted Federal Pollution Prevention Act as any practice which:
  (A) Reduces the amount of any hazardous substance, pollutant, or contaminant entering any waste [pollutant] stream or otherwise released into the environment (including fugitive emissions) prior to recycling, treatment, and disposal; and
  (B) Reduces the hazards to public health and the environment associated with the release of such substances, pollutants, or contaminants.
The term includes equipment or technology modifications, process or procedure modifications, reformulation or redesign of products, substitution of raw materials, and improvements in housekeeping, maintenance, training, or

inventory control. Source reduction does not entail any form of waste management (e.g., recycling and treatment). The Act excludes from the definition of source reduction any practice which alters the physical, chemical, or biological characteristics or volume of a hazardous substance, pollutant, or contaminant through a process or activity which itself is not integral to and necessary for the production of a product or the providing of a service (EPA-91/10, p6).
(2) Practices that reduce the amount of any hazardous substance, pollutant, or contaminant entering any waste stream or otherwise released into the environment; Also reduces the threat to public health and the environment associated with such releases. Term includes equipment or technology modifications, substitution of raw materials, and improvements in housekeeping, maintenance, training or inventory control (EPA-94/04).

Industrial user:
(1) Any industry that introduces pollutants into public sewer systems and whose wastes are treated by a publicly owned treatment facility (EPA-83/06a).
(2) Those industries identified in the Standard Industrial Classification Manual, Bureau of the Budget, 1967, as amended and supplemented, under the category Division D -- Manufacturing and such other classes of significant waste products as, by regulation, the Administrator deems appropriate (CWA502, *see also* 40 CFR 35.905; 35.2005; 403.3-91).

Industrial waste:
(1) Unwanted materials from an industrial operation; may be liquid, sludge, solid, or hazardous waste (EPA-94/04).
(2) *See also* synonym, industrial solid waste.

Industrial waste pollution: A broad category of wastes from manufacturing operations or processes defined by government as noxious. They include floating matter, settleable solids, colloidal matter, dissolved solids, toxic substances, and sludge (*see* pollution for more related terms) (DOI-70/04).

Industrial wastewater (or industrial effluent): The wastewater discharged by an industry (*see* wastewater for more related terms).

Industrial wiper: Paper towels especially made for industrial cleaning and wiping (40 CFR 250.4-91).

Inert atmosphere: A gas incapable of supporting combustion (EPA-83).

Inert contaminant:
(1) A gaseous or liquid contaminant that does not react with surrounding substances under ordinary circumstances.
(2) A contaminant that does not influence the fluid motions by which it is dispersed (NATO-78/10).

Inert gas (noble gas or rare gas): A gas that does not react with other substances under ordinary conditions (EPA-84/09). All inert gases belong to group VIII (group 0) in the periodic table and are all monatomic elements such as helium and neon.

Inert ingredient:

(1) An ingredient which is not active (FIFRA2, *see also* 40 CFR 152.3; 158.153-91).
(2) Pesticide components such as solvents, carriers, dispersants, and surfactants that are not active against target pests. Not all inert ingredients are innocuous (EPA-94/04).

Inert material: Materials lacking active thermal, chemical or biological properties (EPA-83).

Inertia weight class: The class, which is a group of test weights, into which a vehicle is grouped based on its loaded vehicle weight in accordance with the provisions of Part 86 (40 CFR 86.082.2; 600.002.85-91).

Inertial coordinate system: A coordinate system, in which the momentum of a body is conserved when no external forces are applied (NATO-78/10).

Inertial force: The resistance which a body experience when its motion changes in a given coordinate system (NATO-78/10).

Inertial grate stoker: A stoker consisting of a fixed bed of plates that are carried on rollers and activated by an electrically driven mechanism. It draws the bed slowly back against a spring and then releases it so that the entire bed moves forward until stopped abruptly by another spring. The inertia of the solid waste carries it a small distance forward along the stoker surface, and then the cycle is repeated (*see* stoker for more related terms) (SW-108ts).

Inertial separator:
(1) A device that uses centrifugal force to separate waste particles (EPA-94/04).
(2) *See also* synonym, mechanical separator.

Inertial sub-range: The region in the energy spectrum of turbulence for which the wavelengths are small compared to the largest scales of turbulence where turbulent energy is generated, and large compared to the small scales, where turbulent energy is dissipated by viscous forces. The spectrum in this region is only determined by the wave number and the dissipation rate, which leads to a -5/3 power law for the spectrum (NATO-78/10).

Infection: The entry and development or multiplication of an infectious agent in the body of man or animals. Infection is not synonymous with infectious disease; the result may be inapparent. The presence of living infectious agents on exterior surfaces of the body, or upon articles of apparel or soiled articles, is not infection, but contamination of such surfaces and articles. In addition, it should be pointed out that antibody production, i.e., seroconversion, does not necessarily mean that "infection" has occurred (EPA-5/90).

Infection process (or chain of infection): The relationships between the infectious agent, the host, and the mechanism by which the agent is transmitted. The entire chain has four requirements (p2.7-ATSDR-9/90):
(1) The presence of an infectious agent;
(2) A sufficient number of infectious agents to cause infection;
(3) The availability of a susceptible host; and
(4) An appropriate portal of entry into that susceptible host.

Infectious agent:
(1) Any organism (such as a virus or a bacteria) that is capable of being communicated by invasion and multiplication in body tissues and capable of causing disease or adverse health impacts in humans (40 CFR 259.10-91).
(2) Any organism, such as a virus or bacterium, that is pathogenic and capable of being communicated by invasion and multiplication in body tissues (EPA-94/04).

Infectious waste:
(1) (A) Equipment, instruments, utensils, and fomites of a disposable nature from the rooms of patients who are suspected to have or have been diagnosed as having a communicable disease and must, therefore, be isolated as required by public health agencies;
(B) Laboratory wastes such as pathological specimens (e.g., all tissues, specimens of blood elements, excreta, and secretions obtained from patients or laboratory animals) and disposable fomites (any substance that may harbor or transmit pathogenic organisms) attendant thereto;
(C) Surgical operating room pathologic specimens and disposable fomites attendant thereto and similar disposable materials from outpatient areas and emergency rooms (40 CFR 240.101-91, see also 40 CFR 241.101; 243.101; 245.101; 246.101-91).
(2) Hazardous waste with infectious characteristics, including: contaminated animal waste; human blood and blood products; isolation waste, pathological waste; and discarded sharps (needles, scalpels or broken medical instruments) (EPA-94/04).
(3) See waste for more related terms.

Infectious dose 50 (ID50): A dose of microorganisms that is required to infect 50 percent of the exposed population (see dose response for more related terms) (EPA-88/09a).

Infective dose: The infective dose is the number of microorganisms required to produce infection in humans. A great deal of uncertainty is associated with infective dose estimates, due to the variable contribution of a number of factors, including host sensitivity, pathogen virulence, assay technique, etc. (see dose response for more related terms) (EPA-88/12).

Infectivity: The capacity of the organism to proliferate at the site of invasion (EPA-88/09a).

Inference guideline: An explicit statement of a predetermined choice among alternative methods (inference options) that might be used to infer human risk from data that are not fully adequate or are not drawn directly from human experience, e.g., a guideline might specify the mathematical model to be used to estimate the effects of exposure at low doses on the basis of the effects of exposure at high doses (NAC-83).

Infiltration:
(1) Water other than wastewater that enters a sewerage system (including sewer service connections) from the ground through such means as defective pipes, pipe joints, connections, or manholes. Infiltration does not include, and is distinguished tom, inflow (40 CFR 35.905-91, see also 40 CFR 35.2005-91).
(2) (A) The penetration of water through the ground surface into sub-surface soil or the penetration of water from the soil into sewer or other pipes through defective joints, connections, or manhole walls.
(B) A land application technique where large volumes of waste water are applied to land, allowed to penetrate the surface and percolate through the underlying soil (see percolation) (EPA-94/04).
(3) cf. non-excessive infiltration

Infiltration air: The air that leaks into the chambers or ducts of an incinerator (see air for more related terms) (SW-108ts).

Infiltration bed: A device for removing suspended solids from water consisting of natural deposits of granular material under which a system of pipes collect the water after passage through the bed (EPA-76/04).

Infiltration gallery: A subsurface groundwater collection system, typically shallow in depth, constructed with open-jointed or perforated pipes that discharge collected water into a water-tight chamber from which the water is pumped to treatment facilities and into the distribution system. Usually located close to streams or ponds (EPA-94/04).

Infiltration/inflow (I/I):
(1) The total quantity of water from both infiltration and inflow without distinguishing the source (40 CFR 35.905-91).
(2) The quantity of water entering a sewer system. Infiltration means entry through such sources as defective pipes, pipe joints, connections, or manhole walls. Inflow signifies discharge into the sewer system through service connections from such sources as area or foundation drainage, springs and swamps, storm waters, street wash waters or sewers (EPA-74/11).
(3) cf. excessive infiltration/inflow.

Infiltration rate: The quantity of water than can enter the soil in a specified time interval (EPA-94/04).

Infiltration water: The water which permeates through the earth into the plant site (see water for more related terms) (40 CFR 440.141-91).

Inflow:
(1) Water other than wastewater that enters a sewerage system (including sewer service collections) from sources such as roof leaders, cellar drains, yard drains, area drains, foundation drains, drains from springs and swampy areas, manhole covers, cross connections between storm sewers and sanitary sewers, catch basins, cooling towers, storm waters, surface runoff, street wash waters, or drainage. Inflow does not include, and is distinguished from, infiltration (40 CFR 35.905-91, see also 40 CFR 35.2005-91).
(2) Entry of extraneous rain water into a sewer system from sources other than infiltration, such as basement drains, manholes, storm drains, and street washing (EPA-89/12).
(3) cf. non-excessive inflow.

Influent: Water, wastewater, or other liquid flowing into a reservoir, basin, or treatment plant (EPA-94/04).

Influent stream: Stream or portion of stream that contributes water to the groundwater supply (EPA-83).

Influent water: The water contributing to the zone of saturation and thereby sustaining or raising the water table (*see* water for more related terms) (DOI-70/04).

Information file: In the Superfund program, a file that contains accurate, up-to-date documents on a Superfund site. The file is usually located in a public building (school, library, or city hall) convenient for local residents (EPA-94/04).

Infrared: Pertaining to the region of the electromagnetic spectrum from approximately 0.78-1000 um (micrometers). Near infrared is below 10 um and far infrared above 10 um (LBL-76/07-air).

Infrared absorption: One of continuous emission monitors (*see* continuous emission monitor or radiation emission-absorption instrumentation for various types). The infrared portion (IR) of the spectrum ranges from wavelengths of 2.4 to 14 microns and the near infrared (NIR) ranges from 1 to 2.5 microns. Most organic species absorb radiation in the IR region and the concentration of species can be determined by the nature of the transmitted radiation. The concentration of an absorbing compound is inversely proportional to the sample cell path length and proportional to the energy transmitted through the sample divided by the logarithm of the energy entering the sample (Beers law). Sensitivity can be increased through increasing the cell path length and by including multiple radiation passes through the sample. High radiation energy sources such as CO_2 lasers are finding increased use (EPA-84/03a).

Infrared absorption spectrometer: *See* definition under GC/IR.

Infrared incinerator: Any enclosed device that uses electric powered resistance heaters as a source of radiant heat and which is not listed as an industrial furnace (40 CFR 260.10-91).

In general, an infrared system has the following major components:
(1) **Primary combustion chamber (PCC):** The electric powered PCC utilizes a high temperature alloy wire mesh belt for waste feed conveyance and is capable of achieving temperatures up to 1850° F by exposure to infrared radiant heat provided by horizontal rows of electric-powered silicon carbide rods located above the conveyer belt.
(2) **Secondary combustion chamber (SCC):** The gas-fired SCC is capable of reaching temperatures of 2300° F; it provides residence time, turbulence and supplemental energy, if required, to destroy gaseous volatiles emanating from the waste (Lee-88/08)
(3) *See* incinerator for more related terms.

Infrared radiation: The electromagnetic radiation whose wavelengths lie between red light (approximately 0.75 micrometer) and radio waves (approximately 1000 micrometer) (*see* radiation for more related terms).

Infrared spectrometer: An instrument used to measure the concentration of compounds. It is based on the principles that molecular vibrations occur in the infrared region.

Infrared spectrophotometer (IS): One of continuous emission monitors (*see* continuous emission monitor for various types). IS a compound-specific instrument. Each compound being analyzed will absorb radiation at a discrete infrared wavelength. The unit measures how much of the infrared energy (IR) is absorbed and gives readings of percent IR absorbed or ppm of chemical.

Molecules are composed of atoms which are held together by bonds of various types and lengths. These arrangements, as in the ball and spring configurations establish finite locations and discrete movements for each atom (ball) and bond (spring). These movements can be either vibrational, rotational, stretching, or bending of the chemical bonds. The frequencies of these movements are on the order of infrared radiation. A given bond movement can be initiated by stimulating the molecule with IR of varying frequency. As the bond moves, it absorbs the characteristic energy associated with that movement. The frequencies and intensity of IR absorbed are specific for a compound and its concentration, providing a "fingerprint" which can be used as an analytical tool (Course 165.5).

Infrared spectroscopy: The determination of material structure by means of their interaction with infrared radiation (*see* spectroscopy for more related terms).

Infrared spectrum analysis: Analysis of infrared wavelengths.

Ingot: A mass of metals or metal alloys shaped for convenience in storage and handling. Sizes according to weight are 15, 30, 50, and 1000 pounds (EPA-76/12).

Ingredient statement: A statement which contains:
(1) The name and percentage of each active ingredient, and the total percentage of all inert ingredients, in the pesticide, and
(2) If the pesticide contains arsenic in any form, a statement of the percentages of total and water soluble arsenic, calculated as elementary arsenic (FIFRA2-7USC136-91).

Inground tank: A device meeting the definition of tank in 40 CFR 260.10 whereby a portion of the tank wall is situated to any degree within the ground, thereby preventing visual inspection of that external surface area of the tank that is in the ground (*see* tank for more related terms) (40 CFR 260.10-91).

Inhalable diameter: Refers to that aerodynamic diameter of a particle which is considered to be inhalable for the organism. It is used to refer to particles which are capable of being inhaled and may be deposited an)here within the respiratory tract from the trachea to the alveoli. For man, the inhalable diameter is considered as 15 micrometers or less (40 CFR 798.1150-91, *see also* 40 CFR 798.2450; 798.4350-91).

Inhalation toxicity: Gross anatomy, microscopic and ultrastructural anatomy, changes in function (*see* endpoint for more related terms) (Course 165.6; EPA-92/12).

Inherent ash: The portion of the ash of a material found after combustion which is chemically bound to the molecules of the combustible, as distinguished from extraneous non-combustible materials from other sources or which may be mechanically entrained with the combustible (*see* ash for more related terms) (EPA-83).

Inhibition: The slowing down or stoppage of chemical or biological reactions by certain compounds or ions (EPA-83/06a).

Inhibition concentration (IC): A point estimate of the toxicant; concentration that would cause a given percent reduction (e.g., IC_{25}) in a nonlethal biological measurement of the test organisms, such as reproduction or growth (EPA-91/03).

Inhibitor:
(1) A coating on a propellant grain which prevents burning at that point (EPA-76/03).
(2) A chemical which can inhibit or retard a chemical reaction.

Initial calibration verification standard: *See* definition under calibration.

Initial compliance period: The first full three-year compliance period which begins at least 18 months after promulgation (40 CFR 141.2-91).

Initial compliance period (water): The first full three-year compliance period which begins at least 18 months after promulgation (EPA-94/04).

Initial condition: The prescribing of all dependent variables in a differential equation at the initial time (NATO-78/10).

Initial crusher: Any crusher into which nonmetallic minerals can be fed without prior crushing in the plant (40 CFR 60.671-91).

Initial decision: The decision issued by the Presiding Officer based upon the record of the proceedings out of which it arises (40 CFR 22.03-91, *see also* 40 CFR 27.2; 164.2-91).

Initial failure rate: The percentage of vehicles rejected because of excessive emissions of a single pollutant during the first inspection cycle of an inspection/maintenance program. (If inspection is conducted for more than one pollutant, the total failure rate may be higher than the failure rates for each single pollutant) (40 CFR 51-App/N-91).

Initial mixing: Is defined to be that dispersion or diffusion of liquid, suspended particulate, and solid phases of a waste which occurs within four hours after dumping. The limiting permissible concentration shall not be exceeded beyond the boundaries of the disposal site during initial mixing, and shall not be exceeded at any point in the marine environment after initial mixing. The maximum concentration of the liquid, suspended particulate, and solid phases of a dumped material after initial mixing shall be estimated by one of these methods, in order of preference:
(1) When field data on the proposed dumping are adequate to predict initial dispersion and diffusion of the waste, these shall be used, if necessary, in conjunction with an appropriate mathematical model acceptable to EPA or the District Engineer, as appropriate.
(2) When field data on the dispersion and diffusion of a waste of characteristics similar to that proposed for discharge are available, these shall be used in conjunction with an appropriate mathematical model acceptable to EPA or the District Engineer, as appropriate.
(3) When no field data are available, theoretical oceanic turbulent diffusion relationships may be applied to known characteristics of the waste and the disposal site (40 CFR 227.29-91).

Initial pressure (P_I): The pressure applied to the delivery tank at the beginning of the static pressure test, as specified in the appropriate regulation, in mm H_2O (40 CFR 60-App/A(method 27)-91).

Initial rate of absorption (IRA): The gain in weight of a brick whose surface was in contact with water for sixty seconds (ASTM C-6; EPA-83).

Initial vacuum (V_I): The vacuum applied to the delivery tank at the beginning of the static vacuum test, as specified in the appropriate regulation, in mm H_2O (40 CFR 60-App/A(method 27)-91).

Initiation: The ability of an agent to induce a change in a tissue which leads to the induction of tumors after a second agent, called a promoter, is administered to the tissue repeatedly. *See also* Promoter (EPA-92/12).

Initiation of operation date: Specified by the grantee means the on which use of the project begins or the purpose for which it was planned, designed, and built (40 CFR 35.2005-91).

Injectables: Medicinals prepared in a sterile (buffered) form suitable for administration by injection (EPA-83/09).

Injection aerator: An aerator in which air is directly injected into wastewater treatment ponds to avoid pressure loss that occurs with a spray aerator (*see* aerator for more related terms).

Injection interval: That part of the injection zone in which the well is screened, or in which the waste is otherwise directly emplaced (40 CFR 146.61; 148.2-91).

Injection well:
(1) A well into which fluids are being injected (40 CFR 144.3-91).
(2) A well into which fluids are injected for purposes such as waste disposal, improving the recovery of crude oil, or solution mining (*see* well for more related terms) (EPA-94/04).
Other injection well related terms include:
- New underground injection well
- Underground injection

Injection zone:
(1) A geological formation, group of formations, or part of a formation receiving fluids through a well (40 CFR 144.3; 146.3; 147.2902-91).
(2) A geological formation receiving fluids through a well (EPA-94/04).

Ink: A coating used in printing, impressing, or transferring an image onto a substrate (40 CFR 52.741-91, *see also* 40 CFR - 60.581-91). Other ink related terms include:
- Flexographic ink
- Gravure ink
- Heat set ink

- Letterpress ink
- Lithographic ink
- Metallic ink
- Moisture set ink
- News ink
- Oil base ink
- Opaque ink
- Printing ink
- Silk screen ink
- Solvent base ink
- Thermosetting ink
- Water base ink (paint)

Ink receptivity: Ability to absorb ink (EPA-83).

Ink solids: The solids content of an ink as determined by Reference Method 24, ink manufacturer's formulation data, or plant blending records (40 CFR 60.581-91).

Inland oil barge: A non-self-propelled vessel carrying oil in bulk as cargo and certificated to operate only in the inland waters of the United States, while operating in such waters (CWA311-33USC1321-91).

Inland water: For the purposes of classifying the size of discharges, inland water means those waters of the United States in the inland zone, waters of the Great Lakes, and specified ports and harbors on inland rivers (*see* water for more related terms) (40 CFR 300.5-91).

Inland waters of the United States: Those waters of the United States lying inside the baseline from which the territorial sea is measured and those waters outside such baseline which are a part of the Gulf Intracoastal Waterway (CWA311-33USC1321-91).

Inland zone: The environment inland of the coastal zone excluding the Great Lakes and specified ports and harbors of inland rivers. The term inland zone delineates the area of Federal responsibility for response action. Precise boundaries are determined by EPA/USCG agreement and identified in federal regional contingency plans (40 CFR 300.5-91).

Inner liner: A continuous layer of material placed inside a tank or container which protects the construction materials of the tank or container from the contained waste or reagents used to treat the waste (*see* liner for more related terms) (40 CFR 260.10-91).

Innovative control technology: Any system of air pollution control that has not been adequately demonstrated in practice, but would have a substantial likelihood of achieving greater continuous emissions reduction than any control system in current practice or of achieving at least comparable reductions at lower cost in terms of energy, economics, or nonair quality environmental impacts (40 CFR 51.166-91, *see also* 40 CFR 52.21-91).

Innovative technology: The developed wastewater treatment processes and techniques which have not been fully proven under the circumstances of their contemplated use and which represent a significant advancement over the state of the art in terms of significant reduction in life cycle cost or significant environmental benefits through the reclaiming and reuse of water, otherwise eliminating the discharge of pollutants, utilizing recycling techniques such as land treatment, more efficient use of energy and resources, improved or new methods of waste treatment management for combined municipal and industrial systems, or the confined disposal of pollutants so that they will not migrate to cause water or other environmental pollution (40 CFR 35.2005-91, *see also* 40 CFR 125.22-91).

Innovative technologies: New or inventive methods to treat effectively hazardous waste and reduce risks to human health and the environment (EPA-94/04).

Inoculum:
(1) Bacterium placed in compost to start biological action.
(2) A medium containing organisms which is introduced into cultures or living organisms (EPA-94/04).

Inorganic acid: A chemical containing hydrogen and nonmetal elements or nonmetal radicals such as hydrogen chloride HCl, sulfuric acid H_2SO_4 and carbonic acid H_2CO_3 (cf. organic acid and *see* acid for more related terms).

Inorganic arsenic: The oxides and other noncarbon compounds of the element arsenic included in particulate matter, vapors, and aerosols (*see* arsenic for more related terms) (40 CFR 61.171-91).

Inorganic chemical (inorganic material or inorganic matter):
(1) Chemical substances of mineral origin, not of basically carbon structure (EPA-94/04).
(2) A chemical that does not contain carbon. However, some carbon containing compounds such as CO, CO_2, CS_2, carbonates and cyanides are usually considered as inorganic chemicals.

Inorganic chemistry: A branch of chemistry dealing with the chemical reactions and properties of all inorganic matter.

Inorganic material: *See* synonym, inorganic chemical.

Inorganic matter: *See* synonym, inorganic chemical.

Inorganic pesticide: *See* synonym, pesticide.

Inorganic pigment: A class of pigments used in printing inks consisting of compounds of the various metals, e.g., chrome yellow (EPA-79/12a).

Inorganic refuse: Solid waste composed of matter other than plant, animal, and certain carbon compounds, e.g., metals and glass.

Inorganic solid debris: The nonfriable inorganic solids contaminaated with D004-D011 hazardous wastes that are incapable of passing through a 9.5 mm standard sieve; and that require cutting, or crushing and grinding in mechanical sizing equipment prior to stabilization; and are limited to the following inorganic or metal materials:
(1) Metal slags (either dross or scoria)
(2) Glassified slag
(3) Glass
(4) Concrete (excluding cementitious or pozzolanic stabilized hazardous wastes)

(5) Masonry and refractory bricks
(6) Metal cans, containers, drums, or tanks
(7) Metal nuts, bolts, pipes, pumps, valves, appliances, or industrial equipment; and
(8) Scrap metal as defined in 40 CFR 261.1(c)(6) (40 CFR 268.2-91).

Inorganic waste: The waste composed of matter other than plant or animal (i.e., contains no carbon) (*see* waste for more related terms) (EPA-89/11).

Inprocess wastewater: Any water which during manufacturing or processing, comes into direct contact with vinyl chloride or polyvinyl chloride or results from the production or use of any raw material, intermediate product, finished product, by-product, or waste product containing vinyl chloride or polyvinyl chloride but which has not been discharged to a wastewater treatment process or discharged untreated as wastewater. Gasholder seal water is not inprocess wastewater until it is removed from the gasholder (*see* wastewater for more related terms) (40 CFR 61.61-91).

Insect: Any of the numerous small invertebrate animals generally having the body more or less obviously segmented, for the most part belonging to the class insecta, comprising six-legged, usually winged forms, as for example, beetles, bugs, bees, flies, and to other allied classes of arthropods whose members are wingless and usually have more than six legs, as for example, spiders, mites, ticks, centipedes, and wood lice (FIFRA2-7USC136-91).

Insect attractant: A substance that lures insects to trap or poison-bait stations. Usually classed as food, oviposition, and sex attractants (EPA-85/10).

Insect growth regulator (IGR): A chemical substance that disrupts the action of insect hormones controlling molting, maturity from pupal stage to adult, and others (EPA-85/10).

Insecticide:
(1) In according to method of action, insecticide can be classified:
    (A) Stomach poisons, which act in the digestive system;
    (B) Contact poisons, which act by direct external contact with the insect at some stage of its life cycle; and
    (C) Fumigants, which attack the respiratory system (AP-40, p832).
(2) A pesticide compound specifically used to kill or prevent the growth of insects (EPA-94/04).

Inside spray coating operation: The system on each beverage can surface coating line used to apply a coating to the interior of a two-piece beverage can body. This coating provides a protective film between the contents of the beverage can and the metal can body. The inside spray coating operation consists of the coating application station, flashoff area, and curing oven. Multiple applications of an inside spray coating are considered to be a single coating operation (40 CFR 60.491-91).

Inspection/maintenance:
(1) A program to reduce emissions from in-use vehicles through identifying vehicles that need emissions control related maintenance and requiring that maintenance be performed (40 CFR 51-App/N-91).
(2) (A) Activities to assure proper emissions-related operation of mobile sources of air pollutants, particularly automobile emissions controls.
    (B) Also applies to wastewater treatment plants and other anti-pollution facilities and processes (EPA-94/04).

Inspection and maintenance program: A program to reduce emissions from in-use vehicles through identifying vehicles that need emission control related maintenance and requiring and such maintenance be performed (40 CFR 51.731; 52.1161; 52.2485-91).

Inspection criteria: The pass and fail numbers associated with a particular sampling (40 CFR 86.602.84-91, *see also* 40 CFR 86.1002.84; 204.51; 205.51-91).

Insolation: The solar radiation at the earth surface (cf. global radiation) (NATO-78/10).

Insoluble: Unable to be dissolved in a different material.

Instability: The property of a system whereby small disturbances introduced into it increase in magnitude (cf. numerical instability) (NATO-78/10).

Instant photographic film article: A self-developing photographic film article designed so that all the chemical substances contained in the article, including the chemical substances required to process the film, remain sealed during distribution and use (40 CFR 723.175-91).

Instantaneous sampling: Obtaining a sample in a very short period of time such that this sampling time is insignificant in comparison with the duration of the operation or the period being studied (*see* sampling for more related terms) (EPA-83/06).

Instantaneous electric peak demand: The maximum demand at the instant of greatest load, usually determined from the readings of indicating or graphic meters (*see* electric demand for more related terms) (EPA-83).

Instantaneous source: A source which emits pollution over a time period much shorter than the travel time of the emission to a point where its concentration is considered (*see* source for more related terms) (NATO-78/10).

Institutional factor: The social, economic and political impact that can be exerted by groups of persons, industries, political bodies, or public organizations to further their respective goals (EPA-83).

Institutional solid waste: *See* synonym, institutional waste.

Institutional waste (or institutional solid waste): The solid wastes generated by educational, health care, correctional, and other institutional facilities (*see* waste for more related terms) (40 CFR 243.101-91, *see also* 40 CFR 245.101; 246.101-91).

Institutional use: Any application of a pesticide in or round any property or facility that functions to provide service to the general public or to public or private organizations, including but not limited to:

(1) Hospitals and nursing homes.
(2) Schools other than preschool and day care facilities.
(3) Museums and libraries.
(4) Sports facilities.
(5) Office buildings (40 CFR 152.3-91).

Instream use: Water use taking place within a stream channel, e.g., hydroelectric power generation, navigation, water quality improvement, fish propagation, recreation (EPA-94/04).

Instrument check standard: A multi-element standard of known concentrations prepared by the analyst to monitor and verify instrument performance on a daily basis (*see* standard for more related terms) (40 CFR 136-App/C-91).

Instrument class:
(1) **Portable:** Instrument can be readily hand carried for field use (LBL-76/07-bio).
(2) **Mobile:** Instrument can be readily transported in a van, ruggedized, 12 V d.c. operation (LBL-76/07-bio).
(3) **Stationary:** Instrument cannot be readily transported. This may be because of size, weight, the need to operate in a protected environment, fragility, or high maintenance requirements (LBL-76/07-bio).
(4) **Continuous-sensor type:** Instruments measure a constituent or parameter on an uninterrupted basis. These instruments may be used in the field or laboratory (LBL-76/07-water).
(5) **Batch-sampling type:** Instruments measure a constituent or parameter on an interrupted or discrete sampled basis; the analysis is then performed on this sample. These instruments are usually used in the laboratory, but may be used in the field (LBL-76/07-water).
(6) **Laboratory analysis:** Instrumentation ordinarily operates in the laboratory due to the constraints of operator intervention, operational environment, fragility or high maintenance requirements (LBL-76/07-water).

Instrument detection limit (IDL):
(1) The smallest signal an instrument can reliablely detect (ACS-87/11).
(2) The concentration equivalent to a signal, due to the analyte, which is reliably differentiated from zero (or background). Often defined as three times background noise (40 CFR 136-App/C).
(3) *See* detection limit for more related terms.

Instrument range: The maximum and minimum concentration that can be measured by a specific instrument. The minimum is often stated or assumed to be zero (0) and the range expressed only as the maximum. If a single analyzer is used for measuring multiple ranges (either manually or automatically), the performance standards expressed as a percentage of full scale apply to all ranges (EPA-90/04).

Instrument use: Instrument use related terms include:
- Field operation
- Laboratory operation

Instrumental detection limit: The concentration equivalent to a signal, due to the analyte, which is equal to three times the standard deviation of a series of ten replicate measurements of a reagent blank signal at the same wavelength (40 CFR 136-App/C-91).

Insulated furnace wall: A furnace wall behind which insulation material is installed (*see* furnace wall for more related terms) (SW-108ts).

Insulating block: A shaped product having a very low thermal conductivity, suitable for backing up the furnace lining (EPA-83).

Insulating brick: A firebrick having a low thermal conductivity and a bulk density of less than 70 pounds per cubic foot, suitable for lining industrial furnaces. Also called insulating block (*see* brick for more related terms) (SW-108ts).

Insulating paper: A standard material for insulating electrical equipment, usually consisting of bond or kraft paper coated with black or yellow insulating varnish on both sides (*see* paper for more related terms) (EPA-83/03).

Insulation:
(1) A material used to prevent a heat loss or gain between two different temperature systems.
(2) A material having high electrical resistivity and therefore suitable for separating adjacent conductors in an electric circuit or preventing possible future contact between conductors (EPA-83/03).

Insulation board: A panel manufactured from interfelted ligno-cellulosic fibers consolidated to a density of less than 0.5 g/cu cm (less than 31 lb/cu ft) (40 CFR 429.11-91).

Insulator: A non-conducting support for an electric conductor (EPA-83/03).

Insurance: The primary insurance, excess insurance, reinsurance, surplus lines insurance, and any other arrangement for shifting and disturbing risk which is determined to be insurance under applicable State or Federal law (SF401; 42USC9671-91).

Intake: The amount of materials inhaled, ingested, or absorbed dermally during a specified period of time (Course 165.6).

Intake water: The gross water minus reuse water (*see* water for more related terms) (EPA-83/06a).

Integral vista: A view perceived from within the mandatory Class I Federal area of a specific landmark or panorama located outside the boundary of the mandatory Class I Federal area (40 CFR 51.301-91).

Integrated chemical treatment: A waste treatment method in which a chemical rinse tank is inserted in the plating line between the process tank and the water rinse tank. The chemical rinse solution is continuously circulated through the tank and removes the dragout (EPA-83/06a).

Integrated circuit (IC):
(1) A combination of interconnected circuit elements inseparably associated on or within a continuous substrate.
(2) Any electronic device in which both active and passive elements are contained in a single package. Methods of

making an integrated circuit are by masking process, screening and chemical deposition (EPA-83/06a).

Integrated electric demand: The demand averaged over a specified period, usually determined by an integrating demand meter or by the integration of a load curve. It is the average of the continuously varying instantaneous demands during a specified demand interval (*see* electric demand for more related terms) (EPA-83).

Integrated emergency management system (IEMS): Developed by FEMA in recognition of the economies realized in planning for all hazards on a generic functional basis as opposed to developing independent structures and resources to deal with each type of hazard (NRT-87/03).

Integrated exposure assessment:
(1) A summation over time, in all media, of the magnitude of exposure to a toxic chemical (Course 165.6).
(2) Cumulative summation (over time) of the magnitude of exposure to a toxic chemical in all media (EPA-94/04).

Integrated facility: A facility that performs electroplating as only one of several operations necessary for manufacture of a product at a single physical location and has significant quantities of process wastewater from non-electroplating manufacturing operations. In addition, to qualify as an integrated facility one or more plant electroplating process wastewater lines must be combined prior to or at the point of treatment (or proposed treatment) with one or more plant sewers carrying process wastewater from non-electroplating manufacturing operations (40 CFR 413.02-91).

Integrated pest management (IPM):
(1) A system of managing pests by using biological, cultural and chemical means (EPA-74/11).
(2) A mixture of chemical and other, non-pesticide, methods to control pests (EPA-94/04).

Integrated sample:
(1) A sample obtained over a period of time with:
   (A) The collected atmosphere being retained in a single vessel; or
   (B) With a separated component accumulating into a single whole. Examples are dust sampling in which all the dust separated from the air is accumulated in one mass of fluid; the absorption of acid gas in an alkaline solution; and collection of air in a plastic bag or gasometer. Such a sample does not reflect variations in concentration during the period of sampling (EPA-83/06).
(2) A composite of a series of samples or a continuous flow of samples collected over a finite time period and representing an average sample for that period (LBL-76/07-air).
(3) *See* sample for more related terms.

Integrated solid waste management: A practice of using several alternative waste management techniques to manage and dispose of specific components of the municipal solid waste stream. Waste management alternatives include source reduction, recycling, composting, energy recovery and landfilling (EPA-89/11).

Integrated system: A process for producing a pesticide product that: (1) Contains any active ingredient derived from a source that is not an EPA-registered product; or (2) Contains any active ingredient that was produced or acquired in a manner that does not permit its inspection by the Agency under FIFRA section 9(a) prior to its use in the process (40 CFR 158.153-91).

Integrated waste management: Using a variety of practices to handle municipal solid waste; can include source reduction, recycling, incineration, and landfilling (EPA-94/04).

Integrated woolen mill: One in which all functions of wool processing are carried out from beginning to end - opening and scouring, spinning, dyeing, and finishing (DOI-70/04).

Intelligence: Information obtained from existing records or documents, placards, labels, signs, special configuration of containers, visual observations, technical records, eye witnesses, and others (Course 165.5).

Intensity of turbulence: A measure of the strength of the turbulence in a flow. Formally defined as the ratio of the standard deviation of a turbulent velocity component to the mean flow velocity (NATO-78/10).

Intensive property:
(1) A thermodynamic property which is independent of the quantity or the shape of the system under consideration, e.g., temperature, pressure or composition.
(2) Also known as the **intrinsic property**. It is a property whose value is the same for any part of a homogeneous system as it does for the whole system. Examples of the intensive property include pressure, temperature, density, etc. (Jones-p12; Wark-p5).
(3) *See* property for more related terms.

Inter route relief method: A method in which regular crews help collect on other routes when they finish their own (*see* waste collection method for more related terms) (SW-108ts).

Interception: Temporary retention of water (e.g, rainfall) by tree leaves or branches.

Interceptor sewer:
(1) A sewer whose primary purpose is to transport wastewaters from collector sewers to a treatment facility (40 CFR 21.2-91, *see also* 40 CFR 35.905; 35.2005-91; EPA-89/12; EPA-74/11). As the name implies, it means that the sewers intercept polluted drains flowing into any uncontrolled areas such as a river (*see* sewer for more related terms).
(2) Large sewer lines that, in a combined system, control the flow of sewage to the treatment plant. In a storm, they allow some of the sewage to flow directly into a receiving stream, thus keeping it from overflowing onto the streets. Also used in separate systems to collect the flows from main and trunk sewers and carry them to treatment points (EPA-94/04).

Interchange energy: Kilowatt hours delivered to or received by one electric utility system from another for economy purposes. They may be returned in kind at a later time or may be accumulated as energy balances until the end of a stated period. Settlement may be by payment or on a pooling basis (*see* electric energy for more related terms) (EPA-83).

Interconnected: That two or more electric generating units are electrically tied together by a network of power transmission lines, and other power transmission equipment (40 CFR 60.41a-91).

Interface:
(1) In hydrology, the contact zone between two different fluids (Course, 165.7). In air pollution control, the area where the gas phase and the absorbent contact each other (EPA-84/03b).
(2) The common boundary between two substances such as water and a solid, water and a gas, or two liquids such as water and oil (EPA-94/04).

Interference: A discharge which, alone or in conjunction with a discharge or discharges from other sources, both:
(1) Inhibits or disrupts the POTW, its treatment processes or operations, or its sludge processes or operations, use or disposal; and
(2) Therefore is a cause of a violation of any requirement of the POTW's NPDES permit (including an increase in the magnitude or duration of a violation) or of the prevention of sewage sludge use or disposal in compliance with the following statutory provisions and regulations or permits issued thereunder (or more stringent State or local regulations): section 405 of the Clean Water Act, the Solid Waste Disposal Act (SWDA) (including Title II, more commonly referred to as the Resource Conservation and Recovery Act (RCRA), and including State regulations contained in any State sludge management plan prepared pursuant to Subtitle D of the SWDA), the Clean Air Act, the Toxic Substances Control Act, and the Marine Protection, Research and Sanctuaries Act (40 CFR 403.3-i-91).

Interference: Any substance or species which causes a deviation of instrument output from the value which would result from the presence of only the desired constituent (e.g., pollutant of concern) under test.
- Appreciable interferences are those greater than 10%;
- Moderate interferences are those between 5% and 10%; and
- Slight or no interferences are those less than 5%.

Where the interference percent is: $[(X) - (C)]/(i) \times 100$ = % interference, (X) = total response due to constituent and interference, (C) = response due to constiuent under test, (i) = concentration of interference.

The unit of response or concentration must be common to all parameters and should be indicated (LBL-76/07-water).

Interference check: A method for detecting analytical interferences and excessive biases through direct comparison of gas concentrations provided by the measure ment system and by a modified Method 6 procedure. For this check, the modified Method 6 samples are acquired at the sample bypass discharge vent (40 CFR 60-App/A(method 6C)-91).

Interference check sample: A solution containing both interfering and analyte elements of known concentration that can be used to verify background an interelement correction factors (40 CFR 136-App/C-91).

Interference equivalent: Positive or negative response caused by a substance other than the one being measured (40 CFR 53.23-91).

Interference percentages: The positive or negative percentages indicating the interferences caused by substances other than the one being measured. The interference percent is defined as: $[(x) - (y)]/(z) \times 100$ = % interference, where: (x) = Total response due to the pollutant of concern plus interference in ppm; (y) = response due to the pollutant of concern in ppm; (z) = concentration of interference in ppm (LBL-76/07-bio).

Interference response: The output response of the measurement system to a component in the sample gas, other than the gas component being measured (40 CFR 60-App/A(method 7E)-91).

Interflow: That portion of precipitation that infiltrates into the soil and moves laterally under its surface until intercepted by a stream channel or until it resurfaces down slope from its point of infiltration (SW-108ts).

Interfluve: The ridge between two adjacent river valleys (DOI-70/04).

Interim approval: The approval received by a state program that meets the requirements in 40 CFR 281.11(c) (1) and (2) for the time period defined in 40 CFR 281.11(c)(3) (*see* approval for more related terms) (40 CFR 281.12-91).

Interim authorization: The approval by EPA of a State hazardous waste program which has met the requirements of section 3006(c) of RCRA and applicable requirements of Part 271, Subpart B (40 CFR 270.2-91).

Interim (permit) status: Period during which treatment, storage and disposal facilities coming under RCRA in 1980 are temporarily permitted to operate while awaiting a permanent permit. Permits issued under these circumstances are usually called "Part A" or "Part B" permits (EPA-94/04).

Interim status: In RCRA, a temporary permit that allows owners and operators of TSDs (treatment, storage, and disposal facilities) that were in existence, or for which construction had commenced, prior to November 19, 1980 to continue to operate without a permit after this date. Owners and operators of TSDs are eligible for interim status on ongoing basis if the TSD is in existence on the effective date of regulatory changes under RCRA that cause the facility to be subject to Subtitle C regulation. Owners and operators in interim status are subject to and must comply with the applicable standards in 40 CFR 265. Interim status is gained through the notification process and by submitting Part A of the permit application (EPA-86/01; RCRA3005).

Interior body spray coat: A coating applied by spray to the interior of a can body (40 CFR 52.741-91).

Interior drainage: *See* synonym, internal drainage.

Interior paint: A coating for the inside surfaces of a structure (EPA-79/12b).

Interleaved winding: An arrangement of winding coils around a transformer core in which the coils are wound in the form of a disk, with a group of disks for the low-voltage windings stacked alternately with a group of disks for the high-voltage windings (EPA-83/03).

Interlock: A part of an automatic control system which ties together operation of different incinerator components (EPA-89/03b).

Intermediate: Any chemical substance that is consumed, in whole or in part, in chemical reactions used for the intentional manufacture of other chemical substances or mixtures, or that is intentionally present for the purpose of altering the rates of such chemical reactions (40 CFR 704.3-91, *see also* 40 CFR 710.2; 712.3; 720.3; 723.175-91).

Intermediate cover (or middle cover): The cover material that serves the same functions as daily cover, but must resist erosion for a longer period of time, because it is applied on areas where additional cells are not to be constructed for extended periods of time (*see* cover for more related terms) (40 CFR 241.101-91).

Intermediate cover soil: The soil materials placed on completed lifts in areas where there is clear intention to place another lift on top within one year (*see* soil for more related terms) (OME-88/12).

Intermediate duty fireclay brick: A fireclay brick that has a PCE above Cone 29 or does not deform more than 3% at 2460° F in the standard local test (*see* brick for more related terms) (SW-108ts).

Intermediate handler: A facility that either treats regulated medical waste or destroys regulated medical waste but does not do both. The term, as used in this Part, does not include transporters (40 CFR 259.10-91).

Intermediate premanufacture notice or intermediate PMN: Any PMN submitted to EPA for a chemical substance which is an intermediate (as intermediate is defined in 40 CFR 720.3 of this chapter) in the production of a final product, provided that the PMN for the intermediate is submitted to EPA at the same time as, and together with, the PMN for the final product and that the PMN for the intermediate identifies the final product and describes the chemical reactions leading from the intermediate to the final product. If PMNs are submitted to EPA at the same time for several intermediates used in the production of a final product, each of those is an intermediate PMN if they all identify the final product and every other associated intermediate PMN and are submitted to EPA at the same time as, and together with, the PMN for the final product (40 CFR 700.43-91).

Intermediate processing center (IPC): Usually refers to the type of materials recovery facility (MRF) that processes residentially collected mixed recyclables into new products available for market; often used interchangeably with MRF (EPA-89/11).

Intermediate speed: The peak torque speed if peak torque speed occurs between 60 and 75 percent of rated speed. If the peak torque speed is less than 60 percent of rated speed, intermediate speed means 60 percent of rated speed. If the peak torque speed is greater than 75 percent of rated speed, intermediate speed means 75 percent of rated speed (40 CFR 86.082.2-91).

Intermediate standard: For calibration application, those standards cannot be calibrated by measuring physical dimensions, but accuracies of +/- 1 to 2% can be reached. Intermediate standards are calibrated against primary standards (*see* calibration of air flow for more related terms) (EPA-83/06).

Intermediate treatment: *See* synonym, intermediate wastewater treatment.

Intermediate wastewater treatment (or intermediate wastewater treatment): The wastewater treatment such as aeration or chemical treatment, supplementary to primary treatment. Such treatment removes substantial percentages of very finely divided particulate matter, in addition to the suspended solids removed by primary treatment. Supplementary processing improves the efficiency of treatment so that about 60 percent of both BOD and suspended solids are removed (*see* wastewater treatment for more related terms) (DOI-70/04).

Intermittent: Not continuous.

Intermittent control system (ICS): A dispersion technique which varies the rate at which pollutants are emitted to the atmosphere according to meteorological conditions and/or ambient concentrations of the pollutant, in order to prevent ground-level concentrations in excess of applicable ambient air quality standards. Such a dispersion technique is an ICS whether used alone, used with other dispersion techniques, or used as a supplement to continuous emission controls (i.e., used as a supplemental control system) (40 CFR 51.100-91).

Intermittent emission: Those gas streams containing VOC that are generated at intervals during process line operation and includes both planned and emergency releases (*see* emission for more related terms) (40 CFR 60.561-91).

Intermittent filter: A natural or artificial bed of sand or other fine-grained material onto which sewage is intermittently flooded and through which it passes, with time allowed for filtration and the maintenance of aerobic conditions (*see* filter for more related terms) (EPA-83/03).

Intermittent operations: The industrial users does not have a continuous operation (40 CFR 471.02-91).

Intermittent sampling: Sampling successively for a limited period of time throughout an operation or for a predetermined period of time. The duration of sampling periods and of the intervals between are not necessarily regular and are not specified (*see* sampling for more related terms) (EPA-83/06).

Intermittent stream (or ephemeral stream):
(1) A channel that carries water run-off only in times of rainfall and remains as a dry channel during the rest of the year (DOI-70/04).
(2) A stream whose flow is intermittent (DOE-91/04).

Intermittent vapor processing system: A vapor processing system that employs an intermediate vapor holder to accumulate total organic compounds vapors collected from gasoline tank trucks, and treats the accumulated vapors only during automatically controlled cycles (40 CFR 60.501-91).

Intermunicipal agency: An agency established by two or more municipalities with responsibility for planning or administration of solid waste (RCRA1004; 40 CFR 40.115.2-91).

Internal: Any structure or component within a reactor vessel that is not part of the reactor core (DOE-91/04).

Internal boundary layer: In boundary layer meteorology, an interface in the atmospheric boundary layer between flows with different turbulence characteristics. The interface usually starts at a discontinuous change of the surface characteristics. The height of the interface, which defines the internal boundary layer height, grows with the distance down wind of the discontinuity. On both sides of the interface, different dispersion conditions are generally present (NATO-78/10).

Internal combustion engine: An engine in which fuel is combusted in cylinders within the engine and the combustion gas is used to convert combustion heat into work (*see* engine for more related terms).

Internal compaction transfer system: A reciprocating action of a hydraulically powered bulkhead contained within an enclosed trailer packs solid waste against the rear doors (EPA-83).

Internal control: *See* synonym, in plant measure.

Internal drainage (or interior drainage): Drainage in which the waters have no outlet and so do not reach the sea (DOI-70/04).

Internal floating roof: A cover or roof in a fixed-roof tank which rests upon and is supported by the volatile organic liquid being contained and is equipped with a closure seal or seals to close the space between the roof edge and tank shell (40 CFR 52.741-91, *see also* 40 CFR 61.341-91).

Internal measure: *See* synonym, in plant measure.

Internal radiation: The radiation originating from a source within the body as a result of the inhalation, ingestion, or implantation of natural or man-made radionuclides in body tissues (*see* radiation for more related terms) (EPA-88/08a).

Internal redistribution of energy: A dynamic process which occurs during a unimolecular reaction where the chemical structure of a molecule is altered to compensate for a thermal effect (EPA-88/12).

Internal reference material: *See* definition under reference material.

Internal standard spike: An analyte which has the same characteristics as the surrogate, but is added to a sample just prior to analysis. It provides a short term indication of instrument performance, but it may also be an integral part of the analytical method in a non-quality control sense, i.e., to normalize data for quantitation purposes (*see* spike for more related terms) (EPA-84/03).

Internal subunit: A subunit that is covalently linked to at least two other subunits. Internal subunits of polymer molecules are chemically derived from monomer molecules that have formed covalent links between two or more other molecules (40 CFR -704.25-91, *see also* 40 CFR 721.350; 723.250-91).

International shipment: The transportation of hazardous waste into or out of the jurisdiction of the United States (40 CFR 260.10-91).

Intersection: The nonparallel touching or crossing of fibers, with the projection having an aspect ratio of 5:1 or greater (40 CFR 763-App/A-91).

Interspecies dose conversion: The process of extrapolating from animal doses to equivalent human doses (EPA-92/12).

Interstate agency: An agency of two or more States established under an agreement or compact approved by the Congress, or any other agency of two or more States, having substantial powers or duties pertaining to the control of water (40 CFR 35.905-91).

Interstate air pollution control agency:
(1) An air pollution control agency established by two or more States; or
(2) An air pollution control agency of two or more municipalities located in different States (cf. air pollution control agency) (CAA302-42USC7602-91).

Interstate carrier water supply: A source of water for drinking and sanitary use on planes, buses, trains, and ships operating in more than one state. These sources are federally regulated (EPA-94/04).

Interstate commerce: The commerce between any place in a State and any place in another State, or between places in the same State through another State, whether such commerce moves wholly by rail or partly by rail and partly by motor vehicle, express, or water. This definition of interstate commerce for purposes of this regulation is similar to the definition of interstate commerce in section 203(a) of the Interstate Commerce Act (4USC303(a)) (40 CFR 201.1-n-91).

Interstate commerce clause: A clause of the U.S. Constitution which reserves to the federal government the right to regulate the conduct of business across state lines. Under this clause, for example, the U.S. Supreme Court has ruled that states may not inequitably restrict the disposal out-of-state wastes in their jurisdictions (EPA-94/04).

Interstate pipeline: A pipeline or that part of a pipeline that is used in the transportation of hazardous liquids in interstate or foreign commerce (40 CFR 195.2-91).

Interstate waters: Waters that flow across or form part of state or international boundaries, e.g., the Great Lakes, the Mississippi River, or coastal waters (EPA-94/04).

Interstices: The openings or pore spaces in rock; in an aquifer, they are filled with water (DOI-70/04).

Interstitial fibrosis: A progressive formation of fibrous tissue in the interstices in any structure; in the lungs, it reduces aeration of the blood (LBL-76/07-bio).

Interstitial monitoring: The continuous surveillance of the space between the walls of an underground storage tank (*see* monitoring for more related terms) (EPA-94/04).

Interstitial water: The water contained in pore spaces of rocks or soils (*see* water for more related terms).

Intervener: The person who files a motion to be made party under 40 CFR 209.15 or 40 CFR 209.16, and whose motion is approved (40 CFR 85.1807; 209.3-91).

Intranuclear inclusion bodies: Round, oval, or irregularly shaped bodies appearing in the nuclei of cells (LBL-76/07-bio).

Intrastate pipeline: A pipeline or that part of a pipeline to which this part applies that is not an interstate pipeline (40 CFR 195.2-91).

Intrusive: An intruding or perturbing force in a chemical reaction system (EPA-88/12).

Invasiveness: The ability of a microorganism to penetrate the body tissue (EPA-88/09a).

Inventory:
(1) The list of chemical substances manufactured or processed in the United States that EPA compiled and keeps current under section 8(b) of the Act (40 CFR 720.3-91, *see also* 40 CFR 723.250; 747.115; 747.195; 747.200-91).
(2) The TSCA inventory of chemicals produced pursuant to Section 8 (b) of the Toxic Substances Control Act (EPA-94/04).

Inventory form: The Tier I and Tier II emergency and hazardous chemical inventory forms set forth in Subpart D of this Part Material Safety Data Sheet or MSDS means the sheet required to be developed under 40 CFR 900.1200(g) of Title 29 of the Code of Federal Regulations (40 CFR 370.2-91).

Inventory of open dump: The inventory required under section 4005(b) and is defined as the list published by EPA of those disposal facilities which do not meet the criteria (40 CFR 256.06-91).

Inventory system: A method of physically accounting for the quantity of ink, solvent, and solids used at one or more affected facilities during a time period. The system is based on plant purchase or inventory records (40 CFR 60.581-91).

Inversion: A layer of warm air preventing the rise of cooling air and pollutants trapped beneath it. Can cause an air pollution episode (EPA-94/04). Other inversion related terms include:
- Radiation inversion
- Subsidence inversion

Invertebrate: Animals without backbones (LBL-76/07-water).

Inverted siphon: In a sewer design, any dip or sag introduced into a sewer to pass under structures encountered, such as conduits or subways, or under a stream or across a valley.

Investment cost: The capital expenditure required to bring the treatment or control technology into operation (EPA-83/06a).

Investment tax credit: A tax credit that allows businesses to subtract a portion of the cost of qualifying capital purchases from their Federal or State tax liability, thus reducing the net after-tax cost of capital (OTA-89/10). In pollution abatement, reduction in a company's tax by a given percent of the sum invested in pollution abatement equipment and facilities (DOI-70/04).

Invitation for bid: The solicitation for prospective suppliers by a purchaser requesting their competitive price quotations (40 CFR 248.4-91).

Iodide: A chemical containing iodine.

Ion:
(1) An electrically charged atom that can be drawn from waste water during electrodialysis (EPA-94/04).
(2) An electrically charged atom or group of atoms, when a neutral atom or group of atoms loses or gains one or more electrons during chemical reactions by the action of certain forms of radiant energy, etc. The loss of an electron results in a **positive ion** (cation or positive charge), the gain of an electron results in a **negative ion** (anion or negative charge).
(3) *See* atom for more related terms.

Ion concentration (or ion density): Number of ions per unit volume.

Ion density: *See* synonym, ion concentration.

Ion detector: An instrument for measuring ion concentration in a solution.

Ion exchange:
(1) In radiation, the reversible exchange of ions contained in a crystal for different ions in solution, without destroying the crystal structure or disturbing the electrical neutrality (EPA-88/08a).
(2) In waste treatment, ion exchange involves the interchange of ions between an aqueous solution and a solid material (the ion exchanger). After removal of the solution, the exchanger is then exposed to a second aqueous solution of different composition which removes the ions picked up by the exchanger. The process is most frequently carried out by pumping the solutions through one or more fixed beds of exchanger which involves the following four steps:
   (A) **Service (or exhaustion):** The solution containing the ions (polluting ions such as the ions in chromic acid) to be removed is contacted with ion exchanger (R^-). The ions are removed from the feed during the service step by passing the solution downflow the resin column. To recover chrome from chromic acid from plating rinsewater, ion exchange is very suitable for this purpose as follows: $2R^-(OH)^+ + H_2CrO_4 \longrightarrow R_2CrO_4 + 2H_2O$.
   (B) **Backwash:** The reaction bed is washed (generally with water) in a reverse direction to the service cycle in

order to expand and re-settle the resin bed. The flow is adjusted to the amount sufficient to fluidize the resin bed 50 to 100%.

(C) **Regeneration:** Passing a dilute (i.e., 1 to 5 normal) solution of either acid (for cation) or caustic (for anion) down through the reaction bed to convert (regenerate) ion exchanger (R') back to its original form: $R_2CrO_4 + 2NaOH \longrightarrow 2ROH + Na_2CrO_4$.

This equation shows that ROH is in its original form and the chromic acid in the service stage becomes now a chromic salt which can be further processed for chrome recovery.

(D) **Rinse:** This step removes the excess regeneration solution prior to the next service step.

(3) cf. Liquid ion exchange.

Ion exchange chromatography: A chromatographic technique in which the stationary phase materials are ion exchange resins (*see* chromatography for more related terms).

Ion exchange resin: A synthetic resin containing active groups (usually sulfonic, carboxylic, phenol, or substituted amino groups) that gives the resin the ability to combine with or exchange ions with a solution (*see* resin for more related terms) (EPA-83/03).

Ion exchange treatment: A common water-softening method often found on a large scale at water purification plants that remove some organics and radium by adding calcium oxide or calcium hydroxide to increase the ph to a level where the metals will precipitate out (EPA-94/04).

Ion flotation technique: A treatment for electroplating rinse waters (containing chromium and cyanide) in which ions are separated from solutions by flotation (EPA-83/06a).

Ionic bond (or electrovalent bond): A chemical bond between two atoms. The bond is established by the transfer of electrons from one atom to another atom (*see* chemical bond for more related terms).

Ionic exchange: In mining, the replacement of ions on the surface, or sometimes within the lattice, of materials such as clay (EPA-82/05).

Ionic strength: The strength can be expressed as the summation of $m_i c_i^2$. Where m_i is the molality of different ions, and c_i is the charge of each m_i. It is a measure of electrostatic interactions among ions in an electrolyte.

Ionization: The process by which, at the molecular level, atoms or groups of atoms acquire a charge by the loss or gain of one or more electrons (EPA-82/11).

Ionization chamber: A device that measures the intensity of ionizing radiation (EPA-94/04).

Ionization radiation: The energy that breaks a compound into ions. It is an electromagnetic radiation, such as gamma or X-radiation, having wavelengths less than approximately 10 Angstroms (*see* radiation for more related terms) (EPA-88/09a).

Ionizing radiation:

(1) Radiation that can strip electrons from atoms, i.e., alpha, beta, and gamma radiation (EPA-94/04).
(2) Radiation that can displace electrons from atoms or molecules, thereby producing ions (DOE-91/04).
(3) *See* radiation for more related terms.

Ionizing wet scrubber (IWS): One of air pollution control devices. IWS is a device for removing very fine solids from gaseous emissions by attracting them to electrically charged surfaces that are flushed with water (*see* scrubber for more related terms) (CRWI-89/05).

Iridium (Ir): A hard transition metal with a.n. 77; a.w. 192.2; d. 22.5 g/cc; m.p. 2454° C and b.p. 5300° C. The element belongs to group VIII of the periodic table.

Iron (Fe): The product made by the reduction of iron ore. Iron in the steel mill sense is impure and contains up to 4% dissolved carbon along with other impurities (cf. steel) (EPA-74/06a). Iron with a.n. 26; a.w. 55.847; d. 7.86 g/cc; m.p. 1536° C and b.p. 3000° C is a hard transition metal and belongs to group VIII of the periodic table (cf. malleable iron).

Iron and steel: Those byproduct cokemaking operations other than merchant cokemaking operations (40 CFR 420.11-91).

Iron bacteria (or crenothrix): Bacteria that either utilize iron as a source of energy or cause its dissolution or deposition. The former obtain energy by oxidizing ferrous iron to ferric iron, which is precipitated as ferric hydrate; the latter, without oxidizing ferrous iron, alter environmental conditions in such a way as to cause it to be dissolved or deposited (DOI-70/04). Iron bacteria grow well in a water with high iron content.

Iron blast furnace: All blast furnaces except ferromanganese blast furnaces (40 CFR 420.31-b-91).

Iron formation: A sedimentary low grade, iron ore body consisting mainly of chert and fine-grained quartz and ferric oxide segregated in bands or sheets irregularly mingled (cf. taconite) (EPA-82/05).

Iron ore: The raw material from which iron is made. It is primarily iron oxide with impurities such as silica (EPA-74/06a).

Iron oxide: Fe_2O_3; one of the necessary ingredients in the production of cement (ETI-92).

Irony aluminum: A high iron content aluminum alloy recovered from old scrap containing iron. Prepared in the sweating furnace operating at temperatures sufficiently high to melt only the aluminum (EPA-76/12).

Irradiated food: Food subject to brief radioactivity, usually gamma rays, to kill insects, bacteria, and mold, and to permit storage without refrigeration (EPA-94/04).

Irradiation: Exposure to radiation of wavelengths shorter than those of visible light (gamma, x-ray, or ultraviolet), for medical purposes, to sterilize milk or other foodstuffs, or to induce polymerization of monomers or vulcanization of rubber (EPA-94/04).

Irreparable harm: The significant undesirable effects occurring after the date of permit issuance which will not be reversed after cessation or modification of the discharge (40 CFR 125.121-91).

Irreversible effect:
(1) Effect characterized by the inability of the body to partially or fully repair injury caused by a toxic agent (EPA-94/04).
(2) The effect characterized by the inability of the body to partially or fully repair injury caused by a toxic agent (Course 165.6).

Irreversible reaction: A chemical process which can only progress in one direction and is incapable of progressing in a reverse direction.

Irrigation: Applying water or wastewater to land areas to supply the water and nutrient needs of plants (EPA-94/04).

Irrigation efficiency: The amount of water stored in the crop root zone compared to the amount of irrigation water applied (EPA-94/04).

Irrigation return flow: Surface and subsurface water which leaves the field following application of irrigation water (EPA-94/04).

Irritant:
(1) A substance that can cause irritation of the skin, eyes, or respiratory system. Effects may be acute from a single high level exposure, or chronic from repeated low-level exposures to such compounds as chlorine, nitrogen dioxide, and nitric acid (EPA-94/04).
(2) *See* definition under toxicant and effect.

Irrotational: A flow field in which no vorticity is present throughout the whole field (NATO-78/10).

Isentropic process: A thermodynamic process which is reversible and adiabatic (no heat transfer through a system boundary). Its pressure volume relation can be expressed as: PV^k = constant, where: k = Cp/Cv; specific heat ratio; P = pressure; V = volume; Cp = constant pressure specific heat; and Cv = constant volume specific heat. For PV^n = constant, where: n = 0, the process is isobaric; n = 1, the process is isothermal; n = k, the process is isentropic; n = infinite, the process is isometric.

iso-: A prefix indicating equal or same.

ISO standard day conditions: 288 degrees Kelvin, 60 percent relative humidity and 101.3 kilopascals pressure (40 CFR 60.331-91).

Isoabsorptive point: A wavelength at which the absorptivities of two or more substances are equal (LBL-76/07-air).

Isobar: A graphical representation (plot) of the amount of vapor adsorbed versus temperature at a particular (constant) pressure. Also used in absorption studies (EPA-84/09).

Isobestic point: Wavelength at which the % transmission or optical density of a substance is independent of the isomeric form or oxidation state: a point of intersection for all spectral curves of the compound independent of the concentrations of its forms. For each pair of forms there will be one such isobestic point (LBL-76/07-water).

Isodrin: The pesticide 1,4:5,8-Dimethanonaphthalene, 1,2,3,4,10,10-hexacholoro-1,4,4a,5,8,8a-hexahydro-, (1alpha, 4alpha, 4abeta, 5beta, 8 beta, 8abeta)-, CAS Number 465-73-6 (40 CFR 704.102-91, *see also* 40 CFR 721.1150-91).

Isoelectric point:
(1) The pH value of a solution at which the soluble protein becomes insoluble and precipitates out (EPA-75/04), or at which acidic ionization balances basic ionization so that an electrolyte will not migrate in an electrical field (EPA-82/11).
(2) The pH value at which zero charge occurs on a colloidal particle (e.g., bacteria or algae).

Isoelectric precipitation: Precipitation occurs at the isoelectric point.

Isokinetic: A term describing a condition of sampling, in which the flow of gas into the sampling device (at the opening or face of the inlet) has the same flow rate and direction as the ambient atmosphere being sampled (LBL-76/07-air).

Isokinetic sampling: The sampling in which the linear velocity of the gas entering the sampling nozzle is equal to that of the undisturbed gas stream at the sample point (40 CFR 60.2-91).

Isolated system: In thermodynamics, a system which does not interacts with its surroundings (or crosses its boundaries). An isolated system must be a closed system (*see* thermodynamic first law for more related terms).

Isolation: The segregation of waste for separate treatment and/or disposal (EPA-83/06a).

Isomer:
(1) Chemical substances with the same number of characteristic atoms which are electronically arranged differently, producing compounds with distinguishable physical and chemical properties; e.g., the atoms in dichloroethane can be arranged as 1,1-dichloroethane or 1,2-dichloroethane (cf. homologous group) (EPA-88/12).
(2) Two hydrocarbons with the same number of carbon and hydrogen atoms but different structures are called isomers, e.g., there are several different octanes, the C_8H_{18} series, each having 8 carbon atoms and 18 hydrogen atoms, but each has a different structure; these different structures are called the isomers of octane.
(3) *See* hydrocarbon family for more related terms.

Isomeric ratio: The relative amounts of each isomeric chlorinated naphthalene that composes the chemical substance; and for each isomer the relative amounts of each chlorinated naphthalene designated by the position of the chlorine atom(s) on the naphthalene (40 CFR 704.43; 704.45-91).

Isopleth: On a map, a line of constant value of a certain property of quantity, e.g., in air pollution, modeling frequently used with respect to maps on which lines of constant concentration values are shown (NATO-78/10).

Isosbestic point: The wavelength at which the absorptivities of two substances, one of which can be converted into the other, are equal (LBL-76/07-air).

Isostere: A graphical representation of the log (or natural log, ln) of the partial pressure of the solute of adsorbate (pollutant) in the gas phase versus the reciprocal of the absolute temperature (1.0/T) for a fixed quantity of adsorbed vapor (EPA-84/09).

Isotherm: A graphical representation of adsorbent capacity versus the partial pressure of the adsorbate (pollutant) at a particular temperature. Also used in absorption studies (EPA-84/09).

Isotope:
(1) An atom of a chemical element with a specific atomic number and atomic mass. Isotopes of the same element have the same number of protons but different numbers of neutrons and different atomic masses (DOE-91/04).
(2) A variation of an element that has the same atomic number of protons but a different weight because of the number of neutrons. Various isotopes of the same element may have different radioactive behaviors, some are highly unstable.

Isotropic: A condition or process of which the properties are independent of direction (NATO-78/10) (EPA-94/04).

Isotropic soil: The soils that have the same property or properties, e.g., permeability, in all directions (*see* soil for more related terms) (EPA-83).

Isotropic turbulence: The turbulence for which the statistical properties are independent of the orientation of the coordinate axes (*see* turbulence for more related terms) (NATO-78/10).

Issuance of an NSO (Nonferrous Smelter Orders): The final transmittal of the NSO pursuant to 40 CFR 57.107(a) by an issuing agency (other than EPA) to EPA for approval, or the publication of an NSO issued by EPA in the Federal Register (40 CFR 57.103-n-91).

Isotropic tracer: An isotope of an element which is incorporated into the element and is used to follow the change of the element throughout the desired processes.

Jj

Jacket water: The water in the engine jacket space for cooling purposes (*see* water for more related terms).

Jackson turbidimeter: A meter to measure the turbidity in water.

Jar test:
(1) A laboratory procedure that simulates a water treatment plant's coagulation/flocculation units with differing chemical doses, mix speeds, and settling times to estimate the minimum or ideal coagulant dose required to achieve certain water quality goals (EPA-94/04).
(2) A technique used for testing the correct chemical dose in wastewater.

Jaw crusher:
(1) A primary crusher designed to reduce large rocks or ores to sizes capable of being handled by any of the secondary crushers (EPA-82/05).
(2) A primary crusher designed to reduce the size of materials by impact or crushing between a fixed plate and an oscillating plate or between two oscillating plates, forming a tapered jaw (EPA-82/10).
(3) *See* crusher for more related terms.

Jet condenser (or wet condenser): A condenser which makes use of high-velocity jets of water to both condense the vapor and force the non-condensable gases out of the system (*see* condenser for more related terms).

Jig: A (mining) machine in which the feed is stratified in water by means of a pulsating motion and from which the stratified products are separately removed, the pulsating motion being usually obtained by alternate upward and downward currents of the water (EPA-82/05).

Jigging: The separation of the heavy fractions of an ore from the light fractions by means of a jig (EPA-82/05).

Job shop: A facility which owns not more than 50% (annual area basis) of the materials undergoing metal finishing (40 CFR 433.11-91).

Johnstone equation: An equation used to describe collection efficiency in a venturi scrubber (EPA-84/09).

Joint treatment: A treatment, in publicly owned treatment works, of combined municipal wastewaters of domestic origin and wastewaters from other sources (*see* treatment for more related terms) (EPA-74/03d).

Joist: Any of the parallel horizontal beams set from wall to wall to support the floor or ceiling (EPA-88/08).

Joule: An energy unit.

Judicial officer: An officer or employee of the Agency appointed as a Judicial Officer by the Administrator pursuant to this section who shall meet the qualifications and perform functions as follows:
(i) Officer. There may be designated for the purposes of this section one or more Judicial Officers. As work requires, there may be a Judicial Officer designated to act for the purposes of a particular case.
(ii) Qualifications. A Judicial Officer my be a permanent or temporary employee of the Agency who performs other duties for the Agency. Such Judicial Officer shall not be employed by the Office of Enforcement or have any connection with the preparation or presentation of evidence for a hearing held pursuant to this subpart.
(iii) Functions. The Administrator may consult with a Judicial Officer or delegate all or part of his authority to act in a given case under this section to a Judicial Officer: Provided, that this delegation shall not preclude the Judicial Officer from referring any motion or case to the Administrator when the Judicial Officer determines such referral to be appropriate (40 CFR 164.2-I-91).

Jumbo roll: Roll of paper greater than 25" in diameter (EPA-83).

Junction: A region of transition between two different semi-conducting regions in a semiconductor device such as a p-n junction, or between a metal and a semiconductor (EPA-83/03).

Junction box:
(1) A manhole or access point to a wastewater sewer system line (40 CFR 60.691-91).
(2) A protective enclosure into which wires or cables are led and connected to form joints (EPA-83/03).

Junk:
(1) Unprocessed or processed materials suitable for reuse or recycling (SW-108ts).
(2) Old or scrap copper, brass, rope, rags, batteries, gaper,

rubber, junked, dismantled or wrecked automobiles or parts thereof; iron, steel and other old or scrap ferrous or materials which are not held for sale for remelting purposes (EPA-83).

Junk collector: Accepts discarded materials but offers them for sale to processors for reuse or recycling (EPA-83).

Jurisdiction by law: The agency authority to approve, veto, or finance all or part of the proposal (40 CFR 1508.15-91).

Jute: The glossy fibers of two types of east indian plants which are used to make sacking, burlap, and twine. In the paper industry, this largely obsolete term has been used to refer to linerboard made of waste paper - jute linerboard (EPA-83).

Kk

k cu m: 1000 cubic meter(s) (40 CFR 401.11-91).

K factor:
(1) A dimensionless term used to determine the proper flow regime for a settling particle in a gas (EPA-84/09).
(2) The thermal conductivity of a material, expressed as Btu per sq ft per hour in degrees Fahrenheit and inches (cf. thermal conductivity) (SW-108ts).

K_1: *See* definition under uptake rate constant.

K_2: *See* definition under depuration rate constant.

Kaolin: A type of clay minerals.

Karst:
(1) The terrain with characteristics of relief and drainage arising from a high degree of rock solubility in natural waters. The majority of karst occurs in limestones, but karst may also form in dolomite, gypsum, and salt deposits. Features associated with karst terrains typically include irregular topography, sinkholes, vertical shafts, abrupt ridges, caverns, abundant springs, and/or disappearing streams. Karst aquifers are associated with karst terrain (40 CFR 300-App/A-91).
(2) A geologic formation of irregular limestone deposits with sinks, underground streams, and caverns (EPA-94/04).

Karbutilate ($C_{14}H_{21}N_3O_3$): A white solid used as a herbicide.

K_d: *See* definition under adsorption ratio.

Kelvin: *See* definition under absolute temperature.

Kerma (K): The unit used to quantify neutral particle effects in matter with units of (ergs/g). Kerma (K) is the sum of the initial kinetic energies (E) of all charged particles liberated by indirectly ionizing particles in a volume element of the specified material (E in ergs/cm^3) divided by the mass (m) of the matter in that volume element (m in grams/cm^3): K = E/m, in ergs per gram (*see* radiation unit for more related terms) (LBL-76/07-rad).

Ketene (C_2H_2O): A colorless toxic gas used in the manufacturing of cellulose acetate and aspirin.

Keto-: A prefix indicating a keto or a carbonyl radical (C:O) in a compound.

Ketone: A solvent which contains -CO- group and is linked to two hydrocarbon groups.

Kid finish paper: The vellum finish resembling kid leather on soft paper (*see* paper for more related terms) (EPA-83).

Kill tank: A contact chamber in which wastewater or sludge is treated with an inactivation agent to decontaminate waste materials (EPA-88/09a).

Kiln:
(1) A large cylindrical mechanized type of furnace (EPA-83/06a). Kilns have multiple usages, e.g., it is used in the pulp and paper industry to burn lime and calcium carbonate to produce CaO, which is used again with green liquor to form white liquor (cf. incinerator, rotary kiln) (EPA-87/10).
(2) The manufacturing unit in which clinker is formed (ETI-92).

Kiln drying: A method of preparing wood for treatment in which the green stock is dried in a kiln under controlled conditions of temperature and humidity (EPA-74/04).

Kiln scum: A deposit formed during kiln firing, either from soluble salts in the clay or by reaction between the sulphur gases in the kiln atmosphere and minerals in the clay (EPA-83).

Kilogram equivalent weight: 1000 x gram equivalent weight.

Kinematic viscosity: Kinematic viscosity = Viscosity/density, (*see* viscosity for more related terms).

Kinetic energy:
(1) The energy which is possessed by a body as a consequence of its motion (NATO-78/10).
(2) Energy possessed by a moving body of water as a result of its motion (EPA-94/04).
(3) *See* energy for more related terms.

Kinetic rate coefficient: A number that describes the rate at which a water constituent such as a biochemical oxygen demand or dissolved oxygen rises or falls (EPA-94/04).

Kjeldahl process: *See* synonym, total Kjeldahl nitrogen.

Klystron: An evacuated electron-beam tube in which an initial velocity modulation imparted to electrons in the beam results subsequently in density modulation of the beam; used as an amplifier in the microwave region or as an oscillator (EPA-83/03).

Knife hog: A size-reduction device relying primarily on the shearing, cutting or chipping action produced by sharp-edged blades attached to a rotating shaft (mandrel) to shave or chip off pieces of the charged object (*see* size reduction machine for more related terms) (EPA-83).

Knot:
(1) A unit of wind velocity, usually applied in meteorology, One knot is defined as one nautical mile (1853 meter) per hour (NATO-78/10).
(2) An imperfection in paper or lumps in paper stock resulting from incompletely defibered textile materials; the term applies especially to rag paper manufacture (EPA-87/10).
(3) Small undefibered clusters of wood pulp (EPA-87/10).
(4) The basal portion of a branch or limb which has become incorporated in the body of the tree (EPA-87/10).

Known human effects:
(1) Means a commonly recognized human health effect of a particular substances or mixture as described either in:
    (i) Scientific articles or publication abstract in standard reference sources.
    (ii) The firm's product labeling or material safety data sheets (MSDS).
(2) However, an effect is not a "known human effect" if it:
    (i) Was a significantly more severe toxic effect than previously described.
    (ii) Was a manifestation of a toxic effect after a significantly shorter exposure period or lower exposure level than described.
    (iii) Was a manifestation of a toxic effect by an exposure route different from that described (40 CFR 717.3-c-91).

Kodel fiber: A trademark of Eastman for polyester yarns and fibers (*see* fiber for more related terms) (EPA-74/06b).

Kraft: A descriptive term for the (alkaline) sulfate pulping process, the resulting pulp, and paper or paperboard made therefrom (EPA-87/10).

Kraft paper:
(1) A comparatively coarse paper noted for its strength and used primarily as a wrapper or packaging material (OME-88/12).
(2) A paper made predominantly from wood pulp produced by a modified sulfate pulping process. It is a comparatively coarse paper particularly noted for its strength, and in unbleached grades is used primarily as a wrapper or packaging material (EPA-83).
(3) *See* paper for more related terms.

Kraft pulp: *See* synonym, sulfate pulp.

Kraft pulp mill: Any stationary source which produces pulp from wood by cooking (digesting) wood chips in a water solution of sodium hydroxide and sodium sulfide (white liquor) at high temperature and pressure. Regeneration of the cooking chemicals through a recovery process is also considered part of the kraft pulp mill (40 CFR 60.281-91).

Krypton (Kr): A noble gaseous element with a.n. 36; a.w. 83.80; d. 2.6 g/cc; m.p. -157.3° C and b.p. -152° C. The element belongs to group VIIIA of the periodic table.

Kyanize: A process that mercuric chloride is used to treat wood to prevent its decay.

Ll

Label:
(1) (In noise control) means that item, as described in this regulation, which is inscribed on, affixed to or appended to a product, its packaging, or both for the purpose of giving noise reduction effectiveness information appropriate to the product (40 CFR 211.203-91).
(2) (In fuel economy) means a sticker that contains fuel economy information and is affixed to new automobiles in accordance with Subpart D of this part (40 CFR 600.002.85-91).
(3) (In pesticide) means the written, printed, or graphic matter on, or attached to, the pesticide or device or any of its containers or wrappers (FIFRA2-7USC136-91).

Labeling: All labels and all other written, printed, or graphic matter:
(1) Accompanying the pesticide or device at any time; or
(2) To which reference is made on the label or in literature accompanying the pesticide or device, except to current official publications of the environmental Protection Agency, the United States Department of Agricultural and Interior, the Department of Health and Human Services, State experiment stations, State agricultural colleges, and other similar Federal or State institutions or agencies authorized by law to conduct research in the field of pesticides (FIFRA2-7USC136-91).

Laboratory: Any research, analytical, or clinical facility that performs health care related analysis or service. This includes medical, pathological, pharmaceutical, and other research, commercial, or industrial laboratories (40 CFR 259.10-91, *see also* 40 CFR 761.3-91).

Laboratory blank: A blank for evaluating contamination of laboratory measurement (*see* blank for more related terms) (ACS-87/11).

Laboratory matrix spike: A sample created by spiking target analytes into a portion of a sample when it is received in the laboratory. It provides bias information regarding sample preparation and analysis and is the most common type of matrix spike (*see* spike for more related terms) (EPA-84/03).

Laboratory operation: Usually denotes use of fixed equipment capable of analysis only under restricted environmental conditions, this may include mobile laboratories used in the field (*see* instrument use for more related terms) (LBL-76/07-water).

Laboratory sample: A representative portion of the gross sample recieved by the laboratory for further analysis (*see* analytical parameters--laboratory for more related terms) (EPA-83).

Laboratory sample coordinator: That person responsible for the conduct of sample handling and the certification of the testing procedures (40 CFR 763-App/A-91).

Laboratory screening: A treatability study designed to establish the validity of an alternative for treating an operable unit and to identify parame-ters for investigation in later bench- and pilot-scale testing (EPA-89/12a).

Lachrymator: A substance which increases the flow of tears (EPA-87/07a).

Lacquers: Any clear wood finishes formulated with nitrocellulose or synthetic resins to dry by evaporation without chemical reaction, including clear lacquer sanding sealers (40 CFR 52.741-91).

Lacustrine: Found or formed in lakes (DOE-91/04).

Ladle: A vessel used to hold or pour molten metal (EPA-85/10a).

Lag time:
(1) The time interval between a step change in input concentration and the first observable corresponding change in response (40 CFR 53.23-91).
(2) The time interval from a step change in the input concentration at the instrument inlet to the first corresponding change in the instrument output (LBL-76/07-water).
(3) *See* time for more related terms.

Lagoon (stabilization lagoon, stabilization pond or waste stabilization pond):
(1) A shallow pond where sunlight, bacterial action, and oxygen work to purify wastewater; also used for storage of wastewater or spent nuclear fuel rods (EPA-94/04).
(2) Shallow body of water, often separated from the sea by coral reefs or sandbars (EPA-94/04).
(3) A man-made or natural pond (or lake) for holding wastewater where sunlight, bacterial action and oxygen work to remove suspended solids for stabilzation of organic matter by biological oxidation (cf. sewage lagoon) (EPA-83/03).

Other lagoon related terms include:
- Aerated aerobic lagoon (*see* synonym, aerated lagoon)
- Aerated lagoon
- Aerated pond (*see* synonym, aerated lagoon)
- Aerobic lagoon
- Anaerobic lagoon
- Recycle lagoon

Lagrangian frame of reference: A coordinate system in which the properties of a flow are described as a function of time and space where the coordinate system is attached to a fluid particle (cf. eulerian frame for comparison) (NATO-78/10).

Lake: A large fresh or salt water body completely surrounded by land. Other lake related terms include:
- Dry lake
- Dystrophic lake
- Eutrophic lake
- Extinction lake
- Meromictic lake
- Oligotrophic lake
- Publicly owned freshwater lake
- Senescent lake (*see* synonym, extinction lake)
- Storm water lake

Lakewide management plan: A written document which embodies a systematic and comprehensive ecosystem approach to restoring and protecting the beneficial uses of the open waters of each of the Great lakes in accordance with article VI and Annex 2 of the Great lakes Water Quality Agreement (CWA118-33USC1268-91).

Laminar diffusion: An aerosol moving mechanism due to the random motion of fluid particles (EPA-88/09a).

Laminar flame: A flame in which the transport of heat, mass and momentum occurs by molecular conductivity, density, and viscosity gradients. Under these relatively quiescent conditions, diffusion occurs only by the driving forces of the concentration gradients (*see* flame for more related terms).

Laminar flow (Stokes) flow: In a laminar flow, the fluid is constrained to motion in layers (or laminae) by the action of viscosity. The layers of fluid move in parallel paths that remain distinct from one another; any agitation is of a molecular nature only. A laminar flow (compressible flow) occurs when Reynolds Number is usually < 2100 in a tube, pipe, stack, equipment, room, etc., that is characterized by the absence of eddies, i.e., the flow is smooth. For an in compressible flow, Reynolds number less than 2.0 (*see* flow regime for more related terms) (EPA-84/09).

Laminated glass: Two or more pieces of glass held together by an intervening layer or layers of plastic materials. It will crack and break under sufficient impact but the pieces of glass tend to adhere to the plastic and not to fly. If a hole is produced, the edges are likely to be less jagged than would be the case with ordinary glass (*see* safety glass for more related terms) (EPA-83).

Lampworking: Forming glass articles from tubing and cane by heating in a gas flame and shaping (EPA-83).

Land: Any surface or subsurface land that is not part of a disposal site and is not covered by an occupiable building (40 CFR 192.11-91).

Land application: Discharge of wastewater onto the ground for treatment or reuse (*see* land treatment for more related terms) (EPA-94/04).

Land ban: Phasing out of land disposal of most untreated hazardous wastes, as mandated by the 1984 RCRA amendments (EPA-94/04).

Land breeze: The breeze due to a local circulation near a shore line, which blows from the land towards the water. This local circulation is caused by a temperature difference between the water (warm) and the land surface (cold). The land breeze usually blows at night and alternates with the sea breeze (*see* breeze for more related terms) (NATO-78/10).

Land disposal: When used with respect to a specified hazardous waste, land disposal shall be deemed to include, but not be limited to, any placement of such hazardous waste in a landfill, surface impoundment, waste pile, injection well, land treatment facility, salt dome formation, salt bed formation, or under ground mine or cave (RCRA3004.k, *see also* 40 CFR 268.2-91). Land disposal and clean-up of hazardous waste is governed primarily by RCRA (42USC6901-6991i), the Hazardous Materials Transportation Act (49USC1801-1813) and Superfund (42USC9601-9675). Methods of land disposal include:
(1) Co-disposal
(2) Controlled burning dump
(3) Dump
(4) Dumping
(5) Land treatment
(6) Open dump (*see* synonym, dump)
(7) Open dumping (*see* synonym, dump)
(8) Surface dump (*see* synonym, dump)
(9) Water dumping

Land disposal restriction policy: In HSWA, this policy prohibits the land disposal of certain hazardous wastes unless the wastes are treated or they can be demonstrated that there will be "no migration as long as the wastes remain hazardous." Also prohibited from land disposal are bulk or non-containerized liquid hazardous wastes, certain dioxin-containing hazardous wastes, and some solvent wastes (OSWER-87).

Land farming (of waste): A disposal process in which hazardous waste deposited on or in the soil is naturally degraded by microbes (*see* land treatment for more related terms) (EPA-89/12).

Land spreading: *See* synonym, land treatment.

Land subsidence: *See* synonym, land subsidence.

Land treatment (land spreading or sludge farming): The process is a biological treatment in which the large population of microorganisms present in soil is employed to degrade organic wastes. In addition, the large surface area and the sorptive characteristics of the soil particles serve to retain biodegradable organics for microbial assimilation and to attenuate the movement of inorganics such as heavy metals and refractory organics.

Methods of land treatment include:
(1) Land application
(2) Land disposal
(3) Land farming (of waste)
(4) Soil irrigation (*see* synonym, spray irrigation)
(5) Soil percolation (*see* synonym, spray irrigation)
(6) Spray irrigation
(7) Surface irrigation

Land treatment facility: A facility or part of a facility at which hazardous waste is applied onto or incorporated into the soil surface; such facilities are disposal facilities if the waste will remain after closure (40 CFR 260.10-91).

Land use: The activities which are conducted in, or on the shorelands within, the coastal zone, subject to the requirements outlined in section 1456(g) of this title (CZMA304-16USC1453-90).

Land use classification (EPA-90/04):
- Type I1 - urban: Heavy industrial;
- Type I2 - urban: Light/moderate industrial;
- type C1 - urban: Commercial;
- Type R1 - urban: Compact residential (single family);
- Type R2 - urban: Compact residential (multi-family);
- Type R3 - rural: Common residential (Normal easements);
- Type R4 - rural: Estate residential (multi-acre plots);
- Type A1 - rural: Metropolitan natural;
- Type A2 - rural: Agricultural;
- Type A3 - rural: Undeveloped (grasses/woods);
- Type A4 - rural: Undeveloped (heavily wooded); and
- Type A5 - rural: Water surfaces.

Landfill:
(1) A disposal facility or part of a facility where regulated medical waste is placed in or on the land and which is not a land treatment facility, a surface impoundment, or an injection well (40 CFR 259.10-91).
(2) A disposal facility or part of a facility where hazardous waste is placed in or on land and which is not a pile, a land treatment facility, a surface impoundment, an underground injection well, a salt dome formation, a salt bed formation, an underground mine or a cave (40 CFR 260.10-91).
(3) **Sanitary landfills** are disposal sites for non-hazardous solid wastes spread in layers, compacted to the smallest practical volume, and covered by material applied at the end of each operating day (EPA-94/04).
(4) **Secure chemical landfills** are disposal sites for hazardous waste, selected and designed to minimize the chance of release of hazardous substances into the environment (EPA-94/04).

Other landfill related terms include:
- Class I landfill disposal site
- Class II landfill disposal site
- Class III landfill disposal site
- Sanitary landfill
- Secure chemical landfill (*see* synonym, secure landfill)
- Secure landfill
- Sepecially designated landfill

Landfill blade: A U-blade with an extension on top that increases the volume of solid wastes that can be pushed and spread, and protects the operator from any debris thrown out of the solid waste (*see* blade for more related terms) (SW-108ts).

Landfill cell: A discrete volume of a hazardous waste landfill which uses a liner to provide isolation of wastes from adjacent cells or wastes. Examples of landfill cells are trenches and pits (40 CFR 260.10-91).

Landfill gas: The gas that emits from a landfill site. The gas mainly includes CH_4 and CO_2.

Landfill machine: Any machine that is used on a sanitary landfill; generally considered to be dozers, tractors, loaders, compactors, and/or scrapers (EPA-83).

Landscape: The traits, patterns, and structure of a specific geographic area, including its biological composition, its physical environment, and its anthropogenic or social patterns. An area where interacting ecosystems are grouped and repeated in similar form (EPA-94/04).

Landscape characterization: Documentation of the traits and patterns of the essential elements of the landscape (EPA-94/04).

Landscape ecology: The study of the distribution patterns of communities and ecosystems, the ecological processes that affect those patterns, and changes in pattern and process over time (EPA-94/04).

Landscape indicator: A measurement of the landscape, calculated from mapped or remotely sensed data, used to describe spatial patterns of land use and land cover across a geographic area. Landscape indicators may be useful as measures of certain kinds of environmental degradation such as forest fragmentation (EPA-94/04).

Langelier index (LI): An index reflecting the equilibrium pH of a water with respect to calcium and alkalinity; used in stabilizing water to control both corrosion and scale deposition (EPA-94/04).

Lanthanum (La): A rare earth metal with a.n. 57; a.w. 138.91; d. 6.17 g/cc; m.p. 920° C and b.p. 3470° C. The element belongs to group IIIB of the periodic table.

Lantz process: A destructive distillation technique in which the combustible components of waste are converted into combustible gases, charcoal, and a variety of distillates (cf. destructive distillation) (SW-108ts).

Laparotomy: Surgical incision through the abdominal wall (LBL-76/07-bio).

Lapple's method: A method used to calculate cyclone efficiency for a given particle size (EPA-84/09).

Lapse rate: In meteorology, the change of temperature with height (NATO-78/10).

Large: A point source that processes a total annual raw material production of fruits, vegetables, specialties and other products that exceeds 9,080 kkg (10,000 tons) per year (40 CFR 407.61-91, *see* *also* 40 CFR 407.71; 407.81-91).

Large appliance: Any residential and commercial washers, dryers, ranges, refrigerators, freezers, water heaters, dish washers, trash compactors, air conditioners, and other similar products (40 CFR 52.741-91).

Large appliance coating: Any coating applied to the component metal parts (including, but not limited to, doors, cases, lids, panels, and interior support parts) of residential and commercial washers, dryers, ranges, refrigerators, freezers, water heaters, dish washers, trash compactors, air conditioners, and other similar products (40 CFR 52.741-91).

Large appliance coating facility: A facility that includes one or more large appliance coating line(s) (40 CFR 52.741-91).

Large appliance coating line: A coating line in which any protective, decorative, or functional coating is applied onto the surface of large appliances (40 CFR 52.741-91).

Large appliance part: Any organic surface-coated metal lid, door, casing, panel, or other interior or exterior metal part or accessory that is assembled to form a large appliance product. Parts subject to in-use temperatures in excess of 250° F are not included in this definition (40 CFR 60.451-91).

Large appliance product: Any organic surface-coated metal range, oven, microwave oven, refrigerator, freezer, washer, dryer, dishwasher, water heater, or trash compactor manufactured for household, commercial, or recreational use (40 CFR 60.451-91).

Large appliance surface coating line: That portion of a large appliance assembly plant engaged in the application and curing of organic surface coatings on large appliance parts or products (40 CFR 60.451-91).

Large high voltage capacitor: A capacitor which contains 1.36 kg (3 lbs.) or more of dielectric fluid and which operates at 2,000 volts (a.c. or d.c.) or above (*see* capacitor for more related terms) (40 CFR 761.3-91).

Large low voltage capacitor: A capacitor which contains 1.36 kg (3 lbs.) or more of dielectric fluid and which operates below 2,000 volts (a.c. or d.c.) (*see* capacitor for more related terms) (40 CFR 761.3-91).

Large municipal separate storm sewer system: All municipal separate storm sewers that are either:
(1) Located in an incorporated place with a population of 250,000 or more as determined by the latest Decennial Census by the Bureau of Census (appendix F); or
(2) Located in the counties listed in appendix H, except municipal separate storm sewers that are located in the incorporated places, townships or towns within such counties; or
(3) Owned or operated by a municipality other than those described in paragraph (b)(4)(i) or (ii) of this section and that are designated by the Director as part of the large or medium municipal separate storm sewer system due to the interrelationship between the discharges of the designated storm sewer and the discharges from municipal separate storm sewers described under paragraph (b)(4)(i) or (ii) of this section. In making this determination the Director may consider the following factors:
(A) Physical interconnections between the municipal separate storm sewers;
(B) The location of discharges from the designated municipal separate storm sewer relative to discharges from municipal separate storm sewers described in paragraph (b)(4)(i) of this section;
(C) The quantity and nature of pollutants discharged to waters of the United States;
(D) The nature of the receiving waters; and
(E) Other relevant factors; or
(4) The Director may, upon petition, designate as a large municipal separate storm sewer system, municipal separate storm sewers located within the boundaries of a region defined by a storm water management regional authority based on a jurisdictional, watershed, or other appropriate basis that includes one or more of the systems described in paragraph (b)(4)(i), (ii), (iii) of this section (40 CFR 122.26-4-91).

Large MWC plant: An MWC plant with an MWC plant capacity greatr than 225 megagrams per day (250 tons per day) but less than or equal to 1,000 megagrams per day (1,100 tons per day) of MSW (cf. MWC plant) (40 CFR 60.31a-91, *see also* 40 CFR 60.51a-91).

Large quantity generator: Person or facility generating more than 2200 pounds of hazardous waste per month. Such generators produce about 90 percent of the nation's hazardous waste, and are subject to all RCRA requirements (EPA-94/04).

Large release: A release of radioactive material that would result in doses greater than 25 rem to the whole body or 300 rem to the thyroid at 1.6 km from the control perimeter (security fence) of a reactor facility (DOE-91/04).

Large route method: A method in which each crew is assigned a weekly route. The crew works each day without a fixed stopping point or work time, but it completes the route within the working week (*see* waste collection method for more related terms) (SW-108ts).

Large sized plants: The plants which process more than 10,430 kg/day (23,000 lbs/day) raw materials (40 CFR 428.71-91).

Large water system: A water system that serves more than 50,000 persons (EPA-94/04).

Laser beam machining: The use of a highly focused monofrequency collimated beam of light to melt or sublime material at the point of impingement on a workpiece (EPA-83/06a).

Latency: Time from the first exposure of a chemical until the appearance of a toxic effect (EPA-94/04).

Latency of response: The time between the application of a stimulus and the beginning of the response to that stimulus (LBL-76/07-bio).

Latency period: The time between the initial induction of a health effect and the manifestation (or detection) of the health effect; crudely estimated as the time (or some fraction of the time) from first exposure to detection of the effect (EPA-92/12).

Latent heat (Btu/lb or Btu/(lb mole)):
(1) The heat released or absorbed by a change of phase (cf. sensible heat) (EPA-84/09).
(2) Change of enthalpy during a change of state at the same pressure and temperature.
(3) *See* heat for more related terms.
Other latent heat related terms include:
- Condensation: A change of phase from vapor to liquid
- Fusion: A change of phase from solid to liquid
- Solidification: A change of phase from liquid to solid
- Sublimation: A change of phase from solid to vapor
- Vaporization: A change of phase from liquid to vapor

Latent heat of condensation or evaporation (specific latent heat of condensation or evaporation): Thermodynamically, latent heat of condensation or evaporation is the enthalpy difference of a pure condensable fluid between its dry saturated vapor state and its saturated liquid state at the same pressure and temperature.

Lateral sewer:
(1) A sewer which connects the collector sewer to the interceptor sewer (40 CFR 21.2-91).
(2) Pipes that run under city streets and receive the sewage from homes and businesses, as opposed to domestic feeders and main trunk lines (EPA-94/04).
(3) *See* sewer for more related terms.

Laterite: The red residual soil developed in humid, tropical, and subtropical regions of good drainage. It is iron oxides and hydroxides and aluminum, manganese, or nickel (EPA-82/05).

Latex: A suspension of rubber particles in a water solution. Coagulation of the rubber is prevented by protective colloids. A protective colloid is a surface-active substance that prevents a dispersed phase of a suspension from coalescing by forming a thin layer on the surface of each particle (EPA-74/12a).

Latex resin: A resin which is produced by a polymerization process which initiates from free radical catalyst sites and is sold undried (*see* resin for more related terms) (40 CFR 61.61-91).

Latex paint: A paint containing a stable aqueous dispersion of synthetic resin, produced by emulsion polymerization, as the principal constituent of the binder (EPA-79/12b).

Launder:
(1) A trough, channel, or gutter usually of wood, by which water is conveyed, specially in mining, a chute or trough for conveying powdered ore, or for carrying water to or from the crushing apparatus.
(2) A flume (EPA-82/05).

Laundering weir: Sedimentation basin overflow weir (EPA-94/04).

Law: Law related terms include:
- Electrostatic related laws
  - Coulomb's law (or law of electrostatic attraction)
  - Faraday's law of electrolysis
- Gas related laws
  - Beer Lambert law
  - Beer's law
  - Boyle's law
  - Charles' law
  - Dalton's law
  - Darcy's law
  - Ideal gas law
- Fluid related laws
  - Fan law
  - Stokes law
  - Newton's viscosity law
- Photochemistry related laws
  - Grotthus Draper law (the first law of photochemistry)
  - Stark Einstein law (the second law of photochemistry)
- Solution related laws
  - Henry's law
  - Raoult's law
- Statute related laws: *See* Act, environmental law, etc.
- Thermodynamic laws
  - Thermodynamic zeroth law
  - Thermodynamic first law
  - Thermodynamic second law
  - Thermodynamic third law

Law development: A federal law is also known as the **statute**. It is enacted by the United States Congress by three major steps, namely, Bill; Act; and Law:
(1) A bill is introduced in either of the two houses, the Senate or House of Representatives. The bill is debated, read a third time and then voted on. If it passes in one house, it goes to the other house. After passage in one house, the legislation is called an Act.
(2) The Act goes through the same steps in the second house. Often the Act is amended which means the Act is returned to the originating house for study and a vote on the amendments. If the second house passes the Act, it is signed by the Vice President (head of the Senate) and the House Speaker (head of the House of Representative) and then sent to the President for consideration.
(3) The Act becomes law, if it is signed by the President or it is not vetoed within ten days. At this stage, the law is generally referred to as a public law or statute (Coco-85).

Law of electrostatic attraction: *See* synonym, Coulomb's law.

Law related terms:
(1) **Act:** Act is a law (statute) which describes the kind of waste management program that the Congress wants to establish. It also provides the EPA Administrator or his designee with the authority to implement the Act.
(2) **Regulation:** The legal mechanism that spells out how a statute's broad policy directives are to be carried out. Regulations are published in the Federal Register and then codified in the Code of Federal Regulations (CFR) (EPA-86/01).
(3) **Policy (must do):** A document that specifies operating policies that must be followed. It is used by program offices (such as EPA's Office of Solid Waste) to outline the manner

in which pieces of an environmental program are to be carried out (EPA-86/01).

(4) **Guidance (how to):** A document developed by EPA or States outlining a position on a topic or giving instructions on how a procedure must be conducted. It explains how to do something and provides EPA's interpretations on an Act (EPA-86/01)

Lawrencium (Lw): A radioactive metal with a.n. 103 and a.w. 257. The element belongs to group IIIB of the periodic table.

LC50/lethal concentration: Median level concentration, a standard measure of toxicity. It tells how much of a substance is needed to kill half of a group of experimental organisms in a given time. (*see* LD50.) (EPA-94/04)

LD50/lethal dose: The dose of a toxicant that will kill 50 percent of the test organisms within a designated period. The lower the LD 50, the more toxic the compound (EPA-94/04).

Leach ion-exchange flotation process: A mixed method of extraction developed for treatment of copper ores not amenable to direct flotation. The metal is dissolved by leaching, e.g., with sulfuric acid, in the presence of an ion exchange resin. The resin recaptures the dissolved metal and is then recovered in a mineralized froth by the flotation process (EPA-82/05).

Leach precipitation float: A mixed method of chemical reaction plus flotation developed for such copper ores as chrysocolla and the oxidized minerals. The value is dissolved by leaching with acid, and the copper is re-precipitated on finely divided particles of iron, which are then recovered by flotation, yielding an impure concentrate in which metallic copper predominates (EPA-82/05).

Leachate:
(1) Liquid that has percolated through solid waste and has extracted dissolved or suspended materials from it (cf. percolate) (40 CFR 241.101-91, *see also* 40 CFR 257.2; 260.10-91).
(2) Water that collects contaminants as it trickles through wastes, pesticides or fertilizers. Leaching may occur in farming areas, feedlots, and landfills, and may result in hazardous substances entering surface water, ground water, or soil (EPA-94/04).
(3) cf. hazardous waste leachate.

Leachate collection system: A system that gathers leachate and pumps it to the surface for treatment (EPA-94/04).

Leached: Subjected to the action of percolating water or other liquid that removes the soluble parts (LBL-76/07-bio).

Leaching:
(1) The process by which soluble constituents are dissolved and carried down through the soil by a percolating fluid. (*see* leachate) (EPA-89/12).
(2) The process by which a soluble component is "washed" from a solid material typically by percolation of a liquid through the solid material. The solution which contains the soluble components is known as the leachate (ETI-92).
(3) In mining:

(A) The removal in solution of the more soluble minerals by percolating waters.
(B) Extracting a soluble metallic compound from an ore by selectively dissolving it in a suitable solvent and is usually recovered by precipitation of the metal or by other methods (EPA-82/05).

Leaching field:
(1) An area of ground to which wastewater is discharged. Not considered an acceptable treatment method for industrial wastes (EPA-83/06a).
(2) The most commonly used on site disposal technique consisting of tiles which distribute septic tank effluent for subsurface land application. This is a type of soil absorption system (EPA-80/08).

Lead (Pb):
(1) elemental lead or alloys in which the predominant component is lead (40 CFR 60.121-91, *see also* 40 CFR 415.61; 415.671; 420.02-91). Lead is a soft metallic element with a.n. 82; a.w. 207.19; d. 11.4 g/cc; m.p. 327.4° C and b.p. 1725° C. The element belongs to group IVA of the periodic table.
(2) A heavy metal that is hazardous to health if breathed or swallowed. Its use in gasoline, paints, and plumbing compounds has been sharply restricted or eliminated by federal laws and regulations (cf. heavy metals) (EPA-94/04).
(3) Major lead compounds include lead arsenate ($Pb_3(AsO_4)_2$): A toxic white crystal used as an insecticide.

Other lead related terms include:
- Endogenous lead
- Endogenous urinary lead
- Hard lead
- Soft lead

Lead acid battery manufacturing plant: Any plant that produces a storage battery using lead and lead compounds for the plates and sulfuric acid for the electrolyte (40 CFR 60.371-91).

Lead additive: Any substance containing lead or lead compounds (40 CFR 80.2-91).

Lead additive manufacturer: Any person who produces a lead additive or sells a lead additive under his own name (40 CFR 80.2-91).

Lead agency: The Federal agency, State agency, political subdivision, or Indian Tribe that has primary responsibility for planning and implementing a response action under CERCLA (40 CFR 35.6015-91, *see also* 40 CFR 300.5; 1508.16-91).

Lead free:
(1) When used with respect to solders and flux refers to solders and flux containing not more than 0.2 percent lead; and
(2) When used with respect to pipes and pipe fittings refers to pipes and pipe fittings containing not more than 8.0 percent lead (SDWA1417, *see also* 40 CFR 141.43-91).

Lead glass: A glass containing a substantial proportion of lead oxide, PbO (*see* glass for more related terms) (EPA-83).

Lead matte: Any molten solution of copper and other metal sulfides produced by reduction of sinter product from the oxidation of lead sulfide ore concentrates (40 CFR 61.171-91).

Lead mineral: The most important industrial one is galena (PbS), which is usually argentiferous. In the upper parts of deposits the mineral may be altered by oxidation to cerussite (PbCO3) or anglesite (PbSO4). Usually galena occurs in intimate association with sphalerite (ZnS) (EPA-82/05).

Lead oxide: An active material used for manufacture of lead-acid battery plates consisting of a mixture of lead oxides and finely divided elemental lead (EPA-84/08).

Lead oxide manufacturing facility: A facility that produces lead oxide from lead, including product recovery (40 CFR 60.371-91).

Lead recipe: The glass product composition of the following ranges of weight proportions: 50 to 60 percent silicon dioxide, 18 to 35 percent lead oxides, 5 to 20 percent total R_2O (e.g., Na_2M and K_2O), 0 to 8 percent total R_2O_3 (e.g., Al_2O_3), 0 to 15 percent total RO (e.g., CaO, MgO), other than lead oxide, and 5 to 10 percent other oxides (40 CFR 60.291-91).

Lead reclamation facility: The facility that remelts lead scrap and casts it into lead ingots for use in the battery manufacturing process, and which is not a furnace affected under Subpart L of this part (40 CFR 60.371-91).

Lead pigment: Lead compounds (e.g., white lead $2PbCO_3 \bullet Pb(OH)_2$ and lead carbonate $PbCO_3$) used for paint color materials.

Lead poisoning: Poisoning caused by the accumulation of lead over a long period of time. The poisoning may result in brain disease, anemia, etc.

Lead service line: A service line made of lead which connects the water main to the building inlet and any lead fitting connected to it (EPA-94/04).

Leaded gasoline:
(1) Gasoline which is produced with the use of any lead additive or which contains more than 0.05 gram of lead per gallon or more than 0.005 gram of phosphorus per gallon (40 CFR 80.2-91).
(2) Gasoline to which lead (tetraethyllead, $(Pb(C_2H_5)_4)$) has been added to raise the octane level (EPA-89/12).
(3) *See* gasoline for more related terms.

Leaf fiber: A natural fiber, usually sisal, used in twine and rope. A contaminant in paper products (*see* fiber for more related terms) (EPA-83).

Leak: Any of several events that indicate interruption of confinement of vinyl chloride within process equipment. Leaks include events regulated under Subpart V of this part such as:
(1) An instrument reading of 10,000 ppm or greater measured according to Method 21 (*see* Appendix A of 40 CFR 60-91);
(2) A sensor detection of failure of a seal system, failure of a barrier fluid system, or both;
(3) Detectable emissions as indicated by an instrument reading of greater than 500 ppm above background for equipment designated for no detectable emissions measured according to Test Method 21 (*see* appendix A of 40 CFR 60); and
(4) In the case of pump seals regulated under 40 CFR 61.242-2, indications of liquid dripping constituting a leak under 40 CFR 61.242-2. Leaks also include events regulated under 40 CFR 61.65(b)(8)(i) for detection of ambient concentrations in excess of background concentration. A relief valve discharge is not a leak (40 CFR 61.61-91, *see also* 40 CFR 61.301-91).

Leak definition concentration: The local VOC concentration at the surface of a leak source that indicates that a VOC emission (leak) is present. The leak definition is an instrument meter reading based on a reference compound (40 CFR 60-App/A(method 21)-91).

Leak detection system: A system capable of detecting the failure of either the primary or secondary containment structure or the presence of a release of hazardous waste or accumulated liquid in the secondary containment structure. Such a system must employ operational controls (e.g., daily visual inspections for releases into the secondary containment system of aboveground tanks) or consist of an interstitial monitoring device designed to detect continuously and automatically the failure of the primary or secondary containment structure or the presence of a release of hazardous waste into the secondary containment structure (40 CFR 260.10-91).

Leak or leaking: Any instance in which a PCB Article, PCB Container, or PCB Equipment has any PCBs on any portion of its external surface (40 CFR 761.3-91).

Leak tight: That solids or liquids cannot escape or spill out. It also means dust-tight (40 CFR 61.141-91).

Lean combustion: A combustion condition that a combustible mixture contains excess oxidant (oxygen or combustion air) for the complete combustion of the mixture (*see* combustion for more related terms).

Least detectable quantity: The smallest amount of a chemical that will produce a reliable instrument reading greater than zero or background. The terms sensitivity and the least detectable quantity are frequently used interchangeably and often they are numerically equal (cf. instrumental detection limit).

Least square method: A technique of fitting a curve to a set of given points by minimizing the sum of the squares of the deviations of the given points from the curve (NATO-78/10).

Ledge plate: A plate that is adjacent to or overlaps the edge of a stoker (SW-108ts).

Ledger paper:
(1) A type of paper generally used in a broad variety of recordkeeping type applications such as in accounting machines (40 CFR 250.4-91).
(2) Strong, smooth, writing paper used for accounting and bookkeeping records (EPA-83).
(3) *See* paper for more related terms.

Legionella:
(1) A genus of bacteria, some species of which have caused a type of pneumonia called Legionnaires Disease (40 CFR 141.2-91).
(2) A genus of bacteria, some species of which have caused a type of pneumonia called Legionnaires Disease (EPA-94/04).

Legislation: Includes a bill or legislative proposal to Congress developed by or with the significant cooperation and support of a Federal agency, but does not include requests for appropriations. The test for significant cooperation is whether the proposal is in fact predominantly that of the agency rather than another source. Drafting does not by itself constitute significant cooperation. Proposals for legislation include requests for ratification of treaties. Only the agency which has primary responsibility for the subject matter involved will prepare legislative environmental impact statement (40 CFR 1508.17-91).

Lehr (LEER): A long, tunnel-shaped oven for annealing glass by continuous passage (EPA-83).

Lessee: A person holding a leasehold interest in an oil or gas lease on lands beneath navigable waters (as that term is defined in section 2(a) of the Submerged Lands Act (43USC1301(a)) or on submerged lands of the outer Continental Shelf, granted or maintained under applicable state law or the Outer Continental Self Lands Act (43USC1331 wt seq.) (OPA1001-91).

Lethal concentration: The point estimate of the toxicant concentration that would be lethal to a given percentage of the test organisms during a specific period (*see* dose response for more related terms) (EPA-91/03).

Lethal concentration fifty (LC50):
(1) That concentration of material which is lethal to one-half (50%) of the test population of aquatic animals upon continuous exposure for 96 hours or less (40 CFR 116.3; 795.120; 797.1350; 797.1440; 797.1400; 797.1970; 797.2050; 797.1930; 797.1950 & 300-App/A-91).
(2) A median level concentration, a standard measure of toxicity. It tells how much of a substance is needed to kill half of a group of experimental organisms at a specific time of observation (*see* LD50) (EPA-89/12).
(3) A calculated concentration of a substance in air, exposure to which for a specified length of time is expected to cause the death of 50 percent of an entire defined experimental animal population.
(4) *See* dose response for more related terms.

Lethal concentration low (LC LO): The lowest concentration of a substance in air, other than lethal concentration fifty, which has been reported to have caused death in humans or animals (*see* dose response for more related terms).

Lethal dose (LD): A measured quantity administered to test fish or animals. For direct feeding experiments or injections, the median toxic dosage is noted as LD50, or the lethal dose for 50% of the animals. The time of observation may or may not be a criterion for single feedings or injections (*see* dose response for more related terms) (LBL-76/07-water).

Lethal dose fifty (LD50):
(1) The empirically derived dose of the test substance that is expected to result in mortality of 5 percent of a population of birds which is treated with a single oral dose under the conditions of the test (40 CFR 797.2175-91, *see also* 40 CFR 300-App/A-91).
(2) The dose of a toxicant that will kill 50 percent of the test organisms within a designated period of time. The lower the LD 50, the more toxic the compound (EPA-89/12).
(3) A dose of a substance in ppm of medium which is expected to cause the death of 50 percent of an entire defined experimental animal population. The substance refers to a pure chemical, to a chemical containing impurities, or to a mixture of chemicals. It is determined from the exposure to the substance by any route other than inhalation. A group of well known chemicals and their LD(50)s values (concentration mg/Kg) are listed in the following list. The list shows approximate oral LD50 in rats in the unit of mg/kg. The lower LD(50)s are more acutely toxic than those with higher values.
- Sucrose (table sugar) 29,700
- Vitamin A 2,000
- Aspirin 1,000
- Caffeine 192
- DDT 113
- Sodium cyanide 6.4

(4) *See* dose response for more related terms.

Lethal dose low (LD LO):
(1) The lowest dose, other than lethal dose fifty, of a substance introduced by any route, other than inhalation, which has been reported to have caused death in humans or animals (Course 165.5).
(2) The lowest dose at which death occurred (EPA-92/12).
(3) *See* dose response for more related terms.

Lethal dose zero (LD 0): The highest concentration of a toxic substance at which none of the test organisms die (*see* dose response for more related terms) (EPA-89/12).

Lethal mutation: A change in the genome which, when expressed, causes death to the carrier (40 CFR 798.5275-91).

Letterpress ink: An ink used for typographic (raised type) printing which is a viscous, tacky ink which cures by oxidation (*see* ink for more related terms) (EPA-79/12a).

Levee: A bank of material, usually earth, constructed to form a barrier. Also called a dike (EPA-83).

Level indicator: One of solid or sludge flow rate meters. This includes methods based upon mechanical, ultrasonic, nuclear, and radio frequency principles of operation. Nearly all tank level indicators will perform better with somewhat uniform (free-flowing) particles. This will aid in distributing the level of material evenly within the vessel, allowing for more accuracy in whatever monitoring system is used. Typically, these methods can monitor tank levels to within +/- 1% (*see* flow rate meter for more related terms) (EPA-89/06).

Level of concern (LOC): The concentration in air of an extremely hazardous substance above which there may be serious immediate health effects to anyone exposed to it for short periods of time (EPA-94/04).

Level of detection: *See* synonym, quantifiable level.

Level of protection: Equipment to protect the body against contact with known or anticipated toxic chemicals. *see* action level for four levels of protection (Course 165.5).

Level of quantitation or LOQ: The lowest concentration at which HDDs/HDFs can be reproducibly measured in a specific chemical substance within specified confidence limits, as described in this part (40 CFR 766.3-91).

Lewis acid: A chemical that can accept a pair of electrons from a base (e.g., SO_3) (*see* acid for more related terms).

Lewis base: A chemical that can donate a pair of electrons (e.g., OH) (*see* base for more related terms).

Liability: The probable future sacrifices of economic benefits arising from present obligations to transfer assets or provide services to other entities in the future as a result of past transactions or events (cf. current liability) (40 CFR 144.61-91, *see also* 40 CFR 264.141; 265.141-91).

Liable or liability: Shall be construed to be the standard of liability which obtains under section 311 of the Federal Water Pollution Control Act (33USC1321) (OPA1001, *see also* SF101-42USC9601-91).

License: For nuclear reactor operation, *see* 10 CFR 20.3; 30.4; 40.4; 70.4-91.

License or permit: Any license or permit granted by an agency of the Federal Government to conduct any activity which may result in any discharge into the navigable waters of the United States (40 CFR 121.1-91).

Licensed material: The ource material, special nuclear material, or by-product material received, possessed, used, or transferred under a general or specific License issued by the Commission pursuant to the regulations in this chapter (10 CFR 20.3-91).

Licensed site: The area contained within the boundary of a location under the control of persons generating or storing uranium byproduct materials under a license issued pursuant to section 84 of the Act. For purposes of this subpart, "licensed site" is equivalent to "regulated unit" in Subpart F of Part 264 of this chapter (*see* dump for more related terms) (40 CFR 192.31-91).

Licensing or permitting agency: Any agency of the Federal Government to which application is made for a license or permit (40 CFR 121.1-91).

Lidar: Acronym for light detection and ranging (40 CFR 60-App/A(alt. method 1)-91).

Lidar range: The range or distance from the lidar to a point of interest along the lidar line-of-sight (40 CFR 60-App/A(alt. method 1)-91).

Life cycle:
(1) The phases, changes or stages an organism passes through during its lifetime (from the fertilized egg to death of the mature organism) (EPA-74/11).
(2) The series of stages in the form and mode of life of an organism; i.e., the stages between successive recurrences of a certain primary stage such as the spore, fertilized egg, seed, or resting cell (LBL-76/07-water).

Life cycle assessment: The concept of life cycle assessment is to evaluate the environmental effects associated with any given activity from the initial gathering of raw material from the earth until the point at which all residuals are returned to as the earth. Major concepts of life cycle assessment include:
(1) Life cycle assessment is a tool to evaluate the environmental consequences of a product or activity holistically, across its entire life.
(2) There is a trend in many countries toward more environmentally benign products and processes.
(3) A complete life cycle assessment consists of three complementary components: Inventory, Impact, and Improvement Analysis.
(4) Life cycle inventories can be used both internally to an organization and externally, with external applicators requiring a higher standard of accountability.
(5) Life cycle inventory analyses can be used in process analyses, material selection, product evaluation, product comparison, and policy making (EPA-93/02, p1).

Life of the unit, firm power contractual arrangement: A unit participation power sales agreement under which a utility or industrial customer reserves, or is entitled to receive, a specified amount or percentage of nameplate capacity and associated energy generated by any specified generating unit and pays its proportional amount of such unit's total costs, pursuant to a contract:
(1) For the life of the unit;
(2) For a cumulative term of no less than 30 years, including contracts that permit an election for early termination; or
(3) For a period equal to or greater than 25 years or 70 percent of the economic useful life of the unit determined as of the time the unit was built, with option rights to purchase or release some portion of the nameplate capacity and associated energy generated by the unit at the end of the period (40 CFR 72.2-91).

Lifetime exposure: Total amount of exposure to a substance that a human would receive in a lifetime (usually assumed to be 70 years) (EPA-94/04).

Lift: In a sanitary landfill, a compacted layer of solid waste and the top layer of cover material (cf. cell) (EPA-94/04).

Lift and carry container: A large container that can be lifted onto a service vehicle and transported to a disposal site for emptying; also called a detachable container or drop-off box (*see* container for more related terms) (SW-108ts).

Lift depth: Vertical thickness of a compacted volume of solid wastes and the cover material immediately above it in a sanitary landfill (EPA-83).

Lifting station: *See* synonym, pumping station (EPA-94/04).

Ligand:
(1) The molecules attached to a central atom by coordinate covalent bonds (EPA-83/06a).
(2) A molecule, ion, or atom that is attached to the central atom of a coordination compound, a chelate, or other complex (LBL-76/07-bio).

Light duty truck (LDT): Any motor vehicle rated at 3,850 kg gross vehicle weight or less, designed mainly to transport property (*see* truck for more related terms) (40 CFR 52.741-91, *see also* 40 CFR 60.391; 86.082.2-91).

Light duty truck 1: Any light light duty truck up through 3750 lbs loaded vehicle weight (*see* truck for more related terms) (40 CFR 86.094.2-91).

Light duty truck 2: Any light light duty truck greater than 3750 lbs loaded vehicle weight (*see* truck for more related terms) (40 CFR 86.094.2-91).

Light duty truck 3: Any heavy light duty truck up through 5750 lbs adjusted loaded vehicle weight (*see* truck for more related terms) (40 CFR 86.094.2-91).

Light duty truck 4: Any heavy light duty truck greater than 5750 lbs adjusted loaded vehicle weight (*see* truck for more related terms) (40 CFR 86.094.2-91).

Light duty vehicle: A gasoline powered motor vehicle rated at 6,000 lb. gross vehicle weight (GVW) or less (*see* vehicle for more related terms) (40 CFR 51.731-91, *see also* 40 CFR 52.1161; 52.2039; 52.2485, 52.2490; 52.2491; 86.082.2-91).

Light duty vehicle and engine: The new light duty motor vehicles and new light duty motor vehicle engines, as determined under regulations of the Administrator (*see* engine for more related terms) (CAA202-42USC7521-91).

Light light duty truck: Any light-duty truck rated up through 6000 lbs GVW (40 CFR 86.094.2-91).

Light liquid: The VOM in the liquid state which is not defined as heavy liquid (*see* liquid for more related terms) (40 CFR 52.741-91).

Light off time (LOT): The time required for a catalytic converter (at ambient temperature 68-86° F) to warm up sufficiently to convert 50% of the incoming HC and CO to CO_2 and H_2O (40 CFR 85.2122(a)(15)(ii)(C)-91)

Light oil condenser: Any unit in the light oil recovery operation that functions to condense benzene containing vapors (40 CFR 61.131-91).

Light oil decanter: Any vessel, tank, or other type of device in the light oil recovery operation that functions to separate light oil from water downstream of the light oil condenser. A light oil decanter also may be known as a light oil separator (40 CFR 61.131-91).

Light oil storage tank: Any tank, reservoir, or container used to collect or store crude or refined light oil (40 CFR 61.131-91).

Light oil sump: Any tank, pit, enclosure, or slop tank in light oil recovery operations that functions as a wastewater separation device for hydrocarbon liquids on the surface of the water (40 CFR 61.131-91).

Light water: The common form of water (a molecule with two hydrogen atoms and one oxygen atom) in which the hydrogen atom consist largely or completely of the normal hydrogen isotope (one proton and one neutron) (*see* water for more related terms) (DOE-91/04).

Light water reactor: A nuclear reactor in which circulating light water is used to cool the reactor core and to moderate (reduce the energy of) the neutrons created in the core by the fission reactions (DOE-91/04).

Lignin: An organic polymer that cements celluloses together to form plant cell walls.

Lignite coal:
(1) Consolidated lignitic coal having less than 8,300 British thermal units per pound, moist and mineral matter free (SMCRA701-30USC1291-90).
(2) **Lignite:** The coal that is classified as lignite A or B according to the American Society of Testing and Materials' (ASTM) Standard Specification for Classification of Coals by Rank D388-77 (incorporated by reference-*see* 40 CFR 60.17) (40 CFR 60.41a-91, *see also* 40 CFR 60.41b-91).
(3) *See* coal for more related terms.

Lime:
(1) Calcium oxide (CaO), or a mixture of calcium oxide (CaO) and magnesium oxide (MgO). All classes of quicklime and hydrated lime both calcitic (high calcium) and dolomitic (EPA-83). In general, lime is calcium oxide (CaO), but often it means calcium hydroxide ($Ca(OH)_2$).
(2) The common name for calcium oxide (CaO); formed in the calcining zone of the cement kiln (ETI-92).
(3) Quicklime (calcium hydroxide ($Ca(OH)_2$)) obtained by calcining limestone or other forms of calcium carbonate. Loosely used for hydrated or ground calcium carbonate in agricultural lime and for calcium in such expressions as carbonate of lime, chloride of lime, and lime feldspar (EPA-82/05).
(4) cf. settle lime.

Lime based process: Waste fixation techniques based on lime products usually depend on the reaction of lime with fine-grained siliceous (pozzolNic) material and water to produce a concrete-like solid. The most common pozzolanic materials used in waste treatment are fly ash, ground blast-furnace slag, and cement-kiln dust (*see* solidification and stabilization for more related terms).

Lime kiln: A unit used to calcine lime mud, which consists primarily of calcium carbonate, into quicklime, which is calcium oxide (40 CFR 60.281-91).

Lime leg: In the steel industry, the fixed leg of the ammonia still to which milk of lime is added to decompose ammonium salts; the liberated ammonia is steam distilled and returned to the gas stream (cf. free leg) (EPA-74/06a).

Lime manufacturing plant: Any plant which uses a rotary lime kiln to produce lime product from limestone by calcination (40 CFR 60.341-91).

Lime mud: In the wood industry, a solid residue generated from the white liquor clarifier in the lime recovery/white liquor preparation process (EPA-87/10).

Lime mud slurry: The product resulting from the addition of water to lime cake to facilitate pumping of the material for further handling and/or disposal (EPA-74/01a).

Lime pond: An earthen diked area to which the lime mud slurry or waste filter cake is transported and held (EPA-74/01a).

Lime product: The product of the calcination process including, but not limited to, calcitic lime, dolomitic lime, and dead-burned dolomite (40 CFR 60.341-91).

Lime slurry: A form of calcium hydroxide in aqueous suspension that contains considerable free water (EPA-82/05).

Lime softening: A water softening method in which lime is used to react with calcium bicarbonate to form calcium carbonate for precipitation.

Limestone: Either calcitic limestone ($CaCO_3$) or dolomitic limestone ($CaCO_3$, $MgCO_3$) (EPA-83).

Limestone coal pellets as fuel: One of SO_2 emission reduction techniques (see sulfur oxide emission control for control structure). Limestone coal pellets as fuel is an another way to reduce SO_2 emissions from combustion processes by burning pellets made of limestone and coal. Pellets are made by pulverizing coal and limestone and adding a binder forming small, cylinder shaped pellets which are about half the size of a charcoal briquette. These consist of approximately two-thirds coal, and one-third limestone. As the pellet burns, the calcium in the limestone absorbs the SO_2 generated by coal combustion and forms calcium sulfate. Calcium sulfate emissions are subsequently collected in a baghouse or electrostatic precipitator (EPA-81/12, p8-5).

Limestone scrubbing: Use of a limestone and water solution to remove gaseous stack-pipe sulfur before it reaches the atmosphere (EPA-94/04).

Liming: An operation in the beam-house where a lime solution comes in contact with the hide. Liming in conjunction with the use of sharpeners such as sodium sulfhydrate is used to either chemically burn hair from the hide or to loosen it for easier mechanical removal. Hair pulping normally utilizes higher chemical concentrations (EPA-82/11).

Limit of detection (LOD): The smallest concentration-amount of analyte that can be reliably reported as found-detected in a material-sample (see detection limit for more related terms) (ACS-87/11).

Limitation: See synonym, standard.

Limited degradation: An environmental policy permitting some degradation of natural systems but terminating at a level well be neath an established health standard (EPA-94/04).

Limited evidence: According to the U.S. EPA's Guidelines for Carcinogen Risk Assessment, limited evidence is a collection of facts and accepted scientific inferences which suggests that the agent may be causing an effect, but this suggestion is not strong enough to be considered established fact (EPA-92/12).

Limited water-soluble substance: The chemicals which are soluble in water at less than 1,000 mg/L (cf. readily water-soluble substances) (40 CFR 797.1060-91).

Limiting chemical: The chemicals that are the last to be removed (or treated) from a medium by a given Technology. In theory, the cumulative residual risk for a medium may approximately equal the risk associated with he limiting chemicals (EPA-91/12).

Limiting factor: A condition whose absence or excessive concentration, is incompatible with the needs or tolerance of a species or population and which may have a negative influence on their ability to thrive or survive (EPA-94/04).

Limiting orifice: A device that limits flow by constriction to a relatively small area. A constant flow can be obtained over a wide range of upstream pressures (EPA-83/03).

Limiting permissible concentration (LPC) of the liquid phase of a material:
(1) That concentration of a constituent which, after allowance for initial mixing as provided in 40 CFR 227.29, does not exceed applicable marine water quality criteria; or, when there are no applicable marine water quality criteria,
(2) That concentration of waste or dredged material in the receiving water which, after allowance for initial mixing, as specified in 40 CFR 227.29, will not exceed a toxicity threshold defined as 0.01 of a concentration shown to be acutely toxic to appropriate sensitive marine organisms in a bioassay carried out in accordance with approved EPA procedures.
(3) When there is reasonable scientific evidence on a specific waste material to justify the use of an application factor other than 0.01 as specified in paragraph (a)(2) of this section, such alternative application factor shall be used in calculating the LPC (40 CFR 227.27-a-91).

Limiting permissible concentration of the suspended particulate and solid phases of a material: That concentration which will not cause unreasonable acute or chronic toxicity or other sublethal adverse effects based on bioassay results using appropriate sensitive marine organisms in the case of the suspended particulate phase, or appropriate sensitive benthic marine organisms in the case of the solid phase; and which will not cause accumulation of toxic materials in the human food chain. These bioassays are to be conducted in accordance with procedures approved by EPA, or, in the case of dredged material, approved by EPA and the Corps of Engineers (40 CFR 227.27-91).

Limnetic zone: The open-water region of a lake. This region supports plankton and fish as the principal plants and animals (LBL-76/07-water).

Limnology: The study of the physical, chemical, hydrological, and biological aspects of fresh water bodies (EPA-94/04).

Lindane: A pesticide that causes adverse health effects in domestic water supplies and is toxic to freshwater fish and aquatic life (EPA-94/04).

Linde process: An alternative technology to produce high-purity oxygen by compressing air and then purifying and separating it through a series of vessels containing granular adsorbent (molecular sieve) (EPA-77/07).

Line losses: Kilowatt hours and kilowatts lost in transmission and distribution lines under specified conditions (*see* electric loss for more related terms) (EPA-83).

Line section: A continuous run of pipe between adjacent pressure pump stations, between a pressure pump stations, between a pressure pump station and terminal or breakout tanks, between a pressure pump station and a block valve, or between adjacent block valves (40 CFR 195.2-91).

Line source: A one-dimensional source. An example of a line source is the particular emissions from a dirt road (*see* source for more related terms) (EPA-88/09).

Linear alkylate sulfonate (LAS): A surface-active compound in synthetic detergents that decomposes readily by bacterial action where oxygen is present (DOI-70/04).

Linear dynamic range: The concentration range over which the analytical curve remains linear (40 CFR 136-App/C-91).

Linear regression: A method to fit a line through a set of points such that the sum of squared vertical deviations of the point values from the fitted line is a minimum, i.e., no other line, no matter how it is computed, will have a smaller sum of squared distances between the actual and predicted values of the dependent variable. Linear regression can be expressed as $y = a + bx$ (EPA-79/12c).

Linearity:
(1) Expresses the degree to which a plot of instrument response versus known pollutant concentration falls on a straight line. A quantitative measure of linearity may be obtained by performing a regression analysis on several calibration points (LBL-76/07-bio).
(2) The maximum deviation between an actual instrument reading and the reading predicted by drawing a straight line between the upper and lower calibration points (LBL-76/07-water).

Linearized multistage procedure: The modified form of the multistage mode (*see* Multistage Model). The constant q1 is forced to be positive (>0) in the estimation algorithm and is also the slope of the dose-response curve at low doses. The upper confidence limit of q1 (called q1*) is called the slope factor (EPA-92/12).

Liner (or lining):
(1) a continuous layer of natural or man-made materials, beneath or on the sides of a surface impoundment, landfill, or landfill cell, which restricts the downward or lateral escape of hazardous waste, hazardous waste constituents, or leachate (40 CFR 260.10-91).
(2) (A) A relatively impermeable barrier designed to keep leachate inside a landfill. Liner materials include plastic and dense clay (EPA-94/04).
 (B) An insert or sleeve for sewer pipes to prevent leakage or infiltration (EPA-94/04).
(3) The material used on the inside of a furnace wall or stack (chimney); usually of high-grade refractory tiles or bricks or a plastic refractory material. It may have insulating, abrasion, corrosion, water resistant, and/or temperature tolerant properties (EPA-83).

Other liner related terms include:
- Flexible membrane
- Geomembrane
- Geotextile
- Inner liner
- Missile liner
- Sanitary landfill liner
- Soil liner

Lipid: A large group of organic compounds which are insoluble in water but are soluble in organic solvents.

Lipid solubility: The maximum concentration of a chemical that will dissolve in fatty substances. Lipid soluble substances are insoluble in water. They will very selectively disperse through the environment via uptake in living tissue (EPA-94/04).

Lipid solubility: The maximum concentration of a chemical that will dissolve in fatty substances; lipid soluble substances are insoluble in water. If a substance is lipid soluble it will very selectively disperse through the environment via living tissue (EPA-89/12).

Lipophilic: A high affinity for lipids (fats) (EPA-91/03).

Lipoprotein: A weak combination of lipid and protein (EPA-88/09a).

Liquefaction: Changing a solid into a liquid (cf. gasification) (EPA-94/04).

Liquefied petroleum gas (LP gas or LPG): A hydrocarbon mixture of propane, butane, isobutane, propylene, or butylene. The most common commercial products are propane, butane or some mixture of the two (Markes-67).

Liquid: A pumpable material of naturally occurring or man-made origin possessing a relatively fixed volume and a solids content of less than 10 percent (EPA-89/12a). Other liquid related terms include:
- Heavy liquid
- Light liquid

- Petroleum liquid
- Saturated liquid
- Saturated liquid state

Liquid chromatography: A chromatographic technique in which the stationary phase materials are liquids (*see* chromatography for more related terms).

Liquid dripping: Any visible leakage from the seal including spraying, misting, clouding, and ice formation (40 CFR 60.481-91).

Liquid filter: A filter used to remove solids from a liquid stream (*see* filter for more related terms).

Liquid flow rate meter: *See* definition under flow rate meter.

Liquid/gas method: Either of two methods for determining capture which require both gas phase and liquid phase measurements and analysis. The first method requires construction of a TTE. The second method uses the building or room which houses the facility as an enclosure. The second method requires that all other VOM sources within the room be shut down while the test is performed, but all fans and blowers within the room must be operated according to normal procedures (40 CFR 52.741-91).

Liquid injection incinerator:
(1) As the name implies, liquid injection incinerators are applicable almost exclusively for pumpable liquid waste. These incinerators are usually simple, refractory-lined cylinders (either horizontally or vertically aligned) equipped with one or more waste burners. Liquid wastes are injected through the burner(s), atomized to fine droplets, and burned in suspension. Burners as well as separate waste injection nozzles may be oriented for axial, radial, or tangential firing. Improved utilization of combustion space and higher heat release rates, however, can be achieved with the utilization of swirl or vortex burners or designs involving tangential entry (*see* incinerator for more related terms) (Oppelt-87/05).
(2) Commonly used system that relies on high pressure to prepare liquid wastes for incineration breaking them up into tiny droplets to allow easier combustion (EPA-94/04).

Liquid ion exchange (LIE): The basic principle underlying the liquid ion exchange (LIE) reactions is the concept of distribution or partition between phases. When two immiscible phases are placed in contact, a substance that is soluble in both will be partitioned between those phases in a definite proportion. If the substance is much more soluble in one phase than the other, then the partition will favor the first phase. In the liquid ion exchange process, a water-soluble, ionic species is caused to become more soluble in an organic solvent (by salt formation, complexation etc.) thus promoting the partition or extraction of that species into the solvent. LIE involves:
(1) **Loading:** This is to transfer a species of interest from the aqueous phase into a miscible organic phase. The extracted aqueous phase is then termed the raffinate.
(2) **Stripping:** This is to transfer the species of interest from the loaded organic phase into a second aqueous phase, whose composition is such that the species of interest will now transfer to this new aqueous phase. The second aqueous is often referred as the stripping liquor.

Thus the LIE produces two aqueous streams, the raffinate or cleaned stream and the loaded (or pregnant) stripping liquor. The organic phase is virtually always run in a closed loop (cf. ion exchange).

Liquid liquid extraction (or solvent extraction): Liquid liquid extraction is the removal of a solute from another liquid by mixing that combination with a solvent preferential to the substance to be removed (EPA-87/10a). The process involves:
(1) **Extraction:** Feed and solvent, two immiscible liquids, are mixed to allow the solute to transfer from the feed to the solvent. The mixing can be accomplished either by force or by a countercurrent flow caused by density differences.
(2) **Solute removal:** This process can be via a second solvent extraction step, distillation, or some other methods.
(3) **Solvent recovery:** This can be accomplished by stripping, distillation, adsorption or other methods.

Liquid metal reactor: A nuclear reactor in which liquid metal, usually sodium or sodium potassium, is used to cool the reactor core (DOE-91/04).

Liquid mounted seal: A foam or liquid-filled primary seal mounted in contact with the liquid between the tank wall and the floating roof continuously around the circumference of the tank (40 CFR 60.111a-91, *see also* 40 CFR 61.341-91).

Liquid nitriding: A process of case hardening a metal in a molten cyanide bath (EPA-83/06a).

Liquid phase of a material: Subject to the exclusions of paragraph (b) of this section, is the supernatant remaining after one hour undisturbed settling, after centrifugation and filtration through a 0.45 micron filter. The suspended particulate phase is the supernatant as obtained above prior to centrifugation and filtration. The solid phase includes all material settling to the bottom in one hour. Settling shall be conducted according to procedures approved by EPA (40 CFR 227.32-91).

Liquid phase process: A polymerization process in which the polymerization reaction is carried out in the liquid phase; i.e., the monomer(s) and any catalyst are dissolved, or suspended in a liquid solvent (40 CFR 60.561-91).

Liquid phase refining: A metal with an impurity is refined by heating the metal to the point of melting of the low temperature metal. It is separated by sweating out (EPA-83/06a).

Liquid phase slurry process: A liquid phase polymerization process in which the monomer(s) are in solution (completely dissolved) in a liquid solvent, but the polymer is in the form of solid particles suspended in the liquid reaction mixture during the polymerization reaction; sometimes called a particle form process (40 CFR 60.561-91).

Liquid phase solution process: A liquid phase polymerization process in which both the monomer(s) and polymer are in solution (completely dissolved) in the liquid reaction mixture (40 CFR 60.561-91).

Liquid saturation line: The curve line which separates the liquid region from the liquid-vapor region (Wark-p56).

Liquid service: That the equipment or component contains process fluid that is in a liquid state at operating conditions (40 CFR 52.741-91).

Liquid to gas ratio:
(1) In wet scrubbers (absorbers), the ratio of scrubbing liquid (in gallons per minute) to the inlet gas flow rate (in acfm) (EPA-89/03b).
(2) The amount of liquid injected into a wet scrubber per given volume of exhaust flow (with units of gal/1000 acf or gpm/1000 acfm) (EPA-84/03b-p2.1).
(3) In flue gas desulfurization systems, the ratio of scrubber slurry to gas flow (L/G ratio). For a given set of system variables, a minimum L/G ratio is required to achieve the desired SO_2 absorption, based on the solubility of SO_2 in the liquid. High L/G ratios require more piping and structural design considerations, resulting in higher costs (EPA-84/03b-p8.3).

Liquid trap: The sumps, well cellars, and other traps used in association with oil and gas production, gathering, and extraction operations (including gas production plants), for the purpose of collecting oil, water, and other liquids. These liquid traps may temporarily collect liquids for subsequent disposition or reinjection into a production or pipeline stream, or may collect and separate liquids from a gas stream (40 CFR 280.12-91).

Liquid viscosity: In a liquid, transfer of momentum between strata having different velocities is small, compared to the cohesive forces between the molecules. Hence, shear stress is predominantly the result of intermolecular cohesion. Because forces of cohesion decrease with an increase in temperature, the shear stress decreases with an increase in temperature. Therefore, liquid viscosity decreases when the temperature increases (*see* viscosity for more related terms) (EPA-81/12, p2-6).

Liquidus temperature: The maximum temperature at which equilibrium exists between the molten glass and its primary crystalline phase (*see* temperature for more related terms) (EPA-83).

Liquor: A solution of dissolved substance in a liquid (as opposed to a slurry, in which the materials are insoluble) (EPA-89/03b). Other liquor related terms include:
- Ammonia liquor
- Black liquor
- Cooking liquor
- Cotton kier liquor
- Flushing liquor
- Flushing liquor circulation tank
- Green liquor
- Malt liquor
- Mixed liquor
- Mother liquor
- Pickle liquor
- Pregnant liquor
- Press liquor
- Scrubber liquor
- Spent cooking liquor
- Sulfite cooking liquor
- White liquor

List: Shorthand term for EPA list of violating facilities or firms debarred from obtaining government contracts because they violated certain sections of the Clean Air or Clean Water Acts. The list is maintained by The Office of Enforcement and Compliance Monitoring (EPA-94/04).

List of violating facilities: A list of facilities which are ineligible for any agency contract, grant or loan (40 CFR 15.4-91).

List official: An EPA official designated by the Assistant Administrator to maintain the List of Violating Facilities (40 CFR 15.4-91).

Listed mixture: Any mixture listed in 40 CFR 716.120 (40 CFR 716.3-91).

Listed specification: A specification listed in section I of Appendix B of this part (*see* specification for more related terms) (40 CFR 192.3-91).

Listed waste:
(1) Wastes listed as hazardous under RCRA but which have not been subjected to the Toxic Characteristics Listing Process because the dangers they present are considered self-evident (EPA-94/04).
(2) In RCRA, hazardous wastes that have been placed on one of three lists developed by EPA: Non-specific source wastes; Specific source wastes; Commercial chemical products. These lists were developed by examining different types of waste and chemical products to *see* if they exhibit one of the four characteristics, meet the statutory definition of hazardous waste, are acutely toxic or acutely hazardous, or are otherwise toxic (EPA-86/01).
(3) *See* waste for more related terms.

Listing proceeding: An informal hearing, conducted by the Case Examiner, held to determine whether a facility should be placed on the List of Violating Facilities (40 CFR 15.4-91).

Listwise and pairwise deletion of missing data:
(1) Listwise deletion of missing data is a procedure to handle the absence of some of the information in a regression analysis by eliminating any data point with absent information from all calculations.
(2) Pairwise deletion of missing data is a procedure to handle the absence of some of the information in a regression analysis by eliminating the data point from only those calculations which use the absent figures (EPA-79/12c).

Lithic: Pertaining to stone or a stone tool (DOE-91/04).

Lithium (Li): A alkali metal with a.n. 3; a.w. 6.939; d. 0.53 g/cc; m.p. 180.5° C and b.p. 1330° C. The element belongs to group IA of the periodic table.

Lithologic: Structure and composition of a rock (DOE-91/04).

Lithographic ink: An ink used in the lithographic process. The principal characteristic of a good lithographic ink is its ability to resist excessive emulsification by a reservoir of dampering solution (*see* ink for more related terms) (EPA-79/12a).

Lithographic printing line: A printing line, except that the substrate is not necessarily fed from an unwinding roll, in which each roll printer uses a roll where both the image and non-image areas are essentially in the same plane (planographic) (40 CFR 52.741-91).

Lithology: The description of rocks on the basis of their physical and chemical characteristics (40 CFR 146.3-91, *see also* 40 CFR 147.2902; 147.2902-91).

Lithosphere:
(1) The solid part of the Earth below the surface, including any ground water contained within it (40 CFR 191.12-91).
(2) The outer part of the solid earth composed of rock essentially like that explored at the surface and believed to be about 50 miles in thickness (LBL-76/07-water).

Litter:
(1) Carelessly discarded materials (SW-108ts).
(2) The highly visible portion of solid waste carelessly discarded outside the regular garbage and trash collection and disposal system (EPA-94/04).

Littoral: The shallow waters that extend along the edge of a lake or sea (DOI-70/04).

Littoral zone:
(1) The shoreward region of a body of water (LBL-76/07-water).
(2) That portion of a body of fresh water extending from the shoreline lakeward to the limit of occupancy of rooted plants (EPA-94/04).
(3) The strip of land along the shoreline between the high and low water levels (EPA-94/04).

Livable space: Any enclosed space that residents now use or could reasonably adapt for use as living space (EPA-88/08).

Live 18 day embryos: Embryos that are developing normally after 1 days of incubation. This is determined by candling the eggs. Values are expressed as a percentage of viable embryos (fertile eggs) (40 CFR 797.2130-91).

Live 21 day embryos: Embryos that are developing normally after 21 days of incubation. This is determined by candling the eggs. Values are expressed as a percentage of viable embryos (fertile eggs) (40 CFR 797.2150-91).

Live bottom bin: A storage bin for shredded or granular material whereby controlled discharge is by a mechanical or vibrating device across the bin bottom (EPA-83).

Live bottom pit: A storage pit, usually rectangular, receiving truck unloaded material, utilizing a push platen or bulkhead, reciprocating rams or mechanical conveyor across the pit floor for controlled discharge (retrieval) of the material (EPA-83).

Live tank: A metal, wood, or plastic tank with circulating seawater for the purpose of keeping a fish or shellfish alive until processed (*see* tank for more related terms) (EPA-9/75).

Live weight killed (LWK): The total weight of the total number of animals slaughtered during the time to which the effluent limitations apply; i.e., during any one day or any period of thirty consecutive days (40 CFR 432.11; 432.21; 432.31; 432.41-91).

Load (or loading):
(1) An amount of matter or thermal energy that is introduced into a receiving water; to introduce matter or thermal energy into a receiving water. Loading may be either man caused (pollutant loading) or natural (natural background loading) (40 CFR 130.2-91).
(2) The concentration of pollutant, usually expressed in grains of pollutant per cubic foot of contaminated gas stream (EPA-84/09).
(3) *See* pull.

Other load related terms include:
- Base load
- Connected load
- Electric load
- Peak load
- Pollution load
- System load

Load allocation (LA): The portion of a receiving water's loading capacity that is attributed either to one of its existing or future nonpoint sources of pollution or to natural background sources. Load allocations are best estimates of the loading, which may range from reasonably accurate estimates to gross allotments, depending on the availability of data and appropriate techniques for predicting the loading. Wherever possible, natural and nonpoint source loads should be distinguished (cf. waste load allocation) (40 CFR 130.2-91).

Load bearing resistance: The degree to which a refractory resists deformation when subjected to a specified compressive load at a specified temperature and time (SW-108ts).

Load cell: A device external to the locomotive, of high electrical resistance, used in locomotive testing to simulate engine loading while the locomotive is stationary. (Electrical energy produced by the diesel generator is dissipated in the load cell resistors instead of the traction motors) (40 CFR 201.1-91).

Load curve: A curve on a chart showing power (kilowatts) supplied, plotted against time of occurrence, and illustrating the varying magnitude of the load during the period covered (EPA-83).

Load diversity: Load diversity is the difference between the sum of the maxima of two or more individual loads and the coincident or combined maximum load, usually measured in kilowatts (EPA-83).

Load factor: The ratio of the average load in kilowatts supplied during a designated period to the peak or maximum load occurring tn that period (EPA-83).

Loaded emissions test: A sampling procedure for exhaust emissions which requires exercising the engine under stress (i.e., loading) by use of a chassis dynamometer to stimulate actual driving conditions. As a minimum requirement, the loaded emission test must include running the vehicle and measuring exhaust emissions at two speeds and loads other than idle (40 CFR 51-App/N-91).

Loaded vehicle mass: The curb mass plus 80 kg (176 lb.), average driver mass (40 CFR 86.402.78-91).

Loaded vehicle weight: The vehicle curb weight plus 300 Pounds (40 CFR 86.082.2-91).

Loading (*see also* **load**):
(1) The introduction of waste into a waste management unit but not necessarily to complete capacity (or referred to as filling) (40 CFR 61.341-91).
(2) The ratio of the biomass of gammarids (grams, wet weight) to the volume (liters) of test solution in either a test chamber or passing through it in a 24-hour period (40 CFR 795.120-91, *see also* 40 CFR 797.1400; 797.1520; 797.1600; 797.1830; 797.1300; 797.1330; 797.1930; 797.1950; 797.1970-91).

Other loading related terms include:
- Dust loading
- Filter loading
- Grain loading
- Hydraulic loading
- Organic loading
- Particulate loading
- Pollutant loading
- Splash loading
- Waste loading

Loading capacity: The greatest amount of loading that a water can receive without violating water quality standards (40 CFR 130.2-91).

Loading cycle: The time period from the beginning of filling a tank truck, railcar, or marine vessel until flow to the control device ceases, as measured by the flow indicator (40 CFR 61.301-91).

Loading rack: The loading arms, pumps, meters, shutoff valves, relief valves, and other piping and valves necessary to fill delivery tank trucks (40 CFR 60.501-91, *see also* 40 CFR 61.301-91).

Loam: A soft, easily worked soil containing sand, silt, and clay (SW-108ts).

Loan: An agreement or other arrangement under which any portion of a business, activity, or program is assisted under a loan issued by an agency and includes any subloan issued under a loan issued by an agency (cf. low interest loan) (40 CFR 15.4-91).

Loan guarantee: A government funded insurance that protects lenders against the failure of a project to pay back the principal and interest on a loan (OTA-89/10).

Local comprehensive emergency response plan: The emergency plan prepared by the local emergency planning committee as required by section 303 of the Emergency Planning and Community Right-To-Know Act of 1986 (SARA Title III) (40 CFR 310.11-91).

Local education agency (LEA): In the asbestos program, an educational agency at the local level that exists primarily to operate schools or to contract for educational services, including primary and secondary public and private schools. A single, unaffiliated school can be considered an LEA for AHERA purposes (EPA-94/04).

Local emergency planning committee (LEPC): A committee appointed by the state emergency response commission, as required by SARA Title III, to formulate a comprehensive emergency plan for its jurisdiction (EPA-94/04; *see also* 40 CFR 350.1-91).

Local free convection: A condition in the atmospheric boundary layer when forced convection should be assumed near the surface where mechanical turbulence is important and free convection at higher altitudes where thermal turbulence is important (*see* convection for more related terms) (NATO-78/10).

Locomotive: For the purpose of this regulation, a self-propelled vehicle designed for and used on railroad tracks in the transport of rail cars, including self-propelled rail passenger vehicles (40 CFR 201.1-91).

Locomotive load cell test stand: The load cell 40 CFR 201.1(o) and associated structure, equipment, trackage and locomotive being tested (40 CFR 201.1-91).

Lode: A tabular deposit of valuable minerals between definite boundaries. Lode, as used by miners, is nearly synonymous with the term vein as employed by geologists (EPA-82/05).

Loft drying: Form of air drying (EPA-83).

Lofting: The situation when air pollution is prevented from reaching the surface by a stable layer near ground level. Lofting can occur during the transition from unstable to inversion conditions and is mostly observed near sunset (NATO-78/10).

Log normal distribution:
(1) Particle size distribution when a linear relationship exists on a logarithmic/probability (normal) plot of particle size/cumulative distribution (cf. cumulative distribution) (EPA-84/09).
(2) A frequency distribution in which the logarithm of the variable in question is normally distributed (NATO-78/10).

Log normal probabilistic dilution model: A dilution model that calculates the probability distribution of receiving water quality concentrations from the log-normal probability distributions of the input variables (EPA-85/09).

Log P (also expressed as $\log k_{ow}$ or as n-octanal/water partition coefficient): The ratio, in a two-phase system of n-octanol and water at equilibrium, of the concentration of a chemical in the n-octanol phase to that in the water phase (EPA-91/03).

Log sorting and log storage facilities: The facilities whose discharges result from the holding of unprocessed wood, for example, logs or roundwood with bark or after removal of bark held in self-contained bodies of water (mill ponds or log ponds) or stored on land where water is applied intentionally on the logs (wet decking). (*see* 40 CFR 429, Subpart I, including the effluent limitations guidelines) (40 CFR 122.27-91).

Logarithmic wind profile: The variation of the wind speed (u) with height (z) following a logarithmic equation of the form: u = (U/k)ln((z-d)/Z), where: U = friction velocity; k = Von Karman constant; d = zero-plane displacement; and Z = the roughness length. The logarithmic velocity profile occurs in the turbulent flow near the earth's surface in neutral conditions (*see* wind for more related terms) (NATO-78/10).

Logarithm: The common logarithm of a number is equal to the exponential to which 10 must be raised to get the number. Logarithms which are based on 10 are called common, denary, or Briggsian Logarithms. For example:
- $10^2 = 100$ can be expressed as log(100) = 2
- $10^x = y$ can be expressed as log(y) = x

Logarithms which are based on "e" are called natural or Napierian Logarithms. "e" is further defined as:

$e = 1 + 1 + 1/2! + 1/3! + 1/4! + \ldots = 2.718281828459$

If the subscript 10 or e is omitted, the base must be inferred from the context, the base 10 being used in numerical computation, and the base e in theoretical work. In either system,

log (xy) = log (x) + log (y)
log (x/y) = log (x) - log (y)
log (x^y) = ylog (x)
log (1/x) = -log (x)

The two systems are related as follows:
$\log_{10}e = M = 0.4343$
$\log_{10}x = 0.4343\log_e x$
$\ln 10 = \log_e 10 = 1/M = 2.3026$
$\ln x = \log_e x = 2.3026\log_{10}x$
(Marks-p.2-26)

Logging residues: The unused portions of pole and sawtimber trees felled in land clearing (EPA-83).

Logit model: A dose-response model of the form: P(d) = 1/[1 + exp -(a + b log d)]; where: P(d) is the probability of toxic effects from a continuous dose rate d, and a and b are constants (EPA-92/12).

Lognormal probabilistic dilution model: A model that calculates the probability distribution of receiving water quality concentrations from the lognormal probability distributions of the input variables (EPA-91/03).

Long lead time procurement items: Equipment and/or construction materials that must be ordered before the estimated start of construction to ensure their availability at the time needed (DOE-91/04).

Long range transport: The transport of air pollutants by the wind field over distances of the order of a 1000 km (NATO-78/10).

Long stock: Paper with relatively long fibers (EPA-83).

Long term average: The concentration value calculated or measured as an average over a period greater than one month (usually one year) (NATO-78/10).

Long term stabilization: The addition of material on a uranium mill tailings pile for purpose of ensuring compliance with the requirements of 40 CFR 192.02(a) or 192.32(b)(i). These actions shall be considered complete when the Nuclear Regulatory Commission determines that the requirements of 40 CFR 192.02(a) or 192.32(b)(i) have been met (40 CFR 61.221-91).

Longitudinal study: An effect study over a long period of time (EPA-165.5).

Looping: The vertical meandering of a plume during very unstable conditions caused by large turbulent eddies. The plume may reach the ground close to the source resulting in very high concentration there (NATO-78/10).

Loose fill insulation: The insulation in granular, nodular, fibrous, powdery, or similar form, designed to be installed by pouring, blowing or hand placement (40 CFR 248.4-91).

Loss of coolant accidents: A postulated accident that results from the loss of reactor coolant (at a rate that exceeds the capability of the reactor coolant makeup system) from breaks in the reactor coolant pressure boundary, up to and including a break equivalent in size to the double-ended rupture of the largest pipe of the reactor coolant system (DOE-91/04).

Lot: A definite quantity of a material derived under similar conditions of production. *See* gross sample under analytical parameters - laboratory (EPA-83).

Lot size: In quality control, the number of units in a particular lot (EPA-84/03).

Louvre damper: A damper consisting of several blades each pivoted about its center and linked together for simultaneous operation (*see* damper for more related terms).

Low-, mid-, and high-level calibration gas: For calibration gas application, the corresponding low-, mid-, and high-level of propane calibration gas (in air or nitrogen) is a concentration equivalent to 20-30, 45-55 and 80-90% of the applicable span value (*see* calibration gas for more related terms) (EPA-90/04).

Low altitude: Any elevation less than or equal to 1,219 meters (4,000 feet) (40 CFR 86.082.2-91, *see also* 40 CFR 86.1602-91).

Low altitude condition: A test altitude less than 549 meters (1,800 feet) (40 CFR 86.082.2-91).

Low concentration PCB: The PCBs that are tested and found to contain less than 500 ppm PCBs, or those PCB- containing materials which EPA requires to be assumed to be at concentrations below 500 ppm (i.e., untested mineral oil dielectric fluid) (*see* PCB for more related terms) (40 CFR 761.123-91).

Low density polyethylene (LDPE): A thermoplastic polymer or copolymer comprised of at least 50 percent ethylene by weight and having a density of 0.940 g/cm^3 or less (*see* polyethylene for more related terms) (40 CFR 60.561-91).

Low dose: Should correspond to 1/10 of the high dose (cf. high dose) (40 CFR 795.232-91).

Low excess air: One of NO_x emission reduction techniques (*see* nitrogen oxide emission control for control structure). In a combustion system, a certain amount of excess air is required to ensure complete combustion. The more efficient the burners are for air and fuel mixing, the less amount of excess air is required for complete combustion. Because the local flame zone concentration of oxygen is reduced, thus reducing both thermal and fuel NO_x (EPA-81/12, p7-9).

Low NO_x burner: One of NO_x emission reduction techniques (*see* nitrogen oxide emission control for control structure). This technique includes:
(1) Burners are designed to control mixing of fuel and air in a pattern to keep flame temperature low and dissipate the heat quickly.
(2) Burners are designed to control the flame shape to minimize the reaction of nitrogen and oxygen at peak flame temperature.
(3) Burners are designed to have fuel rich and air rich regions to reduce flame temperature and oxygen availability (EPA-81/12, p7-12).
(4) One of several combustion technologies used to reduce emissions of Nitrogen Oxides (NOx) (EPA-94/04).

Low rate filter: A trickling filter designed to receive a small load of BOD per unit volume of filtering material and to have a low dosage rate per unit of surface area (usually 1 to 4 mgd/acre). Also called standard rate filter (*see* filter for more related terms) (EPA-82/11f).

Low flow prewash: A system which concentrates most of the fix carryout in a low volume after-fix prewash tank. The system consists of segmenting the after-fix prewash tank to provide a small prewash section with separate wash water make-up and overflow (EPA-80/10).

Low hazard content: Low hazard contents shall be classified as those of such low combustibility that no self-propagating fire therein can occur and that consequently the only probable danger requiring the use of emergency exits will be from panic, fumes, or smoke, or fire from some external source (*see* hazard for more related terms) (40 CFR 1910.35-91).

Low heat release rate: A heat release rate of 730,000 J/sec-m3 (70,000 Btu/hour-ft3) or less (40 CFR 60.41b-91).

Low heating value (net heating value or lower heating value): Similar to HHV, except that the water produced by the combustion is not condensed but retained as vapor (*see* heating value for more related terms) (EPA-84/09).

Low interest loan: A government subsidy that allows a loan for a specific purpose to be offered at below a market rate (*see* loan for more related terms) (OTA-89/10).

Low level radioactive waste (LLRW): Wastes less hazardous than most of those generated by a nuclear reactor. Usually generated by hospitals, research laboratories, and certain industries. The Department of Energy, Nuclear Regulatory Commission, and EPA share responsibilities for managing them (*see* waste for more related terms) (EPA-94/04).

Low level waste: The waste that contains radioactivity but is not classified as high level waste, transuranic waste, spent nuclear fuel, or "11e(2) by-product material" as defined by DOE Order 5820.2A. Test specimens of fissionable material irradiated for research and development only, and not for the production of power or plutonium, may be classified as low-level waste, provided the concentration of transuranic waste is less than 100 nCi/g (*see* waste for more related terms) (DOE-91/04).

Low noise emission product: Any product which emits noise in amounts significantly below the levels specified in noise emission standards under regulations applicable under section 6 at the time of procurement to that type of product (NCA15-42USC4914-87).

Low noise emission product determination: The Administrators determination whether or not a product, for which a properly filed application has been received, meets the low-noise-emission product criterion (40 CFR 203.1-91).

Low pressure distribution system: A distribution system in which the gas pressure in the main is substantially the same as the pressure provided to the customer (40 CFR 192.3-91).

Low pressure process: A production process for the manufacture of low density polyethylene in which a reaction pressure markedly below that used in a high pressure process is used. Reaction pressure of current low pressure processes typically go up to about 300 psig (40 CFR 60.561-91).

Low processing packinghouse: A packinghouse that processes no more than the total animals killed at that plant, normally processing less than the total kill (cf. high processing packinghouse) (40 CFR 432.31-91).

Low stream flow augmentation: The release of water from dam-controlled reservoirs when the stream level is low (DOI-70/04).

Low temperature rendering: A rendering process in which the cooking is conducted at a low temperature which does not evaporate the raw material moisture. Normally used to produce a high-quality edible product such as lard (*see* rendering for more related terms) (EPA-75/01).

Low terrain: Any area other than high terrain (*see* terrain for more related terms) (40 CFR 51.166-91, *see also* 40 CFR 52.21-91).

Low viscosity poly(ethylene terephthalate): A poly(ethylene terephthalate) that has an intrinsic viscosity of less than 0.75 and is used in such applications as clothing, bottle, and film production (40 CFR 60.561-91).

Low volume waste source: Taken collectively as if from one source, wastewater from all sources except those for which specific limitations are otherwise established in this part. Low volume wastes sources include, but are not limited to: wastewaters from wet scrubber air pollution control systems, ion exchange water treatment system, water treatment evaporator blowdown, laboratory and sampling streams, boiler blowdown, floor drains, cooling tower basin cleaning wastes, and recirculating house service water systems. Sanitary and air conditioning wastes are not included (40 CFR 423.11-91).

Lower detectable limit:
(1) The minimum pollutant concentration which produces a signal of twice the noise level (40 CFR 53.23-91).
(2) The smallest quantity or concentration of sample which causes a response equal to twice the noise level. (Not to be confused with sensitivity, which is response per unit of concentration) (cf. detection limit) (LBL-76/07-bio).

Lower explosive limit:
(1) The lowest percent by volume of a mixture of explosive gases which will propagate a flame in air at 25° C and atmospheric pressure (40 CFR 257.3.8-91).
(2) The concentration of fuel (organic) below which combustion will not occur in the presence of a flame or spark (*see* combustion limit for more related terms) (EPA-81/12, p3-6; 84/09).
(3) The concentration of a compound in air below which the mixture will not catch on fire (EPA-94/04).

Lower heating value: *see* synonym, low heating value.

Lowest achievable emission rate (LAER):
(1) For any source, that rate of emissions which reflects:
  (A) The most stringent emission limitation which is contained in the implementation plan of any State for such class or category of source, unless the owner or operator of the proposed source demonstrates that such limitations are not achievable; or
  (B) The most stringent emission limitation which is achieved in practice by such class or category of source, whichever is more stringent. In no event shall the application of this term permit a proposed new or modified source to emit any pollutant in excess of the amount allowable under applicable new source standards of performance (CAA171, *see also* 40 CFR 51.165; 51-App/S-91).
(2) Under the Clean Air Act, this is the rate of emissions which reflects:
  (A) The most stringent emission limitation which is contained in the implementation plan of any state for such source unless the owner or operator of the proposed source demonstrates such limitations are not achievable; or
  (B) The most stringent emissions limitation achieved in practice, whichever is more stringent. Application of this term does not permit a proposed new or modified source to emit pollutants in excess of existing new source standards (EPA-94/04).
The best available control technology (BACT), as defined in CAA169, shall be substituted for the lowest achievable emission rate (LAER) [CAAA182(c)(7)].

Lowest observed adverse effect (LOAEL): The lowest dose in an experiment which produced an observable effect (Course 165.6).

Lowest observed adverse effect level (or lowest effect level):
(1) The lowest concentration of an effluent or toxicant that results in statistically significant adverse health effects as observed in chronic or subchronic human epidemiology studies or animal exposure (EPA-91/03).
(2) The lowest exposure level at which there are statistically or biologically significant increases in frequency or severity of adverse effects between the exposed population and its appropriate control group (EPA-90/08).
(3) The lowest dose in an experiment which produced an observable adverse effect (EPA-94/04).

Lowest effect level (LEL): *See* synonym, lowest observed adverse effect level (EPA-90/08b).

Lowest temperature in the surroundings: *See* synonym, sink temperature.

Lubricant: A substance added to resin-sand mixtures to permit the easy release of molds from patterns. Calcium stearate, zinc stearate and carnauba wax are common lubricating agents (cf. spent lubricant) (EPA-85/10a).

Lubricanting oil: The fraction of crude oil which is sold for purposes of reducing friction in any industrial or mechanical device. Such term includes re-refined oil (*see* oil for more related terms) (RCRA1004-42USC6903-91).

Luminescence analyzer: One of continuous emission monitors (*see* continuous emission monitor for various types). Luminescence is the emission of light which results from a molecule returning from an elevated energy state to the ground state. Types of luminescence analyzers include:
(1) Chemiluminescence analyzer (CA)
(2) Fluorescence analyzer
(3) Photoluminescence (or flame photometric)

Luminescent material: The materials that emit light upon excitation by such energy sources as photons, electrons, applied voltage, chemical reactions or mechanical energy and which are specifically used as coatings in fluorescent lamps and cathode ray tubes. Luminescent materials include, but are not limited to, calcium halophosphate, yttrium oxide, zinc sulfide, and zinc-cadmium sulfide (40 CFR 469.41-91, *see also* 40 CFR 469.42-91).

Luminous flame: *See* definition under flame combustion.

Lutetium (Lu): A rare earth metal with a.n. 71; a.w. 174.97; d. 9.84 g/cc; m.p. 1652° C and b.p. 3327° C. The element belongs to group IIIB of the periodic table.

Lycra fiber: A trademark of Du Pont for polyurethane multifilament elastic yarns (*see* fiber for more related terms) (EPA-74/06b).

Lye: A strong alkaline solution. Caustic soda (sodium hydroxide) is the most common lye (EPA-74/03).

Lye dump: Spent water from the lye bath that is used to remove the inner membrane of sectionizing fruit. The spent lye solution is discharged periodically (EPA-74/03).

Lye rinse: Rinse water used to remove from the fruit, lye solution carried out of the lye bath in sectionizing operations (EPA-74/03).

Lysimeter:
(1) A device used to measure the quantity or rate of water

movement through or from a block of soil or other material, such as solid waste, or used to collect percolated water for quality analysis (*see* flow velocity meter for more related terms) (SW-108ts).

(2) A device used to measure the quantity or rate of water movement through or from a block of soil or other material, such as used to collect percolated water for quality analysis (EPA-83).

Mm

M5: *See* definition under method 5.

M9IM: Those offshore facilities continuously manned by nine (9) or fewer persons or only intermittently manned by any number of persons (40 CFR 435.11-91, *see also* 40 CFR 435.41-91).

M10: Those offshore facilities continuously manned by ten (10) or more persons (40 CFR 435.11-91, *see also* 40 CFR 435.41-91).

Maceration: *See* synonym, imbibition.

Maceration water: The water applied to the bagasse during the milling process to dissolve sucrose, which is later reclaimed (*see* water for more related terms) (EPA-75/02d).

Mach number: The ratio of the speed of an object to the speed of sound under the same temperature and pressure conditions.

Machine finish: Any finish produced on the papermaking machine vs off machine coating or finish (EPA-83).

Machine glazed: High gloss finish produced mechanically on the papermaking machine (EPA-83).

Machine shop: A facility performing cutting, grinding, turning, honing, milling, deburring, lapping, electrochemical machining, etching, or other similar operations (40 CFR 61.31-91).

Machining: Machining operations include turning, milling, drilling, boring, tapping, planning, broaching, sawing and filing, and chamfering (EPA-83/06a).

Macro encapsulation: The isolation of a waste by embedding it in or surrounding it with a material which acts as a barrier to water or air (OME-88/12).

Macro method: A method requiring more than milligram amounts of sample (*see* method for more related terms) (ACS-87/11).

Macro nutrient: The essential elements (such as nitrogen and potassium) needed in a large quantity for the growth of organisms (*see* nutrient for more related terms).

Macro organism:
(1) Plants, animals, or fungal organisms visible to the unaided eye (LBL-76/07-water).
(2) Those invertebrates visible to the unaided eye and which are retained on a U.S. standard sieve No. 30 (openings of 0.589 mm) (cf. benthic macroorganism) (DOD-78/01).
(3) *See* organism for more related terms.

Macro scale: In meteorology, a length scale of the order of ten thousand kilometers. Applicable to large scale phenomena in the atmosphere, e.g., tidal waves (cf. meso scale or micro scale) (NATO-78/10).

Macroscopic thermodynamics (or classical thermodynamic): The consideration of a matter is in the level of a large scale (cf. microscopic thermodynamic).

Magazine paper: A variety of coated and uncoated papers used in magazines and similar periodicals (*see* paper for more related terms) (EPA-83).

Magnaflux inspection: The trade name for magnetic particle test (EPA-83/03).

Magnesium (Mg): An alkaline earth metal with a.n. 12; a.w. 24.312; d. 1.74 g/cc; m.p. 650° C and b.p. 1107° C. The element belongs to group IIA of the periodic table.

Magnet wire: The aluminum or copper wire formed into an electromagnetic coil (40 CFR 52.741-91).

Magnet wire coating: Any coating or electrically insulating varnish or enamel applied to magnet wire (40 CFR 52.741-91).

Magnet wire coating facility: A facility that includes one or more magnet wire coating line(s) (40 CFR 52.741-91).

Magnet wire coating line: A coating line in which any protective, decorative, or functional coating is applied onto the surface of a magnet wire (40 CFR 52.741-91).

Magnetic fraction: That portion of municipal ferrous scrap remaining after the non-magnetic contaminants have been manually removed and the magnetic fraction washed with water and dried at ambient temperature or as required by ASTM C29 (EPA-83).

Magnetic separation: Use of magnets to separate ferrous materials from mixed municipal waste stream (EPA-94/04).

Magnetic separator:
(1) A device that removes ferrous metals by means of magnets (SW-108ts).
(2) A technique for separating magnet or weakly paramagnetic particles and other non-magnetic materials (down to colloidal particle size) from slurries, sludges, and (after chemical treatment) from solutions. The feed stream is passed through a fine ferromagnetic filter which, when magnetized, collects the magnetic material. The filter is periodically cleaned, with the magnetic material then recovered by a simple wash procedure. The removal of non-magnetic material requires the feed to be treated with a magnetic seed (e.g., magnetite).
(3) A separation process by which a permanent magnet or electromagnet is used to attract magnet materials away from mixed waste. In mining, it is an important process in the beneficiation of iron ores in which the magnetic mineral is separated from non-magnetic material, e.g., magnetite from other minerals, roasted pyrite from sphalerite (EPA-82/05).
(4) *See* separator for more related terms.

Magnetic wind instrument: One of continuous emission monitors (*see* continuous emission monitor or paramagnetic oxygen analyzer for various types). This instrument is based on the principle that paramagnetic attraction of the oxygen molecule decreases as the temperature increases. A typical analyzer utilizes a cross-tube wound with filament wire heated to 200° C. A strong magnetic field covers one half of the coil. Oxygen contained in the sample gas is attracted to the applied field and enters the cross-tube. The oxygen then heats up and its paramagnetic susceptibility is reduced. This heated oxygen then is pushed out by the colder gas jet entering the cross-tube. A wind or flow of gas, therefore, continuously passes through the cross-tube. This gas flow, however, effectively cools the heated filament coil and changes its resistance. The change in resistance detected in the Wheatstone bridge circuit is proportional to the oxygen concentration (EPA-84/03a).

Magnetic tape: Any flexible substrate that is covered on one or both sides with a coating containing magnetic particles and that is used for audio or video recording or information storage (40 CFR 60.711-91).

Magnetite (or magnetic iron ore): The natural black oxide of iron (Fe_3O_4). As black sand, magnetite occurs in placer deposits, and also as lenticular bands. Magnetite is used widely as a suspension solid in dense-medium washing of coal and ores (EPA-82/05).

Magneto-dynamic instrument: One of continuous emission monitors (*see* continuous emission monitor or paramagnetic oxygen analyzer for various types). This instrument utilizes the paramagnetic property of the oxygen molecule by suspending a specially constructed torsion balance in a magnetic field. Here, a dumbbell will be slightly repelled from the magnetic field. When a sample containing oxygen is added, the magnet attracts the oxygen and field lines surrounding the dumbbell are changed. The dumbbell swings to realign itself with the new field. Light reflected from a small mirror placed on the dumbbell is used to indicate the degree of swing of the dumbbell, and hence, the oxygen concentration (EPA-84/03a).

Magnetohydrodynamics (MHD): The study of the interactions of a magnetic field and a conducting fluid for electrical power generation.

Magnetopneumatic instrument: One of continuous emission monitors (*see* continuous emission monitor or paramagnetic oxygen analyzer for various types). This instrument utilizes the paramagnetic property of the oxygen molecule by introducing oxygen into a disproportional magnetic field. The paramagnetic gaseous substance is attracted to the magnetic field to cause the pressure in that field to rise. The pressure elevation is picked out of the magnetic field by using a non-paramagnetic gas (nitrogen). The electromagnet is alternately magnetized and the pressure change is converted into an electrical signal by a condenser microphone. All of the commercial paramagnetic analyzers are extractive systems. Water and particulate matter have to be removed before the sample enters the monitoring system. It should be noted that NO and NO_2 are also paramagnetic and may cause some interferences in the monitoring method if high concentrations are present (EPA-84/03a).

Magnitude: The amount of a pollutant (or pollutant parameter such as toxicity), expressed as a concentration or toxic unit is allowable (EPA-85/09; 91/03).

Magnitude of entry editing: The determination of a value's validity based solely on the size of the figure, and is a test typically performed on raw input data (cf. cross check editing) (EPA-79/12c).

Main: A distribution line that serves as a common source of supply for more than one service line (40 CFR 192.3-91).

Major: Larger publicly owned treatment works (POTWs) with flows equal to at least one million gallons per day (mgd) or servicing population equivalent to 10,000 persons; certain other POTWs having significant water quality impacts (*see* minors) (EPA-94/04).

Major disaster: Any hurricane, tornado, storm, flood, high water, wind-driven water, tidal wave, earthquake, drought, fire, or other catastrophe in any part of the United States which, in the determination of the President, is or threatens to become of sufficient severity and magnitude to warrant disaster assistance by the Federal Government to supplement the efforts and available resources of States and local governments and relief organizations in alleviating the damage, loss, hardship, or suffering caused thereby (40 CFR 109.2-d-91).

Major emitting facility: Any of the following stationary sources of air pollutants which emit, or have the potential to emit, one hundred tons per year or more of any air pollutant from the following types of stationary sources: fossil-fuel fired steam electric plants of more than two hundred and fifty million British thermal units per hour heat input, coal cleaning plants (thermal dryers), kraft pulp mills, Portland Cement plants, primary zinc smelters, iron and steel mill plants, primary aluminum ore reduction plants, primary copper smelters, municipal incinerators capable of charging more than two hundred and fifty tons of refuse per day, hydrofluoric, sulfuric, and nitric acid plants, petroleum refineries, lime plants, phosphate rock processing plants, coke oven batteries, sulfur recovery plants, carbon black plants (furnace process) primary lead smelters, fuel conversion plants, sintering plants, secondary metal production facilities, chemical process plants, fossil-fuel boilers of more than two hundred and fifty million British thermal units per hour heat input, petroleum storage and transfer facilities with a capacity exceeding three hundred thousand barrels, taconite ore processing facilities, glass fiber processing plants, charcoal production facilities. Such term also includes any other source with the potential to emit two hundred and fifty tons per year or more of any air pollutant. This term shall not include new or modified facilities which are nonprofit health or education institutions which have been exempted by the State (CAA169-42USC7479).

Major facility: Any NPDES facility or activity classified as such by the Regional Administrator, or, in the case of approved State programs, the Regional Administrator in conjunction with the State Director (40 CFR 122.2-91, *see also* 40 CFR 124.2; 144.3; 270.2-91).

Major federal action: Includes actions with effects that may be major and which are potentially subject to Federal control and responsibility. Major reinforces but does not have a meaning independent of significantly (40 CFR 1508.27). Actions include the circumstance where the responsible officials fail to act and that failure to act is reviewable by courts or administrative tribunals under the Administrative Procedure Act or other applicable law as agency action:

(1) Actions include new and continuing activities, including projects and programs entirely or partly financed, assisted, conducted, regulated, or approved by federal agencies; new or revised agency rules, regulations, plans, policies, or procedures; and legislative proposals (40 CFR 1506.8, 1508.17). Actions do not include funding assistance solely in the form of general revenue sharing funds, distributed under the State and Local Fiscal Assistance Act of 1972, 31 U.S.C. 1221 et seq., with no Federal agency control over the subsequent use of such funds. Actions do not include bringing judicial or administrative civil or criminal enforcement actions.

(2) Federal actions tend to fall within one of the following categories:
   (A) Adoption of official policy, such as rules, regulations, and interpretations adopted pursuant to the Administrative Procedure Act, 5 U.S.C. 551 et seq.; treaties and international conventions or agreements; formal documents establishing an agency's policies which will result in or substantially alter agency programs.
   (B) Adoption of formal plans, such as official documents prepared or approved by federal agencies which guide or prescribe alternative uses of Federal resources, upon which future agency actions will be based.
   (C) Adoption of programs, such as a group of concerted actions to implement a specific policy or plan; systematic and connected agency decisions allocating agency resources to implement a specific statutory program or executive directive.
   (D) Approval of specific projects, such as construction or management activities located in a defined geographic area. Projects include actions approved by permit or other regulatory decision as well as federal and federally assisted activities (40 CFR 1508.18-91).

Major modification:
(1) Any physical change in or change in the method of operation of a major stationary source that would result in a significant net emissions increase of any pollutant subject to regulation under the Act.
(2) Any net emissions increase that is considered significant for volatile organic compounds shall be considered significant for ozone.
(3) A physical change or change in the method of operation shall not include:
   (A) Routine maintenance, repair and replacement;
   (B) Use of an alternative fuel or raw material by reason of an order under sections 2 (a) and (b) of the Energy Supply and Environmental Coordination Act of 1974 (or any superseding legislation) or by reason of a natural gas curtailment plan pursuant to the Federal Power Act;
   (C) Use of an alternative fuel by reason of an order or rule section 125 of the Act;
   (D) Use of an alternative fuel at a steam generating unit to the extent that the fuel is generated from municipal solid waste;
   (E) Use of an alternative fuel or raw material by a stationary source which:
      (i) The source was capable of accommodating before December 21, 1976, unless such change would be prohibited under any federally enforceable permit condition which was established after December 12, 1976 pursuant to 40 CFR 52.21 or under regulations approved pursuant to 40 CFR subpart I or 40 CFR 51.166, or
      (ii) The source is approved to use under any permit issued under regulations approved pursuant to this section;
   (F) An increase in the hours of operation or in the production rate, unless such change is prohibited under any federally enforceable permit condition which was established after December 21, 1976 pursuant to 40 CFR 52.21 or regulations approved pursuant to 40 CFR part 51 subpart I or 40 CFR 51.166.
   (G) Any change in ownership at a stationary source.
   (H) The addition, replacement or use of a pollution control project at an existing electric utility steam generating unit, unless the reviewing authority determines that such addition, replacement, or use renders the unit less environmentally beneficial, or except:
      (i) When the reviewing authority has reason to believe that the pollution control project would result in a significant net increase in representative actual annual emissions of any criteria pollutant over levels used for that source in the most recent air quality impact analysis in the area conducted for the purpose of title I, if any, and
      (ii) The reviewing authority determines that the increase will cause or contribute to a violation of any national ambient air quality standard or PSD increment, or visibility limitation.

(i) The installation, operation, cessation, or removal of a temporary clean coal technology demonstration project, provided that the project complies with:
 (i) The State Implementation Plan for the State in which the project is located, and
 (ii) Other requirements necessary to attain and maintain the national ambient air quality standard during the project and after it is terminated (40 CFR 51.165-v-91).

Major modification: This term is used to define modifications of major stationary sources of emissions with respect to Prevention of Significant Deterioration and New Source Review under the Clean Air Act (EPA-94/04).

Major municipal separate storm sewer outfall (or major outfall): A municipal separate storm sewer outfall that discharges from a single pipe with an inside diameter of 36 inches or more or its equivalent (discharge from a single conveyance other than circular pipe which is associated with a drainage area of more than 50 acres); or for municipal separate storm sewers that receive storm water from lands zoned for industrial activity (based on comprehensive zoning plans or the equivalent), an outfall that discharges from a single pipe with an inside diameter of 12 inches or more or from its equivalent (discharge from other than a circular pipe associated with a drainage area of 2 acres or more) (cf. outfall or sewer outfall) (40 CFR 122.26-91).

Major outfall: *See* synonym, major municipal separate storm sewer outfall.

Major PSD modification: A major modification as defined in 40 CFR 52.21 (40 CFR 124.41-91).

Major PSD stationary source: A major stationary source as defined in 40 CFR 52.21(b)(1) (40 CFR 124.41-91).

Major source: Any stationary source (or any group of stationary sources located within a contiguous area and under common control) that is either of the following:
(1) A major source as defined in section 112.
(2) A major stationary source as defined in section 302 or part D of title I (CAA501, *see also* CAA112).
(3) *See* source for more related terms.

Major source baseline date:
(1) (A) In the case of particulate matter and sulfur dioxide, January 6, 1975, and
 (B) In the case of nitrogen dioxide, February 8, 1988.
(2) **Minor source baseline date:** The earliest date after the trigger date on which a major stationary source or a major modification subject to 40 CFR 52.21 or to regulations approved pursuant to 40 CFR 51.166 submits a complete application under the relevant regulations. The trigger date is:
 (A) In the case of particulate matter and sulfur dioxide, August 7, 1977, and
 (B) In the case of nitrogen dioxide, February 8, 1988.
(3) The baseline date is established for each pollutant for which increments or other equivalent measures have been established if:
 (A) The area in which the proposed source or modification would construct is designated as attainment or unclassifiable under section 107(d)(i) (D) or (E) of the Act for the pollutant on the date of its complete application under 40 CFR 52.21 or under regulations approved pursuant to 40 CFR 51.166; and
 (B) In the case of a major stationary source, the pollutant would be emitted in significant amounts, or, in the case of a major modification, there would be a significant net emissions increase of the pollutant.
(4) Any minor source baseline date established originally for the TSP increments shall remain in effect and shall apply for purposes of determining the amount of available PM-10 increments, except that the reviewing authority may rescind any such minor source baseline date where it can be shown, to the satisfaction of the reviewing authority, that the emissions increase from the major stationary source, or the net emissions increase from the major modification, responsible for triggering that date did not result in a significant amount of PM-10 emissions (40 CFR 51.166-14).

Major stationary source:
(1) The following types of stationary sources with the potential to emit 250 tons or more of any pollutant; fossil-fuel fired steam electric plants of more than 250 million British thermal units per hour heat input, coal cleaning plants (thermal dryers), kraft pulp mills, Portland Cement plants, primary zinc smelters, iron and steel mill plants, primary aluminum ore reduction plants, primary copper smelters, municipal incinerators capable of charging more than 250 tons of refuse per day, hydrofluoric, sulfuric, and nitric acid plants, petroleum refineries, lime plants, phosphate rock processing plants, coke oven batteries, sulfur recovery plants, carbon black plants (furnace process), primary lead smelters, fuel conversion plants, sintering plants, secondary metal production facilities, chemical process plants, fossil-fuel boilers of more than 250 million British thermal units per hour heat input, petroleum storage and transfer facilities with a capacity exceeding 300,000 barrels, taconite ore processing facilities, glass fiber processing plants, charcoal production facilities (CAA169A-42USC7491).
(2) Term used to determine the applicability of Prevention of Significant Deterioration and new source regulations. In a nonattainment area, any stationary pollutant source with potential to emit more than 100 tons per year is considered a major stationary source. In PSD areas the cutoff level may be either 100 or 250 tons, depending upon the source (EPA-94/04).

Major stationary source and major emitting facility: Except as otherwise expressly provided, the terms major stationary source and major emitting facility mean any stationary facility or source of air pollutants which directly emits, or has the potential to emit, one hundred tons per year or more of any air pollutant (including any major emitting facility or source of fugitive emissions of any such pollutant, as determined by rule by the Administrator) (CAA302-42USC7602-91).

Major stationary source and major modification: The major stationary source and major modification, respectively, as defined in 40 CFR 51.166 (40 CFR 51.301-91).

Makeup solvent: The solvent introduced into the affected facility that compensates for solvent lost from the affected facility during the manufacturing process (*see* solvent for more related terms) (40 CFR 60.601-91).

Makeup water: The water added to compensate for water losses resulting from evaporation and water disposal (*see* water for more related terms) (EPA-89/03b).

Malathion ($C_{10}H_{19}O_6PS_2$): A light yellow liquid used as an insecticide.

Malfunction: Any sudden and unavoidable failure of air pollution control equipment, process equipment, or a process to operate in a normal or usual manner. Failures that are caused entirely or in part by poor maintenance, careless operation, or any other preventable upset condition or preventable equipment breakdown shall not be considered malfunctions (40 CFR 52.741-91, *see also* 40 CFR 57.103; 60.2; 61.141; 61.161; 61.171; 61.181; 86.082.2; 264.1031-91).

Malignant: Tending to become progressively worse and to result in death if not treated; having the properties of anaplasia, invasiveness, and metastasis (EPA-92/12).

Malleable iron: A cast iron made by a prolonged anneal of white cast iron in which decarburization or graphitization, or both, take place to eliminate some or all of the cementite. Graphite is present in the form of temper carbon (*see* iron for more related terms) (40 CFR 464.31-91).

Malt liquor: The molten sugar which has been added a small amount of water (half the weight of the sugar) (*See* liquor for more related terms) (EPA-75/02d).

Man made beta particle and photon emitters:
(1) All radionuclides emitting beta particles and/or photons listed in Maximum Permissible Body Burdens and Maximum Permissible Concentration of Radionuclides in Air or Water for Occupational Exposure, NBS Handbook 69, except the daughter products of thorium-232, uranium-235 and uranium-238 (40 CFR 141.2-91).
(2) All radionuclides emitting beta particles and/or photos listed in Maximum Permissible Body Burdens and Maximum Permissible Concentrations of Radionuclides in Air and Water for Occupational Exposure (EPA-94/04).

Management:
(1) (In AEA), any activity, operation, or process (except for transportation) conducted to prepare spent nuclear fuel or radioactive waste for storage or disposal, or the activities associated with placing such fuel or waste in a disposal system (40 CFR 191.02-91).
(2) In RCRA, *see* hazardous waste management (40 CFR 260.10-91).

Management of migration: The actions that are taken to minimize and mitigate the migration of hazardous substances or pollutants or contaminants and the effects of such migration. Measures may include, but are not limited to, management of a plume of contamination, restoration of a drinking water aquifer, or surface water restoration (40 CFR 300.5-91).

Management plan: Under the Asbestos Hazard Emergency Response Act (AHERA), a document that each Local Education Agency is required to prepare, describing all activities planned and undertaken by a school to comply with AHERA regulations, including building inspections to identify asbestos-containing materials, response actions, and operations and maintenance programs to minimize the risk of exposure (EPA-94/04).

Management system audit: An on-site audit of an organization's quality assurance management system. The audit is used to verify the existence and evaluate the adequacy of the internal management systems necessary for the successful implementation of quality assurance program (EPA-85/08).

Managerial controls: Methods of nonpoint source pollution control based on decisions about managing agricultural wastes or application times or rates for agrochemicals (EPA-94/04).

Mandatory class I Federal areas: The Federal areas which may not be designated as other than class I under this part (CAA169A, *see also* 40 CFR 51.301-91).

Mandatory recycling:
(1) Programs which by law require consumers to separate trash so that some or all recyclable materials are not burned or dumped in landfills (EPA-89/11).
(2) Programs which by law require consumers to separate trash so that some or all recyclable materials are recovered for recycling rather than going to landfills (EPA-94/04).

Mandatory standards: Standards adopted by the U.S. Department of Energy (DOE) that define the minimum requirements that DOE and its contractors must comply with. standards may be classified as mandatory because of applicable federal or state statutes or implementing requirements, or as a matter of DOE policy (DOE-91/04).

Mandrel: A form used as a cathode in electroforming; a mold or matrix (EPA-74/03d).

Manganese (Mn): A brittle transition metal with a.n. 25; a.w. 54.938; d. 7.43 g/cc; m.p. 1245° C and b.p. 2150° C. The element belongs to group VIIB of the periodic table.

Manganese minerals: Those in principal production are pyrolusite, some psilomelane, and wad (impure mixture of manganese and other oxides) (EPA-82/05).

Manganese nodules: The concretions, primarily of manganese salts, covering extensive areas of the ocean floor. They have a layer configuration and may prove to be an important source of manganese (EPA-82/05).

Manganese ore: A term used by the Bureau of Mines for ore containing 35% or more manganese and may include concentrate, nodules, or synthetic ore (EPA-82/05).

Manganiferous iron ore: A term used by the Bureau of Mines for ores containing 5 to 10 % manganese (EPA-82/05).

Manganiferous ore: A term used by the Bureau of Mines for any ore of importance for its manganese content containing less than 35 % percent manganese but not less than 5 % manganese (EPA-82/05).

Manhole (drop manhole or sewer manhole): A hole in a tank, boiler, or sewer system which is big enough for a person to pass through.

Manifest: The form (shipping document) used for identifying the quantity, composition, and the origin, routing, and destination of hazardous waste during its transportation from the point of generation to the point of disposal, treatment, or storage (RCRA1004-91, *see also* 40 CFR 144.3; 260.10; 270.2; 761.3-91).

Manifest document number: The U.S. EPA twelve digit identification number assigned to the generator plus a unique five digit document number assigned to the Manifest by the generator for recording and reporting purposes (40 CFR 260.10-91).

Manifest system: Tracking of hazardous waste from "cradle to grave" (generation through disposal) with accompanying documents known as manifests (*see* cradle to grave) (EPA-94/04).

Manifold business forms: A type of product manufactured by business forms manufacturers that is commonly produced as marginally punched continuous forms in small rolls or fan folded sets with or without carbon paper interleaving. It has a wide variety of uses such as invoices, purchase orders, office memoranda, shipping orders, and computer printout (40 CFR 250.4-91).

Manila paper: Manila paper indicates color and finish and not the use of manila hemp (*see* paper for more related terms) (EPA-83).

Manmade air pollution: The air pollution which results directly or indirectly from human activities (*see* pollution for more related terms) (CAA169A-42USC7491-91).

Manned control center: An electrical power distribution control room where the operating conditions of a PCB Transformer are continuously monitored during the normal hours of operation (of the facility), and, where the duty engineers, electricians, or other trained personnel have the capability to deenergize a PCB Transformer completely within 1 minute of the receipt of a signal indicating abnormal operating conditions such as an overtemperature condition or overpressure condition in a PCB Transformer (40 CFR 761.3-91).

Manning equation: An equation used to calculate the average open channel flow $u = (1.486 MG^{0.5})/n$; where: u = average velocity; M = hydraulic mean depth; G = hydraulic gradient and n = Manning's coefficient of roughness.

Manometer:
(1) A pressure gauge which measures pressure according the relationship. p = density x height of liquid. It usually consists of a U-shaped tube containing a liquid, the surface of which moves proportionally with changed in pressure on the liquid in the other end. Also, a tube type of differential pressure gauge (EPA-82/11f).
(2) A device for measuring the liquid pressure in a system. A manometer measures a pressure difference between a measurable length of a fluid column against the pressure difference. For a fluid in static equilibrium, the relationship between the pressure and elevation within the fluid is expressed as:
pressure difference = (specific weight of a fluid)
              x (height of a fluid column)
The unit is usually denoted by psig (pound force per square inch gage) (Jones-60, p15).
(3) *See* pressure gauge for more related terms.

Mantlerock: The layer of loose rock fragments, the surface part of which is called soil, that covers most of the earth's land area and varies in thickness from place to place (DOI-70/04).

Manual burner: A burner which is purged, started, ignited, modulated and stopped manually (*see* burner for more related terms) (EPA-83).

Manual method: A method for measuring concentrations of an ambient air pollutant in which sample collection, analysis, or measurement, or some combination thereof, is performed manually (40 CFR 53.1-91).

Manual plating: The plating in which the workpieces are conveyed manually through successive cleaning and plating tanks (EPA-83/06a).

Manual separation:
(1) The separation of recyclable or compostible materials from waste by hand sorting (EPA-89/11).
(2) Hand sorting of reyclable or compostable materials in waste (EPA-94/04).

Manufacture: Any person engaged in the manufacturing, assembling, or importation of marine sanitation devices or of vessels subject to standards and regulations promulgated under this section (CWA312).

Manufacture of electronic crystals: The growing of crystals and/or the production of crystals wafers for use in the manufacture of electronic devices (40 CFR 469.22-91).

Manufacture of semiconductors: Those processes, beginning with the use of crystal wafers, which lead to or are associated with the manufacture of semiconductor devices (40 CFR 469.12-91).

Manufacturer: Includes an importer and the term manufacture includes importation (CAA211.o).

Manufacturers formulation: A list of substances or component parts as described by the maker of a coating, pesticide, or other product containing chemicals or other substances (EPA-94/04).

Manufacturer of pesticide active ingredients: The chemical and/or physical conversion of raw materials to technical grade ingredients intended to prevent, destroy, repel, or mitigate any pest (EPA-85/10).

Manufacturer of pesticide intermediates: The manufacture of materials resulting from each reaction step in the creation of pesticide active ingredients, except for the final synthesis step. According to this definition an excess of materials need not be produced (EPA-85/10).

Manufacturer of products other than pesticides: The manufacture of products not specifically defined in the scope of coverage (e.g., organic chemicals, plastics and synthetics, pharmaceuticals, etc.) (EPA-85/10).

Manufacturer's proving ground: A facility whose sole purpose is to develop complete advanced vehicles for an automotive manufacturer (40 CFR 52.136; 52.137-91).

Manufacturer's rated dryer capacity: The dryer's rated capacity of articles, in pounds or kilograms of clothing articles per load, dry basis, that is typically found on each dryer on the manufacturer's name-plate or in the manufacturer's equipment specifications (40 CFR 60.621-91).

Manufacturing: The combining of commercial asbestos - or, in the case of woven friction products, the combining of textiles containing commercial asbestos - with any other material(s), including commercial asbestos, and the processing of this combination into a product. Chlorine production is considered a part of manufacturing (40 CFR 61.141-91).

Manufacturing activities: All those activities at one site which are necessary to produce a substance identified in Subpart D of this Part and make it ready for sale or use as the listed substance, including purifying or importing the substance (40 CFR 704.203-91).

Manufacturing line: The manufacturing equipment comprising the forming section, where molten glass is fiberized and a fiberglass mat is formed; the curing section, where the binder resin in the mat is thermally set; and the cooling section, where the mat is cooled (40 CFR 60.681-91).

Manufacturing process: A method whereby a process emission source or series of process emission sources is used to convert raw materials, feed stocks, subassemblies, or other components into a product, either for sale or for use as a component in a subsequent manufacturing process (40 CFR 52.741-91, *see also* 40 CFR 761.3-91).

Manufacturing stream: All reasonable anticipated transfer, flow, or disposal of a chemical substance, regardless of physical state or concentration, through all intended operations of manufacture, including the cleaning of equipment (40 CFR 721.3-91).

Manufacturing use product: Any pesticide product that is not an end-use product (40 CFR 152.3-91, *see also* 40 CFR 158.153; 162.151-91).

Manure: Primarily the excreta of animals; may contain some spilled feed or bedding (SW-108ts).

Margin of exposure (MOE): The ratio of the no observed adverse effect level (NOAEL) to the estimated exposure dose (EED) (EPA-92/12).

Margin of safety (MOS):
(1) The older term used to describe the margin of exposure (EPA-92/12).
(2) Maximum amount of exposure producing no measurable effect in animals (or studied humans) divided by the actual amount of human exposure in a population (EPA-94/04).

Mariculture: Growth of marine organisms including marine animals and plants.

Marine bays and estuaries: The semi-enclosed coastal waters which have a free connection to the territorial sea (40 CFR 35.2005-91).

Marine environment: That territorial seas, the contiguous zone and the oceans (*see* environment for more related terms) (40 CFR 125.121-91).

Marine mammal: Any mammal which:
(1) Is morphologically adapted to the marine environment (including sea otters and members of the orders Sirenia, Pinnipedia and Cetacea); or
(2) Primarily inhabits the marine environment (such as the polar bear); and, for the purposes of this chapter, includes any part of any such marine mammal, including its raw, dressed, or dyed fur or skin (MMPA3-16USC1362-90).

Marine mammal product: Any item of merchandise which consists, or is composed in whole or in part, of any marine mammal (MMPA3-16USC1362-90).

Marine paint: A varnish specially designed to withstand immersion in water and exposure to marine atmosphere (EPA-79/12b).

Marine Protection, Research, and Sanctuaries Act (MPRSA) of 1972: *See* the term of Act or MPRSA.

Marine sanitation device:
(1) Includes any equipment for installation on board a vessel which is designed to receive, retain, treat, or discharge sewage, and any process to treat such sewage (CWA312, *see also* 40 CFR 140.1-91).
(2) Any equipment or process installed on board a vessel to receive, retain, treat, or discharge sewage (EPA-94/04).

Marine vessel: Any tank ship or tank barge which transports liquid product such as benzene (40 CFR 61.301-91).

Market pulp: A pulp manufactured explicitly for purchase (EPA-87/10).

Markov process: A process in which the future state of the process is only dependent on the present state and not any past states (NATO-78/10).

MARPOL 73/78: The International Convention for the Prevention of Pollution from Ships, 1973, as modified by the Protocol of 1978 relating thereto, Annex 1, which regulates pollution from oil and which entered into force on October 2, 1983 (40 CFR 110.1-91).

Marsh:
(1) An area of low-lying wetland, dominated by grasslike plants (DOE-91/04).
(2) A type of wetland that does not accumulate appreciable peat deposits and is dominated by herbaceous vegetation. Marshes may be either fresh or saltwater, tidal or non-tidal (*see* wetlands) (EPA-94/04).

Masking: Blocking out one sight, sound, or smell with another (OME-88/12).

Mass and weight meter: Mass is the measure of the quantity of matter and weight is the measure of the gravity force acting on a mass. Since this force is in fixed proportion to the mass, the two terms are not often distinguished in engineering and physical calculations. Indeed, almost all practical measures of mass are based on weight. The meter includes various scales and weighers (cf. solid rate meter).

Mass balance (or material balance):
(1) An accumulation of the annual quantities of chemicals transported to a facility, produced at a facility, consumed at a facility, used at a facility, accumulated at a facility, released from a facility, and transported from a facility as a waste or as a commercial product or byproduct or component of a commercial product or byproduct (SF311.1, *see also* 40 CFR 797.2850-91).
(2) A concept based on a fundamental law of physical science (conservation of mass) which says that matter can not be created or destroyed. It is used to calculate all input and output streams of a given substance in a system (cf. energy balance) (EPA-84/09).
(3) An accounting of the weights of materials entering and leaving a processing unit, such as an incinerator, usually on an hourly basis (SW-108ts).

Mass burn incinerator: An incinerator which burns unprocessed, mixed municipal solid waste (MSW) in a single combustion chamber under conditions of excess air. MSW is stored in a pit and is fed into the incinerator by a crane, which also can remove oversized items. It is burned in a sloping, moving grate. The movement (e.g., vibrating, reciprocating, or pulsing) helps agitate the MSW and mix it with air, and causes it to tumble down the slope. Many proprietary grates have been designed. Some systems use a rotating (or rotary) kiln rather than grates to agitate the waste and mix it with air. Many new mass burn incinerators use computer systems to precisely control grate movement, underfire air, and overfire air (*see* incinerator for more related terms) (OTA-89/10).

Mass burn refractory MWC: A combustor that combusts MSW in a refractory wall furnace. This does not include rotary combustors without waterwalls (40 CFR 60.51a-91).

Mass burn rotary waterwall MWC: A combustor that combusts MSW in a cylindrical rotary waterwall furnace. This does not include rotary combustors without waterwalls (40 CFR 60.51a-91).

Mass burn waterwall MWC: A combustor that combusts MSW in a conventional waterwall furnace (40 CFR 60.51a-91).

Mass concentration: Concentration expressed in terms of mass of substance per unit volume of gas or liquid (EPA-83/06).

Mass feed stoker steam generating unit: A steam generating unit where solid fuel is introduced directly into a retort or is fed directly onto a grate where it is combusted (40 CFR 60.41b-91).

Mass flowmeter (or coriolis flowmeter): One of liquid flow rate meters. This instrument applies to liquids of widely varying viscosity and density and most slurries. It has been advertised for use with gases, but that application may be rare. The reported accuracy is within +/- 1% (EPA-89/06) (*see* flow rate meter for more related terms).

Mass flux: *See* synonym, mass velocity.

Mass median aerodynamic diameter (MMAD): Mass median of the distribution of mass with respect to aerodynamic diameter. Graphs for these distributions are constructed by plotting frequency against aerodynamic diameters (EPA-90/08; 92/12).

Mass median size: A measurement of particle size for samples of particulate matter, consisting of that diameter such that the mass of all larger particles is equal to the mass of all smaller particles (EPA-83/06).

Mass of pollutant that can be discharged: The pollutant mass calculated by multiplying the pollutant concentration times the average process water usage flow rate (40 CFR 463.2-91).

Mass selective detector: A bench top, low resolution mass spectrometer; the basic principle underlying a mass spectrometer is that substances in the gaseous or vapor state, when subjected to high voltage electric current, lose electrons and form positively charged ions (cations); these cations are accelerated and deflected by magnetic and/or electric fields; the deflection of an ion depends on its mass, charge, and velocity; if charge, velocity, and deflection force are constant, the deflection is less for a heavy particle and more for a light one; mass spectrum can be created and recorded according to the mass to change ratio of the ions (EPA-88/12).

Mass spectrometer: An instrument that analyzes samples by sorting molecular or atomic ions according to their masses and electrical charges. A MS can be used to detect very low concentrations of a chemical (cf. chromatography) (EPA-81/09).

Mass spectrometry: The use of mass spectrometer to analyze chemical elements and their structures (cf. spectrometry or GC/MS).

Mass spectroscope: A technique to analyze the masses of atoms or molecules based on the variation of ion deflections from different masses (cf. spectroscope).

Mass specroscopy: A technique used to analyze relative atomic masses, isotopes, and ion reactions (*see* spectroscopy for more related terms).

Mass transfer: Mass transfer related terms include:
- Convective mass transfer; and
- Diffusional mass transfer which is also known as diffusion. *See* diffusion.

Mass transfer zone (MTZ): An adsorbent bed where the concentration gradient is present (a concentration gradient zone across which mass transfer takes place). It exists between the location where the concentration is saturated (100%) and a value approaching zero (0%) (i.e., the concentration gradient ranges from a bulk concentration to a value approaching to zero across a very small distance). MTZ in activated carbon applications are in the 2-4 inch range. The MTZ is dependent on the adsorbent, packing size, bed depth, gas velocity, temperature, and total pressure of the gas stream (EPA-84/09).

Mass velocity (or mass flux): The mass flow rate divided by the cross-sectional area through which the mass is flowing. In absorption practice, the mass velocity is expressed in pounds per seconds per square foot, pounds per hour per square foot, or kilograms per second per square meter. An advantage of using mass velocity is that it is independent of temperature and pressure when the flow is steady (constant mass flow rate) and the cross section is unchanged (EPA-84/09).

Master inventory file: EPA's comprehensive list of chemical substances which constitute the Chemical Substances Inventory compiled under section 8(b) of the Act. It includes substances reported under Subpart A of this part and substances reported under Part 720 of this chapter for which a Notice of Commencement of Manufacture or Import has been received under 40 CFR 720.120 of this chapter (40 CFR 710.23-91).

Mat: A glass-fiber product of feltlike nature (EPA-83).

Matching funds: The portion of allowable project costs that a recipient contributes toward completing the technical assistance grant project using non-Federal funds or Federal funds if expressly authorized by statute. The match may include in-kind as well as cash contributions (40 CFR 35.4010-91).

Material: Matter of any kind or description, including, but not limited to, dredged material, solid waste, incinerator residue, garbage, sewage, sewage sludge, munitions, radiological, chemical, and biological warfare agents, radioactive materials, chemicals, biological and laboratory waste, wreck or discarded equipment, rock, sand, excavation debris, industrial, municipal, agricultural, and other waste, but such term does not mean sewage from vessels within the meaning of section 312 of the FWPCA. Oil within the meaning of section 311 of the FWPCA shall constitute "material" for purposes of this Subchapter H only to the extent that it is taken on board a vessel or aircraft for the primary purpose of dumping (40 CFR 220.2-d-91). Other material related terms include:
- Primary material
- Secondary material
- Spent material
- Virgin material

Material balance: *See* synonym, mass balance.

Material category: In the asbestos program, broad classification of materials into thermal surfacing insulation, surfacing material, and miscellaneous material (EPA-94/04).

Material handling: Conveying, elevating, pumping, packing, and shipping (AP-40, p790).

Material management: An municipal solid waste (MSW) management approach that would:
(1) Coordinate product manufacturing with different management methods (e.g., design products for recyclability); and
(2) Manage MSW on a material-by-material basis, by diverting discarded materials to a most appropriate management method based on their physical and chemical characteristics (OTA-89/10).

Material recovery:
(1) Retrieval of materials from waste (OTA-89/10).
(2) The concept of resource recovery, emphasis is on separating and processing waste materials for beneficial use or reuse. Materials recovery usually refers to paper, glass, metals, rubber, plastics, or textiles (cf. front-end system) (EPA-83).

Material recovery facility:
(1) A facility for separating recyclables from mixed waste or for separating commingled recyclables (OTA-89/10).
(2) A facility that processes residentially collected mixed recyclables into new products available for market (EPA-94/04).

Material Type: Classification of suspect material by its specific use or application, e.g., pipe insulation, fireproofing, and floor tile (EPA-94/04).

Material recovery section: The equipment that recovers unreacted or byproduct materials from any process section for return to the process line, off-site purification or treatment, or sale. Equipment designed to separate unreacted or byproduct material from the polymer product are to be included in this process section, provided at least some of the material is recovered for reuse in the process, off-site purification or treatment, or sale, at the time the process section becomes an affected facility. Otherwise, such equipment are to be assigned to one of the other process sections, as appropriate. Equipment that treats recovered materials are to be included in this process section, but equipment that also treats raw materials are not to be included in this process section. The latter equipment are to be included in the raw materials preparation section. If equipment is used to return unreacted or byproduct material directly to the same piece of process equipment from which it was emitted, then that equipment is considered part of the process section that contains the process equipment. If equipment is used to recover unreacted or byproduct material from a process section and return it to another process section or a different piece of process equipment in the same process section or sends it off-site for purification, treatment, or sale, then such equipment are considered part of a material recovery section. Equipment used for the on-site recovery of ethylene glycol from poly(ethylene terephthalate) plants, however, are not included in the material recovery section, but are covered under the standards applicable to the polymerization reaction section (40 CFR 60.562-1(c)(1)(ii)(A) or (2)(ii)(A)) (40 CFR 60.561-91).

Material safety data sheet (MSDS):
(1) The sheet required to be developed under section 1901.1200(g) of title 29 of the Code of Federal Regulations, as that section may be amended from time to time (SF329, *see also* 40 CFR 370.2; 721.3-91).
(2) A compilation of information required under the OSHA

Communication Standard on the identity of hazardous chemicals, health, and physical hazards, exposure limits, and precautions. Section 311 of SARA requires facilities to submit MSDSs under certain circumstances (EPA-94/04).

Material specification: A specification that stipulates the use of certain materials to meet the necessary performance requirements (*see* specification for more related terms) (40 CFR 247.101-91, *see also* 40 CFR 249.04-91).

Mathanoic acid: *See* synonym, formic acid.

Mathematical model:
(1) A mathematical simulation of a real process, expressed as a set of equations (NATO-78/10).
(2) A quantitative equation or system of equations formulated in such a way as to reasonably depict the structure of a situation and the relationships among the relevant variables (EPA-82/10).

Mathieson process: A process of producing chlorine dioxide, using SO_2 as a reducing agent (EPA-87/10).

Matrix:
(1) A fiber of fibers with one end free and the other end embedded in or hidden by a particulate. The exposed fiber must meet the fiber definition (40 CFR 763-App/A-91).
(2) In quality control, the material in which the analyte(s) of primary interest is embedded (EPA-84/03).

Matte: A metal sulfide mixture produced by smelting sulfide ores (EPA-83/03a).

Maximum 30 day average: The maximum average of daily values for 30 consecutive days (40 CFR 439.1-91).

Maximum acceptable toxicant concentration: The maximum concentration at which a chemical can be present and not be toxic to the test organism (4 CFR 797.1330-91, *see also* 40 CFR 797.1950-91).

Maximum actual operating pressure: The maximum pressure that occurs during normal operations over a period of 1 year (40 CFR 192.3-91).

Maximum allowable operating pressure (MAOP): The maximum pressure at which a pipeline or segment of a pipeline may be operated under this part (40 CFR 192.3-91).

Maximum capability: *See* synonym, firm capacity.

Maximum capacity: In wastewater treatment, either the maximum wastewater can be put through a plant hydraulically, or some lower rate established by the management (such as the maximum rate at which wastewater can be treated without seriously interrupting the treatment process) (DOI-70/04).

Maximum contaminant levels (MCLs):
(1) The maximum permissible level of a contaminant in water which is delivered to any user of a public water system (SDWA1401-91, *see also* 40 CFR 142.2; 300-App/A-91).
(2) The maximum permissible level of a contaminant in water delivered to any user of a public water system. MCLs are enforceable standards (EPA-94/04).
(3) *See* the term of Act or SDWA for more related terms.

Maximum contaminant level goals (MCLGs):
(1) Under the Safe Drinking Water Act, a non-enforceable concentration of a drinking water contaminant, set at the level at which no known or anticipated adverse effects on human health occur and which allows an adequate safety margin. The MCLG is usually the starting point for determining the regulated Maximum Contaminant Level (*see* maximum contaminant level (EPA-94/04).
(2) The maximum level of a contaminant in drinking water at which no known or anticipated adverse effect on the health or persons would occur, and which allows an adequate margin of safety. Maximum contaminant level goals are nonenforceable health goals.

MCLGs include two regulations:
(1) Primary drinking water regulation
(2) Secondary drinking water regulation
(3) *See* the term of Act or SDWA for more related terms.

Maximum credible accident: The most serious accident that can occur as a result of a feasible sequence of equipment failures. The design should be aimed at minimizing the risks resulting from such an accident (NATO-78/10).

Maximum day limitation: The effluent limitation value equal to the maximum for one day and is the value to be published by the EPA in the Federal Register (cf. maximum thirty limitation) (EPA-76/03).

Maximum daily discharge limitation: The highest allowable daily discharge.

Maximum demonstrated MWC unit load: The maximum 4-hour block average MWC unit load achieved during the most recent dioxin/furan test demonstrating compliance with the applicable standard for MWC organics specified under 40 CFR 60.53a (40 CFR 60.51a-91).

Maximum demonstrated particulate matter control device temperature: The maximum 4-hour block average temperature measured at the final particulate matter control device inlet during the most recent dioxin/furan test demonstrating compliance with the applicable standard for MWC organics specified under 40 CFR 60.53a. If more than one particulate matter control device is used in series at the affected facility, the maximum 4-hour block average temperature is measured at the final particulate matter control device (40 CFR 60.51a-91).

Maximum design heat input capacity: The ability of a steam generating unit to combust a stated maximum amount of fuel (or combination of fuels) on a steady state basis as determined by the physical design and characteristics of the steam generating unit (40 CFR 60.41c-91).

Maximum electric demand (peak demand or peak load): The greatest of all of the demands of the load under consideration which has occurred during a specified period of time (*see* electric demand for more related terms) (EPA-83).

Maximum emission concentration: Regulatory standard(s) for maximum concentration of air pollutant emission from stationary or moving sources, including opacity, gravimetric emission, standards (expressed as weight of emitted pollutant or percent of gaseous pollutant by volume in the emitted gas or as parts per volume tn specific parts of effluent gas). *See* Emissions (EPA-83).

Maximum exposed individual (MEI):
(1) EPA's hypothetical limit to ensure that emissions of individual metals and HCL do not exceed ambient health based levels (EPA-89/09).
(2) A hypothetical individual that receives the maximum possible exposure to a chemical from all possible sources and pathways (ETI-92).

Maximum exposed individual (MEI) risk: The risk at the point where the maximum concentration occurs regardless of the actual population distribution. EPA is proposing that MEI risk of less than 10^{-5} (1 cancer per 100,000 people exposed to any metal emissions) (EPA-89/09).

Maximum for any one day and average of daily values for thirty consecutive days: Shall be based on the daily average mass of material processed during the peak thirty consecutive day production period (40 CFR 407.61-91, *see also* 40 CFR 407.71; 407.81-91).

Maximum for any one day limitations: Effluent limitations determined by multiplying long-term average effluent concentration by appropriate variability factors (EPA-87/10a).

Maximum heat input capacity: The ability of a steam generating unit to combust a stated maximum amount of fuel on a steady state basis, as determined by the physical design and characteristics of the steam generating unit (40 CFR 60.41b-91).

Maximum monthly for average limitations: Effluent limitations determined by multiplying long term average effluent concentration by appropriate variability factors (EPA-87/10a).

Maximum mixing height: The maximum height of the mixing layer, usually attained over a land surface near sunset. In practice, it is calculated from the temperature profile by determining the height of the intersection of this profile with the dry adiabatic starting from the maximum ground level temperature (NATO-78/10).

Maximum production capacity: Either the maximum demonstrated rate at which a smelter has produced its principal metallic final product under the process equipment configuration and operating procedures prevailing on or before August 7, 1977, or a rate which the smelter is able to demonstrate by calculation is attainable with process equipment existing on August 7, 1977. The rate may be expressed as a concentrate feed rate to the smelter (40 CFR 57.103-91).

Maximum rated capacity: That the portable air compressor, operating at the design full speed with the compressor on load, delivers its rated cfm output and pressure, as defined by the manufacturer (40 CFR 204.51-91).

Maximum rated horsepower: The maximum brake horsepower output of an engine as stated by the manufacturer in his sales and service literature and his application for certification under 40 CFR 86.082.21 (40 CFR 86.082.2-91).

Maximum rated RPM: The engine speed measured in revolutions per minute (RPM) at which peak net brake power (SAE J-245) is developed for motorcycles of a given configuration (40 CFR 205.151-91).

Maximum rated torque: The maximum torque produced by an engine as stated by the manufacturer in his sales and service literature and his application for certification under 40 CFR 86.082.21 (*see* torque for more related terms) (40 CFR 86.082.2-91).

Maximum sound level: The greatest A-weighted sound level in decibels measured during the designated time interval or during the event, with either fast meter response 40 CFR 201.1(1) or slow meter response 40 CFR 201.l(ii) as specified. It is abbreviated as L_{max} (40 CFR 201.1-91).

Maximum theoretical emissions: The quantity of volatile organic material emissions that theoretically could be emitted by a stationary source before add on controls based on the design capacity or maximum production capacity of the source and 8760 hours per year. The design capacity or maximum production capacity includes use of coating(s) or ink(s) with the highest volatile organic material content actually used in practice by the source (40 CFR 52.741-91).

Maximum thirty day limitation: The effluent limitation value for which the average of daily values for thirty consecutive days shall not exceed and is the value to be published by the EPA in the Federal Register (cf. maximum day limitation) (EPA-76/03).

Maximum tolerated dose (MTD):
(1) The maximum dose that an animal species can tolerate for a major portion of its lifetime without significant impairment of growth or observable toxic effect other than carcinogenicity (Course 165.6).
(2) The maximum doses that an animal species can tolerate for a major portion of its lifetime without significant impairment or toxic effect other than carcinogenicity (EPA-94/04).
(3) *See* dose for more related terms.

Maximum total trihalomethane potential (MTP): The maximum concentration of total trihalomethanes produced in a given water containing a disinfectant residual after 7 days at a temperature of 25 C or above (40 CFR 141.2-91).

Maximum true vapor pressure: The equilibrium partial pressure exerted by the stored VOL at the temperature equal to the highest calendar-month average of the VOL storage temperature for VOL's stored above or below the ambient temperature or at the local maximum monthly average temperature as reported by the National Weather Service for VOL's stored at the ambient temperature, as determined:
(1) In accordance with methods described in American

Petroleum Institute Bulletin 2517, Evaporation Loss From External Floating Roof Tanks (incorporated by reference, 40 CFR 0.17); or

(2) As obtained from standard reference texts; or
(3) As determined by ASTM Method D2879-83 (incorporated by reference, 40 CFR 0.17);
(4) Any other method approved by the Administrator (40 CFR 60.111b-f-91).

Mean: The arithmetic average of the individual sample values (EPA-76/03). Other mean related terms include:
- Arithmetic mean
- Geometric mean
- Harmonic mean

Mean fork length: The length of a fish measured from the head to the point where the tail begins to fork (EPA-76/04).

Mean of emission limitation: A system of continuous emission reduction (including the use of specific technology or fuels with specified pollution characteristics) (CAA302-42USC7602-91).

Mean particle diameter (or average particle diameter): In air pollution control, the mean is always based on mass. Thus, if a particle size distribution has a mean of 7.2 microns, 50% of the particles by mass have a size equal to or less than 7.2 microns (EPA-84/09).

Mean spectral response: The wavelength that is the arithmetic mean value of the wavelength distribution for the effective spectral response curve of the transmissometer (40 CFR 60-App/B-91).

Mean retention time: The time obtained by dividing a reservoir's mean annual minimum total storage by the non-zero 30-day, ten-year low-flow from the reservoir (40 CFR 131.35-91).

Mean value: With respect to a set of data varying within a certain range, the mean value is representative for the center value of the set (NATO-78/10).

Mean velocity: The average velocity of a stream flowing in a channel or conduit at a given cross section or in a given reach. It is equal to the discharge divided by the cross sectional area of the reach. Also called average velocity (EPA-82/11f).

Meander: One of the curves in a river course that swings from side to side in wide loops as it progresses across flat country. A meander is continually being accentuated by the river itself, the concave bank being worn away by the current while solid material is being deposited at the convex bank (DOI-70/04).

Measure: To ascertain the extent, degree, quantity, dimensions, or capability with respect to a standard, hence to estimate (ACS-87/11).

Measurement: Refers to the ability of the analytical method or protocol to quantify as well as identify the presence of the substance in question (40 CFR 403.7-91). Other measurement related terms include:
- Gas concentration measurement system
- Total hydrocarbon measurement

Measurement period: A continuous period of time during which noise of railroad yard operations is assessed, the beginning and finishing times of which may be selected after completion of the measurements (40 CFR 201.1-91).

Measurement system performance specifications: For hazardous waste incineration, the performance of CO_2, O_2 and THC measurement is recommended below (EPA-90/04):
(1) For CO_2 measurement
  (A) Zero Drift (less than span value): --
  (B) Calibration Drift (less than span value): +/- 5%
  (C) Calibration Error (less than calibration gas value): +/- 5%
(2) For O_2 measurement
  (A) Zero Drift (less than span value): --
  (B) Calibration Drift (less than span value): +/- 0.5
  (C) Calibration Error (less than calibration gas value): +/- 0.5%
(3) For THC measurement
  (A) Zero Drift (less than span value): +/- 3%
  (B) Calibration Drift (less than span value): +/- 3%
  (C) Calibration Error (less than calibration gas value): +/- 5%

Meat cutter: An operation which fabricates, cuts, or otherwise produces fresh meat cuts and related finished products from livestock carcasses, at rates greater than 2730 kg (6000 lb) per day (40 CFR 432.61-91).

Mechanical aeration: Use of mechanical energy to inject air into water to cause a waste stream to absorb oxygen (see aeration for more related terms) (EPA-94/04).

Mechanical agitation: The agitation of a liquid medium through the use of mechanical equipment such as impellers or paddles (EPA-83/06a).

Mechanical and thermal integrity: The ability of a converter to continue to operate at its previously determined efficiency and light-off time and be free from exhaust leaks when subject to thermal and mechanical stresses representative of the intended application (40 CFR 85.2122(a)(15)(ii)(G)-91).

Mechanical clarifier: A man-made device designed specifically for the detention of waste water for the purpose of removal of the settleable solids from the waste water under controlled operating conditions (see clarifier for more related terms) (EPA-74/01a).

Mechanical collector: A device which uses inertial and gravitational forces to separate dry dust from gas (see collector for more related terms) (EPA-83).

Mechanical composting method: A composting procedure characterized by continuous mechanical mixing and forced aeration. Also called high rate composting and continuous-mix composting (EPA-83).

Mechanical draft tower: A cooling tower in which the air flow through the tower is maintained by fans. In **forced draft towers**, the air is forced through the tower by fans located at its base; whereas in **induced draft towers**, the air is pulled through the tower by fans mounted on top of the tower (cf. natural draft cooling tower) (EPA-82/11f).

Mechanical finish: Final operations on a product performed by a machine or tool. They include polishing, buffing, barrel finishing, shot peening, power brush finishing (EPA-83/06a).

Mechanical plating: Providing a coating wherein fine metal powders are peened onto the part by tumbling or other means (*see* plating for more related terms) (EPA-83/06a).

Mechanical process composting: A method in which the compost is continuously and mechanically mixed and aerated (*see* composting for more related terms) (SW-108ts).

Mechanical pulp: The pulp produced by physical means without the use of chemicals or heat, often referred to as groundwood (*see* pulp for more related terms) (EPA-87/10).

Mechanical scrubber: One of air pollution control devices. A mechanical scrubber is a mechanically aided scrubber which incorporates a motor-driven device such as fan blades which cause gas flows and on which particles collect by impaction. When liquid is introduced at the hub of the rotating fan blades, some atomizes upon impact with the fan and some runs over, and washes the blades. The liquid is recaptured by the fan housing, which drains into a sump. As with the orifice types, the water is usually recirculated (*see* scrubber for more related terms) (AP-40; Calvert-84).

Mechanical separation:
(1) Separation of waste into various components using mechanical means, such as cyclones, trommels, and screens (*see* separation for more separation methods) (EPA-89/11).
(2) Using mechanical means to separate waste into various components (EPA-94/04).

Mechanical separator (centrifugal collector, centrifugation, or inertial separator):
(1) A physical process whereby the components of a fluid mixture are prepared mechanically by the application of centrifugal force, which is applied by rapidly rotating the mass of fluid within a confined rigid vessel. Centrifugal forces acting on the revolving mass of fluid cause the solids suspended in the fluid to migrate to the periphery of the vessel where they can be separated. The particles are removed as a liquid/solid mixture significantly more concentrated than the original liquid.
(2) A mechanical system using centrifugal and/or gravitational force to remove particulate matter or aerosols from a gas stream or to dewater sludge. This includes those designed with no moving parts, such as settling chambers and cyclones where the collection energy comes from the inlet gas stream. An inertial separator operates by the principle of imparting centrifugal force to the particle to be removed from the carrier gas stream. This force is produced by directing the gas in a circular path or effecting an abrupt change in direction. Although an inertial separator is suitable for medium-sized particulates (15 to 40 microns), it is generally unsuitable for fine dusts or metallurgical fumes. Dusts with a particle size ranging from 5 to 10 microns are normally too fine to be collected efficiently (AP-40, p91).
(3) *See* separator for more related terms. Types of mechanical separators include:
(A) Cyclone
(B) Cyclone collector (*see* synonym, cyclone)
(C) Cyclone separator (*see* synonym, cyclone)
(D) Multi-cyclone

Mechanical torque rate: A term applied to a thermostatic coil, defined as the torque accumulation per angular degree of deflection of a thermostatic coil (*see* torque for more related terms) (40 CFR 85.2122(a)(2)(iii)(F)-91).

Mechanical turbulence: Random irregularities of fluid motion in air caused by buildings or other non-thermal, processes (*see* turbulence for more related terms) (EPA-94/04).

Mechanistic model: A mathematical description of a process based on understanding relevant phenomena as opposed to empirical modeling (EPA-88/09a).

Mechanism: An administrative procedure, guideline, manual, or written statement (40 CFR 56.1-91).

Media: Specific environments-air, water, soil-which are the subject of regulatory concern and activities (EPA-94/04).

Median: In a statistical array, the value lying in the middle of an increasing or decreasing series of numbers such that the same number of values appears above the median as do below it (EPA-87/10a).

Median diameter: *See* synonym, geometric mean diameter.

Median lethal dose (LD50): The dose lethal to 50 percent of a group of test organisms for a specified period. The dose material may be ingested or injected (EPA-76/03).

Median tolerance limit (TLM): In toxicological studies, the concentration of pollutants at which 50 percent of the test animals can survive for a specified period of exposure (EPA-76/03).

Median value: A data observation located at the 50th percentile or the mid-range (EPA-82/10).

Medical device: Any device (as defined in the Federal Food, Drug, and Cosmetic Act (21USC321)), diagnostic product, drug (as defined in the Federal Food, Drug, and Cosmetic Act), and drug delivery system - (A) if such device, product, drug, or drug delivery system utilizes a class I or class II substance for which no safe and effective alternative has been developed, and where necessary, approved by the Commissioner; and (B) if such device, product, drug, or drug delivery system, has, after notice and opportunity for public comment, been approved and determined to be essential by the Commissioner in consultation with the Administrator (CAA601-42USC7671-91).

Medical emergency: Any unforeseen condition which a health professional would judge to require urgent and unscheduled medical attention. Such a condition is one which results in sudden and/or serious symptom(s) constituting a threat to a person's physical or psychological well-being and which requires immediate medical attention to prevent possible deterioration, disability, or death (40 CFR 350.40-91).

Medical surveillance: A periodic comprehensive review of a worker's health status; acceptable elements of such surveillance program are listed in the Occupational Safety and Health Administration standards for asbestos (EPA-94/04).

Medical use: The intentional internal or external administration of byproduct material, or the radiation therefrom, to human beings in the practice of medicine in accordance with a license issued by a State or Territory of the United States, the District of Columbia, or the Commonwealth of Puerto Rico (10 CFR 30.4-91).

Medical waste:
(1) Except as otherwise provided in this paragraph, the term, medical waste, means any solid waste which is generated in the diagnosis, treatment, or immunization of human beings or animals, in research pertaining thereto, or in the production or testing of biologicals. Such term does not include any hazardous waste identified or listed under subtitle C or any household waste as defined in regulations under subtitle C (CWA502, *see also* RCRA1004; 40 CFR 60.51a; 259.10-91).
(2) Any solid waste generated in the diagnosis, treatment, or immunization of human beings or animals, in research pertaining thereto, or in the production or testing of biologicals, excluding hazardous waste identified or listed under 40 CFR Part 261 or any household waste as defined in 40 CFR Sub-section 261.4 (b)(1)) (EPA-94/04).
(3) *See* waste for more related terms.
Other medical waste related terms include:
- Biomedical waste
- Destroyed regulated medical waste
- Hospital waste
- Oversized regulated medical waste
- Regulated medical waste
- Treated regulated medical waste
- Untreated regulated medical waste

Medical waste incinerator: An incinerator operated for the purpose of burning hospital and medical waste. There are three basic types of incinerators used for medical waste incineration. They are:
(1) Controlled air
(2) Multiple chamber
(3) Rotary kiln incinerators (EPA-89/03b)
(4) *See* incinerator for more related terms

Medical Waste Tracking Act (MWTA) of 1988: *See* the term of Act or MWTA.

Medium: A point source that processes a total annual raw material production of fruits, vegetables, specialties and other products that is between 1,816 kkg (2,000 tons) per year and 9,080 kkg (10,000 tons) per year (cf. media) (40 CFR 407.61-91, *see also* 40 CFR 407.71; 407.81-91).

Medium density polyethylene (MDPE): A polymer prepared by low-pressure polymerization of ethylene as the principal monomer and having the characteristics of ASTM D134 Type II polyethylene. Such polymer resins have density less than 0.41 g/cc as noted in ASTM D124 (EPA-91/05).

Medium duty vehicle: A gasoline powered motor vehicle rated at more than 6,000 lb GVW and less than 10,000 lb GVW (*see* vehicle for more related terms) (40 CFR 52.1161-91, *see also* 40 CFR 52.2485; 52.2490-91).

Medium municipal separate storm sewer system: All municipal separate storm sewers that are either:
(1) Located in an incorporated place with a population of 100,000 or more but less than 250,000, as determined by the latest Decennial Census by the Bureau of Census (appendix G); or
(2) Located in the counties listed in appendix I, except municipal separate storm sewers that are located in the incorporated places, townships or towns within such counties; or
(3) Owned or operated by a municipality other than those described in paragraph (b)(4)(i) or (ii) of this section and that are designated by the Director as part of the large or medium municipal separate storm sewer system due to the interrelationship between the discharges of the designated storm sewer and the discharges from municipal separate storm sewers described under paragraph (b)(4)(i) or (ii) of this section. In making this determination the Director may consider the following factors:
 (A) Physical interconnections between the municipal separate storm sewers;
 (B) The location of discharges from the designated municipal separate storm sewer relative to discharges from municipal separate storm sewers described in paragraph (b)(7)(i) of this section;
 (C) The quantity and nature of pollutants discharged to waters of the United States;
 (D) The nature of the receiving waters; or
 (E) Other relevant factors; or
(4) The Director may, upon petition, designate as a medium municipal separate storm sewer system, municipal separate storm sewers located within the boundaries of a region defined by a storm water management regional authority based on a jurisdictional, watershed, or other appropriate basis that includes one or more of the systems described in paragraphs (b)(7)(i), (ii), (iii) of this section (40 CFR 122.26-7-91).

Medium-size water system: A water system that serves than 3,300 to 50,000 persons (EPA-94/04).

Medium sized plant: The plants which process between 3,720 kg/day (8,200 lbs/day) and 10,430 kg/day (23,000 lbs/day) of raw materials (40 CFR 428.61-91).

Megawatt (MW): A unit of power equal to or 1 million watts (1000 kilowatts). Megawatt thermal (MWt) is commonly used to define heat produced, while megawatt electric (MWe) defines electricity produced (DOE-91/04).

Megawatt hour (Mwh): The megawatt hour(s) of electrical energy consumed in the smelting process (furnace power consumption) (40 CFR 424.11-91, *see also* 40 CFR 424.21-91).

Melt (or melting):
(1) That amount of raw material (raw sugar) contained within aqueous solution at the beginning of the process for production of refined cane sugar (40 CFR 409.21-91, *see also* 40 CFR 409.31-91).
(2) A specific quantity of glass made at one time (EPA-83).

Melt test: Technique of melting sample in an induction furnace under a blanket of argon gas as the first step tn determining chemical analysis of municipal ferrous scrap (EPA-83).

Meltdown: The melting of the scrap and other solid metallic elements of the charge (EPA-74/06a).

Meltdown and refining: That phase of the steel production cycle when charge material is melted and undesirable elements are removed from the metal (40 CFR 60.271-91).

Meltdown and refining period: The time period commencing at the termination of the initial charging period and ending at the initiation of the tapping period, excluding any intermediate charging periods (40 CFR 60.271-91).

Melting (*see also* melt):
(1) That phase of steel production cycle during which the iron and steel scrap is heated to the molten state (40 CFR 60.271a-91).
(2) The thermal process by which the charge is completely converted into molten glass free from undissolved batch (EPA-83).

Melting point: The range of furnace temperatures within which melting takes place at a commercially desirable rate, and at which the resulting glass generally has a viscosity of 10 to 25 poises (*see* temperature for more related terms) (EPA-83).

Melting point temperature: For a given pressure, the temperature at which a solid changes phase to a liquid phase. That is the solid and liquid phases of the substance are in equilibrium. The opposite change in phases is the freezing point (*see* temperature for more related terms) (Course 165.5).

Membrane: A thin sheet of synthetic polymer through the apertures of which small molecules can pass, while larger ones are retained (EPA-83/06a).

Membrane barrier: A thin layer of a material impermeable to the flow of gas or water (SW-108ts).

Membrane filtration:
(1) A filtration at pressures ranging from 50 to 100 psig with the use of membranes or thin film. The membranes have accurately controlled pore sites and typically low flux rates (EPA-83/06a).
(2) A method of quantitative or qualitative analysis of bacterial or particulate matter in a water sample filtration through a membrane capable of retaining bacteria (LBL-76/07-water).
(3) *See* filtration for more related terms.

Membrane process: Such as reverse osmosis and ultrafiltration. They are used primarily in the metal industry to remove solutes from wastewater (EPA-85/10).

Mendelevium (Md): A radioactive metal with a.n. 101 and a.w. 256. The element belongs to group IIIB of the periodic table.

Meniscus:
(1) The curved edge of a liquid at its plane of contact with a perpendicular surface (usually a glass tube: (EPA-83))
(2) The curved top of a column of liquid in a small tube (EPA-94/04).

Mercaptan: Various compounds with the general formula R-SH that are analogous to the alcohols and phenols but contain sulfur in place of oxygen and often have disagreeable odors (EPA-85/10).

Merchant: Those by-product cokemaking operations which provide more than fifty percent of the coke produced to operations, industries, or processes other than iron making blast furnaces associated with steel production (40 CFR 420.11-91).

Mercury:
(1) The element mercury, excluding any associated elements, and includes mercury in particulates, vapors, aerosols, and compounds (40 CFR 61.51-91).
(2) The total mercury present in the process wastewater stream exiting the mercury treatment system (40 CFR 415.61-91).
(3) A heavy metal that can accumulate in the environment and is highly toxic if breathed or swallowed (*see* heavy metals) (EPA-94/04).
(4) A liquid metallic element with a.n. 80; a.w. 200.59; d. 13.6 g/cc; m.p. -38.4° C and b.p. 357° C. The element belongs to group IIB of the periodic table.

Major mercury compounds include:
(1) Mercuric chloride ($HgCl_2$): A toxic crystal used in preserving wood and disinfecting.
(2) Mercuric sulfide (HgS):
   (A) A toxic black powder used in pigment for rubber, etc.
   (B) A toxic red powder (Chinese red) used in plastics coloring and pigment.

Mercury chlor alkali cell: A device which is basically composed of an electrolyzer section and a denuder (decomposer) section and utilizes mercury to produce chlorine gas, hydrogen gas, and alkali metal hydroxide (40 CFR 61.51-91).

Mercury chlor alkali electrolyzer: An electrolytic device which is part of a mercury chlor-alkali cell and utilizes a flowing mercury cathode to produce chlorine gas and alkali metal amalgam (40 CFR 61.51-91).

Mercury minerals: The main source is cinnabar (HgS) (EPA-82/05).

Mercury ore: A mineral mined specifically for its mercury content (40 CFR 61.51-91).

Mercury ore processing facility: A facility processing mercury ore to obtain mercury (40 CFR 61.51-91).

Meromictic lake: The lakes in which dissolved substances create a gradient of density differences in depth, preventing complete mixing or circulation of the water (*see* lake for more related terms) (LBL-76/07-water).

Mesh: Number of wires per inch in a screen (EPA-88/08a).

Mesh size: The particle size of granular activated carbon as determined by the U.S. Sieve series. Particle size distribution within a mesh series is given in the specification of the particular carbon (EPA-82/11f).

Meso scale: In meteorology, a length scale of the order of a hundred kilometers. Applicable to phenomena such as land sea breeze, mountain valley wind etc. (cf. macro scale) (NATO-78/10).

Mesophiles: Grow best at medium temperatures, 25 to 40 degrees C (EPA-83). Mesophilic digestion means that the anaerobic digestion temperature is maintained this temperature (*see* thermophile for more related terms).

Mesosphere: The atmospheric region between the stratosphere and the thermosphere.

Mesotrophic: Reservoirs and lakes which contain moderate quantities of nutrients and are moderately productive in terms of aquatic animal and plant life (EPA-94/04).

Metastasis: The transfer of disease from one organ or part to another not directly connected with it; adj., metastatic (EPA-92/12).

Metabolism:
(1) The study of the sum of the processes by which a particular substance is handled in the body and includes absorption, tissue distribution, biotransformation and excretion (40 CFR 795.228-91, *see also* 40 CFR 795.228; 795.231; 795.232-91).
(2) The sum of the chemical reactions occurring within a cell or a whole organism; includes the energy-releasing breakdown of molecules (catabolism) and the synthesis of new molecules (anabolism) (Course 165.6).
(3) The sum of the processes concerned in the building up of protoplasma and its destructive incidental to life: The chemical changes in living cells by which energy is provided for the vital process and activities and new material is assimilated to repair the waste (LBL-76/07-bio).

Metabolite:
(1) A chemical entity produced by one or more enzymatic or nonenzymatic reactions as a result of exposure of an organism to a chemical substance (40 CFR 723.50-91).
(2) Any substances produced by biological processes, such as those from pesticides (EPA-94/04).

Metabolizable: May be utilized by organisms in respiratory and/or growth processes (LBL-76/07-water).

Metal: Metal includes all ferrous, nonferrous and alloy materials (EPA-83). In the secondary materials industry, it includes all nonferrous materials, copper, brass, aluminum, zinc, lead, but not iron and steel (EPA-83). Other metal related terms include:
- Co-precipitation of metal
- Combined metal
- Common metal
- Dissolved metal
- extractable metal
- Ferrous metal
- Heavy metal
- MWC (municipal waste combustion) metal
- Non-ferrous metal
- Precious metal
- Rare earth
- Trace metal
- Refractory metal
- Scrap metal
- Silicon metal
- Suspended metal
- Total metal
- Total recoverable metal
- Toxic metal

Metal analyzer: Types of metal analyzers include:
(1) Cold vapor atomic absorption spectrometry (CVAAS)
(2) Flame atomic absorption spectrometry (FAAS)
(3) Graphite furnace atomic absorption spectrometry (GFAAS)
(4) Hydrite atomic absorption spectrometry (HAAS)
(5) Inductively coupled argon plasma emission spectrometry (ICP)

Metal and hydrogen chloride controls from hazardous waste incineration: Under RCRA activities, EPA issued regulations to control metals, HCL, and products of incomplete combustion (PICs). The regulations entitled, "Burning of Hazardous Waste in Boilers and Industrial Furnaces; Final Rule," were codified in 40 CFR 260, 261, 264, 265, 266, 270 and 271 and were published in Federal Register Vol.56, No.35, February 21, 1991). Some key features of the regulations are provided below:
(1) The risk specific dose (RSD) for carcinogen and reference air concentrations (RAC) for health effects are as follows:
   (A) RSD[1] FOR CARCINOGENIC METALS (40 CFR 266-App/V)
- Arsenic (As)   0.00230 ($\mu g/m^3$)
- Beryllium (Be)   0.00420
- Chromium (Cr)   0.00083
- Cadmium (Cd)   0.00560

   (B) RAC[2] FOR TOXIC METALS (40 CFR 266-App/IV)
- Antimony (Sb)   0.30 ($\mu g/m^3$)
- Barium (Ba)   50.0
- Lead (Pb)   0.09
- Mercury (Hg)   0.08
- Silver (Ag)   3.00
- Thallium (Ti)   0.30

RSD[1]: The dose in which one out of 100,000 population will have a chance to obtain a cancer, if they continuously breath carcinogen metals at the designated RSD concentrations (e.g., chromium: 00230 ug/m^3) for 70 years.

RAC[2]: The concentration in which one out of 100,000 population will have a chance to obtain a health effect, if they continuously breath carcinogen metals at the designated RAC concentrations (e.g., lead: 0.09 $\mu g/m^3$) for 70 years.

(2) The approaches to regulate the above metals and hydrogen

chloride are as follows:

(A) **Tier I:** Tier I sets limits on feed rates. It would be back-calculated from the Tier II emission limits assuming no credit for partitioning of metals to bottom ash or for removal of metals or HCL from stack gases by air pollution control equipment. The Tier I feed rate limits and the Tier II emission limits are numerically equal but expressed in different units: lb/hr feed rate versus g/sec. emission rate. Compliance with Tier I could be demonstrated simply by analysis of waste feeds.

(B) **Tier II:** Tier II sets limits on emissions. It can be derived by back-calculating from ambient levels posing acceptable health risks using dispersion coefficients for reasonable, worst-case facilities. Compliance must be demonstrated by stack emissions tests, thus, partitioning and air pollution control equipment removal efficiency would be considered.

(C) **Tier III:** Tier III allows the applicant to demonstrate by site-specific dispersion modeling that emissions higher than the Tier II limits will nonetheless not result in an exceedance of ambient levels that pose unacceptable health risks. In effect, the applicant would be demonstrating that dispersion of emissions from the facility being permitted is not as severe (i.e., it has better dispersion) as for the reasonable, worst-case facilities used to derive the Tier II limits.

Metal cladding: *See* synonym, cladding

Metal cleaning waste: Any wastewater resulting from cleaning [with or without chemical cleaning compounds] any metal process equipment including, but not limited to, boiler tube cleaning, boiler fireside cleaning, and air preheater cleaning (*see* waste for more related terms) (40 CFR 423.11-91).

Metal coil surface coating operation: The application system used to apply an organic coating to the surface of any continuous metal strip with thickness of 0.15 millimeter (mm) (0.006 in.) or more that is packaged in a roll or coil (40 CFR 60.461-91).

Metal furniture: A furniture piece including, but not limited to, tables, chairs, waste baskets, beds, desks, lockers, benches, shelving, file cabinets, lamps, and room dividers (40 CFR 52.741-91).

Metal furniture coating: Any non-adhesive coating applied to any furniture piece made of metal or any metal part which is or will be assembled with other metal, wood, fabric, plastic or glass parts to form a furniture piece including, but not limited to, tables, chairs, waste baskets, beds, desks, lockers, benches, shelving, file cabinets, lamps, and room dividers. This definition shall not apply to any coating line coating miscellaneous metal parts or products (40 CFR 52.741-91).

Metal furniture coating facility: A facility that includes one or more metal furniture coating line(s) (40 CFR 52.741-91).

Metal furniture coating line: A coating line in which any protective, decorative, or functional coating is applied onto the surface of metal furniture (40 CFR 52.741-91).

Metal ion: An atom or radical that has lost or gained one or more electrons and has thus acquired an electric charge. Positively charged ions are cations, and those having a negative charge are anions. An ion often has entirely different properties from the element (atom) from which it was formed (EPA-83/06a).

Metal modifiers: Metals used in explosives or propellants to modify their property, e.g., aluminum increases the energy of an explosion (EPA-76/03).

Metal oxidation refining: A refining technique that removes impurities from the base metal because the impurity oxidizes more readily than the base. The metal is heated and oxygen supplied. The impurity upon oxidizing separates by gravity or volatilizes (EPA-83/06a).

Metal oxide semiconductor (MOS) device: A metal insulator semiconductor structure in which the insulating layer is an oxide of the substrate material; for a silicon substrate, the insulating layer is silicon dioxide (SiO_2) (EPA-83/03).

Metal paste production: The manufacture of metal pastes for use as pigments by mixing metal powders with mineral spirits, fatty acids and solvents. Grinding and filtration are steps in the process (EPA-83/06a).

Metal powder production:
(1) Operations are mechanical process operations which convert metal to a finely divided form (40 CFR 471.02-91).
(2) The production of metal particles for such use as pigments either by milling and grinding of scrap or by atomization of molten metal (EPA-83/06a).

Metal preparation: Any and all of the metal processing steps preparatory to applying the enamel slip. Usually this includes cleaning, picking and applying a nickel flash or chemical coating (40 CFR 466.02-91).

Metal separation: A metallic ion removal method from wastewater by conversion to an insoluble form using such agents as lime, soda ash, or caustic followed by a separation process, usually clarification of filtration (*see* separation for more separation methods) (EPA-85/10).

Metal spraying: Coating metal objects by spraying molten metal upon the surface with gas pressure (EPA-83/06a).

Metalimnion: Methoxychlor: Pesticide which causes adverse health effects in domestic water supplies and is toxic to freshwater and marine aquatic life.: The middle layer of a thermally stratified lake or reservoir. In this layer there is a rapid decrease in temperature with depth. Also called thermocline (EPA-94/04).

Metallic bond: A chemical bond that holds metal or alloy atoms together (*see* chemical bond for more related terms).

Metallic element: An element whose atoms are held together by metallic bonds. It shows luster, electrical conductivity and malleability.

Metallic ink: An ink composed of aluminum or bronze powders in varnish to produce gold or silver color effects (*see* ink for more related terms) (EPA-79/12a).

Metallic mineral concentrate: A material containing metallic compounds in concentrations higher than naturally occurring in ore but requiring additional processing if pure metal is to be isolated. A metallic mineral concentrate contains at least one of the following metals in any of its oxidation states and at a concentration that contributes to the concentrate's commercial value: Aluminum, copper, gold, iron, lead, molybdenum, silver, titanium, tungsten, uranium, zinc, and zirconium. This definition shall not be construed as requiring that material containing metallic compounds be refined to a pure metal in order for the material to be considered a metallic mineral concentrate to be covered by the standards (40 CFR 60.381-91).

Metallic mineral processing plant: Any combination of equipment that produces metallic mineral concentrates from ore. Metallic mineral processing commences with the mining of ore and includes all operations either up to and including the loading of wet or dry concentrates or solutions of metallic minerals for transfer to facilities at non-adjacent locations that will subsequently process metallic concentrates into purified metals (or other products), or up to and including all material transfer and storage operations that precede the operations that produce refined metals (or other products) from metallic mineral concentrates at facilities adjacent to the metallic mineral processing plant. This definition shall not be construed as requiring that mining of ore be conducted in order for the combination of equipment to be considered a metallic mineral processing plant (cf. metallic mineral concentrate) (40 CFR 60.381-91).

Metallic replacement: This occurs when a metal such as iron comes in contact with a solution containing dissolved ions of a less active metal such as silver. The dissolved silver ions react with solid metal (iron). The more active metal (iron) goes into solution as an ion and ions of the less active metal become solid metal (silver) (EPA-80/10).

Metallic shoe seal: Includes but is not limited to a metal sheet held vertically against the tank wall by springs or weighted levers and is connected by braces to the floating roof. A flexible coated fabric (envelope) spans the annular space between the metal sheet and the floating roof (40 CFR 60.111a-91).

Metallic shoe type seal: A primary or secondary seal constructed of metal sheets (shoes) which are joined together to form a ring, springs, or levels which attach the shoes to the floating roof and hold the shoes against the tank wall, and a coated fabric which is suspended rom the shoes to the floating roof (40 CFR 52.741-91).

Metallic yield: The weight percent of municipal ferrous scrap which s generally recoverable as metal or alloy (EPA-83).

Metallo organic active ingredient: The carbon containing active ingredients containing one or more metallic atoms in the structure (40 CFR 455.31-91).

Metallo organic pesticide: *See* synonym, pesticide.

Metalloid (semi metal): A chemical whose properties are between metals and nonmetals and can be used as electric semiconductors. Typical metalloids include boron and silicon.

Metalworking fluid: A liquid of any viscosity or color containing intentionally added water used in metal machining operations for the purpose of cooling, lubricating, or rust inhibition (40 CFR 721.3; 747.115; 747.195; 747.200-91).

Metamorphosis: A structural transformation during the life cycle of many animals (e.g., the change from a tadpole to a frog and from a pupa to an insect).

Metaphysis (plural, metaphyses): The wider part at the end of the shaft of a long bone, which during development contains the growth zone and consists of spongy bone (LBL-76/07-bio).

Metathesis: *See* synonym, abstraction.

Meteorological conditions: The complex of meteorological parameters which are relevant when studying a phenomenon in the atmosphere (NATO-78/10).

Meteorological data: The data required for input to a dispersion model that characterizes the transport and turbulent dispersion properties of the atmosphere and mixing depth. The parameters that are commonly used to characterize these processes are wind direction, wind speed, atmospheric stability, and mixing height. Wind direction determines the direction of the plume movement, i.e., its advection. Wind speed affects the speed of transport, and therefore the initial dilution of the pollutant as it is emitted form the stack. Atmospheric stability determines the rate of turbulent dispersion of the plume as it moves downward. Mixing height determines the depth of the atmosphere through which pollutants can be dispersed in the vertical (EPA-88/09).

Meteorological precipitation: The precipitation of water from the atmosphere in the form of hail, mist, rain, sleet, and snow. Deposits of dew, fog, and frost are excluded (*see* precipitation for more related terms) (EPA-83/06).

Metering rod: In the paper industry, a rod used to apply coating to the surface of a sheet, metering even thickness coating layers on the surface (EPA-87/10).

Methane (or marsh gas, CH_4): A colorless, nonpoisonous, flammable gas created by anaerobic decomposition of organic compounds (EPA-94/04, *see also* EPA-84/09).

Methane fermentation: The bacteria action in which CH_4 and CO_2 are formed during the anaerobic digestion of organic materials. The rate is approximately 1 m^3 (CH_4 ((70%) + CO_2 (30%)) gas produced per 1 kg solid digested.

Methanol: An alcohol that can be used as an alternative fuel or as a gasoline additive. It is less volatile than gasoline; when blended with gasoline it lowers the carbon monoxide emissions but increases hydrocarbon emissions. Used as pure fuel, its emissions are less ozone-forming that those from gasoline (EPA-94/04).

Methanol fueled: Any motor vehicle or motor vehicle engine that is engineered and designed to be operated using methanol fuel (i.e., a fuel that contains at least 50 percent methanol (CH_3OH) by volume) as fuel. Flexible fuel vehicles are methanol-fueled vehicles (40 CFR 86.090.2-91).

Method: In air pollution control, a method means mostly sampling and analyzing methods. Other method related terms include:
- Absolute method
- Alternative method
- Comparative method
- Candidate method
- Equivalent method
- Field method
- Macro method
- Micro method
- Reference method
- Routine method
- Standard method
- Standard reference
- Trace method
- Ultra trace method

Method 5 (M5): M5 is used to capture particulates from combustion stack gases (SW-846).

Method 18: An EPA test method which uses gas chromatographic techniques to measure the concentration of volatile organic compounds in a gas stream (EPA-94/04).

Method 24: An EPA reference method to determine density, water content and total volatile content (water and VOC) of coatings (EPA-94/04).

Method 25: An EPA reference method to determine the VOC concentration in a gas stream (EPA-94/04).

Method detection limit (MDL):
(1) The lowest concentration of analyte that a method can detect reliably in either a sample or blank (40 CFR 300-App/A-91).
(2) The minimum concentration of a substance that can be measured and reported with 99% confidence that the analyte concentration is greater than zero and is determined from analysis of a sample in a given matrix containing the analyte (40 CFR 136-App/B).
(3) The minimum concentration of a substance (analyte) that a measurement system can consistently detect and/or measure in replicated field samples. There have been many terms used to designate detection limits and they have been defined in various ways such as (a) detection sensitivity, (b) lower limits of detection (LLD), (c) minimum detection amount (EPA-84/03)
(4) *See* detection limit for more related terms.

Method of standard addition: The standard addition technique involves the use of the unknown and the unknown plus a known amount of standard (40 CFR 136-App/C-91, *see also* Method 7000 in SW-846).

Method quantification limit (MQL): The minimum concentration of a substance that can be measured and reported.

Method of standard addition: The standard addition technique involves the use of the unknown and the unknown plus a known amount of standard (40 CFR 136-App/C-91).

Methoxychlor: Pesticide that causes adverse health effects in domestic water supplies and is toxic to freshwater and marine aquatic life (EPA-94/04).

Methyl alcohol (CH_3OH): A toxic liquid used in antifreeze for autos.

Methyl blue ($C_{37}H_{27}N_3O_9S$): A deep blue powder used in coloring, dye for cotton and silk.

Methyl orange alkalinity: A measure of the total alkalinity in a water sample in which to color of methyl orange reflects the change in level (EPA-94/04).

mho (or Siemens): A reciprocal of ohm. Ohm is the unit or electrical resistance and mho is a former unit of electrical conductance (now also known as Siemens).

Mica: A group of aluminum silicate minerals that are characterized by their ability to split into thin, flexible flakes because of their basal cleavage (EPA-83/03).

Micelle: An agglomeration of molecules.

Micro (μ):
(1) 10^{-6}.
(2) A prefix indicating a very small size.

Micro aerophile: A bacteria that grows in a low level of oxygen.

Micro capsule: Capsules in the size approximately from 10^{-6} to 0.02 meter.

Micro climate: The local climate (cf. micro-scale).

Micro encapsulation: The isolation of a waste from external effects by mixing it with a material which then cures or converts it to a solid, non-leaching barrier (OME-88/12).

Micro densitomer: An instrument used in spectroscopy to detect spectrum lines so that the spectrum lines can be seen by human eyes.

Micro filtration: *See* synonym, ultra filtration.

Micro method: A method requiring milligram or smaller amounts of sample (*see* method for more related terms) (ACS-87/11).

Micro organism:
(1) Living organisms so small that individually they can usually only be seen through a microscope (EPA-89/12).
(2) Generally any living things microscopic in size, including the bacteria, actinomycetes, yeasts, simple fungi, some algae, rickettsiae, spirochetes, slime molds, protozoans and some of the simpler multicellular organisms (Benthic microorganism) (EPA-83).
(3) *See* organism for more related terms.

Micro nutrient: A relatively small amount of nutrient required by organisms to maintain their lives (*see* nutrient for more related terms).

Micro scale: In meteorology, a length scale of the order of the predominant scale of turbulence in the atmospheric boundary layer. Values of this scale vary between a few meters and several kilometers (cf. macro scale or micro climate) (NATO-78/10).

Micro straining: A process for removing solids from water, which consists of passing the water stream through a micro screen (the openings are about 15 to 65 micrometer) with the solids being retained on the screen (EPA-83/06a).

Microscopic thermodynamics: The consideration of a matter is in the level of a molecular or atomic scale (cf. macroscopic thermodynamic).

Microbe (microbiota or microorganism): Microscopic organisms such as algae, animals, viruses, bacteria, fungus, and protozoa, some of which cause diseases (cf. microorganism) (EPA-89/12).

Microbead: Beads with a diameter of approximately 100 um (micro meter) (polystyrene, sephadex, or polyacrylamide) used to maximize surface for culturing monolayer cells (EPA-88/09a).

Microbial: Of or pertaining to microbes, single-celled organisms (e.g., bacteria) (EPA-87/10a).

Microbial film: *See* synonym, biological film.

Microbial growth: The activity and growth of microorganisms such as bacteria, algae, diatoms, plankton, and fungi (EPA-94/04).

Microbial pesticide: A microorganism that is used to control a pest, but of minimum toxicity to man (EPA-94/04).

Microbiological aerosol: An airborne particle either exclusively or partially composed of micro-organisms including bacteria, fungi and viruses (EPA-83).

Microbiology: The science of microorganisms including bacteriology, cytology, enzymology, mycology, and virology.

Microbiota: *See* synonym, microbe.

Microclimate: The localized climate conditions with in an urban area or neighborhood (EPA-94/04).

Microcurie (μCi): means that amount of radioactive material which disintegrates at the rate of 37 thousand atoms per second (10 CFR 30.4-91). 1 μCi = $(10^{-6}) \times (37 \times 10^9)$ = 37×10^3 transformations per second (*see* curie for more related terms).

Microcytic anemia: A condition in which the majority of the red cells are smaller than normal (LBL-76/07-bio).

Micrometer: 10^{-6} meter, also known as **micron**.

Micron: *See* definition under micrometer.

Micronuclei: The small particles consisting of acentric fragments of chromosomes or entire chromosomes, which lag behind at anaphase of cell division. After telophase, these fragments may not be included in the nuclei of daughter cells and form single or multiple micronuclei in the cytoplasm (40 CFR 798.5395-91)

Micron efficiency curve: A curve showing how well a collector (air pollution control device) traps micron-size particles (SW-108ts).

Microorganism: *See* synonym, microbe.

Microrem: *See* definition under radiation unit.

Microrem per hour: *See* definition under radiation unit.

Microsome: One of the finer granular elements of protoplasm (LBL-76/07-bio).

Microwave desulfurization: One of SO_2 emission reduction techniques (*see* sulfur oxide emission control for control structure). In microwave desulfurization, coal is crushed, then heated for 30 to 60 seconds by exposure to microwaves. Mineral sulfur selectively absorbs this radiation forming hydrogen sulfide gas (H_2S). The H_2S is usually reduced to elemental sulfur by the Claus process. Another microwave process adds calcium hydroxide [$Ca(OH)_2$] to crushed coal. The organic sulfur converts calcium sulfite ($CaSO_3$) when exposed to this radiation. The coal is washed with water to remove the $CaSO_3$ and other impurities. As much as 70% of the sulfur can be removed by the microwave process (EPA-81/12, p8-4).

Microwave discharge: A microwave discharge, also known as **microwave plasma**, in the treatment of hazardous wastes is a special application in the general field of plasma chemistry. Microwave is an electro-magnetic wave that has a wavelength between about 0.3 to 30 cm, corresponding to a frequency of 1 to 100 GHz. The technology uses microwave energy to excite the molecules of a carrier gas, such as helium or air, thus raising electron energy levels and essentially forming very reactive free radicals. The gas in this high energy condition is called plasma. The excited electrons transfer this energy to break chemical bonds of materials. Carbon-carbon bonds are among those most susceptible. Thus, theoretically, any organic waste (liquid, solid, or gas) placed into the plasma can be degraded to intermediate or ultimate products, perhaps destroying their toxic properties. Residence time within the plasma varies from 0.1 to 1.0 second. The temperature for plasma destruction can be as low as 150° C (300° F) (*see* plasma arc for more related terms) (EPA-76/11; 81/09; 88/09a).

Microwave plasma: *See* definition under microwave discharge.

Middle cover: *See* synonym, intermediate cover.

Midget impinger: A specific instrument employing wet impingement, typically using a liquid volume of 10 ml and a gas flow of 0.1 ft^3 per min. (*see* impinger for more related terms) (EPA-83/06).

Migration velocity: The velocity which a particle migrates (drifts) toward the collection plate in an electrostatic precipitator (EPA-84/09).

Mil: A unit of thickness. 0.001 inch (EPA-82/11e).

Mild steel: Carbon steel containing a maximum of about 0.25% carbon. Mild steel is satisfactory for use where severe corrodants are not encountered or where protective coatings can be used to prevent or reduce corrosion rates to acceptable levels (EPA-86/12).

Milestone: An important or critical event that must occur in a project cycle in order to achieve the project objectives. It must be clearly defined and easily measurable, leaving no doubt as to when it is achieved (DOE-91/04).

Military engine: Any engine manufactured solely for the Department of Defense to meet military specifications (*see* engine for more related terms) (40 CFR 86.082.2-91).

Milk equivalent: The quantity of milk (in pounds) to produce one pound of product (cf. whey and raw, skim, or whole milk) (EPA-74/05).

Milkiness: A condition of pronounced cloudiness in glass (EPA-83).

Milking center: A separate milking area with storage and cooling facilities adjacent to a free stall barn or cowyard dairy operation (40 CFR 412.11-91).

Milkroom: The milk storage and cooling rooms normally used for stall barn dairies (40 CFR 412.11-91).

Mill:
(1) A preparation facility within which the metal ore is cleaned, concentrated, or otherwise processed before it is shipped to the customer, refiner, smelter, or manufacturer. A mill includes all ancillary operations and structures necessary to clean, concentrate, or otherwise process metal ore, such as ore and gangue storage areas and loading facilities (40 CFR 440.132-91).
(2) A monetary unit equal to one-tenth of a cent (DOE-91/04).

Mill broke: Any paper waste generated in a paper mill prior to completion of the papermaking process. It is usually returned directly to the pulping process. Mill broke is excluded from the definition of recovered materials (40 CFR 250.4-91).

Milled refuse: The solid waste that has been mechanically reduced in size (*see* refuse for more related terms) (SW-108ts).

Millboard: An asbestos-containing product made of paper and similar in consistency to cardboard produced in sections rather than as a continuous sheet. Major applications of this product include: thermal protection for large circuit breakers; barriers from flame or heat; linings in floors, partitions, and fire doors; linings for stoves and heaters; gaskets; table pads; trough liners; covers for operations involving molten metal; and stove mats (40 CFR 763.163-91).

Millicurie (mCi): means that amount of radioactive material which disintegrates at the rate of 37 million atoms per second (10 CFR 30.4-91). 1 mCi = $(10^{-3}) \times (37 \times 10^9) = 37 \times 10^6$ transformations per second (*see* curie for more related terms).

Milling:
(1) The mechanical treatment of a nonferrous metal to produce powder, or to coat one component of a powder mixture with another (40 CFR 471.02-91).
(2) Using a rotary tool with one or more teeth which engage the workpiece and remove material as the workpiece moves past the rotating cutter.
   (A) **Face milling:** Milling a surface perpendicular cutting edges remove the bulk of the material while the face cutting edges provide the finish of the surface being generated.
   (B) **End milling:** Milling accomplished with a tool having cutting edges on its cylindrical surfaces as well as on its end. In end milling - peripheral, the peripheral cutting edges on the cylindrical surface are used; while in end milling slotting, both end and peripheral cutting edges remove metal.
   (C) **Slide and slot milling:** Milling of the side or slot of a workpiece using a peripheral cutter.
   (D) **Slab milling:** Milling of a surface parallel to the axis of a helical, multiple-toothed cutter mounted on an arbor.
   (E) **Straddle milling:** Peripheral milling a workpiece on both sides at once using two cutters spaced as required (EPA-83/06a).

Million-gallons per day (MGD): A measure of water flow (EPA-94/04).

Mimeo paper: A grade of writing paper used for making copies on stencil duplicating machines (*see* paper for more related terms) (40 CFR 250.4-91).

Minable:
(1) Capable of being mined.
(2) Material that can be mined under present day mining technology and economics (EPA-82/05).

Mine: An area of land, surface or underground, actively mined from the production of crushed and broken tone from natural deposits (40 CFR 436.21-91, *see also* 40 CFR 436.31; 436.41; 436.181; 440.132; 440.141-91). Other mine related terms include:
- Abandoned mine
- Active mine
- Active mining area
- Conventional mine
- Deep mine

Mine area: The land area from which overburden is stripped and ore is removed prior to moving the ore to the beneficiation area (40 CFR 440.141-91).

Mine dewatering: Any water that is impounded or that collects in the mine and is pumped, drained or otherwise removed from the mine through the efforts of the mine operator. However, if a mine is also used for treatment of process generated waste water, discharges of commingled water from the facilities shall be deemed discharges of process generated waste water (40 CFR 436.21-91, *see also* 40 CFR 436.31; 436.41; 436.181-91).

Mine drainage: Any drainage, and any water pumped or siphoned, from an active mining area or a post-mining area (40 CFR 434.11-91, *see also* 40 CFR 440.132; 440.141-91). Other mine drainage related terms include:
- Acid mine drainage
- Ferruginous mine drainage (*see* synonym, acid mine drainage)
- Alkaline mine drainage
- Controlled surface mine drainage

Mine mouth plant: A steam electric powerplant located within a short distance of a coal mine and to which the coal is transported from the mine by a conveyer system, slurry pipeline, or truck (EPA-82/11f).

Miner of asbestos: A person who produces asbestos by mining or extracting asbestos-containing ore so that it may be further milled to produce bulk asbestos for distribution in commerce, and includes persons who conduct milling operations to produce bulk asbestos by processing asbestos-containing ore. Milling involves the separation of the fibers from the ore, grading and sorting the fibers, or fiberizing crude asbestos ore. To mine or mill is to "manufacture" for commercial purposes under TSCA (*see* asbestos for more related terms) (40 CFR 763.63-91).

Mineral: An inorganic substance occurring in nature, though not necessarily of inorganic origin, which has (1) a definite chemical composition, or commonly a characteristic range of chemical composition, and (2) distinctive physical properties, or molecular structure. With few exceptions, such as opal (amorphous) and mercury (liquid), minerals are crystalline solids (cf. other mineral) (EPA-82/05).

Mineral dressing: *See* synonym, mineral processing.

Mineral fiber insulation: The insulation (rock wool or fiberglass) which is composed principally of fibers manufactured from rock, slag or glass, with or without binders (40 CFR 248.4-91).

Mineral handling and storage facility: The areas in asphalt roofing plants in which minerals are unloaded from a carrier, the conveyor transfer points between the carrier and the storage silos, and the storage silos (40 CFR 60.471-91).

Mineral oil PCB transformer: Any transformer originally designed to contain mineral oil as the dielectric fluid and which has been tested and found to contain 500 ppm or greater PCBs (40 CFR 761.3-91).

Mineral processing (ore processing or mineral dressing): The dry and wet crushing and grinding of ore or other mineral-bearing products for the purpose of raising concentrate grade, removal of waste and unwanted or deleterious substances from an otherwise useful products, separation into distinct species of mixed minerals, chemical attack and dissolution of selected values (EPA-82/05).

Mineral spirit: A petroleum derivative used as a vehicle for inks and varnishes. It usually boils in the range of 149 to 204° C and has a flash point just about 27° C (EPA-79/12a).

Mineral water: The water containing either artificially or naturally supplied minerals required by living organisms (*see* water for more related terms).

Mineralization: In wastewater treatment, it refers to the relative increase in the ratio of minerals to organic materials, because organic materials are continuously digested by bacteria and are converted to CH_4 or CO_2.

Mineralogy: A science (branch of geology) dealing with the study of minerals.

Minimal risk: That the probability and magnitude of harm or discomfort anticipated in the research are not greater in and of themselves than those ordinarily encountered in daily life or during the performance of routine physical or psychological examinations or tests (40 CFR 26.102-91).

Minimization: A comprehensive program to minimize or eliminate wastes, usually applied to wastes at their point of origin (*see* waste minimization) (EPA-94/04).

Minimize: To reduce to the smallest possible amount or degree (40 CFR 6-App/A-91).

Minimum detectable sensitivity: *See* synonym, detection limit.

Minimum level (ML): Refers to the level at which the entire analytical system gives recognizable mass spectra and acceptable calibration points when analyzing for pollutants of concern. This level corresponds to the lowest point at which the calibration curve is determined (EPA-91/03).

Mining: The depletion of a resource without making any provision for replenishment (DOI-70/04). In reference ground water, withdrawals in excess of natural replenishment and capture. Commonly applied to heavily pumped areas in semiarid and arid regions, where opportunity for natural replenishment and capture is small. The term is hydrologic and excludes any connotation of unsatisfactory water-management practice (cf. active mining area) (Course, 165.7).

Mining of an aquifer: Withdrawal of ground water over a period of time that exceeds the rate of recharge of the aquifer (EPA-94/04).

Mining overburden returned to the mine site: Any material overlying an economic mineral deposit which is removed to gain access to that deposit and is then used for reclamation of a surface mine (40 CFR 260.10-91).

Mining runoff: The water drained from mining.

Mining waste: The residues which result from the extraction of raw materials from the earth (or residues left after ore beneficiation) (*see* waste for more related terms) (40 CFR 243.101-91, *see also* 40 CFR 246.101-91; EPA-83).

Minor: Publicly owned treatment works with flows less than 1 million gallons per day (*see* majors) (EPA-94/04).

Minor source baseline date: *See* definition under major source baseline date.

Misbranded:
(1) A pesticide is misbranded if:
- (A) Its labeling bears any statement, design, or graphic representation relative thereto or to its ingredients which is false or misleading in any particular;
- (B) It is contained in a package or other container or wrapping which does not conform to the standards established by the Administrator pursuant to section 25(c)(3);
- (C) It is an imitation of, or is offered for sale under the name of, another pesticide;
- (D) Its label does not bear the registration number assigned under section 7 to each establishment in which it was produced;
- (E) Any work, statement, or other information required by or under authority of this Act to appear on the label or labeling is not prominently placed thereon with such conspicuousness (as compared with other words, statements, designs, or graphic matter in the labeling) and in such terms as to render it likely to be read and understood by the ordinary individual under customary conditions of purchase and use;
- (F) The labeling accompanying it does not contain directions for use which are necessary for effecting the purpose for which the product is intended and if complied with, together with any requirements imposed under section 3(d) of this Act, are adequate to protect health and the environment;
- (G) The label does not contain a warning or caution statement which may be necessary and if complied with, together with any requirements imposed under section 3(d) of this Act, is adequate to protect health and the environment; or
- (H) In the case of a pesticide not registered in accordance with section 3 of this Act and intended for export, the label does not contain, in words prominently placed thereon with such conspicuousness (as compared with other words, statements, designs, or graphic matter in the labeling) as to render it likely to be noted by the ordinary individual under customary conditions of purchase and use, the following: "Not Registered for Use in the United States of America."

(2) A pesticide is misbranded if:
- (A) The label does not bear an ingredient statement on that part of the immediate container (and on the outside container or wrapper of the retail package, if there be one, through which the ingredient statement on the immediate container cannot be clearly read) which is presented or displayed under customary conditions of purchase, except that a pesticide is not misbranded under this subparagraph if:
  - (i) The size or form of the immediate container, or the outside container or wrapper of the retail package, makes it impracticable to place the ingredient statement on the part which is presented or displayed under customary conditions of purchase: and
  - (ii) The ingredient statement appears prominently on another part of the immediate container, or outside container or wrapper, permitted by the Administrator;
- (B) The labeling does not contain a statement of the use classification under which the product is registered;
- (C) There is not affixed to its container, and to the outside container or wrapper of the retail package, if there be one, through which the required information on the immediate container cannot be clearly read, a label bearing:
  - (i) The name and address of the producer, registrant, or person for whom produced;
  - (ii) The name, brand, or trademark under which the pesticide is sold;
  - (iii) The net weight or measure of the content, except that the Administrator may permit reasonable variations; and
  - (iv) When required by regulation of the Administrator to effectuate the purposes of the Act, the registration number assigned to the pesticide under this Act, and the use classification; and
- (D) The pesticide contains any substances in quantities highly toxic to man, unless the label shall bear, in addition to any other matter required by this Act:
  - (i) The skull and crossbones;
  - (ii) The word "poison" prominently in red on a background of distinctly contrasting color; and
  - (iii) A statement of a practical treatment (first aid or otherwise) in case of poisoning by the pesticide (FIFRA2-7USC136-91).

Miscellaneous ACM:
(1) The miscellaneous material that is ACM in a school building (40 CFR 763.83-91).
(2) Interior asbestos-containing building material or structural components, members or fixtures, such as floor and ceiling tiles; does not include surfacing materials or thermal system insulation (EPA-94/04).

Miscellaneous fabric product manufacturing process:
(1) A manufacturing process involving one or more of the following applications, including any drying and curing of formulations, and capable of emitting VOM:
- (A) Adhesives to fabricate or assemble components or products.
- (B) Asphalt solutions to paper or fiberboard.
- (C) Asphalt to paper or felt.
- (D) Coatings or dye to leather.
- (E) Coatings to plastic.
- (F) Coatings to rubber or glass.
- (G) Disinfectant material to manufactured items.
- (H) Plastic foam scrap or "fluff" from the manufacture of foam containers and packaging material to form resin pallets.
- (i) Resin solutions to fiber substances.
- (J) Viscose solutions for food casings.

(2) The storage and handling of formulations associated with the process described above, and the use and handling of organic liquids and other substances for clean-up operations associated with the process described in this definition (40 CFR 52.741-91).

Miscellaneous formulation manufacturing process:
(1) A manufacturing process which compounds one or more of the following and is capable of emitting VOM:
  (A) Adhesives.
  (B) Asphalt solutions.
  (C) Caulks, sealants, or waterproofing agents.
  (D) Coatings, other than paint and ink.
  (E) Concrete curing compounds.
  (F) Dyes.
  (G) Friction materials and compounds.
  (H) Resin solutions.
  (i) Rubber solutions.
  (J) Viscose solutions.
(2) The storage and handling of formulations associated with the process described above, and the use and handling of organic liquids and other substances for clean-up operations associated with the process described in this definition (40 CFR 52.741-91).

Miscellaneous material:
(1) The building material on structural components, structural members or fixtures, such as floor and ceiling tiles. The term does not include surfacing material or thermal system insulation (TSCA-AIA1, *see also* 40 CFR 763.83-91).
(2) Interior building materials on structural components, such as floor or ceiling tiles (EPA-94/04).

Miscellaneous metal parts or products: Any metal part or metal product, even if attached to or combined with a nonmetal part or product, except cans, coils, metal furniture, large appliances, magnet wire, automobiles, ships, and airplane bodies (40 CFR 52.741-91).

Miscellaneous metal parts and products coating: Any coating applied to any metal part or metal product, even if attached to or combined with a nonmetal part or product, except cans, coils, metal furniture, large appliances, and magnet wire. Prime coat, prime surfacer coat, topcoat, and final repair coat for automobiles and light-duty trucks are not miscellaneous metal parts and products coatings. However, underbody anti-chip (e.g., underbody plastisol) automobile, and light-duty truck coatings are miscellaneous metal parts and products coatings. Also, automobile or light-duty truck refinishing coatings, coatings applied to the exterior of marine vessels, coatings applied to the exterior of airplanes, and the customized topcoating of automobiles and trucks if production is less than 35 vehicles per day are not miscellaneous metal parts and products coatings (40 CFR 52.741-91).

Miscellaneous metal parts or products coating facility: A facility that includes one or more miscellaneous metal parts or products coating lines (40 CFR 52.741-91).

Miscellaneous metal parts or products coating line: A coating line in which any protective, decorative, or functional coating is applied onto the surface of miscellaneous metal parts of products (40 CFR 52.741-91).

Miscellaneous oil spill control agent: Any product, other than a dispersant, sinking agent, surface collecting agent, biological additive, or burning agent, that can be used to enhance oil sill cleanup, removal, treatment, or mitigation (40 CFR 300.5-91).

Miscellaneous organic chemical manufacturing process:
(1) A manufacturing process which produces by chemical reaction, one or more of the following organic compounds or mixtures of organic compounds and which is capable of emitting VOM:
  (A) Chemicals listed in appendix A of this section
  (B) Chlorinated and sulfonated compounds
  (C) Cosmetic, detergent, soap, or surfactant intermediaries or specialties and products
  (D) Disinfectants
  (E) Food additives
  (F) Oil and petroleum product additives
  (G) Plasticizers
  (H) Resins or polymers
  (i) Rubber additives
  (J) Sweeteners
  (K) Varnishes
(2) The storage and handling of formulations associated with the process described above and the use and handling of organic liquids and other substances for clean-up operations associated with the process described in this definition (40 CFR 52.741-91).

Miscellaneous unit: A hazardous waste management unit where hazardous waste is treated, stored, or disposed of and that is not a container, tank, surface impoundment, pile, land treatment unit, landfill, incinerator, boiler, industrial furnace, underground injection well with appropriate technical standards under 40 CFR 146, or unit eligible for a research, development, and demonstration permit under 40 CFR 270.65 (40 CFR 260.10-91).

Miscellaneous waste stream: The following additional waste streams related to forming copper: hydrotesting, sawing, surface milling, and maintenance (40 CFR 468.02-91).

Miscellaneous wastewater stream: The combined wastewater streams from the process operations listed below for each subcategory. If a plant has one of these streams then the plant receives the entire miscellaneous waste stream allowance:
(1) Cadmium subcategory. Cell wash, electrolyte preparation, floor and equipment wash, and employee wash.
(2) Lead subcategory. Floor wash, wet air pollution control, battery repair, laboratory, hand wash, and respirator wash.
(3) Lithium subcategory. Floor and equipment wash, cell testing, and lithium scrap disposal.
(4) Zinc subcategory. Cell wash, electrolyte preparation, employee wash, reject cell handling, floor and equipment wash (40 CFR 461.2-91).

Mischmetal: Refers to a rare earth metal alloy comprised of the natural mixture of rare earths to about 94-99 percent. The balance of the alloy includes traces of other elements and one to tow percent iron (40 CFR 421.271-91).

Miscible liquids: Two or more liquids that can be mixed and will remain mixed under normal conditions (EPA-94/04).

Missed detection: The situation that occurs when a test indicates that a tank is "tight" when in fact it is leaking (EPA-94/04).

Missile liner: An asbestos containing product used as a liner for coating the interior surfaces of rocket motors (*see* liner for more related terms) (40 CFR 763.163-91).

Mist: Liquid particles measuring 40 to 500 microns, are formed by condensation of vapor. By comparison, fog particles are smaller than 40 microns (EPA-94/04).

Mist coating: *See* synonym, fog coating.

Mist eliminator (or entrainment separator): A control device used to remove entrained water droplets downstream from a scrubber (EPA-81/09. Other mist eliminator related terms include:
- Cyclone mist eliminator
- Fiber bed mist eliminator
- Wire mesh eliminator

Miticide: A chemical substance used to destroy mites, acaricides (EPA-85/10).

Mitigation:
(1) Includes:
 (A) Avoiding the impact altogether by not taking certain action or parts of an action.
 (B) Minimizing impacts by limiting the degree or magnitude of the action and its implementation.
 (C) Rectifying the impact by repairing, rehabilitating, or restoring the affected environment.
 (D) Reducing or eliminating the impact over time by preservation and maintenance operations during the life of the action.
 (E) Compensating for the impact by replacing or providing substitute resources or environments (40 CFR 1508.20-91).
(2) Any measure taken to reduce the severity of the adverse effects associated with the accidental release of a hazardous chemical (EPA-86/12).
(3) Measures taken to reduce adverse impacts on the environment (EPA-94/04).

Mitigator: A building trades professional who works for profit to correct radon problems, a person experienced in radon remediation. At present, training programs are underway to provide working professionals with the knowledge and experience necessary to control radon exposure problems. Some State radiological health offices have lists of qualified professionals (EPA-88/08).

Mitin fiber: A trademark of Geigy for a moth-repellent finish for woolens (*see* fiber for more related terms) (EPA-74/06b).

Mitochondria: Small granules or rod-shaped structures seen by differential staining in the cytoplasm of cells (LBL-76/07-bio).

Mitosis: A cell division involving exact separation of chromosomes so that each of the two daughter cells carries the same chromosome from the mother cell.

Mitotic gene conversion: Is detected by the change of inactive alleles of the same gene to wild-type alleles through intragenic recombination in mitotic cells (40 CFR 798.5575-91).

Mitsubishi process: A process used in primary copper refining which incorporates three furnaces to combine roasting, smelting, and converting into one continuous process. The Mitsubishi process results in reduced smelting rates and heating costs (EPA-83/03a).

Mixed fertilizer: A mixture of wet and/or dry straight fertilizer materials, mixed fertilizer materials, fillers and additives prepared through chemical reaction to a given formulation (cf. blend fertilizer) (40 CFR 418.71-91).

Mixed funding: Settlements in which potentially responsible parties and EPA share the cost of a response action (EPA-94/04).

Mixed kraft bag: The bag consists of used kraft bags free from twisted or woven stock and other similar objectionable materials (EPA-83).

Mixed liquor: A mixture of activated sludge and water containing organic matter undergoing activated sludge treatment in an aeration tank (*see* liquor for more related terms) (EPA-94/04).

Mixed liquor suspended solids (MLSS): The dry suspended solids in mixed liquor (mg/liter). It is a measure of the concentration of matter in a biological treatment process (EPA-87/10a).

Mixed liquor volatile suspended solids (MLVSS): The dry volatile suspended solids in mg/liter of mixed liquor.

Mixed media filtration: A filter which uses two or more filter materials of differing specific gravities selected so as to produce a filter uniformly grade from coarse to fine (EPA-83/03).

Mixed municipal refuse: *See* synonym, mixed municipal solid waste.

Mixed municipal solid waste (or mixed municipal refuse): The trash that is not sorted into categories of materials (*see* waste for more related terms) (OTA-89/10).

Mixed radioactive and other hazardous substances: The material containing both radioactive hazardous substances and nonradioactive hazardous substances, regardless of whether these types of substances are physically separated, combined chemically, or simply mixed together (*see also* waste, mixed) (40 CFR 300-App/A-91).

Mixed sample: *See* synonym, composite sample.

Mixed waste: The waste that is both hazardous waste and radioactive waste as defined in this glossary (*see also* mixed radioactive and other hazardous substances) (*see* waste for more related terms) (DOE-91/04).

Mixing:
(1) Agitating, beating, blending, diffusing, dispersing, emulsifying, homogenizing, kneading, stirring, whipping, working, and so forth (AP-40, p790).
(2) The incorporation of ingredients into a coating with the use of little or no shearing energy (EPA-79/12a).
(3) One of very important factors in promoting the contact between oxygen and organic compounds for incineration (cf. three T's).

Mixing chamber: A chamber or zone in an excess air incinerator between the primary and secondary chamber where combustion gases and secondary combustion air are mixed (EPA-89/03b).

Mixing depth (or mixing height):
(1) The thickness of the atmospheric layer near the surface in which air pollutants are dispersed by turbulence. At the top of the mixing layer, a barrier to the dispersion is often formed by a temperature inversion (NATO-78/10).
(2) The turbulent air near the ground is frequently bounded above by a layer of stable air. The distance from the ground to the bottom of the stable layer is the mixing height. The stable layer prevents substantial vertical dispersion above the mixing height (EPA-88/09).
(3) cf. mixing layer.

Mixing height: *See* synonym, mixing depth.

Mixing layer: The layer above the surface through which relatively vigorous vertical mixing occurs (cf. mixing depth) (DOE-91/04).

Mixing zone (or dilution zone):
(1) a limited area or volume of water where initial dilution of a discharge takes place; and where numeric water quality criteria can be exceeded but acutely toxic conditions are prevented from occurring (40 CFR 131.35-91, *see also* 40 CFR 125.121; 230.3-91).
(2) An area where an effluent discharge undergoes initial dilution and is extended to cover the secondary mixing in the ambient waterbody. A mixing zone is an allocated impact zone where water quality criteria can be exceeded as long as acutely toxic conditions are prevented (EPA-91/03).

Mixolimnion: The upper layer of a lake. It has a lower density and free circulation because of wind effects (cf. monimolimnion).

Mixture: Any combination of two or more chemical substances if the combination does not occur in nature and is not, in whole or in part, the result of a chemical reaction; except that such term does include any combination which occurs, in whole or in part, as a result of a chemical reaction if none of the chemical substances comprising the combination is a new chemical substance and if the combination could have been manufactured for commercial purposes without a chemical reaction at the time the chemical substances comprising the combination were combined (TSCA3, *see also* 40 CFR 2.306; 116.3; 355.20; 372.3; 710.2; 720.3; 723.250; 712.3; 761.3-91).

Mixture property:
(1) The mass of a mixture is equal to the sum of the masses of its constituents: $m = SUM\ m_i$. Where: m = total mass of a mixture; m_i = mass of constituent I; SUM = summation of constituents.
(2) The total number of moles of a mixture is equal to the sum of the number of moles of its constituents: $n = SUM\ n_i$, where n = total moles of a mixture, n_i = moles of constituent I.
(3) The mole fraction is defined as: $y_i = n_i/n$. Where: y_i = mole fraction of constituent I; n_i = number of moles of constituent I; n = total moles of a mixture.
(4) **Dalton model:** When gases or vapors (having no chemical interaction) are present as a mixture in a given space or a given volume, the properties (e.g., pressure) of each component are considered as though each component existed separately at the same volume and temperature of the mixture (EPA-81/12, p2-9; Sonntag-71, p362). The pressure exerted by one component of a gas-mixture is called its partial pressure. For an ideal gas mixture, the total pressure of the gas-mixture is the sum of the partial pressures, namely: $P = SUM\ P_i$. Where: P = total pressure; P_i = pressure of each component of the system. For this case, the partial pressure of component I is equal to the mole fraction of component I in the mixture, i.e., $y_i = n_i/n = P_i/P$.
(5) **Amagat model:** When gases or vapors (having no chemical interaction) are present as a mixture in a given space or a given volume, the properties of each component are considered as though each component existed separately at the same pressure and temperature of the mixture (Sonntag-71, p363). For an ideal mixture, the total volume of the gas-mixture is the sum of the each component volume, namely: $V = SUM\ V_i$. Where: V = total pressure; V_i = volume of each component of the system. For this case, the volume fraction of component I is equal to the mole fraction of component I in the mixture, i.e., $y_i = n_i/n = V_i/V$.
(6) A mixture of ideal gases is also an ideal gas.

Mobile incinerator systems: Hazardous waste incinerators that can be transported from one site to another (*see* incinerator for more related terms) (EPA-94/04).

Mobile offshore drilling unit: A vessel (other than a self-elevating lift vessel) capable of use as an offshore facility (OPA1001-91).

Mobile source:
(1) Any vehicle, rolling stock, or other means of transportation which contains or carries a reportable quantity of a hazardous substance (or other pollutant) (*see* source for more related terms) (40 CFR 117.1-91).
(2) Any non-stationary source of air pollution such as cars, trucks, motorcycles, buses, airplanes, locomotives (EPA-94/04).

Mobile treatment unit: Any equipment or device for the treatment of solid or hazardous waste which is designed or fabricated so as to be moved from one location to another.

Mobility: The ability of a contaminant to migrate from its source (EPA-89/12a).

Model:
(1) A specific combination of car line, body style, and drive-train configuration (40 CFR 86.082.2-91).
(2) A quantitative or mathematical representation or computer simulation which attempts to describe the characteristics or relationships of physical events (EPA-88/09).
(3) A mathematical function with parameters which can be adjusted so that the function closely describes a set of empirical data. A mathematical or mechanistic model is usually based on biological or physical mechanisms, and has model parameters that have real world interpretation. In contrast, statistical or empirical models are curve-fitting to data where the math function used is selected for its

numerical properties. Extrapolation from mechanistic models (e.g., pharmacokinetic equations) usually carries higher confidence than extrapolation using empirical models (e.g., logit) (EPA-92/12).

Model calibration: An adjustment of the results of an air pollution simulation model to fit measured concentration data. This is usually done by specifying some free constants in the model (NATO-78/10).

Model plant: A hypothetical plant design used for developing economic, environmental, and energy impact analyses as support for regulations or regulatory guidelines; first step in exploring the economic impact of a potential NSPS (EPA-94/04).

Model sensitivity: In biosafety, an extent to which model predictions are affected by a small change in input information, unknown parameter, or by an assumption (EPA-88/09a).

Model specific code: The designation used for labeling purposes in Sections 205.15 and 205.16 for identifying the motorcycle manufacturer, class, and advertised engine displacement, respectively (40 CFR 205.151-91).

Model type: The meaning given it in section 501(11) of the Act, 15USC2001(11) (40 CFR 2.311-91, *see also* 40 CFR 86.082-2; 600.002.85-91).

Model year: The manufacturers annual production period (as determined by the Administrator) which includes January 1 of such calendar year; however, if the manufacturer has no annual production period, the term model year shall mean the calendar year (40 CFR 85.2102-91, *see also* CAA202; 40 CFR 85.1502; 86.082.2; 86.402.78; 204.51; 205.51; 205.151; 600.002.85; 763.163-91).

Modeling: An investigative technique using a mathematical or physical representation of a system or theory that accounts for all or some its known properties. Models are often used to test the effect of changes of system components on the overall performance of the system (EPA-89/12).

Modification:
(1) Any alteration in the terms and conditions of a contract, including supplemental agreements, amendments and extensions (40 CFR 8.2-91).
(2) Any change to a parking facility that increases or may increase the motor vehicle capacity of, or the motor vehicle activity associated with, such parking facility (40 CFR 52.2486-91).
(3) Any physical change in, or change in the method of operation of, an existing facility which increases the amount of any air pollutant (to which a standard applies) emitted into the atmosphere by that facility or which results in the emission of any air pollutant (to which a standard applies) into the atmosphere not previously emitted (40 CFR 60.2-91, *see also* CAA111; CAA112-91).
(4) Any change to the original design and construction of a cargo tank or a cargo tank motor vehicle which affects its structural integrity or lading retention capability. Excluded from this category are the following:

(A) A change to motor vehicle equipment such as lights, truck or tractor power train components, steering and brake systems, and suspension parts, and changes to appurtenances, such as fender attachments, lighting brackets, ladder brackets; and
(B) Replacement of components such as valves, vents, and fittings with a component of a similar design and of the same size (40 CFR 180.403-91).

Modified discharge: The volume, composition and location of the discharge proposed by the applicant for which a modification under section 301(h) of the Act is requested. A modified discharge may be a current discharge, improved discharge, or altered discharge (*see* discharge for more related terms) (40 CFR 125.58-91).

Modified method 5 (MM5) sampling train (or semi-VOST): MM5 is based on the sampling unit under EPA method 5. It is used to capture both particulates and semi-volatile compounds during sampling. During operation, the MM5 collects a 4-6 dry standard cubic meter sample over three to four hours. Its major components include (cf. source assessment sampling system) (EPA-82/02; SW-846):
(1) **A probe:** Either medium wall Pyrex glass (for probes less than 7 feet in length) or 1.6 cm OD Incology 825 tubing (for probes greater than 7 ft long) are wrapped with heating wire and a stainless steel jacket. Samples are collected while the probe is heated to a gas temperature above the dew point of the stack gas. If sampling of combustion zone vapors is necessary, a water cooled, quarts-lined sampling probe is used.
(2) **An optional glass cyclone with an attached collection flask:** The cyclone is to remove large quantities of particulates to prevent plugging of the downstream filter.
(3) **Filter:** The filter is to further remove particulates in the collected gas. The cyclone, flask, and filter holder are contained in an electrically heated enclosed box at a temperature of approximately 120° C.
(4) **Sorbent module:** Downstream of the filter, the sampled gas passes through a water-cooled module and then to a sorbent module that is filled with XAD-2 resin which is a porous polymer resin with the capability of adsorbing a broad range of organic species with boiling points $>/ = 100°$ C.
(5) **Four impingers:** The four impingers are connected in series and immersed in an ice bath. The first impinger, connected to the outlet of the sorbent module, is empty and is used to collect the condensate which percolates through the sorbent resin module. The second and third impingers have long stems which are immersed in the appropriate scrubbering solution. The sampled gas bubbled through the solution. The selection of the solution is contingent upon the type of vapors that are suspected of being contained in the sampled gas. A caustic solution such as sodium hydroxide or sodium acetate is used to collect acid gases such as HCL. The fourth impinger is typically filled with silica gel to absorb any moisture in the sampled gas. Moisture removal is important to ensure accurate gas flow measurements and to prevent damage to the pumping system.
(6) The downstream of the four impingers are an air tight pump, and a dry gas meter.

Modified rosin: A rosin that has been treated with heat or catalysts, or both; with or without added chemical substances, so as to cause substantial change in the structure of the rosin acids (*see* rosin for more related terms) (EPA-79/12).

Modified solid waste incineration unit: A solid waste incineration unit at which modifications have occurred after the effective date of a standard under subsection (a) if:
(1) The cumulative cost of the modifications, over the life of the unit, exceed 50 per centum of the original cost of construction and installation of the unit (not including the cost of any land purchased in connection with such construction or installation) updated to current costs, or
(2) The modification is a physical change in or change in the method of operation of the unit which increases the amount of any air pollutant emitted by the unit for which standards have been established under this section or section 111 (CAA129.g-42USC7429).
(3) *See* solid waste incineration unit for more related terms.

Modified source:
(1) The enlargement of a major stationary source is often referred to as modification, implying that more emissions will occur (EPA-94/04).
(2) *See also* synonym, modification source.

Modified steaming: A technique for conditioning (wood) logs which is a variety of the steam vat process in that steam is produced by heating water with coils set in the bottom of the vat (EPA-74/04).

Modifier:
(1) In froth flotation, reagents used to control alkalinity and to eliminate harmful effects of colloidal material and soluble salts.
(2) Chemicals which increase the specific attractive between collector agents and particle surfaces or conversely which increase the wettability of those surfaces (EPA-82/05). A substance added to a propellant to reduce the dependence of burning rate on pressure (EPA-76/03).

Modifying factor (MF): An uncertainty factor that is greater than zero and less than or equal to 10; its magnitude reflects professional judgement regarding scientific uncertainties of the data base or study design not explicitly treated by the uncertainty factors (e.g., the number of animals tested). The default value for the MF is 1 (EPA-90/08; 92/12).

Modular concept: Ability to add to or close down separate subsystems of the process without interruption to the whole operation. The assembly of a system from component subassemblies such that the resulting system is regarded as a unit (EPA-83).

Modular combustion unit: *See* synonym, modular incinerator.

Modular excess air MWC: A combustor that combusts MSW and that is not field-erected and has multiple combustion chambers, all of which are designed to operate at conditions with combustion air amounts in excess of theoretical air requirements (40 CFR 60.51a-91).

Modular high-temperature gas-cooled reactor (MHTGR): A relatively small nuclear reactor of standardized design in which graphite (a compound of electrical carbon) is used to moderate (reduce the energy of) the neutrons created in the core by fission reactions and a gas (helium) is used to cool the reactor core. The MHTGR concept proposed for the NPR consists of a complex of eight separate reactor modules, each with its own core, vessel, cooling system, and containment structure (DOE-91/04).

Modular incinerator (or modular combustion unit):
(1) Smaller-scale waste combustion units prefabricated at a manufacturing facility and transported to the waste combustion sites (EPA-89/11).
(2) One of a series of incinerator units designed to operate independently (OME-88/12). When applied to municipal waste incineration, it is similar to a mass burn incinerator in that it burns unprocessed municipal solid waste (MSW). However, it features two combustion chambers and feeds MSW with a hydraulic ram. Its primary chamber is operated in a slightly oxygen deficient condition. Wastes are vaporized in this chamber and the resulting gases are sent to the secondary chamber, which is operated in an excess condition, for complete combustion. One disadvantage of the two chamber system is that waste burnout is not always complete which increases ash quantities and reduces the efficiency of energy recovery. In general, a modular incinerator is a small, factory fabricated plant, usually custom designed to fit a particular application. It is also used for industrial and medical waste combustion (OTA-89/10).
(3) *See* incinerator for more related terms.

Modular starved air MWC: A combustor that combusts MSW (municipal solid waste) and that is not field-erected and has multiple combustion chambers in which the primary combustion chamber is designed to operate at substoichiometric conditions (40 CFR 60.51a-91).

Modulated stem: The stem attached to the vacuum break diaphragm in such a manner as to allow stem displacement independent of diaphragm displacement (40 CFR 85.2122(a)(1)(ii)(G)-91).

Modulated stem displacement: The distance through which the modulated stem may move when actuated independent of diaphragm displacement (40 CFR 85.2122(a)(1)(ii)(C)-91).

Modulated stem displacement force: The amount of force required t start and finish of a modulated stem displacement (40 CFR 85.2122(a)(1)(ii)(D)-91).

Moisture: The water content of substances or matters such as air, wood or a timber product expressed as a percentage of total weight.

Moisture content: The weight loss (expressed in percent) when a sample of material is dried to a constant weight at a temperature of $105°$ C. Conventionally expressed as a percentage of the total mass of the wet material (EPA-83).

Moisture content of solid waste: The weight loss (expressed in percent) when a sample of solid waste is dried to a constant weight at a temperature of $100°$ C to $105°$ C. The percent of moisture contained in a solid waste sample can be calculated on a dry or wet

basis as follows (SW-108ts):
- wet % = (water content)/(dry weight + water content)
- dry % = (water content)/(dry weight of sample)

Moisture holding capacity: *See* synonym, field capacity of solid waste.

Moisture penetration: The depth to which irrigation water or rain penetrates soil before the rate of downward (penetration) movement becomes negligible (SW-108ts).

Moisture set ink: An ink that dries or sets principally by precipitation. The vehicle consists of a water insoluble resin dissolved in a hygroscopic solvent. Drying occurs when the hygroscopic solvent has absorbed sufficient moisture either form the atmosphere, substrate or external application to precipitate the binder. An important characteristic of these inks is their low odor (*see* ink for more related terms) (EPA-79/12a).

Molal concentration: *See* synonym, molarity.

Molal solution: The number of moles of solute per 1000 grams of solvent (*see* concentration for more related terms).

Molal unit: The molecular weight of a compound or an element is simply the sum of all the atomic weights (from the Periodic Chart) of all the atoms in the molecule. A mole of any pure substance is defined as the amount of substance that numerically equals the molecular weight of that compound. Therefore, by definition, a gram mole or pound mole of oxygen weighs 32 grams or pounds depending on the units used (EPA-81/12, p2-5).

Molality: The number of moles per 1000 grams of solvent (*see* concentration for more related terms).

Molar absorptivity (e_r):
(1) Is defined as the proportionality constant in the Beer-Lambert law when the concentration is given in terms of moles per liter (i.e., molar concentration). Thus, $A = e_r Cl$, where A and e_r represent the absorbance and molar absorptivity at wavelength r and l and C are defined in (3). The units of e_r are molar^{-1} cm^{-1}. Numerical values of molar absorptivity depend upon the nature of the absorbing species (other identical or similar definitions are provided in 40 CFR 796.3780-iv) (40 CFR 796.3700-iv-91).
(2) The product of the absorptivity and the molecular weight of the substance (*see* absorptivity for more related terms) (LBL-76/07-bio).

Molar analysis: The analysis of a gas mixture is based on the number of moles of each component in the system (*see* property for more related terms).

Molar solution: A solution containing 1 mole (gram-molecular weight) of solute in 1 liter of the solution (*see* concentration for more related terms).

Molarity (mole fraction or molal concentration): The number of grams- molecular weights of a compound dissolved in 1 liter of solvent (*see* concentration for more related terms).

Mold: A form made of sand, metal, or refractory material that contains the cavity into which molten metal is poured to produce a casting of definite shape and outline (EPA-85/10a).

Molded glass: A glass which is formed in a mold, as distinct from cast, rolled, drawn, or offhand ware (*see* glass for more related terms) (EPA-83).

Molded pulp product: Contoured products, such as egg packaging items, food trays, plates, and bottle protectors, made by depositing fibers from a pulp slurry onto a forming mold of the contour and shape desired in the product (EPA-87/10).

Mole (n): The quantity of a pure substance whose weight (any unit of mass) is numerically equal to its molecular weight (e.g., one g-mole of oxygen would be 32 g and 1 lb-mole of oxygen would be 32 lb. In engineering calculations, the pound mole is the most commonly used. A pound mole of hydrogen (H_2, molecular weight 2) is 2 pounds and a pound mole of oxygen (O_2, molecular weight 32) 32 pounds. Number of moles can be calculated by n=m/M, where m = weight of a substance and M = molecular weight of the substance. For Example: for 64 pounds of O_2, n = 64 #/(32 #/(#-mole)) = 2 #-mole.

The measure applies to mixtures of gases, e.g., a pound mole of air is approximately 29 pounds. A pound mole of the gases just mentioned or any gas or mixture occupies the same volume under standard conditions at 60° F and 1 atm, i.e., 379 cu ft. This important fact, based on Avogadro's principle (The number of molecules per mole is 6.025 x 10^{23}; it is known as Avogdro's Number), establishes the basis for calculation of gas flow rates and other factors which are a part of control equipment design (EPA-84/09).

Mole fraction: A ratio employed in expressing concentrations of solutions and mixtures. The mole fraction of any component of a mixture or solution is defined as the number of moles of that component divided by the sum of the number of moles of all components (*see* concentration for more related terms) (EPA-84/09).

Molecular absorption: Refers to the measurement of the absorption of specific wavelengths of electromagnetic radiation by molecules (LBL-76/07-bio).

Molecular absorption spectrometry (or colorimetry): The measurement of color naturally present in samples or developed therein by the addition of reagents (LBL-76/07-water).

Molecular diffusion:
(1) The diffusion of a property within a fluid by molecular processes. In a turbulent flow, molecular diffusion is often negligible compared to diffusion by turbulence (NATO-78/10).
(2) A process of spontaneous intermixing of different substances, attributable to molecular motion and tending to produce uniformity of concentration (LBL-76/07-air).

Molecular formula: A formula representing the numbers of atoms and the composition of a chemical compound by a specific symbols (e.g., CH_4 represents 4 atoms of hydrogen and one atom of carbon for a methane compound).

Molecular weight: The sum of the weight of the atoms which comprise the molecular. Therefore, the molecular weight of a substance would be equal to the sum of the atomic weight of each atom in its molecule, e.g., a molecule of hydrogen (2 atoms, H_2) has a molecular weight of 2, a molecular of oxygen (2 atoms, O_2) has a molecular weight of 32, and water has a molecular weight of 18 (cf. atomic weight) (EPA-84/09).

Molecule:
(1) A chemical unit composed of one or more atoms (EPA-83/06a).
(2) The smallest division of a compound that still retains or exhibits all the properties of the substance (EPA-94/04).

Molluscicide: A chemical used to kill or control snails and slugs (EPA-85/10).

Mollusk (mollusca): A large animal group including those forms popularly called shellfish (but not including crustaceans). All have a soft unsegmented body protected in most instances by a calcareous shell. Examples are snails, mussels, clams, an oysters (LBL-76/07-water).

Molten salt combustion (MSC): *see* synonym, molten salt incinerator

Molten salt incinerator (or molten salt combustion): Molten salt destruction is a method of combusting organic material while, at the same time, scrubbing in-situ objectionable by-products of the combustion and thus preventing their emission in the effluent gas stream. This process of simultaneous combustion and scrubbing is accomplished by injecting the material to be burned with air, or oxygen enriched air, under the surface of a pool of (generally) molten carbonate salts. The melt is maintained at temperatures on the order of 1650° F (900° C) to promote rapid oxidation of organic waste. Halogen species in the waste form halide salts and phosphorus, sulfur, arsenic or silicon (from glass or ash in the waste) form the oxygenated salts such as sodium phosphate, sulfate, arsenate or silicate, respectively. These products are retained in the melt as inorganic salts rather than released to the atmosphere as volatile gases (*see* incinerator for more related terms) (Lee-83/07).

Molten salt reactor: A thermal treatment unit that rapidly heats waste in a heat-conducting fluid bath of carbonate salt (*see* incinerator for more related terms) (EPA-94/04).

Molybdenum (Mo): A hard transition metal element with a.n. 42; a.w. 95.94; d. 10.2 g/cc; m.p. 2610° C and b.p. 5560° C. The element belongs to group VIB of the periodic table.

Molybdenite: The most common ore of molybdenum (MoS_2) (EPA-82/05).

Molybdenite concentrate: A commercial molybdenite ore after the first processing operations. It contains about 90% (MoS_2) along with quartz, feldspar, water, and processing oil (EPA-82/05).

Momentary flood peak: The maximum rate of flow during a flood event; usually this is the flow at the time flood crest is reached (*see* flood for more related terms) (DOI-70/04).

Momentum: The quantity of motion measured by the product of the mass and the velocity of a body (NATO-78/10).

Momentum flowmeter: One of solid or sludge flow rate meters. Two types of these solid flowmeters are available, based upon either impact or torque. This device works fairly well with dry, flowable materials but is less accurate if feed particles are very large, nonuniform, or viscous. Typical accuracies are within +/- 2%. Sludges are not recommended because of their viscosity and splashing effects (*see* flow rate meter for more related terms) (EPA-89/06).

Monimolimnion: The bottom layer of a lake. It has a higher density and no circulation because of no wind effects (cf. mixolimnion).

Monitor (n): *See* synonym, analyzer.

Monitor (v): To measure and record (40 CFR 52.741-91).

Monitoring (or air monitoring):
(1) Periodic or continuous surveillance or testing to determine the level of compliance with statutory requirements and/or pollutant levels in various media or in humans, plants, and animals (EPA-94/04).
(2) The process of measuring and/or sampling the amount of pollutants or radioactive contamination present in the environment on a continuous, periodical or random basis for spatial and time variations (EPA-84/09; 74/11).
(3) cf. interstitial monitoring.
Types of monitoring include:
(1) Continuous monitoring
(2) Periodical monitoring
(3) Random monitoring

Monitoring and analysis order: *See* definition under administrative order.

Monitoring device: The total equipment, required under the monitoring of operations sections in applicable subparts, used to measure and record (if applicable) process parameters (40 CFR 60.2-91).

Monitoring system: Any system, required under the monitoring sections in applicable subparts, used to sample and condition (if applicable), to analyze, and to provide a record of emissions or process parameters (40 CFR 61.02-91).

Monitoring system types: There are three basic types of monitoring systems: cross stack, extractive, and in-situ. Carbon monoxide monitoring systems generally are extractive or cross stack, while oxygen monitors are either extractive or in-situ (EPA-90/04):
(1) **Cross stack analyzer:** An analyzer which measures the parameter of interest by placing a source beam on one side of the stack and either the detector (in single pass instruments) or a retro-reflector (in double pass instruments) on the other side and measuring the parameter of interest (e.g., CO) by the attenuation of the beam by the gas in its path.

(2) **Extractive analyzer:** An analyzer which uses a pump or other mechanical, pneumatic, or hydraulic means to draw a small portion of the stack or flue gas and convey it to the remotely located analyzer.

(3) **In-situ analyzer:** An analyzer which places the sensing or detecting element directly in the flue gas stream and thus perform the analysis without removing a sample from the stack.

Monitoring well:
(1) A well used to obtain water quality samples or measure groundwater levels (EPA-94/04).
(2) Well drilled at a hazardous waste management facility or Superfund site to collect ground-water samples for the purpose of physical, chemical, or biological analysis to determine the amounts, types, and distribution of contaminants in the ground water beneath the site (EPA-94/04).

Monochromator:
(1) A device or instrument that, with an appropriate energy source, may be used to provide a continuous calibrated series of electro-magnetic energy bands of determinable wavelength or frequency range (LBL-76/07-air).
(2) A device that combines polychromatic radiations (or wavelengths) into a monochromatic radiation or isolates a particular narrow band of wavelengths for focusing on a detector.

Monoclonal antibodies (also called MABs or MCAs): Molecules of living organisms that selectively find and attach to other molecules to which their structure conforms exactly. This could also apply to equivalent activity by chemical molecules (EPA-89/12).

Monoclonal antibodies: (Also called MABs and MCAs) 1. Man-made clones of a molecule, produced in quantity for medical or research purposes. 2. Molecules of living organisms that selectively find and attach to other molecules to which their structure conforms exactly. This could also apply to equivalent activity by chemical molecules (EPA-94/04).

Monofill: A sanitary landfill method for one type of waste only (*see* sanitary landfill for more related terms) (OTA-89/10).

Monolithic: Describing a structure which is without cracks or seams, self-supporting, and essentially homogeneous (OME-88/12).

Monolithic lining: A refractory lining, made on-site in large sections without conventional layers and joints, may be formed by pouring or casting, gunniting, ramming or sintering a fine granular material into place (EPA-83).

Monomer: A chemical substance that has the capacity to form links between two or more other molecules (cf. polymer) (40 CFR 704.25; 721.350; 723.250-91).

Mononitrotoluene (MNT) and dinitrotoluene (DNT): The intermediate products formed during the manufacture of TNT. DNT is also used in the formulation of single-base propellants (EPA-76/03).

Monomictic: Lakes and reservoirs which are relatively deep, do not freeze over during the winter months, and undergo a single stratification and mixing cycle during the year (usually in the fall) (EPA-94/04).

Monorail crane: A crane consisting of a lifting unit that hangs from a single suspended, horizontal rail in such a way that the unit can travel the length of the rail (EPA-83).

Monte Carlo simulation: A statistical modeling technique that involves the random selection of sets of input data for use in repetitive model runs in order to predict the probability distributions of receiving water quality concentrations (EPA-85/09; 91/03).

Month: A calendar month or a prespecified period of 28 days or 35 days (utilizing a 4-4-5-week recordkeeping and reporting schedule) (40 CFR 60.541-91).

Monthly average: The arithmetic average of eight (8) individual data points from effluent sampling and analysis during any calendar month (40 CFR 425.02-91).

Monthly average regulatory values: The basis for the monthly average discharge in direct discharge permits nd for pretreatment standards. Compliance with the monthly discharge limit is required regardless of the number of samples analyzed and averaged (40 CFR 461.3-91).

Montreal Protocol: The Montreal Protocol on Substances that Deplete the Ozone Layer, a protocol to the Vienna Convention for the Protection of the Ozone Layer, including adjustments adopted by Parties thereto and amendments that have entered into force (40 CFR 82.3-91).

Montreal Protocol and the Protocol: The Montreal Protocol on Substances that Deplete the Ozone Layer, a protocol to the Vienna Convention for the Protection of the Ozone Layer, including adjustments adopted by Parties thereto and amendments that have entered into force (CAA601-42USC7671-91).

Moratorium:
(1) A complete cessation of the taking of marine mammals and a complete ban on the importation into the United States of marine mammals and marine mammal products, except as provided in this chapter (MMPA3-16USC1362-90).
(2) During the negotiation process, a period of 60 to 90 days during which EPA and potentially responsible parties may reach settlement but no site response activities can be conducted (EPA-94/04).

Morbidity:
(1) Disease state (EPA-90/05).
(2) Rate of disease incidence (EPA-94/04).

Morphologic transformation: The acquisition of certain phenotypic, characteristics most notably loss of contact inhibition and loss of anchorage dependence which are often but not always associated with the ability to induce tumors in appropriate hosts (40 CFR 795.285-91).

Mortality: Death state (EPA-5/90).

Mortar: A mixture of gypsum plaster with aggregate or hydrated lime, or both, and water to produce a trowelable slurry (EPA-83). Other mortar related terms include:
- Air setting refractory mortar
- Cold setting refractory mortar (*see* definition under air setting refractory mortar)
- Fireclay mortar
- Heat setting refractory mortor
- Hot setting refractory mortar (*see* definition under heat setting refractory mortar)
- Hydraulic setting mortar

Moss: Any bryophytic plant characterized by small, leafy, often tufted stems bearing sex organs at the tips (LBL-76/07-water).

Most probable number (MPN):
(1) In the testing of bacterial density by the dilution method, number of organisms volume which, in accordance with statistical theory, would be more likely than any other possible number to yield the observed test result or which would yield the observed test result with the greatest frequency. Expressed as density of organisms per 100 mL (EPA-74/01a).
(2) The most probable number of coliform-group organisms per unit volume of sample water (EPA-94/04).

Mother liquor: A concentrated solution substantially freed from undissolved matter by filtration, centrifugation, or decantation. Crystals are formed from the mother liquor (*see* liquor for more related terms) (EPA-77/07).

Motile: Exhibiting or capable of spontaneous movement (LBL-76/07-water).

Motor activity: Any movement of the experimental animal (40 CFR 798.6200-91).

Motor carrier: A common carrier by motor vehicle, a contract carrier by motor vehicle, or a private carrier of property by motor vehicle as those terms are defined by paragraphs (14), (15), and (1) of section 203(a) of the Interstate Commerce Act (49USC303(a)) (40 CFR 202.10-91).

Motor controller: An electronic or electro-mechanical device to convert energy stored in an energy storage device into a form suitable to power the traction motor (40 CFR 600.002.85-91).

Motor fuel: Petroleum or a petroleum based substance that is motor gasoline, aviation gasoline, No. 1 or No. 2 diesel fuel, or any grade of gasohol, and is typically used in the operation of a motor engine (*see* fuel for more related terms) (40 CFR 280.12-91).

Motor vehicle: Any self-propelled vehicle designed for transportation of persons or property on a street or highway (*see* vehicle for more related terms) (40 CFR 52.136-91, *see also* CAA216; 52.137; 202.10-91).

Motor Vehicle Information and Cost Savings Act (MVICSA): *See* the term of Act or MVICSA.

Motor vehicle or engine part manufacturer: As used in sections 207 and 208 means any person engaged in the manufacturing, assembling or rebuilding of any device, system, part, component or element of design which is installed in or on motor vehicles or motor vehicle engines (*see* vehicle for more related terms) (CAA216-42USC7550-91).

Motorcycle: Any motor vehicle with a headlight, taillight, and stoplight and having: two wheels, or three wheels and a curb mass less than or equal to 680 kilograms (1499 pounds) (cf. federally regulated motorcycle) (40 CFR 86.402.78-91, *see also* 40 CFR 205.151-91).

Motorcycle noise level: The A-weighted noise level of a motorcycle as measured by the acceleration test procedure (40 CFR 205.151-91).

Mountain valley wind: A diurnal variation of the wind, in a mountain valley blowing uphill by day and downhill by night forming a meso scale circulation system (*see* wind for more related terms) (NATO-78/10).

Mouse: In seaming a landfill liner, a synonymous term for hot wedge, or hot shoe, seaming device (EPA-89/09, *see also* EPA-91/05).

Movable bulkhead loader: A type of side-loading, enclosed compactor truck equipped with a movable bulkhead that pushes the solid wastes from the front loading area to the rear of the vehicle (EPA-83).

Movable grate: A grate with moving parts. A movable grate designed to feed solid fuel or solid waste to furnace is called a stoker (*see* grate for more related terms).

Movement:
(1) An exhaust configuration of a building or emission control device (e.g., positive-pressure fabric filter) that extends the length of the structure and has a width very small in relation to its length (i.e., length to width ratio is typically greater than 5:1). The exhaust may be an open vent with or without a roof, louvered vents, or a combination of such features (40 CFR 60.61-91).
(2) That hazardous waste transported to a facility in an individual vehicle (40 CFR 260.10-91).

Muck: The usually blackish, fine-textured, and largely organic deposits at the bottom of a water body (DOI-70/04).

Muck soils: Earth made from decaying plant materials (*see* soil for more related terms) (EPA-94/04).

Mud filter: In the paper industry, a mud filter means a piece of equipment used to thicken and wash lime mud prior to burning it in the lime kiln (*see* filter for more related terms) (EPA-87/10).

Mud washer: In the paper industry, a piece of equipment used to wash the sodium base chemicals from the lime mud prior to burning it in the lime kiln (EPA-87/10).

Mudballs: Round material that forms in filters and gradually increases in size when not removed by backwashing (EPA-94/04).

Muffle furnace: A furnace heated through the outside of a refractory chamber containing the hearth (*see* furnace for more related terms) (EPA-83).

Muffler: A device for abating the sound of escaping gases of an internal combustion engine (40 CFR 202.10-91).

Mulch: A layer of material (wood chips, straw, leaves, etc.) placed around plants to hold moisture, prevent weed growth, and enrich or sterilize the soil (EPA-94/04).

Mullen: Measurement of the resistance of paper to a bursting pressure (EPA-83).

Multi-collinearity: A regression analysis situation in which some or all of the independent variables are very highly intercorrelated. The presence of this situation may cause model development to be extremelu difficult or impossible (EPA-79/12c).

Multi-cyclone (or multiple cyclone): A dust collector consisting of a number of cyclone collectors that operate in parallel. The volume and velocity of combustion gas can be regulated by dampers to maintain efficiency over a given load range (*see* mechanical separator for more related terms) (SW-108ts).

Multi-cyclone collector: An assembly of cyclone tubes operating in parallel (*see* collector for more related terms) (EPA-83).

Multi-effect evaporator (or multiple effect evaporation): A series of evaporations and condensations with the individual units set up in series and the latent heat of vaporization from one unit used to supply energy for the next (cf. multiple effect evaporator system) (EPA-83/06a; EPA-75/02d).

Multi-hit: A sensitive target which requires several hits for inactivation (cf. multi-target) (EPA-88/09a).

Multi-media: Water, air, and land (PPA6603-91).

Multi-media environmental goal (MEG): The level of contaminants (in ambient air, water, or land, or in emissions or effluents conveyed to ambient media) that:
(1) Will not produce negative effects in the surrounding populations or ecosystems; or
(2) Represent control limits demonstrated to be achievable through technology (EPA-81/09).

Multi-media filter: A filtration device designed to remove suspended solids from wastewater by trapping the solids in a porous medium. The multimedia filter is characterized by fill material ranging from large particles with low specific gravities to small particles with a higher specific gravity. Gradation from large to small media size is in the direction of normal flow (*see* filter for more related terms) (EPA-82/11).

Multi-parameter capability: Ability to measure other constituents (e.g., pollutants) or parameters (LBL-76/07-water).

Multi-purpose bucket: A two-piece, hinged container affixed to movable arms of a loader used to haul, excavate and spread cover material and crush and spread solid wastes (EPA-83).

Multi-stage separator: An apparatus used to physically separate several constituents or components of a solution based on density differences (*see* separator for more related terms) (EPA-77/07).

Multi-target: A microorganism possesses several vital sites, each of which must be hit once before inactivation (cf. multi-hit) (EPA-88/09a).

Multiple bond: A chemical bond that contains more than one pair of electrons (*see* chemical bond for more related terms).

Multiple chamber incinerator: A multiple chamber incinerator consists of two or more chambers, arranged as in-line or with retorts, interconnected by gas passage ports or ducts (*see* incinerator for more related terms) (EPA-83).

Multiple hearth incinerator: A multiple hearth incinerator is a flexible incinerator unit, composed of a series of vertical solid flat hearths (usually six to nine), which can be utilized to dispose of sewage sludges, tars, solids, gases, and liquid combustible wastes. The waste is fed through the top of the incinerator and spirals from the top hearth to bottom hearth through a series of drop holes. This type of incinerator was originally designed to incinerate sewage plant sludges (*see* incinerator for more related terms) (EPA-81/09).

Multiple cyclone: *See* synonym, multi-cyclone.

Multiple dosing tank: *See* definition under dosing tank.

Multiple effect evaporation: *See* synonym, multi-effect evaporator.

Multiple effect evaporator system: The multiple-effect evaporators and associated condenser(s) and hotwell(s) used to concentrate the spent cooking liquid that is separated from the pulp (black liquor) (cf. multi-effect evaporator) (40 CFR 60.281-91).

Multiple ferrous melting furnace scrubber configuration: A configuration where two or more discrete wet scrubbing devices are employed in series in a single melting furnace exhaust gas stream. The ferrous melting furnace scrubber mass allowance shall be given to each discrete wet scrubbing device that has an associated wastewater discharge in a multiple ferrous melting furnace scrubber configuration. The mass allowance for each discrete wet scrubber shall be identical and based on the air flow of the exhaust gas stream that passes through the multiple scrubber configuration (40 CFR 464.31-91).

Multiple hearth incinerator: A multiple hearth incinerator is a flexible incinerator unit, composed of a series of vertical solid flat hearths (usually six to nine), which can be utilized to dispose of sewage sludges, tars, solids, gases, and liquid combustible wastes. The waste is fed through the top of the incinerator and spirals from the top hearth to bottom hearth through a series of drop holes. This type of incinerator was originally designed to incinerate sewage plant sludges (*see* incinerator for more related terms) (EPA-9/81).

Multiple linear regression: A method to fit a plane through a set of points such that the sum of squared distances between the individual observations and the estimated plane is a minimum. This statistical technique is an extension of linear regression in that more than one independent variable is used in the least squares equation

(EPA-79/12c). Multiple linear regression can be expressed as y = a + b₁x₁ + b₂x₂ + b₃x₃ + ...

Multiple operation machinery: Two or more tools are used to perform simultaneous or consecutive operations (EPA-83/06a).

Multiple package coating: A coating made from more than one different ingredient which must be mixed prior to using and has a limited pot life due to the chemical reaction which occurs upon mixing (40 CFR 52.741-91).

Multiple purpose development: In water projects, the development that takes into account the use and control of water in all possible aspects: irrigation, power, flood control, domestic and industrial water supply, pollution control, navigation, recreation, fish and wildlife. The first multiple-purpose project authorized and designed as such was the Boulder Canyon Project (Hoover Dam) in 1928 (DOI-70/04).

Multiple stack: The closely grouped point sources where the plumes may mutually influence each other, e.g., with respect to plume rise (*see* stack for more related terms) (NATO-78/10).

Multiple stand: Those recirculation or direct application cold rolling mills which include more than one stand of work rolls (cf. single stand) (40 CFR 420.101-91).

Multiple subcategory plant: A plant discharging process wastewater from more than one manufacturing process subcategory (EPA-83/06a). Multiple use: The use of land for more than one purpose; i.e., grazing of livestock, wildlife production, recreation, watershed, and timber production. Could also apply to use of bodies of water for recreational purposes, fishing, and water supply (EPA-89/12).

Multiple use: Use of land for more than one purpose; i.e., grazing of livestock, watershed and wildlife protection, recreation, and timber production. Also applies to use of bodies of water for recreational purposes, fishing, and water supply (EPA-94/04, *see also* FLPMA103).

Multistage model: A dose-response model often expressed in the form: P(d) = 1 - exp (-[q(0) + q(l)d + q(2)d**2 + ... + q(k)d**k]); where P(d) is the probability of cancer from a continuous dose rate d, the q(i) are the constants, and k is the number of dose groups (or, if less, k is the number of biological stages believed to be required in the carcinogenesis process). Under the multistage model, it is assumed that cancer is initiated by cell mutations in a finite series of steps. A one-stage model is equivalent to a one-hit model (EPA-92/12).

Multistage remote sensing: A strategy for landscape characterization that involves gathering and analyzing information at several geographic scales, ranging from generalized levels of detail at the national level through high levels of detail at the local scale (EPA-94/04).

Mundo garbage: The world of garbage (EPA-83).

Mungo: The waste of milled wool that is combined with other fibers to make low-quality cloth (SW-108ts).

Municipal aluminum scrap (or aluminum scrap): The aluminum alloy product originating from municipal solid waste, that is recovered from industrial, commercial or household wastes destined for disposal facilities (EPA-83). It includes:
(1) **Combustible material--organic:** Material that is measured by weight loss of a dried sample input after heating to a red heat in an open crucible in a vented furnace. Combustibles include both loose organics and organic coatings (EPA-83).
(2) **Loose combustible material--organic:** Materials that consist of, but are not limited to, nonmetallic materials such as paper, rags, plastic, rubber, wood, food wastes, and yard or lawn wastes, etc., which are not permanently attached to noncombustible objects. The LCO's are defined as material larger than 12 mesh (U.S. Standard Sieve). A determination of LCO's is best done by sampling the material and handpicking, hand cleaning, and visually identifying the materials described previously (EPA-83).
(3) **Recovery aluminum:** The percent material recovered after an assay using the procedures prescribed in ASTM E-38 specification (EPA-83).
(4) *See* scrap for more related terms.

Municipal collection:
(1) Collection of solid wastes by a city-operated agency (*see* collection for more related terms) (EPA-83).
(2) The collection of solid waste by public employees and equipment under the supervision and direction of a municipal department or official (SW-108ts).
(3) *See* waste collection for more related terms.

Municipal discharge: Discharge of effluent from waste water treatment plants which receive waste water from households, commercial establishments, and industries in the coastal drainage basin. Combined sewer/separate storm overflows are included in this category (EPA-94/04).

Municipal ferrous scrap: The ferrous waste that is collected from industrial, commercial or household sources and destined for disposal facilities. Typically, municipal ferrous scrap consists of a metal or alloy fraction, a combustible fraction and an inorganic noncombustible fraction which includes metal oxide (EPA-83). It includes:
(1) **Total combustibles:** Materials that include paints, lacquers, coatings, plastics, etc., associated with the original ferrous product, as well as combustible materials (paper, plastics, textiles, etc.) which become associated with the ferrous product after it is manufactured (EPA-83).
(2) **Metallic yield:** The weight percent of the municipal ferrous scrap that is generally recoverable as metal or alloy (EPA-83).
(3) *See* scrap for more related terms.

Municipal incinerator: *See* synonym, municipal waste incinerator.

Municipal separate storm sewer: A conveyance or system of conveyances (including roads with drainage systems, municipal streets, catch basins, curbs, gutters, ditches, man-made channels, or storm drains):
(1) Owned or operated by a State, city, town, borough, county, parish, district, association, or other public body (created by or pursuant to State law) having jurisdiction over disposal of

sewage, industrial wastes, storm water, or other wastes, including special districts under State law such as a sewer district, flood control district or drainage district, or similar entity, or an Indian tribe or an authorized Indian tribal organization, or a designated and approved management agency under section 208 of the CWA that discharges to waters of the United States;
(2) Designed or used for collecting or conveying storm water;
(3) Which is not a combined sewer; and
(4) Which is not part of a Publicly Owned Treatment Works (POTW) as defined at 40 CFR 122.2 (40 CFR 122.26-8-91).
(5) *See* sewer for more related terms.

Municipal sewage: Wastes (mostly liquid) originating from a community; may be composed of domestic wastewaters and/or industria discharges (EPA-94/04).

Municipal solid waste: *See* synonym, municipal waste.

Municipal solid waste (MSW) composting: The controlled degradation of municipal solid waste including after some form of preprocessing to remove non-compostible inorganic materials (*see* composting for more related terms) (EPA-89/11).

Municipal type solid waste: *See* synonym, municipal waste.

Municipal waste (municipal solid waste or municipal type solid waste): Normally, residential and commercial solid wastes generated within a community (40 CFR 240.101-91). Term related municipal waste include refuse derived fuel; refuse; rubbish; rubble; trash; swill; waste, residential; waste, yard; white goods, etc.) (*see* waste for more related terms).

Municipal waste combustor or MWC or MWC unit: Any device that combusts, solid, liquid, or gasified MSW including, but not limited to, field-erected incinerators (with or without heat recovery), modular incinerators (starved air or excess air), boilers (i.e., steam generating units), furnaces (whether suspension-fired, grate-fired, mass-fired, or fluidized bed-fired) and gasification/combustion units. This does not include combustion units, engines, or other devices that combust landfill gases collected by landfill gas collection systems (40 CFR 60.51a-91).

Municipal waste incinerator (or municipal incinerator): A privately or publicly owned incinerator primarily designed and used to burn residential and commercial solid waste (SW-108ts). The main technologies are:
(1) Mass burn incinerator
(2) Refuse derived fuel incinerator
(3) Modular incinerator
Other technologies include:
(1) Fluidized bed
(2) Pyrolysis
(3) Ocean incineration (OTA-89/10)
(4) *See* incinerator for more related terms

Municiple wastewater treatment: *See* synonym, publicly owned treatment works.

Municipality: A city, town, borough, county, parish, district, association, or other public body created by or under State law and having jurisdiction over disposal of sewage, industrial wastes, or other wastes, or an Indian tribe or an authorized Indian tribal organization, or a designated and approved management agency under section 208 of CWA (40 CFR 122.2-91).

Mustard gas (($ClCH_2CH_2)_2S$): A toxic (poisonous) gas.

Mutagen:
(1) Any substance that can cause a change in genetic material (EPA-89/12).
(2) An agent that causes biological mutation (EPA-87/07a).

Mutagen/mutagenicity: An agent that causes a permanent genetic change in a cell other than that which occurs during normal genetic recombination. Mutagenicity is the capacity of a chemical or physical agent to cause such permanent alternation (EPA-94/04).

Mutagenic:
(1) Capable of inducing mutation (LBL-76/07-bio).
(2) The property of a substance or mixture of substances which, when it interacts with a living organism, causes the genetic characteristics of the organism to change and its offspring to have a decreased life expectancy (OME-88/12).

Mutagenicity:
(1) The capacity of a chemical or physical agent to cause permanent alteration of the genetic material within living cells (Course 165.6).
(2) The potential of a substance to cause a mutation or an adverse change in DNA (i.e., genetic material) (ETI-92).
(3) Chromosome alterations, bacterial mutations, DNA damage (*see* endpoint for more related terms) (EPA-92/12).

Mutate: To bring about a change in the genetic constitution of a cell by altering its DNA. In turn, mutagenesis is any process by which cells are mutated (EPA-89/12).

Mutualism: Beneficial mutual-interactions between two species.

MWC acid gases: All acid gases emitted in the exhaust gases from MWC units including, but not limited to, sulfur dioxide and hydrogen chloride gases (40 CFR 60.51a-91).

MWC metals: The metals and metal compounds emitted in the exhaust gases from MWC units (40 CFR 60.51a-91).

MWC organics: The organic compounds emitted in the exhaust gases from MWC units and includes total tetra- through octa-chlorinated dibenzo-p-dioxins and dibenzofurans (40 CFR 60.51a-91).

MWC plant: One or more MWC units at the same location for which construction, modification, or reconstruction is commenced after December 20, 1989 (40 CFR 60.51a-91, *see also* 40 CFR 60.31a-91).

MWC plant capacity: The aggregate MWC unit capacity of all MWC units at an MWC plant for which construction, modification, or reconstruction commenced after December 20, 1989. Any MWC units for which construction, modification, or reconstruction is commenced on or before December 20, 1989, are not included for determining applicability under this subpart (40 CFR 60.51a-91, *see also* 40 CFR 60.31a; 60.51a-91).

Mycelia: The filamentous material which makes up the vegetative body of a fungus (EPA-83/09).

Mycology: The study of fungi (LBL-76/07-water).

Mycoplasma: A bacteria group, also known as pleuropneumonia-like cells or PPLO, that are dedicate, flexible fluid-like cells without cells. Their form depends on the methods of cultivation and handling (EPA-88/09a).

Myelopathy: Pathology of the muscle fiber (LBL-76/07-bio).

Myocarditis: Inflammation of the heart muscle (LBL-76/07-bio).

Nn

N-compartment model: In biosafety analysis, a model that assumes uniform but different conditions in each of the n-compartments identified in a closed facility (EPA-88/09a).

Name plate rating: The full load continuous rating of an electric generator and its prime mover or other electrical equipment under specific conditions as designated by the manufacturer. Usually indicated on a name plate attached mechanically to the individual machine or device. Name plate rating is generally less than, but for older equipment may be greater than, demonstrated capability of the installed machine. *See* capacity (EPA-83).

Nameplate capacity: *See* synonym, design capacity (EPA-83).

Nanno-plankton: The small plankton including algae, bacteria and protozoans (nano = 10^{-9}) (cf. plankton) (EPA-89/12).

Nanometer: Nano- is a prefix indicating 10^{-9}. Nanometer means a unit of length equal to 10^{-9} meter (LBL-76/07-bio).

Naphthalene ($C_{10}H_8$): A white crystal used in fungicides, moth repellents, lubricants, resins and solvents.

Naphthalene processing: Any operations required to recover naphthalene including the separation, refining, and drying of crude or refined naphthalene (40 CFR 61.131-91).

Narrow spectrum antibiotic: An antibiotic only effective for a limited number of organisms.

National air monitoring station (NAMS): Collectively the NAMS are a subset of the SLAMS ambient air quality monitoring network (40 CFR 58.1-91).

National ambient air quality standards (NAAQSs): Air quality standards established by the Clean Air Act. The primary NAAQS are intended to protect the public health with an adequate margin of safety, and the secondary NAAQS are intended to protect the public welfare from any known or anticipated adverse effects of a pollutant (*see* criteria pollutants, state implementation plans, emissions trading) (DOE-91/04; EPA-89/12).

The main CAA regulations are the provisions of Sections 108 and 109 under which EPA has set forth for:
(1) the six most common pollutants (criteria pollutants, *see* pollutants, criteria); and
(2) The maximum level of pollution permitted in the ambient air in designated areas called air quality control regions. Such standards are called ambient standards, officially called National Ambient Air Quality Standards (cf. air quality control region, state implementation plans, or emissions trading) (40 CFR 50.4-50.12). The standards include two major items:
(A) **National primary standards:** The levels of air quality necessary, with an adequate margin of safety, to protect the public health.
(B) **National secondary standards:** The levels of air quality necessary to protect the public welfare (such as agricultural crops, livestock, and deterioration of materials and property) from any known or anticipated adverse effects of a pollutant. (*see also* 40 CFR 57.103; 300-App/A-91).

National consensus standard: Any occupational safety and health standard or modification thereof which:
(1) Has been adopted and promulgated by a nationally recognized standards-producing organization under procedures whereby it can be determined by the Secretary that persons interested and affected by the scope or provisions of the standard have reached substantial agreement on its adoption;
(2) Was formulated in a manner which afforded an opportunity for diverse views to be considered; and
(3) Has been designated as such a standard by the Secretary, after consultation with other appropriate Federal agencies (OSHA3, *see also* 29 CFR 1910.2-91).

National contingency plan: The National Contingency Plan prepared and published under subsection (d) (CWA311, *see also* OPA1001; SF101; 40 CFR 304.12; 310.11-91).

National emissions standards for hazardous air pollutants (NESHAPS):
(1) Standards established for substances listed under section 112 of the Clean Air Act, as amended. Only those NESHAPs promulgated in ambient concentration units apply in the HRS (40 CFR 300-App/A-91).
(2) Emissions standards set by EPA for an air pollutant not covered by NAAQS that may cause an increase in deaths or in serious, irreversible, or incapacitating illness. Primary standards are designed to protect human health, secondary standards to protect public welfare (EPA-89/12).

The decision to list a pollutant as a hazardous pollutant triggers a mandatory duty on EPA's part to propose emission standards for the pollutant within 180 days, hold a public hearing 30 days thereafter, and then either promulgate the standards or publish a statement explaining why information at the hearing supports a conclusion that the pollutant clearly is not a hazardous air pollutant. Thus far, EPA has not met these deadlines and has listed only six pollutants as hazardous under Section 112: asbestos, beryllium, mercury, vinyl chloride, benzene and radionuclides. NESHAPS are applicable to both new and existing sources of pollutants (Winthrop-89/09).

National Environmental Policy Act (NEPA) of 1969: *see* the term of Act or NEPA.

National environmental research park: Outdoor laboratories set aside for ecological research to study the environmental impacts of energy developments and for informing the public of environmental and land use options. The parks were established under the Department of Energy to provide protected land areas for research and education in the environmental sciences and to demonstrate the environmental compatibility of energy technology development and use (DOE-91/04).

National fire academy (NFA): NFA is a component of FEMA's National Emergency Training Center located in Emmitsburg, Maryland. It provides fire prevention and control training for the fire service and allied services. Courses on campus are offered in technical, management, and prevention subject areas. A growing off-campus course delivery system is operated in conjunction with State fire training program offices (NRT-87/03).

National hazardous materials information exchange (NHMIE): NHMIE provides information on HAZMAT training courses, planning techniques, events and conferences, and emergency response experiences and lessons learned. Call toll-free 1-800-752-6367 (in Illinois, 1-800-367-9592). Planners with personal computer capabilities can access NHMIE by dialing FTS 972-3275 or (312) 972-3275 (NRT-87/03).

National oil and hazardous substances contingency plan (NOHSCP): The federal regulation that guides determination of the sites to be corrected under the Superfund program and the program to prevent or control spills into surface waters or other portions of the environment (EPA-89/12).

National oil and hazardous substances pollution contingency plan (NCP): (40 CFR 300), prepared by EPA to put into effect the response powers and responsibilities created by CERCLA and the authorities established by Section 311 of the Clean Water Act (NRT-87/03).

National panel of environmental arbitrators or panel: A panel of environmental arbitrators selected and maintained by the Association to arbitrate cost recovery claims under this part (40 CFR 304.12-91).

National pollutant discharge elimination system (NPDES):
(1) The national program for issuing, modifying, revoking and reissuing, terminating, monitoring and enforcing permits, and imposing and enforcing pretreatment requirements, under sections 307, 402, 318, and 405 of CWA. The term includes an approved program (40 CFR 122.2-91, *see also* 40 CFR 136.2; 270.2-91).
(2) A provision of the Clean Water Act which prohibits discharge of pollutants into waters of the United States unless a special permit is issued by EPA, State, or (where delegated) a tribal government on an Indian reservation (EPA-89/12).

National pretreatment standard or pretreatment standard: Any regulation containing pollutant discharge limits promulgated by the EPA in accordance with section 307 (b) and (c) of the Act, which applies to industrial users of a publicly owned treatment works. It further means any State or local pretreatment requirement applicable to a discharge and which is incorporated into a permit issued to a publicly owned treatment works under section 402 of the Act (40 CFR 117.1-91, *see also* 40 CFR 403.3-91).

National primary and secondary ambient air quality standards: In CAA:
(1) The national primary ambient air quality standards (NPAAQSs) define levels of air quality which the EPA Administrator judges are necessary, with an adequate margin of safety, to protect the public health.
(2) The national secondary ambient air quality standards (NSAAQSs) define levels of air quality which the Administrator judges are necessary to protect the public welfare from any known or anticipated adverse effects of a pollutant (40 CFR 50.2). *see* Appendix 1A for the CAA NPAAQS and NSAAQS standards.

National primary drinking water regulation: Any primary drinking water regulation contained in Part 141 of this chapter (40 CFR 142.2-91).

National priorities list (NPL): EPA's list of the most serious uncontrolled or abandoned hazardous waste sites identified for possible long-term remedial action under Superfund. A site must be on NPL to receive money from the Trust Fund for remedial action. The list is based primarily on the score a site receives from the Hazard Ranking System. (EPA is required to update the NPL at least once a year) (EPA-89/12, *see also* 40 CFR 35.6015; 300.5-91).

National program assistance agreements: The assistance agreements approved by the EPA Assistant Administrator for Water for work undertaken to accomplish broad NEP goals and objectives (40 CFR 35.9010-91).

National response center:
(1) The national communications center located in Washington, DC, that receives and relays notice of oil discharge or releases of hazardous substances to appropriate Federal officials (40 CFR 310.11-91).
(2) The federal operations center that receives notifications of all releases of oil and hazardous substances into the environment. The Center, open 24 hours a day, is operated by the U.S. Coast Guard, which evaluates all reports and notifies the appropriate agency (EPA-89/12).
(3) A communications center for activities related to response actions, is located at Coast Guard headquarters in Washington, DC. The NRC receives and relays notices of

discharges or releases to the appropriate OSC, disseminates osc and NRT reports to the NRT when appropriate, and provides facilities for the NRT to use in coordinating a national response action when required. The toll-free number (800-424-8802, or 202-426-2675 or 202-267-2675 in the Washington, DC area) can be reached 24 hours a day for reporting actual or potential pollution incidents (NRT-87/03).

National response team (NRT):
(1) Representatives of 14 federal agencies that, as a team, coordinate federal responses to nationally significant incidents of pollution and provide advice and technical assistance to the responding agency(ies) before and during a response action (EPA-89/12).
(2) NRT consisting of representatives of 14 government agencies (DOD, DOI, DOT/RSPA, DOT/USCG, EPA, DOC, FEMA, DOS, USDA, DOJ, HHS, DOL, Nuclear Regulatory Commission, and DOE), is the principal organization for implementing the NCP. When the NRT is not activated for a response action, it serves as a standing committee to develop and maintain preparedness, to evaluate methods of responding to discharges or releases, to recommend needed changes in the response organization, and to recommend revisions to the NCP. The NRT may consider and make recommendations to appropriate agencies on the training, equipping, and protection of response teams; and necessary research, development, demonstration, and evaluation to improve response capabilities (NRT-87/03).

National response unit: The National Response Unit established under subsection (j) (CWA311-33USC1321-91).

National register of historic places: A list maintained by the National Park Service of architectural, historic, archaeological, and cultural sites of local, state, or national significance (DOE-91/04).

National security exemption: An exemption from the prohibitions of section 10(a) (1), (2), (3), and (5) of the Act, which may be granted under section 10(b)(l) of the Act for the purpose of national (other identical or similar definitions are provided in 40 CFR 205.2-5) (40 CFR 204.2-5-91).

National standard: Either a primary or secondary standard (40 CFR 51.100-91, *see also* 40 CFR 51.301-91).

National strike force (NSF): NSF is made up of three Strike Teams. The USCG counterpart to the EPA ERTs (NRT-87/03).

Natural barrier: A natural object that effectively precludes or deters access. Natural barriers include physical obstacles such as cliffs, lakes or other large bodies of water, deep and wide ravines, and mountains. Remoteness by itself is not a natural barrier (40 CFR 61.141-91).

Natural condition: Includes naturally occurring phenomena that reduce visibility as measured in terms of visual range, contrast, or coloration (40 CFR 51.301-91).

Natural convection: A fluid flow caused by different fluid density within compartments (*see* convection for more related terms) (EPA-9/88a).

Natural draft: A negative pressure created by the height of a stack or chimney and the difference in temperature between flue gases and the atmosphere (*see* draft for more related terms) (OME-88/12).

Natural draft cooling tower: A cooling tower through which air is circulated by a natural or chimney effect. A hyperbolic tower is a natural draft tower that is hyperbolic in shape (cf. mechanical draft tower) (EPA-82/11f).

Natural draft opening: Any opening in a room, building, or total enclosure that remains open during operation of the facility and that is not connected to a duct in which a fan is installed. The rate and direction of the natural draft across such an opening is a consequence of the difference in pressures on either side of the wall containing the opening (40 CFR 60.711-91, *see also* 40 CFR 60.741-91).

Natural emission: The emission caused by natural processes, e.g., by volcanoes, forest, fires, wind blown sand, swamps, sea-spray, etc. (*see* emission for more related terms) (NATO-78/10).

Natural gas:
(1) A naturally occurring mixture of hydrocarbon and nonhydrocarbon gases found in geologic formations beneath the earth's surface, of which the principal constituent is methane; or
(2) Liquid petroleum gas, as defined by the American Society for Testing and Materials in ASTM D1835-82, Standard Specification for Liquid Petroleum Gases (IBR-*see* 40 CFR 60.17) (40 CFR 60.41b-91, *see also* 40 CFR 60.41c; 60.641-91).
(3) A natural fuel containing primarily methane and ethane that occurs in certain geologic formations (EPA-89/12).

Natural gas liquid: The hydrocarbons, such as ethane, propane, butane, and pentane, that are extracted from field gas (40 CFR 60.631-91).

Natural gas processing plant (gas plant): Any processing site engaged in the extraction of natural gas liquids from field gas, fractionation of mixed natural gas liquids to natural gas products, or both (40 CFR 60.631-91).

Natural pollution: The soil, mineral, or bacterial impurities picked up by water from the earth's surface, apart from any human activity (*see* pollution for more related terms) (DOI-70/04).

Natural pozzolan: The materials that in the natural state, exhibit pozzolanic properties, such as some volcanic ash and lava deposits (*see* pozzolan for more related terms) (EPA-83).

Natural radiation: The radiation from sources beyond the earth (cosmic radiation) and naturally occurring radiation in rocks, soils, plants, and animals (*see* radiation for more related terms) (DOE-91/04).

Natural resources: The land, fish, wildlife, biota, air, water, ground water, drinking water supplies, and other such resources belonging to, managed by, held in trust by, appertaining to, or otherwise controlled by the United States (including the resources of the fishery conservation zone established by the Fishery Conservation and Management Act of 1976), any State, local government, or any foreign government, and Indian tribe, or, if such resources are subject to a trust restriction or alienation, any member of an Indian tribe (SF101, *see also* 300.5-91).

Natural resource limitation area: A land which is particularly sensitive to development, such as steep hillsides, wetlands and floodplans. Also referred to as sensitive areas (EPA-80/08).

Natural rosin (colophony rosin, common rosin, gum rosin, or pine resin): A resin obtained as a residue from distillation of turpentine oils from crude turpentine. Rosin is primarily an isomeric form of the anhydride of abietic acid. It is one of the more common binders in the foundry industry (*see* natural rosin for more related terms) (EPA-85/10a).

Natural selection: The process of survival of the fittest, by which organisms that adapt to their environment survive and those that do not disappear (EPA-89/12).

Naturally occurring or accelerator produced radioactive material (NARM): Any radioactive material except for material classified as source, byproducts, or special nuclear material under the Atomic Energy Act, as amended (EPA-88/08a).

Navigable water:
(1) The waters of the United States, including the territorial seas (CWA502, *see also* OPA1001; SP101; 40 CFR 110.1; 112.2; 116.3; 117.1; 300.5; 302.3; 401.11-91).
(2) Traditionally, waters sufficiently deep and wide for navigation by all, or specified sizes of vessels; such waters in the United States come under federal jurisdiction and are included in certain provisions of the Clean Water Act (EPA-89/12).
(3) *see* water for more related terms.

NC fines: Fine nitrocellulose particles as a result of the purification of nitrocellulose (EPA-76/03).

Near Gaussian residence time distribution: A temporal field in UDRI's reactor characterized by very small time duration between maximum and minimum residence time (< 0.15 second for $time_{max} = 2.0$ s); axial distribution well described by Gaussian function (EPA-88/12).

Near the first service connection: At one of the 20 percent of all service connections in the entire system that are nearest the water supply treatment facility, as measured by water transport facility, as measured by water transport time within the distribution system (40 CFR 141.2-91).

Near region: The region of the atmospheric path along the lidar line-of-sight between the lidar's convergence distance and the plume being measured (40 CFR 60-App/A(alt. method 1)-91).

Nearby: As used in 40 CFR 51.100(ii) of this part is defined for a specific structure or terrain feature and:
(1) For purposes of applying the formulae provided in 40 CFR 51.100(ii)(2) means that distance up to five times the lesser of the height or the width dimension of a structure, but not greater than 0.8 km (1/2 mile), and
(2) For conducting demonstrations under 40 CFR 51.100(ii)(3) means not greater than 0.8 km (1/2 mile), except that the portion of a terrain feature may be considered to be nearby which falls within a distance of up to 10 times the maximum height (Ht) of the feature, not to exceed 2 miles if such feature achieves a height (Ht) 0.8 km from the stack that is at least 40 percent of the GEP stack height determined by the formulae provided in 40 CFR 51.100(ii)(2)(ii) of this part or 26 meters, whichever is greater, as measured from the ground-level elevation at the base of the stack. The height of the structure or terrain feature is measured from the ground-level elevation at the base of the stack (40 CFR 51.100-jj-91).

Neat oil: A pure oil with no or few impurities added. In aluminum forming its use is mostly as a lubricant (*see* oil for more related terms) (40 CFR 467.02-91, *see also* 40 CFR 471.02-91).

Neat soap: The solution of completely saponified and purified soap containing about 2030 percent water which is ready for final formulation into a finished product (*see* soap for more related terms) (40 CFR 417.11-91, *see also* 40 CFR 417.31; 417.61; 417.71-91)).

Necessary preconstruction approvals or permits: Those permits or approvals required by the permitting authority as a precondition to undertaking any activity under clauses (i) or (ii) of subparagraph (A) of this paragraph (CAA169, *see also* 40 CFR 51-App/S; 51.165; 51.166; 52.21; 52.24-91).

Necrosis: The death cells of plants or animals. In plants, necrosis can discolor areas on the plant or kill it entirely (EPA-89/12).

Necrosis producer: *See* definition under toxicant and effect.

Negative electrode: *See* synonym, cathode.

Negative ion: *See* definition under ion.

Negative pressure fabric filter: A fabric filter with the fans on the downstream side of the filter bags (*see* filter for more related terms) (40 CFR 60.271a-91).

Negligence: The omission to do something which a reasonable man, guided by those ordinary considerations which ordinarily regulate human affairs, would do, or the doing of something which a reasonable and prudent man would not do. Negligence is that part of the law of torts which deals with acts not intended to inflict injury (*see* common law for more related terms) (Sullivan-95/04, p13).

Negligible residue: Any amount of a pesticide chemical remaining in or on a raw agricultural commodity or group of raw agricultural commodities that would result in a daily intake regarded as toxicologically insignificant on the basis of scientific judgment of adequate safety data. Ordinarily this will add to the diet an amount which will be less than 1/2,000th of the amount that has been demonstrated to have no effect from feeding studies on the most sensitive animal species tested. Such toxicity studies shall usually include at least 90-day feeding studies in two species of mammals (40 CFR 180.1-91).

Nematocide: A chemical agent which is destructive to nematodes (round worms or threadworms) (EPA-89/12).

Nematoda: Unsegmented roundworms or threadworms. Some are free living in soil, fresh water, and salt water; some are found living in plant tissue; others live in animal tissue as parasites (LBL-76/07-water).

Nematode: The invertebrate animals of the phylum nemathelminthes and class nematoda, that is, unsegmented round worms with elongated, fusiform, or saclike bodies covered with cuticle, and inhabiting soil, water, plants, or plant parts; may also nemas or eelworms (FIFRA2-7USC136-91).

Neodymium (Nd): A rare earth metal with a.n. 60; a.w. 144.24; d. 7.00 g/cc; m.p. 1024° C and b.p. 3027° C. The element belongs to group IIIB of the periodic table.

Neon (Ne): A noble gaseous element with a.n. 10; a.w. 20.183; d. 1.2 g/cc; m.p. -248.6° C and b.p. -246° C. The element belongs to group VIIIA of the periodic table.

NEPA process: All measures necessary for compliance with the requirements of section 2 and Title I of NEPA (40 CFR 1508.21-91).

Nephelometer: An instrument for comparing turbidities of solutions by passing a beam of light through a transparent tube and measuring the ratio of the intensity of the scattered light to that of the incident light (LBL-76/07-water).

Nephelometric turbidity unit (NTU): A unit of measuring turbidity.

Nephelometry: The use of nephelometer to study the properties of suspended solids.

Neptunium (Np): A radioactive metal with a.n. 93; a.w. 237; d. 19.5 g/cc and m.p. 637° C. The element belongs to group IIIB of the periodic table.

Nesslerization: A method of measuring the ammonia content in water.

Nessler tube: A method of comparing sample colors with a standard color solution contained in the Nessler tube.

Net: The addition of pollutants (40 CFR 409.21-91, *see also* 40 CFR 409.31-91).

Net cane: That amount of gross cane less the weight of extraneous material (40 CFR 409.61-91).

Net capacity: *See* synonym, design capacity.

Net emissions increase:
(1) The amount by which the sum of the following exceeds zero:
 (A) Any increase in actual emissions from a particular physical change or change in the method of operation at a stationary source; and
 (B) Any other increases and decreases in actual emissions at the source that are contemporaneous with the particular change and are otherwise creditable.
(2) An increase or decrease in actual emissions is contemporaneous with the increase from the particular change only if it occurs before the date that the increase from the particular change occurs.
(3) An increase or decrease in actual emissions is creditable only if:
 (A) It occurs within a reasonable period to be specified by the reviewing authority; and
 (B) The reviewing authority has not relied on it in issuing a permit for the source under regulations approved pursuant to this section which permit is in effect when the increase in actual emissions from the particular change occurs.
(4) An increase in actual emissions is creditable only to the extent that the new level of actual emissions exceeds the old level.
(5) A decrease in actual emissions is creditable only to the extent that:
 (A) The old level of actual emission or the old level of allowable emissions whichever is lower, exceeds the new level of actual emissions;
 (B) It is federally enforceable at and after the time that actual construction on the particular change begins; and
 (C) The reviewing authority has not relied on it in issuing any permit under regulations approved pursuant to 40 CFR part 51 subpart I or the state has not relied on it in demonstrating attainment or reasonable further progress;
 (D) It has approximately the same qualitative significance for public health and welfare as that attributed to the increase from the particular change.
(6) An increase that results from a physical change at a source occurs when the emissions unit on which construction occurred becomes operational and begins to emit a particular pollutant. Any replacement unit that requires shakedown becomes operational only after a reasonable shakedown period, not to exceed 180 days (40 CFR 51.165-vi-91).

Net for distribution energy: On an electric system or company basis, this means the kilowatt hours available for total system or company load. Specifically it is the sum of net generation by the system's own plants, purchased energy and net interchange (in less out) (*see* electric energy for more related terms) (EPA-83).

Net heating value: *See* synonym, low heating value.

Net system capacity: The sum of the net electric generating capability (not necessarily equal to rated capacity) of all electric generating equipment owned by an electric utility company (including steam generating units, internal combustion engines, gas turbines, nuclear units, hydroelectric units, and all other electric generating equipment) plus firm contractual purchases that are interconnected to the affected facility that has the malfunctioning flue gas desulfurization system. The electric generating capability of equipment under multiple ownership is prorated based on ownership unless the proportional entitlement to electric output is otherwise established by contractual arrangement (40 CFR 60.41a-91).

Net ton: *See* definition under ton.

Net working capital: The current assets minus current liabilities (40 CFR 144.61-91).

Net worth: The total assets minus total liabilities and is equivalent to owner's equity (cf. tangible net worth) (40 CFR 144.61-91, *see also* 40 CFR 264.141; 265.141-91).

Network: A system of transmission or distribution lines so cross connected and operated as to permit multiple energy supply to principal points on it (EPA-83).

Neurons: Cells that carry nervous impulses (NJIT-5/88).

Neurotoxicity: Gross observation, clinical evaluation, neurohistopathological tests, neurochemical tests (*see* endpoint for more related terms) (Course 165.6; EPA-92/12).

Neuston: The community of minute organisms living in the surface film of water (DOI-70/04).

Neurotoxic target esterase (NTE): A membrane-bound neural protein that hydrolyze phenyl valerate and is highly correlated with the initiation of OPIDN. NTE activity is operationally defined as the phenyl valerate hydrolytic activity resistant to paraoxon but sensitive to mipafox or neuropathic O-P ester inhibition (40 CFR 798.6450-91).

Neurotoxicity: Any adverse effect on the structure or function of the central and/or peripheral nervous system related to exposure to a chemical substance (40 CFR 798.6050-91, *see also* 40 CFR 798.6200-91).

Neurotoxicity or a neurotoxic effect: Any adverse change in the structure or function of the nervous system following exposure to a chemical substance (40 CFR 798.6400-91, *see also* 40 CFR 798.6500; 798.6850-91).

Neutral: The property of a system whereby small disturbances introduced into it, neither grow nor die out (NATO-78/10).

Neutral atmosphere: The atmosphere condition for which the vertical temperature profile is equal to the adiabatic lapse rate over the whole boundary layer. Vertical air motions are neither enhanced nor suppressed. The turbulence intensity is moderate (NATO-78/10).

Neutral plane: A roughly horizontal plane through a house defining the level at which the pressure indoors equals the pressure outdoors. During cold weather, when the thermal stack effect is occurring, indoor pressures below the neutral plane will be lower than outdoors, so that outdoor air and soil gas will infiltrate. Above the neutral plane, indoor pressures will be higher than outdoors, so that house air will exfiltrate (EPA-88/08).

Neutral refractory: The resistant to chemical attack by both acid and basic slags, or fluxes at high temperatures (*see* refractory for more related terms) (EPA-83).

Neutral sulfite semi-chemical (NSSC): The virgin fiber pulp made using the semi-chemical process; utilized principally for making corrugating medium. (*see also* Semi-chemical Paperboard) (EPA-83).

Neutral sulfite semichemical pulping operation: Any operation in which pulp is produced from wood by cooking (digesting) wood chips in a solution of sodium sulfite and sodium bicarbonate, followed by mechanical defibrating (grinding) (40 CFR 60.281-91).

Neutralization:
(1) Those acid pickling operations that do not include acid recovery or acid regeneration processes (40 CFR 420.91-91).
(2) Decreasing the acidity or alkalinity of a substance by adding to it alkaline or acidic materials, respectively (EPA-89/12).

The process of neutralization is the interaction of an acid with a base. The typical properties exhibited by acids in solution are due to the hydrogen ion, [H+]. Similarly, alkaline (or basic) properties are a result of the hydroxyl ion, [OH-]. In aqueous solutions, acidity and alkalinity are defined with respect to pH, where pH = $-\log[H^+]$ and, at room temperature, pH = $14 + \log[OH^-]$. In the strict sense, neutralization is the adjustment of pH to 7, at which level the concentrations of hydrogen and hydroxyl ions are equal. Solutions with excess hydroxyl ion concentration (pH > 7) are said to be basic; solutions with excess hydrogen ions (pH < 7) are acidic. Since adjustment of the pH to 7 is not often practical or even desirable in waste treatment, the term neutralization is sometimes used to describe adjustment of pH to values other than 7.

The actual process of neutralization is accomplished by the addition of an alkaline to an acidic material or by adding an acidic to an alkaline material, as determined by the required final pH. The primary products of the reaction are a salt and water. A simple example of acid-base neutralization is the reaction between hydrochloric acid and sodium hydroxide: $HCl + NaOH \longrightarrow H_2O + NaCl$, where the product, sodium chloride, in aqueous solution is neutral with pH = 7.0.

Neutron: A no charged particle and has a mass approximately equal to a proton (*see* atom for more related terms).

Neutron activation analysis: An instrument in which neutrons are used to bombard a specimen. The resulting isotopes are used to identify the specimen.

New aircraft turbine engine: An aircraft gas turbine engine which has never been in service (*see* engine for more related terms) (40 CFR 87.1-91).

New biochemical and microbial registration review: The review of an application or registration of a biochemical or microbial pesticide product containing a biochemical or microbial active ingredient not contained in any other pesticide product that is registered under FIFRA at the time the application is made. For purposes of this Subpart, the definitions of biochemical and microbial pesticides contained in Section 15.6 and (b) of this chapter shall apply (40 CFR 152.403-91).

New chemical: An active ingredient not contained in any currently registered pesticide (40 CFR 166.3-91).

New chemical registration review: The review of an application for registration of a pesticide product containing a chemical active ingredient which is not contained as an active ingredient in any other pesticide product that is registered under FIFRA at the time the application is made (40 CFR 152.403-91).

New chemical substance: Any chemical substance which is not included in the chemical substance list compiled and published under section 8(b) (cf. chemical substance) (TSCA3).

New class II well: The wells constructed or converted after the effective date of this program, or which are under construction on the effective date of this program (*see* well for more related terms) (40 CFR 147.2902-91).

New corrugated cutting: The cutting consists of baled corrugated cuttings having two or more liners of either jute or kraft. Non-soluble adhesives, butt rolls, slabbed or hogged medium, and treated medium or liners are not acceptable in this grade (EPA-83).

New corrugated cutting wastepaper: *See* definition under waste paper.

New discharger: Any building, structure, facility, or installation:
(1) From which there is or may be a "discharge of pollutants";
(2) That did not commence the "discharge of pollutants" at a particular "site" prior to August 13, 1979;
(3) Which is not a "new source"; and
(4) Which has never received a finally effective NPDES permit for discharges at that "site."
This definition includes an "indirect discharger" which commences discharging into "waters of the United States" after August 13, 1979. It also includes any existing mobile point source (other than an offshore or coastal oil and gas exploratory drilling rig or a coastal oil and gas developmental drilling rig) such as a seafood processing rig, seafood processing vessel, or aggregate plant, that begins discharging at a "site" for which it does not have a permit; and any offshore or coastal mobile oil and gas exploratory drilling rig or coastal mobile oil and gas developmental drilling rig that commences the discharge of pollutants after August 13, 1979, at a "site" under EPA's permitting jurisdiction for which it is not covered by an individual or general permit and which is located in an area determined by the Regional Administrator in the issuance of a final permit to be an area or biological concern. In determining whether an area is an area of biological concern, the Regional Administrator shall consider the factors specified in 40 CFR 125.122(a) (1) through (10). An offshore or coastal mobile exploratory drilling rig or coastal mobile developmental drilling rig will be considered a "new discharger" only for the duration of its discharge in an area of biological concern (40 CFR 122.2-91).

New hazardous waste management facility or new facility: A facility which began operation, or for which construction commenced after October 21, 1976. (*see also* Existing hazardous waste management facility) (40 CFR 260.10-91).

New HWM facility: A Hazardous Waste Management facility which began operation or for which construction commenced after November 19, 1980 (40 CFR 270.2-91).

New injection well: An injection well which began injection after a UIC program for the State applicable to the well is approved or prescribed (*see* well for more related terms) (40 CFR 144.3-91).

New kraft lined corrugated cuttings: Consists of baled corrugated cuttings having all liners of kraft. Non-soluble adhesives, butt rolls, slabbed or hogged medium, and treated medium or liners are not acceptable in this grade (EPA-83).

New motor vehicle: Except with respect to vehicles or engines imported or offered for importation, the term "new motor vehicle" means a motor vehicle the equitable or legal title to which has never been transferred to an ultimate purchaser; and the term "new motor vehicle engine" means an engine in a new motor vehicle or a motor vehicle engine the equitable or legal title to which has never been transferred to the ultimate purchaser; and with respect to imported vehicles or engines, such terms mean a motor vehicle and engine, respectively, manufactured after the effective date of a regulation issued under section 202 which is applicable to such vehicle or engine (or which would be applicable to such vehicle or engine had it been manufactured for importation into the United States) (CAA216-42USC7550).

New OCS source: An OCS (outer continebtal shelf) source which is a new source within the meaning of section 111(a) (*see* source for more related terms) (CAA328-42USC7627-91).

New product:
(1) A product the equitable or legal title of which has never been transferred to an ultimate purchaser; or
(B) A product which is imported or offered for importation into the United States and which is manufactured after the effective date of a regulation under section 6 or section 8 which would have been applicable to such product had it been manufactured in the United States (NCA3, *see also* 40 CFR 162.151; 205.2-91).

New solid waste incineration unit: A solid waste incineration unit the construction of which is commenced after the Administrator proposes requirements under this section establishing emissions standards or other requirements which would be applicable to such unit or a modified solid waste incineration unit (*see* solid waste incineration unit for more related terms) (CAA129.g-42USC7429-91).

New source: Any stationary source, the construction or modification of which is commended after the publication of regulations (or, if earlier, proposed regulations) prescribing a standard of performance under this section which will be applicable to such source (*see* source for more related terms) (CAA111).

New source coal mine:
(1) A coal mine (excluding coal preparation plants and coal preparation plant associated areas) including an abandoned mine which is being re-mined:
(i) The construction of which is commenced after May 4, 1984; or
(ii) Which is determined by the EPA Regional Administrator to constitute a "major alteration." In

making this determination, the Regional Administrator shall take into account whether one or more of the following events resulting in a new, altered or increased discharge or pollutants has occurred after May 4, 1984 in connection with the mine for which the NPDES permit is being considered:

(A) Extraction of a coal seam not previously extracted by that mine;
(B) Discharge into a drainage area not previously affected by wastewater discharge from the mine;
(C) Extensive new surface disruption at the mining operation;
(D) A construction of a new shaft, slope, or drift; and
(E) Such other factors as the Regional Administrator deems relevant.

(2) No provision in this part shall be deemed to affect the classification as a new source of a facility which was classified as a new source coal mine under previous EPA regulations, but would not be classified as a new source under this section, as modified. Nor shall any provision in this part be deemed to affect the standards applicable to such facilities, except as provided in 40 CFR 434.65 of this chapter (40 CFR 434.11-j-91).

New source performance standards (NSPS): Uniform national EPA air emission and water effluent standards which limit the amount of pollution allowed from new sources or from existing sources that have been modified (EPA-89/12).

New tank system: A tank system that will be used to contain an accumulation of regulated substances and for which installation has commenced after December 22, 1988. (*see also* existing tank system) (40 CFR 280.12-91).

New tank system or new tank component: A tank system or component that will be used for the storage or treatment of hazardous waste and for which installation has commenced after July 14, 1986; except, however, for purposes of 40 CFR 264.193(g)(2) and J 265.193(g)(2), a new tank system is one for which construction commences after July 14, 1986. (*see also* existing tank system) (40 CFR 260.10-91).

New unit: A unit that commences commercial operation on or after the date of enactment of the Clean Air Act Amendments of 1990 (CAA402-42USC7651a-91).

New use: When used with respect to a product containing a particular active ingredient, means:
(1) Any proposed use pattern that would require the establishment of, the increase in, or the exemption from the requirement of, a tolerance or food additive regulation under section 408 or 409 of the Federal Food, Drug and Cosmetic Act;
(2) Any aquatic, terrestrial, outdoor, or forestry use pattern, if no product containing the active ingredient is currently registered for that use pattern; and
(3) Any additional use pattern that would result in a significant increase in the level of exposure, or a change in the route of exposure, to the active ingredient of man or other organisms (40 CFR 152.3-p-91).

New use pattern registration review: The review of an application for registration, or for amendment of a registration entailing major change to the use pattern of an active ingredient contained in a product registered under FIFRA or pending Agency decision on a prior application at the time of application. For purposes of this paragraph, examples of major changes include but are not limited to, changes from non-food to food use, outdoor to indoor use, ground to aerial application, terrestrial to aquatic use, and nonresidential to residential use (40 CFR 152.403-c-91).

New uses of asbestos: The commercial uses of asbestos not identified in 40 CFR 763.165 the manufacture, importation or processing of which would be initiated for the first time after August 25, 1989 (under the Toxic Substances Control Act) (40 CFR 763.163-91).

New vessel: Includes every description of watercraft or other artificial contrivance used, or capable of being used, as a means of transportation on the navigable waters, the construction of which is initiated after promulgation of standards and regulation under this section (CWA312, *see also* 40 CFR 140.1-91).

New underground injection well: An underground injection well whose operation was not approved by appropriate State and Federal agencies before the date of the enactment of this title (*see* injection well for more related terms) (SDWA1424-42USC300h.3-91).

New water: The water from any discrete source such as a river, creek, lake or well which is deliberately allowed or brought into the plant site (*see* water for more related terms) (40 CFR 440.141-91).

New well: Any Class I hazardous waste injection well which is not an existing well (*see* well for more related terms) (40 CFR 146.61-91).

News ink: A printing ink designed to run on newsprint, consisting basically of carbon black or colored pigments dispersed in mineral oil vehicles, which dry by absorption. Recent developments utilize emulsion, oxidation, and heat set systems (*see* ink for more related terms) (EPA-79/12a).

News wastepaper: *see* definition under waste paper.

Newsprint: *see* synonym, newsprint paper

Newsprint paper: The paper of the type generally used in the publication of newspapers or special publications like the Congressional Record. It is made primarily from mechanical wood pulps combined with some chemical wood pulp (*see* paper for more related terms) (40 CFR 250.4-91).

Newton: A force unit (1 kg-m/s^2).

Newtonian fluid: A fluid whose velocity gradient is directly proportional to its shear force; or a fluid whose applied shear stress has a linear relation with the rate of deformation. A non-Newtonian fluid, the relation is non-linear.

Newton's viscosity law: In fluid mechanics, the applied shear stress

(S_s) is proportional to the rate of deformation or to the velocity gradient normal to the velocity. Mathematically, it can be expressed as:

$$S_s = \mu(dV/dy)$$

where:
- S_s = shear stress
- V = velocity in differential amount
- y = distance normal to the velocity direction in differential amount
- μ = absolute or dynamic viscosity (*see* law for more related terms).

Niccolite: A copper-red arsenide of nickel which usually contains a little iron, cobalt, and sulfur. It is one of the chief ores of metallic nickel (EPA-82/05).

Nickel (Ni): The total nickel present in the process wastewater stream exiting the wastewater treatment system (40 CFR 415.361-91, *see also* 40 CFR 415.471; 415.651; 420.02-91). Nickel is a transition metal with a.n. 28; a.w. 58.71; d. 8.9 g/cc; m.p. 1453° C and b.p. 2730° C and belongs to group VIII of the periodic table.

Nickel cyanide ($Ni(CN)_2 \cdot 4H_2O$): A toxic powder used for nickel electroplating and metallurgy.

Nickel flash: A chemical preparation process in which nickel compounds are reduced to metallic nickel and deposited on the surface of the treated item, while iron is oxidized to the ferrous ion (EPA-82/11e).

Nickel mineral: The nickel-iron sulfide, pentlandite is the principal present economic source of nickel and garnierite (nickel magnesium hydrosilicate) is next in economic importance (EPA-82/05).

Night soil: The human waste (night implies that the waste is often removed in the evening (*see* soil for more related terms).

Niobium (Nb): A transition metal with a.n. 41; a.w. 92.906; d. 8.4 g/cc; m.p. 2468° C and b.p. 3300° C. The element belongs to group VB of the periodic table.

Nitrate: A salt or a compound containing $(NO_3)^-$ radical which can exist in the atmosphere or as a dissolved gas in water and which can have harmful effects on humans and animals. Nitrates in water can cause severe illness in infants and cows (EPA-89/12).

Nitrate nitrogen: The final decomposition product of the organic nitrogen compounds. Determination of this parameter indicates the degree of waste treatment (EPA-87/10a).

Nitration: The replacement of hydrogen on a carbon atom with a nitro group (-NO) through the use of nitric acid or mixed acid (EPA-87/10a).

Nitric acid (HNO_3): A poisonous liquid oxidant used for fertilizer manufacture.

Nitric acid plant: Any facility producing nitric acid 30 to 70 percent in strength by either the pressure or atmospheric pressure process (40 CFR 51.100-91).

Nitric acid concentrator (NAC): A distillation process which concentrates weak nitric acid (sixty percent) to strong nitric acid (ninety-eight percent) (EPA-76/03).

Nitric acid production unit: Any facility producing weak nitric acid by either the pressure or atmospheric pressure process. (*see also* weak nitric acid) (40 CFR 60.71-91).

Nitric oxide (NO): A component of nitrogen oxide. For more details, *see* nitrogen oxide.

Nitrification:
(1) The process whereby ammonia in wastewater is oxidized to nitrite and then to nitrate by bacterial or chemical reactions (EPA-89/12).
(2) Bacterial oxidation of ammonia and nitrites to nitrates (LBL-76/07-water).
(3) Oxidative process that converts ammonium salts to nitrites and nitrites to nitrates (LBL-76/07-bio).

Nitrifiers: Bacteria which cause the oxidation of ammonia to nitrites and nitrates (EPA-76/03).

Nitrile rubber: A synthetic rubber made by emulsion polymerization of acrylonitrile with butadiene (*see* rubber for more related terms) (EPA-74/12a).

Nitrilotriacetic acid (NTA): A compound being used to replace phosphates in detergents (EPA-89/12).

Nitrite:
(1) A compound containing $(NO_2)^-$ radical.
(2) Nitrous oxide salts used in food preservation (EPA-89/12).

Nitrite nitrogen: An intermediate stage in the decomposition of organic nitrogen to the nitrate form. Tests for nitrite nitrogen can determine whether the applied treatment is sufficient (EPA-87/10a).

Nitrobacteria: Bacteria (an autotrophic genus) that oxidize nitrite nitrogen to nitrate nitrogen (EPA-76/03).

Nitrocellulose: A basic ingredient used in propellant manufacturing, made by nitrating woodpulp or cotton fibers with mixed acid (EPA-76/03).

Nitrogen (N): A gaseous element with a.n. 7; a.w. 14.0067; d. 0.81 g/cc; m.p. -210° C and b.p. -195.8° C. The element belongs to group VA of the periodic table (cf. dry nitrogen).

Nitrogen balance: The net nitrogen intake as protein and nitrogen excretion for living organisms.

Nitrogen cycle: Organic nitrogen in waste is oxidized by bacteria into ammonia. If oxygen is present, ammonia is bacterially oxidized first into nitrite and then into nitrate. If oxygen is not present, nitrite and nitrate are bacterially reduced to nitrogen gas. The second step is called "denitrification." (EPA-76/03).

Nitrogen dioxide (NO_2): A component of nitrogen oxide. For more details, *see* nitrogen oxide.

Nitrogen fixation: Biological nitrogen fixation is carried on by a selected group of bacteria which take up atmospheric nitrogen and convert it to amine groups or for amino acid synthesis (EPA-76/03).

Nitrogen oxide (or oxide of nitrogen):
(1) All oxides of nitrogen except nitrous oxide, as measured by test methods set forth in this part (40 CFR 60.2-91, *see also* 40 CFR 86.082.2-91, 86.402.78-91). It includes N_2O, NO, N_2O_3, N_2O_4, NO_2, N_2O_5 and NO_3.
(2) Product of combustion from transportation and stationary sources and a major contributor to the formation of ozone in the troposphere and acid deposition (EPA-89/12).
(3) Nitrogen oxides have far greater significance in photochemical smog than any of the other air contaminants, e.g., NO_2 causes a reddish brown plume (EPA-84/09).
(4) cf. family mitrogen oxide (NO_x) emission limit.

Nitrogen oxide emission control: The techniques of nitrogen oxide emission control include:
(1) Fuel denitrogenation (EPA-81/12, p7-4).
(2) Combustion modification (EPA-84/03b, p7-5).
  (A) Staged combustion (or off-stoichiometric combustion) (EPA-84/03b, p7-8)
  (B) Low excess air (EPA-84/03b, p7-9)
  (C) Flue gas recirculation (FGR) (EPA-84/03b, p7-10)
  (D) Low NO_x burner (EPA-84/03b, p7-12)
  (E) Reduced air preheat and load reduction (EPA-84/03b, p7-13)
  (F) Steam and water injection (EPA-84/03b, p7-13)
  (G) Catalytic combustion (EPA-81/12, p7-8)
(3) Flue gas treatment: Nitrogen oxide emissions can be reduced by treating the flue gas after it leaves the combustion zone. The techniques include:
  (A) Exxon thermal $DeNO_x$ process (EPA-84/03b, p7-13).
  (B) Selective catalytic reduction (SCR) (EPA-84/03b, p7-15)
  (C) Shell UOP (EPA-84/03b, p7-20)
  (D) Wet simultaneous NO_x and SO_2 reduction (EPA-81/12, p7-24).

Nitrogen oxide formation: When fossil fuels are burned with air in a furnace, some of the oxygen (O_2) and nitrogen (N_2) present combine to form oxides of nitrogen (in the intense heat environment of any combustion process.) Most of the oxides form according to the following reaction; $N_2 + O_2 ---> 2NO$. Once the NO forms, the rate of decomposition is very slow and NO does not dissociate into N_2 and O_2 in any appreciate amounts. The NO formed can react with more oxygen to form NO_2 by: $NO + 0.5O_2 ---> NO_2$ (EPA-81/12, p7-4).

Two major nitrogen oxides, nitric oxide and nitrogen dioxide are important as air contaminants. NO is considerably less toxic than the NO_2. NO acts as an asphyxiant when in concentrations great enough to reduce the normal oxygen supply from the air. NO_2, on the other hand, in concentrations of approximately 10 ppm for 8 hours, can produce lung injury and edema, and in greater concentrations, i.e., 20 to 30 ppm for 8 hours can produce fatal lung damage. A typical flue gas contains 90-95% nitric oxide and 5-10% nitrogen dioxide (AP-40, p15).

Nitrogenous bacteria oxygen demand (nitrogenous BOD): The oxygen demand by bacteria to oxidize ammonia to nitrite or nitrate.

Nitrogenous waste: The animal or vegetable residues that contain significant amounts of nitrogen (*see* waste for more related terms) (EPA-89/12).

Nitroglycerin ($CH_2NO_3CHNO_3CH_2NO_3$, NG): A colorless highly explosive oil which is a nitration product of glycerin. NG, as it is frequently called, is a principal constituent of dynamite and certain propellants (rocket grains). NG is extremely sensitive to impact and freezes at 56° F. A basic ingredient used in propellant manufacturing, made by nitrating glycerin with mixed acid (EPA-76/03).

Nitroguanidine: The third base raw material used in the manufacture of triple-base propellant. The other two are nitrocellulose and nitroglycerin (EPA-76/03).

Nitrosating agent: Any substance that has the potential to transfer a nitrosyl group (-NO) to a primary, secondary, or tertiary amine to form the corresponding nitrosamine (40 CFR 747.115-91, *see also* 40 CFR 747.195; 747.200).

Nitrosomonas: Bacteria which oxidize ammonia nitrogen into nitrite nitrogen; an aerobic autotrophic life form (EPA-76/03).

No adverse effect level: *See* synonym, no effect level.

No bake binder: Sand binders that set without the addition of heat. Furan resins and alkyd-isocyanate compounds are the two predominant no-bake binders. Furan resins are cyclic compounds which use phosphoric acid or toluenesulfonic acid as the setting agents (EPA-85/10a).

No data: According to the U.S. EPA Guidelines for Carcinogen Risk Assessment, no data describes a category of human and animal evidence in which no studies are available to permit one to draw conclusions as to the induction of a carcinogenic effect (EPA-92/12).

No detectable emission: Less than 500 ppm above background levels, as measured by a detection instrument in accordance with Method 21 in Appendix A of 40 CFR 60 (40 CFR 60.691-91, *see also* 40 CFR 60-App/A(method 21); 61.341-91).

No discharge of free oil: That a discharge does not cause a film or sheen upon or a discoloration on the surface of the water or adjoining shorelines or cause a sludge or emulsion to be deposited beneath the surface of the water or upon adjoining shorelines (40 CFR 435.11-91, *see also* 40 CFR 435.41-91).

No discharge of pollutant: No net increase of any parameters designated as a pollutants to the accuracy that can be determined from the designated analytical methods (*see* pollutant for more related terms) (EPA-74/04c).

No effect level (no adverse effect level, no observed effect level or no toxic effect level):
(1) The maximum dose used in a test which produces no observed adverse effects. A no-observed-effect level is expressed in terms of the weight of a substance given daily per unit weight of test animal (mg/kg). When administered to animals in food or drinking water, the no-observed-effect

level is expressed as mg/kg of food of mg/mL of water (40 CFR 795.260-91, *see also* 40 CFR 798.2250; 798.2450; 798.2650; 798.2675; 798.4350; 798.4900-91).
(2) An exposure level at which there are no statistically or biologically significant increases in the frequency or severity of any effect between the exposed population and its appropriate control (EPA-90/08; 92/12).

No evidence of carcinogenicity: According to the U.S. EPA Guidelines for Carcinogen Risk Assessment, a situation in which there is no increased incidence of neoplasms in at least two well-designed and well-conducted animal studies of adequate power and dose in different species (EPA-92/12).

No observed adverse effect level (NOAEL):
(1) The highest dose in an experiment which did not produce an observable adverse effect (same as no-effect level) (Course 165.6, *see also* EPA-85/09).
(2) A tested dose of an effluent or a toxicant below which no adverse biological effects are observed, as identified from chronic or subchronic human epidemiology studies or animal exposure studies (EPA-91/03).
(3) An exposure level at which there are no statistically or biologically significant increases in the frequency or severity of adverse effects between the exposed population and its appropriate control. Some effects may be produced at this level, but they are not considered as adverse, nor precursors to specific adverse effects. In an experiment with several NOAELs, the regulatory focus is primarily on the highest one, leading to the common usage of the term NOAEL as the highest exposure without adverse effect (EPA-90/08b; 92/12).

No observed effect concentration (NOEC):
(1) The highest tested concentration in an acceptable early life stage test:
  (A) Which did not cause the occurrence of any specified adverse effect (statistically different from the control at the 5 percent level); and
  (B) Below which no tested concentration caused such an occurrence (40 CFR 797.1600-91).
(2) The highest tested concentration of an effluent or a toxicant at which no adverse effects are observed on the aquatic test organisms at a specific time of observation. Determined using hypothesis testing (EPA-91/03).

No observed effect level (NOEL): *See* synonym, no effect level.

No reasonable alternative:
(1) No land-based disposal sites, discharge point(s) within internal waters, or approved ocean dumping sites within a reasonable distance of the site of the proposed discharge the use of which would not cause unwarranted economic impacts on the discharger, or, notwithstanding the availability of such sites;
(2) On-site disposal is environmentally preferable to other alternative means of disposal after consideration of:
  (A) The relative environmental harm of disposal on-site, in disposal sites located on land, from discharge point(s) within internal waters, or in approved ocean dumping sites; and
  (B) The risk to the environment and human safety posed by the transportation of the pollutants (40 CFR 125.121-91).

No toxic effect level: *See* synonym, no effect level.

No. 1 mixed wastepaper: *See* definition under waste paper.

No. 1 sorted colored ledger wastepaper: *See* definition under waste paper.

No. 1 sorted white ledger wastepaper: *See* definition under waste paper.

No. 2 mixed wastepaper: *See* definition under waste paper.

Nobelium (No): A radioactive metal with a.n. 102 and a.w. 254. The element belongs to group IIIB of the periodic table.

Noble gas: *See* synonym, inert gas.

Noble metal: The metals that are below hydrogen in the electromotive force series; includes antimony, copper, rhodium, silver, gold, bismuth (cf. precious metals) (EPA-83/06a).

Noise:
(1) Spontaneous, short duration deviations in output, about the mean output, which are not caused by input concentration changes. Noise is determined as the standard deviation about the mean and is expressed in concentration units (40 CFR 53.23-91).
(2) Spontaneous deviation from a mean output not caused by input concentration changes (exressed as percent of full scale) (LBL-76/07-bio).
(3) Any undesired audible signal thus, in acoustics, noise is any undesired sound (cf. sound) (EPA-74/11).
Other noise related terms include:
- Aircraft noise
- Environmental moise

Noise control: The process to control the audible sound to an acceptable level.

Noise Control Act (NCA) of 1972: *See* the term of Act or NCA.

Noise control system: Includes any vehicle part, component or system the primary purpose of which is to control or cause the reduction of noise emitted from a vehicle (40 CFR 205.51-91, *see also* 40 CFR 205.151-91).

Noise emission standard: The noise levels in 40 CFR 205.152 or 40 CFR 205.166 (40 CFR 205.151-91).

Noise emission test: A test conducted pursuant to the measurement methodology specified in 40 CFR 204.54 (40 CFR 204.51-91, *see also* 40 CFR 205.51; 205.151-91).

Noise pollution: The unwanted sound or excessive noise in the human environment (*see* pollution for more related terms).

Noise reduction rating (NRR): A single number noise reduction factor in decibels, determined by an empirically derived technique which takes into account performance variation of protectors in noise reducing effectiveness due to differing noise spectra, fit variability and the mean attenuation of test stimuli at the one-third octave band test frequencies (40 CFR 211.203-91).

Noiseless paper: The paper used in any place where rattle or rustle would be objectionable, such as in theater programs (*see* paper for more related terms) (EPA-83).

Nominal 1-month period: A calendar month or, if established prior to the performance test in a statement submitted with notification of anticipated startup pursuant to 40 CFR 60.7(a)(2), a similar monthly time period (e.g., 30-day month or accounting month) (40 CFR 60.711-91, *see also* 40 CFR 60.721; 60.741-91).

Nominal concentration: The amount of an ingredient which is expected to be present in a typical sample of a pesticide product at the time the product is produced, expressed as a percentage by weight (40 CFR 158.153-91).

Nominal fuel tank capacity: The volume of the fuel tank(s), specified by the manufacturer to the nearest tenth of a U.S. gallon, which may be filled with fuel from the fuel tank filler inlet (40 CFR 86.082.2-91).

Nominal wall thickness: The wall thickness listed in the pipe specifications (40 CFR 195.2-91).

Non-agreement State: *See* definition under agreement State.

Non-attainment area:
(1) For any air pollutant, an area which is designated nonattainment with respect to that pollutant within the meaning of section 107(d) (CAA171-91, *see also* 40 CFR 51.138-91).
(2) The geographic area which does not meet one or more of the National Ambient Air Quality Standards for the criteria pollutants designated in the Clean Air Act (EPA-94/04).
(3) An air quality control region (or portion thereof) in which the U.S. Environmental Protection Agency has determined that ambient air concentrations exceed national ambient air quality standards for one or more criteria pollutants (DOE-91/04).

Non-biodegradable: Materials that can not be decomposed by bacteria.

Non-coincident electric demand: The sum of two or more individual demands which do not occur in the same period interval. Meaningful only when considering demands within a limited period of time, such as a day, week, month, a heating or cooling season, and usually for not more than one year (*see* electric demand for more related terms) (EPA-83).

Non-combustible: The components of a material which remain after combustion of all combustible matter including inert materials such as glass, dirt and sand and wholly oxidized metals (cf. ash) (EPA-83).

Non-combustible rubbish: Miscellaneous bulkier refuse materials that are unburnable at ordinary incinerator temperatures (1300° F to 2000° F). *See* Solid Waste (EPA-83).

Non-combustible waste: Includes metals, tin cans, foils, dirt, gravel, bricks, ceramics, glass, crockery, ashes (*see* waste for more related terms) (EPA-83).

Non-community water system:
(1) A public water system that is not a community water system (40 CFR 141.2-91).
(2) A public water system that is not a community water system, e.g., the water supply at a camp site or national park (EPA-89/12).

Non-conformance penalty (NCP): as described in section 206(g) of the Clean Air Act and in this subpart (40 CFR 86.1102.87-91).

Non-conforming vehicle or engine: A motor vehicle or motor vehicle engine which is not covered by a certificate of conformity prior to final or conditional importation and which has not been finally admitted into the United States under the provisions of 40 CFR 85.1505, 40 CFR 85.1509, or the applicable Provisions of 40 CFR 85.1512. Excluded from this definition are vehicles admitted under provisions of 40 CFR 85.1512 covering EPA approved manufacturer and U.S. Government Agency catalyst and O_2 sensor control programs (*see* vehicle for more related terms) (40 CFR 85.1502-91).

Non-conservative substance: Organic substances that can be decomposed by bacteria.

Non-consumer article: Any article subject to TSCA which is not a consumer product within the meaning of the Consumer Product Safety Act (CPSA), 15USC2052 (40 CFR 762.3-91).

Non-consumptive use of water: The water use in which only a small portion is lost to the atmosphere by evapotranspiration or by being combined with a manufactured product. Non-consumptive use returns to the stream or ground approximately the same amount of water as is diverted or used (*see* water for more related terms) (DOI-70/04).

Non-contact cooling water: The water used for cooling which does not come into direct contact with any raw material, intermediate product, waste product or finished product (*see* cooling water for more related terms) (40 CFR 401.11-91, *see also* 40 CFR 418.11; 418.21; 418.51; 60.691-91).

Non-contact cooling water pollutant: Pollutants present in non-contact cooling waters (*see* pollutant for more related terms) (40 CFR 401.11-91).

Non-contact cooling water system: A once-through drain, collection and treatment system designed and operated for collecting cooling water which does not come into contact with hydrocarbons or oily wastewater and which is not recirculated through a cooling tower (*see* cooling water for more related terms) (40 CFR 60.691-91).

Non-contact process wastewater (or non-process wastewater): The wastewater generated by a manufacturing process which has not come in direct contact with the reactants or products used in the process. These include such streams as non-contact cooling water, cooling tower blowdown, boiler blowdown, etc. (*see* wastewater for more related terms) (EPA-76/03; 87/10a).

Non-contact wastewater: The wastewater which is not contaminated by the process or related materials. Examples include boiler blowdown, cooling water, sanitary sewage. Storm water from outside the immediate manufacturing area may be included in this definition if it is not contaminated from product spills, etc. (*see* wastewater for more related terms) (EPA-85/10).

Non-continental area: The State of Hawaii, the Virgin Islands, Guam, American Samoa, the Commonwealth of Puerto Rico, or the Northern Mariana Islands (40 CFR 60.41a-91, *see also* 40 CFR 60.41b; 60.41c-91).

Non-continuous discharger: A mill which is prohibited by the NPDES authority from discharging pollutants during specific periods of time for reasons other than treatment plant upset control, such periods being at least 24 hours in duration. A mill shall not be deemed a non-continuous discharger unless its permit, in addition to setting forth the prohibition described above, requires compliance with the effluent limitations established for non-continuous dischargers and also requires compliance with maximum day and average of 30 consecutive days effluent limitations. Such maximum day and average of 30 consecutive days effluent limitations for non-continuous dischargers shall be established by the NPDES authority in the form of concentrations which reflect wastewater treatment levels that are representative of the application of the best practicable control technology currently available, the best conventional pollutant control technology, or new source performance standards in lieu of the maximum day and average of 30 consecutive days effluent limitations for conventional pollutants set forth in each subpart (40 CFR 430.01-c-91).

Non-conventional pesticide pollutant: All pesticide active ingredients that are not toxic pollutants under 40 CFR 401.15 or conventional pollutants under 40 CFR 401.16 and for which BAT and effluent limitations guidelines, NSPS and pretreatment standards are established in this part (*see* pollutant for more related terms) (40 CFR 455.10-91).

Non-conventional pollutant:
(1) Any pollutant which is not a statutorily listed or which is poorly understood by the scientific community (EPA-89/12).
(2) Parameters selected for consideration in performance standards that have not been previously designated as either conventional or toxic pollutants (EPA-85/10a).
(3) For the pesticide industry, non-conventional pollutants are defined as nonpriority pollutant pesticides, COD, ammonia, and manganese (EPA-85/10).
(4) *See* pollutant for more related terms.

Non-corrugating medium furnish subdivision mills: The mills where recycled corrugating medium is not used in the production of paperboard (cf. corrugating medium furnish subdivision mills) (40 CFR 430.51-91).

Non-criteria pollutant: Any identified potential pollutants exclusive of criteria pollutants which may include:

(1) Non-carcinogenic pollutants such as HCl and Cl;
(2) Suspect carcinogenic elements such as As, Be, Cd, Cr(+6), and NI; and
(3) Organic carcinogens such as PCDDs, PCDFs and PCBs (CRWI-89/05).
(4) *See* pollutant for more related terms.

Non-degradation: A water quality classification used to denote a water body to which no pollutants may be legally added (EPA-80/08).

Non-degradation principle: A principle of avoiding deterioration of the present state of ambient air quality (also known as stand-still principle) (NATO-78/10).

Non-destructive detector: The instruments that can measure a variable without sampling or destruction of the material in which the variable occurs (LBL-76/07-bio).

Non-destructive test: A test method which does not require the removal of samples from, nor damage to, the installed liner system. The evaluation is done in an in-situ manner as with a vacuum box test (*see* test for more related terms) (EPA-89/09, *see also* EPA-91/05).

Non-deterministic model: A model which is based on empirical relationships between air quality data and other parameters which are given for a certain period of time. It is then used to calculate the air quality in another period. In this model no direct relation is established between air quality and emissions (NATO-78/10).

Non-dispersion infrared Analyzer (NDIR): One of continuous emission monitors (*see* continuous emission monitor for various types). NDIR analyzer monitors CO, CO_2, NO_x, SO_2, hydrocarbons, and other gases that absorb light in the infrared region of the spectrum. In a typical NDIR analyzer, infrared light from a lamp or glower passes through two gas cells, a reference cell and a sample cell. The reference cell generally contains dry nitrogen gas, which does not absorb light at the wave-length used in the instrument. As the light passes through the sample cell, pollutant molecules absorb some of the infrared light. As a result, when the light emerges from the end of the sample cell, it has less energy than when it entered. It also has less energy than the light emerging from the reference cell. The energy difference is then sensed by some type of detector, such as a thermistor, a thermocouple, or microphone arrangement (EPA-81/09).

Non-dispersion ultraviolet analyzer (NDUV): One of continuous emission monitors (*see* continuous emission monitor for various types). NDUV monitors gases that absorb light in the ultraviolet and visible region of the spectrum. Essentially, it measures the degree of absorption at a wavelength in the absorption band of the molecule of interest. This is similar to the NDIR method, but the major different is that a reference cell is not used. Instead, a reference wavelength, in a region where the pollutant has minimal absorption, is utilized. This method is often differential absorption, since measurements are performed at two different frequencies. This method is not limited to extractive monitoring systems, but it also is used in both in-situ analyzers and remote sensors. As with all extractive monitoring systems, particulate matter is removed before entering the analyzer. It is not necessary, however, to remove water vapor in some of these systems. A heated sample line and heated cell prevent condensation in the analyzer. Since water does not absorb light in this region of the ultraviolet spectrum, no interference occurs (EPA-81/09).

Non-domestic construction material: A construction material other than a domestic construction material (40 CFR 35.936.13-91).

Non-emission related maintenance: That maintenance which does not substantially affect emissions and which does not have a lasting effect on the deterioration of the vehicle or engine with respect to emissions once the maintenance is performed at any particular date (40 CFR 86.084.2-91, *see also* 40 CFR 86.088.2-91).

Non-enclosed process: Any equipment system (such as an open-top reactor, storage tank, or mixing vessel) in which a chemical substance is manufactured, processed, or otherwise used where significant direct contact of the bulk chemical substance and the workplace air may occur (40 CFR 721.3-91).

Non-excessive infiltration: The quantity of flow which is less than 120 gallons per capita per day (domestic base flow and infiltration) or the quantity of infiltration which cannot be economically and effectively eliminated from a sewer system as determined in a cost effectiveness analysis. (*see* 40 CFR 35.2005(b)(16) and 35.2120) (40 CFR 35.2005-91).

Non-excessive inflow: The maximum total flow rate during storm events which does not result in chronic operational problems related to hydraulic overloading of the treatment works or which does not result in a total flow of more than 25 gallons per capita per day (domestic base flow plus infiltration plus inflow). Chronic operational problems may include surcharging, backups, bypasses, and overflows (*see* inflow for more related terms) (*see* 40 CFR - 35.2005(b)(16) and 35.2120) (40 CFR 35.2005-91).

Non-expendable personal property: The personal property with a useful life of at lest two years and an acquisition cost of $500 or more (40 CFR 30.200-91).

Non-ferrous metals:
(1) Any pure metal other than iron or any metal alloy for which a metal other than iron is its major constituent in percent by weight (40 CFR 471.02-91).
(2) Metals that contain no iron, e.g., aluminum, copper, brass and bronze (EPA-83/06a).
(3) *See* metal for more related terms.

Non-filterable (suspended) metals: Those metals which are retained by a 0.45 um membrane filter (LBL-76/07-water).

Non-fractionating plant: Any gas plant that does not fractionate mixed natural gas liquids into natural gas products (40 CFR 60.631-91).

Non-friable: The material in a school building which when dry may not be crumbled, pulverized , or reduced to powder by hand pressure (40 CFR 763.83-91).

Non-friable asbestos-containing material: Any material containing more than 1 percent asbestos as determined using the method specified in appendix A, subpart F, 40 CFR 763, section 1, Polarized Light Microscopy, that, when dry, cannot be crumbled, pulverized, or reduced to powder by hand pressure (40 CFR 61.141-91).

Non-gaseous losses: The solvent that is not volatilized during fiber production, and that escapes the process and is unavailable for recovery, or is in a form or concentration unsuitable for economical recovery (40 CFR 60.601-91).

Non-impervious solid surfaces: The solid surfaces which are porous and are more likely to absorb spilled PCBs prior to completion of the cleanup requirements prescribed in this policy. Nonimpervious solid surfaces include, but are not limited to, wood, concrete, asphalt, and plasterboard (40 CFR 761.123-91).

Non-industrial source: Any source of pollutants which is not an industrial source (*see* source for more related terms) (40 CFR 125.58-91).

Non-industrial use: The use other than at a facility where chemical substances or mixtures are manufactured, imported, or processed (40 CFR 721.3-91).

Non-ionization radiation (or radio frequency radiation): The electromagnetic radiation, such as ultraviolet rays, having wavelength between approximately 1 to 1000 Angstroms (*see* radiation for more related terms) (EPA-88/09a; 89/12).

Non-ionizing electromagnetic radiation:
(1) Radiation that does not change the structure of atoms but does heat tissue and may cause harmful biological effects.
(2) Microwaves, radio waves, and low-frequency electromagnetic fields from high-voltage transmission lines (EPA-89/12).
(3) *See* radiation for more related terms.

Non-isolated intermediate: Any intermediate that is not intentionally removed from the equipment in which it is manufactured, including the reaction vessel in which it is manufactured, equipment which is ancillary to the reaction vessel, and any equipment through which the substance passes during a continuous flow process, but not including tanks or other vessels in which the substance is stored after its manufacture. Mechanical or gravity transfer through a closed system is not considered to be intentional removal, but storage or transfer to shipping containers "isolates" the substance by removing it from process equipment in which it is manufactured (40 CFR 704.3-91).

Non-leaching: Generally applied to a landfill structure or material, describing some materials which do not permit fluids to enter or to leave (OME-88/12).

Non-metallic mineral: Any of the following minerals or any mixture of which the majority is any of the following minerals:
(1) Crushed and Broken Stone, including Limestone, Dolomite, Granite, Traprock, Sandstone, Quartz, Quartzite, Marl, Marble, Slate, Shale, Oil Shale, and Shell;
(2) Sand and Gravel;
(3) Clay including Kaolin, Fireclay, Bentonite, Fuller's Earth,

Ball Clay, and Common Clay;
(4) Rock Salt;
(5) Gypsum;
(6) Sodium Compounds, including Sodium Carbonate, Sodium Chloride, and Sodium Sulfate;
(7) Pumice;
(8) Gilsonite;
(9) Talc and Pyrophyllite;
(10) Boron, including Borax, Kernite, and Colemanite;
(11) Barite;
(12) Fluorospar;
(13) Feldspar;
(14) Diatomite;
(15) Perlite;
(16) Vermiculite;
(16) Mica; and
(17) Kyanite, including Andalusite, Sillimanite, Topaz, and Dumortierite (40FR60.671-91).

Non-metallic mineral processing plant: Any combination of equipment that is used to crush or grind any nonmetallic mineral wherever located, including lime plants, power plants, steel mills, asphalt concrete plants, portland cement plants, or any other facility processing nonmetallic minerals except as provided in 40 CFR 60.670 (b) and (c) (40 CFR 60.671-91).

Non-methane organic gas (NMOG): The sum of nonoxygenated and oxygenated hydrocarbons contained in a gas sample, including, at a minimum, all oxygenated organic gases containing 5 or fewer carbon atoms (i.e., aldehydes, ketones, alcohols, ethers, etc.), and all known alkanes, alkenes, alkynes, and aromatics containing 12 or fewer carbon atoms. To demonstrate compliance with a NMOG standard, NMOG emissions shall be measured in accordance with the "California Non-Methane Organic Gas Test Procedures." In the case of vehicles using fuels other than base gasoline, the level of NMOG emissions shall be adjusted based on the reactivity of the emissions relative to vehicles using base gasoline (CAA241-42USC7581).

Non-operational storage tank: Any underground storage tank in which regulated substances will not be deposited or from which regulated substances will not be dispensed after the date of the enactment of the Hazardous and Solid Waste Amendments of 1984 (RCRA9001-42USC6991-91).

Non-oxygenated hydrocarbon: The organic emissions measured by a flame ionization detector excluding methanol (40 CFR 86.090.2-91).

Non-parametic method: A statistical method which does not require the assumption of a distributional form, such as a normal distribution (EPA-87/10).

Non-pass through pollutant: The pollutants that are biodegradable, and do not pass through biological oxidation treatment systems (*see* pollutant for more related terms) (EPA-85/10).

Non-passenger automobile: An automobile that is not a passenger automobile, as defined by the Secretary of Transportation at 49 CFR 523.5 (40 CFR 600.002.85-91).

Non-PCB transformer: Any transformer that contains less than 50 ppm PCB; except that any transformer that has been converted from a PCB Transformer or a PCB-Contaminated transformer cannot be classified as a non-PCB Transformer until reclassification has occurred, in accordance with the requirements of 40 CFR 761.30(a)(2)(v) (*see* transformer for more related terms) (40 CFR 761.3-91).

Non-perishable raw agricultural commodity: Any raw agricultural commodity not subject to rapid decay or deterioration that would render it unfit for consumption. Examples are cocoa beans, coffee beans, field-dried beans, field-dried peas, grains, and nuts. Not included are eggs, milk meat, poultry, fresh fruits, and vegetables such as onions, parsnips, potatoes, and carrots (40 CFR 180.1-91).

Non-point source: The pollution sources which generally are not controlled by establishing effluent limitations under sections 301, 302, and 402 of the Act. Nonpoint source pollutants are not traceable to a discrete identifiable origin, but generally result from land runoff, precipitation, drainage. or seepage (*see* source for more related terms) (40 CFR 35.1605.4-91, *see also* EPA-89/12).

Non-pressure process: A method of treating wood at atmospheric pressure in which the wood is simply soaked in hot or cold preservative (EPA-74/04).

Non-process wastewater: *See* synonym, non-contact process wastewater.

Non-process wastewater cooling tower: The water used for cooling purposes which has no direct contact with any process raw materials, intermediates, or final products (*see* cooling water for more related terms) (EPA-87/10a).

Non-process water: The water used for the heating and cooling of process solutions to maintain proper operating conditions or for the make-up water in cooling towers, boilers and lawn sprinkling systems. This water is not process water as it does not come in contact with raw materials or the product (*see* water for more related terms) (EPA-80/10).

Non-procurement list: The portion of the list of parties excluded from Federal Procurement or Nonprocurement Programs complied, maintained and distributed by the General Services Administration (GSA) containing the names and other information about persons who have been debarred, suspended, or voluntarily excluded under Executive Order 12549 and these regulation, and those who have been determined to be ineligible (40 CFR 32.105-91).

Non-productive rubber stock (or non-reactive rubber stock): A rubber stock which has been compounded but which contains no curing agents (cf. productive rubber stock) (EPA-74/12a).

Non-putrescible: Incapable of organic decomposition or decay (EPA-76/03).

Non-reactive rubber stock: *See* synonym, non-productive rubber stock.

Non-road engine: An internal combustion engine (including the

Non-road vehicle: A vehicle that is powered by a nonroad engine and that is not a motor vehicle or a vehicle used solely for competition (*see* vehicle for more related terms) (CAA216-42USC7550-91).

fuel system) that is not used in a motor vehicle or a vehicle used solely for competition, or that is not subject to standards promulgated under section 111 or section 202 (*see* engine for more related terms) (CAA216-42USC7550-91).

Non-roof coating: An asbestos-containing product intended for use as a coating, cement, adhesive, or sealant and not intended for use on roofs. Major applications of this product include: liquid sealants; semi-liquid glazing, caulking and patching compounds; asphalt-based compounds; epoxy adhesives; butyl rubber sealants; vehicle undercoatings; vinyl sealants; and compounds containing asbestos fibers that are used for bonding, weather proofing, sound deadening, sealing, coating; and other such applications (40 CFR 763.163-91).

Non-scheduled renovation operation: A renovation operation necessitated by the routine failure of equipment, which is expected to occur within a given period based on past operating experience, but for which an exact date cannot be predicted (40 CFR 61.141-91).

Non-sudden accidental occurrence: An occurrence which takes place over time and involves continuous or repeated exposure (*see* accidental occurrence for more related terms) (40 CFR 264.141-91, *see also* 40 CFR 265.141-91).

Non-target analyte spike: The spiking of surrogate analytes into the sample. A surrogate analyte is one which mimics the behavior of target analytes in terms of stability, preparation losses, measurement artifacts, etc., but does not interfere with target analyte measurement. Surrogates, like matrix spikes, can be added in the laboratory or in the field; results are interpreted in a fashion similar to matrix spikes (*see* spike for more related terms) (EPA-84/03).

Non-target organism: A plant or animal other than the one against which the pesticide is applied (*see* organism for more related terms) (40 CFR 171.2-91).

Non-threshold effect: Associated with exposure to chemicals that have no safe exposure levels (i.e., cancer) (EPA-91/03).

Non-transferred plasma arc: An arc established between two internal electrodes in the arc-producing device. A small column of injected gas is heated by the electric arc creating a plasma flow which extends beyond the tip of the arc-forming device. Nontransferred arcs heat only by way of conduction (*see* plasma arc for more related terms) (Wittle-93/07).

Non-transient non-community water system or NTNCWS: A public water system that is not a community water system and that regularly serves at least 25 of the same persons over 6 months per year (40 CFR 141.2-91).

Non-transportation-related onshore and offshore facilities:
(1) Mixed onshore and offshore oil well drilling facilities including all equipment and appurtenances related thereto used in drilling operations for exploratory or development wells, but excluding any terminal facility, unit or process integrally associated with the handling or transferring of oil in bulk to or from a vessel.
(2) Mobile onshore and offshore oil well drilling platforms, barges, trucks, or other mobile facilities including all equipment and appurtenances related thereto when such mobile facilities are fixed in position for the purpose of drilling operations for exploratory or development wells, but excluding any terminal facility, unit or process integrally associated with the handling or transferring of oil in bulk to or from a vessel.
(3) Fixed onshore and offshore oil production structures, platforms, derricks, and rigs including all equipment and appurtenances related thereto, as well as completed wells and the wellhead separators, oil separators, and storage facilities used in the production of oil, but excluding any terminal facility, unit or process integrally associated with the handling or transferring of oil in bulk to or from a vessel.
(4) Mobile onshore and offshore oil production facilities including all equipment and appurtenances related thereto as well as completed wells and wellhead equipment, piping from wellheads to oil separators, oil separators, and storage facilities used in the production of oil when such mobile facilities are fixed in position for the purpose of oil production operations, but excluding any terminal facility, unit or process integrally associated with the handling or transferring of oil in bulk to or from a vessel.
(5) Oil refining facilities including all equipment and appurtenances related thereto as well as in-plant processing units, storage units, piping, drainage systems and waste treatment units used in the refining of oil, but excluding any terminal facility, unit or process integrally associated with the handling or transferring of oil in bulk to or from a vessel.
(6) Oil storage facilities including all equipment and appurtenances related thereto as well as fixed bulk plant storage, terminal oil storage facilities, consumer storage, pumps and drainage systems used in the storage of oil, but excluding inline or breakout storage tanks needed for the continuous operation of a pipeline system and any terminal facility, unit or process integrally associated with the handling or transferring of oil in bulk to or from a vessel.
(7) Industrial, commercial, agricultural or public facilities which use and store oil, but excluding any terminal facility, unit or process integrally associated with the handling or transferring of oil in bulk to or from a vessel.
(8) Waste treatment facilities including in-plant pipelines, effluent discharge lines, and storage tanks, but excluding waste treatment facilities located on vessels and terminal storage tanks and appurtenances for the reception of oily ballast water or tank washings from vessels and associated systems used for off-loading vessels.
(9) Loading racks, transfer hoses, loading arms and other equipment which are appurtenant to a nontransportation-related facility or terminal facility and which are used to transfer oil in bulk to or from highway vehicles or railroad cars.
(10) Highway vehicles and railroad cars which are used for the transport of oil exclusively within the confines of a nontransportation-related facility and which are not intended

to transport oil in interstate or intrastate commerce.
(11) Pipeline systems which are used for the transport of oil exclusively within the confines of a nontransportation-related facility or terminal facility and which are not intended to transport oil in interstate or intrastate commerce, but excluding pipeline systems used to transfer oil in bulk to or from a vessel (40 CFR 112-AA-91).

Non-utility unit: A unit other than a utility unit (CAA402-42USC7651a-91).

Non-volatile organic chemical[1]: A chemical compound with the following properties:
(1) Boiling point (degree C): >300
(2) Sampling method: M5 (MM5)
(3) Capture method: filter
(4) Analytical method: GC/MS, ICP, AAS
(5) see organic chemical for more related terms
Note:[1] Non-VOC includes inorganics and metals
AAS=atomic absorption spectroscopy
GC/MS=gas chromatographic/mass spectroscopy
ICP=inductively coupled argon plasma emission spectroscopy
M5=Method 5
MM5: Modified Method 5 (sampling method)

Non-vapor tight: Any tank truck, railcar, or marine vessel that does not pass the required vapor-tightness test (40 CFR 61.301-91).

Non-wastewater: The wastes that do not meet the criteria for wastewaters in paragraph (f) of this section (see wastewater for more related terms) (40 CFR 268.2-91).

Non-water quality: Thermal, air, noise and all other environmental parameters except water (EPA-75/04).

Non-water quality environmental impact: The ecological impact as a result of solid, air, or thermal pollution due to the application of various wastewater technologies to achieve the effluent guidelines limitation. Also associated with the non-water quality aspect is the energy impact of wastewater treatment (EPA-85/10a).

Non-wood fibers: Fibers not of the wood family used to produce pulp, paper, and paperboard. Such as vegetable fibers (cotton, flax, jute, hemp, cereal straw, bagasse, bamboo, esparto, abaca, sisal, pineapple), animal fiber (wool), mineral fiber (asbestos, glass), and man-made or artificial fiber (rayon, nylon, orlon, dacron) (EPA-87/10).

Normal ambient value: That concentration of a chemical species reasonably anticipated to be present in the water column, sediments, or biota in the absence of disposal activities at the disposal site in question (40 CFR 228.2-91).

Normal cubic meter (Nm³): A volume unit measured under standard pressure and temperature (cf. scf).

Normal distribution: A statistical distribution identified by a bell shaped curve which is the most important of all continuous distributions. This distribution curve is symmetrical about the mean (EPA-87/10).

Normal liquid detergent operations: All such operations except those defined as fast turnaround operation of automated fill lines (40 CFR 417.161-91).

Normal operation of a spray tower:
(1) Operation utilizing formulations that present limited air quality problems from stack gases and associated need for extensive wet scrubbing, and without more than 6 turnarounds in a 30 consecutive day period, thus permitting essentially complete recycle of waste water (40 CFR 417.151-91).
(2) The range of full-power operating conditions that can be achieved when seasonal variations in ambient conditions are taken into account (DOE-91/04).

Normal range of a release: All releases (in pounds or kilograms) of a hazardous substance reported or occurring over any 24-hour period under normal operating conditions during the preceding year. Only releases that are both continuous and stable in quantity and rate may be included in the normal range (see release for more related terms) (40 CFR 302.3-91).

Normal solution: A solution that contains 1 gm molecular weight of the dissolved substance divided by the hydrogen equivalent of the substance (that is, one gram equivalent) per liter of solution. Thus, a one normal solution of sulfuric acid (H_2SO_4, mol. wt. 98) contains 98/2 or 49 grams of H_2SO_4, per liter (EPA-87/10a).

Normality: A unit to express the number of gram-equivalent weights of a compound per liter of solution.

Normalized coefficients: In statistics, the ratio of the differences between the regression constant and its expected value to its standard error.

Normalizing: The heat treatment of iron-base alloys above the critical temperature, followed by cooling in still air. (This is often done to refine or homogenize the grain structure of castings, forgings and wrought steel products) (EPA-83/06a).

Notice of deficiency (NOD): In RCRA, a written notice from a permit writer to a permit applicant relative to the deficient information submitted (Lee).

Notice of intent (NOI): A notice that an environmental impact statement will be prepared and considered. The notice shall briefly:
(1) Describe the proposed action and possible alternatives.
(2) Describe the agency's proposed scoping process including whether, when, and where any scoping meeting will be held.
(3) State the name and address of person within the agency who can answer questions about the proposed action and the environmental impact statement (40 CFR 1508.22-91, see also 40 CFR 6.105-91).

Novel paper: A grade used in pulp magazines (thrillers, romance, etc.). Also called pulp paper (see paper for more related terms) (EPA-83).

Nozzle: See synonym, spray nozzle.

NPDES State: A State (as defined in 40 CFR 122.2) or Interstate water pollution control agency with an NPDES permit program approved pursuant to section 402(b) of the Act (40 CFR 403.3-91).

NPR site: The primary location for the NPR at each alternative DOE-owned site (Hanford, INEL, and SRS) (DOE-91/04).

NRC licensed facility: Any facility licensed by the Nuclear Regulatory Commission or any Agreement State to receive title to, receive, possess, use, transfer, or deliver any source, by-product, or special nuclear material (40 CFR 61.101-91).

Nuclear absorption meter: One of solid or sludge flow rate meters. This Method, based upon absorption of gamma radiation, includes nuclear level meters, nuclear belt or auger scales, and a combination of nuclear density meters and ultrasonic flowmeters. It only measures density; therefore, another instrument must also be used to measure volume, speed, or another parameter to obtain feed rate. Nuclear instruments can be used on nearly any materials including sludges. Radiation absorption is proportional to the mass present, so particle size and configuration will not greatly hinder accuracy. Sludge operations will work best with a nuclear density detector/ultrasonic flowmeter combination, enabling the process material to be fed through conventional piping. Accuracies of nuclear devices may not be as high as gravimeteric systems but may be sufficient on a practical basis (*see* flow rate meter for more related terms) (EPA-89/06).

Nuclear criticality: A self-sustaining nuclear chain reaction (DOE-91/04).

Nuclear criticality safety: The prevention or termination of inadvertent nuclear criticality, mitigation of consequences, and protection against injury or damage due to accidental nuclear criticality (DOE-91/04).

Nuclear energy: The energy derived from the fission (splitting) of nuclei of heavy elements such as uranium or thorium or from the fusion (combining) of the nuclei of light elements such as deuterium or tritium (EPA-82/11f).

Nuclear facility: A facility whose operations involve radioactive materials in such form and quantity that a significant nuclear hazard potentially exists to the employees or the general public. Included are facilities that:
(1) Produce, process, or store radioactive liquid or solid waste, fissionable materials, or tritium;
(2) Conduct separations operations;
(3) Conduct irradiated materials inspection, fuel fabrication, decontamination, or recovery operations; or
(4) Conduct fuel enrichment operations. Incidental use of radioactive materials in a facility operation (e.g., check sources, radioactive sources, and X-ray machines) does not necessarily require a facility to be included in this definition (DOE-91/04).

Nuclear fuel: Nuclear energy used as a source of energy for primarily producing electricity (cf. spent muclear fuel).

Nuclear fusion: A nuclear reaction in which two light nuclei fuse to form a heavier nucleus with a release of large amounts of energy (*see* fusion for more related terms).

Nuclear fuel cycle: The operations defined to be associated with the production of electrical power for public use by any fuel cycle through utilization of nuclear energy (40 CFR 190.02-91).

Nuclear grade: Material of a quality adequate for use in a nuclear application (DOE-91/04).

Nuclear island: A nuclear island consists of a set of four reactor modules, their support buildings, and operation center (DOE-91/04).

Nuclear material: Composite term applied to:
(1) By-product material
(2) Other nuclear material
(3) Other nuclear material
(4) Source material
(5) Special nuclear material
(6) Special nuclear material of low strategic significance
(7) Special nuclear material of moderate strategic significance
(8) Special nuclear material scrap

Nuclear power plant: A facility that converts atomic energy into usable power; heat produced by a reactor makes steam to drive turbines which produce electricity (EPA-89/12).

Nuclear radiation: The radiation of alpha, beta or gamma particles resulting from a change in atomic nucleus such as fission or fusion (*see* radiation for more related terms).

Nuclear winter: Prediction by some scientists that smoke and debris rising from massive fires resulting from a nuclear war could enter the atmosphere and block out sunlight for weeks or months. The scientists making this prediction project a cooling of the earth's surface, and changes in climate which could, for example, negatively effect world agricultural and weather patterns (EPA-89/12).

Nucleic acid transition: A mutational change in nucleic acid in which a purine (adenine or guanine) is replaced by another purine, and a pyrimidine (thymine, cytosine, or uracil) is replaced by another pyrimidine, e.g., an adenine: thymine base pair might be replaced by a guanine: cytosine base pair (EPA-88/09a).

Nucleus: *See* definition under atom.

Nuclide: An atomic nucleus specified by its atomic weight, atomic number, and energy state (DOE-91/04).

Nuggetized: *See* synonym, balled (EPA-83).

Nuisance: That activity which arises from the unreasonable, unwarrantable or unlawful use by a person of his own property, working an obstruction or injury to the right of another or to the public, and producing such material annoyance, inconvenience, and discomfort that the law will presume resulting damage (*see* common law for more related terms) (Sullivan-95/04, p7).

Number average molecular weight: The arithmetic average (mean) of the molecular weight of all molecules in a polymer (40 CFR

723.250-91).

Numerical modeling: A solution by numerical techniques of a set of equations resulting from a mathematical model (NATO-78/10).

Numerical instability: A property of numerical algorithms for differential equations, whereby the numerical solution will grow without bound, due to amplification of small errors in the computation, whereas the exact solution of the problem is bounded (*see* instability for more related terms) (NATO-78/10).

NUREG 0654/FEMA-REP-1: Criteria for Preparation and Evaluation of Radiological Emergency Response Plans and Preparedness in Support of Nuclear Power Plants, prepared by NRC and FEMA. Provides a basis for State and local government and nuclear facility operators to develop radiological emergency plans and improve emergency preparedness. The criteria also will be used by Federal agency reviewers in determining the adequacy of State, local, and nuclear facility emergency plans and preparedness (NRT-87/03).

Nutrient: Any substance assimilated by living things that promotes growth. The term is generally applied to nitrogen and phosphorus in wastewater, but is also applied to other essential and trace elements (EPA-89/12). The nutrients in contaminated water are routinely analyzed to characterize the food available for microorganisms to promote organic decomposition and are provided below (unit in mg/L) (EPA-74/04):
- Nitrogen, ammonia (NH_3) as N
- Nitrogen, total nitrogen measured by the Kjeldahl process, (NH_3 and Organic N) as N
- Nitrogen nitrate (NO_3) as N
- Total phosphate as P
- Ortho phosphate as P

Other nutrient related terms include:
- Macro nutrient
- Micro nutrient

Nutrient addition: The process of adding nitrogen or phosphorous in a chemically combined form to a waste stream (EPA-87/10a).

Nylon fiber: A generic name for a fiber in which the fiber forming substance is any long-chain synthetic polyamide having recurring amide groups as an integral part of the polymer chain (*see* fiber for more related terms) (EPA-74/06b).

Oo

O_3: Ozone (40 CFR 58.1-91).

Observation: *See* synonym, observed value.

Observed value (observation or variate): In quality control, the particular value of a characteristic and designated parameter such as X_1, X_2, X_3, and so on (EPA-84/03).

Obligate aerobe: *See* synonym, aerobe.

Obligate anaerobe: *See* synonym, anerobe.

Obligate parasite: An organism (e.g., viruses) which can only be a parasite (live in or on another species) (cf. parasitic organism).

Obligation: The amounts of orders placed, contracts and subgrants awarded, goods and services received, and similar transactions during a given period that will require payment by the grantee during the same or a future period (*see* cost for more related terms) (40 CFR 31.3-91).

Observation period: The accumulated time period during which observations are conducted, not to be less than the period specified in the applicable regulation (40 CFR 60-App/A(method 22)-91, *see also* 40 CFR 797.2175-91).

Observed effect concentration (OEC): The lowest tested concentration in an acceptable early life stage test:
(1) Which caused the occurrence of any specified adverse effect (statistically different from the control at the 5 percent level); and
(2) Above which all tested concentrations caused such an occurrence (40 CFR 797.1600-91).

Obsolete scrap: The scrap that results when a material becomes worn or otherwise unusable for its original purpose (*see* scrap for more related terms) (SW-108ts).

Obstipation: Extreme constipation (LBL-76/07-bio).

Occupational dose: Includes exposure of an individual to radiation:
(1) In a restricted area; or
(2) In the course of employment in which the individual's duties involve exposure to radiation, provided, that occupational dose shall not be deemed to include any exposure of an individual to radiation for the purpose of medical diagnosis or medical therapy of such individual (10 CFR 20.3-91).

Occupational exposure: Reasonably anticipated skin, eye, mucous membrane, or parenteral contact with blood or other potentially infectious materials that may result from the performance of an employee's duties. This definitions excludes incidental exposures that may take place on the job, and that are neither reasonably nor routinely expected and that the worker is not required to incur in the normal course of employment (*see* exposure for more related terms) (29 CFR 1910).

Occupational Safety and Health Act (OSHA) of 1970: *see* the term of Act or OSHA.

Occurrence: An accident, including continuous or repeated exposure to conditions, which results in a release from an underground storage tank (40 CFR 280.92-91).

Ocean discharge: The discharge of wastewater into an ocean (EPA-85/10).

Ocean disposal: Dumping of waste materials at sea (EPA-83).

Ocean dumping:
(1) The disposition of waste into an ocean or estuarine body of water (SW-108ts).
(2) The disposal of pesticides in or on the oceans and seas, as defined in Pub. L. 92-532 (40 CFR 165.1-91).

Ocean Dumping Act: *See* the term of Act or MPRSA.

Ocean incineration (or at sea incineration):
(1) Disposal of waste by burning at sea on specially-designed incinerator ships (EPA-89/12).
(2) The thermal destruction of waste at sea in a specially designed tanker vessel equipped with high temperature incinerators. The principle of operation of these incinerators is identical to that of land-based incinerators with the exception that current ocean incinerators are not equipped with air pollution controlled systems. Acid gas produced from incinerating chlorinated wastes is discharged to the air without treatment to be neutralized by contact with sea water, which has a naturally high buffering capacity (Oppelt-87/05).

The extent to which ocean incineration is used will depend upon need, cost and risk considerations. The availability and cost of land based incineration units will affect demand for ocean incineration. Other major factors affecting ocean incineration are the ability to designate sites, to obtain permits for portside support facilities, and to permit ocean incinerator ships. Additionally, the cost of transporting and incinerating wastes in land based versus ocean incinerators would affect utilization rates for each technology. Finally, managing risk to satisfy the public is an issue which affects both land and sea incineration of hazardous wastes.

(3) See incineration for more related terms.

Ocean thermal energy conversion (OTEC): An energy conversion process in which the different content of energy associated with the high temperature surface ocean water and with the low temperature deep ocean water is converted into electrical energy or other useful form of energy.

Ocean water: Those coastal waters landward of the baseline of the territorial seas, the deep waters of the territorial seas, or the waters of the contiguous zone (see water for more related terms) (40 CFR 125.58-91, see also 40 CFR 220.2-91).

Oceanography: The study of the physical, chemical, geological, and biological aspects of the sea (LBL-76/07-water).

Octane ($C_{18}H_{18}$): A flammable liquid used as a solvent.

Octane number: A number to rate the tendency of a fuel to resist knocking when it is combusted in an internal combustion engine under a standard condition (e.g., n-heptane is 0 and isooctane is 100) (cf. cetane number).

Octanol water partition coefficient: The measure of the extent of partitioning of a substance between water and octanol at equilibrium. The K(ow) is determined by the ratio between the concentration in octanol divided by the concentration in water at equilibrium. [unitless] (40 CFR 300-App/A-91, see also 40 CFR 796.1720-91).

Octave band attenuation: The amount of sound reduction determined according to the measurement procedure of 40 CFR 21.20 for one-third octave bands of noise (40 CFR 211.203-91).

Odometer: A distance measurement instrument as used in a vehicle.

Odor: The property of a substance affecting the sense of smell; any smell; scent; perfume (EPA-83/06).

Odor concentration: The number of unit volumes that a unit volume of sample will occupy when diluted to the odor threshold (EPA-83/06).

Odor control: Methods to control odor include:
(1) Aeration to control odor from treatment ponds;
(2) Ozonation, chlorination or activated carbon to control odor from water supply;
(3) Scrubber containing alkali or activated carbon to control odor from gases.

Odor threshold:
(1) The lowest concentration of an airborne odor that a human being can detect (without the aid of mechanical instruments). The unit volume of air at the odor threshold (EPA-83/06, see also EPA-83).
(2) The concentration of an odorous compound at which the physiological effect elicits a response 50 percent of the time (LBL-76/07-air).

Odorant: Odorous substance (such as an aromatic gas) (EPA-83/06, see also EPA-83).

Off-gas:
(1) Gaseous products from chemical decomposition of a material (EPA-83).
(2) Gases, vapors, and fumes produced as a result of metal molding and casting operations (EPA-85/10a).

Off-kg (off-lb or off-pound): The mass of metal or metal alloy removed from a forming operation at the end of a process cycle for transfer to a different machine or process (40 CFR 471.02-91, see also 40 CFR 467.02; 468.02-91).

Off-lb: See synonym, off-kg.

Off-peak energy: Energy supplied during periods of relatively low system demands as specified by the supplier (see electric energy for more related terms) (EPA-83).

Off-pound: See synonym, off-kg.

Off-road motorcycle: Any motorcycle that is not a street motorcycle or competition motorcycle (40 CFR 205.151-91).

Off-scene support: Assistance (via telephone, radio, or computer) from technical persons, agencies, shippers, responders, etc. not at the accident site (EPA-85/11).

Off-shore facility: Any facility of any kind located in, on, or under any of the navigable waters of the United States, which is not a transportation-related facility (see facility for more related terms) (40 CFR 112.2-91, see also CWA311; SF101; 40 CFR 110.1; 116.3; 300.5; 302.3-91).

Off-site: Any site which is not on-site (cf. on-site) (40 CFR 270.2-91).

Off-site facility: A hazardous waste treatment, storage or disposal area that is located at a place away from the generating site (see facility for more related terms) (EPA-89/12).

Offal: Intestines and discarded parts, including paunch manure, of slaughtered animals (SW-108ts).

Office Director--approved emission test or emission short test: Any test prescribed under 40 CFR 85.2201 et seq., and meeting all of the requirements thereunder(40FR85.2102-91).

Office of Ombudsman: An ombudsman is a public official to investigate citizen's complaints against local or national government agencies that may be infringing on the rights of individuals. EPA'S

Office of Ombusdsman was established to receive individual complaints, grievances, request for information submitted by any person with respect to any program or requirement under the RCRA (RCRA2008).

Office papers: The note pads, loose-leaf fillers, tablets, and other papers commonly used in offices, but not defined elsewhere (*see* paper for more related terms) (40 CFR 250.4-91).

Offset:
(1) With respect to printing and publishing operations, use of a blanket cylinder to transfer ink from the plate cylinder to the surface to be printed (40 CFR 52.741-91).
(2) Paper made especially for lithographic printing process. Also refers to the transfer of wet ink to succeeding sheets in printing.
(3) An imperfection resulting from mold parts not properly matched, that is, a finish or base offset from body or neck (EPA-83).

Offset printing paper: An uncoated or coated paper designed for offset lithography (*see* paper for more related terms) (40 CFR 250.4-91).

Offshore: Beyond the line of ordinary low water along that portion of the coast of the United States that is in direct contact with the open seas and beyond the line marking the seaward limit of inland waters (40 CFR 192.3-91, *see also* 40 CFR 195.2-91).

Offshore facility: Any facility of any kind located in, on, or under any of the navigable waters of the United States, and any facility of any kind that is subject to the jurisdiction of the United States and is located in, on, or under any other waters, other than a vessel or a public vessel (40 CFR 110.1-91, *see also* CWA311; OPA1001; SF101; 40 CFR 112.2; 116.3; 300.5; 302.3-91).

Offshore platform gas turbines: Any stationary gas turbine located on a platform in an ocean (40 CFR 60.331-91).

Ohm: The unit of electrical resistance (EPA-74/03d).

Oil: Oil of any kind or in any form, including, but not limited to, petroleum, fuel oil, sludge, oil refuse, and oil mixed with wastes other than dredged spoil (40 CFR 109.2-91, *see also* CWA311; OPA1001; 40 CFR 60.41b; 60.41c; 110.1; 112.2; 113.3; 300.5; 426.81; 426.111; 426.121-91). Other oil related terms include:
- Core oil
- Cruide oil
- Crude tall oil
- Distillate oil
- Drying oil
- Essential oil
- Fatty oil
- Heating oil
- Lubricating oil
- Neat oil
- Quenching oil
- Re-refined oil
- Recycled oil
- Residual oil
- Slop oil
- Tall oil
- Used oil
- Very low sulfur oil

Oil and grease: The total recoverable oil and grease as measured by the procedure listed in 40 CFR 136 (40 CFR 410.11-91).

Oil and hazardous materials technical assistance data system (OHMTADS): A computerized data base containing chemical, biological, and toxicological information about hazardous substances. OSCs use OHMTADS to identify unknown chemicals and to learn how to best handle known chemicals (NRT-87/03).

Oil base ink (paint): An ink (paint) that uses oils or resins as the prime vehicle ingredient (*see* ink for more related terms) (EPA-75/07).

Oil burner: A burner which is used to burn oil (*see* burner for more related terms) (AP-40).

Oil cooker: An open-topped vessel containing a heat source and typically maintained at 68° C for the purpose of driving off excess water from waste oils (EPA-83/06a).

Oil filled capacitor: A capacitor whose conductor and insulating elements are immersed in an insulating fluid that is usually but not necessarily, oil (EPA-83/03).

Oil fingerprinting: A method that identifies sources of oil and allows spills to be traced back to their source (EPA-89/12).

Oil hole or pressurized coolant drilling: Using a drill with one or more continuous holes through its body and shank to permit the passage of a high pressure cutting fluid which emerges at the drill point and ejects chips (*see* drilling for more related terms) (EPA-83/06a).

Oil pollution: The pollution caused by the discharge of oily materials (*see* pollution for more related terms).

Oil Pollution Act (OPA) of 1990: *See* the term of Act or OPA.

Oil pollution fund: The fund established by section 311(k) of the CWA (cf. fund) (40 CFR 300.5-91).

Oil recovery system: The equipment used to reclaim oil from wastewater (EPA-74/04).

Oil shale: The rock or kerogen that contains carbon or hydrocarbons from which oil can be extracted.

Oil spill: An accidental or intentional discharge of oil which reaches bodies of water. Can be controlled by chemical dispersion, combustion, mechanical containment, and/or adsorption (EPA-89/12).

Oil water separator: A waste management unit, generally a tank or surface impoundment, used to separate oil from water. An oil-water separator consists of not only the separation unit but also the forebay and other separator basins, skimmers, weirs, grit chambers, sludge hoppers, and bar screens that are located directly after the

individual drain system and prior to additional treatment units such as an air flotation unit, clarifier, or biological treatment unit. Examples of an oil-water separator include an API separator, parallel-plate interceptor, and corrugated-plate interceptor with the associated ancillary equipment (*see* separator for more related terms) (40 CFR 61.341).

Oily preservative: Pentachlorophenol-petroleum solutions and creosote in the various forms in which it is used (EPA-74/04).

Oily wastewater: The wastewater generated during the refinery process which contains oil, emulsified oil, or other hydrocarbons. Oily wastewater originates from a variety of refinery processes including cooling water, condensed stripping steam, tank draw-off, and contact process water (*see* wastewater for more related terms) (40 CFR 60.691-91).

Old chemical registration review: The review of an application for registration of a new product containing active ingredients and uses which are substantially similar or identical to those currently registered or for which an application is pending Agency decision (40 CFR 152.403-91).

Old scrap: *See* synonym, post consumer waste.

Olefiant gas: *See* synonym, ethylene.

Olefin (C_nH_{2n}): Unsaturated straight-chain hydrocarbon compounds seldom present in crude oil, but frequently in cracking processes (EPA-74/04b).

Olefinic hydrocarbons: A specific subgroup of aliphatic hydrocarbons sharing the common characteristic of at least one unsaturated carbon-to-carbon atomic bond in the hydrocarbon molecule, and with straight or branched chain structure (EPA-86/12).

Oleic acid ($C_{17}H_{33}COOH$): A mono-saturated fatty acid. It is a common component of almost all naturally occurring fats as well as tall oil. Most commercial oleic acid is derived from animal tallow or natural vegetable oils (EPA-82/05).

Oleum: A solution of SO_3 in sulfuric acid (EPA-74/04c).

Oligotrophic lake: The deep clear lakes with low nutrient supplies. They contain little organic matter and have a high dissolved oxygen level (*see* lake for more related terms) (EPA-89/12).

Oligotrophic water: The water with a small supply of nutrients; thus, they support little organic production (*see* water for more related terms) (LBL-76/07-water).

Oliguria: Deficiency in the formation and excretion of urine (LBL-76/07-bio).

Ombusdsman: *See* definition under office of ombusdsman.

Omnibus authority: An authority, under Section 3005(c) of the Resource Conservation and Recovery Act (RCRA) as amended by the Hazardous and solid Waste Amendments of 1984 (HSWA), for EPA to establish permit conditions for hazardous waste facilities beyond the scope of existing regulations. It states, "each permit...shall contain such items and conditions as the Administrator or State determines necessary to protect human health and the environment." This language has been added verbatim to EPA's hazardous waste regulations at 40 CFR 270.32(b)(2) by the Codification Rule published at 50 FR 28701-28755 on July 15, 1985. It is also listed as a self-implementing HSWA provision at 40 CFR 271.1(j) in 51 FR 22712-23 (September 22, 1986) (EPA-90/04).

Omnivore: Animals or organisms which eat both meat or vegetable for living.

On-peak energy: Energy supplied during periods of relatively high system demands as specified by the supplier (*see* electric energy for more related terms) (EPA-83).

On-scene coordinator (OSC):
(1) The single Federal representative designated pursuant to the National Oil and Hazardous Substances Pollution Contingency Plan and identified in approved Regional Oil and Hazardous Substances Pollution Contingency Plans (40 CFR 113.3-91, *see also* 40 CFR 300.5-91).
(2) The predesignated EPA, Coast Guard, or Department of Defense official who coordinates and directs Superfund removal actions or Clean Water Act oil-or hazardous-spill corrective actions (EPA-89/12; 85/11; NRT-87/03).

On-shore: All facilities except those that are located in the territorial seas or on the outer continental shelf (40 CFR 60.631-91, *see also* 40 CFR 60.641; 435.51; 435.61-91).

On-shore facility: Any facility of any kind located in, on, or under any land within the United States, other than submerged lands, which is not a transportation-related facility (*see* facility for more related terms) (40 CFR 112.2-91, *see also* CWA311; SF101; 40 CFR 110.1; 116.3; 300.5; 302.3-91).

On-shore oil storage facility: Any facility (excluding motor vehicles and rolling stock) of any kind located in, on, or under, any land within the United States, other than submerged land (40 CFR 113.3-91).

On-site:
(1) Within the boundaries of a contiguous property unit (40 CFR 761.3-91, *see also* 40 CFR 260.10; 270.2; 300.5; 761.3-91).
(2) Presence within the boundaries of the work-site.

On-site coating mix preparation equipment: Those pieces of coating mix preparation equipment located at the same plant as the coating operation they serve (40 CFR 60.741-91).

On-site disposal: The utilization of methods or processes to eliminate or reduce the volume or weight of solid waste on the property of the waste generator (SW-108ts).

On-site facility: A hazardous waste treatment, storage or disposal area that is located on the generating site (*see* facility for more related terms) (EPA-89/12).

On-site incinerator: An incinerator that burns solid waste on the property utilized by the generator thereof (*see* incinerator for more related terms) (SW-108ts).

On-site system: A self-contained system which provides both treatment and disposal of wastewater on an individual lot (EPA-80/08).

On the weight of the fiber (OWF): In the textile industry, a basis for calculating the polluting characteristics of different operations and different fibers (DOI-70/04).

Once through cooling water: The water passed through the main cooling condensers in one or two passes for the purpose of removing waste heat (*see* cooling water for more related terms) (40 CFR 423.11-91, *see also* 40 CFR 419.11-91).

Oncogene: A gene that can cause cancer.

Oncogenic: A substance that causes tumors, whether benign or malignant (EPA-89/12).

Oncology:
(1) The branch of medicine dealing with tumors.
(2) The study of cancer (Course 165.6).

One hit model: A dose-response mode of the form: $P(d) = a - \exp(-b\,d)$; where $P(d)$ is the probability of cancer from a continuous dose rate d, and b is a constant. The one-hit model is based on the concept that a tumor can be induced after a single susceptible target or receptor has been exposed to a single effective dose unit of a substance (EPA-92/12).

One hour period: Any 60-minute period commencing on the hour (40 CFR 60.2-91).

One hundred year (100-year) flood: As used in paragraph (b)(l) of this section, means a flood that has a one percent chance of being equaled or exceeded in any given year (40 CFR 264.18-91)

One hundred year (100-year) floodplain: As used in paragraph (b)(1) of this section, means any land area which is subject to a one percent or greater chance of flooding in any given year from any source (40 CFR 264.18-91).

One (1)-year, 2-year, and 10-year, 24-hour precipitation events: The maximum 24-hour precipitation event with a probable recurrence interval of once in one, two, and ten years respectively as defined by the National Weather Service and Technical Paper No. 40, Rainfall Frequency Atlas of the U.S., May 1961, or equivalent regional or rainfall probability information developed therefrom (40 CFR 434.11-91).

One (1)-year, 24-hour precipitation event: The maximum 24-hour precipitation event with a probably recurrence interval or once in one year as defined by the National Weather Service and Technical Paper No. 40, Rainfall Frequency Atlas of the U.S., May 1961, or equivalent regional or rainfall probability information developed therefrom (40 CFR 434.11-91).

Onground tank: A device meeting the definition of tank in 40 CFR 260.10 and that is situated in such a way that the bottom of the tank is on the same level as the adjacent surrounding surface so that the external tank bottom cannot be visually inspected (*see* tank for more related terms) (40 CFR 260.10-91).

Onionskin paper: A thin paper, somewhat transparent used when small bulk is desired (*see* paper for more related terms) (EPA-83).

Opacity:
(1) The fraction of incident light that is attenuated by an optical medium. Opacity (Op) and transmittance (Tr) are related by: $Op = 1 - Tr$ (40 CFR 60-App/B-91, *see also* 40 CFR 60.2; 60-App/A(alt. method 1); 61.171; 61.181; 86.082.2-91).
(2) The amount of light obscured by particulate pollution in the air; clear window glass has a zero opacity, a brick wall has 100 percent opacity. Opacity is used as an indicator of changes in performance of particulate matter pollution control systems (EPA-89/12).
(3) The amount of light obscured by particulate pollution in the air; clear window glass has a zero opacity, a brick wall has 100 percent opacity. It is a measure of the degree to which the stack gas plume blocks light. Opacity is primarily caused by unburned ash or unburned carbon (soot) in the flue gas. High opacities can indicate poor mixing, low level of combustion air, high level of HCl emissions or poor burner operation in the secondary chamber. If a large amount of water vapor is present in the combustion gas, the water can condense when it cools as it leaves the stock forming a dense white steam plume. This is not an indicator of poor combustion and should not be confused with a black or white smoke caused by soot or acid gases. Opacity can be measured by a transmissometer (EPA-89/03b).
(4) *See also* combustion indicator or shop opacity.

Opacity rating:
(1) The apparent obscuration of an observer's vision that equals the apparent obscuration of smoke of a given rating on the Ringelmann Chart (SW-108ts).
(2) A measurement of the opacity of emissions, defined as the apparent obscuration of an observer's vision to a degree equal to the apparent obscuration of smoke of a given rating on the Ringelmann Chart (LBL-76/07-air).

Opal glass: Any translucent glass (*see* glass for more related terms) (EPA-83).

Opaque ink: An ink that does not allow the light to pass through it and has good hiding power. It does not permit the paper or previous printing to show through (*see* ink for more related terms) (EPA-79/12a).

Opaque stain: All stains that are not semi-transparent stains (40 CFR 52.741-91).

Open air: *See* synonym, ambient air.

Open arc furnace: A furnace that heat is generated in an open arc furnace by the passage of an electric arc, either between two electrodes or between one or more electrodes and the charge. The arc furnace consists of a furnace chamber and two or more electrodes. The furnace chamber has a lining which can withstand the operating temperatures and which is suitable for the material to be heated. The lining is contained within a steel shell which, in most cases, can be tilted or moved (*see* furnace for more related terms) (EPA-74/02a).

Open burning:
(1) The combustion of solid waste without:
   (A) Control of combustion air to maintain adequate temperature for efficient combustion;
   (B) Containment of the combustion reaction in an enclosed device to provide sufficient residence time and mixing for complete combustion; and
   (C) Control of the emission of the combustion products (40 CFR 257.3.7-91, *see also* 40 CFR 165.1; 240.101; 241.101; 260.10-91).
(2) Uncontrolled burning of wastes in an open space (EPA-83, *see also* EPA-89/12).

Open caustic system: A tank or tub cleaning system that does not reuse any part of a secondary water rinse following caustic washing (*see* caustic system for more related terms) (EPA-79/12b).

Open chain: *See* synonym, chain.

Open channel flow: A flow in an open channel whose liquid surface that is subject to atmospheric pressure must exist (*see* flow for more related terms) (M&EI-72).

Open circuit apparatus: An apparatus of the following types from which exhalation is vented to the atmosphere and not rebreathed (NIOSH-84/10).

Open combustion: Those basic oxygen furnace steel making wet air cleaning systems which are designed to allow excess air enter the air pollution control system for the purpose of combusting the carbon monoxide in furnace gases (*see* combustion for more related terms) (40 CFR 420.41-91).

Open cut mine: Any form of recovery of ore from the earth except by a dredge(40 CFR 440.141-91).

Open dump: *See* synonym, dump.

Open dumping: *See* synonym, dump.

Open ended valve: Any valve, except pressure relief devices, having one side of the valve in contact with process fluid and one side open to the atmosphere, either directly or through open piping (40 CFR 52.741-91, *see also* 40 CFR 60.481; 61.241; 264.1031-91).

Open formation: The formation of lead battery plates done with the plates in open tanks of sulfuric acid. Following formation plates are placed in the battery cases (EPA-84/08).

Open furnace: An electric submerged-arc furnace with the surface of the charge exposed to the atmosphere, whereby the reaction gases are burned by the inrushing air (*see* furnace for more related terms) (EPA-75/02).

Open hearth furnace: A furnace used for making steel. It has a large flat saucer shaped hearth to hold the melted steel. Flames play over top of the steel and melt is primarily by radiation (*see* furnace for more related terms) (EPA-74/06a).

Open hearth furnace steelmaking: The production of steel from molten iron, steel scrap, fluxes, and various combinations thereof, in refractory lined fuel-fired furnaces equipped with regenerative chambers to recover heat from the flue and combustion gases (*see* furnace for more related terms) (40 CFR 420.41-91).

Open pit incinerator: A burning device that has an open top and a system of closely spaced nozzles that place a stream of high velocity air over the burning zone (*see* incinerator for more related terms) (SW-108ts).

Open pit mining (or open cut mining): A form of operation designed to extract minerals that lie near the surface. Waste, or overburden is first removed, and the mineral is broken and loaded. Important chiefly in the mining of ores of iron and copper (EPA-82/05).

Open pot: A pot open to the flames and gases of combustion (EPA-83).

Open pile method: Open-air composting, either anaerobic or aerobic, accomplished by placing compostable material in windrows, piles, ventilated bins or pits and turning it occasionally. Also called windrow method (EPA-83).

Open site: An area that is essentially free of large sound-reflecting objects, such as barriers, walls, Board fences, signboards, parked vehicles, bridges, or buildings (40 CFR 202.10-91).

Open system: A thermodynamic system that admits the transfer of mass across its boundaries (*see* thermodynamic system for more related terms).

Open top vapor depressing: The batch process of cleaning and removing soils from surfaces by condensing hot solvent vapor on the colder metal parts (40 CFR 52.741-91).

Operable: For a nuclear facility, a situation wherein a reactor and fuel/target cycle facilities re being operated or have the potential for being operated. A reactor and fuel/target cycle facility that cannot be operated on a day-to-day basis because of refueling, extensive modifications, or technical problems is still considered operable (DOE-91/04).

Operable unit:
(1) A response action taken as one part of an overall site response. A number of operable units may occur in the course of a site response (40 CFR 35.4010-91, *see also* 40 CFR 300.5-91).
(2) A term for each of a number of separate activities undertaken as part of a Superfund site cleanup. A typical operable unit would be removing drums and tanks from the surface of a site (EPA-89/12).
(3) Defined in 40 CFR 300.6 as a discrete part of an entire response that decreases a release, threat of release, or pathway of exposure (DOE-91/04).

Operant, operant behavior, operant conditioning:
(1) **An operant:** A class of behavioral responses which change or operates on the environment in the same way.
(2) **An operant behavior:** Further distinguished as behavior which is modified by its consequences.
(3) **An operant conditioning:** The experimental procedure used to modify some class of behavior by reinforcement or punishment (40 CFR 798.6500-91).

Operating day: For the purposes of these standards, any calendar day during which equipment used in the manufacture of polymer was operating for at least 8 hours or one labor shift, whichever is shorter. Only operating days shall be used in determining compliance with the standards specified in 40 CFR 60562-1(c)(1)(ii)(B), (1)(ii)(C), (2)(ii)(B), and (2)(ii)(C). Any calendar day in which equipment is used for less than 8 hours or one labor shift, whichever is less, is not an "operating day" and shall not be used as part of the rolling 14-day period for determining compliance with the standards specified in 40 CFR 60.562-1(c)(1)(ii)(B), (1)(ii)(C), (2)(ii)(B), and (2)(ii)(C) (40 CFR 60.561-91).

Operating line: A term used to describe operating conditions within an absorber. It generally appears on an equilibrium diagram as a straight line for dilute solutions. The slope of this line is given by the ratio of the liquid to gas molar flow rates (EPA-84/09).

Operation:
(1) That an impoundment is being used for the continued placement of new tailings or is in standby status for such placement. An impoundment is in operation from the day that tailings are first placed in the impoundment until the day that final closure begins (40 CFR 61.251-91).
(2) Any step in the electroplating process in which a metal is electrodeposited on a basis material and which is followed by a rinse; this includes the related operations of alkaline cleaning, acid pickle, stripping, and coloring when each operation is followed by a rinse (40 CFR 413.11-91, *see also* 40 CFR 413.21; 413.41; 413.51; 413.61; 413.71; 413.81-91).

Operation and maintenance: The activities required to assure the dependable and economical function of treatment works:
(1) Maintenance: Preservation of functional integrity and efficiency of equipment and structures. This includes preventive maintenance, corrective maintenance and replacement of equipment (*see* 40 CFR 35.2005(b)(36)) as needed.
(2) Operation: Control of the unit processes and equipment which make up the treatment works. This includes financial and personnel management; records, laboratory control, process control, safety and emergency operation planning (40 CFR 35.2005-30-91).

Operation and maintenance cost: In wastewater treatment, the cost of running the wastewater treatment equipment. This includes labor costs, material and supply costs, and energy and power costs (EPA-83/06a).

Operational life: Refers to the period beginning when installation of the tank system has commenced until the time the tank system is properly closed under subpart G (40 CFR 280.12-91).

Operational readiness review: A structured method for determining that a project, process, or facility is ready to operate or occupy. It includes, as a minimum, a review of the readiness of the plant and hardware, personnel, and procedures. The review includes a determination of compliance with applicable environmental, health, and safety requirements (DOE-91/04).

Operational test period: A period of time (168 hours) during which the CEMS is expected to operate within the established performance specifications without any unscheduled maintenance, repair, or adjustment (40 CFR 60-App/B-91).

Operon: A group of linked genes controlled as a single unit (EPA-88/09a).

Ophthalmic glass: A glass used in spectacles, generally having specified optical and physical properties and quality (*see* glass for more related terms) (EPA-83).

Opportunity cost: The cost of foregoing alternative uses of a resource (*see* cost for more related terms) (OTA-89/10).

Optic: An optical instrument with a lens or prism having variations in wall thickness, producing refractive effects (EPA-83).

Optical crown glass: An optical glass with a low dispersion and low index of refraction, usually forming the converging element of an optical system. Any optical glass possessing a Nu-value of at least 55.0; or any optical glass with a Nu-value between 50.0 and 55.0 having a refractive index greater than 1.60 (*see* glass for more related terms) (EPA-83).

Optical density: A logarithmic measure of the amount of incident light attenuated. Optical density (D) is related to the transmittance (Tr) and opacity (Op) as follows: $D = -\log(10) \, Tr = -\log(10)(1 - Op)$ (40 CFR 60-App/B-91).

Optical flint glass: An optical glass with a high dispersion and high index of refraction, usually forming the diverging elements of an optical system. Any optical glass possessing a Nu-value less than 50.0; or any optical glass with a Nu-value between 50.0 and 55.0 having a refractive index less than 1.60 (*see* glass for more related terms) (EPA-83).

Optical glass numerical designation: Based on the index of refraction for sodium line (nd) and the Nu-value (v). the unity factor for the index is dropped (that is, 1.496 becomes 496), and the decimal point for the Nu-value is also dropPed (Nu=61.1 becomes 644). The glass is specified 496/644 (EPA-83).

Optical pyrometer: A temperature measuring instrument that matches the intensity of radiation at a single wavelength from a tungsten filament with the intensity of the radiation at the same wavelength emitted by a heat source (*see* temperature for more related terms) (SW-108ts).

Optical sorting: The use of photo cells to individually measure the reflectance of passing particles (EPA-83).

Optical spectrometer: An instrument with an entrance slit, a dispersing device, and with one or more exit slits, with which measurements are made at selected wavelengths within the spectral range, or by scanning over the range. The quantity detected is a function of radiant power (LBL-76/07-bio).

Optimum concentration range: A range, defined by limits expressed in concentration, below which scale expansion must be used and above which curve correction should be considered. This range will vary with the sensitivity of the instrument and the operating conditions employed (Course 165.6).

Optimum sustainable population: With respect to any population stock, the number of animals which will result in the maximum productivity of the population or the species, keeping in mind the carrying capacity of the habitat and the health of the ecosystem of which they form a constituent element (MMPA3-16USC1362-90).

Optimum yield: The best use of ground water that can be made under the circumstances; a use dependent not only upon hydrologic factors but also legal, social, and economic factors (Course 165.7).

Order of magnitude: The power of 10. This term is often used to express the major differences of two values.

Ordinary hazard content: Ordinary hazard contents shall be classified as those which are liable to burn with moderate rapidity and to give off a considerable volume of smoke but from which neither poisonous fumes nor explosions are to be feared in case of fire (*see* hazard for more related terms) (40 CFR 1910.35-91).

Ore: A mineral of sufficient value as to quality and quantity which may be mined with profit (EPA-82/05, *see also* 40 CFR 440.141-91). Other ore related terms include:
- Oxidized ore
- Unrefined and unprocessed ore

Ore dressing: *See* synonym, mineral processing.

Ore reserve: The term usually restricted to ore which the grade and tonnage have been established with reasonable assurance by drilling and other means (EPA-82/05).

Organ: Any human organ exclusive of the dermis, the epidermis, or the cornea (cf. critical organ) (40 CFR 190.02-91).

Organelle: A specific particle of organized living substance in most cells (LBL-76/07-bio).

Organic:
(1) Referring to or derived from living organisms.
(2) In chemistry, any compound containing carbon (EPA-89/12).

Organic acid: A product of biochemical activity containing the carboxyl radical (COOH) in its structure which readily reacts with other compounds (EPA-83). Examples include butyric acid ($CH_3(CH_2)_2COOH$) and benzoic acid (C_6H_5COOH) (cf. inorganic acid).

Organic active ingredients: The carbon-containing active ingredients used in pesticides, excluding metalloorganic active ingredients (40 CFR 455.21-91).

Organic analyzer: That portion of the system that senses organic concentration and generates an output proportional to the organic concentration (40 CFR 60-App/A(method 25A)-91).

Organic brain damage: Structural impairment or change in the brain (LBL-76/07-bio).

Organic carbon partition coefficient (K_{oc}): The measure of the extent of partitioning of a substance, at equilibrium, between organic carbon in geologic materials and water. The higher the K_{oc}, the more likely a substance is to bind to geologic materials than to remain in water. [ml/g] (40 CFR 300-App/A-91).

Organic chemical (organic compound, organic matter or organic material): A chemical compound containing carbon. Some regulatory definitions include carbon monoxide, carbon dioxide, etc. as organic compounds, some definitions exclude them as organic compounds. *see* specific definitions under 40 CFR 52.741; 52.1145; 796.2750-91). It can be grouped into:
(1) Volatile organic chemical (VOC)
(2) Semi-VOC
(3) Non-VOC

Organic chemistry: A branch of chemistry dealing with the study of composition, reaction, properties, etc. of organic compounds.

Organic chlorine: The chlorine associated with all chlorine containing compounds that elute just before lindane to just after mirex during gas chromatographic analysis using a halogen detector (40 CFR 797.1520-91).

Organic coating: Any coating used in a surface coating operation, including dilution solvents, from which volatile organic compound emissions occur during the application or the curing process. For the purpose of this regulation, powder coatings are not included in this definition (40 CFR 60.311-91, *see also* 40 CFR 60.451-91).

Organic compound: *See* synonym, organic chemical.

Organic content: Synonymous with volatile solids, except for small traces of some inorganic materials such as calcium carbonate, that lose weight at temperatures used in determining volatile solids (SW-108ts).

Organic detritus: The particulate remains of disintegrated plants and animals (LBL-76/07-water).

Organic heterocyclic compounds: Organic ring structures containing atom(s) other than carbon (e.g., N.O.S.) associated with the ring carbons (LBL-76/07-water).

Organic loading: In the activated sludge process, the food to microorganisms (F/M) ratio defined as the amount of biodegradable material available to a given amount of microorganisms per unit of time (*see* loading for more related terms) (EPA-10/87a).

Organic material: *See* synonym, organic chemical.

Organic material hydrocarbon equivalent: The sum of the carbon mass contributions of non-oxygenated hydrocarbons, methanol and formaldehyde as contained in a gas sample, expressed as gasoline fueled vehicle hydrocarbons. In the case of exhaust emissions, the hydrogen to carbon ratio of the equivalent hydrocarbon is 1.85:1. In the case of diurnal and hot soak emissions, the hydrogen to carbon ratios of the equivalent hydrocarbons are 2.33:1 and 2.2:1, respectively (40 CFR 86.090.2-91).

Organic matter: *See* synonym, organic chemical.

Organic nitrogen: Nitrogen combined in an organic molecules such as protein, amines, and amino acids.

Organic pesticide: *See* synonym, pesticide.

Organic pesticide chemical: The sum of all organic active ingredients listed in 40 CFR 455.20(b) which are manufactured at a facility subject to this subpart (40 CFR 455.21-91).

Organic pigment: General classification of pigments which are manufactured from coal tar and its derivatives. Compare with inorganic pigments, as a class, it is generally stronger and brighter, e.g., lithol rubine (EPA-79/12a).

Organic polymer process: The most thoroughly tested organic polymer solidification technique is the urea-formaldehyde system. The polymer is generally formed in a batch process where the wet or dry wastes are blended with a propolymer in a waste receptacle (steel drum) or a specially designed mixer. When these two components are thoroughly mixed, a catalyst is added and mixing is continued until the catalyst is thoroughly dispersed. Mixing is terminated before the polymer has formed, and the resin-waste mixture is transferred to a waste container if necessary. The polymerized material does not chemically combine with the waste but, instead, forms a spongy mass that traps the solid particles. Any liquid associated with the waste will remain after polymerization. The polymerized mass must often be dried before disposal (*see* solidification and stabilization for more related terms).

Organic solvent: The organic materials, including diluents and thinners, which are liquids at standard conditions and which are used as dissolvers, viscosity reducers, or cleaning agents (*see* solvent for more related terms) (40 CFR 52.1088-91, *see also* 40 CFR 52.1107; 52.1145; 52.2440-91).

Organic solvent based green tire spray: Any mold release agent and lubricant applied to the inside or outside of green tires that contains more than 12 percent, by weight, of VOC as sprayed (40 CFR 60.541-91).

Organic vapor: The gaseous phase of an organic material or a mixture of organic materials present in the atmosphere (40 CFR 52.741-91).

Organic waste: The waste material containing carbon. The organic fraction of municipal solid waste includes paper, wood, food wastes, plastics, and yard wastes (*see* waste for more related terms) (EPA-89/11).

Organism: Any living thing (EPA-89/12). Other organism related terms include:.
- Botulinus organism
- Critical aquatic organism
- Non-target organism

Organo-: A prefix indicating that a compound contains an organic group(s).

Organochlorine pesticides: Those pesticides which contain carbon and chlorine such as aldrin, DDD, DDE, DDT, dieldrin, endrin, and heptachlor (40 CFR 797.1520-91).

Organoleptic: Affecting or involving a sense organ as of taste, smell, or sight (EPA-92/12).

Organometallic compound: Any compound that contains both metal and carbon elements. Zineb ($C_4H_6N_2S_4Zn$) is an organometallic compound which is one of fungicide substances. It contains both metal (zinc) and carbon. Lead arsenate ($PbHAsO_4$) is an inorganometallic compound which is one of insecticide substances. It contains only metal (As) and has no carbon element in the compound.

Organophosphates:
(1) Pesticide chemicals that contain phosphorus; used to control insects. They are short-lived, but some can be toxic when first applied (EPA-89/12).
(2) A group of pesticide chemicals containing phosphorus, such as malathion and parathion, intended to control insects. These compounds are short-lived and, therefore, do not normally contaminate the environment. However, some organophosphates, such as parathion, are extremely toxic when initially applied and exposure to them can interfere with the normal processes of the nervous system, causing convulsions and eventually death. Malathion, on the other hand, is low in toxicity and relatively safe for humans and animals; it is a common ingredient in household insecticide products (EPA-74/11).

Organophosphorus induced delayed neurotoxicity (OPIDN): A neurological syndrome in which limb weakness and upper motor neuron spasticity are the predominant clinical signs and distal axonopathy of peripheral nerve and spinal cord are the correlative pathological signs. Clinical signs and pathology first appear between 1 and 2 weeks following exposure which normally inhibits greater than 80 percent of NTE (for O-Ps that age) (40 CFR 798.6450-91).

Organotins: Chemical compounds used in anti-foulant paints to protect the hulls of boats and ships, buoys, and dock pilings from marine organisms such as barnacles (EPA-89/12).

Orifice: An opening of regulated size and edge characteristics designed to direct, shape or impart flow parameters to fluids; to impart a measurable pressure difference for determining flow rate (EPA-83).

Orifice meter: One of liquid flow rate meters. This instrument is used with gases and low viscosity fluids. Typical accuracies are +/- 1% full-scale, which is the accuracy of the differential pressure measuring device on a clean fluid. When used with dirty or viscous fluids, both accuracy and life of the instrument are sacrificed. An accuracy of +/- 5% may be more realistic in these cases (*see* flow rate meter for more related terms) (EPA-89/06).

Orifice ring: That ring that forms the hole through which glass flows in the feeder process. *see also* Bushing (EPA-83).

Orifice type scrubber: One of air pollution control devices. The scrubber uses air velocity to promote liquid contact. The air flow through a restricted passage (usually curved) partially filled with water causes the dispersion of the water. In turn, centrifugal forces, impingement, and turbulence cause wetting of the particles and their collection. Water quantities in motion are relatively high, but most of the water can be recirculated without pumps or spray nozzles. Recirculation rates are as high as 20 gpm per 1000 cfm gas handled (*see* scrubber for more related terms) (AP-40).

Original equipment manufacturer (OEM): The entity which originally manufactured the motor vehicle or motor vehicle engine prior to conditional importation (40 CFR 85.1502-91).

Original equipment part: A part present in or on a vehicle at the time the vehicle is sold to the ultimate purchaser, except for components installed by a dealer which are not manufactured by the vehicle manufacturer or are not installed at the direction of the vehicle manufacturer (40 CFR 85.2102-91).

Original generation point: The location where regulated medical waste is generated. Waste may be taken from original generation points to a central collection point prior to off-site transport or on-site treatment (40 CFR 259.10-91).

Original production (OP) year: The calendar year in which the motor vehicle or motor vehicle engine was originally produced by the OEM (40 CFR 85.1502-91).

Original production (OP) years old: The age of a vehicle as determined by subtracting the original production year of the vehicle from the calendar year of importation (40 CFR 85.1502-91).

Orlon fiber: A trademark of Du Pont for acrylic fibers (*see* fiber for more related terms) (EPA-74/06b).

Ornamental: Trees, shrubs, and other plantings in and around habitations generally, but not necessarily located in urban and suburban areas, including residences, parks, streets, retail outlets, industrial and institutional buildings (40 CFR 171.2-91).

Orographic:
(1) The nature of a terrain with respect to differences in its elevation (NATO-78/10).
(2) Pertaining to the physical geography of mountains or mountain ranges (DOE-91/04).

ORP recorders: Oxidation-reduction potential recorders (EPA-83/06a).

Orsat (or orsat analyzer): An apparatus used to volumetrically analyze O_2, CO and CO_2 by passing the mixture gases through various solvents that absorb them (SW-108ts). A gas sample, on a dry basis, is first contained in an orsat analyzer at a known temperature and pressure. A liquid which can absorbs O_2 is then brought into constant with the sample gas. Because the temperature and pressure are maintained constant, the volume of the gas decreases as O_2 is absorbed. The reduced volume is recorded. Similarly, different liquids which can absorb CO_2 and CO selectively are sequentially brought into contact with the sample gas. For each test, the change in volume is recorded. Based on the property of a gas mixture, the change in volume is the measurement of volumetric fraction of each gas species. Also based on the property of a gas mixture, the volumetric fraction is equal to the mole fraction of each gas species. The remaining gas in the apparatus is assumed to be nitrogen.

Orthophosphate: An acid or salt containing phosphorus such as potassium orthophosphate $K_3(PO_4)$.

Orthotolidine residual: A measure of chlorine residual left in treated water after application of chlorine (EPA-75/10).

Osborne separator: A device that utilizes a pulsed, rising column of air to separate small particles of glass, metal, and other dense items from compost (*see* separator for more related terms) (SW-108ts).

Oscillating grate: The grate surface oscillates to move the fuel and residue from feed end to discharge (*see* grate for more related terms) (EPA-83).

Oscillating grate stoker: A stoker whose entire grate surface oscillates to move the solid waste and residue over the grate surface (*see* stoker for more related terms) (SW-108ts).

Osmium (Os): A hard transition metal with a.n. 76; a.w. 190.2; d. 22.6 g/cc; m.p. 3000° C and b.p. 5500° C. The element belongs to group VIII of the periodic table.

Osmole: The standard unit for expressing osmotic pressure. One osmole is the osmotic pressure exerted by a one-molar solution of an ideal solute (LBL-76/07-water).

Osmoregulation: The process of regulating fluids in and around living cells.

Osmosis: The diffusion of a solvent through a semi-permeable membrane into a more concentrated solution (EPA-76/03, *see also* EPA-89/12).

Osmotic pressure: The equilibrium pressure differential across a semipermeable membrane which separates a solution of lower to one of higher concentration (EPA-82/11f).

Osmotic shock treatment: The use of hypotonic buffers to lyse cells releasing cellular products in the periplasm (EPA-88/09a).

Other coatings: The coating steel products with metals other than zinc or terne metal by the hot dip process including the immersion of the steel product in a molten bath of metal, and the related operations preceding the subsequent to the immersion phase (40 CFR 420.121-91).

Other lead emitting operation: Any lead-acid battery manufacturing plant operation from which lead emissions are collected and ducted to the atmosphere and which is not part of a grid casting, lead oxide manufacturing, lead reclamation, paste mixing, or three-process operation facility, or a furnace affected under Subpart L of this part (40 CFR 60.371-91).

Other mineral: The clay, stone, sand, gravel, metalliferous and nonmetalliferous ores, and any other solid material or substances of commercial value excavated in solid form from natural deposits on or in the earth, exclusive of coal and those minerals which occur naturally in liquid or gaseous form (*see* mineral for more related terms) (SMCRA701-30USC1291-90).

Other nuclear material: *See* Figure I-1 of DOE Order 5633.3. This figure defines other nuclear materials, which include tritium, deuterium, lithium-6, and neptunium-237 (*see* nuclear material for more related terms) (DOE-91/04).

Other potentially infectious materials:
(1) The following body fluids: semen, vaginal secretions, cerebrospinal fluid, synovial fluid, pleural fluid, pericardial fluid, peritoneal fluid, amniotic fluid, saliva in dental procedures, and any body fluid that is visibly contaminated with blood;
(2) Any unfixed tissue or organ (other than intact skin) from a human (living or dead); and
(3) Hepatitis B virus (HBV) or human immunodeficiency virus (HIV) containing cell or tissue cultures, organ cultures, and culture medium or other solutions: and blood, organs or other tissues from experimental animals infected with HIV or HBV (29 CFR 1910).

Otto cycle: The type of engine with operating characteristics significantly similar to the theoretical Otto combustion cycle. The use of a throttle during normal operation is indicative of an Otto-cycle engine. This definition applies beginning with the 10 model year (40 CFR 86.090.2-91).

Outcrop: The exposing of bedrock or strata projecting through the overlying cover of detritus and soil (EPA-82/10).

Outdoor electrical substations: Outdoor, fenced-off, and restricted access areas used in the transmission and/or distribution of electrical power. Outdoor electrical substations restrict public access by being fenced or walled off as defined under 40 CFR 61.30(l)(1)(ii). For purposes of this TSCA policy, outdoor electrical substations are defined as being located at least 0.1 km from a residential/commercial area. Outdoor fenced-off and restricted access areas used in the transmission and/or distribution of electrical power which are located less than 0.1 km from a residential/commercial areas (40 CFR 761.123-91).

Outer continental shelf: The meaning provided by section 2 of the Outer Continental Shelf Lands Act (43USC1331) (CAA328, *see also* OPA1001-91).

Outer continental shelf energy activity: Any exploration for, or any development or production of, oil or natural gas from the outer Continental Shelf (as defined in section 1331(a) of title 43) or the siting, construction, expansion, or operation of any new or expanded energy facilities directly required by such exploration, development, or production (CZMA304-16USC1453-90).

Outer continental shelf source and OCS source: Include any equipment, activity, or facility which:
(1) Emits or has the potential to emit any air pollutant,
(2) Is regulated or authorized under the Outer Continental Shelf Lands Act, and
(3) Is located on the Outer Continental Shelf or in or on waters above the Outer Continental Shelf. Such activities include, but are not limited to, platform and drill ship exploration, construction, development, production, processing, and transportation. For purposes of this subsection, emissions from any vessel servicing or associated with an OCS source, including emissions while at the OCS source or en route to or from the OCS source within 25 miles of the OCS source, shall be considered direct emissions from the OCS source (CAA328-42USC7627).

Outfall:
(1) A point source as defined by 40 CFR 122.2 at the point where a municipal separate storm sewer discharges to waters of the United States and does not include open conveyances connecting two municipal separate storm sewers, or pipes, tunnels or other conveyances which connect segments of the same stream or other waters of the United States and are used to convey waters of the United States (40 CFR 122.26-91).
(2) The point where an effluent (sewer or wastewater) is discharged into receiving waters (cf. major outfall or sewer outfall) (EPA-89/12).

Outlays (expenditures): The charges made to the project or program. They may be reported on a cash or accrual basis. For reports prepared on a cash basis, outlays are the sum of actual cash disbursement for direct charges for goods and services, the amount of indirect expense incurred, the value of in-kind contributions applied, and the amount of cash advances and payments made to contractors and subgrantees. For reports prepared on an accrued expenditure basis, outlays are the sum of actual cash disbursements, the amount of indirect expense incurred, the value of in-kind contributions applied, and the new increase (or decrease) in the amounts owed by the grantee for goods and other property received, for services performed by employees, contractors, subgrantees, subcontractors, and other payees, and other amounts becoming owed under programs for which no current services or performance are required, such as annuities, insurance claims, and other benefit payments (40 CFR 31.3-91).

Outokumpu furnace: A furnace used for flash smelting, in which hot sulfide concentrate is fed into a reaction shaft along with preheated air and fluxes. The concentrate roasts and smelts itself in a single autogeneous process (*see* furnace for more related terms) (EPA-83/03a).

Output: An activity or product which the applicant agrees to complete during the budget period (40 CFR 35.105-91).

Outside air: The air outside buildings and structures, including, but not limited to, the air under a bridge or in an open air ferry dock (40 CFR 61.141-91).

Outthrows: All papers that are so manufactured or treated or are in such form as to be unsuitable for consumption as the grade specified (EPA-83).

Oven: A chamber within which heat is used for one or more of the following purposes: dry, bake, cure, or polymerize a coating or ink (40 CFR 52.741-91, *see also* 40 CFR 60.441-91).

Oven glass: A glass suitable for manufacture of articles to be used in baking and roasting foods (*see* glass for more related terms) (EPA-83).

Over issue news: Consists of unused over-run regular newspapers printed on newsprint, baled or securely tied in bundles, containing not more than the normal percentage of rotogravure and colored sections (EPA-83).

Over the head position: The mode of use of a device with a headband, in which the headband is worn such that it passes over the users head. This is contrast to the behind-the-head and under the chin positions (40 CFR 211.203-91).

Overall control: The product of the capture efficiency and the control device efficiency (40 CFR 52.741-91).

Overburden:
(1) Any material of any nature, consolidated or unconsolidated, that overlies a mineral deposit, excluding topsoil or similar naturally-occurring surface materials that are not disturbed by mining operations (40 CFR 122.26-91).
(2) Rock and soil cleared away before mining (EPA-89/12).

Overdraft: The withdrawal of groundwater at rates perceived to be excessive, and therefore an unsatisfactory water management practice. *See also* mining (Course 165.7).

Overfill release: A release that occurs when a tank is filled beyond its capacity, resulting in a discharge of the regulated substance to the environment (*see* release for more related terms) (40 CFR 280.12-91).

Overfire air:
(1) Air, under control as to quantity and direction, introduced above or beyond a fuel bed by induced or forced draft.
(2) Air forced into the top of an incinerator or boiler to fan the flames (EPA-89/12).
(3) *See* air for more related terms.

Overfire air fan (or overfire fan): A fan used to provide air above a fuel bed (*see* fan for more related terms) (SW-108ts).

Overfire air port: An entry point of combustion air introduced above and beyond the fuel or waste bed (*see* port for more related terms) (EPA-89/03b).

Overfire fan: *See* synonym, overfire air fan.

Overflow: Excess water discharged from the treatment system (EPA-82/10, *see also* 40 CFR 403.7-91).

Overland flow: A land application technique that cleanses waste water by allowing it to flow over a sloped surface. As the water flows over the surface, the contaminants are removed and the water is collected at the bottom of the slope for reuse (EPA-89/12).

Overpress: An imperfection; projecting excess glass resulting from imperfect closing of mold joints (EPA-83).

Oversaturated solution (or supersaturated solution): A solution that contains a greater concentration of a solute than is possible at equilibrium under fixed conditions of temperature and pressure (40 CFR 796.1840-91).

Oversize waste: *See* synonym, bulky waste.

Oversized regulated medical waste: The medical waste that is too large to be placed in a plastic bag or standard container (*see* medical waste or waste for more related terms) (40 CFR 259.10-91).

Overturn: The period of mixing (turnover), by top to bottom circulation, of previously stratified water masses. This phenomenon may occur in spring and/or fall, or after storms. It results in a uniformity of chemical and physical properties of the water at all depths (EPA-89/12).

Overvarnish: A transparent coating applied directly over ink or coating (40 CFR 52.741-91).

Overvarnish coating operation: The system on each beverage can surface coating line used to apply a coating over ink which reduces friction for automated beverage can filling equipment, provides gloss, and protects the finished beverage can body from abrasion and corrosion. The overvarnish coating is applied to two-piece beverage can bodies. The overvarnish coating operation consists of the coating application station, flashoff area, and curing oven (40 CFR 60.491-91).

Ovicide: A chemical that destroys an organism's eggs (EPA-85/10).

Oxalic acid ($(COOH)_2$): A poisonous crystalline solid used as a bleach or rust remover.

Oxbow lake: A lake formed when a meandering river, having bent in almost a complete circle, cuts across the narrow neck of land between the two stretches and leaves a backwater; silt is gradually deposited by the river at the entrance to this backwater till it is finally separated from the river and becomes a lake (DOI-70/04).

Oxidant: A substance containing oxygen that reacts chemically in air to produce a new substance. The primary ingredient of photochemical smog (EPA-89/12).

Oxidation (or chemical oxidation):
(1) The combination of a reactant with oxygen (EPA-83).
(2) The process in chemistry whereby electrons are removed from a molecule.
(3) The addition of oxygen which breaks down organic waste or chemicals such as cyanides, phenols, and organic sulfur compounds in sewage by bacterial and chemical means (EPA-89/12).
(4) Oxidation is a chemical reaction process of converting a substance to another form by combination with oxygen. In rapid oxidation, heat is released. For some types of oxidation, heat is required.

Oxidation control system: An emission control system which reduces emissions from sulfur recovery plants by converting these emissions to sulfur dioxide (40 CFR 60.101-91).

Oxidation number (or oxidation state): The number of electrons to be added to or to be subtracted from an atom, when the atom is converted from a combined state (a compound) to its original elemental form.

Oxidation pond: A man made lake or body of water in which waste is consumed by bacteria. It is used most frequently with other waste treatment processes. An oxidation pond is basically the same as a sewage lagoon (*see* pond for more related terms) (EPA-89/12).

Oxidation reduction indicator: An indicator (compound) whose color in the oxidized state is different from that in the reduced state. This different color property is used to test if a substance is in the oxidized state or reduced state.

Oxidation reduction potential (ORP):
(1) A measurement that indicates the activity ratio of the oxidizing and reducing species present (EPA-76/03).
(2) The emf developed by a platinum electrode immersed in water, referred to the Standard Hydrogen Electrode (LBL-76/07-water).

Oxidation reduction (redox) reaction:
(1) An atom or group of atoms loses electrons; or
(2) The introduction of one or more oxygen atoms into a molecule, accompanied by the release of energy (EPA-87/10a), e.g., the reaction in alkaline solution: $2MnO4^- + CN^- + 2OH^- ---> 2MnO4^{-2} + CNO^- + H_2O$.

The oxidation state of the cyanide ion is raised from -1 to +1 (the cyanide is oxidized as it combines with an atom of oxygen to form cyanate); the oxidation state of the permanganate decreases from -1 to -2 (permanganate is reduced to manganate). This change in oxidation state implies that an electron was transferred from the cyanide ion to the permanganate. The increase in the positive valence (or decrease in the negative valence) with oxidation take place simultaneously with reduction in chemically equivalent ratios.

Oxidation state: *See* synonym, oxidation number.

Oxide: A compound of two elements, one of which is oxygen.

Oxide layer: In seaming a landfill liner, the taking of atmospheric oxygen in the form of a surface film after a polyethylene sheet is extruded or otherwise manufactured (EPA-89/09, *see also* EPA-91/05).

Oxide mask: An oxidized layer of silicon wafer through which windows are formed which will allow for dopants to be introduced into the silicon (EPA-83/03).

Oxide of nitrogen: *See* synonym, nitrogen oxide.

Oxidizable cyanide: The cyanide amenable to oxidation (*see* cyanide for more related terms) (EPA-83/06a).

Oxidized ore: The alteration of metalliferous minerals by weathering and the action of surface waters, and the conversion of the minerals into oxides, carbonates, or sulfates (*see* ore for more related terms) (EPA-82/05).

Oxidized sludge: The stable sludge. Raw sludge that has been treated by either biological or chemical means and becomes a stable sludge (*see* sludge for more related terms).

Oxidized zone: The portion of an ore body near the surface, which has been leached by percolating water carrying oxygen, carbon dioxide or other gases (EPA-82/05).

Oxidizing: Combining the material concerned with oxygen (EPA-83/06a).

Oxidizing agent: An agent (or a compound) which is (1) providing oxygen, (2) removing hydrogen from another compound or (3) attracting negative electrons during oxidation reactions. As a result, the agent is reduced.

Oxidizing bleach: Bleaching of pulp by oxidizing chemicals such as hypochlorite, peroxide, chlorine dioxide, etc (EPA-83).

Oxidizing salt bath descaling: The removal of scale from semi-finished steel products by the action of molten salt baths other than those containing sodium hydride (*see* salt bath descaling for more related terms) (40 CFR 420.81-91).

Oxidizing slag: In the steel industry, a fluxing agent that is used to remove certain oxides such as silicon dioxide, manganese oxide, phosphorus pentoxide and iron oxide from hot metals (*see* slag for more related terms) (EPA-74/06a).

OXO process (hydrocarbonylation or hydroformylation): A process wherein olefinic hydrocarbon vapors are passed over cobalt catalysts in the presence of carbon monoxide and hydrogen to produce alcohols, aldehydes, and other oxygenated organic compounds (EPA-87/10a).

Oxyacetylation: A process using ethylene, acetic acid, and oxygen commonly used to produce vinyl acetate (EPA-87/10a).

Oxy-: A prefix indicating that a compound contains the oxygen radical (-O-).

Oxygen (O): A gaseous element with a.n. 8; a.w. 15.9994; d. 1.14 g/cc; m.p. -218.8° C and b.p. -183° C. The element belongs to group VIA of the periodic table. Other oxygen related terms include:
- Available oxygen
- Biological oxygen demand (*see* synonym, biochemical oxygen demand)
- Biochemical oxygen demand
- Chemical oxygen demand (COD)
- Dissolved oxygen (DO)
- Dissolved oxygen unit
- Dry oxygen
- Five day BOD (BOD5) (*see* definition under biochemical oxygen demand)
- Immediate dissolved oxygen demand
- Oxygen depleting effect (*see* synonym, biochemical oxygen demand)
- Theoretical oxygen

- Total oxygen demand
- Ultimate oxygen demand
- Wet oxygen

Oxygen activated sludge: An activated sludge process using pure oxygen as an aeration gas (rather than air). This is a patented process marketed by Union Carbide under the trade name UNOX (*see* sludge for more related terms) (EPA-87/10a).

Oxygen debt: A phenomenon that occurs in an organism when available oxygen is inadequate to supply the respiratory demand. During such a period the metabolic processes result in the accumulation of breakdown products that are not oxidized until sufficient oxygen becomes available (LBL-76/07-water).

Oxygen deficient atmosphere: An atmosphere which contains an oxygen partial pressure of less than 148 millimeters of mercury (19.5 percent by volume at sea level) (NIOSH-84/10).

Oxygen deficit: The amount of oxygen difference between the dissolved oxygen and the oxygen saturation value in a water body.

Oxygen depleted pathway: Chemical transformations which occur under oxygen deficient reaction atmospheres (EPA-88/12).

Oxygen enriched incineration: An incineration process whose combustion air is enriched with oxygen to the level higher than that of the regular air oxygen content. Incineration requires an intensive, complete destruction oxidation of waste molecules with oxygen. Current incinerators require up to about 50 - 150% excess air to provide enough oxygen for oxidation, and require that the so-called 3-T's factors (temperature, turbulence, and residence time) be adequate to insure efficient destruction. Because 79% of air is nitrogen, the majority of any excess air used will not contribute to the effectiveness of incineration and will only result in extra energy required to raise the nitrogen to combustion temperature, and additional product gas handling and cleaning requirements. As a matter of fact, two of the 3-T's (turbulence and residence time) are essentially the physical parameters used to promote the contact of hazardous waste particles with oxygen. Therefore, it is logical that increased oxygen concentration should improve incineration or destruction efficiency (*see* incineration for more related terms).

Oxygen indicator: An oxygen indicator, which is used to detect the oxygen concentration in air, has two principal components for operation. These are the oxygen sensor and the meter read-out. In some units, air is drawn into the oxygen detector with a pump; in other units, the ambient air is allowed to diffuse to the sensor. The oxygen detector uses either a paramagnetic or an electrochemical sensor to determine the oxygen concentration in air. A typical sensor consists of two electrodes; a housing containing a basic electrolytic solution (KOH electrolyte) and a semi-permeable teflon membrane.

Oxygen molecules (O_2) diffuse through the membrane into the solution. Reactions between the oxygen, the solution and the electrodes produce an electric current proportional to the oxygen content. The current passes through the electronic circuit. The resulting signal is shown as a needle deflection on a meter or digital reading (*see* air analyzer for more related terms) (Course 165.5).

Oxygen recorder: An instrument for continuously monitoring the percentage of oxygen content in a gas (EPA-83).

Oxygen sag curve: A graph to show dissolved oxygen consumption against time (or distance downstream) in a water course from the point of pollution.

Oxygen saturation value: The value of oxygen in water and in air is in equilibrium.

Oxygen tension: Partial pressure of oxygen in solution (LBL-76/07-water).

Oxygenated solvent: An organic solvent containing oxygen as part of the molecular structure. Alcohols and ketones are oxygenated compounds often used as paint solvents (*see* solvent for more related terms) (EPA-89/12).

Oxygenation:
(1) Impregnation or combination with oxygen (DOI-70/04).
(2) In wastewater treatment, it means to increase the content of dissolved oxygen in wastewater.

Oxygenation capacity: The rate of absorbing oxygen per unit volume of de-oxygenated liquid.

Ozonation: A water or wastewater treatment process involving the use of ozone as an oxidizing agent (EPA-87/10a). Since ozone (O_3) is a powerful oxidizing agent and an extremely reactive gas that cannot be shipped or stored, it must be generated on site immediately prior to use. Ozone, beside being a powerful oxidizing agent, has anti-bacterial and anti-viral properties.

Ozonator: A device that adds ozone to water (EPA-89/12).

Ozone (O_3):
(1) The molecular oxygen with three atoms of oxygen forming each molecule. The third atom of oxygen in each molecule of ozone is loosely bound and easily released. Ozone is used sometimes for the disinfection of water but more frequently for the oxidation of taste-producing substances, such as phenol, in water and for the neutralization of odors in gases or air (EPA-87/10a).
(2) (Ozone can be) found in two layers of the atmosphere, the stratosphere and the troposphere. In the stratosphere (the atmospheric layer beginning 7 to 10 miles above the earth's surface) ozone is a form of oxygen found naturally which provides a protective layer shielding the earth from ultraviolet radiation's harmful health effects on humans and the environment. In the troposphere (the layer extending up 7 to 10 miles from the earth's surface), ozone is a chemical oxidant and major component of photochemical smog. Ozone can seriously affect the human respiratory system and is one of the most prevalent and widespread of all the criteria pollutants for which the Clean Air Act required EPA to set standards. Ozone in the troposphere is produced through complex chemical reactions of nitrogen oxides, which are among the primary pollutants emitted by combustion sources; hydrocarbons, released into the atmosphere through the combustion, handling and processing of petroleum products; and sunlight (EPA-89/12).

Ozone depletion: Destruction of the stratospheric ozone layer which shields the earth from ultraviolet radiation harmful to biological life. This destruction of ozone is caused by the breakdown of certain chlorine and/or-bromine containing compounds (chlorofluorocarbons or halons) which break down when they reach the stratosphere and catalytically destroy ozone molecules (EPA-89/12).

Ozone depletion potential: A factor established by the Administrator to reflect the ozone depletion potential of a substance, on a mass per kilogram basis, as compared to chlorofluorocarbon-11 (CFC-11). Such factor shall be based upon the substance's atmospheric lifetime, the molecular weight of bromine and chlorine, and the substance's ability to be photolytically disassociated, and upon other factors determined to be an accurate measure of relative ozone depletion potential (CAA601-42USC7671-91).

Ozone layer (or ozonosphere): The stratum of the upper atmosphere (about 15-50 km above the earth surface) in which most of the atmosphere's ozone is concentrated.

Ozone protection policy: This policy involves providing information on the effects of human activities on ozone, the effects of ozone change on the public health and welfare, and the need for additional legislation (CAA150-42USC7450-91).

Ozonide: A compound containing the ion $(O_3)^-$ after reaction with ozone.

Ozonosphere: *See* synonym, ozone layer.

Pp

Package plant: A prefabricated or pre-built wastewater treatment plant (EPA-74/11).

Package sewage treatment plant: A sewage treatment facility contained in a small area and generally prefabricated in a complete package (EPA-82/11f).

Packaged boiler: A boiler equipped and shipped complete with fuel burning equipment, mechanical draft equipment, automatic controls and accessories (*see* boiler for more related terms).

Packaging: Capping, closing, filling, labeling, packing, wrapping, and so forth (AP-40, p790).

Packaging rotogravure printing: A rotogravure printing upon paper, paper board, metal foil, plastic film, and other substrates, which are, in subsequent operations, formed into packaging products or labels for articles to be sold (40 CFR 52.741-91).

Packaging rotogravure printing line: A rotogravure printing line is which surface coatings are applied to paper, paperboard, foil, film, or other substrates which are to be used to produce containers, packaging products, or labels for articles (40 CFR 52.741-91).

Packed absorber: *See* synonym, packed tower.

Packed bed scrubber: *See* synonym, packed tower.

Packed column: *See* synonym, packed tower.

Packed tower (packed absorber, packed column or packed bed scrubber):
(1) One of air pollution control devices. A packed tower is a pollution control device that forces dirty air through a tower packed with crushed rock or wood chips while liquid is sprayed over the packing material. The pollutants in the air stream either dissolve or chemically react with the liquid (EPA-89/12).
(2) An air pollution control device that forces dirty gas through the bottom of a tower packed with 1- to 3-inch (2.5- to 7.6-cm) diameter plastic shapes, crushed rock, wood chips, or other packing materials while scrubber liquid is sprayed over the packing materials from the top of the tower. The liquid either absorbs (dissolves) the pollutants in the gas stream or chemically react with them. The collection of the pollutants depends upon the length of contact time of the gas stream on the collecting surface. The packing materials that are intended to maximize the surface area are to provide high surface area to volume ratios for intimate gas/liquid contact for mass transfer. The scrubber, which is used primarily for acid gas control, removes some particulate matter, however, it has a low collection efficiency for fine particulates.
(3) The major components of a packed-bed scrubber include:
 (A) A cylindrical shell to house the scrubbing media;
 (B) Packing media and supporting plates;
 (C) Liquid spray nozzles to distribute the scrubbing liquid;
 (D) Demister pads to remove liquid droplets from the clean flue gas; and
 (E) An induced draft fan for moving the flue gas through the scrubber (EPA-81/09, 89/03b; AP-40).
(4) *See* scrubber for more related terms.

Packer:
(1) A device lowered into a well to produce a fluid tight seal (40 CFR 146.3-91, *see also* 40 CFR 147.2902-91).
(2) cf. packer vehicle

Packer vehicle (or packer): There are two types:
(1) A compactor collection truck which is an enclosed vehicle provided with special mechanical devices for loading, compressing and distributing refuse within the body.
(2) A stationary compactor which is an adjunct of a refuse collection system which compacts refuse into a pull on detachable container at the site of generation (cf. compactor) (EPA-83).
(3) *See* vehicle for more related terms.

Packing:
(1) An asbestos containing product intended for use as a mechanical seal in circumstances involving rotary, reciprocating, and helical motions, and which are intended to restrict fluid or gas leakage between moving and stationary surfaces. Major applications of this product include: seals in pumps; seals in valves; seals in compressors; seals in mixers; seals in swing joints; and seals in hydraulic cylinders (40 CFR 763.163-91).
(2) Also known as fill, it is the material which forms the heat transfer surface within a cooling tower, and over which the water is distributed in its passage down the tower (Gurney-66).

Packinghouse: A plant that both slaughters animals and

subsequently processes carcasses into cured, smoked, canned or other prepared meat products (40 CFR 432.31-91, *see also* 40 CFR 432.41-91).

Paint: A combination of a pigment, extender and vehicle, and frequently other additives, in a liquid composition, which is converted to an opaque solid film after application (EPA-79/12b). Other paint related terms include:
- Industrial sales paint
- Solvent base paint
- Tint base paint
- Trade sales paint
- Water base paint

Paint baking: The process of both drying and baking, curing, or polymerizing coatings (AP-40, p865).

Paint manufacturing plant: A plant that mixes, blends, or compounds enamels, lacquers, sealers, shellacs, stains, varnishes, or pigmented surface coatings (40 CFR 52.741-91).

Paint sludge: Sticky mass of paint wastes, usually spray-booth residue. Its consistency is that of modeling clay (*see* sludge for more related terms) (EPA-83).

Paint stripping: The process of removing an organic coating from a workpiece or painting fixture. The removal of such coatings using processes such as caustic, acid, solvent and molten salt stripping are included (EPA-83/06a).

Pairwise deletion of missing data: *See* definition under listwise and pairwise deletion of missing data.

Paleontology: The study of fossils (DOE-91/04).

Palladium (Pd): A transition metal with a.n. 46; a.w. 106.4; d. 12.0 g/cc; m.p. 1552° C and b.p. 3980° C. The element belongs to group VIII of the periodic table.

Pandemic: Widespread throughout an area, nation or the world (EPA-89/12).

Pan body: Tractor drawn scraper body capable of carrying a load (EPA-83).

Panel:
(1) In electronics, the front, screen portion of the glass enclosure of a cathode ray tube (EPA-83/03).
(2) *See* national panel of environmental arbitrators (40 CFR 304.12-91).

Panel spalling test: A standardized test to provide an index to the spalling behavior of refractories (EPA-83).

Paper: One of two broad subdivisions of paper products, the other being paperboard. Paper is generally lighter in basis weight, thinner, and more flexible than paperboard. Sheets 0.012 inch or less in thickness are generally classified as paper. Its primary uses are for printing, writing, wrapping, and sanitary purposes. However, in this guideline, the term paper is also used as a generic term that includes both paper and paperboard. It includes the following types of papers: bleached paper, bond paper, book paper, brown paper, coarse paper, computer paper, cotton fiber content paper, cover stock or cover paper, duplicator paper, form bond, ledger paper, manifold business forms, mimeo paper, newsprint, office papers, offset printing paper, printing paper, stationery, tabulating paper, unbleached papers, writing paper, and xerographic/copy paper (40 CFR 250.4-aa-91). Other paper related terms include:
- Bleached paper
- Bond paper
- Book paper
- Braille paper
- Brown paper
- Coarse paper
- Coated paper
- Commercial paper
- Computer paper
- Computer printout paper
- Corrugated paper
- Cotton fiber content paper
- Currency paper
- Developing paper (*see* synonym, sensitized paper)
- Duplicator paper
- Fine paper
- Glassine paper
- Glazed paper
- Grain paper
- Green paper
- Groundwood paper
- Hardness paper
- High grade electrical paper
- High grade paper
- Insulating paper
- Kid finish paper
- Kraft paper
- Ledger paper
- Magazine paper
- Manila paper
- Mimeo paper
- Newsprint (*see* synonym, newsprint paper)
- Newsprint paper
- Noiseless paper
- Novel paper
- Office paper
- Offset printing paper
- Onionskin paper
- Paraffin paper
- Photographic paper (*see* synonym, sensitized paper)
- Post consumer recovered paper
- Printing paper
- Rag paper
- Recyclable paper
- Register paper
- Sensitized paper
- Specialty paper
- Starch coated paper
- Super mixed paper
- Tabulating paper
- Unbleached paper
- Waxed paper
- White paper

- Writing paper
- Xerographic/copy paper

Paper chromatography: A chromatography method for analyzing a chemical mixture in which the stationary phase is a special grade of absorbent paper. A sample mixture can be identified, based on the progress of the absorption of the sample by the paper.

Paper coating: Any coating applied on paper, plastic film, or metallic foil to make certain products, including (but not limited to) adhesive tapes and labels, book covers, post cards, office copier paper, drafting paper, or pressure sensitive tapes. Paper coating includes the application of coatings by impregnation and/or saturation (40 CFR 52.741-91).

Paper coating facility: A facility that includes one or more paper coating lines (40 CFR 52.741-91).

Paper coating line: A coating line in which any protective, decorative, or functional coating is applied on, saturated into, or impregnated into paper, plastic film, or metallic foil to make certain products, including (but not limited to) adhesive tapes and labels, book covers, post cards, office copier paper, drafting paper, and pressure sensitive tapes (40 CFR 52.741-91).

Paper converting operation: A manufacturing facility that transforms paper into products such as envelopes or boxes (OTA-89/10).

Paper napkin: The special tissues, white or colored, plain or printed, usually folded, and made in a variety of sizes for use during meals or with beverages (40 CFR 250.4-91).

Paper packer: Plant facilities for converting wastepaper to paper stock. There are many that are both packers and dealers (EPA-83).

Paper producer: An industrial or commercial establishment that produces wastepaper as a result of its operations, and for reasons of economy or volume involved considers it practical to convert that wastepaper to paper stock (EPA-83).

Paper product: Any item manufactured from paper or paperboard. The term "paper product" is used in this guideline to distinguish such items as boxes, doilies, and paper towels from printing and writing papers. It includes the following types of products: corrugated boxes, doilies, envelopes, facial tissue, fiber of fiberboard boxes, folding boxboard, industrial wipers, paper napkins, paper towels, tabulating cards, and toilet tissue (40 CFR 250.4-cc-91).

Paper stock: Paper which has been collected, sorted, and graded to meet specifications. It is important not to use this term interchangeably with wastepaper (EPA-83).

Paper supplier: Generally refers to the dealers, packers, and brokers (EPA-83).

Paper towel: Paper toweling in folded sheets, or in raw form, for use in drying or cleaning, or where quick absorption is required (40 CFR 250.4-91).

Paperboard: One of the two broad subdivisions of paper, the other being paper itself. Paperboard is usually heavier in basis weight and thicker than paper. Sheets 0.012 inch or more in thickness are generally classified as paperboard. The broad classes of paperboard are containerboard, which is used for corrugated boxes, boxboard, which is principally used to make cartons, and all other paperboard (40 CFR 250.4-91).

para-: A prefix indicating that two substituents are in the 1,4 positions of a benzene ring compound (directly opposite each other).

Paraffin (C_nH_{2n+2}): A methane series which is a saturated aliphatic hydrocarbon.

Paraffin paper: The paper treated with a wax to make it waterproof (*see* paper for more related terms) (EPA-83).

Paramagnetic oxygen analyzer (POA): One of continuous emission monitors (*see* continuous emission monitor for various types). Molecules behave in different ways when placed in a magnetic field. This magnetic behavior is either diamagnetic or paramagnetic. Most materials are diamagnetic and, when placed in a magnetic field, is repelled by it. Paramagnetism arises when a molecule has one or more electrons spinning in the same direction. Most materials have paired electrons; the same number of electrons spinning counterclockwise as spinning clockwise. Oxygen, however, has two unpaired electrons that spin in the same direction. These two electrons give the oxygen molecule a permanent magnetic moment. When an oxygen molecule is placed near a magnetic field, the molecule is drawn to the field and the magnetic moments of the electrons become aligned with it. This striking phenomenon was first discovered by Faraday and forms the basis of the paramagnetic method for measuring oxygen concentrations (EPA-81/09). Types of paramagnetic oxygen analyzers include:
(1) Magnetic wind instrument
(2) Magneto-dynamic instrument
(3) Magnetopneumatic instrument

Parameter:
(1) A quantity in an equation which must be specified beside the independent variables to obtain the solution for the dependent variables.
(2) In quality control, a constant or coefficient that describes some characteristics of population (e.g., standard deviation, mean, regression coefficient (EPA-84/03).

Parametric method: Classical statistical methods which are effective for samples taken from normally distributed populations (EPA-87/10).

Paraquat ($CH_3(C_5H_4N)_2CH_3 \cdot 2CH_3SO_4$): A standard herbicide used to kill various types of crops, including marijuana (EPA-89/12).

Parasitic organism: An organism that lives in or on another species (host) (cf. obligate parasite).

Parent company: A company that owns or controls another company (40 CFR 704.3-91).

Parent corporation: A corporation which directly owns at least 50 percent of the voting stock of the corporation which is the facility owner or operator; the latter corporation is deemed a "subsidiary" of the parent corporation (40 CFR 144.61-c-91).

Paresis: Incomplete paralysis (LBL-76/07-bio).

Parke's process: A process in which zinc is added to molten lead to form insoluble zinc-gold and zinc-silver compounds. The compounds are skimmed and the zinc is removed through vacuum de-zincing (EPA-83/03a).

Parison: A preliminary shape or blank from which a glass article is to be formed (EPA-83).

Parison mold: *See* synonym, blank mold

Parshall flume: A calibrated device developed by Parshall for measuring the flow of liquid in an open conduit. It consists essentially of a contracting length, a throat, and an expanding length. At the throat is a sill over which the flow passes at critical depth. The upper and lower heads are each measured at a definite distance from the sill. The lower head cannot be measured unless the sill is submerged more than about 67 percent (EPA-82/11e).

Part A or part B permit: *see* definition under interim (permit) Status.

Partial closure: The closure of a hazardous waste management unit in accordance with the applicable closure requirements of Parts 264 and 265 of this Chapter at a facility that contains other active hazardous waste management units. For example, partial closure may include the closure of a tank (including its associated piping and underlying containment systems), landfill cell, surface impoundment, waste pile, or other hazardous waste management unit, while other units of the same facility continue to operate (*see* closure for more related terms) (40 CFR 260.10-91).

Partial molal free energy: Free energy of a system expressed as the summation of component free energies (LBL-76/07).

Partial pressure: The pressure of one gas component in a mixture of gaseous components, if it alone occupies the mixture's container. If the gas mixture behaves approximately as an ideal gas, the gas mixture exerts a total pressure in the container which is equal to the sum of the individual partial pressures (cf. partial pressure under mixture property and *see* pressure for more related terms) (EPA-9/84).

Partial recycle caustic system: Any tank or tub cleaning operation which recycles a primary caustic rinse and uses only a portion of secondary water rinse as make-up water for the caustic (*see* caustic system for more related terms) (EPA-79/12b).

Participating PRP: Any potentially responsible party (PRP) who has agreed, pursuant to 40 CFR 304.21 of this part, to submit one or more issues arising in an EPA claim for resolution pursuant to the procedures established by this part (40 CFR 304.12-91).

Participation rate:
(1) The portion of a population participating in a recycling program (OTA-89/10).

(2) A measure of the number of people participating in a recycling program compared to the total number that could be participating (EPA-89/11).

Particle (or particulate): Fine liquid or solid particles such as dust, smoke, mist, fumes, or smog, found in air or emissions (EPA-89/12). Other particle related terms include:
- Alpha particle
- Beta particle

Particle collection mechanisms for wet scrubbing systems: Wet scrubbers capture relatively small dust particles with large liquid droplets. Droplets are produced by injecting liquid at high pressure through specially designed nozzles, by aspirating the particle-laden gas stream through a liquid pool, or by submerging a whirling rotor in a liquid pool. These droplets collect particles by using one or more of several collection mechanisms. These mechanisms include (impaction and diffusion are the two primary ones):
(1) **Impaction:** Particles are too large to follow gas streamlines around a droplet collide with it.
(2) **Diffusion:** Very tiny particles move randomly, colliding with droplets because they are confined in a limited space.
(3) **Direct interception:** An extension of the impaction mechanism. The center of a particle follows the streamlines around the droplet, but a collision occurs if the distance between the particle and droplet is less than the radius of the particle.
(4) **Electrostatic attraction:** Particles and droplets become oppositely charged and attract each other.
(5) **Condensation:** When hot gas cools rapidly, particles in the gas stream can act as condensation nuclei and, as a result, become larger.
(6) **Centrifugal force:** The shape or curvature of a collector causes the gas stream to rotate in a spiral motion, throwing larger particles toward the wall.
(7) **Gravity:** Large particles moving slowly enough will fall from the gas stream and be collected (EPA-84/03b, p1-4)

Particle collector: *See* definition under particulate matter control in the term of air pollution control equipment.

Particle concentration: The number of particles per unit volume of air or other gas. (Note: on expressing particle concentration, the method of determining the concentration should be stated) (EPA-83/06).

Particle fall: A measurement of air contamination consisting of the mass rate at which solid particles deposit from the atmosphere. A term used in the same sense as the terms Dust Fall and Soot Fall but without any implication as to nature and source of the particles (EPA-83/06).

Particle size: An expression for the size of liquid or solid particle usually expressed in microns (EPA-89/03b).

Particle size analysis: The determination of the various amounts of the different particle sizes in a soil sample (i.e., sand, silt, clay) usually by sedimentation, sieving, micrometry, or combinations of these methods. The names and size limits of these particles as widely used in the United States are set forth in paragraph (a)(2)(ii) of this section (40 CFR 796.2700-91, *see also* 40 CFR 796.2750-

91).

Particle size discretization: A representation of a continuous range of particle sizes by a finite set of size channels; each channel is represented by a single characteristic size (EPA-88/09a).

Particle size distribution:
(1) The relative percentage by weight or number of each of the different size fractions of particulate matter (EPA-83/06).
(2) Distribution of particles of different sizes within a matrix of aerosols; numbers of particles of specified sizes or size ranges, usually in micrometers (EPA-89/03b).

Particle size measurement: Particle size measurement related terms include:
- Anderson sampler
- Cascade impactor

Particulate: *See* synonym, particle. Other particulate related terms include:
- Airborne particulate
- Total suspended particulate

Particulate asbestos material: Finely divided particles of asbestos or material containing asbestos (40 CFR 61.141-91).

Particulate emission: *See* synonym, PM emission.

Particulate filter respirator: An air purifying respirator, commonly referred to as a dust or a fume respirator, which removes most of the dust or fume from the air passing through the device (29 CFR 1910.94a-91).

Particulate loading: The mass of particulates per unit volume of air or water (*see* loading for more related terms) (EPA-89/12).

Particulate matter: *See* definition under PM.

Particulate matter emission: *See* definition under PM emission.

Particulate matter, 10 micron (PM10): *See* definition under PM10.

Parting line: Line or seam resulting from joint of two mold parts (EPA-83).

Parts per million: *See* definition under ppm.

Pascal (Pa): The standard international unit of vapor pressure and is defined as newtons per square meter (N/m^2). A newton is the force necessary to give acceleration of one meter per second squared to one kilogram of mass (40 CFR 796.1950-91).

Pass through: A discharge which exits the POTW into waters of the United States in quantities or concentrations which, alone or in conjunction with a discharge or discharges from other sources, is a cause of a violation of any requirement of the POTW's NPDES permit (including an increase in the magnitude or duration of a violation) (40 CFR 403.3-91).

Pass through pollutant: The pollutants that are not readily biodegradable and pass through biological oxidation treatment systems (*see* pollutant for more related terms) (EPA-85/10).

Passivation: The changing of the chemically active surface of a metal to a much less reactive state by means of an acid dip (EPA-83/06a).

Passivation film: A layer of oxide or other chemical compound of a metal on its surface that acts as a protective barrier against corrosion or further chemical reaction (EPA-86/12).

Passive institutional control:
(1) Permanent markers placed at a disposal site;
(2) Public records and archives;
(3) Government ownership and regulations regarding land or resource use; and
(4) Other methods of preserving knowledge about the location, design, and contents of a disposal system (40 CFR 191.12-91).

Paste mixing facility: The facility including lead oxide storage, conveying, weighing, metering, and charging operations; paste blending, handling, and cooling operations; and plate pasting, takeoff, cooling, and drying operations (40 CFR 60.371-91).

Pasteurize: To heat a substance to a temperature and for a time necessary to destroy pathogens but not necessarily all microbial life. Pasteurization is a form of disinfection (cf. disinfect or sterilize) (EPA-83).

Pastures crops: Crops such as legumes, grasses, grain stubble and stover which are consumed by animals while grazing (40 CFR 257.3.5-91).

Patent: An official document issued by the U.S. Office of Patents conferring an exclusive right or privilege to produce, use, or sell a pesticide for a specified period of time.

Path function: A thermodynamic function whose results depend on the path or the process between the two states. Heat and work are path functions. They can be calculated, only when the path of the process is known (*see* thermodynamic process for more related terms) (Jones-p50; Wark-p5).

Path length: The depth of effluent in the light beam between the receiver and the transmitter of a single-pass transmissometer, or the depth of effluent between the transceiver and reflector of a double-pass transmissometer. Two path lengths are referenced by this specification as follows:
(1) **Monitor path length:** The path length (depth of effluent) at the installed location of the CEMS.
(2) **Emission outlet path length:** The path length (depth of effluent) at the location where emissions are released to the atmosphere. For noncircular outlets, $D(e) = (2LW)/(L + W)$, where L is the length out the outlet and W is the width of the outlet. Note that this definition does not apply to pressure baghouse outlets with multiple stacks, side discharge vents, ridge roof monitors, etc (40 CFR 60-App/B-91).

Pathogen:
(1) Microorganisms that can cause disease in other organisms or

in humans, animals and plants. They may be bacteria, viruses, or parasites and are found in sewage, in runoff from animal farms or rural areas populated with domestic and/or wild animals, and in water used for swimming. Fish and shellfish contaminated by pathogens, or the contaminated water itself, can cause serious illnesses (EPA-89/12).
(2) Any microorganism capable of causing disease (EPA-5/90).

Pathogenic: Capable of causing disease (EPA-89/12).

Pathogenic bacteria: Bacteria inimical to man's welfare (LBL-76/07-water).

Pathogenicity: The capability of an infectious agent to cause disease in a susceptible host (EPA-5/90).

Pathological: Relating to the study of the essential nature of disease and generally altered or caused by disease.

Pathological waste:
(1) Waste material capable of causing disease.
(2) Waste material consisting of anatomical parts (cf. waste, infectious) (EPA-89/03b).
(3) *see* medical waste for more related terms.

Pathology: The study of disease.

Pathway of dispersion: The mode (water, groundwater, soil, and air) by which a chemical moves through the environment (Course 165.5).

PCB: Polychlorinated biphenyls (PCBs)
(1) A mixture of compounds composed of the biphenyl molecule which has been chlorinated to varying degrees (40 CFR 129.4-91, *see also* 40 CFR 268.2; 704.43; 761.3; 761.123-91).
(2) Chlorinated aromatic compounds that were banned in the 1970s because of their toxicity to laboratory animals and long-term stability in the environment (ETI-92).
(3) A group of toxic, persistent chemicals used in transformers and capacitors for insulating purposes and in gas pipeline systems as a lubricant. Further sale of new use was banned by law in 1979 (EPA-89/12).
PCB is a colorless liquid, used as an insulating fluid in electrical equipment. (The future use of PCB for new transformers was banned by the Toxic Substances control Act of October 1976). In the environment, PCBs exhibit many of the same characteristics as DDT and may, therefore, be confused with that pesticide. PCBs are highly toxic to aquatic life, they persist in the environment for long periods of time, and they are biologically accumulative. (*see* di-n-octyl-phthalate which replaces PCB as a dielectric fluid) (EPA-74/11; 83/03). Other PCB related terms include:
- Commercial storer of PCB waste
- Disposer of PCB waste
- Generator of PCB waste
- High concentration PCB
- Low concentration PCB
- Recycled PCB

PCB article: Any manufactured article, other than a PCB container, that contains PCBs and whose surface(s) has been in direct contact with PCBs. "PCB article" includes capacitors, transformers, electric motors, pumps, pipes and any other manufactured item:
(1) Which is formed to a specific shape or design during manufacture,
(2) Which has end use function(s) dependent in whole or in part upon its shape or design during end use, and
(3) Which has either no change of chemical composition during its end use or only those changes of composition which have no commercial purpose separate from that of the PCB Article (40 CFR 761.3-91).

PCB article container: Any package, can, bottle, bag, barrel, drum, tank, or other device used to contain PCB Articles or PCB Equipment, and whose surface(s) has not been in direct contact with PCBs (40 CFR 761.3-91).

PCB container: Any package, can, bottle, bag, barrel, drum, tank, or other device that contains PCBs or PCB Articles and whose surface(s) has been in direct contact with PCBs (40 CFR 761.3-91).

PCB contaminated electrical equipment: Any electrical equipment, including but not limited to transformers (including those used in railway locomotives and self-propelled cars), capacitors, circuit breakers, reclosers, voltage regulators, switches (including sectionalizers and motor starters), electromagnets, and cable, that contain 50 ppm or greater PCB, but less than 500 ppm PCB.
Oil filled electrical equipment other than circuit breakers, reclosers, and cable whose PCB concentration is unknown must be assumed to be PCB-Contaminated Electrical Equipment. (*see* 40 CFR 761.30 (a) and (h) for provisions permitting reclassification of electrical equipment containing 500 ppm or greater PCBs to PCB-Contaminated Electrical Equipment) (40 CFR 761.3-91).

PCB equipment: Any manufactured item, other than a PCB Container or a PCB Article Container, which contains a PCB Article or other PCB Equipment, and includes microwave ovens, electronic equipment, and fluorescent light ballasts and fixtures (40 CFR 761.3-91).

PCB item: Any PCB Article, PCB Article Container, PCB Container, or PCB Equipment, that deliberately or unintentionally contains or has a part of it any PCB or PCBs (40 CFR 761.3-91).

PCB manufacturer: A manufacturer who produces polychlorinated biphenyls (40 CFR 129.105-91).

PCB transformer: Any transformer that contains 500 ppm PCB or greater (cf. non-PCB transformer) (*see* transformer for more related terms) (40 CFR 761.3-91).

PCB waste: Those PCBs and PCB Items that are subject to the disposal requirements of subpart D of this part (*see* waste for more related terms) (40 CFR 761.3-91).

Peak air flow: The maximum engine intake mass air flow rate measure during the 195 second to 202 second time interval of the Federal Test Procedure (40 CFR 85.2122(a)(15)(ii)(D)-91).

Peak capability: *See* synonym, firm capacity.

Peak demand: *See* synonym, maximum electric demand.

Peak load:
(1) Hundred (100) percent of the manufacturer's design capacity of the gas turbine at ISO standard day conditions (*see* load for more related terms) (40 CFR 60.331-91).
(2) *See* synonym, maximum electric demand.

Peak load plant (or peak load station): An electrical generating facility operated only during periods at maximum demand (EPA-82/11f).

Peak load station: *See* synonym, peak load plant.

Peak per cent losses: The difference between the power input and output, as a result of losses due to the transfer of power between two or more points on a system at the time of maximum load, divided by the power input (*see* electric loss for more related terms) (EPA-83).

Peak spectral response: The wavelength of maximum sensitivity of the transmissometer (40 CFR 60-App/B-91).

Peak torque speed: The speed at which an engine develops maximum torque (40 CFR 86.082.2-91).

Pearl Benson Index: A measure of color producing substances (EPA-74/04).

Peat: Partially decomposed organic materials (SW-108ts).

Peel apart film article: A self developing photographic film article consisting of a positive image receiving sheet, a light sensitive negative sheet, and a sealed reagent pod containing a developer reagent and designed so that all the chemical substances required to develop or process the film will not remain sealed within the article during and after the development of the film (40 CFR 723.175-91).

Peep door: A small door or hole in an incinerator through which combustion can be observed (SW-108ts).

Peephole: A small opening in a furnace wall for observation (EPA-83).

Pelagic zone: The free-water region of a sea. (Pelagic refers to the sea, and limnetic refers to bodies of fresh water (LBL-76/07-water).

Pellet burning wood heater: A wood heater which meets the following criteria:
(1) The manufacturer makes no reference to burning cord wood in advertising or other literature;
(2) The unit is safety listed for pellet fuel only;
(3) The unit operating and instruction manual must state that the use of cordwood is prohibited by law; and
(4) The unit must be manufactured and sold including the hopper and auger combination as integral parts (40 CFR 60-App/A(method 28)-91).

Pelletized: An agglomeration process in which an unbaked pellet is heat hardened. The pellets increase the reduction rate in a blast furnace by improving permeability and gas-solid contact (EPA-83/03a).

Penetration:
(1) A fraction of suspended particulate that passes through a collection device (EPA-89/03b).
(2) The flow of a hazardous liquid chemical through zippers, stitched seams, and pinholes or other imperfections in a protective clothing material (NIOSH-84/10).

Pentachlorophenol: A chlorinated phenol with the formula C_6Cl_5OH and formula weight of 266.35 that is used as a wood preservative. Commercial grades of this chemical are usually adulterated with tetrachlorophenol to improve its solubility (EPA-74/04).

Peptide: A chemical compound joined by a peptide bond (-CO-NH-).

Peptide bond: A chemical bond (-CO-NH-) formed by a carboxyl group and an amino acid (*see* chemical bond for more related terms).

Peptidoglycan: The rigid backbone of the bacteria cell wall consisting of two major sub-units, N-acetyl muramic acid and N-acetyl glucosamine, and a number of amino acids (EPA-88/09a).

Percability: Permeability (DOI-70/04).

Perched water: A water table, usually of limited area, maintained above the normal free-water elevation by the presence of an intervening, relatively impervious stratum (cf. perched water body and *see* water for more related terms) (SW-108ts).

Perched water body: A suspended, isolated body of ground-water occurring in a saturated zone, and separated from the main body of groundwater by unsaturated, impermeable rock. The isolated body has its own local water table, a perched water table, below which shallow wells can obtain water (DOI-70/04).

Perched water table: The top of a zone of saturation that lies on an impermeable horizon above the level of the general water table in the area. It is generally near the surface and frequently supplies a hillside spring (EPA-83).

Perchlorate: A compound (salt) containing the $(ClO_4)^-$ radical (e.g., sodium perchlorate $NaClO_4$).

Perchlorate acid ($HClO_4$): A strong, corrosive oxidant used in electrolytic baths and oxidizers.

Percent absorption: 100 times the ratio between total excretion of radioactivity following oral or dermal administration and total excretion following intravenous administration of test substance (40 CFR 795.228-91).

Percent load: The fraction of the maximum available torque at a specified engine speed (40 CFR 86.082.2-91).

Percent removal: A percentage expression of the removal efficiency across a treatment plant for a given pollutant parameter, as determined from the 30-day average values of the raw wastewater influent pollutant concentrations to the facility and the 30-day average values of the effluent pollutant concentrations for a given time period (40 CFR 133.101-91).

Percent strength: For a gas mixture, it is the number of moles of the gas of interest divided by the total number of moles of gas; this will equal the partial pressure of the gas of interest divided by the total pressure of the gases. For a liquid solution, the percentage concentration is the number of grams of compound per 100 grams of solution (EPA-88/09a).

Percentage of completion method: Refers to a system under which payments are made for construction work according to the percentage of completion of the work, rather than to the grantee's cost incurred (40 CFR 31.3-91).

Percentage reduction: The ratio of material removed from water or sewage by treatment to the material originally present (expressed as a percentage) (EPA-74/01a).

Perceptible leaks: Any petroleum solvent vapor or liquid leaks that are conspicuous from visual observation or that bubble after application of a soap solution, such as pools or droplets of liquid, open containers or solvent, or solvent laden waste standing open to the atmosphere (40 CFR 60.621-91).

Perched water: A water table, usually of limited area, maintained above the normal free-water elevation by the presence of an intervening, relatively impervious stratum (*see* water for more related terms) (SW-108ts).

Perched water body: A suspended, isolated body of ground-water occurring in a saturated zone, and separated from the main body of groundwater by unsaturated, impermeable rock. The isolated body has its own local water table, a perched water table, below which shallow wells can obtain water (*see* water for more related terms) (DOI-4/70).

Perched water table: The top of a zone of saturation that lies on an impermeable horizon above the level of the general water table in the area. It is generally near the surface and frequently supplies a hillside spring (*see* water for more related terms) (EPA-83).

Percolate: To ooze or trickle through a permeable substance. Ground water may percolate into the bottom of an unlined landfill (cf. leachate) (EPA-89/11).

Percolating filter: A trickling filter (*see* filter for more related terms).

Percolation: The movement of water downward and radially through the subsurface soil layers, usually continuing downward to the ground water (cf. seepage) (EPA-89/12).

Perennial: A plant that can live for many years.

Perennial stream: A stream that carries water at all times (DOI-70/04).

Performance assessment: An analysis that:
(1) Identifies the processes and events that might affect the disposal system;
(2) Examines the effects of these processes and events on the performance of the disposal system; and
(3) Estimates the cumulative releases of radionuclides, considering the associated uncertainties, caused by all significant processes and events. These estimates shall be incorporated into an overall probability distribution of cumulative release to the extent practicable (29 CFR 191.12-91).

Performance audit (or performance evaluation audit): A quantitative evaluation of the measurement systems of a program. It requires testing the measurement systems with samples of known composition or behavior to evaluate precision and accuracy. The performance audit is carried out by or under the auspices of the QA Officer without the knowledge of the analysts. Since this is seldom achievable, many variations are used that increase the awareness of the analyst as to the nature of the audit material (*see* audit for more related terms) (EPA-85/08; EPA-86/10a).

Performance averaging period: 30 calendar days, one calendar month, or four consecutive weeks as specified in sections of this subpart (40 CFR 60.431-91).

Performance curve: A graph describing collection efficiency of a control device as a function of particle size (EPA-84/09).

Performance evaluation audit: *See* synonym, performance audit.

Performance evaluation sample: A reference sample provided to a laboratory for the purpose of demonstrating that the laboratory can successfully analyze the sample within limits of performance specified by the Agency. The true value of the concentration of the reference material is unknown to the laboratory at the time of the analysis (40 CFR 141.2-91).

Performance goal: A predetermined level of effectiveness that a treatment technology seeks to attain. Performance goals are set in terms of the percentage reduction in toxicity, mobility, or volume of a waste and its contaminants (EPA-89/12a).

Performance specification: A specification that states the desired operation or function of a product but does not specify the materials from which the product must be constructed (*see* specification for more related terms) (40 CFR 247.101-91).

Performance test: A test devised to permit rigorous observation and measurement of the performance of a unit of equipment or a system under prescribed operating conditions (EPA-83).

Periodic application of cover material: The application and compaction of soil and other suitable material over disposed solid waste at the end of each operating day or at such frequencies and in such a manner as to reduce the risk of fire and to impede vectors access to the waste (40 CFR 257.3.6-91, *see also* 40 CFR 257.3.8-91).

Periodic table: A chemical element table in which elements are organized in accordance with their atomic numbers. The table includes horizontal rows (periods) and vertical columns (groups). Elements in the same row or in the same column share some similar chemical properties.

Periodical monitoring: Monitoring with fixed time intervals between consecutive observations (*see* monitoring for more related terms).

Periphyton: Organisms attached to underwater surfaces (DOE-91/04).

Perlite composite board: The insulation board composed of expanded perlite and fibers formed into rigid, flat, rectangular units with a suitable sizing material incorporated in the product. It may have on one or both surfaces a facing or coating to prevent excessive hot bitumen strike-in during roofing installation (40 CFR 248.4-91).

Periphyton: The association of aquatic organisms attached or clinging to stems and leaves of rooted plants or other surfaces projecting above the bottom (LBL-76/07-water).

Perimeter drain: *See* synonym, French drain.

Periphyton: Organisms (including both plants and animals) that commonly grow on submerged surfaces such as stones, wood, aquatic plants, or other objects, forming more or less continuous slimy or woolly felted coatings on these objects (DOI-70/04).

Permanent expansion (or secondary expansion): The ability of some refractories to increase in size permanently at temperature within their useful range (SW-108ts).

Permanent hardness: The none-carbonate hardness such as calcium sulphate ($CaSO_4$) and calcium fluoride (CaF_2). These compounds can not be removed by boiling (*see* hardness for more related terms).

Permanent opening: An opening designed into an enclosure to allow tire components to pass through the enclosure by conveyor or other mechanical means, to provide access for permanent mechanical or electrical equipment, or to direct air flow into the enclosure. A permanent opening is not equipped with a door or other means of obstruction of air flow (40 CFR 60.541-91).

Permanent storage capacity: The grain storage capacity which is inside a building, bin, or silo (40 CFR 60.301-91).

Permanganate number (K NO.): A method (T-214-TAPPI Std.) used to determine the relative hardness or bleach requirements (bleach ability) of paper pulp. By definition, it is the number of milliliters of 0.1 N potassium permanganate solution absorbed by 1 gram of moisture-free pulp under specified control conditions (EPA-87/10).

Permeability:
(1) The rate at which liquids pass through soil or other materials in a specified direction (EPA-89/12).
(2) The capacity of a porous medium to conduct or transmit fluids (SW-108ts).
(3) In ground water, a qualitative sense, while hydraulic conductivity which considers both permeability and hydraulic head is a quantitative term. They refer to the degree of ease with which water can pass through a rock unit. They often are expressed in units of gallons per day per square foot (cf. hydraulic conductivity) (EPA-87/03).

Permeability coefficient: A water flow rate in which the amount of water passes through a cross section of 1 square foot under a unit hydraulic gradient at 60° F temperature.

Permeable: Having pores or openings that permit liquids or gases to pass through (EPA-89/11).

Permeable rock (or pervious rock): A rock, either porous (such as sandstone) or fissured, that allows water to soak into it and pass through it freely (DOI-70/04).

Permeation: The process by which a hazardous liquid chemical moves through a protective clothing material on a molecular level. Permeation involves: (1) sorption of molecules of the liquid into the contacted (outside) surface of a material; (2) diffusion of the sorbed molecules in the material; and (3) desorption of the molecules from the opposite (inside) surface of the material into the collecting medium (NIOSH-84/10).

Permissible dose: The dose of a chemical that may be received by an individual without the expectation of significantly harmful result (Course 165.6).

Permissible exposure limit (PEL): The employer shall ensure that no employee is exposed to an airborne concentration in excess of 0.2 fiber per cubic centimeter of air as an 8-hour time-weighted average (TWA), as determined by the method described in Appendix A of this section or by an equivalent method (40 CFR 763.121-91).

Permit: An authorization, license, or equivalent control document issued by EPA or an approved state agency to implement the requirements of an environmental regulation; e.g., a permit to operate a wastewater treatment plant or to operate a facility that may generate harmful emissions (EPA-89/12, *see also* SMCRA701; 40 CFR 2.309; 21.2; 22.03; 122.2; 124.2; 124.41; 129.2; 144.3; 146.3; 147.2902; 232.2; 256.06; 403.3; 501.2-91).

40 CFR 124.1 describes the general permit procedures for issuing, modifying, revoking and reissuing or terminating all RCRA, UIC, PSD, and NPDES permits. Other permit related terms include:
- Draft permit
- Emergency permit
- General permit
- Proposed permit

Permit applicant or applicant: A person applying for a permit (SMCRA701-30USC1291-90).

Permit area: The area of land specified or referred to in an NPDES permit in which active mining and related activities may occur that result in the discharge regulated under the terms of the permit. Usually this is specifically delineated in an NPDES permit or permit application, but in other cases may be ascertainable from an Alaska Tri-agency permit application or similar document specifying the mine location, mining plan and similar data (40 CFR 440.141-91, *see also* SMCRA701).

Permit averaging period: The duration of time over which a permit limit is calculated - day(s), week, or month (EPA-85/09; 91/03).

Permit-by-rule: A provision of these regulations stating that a facility or activity is deemed to have a RCRA permit if it meets the requirements of the provision (40 CFR 270.2-91).

Permit for treatment, storage, or disposal of hazardous waste: In RCRA, the permit procedure includes (RCRA Sec. 3005; EPA-1/89):
(1) **Part A:** The first part of the two part permit application that must be submitted by a TSD (treatment, storage, and disposal) facility to receive a permit. It contains general facility information (such as the name and location of the facility, its owner, the type of waste accepted, the maximum capacity, and other environmental permits that the facility has obtained). There is a standard form for the Part A (EPA-86/01).
(2) **Part B:** Trial burn is probably the most important and difficult part of entire permit application. It includes detailed and highly technical information concerning the TSD in question. There is no standard form for the Part B. *see* six (6) Hazardous Waste Incineration Guidance Series for permit procedures.

Permit or license applicant: When used with respect to an action of a Federal agency for which exemption is sought under section 1536 of this title, any person whose application to such agency for a permit or license has been denied primarily because of the application of section 1536(a) of this title to such agency action (ESA3-16USC1531-90).

Permit or PSD permit: A permit issued under 40 CFR 52.21 or by an approved State (40 CFR 124.41-91).

Permittee: A person holding a permit (SMCRA701, *see also* OPA1001; 40 CFR 172.1-91).

Permitting authority: The Administrator or the air pollution control agency authorized by the Administrator to carry out a permit program under this title (CAA501, *see also* CAA402; 40 CFR 230.3-91).

Peroxide:
(1) A chemical compound which contains the peroxy -O-O- radical (e.g., hydrogen peroxide H_2O_2).
(2) A chemical used in bleaching of wood pulps, usually groundwood pulps (EPA-87/10).

Peroxyacetyl nitrate (PAN): A pollutant created by the action of sunlight on hydrocarbons and nitrogen oxides in the air. PANs are an integral part of photochemical smog (EPA-84/09).

Persistence: Refers
(1) To the length of time a compound, once introduced into the environment, stays there. A compound may persist for less than a second or indefinitely (EPA-89/12).
(2) In meteorology, the continuation of the same situation during a specified period of time. For a wind direction, it is also called steadiness or constancy (NATO-78/10).

Persistent chemical: A substance which resists biodegradation and/or chemical oxidation when released into the environment and tends to accumulate on land, in air, in water, or in organic matter (Course 165.5).

Persistent pesticides: Pesticides that do not break down chemically or break down very slowly and that remain in the environment after a growing season (EPA-89/12).

Persistent pollutant: A pollutant which is not subject to decay, degradation, transformation, volatilization, hydrolysis, or photolysis (*see* pollutant for more related terms) (EPA-91/03).

Person rem: The unit of collective radiation dose commitment to a given population; the sum of the individual doses received by a population segment (DOE-91/04).

Personal protective equipment: Any chemical protective clothing or device placed on the body to prevent contact with, and exposure to, an identified chemical substance or substances in the work area. Examples include, but are not limited to, chemical protective clothing, aprons, hoods, chemical goggles, face splash shields, or equivalent eye protection, and various types of respirators. Barrier creams are not included in this definition (40 CFR 721.3-91).

Persulfate: A strong oxidizing agent used to regenerate ferrocyanide to ferricyanide in bleaches (EPA-80/10).

Pest: An insect, rodent, nematode, fungus, weed or other form of terrestrial or aquatic plant or animal life or virus, bacterial or microorganism that is injurious to health or the environment (EPA-89/12, *see also* FIFRA2; 40 CFR 455.10-91).

Pest problem:
(1) A pest infestation and its consequences; or
(2) Any condition for which the use of plant regulators, defoliants, or desiccants would be appropriate (40 CFR 162.151-91).

Pesticidal product report: The information showing the types and amounts of pesticidal products which were:
(1) Produced in the past calendar year;
(2) Produced in the current calendar year; and
(3) Sold or distributed in the past calendar year.
For active ingredients, the pesticidal product report must include information on the types and amounts of an active ingredient for which there is actual or constructive knowledge of its use or intended use as a pesticide. This pesticidal product report also pertains to those products produced for export only which must also be reported. A positive or a negative annual report is required in order to maintain registration for the establishment (40 CFR 167.3-91).

Pesticidal report: A pesticide, active ingredient, or device (40 CFR 167.3-91).

Pesticide:
(1) Any substance or mixture of substances intended for preventing, destroying, repelling, or mitigating any pest; or
(2) Any substance or mixture of substances intended for use as a plant regulator, defoliant, or desiccant (40 CFR 165.1-91, *see also* FIFRA2; 40 CFR 125.58; 152.3; 455.10; 710.2; 720.3-91; EPA-89/12). Pesticide includes:
  (A) **Excess pesticide:** All pesticides which cannot be legally sold pursuant to the Act or which are to be discarded (40 CFR 165.1-91).
  (B) **Organic pesticide:** Carbon-containing substances used as pesticides, excluding metallo-organic compounds (40 CFR 165.1-91).
  (C) **Inorganic pesticide:** Non-carbon-containing substances used as pesticides (40 CFR 165.1-91).
  (D) **Metallo organic pesticide:** A class of organic pesticides containing one or more metal or metalloid atoms in the structure (40 CFR 165.1-91).

Other pesticide related terms include:
- Botanical pesticide
- Contact insecticide (*see* synonym, contact pesticide)
- Contact pesticide
- Restricted use pesticide
- Restricted use pesticide retail dealer
- Selective pesticide
- Systemic pesticide

Pesticide chemical: Any substance which alone, or in chemical combination with or in formulation with one or more other substances, is a pesticide within the meaning of FIFRA and which is used in the production, storage, or transportation of any raw agricultural commodity or processed food. The term includes any substance that is an active ingredient, intentionally added inert ingredient, or impurity of such a pesticide (40 CFR 177.3-91, *see also* 40 CFR 163.2; 180.1; 455.10-91).

Pesticide incinerator: Any installation capable of the controlled combustion of pesticides, at a temperature of 1000° C (1832° F) for two second dwell time in the combustion zone, or lower temperatures and related dwell times that will assure complete conversion of the specific pesticide to inorganic gases and solid ash residues. Such installation complies with the Agency Guidelines for the Thermal Processing of Solid Wastes as prescribed in 40 CFR 240 (*see* incinerator for more related terms) (40 CFR 165.1-91).

Pesticide product: A pesticide in the particular form (including composition, packaging, and labeling) in which the pesticide is, or is intended to be, distributed or sold. The term includes any physical apparatus used to deliver or apply the pesticide if distributed or sold with the pesticide (40 CFR 152.3-91, *see also* 40 CFR 162.151-91).

Pesticide related waste: All pesticide-containing wastes or by-products which are produced in the manufacturing or processing of a pesticide and which are to b discarded, but which, pursuant to acceptable pesticide manufacturing or processing operations, are not ordinarily a part of or contained within an industrial waste stream discharged into a sewer or the waters of a state (40 CFR 165.1-91).

Pesticide residue: A residue of a pesticide chemical or of any metabolite or degradation product of a pesticide chemical (40 CFR 177.3-91).

Pesticide safety team network (PSTN): PSTN is operated by the National Agricultural Chemicals Association to minimize environmental damage and injury arising from accidental pesticide spills or leaks. PSTN area coordinators in ten regions nationwide are available 24 hours a day to receive pesticide incident notifications from CHEMTREC (NRT-87/03).

Pesticide solid waste: *See* synonym, pesticide waste.

Pesticide tolerance: The amount of pesticide residue allowed by law to remain in or on a harvested crop. By using various safety factors, EPA sets these levels well below the point where the chemicals might be harmful to consumers (EPA-89/12).

Pesticide use: A use of a pesticide (described in terms of the application site and other applicable identifying factors) that i included in the labeling of a pesticide product which is registered, or for which an application for registration is pending, and the terms and conditions (or proposed terms and conditions) of registration for the use (40 CFR 154.3-91).

Pesticide waste (or pesticide solid waste): The residue resulting from the manufacturing, handling, or use of chemicals for killing plants and animals pests (*see* waste for more related terms) (SW-108ts).

Pesticide related wastes: All pesticide-containing wastes or by-products which are produced in the manufacturing or processing of a pesticide and which are to b discarded, but which, pursuant to acceptable pesticide manufacturing or processing operations, are not ordinarily a part of or contained within an industrial waste stream discharged into a sewer or the waters of a state (*see* waste for more related terms) (40 CFR 165.1-91).

Petrochemical: Chemicals made from petroleum-derived feedstocks.

Petrochemical operations: The production of second-generation petrochemicals (i.e., alcohols, ketones, cumene, styrene, etc.) or first generation petrochemicals and isomerization products (i.e. BTX, olefins, cyclohexane, etc.) when 15 percent or more of refinery production is as first-generation petrochemicals and isomerization products (40 CFR 419.31-91).

Petrochemistry: A science studying the chemical reactions associated with petroleum, natural gas or asphalt deposits.

Petroleum: The crude oil removed from the earth and the oils derived from tar sands, shale, and coal (40 CFR 52.741-91, *see also* RCRA9001; 40 CFR 60.101; 60.111; 60.111a; 60.111b; 60.591; 60.691; 61.341-91).

Petroleum coke: The coke formed by the destructive distillation of petroleum. Like foundry coke, petroleum coke can also be used for making cast iron in a cupola (*see* coke for more related terms) (EPA-85/10a).

Petroleum dry cleaner: A dry cleaning facility that uses petroleum solvent in a combination of washers, dryers, filters, stills, and settling tanks (40 CFR 60.621-91).

Petroleum liquid: Petroleum, condensate and any finished or intermediate products manufactured in a petroleum refinery (*see* liquid for more related terms) (40 CFR 60.111b-91, *see also* 40 CFR 60.111; 60.111a-91).

Petroleum marketing facilities: Include all facilities at which petroleum is produced or refined and all facilities from which petroleum is sold or transferred to other petroleum marketers or to the public (40 CFR 280.92-91).

Petroleum marketing firms: All firms owning petroleum marketing facilities. Firms owning other types of facilities with USTs as well as petroleum marketing facilities are considered to be petroleum marketing firms (40 CFR 280.92-91).

Petroleum refinery: Any facility engaged in producing gasoline, kerosene, distillate fuel oils, residual fuel oils, lubricants, or other products through distillation of petroleum, or through redistillation, cracking, or reforming of unfinished petroleum derivatives (40 CFR 52.741-91, *see also* 40 CFR 60.41b; 60.101; 60.111; 60.111a; 60.591; 60.691; 61.341-91).

Petroleum UST system: An underground storage tank system that contains petroleum or a mixture or petroleum with de minimis quantities of other regulated substances. Such systems include those containing motor fuels, jet fuels, distillate fuel oils, residual fuel oils, lubricants, petroleum solvents, and used oils (40 CFR 280.12-91).

pH:
(1) The negative logarithm of the hydrogen ion concentration (40 CFR 131.35-91, *see also* 40 CFR 257.3.5; 420.02; 796.2750-91; EPA-89/12).
(2) A measure of the acidity (0, most acid) or alkalinity (14, most alkaline) of a liquid or solid on a scale of 0 to 14 as follows: $pH = -\log (H^+)$.
(3) **Acid:** Gases or liquids with pH < 7, i.e., H^+ donors.
(4) **Base:** Gases or liquids with pH > 7, i.e., H^+ acceptors.
(5) **Neutral:** pH = 7.

pH shows hydrogen ion concentration which reflects the balance between acids and alkalis. The pH of most natural waters falls within the range 4 to 9. A pH of 7.0 indicates neutral water. A 6.5 reading is slightly acid; and 8.5 reading is alkaline. Slight decrease in pH may greatly increase the toxicity of substances such as cyanides, sulfides, and most metals. Slight increase in pH may greatly increase the toxicity of pollutants such as ammonia. Alkaline water will tend to form a scale; acid water is corrosive; good water should be nearly neutral (DOI-70/04).

pH adjustment: A means of maintaining the optimum pH through the use of chemical additives. Can be manual, automatic, or automatic with flow corrections (EPA-83/03).

pH buffer: A substance used to stabilize the acidity or alkalinity in a solution (EPA-83/06a).

pH electrode: Similar to the glass electrode structure, the pH electrode is used as the hydrogen-ion sensor of most pH meters (*see* electrode for more related terms).

pH modifier: An agent to control pH for proper functioning of a cationic or anionic flotation. Modifying agents used include soda ash, sodium hydroxide, sodium silicate, sodium phosphates, lime, sulfuric acid, and hydrofluoric acid (EPA-82/02).

Pharmaceutical: Any compound or mixture, other than food, used in the prevention, diagnosis, alleviation, treatment, or cure of disease in man and animal (40 CFR 52.741-91).

Pharmaceutical coating operation: A device in which a coating is applied to a pharmaceutical, including air drying or curing of the coating (40 CFR 52.741-91).

Pharmacokinetics: The study of the rates of absorption, tissue distribution, biotransformation, and excretion (40 CFR 795.228-91, *see also* 40 CFR 795.231; 795.232-91).

Pharmacology: The science of medicine and drugs.

Phase:
(1) In physical chemistry, the uniform appearance of substances. Three phases, gas, liquid, and solid, usually can be identified from the appearance. A substance may exist in several phases:
- A pure solid phase known as ice
- A pure liquid phase
- A pure vapor phase known as steam
- An equilibrium mixture of liquid and vapor phases
- An equilibrium mixture of liquid and solid phases
- An equilibrium mixture of solid and vapor phases

(2) In electronics, one of the separate circuits or windings of a poly-phase system, machine or other apparatus (EPA-83/03).

Phase assembly: The coil-core assembly of a single phase of a transformer (EPA-83/03).

Phase diagram: A projection of a pressure-specific volume-temperature surface upon the pressure-temperature plane. The diagram shows more than one of the saturation lines (liquid-vapor, liquid-solid, etc.) of a pure substance (Jones-p115; Wark-p60).

Phase equilibria: The equilibrium relationships among various phases, e.g., gas, liquid or solid of substances under different conditions such as temperature or pressure.

Phase of super saturation: The metastable phase in which existing sugar crystals grow but new crystals do not form; the intermediate phase in which existing crystals grow and new crystals do form; and the labile phase in which new crystals form spontaneously without the presence of others (EPA-75/02d).

Phased disposal: A method of tailings management and disposal which uses lined impoundments which are filled and then immediately dried and covered to meet all applicable Federal standards (*see* disposal for more related terms) (40 CFR 61.251-91).

Phenol (C_6H_5OH):
(1) Total phenols as measured by the procedure listed in 40 CFR 136 (40 CFR 410.01-91).
(2) Organic compounds that are by products of petroleum refining, tanning, and textile, dye, and resin manufacturing. Low concentrations cause taste and odor problems in water; higher concentrations can kill aquatic life and humans (EPA-89/12).
(3) cf. total phenol.

Phenol 4AAP (or phenolic compound): The value obtained by the method specified in 40 CFR 136.3 (cf. four (4)-AAP colorimeteric method) (40 CFR 420.02-91).

Phenol coefficient: A device for indicating the efficiency of antiseptics.

Phenol formaldehyde resin: *See* synonym, phenolic resin.

Phenolic compound: *See* synonym, phenol 4AAP.

Phenolic insulation: The insulation made with phenolic plastics which are plastics based on resins made by the condensation of phenols, such as phenol or cresol, with aldehydes (40 CFR 248.4-91).

Phenolic resin (or phenol formaldehyde resin): A group of varied and versatile synthetic resins. They are made by reacting almost any phenolic and an aldehyde. In some cases, hexamethylenetetramine is added to increase the aldehyde content. Both types of materials are used separately or in combination in the blending of commercial molding materials. Due to the thermal degradation of phenolic resins that may occur during metal pouring, phenol and formaldehyde may be generated (*see* resin for more related terms) (EPA-85/10a).

Phenotypic expression time: A period during which unaltered gene products are depleted from newly mutated cells (40 CFR 798.5300-91).

Pheromone: Hormonal chemical produced by female of a species to attract a mate (EPA-89/12).

Phosphate: Certain chemical compounds containing phosphorus (EPA-89/12).

Phosphate coating: A process of forming a conversion coating on iron or steel by immersing in a hot solution of manganese, iron or zinc phosphate. Often used on a metal part prior to painting or porcelainizing (EPA-83/06a).

Phosphate glass: A glass in which the essential glass former is phosphorus pentoxide instead of silica (*see* glass for more related terms) (EPA-83).

Phosphate rock feed: All material entering the process unit including, moisture and extraneous material as well as the following ore minerals: Fluorapatite, hydroxylapatite, chlorapatite, and carbonateapatite (40 CFR 60.401-91).

Phosphate rock plant: Any plant which produces or prepares phosphate rock product by any or all of the following processes: Mining, beneficiation, crushing, screening, cleaning, drying, calcining, and grinding (40 CFR 60.401-91).

Phosphatizing: A process of forming rust-resistant coating on iron or steel by immersing in a hot solution of acid manganese, iron or zinc phosphates (EPA-83/06a).

Phosphodiester bond: In nucleic acid (DNA or RNA), the phosphorus bond between the 5 position of one sugar unit of a nucleotide and the 3 position of a sugar of a second nucleotide. Thus, Phosphodiester bonds are the bonds that form a single strand of nucleic acid (EPA-88/09a).

Phosphogypsum: The waste or other form of byproduct which results from the process of wet acid phosphorus production and which contains greater than [up to 10] pCi/g radium (40 CFR 61.201-91).

Phosphogypsum stacks or stacks:
(1) Piles of waste from phosphorus fertilizer production containing phosphogypsum. Stacks shall also include phosphate mines that are used for the disposal of phosphogypsum; or
(2) Piles of waste or other form of byproduct which results from wet acid phosphorus production containing phosphogypsum. Stacks shall also include phosphate mines that are used for the disposal of phosphogypsum (40 CFR 61.201-91).

Phosphor: A crystalline inorganic compound that produces light when excited by ultraviolet radiation (EPA-83/03).

Phosphorus (P): A nonmetallic element with a.n. 15; a.w. 30.9738; d. 1.82 g/cc; m.p. 44.2° C and b.p. 280° C. The element belongs to group VA of the periodic table. It is an essential chemical food element that can contribute to the eutrophication of lakes and other water bodies. Increased phosphorus levels result from discharge of phosphorus-containing materials into surface waters (EPA-89/12). Major phosphorous compounds include:

Phosphorus precipitation: The addition of the multivalent metallic ions of calcium, iron and aluminum to wastewater to form insoluble precipitates with phosphorus (EPA-87/10a).

Phosphorylation: Introduction of the phosphoryl group into an organic compound (LBL-76/07-bio).

Photic zone: The upper zone of a water body in which sufficient light is available for photosynthesis (cf. profundal zone) (DOI-70/04).

Photo resist:
(1) A light-sensitive coating that is applied to a substrate or board, exposed, and developed prior to chemical etching, the exposed areas serve as a mask for selective etching (EPA-83/03).
(2) Thin coatings produced from organic solutions which when exposed to light of the proper wave length are chemically changed in their solubility to certain solvents (developers). This substance is placed over a surface which is to be protected during processing such as in the etching of printer

circuit boards (EPA-83/06a).

Photochemical oxidant: Air pollutants formed by the action of sunlight on oxides of nitrogen and hydrocarbons (EPA-89/12).

Photochemical smog:
(1) Air pollution caused by chemical reactions of various pollutants emitted from different sources (EPA-89/12).
(2) A type of air pollution resulting from photochemical reactions (LBL-76/07-air).
(3) A mixture of secondary air pollutants including ozone, organic nitrates etc., which are produced from primary pollutants such as nitrogen oxides, hydrocarbons etc. by photochemical reactions (NATO-78/10).

Photochemically reactive solvent:
(1) Any solvent with an aggregate of more than 20 percent of its total volume composed of the chemical compounds classified below or which exceeds any of the following individual percentage composition limitations, as applied to the total volume of solvent.
(2) A combination of hydrocarbons, alcohols, aldehydes, esters, ethers, or ketones having an olefinic or cycloolefinic type of unsaturation: 5 percent.
(3) A combination of aromatic compounds with 8 or more carbon atoms to the molecule except ethylbenyene: 8 percent.
(4) A combination of ethylbenyene or ketones having branched hydrocarbon structures, trichloroethylene or toluene: 20 percent (40 CFR 52.1088-91, *see also* 40 CFR 52.1107; 52.2440-91).
(5) *See* solvent for more related terms.

Photochemistry: A science studying the light effects on chemical reactions.

Photoconductive effect: An increase in electrical conductivity of a semiconductor due to the incident of photons (cf. photoelectric effect).

Photodegradable plastic: A plastic that breaks down in the presence of ultraviolet (UV) light (OTA-89/10).

Photodegradation: Decomposition or break-down of a chemical compound by photo (radiant) energy (cf. biodegradation).

Photoelectric colorimetry: A device that uses a phototube or a photocell for the determination of color.

Photoelectric effect: The liberation of electric charges from a substance due to the irradiation of electromagnetic energy. Other phenomena related to the photoelectric effect include photoconductive effect and photovoltaic effect.

Photoelectron spectroscopy: A method for measuring the ionization potentials of molecules (*see* spectroscopy for more related terms).

Photographic article: Any article which will become a component of an instant photographic or peel apart film article (40 CFR 723.175-91).

Photographic paper: *See* synonym, sensitized paper (EPA-83).

Photoionization: Ionization of an atom or molecule due to the absorption of a photon of light.

Photoionization detector (PID): *See* definition under GC/FID.

Photolithography: The process by which a microscopic pattern is transferred from a photomask to a material layer (e.g., SiO) in an actual circuit (EPA-83/03).

Photoluminescence (or flame photometric analyzer): One of continuous emission monitors (*see* continuous emission monitor or luminescence analyzer for various types). Flame photometry is a branch of spectrochemical analysis in which a sample is excited to luminescence by introduction into a flame. It is primarily used in ambient air sampling, but has been applied to stationary source sampling by using sample dilution techniques (EPA-84/03a).

Photolysis: The chemical reaction of a substance caused by direct absorption of solar energy (direct photolysis) or caused by other substances that absorb solar energy (indirect photolysis) (40 CFR 300-App/A-91). It is a process which chemical bonds are broken under the influence of ultraviolet (UV) or visible light. Products of photo-degradation vary according to the matrix in which the process occurs, but the complete conversion of an organic contaminant to CO_2, H_2O, etc., is not probable (cf. direct photolysis).

Photomask: A film or glass negative that has many high-resolution images, used in the production of semiconductor devices and integrated circuits (EPA-83/03).

Photometer:
(1) A device having the capability of converting radiant energy into electrical energy (LBL-76/07-air).
(2) An instrument that measures the intensity of light or the degree of light absorption (LBL-76/07-water).

Photon: A quantum of electromagnetic energy (EPA-83/03).

Photoperiod: The light and dark periods in a 24 hour day. This is usually expressed in a form such as 17 hours light/7 hours dark or 17L/7D (40 CFR 797.2130-91, *see also* 40 CFR 797.2150-91).

Photoperiodism: The response of an organism to the light and dark periods of a day.

Photosensitive coating: A chemical layer that is receptive to the action of radiant energy (EPA-83/06a).

Photosynthesis:
(1) The manufacture by plants of carbohydrates and oxygen from carbon dioxide and water in the presence of chlorophyll, using sunlight as an energy source (EPA-89/12).
(2) The metabolic process by which simple sugars are manufactured from carbon dioxide and water by plant cells using light as an energy source (DOD-78/01).

Photovoltaic effect: The generation of a voltage between two different contact materials due to the incident of photons of light (cf. photoelectric effect).

Physical adsorption: An adsorption process in which the adsorbate is held together by weak physical forces (*see* adsorption for more related terms).

Physical and chemical treatment: The processes generally used in large scale wastewater treatment facilities. Physical processes may involve air stripping or filtration. Chemical treatment includes coagulation, chlorination, or ozone addition. The term can also refer to treatment processes, treatment of toxic materials in surface waters and ground waters, oil spills, and some methods of dealing with hazardous materials on or in the ground (*see* treatment for more related terms) (EPA-89/12).

Physical and thermal integrity: The ability of the material of the cap and/or rotor to resist physical and thermal breakdown (40 CFR 85.2122(a)(7)(ii)(C)-91).

Physical chemical treatment system: Those full scale coke plant wastewater treatment systems incorporating full scale granular activated carbon adsorption units which were in operation prior to January 7, 1981, the date of proposal of this regulation (40 CFR 420.11-91).

Physical chemistry: A science dealing with the effects of physical properties on chemical phenomena.

Physical coal cleaning: One of SO_2 emission reduction techniques (*see* sulfur oxide emission control for control structure). Physical coal cleaning depends on the differences in density of both coal and the impurities. Coal is crushed, washed, and then separated by settling processes using cyclones, air classifiers or magnetic separators. Approximately 40 to 90% of the pyritic sulfur content can be removed by physical coal cleaning. Its effectiveness depends on the size of pyritic sulfur particles and the amount of pyritic sulfur contained in the coal (EPA-81/12, p8-4).

Physical construction: The excavation, movement of earth, erection of forms or structures, or similar activity to prepare an HWM facility to accept hazardous waste (40 CFR 270.2-91).

Physical model: The simulation of a real process by a physical experiment which models the important features of the original process that are the object of study, e.g., the simulation of dispersion problem in a wind tunnel (NATO-78/10).

Physical or mental impairment: *See* definition under handicapped person (40 CFR 7.25-91).

Physical treatment: A treatment process that alters the physical structure of a toxic waste contaminant to reduce the waste's toxicity, mobility, or volume (*see* treatment for more related terms) (EPA-89/12a).

Physiologically based pharmacokinetic (PBPK) model: Physiologically based compartmental model used to quantitatively describe pharmacokinetic behavior (EPA-92/12).

Phytoplankton: That portion of the plankton community comprised of tiny plants, e.g., algae, diatoms (EPA-89/12).

Phytotoxic: Something that harms plants (EPA-89/12).

pi: 3.141592... It is the ratio of the circumference of a circle to its diameter.

Pica: Ingestion of nonfood items (LBL-76/07-bio).

Pick interval: The time or range intervals in the lidar backscatter signal whose minimum average amplitude is used to calculate opacity. Two pick intervals are required, one in the near region and one in the far region (40 CFR 60-App/A(alt. method 1)-91).

Picking table or belt: A table or belt on which solid waste is manually sorted and certain items are removed. It is normally used in composting and salvage operations (SW-108ts).

Pickle liquor: In steel manufacture, a dilute acid solution used to clean steel (*see* liquor for more related terms) (DOI-70/04).

Pickling:
(1) An immersion of all or part of a workpiece in a corrosive media such as acid to remove scale and related surface coatings (EPA-83/06a).
(2) The process of immersing hot steel in a bath of hot dilute acid to prepare it for being cold-rolled and finished by galvanizing and tin-plating (DOI-70/04).
(3) The process that follows bating whereby the hide is immersed in a brine and acid solution to bring the skin or hide to an acid condition; prevents precipitation of chromium salts on the hide (EPA-82/11).

Pickling bath: Any chemical bath (other than alkaline cleaning) through which a workpiece is processed (40 CFR 468.02-91).

Pickling fume scrubber: The process of using an air pollution control device to remove particulates and fumes from air above a pickling bath by entraining the pollutants in water (40 CFR 468.02-91).

Pickling rinse: A rinse, other than an alkaline cleaning rinse, through which a workpiece is processed. A rinse consisting of a series of rinse tanks is considered as a single rinse (40 CFR 468.02-91).

Pickling rinse for forged parts: A rinse, other than an alkaline cleaning rinse, through which forged parts are processed. A rinse consisting of a series of rinse tanks is considered as a single rinse (40 CFR 468.02-91).

pico-: A prefix denoting 10^{-12}.

Picocurie (pCi): The quantity of radioactive material producing 2.22 nuclear transformations per minute (40 CFR 141.2-91). 1 pCi = $(10^{-12}) \times (37 \times 10^9) \times (60)$ = 2.22 transformations per minute (*see* radiation unit for more related terms).

Picocurie per liter (pCi/L): A common unit of measurement of the concentration of radioactivity in a gas or liquid. A picocurie per liter corresponds to 0.03 radioactive disintegrations per second in every liter (*see* radiation unit for more related terms) (EPA-88/08a, *see also* EPA-88/08).

Picric acid ($C_6H_2(NO_2)_3$): 2,4,6-trinitrophenol, a yellow explosive compound used for making matches, batteries, etc.

Pieces: The floor tile measured in the standard size of 12 X 12 X 3/32 (40 CFR 427.71-91).

Piezometer: A gage for measuring liquid pressure.

Piezometer tube: An open topped tube used to measure the pressure of confined groundwater. The lower end of the tube is submerged into the confined groundwater, the pressure is indicated by the risen water level.

Piezometric surface: The surface of standing water level shown by a piezometer tube.

Piezoelectric monitor: A type of particle monitor which measures mass concentration by utilization of a vibrating piezoelectric crystal driven by a standard oscillation circuit (EPA-81/09).

Pig:
(1) A container, usually lead, used to ship or store radioactive materials (EPA-89/12).
(2) In metallurgy, an ingot of aluminum alloy weighing 15 to 50 pounds (EPA-76/12).

Pigment:
(1) A general term used to describe any of a wide variety of organic, inorganic, natural, or synthetic insoluble material which are added to coatings to produce a desired color, viscosity, solids level or gloss (EPA-79/12b).
(2) The colorant used to give printing inks the desired hue and color (EPA-79/12a).

Pigmented coatings: The opaque coatings containing binders and colored pigments which are formulated to conceal the wood surface either as an undercoat or topcoat (40 CFR 52.741-91).

Pile:
(1) Any non-containerized accumulation of solid, nonflowing hazardous waste that is used for treatment or storage (40 CFR 260.10-91).
(2) (A) The fuel element in a nuclear reactor.
    (B) A heap of waste (EPA-89/12).

Pili (singular, pilus): A proteinaceous surface structure of some bacteria (EPA-88/09a).

Pilkington process: A process for making flat glass in which the glass continuously pours from a tank onto a spout and thence between forming rolls and is subsequently annealed as one continuous sheet (EPA-83).

Pilot: A burner that is used to ignite waste and auxiliary fuel during startup (*see* burner for more related terms) (EPA-89/03b).

Pilot and main burner: One of main burner components. The pilot is lit first, and, once a flame is detected, the fuel supply to the main burner is opened allowing the pilot to light the main burner (*see* burner component for more related terms) (EPA-89/03b).

Pilot scale testing: A treatability study designed to provide the detailed cost and design data required to optimize a treatment technology's performance and to provide information in support of remedy implementation (EPA-89/12a).

Pinholes:
(1) In constructing a landfill liner, a small imperfections in sheet or seamed flexible membrane liners which allow for escape of the contained materials, i.e., leaks (EPA-89/09, *see also* EPA-91/05).
(2) Imperfections in paper which appear as very small holes (EPA-83).

Pink water: After loading TNT into munitions, the loading bays are washed. TNT particles in concentrations of 100-150 mg/l produce in sunlight an orange or light rust colored effluent termed "pink water" (*see* water for more related terms) (EPA-76/03).

Pine rosin: *See* synonym, natural rosin.

Pipe: Any pipe or tubing used in the transportation of gas, including pipe-type holders (40 CFR 192.3-91).

Pipe, tube and other: Those acid pickling operations that pickle pipes, tubes or any steel product other than those included in paragraphs (k), (l) and (m) of this section (40 CFR 420.91-91).

Pipe and tube mill: Those steel hot forming operations that produce butt welded or seamless tubular steel products (40 CFR 420.71-91).

Pipe flow: The flow of liquid is full in a pipe (*see* flow for more related terms) (M&E-72).

Pipe or line pipe: A tube, usually cylindrical, through which a hazardous liquid flows from one point to another (40 CFR 195.2-91).

Pipe or piping: A hollow cylinder or tubular conduit that is constructed of non-earthen materials (40 CFR 280.12-91).

Pipeline: All parts of those physical facilities through which gas moves in transportation, including pipe, valves, and other appurtenance attached to pipe, compressor units, metering stations, regulator stations, delivery stations, holders, and fabricated assemblies (40 CFR 192.3-91).

Pipeline or pipeline system: All parts of a pipeline facility through which a hazardous liquid moves in transportation, including, but not limited to, line pipe, valves and other appurtenances connected to line pipe, pumping units, fabricated assemblies associated with pumping units, metering and delivery stations and fabricated assemblies therein, and breakout tanks (40 CFR 195.2-91).

Pipeline facility: The new and existing pipelines, rights-of-way, and any equipment, facility, or building used in the transportation of gas or in the treatment of gas during the course of transportation (40 CFR 192.3-91, *See also* 40 CFR 195.2-91).

Pipeline wrap: An asbestos-containing product made of paper felt intended for use in wrapping or coating pipes for insulation purposes (40 CFR 763.163-91).

Piping: A hollow cylinder or tubular conduit that is constructed of non-earthen materials (40 CFR 280.12-91) (40 CFR 280.12-91).

Piscicide: Chemicals used to kill fish (EPA-85/10).

Pit crucible furnace: Pit crucible furnace derives its name from its location. The top of the furnace is near floor level, which facilitates charging of the metal to the furnace and removing of the crucible for pouring (*see* furnace for more related terms) (AP-40, p238).

Pitch:
(1) In chemistry, synonymous with asphalt.
(2) In physics, a cycle (or frequency) of a sound.

Pitch binder: Thermosetting binders used in core making. Baking of the sand-binder mixture is required for evaporation-oxidation and polymerization to take place (EPA-85/10a).

Pitch coke: The coke formed by the destructive distillation of petroleum pitch. Used in foundry operations as sand binders (*see* coke for more related terms) (EPA-85/10a).

Pitch polishing: Polishing operation in which pitch rather than felt is the resilient carrier for the polishing agent (EPA-83).

Pitot tube:
(1) A device for measuring the velocity of a flowing fluid by using the velocity head of the stream as an index velocity (AP-40, p72; OME-88/12).
(2) An instrument which will sense the total pressure and static pressure in a gas stream. Used to determine gas velocity (EPA-83).

Pituita-adrenal axis: The interrelation of the anterior pituitary and adrenal glands whereby the activity of one is stimulated or inhibited by a hormone of the other, e.g., regulation of ACTH secretion by the concentrations of adrenal corticoids in the blood (LBL-76/07-bio).

Pituitary gonadotropic hormones: Hormones secreted in the anterior pituitary that control the gonads (LBL-76/07-bio).

Pituitary thyroid axis: The interrelated activities of the anterior pituitary and thyroid gland (LBL-76/07-bio).

PK value: $PK = \log_{10}(1/K)$, where K is the acid dissociation constant. The term is used to measure the strength of an acid on a logarithmic scale.

Placer mine:
(1) A deposit of sand, gravel, or talus from which some valuable mineral is extracted.
(2) To mine gold, platinum, tin or other valuable minerals by washing the sand, gravel, etc. (EPA-82/05).

Placer mining: The extraction of heavy mineral from a placer deposit by concentration in running water. It includes ground sluicing, panning, shoveling gravel into a sluice, scraping by power scraper, excavation by dragline or extraction by means of various types of dredging activities (EPA-82/05).

Plain: Relatively free from bubbles and seed during the melting (EPA-83).

Plain sedimentation: The sedimentation of suspended matter in a liquid unaided by chemicals or other special means and without any provision for the decomposition of the deposited solids in contact with the sewage. (The term plain settling is preferred) (EPA-76/03).

Planck's constant (h): 6.62620×10^{-34} joule second.

Planetary boundary layer: The layer in the atmosphere near the earth's surface, where the influence of the surface is present primarily through frictional forces. It is usually subdivided into two sections; the surface layer and the Ekman layer (NATO-78/10).

Plankton: Tiny plants and animals that live in water (cf. nanno-plankton) (EPA-89/12).

Planned renovation operation: A renovation operation, or a number of such operations, in which some RACM will be removed or stripped within a given period of time and that can be predicted. Individual nonscheduled operations are included if a number of such operations can be predicted to occur during a given period of time based on operating experience (40 CFR 61.141-91).

Plant:
(1) Any member of the plant kingdom, including seeds, roots and other parts thereof (ESA3-16USC1531-90).
(2) One or more facilities at the same location owned by or under common control of the same person (40 CFR 82.3-91, *see also* 40 CFR 52.741-91).
(3) A location at which a process or set of processes are used to produce, refine, or repackage, chemicals (EPA-87/07a).

Plant blending records: Those records which document the weight fraction of organic solvents and solids used in the formulation or preparation of inks at the vinyl or urethane printing plant where they are used (40 CFR 60.581-91).

Plant effluent or discharge after treatment: The wastewater discharged from the industrial plant. In this definition, any waste treatment device (pond, trickling filter, etc.) is considered part of the industrial plant (EPA-83/06a).

Plant regulator: Any substance or mixture of substances intended, through physiological action, for accelerating or retarding the rate of growth or rate of maturation, or for otherwise altering the behavior of plants or the produce thereof, but shall not include substances to the extent that they are intended as plant nutrients, trace elements, nutritional chemicals, plant inoculants, and soil amendments. Also, the term, plant regulator shall not be required to include any of such of those nutrient mixtures or soil amendments as are commonly known as vitamin-hormone horticultural products, intended for improvement, maintenance, survival, health, and propagation of plants, and as are not for pest destruction and are non-toxic, nonpoisonous in the undiluted packaged concentration (FIFRA2-7USC136-91).

Plant sanitation: The aspects of plant housekeeping which reduce the incidence of water contamination resulting from equipment leaks, spillage of preservative etc. (EPA-74/04).

Plant site:
(1) The area occupied by the mine, necessary haulage ways from the mine to the beneficiation process, the beneficiation area, the area occupied by the wastewater treatment facilities and the storage areas for waste materials and solids removed from the wastewaters during treatment (40 CFR 440.141-15-91).
(2) All contiguous or adjoining property that is under common control, including properties that are separated only by a road or other public right-of-way. Common control includes properties that are owned, leased, or operated by the same entity, parent entity, subsidiary, or any combination thereof (40 CFR 63.101-91).

Plant waste: Dunnage, shipping, packaging, storage, and general office waste. Not production or process wastes (*see* waste for more related terms) (EPA-83).

Plaque: A porous body of sintered metal on a metal grid used as a current collector and holder of electrode active materials, especially for nickel-cadmium batteries (EPA-84/08).

Plasma: The fluid part of blood, lymph, or intramuscular fluid in which cells are suspended (EPA-83/09).

Plasma arc: Plasma arc related terms include:
- Microwave discharge
- Microwave plasma (*see* definition under microwave discharge)
- Transferred plasma arc
- Nontransferred plasma arc

Plasma arc incinerator: Any enclosed device using a high intensity electrical discharge or arc as a source of heat and which is not listed as an industrial furnace (40 CFR 260.10-91).

The plasma arc has been referred to as the fourth state of matter since the arc does not always behave as a solid, liquid or gas. A plasma may be defined as a conductive gas flow consisting of charged and neutral particles, having an overall charge of approximately zero, and all exhibiting collective behavior. The plasma, when applied to waste disposal, can best be understood by thinking of it as an energy conversion and energy transfer device. The electrical energy input is transformed into a plasma with a temperature equivalent of up to 18,000° F at the centerline of the reactor. As the activated components of the plasma decay, their energy is transferred to waste materials exposed to the plasma. The wastes are then broken into atoms, ionized, pyrolyzed and finally destroyed as they interact with the decaying plasma species. The heart of this technology is that the breakdown of the wastes into atoms occurs virtually instantaneously and no large molecular intermediary compounds are produced during the kinetic recombination (*see* incinerator for more related terms) (Lee-83/07).

Plasma arc machining: The process of material removal or shaping of a workpiece by a high velocity jet of high temperature ionized gas (cf. incinerator, plasma arc) (EPA-83/06a).

Plasmid: A circular piece of DNA that exists apart from the chromosome and replicates independently of it. Bacterial plasmids carry information that renders the bacteria resistant to antibiotics. Plasmids are often used in genetic engineering to carry desired genes into organisms (EPA-89/12).

Plastic: Non-metallic compounds that result from a chemical reaction, and are molded or formed into rigid or pliable construction materials or fabrics (EPA-89/12). Other plastic related terms include:
- Biodegradable plastic

Plastic body: An automobile or light-duty truck body constructed of synthetic organic material (40 CFR 60.391-91).

Plastic body component: Any component of an automobile or light-duty truck exterior surface constructed of synthetic organic material (40 CFR 60.391-91).

Plastic insulation: An insulation that is plastic enough when mixed with water that it can adhere to outer furnace walls or be placed over furnace arches (SW-108ts).

Plastic material: A synthetic organic polymer (i.e., a thermoset polymer, a thermoplastic polymer, or a combination of a natural polymer and a thermoset or thermoplastic polymer) that is solid in its final form and that was shaped by flow. The material can either be a homogeneous polymer and a polymer combined with fillers, plasticizers, pigments, stabilizers, or other additives (40 CFR 463.2-91).

Plastic molding and forming: A manufacturing process in which materials are blended, molded, formed or otherwise processed into intermediate or final products (40 CFR 463.2-91).

Plastic part: The panels, housings, bases, covers, and other business machine components formed of synthetic polymers (40 CFR 60.721-91).

Plastic refractory: A blend of ground fireclay materials in a plastic form, that is suitable for ramming into place to form monolithic linings or special shapes. It may be air-setting or heat setting and is available in different qualities of heat resistance (*see* refractory for more related terms) (EPA-83).

Plastic rigid foam: The cellular polyurethane insulation, cellular poly-isocyanurate insulation, glass fiber reinforced polyisocyanurate-polyurethane foam insulation, cellular polystyrene insulation, phenolic foam insulation, spray-in-place foam and foam-in-place insulation (40 CFR 248.4-91).

Plasticity soil: The property of soil which allows it to be deformed without appreciable volume change or cracking (*see* soil for more related terms) (EPA-83).

Plasticizer:
(1) A material, generally an organic liquid, incorporated in a plastic or rubber formulation to soften the resin polymer and improve flexibility, ductility and extensibility (EPA-91/05).
(2) A substance added to printing ink to impart flexibility (EPA-79/12a).
(3) A high boiling liquid which is used in the formulation of a propellant to help make it plastic (EPA-76/03).

Plate: *See* synonym, anode.

Plate column: *See* synonym, tray column.

Plate glass (or polished plate glass): A flat glass formed by a rolling process, ground and polished on both sides, with surfaces essentially plane and parallel (*see* glass for more related terms) (EPA-83).

Plate mill: Those steel hot forming operations that produce flat hot-rolled products which are (1) between 8 and 48 inches wide and over 0.23 inches thick; or (2) greater than 48 inches wide and over 0.18 inches thick (40 CFR 420.71-91).

Plate scrubber: One of air pollution control devices. A plate scrubber is a scrubber that relies on a gas absorption process for the removal of contaminants. The basic design is a vertical cylindrical column with a number of plates or trays inside. The scrubbing liquid is introduced at the top plate and flows successively across each plate as it moves downward to the liquid outlet at the tower bottom. Flue gas comes in at the bottom of the tower and passes through openings in each plate before leaving through the top. Gas absorption is promoted by the breaking up of the gas phase into bubbles which pass through the volume of liquid in each plate (*see* scrubber for more related terms) (Calvert-84).

Plate soak: The process operation of soaking or reacting lead subcategory battery plates, that are more than 2.5 mm (0.100 in) thick, in sulfuric acid (40 CFR 461.2-91).

Plated area: A surface upon which an adherent layer of metal is deposited (EPA-83/06a).

Plating:
(1) Forming an adherent layer of metal upon an object (EPA-83/06a).
(2) The finishing operation where the skin or hide is pressed in order to make it smoother. Plating may be done with an embossing plate which imprints textured effects into the leather surface (EPA-82/11).
Other plating related terms include:
- Automatic plating
- Barrel plating
- Mechanical plating
- Vapor plating

Plating barrel: A container in which parts are placed loosely, so they can tumble as the barrel rotates in the plating or processing solution (EPA-74/03d).

Plating rack: A fixture that permits moving one or more workpieces in and out of a treating or plating tank and transferring electric current to the workpieces when in the tank (EPA-74/03d).

Platinum (Pt): A transition metal with a.n. 78; a.w. 195.09; d. 21.4 g/cc; m.p. 1769° C and b.p. 4530° C. The element belongs to group VIII of the periodic table.

Platinum electrode: An electrode made from a platinum wire. The electrode is used in the voltammetric analysis of electrolytes (*see* electrode for more related terms).

Platinum mineral: Platinum, ruthenium, rhodium, palladium, osmium, and iridium are members of a group characterized by high specific gravity, unusual resistance to oxidizing and acidic attack and high melting point (EPA-82/02).

Playa: Level area at the bottom of a desert basin that at times is temporarily covered with water; a dry lake bed (DOE-91/04).

Pleistocene: Geologic epic beginning approximately 3 to 5 million years ago (DOE-91/04).

Pliocene: Geologic period between the Miocene and the Pleistocene periods (DOE-91/04).

Plug flow: A fluid flow pattern characterized by complete mixing in the radial direction and no mixing in the axial direction (EPA-88/09a).

Plug flow sampling: A monitoring procedure that follows the same slug of wastewater throughout its transport in the receiving water. Water quality samples are collected at receiving water stations, tributary inflows, and point source discharges only when a dye slug or tracer passes that point (EPA-85/09).

Plug flow reactor (PFR): A chemical reactor having a characteristic plug fluid-flow pattern (EPA-88/09a).

Plugging:
(1) The act or process of stopping the flow of water, oil or gas into or out of a formation through a borehole or well penetrating that formation (40 CFR 144.3; 146.3; 147.2902-91).
(2) Stopping a leak or sealing off a pipe or hose (EPA-89/12).

Plugging and abandonment plan: The plan for plugging and abandonment prepared in accordance with the requirements of 40 CFR 144.28 and 40 CFR 144.51 (40 CFR 144.61-91).

Plugging record: A systematic listing of permanent or temporary abandonment of water, oil, gas, test, exploration and waste injection wells, and may contain a well log, description of amounts and types of plugging material used, the method employed for plugging, a description of formations which are sealed and a graphic log of the well showing formation location, formation thickness, and location of plugging structures (40 CFR 146.3-91).

Plume:
(1) A discharge of substances from a given point of origin. It can be visible such as combustion gases exhausted from the stack or invisible such as thermal plume in water (EPA-89/03b).
(2) (A) The area of measurable and potentially harmful radiation leaking from a damaged reactor.
    (B) The distance from a toxic release considered dangerous for those exposed to the leaking fumes (EPA-89/12; 85/11).
(3) The plume being measured by lidar (40 CFR 60-App/A(alt. method 1)-91).

Plume path: The curve along which the plume moves in the atmosphere. The form and direction of this curve is determined by the buoyancy of the plume, its exit speed, the wind profile, the stability of the atmosphere and the turbulence in the atmosphere (NATO-78/10).

Plume rise: The upward motion of a plume after its emission from the source due to its exit velocity and/or its buoyancy. This buoyancy is caused by the higher temperature of the plume with respect to its surroundings (NATO-78/10).

Plume signal: The backscatter signal resulting from the laser light pulse passing through a plume (40 CFR 60-App/A(alt. method 1)-91).

Plunger pump: *See* synonym, accelerator pump.

Plutonium (Pu):
(1) A transuranic, heavy, silvery metal with 15 isotopes that is produced by the neutron irradiation of natural uranium. The isotope plutonium-239 is the most important isotope, used both in nuclear weapons and commercial nuclear power applications (DOE-91/04).
(2) A radioactive metal with a.n. 94; a.w. 242; d. 19.84 g/cc; m.p. 640° C and b.p. 3235° C. The element belongs to group IIIB of the periodic table.

Plutonium processing and fuel fabrication plant: A plant in which the following operations or activities are conducted:
(1) Operations for manufacture of reactor fuel containing plutonium including any of the following:
  (A) Preparation of fuel material;
  (B) Formation of fuel material into desired shapes;
  (C) Application of protective cladding;
  (D) Recovery of scrap material; and
  (E) Storage associated with such operations; or
(2) Research and development activities involving any of the operations described in paragraph (r)(1) of this section, except for research and development activities utilizing unsubstantial amount of plutonium (10 CFR 70.4-91).

Plutonium target processing facility: The land, buildings, equipment, and processes used to extract plutonium from plutonium targets after they have been irradiated in a reactor and to purify the plutonium. The facility may also be used to treat the remaining uranium for recycling and to treat other remaining materials for transfer to waste treatment, storage, or disposal facilities (DOE-91/04).

Plywood: An assembly of a number of layers of wood, or veneers, joined together by means of an adhesive. Plywood consists of two main types (EPA-74/04): It includes:
(1) **Hardwood plywood:** Plywood which has a face ply of hardwood and is generally used for decorative purposes.
(2) **Softwood plywood:** The veneers typically are of softwood and the usage is generally for construction and structural purposes.

Plywood core: The innermost segment of a plywood panel (EPA-74/04).

PM (or particulate matter): Any airborne finely divided solid or liquid material with an aerodynamic diameter smaller than 100 micrometers (40 CFR 51.100-91, *see also* 40 CFR 60.2; 60.51a; 61.171-91).

PM emission (or particulate matter emission):
(1) All finely divided solid or liquid material, other than uncombined water, emitted to the ambient air as measured by applicable reference methods, or an equivalent or alternative method, specified in this chapter, or by a test method specified in an approved State implementation plan (40 CFR 51.100-91).
(2) Fine solid matter suspended in combustion gases carried to the atmosphere. The emission rate is usually expressed as a concentration such as grains per dry standard cubic feet (gr/dscf) corrected to a common base, usually 12 percent CO (EPA-89/03b).

PM10:
(1) The particulate matter with an aerodynamic diameter less than or equal to a nominal ten micrometers, as measured by such method as the Administrator may determine (other identical or similar definitions are provided in 40 CFR 58.1-u) (CAA302-42USC7602).
(2) A new standard for measuring the amount of solid or liquid matter suspended in the atmosphere, i.e., the amount of particulate matter over 10 micrometers in diameter; smaller PM-10 particles penetrate to the deeper portions of the lung, affecting sensitive population groups such as children and individuals with respiratory ailments (EPA-94/04).

PM10 emission: The finely divided solid or liquid material, with an aerodynamic diameter less than or equal to a nominal 10 micrometers emitted to the ambient air as measured by an applicable reference method, or an equivalent or alternative method, specified in this chapter or by a test method specified in an approved State implementation plan (40 CFR 51.100-91).

PM10 sampler: A device, associated with a manual method for measuring PM10, designed to collect PM10 from an ambient air sample, but lacking the ability to automatically analyze or measure the collected sample to determine the mass concentration of PM10 in the sampled air (40 CFR 53.1-91).

Pneumatic ash handling: A system of pipes and cyclone separators that conveys fly ash or floor dust to bin via an air stream (SW-108ts).

Pneumatic coal-cleaning equipment: Any facility which classifies bituminous coal by size or separates bituminous coal from refuse by application of air stream(s) (40 CFR 60.251-91).

Poaching process: A process of boiling nitrocellulose (NC) in soda ash at 96° C for four hours followed by fresh water at 96° C for two hours. The NC will then settle and the water is drained off (EPA-76/03).

Pocket plate: A type of battery construction where the electrode is a perforated metal envelope containing the active material (EPA-84/08).

POHC-PIC surrogate soup: A homogeneous mixture of organic and inorganic compounds used to evaluate combustion efficiency of an incinerator (EPA-88/12).

Point: Thickness of a sheet of paper measured in thousandths of an inch (EPA-83).

Point design: A design for the NPR that comprises a collection of design and component information representing a concept without the benefit of integration criteria, supporting analyses, and optimization studies: design that represents a point on the performance curve (DOE-91/04).

Point function: A thermodynamic function whose results depend only on the values of two end states. For example, temperature is a point function, Therefore, the temperature change from state 1 (T_1) to state 2 (T_2) can be expressed as $T_2 - T_1$, regardless what processes the temperature change follows between the two states. Contrary to the point function, heat and work are path functions (*see* thermodynamic process for more related terms) (Wark-p6).

Point of disinfectant application: The point where the disinfectant is applied and water downstream of that point is not subject to recontamination by surface water runoff (40 CFR 141.2-91).

Point of entry treatment device: A treatment device applied to the drinking water entering a house or building for the purpose of reducing contaminants in the drinking water distributed throughout the house or building (40 CFR 141.2-91).

Point of use treatment device: A treatment device applied to single tap used for the purpose of reducing contaminants in drinking water at that one tap (40 CFR 141.2-91).

Point of waste generation: The location where samples of a waste stream are collected for the purpose of determining the waste flow rate, water content, or benzene concentration in accordance with procedures specified in 40 CFR 1.355 of this subpart. For a chemical manufacturing plant or petroleum refinery, the point of waste generation is a location after the waste stream exits the process unit component, product tank, or waste management unit generating the waste, and before the waste is exposed to the atmosphere or mixed with other wastes. For a coke-byproduct recovery plant subject to and complying with the control requirements of 40 CFR 61.132, 61.133, or 61.134 of this part, the point of waste generation is a location after the waste stream exits the process unit component or waste management unit controlled by that subpart, and before the waste is exposed to the atmosphere. For other facilities subject to this subpart, the point of waste generation is a location after the waste enters the facility, and before the waste is exposed to the atmosphere or placed in a facility waste management unit (40 CFR 61.341-91).

Point source: A stationery location or fixed facility from which pollutants are discharged or emitted. Also, any single identifiable source of pollution, e.g., a pipe, ditch, ship, ore pit, factory smokestack (*see* source for more related terms) (EPA-89/12).

Poison: A substance which is harmful to living organisms (cf. economic poison).

Polar capacitor: An electrolytic capacitor having an oxide film on only one foil or electrode which forms the anode or positive terminal (EPA-83/03).

Polarographic analyzer (voltametric analyzer or electrochemical transducer): One of electroanalytical methods for continuous emission monitoring (*see* continuous emission monitor for various types). With the proper choice of electrodes and electrolytes, the analyzer utilizes the principles of polarography to monitor CO, CO_2, H_2S, NO_2, SO_2, and other gases. The transducers in these instruments is generally a self-contained electrochemical cell in which a chemical reaction takes place involving the pollutant molecule. Two basic techniques are used in the transducers:
(1) The utilization of a selective semi-permeable membrane that allows the pollutant molecule to diffuse to an electrolytic solution; and
(2) The measurement of the current change produced at an electrode by the oxidation or reduction of the dissolved gas at the electrode (EPA-81/09; 84/03a).

Pole type transformer: A transformer suitable for mounting on a pole or similar structure (*see* transformer for more related terms) (EPA-83/03).

Poling: A step in the production of ceramic piezoelectric bodies which orients the oxes of the crystallites in the preferred direction (EPA-83/03).

Policy (must do): *See* definition under law related terms.

Polished plate glass: *See* synonym, plate glass.

Polished wire glass: A wire glass, ground and-polished on both sides (*see* glass for more related terms) (EPA-83).

Polishing:
(1) A final water treatment step used to remove any remaining organics from the water (EPA-87/10a).
(2) The process of removing stock from a workpiece by the action of loose or loosely held abrasive grains carried to the workpiece by a flexible support. Usually, the amount of stock removed in a polishing operation is only incidental to achieving a desired surface finish or appearance (EPA-83/06a).

Polishing and buffing wheels: All power-driven rotatable wheels composed all or in part of textile fabrics, wood, felt, leather, paper, and may be coated with abrasives on the periphery of the wheel for purposes of polishing, buffing, and light grinding (29 CFR 1910.94b-91).

Polishing compounds: Fluid or grease stick lubricants composed of animal tallows, fatty acids, and waxes. Selection depends on surface finish desired (EPA-83/06a).

Polishing pond: A stabilization lagoon used as a final treatment step to remove any remaining organics (*see* pond for more related terms) (EPA-87/10a).

Pollen:
(1) (A) A fine dust produced by plants.
    (B) The fertilizing element of flowering plants.
    (C) A natural or background air pollutant (EPA-89/12).
(2) A fine dust produced by plants; a natural or background air pollutant (LBL-76/07-air).

Pollutant:
(1) Generally, any substance introduced into the environment that adversely affects the usefulness of a resource (EPA-89/12, *see also* CWA502; 40 CFR 21.2; 122.2; 230.3; 257.3.3; 401.11-91).
(2) Any introduced gas, liquid or solid that makes a resource unfit for a specific purpose (LBL-76/07-air).
(3) In CERCLA, any elements, substances, compounds, or mixtures, including disease-causing agents, which after release into the environment and upon exposure, ingestion, inhalation, or assimilation into any organism, either directly from the environment or indirectly by ingesting through food chains, will or may reasonably be anticipated to cause death, disease, behavioral abnormalities, cancer, genetic mutation, physiological malfunctions (including malfunctions in reproduction), or physical deformation in such organisms or their offspring. The term does not include petroleum, including crude oil any or fraction thereof which is not otherwise specifically listed or designated as a hazardous substance under section 101(14)(A) through (F) of CERCLA, nor does it include natural gas, liquefied natural gas, or synthetic gas of pipeline quality (or mixtures of natural gas and synthetic gas) (CERCLA Sec. 101 & 40 CFR 300.6).
(4) In CWA, dredged spoils, solid wastes, incinerator residue, filter backwash, sewage, garbage, sewage sludge, munitions, chemical wastes, biological materials, radioactive material (except those regulated under the Atomic Energy Act of 1954, as amended (42 U.S.C. 2011 et seq.), heat, wrecked or discarded equipment, rock, sand, cellar dirt and industrial, municipal, and agricultural waste discharged into water. It does not mean (40 CFR 233.3):
  (A) Sewage from vessels; or
  (B) Water, gas, or other materials which are injected into a well to facilitate production of oil or gas, or water derived in association with oil and gas production and disposed of in a well, if the well used either to facilitate production or for disposal purposes is approved by authority of the State in which the well is located, and if the State determines that injection or disposal will not result in the degradation of ground or surface water resource.

Other pollutant related terms include:
- Air pollutant
- Compatible pollutant
- Conservative pollutant
- Conventional pollutant
- Criteria pollutant
- Critical pollutant
- Designated pollutant
- Hazardous air pollutant
- Incompatible pollutant
- No discharge of pollutant
- Non-contact cooling water pollutant
- Non-conventional pollutant
- Non-conventional pesticide pollutant
- Non-criteria pollutant
- Non-pass through pollutant
- Pass-through pollutant
- Persistent pollutant
- Primary pollutant
- Primary significance pollutant
- Priority pollutant
- Process wastewater pollutant
- Refractory pollutant
- Secondary pollutant
- Secondary significance pollutant
- Toxic pollutant
- Toxic air pollutant
- Traditional pollutant parameter

Pollutant loading: The ratio of the total daily mass discharge of a particular pollutant to the total daily production expressed in terms of (g pollutant)/(Kg production) (*see* loading for more related terms) (EPA-87/10a).

Pollutant or contaminant: Includes, but not be limited to, any element, substance, compound, or mixture, including disease-causing agents, which after release into the environment and upon exposure, ingestion, inhalation, or assimilation into any organism, either directly from the environment or indirectly by ingestion through food chains, will or may reasonably be anticipated to cause death, disease, behavioral abnormalities, cancer, genetic mutation, physiological malfunctions (including malfunctions in reproduction) or physical deformations, in such organisms or their offspring; except that the term pollutant or contaminant shall not include petroleum, including crude oil or any fraction thereof which is not otherwise specifically listed or designated as a hazardous substance under subparagraphs (A) through (F) of paragraph (14) and shall not include natural gas, liquefied natural gas, or synthetic gas of pipeline quality (or mixtures of natural gas and such synthetic gas) (SF101, *see also* 40 CFR 300.5; 310.11-91).

Pollutant parameter: The constituents of wastewater determined to be detrimental to human health or the environment, thus requiring control (EPA-85/10a).

Pollutant standard index (PSI): A measure of adverse health effects of air pollution levels in major cities (EPA-89/12, *see also* 40 CFR 58-App/G-91). In general, PSI values and their air quality levels are as follows: 0-50, good conditions; 50-100, acceptable conditions; 100-200, unacceptable conditions; 200-300, hazardous conditions. PSI is also known as air quality standard.

Pollutant transport: An array of mechanisms by which a substance may migrate outside the immediate location of the release or discharge of the substance. e.g., pollution of groundwater by the migration of hazardous wastes from landfill (Course 165.5).

Pollution: Generally, the presence of matter or energy whose nature, location or quantity produces undesired environmental effects. Under the Clean Water Act, for example, the term is defined as the man-made or man-induced alteration of the physical, biological, and radiological integrity of water (EPA-89/12, *see also* CWA502; 40 CFR 130.2; 401.11; 230.3-91; SW-108ts). Other pollution related terms include:
- Agricultural pollution
- Air pollution
- Indoor air pollution
- Industrial waste pollution
- Manmade air pollution
- Natural pollution
- Noise pollution

- Oil pollution
- Sewage pollution
- Soil pollution
- Thermal pollution
- Water pollution
- Wood burning stove pollution

Pollution indicator organism: A plant or animal form, such as the rat-tailed maggot or blue-green algae, that thrives in polluted water (cf. biological indicator) (DOI-70/04).

Pollution liability: The liability for injuries arising from the release of hazardous substances or pollutants or contaminants (SF401-42USC9671-91).

Pollution load: A measure of the unit mass of a wastewater in terms of its solids or oxygen demanding characteristics, or in terms of harm to receiving water (see load for more related terms) (EPA-83/03).

Pollution/pollutant: The terms pollution and pollutants refer to all nonproduct outputs, irrespective of and recycling or treatment that may prevent or mitigate releases to the environment (EPA-91/10, p6).

Pollution prevention:
(1) The use of materials, processes, or practices that reduce or eliminate the creation of pollutants or wastes. It includes practices that reduce the use of hazardous materials, energy, water or other resources, and practices that protect natural resources through conservation or more efficient uses (EPA-91/10, p4).
(2) A preventive action or a measure taken to minimize waste generation or waste toxicity, if waste generation is inevitable. In a broad sense, it incorporates clean technology, low-waste technology, non-waste technology, prevention, quantity reduction, recycling, reduction, resource conservation, resource recovery, source reduction, toxicity reduction, waste minimization, etc. (EPA-89/12; OTA-89/10).
(3) cf. industrial pollution prevention.

Pollution Prevention Act (PPA) of 1990: see the term of Act or PPA.

Pollution source: Generators of various pollutants (see source for more related terms).

Polonium (Po): A radioactive metal with a.n. 84; a.w. 210; d. 9.2 g/cc; m.p. 254° C and b.p. 962° C. The element belongs to group VIA of the periodic table.

Population stock or stock: A group of marine mammals of the same species or smaller taxa in a common spatial arrangement, that interbreed when mature (MMPA3-16USC1362-90).

Polybrominated dibenzofurans: Refers to any member of a class of dibenzofurans with two to eight bromine substituents (40 CFR 766.3-91).

Polybrominated dibenzo-p-dioxin (PBDD): Any member of a class of dibenzo-p-dioxins with two to eight bromine substituents (40 CFR 766.3-91).

Polybutadiene rubber: A synthetic rubber made by solution polymerization of butadiene (see rubber for more related terms) (EPA-74/12a).

Polychlorinated biphenyl: See definition under PCB.

Polychlorinated dibenzofuran: Any member of a class of dibenzofurans with two to eight chlorine substituents (40 CFR 766.3-91).

Polychlorinated dibenzo-p-dioxin (PCDD): Any member of a class of dibenzo-p-dioxins with two to eight chlorine substituents (40 CFR 766.3-91).

Polycyclic aromatic hydrocarbon (PAH): A group of organic compounds. Some are known to be potent human carcinogens (EPA-88/09b).

Polycyclic organic matter (POM): Same as polycyclic aromatic hydrocarbon

Polyelectrolytes:
(1) Synthetic chemicals that help solids to clump during sewage treatment (EPA-89/12).
(2) A high polymer substance, either natural or synthetic, containing ionic constituents; they may be either cationic or anionic (EPA-83/06a). Polyelectrolytes such as $Al_2(SO)_3$, $FeCl_3$, etc. form AL^- and SO_4^- ions in a solution. These ions are called coagulants.

Polyester: A chemical substance that meets the definition of polymer and whose polymer molecules contain at least two carboxylic acid ester linkages, at least one of which links internal subunits together (40 CFR 723.250-91).

Polyester fiber: The fiber in which the fiber forming substance is any long chain synthetic polymer composed of at least 85% by weight of an ester of dihydric alcohol and terephthalic acid (see fiber for more related terms) (EPA-74/06b).

Polyethylene:
(1) A thermoplastic polymer or copolymer comprised of at least 50 percent ethylene by weight; see low density polyethylene and high density polyethylene (40 CFR 60.561-91).
(2) A semicrystalline thermoplastic polymer made largely of ethylene, often incorporating lesser amounts of one or more co-monomers. PE has the following classifications in accordance with its density variations (EPA-89/09, see also EPA-91/05):
- Type 0, density under ... 0.910 (g/cc)
- Type I, low density (LDPE and LLDPE) ... 0.910-0.925
- Type II, medium density ... 0.926-0.940
- Type III, High density (copolymer) ... 0.941-0.959
- Type IV, High density (homopolymer) ... 0.960 & higher

Other polyethylene related terms include:
- Low density polyethylene
- High density polyethylene

- Medium density polyethylene
- Very low density polyethylene

Poly(ethylene terephthalate) (PET): A polymer or copolymer comprised of at least 50 percent bis-(2- hydroxyethyl)-terephthalate (BHET) by weight (40 CFR 60.561-91).

Poly(ethylene terephthalate) (PET) manufacture using dimethyl terephthalate: The manufacturing of poly(ethylene terephthalate) based on the esterification of dimethyl terephthalate (DMT) with ethylene glycol to form the intermediate monomer bis-(2-hydroxyethyl)-terephthalate polymerized to form PET (40 CFR 60.561-91).

Poly(ethylene terephthalate) (PET) manufacture using terephthalic acid: The manufacturing of poly(ethylene terephthalate) based on the esterification reaction of terephthalic acid (TPA) with ethylene glycol to form the intermediate monomer bis-(2-hydroxyethyl)-terephthalate (BHET) that is subsequently polymerized to form PET (40 CFR 60.561-91).

Polyhalogenated dibenzofuran or PHDF: Any member of a class of dibenzofurans containing two to eight chlorine, bromine, or a combination of chlorine and bromine substituents (40 CFR 766.3-91).

Polyhalogenated dibenzo-p-dioxin or PHDD: Any member of a class of dibenzo-p-dioxins containing two to eight chlorine substituents or two to eight bromine substituents (40 CFR 766.3-91).

Polymer:
(1) Any of the natural or synthetic compounds of usually high molecular weight that consist of many repeated links, each link being a relatively light and simple molecule (40 CFR 60.601-91, *see also* 40 CFR 704.25; 721.350; 723.250-91).
(2) Basic molecular ingredients in plastic (EPA-89/12).
(3) A carbon based organic chemical material formed by the chemical reaction of monomers having either the same or different chemical structures. Plastics, rubbers and textile fibers are all relatively high molecular weight polymers (EPA-91/05).

Other polymer related terms include:
- Thermoplastic polymer
- Thermoset polymer

Polymer molecule: A molecule which includes at least four covalently linked subunits, at least two of which are internal subunits (40 CFR 704.25-91, *see also* 40 CFR 721.350; 723.250-91).

Polymeric coating of supporting substrates: A web coating process that applies elastomers, polymers, or prepolymers to a supporting web other than paper, plastic film, metallic foil, or metal coil (40 CFR 60.741-91).

Polymeric flocculants:
(1) High molecular weight compounds which, due to their polar charges, aid in particle binding and agglomeration (EPA-85/10a).
(2) A polar polymers which catches the particles due to polarity of polymers and also the physical meshes.

Polymerization: A chemical reaction in which the molecules of a monomer are linked together to form large molecules whose molecular weight is a multiple of that of the original substance. When two or more monomers are involved, the process is called co-polymerization (EPA-75/01a).

Polymerization reaction section: The equipment designed to cause monomer(s) to react to form polymers, including equipment designed primarily to cause the formation of short polymer chains (oligomers or low polymers), but not including equipment designed to prepare raw materials for polymerization, e.g., esterification vessels. For the purposes of these standards, the polymerization reaction section begins with the equipment used to transfer the materials from the raw materials preparation section and ends with the last vessel in which polymerization occurs. Equipment used for the on-site recovery of ethylene glycol from poly(ethylene terephthalate) plants, however, are included in this process section, rather than in the material recovery process section (40 CFR 60.561-91).

Polynuclear aromatic compound (PNA): A class of organic materials with characteristic multiple ring molecular structure (e.g., naphthalene) which is suspected of causing genetic damage (OME-88/12).

Polypropylene (PP): A thermoplastic polymer or copolymer comprised of at least 50 percent propylene by weight (40 CFR 60.561-91).

Polystyrene (PS): A thermoplastic polymer or copolymer comprised of at least 80 percent styrene or para-methylstyrene by weight (40 CFR 60.561-91).

Polyvinyl chloride ((CH_2CHCl)$_x$ PVC): A common plastic material ($CH_2 = CHCl$) which is tasteless, odorless, and generally insoluble; can release hydrochloric acid (HCL) when burned (EPA-83, *see also* EPA-89/12; 91/05).

Polyvinyl chloride plant: Includes any plant where vinyl chloride alone or in combination with other materials is polymerized (40 CFR 61.61-91).

Pond: A natural or man-made water body, generally, smaller than a lake. Other pond related terms include:
- Cooling pond
- Evaporation pond
- Holding pond
- Oxidation pond
- Polishing pond
- Tailing pond

Pond lime: A lime cake after being run into waste ponds (EPA-74/01a).

Pond water surface area: For the purpose of calculating the volume of waste water, shall mean the area within the impoundment for rainfall and the actual water surface area for evaporation (40 CFR 421.11-91, *see also* 40 CFR 421.61-91).

Pondage: The amount of water stored behind a dam of relatively small storage capacity used to control the flow of a river (DOI-70/04).

Population: A group of interbreeding organisms of the same kind occupying a particular space. Generically, the number of humans or other living creatures in a designated area (EPA-89/12).

Population at risk: A population subgroup that is more likely to be exposed to a chemical, or is more sensitive to a chemical, than is the general population (Course 165.6).

Population equivalent (PE): An expression of the relative strength of a waste (usually industrial) in terms of its equivalent in domestic waste, expressed as the population that would produce the equivalent domestic waste. A population equivalent of 160 million persons means the pollution effect equivalent to raw sewage from 160 million persons; 0.17 pounds BOD (the oxygen demand of untreated wastes from one person) = 1 PE (DOI-70/04).

Porcelain enameling: The entire process of applying a fused vitreous enamel coating to a metal basis material. Usually this includes metal preparation and coating operations (40 CFR 466.02-91).

Pore water: The water that is filled in the soil cavities (*see* water for more related terms).

Porosity:
(1) The ratio of the volume in any porous material that is not filled with solid matter to the total volume occupied (cf. void ratio) (SW-108ts).
(2) In soil application, the capacity of rock or soil to contain water. The amount of water that rock can contain depends on the open spaces between the grains or cracks that can fill with water. Well-sorted soil is more porous than poorly-sorted soil. Soil is well sorted if the grains are all about the same size (as in the case of gravel or sand); spaces account for a large proportion of the total volume. Soil is poorly sorted if the grains are not all the same size; spaces between larger grains will fill with small grains instead of with water. Poorly-sorted rock thus holds less water than well-sorted (DOI-4/72).
(3) In groundwater application, the ratio of the volume of pores of a material to the volume of its mass. In ground water, it defines the amount of water a saturated rock volume can contain. If a unit volume of saturated rock is allowed to drain by gravity, not all of the water it contains will be released. The volume drained is the specific yield, a percentage, and the volume retained is the specific retention. It is the specific yield that is available to wells. Therefore, porosity is equal to specific yield plus specific retention (EPA-87/03).
(4) In air application, a measure of time required for 100 cm^3 of air to flow through a sample area. Also termed "air resistance" (in seconds per 100 cm^3) (EPA-87/10).

Porphyria: A disturbance of porphyrin metabolism characterized by marked increase in formation and excretion of porphyrins or their precursors (LBL-76/07-bio).

Port: Any opening in a furnace through which fuel or flame enters or exhaust gases escape (EPA-83). Other port related terms include:
- Overfire air port
- View port

Portable air compressor or compressor: Any wheel, skid, truck, or railroad car mounted, but not self-propelled, equipment designed to activate pneumatic tools. This consists of an air compressor (air end), and a reciprocating rotary or turbine engine rigidly connected in permanent alignment and mounted on a common frame. Also included are all cooling, lubricating, regulating, starting, and fuel systems, and all equipment necessary to constitute a complete, self-contained unit with a rated capacity of 75 cfm or greater which delivers air at pressures greater than 50 psig, but does not include any pneumatic tools themselves (40 CFR 204.51-a-91).

Portable grinder: Any power-driven rotatable grinding, polishing, or buffing wheel mounted in such manner that it may be manually manipulated (29 CFR 1910.94b-91).

Portable plant: Any nonmetallic mineral processing plant that is mounted on any chassis or skids and may be moved by the application of a lifting or pulling force. In addition, there shall be no cable, chain, turn buckle, bolt or other means (except electrical connections) by which any piece of equipment is attached or clamped to any anchor, slab, or structure, including bedrock that must be removed prior to the application of a lifting or pulling force for the purpose of transporting the unit (40 CFR 60.671-91).

Portal of entry effect: A local effect produced at the tissue or organ of first contact between the biological system and the toxicant (EPA-90/08).

Portland cement:
(1) The most common variety of cement (ETI-92).
(2) The product obtained by pulverizing clinker consisting essentially of hydraulic calcium silicates, to which no additions have been made subsequent to calcination other than water and/or untreated calcium sulphate (EPA-74/01b).
(3) cf. cement.

Portland cement plant: Any facility manufacturing portland cement by either the wet or dry process (40 CFR 60.61-91).

Posing an exposure risk to food or feed: Being in any location where human food or animal feed products could be exposed to PCBs released from a PCB item. A PCB item poses an exposure risk to food or feed if PCBs released in any way from the PCB item have a potential pathway to human food or animal feed. EPA considers human food or animal feed to include items regulated by the U.S. Department of Agriculture or the Food and Drug Administration as human food or animal feed; this includes direct additives. Food or feed is excluded from this definition if it is used or stored in private homes (40 CFR 761.3-91).

Positive crankcase ventilation (PCV) valve: A device to control the flow of blow-by gasses and fresh air from the crankcase to the fuel induction system of the engine (40 CFR 85.2122(a)(4)(ii)-91).

Positive displacement meter: One of liquid flow rate meters. This type of flowmeter is more applicable than other types for use with higher viscosity fluids. However, accuracy is highest when used with a clean, moderately viscous fluid. It cannot be used with multiphase liquids, gases, or slurries of varying density (*see* flow rate meter for more related terms) (EPA-89/06).

Positive electrode: *See* synonym, anode.

Positive ion: *See* definition under ion.

Positive pressure fabric filter: A fabric filter with the fans on the upstream side of the filter bags (*see* filter for more related terms) (40 CFR 60.271a; 60.341-91).

Positive temperature coefficient (PTC) type choke heater: A positive temperature coefficient resistant ceramic disc capable of providing heat to the thermostatic coil when electrically energized (40 CFR 85.2122(a)(2)(iii)(i)-91).

Positive test result:
(1) Any resolvable gas chromatographic peak for any 2,3,7,8-HDD or HDF which exceeds the LOQ listed under 40 CFR 766.27 for that congener; or
(2) Exceeds LOQs approved by EPA under 40 CFR 766.28 (40 CFR 766.3-91).

Possession or control: In the possession or control of any person, or of any subsidiary, partnership in which the person is a general partner, parent company, or any company or partnership which the parent company owns or controls, if the subsidiary, parent company, or other company or partnership is associated with the person in the research, development, test marketing, or commercial marketing or the substance in question. Information is in the possession or control of a person if it is:
(1) In the person's own files including files maintained by employees of the person in the course of their employment.
(2) In commercially available data bases to which the person has purchased access.
(3) Maintained in the files in the course of employment by other agents of the person who are associated with research, development, test marketing or commercial marketing of the chemical substance in question (40 CFR 704.3-91).

Post aeration: Adding air into a treated effluent to increase the content of dissolved oxygen (*see* aeration for more related terms).

Post chlorination: Adding chlorine into a treated effluent to further disinfect or to prevent microbial growth in the effluent (cf. prechlorination).

Post closure: The time period following the shutdown of a waste management or manufacturing facility. For monitoring purposes, this is often considered to be 30 years (*see* closure for more related terms) (EPA-89/12).

Post closure plan: The plan for post-closure care prepared in accordance with the requirements of 40 CFR 264.117 through 40 CFR 264.120 (*see* closure for more related terms) (40 CFR 264.141; 265.141-91).

Post consumer recovered paper:
(1) Paper, paperboard and fibrous wastes from retail stores, office buildings, homes and so forth, after they have passed through their end-usage as a consumer item including: used corrugated boxes; old newspapers; old magazines; mixed waste paper; tabulating cards and used cordage; and
(2) All paper, paperboard and fibrous wastes that enter and are collected from municipal solid waste (40 CFR 248.4-91).
(3) *See* paper for more related terms.

Post consumer recycling: The reuse of materials generated from residential and commercial waste, excluding recycling of material from industrial processes that has not reached the consumer, such as glass broken in the manufacturing process (EPA-89/11).

Post consumer waste (or old scrap): A material or product that has served its intended use and has been discarded for disposal or recovery after passing through the hands of a final consumer (*see* waste for more related terms) (40 CFR 246.101-91, *see also* 40 CFR 247.101-91).

Post exposure period: The portion of the test that begins with the test birds being returned from a treated diet to the basal diet. This period is typically 3 days in duration, but may be extended if birds continue to die or demonstrate other toxic effects (40 CFR 797.2050-91).

Post flame zone: A three-dimensional space in an incinerator where high temperature oxidation occurs in absence of emission of visible radiation and large exothermic chemical reactions; temperature gradients; radical concentrations and reaction rate substantially lower than flame zone (EPA-88/12).

Post impregnation suspension process: A manufacturing process in which polystyrene beads are first formed in a suspension process, washed, dried, or otherwise finished and then added with a blowing agent to another reactor in which the beads and blowing agent are reacted to produce expandable polystyrene (40 CFR 60.561-91).

Post mining area:
(1) A reclamation area; or
(2) The underground workings of an underground coal mine after the extraction, removal, or recovery of coal from its natural deposit has ceased and prior to bond release (40 CFR 434.11-91).

Post removal site control: Those activates that are necessary to sustain the integrity of a Fund-financed removal action following its conclusion. Post-removal site control may be a removal or remedial action under CERCLA. The term includes, without being limited to, activities such as relighting gas flares, replacing filters, and collecting leachate (40 CFR 300.5-91).

Post translation modification: An alteration of a polypeptide that occurs after translation of the mRNA (e.g., glycosylation) (EPA-88/09a).

Post treatment: Treatment of treated water or wastewater to improve the water quality.

Postictally: After a stroke or seizure, such as an acute epileptic attack (LBL-76/07-bio).

Pot furnace: A glass melting furnace that contains one or more refractory vessels in which glass is melted by indirect heating. The openings of the vessels are in the outside wall of the furnace and are covered with refractory stoppers during melting (*see* furnace for more related terms) (40 CFR 61.161-91; AP-40, p238).

Potable water (or drinking water): The water that is safe for drinking and cooking (*see* water for more related terms) (EPA-89/12).

Potash: Potassium oxide (K_2O). A carbonate of potassium (EPA-83). Potassium carbonate can be obtained from wood ashes. The mineral potash is usually a muriate (chloride). Caustic potash is its hydrated form (EPA-83/09).

Potassium (K): A alkali metal with a.n. 19; a.w. 39.098; d. 0.86 g/cc; m.p. 63.7° C and b.p. 760° C. The element belongs to group IA of the periodic table. Major potassium compounds include:
- Potassium cyanide (KCN): Toxic crystals used for insecticide, electroplating, etc.
- Potassium dichromate ($K_2Cr_2O_7$): Toxic crystals used for electroplating, matches, etc.
- Potassium manganate (K_2MnO_4): Water soluble crystals used for disinfectant, photography, printing and water purification.
- Potassium permanganate ($KMnO_4$): Water soluble crystals used for disinfectant, dyes, bleaches, etc.

Potential combustion concentration: The theoretical emissions (ng/J, lb/million Btu heat input) that would result from combustion of a fuel in an uncleaned state (without emission control systems) and:
(1) For particulate matter is: (A) 3,000 ng/J (7.0 lb/million Btu) heat input for solid fuel; and (B) 75 ng/J (0.17 lb/million Btu) heat input for liquid fuels.
(2) For sulfur dioxide is determined under 40 CFR 60.48a(b).
(3) For nitrogen oxides is: (A) 290 ng/J (0.67 lb/million Btu) heat input for gaseous fuels; (B) 310 ng/J (0.72 lb/million Btu) heat input for liquid fuels; and (C) 990 ng/J (2.30 lb/million Btu) heat input for solid fuels (40 CFR 60.41a-91).

Potential damage: The circumstances in which:
(1) Friable ACBM is in an area regularly used by building occupants, including maintenance personnel, in the course of their normal activities.
(2) There are indications that there is a reasonable likelihood that the material or its covering will become damaged, deteriorated, or delaminated due to factors such as changes in building use, changes in operations and maintenance practices, changes in occupancy, or recurrent damage (40 CFR 763.83-91).

Potential electrical output capacity: Is defined as 33 percent of the maximum design heat input capacity of the steam generating unit (e.g., a steam generating unit with a 100MW (340 million Btu/hr) fossil-fuel heat input capacity would have a 33MW potential electrical output capacity). For electric utility combined cycle gas turbines the potential electrical output capacity is determined on the basis of the fossil-fuel firing capacity of the steam generator exclusive of the heat input and electrical power contribution by the gas turbine (40 CFR 60.41a-91).

Potential energy: The energy which is possessed by a body as a consequence of its position, shape or configuration (*see* energy for more related terms).

Potential for industry-wide application: That an innovative technology can be applied in two or more facilities which are in one or more industrial categories (40 CFR 125.22-91).

Potential hydrogen chloride emission rate: The hydrogen chloride emission rate that would occur from combustion of MSW in the absence of any hydrogen chloride emissions control (40 CFR 60.51a-91).

Potential production allowances: The production allowances obtained under 40 CFR 82.9(a) (40 CFR 82.3-91).

Potential significant damage: The circumstances in which:
(1) Friable ACBM is in an area regularly used by building occupants, including maintenance personnel, in the course of their normal activities.
(2) There are indications that there is a reasonable likelihood that the material or its covering will become damaged, deteriorated, or delaminated due to factors such as changes in building use, changes in operations and maintenance practices, changes in occupancy, or recurrent damage.
(3) The material is subject to major or continuing disturbance, due to factors including, but not limited to, accessibility or, under certain circumstances, vibration or air erosion (40 CFR 763.83-91).

Potential sulfur dioxide emission rate: The theoretical sulfur dioxide emissions (ng/J, lb/million Btu heat input) that would result from combusting fuel in an uncleaned state and without using emission control systems (40 CFR 60.41b-91, *see also* 40 CFR 60.41c; 60.51a-91).

Potential to emit: The maximum capacity of a stationary source to emit a pollutant under its physical and operational design. Any physical or operational limitation on the capacity of the source to emit a pollutant including air pollution control equipment and restrictions on hours of operation or on the type or amount of material combusted, stored, or processed, shall be treated as part of its design if the limitation or the effect it would have on emissions is federally enforceable. Secondary emissions do not count in determining the potential to emit of a stationary source (40 CFR 51.166-91, *see also* 40 CFR 51.165; 51-App/S; 51.301; 52.21; 52.24; 66.3-91).

Potentially responsible party (PRP): Any individual(s) or company(ies) (such as owner, operators, transporters, or generators) potentially responsible under sections 106 or 107 of CERCLA for the contamination problems at a Superfund site (40 CFR 35.4010-91, *see also* 40 CFR 35.6015; 304.12-91; EPA-89/12).

Potentiation: The effect of one chemical to increase the effect of another chemical (Course 165.6).

Potentiometer: An instrument used to measure or to divide small voltage.

Potentiometric titration: A solution titration in which the end point is determined by measuring the potential on an electrode immersed in the solution (*see* titration for more related terms).

Potentiometric surface:
(1) An imaginary surface representing the static head of ground water, of which the water table is one type (Course 165.7).
(2) Contoured water-level elevations for wells completed in an unconfined aquifer (DOE-91/04).

Potentiometric surface map: A water-level map which is a graphical representation of a water gradient and can be prepared by plotting water-level measurements on a base map and then drawing contours (EPA-87/03).

Potentiometry: A technique of using a potentiometer to measure electromotive forces.

Potroom: A building unit which houses a group of electrolytic cells in which aluminum is produced (40 CFR 60.191-91).

Potroom group: An uncontrolled potroom, a potroom which is controlled individually, or a group of potrooms or potroom segments ducted to a common control system (40 CFR 60.191-91).

POTW treatment plant: That portion of the POTW which is designed to provide treatment (including recycling and reclamation) of municipal sewage and industrial waste (40 CFR 403.3-91).

POTW pretreatment program: *see* synonym, approved POTW pretreatment program.

Pound atomic weight: The atomic weight of an element expressed in pounds (cf. gram atomic weight).

Pour point temperature: The lowest temperature at which an oil will flow or can be poured under specified conditions of tests (*see* temperature for more related terms) (OME-88/12).

Pouring: The removal of blister copper from the copper converter bath (40 CFR 61.171-91).

Powder coating: Any surface coating which is applied as a dry powder and is fused into a continuous coating film through the use of heat (40 CFR 60.311; 60.451-91).

Powder forming: Includes forming and compressing powder into a fully dense finished shape, and is usually done within closed dies (40 CFR 471.02-91).

Powder or dry solid form: A state where all or part of the substance would have the potential to become fine, loose, solid particles (40 CFR 721.3-91).

Power cycle: A device such as a heat engine which is used to produce power continuously (Holman-p397).

Power law wind profile: The empirical equation describing the variation of the wind speed (u) with height (z) in the following form: $u = u_1 (z/z_1)^p$, where: z_1 = reference height with the reference wind speed u_1; and p = constant. In general, the exponent p is a function of stability, roughness and also height (*see* wind for more related terms) (NATO-78/10).

Power regulator: A transformer used to maintain constant output current for changes in temperature output load, line current, and time (EPA-83/03).

Power setting: The power or thrust output of an engine in terms of kilonewtons thrust for turbojet and turbofan engines and shaft power in terms of kilowatts for turboprop engines (40 CFR 87.1-91).

Power transformer: A transformer used at a generating station to step up the initial voltage to high levels for transmission (*see* transformer for more related terms) (EPA-83/03).

Powered spray module: *See* synonym, spray module.

Powerplant: Equipment that produces electrical energy generally by conversion from heat energy produced by chemical or nuclear reaction (EPA-82/11f).

Pozzolan: A finely-divided inorganic material which will react with lime and water to give a hardened product. Pozzolans are frequently used as ingredients of portland cement concrete. Certain fly ashes have useful pozzolanic properties (EPA-83). Other pozzolan related terms include:
- Calcined pozzolan
- Natural pozzolan

ppb (part per billion): A unit of measure of the concentration of gases in air expressed as parts of the gas per billion (10^9) parts of the air gas mixture, normally both by volume (LBL-76/07-air). It is a concentration unit defined as 10^{-3} times the concentration in ppm.

ppm (parts per million):
(1) (By volume) means a volume/volume ratio which expresses the volumetric concentration of gaseous air contaminant in a million unit volume of gas (40 CFR 52.741-91), e.g., if one cubic foot of benzene vapor is mixed with 1 million cubic feet of air, the benzine concentration in air is 1 ppm. The same analysis can be extended to ppb, parts per billion (EPA-84/09). In wastewater treatment, the unit commonly used to represent the degree of pollutant concentration where the concentrations are small. Larger concentrations are given in percentages. Thus BOD is represented in ppm while suspended solids in water are expressed in percentages.
(2) (By weight) ppm is equal to milligrams per liter divided by the specific gravity. It should be noted that in water analysis ppm is always understood to imply a weight/weight ratio, even though in practice a volume may be measured instead of a weight (EPA-76/03).

Practical quantitation limit (PQL): A correction factor, sometimes arbitrarily defined, used to account for uncertainty in measurement precision (EPA-91/03).

Praseodymium (Pr): A rare earth metal with a.n. 59; a.w. 140.907; d. 6.77 g/cc; m.p. 935° C and b.p. 3127° C. The element belongs to group IIIB of the periodic table.

Pre-aeration: A preparatory treatment of sewage consisting of aeration to remove gases and add oxygen or to promote the flotation of grease and aid coagulation (*see* aeration for more related terms) (EPA-76/03).

Pre-application: The phase of the Grants Program where the Grantee--frequently with the assistance of a consulting engineer--prepares an application for Step 1 grant funds (EPA-80/08).

Pre-certification vehicle: An uncertified vehicle which a manufacturer employs in fleets from year to year in the ordinary course of business for product development, production method assessment, and market promotion purposes, but in a manner not involving lease or sale (40 CFR 85.1702-91).

Pre-certification vehicle engine: An uncertified heavy-duty engine owned by a manufacturer and used in a manner not involving lease or sale in a vehicle employed from year to year in the ordinary course of business for Product development, production method assessment, and market promotion purposes (40 CFR 85.1702-91).

Pre-chlorination:
(1) Chlorination of water (adding chlorine into water) prior to filtration.
(2) Chlorination of sewage prior to treatment to restrict microbial growths in the sand filter (cf. post chlorination) (EPA-83/03).

Pre-consumer waste: The waste generated in processing materials or manufacturing them into final products (*see* waste for more related terms) (OTA-89/10).

Pre-existing discharge: Any discharge at the time of permit application under this subsection (CWA301.p-33USC1311-91).

Pre-plating treatment waste: The waste contributed by preplating treatments. The waste is affected by the basis materials, any surface soil on the workpieces, formulation of solutions used for cleaning or activating the materials, solution temperatures, and cycling times (*see* waste for more related terms) (EPA-74/03d).

Precalciner kiln: Identical to a preheater kiln except for a secondary firing in the tower of heat-exchanging cyclones which nearly completes the calcination of the feed material (ETI-92).

Precious metals: Gold, platinum, palladium, rhodium, iridium, osmium, and ruthenium (*see* metal for more related terms) (40 CFR 421.261-91, *see also* 40 CFR 466.02; 471.02; 468.02-91).

Precipitate: A solid that separates from a solution because of some chemical or physical change (EPA-89/12).

Precipitation:
(1) In air pollution control, the process of charging, collecting, and removing of particles. (*see also* electrostatic precipitator) (EPA-84/09).
(2) In wastewater treatment, a physicochemical process whereby some or all of substance in solution is transformed into a solid phase and is subsequently removed from the solution, e.g., zinc chloride is highly soluble in water, as is sodium sulfide. Zinc sulfide, however, has an extremely low solubility in water. Thus if an aqueous solution of zinc chloride is mixed with an aqueous solution of sodium sulfide, zinc ions and sulfide ions will rapidly combine to form solid zinc sulfide particles. Precipitation, flocculation and sedimentation are discussed together because in waste treatment they are most commonly used together, as consecutive treatments to the same stream. Precipitation removes a substance in solution and transforms it into a second phase, often in the form of solid particles that may be small or even colloidal. Flocculation transforms small suspended particles into larger suspended particles so that they can be more easily removed. Sedimentation removes the suspended particles from the liquid.
(3) In meteorology, any form of water, whether liquid or solid, that falls to the ground from the atmosphere; it includes drizzle, rain, snow, snow pellets, snow grains, ice crystals, ice pellets, and bail; the amount of precipitation is usually expressed in inches of equivalent liquid water depth at a given point over a specified period of time (DOI-70/04).

Precipitation can be the result of:
(1) Uneven heating of the ground, which causes the air to rise and expand, vapor to condense, and precipitation to occur (conventional precipitation).
(2) Topographic barriers that force the moisture laden air to rise and cool (orographic precipitation) (EPA-87/03).

Other precipitation related terms include:
- Meteorological precipitation
- Thermal precipitation
- Ultrasonic precipitation

Precipitation bath: The water, solvent, or other chemical bath into which the polymer or prepolymer (partially reacted material) solution is extruded, and that causes physical or chemical changes to occur in the extruded solution to result in a semihardened polymeric fiber (40 CFR 60.601-91).

Precipitation hardening metal: Certain metal compositions which respond to precipitation hardening or aging treatment (EPA-83/06a).

Precipitation supernatant: A liquid or fluid forming a layer above precipitated solids (EPA-83/03a).

Precipitator: Air pollution control devices that collect particles from an emission (EPA-89/12).

Precision:
(1) The capability of a person, an instrument or a method to obtain reproducible results; specifically, a measure of the random error as expressed by the variance, the standard error or a multiple of the standard error (*see* analytical parameters--laboratory for more related terms) (EPA-83).
(2) *See also* definition under uncertainty.

Precoat: A coating operation in which a coating other than an adhesive or release is applied to a surface during the production of a pressure sensitive tape or label product (40 CFR 60.441-91).

Precoat filter: A type of filter in which the media is applied to an existing surface prior to filtration (*see* filter for more related terms) (EPA-75/02d).

Preconditioning: The operation of an automobile through one (1) EPA Urban Dynamometer Driving Schedule, described in 40 CFR 86 (40 CFR 610.11-91).

Precontrolled vehicle: The light duty vehicles sold nationally (except in California) prior to the 1968 model year and light duty vehicles sold in California prior to the 1966 model year (*see* vehicle for more related terms) (40 CFR 51-App/N-91).

Precook: Prehydrolysis (EPA-87/10).

Precursor:
(1) A chemical substance which is not contaminated due to the process conditions under which it is manufactured, but because of its molecular structure, and under favorable process conditions, it may cause or aid the formation of HDDs/HDFs in other chemicals in which it is used as a feedstock or intermediate (40 CFR 766.3-91).
(2) In photochemical terminology, a compound such as a volatile organic compound (VOC) that precedes an oxidant. Precursors react in sunlight to form ozone or other photochemical oxidants (EPA-89/12).

Pregnant liquor: A solution containing the metal values prior to their removal and recovery (*see* liquor for more related terms) (EPA-75/02b).

Pregnant solution: A value bearing solution in a hydrometallurgical operation (EPA-82/05).

Preganglionic transmission block: Blocking of nerve impulse transmission at the synapse before the nerve enters the ganglion (LBL-76/07-bio).

Pregnant solvent: In solvent extraction, the value-bearing solvent produced in the solvent extraction circuit (EPA-82/05).

Preheater: In air preheating, a unit used to heat the air needed for combustion of absorbing heat from the products of combustion (EPA-82/11f).

Preheater kiln: A cement kiln equipped with a tower of heat-exchanging cyclones which preheat and partially calcine the feed material (ETI-92).

Prehydrolysis: The pre-steaming of wood chips in the digester prior to cooking; usually associated with improved bleaching of kraft pulps (EPA-87/10).

Preliminary analysis: The engineering analysis performed by EPA prior to testing prescribed by the Administrator based on data and information submitted by a manufacturer or available from other sources (40 CFR 610.11-91).

Preliminary assessment (PA):
(1) Review of existing information and an off-site reconnaissance, if appropriate, to determine if a release may require additional investigation or action. A PA may include an on-site reconnaissance, if appropriate (40 CFR 300.5-91).
(2) The process of collecting and reviewing available information about a known or suspected waste site or release (EPA-89/12).

Preliminary filter: The use of quick gravity filters or microstainers for raw wastewater prior to the slow sand filters to increase the length of their run (*see* filter for more related terms).

Preliminary remediation goals (PRGs): Initial clean-up goals that:
(1) Are protective of human health and he environment and
(2) Comply with ARARs. They are developed early in the process based on readily available information and are modified to reflect results of the baseline risk assessment. They :so are used during analysis of remedial alternatives in the remedial investigation/feasibility study (RI/FS) (EPA-91/12).

Preliminary treatment: *See* synonym, pretreatment.

Premanufacture notice or PMN: Any notice submitted to EPA pursuant to 40 CFR 720 of this chapter or 40 CFR 723.250 of this chapter (40 CFR 700.43-91).

Under Section 5(a) of TSCA, a manufacturer must notify EPA ninety days before producing a new chemical substance. Within five days of receiving the notice, EPA must publish in the Federal Register an item identifying the chemical substance, listing its intended uses, and a description of the toxicological tests required to demonstrate that there will no unreasonable risk of injury to health or the environment (Arbuckle-89).

Premature dementia: Premature organic brain deterioration (LBL-76/07-bio).

Premises: A tract or parcel of land with or without habitable buildings (EPA-83).

Premixed flame: A flame produced by a combustion process in which the fuel and air are mixed prior to entering the combustion zone. Premixed flame combustion is largely applied to relatively small devices such as automobile engines and laboratory burners, and is usually limited to fuels that are gaseous at ambient temperature or to those that vaporize at relatively low temperatures (*see* flame for more related terms).

Preponderance of evidence: Proof by information that, compared with that opposing it, leads to the conclusion that the fact at issue is more probably true than not (40 CFR 32.105-91).

Present worth: An estimate of the amount of money which would now equal all future costs of the wastewater system. It is the sum of all construction and O&M costs for the design period, discounted to the present (EPA-80/08).

Preservative: Chemicals used to preserve the quality of materials, e.g., the chemicals added to water thinned paints to prevent the growth of bacteria or yeast in the can during paint storage (EPA-79/12b).

Preserve: To prevent modification to the natural floodplain environment or to maintain it as closely as possible to its natural state (40 CFR 6-App/A-91).

Press and blow process: A process of glass manufacture in which the finish and parison are pressed and the parison is subsequently blown to form the final shape (EPA-83).

Press liquor:
(1) Stick water resulting from the compaction of recovered fish solids (EPA-74/06).
(2) The liquor obtained when citrus peel is chopped, treated with lime, and pressed or squeezed (EPA-74/03).
(3) *See* liquor for more related terms.

Pressed and blown glass:
(1) The glass includes the broad classifications of kitchen and tableware, art objects, novelty items and the like, lighting and electronic glassware, and insulation glassware, insulation and manufactured products using glass fibre (EPA-83).
(2) A glass which is pressed, blown, or both, including textile fiberglass, noncontinuous flat glass, noncontainer glass, and other products listed in SIC 3229. It is separated into:
   (A) Glass of borosilicate recipe.
   (B) Glass of soda-lime and lead recipes.
   (C) Glass of opal, fluoride, and other recipes (40 CFR 60.291-91).
(3) *See* glass for more related terms.

Pressed glass: A glassware formed by pressure between a mold and a plunger (*see* glass for more related terms) (EPA-83).

Pressed powder: A method of making an electrode by pressing powdered active material into a metal grid (EPA-84/08).

Pressed wood product: A group of materials used in building and furniture construction that is made from wood veneers, particles, or fibers bonded together with an adhesive under heat and pressure (EPA-88/09b).

Pressure:
(1) The total load or force per unit area acting on a surface (40 CFR 146.3-91, *see also* 40 CFR 146.3; 147.2902; 147.2902-91).
(2) The normal component of force per unit area. Various pressure units are shown below (Course 502, p20, Student Workbook):
1 atm = 14.6959 pounds/inch^2 (psia)
= 2116 pounds/foot^2 absolute (psfa)
= 33.91 ft. water
= 406.9 in. water
= 29.92 in. mercury (in Hg) at 32° F
= 760 millimeters of mercury (mm Hg)
= 760 Torr
= 1.013 E+06 dynes/cm^2
= 1.013x10^5 newtons/m^2
1 inch Hg = 0.491 psi
1 micron Hg = 10^{-6} m = 10^{-3} mm Hg
= 1.933 x 10^{-05} psi
1 microbar = 1 dyne/square cm
1 torr = 1 mm Hg

Other pressure related terms include:
- Absolute pressure
- Atmospheric pressure
- Barometric pressure (*see* synonym, atmospheric pressure)
- Critical pressure
- Dew point pressure
- Gauge (gage) pressure
- Partial pressure
- Saturated pressure (*see* synonym, saturation pressue)
- Saturation pressure
- Standard atmospheric pressure
- Static pressure
- Surge pressure
- Total pressure
- Vacuum pressure
- Velocity pressure

Pressure check: An imperfection; a check or crack in a glass article resulting from too much pressure in forming (EPA-83).

Pressure deformation: The process of applying force, (other than impact force), to permanently deform or shape a workpiece. Pressure deformation operations may include operations such as rolling, drawing, bending, embossing, coining, swaging, sizing, extruding, squeezing, spinning, seaming, piercing, necking, reducing, forming, crimping, coiling, twisting, winding, flaring or weaving (EPA-83/06a).

Pressure demand type apparatus: An apparatus in which the pressure inside the facepiece in relation to the immediate environment is positive during both inhalation and exhalation (NIOSH-84/10).

Pressure drop (or gas-side pressure drop):
(1) A measure, in kilopascals, of the difference in static pressure measured immediately upstream and downstream of the air filter element (40 CFR 85.2122(a)(16)(ii)(B)-91).
(2) The difference in static pressure between two points due to energy losses in a gas stream (EPA-89/03b).
(3) In wet scrubbing, the pressure difference, or pressure drop that occurs as the exhaust gases pushed or pulled through the scrubber, disregarding the pressure that would be used for pumping or spraying the liquid into the scrubber. The terms pressure drop and gas-side pressure drop are used interchangeably (EPA-84/03-p1/9).

Pressure field extension: A spatial extension of a variation in pressure as occurs under a slab when a fan ventilates at one or a few distinct points (EPA-88/08).

Pressure filter: A filter in which the pressure on the input side of the filter medium is greater than atmospheric pressure (*see* filter for more related terms) (DOI-70/04).

Pressure filtration: The process of solid/liquid phase separation effected by passing the more permeable liquid phase through a mesh which is impenetrable the solid phase (*see* filtration for more related terms) (EPA-82/11e).

Pressure gauge: A gauge that measures the pressure difference between a given fluid and that of the atmosphere. Other pressure gauge related terms include:
- Barometer
- Manometer

Pressure head: The height of a column of a fluid required to produce a given pressure at its base. The relationship between pressure and pressure-head is: $p = dhg/g_c$. Where: p = pressure (force/area); d = density of fluid (mass/volume); g = local acceleration due to gravity, length/time^2; g_c = dimensional constant; and h = pressure-head in terms of length. Pressure-head may be expressed in terms of any fluid that is convenient, e.g., Hg or H_2O. **Effective head** = pressure head - head due to friction, etc.

Pressure process: A process in which wood preservatives and fire retardants are forced into wood using air or hydrostatic pressure (EPA-74/04).

Pressure release: The emission of materials resulting from system pressure being greater than set pressure of the pressure relief device (40 CFR 60.481-91, *see also* 40 CFR 61.241; 264.1031-91).

Pressure roller: In seaming a landfill liner, rollers accompanying a seaming technique which apply pressure to the opposing FML (flexible membrane liner) sheets to be joined. They closely follow the actual melting process and are self-contained within the seaming device (EPA-89/09, *see also* EPA-91/05).

Pressure sewer: A system of pipes in which water, wastewater, or other liquid is transported to a higher elevation by use of pumping force (*see* sewer for more related terms) (EPA-89/12, *see also* EPA-88/08).

Pretreatment (or preliminary treatment):
(1) The reduction of the amount of pollutants, the elimination of pollutants, or the alteration of the nature of pollutant properties in wastewater prior to or in lieu of discharging or otherwise introducing such pollutants into a POTW (for complete definition, *see* 40 CFR 403.3-91).
(2) Processes used to reduce, eliminate, or alter the nature of wastewater pollutants from non-domestic sources before they are discharged into publicly owned treatment works (EPA-89/12).

Pretreatment (or preliminary treatment):
(1) The reduction of the amount of pollutants, the elimination of pollutants, or the alteration of the nature of pollutant properties in wastewater prior to or in lieu of discharging or otherwise introducing such pollutants into a POTW. The reduction or alteration may be obtained by physical, chemical or biological processes, process changes or by other means, except as prohibited by 40 CFR 403.6(d). Appropriate pretreatment technology includes control equipment, such as equalization tanks or facilities, for protection against surges or slug loadings that might interfere with or otherwise be incompatible with the POTW. However, where wastewater from a regulated process is mixed in an equalization facility with unregulated wastewater or with wastewater from another regulated process, the effluent from the equalization facility must meet an adjusted pretreatment limit calculated in accordance with 40 CFR 403.6(e) (40 CFR 403.3-q-91).
(2) Processes used to reduce, eliminate, or alter the nature of wastewater pollutants from non-domestic sources before they are discharged into publicly owned treatment works (POTWs) (EPA-94/04).

Pretreatment control authority:
(1) The POTW if the POTW's submission for its pretreatment program has been approved in accordance with the requirements of 40 CFR 403.11; or
(2) The Approval Authority if the submission has not been approved (40 CFR 414.10-91).

Pretreatment requirements: Any substantive or procedural requirement related to Pretreatment, other than a National Pretreatment Standard, imposed on an Industrial User (40 CFR 403.3-91).

Pretreatment standard: *See* synonym, national pretreatment standard.

Prevalence study: An epidemiological study which examines the relationships between diseases and exposures as they exist in a defined population at a particular point in time (Course 165.6).

Prevention:
(1) Measures taken to minimize the release of wastes to the environment (cf. waste minimization) (EPA-89/12).
(2) Design and operating measures applied to a process to ensure that primary containment of toxic chemicals is maintained. Primary containment means confinement of toxic chemicals within the equipment intended for normal operating conditions (EPA-87/07a).

Prevention of significant deterioration (PSD):
(1) EPA program in which state and/or federal permits are required that are intended to restrict emissions for new or modified sources in places where air quality is already better than required to meet primary and secondary ambient air quality standards (EPA-89/12).
(2) PSD means that sources in "clean air" areas do not have the right to pollute the air even if the National Ambient Air Quality Standards would not be violated. Sources must meet permit standards that ensure that there is no significant deterioration of air quality in a region. PSD requirements must be part of all SIPs (State Implementation Plans) (Winthrop-89/09).

Preventive measures: The actions taken to reduce disturbance of ACBM or otherwise eliminate the reasonable likelihood of the material's becoming damaged or significantly damaged (40 CFR 763.83-91).

Price analysis: The process of evaluating a prospective price without regard to the contractor's separate cost elements and proposed profit. Price analysis determines the reasonableness of the proposed subagreement price based on adequate price competition, previous experience with similar work, established catalog or market price, law, or regulation (40 CFR 33.005-91, *See also* 40 CFR 35.6015-91).

Primary air: *See* synonym, primary combustion air.

Primary air fan: A fan for transport, ignition and combustion stabilization air (*See* fan for more related terms) (EPA-83).

Primary alcohol: An alcohol whose OH attached carbon joins at least with two hydrogen atoms. Its molecular structure can be expressed as RCH$_2$OH, e.g., methanol (HCH$_2$OH) (*See* alcohol for more related terms).

Primary aluminum reduction plant: Any facility manufacturing aluminum by electrolytic reduction (40 CFR 60.191-91).

Primary amine: An amine whose molecular structure can be expressed as RNH$_2$, e.g., methylamine (*See* amine for more related terms) (CH$_3$NH$_2$).

Primary battery: A battery which must usually be replaced after one discharge; i.e., the battery cannot be recharged (*See* battery for more related terms) (EPA-84/08).

Primary burner (or primary combustion chamber burner): A fuel burner for preheating combustion chamber, igniting waste, and maintaining temperature in the primary chamber (*See* burner for more related terms) (EPA-89/03b).

Primary clarifier (primary settling tank or primary sedimentation tank): The settling tank into which the wastewater (sewage) first enters and from which the solids are removed as raw sludge (*See* clarifier for more related terms) (EPA-76/03).

Primary chamber: *See* synonym, primary combustion chamber.

Primary combustion air (or primary air):
(1) Air admitted to a combustion system at the point where the fuel is first oxidized (SW-108ts).
(2) Air for combustion supplied to combustion devices such as furnaces and incinerators to atomize fuel liquid droplets.
(3) *See* combustion air for more related terms.

Primary combustion chamber (or primary chamber):
(1) A combustion chamber where waste is fed and combustion begins. It is the first chamber of multi-combustion chamber incinerators, e.g., a rotary kiln is the primary combustion chamber of an incineration system (EPA-89/03b).
(2) Chamber wherein primary ignition and burning occurs (EPA-83).
(3) *See* combustion chamber for more related terms.

Primary combustion chamber air blower: A forced air blower for providing underfire combustion air to the primary combustion chamber (*See* blower for more related terms) (EPA-89/03b).

Primary combustion chamber burner: *See* synonym, primary burner.

Primary combustion chamber water spray: A spray system to inject a fine water mist into the primary chamber to assist in temperature control (*See* combustion chamber for more related terms) (EPA-89/03b).

Primary combustion chamber underfire steam injection: A steam injection system to inject steam into ash to assist in temperature control (*See* combustion chamber for more related terms) (EPA-89/03b).

Primary containment: The containment provided by the piping, vessels and machinery used in a facility for handling chemicals under normal operating conditions (*See* containment for more related terms) (EPA-87/07a).

Primary control system: An air pollution control system designed to remove gaseous and particulate flourides from exhaust gases which are captured at the cell (40 CFR 60.191-91).

Primary coolant system boundary: In a modular high-temperature gas-cooled reactor, those systems and components that contain the primary coolant. The primary coolant system boundary includes the reactor vessel, circulators, steam generators, heat exchangers, associated primary coolant system piping and valves, and any other system that connects with the primary coolant system to the second isolation valve (DOE-91/04).

Primary copper smelter: Any installation or any intermediate process engaged in the production of copper from copper sulfide ore concentrates through the use of pyrometallurgical techniques (*See* smelter for more related terms) (40 CFR 60.161-91, *See also* 40 CFR 61.171-91).

Primary drinking water regulation: A regulation which:
(1) Applies to public water systems;
(2) Specifies contaminants which, in the judgement of the Administrator, may have any adverse effect on the health of persons;
(3) Specifies for each such contaminant either:
   (A) A maximum contaminant level, if, in the judgment of the Administrator, it is economically and technologically feasible to ascertain the level of such contaminant in water in public water systems; or
   (B) If, in the judgment of the Administrator, it is not economically or technologically feasible to so ascertain the level of such contaminant, each treatment technique known to the Administrator which leads to a reduction in the level of such contaminant sufficient to satisfy the requirements of section 1412; and
(4) Contains criteria and procedures to assure a supply of drinking water which dependably complies with such maximum contaminant levels; including quality control and testing procedures to insure compliance with such levels and to insure proper operation and maintenance of the system, and requirements as to:
   (A) The minimum quality of water which may be taken into the system; and
   (B) Siting for new facilities for public water systems (cf. secondary drinking water regulation) (40 CFR 141; SDWA1401-42USC300f).
(5) *See* the term of Act or SDWA for more related terms.

Primary effluent: The effluent from the primary clarifiers (*See* effluent for more related terms).

Primary emission:
(1) Particulate matter emissions from the BOPF generated during the steel production cycle and captured by the BOPF primary control system (40 CFR 60.141-91, *See also* 40 CFR 60.141a-91).

(2) The material that is released into the air by a discrete source (SW-108ts).
(3) *See* emission for more related terms.

Primary emission control system: The combination of equipment used for the capture and collection of primary emissions (e.g., an open hood capture system used in conjunction with a particulate matter cleaning device such as an electrostatic precipitator or a closed hood capture used in conjunction with matter cleaning device such as a scrubber) (40 CFR 60.141a-91, *See also* 40 CFR 61.171; 61.181-91).

Primary feeder circuit (substation) transformer: A transformer used to reduce the voltage from the subtransmission level to the primary feeder level (*See* transformer for more related terms) (EPA-83/03).

Primary gout: A condition characterized by abnormal purine metabolism producing an excess of uric acid in the blood, chalky deposits chiefly urates) in the joints, and attacks of acute arthritis (LBL-76/07-bio).

Primary industry category: Any industry category listed in the NRDC settlement agreement (Natural Resources Defense Council et al. v. Train, 8 E.R.C. 2120 (D.D.C. 1976), modified 12 E.R.C. 1833 (D.D.C. 1979)); also listed in Appendix A of Part 122 (cf. secondary industry category) (40 CFR 122.2-91).

Primary lead smelter: Any installation or any intermediate process engaged in the production of lead from lead sulfide ore concentrates through the use of pyrometallurgical techniques (*See* smelter for more related terms) (40 CFR 60.181-91).

Primary manufacturing residues: Sawdust, chips, slabs and the like created from the basic conversion of roundwood into a lumber product. Sawmills and plywood and veneer mills are the principal operations creating primary manufacturing residues from their processing operations (EPA-83).

Primary material: The virgin or new materials used for manufacturing basic products. Examples include wood pulp, iron ore and silica sand (*See* material for more related terms) (EPA-83).

Primary maximum contaminant levels: The levels of water quality necessary, with an adequate margin of safety, to protect the public health (*See* the term of Act or SDWA for more related terms). *See* 40 CFR 141.11-141.16 for standards.

Primary metal cast: The metal that is poured in the greatest quantity at an individual plant (40 CFR 464.31-91).

Primary mill: Whose steel hot forming operations at reduce ingots to blooms or slabs by passing the ingots between rotating steel rolls. The first hot forming operation performed on solidified steel after it is removed from the ingot molds is carried out on a primary mill (40 CFR 420.71-91).

Primary oxygen blow: The period in the steel production cycle of a BOPF during which a high volume of oxygen-rich gas is introduced to the bath of molten iron by means of a lance inserted from the top of the vessel or though tuyeres in the bottom or through the bottom and sides of the vessel. This definition does not include any additional or secondary oxygen blows made after the primary blow or the introduction of nitrogen or other inert gas through tuyeres in the bottom or bottom and the sides of the vessel (40 CFR 60.141-91, *See also* 40 CFR 60.141a-91).

Primary panel: The surface that is considered to be the front surface or that surface which is intended for initial viewing at the point of ultimate sale or the point of distribution for use (40 CFR 211.203-91).

Primary physical containment: The set of operating practices and equipment characteristics that prevent the escape of viable recombinant organisms from a bioprocess (EPA-88/09a).

Primary pollutant:
(1) Pollutants emitted directly from a polluting stack.
(2) Pollutants remaining in the form as they are when emitted from the source (NATO-78/10).
(3) *See* pollutant for more related terms.

Primary processor of asbestos: A person who processes for commercial purposes bulk asbestos (*See* asbestos for more related terms) (40 CFR 763.63-91).

Primary recycling: The return of a secondary material to the same industry from which it came and processing of the secondary material so that it will yield the same or similar product which it was as a secondary material. Examples are the return of broken glass containers to glass container manufacturing plants for making new containers and the recycling of sheet steel scrap to steel furnaces for the manufacture of new sheet steel (EPA-83).

Primary resistor: A device used in the primary circuit of an inductive ignition system to limit the flow of current (40 CFR 85.2122(a)(10)(ii)-91).

Primary sedimentation tank: *See* synonym, primary clarifier.

Primary settling: The first settling unit for the removal of settleable solids through which wastewater is passed in a treatment works (*See* settling for more related terms) (EPA-84/08).

Primary settling tank: *See* synonym, primary clarifier.

Primary significance pollutant: The pollutants the are of primary significance if they are recommended for regulation due to their deleterious effects on humans and the environment (*See* pollutant for more related terms) (EPA-85/10).

Primary sludge: The sludge from the primary clarifiers (*See* sludge for more related terms) (EPA-76/03).

Primary standard: A national primary ambient air quality standard promulgated pursuant to section 109 of the Act (*See* standard for more related terms) (40 CFR 51.100-91).

Primary standard: For calibration application, those whose volumes can be calibrated simply by measuring dimensions alone. The measured internal dimensions are regular, and accuracies better than +/- 0.30% can be reached (*See* calibration of air flow for more related terms) (EPA-83/06).

Primary standard attainment date: The date specified in the applicable implementation plan for the attainment of a national primary ambient air quality standard for any air pollutant (CAA302-42USC7602-91).

Primary treatment: *See* synonym, primary wastewater treatment.

Primary wastewater treatment (or primary treatment): The first steps in wastewater treatment; screens and sedimentation tanks are used to remove most materials that floats or will settle. Primary treatment results in the removal of about 30 percent of carbonaceous biochemical oxygen demand from domestic sewage (*See* treatment or wastewater treatment for more related terms) (EPA-89/12).

Primary winding: The winding on the supply (i.e., input) side of a transformer (EPA-83/03).

Primary zinc smelter: Any installation engaged in the production, or any intermediate process in the production, of zinc or zinc oxide from zinc sulfide ore concentrates through the use of pyrometallurgical techniques (*See* smelter for more related terms) (40 CFR 60.171-91).

Prime coat: The first of two or more coatings applied to a surface (40 CFR 52.741-91, *See also* 40 CFR 60.721-91).

Prime coat operation: The prime coat spray booth or dip tank, flash-off area, and bake oven(s) which are used to apply and dry or cure the initial coating on components of automobile or light-duty truck bodies (40 CFR 60.391-91, *See also* 40 CFR 60.461-91).

Prime contractor: Any person holding a contract, and for the purposes of subpart B (General Enforcement, compliance Review, and complaint Procedure) of the rules, regulations, and relevant orders of the Secretary of labor, any person who has held a contract subject to the order (cf. contractor) (40 CFR 8.2-91).

Prime farmland: The same meaning as that previously prescribed by the Secretary of Agriculture on the basis of such factors as moisture availability, temperature regime, chemical balance, permeability, surface layer composition, susceptibility to flooding, and erosion characteristics, and which historically have been used for intensive agricultural purposes, and as published in the Federal Register (SMCRA701-30USC1291-90).

Prime mover: The engine, turbine, water wheel or similar machine which drives an electric generator (EPA-83).

Prime surfacer coat: A coating used to touch up areas on the surface of automobile or light-duty truck bodies not adequately covered by the prime coat before application of the top coat. The prime surfacer coat is applied between the prime coat and topcoat. An anti-chip coating applied to main body parts (e.g., rocker panels, bottom of doors and fenders, and leading edge of roof) is a prime surfacer coat (40 CFR 52.741-91).

Primers: Any coatings formulated and applied to substrates to provide a firm bond between the substrate and subsequent coats (40 CFR 52.741-91).

Principal organic hazardous constituents (POHCs):
(1) Organic compounds chosen to determine the DRE of a combustion device in a trial burn. POHCs are representative of the compounds in the waste stream that are the most abundant and the most difficult to destroy (ETI-92).
(2) Specific hazardous waste compounds in Appendix 8 of 40 CFR 261 selected for monitoring during the trial burn of a hazardous waste incinerator. They must represent the broad range of physical and chemical characteristics of the incinerator's normal waste feed. At least one of the POHCs must be recognized as being equally or more difficult to incinerate than any waste constituent which will be burned during normal operations.

Principal source aquifer: *See* synonym, sole source aquifer.

Principal study: The study that contributes most significantly to the qualitative and quantitative risk assessment (EPA-92/12).

Principle of (instrument) operation: The technique used to detect and measure the pollutant or parameter, and/or a description of the major components of the instrument (LBL-76/07-bio).

Printability: The ability of a paper surface to accept printing ink (EPA-87/10).

Printed circuit board: A circuit in which the interconnecting wires have been replaced by conductive strips printed, etched, etc., onto an insulating board. Methods of fabrication include etched circuit, electroplating, and stamping (EPA-83/06a).

Printing: The application of words, designs, and pictures to a substrate using ink (40 CFR 52.741-91).

Printing ink: An ink used in printing, impressing, stamping, or transferring on paper or paper-like substances, wood, fabrics, plastics, films or metals, by the recognized mechanical reproductive processes employed in printing, publishing and related services (*See* ink for more related terms) (EPA-79/12a).

Printing line: An operation consisting of a series of one or more roll printers and any associated roll coaters, drying areas, and ovens wherein one or more coatings are applied, dried, and/or cured (40 CFR 52.741-91).

Printing paper: The paper designed for printing, other than newsprint, such as offset and book paper (*See* paper for more related terms) (40 CFR 250.4-91).

Priority list for chemical testing: Under TSCA, a list of chemicals selected by a committee of eight members for testing if unreasonable risk to health or the environment is suspected. The priority list is supposed to be updated every six months under Section 4(e)(2) of TSCA. Forty two compounds were selected by the committee as of 1985 (Arbuckle-89).

Priority pollutant:
(1) In CWA, priority pollutants means the 126 pollutants listed in 40 CFR 423 Appendix A (40 CFR -App/A-92).

(2) Those pollutants listed by the Administrator under CWA Section 307(a) (EPA-91/03).
(3) *See* pollutant for more related terms.

Priority water quality areas: For the purposes of Section 35.2015, specific stream segments or bodies of water, as determined by the State, where municipal discharges has resulted in the impairment of a designated use or significant public health risks, and where the reduction of pollution from such discharges will substantially restore surface or groundwater uses (40 CFR 35.2005-91).

Private applicator: A certified applicator who uses or supervises the use of any pesticide which is classified for restricted use for purposes of producing any agricultural commodity on property owned or rented by him or his employer or (if applied without compensation other than trading of personal services between producers of agricultural commodities) on the property of another person (40 CFR 171.2-91).

Private collection:
(1) The collecting of solid wastes for which citizens or firms, individually or tn limited groups, pay collectors or private operating agencies. Also known as private disposal (EPA-83).
(2) The collection of solid waste by individuals or companies from residential, commercial, or industrial premises. The arrangements for the service are made directly between the owner or occupier of the premises and the collector (SW-108ts).
(3) *See* waste collection for more related terms.

Private solid waste utility: A private business that collects, processes, and disposes of solid waste under a government license or monopoly franchise (EPA-83).

Privately owned treatment works: Any device or system which is (a) used to treat wastes from any facility whose operator is not the operator of the treatment works and (b) not a POTW (cf. publicly owned treatment works) (40 CFR 122.2-91).

Probabilistic variable: A variable which uses the value, between 0 and 1, to express its occurrence distribution (*See* variable for more related terms).

Probability: A number expressing the likelihood of occurrence of a specific event, such as the ratio of the number of outcomes that will produce a given event to the total number of possible outcomes (EPA-85/09; 91/03).

Probability/potential: A measure, either qualitative or quantitative, that an event will occur within some unit of time (EPA-87/07a).

Probability distribution: A mathematical representation of the probabilities that a given variable will have various values (EPA-85/09; 91/03).

Probable maximum flood: Maximum flood predicted for a scenario having hydrological conditions that maximize the flow of surface waters (DOE-91/04).

Probe:
(1) A tube used for sampling or for measuring pressures at a distance from the actual collection or measuring apparatus. It is commonly used for reaching inside stacks and ducts (EPA-83/06).
(2) One of several devices which are driven into the ground and which allow soil gas to be collected and subsequently analyzed (Course 165.7).
(3) A tube used for sampling or for measuring pressures at a distance from the actual collection or measuring apparatus. It is commonly used for reaching inside stacks and ducts (LBL-76/07-air).

Probit model: A dose-response model of the form: $P(d) = 0.4$(integral from minus infinity to $[\log(d - u)]/s$ of $[\exp{-(y**2)/2}]dy$); where $P(d)$ is the probability of cancer from a continuous dose rate d, and u and s are constants (EPA-92/12).

Problem waste: Bulky wastes, dead animals, abandoned vehicles, construction and demolition wastes, industrial refuse, tree debris, mining spoils, fly ash etc. (*See* waste for more related terms) (EPA-83).

Procaryotic cell: The simple cell type, characterized by the lack of a nuclear membrane, the absence of mitochondria, and a haploid state at all times (EPA-88/09a).

Proceeding: Any rulemaking, adjudication, or licensing conducted by EPA under the Act or under regulations which implement the Act, except for determinations under this part (40 CFR 2.302-91, *See also* 40 CFR 2.304; 2.305; 2.310; 2.301; 2.303; 2.306; 17.2-91).

Process:
(1) The preparation of a chemical substance or mixture, after its manufacture, for distribution in commerce:
   (A) In the same form or physical state as, or in a different form or physical state from, that in which it was received by the person so preparing such substance or mixture; or
   (B) As part of an article containing the chemical substance or mixture (TSCA3, *See also* 40 CFR 52.741; 372.3; 704.3; 710.2; 716.3; 720.3; 723.175; 723.250; 747.115; 747.195; 747.200; 761.3; 762.3; 763.163-91).
(2) The sequence of physical and chemical operations for the production, refining, repackaging or storage of chemicals (EPA-87/07a).
(3) *See* synonym, thermodynamic process.

Other process related terms include:
- Dry process
- Semi-dry process
- Wet process

Process and size factor:
(1) The process factor is a factor, based upon process configuration, for calculating a petroleum refinery's BPT limitations.
(2) The size factor is a factor, based upon a petroleum refinery's size, for calculating a petroleum refinery's BPT limitations (EPA-79/12c).

Process area: The dimensional area directly involved in a particular processing step (expressed in terms of square feet and square meters) (EPA-82/11e).

Process configuration: A numerical measurement of a refinery's process complexity, developed for use in calculating BPT guidelines for this industry (EPA-79/12c).

Process emission: The particulate matter which is collected by a capture system (*See* emission for more related terms) (40 CFR 60.301-91, *See also* 40 CFR 61.171; 61.181; 704.3-91).

Process for commercial purposes: To process (1) for distribution in commerce, including for test marketing purposes, or (2) for use as an intermediate (40 CFR 710.2-91, *See also* 40 CFR 712.3; 716.3; 717.3; 721.3; 763.63-91).

Process fugitive emissions: The particulate matter emissions from an affected facility that are not collected by a capture system (*See* emission for more related terms) (40 CFR 60.381-91).

Process gas: Any gas generated by a petroleum refinery process unit, except fuel gas and process upset gas as defined in this section (40 CFR 60.101-91).

Process generated wastewater: Water directly or indirectly used in the operation of a feedlot for any or all of the following: Spillage or overflow from animal or poultry watering systems; washing, cleaning or flushing pens, barns, manure pits or other feedlot facilities; direct contact swimming, washing or spray cooling of animals; and dust control (*See* wastewater for more related terms) (40 CFR 412.11-91).

Process heater: A device that is primarily used to heat a material to initiate or promote a chemical reaction in which the material participates as a reactant or catalyst (40 CFR 60.41b, *See also* 40 CFR 60.41c; 60.561; 60.611; 60.661; 61.301; 264.1031-91).

Process improvement: The routine changes made for safety and occupational health requirements, for energy savings, for better utility, for ease of maintenance and operation, for correction of design deficiencies, for bottleneck removal, for changing product requirements, or for environmental control (40 CFR 60.481-91).

Process line: A group of equipment assembled that can operate independently if supplied with sufficient raw materials to produce polypropylene, polyethylene, polystyrene (general purpose, crystal, or expandable) or poly(ethylene terephthalate) or one of their copolymers. A process line consists of the equipment in the following process sections (to the extent that these process sections are present at a plant): raw materials preparation, polymerization reaction, product finishing, product storage, and material recovery (40 CFR 60.561-91).

Process machinery: Process equipment, such as pumps, compressors, heaters, or agitators, that would not be categorized as piping and vessels (EPA-87/07a).

Process modification: The reduction of water pollution through process modification (in-plant technology) (EPA-83/06a).

Process section: The equipment designed to accomplish a general but well-defined task in polymer production. Process sections include raw materials preparation, polymerization reaction, material recovery, product finishing, and product storage and may be dedicated to a single process line or common to more than one process line (40 CFR 60.561-91).

Process stream: All reasonably anticipated transfer, flow, or disposal of a chemical substance, regardless of physical state or concentration, through all intended operations of processing, including the cleaning of equipment (40 CFR 721.3-91).

Process unit: The components assembled to produce, as intermediate or final products, one or more of the chemicals listed in 40 CFR 60.489 of this part. A process unit can operate independently if supplied with sufficient feed or raw materials and sufficient storage facilities for the product (40 CFR 60.481-91, *See also* 40 CFR 60.561; 60.591; 60.611; 60.631; 60.661; 61.241; 61.341-91).

Process unit shutdown: A work practice or operational procedure that stops production from a process unit or part of a process unit. An unscheduled work practice or operational procedure that stops production from a process unit or part of a process unit for less than 24 hours is not a process unit shutdown. The use of spare equipment and technically feasible bypassing of equipment without stopping production are not process unit shutdowns (40 CFR 60.481-91, *See also* 40 CFR 61.241-91).

Process upset gas: Any gas generated by a petroleum refinery process unit as a result of start-up, shut-down, upset or malfunction (40 CFR 60.101-91).

Process vent: Any open-ended pipe or stack that is vented to the atmosphere either directly, through a vacuum-producing system, or through a tank (e.g., distillate receiver, condenser, bottoms receiver, surge control tank, separator tank, or hot well) associated with hazardous waste distillation, fractionation, thin-film evaporation, solvent extraction, or air or steam stripping operations (40 CFR 264.1031-91).

Process vessel: Each tar decanter, flushing-liquor circulation tank, light-oil condenser, light-oil decanter, wash-oil decanter, or wash-oil circulation tank (40 CFR 61.131-91).

Process waste: The waste material from an industrial process. Examples of process wastes are flue gas scrubber sludges, cement kiln dusts, sawmill dust and powder, spent solvents, contaminated oils, etc. (*See* waste for more related terms) (EPA-83, *See also* 40 CFR 129.2-91)

Process wastewater: Any water which, during manufacturing or processing, comes into direct contact with or results from the production or use of any raw material, intermediate product, finished product, byproduct, or waste product (*See* wastewater for more related terms) (40 CFR 117.1-91).

Process wastewater cooling tower: The water used for cooling purposes which may become contaminated through contact with process raw materials, intermediates, or final products (*See* cooling water for more related terms) (EPA-87/10a).

Process wastewater pollutant: The pollutants present in process wastewater (*See* pollutant for more related terms) (40 CFR 401.11-91).

Process wastewater stream: A waste stream that contains only process wastewater (40 CFR 61.341-91).

Process water: Any raw, service, recycled, or reused water that contracts the plastic product or contacts the shaping equipment surfaces such as molds and mandrels that are, or have been, in contract with the plastic product (*See* water for more related terms) (40 CFR 463.2-91).

Process weight: Total weight of all materials, including fuel, used in a manufacturing process. It is used to calculate the allowable particulate emission rate from the process (EPA-89/12).

Processing: Any method, system, or other treatment designed to change the physical form or chemical content of solid waste (SW-108ts).

Processing activities: All those activities which include:
(1) Preparation of a substance identified in Subpart D of this Part after its manufacture to make another substance for sale or use;
(2) Repackaging of the identified substance; or
(3) Purchasing and preparing the identified substance for use or distribution in commerce (40 CFR 704.203-91).

Processing site: Any site, including the mill, containing residual radioactive materials at which all or substantially all of the uranium was produced for sale to any Federal agency prior to January 1, 1971, under a contract with any Federal agency (for complete definition, *See* 40 CFR 192.10-91).

Processing site:
(1) Any site, including the mill, containing residual radioactive materials at which all or substantially all of the uranium was produced for sale to any Federal agency prior to January 1, 1971, under a contract with any Federal agency, except in the case of a site at or near Slick Rock, Colorado, unless:
  (A) Such site was owned or controlled as of January 1, 1978, or is thereafter owned or controlled, by any Federal agency, or
  (B) A license (issued by the (Nuclear Regulatory) Commission or its predecessor agency under the Atomic Energy Act of 1954 or by a State as permitted under section 274 of such Act) for the production at site of any uranium or thorium product derived from ores is in effect on January 1, 1978, or is issued or renewed after such date; and
(2) Any other real property or improvement thereon which:
  (A) Is in the vicinity of such site, and
  (B) Is determined by the Secretary, in consultation with the Commission, to be contaminated with residual radioactive materials derived from such site (40 CFR 192.10-91).

Processor: Any person who processes a chemical substance or mixture (40 CFR 704.3-91, *See also* TSCA3; 40 CFR 710.2; 720.3; 762.3; 747.115; 747.195; 747.200; 763.163-91).

Procurement: The purchase of materials and services, usually, in the case of government procurement, through awarding contracts to low bidders (OTA-89/10).

Prodromal manifestations: Premonitory signs, indicating the approach of a disease or other morbid state (LBL-76/07-bio).

Produce:
(1) When used in relation to special nuclear material,
  (A) To manufacture, make, produce, or refine special nuclear material;
  (B) To separate special nuclear material from other substances in which such material may be contained; or
  (C) To make or to produce new special nuclear material (10 CFR 70.4-91).
(2) Manufacture, prepare, propagate, compound, or process any pesticide, including any pesticide produced pursuant to section 5 of the Act, ny active ingredient or device, or to package, repackage, label, relabel, or otherwise change the container of any pesticide or device (40 CFR 167.3-91).

Produce, produced, and production, refer to the manufacture of a substance from any raw material or feedstock chemical, but such terms do not include:
(A) The manufacture of a substance that is used and entirely consumed (except for trace quantities) in the manufacture of other chemicals; or
(B) The reuse or recycling of a substance (CAA601-42USC7671-91).

Producer and produce:
(1) The term producer means the person who manufacturers, prepares, compounds, propagates, or processes any pesticide or device or active ingredient used in producing a pesticide.
(2) The term produce means to manufacture, prepare, compound, propagate, or process any pesticide or device, or active ingredient used in producing a pesticide. The dilution by individuals of formulated pesticides for their own use and according to the directions on registered labels shall not of itself result in such individuals being included in the definition of this Act (FIFRA2-7USC136-91).

Product accumulator vessel: Any distillate receiver, bottoms receiver, surge control vessel, or product separator in VHAP service that is vented to atmosphere either directly or through a vacuum-producing system. A product accumulator vessel is in VHAP service if the liquid or the vapor in the vessel is at least 10 percent by weight VHAP (40 CFR 61.241-91).

Product change: Any change in the composition of the furnace charge that would cause the electric submerged arc furnace to become subject to a different mass standard applicable under this subpart (40 CFR 60.261-91).

Product fee: A tax or fee on materials or products that can be designed to add the cost of their disposal to the purchase price (*See* fee for more related terms) (OTA-89/10).

Product frosted: That portion of the furnace pull associated with the fraction of finished incandescent lamp envelopes which is frosted; this quantity shall be calculated by multiplying furnace pull by the fraction of finished incandescent lamp envelopes which is frosted (40 CFR 426.121-91).

Product of combustion: The gases, vapors, and solids that result from the combustion of a material (EPA-83).

Product of complete combustion: An terminal product of chemical oxidation of a combustible constituent that is formed in a combustion unit, e.g., carbon dioxide, water and HCl (CRWI-89/05).

Product of incomplete combustion (PICs):
(1) An ideal waste incineration process is all carbon, hydrogen and halogen elements in the waste convert to carbon dioxide, water and hydrogen halide (HBr, HCl, and HF) or halogen (Br_2, Cl_2 and F_2) respectively. Any other organic compounds emitted from an incinerator are PICs which include carbon monoxide and thermal decomposition products of the original compounds (Huffman-91/10).
(2) Organic compounds formed by combustion. Usually generated in small amounts and sometimes toxic, PICs are heat-altered versions of the original material fed into the incinerator (e.g., charcoal is a P.C. from burning wood) (EPA-94/04).
(2) "Unburned organic compounds that were present in the waste, thermal decomposition products resulting from organic constituents in the waste, or compounds synthesized during or immediately after combustion" (ETI-92).

Product tank: A stationary unit that is designed to contain an accumulation of materials that are fed to or produced by a process unit, and is constructed primarily of non-earthen materials (e.g., wood, concrete, steel, plastic) which provide structural support (*See* tank for more related terms) (40 CFR 61.341-91).

Product tank drawdown: Any material or mixture of materials discharged from a product tank for the purpose of removing water or other contaminants from the product tank (40 CFR 61.341-91).

Production equipment exhaust system: A system for collecting and directing into the atmosphere emissions of volatile organic material from reactors, centrifuges, and other process emission sources (40 CFR 52.741-91).

Production normalizing mass (/kkg) for each core or ancillary operation: The mass (off-kkg or off-lb) processed through that operation (40 CFR 467.02-91).

Production normalizing parameter (PNP): The unit of production specified in the regulations used to determine the mass of pollution a production facility may discharge (EPA-83/03a).

Production weighted average: The manufacturer's production-weighted average particulate emission level, for certification purposes, of all of its diesel engine families included in the particulate averaging program. It is calculated at the end of the model year by multiplying each family particulate emission limit by its respective production, summing these terms, and dividing the sum by the total production of the effected families. Those vehicles produced for sale in California or at high altitude shall each be averaged separately from those produced for sale in any other area (40 CFR 86.085.2-91).

Production weighted NO_x average: The manufacturer's production-weighted average NO_x emission level, for certification purposes, of all of its light-duty truck engine families included in the NO_x averaging program. It is calculated at the end of the model year by multiplying each family NO_x emission limit by its respective production, summing those terms, and dividing the sum by the total production of the effected families. Those vehicles produced for sale in California or at high altitude shall each be averaged separately from those produced for sale in any other area (40 CFR 86.088.2-91).

Production weighted particulate average: The manufacturer's production-weighted average particulate emission level, for certification purposes, of all of its diesel engine families included in the light-duty particulate averaging program. It is calculated at the end of the model year by multiplying each family particulate emission limit by its respective production, summing those terms, and dividing the sum by the total production of the effected families. Those vehicles produced for sale in California or at high altitude shall each be averaged separately from those produced for sale in any other area (40 CFR 86.090.2-91).

Productive (reactive) rubber stock: A compounded rubber which contains curing agents and which can be vulcanized (EPA-74/12a).

Profundal zone: The deep region of a water body that lies below the light-controlled limit of plant growth (cf. photic zone) (DOI-70/04).

Prohibit specification: To prevent the designation of an area as a present or future disposal site (*See* specification for more related terms) (40 CFR 231.2-91).

Prohibitive material: Any materials which by their presence in a packing of paper stock, in excess of the amount allowed, will make the packing unusable as the grade specified. Any materials that may be damaging to equipment (EPA-83).

Project performance standards: The performance and operations requirements applicable to a project including the enforceable requirements of the Act and the specifications, including the quantity of excessive infiltration and inflow proposed to be eliminated, which the project is planned and designed to meet (40 CFR 35.2005-91).

Promethium (Pm): A rare earth metal with a.n. 61; a.w. 147; d. 7.26 g/cc; m.p. 1042° C and b.p. 3000° C. The element belongs to group IIIB of the periodic table.

Promoter:
(1) A reagent used in froth-flotation process, usually called the collector (EPA-82/05).
(2) In studies of skin cancer in mice, an agent which results in an increase in cancer induction when administered after the animal has been exposed to an initiator, which is generally given at a dose which would not result in tumor induction if given alone. A co-carcinogen differs from a promoter in that it is administered at the same time as the initiator. Co-

carcinogens and promoters do not usually induce tumors when administered separately. Complete carcinogens act as both initiator and promoter. Some known promoters also have weak tumorigenic activity, and some also are initiators. Carcinogens may act as promoters in some tissue sites and as initiators in others (EPA-92/12).

Prompt industrial scrap:
(1) Scrap that is left over from the fabrication of iron and steel products (SW-108ts).
(2) Waste produced in an intermediate stage of processing and returned to the basic production facility for reuse (OTA-89/10).
(3) *See* scrap for more related terms.

Proof press: Any device used only to check the quality of the image formation of newly engraved or etched gravure cylinders and prints only non-saleable items (40 CFR 60.431-91).

Propagation rate: Speed of flame front or pressure wave progress (cf. deflagration) (EPA-83).

Propellant: A fuel and oxidizer physically or chemically combined which undergoes combustion to provide rocket propulsion (40 CFR 61.31-91).

Propellant plant: Any facility engaged in the mixing, casting, or machining of propellant (40 CFR 61.31-91).

Property: *See* synonym, thermodynamic property.

Property (or thermodynamic property): A parameter which is used to define the state, or condition, of a substance in a system. Examples of thermodynamic properties include: temperature, pressure, mass, volume, density, etc. Other property related terms include (Wark-p5):
- Extensive property
- Intensive property
- Specific property

Property of gas mixture:
(1) **Mass:** The mass of a mixture is equal to the sum of the masses of its components. Mathematically, it can be expressed as
$$m = x_1 + x_2 + ... m_i + ...$$
where:
- m = total mass of the mixture
- m_i = mass of component i

(2) **Mass fraction:** The mass of component i divided by the total mass of the system. Mathematically, mole fraction (x_i) is defined as
$$x_i = m_i/m$$
where
- x_i = mass fraction component i
- m_i = mass of component i
- m = total mass of the mixture

(3) **Mole:** The total number of moles of a mixture is equal to the sum of the number of moles of its components, namely
$$n = n_1 + n_2 + ...n_i + ...$$
where:
- n = total number of moles of the mixture
- n_i = moles of component i.

(4) **Mole fraction:** The moles of component i divided by the total number of moles. Mathematically, mole fraction (y_i) is defined as
$$y_i = n_i/n$$
where
- y_i = mole fraction
- n_i = moles of component i
- n = total number of moles

The number of moles, the mass, and the molecular weight of a substance are related by
$$m = nM$$
where
- m = mass
- n = number of moles
- M is the molecular weight

(5) **Pressure:** The total pressure of a mixture is the sum of the partial pressures, namely:
$$p = p_1 + p_2 + p_i + ...$$
where
- p = total pressure
- p_i = pressure of component i

(6) **Partial pressure:** The pressure exerted by one component of a mixture. Partial pressure is defined as
$$p_i = y_i p$$
where
- p_i = partial pressure
- y_i = mole fraction
- p = total pressure of the system

For this case, the partial pressure of component i is equal to the mole fraction of component i in the mixture, i.e.
$$y_i = n_i/n = p_i/p.$$

Property of ideal gas mixture: A mixture of ideal gases is also an ideal gas. Properties of ideal gas mixture include (Jones-p397):
(1) **Gibbs-Dalton law:** Also known as the **Dalton law**. The law states that in a mixture of ideal gases, the properties of each component behaves as if it existed alone in the system at the volume and the temperature of the mixture (Holman-p312; Wark-p323).

(2) **Constant pressure specific heat:** The constant pressure specific heat of an ideal gas mixture is the sum of each component.
$$m_m Cp_m = m_1 Cp_1 + m_2 Cp_2 + m_3 Cp_3 + .$$
$$Cp_m = (m_1/m_m)Cp_1 + (m_2/m_m)Cp_2 + (m_3/m_m)Cp_3 +$$
$$Cp_m = x_1 Cp_1 + x_2 Cp_2 + x_3 Cp_3 + ...$$
where:
- Cp_m = constant pressure specific heat of the mixture
- Cp_1, Cp_2, Cp_3 = constant pressure specific heat of components 1, 2, and 3 respectively
- m_m = total mass of the mixture
- m_1, m_2, m_3 = mass of components 1, 2, and 3 respectively
- Cp_1, Cp_2, Cp_3 = constant pressure specific heat of components 1, 2, and 3 respectively
- x_1, x_2, x_3 = mass fraction of components 1, 2, and 3 respectively

(3) **Constant volume specific heat:** The constant volume specific heat of an ideal gas mixture is the sum of each component.

$$m_m Cv_m = m_1 Cv_1 + m_2 Cv_2 + m_3 Cv_3 + \ldots$$
$$Cv_m = (m_1/m_m)Cv_1 + (m_2/m_m)Cv_2 + (m_3/m_m)Cv_3 + \ldots$$
$$Cv_m = x_1 Cv_1 + x_2 Cv_2 + x_3 Cv_3 + \ldots$$
where:
- Cv_m = constant volume specific heat of the mixture
- Cv_1, Cv_2, Cv_3 = constant volume specific heat of components 1, 2, and 3 respectively
- m_m = total mass of the mixture
- m_1, m_2, m_3 = mass of components 1, 2, and 3 respectively
- Cv_1, Cv_2, Cv_3 = constant volume specific heat of components 1, 2, and 3 respectively
- x_1, x_2, x_3 = mass fraction of components 1, 2, and 3 respectively

(4) **Enthalpy:** The enthalpy of an ideal gas mixture is the sum of each component.
$$H_m = H_1 + H_2 + H_3 + \ldots$$
$$m_m h_m = m_1 h_1 + m_2 h_2 + m_3 h_3 + \ldots$$
$$h_m = (m_1/m_m)h_1 + (m_2/m_m)h_2 + (m_3/m_m)h_3 + \ldots$$
$$h_m = x_1 h_1 + x_2 h_2 + x_3 h_3 + \ldots$$
where:
- H_m = enthalpy of the mixture
- H_1, H_2, H_3 = enthalpy of components 1, 2, and 3 respectively
- m_m = total mass of the mixture
- m_1, m_2, m_3 = mass of components 1, 2, and 3 respectively
- h_1, h_2, h_3 = specific enthalpy of components 1, 2, and 3 respectively
- x_1, x_2, x_3 = mass fraction of components 1, 2, and 3 respectively

(5) **Entropy:** The entropy of an ideal gas mixture is the sum of each component.
$$S_m = S_1 + S_2 + S_3 + \ldots$$
$$m_m s_m = m_1 s_1 + m_2 s_2 + m_3 s_3 + \ldots$$
$$s_m = (m_1/m_m)s_1 + (m_2/m_m)s_2 + (m_3/m_m)s_3 + \ldots$$
$$s_m = x_1 s_1 + x_2 s_2 + x_3 s_3 + \ldots$$
where:
- S_m = entropy of the mixture
- S_1, S_2, S_3 = entropy of components 1, 2, and 3 respectively
- m_m = total mass of the mixture
- m_1, m_2, m_3 = mass of components 1, 2, and 3 respectively
- s_1, s_2, s_3 = specific entropy of components 1, 2, and 3 respectively
- x_1, x_2, x_3 = mass fraction of components 1, 2, and 3 respectively

(6) **Gas constant:** The gas constant of an ideal gas mixture is the sum of each component.
$$m_m R_m = m_1 R_1 + m_2 R_2 + m_3 R_3 + \ldots$$
$$R_m = (m_1/m_m)R_1 + (m_2/m_m)R_2 + (m_3/m_m)R_3 + \ldots$$
$$R_m = x_1 R_1 + x_2 R_2 + x_3 R_3 + \ldots$$
where:
- R_m = gas constant of the mixture
- R_1, R_2, R_3 = gas constant of components 1, 2, and 3 respectively
- m_m = total mass of the mixture
- m_1, m_2, m_3 = mass of components 1, 2, and 3 respectively
- R_1, R_2, R_3 = gas constant of components 1, 2, and 3 respectively
- x_1, x_2, x_3 = mass fraction of components 1, 2, and 3 respectively

For the universal gas constant (Ru), it can be expressed as:
$$Ru = M_m R_m$$
where:
- Ru = universal gas constant
- R_m = gas constant of the mixture
- M_m = molecular weight of the mixture

(7) **Internal energy:** The internal energy of an ideal gas mixture is the sum of each component.
$$U_m = U_1 + U_2 + U_3 + \ldots$$
$$m_m u_m = m_1 u_1 + m_2 u_2 + m_3 u_3 + \ldots$$
$$u_m = (m_1/m_m)u_1 + (m_2/m_m)u_2 + (m_3/m_m)u_3 + \ldots$$
$$u_m = x_1 u_1 + x_2 u_2 + x_3 u_3 + \ldots$$
where:
- U_m = internal energy of the mixture
- U_1, U_2, U_3 = internal energy of components 1, 2, and 3 respectively
- m_m = total mass of the mixture
- m_1, m_2, m_3 = mass of components 1, 2, and 3 respectively
- u_1, u_2, u_3 = specific internal energy of components 1, 2, and 3 respectively
- x_1, x_2, x_3 = mass fraction of components 1, 2, and 3 respectively

(8) **Mass:** The mass of an ideal gas mixture is the sum of each component.
$$m_m = m_1 + m_2 + m_3 + \ldots$$
where:
- m_m = total mass of the mixture
- m_1, m_2, m_3 = mass of components 1, 2, and 3 respectively

(9) **Mole:** The number of moles of an ideal gas mixture is the sum of each component.
$$n_m = n_1 + n_2 + n_3 + \ldots$$
where:
- n_m = number of moles of the mixture
- n_1, n_2, n_3 = number of moles of components 1, 2, and 3 respectively

(10) **Molecular weight:** The molecular weight of an ideal gas mixture can be expressed as
$$M_m = y_1 M_1 + y_2 M_2 + y_3 M_3 + \ldots$$
The above equation can be derived from the mass equation.
$$m_m = m_1 + m_2 + m_3 + \ldots$$
because m = nM, where m = mass, n = number of moles, and M molecular weight, the above equation becomes
$$n_m M_m = (n_1/n_m)M_1 + (n_2/n_m)M_2 + (n_3/n_m)M_3 + \ldots$$
$$M_m = y_1 M_1 + y_2 M_2 + y_3 M_3 + \ldots$$
where:
- M_m = molecular weight of the mixture
- n_m = number of moles of the mixture
- n_1, n_2, n_3 = number of moles of components 1, 2, and 3 respectively
- M_1, M_2, M_3 = molecular weight of components 1, 2, and 3 respectively
- y_1, y_2, y_3 = mole fraction of components 1, 2, and 3 respectively

(11) **Pressure:** Dalton law or the law of additive pressures: In a mixture of ideal gases, the pressure of a mixture of ideal gases equals the sum of the pressures of its components as if each component existed alone at the temperature and volume of the mixture (Jones-p393). Mathematically, it can be derived from the ideal gas law equation as follows:

$$p_m = n_m R u T_m / V_m$$
$$= (n_1 + n_2 + n_3 + ...) R u T_m / V_m$$
$$= n_1 R u T_m / V_m + n_2 R u T_m / V_m + n_3 R u T_m / V_m + ...$$
$$p_m = p_1(T_m, V_m) + p_2(T_m, V_m) + p_3(T_m, V_m)$$

where:
- p_m = pressure of the mixture
- Ru = universal gas constant
- T_m = temperature of the mixture
- V_m = volume of the mixture
- n_m = number of moles of the mixture
- n_1, n_2, n_3 = number of moles of components 1, 2, and 3 respectively
- $p_1(T_m, V_m), p_2(T_m, V_m), p_3(T_m, V_m)$ = pressure of components, 1, 2, and 3 existing at the temperature T_m and the volume V_m

(12) **Pressure fraction:** Pressure fraction is equal to mole fraction (y_i). Using component 1 as an example, mole fraction for component 1 is

$$y_1 = n_1 / n_m$$
$$= [(p_1 V_m)/(R u T_m)][(R u T_m)/p_m V_m]$$
$$= p_1 / p_m$$

where
- y_1 = mole fraction of component 1
- p_1/p_m = definition of pressure fraction
- V_m = volume of mixture
- n_m = number of moles of the mixture
- n_1 = number of moles of component 1
- p_1 = pressure of component 1 existing at the temperature T_m and the pressure V_m

(13) **Temperature:** The temperature of an ideal gas mixture is the same for each component and for the mixture.

$$T_m = T_1 = T_2 = T_3 = ...$$

where:
- T_m = temperature of the mixture
- T_1, T_2, T_3 = temperature of components 1, 2, and 3 respectively

(14) **Volume:** In general, for any gas or gas-vapor mixture, the volume of each component of a mixture is the same as the volume of the mixture because the molecules of each component are free to move throughout the entire space occupied by the mixture. Consequently,

$$V_m = V_1 = V_2 = V_3 = ...$$

where:
- V_m = volume of mixture
- V_1, V_2, V_3 = volume of components 1, 2, and 3 respectively

However, under the Amagat law, Leduc law or the law of additives volumes, in a mixture of ideal gases, the volumes of a mixture of ideal gases equals the sum of the volumes of its components as if each component existed alone at the temperature and pressure of the mixture (Jones-p394). Mathematically, it can be derived from the ideal gas law equation as follows:

$$V_m = n_m R u T_m / p_m$$
$$= (n_1 + n_2 + n_3 + ...) R u T_m / p_m$$
$$= n_1 R u T_m / p_m + n_2 R u T_m / p_m + n_3 R u T_m / p_m + ...$$
$$V_m = V_1(T_m, p_m) + V_2(T_m, p_m) + V_3(T_m, p_m) + ...$$

where:
- p_m = pressure of the mixture
- Ru = universal gas constant
- T_m = temperature of the mixture
- V_m = volume of the mixture
- n_m = number of moles of the mixture
- n_1, n_2, n_3 = number of moles of components 1, 2, and 3 respectively
- $V_1(T_m, p_m), V_2(T_m, p_m), V_3(T_m, p_m)$ = volume of components 1, 2, and 3 respectively existing at the temperature T_m and the pressure p_m

(15) **Volume fraction:** Volume fraction is equal to mole fraction (y_i). Using component 1 as an example, mole fraction for component 1 is

$$y_1 = n_1 / n_m$$
$$= [(p_m V_1)/(R u T_m)][(R u T_m)/p_m V_m]$$
$$= V_1 / V_m$$

where
- y_1 = mole fraction of component 1
- V_1/V_m = definition of pressure fraction
- V_m = volume of mixture
- n_m = number of moles of the mixture
- n_1 = number of moles of component 1
- V_1 = volume of component 1 existing at the temperature T_m and the pressure p_m

Property damage: Shall have the meaning given this term by applicable state law. This term shall not include those liabilities which, consistent with standard insurance industry practices, are excluded from coverage in liability insurance policies for property damage. However, such exclusions for property damage shall not include corrective action associated with releases from tanks which are covered by the policy (40 CFR 280.92-91).

Proportional sampling: The sampling at a rate that produces a constant ration of sampling rate to stack gas flow rate (40 CFR 60.2-91).

Proportionate mortality ratio (PMR): The number of deaths from a specific cause and in a specific period of time per 100 deaths in the same time period (EPA-92/12).

Proposed permit: A State NPDES permit prepared after the close of the public comment period (and, when applicable, any public hearing and administrative appeals) which is sent to EPA for review before final issuance by the State. A proposed permit is not a draft permit (*See* permit for more related terms) (40 CFR 122.2-91).

Prospective study: A study in which subjects are followed forward in time from initiation of the study. This is often called a longitudinal or cohort study (EPA-92/12).

Protactinium (Pa): A radioactive metal with a.n. 91; a.w. 231; d. 15.4 g/cc and m.p. 1230° C. The element belongs to group IIIB of the periodic table.

Protect health and the environment: The protection against any unreasonable adverse effects on the environment (FIFRA2-7USC136-91).

Protection: The measures taken to capture or destroy a toxic chemical that has breached primary containment, but before an uncontrolled release to the environment has occurred (EPA-87/07a).

Protection factors: The respirator protection factor indicates how much protection a respirator provides. It is the ratio of the contaminant concentrations outside and inside the respirator (NIOSH-84/10).

Protective clothing: At least a hat or other suitable head covering, a long sleeved shirt and long legged trousers or a coverall type garment (all of closely woven fabric covering the body, including arms and legs), shoes and socks (40 CFR 170.2-91).

Protective equipment: The clothing or any other materials or devices that shield against unintended exposure to pesticides (40 CFR 171.2-91).

Protocol: The plan and procedures which are to be followed in conducting a test (cf. method) (40 CFR 790.3-91, *See also* TSCA-AIA1-91).

Protein: The complex nitrogenous organic compounds of high molecular weight that contain amino acids as their basic unit and are essential for growth and repair of animal tissue. Many proteins are enzymes (EPA-89/12).

Protista: The kingdom of all simple biological organisms.

Proton: A positively charged particle and has a mass of 1.673×10^{-27} kg which is about 1836 times of an electron (*See* atom for more related terms).

Protoplasm: The colloidal materials which consist of the living contents of a cell.

Protoplast: A membrane-bound cell from which the outer cell wall has been partially or completely removed. The term often is applied to plant cells (EPA-89/12).

Proven emission control systems: The emission control components or systems (and fuel metering systems) that have completed full durability testing evaluation over a vehicle's useful life in some other certified engine family, or have completed bench or road testing demonstrated to be equal or more severe than certification mileage accumulation requirements. Alternatively, proven components or systems are those that are determined by EPA to be of comparable functional quality and manufactured using comparable materials and production techniques as components or systems which have been durability demonstrated in some other certified engine family. In addition, the components or systems must be employed in an operating environment (e.g.. temperature, exhaust flow, etc.,) similar to that experienced by the original or comparable components or systems in the original certified engine family (40 CFR 86.092.2-a-91).

Proximate analysis:
(1) The proximate analysis is to provide data relating to the physical form (parameters) of the waste (such as moisture content, volatile matter, ash, fixed carbon, and heat content) and to provide an approximate mass balance as to the composition of the waste (*See* analysis for more related terms) (EPA-82/02).
(2) The determination by prescribed methods of moisture, volatile matter, fixed carbon (by difference), ash and usually heating value (*See* analytical parameters--fuels for more related terms) (EPA-83).

PSD station: Any station operated for the purpose of establishing the effect on air quality of the emissions from a proposed source for purposes of prevention of significant deterioration as required by 51.24(n) of Part 51 of this chapter (40 CFR 58.1-91).

Pseudo diffusion: The extra diffusion which in the numerical solution of the diffusion equation results from the truncation error related to the finite difference of the advection terms. Its magnitude is dependent on the particular finite difference scheme that is used. It may sometimes completely mask the influence of the other diffusion terms in the equation (NATO-78/10).

Pseudo spectral method: A numerical solution method related to the Galerkin method. The solution is represented by a truncated Fourier series. In the pseudo spectral method the Fourier series are only used to evaluate the space derivatives in the problem, so that these derivatives are calculated in Fourier space, while the time derivatives and the local products are evaluated in real space on grid points (NATO-78/10).

Psychiatric disorders: Includes diseases of the nervous system which affect mental health (PHSA2-l).

Psychrometer: An instrument for measuring relative humidities by means of wet- and dry-bulb temperatures.

Psychrometric chart: A chart that graphically displays the relationship between air-water-vapor mixtures and their properties (EPA-82/11f). An important conversion factor used with the chart is $1\ lb_m = 7000$ grains (*See* humidity for more related terms).

Psychrophiles: Grow best at colder temperatures, below 20° C (*See* thermophile for more related terms) (EPA-83).

Public Health Service Act (PHSA): *See* the term of Act or PHSA.

Public hearing:
(1) A forum conducted by EPA or the State during the permitting process where citizens can voice their comments, concerns, and questions.
(2) An open meeting held by the grant applicant to obtain formal comments on a plan. Notification of the public through newspaper and other media advertisements is required. A transcript of the hearing is maintained (EPA-80/08).

Public lands: Any land and interest in land owned by the United States within the several States and administered by the Secretary of the Interior through the Bureau of Land Management, without regard to how the United States acquired ownership, except:
(1) Lands located on the Outer Continental Shelf; and
(2) Lands held for the benefit of Indians, Aleuts, and Eskimos (FLPMA103-43USC1702-90).

Public participation: *See* synonym, community relations.

Public record: The NPDES permit application or the NPDES permit itself and the "record for final permit" as defined in 40 CFR 124.122 (40 CFR 117.1-91).

Public vessel: A vessel owned or bareboat chartered and operated by the United States, or by a State or political subdivision thereof, or by a foreign nation, except when such vessel is engaged in commerce (40 CFR 110.1-91, *See also* CWA311; OPA1001; 40 CFR 116.3; 300.5-91).

Public water: All navigable waters of the United States and the tributaries thereof; all interstate waters and tributaries thereof; and all intrastate lakes, rivers, streams and tributaries thereof not privately owned (*See* water for more related terms) (EPA-79/12a).

Public water supplies: The water distributed from a public water system (40 CFR 125.58-91).

Public water system: A system that provides piped water for human consumption to at least 15 service connections or regularly serves 25 individuals (EPA-89/12, *See also* SDWA1401; 40 CFR 125.58; 141.2; 142.2; 143.2-91).

Public water system supervision program: A program for the adoption and enforcement of drinking water regulations (with such variances and exemptions from such regulations under conditions and in a manner which is not less stringent than the conditions under, and the manner in, which variances and exemptions may be granted under sections 1415 and 1416) which are no less stringent than the national primary drinking water regulations under section 1412, and for keeping records and making reports required by section 1413(a)(3) (SDWA1443-42USC300j.2).

Publication rotogravure printing line: A rotogravure printing line in which coatings are applied to paper which is subsequently formed into books, magazines, catalogues, brochures, directories, newspaper supplements, or other types of printed material (40 CFR 52.741-91).

Publication rotogravure printing press: Any number of rotogravure printing units capable of printing simultaneously on the same continuous web or substrate and includes any associated device for continuously cutting and folding the printed web, where the following saleable paper products are printed:
- Catalogues, including mail order and premium,
- Direct mail advertisements, including circulars, letters, pamphlets, cards, and printed envelopes,
- Display advertisements, including general posters, outdoor advertisements, car cards, window posters; counter and floor displays; point-of-purchase, and other printed display material,
- Magazines,
- Miscellaneous advertisements, including brochures, pamphlets, catalogue sheets, circular folders, announcements, package inserts, book jackets, market circulars, magazine inserts, and shopping news,
- Newspapers, magazine and comic supplements for newspapers, and preprinted newspaper inserts, including hi-fi and spectacolor rolls and sections, periodicals, and
- Telephone and other directories, including business reference services (40 CFR 60.431-91).

Publicly owned freshwater lake: A freshwater lake that offers public access to the lake through publicly owned contiguous land so that any person has the same opportunity to enjoy non-consumptive privileges and benefits of the lake as any other person (*See* lake for more related terms) (for complete definition, *See* 40 CFR 35.1605.3-91).

Publicly owned freshwater lake: A freshwater lake that offers public access to the lake through publicly owned contiguous land so that any person has the same opportunity to enjoy non-consumptive privileges and benefits of the lake as any other person. If user fees are charged for public use and access through State or substate operated facilities, the fees must be used for maintaining the public access and recreational facilities of this lake or other publicly owned freshwater lakes in the State, or for improving the quality of these lakes (40 CFR 35.1605-3-91).

Publicly owned treatment works (POTW) (or municiple wastewater treatment): The treatment works treating domestic sewage that is owned by a municipality or State (40 CFR 501.2-91, *See also* 40 CFR 117.1; 122.2; 125.58; 260.10; 270.2; 403.3-91; EPA89/12).

Pucker: An uneven quality in a paper due to improper drying (EPA-83).

Puckering: In seaming a landfill liner:
(1) The thermal distortion of the seamed region after completion and cooling of the seam. it is often observed on the under side of the seam (EPA-89/09).
(2) A heat related sign of localized strain caused by improper seaming using extrusion or fusion methods. It often occurs on the bottom of the lower geomembrane and in the shape of a shallow inverted V (EPA-91/05).

Puff model: A mathematical model which describes the transport and dispersion of pollutants in the atmosphere from a source by releasing a quantity of pollutant at each time step (in the case of an instantaneous source one release is sufficient). This quantity of material is transported by the wind field and dispersed by turbulent diffusion (NATO-78/10).

Pull: The quantity of glass delivered by a furnace in 24 hours (EPA-83).

Pull on container: Detachable container system in a large container (around 20-30 cubic yards) is pulled onto a service vehicle or tilt frame or hoist truck by mechanical or hydraulic means and carried to disposal site for emptying (EPA-83).

Pullery: A plant where sheepskin is processed by removing the wool and then pickling before shipment to a tannery (EPA-82/11).

Pulp: (1) In leather industry, pulp is a method of unhairing in which depilatory agents are used to dissolve hair entirely in a few hours (EPA-82/11). (2) In paper industry, pulp is cellulosic fibers after conversion from wood chips (EPA-87/10). Other pulp related terms include:
- Acid pulp (*See* synonym, sulfite pulp)
- Beet pulp
- Chemi-mechanical pulp

- Chemical wood pulp
- Dissolving pulp
- Groundwood pulp
- Kraft pulp (*See* synonym, sulfate pulp)
- Mechanical pulp
- Sulfate pulp
- Sulfite pulp
- Wood pulp

Pulp physicals: Strength properties of the pulp (EPA-83).

Pulp press: A mechanical pressure device which squeezes the exhausted cossetted (pulp) to remove a portion of the inherent water (EPA-74/01a).

Pulp screen water: The water which is drained from the wet insoluble pulp after the diffusion process but before the pulp is pressed to remove extraneous water and sugar (*See* water for more related terms) (EPA-74/01a).

Pulp silo drainage: Drainage water resulting from discharge of pulp from the diffuser with screenings to a silo equipped with channels for drainage water collection (EPA-74/01a).

Pulper:
(1) A mechanical device used to separate fiber bundles in the presence of water prior to papermaking (EPA-87/10).
(2) Machine which converts a mixture of dry pulp or paper and water into a fibre slurry by violent agitation. Also called **hydrapulper** (EPA-83).

Pulping: The operation of reducing a cellulosic raw material, such as pulpwood, rags, straw, reclaimed paper, etc., into a pulp suitable for further processing into paper or paperboard or for chemical conversion (into rayon, cellophane, etc.). pulping may vary from simple mechanical action to rather complex digesting sequences and may be conducted in batch or in continuous process equipment (EPA-83, *See also* EPA-87/10).

Pulpwood: The wood which is suitable for the manufacture of chemical or mechanical wood pulp. The wood may be in the form of logs as they come from the forest or cut into shorter lengths suitable for the chipper or the grinder (EPA-87/10).

Pulverized coal: A coal that has been ground to a powder, usually of a size where 80 percent passes through a #200 U.S.S. sieve (*See* coal for more related terms) (EPA-82/11f).

Pulverized coal fired steam generating unit: A steam generating unit in which pulverized coal is introduced into an air stream that carries the coal to the combustion chamber of the steam generating unit where it is fired in suspension. This includes both conventional pulverized coal fired and micropulverized coal fired steam generating units (40 CFR 60.41b-91).

Pulverization: The crushing or grinding of material into very fine particle size (*See* size reduction machine for more related terms) (EPA-83).

Pumped storage: Water pumped into a storage reservoir during periods of low electric-power demand to be used to generate power during peak demand periods (DOI-70/04).

Pumping station (or lifting station): Mechanical devices installed in sewer or water systems or other liquid-carrying pipelines that move the liquids to a higher level (EPA-89/12).

Punch ware: Handmade, thin, blown glassware, especially tumblers (EPA-83).

Puncture test: The resistance of paper to puncture as measured by special uniform tests (EPA-83).

Punk stick: A small tube used to generate smoke from smoldering materials (EPA-88/08).

Pure substance: A substance which is chemically homogeneous and has the same chemical composition in all phases. Water is a pure substance, because it chemical composition is the same in all phases, even for a mixture of liquid and vapor (Holman-p63; Jones-p143).

Purge or line purge: The coating material expelled from the spray system when clearing it (40 CFR 60.391-91).

Purity: A measure of a pure substance, e.g., the actual sugar content in relation to the total dry substance in sugar beets. Specifically, the percentage of surcrose in total solids (EPA-74/01a).

Push pit: A storage system sometimes used in stationary compactor transfer systems. It is a hydraulically powered bulkhead that traverses the length of the pit periodically pushes the stored waste into the hopper of a compactor (SW-108ts).

Putrefaction:
(1) The decomposition of organic matter by microorganisms and oxidation, resulting in odors (SW-108ts).
(2) The biological decomposition of organic matter accompanied by the production of foul-smelling products associated with anaerobic conditions (EPA-76/03).

Putrescible: Able to rot quickly enough to cause odors and attract flies (EPA-89/12).

Putrescible material: Capable of being decomposed by microorganisms with sufficient rapidity as to cause nuisances from odors and gases. Kitchen wastes, offal and dead animals are examples (EPA-83).

Putrescible waste: The solid waste which contains organic matter capable of being decomposed by microorganisms and of such a character and proportion as to be capable of attracting or providing food for birds (*See* waste for more related terms) (40 CFR 257.3.8-91).

Putrescine: In biochemistry, a colorless, ill-smelling ptomaine (C_4H_2N) resulting from the bacterial decomposition of animal tissues in the presence of moisture and heat, but in the absence of air (EPA-83).

Putty: A white polishing compound (EPA-83).

Pyrazolone colorimetric: A standard method of measuring cyanides in aqueous solutions (EPA-83/06a).

Pyrimidine dimer: Covalent bonding of adjacent pyrimidine residues in a nucleic acid exposed to ultraviolet radiation. A pyrimidine dimer distorts a nucleic acid double helix structure and must be excised so replication and gene expression can proceed (EPA-88/09a).

Pyrite: A combination of iron and sulfur found in coal as FeS_2 (EPA-82/11f).

Pyrogenic lipopolysaccharides: Complex molecules found in the cell walls (*See* endotoxins) of microorganisms that cause inflammatory reactions in the body (EPA-88/09a).

Pyrolysis:
(1) The thermal/chemical decomposition of a compound by heat in an oxygen deficient manner (SW-108ts).
(2) Chemical change caused by heat (DOE-91/04).

Pyrolytic gas and oil: The gas or liquid products that possess usable heating value that is recovered from the heating of organic material (such as that found in solid waste), usually in an essentially oxygen-free atmosphere (40 CFR 245.101-91).

Pyrolytic incinerator: *See* synonym, starved air incinerator.

Pyrolytic thermal decomposition: The thermal decomposition of an organic compound in the complete or partial absence of oxygen (EPA-88/12).

Pyrometallurgical: The use of high-temperature processes to treat metals (EPA-83/03a).

Pyrophoric: Capable of ignition spontaneously (EPA-81/09).

Pyrometer: An instrument to measure high temperature. There are two types:
(1) Optical pyrometer
(2) Radiation pyrometer

Pyrometric cone equivalent (PCE): An index to the refractoriness of a material. It is obtained by a test that provides the number of a standard pyrometric cone that is closest in its bending behavior to that of a pyrometric cone made of the material when both are heated in accordance with ASTM Standard Method of Test for Pyrometric Cone Equivalent of Refractory Materials (SW-108ts).

Pyrophoric: A substance that spontaneously ignites in air at or below room temperature without supply of heat, friction, or shock (EPA-87/07a).

Qq

q1*: Upper bound on the slope of the low-dose linearized multistage procedure (EPA-92/12).

Quad: Quadrillion or 10^5.

Quadrat: A ground area for sampling all biological species of interest.

Qualified incinerator: One of the following:
(1) An incinerator approved under the provisions of 40 CFR 761.70. Any level of PCB concentration can be destroyed in an incinerator approved under 40 CFR 761.70.
(2) A high efficiency boiler which complies with the criteria of 40 CFR 761.60(a)(2)(iii)(A), and for which the operator has given written notice to the appropriate EPA Regional Administrator in accordance with the notification requirements for the burning of mineral oil dielectric fluid under 40 CFR 761.60(a)(2)(iii)(B).
(3) An incinerator approved under section 3005(c) of the Resource Conservation and Recovery Act (42 U.S.C. 6925(c)) (RCRA).
(4) Industrial furnaces and boilers which are identified in 40 CFR 260.10 and 40 CFR 279.61(a)(1) and (2) when operating at their normal operating temperatures (this prohibits feeding fluids, above the level of detection, during either startup or shutdown operations) (40 CFR 761.3).
(5) *See* incinerator for more related terms.

Qualifying phase I technology: A technological system of continuous emission reduction which achieves a 90 percent reduction in emissions of sulfur dioxide from the emissions that would have resulted from the use of fuels which were not subject to treatment prior to combustion (CAA402-42USC7651a-91).

Qualitative: An analysis in which some or all of the components of a sample are identified (cf. quantitative) (EPA-89/12a).

Qualitative evaluation: Assessing the risk of an accidental release at a facility in relative terms; the end result of the assessment being a verbal description of the risk (EPA-87/07a).

Quality:
(1) Quality is the absence of defects. In contrast, deficiency means general inadequacy.
(2) In quality control, quality is the totality of features and characteristics of a product or service that bear on its ability to satisfy given needs (EPA-84/03). It includes:
 (A) **Design quality:** The design developed to meet a need; and
 (B) **Conformance quality:** The actual production that meets the need (ACS-87/11).
(3) In thermodynamics, the fraction of vapor mass in a saturated liquid-vapor mixture. Quality related equations include:
- By definition
 $x = m_g/(m_f + m_g)$; where:
 x = quality
 m_g = vapor mass in a mixture
 m_f = liquid mass in a mixture
- Enthalpy (h_x), (Btu/lb), of a saturated liquid-vapor mixture at a given temperature or pressure
 $h_x = h_f + x(h_g - h_f) = h_f + xh_{fg}$; where:
 h_f = enthalpy of saturated liquid
 h_g = enthalpy of saturated vapor
 h_{fg} = enthalpy difference between saturated vapor and saturated liquid, h_{fg} is also known as latent heat of vaporization
 The values of h_f, h_g, and h_{fg} can be obtained from a steam table
- Entropy (s_x), (Btu/(lb-R)), of a saturated liquid-vapor mixture
 $s_x = s_f + x(s_g - s_f) = s_f + xs_{fg}$; where:
 s_f = entropy of saturated liquid
 s_g = entropy of saturated vapor
 s_{fg} = entropy difference between saturated vapor and saturated liquid
 The values of s_f, s_g, and s_{fg} can be obtained from a steam table
- Internal energy (u_x), (Btu/lb), of a saturated liquid-vapor mixture
 $u_x = u_f + x(u_g - u_f) = u_f + xu_{fg}$; where:
 u_f = internal energy of saturated liquid
 u_g = internal energy of saturated vapor
 u_{fg} = internal energy difference between saturated vapor and saturated liquid
 The values of s_f, s_g, and s_{fg} can be obtained from a steam table
- Specific volume (v_x), (ft^3/lb), of a saturated liquid-vapor mixture
 $v_x = v_f + x(v_g - v_f) = v_f + xv_{fg}$; where:
 v_f = specific volume of saturated liquid
 v_g = specific volume of saturated vapor
 v_{fg} = specific volume difference between

saturated vapor and saturated liquid

The values of v_f, v_g, and v_{fg} can be obtained from a steam table (Holman-p66; Jones-p120; Wark-p66)

Other quality related terms include:
- Quality data
- Data quality objective

Quality assessment: The overall system of activities whose purpose is to provide assurance that the overall quality control job is being done effectively. It involves a continuing evaluation of the products produced and the performance of the production system (ACS-87/11).

Quality assurance (QA):
(1) The total integrated program for assuring the reliability of monitoring and measurement data. A system for integrating the quality planning, quality assessment, and quality improvement efforts to meet user requirements (EPA-86/10a).
(2) A system of activities whose purpose is to provide to the producer or user of a product or service the assurance that it meets defined standards of quality with a stated level of confidence. QA is understanding the measurement process, what needs to be measured, what needs to done, doing what needs to be done, evaluating what was done, reporting evaluated data which is technically sound and legally defensible (ACS-87/11).
(3) cf. construction puality assurance.

Quality assurance narrative statement: A description of how precision, accuracy, representativeness, completeness, and compatibility will be assessed, and which is sufficiently detailed to allow an unambiguous determination of the quality assurance practices to be followed throughout a research project (40 CFR 30.200-91).

Quality assurance plan: A document that contains or references the quality assurance elements established for an activity, group of activities, scientific investigation, or a project, and describes how conformance with such requirements is to be assured for structures, systems, computer software, components, and their operation commensurate with:
(1) The scope, complexity, duration, and importance to satisfactory performance;
(2) The potential impact on environment, safety, and health; and
(3) Requirements for reliability and continuity of operations (DOE-91/04).

Quality assurance program plan: A formal document which describes an orderly assembly of management policies, objectives, principles, organizational responsibilities, and procedures by which an agency or laboratory specifies how it intends to:
(1) Produce data of documented quality; and
(2) Provide for the preparation of quality assurance project plans and standard operating procedures (40 CFR 30.200-91, *See also* EPA-91/02).

Quality assurance project plan (QAPP): An organizations written procedures which delineate how it produces quality data for a specific project or measurement method (40 CFR 30.200-91, *See also* 40 CFR 35.6015; 300.5-91). Because the end use of the data determines the degree of quality assurance that is required, EPA's RREL uses four categories in its QA Program as follow:
(1) Category I Projects require the most rigorous and detailed QA, since the resulting data must both legally and scientifically defensible. Category I projects include enforcement actions and projects of significant national or congressional visibility. Such projects are typically monitored by the Administrator. Category I projects must produce results that are autonomous; that is, results that can prove or disprove a hypothesis without reference to complementary projects.
(2) Category II Projects are those producing results that complement other inputs. These projects are of sufficient scope and substance that their results could be combined with those from other projects of similar scope to produce information for making rules, regulations, or policies. In addition, projects that do not fit this pattern, but have high visibility, would also be included in this category.
(3) Category III Projects are those producing results used to evaluate and select basic options, or to perform feasibility studies or preliminary assessments of unexplored areas which might lead to further work.
(4) Category IV Projects are those producing results for the purpose of assessing suppositions (EPA-2/91).

Quality assurance and quality control (QA/QC): A system of procedures, checks, audits, and corrective actions to ensure that all EPA research design and performance, environmental monitoring and sampling, and other technical and reporting activities are of the highest achievable quality (EPA-89/12). QA/QC related terms include:
(1) **Method:** An assemblage of techniques. It implies reduction to practice.
(2) **Procedure:** A detailed instruction to permit replication of a method, e.g., ASTM-D2914.
(3) **Protocol:** A methodology specified in regulatory, authoritative, or contractual situations, e.g., EPA method 5.
(4) **Technique:** A physical or chemical principle for characterizing materials of chemical systems. A standardization of a method can be expressed as "technique --> method --> procedure --> protocol" (EPA-89/12a, *See also* ACS-87/11).

Quality assurance unit: Any person or organizational element, except the study director, designated by testing facility management to perform the duties relating to quality assurance of the studies (40 CFR 160.3-91, *See also* 40 CFR 792.3-91).

Quality control (QC):
(1) The routine application of procedures for obtaining prescribed standards of performance in the monitoring and measurement process (EPA-86/10a).
(2) The overall system of activities whose purpose is to control the quality of a product or service, so that it meets the needs of users. The aim is to provide quality that is satisfactory, adequate, dependable, and economic (ACS-87/11).
(3) Duplication of a portion of the analytical tests per-formed to estimate the overall quality of the results and to determine what, if any, changes must be made to achieve or maintain the required level of quality (EPA-89/12a).

(4) cf. construction puality control.

Quality control sample: A solution obtained from an outside source having known, concentration values to be used to verify the calibration standards (40 CFR 136-App/C-91).

Quality indicator: Includes:
(1) **Comparability:** In quality control, the confidence with which one data set can be compared to another (ACS-87/11; EPA-85/08).
(2) **Completeness:** In QA, a quality indicator. It is a measure of the amount of valid data obtained from a measurement system compared to the amount of valid data that were intended to be obtained under correct normal conditions (or compared to the amount of data collected that was expected, usually expressed as percent) (ACS-87/11; EPA-85/08; 89/03).
(3) **Representativeness:** The degree to which the data accuratively and precisely represent a characteristic of a population parameter, variation of a property, a process characteristic, or an operational condition (ACS-87/11; EPA-86/10a).

Quality of wet vapor: See quality definition under thermodynamics.

Quantifiable level/level of detection: 2 micrograms per gram from any resolvable gas chromatographic peak, i.e., 2 ppm (40 CFR 761.3-91).

Quantitation limit (QL): The lowest level at which a chemical can be accurately and producibly quantitated. Usually equal to the method detection limit multiplied by a factor of three to five, but varies for different chemicals and different samples (EPA-91/12).

Quantitative: An analysis in which the amount of one or more components of a sample is determined (cf. qualitative) (EPA-89/12a).

Quantitative evaluation: Assessing the risk of an accidental release at a facility in numerical terms; the end result of the assessment being some type of number reflects risk, such as faults per year or mean time between failure (EPA-87/07a).

Quantity reduction: In municipal waste, changing the design of a product so that less municipal waste is generated when the product or its residuals are discarded, or so that the product is more durable or repairable (cf. pollution prevention) (OTA-89/10).

Quantum chemistry: A science of using quantum mechanics to interprete chemical phenomena.

Quarry method: A variation of the area method in which the wastes are spread and compacted in a depression and cover materials are generally obtained elsewhere (See sanitary landfill for more related terms) (SW-108ts).

Quarter: A 3-month period; the first quarter concludes on the last day of the last full month during the 180 days following initial startup (40 CFR 60.481-91).

Quartz glass: A term applied to silica glass made by fusing vein quartz (See glass for more related terms) (EPA-83).

Quasi-static process: See synonym, reversible process.

Quaternary ammonium salt: Chemical compound having a chlorine or bromine ion attached to a nitrogen atom with four carbon-nitrogen bonds. May be used as algicides, bactericides, piscicides, etc. (EPA-85/10).

Quench:
(1) To cool rapidly (EPA-81/09).
(2) Cooling of hot gases by rapid evaporation of water (EPA-89/03b).

Quench station: That portion of the metal coil surface coating operation where the coated metal coil is cooled, usually by a water spray, after baking or curing (40 CFR 60.461-91).

Quench tank: A water-filled tank used to cool incinerator residues or hot materials during industrial processes (See tank for more related terms) (EPA-89/12).

Quench trough: A water-filled trough into which burning residue drops from an incinerator furnace (SW-108ts).

Quenching: Shock cooling by immersion of liquid or molten material in a cooling medium (liquid or gas). Used in metallurgy, plastics forming, and petroleum refining (EPA-83/03).

Quenching oil: The medium to heavy grade mineral oils used in the cooling of metals. Standard weight or grade of oils would be similar to standard SAE 60 (See oil for more related terms) (EPA-85/10a).

Quick setting ink: An ink for letterpress and offset dry by either filtration, coagulation, selective absorption or often a combination of these with some of the other drying methods. The vehicles are generally special resin-oil combinations which, after the ink has been printed, separate into a solid material which remains on the surface as a dry film and an oily material which penetrates rapidly into the stock. This rapid separation gives the effect of very quick setting or drying (See ink for more related terms) (EPA-79/12a).

Quicklime: See definition under calcium.

Quinhydrone electrode: A half cell with an electrode made from a platinum wire in an equal molar solution of quinone and hydroquinone. The electrode is used as a reversible electrode standard in pH determinations (See electrode for more related terms).

Quinone (or cyclohexadiene-1,4-dione, $CO(CHCH)_2CO$): A yellow crystal used to make dyes and hydroquinone.

Rr

1/R² correction: The correction made for the systematic decrease in lidar backscatter signal amplitude with range (40 CFR 60-App/A(alt. method 1)-91).

Rack: A bar screen used to remove large suspended solids.

Rack dryer: Any equipment used to reduce the moisture content of grain in which the grain flows from the top to the bottom in a cascading flow around rows of baffles (racks) (40 CFR 60.301-91).

Rad: *See* definition under radiation absorbed dose (RAD).

Radiant energy or radiation: The energy traveling as a wave unaccompanied by transfer of matter. Examples include x-rays, visible light, ultraviolet light, radio waves, etc (40 CFR 796.3700-91).

Radiation:
(1) Any or all of the following: alpha, beta, gamma, or X-rays; neutrons; and high energy electrons, protons, or other atomic particles; but not sound or radio waves, nor visible, infrared, or ultra-violet light (40 CFR 190.02-91, *See also* 10 CFR 20.3; 300-App/A; 796.3700-91).
(2) Any form of energy propagated as rays, waves, or streams of energetic particles. The term is frequently used in relation to the emission of rays from the nucleus of an atom (cf. radiation under heat transfer) (EPA-88/08a).
(3) The emission of energy in the form of electromagnetic waves. All bodies above absolute zero temperature radiate. Radiation incident on a body may be absorbed, reflected and transmitted (NATO-10/78).
(4) Transmission of energy through space or any medium. Also known as radiant energy (EPA-94/04).

Other radiation related terms include:
- Alpha radiation
- Background radiation
- Beta radiation
- External radiation
- Gamma radiation
- Gamma ray (*See* synonym, gamma radiation)
- Global radiation
- Infrared radiation
- Internal radiation
- Ionization radiation
- Ionizing radiation
- Natural radiation
- Non-ionization radiation
- Non-ionizing electromagnetic radiation
- Nuclear radiation
- Solar radiation
- Ultraviolet radiation

Radiation absorbed dose (rad):
(1) The basic concept in radiation dosimetry. It is the absorption of energy in living tissue and is defined by the International Commission on Radiation Units (ICRU): 1 rad = absorbed energy of 100 ergs per gram of tissue, (0.01 Joules per kilogram of tissue) (LBL-76/07-rad, *See also* EPA-89/12).
(2) The basic unit of absorbed dose equal to the absorption of 0.01 joule per kilogram of absorbing material (DOE-91/04).
(3) *See* radiation unit for more related terms.

Radiation absorbed dose (RAD): The basic unit of absorbed dose equal to the absorption of 0.01 joule per kilogram of absorbing material (DOE-91/04).

Radiation biochemistry: A science dealing with the response of living organisms to radiation.

Radiation catalysis: A catalysis process in which radiation is used to activate or speed up chemical or physical change (*See* catalysis for more related terms).

Radiation cooling: In meteorology, usually employed with respect to the cooling of the earth's surface, particularly during the night caused by radiation heat losses by the surface (NATO-78/10).

Radiation emission-absorption instrumentation: One of continuous emission monitors (*See* continuous emission monitor for various types). All organic compounds absorb electromagnetic radiation because all contain valence electrons that can be excited to higher energy levels. Absorption measurements in the visible and ultraviolet wavelength regions are useful for detecting the presence of certain functional groups that contain valence electrons with relatively low excitation energies. Absorption measurements in the infrared wavelength regions are useful for identifying specific compounds since no two organic compounds (except for optical isomers) have identical infrared absorption curves. The applicability of absorption measurements to characterize incinerator effluents involves the choice of the appropriate energies and wavelength. Types of radiation emission-absorption instrumentation include:
(1) Infrared absorption

(2) Ultraviolet absorption

Radiation inversion: The inversion which develops during the night over a land surface, caused by the strong cooling of the surface due to radiative losses. A favorable condition for the formation of a radiation inversion is a virtually cloudless sky during the night (*See* inversion for more related terms) (NATO-78/10).

Radiation pyrometer:
(1) A device that determines temperature by measuring the intensity of radiation at all wavelengths emitted by a material having a high temperature (SW-108ts).
(2) A device which determines temperature by measuring the intensity of radiation from a heat-generating body (EPA-83).
(3) *See* temperature for more related terms

Radiation standard: Regulations that set maximum exposure limits for protection of the public from radioactive materials (EPA-94/04).

Radiation unit: Radiation unit related terms include:
- Becquerel (Bq)
- Curie (Ci)
- Kerma (K)
- Rad (radiation absorbed dose)
- REM (roentgen equivalent man)
- Rep
- Roentgen (R)

Radical: *See* synonym, free radical.

Radical attack: Generally a bimolecular reaction between a stable molecule and a radical species; e.g., H atom metathesis or Cl atom displacement (EPA-88/12).

Radicle: That portion of the plant embryo which develops into the primary root (40 CFR 797.2750-91).

Radio frequency heat: Heat generated by the application of an alternating electric current, oscillating in the radio frequency range, to a dielectric material. In recent years this method has been used to cure synthetic resin glues (EPA-74/04).

Radio frequency radiation: *See* synonym, non-ionization radiation (EPA-94/04).

Radioactive decay:
(1) The process of spontaneous nuclear transformation, whereby an isotope of one element is transformed into an isotope of another element, releasing excess energy in the form of radiation (40 CFR 300-App/A-91).
(2) Spontaneous change in an atom by emission of charged particles and/or gamma rays; also known as radioactive disintegration and radioactivity (EPA-94/04).

Radioactive half life: The time required for one half the atoms in a given quantity of a specific radionuclide to undergo radioactive decay (40 CFR 300-App/A-91).

Radioactive material: Any material which spontaneously emits radiation (40 CFR 190.02-91, *See also* 10 CFR 20.3-91). Other radioactive material related terms include:

- Airborne radioactive material
- Naturally occurring or accelerator produced radioactive material
- Residual radioactive material

Radioactive substance:
(1) Solid, liquid, or gas containing atoms of a single radionuclide or multiple radionuclides (40 CFR 300-App/A-91; EPA-88/03).
(2) Substances that emit ionizing radiation (EPA-94/04).

Radioactive waste (or radwaste):
(1) Any waste which contains radioactive material in concentrations which exceed those listed in 10 CFR 20, Appendix B, Table II, Column 2 (40 CFR 144.3-91, *See also* 40 CFR 146.3-91).
(2) The high level and transuranic radioactive waste covered by this part (40 CFR 191.02-91).
(3) Materials from nuclear operations that are radioactive or are contaminated with radioactive materials, and for which use, reuse, or recovery are impractical (DOE-91/04).

Other radioactive waste related terms include:
- High level radioactive waste
- Low level radioactive waste
- Transuranic radioactive waste
- *See* waste for more related terms

Radioactivity:
(1) The property of those isotopes of elements that exhibit radioactive decay and emit radiation (40 CFR 300-App/A-91).
(2) A property possessed by some elements, such as uranium, whereby alpha, beta, or gamma rays are spontaneously emitted (EPA-88/08a). Radioactivity occurs when an unstable nucleus undergoes atomic disintegration by emitting particles or electromagnetic radiation (Markes-67).
(3) The spontaneous decay or disintegration of unstable atomic nuclei, accompanied by the emission of radiation (DOE-91/04).

Radioassay: A technique to measure the effect of radiation intensity of a radioactive sample (cf. radioisotope assay).

Radiobiology: The study of radiation effects on living things (EPA-89/12).

Radiochemistry: The study of the chemistry of radioactive substances.

Radioecology: The study of the effects of radiation on species of plants and animals in natural communities (EPA-74/11).

Radiograph: Images made on film by ionizing radiation (DOE-91/04).

Radiography: A non-destructive method of internal examination in which metal or other objects are exposed to a beam of x-ray or gamma radiation. Differences in thickness, density or absorption, caused by internal discontinuities, are apparent in the shadow image either on a fluorescent screen or on photographic film placed behind the object (EPA-83/06a, *See also* 10 CFR 30.4-91).

Radioisotope:
(1) A naturally occurring or artificially created radioactive isotope of a chemical element. Some radioisotopes such as cobalt-60 can be used in medical therapy, biological research, etc.
(2) Chemical variants of an element with potentially oncogenic, teratogenic, and mutagenic effects on the human body (EPA-94/04).

Radioisotope assay: A technique to separate and to measure a radioactive tracer (cf. radioassay).

Radiometer: An instrument used to measure radiant power.

Radionuclide:
(1) A type of atom which spontaneously undergoes radioactive decay (40 CFR 61.91-91, *See also* 40 CFR 61.101-91).
(2) Radioactive isotopes of various elements are collectively referred to as radionuclides (EPA-87/03).
(3) Radioactive particle, man-made or natural, with a distinct atomic weight number. Can have a long life as soil or water pollutants (EPA-94/04).
(4) A radioactive nuclide. (DOE-91/04)

Radionuclide/radioisotope: The isotope of an element exhibiting radioactivity. For HRS purposes, radionuclide and radioisotope are used synonymously (40 CFR 300-App/A-91).

Radiosonde:
(1) An instrument which rises through the atmosphere by means of a balloon and which measures and simultaneously transmits meteorological data, usually pressure, temperature and humidity (NATO-78/10).
(2) A miniature radio transmitter with instruments that is carried aloft (as by an unmanned balloon) for broadcasting by means of precise tone signals or other suitable method the humidity, temperature, pressure, or other parameter every few seconds (LBL-76/07-air).

Radium (Ra): An alkaline earth, radioactive metal with a.n. 88; a.w. 226; d. 5.0 g/cc; m.p. 700° C and b.p. 1140° C. The element belongs to group IIA of the periodic table.

Radius of vulnerability zone: The maximum distance from the point of release of a hazardous substance in which the airborne concentration could reach the level of concern under specified weather conditions (EPA-94/04).

Radon (Rn):
(1) A radioactive gaseous element with a.n. 86; a.w. 222; d. 9.73 g/cc; m.p. -71° C and b.p. -61.8° C. The element belongs to group VIIIA of the periodic table.
(2) A colorless naturally occurring, radioactive, inert gas formed by radioactive decay of radium atoms in soil or rocks (EPA-94/04).

Radon (Rn-222): The radioactive gaseous element and its short-lived decay products produced by the disintegration of the element radium occurring in the air, water, soil, or other media (TSCA302, *See also* EPA-89/12; 88/08a; 88/08).

Radon daughter: *See* synonym, radon decay product.

Radon daughters/radon progeny: Short-lived radioactive decay products of radon that decay into longer-lived lead isotopes, The daughter isotopes can attach themselves to airborne dust and other particles and, if inhaled, damage to lining of the lung. Also known as radon decay products (EPA-94/04).

Radon decay product (radon daughter or radon progeny):
(1) The products are ultrafine solids which tend to adhere to other solids, including dust particles in the air and solid surfaces in a room and can be breathed into the lung where they continue to release radiation as they further decay (EPA-88/09b, *See also* EPA-88/08a).
(2) A term used to refer collectively to the immediate products of the radon decay chain. These include Po-218, Pb-214, Bi-214, and Po-214, which have an average combined half-life of about 30 minutes (EPA-94/04).

Radon progeny: *See* synonym, radon decay product.

Radon progeny integrated sampling unit: A radon decay product measurement system consisting of a low flow-rate air pump that pulls air continuously through a detector assembly containing a thermoluminescent dosimeter. The unit is operated for 100 hours or longer and then the detector assembly is returned to the laboratory for analysis (EPA-88/08).

Radwaste: *See* synonym, radioactive waste.

Raffinate: In the petroleum industry, the portion of the oil which remains undissolved and is not removed by solvent extraction (EPA-74/04b). The less soluble residue that remains after extraction.

Rag paper: A paper product manufactured by use of such materials as cotton or linen threads, flax and hemp, raw cotton, and other textile fibers and cotton linters, as well as rags (*See* paper for more related terms) (EPA-87/10).

Rail car: A non-self-propelled vehicle designed for and used on railroad tracks (40 CFR 201.1-91, *See also* 40 CFR 60.301-91).

Rail haul: The hauling of material by rail (cf. barge haul) (EPA-83).

Railcar loading station: That portion of a metallic mineral processing plant where metallic minerals or metallic mineral concentrates are loaded by a conveying system into railcars (40 CFR 60.381-91).

Railcar unloading station: That portion of a metallic mineral processing plant where metallic ore is unloaded from a railcar into a hopper, screen, or crusher (40 CFR 60.381-91).

Railroad: All the roads in use by any common carrier operating a railroad, whether owned or operated under a contract, agreement, or lease (40 CFR 201.1-91).

Rain: Water drops resulting from the precipitation in the atmosphere.

Rain shadow: An area that has a relatively light average rainfall due to its situation on the lee side of a range of mountains or hills where it is sheltered from the prevailing rain-bearing winds. On the windward side the rainfall is heavy, owing to the forced ascent of the moisture-laden air; as the air descends on the lee side it is warmed and dried, so that little rain is deposited there (DOI-70/04).

Rainout: The scavenging of air pollutants in clouds by liquid or solid particles (NATO-78/10).

Rainwash: A thin sheet of water flowing evenly downslope, quickly concentrated by converging slopes into the shortest and steepest routes downward. This is the first step in the formation of a stream (DOI-70/04).

Rake classifier: A mechanical classifier utilizing reciprocal rakes on an inclined plane to separate coarse from fine material contained in a water pulp (EPA-75/10c).

Ramp method: A variation of area method in which cover materials are obtained by excavating in front of the working face. A variation of this method is known as the progressive slope sanitary landfilling method (*See* sanitary landfill for more related terms) (SW-108ts).

Ramsbottom carbon residue: The carbon residue left after evaporation and pyrolysis of an oil. *See* ASTM D254 (EPA-83).

Random incident field: A sound field ir which the angle of arrival of sound at a given point in space is random in time (40 CFR 211.203-91).

Random monitoring: Monitoring with time intervals determined on a random basis between consecutive observation (*See* monitoring for more related terms).

Random process: A procedure which varies according to some probability function (EPA-79/12c).

Random variable: A variable whose values occur according to the distribution of some probability function (*See* variable for more related terms) (EPA-79/12c).

Range:
(1) Nominal minimum and maximum concentration which a method is capable of measuring (40 CFR 53.23-91).
(2) In statisitics, the difference between the largest and smallest numbers in a set of numbers (EPA-84/03).
(3) The lower and upper detectable limits of concentration, absorbance, % transmittance, etc. (The lower limit should not be reported as 0.0 ppm unless it is the true lower detectable limit (LBL-76/07-bio).
(4) The minimum and maximum measurement limits (the minimum limit is usually reported as 0 ug/L; this is somewhat misleading, and it would be better to report it as the true minimum measurement limit) (LBL-76/07-water).
(5) cf. relative range.

Range of concentration: The highest concentration, the lowest concentration, and the average concentration of an additive in a fuel (40 CFR 79.2-91).

Rank of coal: A classification of coal based upon the fixed carbon as a dry weight basis and the heat value (EPA-82/11f).

Rankine: *See* definition under absolute temperature.

Rankine cycle: A thermodynamic cycle which is the basis of the steam electric generating process (EPA-82/11f).

Rapid sand filter: A filter for the purification of water which has been previously treated, usually be coagulation and sedimentation. The water passes downward through a filtering medium consisting of a layer of sand, prepared anthracite coal or other suitable material, usually from 24 to 30 inches thick and resting on a supporting bed of gravel or other porous medium. The filtrate is removed by an under-drain system. The filter is cleaned periodically by reversing the flow of the water upward through the filtering medium; sometimes supplemented by mechanical or air agitation during backwashing to remove mud and other impurities that are lodged in the sand (*See* filter for more related terms) (EPA-82/10).

Rapping: A process to rap the dust off a electrostatic precipitator.

Rare earth deposit: The source of cerium, terbium, yttrium, and related elements of the rare-earth's group, as well as thorium (EPA-82/05).

Rare gas: *See* synonym, inert gas.

Rare earth metal: Refers to the elements scandium, yttrium, and lanthanum to lutetium, inclusive (*See* metal for more related terms) (40 CFR 421.271-91).

Rare species: *See* synonym, endangered species.

RASP (or rasper):
(1) A grinding machine in the form of a large vertical drum containing heavy hinged arms that rotate horizontally over a rasp-and-sieve floor (SW-108ts).
(2) A machine that grinds waste into a manageable material and helps prevent odor (EPA-94/04).
(3) *See* size reduction machine for more related terms.

Rasper: *See* synonym, RASP.

Rasping system: A procedure in which refuse is ground through a screen partly covered with steel pins, that have the effect of a rasp (EPA-83).

Rate coefficient: A parameter which relates reaction rate of a given molecule(s) to their concentration(s) (EPA-88/12).

Rate constant: The constant in an equation for the rate (concentration change per unit time) of a chemical reaction.

Rate determining step: The slowest step of all reaction steps involved in a chemical reaction system. The slowest step determines the overall rate of the reaction.

Rated capacity:
(1) (In a wastewater treatment plant), the rate of a wastewater

flow that a treatment plant is considered capable of treating on a continuous basis with proper disposal of sludge and no loss in efficiency (DOI-70/04).
(2) (In a general system), the quantity of material that the system can process under demonstrated test conditions (EPA-83).
(3) See capacity for more related terms.

Rated load: The maximum load which a piece of equipment designed to handle safely (*See* load for more related terms) (EPA-83).

Rated incinerator capacity: The number of tons of solid waste that can be processed in an incinerator per 24-hour period when specified criteria prevail (*See* capacity for more related terms) (OME-88/12).

Rated output (RO): The maximum power/thrust available for takeoff at standard day conditions as approved for the engine by the Federal Aviation Administration, including reheat contribution where applicable, but excluding any contribution due to water injection (40 CFR 87.1-91).

Rated pressure ratio (rPR): The ratio between the combustor inlet pressure and the engine inlet pressure achieved by an engine operating at rated output (40 CFR 87.1-91).

Rated speed: The speed at which the manufacturer specifies the maximum rated horsepower of an engine (40 CFR 86.082.2-91).

Rational method: A method of estimating the amount of rainfall runoff.

Rattle: Crackling sound produced by shaking a sheet of paper. It indicates rigidity or stiffness of the paper (EPA-83).

Raoult's law: For concentrated solutions where the components do not interact, the resulting vapor pressure (p) of component "A" in equilibrium with other components can be expressed as: $p = x_A P_A$; where:
- x = mole fraction of component "A" in solution
- P_A = vapor pressure of pure component "A" at the same temperature and pressure as the solution.

In comparing with Raoult's law, Henry's law is for dilute solutions (*See* gas absorption for more related terms) (Hesketh-79, p144).

Raw batch: A glass charge without cullet (cf. raw batch) (EPA-83).

Raw cullet: A glass charge made totally of cullet (*See* cullet for more related terms) (EPA-83).

Raw data: Any laboratory worksheets, records, memoranda notes, or exact copies thereof, that are the result of original observations and activities of a study and are necessary for the reconstruction and evaluation of the report of that study. In the event that exact transcripts of raw data have been prepared (e.g., tapes which have been transcribed verbatim, dated, and verified accurate by signature), the exact copy or exact transcript may be substituted for the original source as raw data. Raw data may include photographs, microfilm or microfiche copies, computer printouts, magnetic media, including dictated observations, and recorded data from automated instruments (40 CFR 160.3-91).

Raw ink: All purchased ink (40 CFR 60.431-91).

Raw material: All natural and synthetic rubber, carbon black, oils, chemical compounds, fabric and wire used in the manufacture of pneumatic tires and inner tubes or components thereof (40 CFR 428.11-91, *See also* 40 CFR 425.02; 428.51; 428.61; 428.71; 428.101; 428.111; 432.101-91).

Raw material equivalent: Equal to the raw material usage multiplied by the volume of air scrubbed via wet scrubbers divided by the total volume of air scrubbed (40 CFR 428.51-91, *See also* 40 CFR 428.61; 428.71-91).

Raw mine drainage: Untreated or unprocessed water drained, pumped or siphoned from a mine (EPA-82/05).

Raw sewage:
(1) The untreated sewage (EPA-76/03).
(2) Untreated wastewater and its contents (EPA-94/04).
(3) *See* sewage for more related terms.

Raw sewage sludge: The solids concentrated by various methods in wastewater treatment plants, usually contain 90 to 96% water (*See* sewage for more related terms) (EPA-83).

Raw sludge: The sludge that has not been treated by either aerobic or anaerobic digestion (*See* sludge for more related terms).

Raw sugar: An intermediate sugar product consisting of crystals of high purity covered with a film of low quality syrup (*See* sugar for more related terms) (EPA-74/01a).

Raw sugar juice: The liquid product remaining after extraction of sugar from the sliced beets (cossetees) during the diffusion process (*See* sugar for more related terms) (EPA-74/01a).

Raw waste load: The quantity of pollutant in wastewater prior to treatment (*See* load for more related terms) (EPA-87/10a).

Raw wastewater: The wastewater prior to any treatment (*See* wastewater for more related terms) (EPA-83/03).

Raw water:
(1) The plant intake water prior to any treatment or use (*See* water for more related terms) (EPA-83/06a).
(2) Intake water prior to any treatment or use (EPA-94/04).

Rayon fiber:
(1) A manufactured fiber composed of regenerated cellulose, as well as manufactured fibers composed of regenerated cellulose in which substituents have replaced not more than 15 percent of the hydrogens of the hydroxyl groups (40 CFR 60.601-91).
(2) A generic name for man made fibers, monofilaments, and continuous filaments, made from regenerated cellulose. Fibers produced by both viscose and cuprammonium process are classified as rayon (EPA-74/06b).
(3) *See* fiber for more related terms.

RDF stoker: A steam generating unit that combusts RDF (refuse-derived fuel) in a semi-suspension firing mode using air-fed distributors (40 CFR 60.51a-91).

Re-aeration:
(1) The addition of air to return activated sludge in a second aeration tank.
(2) Introduction of air into the lower layers of a reservoir. As the air bubbles form and rise through the water, the oxygen from the air dissolves into the water and replenishes the dissolved oxygen. The rising bubbles also cause the lower waters to rise to the surface where they take on oxygen from the atmosphere (EPA-94/04).
(3) *See* aeration for more related terms.

Re-refined oil: The used oil from which the physical and chemical contaminants acquired through previous use have been removed through a refining process (*See* oil for more related terms) (RCRA1004, *See also* 40 CFR 252.4-91).

Reactant: A chemical substance that is used intentionally in the manufacture of a polymer to become chemically a part of the polymer composition (cf. chemical reaction) (40 CFR 723.250-91).

Reaction: The force from a body which is equal and opposite in direction against outside forces exerted on the body. Other reaction related terms include:
- Bimolecular reaction
- Unimolecular reaction

Reaction cell: A chamber in which the chemical reactant is rapidly recirculated to prevent chemical depletion, facilitate sludge removal and automatically provide chemical replenishment control (EPA-82/11e).

Reaction quantum yield for an excited state process: The fraction of absorbed light that results in photoreaction at a fixed wavelength. It is the ratio of the number of molecules that photoreact to the number of quanta of light absorbed or the ratio of the number of moles that photoreact to the number of einsteins of light absorbed at a fixed wavelength (40 CFR 796.3700-91).

Reaction spinning process: The fiber-forming process where a prepolymer is extruded into a fluid medium and solidification takes place by chemical reaction to form the final polymeric material (40 CFR 60.601-91).

Reaction temperature: A temperature at which oxidation occurs in a combustion system (*See* temperature for more related terms) (EPA-84/09).

Reaction turbine: A type of water wheel in which water turns the blades of a rotor, which then drives an electrical generator or other machine (DOI-70/04).

Reactive functional group: An atom or associated group of atoms in a chemical substance that is intended or can reasonably be anticipated to undergo facile chemical reaction (40 CFR 723.250-91).

Reactive site: A chemical configuration on a molecule with which a bond is made by other specific molecules (LBL-76/07-bio).

Reactivity:
(1) One of the four U.S. EPA hazardous waste characteristics (ETI-92).
(2) The ability of one chemical to undergo a chemical reaction with another chemical. Reactivity of one chemical is always measured in reference to the potential for reaction with itself or with another chemical. A chemical is sometimes said to be reactive, or have high reactivity, without reference to another chemical. Usually this means that the chemical has the ability to react with common materials such as water, or common materials of construction such as carbon steel (*See also* hazardous waste characteristics) (EPA-87/07a).
(3) The waste which is extremely unstable and has a tendency to react violently or explode during stages of their management.
(4) *See* hazardous waste characteristics for more related terms.

Reactor: A vat, vessel, or other device in which chemical reactions take place (40 CFR 52.741-91, *See also* 40 CFR 61.61-91).

Reactor coolant pressure boundary: In a heavy-water reactor, those systems and components that contain the reactor coolant. The pressure boundary includes the reactor vessel, pressure tubes, inlet and effluent plenums, pressurizer, reactor coolant pumps, heat exchanger tubes and plenums, associated reactor coolant system piping and valves, and any other system that connects with the reactor coolant system up to the second isolation valve. The moderator and control rod coolant system are considered part of the reactor coolant boundary (DOE-91/04).

Reactor core:
(1) In a heavy-water reactor: the fuel assemblies, including the fuel and target tubes, control assemblies, blanket assemblies, safety rods, and coolant/moderator.
(2) In a light-water reactor: the fuel assemblies, including the fuel and target rods, control rods, and coolant/moderator.
(3) In a modular high-temperature gas-cooled reactor: the graphite elements, including the fuel and target elements, control rods, any other reactor shutdown mechanisms, and the graphite reflectors (DOE-91/04).
(4) The uranium-containing heart of a nuclear reactor, where energy is released (EPA-89/12).

Reactor facility: Unless it is modified by words such as containment, vessel, or core, the term reactor facility includes the housing, equipment, and associated areas devoted to the operation and maintenance of one or more reactor cores. Any apparatus that is designed or used to sustain nuclear chain reactions in a controlled manner, including critical and pulsed assemblies and research, test, and power reactors, is defined as a reactor. All assemblies designed to perform subcritical experiments that could potentially reach criticality are also to be considered reactors. Critical assemblies are special nuclear devices designed and used to sustain nuclear reactions. Critical assemblies may be subject to frequent changes in core and lattice configuration, and they may be used frequently as mockups of reactor configurations. Therefore, requirements for modifications do not apply unless the overall assembly room is modified, a new assembly room is proposed, or a new configuration is not covered in previous safety evaluations (DOE-91/04).

Reactor opening loss: The emissions of vinyl chloride occurring when a reactor is vented to the atmosphere for any purpose other than an emergency relief discharge as defined in 40 CFR 61.65(a) (40 CFR 61.61-91).

Reactor operations: All activities involved in operating and using a reactor, beginning with he initial loading of fuel in the reactor vessel and ending with the removal of fuel to officially decommission the reactor or place it in a standby status (DOE-91/04).

Reactor year: A unit of time by which accident frequency and core damage frequency are measured; it assumes that more than one reactor can operate during the year (a calendar year during which three reactors operated would be the experience equivalent of 3 reactor years) and it assumes that a reactor might not operate continuously for the entire year (a reactor operating only 60% of the calendar year would be the equivalent of 0.6 reactor year) (DOE-91/04).

Readily water soluble substances: The chemicals which are soluble in water at a concentration equal to or greater than 1,000 mg/L (cf. limited water-soluble substances) (40 CFR 797.1060-91).

Ready biodegradability: An expression used to describe those substances which, in certain biodegradation test procedures, produce positive results that are unequivocal and which lead to the reasonable assumption that the substance will undergo rapid and ultimate biodegradation in aerobic aquatic environments (40 CFR 796.3100-91).

Reagent:
(1) A reactive material used to remove acid gases from the combustion gases (EPA-89/03b).
(2) A chemical or solution used to produce a desired chemical reaction, a substance used in assaying or in flotation (EPA-82/05).

Reagent blank:
(1) An aliquot of analyte-free water or solvent analyzed with the analytical batch.
(2) A volume of deionized, distilled water containing the same acid matrix as the calibration standards carried through the entire analytical scheme (40 CFR 136. App. C) (*See* blank for more related terms).

Reagent grade: An analytical reagent grade, ACS (American Chemical Society) reagent grade and reagent grade are synonymous terms for reagents which conform to the current specifications of the Committee on Analytical Reagents of the ACS.

Real ear protection at threshold: The mean value in decibels of the occluded threshold of audibility (hearing protector in place) minus the open threshold of audibility (ears open and uncovered) for all listeners on all trials under otherwise identical test conditions (40 CFR 211.203-91).

Real property: The land, including land improvements, and structures and appurtenances, excluding movable machinery and equipment (40 CFR 30.200-91, *See also* 40 CFR 31.3; 35.6015; 247.101-91).

Real time mode: The use of an air quality simulation model for immediate application taking into account current input data (NATO-78/10).

Ream:
(1) Quantity of paper, generally 500 sheets.
(2) An imperfection; nonhomogeneous layers in flat glass (EPA-83).

Ream weight: The weight of one ream of paper. Also called basis weight (EPA-83).

Reasonable further progress: Annual incremental reductions in air pollution emissions as reflected in a State Implementation Plan, that EPA deems sufficient to provide for attainment of the applicable national ambient air quality standard by the statutory deadline (EPA-94/04).

Reasonable maximum exposure: The maximum exposure reasonably expected to occur in a population (EPA-94/04).

Reasonable potential: Is where an effluent is projected or calculated to cause an excursion above a water quality standard based on a number of factors including, a.s a minimum, the four factors listed in 40 CFR 122.44(d)(1)(ii) (EPA-91/03).

Reasonably available control measures (RACM): A broadly defined term referring to technological and other measures for pollution control (EPA-94/04).

Reasonably available control technology (RACT):
(1) Devices, systems process modifications, or other apparatus or techniques that are reasonably available taking into account:
  (A) The necessity of imposing such controls in order to attain and maintain a national ambient air quality standard;
  (B) The social, environmental and economic impact of such controls; and
  (C) Alternative means of providing for attainment and maintenance of such standard. (This provision defines RACT for the purposes of 40 CFR 51.110(c)(2) and 51.341(b) only) (40 CFR 51.100-91).
(2) Control technology that is both reasonably available, and both technologically and economically feasible. Usually applied to existing sources in nonattainment areas; in most cases is less stringent than new source performance standards (EPA-94/04).

Reboil: Reappearance of bubbles in molten glass after it previously appeared plain (EPA-83).

Rebricking: The cold replacement of damaged or worn refractory parts of the glass melting furnace. Rebricking includes replacement of the refractories comprising the bottom, sidewalls, or roof of the melting vessel; replacement of refractory work in the heat exchanger; replacement of refractory portions of the glass conditioning and distribution system (40 CFR 60.291-91, *See also* 40 CFR 61.161-91).

Recarbonation:
(1) Recharge of carbon dioxide to lower the pH value of a fluid.
(2) Process in which carbon dioxide is bubbled into water being treated to lower the pH (EPA-94/04).

Receiving water: A river, lake, ocean, stream or other watercourse into which wastewater or treated effluent is discharged (*See* water for more related terms) (EPA-94/04).

Receiving water concentration (RWC): The concentration of a toxicant or the parameter toxicity in the receiving water after mixing (formerly termed instream waste concentration [IWC]) (EPA-91/03).

Receptor:
(1) In biochemistry, a specialized molecule in a cell that binds a specific chemical with high specificity and high affinity.
(2) In exposure assessment, an organism that receives, may receive, or has received environmental exposure to a chemical (Course 165.6).
(3) In dispersion modeling, a point in space at which the ambient air quality is being estimated. For a given receptor, the data required as input to a dispersion model are its coordinate location relative to that of the pollutant source and elevation (if terrain is being considered in the modeling analysis (EPA-88/09).

Receptor point: The geographical point where an air pollutant concentration is measured or is calculated by means of an air pollution dispersion model (NATO-78/10).

Recessive mutation: A change in the genome which is expressed in the homozygous or homozygous condition (40 CFR 798.5275-91).

Recharge:
(1) A process, natural or artificial, by which water is added to the saturated zone of an aquifer (40 CFR 149.2-91).
(2) Replenishment of water to an aquifer (DOE-91/04).
(3) The process by which water is added to a zone of saturation, usually by percolation from the soil surface, e.g., the recharge of an aquifer (EPA-94/04).

Recharge area:
(1) An area in which water reaches the zone of saturation (ground water) by surface infiltration; in addition, a major recharge area is an area where a major part of the recharge to an aquifer occurs (40 CFR 149.2-91).
(2) A land area in which water reaches the zone of saturation from surface infiltration, e.g., where rainwater soaks through the earth to reach an aquifer (EPA-94/04).

Recharge rate: The quantity of water per unit of time that replenishes or refills an aquifer (EPA-94/04).

Recharge zone: The area through which water enters the Edwards Underground Reservoir as defined in the December 16, 1975, Notice of Determination (40 CFR 149.101-91).

Reciprocal translocation: The chromosomal translocations resulting from reciprocal exchanges between two or more chromosomes (40 CFR 798.5955-91).

Reciprocating compressor: A piece of equipment that increases the pressure of a process gas by positive displacement, employing linear movement of the driveshaft (*See* compressor for more related terms) (40 CFR 60.631-91).

Reciprocating grate: A stoker grate surface having alternate lateral stationary and moving rows which reciprocate continuously and slowly, forward and backward, for the purpose of stirring the combustible bed of material while conveying it and the resulting residue from the feeding end to the discharge end of the furnace (*See* grate for more related terms) (EPA-83).

Reciprocating grate stoker: A stoker with a bed of bars or plates arranged so that alternate pieces, or rows of pieces, reciprocate slowly in a horizontal sliding mode and act to push the solid waste along the stoker surface (*See* stoker for more related terms) (SW-108ts).

Recirculated cooling water: The water which is passed through the main condensers for the purpose of removing waste heat, passed through a cooling device for the purpose of removing such heat from the water and then passed again, except for blowdown, through the main condenser (*See* cooling water for more related terms) (40 CFR 423.11-91, *See also* EPA-82/11a).

Recirculating cooling system: In a manufacturing or processing plant, a system that reduces the temperature of used water in a cooling tower by evaporating a small percent of the recirculating stream; although the evaporated water is permanently removed from the supply, overall water withdrawal is reduced to a small percent of what it would otherwise be; discharge of contaminated water may be reduced to as little as 1 percent (DOI-70/04).

Recirculating spray: A spray rinse in which the drainage is pumped up to the spray and is continually recirculated (EPA-83/06a).

Recirculation: Those cold rolling operations which include recirculation of rolling solutions at all mill stands (40 CFR 420.101-91).

Recirculation system: A system which is specifically designed to divert the major portion of the cooling water discharge back for reuse (EPA-82/11f).

Reclaim rinse: The first step following a plating process to retain as much of the chemicals as possible and to allow return of the dragout solution to the plating tank (EPA-74/03d).

Reclaimed: A material is reclaimed if it is processed to recover a usable product, or if it is regenerated. Examples are recovery of lead values from spent batteries and regeneration of spent solvents (40 CFR 261.1-91).

Reclamation:
(1) The process of deriving usable materials from waste, by-products, etc., through physical or chemical treatment (EPA-80/10).
(2) A procedure by which a disturbed area can be reworked to

make it productive, useful, or aesthetically pleasing (cf. extraction, reccvery or recycle) (EPA-82/05).
(3) The restoration to a better or ore useful state, such as land reclamation by sanitary landfilling, or the obtaining of useful materials from solid waste (EPA-83).
(4) (In recycling) Restoration of materials found in the waste stream to a beneficial use which may be for purposes other than the original use (EPA-94/04).

Reclamation area: The surface area of a coal mine which has been returned to required contour and on which revegetation (specifically, Seeding or planting) work has commenced (40 CFR 434.11-91).

Reclamation plan: A plan submitted by an applicant for a permit under a State program or Federal program which sets forth a plan for reclamation of the proposed surface coal mining operations pursuant to section 1258 of this title (SMCRA701-30USC1291-90).

Recombinant bacteria: A microorganism whose genetic makeup has been altered by deliberate introduction of new genetic elements. The offspring of these altered bacteria also contain these new genetic elements, i.e., they "breed true" (EPA-94/04).

Recombinant DNA: The new DNA that is formed by combining pieces of DNA from different organisms or cells (EPA-94/04).

Recombination: The rearrangement of genes that differs from those of the parents as a result of independent assortment, linkage and crossing over.

Recommencing discharger: A source which recommences discharge after terminating operations. Regional Administrator means the Regional Administrator of the appropriate Regional Office of the Environmental Protection Agency or the authorized representative of the Regional Administrator (40 CFR 122.2-91).

Recommended maximum contaminant level (RMCL:) The maximum level of a contaminant in drinking water at which no known or anticipated adverse affect on human health would occur, and that includes an adequate margin of safety. Recommended levels are nonenforceable health goals (*See* maximum contaminant level) (EPA-94/04).

Reconfigured emission data vehicle: An emission data vehicle obtained by modifying a previously used emission data vehicle to represent another emission data vehicle (*See* vehicle for more related terms) (40 CFR 86.082.2-91).

Reconstructed source: Facility in which components are replaced to such an extent that the fixed capital cost of the new components exceed 50 percent of the capital cost of constructing a comparable brand-new facility. New-source performance standards may be applied to sources reconstructed after the proposal of the standard if it is technologically and economically feasible to meet the standard (*See* source for more related terms) (EPA-94/04).

Reconstruction: Will be presumed to have taken place where the fixed capital cost of the new component exceeds 50 percent of the fixed capital cost of a comparable entirely new source. Any final decision as to whether reconstruction has occurred must be made in accordance with the provisions of 60.15 (f) (1) through (3) of this title (40 CFR 51.301-91).

Record: Any item, collection, or grouping of information regardless of the form or process by which it is maintained (e.g., paper, document, microfiche, microfilm, X-ray film, or automated data processing) (cf. EPA record) (29 CFR 1910.20-91, *See also* 40 CFR 16.2; 1516.2-91).

Record of decision (ROD):
(1) A public document, signed by the lead agency and any RPs, that explains which remedial alternative(s) will be used at a particular CERCLA site. The ROD is based on data generated during the site characterization and treatability study phases of the RI/FS and on consideration given to public comments and State and community concerns (EPA-89/12a).
(2) A public document that ex plains which cleanup alternative(s) will be used at National Priorities List sites where, under CERCLA, Trust Funds pay for the cleanup (EPA-94/04).

Recorded: The written or otherwise registered in some form for preserving information, including such forms as drawings, photographs, videotape, sound recordings, punched cards, and computer tape or disk (40 CFR 2.201-91).

Recoverable: Refers to the capability and likelihood of being recovered from solid waste for a commercial or industrial use (RCRA1004-42USC6903-91).

Recoverable resources: The materials that still have useful physical, chemical, or biological properties after serving their original purpose and can, therefore, be reused or recycled for the same or other purposes (40 CFR 245.101-91, *See also* 40 CFR 246.101-91).

Recovered materials: The waste material and byproducts which have been recovered or diverted from solid waste, but such term does not include those materials and byproducts generated from, and commonly reused within, an original manufacturing process (40 CFR 248.4-91, *See also* RCRA1004; 40 CFR 249.04; 250.4-91).

Recovered resources: The material or energy recovered from solid waste (RCRA1004-42USC6903-91).

Recovered solvent: The solvent captured from liquid and gaseous process streams that is concentrated in a control device and that may be purified for reuse (*See* solvent for more related terms) (40 CFR 60.601-91).

Recovery (or salvage):
(1) The process of obtaining materials or energy resources from solid waste (40 CFR 245.101; 246.101-91).
(2) The process of retrieving materials or energy resources from wastes. Also referred to as extraction, reclamation, recycling, salvage (EPA-83).
(3) cf. salvage.

Recovery device: An individual unit of equipment, such as an absorber, condenser, and carbon adsorber, capable of and used to recover chemicals for use, reuse or sale (40 CFR 60.611-91, *See* *also* 40 CFR 60.661-91).

Recovery furnace (or recovery boiler):
(1) Either a straight kraft recovery furnace or a cross recovery furnace, and includes the direct-contact evaporator for a direct-contact furnace (40 CFR 60.281-91).
(2) A boiler which burns the high heat content materials recovered from waste or processes such as strong black liquor (EPA-87/10).

Recovery rate:
(1) Percentage of usable recycled materials that have been removed from the total amount of municipal solid waste generated in a specific area or by a specific business (EPA-94/04).
(2) *See also* synonym, diversion rate.

Recovery system: An individual unit or series of material recovery units, such as absorbers, condensers, and carbon adsorbers, used for recovering volatile organic compounds (40 CFR 60.561-91, *See also* 40 CFR 60.611; 60.661-91).

Recrystallization:
(1) Formation of new crystals from previously melted sugar liquor. Recrystallization is encouraged by evaporators and accomplished in vacuum pans (EPA-75/02d).
(2) A process of repeated crystallization to purify a product.

Rectangular weir: A weir having a notch that is rectangular in shape (cf. weir) (EPA-82/11e).

Rectifier:
(1) A device for converting alternating currents into direct currents.
(2) A nonlinear circuit component that, ideally, allows current to flow in one direction unimpeded but allows no current to flow in the other direction (EPA-83/03).
(3) a distillation unit.

Recuperator:
(1) A steel or refractory chamber used to reclaim heat from waste gases (EPA-85/10a).
(2) A continuous heat exchanger in which heat is conducted from the products of combustion to incoming air through flue walls (EPA-83).

Recurrence interval: The average number of years within that a variable will be less than or equal to a specified value. This term is synonymous with return period (EPA-91/03).

Recurrent expenditure: Those expenses associated with the activities of a continuing environmental program. All expenditures, except those for equipment purchases with a unit acquisition cost of $5.000 or more, are considered recurrent unless justified by the applicant as unique and approved as such by the Regional Administrator in the assistance award (40 CFR 35.105-91).

Recyclable: That materials that still have useful physical or chemical properties after serving their original purpose and that can, therefore, be reused or remanufactured into additional products (EPA-89/11).

Recyclable paper: Any paper separated at its point of discard or from the solid waste stream for utilization as a raw material in the manufacture of a new product. It is often called waste paper or paper stock. Not all paper in the waste stream is recyclable. It may be heavily contaminated or otherwise unusable (*See* paper for more related terms) (40 CFR 250.4-91).

Recycle/reuse: Minimizing waste generation by recovering and reprocessing usable products that might otherwise become waste (.i.e. recycling of aluminum cans, paper, and bottles, etc.) (EPA-94/04).

Recycle or recycling (or reuse): The process of minimizing the generation of waste by recovering usable products that might otherwise become waste. Examples are the recycling of aluminum cans, wastepaper, and bottles (EPA-89/12). The process by which waste materials are transformed into new products in such a manner that the original products may lose their identity (cf. extraction, reclamation, recovery and reuse) (SW-108ts).

Recycled: A material is recycled if it is used, reused, or reclaimed (40 CFR 261.1-91).

Recycled material: A material that is utilized in place of a primary, raw, or virgin material in manufacturing a product (40 CFR 245.101-91, *See also* 40 CFR 246.101; 247.101-91).

Recycled oil: Any used oil which is reused, following its original use, for any purpose (including the purpose for which the oil was originally used). Such term includes oil which is re-refined, reclaimed, burned, or reprocessed (*See* oil for more related terms) (RCRA1004-42USC6903-91).

Recycled PCBs: Those PCBs which appear in the processing of paper products or asphalt roofing materials from PCB-contaminated raw materials. Processes which recycle PCBs must meet the following requirements:
(1) There are no detectable concentrations of PCBs in asphalt roofing material products leaving the processing site.
(2) The concentration of PCBs in paper products leaving any manufacturing site processing paper products, or in paper products imported into the United States, must have an annual average of less than 25 ppm with a 50 ppm maximum.
(3) The release of PCBs at the point at which emissions are vented to ambient air must be less than 10 ppm.
(4) The amount of Aroclor PCBs added to water discharged from an asphalt roofing processing site must at all times be less than 3 micrograms per liter ($\mu g/L$) for total Aroclors (roughly 3 parts per billion (3 ppb)). Water discharges from the processing of paper products must at all times be less than 3 micrograms per liter ($\mu g/1$) for total Aroclors (roughly 3 ppb), or comply with the equivalent mass-based limitation.
(5) Disposal of any other process wastes at concentrations of 50 ppm or greater must be in accordance with subpart D of this part (40 CFR 761.3-91).

Recycled water: The process wastewater or treatment facility effluent which is recirculated to the same process (*See* water for more related terms) (EPA-83/03).

Recycle lagoon: A pond that collects treated wastewater, most of which is recycled as process water (*See* lagoon for more related terms) (EPA-83/06a).

Recycling: The process by which recovered materials are transformed into new products (40 CFR 244.101-91, *See also* 40 CFR 245.101; 246.101-91).

Recycling rate: *See* synonym, diversion rate.

Red bag waste:
(1) Refers to infectious waste; the name comes from the use of red plastic bags to contain the waste and to clearly identify that the waste should be handled as infectious (*See* waste for more related terms) (EPA-89/03b).
(2) *See* infectious waste (EPA-94/04).

Red border: An EPA document undergoing review before being submitted for final management decision-making (EPA-94/04).

Red brass: An alloy 83% copper, 17% zinc, used for valves, bearings, castings (*See* brass for more related terms) (EPA-83).

Red edge: Numerous rouge pits located around the edges of a large sheet of polished plate glass (EPA-83).

Red tide: A proliferation of a marine plankton toxic and often fatal to fish, perhaps stimulated by the addition of nutrients. A tide can be red, green, or brown, depending on the coloration of the plankton (EPA-94/04).

Red stock: Sulfite pulp after the pulping process, prior to other treatments, such as bleaching (EPA-87/10).

Red tide: A proliferation of a marine plankton that is toxic and often fatal to fish. This natural phenomenon may be stimulated by the addition of nutrients. A tide can be called red, green or brown, depending on the coloration of the plankton (EPA-89/12).

Red water: The effluent coming from the sellite wash of crude TNT. Sellite has a selective affinity for the unsymmetrical, unwanted isomers of TNT. The result is a blood red effluent high in sulfate concentration. A red waste liquid resulting from the purification of TNT, normally incinerated or sold to the paper industry (*See* water for more related terms) (EPA-76/03).

Redox: Reduction-oxidation reaction.

Redox potentiometry: The measurement of a solution potential (voltage) between a neutral electrode probe after the solution is developed by a reduction or oxidation process.

Redox titration: A titration by using the process of transferring electrons from one substance to another (reduction to oxidation). The titration end point is determined by the color comparison or potential measurement (*See* titration for more related terms).

Reduced air preheat and load reduction: One of NO_x emission reduction techniques (*See* nitrogen oxide emission control for control structure). These techniques are used sparingly in large boilers due to the energy penalty involved and the relatively low emission reduction occurring (EPA-81/12, p7-13).

Reduced sulfur compounds: The hydrogen sulfide (H_2S), carbonyl sulfide (COS) and carbon disulfide (CS_2) (40 CFR 60.101-91, *See also* 40 CFR 60.641-91).

Reducing agent: A chemical that lowers the state of oxidation of other chemicals (EPA-83, *See also* EPA-75/02).

Reducing salt bath descaling: The removal of scale from semi-finished steel products by the action of molten salt baths containing sodium hydride (*See* salt bath descaling for more related terms) (40 CFR 420.81-91).

Reducing slag: Reducing slag is used in the electric furnace following the slagging off of an oxidizing slag to minimize the loss of alloys by oxidation (*See* slag for more related terms) (EPA-74/06a).

Reductant: A reducing agent which is used to remove oxygen in a reaction system.

Reduction (in chemical application):
(1) The addition of hydrogen, removal of oxygen, or addition of electrons to an element or compound (EPA-94/04).
(2) Also known as chemical reduction. It is a process in which an atom (or group of atoms) gains electrons (EPA-87/10a).
Chemical reduction can reduce a metal to its elemental form for potential recycle or can convert it to less toxic oxidation states. One such metal is chromium, which, when present as chromium (VI), is a very toxic material. In the reduced state, chromium (III), the hazards are lessened and the chromium can be precipitated for removal. At the present time, chemical reduction is applied primarily to the control of hexavalent chromium in the plating and tanning industries and the removal of mercury from caustic/chlorine electrolysis cell effluents. An example of reduction-oxidation, or "Redox" is as follows: $2H_2CrO_4 + 3SO_2 + 3H_2O \longrightarrow Cr_2(SO_4)_3 + 5H_2O$.

The oxidation state of Cr changes from 6+ to 3+ (Cr is reduced); the oxidation state of S increases from 2+ to 3+ (S is oxidized). This change of oxidation state implies that an electron was transferred from S to Cr(VI). The decrease in the positive valence (or increase in the negative valence) with reduction takes place simultaneously with oxidation in chemically equivalent ratios. Reduction is used to treat wastes in such a way that the reducing agent lowers the oxidation state of a substance in order to reduce its toxicity, reduce its solubility, or transform it into a form that can be more easily handled.

Reduction (in waste application): An activity by manufacturers (e.g., modifying products) and consumers (e.g., modifying pruchasing decisions) that reduces toxicity or quantity of products before they are purchased (cf. pollution prevention) (OTA-89/10). Other reduction related terms include:
- Quantity reduction
- Toxicity reduction

Reduction cell: A vessel for conducting reduction processes.

Reduction control system: An emission control system which reduces emissions from sulfur recovery plants by converting these emissions to hydrogen sulfide (40 CFR 60.101-91).

Reduction plant: A mill or a treatment place for the extraction of values from ore (EPA-82/05).

Reduction potential: The potential to reduce an ion from a positive charge form to neutral form or from a neutral form to a negative charge form.

Redundancy: For control systems, redundancy is the presence of a second piece of control equipment where only one would be required. The second piece of equipment is installed to act as a backup in the event that the primary piece of equipment fails. Redundant equipment can be installed to backup all or selected portions of a control system (EPA-87/07a).

Reentry: The period of time immediately following the application of a pesticide to a field when unprotected workers should not enter as provided for in 40 CFR 170.3(b) (40 CFR 170.2-91).

Reentry interval: The period of time immediately following the application of a pesticide during which unprotected workers should not enter a field (EPA-94/04).

Reference: Published or unpublished information pertaining to instrument specifications and performance characteristics (LBL-76/07-bio).

Reference air concentration: *See* definition under risk specific dose.

Reference ambient concentration (RAC): The concentration of a chemical in water that will not cause adverse impacts to human health. RAC is expressed in units of mg/L (EPA-91/03).

Reference compound: The VOC (volatile organic compound) species selected as an instrument calibration basis for specification of the leak definition concentration. (For example: If a leak definition concentration is 10,000 ppmv as methane, then any source emission that results in a local concentration that yields a meter reading of 10,000 on an instrument calibrated with methane would be classified as a leak. In this example, the leak definition is 10,000 ppmv, and the reference compound is methane (40 CFR 60-AA (method 21-2.2))).

Reference concentration (RfC):
(1) An estimate (with uncertainty spanning perhaps an order of magnitude) of a continuous inhalation exposure to the human population (including sensitive subgroups) that is likely to be without an appreciable risk of deleterious non-cancer effects during a lifetime (EPA-92/12).
(2) An estimate (with uncertainty spanning perhaps an order of magnitude) of a daily exposure to the human population (including sensitive subgroups) that is likely to be without an appreciable risk of deleterious effects during a lifetime. The inhalation reference dose is for continuous inhalation exposures and is appropriately expressed in units of mg/m^3. It may be expressed as mg/kg/day, in order to compare with oral RfD units, utilizing specified conversion assumptions (EPA-90/08).

Reference dose (RfD):
(1) The estimate of a daily exposure level of a substance to a human population below which adverse noncancer health effects are not anticipated. [milligrams toxicant per kilogram body weight per day (mg/kg-day)] (40 CFR 300-App/A-91).
(2) An estimate of the daily exposure to human population that is likely to be without an appreciable risk of deleterious effect during a lifetime; derived from non-observed adverse effect level or lowest observed adverse effect level (EPA-91/03).
(3) The Agency's preferred toxicity value for evaluating potential r carcinogenic effects in humans resulting from contaminant) exposures at CERCLA sites. (*See* RAGS/HHEM Part A for a discussion of different kinds of reference doses and reference concentrations) (EPA-91/12).
(4) An estimate (with uncertainty spanning perhaps an order of magnitude) of a daily exposure to the human population (including sensitive subgroups) that is likely to be without appreciable risk of deleterious effects during a lifetime (EPA-92/12).
(5) The concentration of a chemical known to cause health problems; also be referred to as the ADI, or acceptable daily intake (EPA-94/04).
(6) *See* dose for more related terms

Reference electrode: *See* synonym, standard electrode.

Reference level: The based level used for the comparison of measurements.

Reference material (RM): A material of known or established concentration that is used to calibrate or to assess the bias of a measurement system (EPA-84/03). It includes:
(1) **Internal reference material:** A reference material developed by a laboratory for its own internal use (ACS-87/11).
(2) **External reference material:** A reference material provided by someone other than the end-user laboratory (ACS-87/11).
(3) **Certified reference material:** A reference material accompanied by a certificate issued by an organization generally accepted as technically capable to do so (ACS-87/11).
(4) **Standard reference material:**
   (A) A National Bureau of Standard certified reference material.
   (B) A material produced in quantity, of which certain properties have been certified by the National Bureau of Standards (NBS) or other agencies to the extent possible to satisfy its intended use. The material should be in a matrix similar to actual samples to be measured by a measurement system or be used directly in preparing such a matrix. Intended uses include:
      (a) Standardization of solutions;
      (b) Calibration of equipment; and
      (c) Monitoring the accuracy and precision of measurement systems (LBL-76/07-air).

Reference method (or sampling and analyzing method):
(1) A method of sampling and analyzing the ambient air for an air pollutant that is specified as a reference method in an appendix to this part, or a method that has been designated as

a reference method in accordance with Part 53 of this chapter; it does not include a method for which a reference method designation has been canceled in accordance with 40 CFR 53.11 or 40 CFR 53.16 of this chapter (40 CFR 50.1-91, *See also* 53.1-91). A list of test methods was provided in Appendixes to this part, as follows, (40 CFR 60-App/A-91)):

- Appendix A - reference method for the determination of sulfur dioxide in the atmosphere (pararosaniline method).
- Appendix B - reference method for the determination of suspended particulate matter in the atmosphere (high volume method).
- Appendix C - measurement principle and calibration procedure for the measurement of carbon monoxide in the atmosphere (non-dispersive infrared photometry).
- Appendix D - measurement principle and calibration procedure for the measurement of ozone in the atmosphere.
- Appendix E - reference method for the determination of hydrocarbons corrected methane.
- Appendix F - measurement principle and calibration procedure for the measurement of nitrogen dioxide in the atmosphere (gas phase chemiluminescence).
- Appendix G - reference method for the determination of lead in suspended particulate matter collected from ambient air.
- Appendix H - interpretation of the National Ambient Air Quality Standards for ozone.
- Appendix I - reserved.
- Appendix J - reference method for the determination of particulate matter as PM10 in the atmosphere.
- Appendix K - interpretation of the National Ambient Air Quality Standards for particulate matter.

(2) Any method of sampling and analyzing for an air pollutant as specified in the applicable subpart (40 CFR 60.2-91). (A list of test methods was provided in Appendix A to this part, as follows, (40 CFR 60-App/A-91)):

- Method 1 - Sample and velocity traverses for stationary sources.
- Method 1A - Sample and velocity traverses for sources with small stacks or ducts.
- Method 2 - Determination of stack gas velocity and volumetric flow rate (Type S pitot tube).
- Method 2A - Direct measurement of gas volume through pipes and small ducts.
- Method 2B - Determination of exhaust gas volume flow rate from gasoline vapor incinerators.
- Method 2C - Determination of stack gas velocity and volumetric flow rate in small stacks or ducts (standard pitot tube).
- Method 2D - Measurement of gas volumetric flow rates in small pipes and ducts.
- Method 3 - Gas analysis for carbon dioxide, oxygen, excess air, and dry molecular weight.
- Method 3A - Determination of Oxygen and Carbon Dioxide Concentrations in Emissions From Stationary Sources (Instrumental Analyzer Procedure).
- Method 4 - Determination of moisture content in stack gases Method 5 - Determination of particulate emissions from stationary sources.
- Method 5A - Determination of particulate emissions from the asphalt processing and asphalt roofing industry.
- Method 5B - Determination of nonsulfuric acid particulate matter from stationary sources.
- Method 5C [Reserved].
- Method 5D - Determination of particulate emissions from positive pressure fabric filters.
- Method 5E - Determination of particulate emissions from the wool fiberglass insulation industry.
- Method 5F - Determination of nonsulfate particulate matter from stationary sources.
- Method 5G - Determination of particulate emissions from wood heaters from a dilution tunnel sampling location.
- Method 5H - Determination of particulate emissions from wood heaters from a stack location.
- Method 6 - Determination of sulfur dioxide emissions from stationary sources.
- Method 6A - Determination of sulfur dioxide, moisture, and carbon dioxide emissions from fossil fuel combustion sources.
- Method 6B - Determination of sulfur dioxide and carbon dioxide daily average emissions from fossil fuel combustion sources.
- Method 6C - Determination of Sulfur Dioxide Emissions From Stationary Sources (Instrumental Analyzer Procedure).
- Method 7 - Determination of nitrogen oxide emissions from stationary sources.
- Method 7A - Determination of nitrogen oxide emissions from stationary sources -- Ion chromatographic method.
- Method 7B - Determination of nitrogen oxide emissions from stationary sources (Ultraviolet spectrophotometry).
- Method 7C - Determination of nitrogen oxide emissions from stationary sources -- Alkaline-permanganate/colorimetric method.
- Method 7D - Determination of nitrogen oxide emissions from stationary sources -- Alkaline-permanganate/ion chromatographic method.
- Method 7E - Determination of Nitrogen Oxides Emissions From Stationary Sources (Instrumental Analyzer Procedure).
- Method 8 - Determination of sulfuric acid mist and sulfur dioxide emissions from stationary sources.
- Method 9 - Visual determination of the opacity of emissions from stationary sources
- Alternate method 1 - Determination of the opacity of emissions from stationary sources remotely by lidar.
- Method 10 - Determination of carbon monoxide emissions from stationary sources.
- Method 10A - Determination of carbon monoxide emissions in certifying continuous emission monitoring systems at petroleum refineries.
- Method 10B - Determination of carbon monoxide emissions from stationary sources.
- Method 11 - Determination of hydrogen sulfide content of fuel gas streams in petroleum refineries.
- Method 12 - Determination of inorganic lead

emissions from stationary sources.
- Method 13A - Determination of total fluoride emissions from stationary sources -- SPADNS zirconium lake method.
- Method 13B - Determination of total fluoride emissions from stationary sources -- Specific ion electrode method.
- Method 14 - Determination of fluoride emissions from potroom roof monitors for primary aluminum plants.
- Method 15 - Determination of hydrogen sulfide, carbonyl sulfide, and carbon disulfide emissions from stationary sources.
- Method 15A - Determination of total reduced sulfur emissions from sulfur recovery plants in petroleum refineries.
- Method 16 - Semicontinuous determination of sulfur emissions from stationary sources.
- Method 16A - Determination of total reduced sulfur emissions from stationary sources (impinger technique).
- Method 16B - Determination of total reduced sulfur emissions from stationary sources.
- Method 17 - Determination of particulate emissions from stationary sources (instack filtration method).
- Method 18 - Measurement of gaseous organic compound emissions by gas chromatography.
- Method 19 - Determination of sulfur dioxide removal efficiency and particulate, sulfur dioxide and nitrogen oxides emission rates.
- Method 20 - Determination of nitrogen oxides, sulfur dioxide, and diluent emissions from stationary gas turbines.
- Method 21 - Determination of volatile organic compound leaks.
- Method 22 - Visual determination of fugitive emissions from material sources and smoke emissions from flares.
- Method 24 - Determination of volatile matter content, water content, density, volume solids, and weight solids of surface coatings.
- Method 24A - Determination of volatile matter content and density of printing inks and related coatings.
- Method 25 - Determination of total gaseous nonmethane organic emissions as carbon.
- Method 25A - Determination of total gaseous organic concentration using a flame ionization analyzer.
- Method 25B - Determination of total gaseous organic concentration using a nondispersive infrared analyzer.
- Method 27 - Determination of vapor tightness of gasoline delivery tank using pressure-vacuum test.
- Method 28 - Certification and auditing of wood heaters Method 28A--Measurement of air to fuel ratio and minimum achievable burn rates for wood-fired appliances.

(3) Any method of sampling and analyzing for an air pollutant, as described in Appendix B to this part (40 CFR 61.02-91). (A list of test methods was provided in Appendix B to this part, as follows, (40 CFR 61-App/B-91)):
- Method 101 - Determination of particulate and gaseous mercury emissions from chlor-alkali plants--air streams.
- Method 101A - Determination of particulate and gaseous mercury emissions from sewage sludge incinerators.
- Method 102 - Determination of particulate and gaseous mercury emissions from chlor-alkali plants--hydrogen streams.
- Method 103 - Beryllium screening method.
- Method 104 - Determination of beryllium emissions from stationary sources.
- Method 105 - Determination of mercury in wastewater treatment plant sewage sludges.
- Method 106 - Determination of vinyl chloride from stationary sources.
- Method 107 - Determination of vinyl chloride content of inprocess wastewater samples, and vinyl chloride content of polyvinyl chloride resin, slurry, wet cake, and latex samples.
- Method 107A - Determination of vinyl chloride content of solvents, resin-solvent solution, polyvinyl chloride resin, resin slurry, wet resin, and latex samples.
- Method 108 - Determination of particulate and gaseous arsenic emissions.
- Method 108A - Determination of arsenic content in ore samples from nonferrous smelters.
- Method 108B - Determination of arsenic content in ore samples from nonferrous smelters.
- Method 108C - Determination of arsenic content in ore samples from nonferrous smelters.
- Method 111 - Determination of Polonium-210 emissions from stationary sources.
- Method 114 - Test methods for measuring radionuclide emissions from stationary Sources.
- Method 115 - Monitoring for radon-222 emissions.
- Method 101 - Determination of particulate and gaseous mercury emissions from chlor-alkali plants--air streams.
- Method 101A - Determination of particulate and gaseous mercury emissions from sewage sludge incinerators.
- Method 102 - Determination of particulate and gaseous mercury emissions from chlor-alkali plants--hydrogen streams.
- Method 103 - Beryllium screening method.
- Method 104 - Determination of beryllium emissions from stationary sources.
- Method 105 - Determination of mercury in wastewater treatment plant sewage sludges.
- Method 106 - Determination of vinyl chloride from stationary sources.
- Method 107 - Determination of vinyl chloride content of inprocess wastewater samples, and vinyl chloride content of polyvinyl chloride resin, slurry, wet cake, and latex samples.
- Method 107A - Determination of vinyl chloride content of solvents, resin-solvent solution, polyvinyl chloride resin, resin slurry, wet resin, and latex samples.
- Method 108 - Determination of particulate and gaseous arsenic emissions.
- Method 108A - Determination of arsenic content in

ore samples from nonferrous smelters.
- Method 108B - Determination of arsenic content in ore samples from nonferrous smelters.
- Method 108C - Determination of arsenic content in ore samples from nonferrous smelters.
- Method 111 - Determination of Polonium-210 emissions from stationary sources.
- Method 114 means test methods for measuring radionuclide emissions from stationary Sources.
- Method 115 means monitoring for radon-222 emissions.

(4) Methodology for the determination of metals emissions in exhaust gases from hazardous waste incineration and similar combustion processes (40 CFR 266-A/9-3.1)
- Methods of sampling and analyzing for metals, as described in SW-846, Example metals and their sampling and analysis methods are as follows:
- Arsenic(As), 7060; Barium(Ba), 7080; Cadmium(Cd), 7130; Chromium(Cr), 7190; Lead(Pb), 7420; Mercury(Hg), 7470; Selenium(Se), 7740; Silver(Ag), 7760.

(5) Methods for organics chemical analysis of municipal and industrial wastewater (40 CFR 136-App/A):
- Method 601 - Purgeable halocarbons
- Method 602 - Purgeable aromatics
- Method 603 - Acrolein and acrylonitrile
- Method 604 - Phenols
- Method 605 - Benzidines
- Method 606 - Phthalate
- Method 607 - Nitrosamines
- Method 608 - Organochlorine pesticides and PCBs
- Method 609 - Nitroaromatics and isophorone
- Method 610 - Polynuclear aromatic hydrocarbons
- Method 611 - Haloethers
- Method 612 - Chlorinated hydrocarbons
- Method 613 - 2,3,7,8-Tetrachlorodibenzo-p-dioxins
- Method 624 - Purgeables
- Method 625 - Base/neutrals and acids
- Method 1624 revision B - Volatile organic compounds by isotope dilution GC/MS
- Method 1625 revision B - Semivolatile organic compounds by isotope dilution GC/MS

(6) (Methods for chemical analysis of drinking water - analysis of trihalomethanes) (40 CFR 141-App/C):
- Part I - The analysis of trihalomethanes in drinking water by the purge and trap method
- Part II - The analysis of trihalomethanes in drinking water by liquid/liquid extraction
- Part III - Determination of maximum total trihalomethane potential

(7) The following 47 analytical testing methods are contained in the Third Edition of "Test Methods for Evaluating Solid Waste, Physical/Chemical Methods" EPA Publication SW-846 (November 1986) and its Revision I (December 1987), which are available for the cost of $110.00 from the Government Printing Office, Superintendent of Documents, Washington, DC 20402, (202)783-3238 (document number 955-001-00000-1) (40 CFR 260.11):
- 0010 Modified Method 5 Sampling Train
- 0020 Source Assessment Sampling System (SASS)
- 0030 Volatile Organic Sampling Train
- 1320 Multiple Extraction Procedure
- 1330 Extraction Procedure for Oily Wastes
- 3611 Alumina Column Cleanup and Separation of Petroleum Wastes
- 5040 Protocol for Analysis of Sorbent Cartridges from Volatile Organic Sampling Train
- 6010 Inductively Coupled Plasma Atomic Emission Spectroscopy
- 7090 Beryllium (AA, Direct Aspiration)
- 7091 Beryllium (AA, Furnace Technique)
- 7198 Chromium, Hexavalent (Differential Pulse Polarography)
- 7210 Copper (AA, Direct Aspiration)
- 7211 Copper (AA, Furnace Technique)
- 7380 Iron (AA, Direct Aspiration)
- 7381 Iron (AA, Furnace Technique)
- 7460 Manganese (AA, Direct Aspiration)
- 7461 Manganese (AA, Furnace Technique)
- 7550 Osmium (AA, Direct Aspiration)
- 7770 Sodium (AA, Direct Aspiration)
- 7840 Thallium (AA, Direct Aspiration)
- 7841 Thallium (AA, Furnace Technique)
- 7910 Vanadium (AA, Direct Aspiration)
- 7911 Vanadium (AA, Furnace Technique)
- 7950 Zinc (AA, Direct Aspiration)
- 7951 Zinc (AA, Furnace Technique)
- 9022 Total Organic Halides (TOX) by Neutron Activation Analysis
- 9035 Sulfate (Colorimetric, Automated, Chloranilate)
- 9036 Sulfate (Colorimetric, Automated, Methylthymol Blue, AA II)
- 9038 Sulfate (Turbidimetric)
- 9060 Total Organic Carbon
- 9065 Phenolics (Spectrophotometric, Manual 4-AAP with Distillation)
- 9066 Phenolics (Colorimetric, Automated 4-AAP with Distillation)
- 9067 Phenolics (Spectrophotometric, MBTH with Distillation)
- 9070 Total Recoverable Oil and Grease (Gravimetric, Separatory Funnel Extraction)
- 9071 Oil and Grease Extraction Method for Sludge Samples
- 9080 Cation-Exchange Capacity of Soils (Ammonium Acetate)
- 9081 Cation-Exchange Capacity of Soils (Sodium Acetate)
- 9100 Saturated Hydraulic Conductivity, Saturated Leachate Conductivity, and Intrinsic Permeability
- 9131 Total Coliform: Multiple Tube Fermentation Technique
- 9132 Total Coliform: Membrane Filter Technique
- 9200 Nitrate
- 9250 Chloride (Colorimetric, Automated Ferricyanide (AAI)
- 9251 Chloride (Colorimetric, Automated Ferricyanide (AAII)
- 9252 Chloride (Titrimetric, Mercuric Nitrate)
- 9310 Gross Alpha and Gross Beta
- 9315 Alpha-Emitting Radium Isotopes

- 9320 Radium-228
(8) Identification and listing of hazardous waste (40 CFR 261)
- Representative sampling methods (40 CFR 261-A/1)
- Method 1311, toxicity characteristic leaching procedure (TCLP) (40 CFR 261-A/2; 268-A/1; SW846-method 1311)
- Method 8280, method of analysis for chlorinated dibenzo-p-dioxins and dibenzofurans (40 CFR 261-A/10)
- Extraction procedures (EP) toxicity test method and structural integrity test(40 CFR 268-A/9; SW846-method 1310A)

(9) *See* method for more related terms.

Reference signal: The backscatter signal resulting from the laser light pulse passing through ambient air (40 CFR 60-App/A(alt. method 1)-91).

Reference standards: Guides or standards that DOE and its contractors should consider for guidance, as applicable, in addition to mandatory standards (DOE-91/04).

Reference substance: Any chemical substance or mixture or analytical standard, or material other than a test substance, feed, or water, that is administered to or used in analyzing the test system in the course of a study for the purposes of establishing a basis for comparison with the test substance for known chemical or biological measurements (40 CFR 160.3-91, *See also* 40 CFR 792.3; 797.1350-91).

Reference tissue concentration (RTC): The concentration of a chemical in edible fish or shellfish tissue that will not cause adverse impacts to human health when ingested. RTC is expressed in units of mg/kg (EPA-91/03).

Refined sugar: A high purity sugar normally used for human consumption (*See* sugar for more related terms) (EPA-74/01a).

Refiner:
(1) Any person who owns, leases, operates, controls, or supervises a refinery (40 CFR 52.137-91, *See also* 40 CFR 52.741; 80.2-91).
(2) A machine used to rub, macerate, bruise, and cut fibrous material, usually cellulose, in a water suspension to convert the raw fiber into a form suitable for formation into a web of desired characteristics on a paper-machine (cf. deflaker or disk refiner) (EPA-87/10).
(3) A compartment of a glass tank furnace, for the purpose of conditioning the glass (EPA-83).

Refinery: A plant at which gasoline is produced (40 CFR 80.2-91).

Refinery process unit: Any segment of the petroleum refinery in which a specific processing operation is conducted (40 CFR 60.101-91).

Refinery unit, process unit or unit: A set of components which are a part of a basic process operation such as distillation, hydrotreating, cracking, or reforming of hydrocarbons (40 CFR 52.741-91).

Refining:
(1) That phase of the steel production cycle during which undesirable elements are removed from the molten steel and alloys are added to reach the final metal chemistry (40 CFR 60.271a-91).
(2) In the paper industry, a general term applied to several operations, all of which involve the mechanical treatment of pulp in a water suspension to develop the necessary papermaking properties of the fibers and to cut the fibers to the desired length distribution (EPA-87/10).

Reflection: In air pollution modeling, used when due to the presence of a physical barrier to the diffusion, the pollutant is assumed to be reflected by this boundary. Mathematically this is performed by assumed an image source symmetrically with respect to the boundary (NATO-78/10).

Reflectometer: A photoelectric device that is used to measure the light that is reflected from a surface.

Reflux: A condensation of a vapor and return of the liquid to the zone from which it was removed (EPA-75/01a).

Reforming: A process wherein heat and pressure are used for the rearrangement of the molecular structure of hydrocarbons or low-octane petroleum fractions (EPA-87/10a).

Reformulated gasoline:
(1) Any gasoline which is certified by the Administrator under this section as complying with this subsection (CAA211.k-42USC7545-91).
(2) Gasoline with a different composition from conventional gasoline (e.g., lower aromatics content) that cuts air pollutants (EPA-94/04).
(3) *See* gasoline for more related terms.

Refractive index: The ratio of the speed of electromagnetic radiation in a vacuum space to that in a specified medium.

Refractometer: An instrument that is used to measure the refractory index of a substance.

Refractometry: The measurement of the refractive index.

Refractory (or refractory material): Materials, usually ceramic like substances used to line furnaces because they can endure high temperatures. In addition, they normally resist one or more of the following destructive influences: abrasion, pressure, chemical attack, and rapid changes in temperature (EPA-83). Other refractory related terms include:
- Acid refractory
- Basic refractory
- Castable refractory
- High alumina refractory
- Neutral refractory
- Plastic refractory

Refractory erosion: The wearing away of refractory surfaces by the washing action of moving liquids, such as molten slags or metals, or the action of moving gases (*See* erosion for more related terms) (SW-108ts).

Refractory expansion joint: An open joint left open so that refractories can expand thermally or permanently. Also, small spaces or gaps built into a refractory structure to permit sections of masonry to expand and contract freely and to prevent the distortion or buckling of furnace structures under excessive expansion stresses. These joints are built in such a way that the masonry can move but that little or no air or gas can leak through it (SW-108ts).

Refractory furnace wall: *See* definition under furnace wall.

Refractory material: *See* synonym, refractory.

Refractory metals: Includes the metals of columbium, tantalum, molybdenum, rhenium, tungsten and vanadium and their alloys (40 CFR 471.02-91). These metals all have a very high melting point and high hardness (*See* metal for more related terms).

Refractory BOD5: *See* synonym, refractory organics.

Refractory organics (refractory BOD5): Organic materials that are only partially degraded or entirely non-biodegradable in biological waste treatment processes. Refractory organics include detergents, pesticides, color-and odor-causing agents, tannins, lignins, ethers, olefins, alcohols, amines, aldehydes, ketones, etc. (EPA-76/03).

Refractory furnace wall: A wall made of heat resistant ceramic material (*See* furnace wall for more related terms) (SW-108ts).

Refractory pollutant: The pollutants that resist treatment (*See* pollutant for more related terms) (DOI-70/04).

Refrigerant: A substance used as an agent in cooling or refrigeration. A number of refrigerants have been developed for application. Factors that are important include:
- Chemical, thermodynamic, and physical properties
- System capacity required
- Compressor type
- Desired temperature level
- Safety consideration

Important properties for a refrigerant include:
- Boiling temperature and pressure
- Freezing temperature
- Critical temperature and pressure
- Condenser and evaporator pressure
- Specific volume
- Latent heat
- Specific heat of liquid
- Molecular weight
- Theoretical horsepower per ton
- Discharge temperature
- Miscibility
- Safety aspect

Refrigerated condenser: A surface condenser in which the coolant supplied to the condenser has been cooled by a mechanical device, other than by a cooling tower or evaporative spray cooling, such as refrigeration unit or steam chiller unit (40 CFR 52.741-91).

Refrigeration: A process to keep cool. A ton of refrigeration is the refrigeration produced by melting one ton of ice at a temperature of 32° F in 24 hour. It is a rate of removing heat equivalent to the removal of 12000 Btu/hr or 200 Btu/min.

Refrigerator (or heat pump):
(1) A device which receives heat from a low temperature part of its surroundings and discharges heat to a higher temperature of its surroundings while executing a cycle (Jones-p93).
(2) A Carnot heat engine which is operated in a reversed direction of its cycle (Wark-p276).

Refueling emissions: Emissions released during vehicle refueling (EPA-94/04).

Refund: The sum, equal to the deposit, that is given to the consumer or the dealer or both in exchange for empty returnable beverage containers (40 CFR 244.101-91).

Refurbishment: With reference to a vapor processing system, replacement of components of, or addition of components to, the system within any 2-year period such that the fixed capital cost of the new components required for such component replacement or addition exceeds 50 percent of the cost of a comparable entirely new system (40 CFR 60.501-91).

Refuse: Putrescible and non-putrescible solid wastes, except body wastes and including kitchen discards, rubbish, ashes, incinerator ash, incinerator residue, street cleanings, and market, commercial, office, and industrial wastes (EPA-83). Other refuse related terms include:
- Commercial refuse
- Domestic garbage (*See* synonym, domestic refuse)
- Domestic refuse
- Milled refuse
- Residential refuse
- Shredded refuse
- Street refuse
- *See* solid waste (EPA-94/04)

Refuse burner: A device for either central or on-site volume reduction of solid waste by burning. It is of simple construction and all the factors of combustion can not be controlled (cf. municipal waste incinerator or *See* burner for more related terms) (SW-108ts).

Refuse chute: A pipe, duct, or trough through which solid waste is conveyed pneumatically or by gravity to a central storage area (SW-108ts).

Refuse derived fuel (RDF):
(1) A type of MSW through shredding and size classification. This includes all classes of RDF including low density fluff RDF through densified RDF and RDF fuel pellets (40 CFR 60.51a-91).
(2) Fuel extracted from solid waste to be used for combustion processes or as feedstock to other process systems.
(3) Forms of refuse fuel derived from municipal solid waste (EPA-83, *See also* EPA-89/11).
   - **Rdf-1:** Wastes used as a fuel in as-discarded form with only bulky wastes removed (EPA-83).
   - **Rdf-2:** Wastes processed to coarse particle size with or without ferrous metal separation (EPA-83).
   - **Rdf-3:** Combustible waste fraction processed to

particle sizes - 95 percent passing 2 inch square screening (EPA-83).
- **Rdf-4:** Combustible waste fraction processed into powder form - 95 percent passing 10 mesh screening (EPA-83).
- **Rdf-5:** Combustible waste fraction densified (compressed) into the form of pellets, slugs, cubettes or briquettes (EPA-83).
- **Rdf-6:** Combustible waste fraction processed into liquid fuel. RDF - combustible waste fraction processed into gaseous fuel (EPA-83).
- **Refuse derived fuel (RDF-2, RDF-3):** Shredded refuse fuel, used principally as a supplement in utility or industrial boilers which have ash handling capabilities. Using a separation system, much of metal, glass and other inorganics are first removed. RDF is the remaining organic fraction which has been processed to relatively uniform size particles (EPA-83).
- **Refuse derived fuel three (RDF-3):** A shredded fuel derived from municipal solid waste (MSW) that has been processed so as to remove metal, glass, and other entrained inorganics. The material has a particle size such that 95 weight percent passes through a 2 inch (50 mm) square mesh screen. RDF-3 used as a primary or supplementary fuel in existing or new industrial or utility boilers (EPA-83).

(4) *See* fuel for more related terms.

Refuse derived fuel (RDF) incinerator: An incinerator which burns refuse derived fuel (RDF). RDF is a homogeneous fuel derived from municipal solid waste (MSW) by a mechanical means. RDF can be coarse, fluff, powder, and densified pellets, briquettes, or similar forms. RDF can be burned in two types of boilers. It can be used as the sole or primary fuel in dedicated boilers, or it can be co-fired with conventional fossil fuels (e.g., coal and oil) or even wood in existing industrial or utility boilers. Boilers using RDF can recover energy. In addition, materials such as steel and glass recovered during the initial processing can be sold (*See* incinerator for more related terms) (OTA-89/10).

Refuse handling: What is done to prepare refuse for disposal or for processing which is conversion of wastes into something useful (EPA-83).

Refuse feeder: The mechanism in which refuse is fed into an incinerator.

Refuse generation: Refuse generation means the act or process of producing solid waste (EPA-83).

Refuse reclamation: Conversion of solid waste into useful products, e.g., composting organic wastes to make soil conditioners or separating aluminum and other metals for recycling (EPA-94/04).

Refuse reduction: A process of salvaging fats and oils from animal discards by cooking, followed by extraction with solvents, and separation from the solvents by distillation (cf. rendering) (EPA-83).

Refuse shed: A region or area which for reasons of typography, contiguous population or other common features includes refuse sources which may be considered collectively in general planning. Usually synonymous with a general populated or metropolitan area, and not necessarily limited by lines of political jurisdiction or divisions (EPA-83).

Refuse sorting: Classification of refuse for reuse, combustible, non-combustible, etc.

Refuse train: *See* synonym, container train.

Regenerant: An agent that is used to restore the activities of an ion-exchange resin.

Regenerated water: *See* synonym, return flow.

Regeneration:
(1) Manipulation of cells to cause them to develop into whole plants (EPA-94/04).
(2) *See* synonym, desorption.

Regenerative cycle gas turbine: Any stationary gas turbine that recovers thermal energy from the exhaust gases and utilizes the thermal energy to preheat air prior to entering the combustor (*See* turbine for more related terms) (40 CFR 60.331-91, *See also* 40 CFR 60.331-91).

Regenerator: A cyclic heat interchanger that alternately receives heat from gaseous combustion products and transfers heat to air or gas before combustion (cf. recuperator) (EPA-83).

Region: An area designated as an air quality control region (AQCR) under section 107(c) of the Act (40 CFR 51.100-91, *See also* 40 CFR 60.21-91).

Regional deposited dose (RDD):
(1) The deposited dose (mg/cm^2 of lung region surface area per minute) calculated for the region of interest as related to the observed effect (i.e., calculated for the tracheobronchial region for an effect concerning the conducting airways) (EPA-90/08).
(2) The deposited dose of particles calculated for the region of interest as related to the observed effect. For respiratory effects of particles, the deposited dose is adjusted for ventilatory volumes and the surface area of the respiratory region effected (mg/min-sq.cm). For extra respiratory effects of particles, the deposited dose in the total respiratory system is adjusted for ventilatory volumes and body weight (mg/min-kg) (EPA-92/12).

Regional deposited dose ratio (RDDR):
(1) The ratio of the regional deposited dose calculated for a given exposure in the animal species of interest to the regional deposited dose of the same exposure in a human. This ratio is used to adjust the exposure effect level for inter-species dosimetric differences to derive a human equivalent concentration for particles (EPA-92/12).
(2) The ratio of the regional deposited dose in the animal species of interest (RDDA) to that of humans (RDDH) This ratio is used to adjust the exposure effect level for inter-species

dosimetric differences (EPA-90/08).

Regional gas dose (RGD): The gas dose calculated for the region of interest as related to the observed effect for respiratory effects. The deposited dose is adjusted for ventilatory volumes and the surface area of the respiratory region effected (mg/min-sq.cm) (EPA-92/12).

Regional gas dose ratio (RGDR): The ratio of the regional gas dose calculated for a given exposure in the animal species of interest to the regional gas dose of the same exposure in humans. This ratio is used to adjust the exposure effect level for inter-species dosimetric differences to derive a human equivalent concentration for gases with respiratory effects (EPA-92/12).

Regional response team (RRT):
(1) RRT is composed of representatives of Federal agencies and a representative from each State in the Federal region. During a response to a major hazardous materials incident involving transportation or a fixed facility, the OSC may request that the RRT be convened to provide advice or recommendations in specific issues requiring resolution. Under the NCP, RRTs may be convened by the chairman when a hazardous materials discharge or release exceeds the response capability available to the OSC in the place where it occurs; crosses regional boundaries; or may pose a substantial threat to the public health, welfare, or environment, or to regionally significant amounts of property. Regional contingency plans specify detailed criteria for activation of RRTs. RRTs may review plans developed in compliance with Title III, if the local emergency planning committee so requests (NRT-87/03).
(2) Representatives of federal, local, and state agencies who may assist in coordination of activities at the request of the On-Scene Coordinator before and during a significant pollution incident such as an oil spill, major chemical release, or a Superfund response (EPA-94/04).

Regional scale: In air pollution modeling, a length scale of the order of a hundred kilometers. Usually applied to an area encompassing urban and/or industrial areas (NATO-78/10).

Register paper: The type of paper for multiple form use (*See* paper for more related terms) (EPA-83).

Registrant:
(1) A person who has registered any pesticide pursuant to the provisions of this Act (FIFRA2; 40 CFR 2.307; 153.62; 164.2-91).
(2) Any manufacturer or formulator who obtains registration for a pesticide active ingredient or product (EPA-94/04).

Registration: The formal listing with EPA of a new pesticide before it can be sold or distributed in intra- or inter-state commerce. The product must be registered under the Federal Insecticide, Fungicide, and Rodenticide Act. EPA is responsible for registration (pre-market licensing) of pesticides on the basis of data demonstrating that they will not cause unreasonable adverse effects on human health or the environment when used according to approved label directions (EPA-94/04, *See also* FIFRA2-7USC136-91).

Registration (of a pesticide): Under FIFRA and its amendments, new pesticide products cannot be sold unless they are registered with the U.S. EPA. Registration involves a comprehensive evaluation of risks and benefits based on all relevant data (EPA-92/12).

Registration standards: Published documents which include summary reviews of the data available on a pesticide's active ingredient, data gaps, and the Agency's existing regulatory position on the pesticide (EPA-94/04).

Regression: A relationship of y and x in a function of $y = f(x)$, where: y is the expected value of an independent random variable x. The parameters in the function $f(x)$ are determined by the method of least squares. When $f(x)$ is a linear function of x, the term linear regression is used (NATO-78/10). Other regression related terms include:
- Forward stepping multiple linear regression
- Linear regression
- Multiple linear regression

Regression coefficient: A quantity that describes the slope and intercept of a regression line (EPA-84/03).

Regression line or equation: The function that indicates the regression relationship (EPA-84/03).

Regression model: A mathematical model, usually a single equation, developed through the use of a least squares linear regression analysis (EPA-82/10).

Regression statistics: Values generated during a regression analysis which identify the significance, or reliability, of the regression generated figures (EPA-79/12c).

Regulated area: An area established by the employer to demarcate areas where airborne concentrations of asbestos exceed or can reasonably be expected to exceed the permissible exposure limit. The regulated area may take the form of:
(1) A temporary enclosure, as required by paragraph (e)(6) of this section; or
(2) An area demarcated in any manner that minimizes the number of employees exposed to asbestos (40 CFR 763.121-91).

Regulated asbestos containing material (RACM):
(1) Friable asbestos material;
(2) Category I nonfriable ACM that has become friable;
(3) Category I nonfriable ACM that will be or has been subjected to sanding, grinding, cutting, or abrading; or
(4) Category II nonfriable ACM that has a high probability of becoming or has become crumbled, pulverized, or reduced to powder by the forces expected to act on the material in the course of demolition or renovation operations regulated by this subpart (40 CFR 61.141-91).
(5) Friable asbestos material or nonfriable ACM that will be or has been subjected to sanding, grinding, cutting, or abrading or has crumbled, or been pulverized or reduced to powder in the course of demolition or renovation operations (EPA-94/04).
(6) *See* asbestos for more related terms.

Regulated chemical: Any chemical substance or mixture for which export notice is required under 40 CFR 707.60 (40 CFR 707.63-91).

Regulated entity:
(1) Manufacturers, processors, wholesale distributors, or importers of consumer or commercial products for sale or distribution in interstate commerce in the United States; or
(2) Manufacturers, processors, wholesale distributors, or importers that supply the entities listed under clause (1) with such products for sale or distribution in interstate commerce in the United States (CAA183.e-42USC7511b-91).

Regulated material: A substance or material that is subject to regulations set forth by Federal Agencies (Course 165.5).

Regulated medical waste:
(1) Those medical wastes that have been listed in 40 CFR 259.30(a) of this part and that must be managed in accordance with the requirements of this part (40 CFR 259.10-91).
(2) (A) Regulated medical waste is any solid waste, defined in 40 CFR 259.10(a) of this part, generated in the diagnosis, treatment, (e.g., provision of medical services), or immunization of human beings or animals, in research pertaining thereto, or in the production or testing of biologicals, that is not excluded or exempted under paragraph (b) of this section. (For detailed definition, *See* 40 CFR 259.30-91).
   NOTE to paragraph (a): The term solid waste includes solid, semisolid, or liquid materials, but does not include domestic sewage materials identified in 40 CFR 261.4(a)(1) of this subchapter (*See* medical waste or waste for more related terms).
(3) Under the Medical Waste Tracking Act of 1988, any solid waste generated in the diagnosis, treatment, or immunization of human beings or animals, in research pertaining thereto, or in the production or testing of biologicals. Included are cultures and stocks of infectious agents; human blood and blood products; human pathological body wastes from surgery and autopsy; contaminated animal carcasses from medical research; waste from patients with communicable diseases; and all used sharp implements, such as needles and scalpels, etc., and certain unused sharps. (See; treated medical waste; untreated medical waste; destroyed medical waste. (EPA-94/04))

Regulated pest: A specific organism considered by a State or Federal agency to be a pest requiring regulatory restrictions, regulations or control procedures in order to protect the host, man and/or his environment (40 CFR 171.2-91).

Regulated substance:
(1) Any substance defined in section 101(14) of the Comprehensive Environmental Response, Compensation and Liability Act (CERCLA) of 1980 (but not including any substance regulated as a hazardous waste under subtitle C), and
(2) Petroleum, including crude oil or any fraction thereof that is liquid at standard conditions of temperature and pressure (60° Fahrenheit and 14.7 pounds per square inch absolute).

The term "regulated substance" includes but is not limited to petroleum and petroleum-based substances comprised of a complex blend of hydrocarbons derived from crude oil though processes of separation, conversion, upgrading, and finishing, such as motor fuels, jet fuels, distillate fuel oils, residual fuel oils, lubricants, petroleum solvents, and used oils (40 CFR 280.12).

Regulation: *See* definition under law related terms.

Regulatory dose (RgD): The daily exposure to the human population reflected in the final risk management decision; it is entirely possible and appropriate that a chemical with a specific RfD may be regulated under different statutes and situations through the use of different RgDs (EPA-92/12).

Reheater:
(1) A heat exchange device for adding superheat to steam which has been partially expanded in the turbine (EPA-82/11f).
(2) Heat transfer apparatus for heating steam after it has given up some of its original heat in doing work (EPA-83).

Reid vapor pressure: The absolute vapor pressure of volatile crude oil and volatile nonviscous petroleum liquids, except liquified petroleum gases, as determined by ASTM D323-82 (incorporated by reference - *See* 40 CFR 60.17) (40 CFR 60.111; 60.111a; 60.111b-91).

Reimbursement period: The period that begins when the data from the last test to be completed under this part for a specific chemical substance listed in 766.25 is submitted to EPA, and ends after an amount of time equal to that which had been required to develop that data or 5 years, whichever is later (40 CFR 766.3-91, *See also* 40 CFR 790.3; 791.3; 763.163-91).

Reinjection: Reintroduction of fly ash into a furnace to burn out all the combustibles (cf. fly ash reinjection under ash) (SW-108ts).

Rejection of a batch: That the number of non-complying compressors in the batch sample is greater than or equal to the rejection number as determined by the appropriate sampling plan (40 CFR 204.51-91, *See also* 40 CFR 205.51-91).

Rejection of a batch sequence: That the number of rejected batches in a sequence is greater than or equal to the sequence rejection number as determined by the appropriate sampling Plan (40 CFR 204.51-91), *See also* 40 CFR 205.51-91).

Related coatings: All non-ink purchased liquids and liquid-solid mixtures containing VOC solvent, usually referred to as extenders or varnishes, that are used at publication rotogravure printing presses (40 CFR 60.431-91).

Relative accuracy (RA): The absolute mean difference between the gas concentration or emission rate determined by the CEMS (continuous emission monitoring system) and the value determined by the RM's (reference method's) plus the 2.5 percent error confidence coefficient of a series of tests divided by the mean of the RM tests or the applicable emission limit (40 CFR 60-App/F-91, *See also* EPA-90/04).

Relative diffusion: The dispersion of a cluster of particles considered with respect to each other (NATO-78/10).

Relative humidity (RH): The ratio of the actual amount of water vapor present in the air to the amount which could exist at saturation. When the radio or TV announcer talks about percent humidity, he means percent relative humidity (% RH) (EPA-84/09). RH can be expressed in the following equation formats:

RH = [actual vapor mass (m_v)]/[vapor mass required to produce a saturated mixture (m_g)] (A)

= [vapor partial pressure (p_v) in air-vapor mixture]/[saturation vapor pressure (p_g) at mixture temperature] (B)

= [saturation specific volume (v_g) of air-water vapor mixture]/[specific vapor volume (v_v) in the air-water vapor mixture] (C)

Equations (B) and (C) can be obtained by applying the ideal gas equation to equation (A) (*See* humidity for more related terms).

Relative humidity range: The range of ambient relative humidity over which the instrument will meet stated performance specifications (LBL-76/07-bio).

Relative percent of percutaneous absorption: 100 times the ratio between total urinary excretion of compound following topical administration and total urinary excretion of compound following intravenous injection (40 CFR 795.223-91).

Relative range: The range divided by the mean of a particular set of numbers (*See* range for more related terms) (EPA-84/03).

Relative risk (or risk ratio):
(1) A comparison of the disease rate in an exposed group to that in an unexposed group (EPA-165.5).
(2) The ratio of incidence or risk among exposed individuals to incidence or risk among non-exposed individuals (EPA-92/12).

Relative stability test: A color test in which methylene blue is used to indicate the dissolved oxygen level in the polluted water.

Relative toxicity: The toxicity of the effluent when it is mixed with the receiving water, or a dilution water of similar composition for toxicity testing (EPA-91/03).

Relative utilization factor: The ratio of the utilization efficiency of the fuel under consideration to the utilization efficiency of the base (comparative) fuel provides realistic assessment of equivalency (EPA-83).

Release:
(1) Any spilling, leaking, pumping, pouring, emitting, emptying, discharging, injecting, escaping, leaching, dumping, or disposing into the environment of a hazardous or toxic chemical or extremely hazardous substance (EPA-94/04).
(2) As defined by section 101(22) of CERCLA, means any spilling, leaking, pumping, pouring, emitting, emptying, discharging, injection, escaping, leaching, dumping, or disposing into the environment (including the abandonment or discarding of barrels, containers, and other closed receptacles containing any hazardous substance or pollutant or contaminant), but excludes: Any release which results in exposure to persons solely within a workplace, with respect to a claim which such persons may assert against the employer of such persons; emissions from the engine exhaust of a motor vehicle, rolling stock, aircraft, vessel, or pipeline pumping station engine; release of source, byproduct or special nuclear material from a nuclear incident, as those terms are defined in the Atomic Energy Act of 1954, if such release is subject to requirements with respect to financial protection established by the Nuclear Regulatory Commission under section 170 of such Act, or, for the purpose of section 104 of CERCLA or any other response action, any release of source, byproduct, or special nuclear material from any processing site designated under section 102(a)(1) or 302(a) of the Uranium Mill Tailings Radiation Control Act of 1978; and the normal application of fertilizer. For the purpose of the NCP, release also means substantial **threat of release** (40 CFR 300.5-91).

Other release related terms include:
- Aboveground release
- Accidental release
- Belowground release
- Continuous release
- Federally permitted release
- Normal range of a release
- Overfill release
- Routine release
- Statistically significant increase in a release

Release detection: Determining whether a release of a regulated substance has occurred form the UST system into the environment or into the interstitial space between the UST system and its secondary barrier or secondary containment around it (40 CFR 280.12-91).

Release that is stable in quantity and rate: A release that is predictable and regular in amount and rate of emission (40 CFR 302.3-91).

Release volume: A real or hypothetical volume surrounding a source of airborne contaminants (EPA-88/09a).

Release zone: The area swept out by the locus of points constantly 100 meters from the perimeter of the conveyance engaged in dumping activities, beginning at the first moment in which dumping is scheduled to occur and ending at the last moment in which dumping is scheduled to occur. No release zone shall exceed the total surface area of the dumpsite (40 CFR 227.28-91).

Relevant and appropriate requirements: Those cleanup standards, standards of control, and other substantive requirements, criteria, or limitations promulgated under federal environmental or state environmental or facility siting laws that, while not "applicable" to a hazardous substance, pollutant, contaminant, remedial action, location, or other circumstance at a CERCLA site, address problems or situations sufficiently similar to those encountered at the CERCLA site that their use is well suited to the particular site. Only those state standards that are identified in a timely manner and are more stringent than federal requirements may be relevant and appropriate (40 CFR 300.5-91).

Relief valve: Each pressure relief device including pressure relief valves, rupture disks and other pressure relief systems used to protect process components from overpressure conditions. Relief valve does not include polymerization shortstop systems, refrigerated water systems or control valves or other devices used to control flow to an incinerator or other air pollution control device (40 CFR 61.61-91).

Relief valve discharge: Any nonleak discharge through a relief valve (40 CFR 61.61-91).

REM: *See* definition under roentgen equivalent man (REM).

Rem-jet: A coating on the back of certain films for the reduction of light reflections during exposure. The rem-jet backing is removed during processing by washing or by washing and mechanical buffing (EPA-80/10).

Remedial action (RA): The actual construction or implementation phase of a Superfund site cleanup that follows remedial design (cf. remedy action) (EPA-94/04, *See also* 40 CFR 192.01-91).

Remedial action plan: A written document which embodies a systematic and comprehensive ecosystem approach to restoring and protecting the beneficial uses of areas of concern, in accordance with article VI and Annex 2 of the Great lakes Water Quality Agreement (CWA118-33USC1268-91).

Remedial design (RD):
(1) The technical analysis and procedures which follow the selection of remedy for a site and result in a detailed set of plans and specifications for implementation of the remedial action (40 CFR 300.5-91, *See also* EPA-89/12a).
(2) A phase of remedial action that follows the remedial investigation/feasibility study and includes development of engineering drawings and specifications for a site cleanup (EPA-94/04).

Remedial investigation (RI):
(1) A process undertaken by the lead agency to determine the nature and extent of the problem presented by the release. The RI emphasizes data collection and site characterization, and is generally performed concurrently and in an interactive fashion with the feasibility study. The RI includes sampling and monitoring, as necessary, and includes the gathering of sufficient information to determine the necessity for remedial action and to support the evaluation of remedial alternatives (40 CFR 300.5-91, *See also* EPA-89/12a).
(2) An in-depth study designed to gather data needed to determine the nature and extent of contamination at a Superfund site; establish site cleanup criteria; identify preliminary alternatives for remedial action; and support technical and cost analyses of alternatives. The remedial investigation is usually done with the feasibility study. Together they are usually referred to as the "RI/FS" (EPA-94/04).

Remedial investigation/feasibility study (RI/FS): The Superfund program's methodology for characterizing the nature and extent of risks posed by CERCLA sites and for identifying and evaluating potential remedial alternatives for those sites. The process is divided into two parts (the remedial investigation and the feasibility study), which are conducted concurrently; data collected in one part influence the tasks performed in the other part and visa versa (EPA-89/02).

Remedial project manager (RPM):
(1) The official designated by the lead agency to coordinate, monitor, or direct remedial or other response actions under Subpart E of the NCP (40 CFR 300.5-91).
(2) The EPA or state official responsible for overSeeing on-site remedial action (EPA-94/04).

Remedial response: Long-term action that stops or substantially reduces a release or threat of a release of hazardous substances that is serious but not an immediate threat to public health (EPA-94/04).

Remediation:
(1) Cleanup or other methods used to remove or contain a toxic spill or hazardous materials from a Superfund site (EPA-94/04).
(2) For the Asbestos Hazard Emergency Response program, abatement methods including evaluation, repair, enclosure, encapsulation, or removal of greater than 3 linear feet or square feet of asbestos-containing materials from a building (EPA-94/04).

Remedy or remedial action: Those actions consistent with permanent remedy taken instead of or in addition to removal actions in the event of a release or threatened release of a hazardous substance into the environment, to prevent or minimize the release of hazardous substances so that they do not migrate to cause substantial danger to present or future public health or welfare or the environment. The term includes, but is not limited to, such actions at the location of the release as storage, confinement, perimeter protection using dikes, trenches, or ditches, clay cover, neutralization, cleanup of released hazardous substances and associated contaminated materials, recycling or reuse, diversion, destruction, segregation of reactive wastes, dredging or excavations, repair or replacement of leaking containers, collection of leachate and runoff, onsite treatment or incineration, provision of alternative water supplies, and any monitoring reasonably required to assure that such actions protect the public health and welfare and the environment. The term includes the costs of permanent relocation of residents and businesses and community facilities where the President determines that, alone or in combination with other measures, such relocation is more cost-effective than and environmentally preferable to the transportation, storage, treatment, destruction, or secure disposition offsite of hazardous substances, or may otherwise be necessary to protect the public health or welfare; the term includes offsite transport and offsite storage, treatment, destruction, or secure disposition of hazardous substances and associated contaminated materials (SP101SP101-42USC9601).

Remedy selection: The remedial alternative(s) identified in the ROD for CERCLA site cleanup (EPA-89/12a).

Remelt: A solution of low grade sugar in clarified juice or water (EPA-75/02d).

Remined area: Only that area of any coal remining operation on which coal mining was conducted before the effective date of the Surface Mining Control and Reclamation Act of 1977 (CWA301.p-33USC1311-91).

Remote sensing:
(1) A quantitative or qualitative determination of air pollutants or of meteorological parameters by means of instruments not in physical contact with the sample being examined (NATO-78/10).
(2) The collection and interpretation of information about an object without physical contact with the object; e.g., satellite imaging and aerial photograph (EPA-94/04).

Removal:
(1) A reduction in the amount of a pollutant in the POTW's effluent or alteration of the nature of a pollutant during treatment at the POTW. The reduction or alteration can be obtained by physical, chemical or biological means and may be the result of specifically designed POTW capabilities or may be incidental to the operation of the treatment system. Removal as used in this subpart shall not mean dilution of a pollutant in the POTW (40 CFR 403.7-91).
(2) The taking out or the stripping of substantially all ACBM from a damaged area, a functional space, or a homogeneous area in a school building (40 CFR 763.83-91, *See also* 40 CFR 763.121-91).

Removal action:
(1) An action that removes hazardous substances from the site for proper disposal or destruction in a facility permitted under the Resource Conservation and Recovery Act or the Toxic Substances Control Act or by the Nuclear Regulatory Commission (40 CFR 300-App/A-91).
(2) Short-term immediate actions taken to address releases of hazardous substances that require expedited response (*See* cleanup) (EPA-94/04).

Removal costs: The costs of removal that are incurred after a discharge of oil has occurred or, in any case in which there is a substantial threat of a discharge of oil, the costs to prevent, minimize, or mitigate oil pollution from such an incident (OPA1001-91).

Remove: To take out RACM or facility components that contain or are covered with RACM from any facility (40 CFR 61.141-91).

Remove or removal: Refers to the removal of the oil from the water and shorelines or the taking of such other actions as may be necessary to minimize or mitigate damage to the public health or welfare, including, but not limited to, fish, shellfish, wildlife, and public and private property, shorelines, and beaches (40 CFR 109.2-c).

Removal efficiency: The ratio of the mass flow rate of the contaminants going into a control device minus the mass flow rate of the contaminants going out of the control device to the mass flow rate of the contaminants going into the control device (EPA-81/09).

Removal flux: Mass or number of aerosol particles being retained by unit deposition area per unit time (EPA-88/09a).

Renal toxicity: Urinalysis, function tests (clearance, glomerular filtration rate), gross and microscopic examinations organ weight (*See* endpoint for more related terms) (Course 165.6; EPA-92/12).

Renderer: An independent or off-site rendering operation, conducted separate from a slaughterhouse, packinghouse or poultry dressing or processing plant, which manufactures at rates greater than 75,000 pounds of raw material per day of meat meal, tankage, animal fats or oils, grease, and tallow, and may cure cattle hides, but excluding marine oils, fish meal, and fish oils (40 CFR 432.101-91).

Rendering:
(1) A process of recovering fatty substances from animal parts by heat treatment, extraction, and distillation (SW-108ts).
(2) The separation of fats and water from poultry offal (inedible parts of poultry) by heat or physical energy. The process includes feather hydrolysis and blood processing for animal feeds (EPA-75/04).

Other rendering related terms include:
- Low temperature rendering
- Wet rendering

Renewable energy: The energy from photovoltaic, solar thermal, wind, geothermal, and biomass energy production technologies (CAA808-42USC7171-91).

Renewal system: The technique in which test organisms are periodically transferred to fresh test solution of the same composition (40 CFR 797.1330-91).

Renewal test: A test without continuous flow of solution, but with occasional renewal of test solutions after prolonged periods, e.g., 24 hours (40 CFR 797.1350-91).

Renovation: The modifying of any existing structure, or portion thereof, where exposure to airborne asbestos may result (40 CFR 763.121-91, *See also* 40 CFR 61.141-91).

Reoxygenation: Addition of air or oxygen to water from any sources (cf. deoxygenation).

Rep: A unit of measurement of any kind of radiation absorbed by man (*See* radiation unit for more related terms) (EPA-74/11).

Repackager: A person who buys a substance identified in Subpart D of this Part or mixture, removes the substance or mixture from the container in which it was bought, and transfers this substance, as is, to another container for sale (40 CFR 704.203-91).

Repair coatings: The coatings used to correct imperfections or damage to furniture surface (40 CFR 52.741-91).

Repeat compliance period:
(1) Any subsequent compliance period after the initial compliance period (40 CFR 141.2-91).
(2) Any subsequent compliance period after the initial one (EPA-94/04).

Repeatability:
(1) The degree of variation between repeated measurements of

the same concentration (LBL-76/07-bio).
(2) The precision of a method expressed as the agreement attainable between independent determinations performed by a single analyst. *See* ASTM E180, also reproducibility and analysis parameters: laboratory (EPA-83).

Repellency: Capability to repel water.

Repellent: A substance used to repel ticks, chiggers, gnats, flies, mosquitoes, and fleas (EPA-85/10).

Replacement: As used in this title means those expenditures for obtaining and installing equipment, accessories, or appurtenances during the useful life of the treatment works necessary to maintain the capacity and performance for which such works are designed and constructed (CWA212, *See also* 40 CFR 35.905; 35.2005-91).

Replacement cost: The capital needed to purchase all the depreciable components in a facility (40 CFR 60.481-91).

Replicate: Two or more duplicate tests, samples, organisms, concentrations, or exposure chambers (40 CFR 797.1600-91).

Replicate sample (or duplicate sample): A sample that has been divided into two or more portions at a step in the measurement process. Each portion is then carried through the remaining steps in the measurement process (EPA-84/03). Duplicate samples are considered to be two replicates (*See* sample for more related terms).

Replicon: A unit of replication that possesses the genetic locus for the initiator (protein) and the locus where replication of genetic material is to begin (EPA-88/09a).

Reportable quantity:
(1) That quantity, as set forth in this part, the release of which requires notification pursuant to this part (40 CFR 302.3-91, *See also* 40 CFR 117.1; 355.20-91).
(2) The quantity of a hazardous substance that is considered reportable under CERCLA. Reportable quantities are:
   (A) One pound; or
   (B) For selected substances, an amount established by regulation either under CERCLA or under Section 311 of the Clean Water Act. Quantities are measured over a 24-hour period (EPA-92/12).
(3) Quantity of a hazardous substance that triggers reports under CERCLA. If a substance exceeds its RQ, the release must be reported to the National Response Center, the SERC, and community emergency coordinators for areas likely to be affected (EPA-94/04).

Reporting year: The most recent complete corporate fiscal year during which a person manufactures, imports, or processes the listed substance, and which falls within a coverage period identified with a substance in Subpart D of this Part (40 CFR 704.203-91).

Repowering:
(1) The replacement of an existing coal-fired boiler with one of the following clean coal technologies: atmospheric or pressurized fluidized bed combustion, integrated gasification combined cycle, magnetohydrodynamics, direct and indirect coal-fired turbines, integrated gasification fuel cells, or as determined by the Administrator, in consultation with the Secretary of Energy, a derivative of one or more of these technologies, and any other technology capable of controlling multiple combustion emissions simultaneously with improved boiler or generation efficiency and with significantly greater waste reduction relative to the performance of technology in widespread commercial use as of the date of enactment of the Clean Air Act Amendments of 1990. Notwithstanding the provisions of section 409(a), for the purpose of this title, the term "repowering" shall also include any oil and/or gas-fired unit which has been awarded clean coal technology demonstration funding as of January 1, 1991, by the Department of Energy (CAA402-42USC7651a).
(2) Rebuilding and replacing major components of a power plant instead of building a new one (EPA-94/04).

Repowering: Rebuilding and replacing major components of a power plant instead of building a new one (EPA-94/04).

Representative important species: The species which are representative, in terms of their biological needs, of a balanced, indigenous community of shellfish, fish and wildlife in the body of water into which a discharge of heat is made (40 CFR 125.71-91).

Representative sample:
(1) A sample of a universe or whole (e.g., waste pile, lagoon, ground water) which can be expected to exhibit the average properties of the universe or whole (*See* sample for more related terms) (40 CFR 260.10-91, *See also* EPA-86/01; LBL-76/07-water).
(2) A sample collected in such a manner that it has characteristics equivalent to the lot sample (*See* analytical parameters--laboratory for more related terms) (EPA-83).
(3) A portion of material or water that is as nearly identical in content and consistency as possible to that in the larger body of material or water being sampled (EPA-94/04).

Representativeness: *See* definition under quality indicator.

Reprocessing: Changing the character of secondary materials i.e., minor, such as crushing or shredding; major - such as biochemical conversion of cellulose into yeast (EPA-83).

Reproducibility: The precision of determinations by different analysts in different laboratories (cf. repeatability or ASTM E180) (EPA-83).

Reproductive toxicity: Fertility, litter size and survival, gestation survival, postnatal body weight (*See* endpoint for more related terms) (Course 165.6; EPA-92/12).

Repulping: The operation of rewetting and fiberizing pulp or paper for subsequent sheet formation (EPA-87/10).

Request for proposal: A request for an offer by one party to another of terms and conditions with references to some work or undertaking; the initial overture or preliminary statement for consideration by the other party to a proposed agreement (40 CFR 248.4-91).

Requirements and standards:
(1) Requirements as used in this policy refers to both the procedural responses and numerical decontamination levels set forth in this policy as constituting adequate cleanup of PCBs.
(2) Standards refers to the numerical decontamination levels set forth in this policy (40 CFR 761.123-91).

Reregistration: The reevaluation and relicensing of existing pesticides originally registered prior to current scientific and regulatory standards. EPA reregisters pesticides through its Registration Standards Program (EPA-94/04).

Research: A systematic investigation, including research development, testing and evaluation, designed to develop or contribute to generalizable knowledge. Activities which meet this definition constitute research for purposes of this policy, whether or not they are conducted or supported under a program which is considered research for other purposes. For example, some demonstration and service programs may include research activities (40 CFR 26.102-91).

Research and development:
(1) Theoretical analysis, exploration, or experimentation; or
(2) The extension of investigative findings and theories of a scientific or technical nature into practical application for experimental and demonstration purposes, including the experimental production and testing of models, devices, equipment, materials and processes. Research and development as used in this part and Parts 31 through 35 does not include the internal or external administration of byproduct material, or the radiation therefrom, to human beings (10 CFR 30.4-91, See also 10 CFR 70.4-91).

Reserve capacity: Extra treatment capacity built into solid waste and wastewater treatment plants and interceptor sewers to accommodate flow increases due to future population growth (EPA-94/04).

Reserve cell: A class of cells which is designated as a reserve cell, because it is supplied to the user in a non-activated state. Typical of this class of cells is the carbon-zinc air reserve cell, which is produced with all the components in a dry or non-activated state, and is activated with water when it is ready to be used (EPA-84/08).

Reserve volume: Volume of air remaining in the lungs after a maximal expiration (EPA-90/08).

Reservoir:
(1) In risk assessment, a tissue in an organism or a place in the environment where a chemical accumulates, from which it may be released at a later time (Course 165.6).
(2) In wastewater treatment, a pond, lake, tank or basin, natural or man-made, used for the storage, regulation and control of water (EPA-74/11).
(3) Any natural or artificial holding area used to store, regulate, or control water (EPA-94/04).

Reservoir of infection: Man, animal, plants, soil or inanimate organic matter tn which an infectious agent lives and multiplies and depends primarily for survival, reproducing itself in such manner that it can be transmitted to man. Man himself is the most frequent reservoir of infectious agents pathogenic for man (EPA-83).

Reservoir route method: A method in which several crews are used to pick up on a centrally located route after having collected on peripheral routes (See waste collection method for more related terms) (SW-108ts).

Reservoir sediment storage: The natural accumulation of sediment in a reservoir that must be taken into account when calculating reservoir capacity (DOI-70/04).

Residence time:
(1) The amount of time that a gas spends in a particular location or at a particular temperature (ETI-92).
(2) The length of time that the combustion gas is exposed to the combustion temperature in an incinerator. It can be expressed as:

$t = V/q$

where:
t = residence time
V = combustion chamber volume
q = combustion gas flow rate (EPA-81/09, p4-98).
(3) The characteristic time during which a substance remains in the atmosphere after its emission, taking into account all the possible sinks of this substance (NATO-78/10).
(4) See time for more related terms.

Residential burner: A device used to burn the solid wastes generated in an individual dwelling (cf. incinerator, municipal waste) (See burner for more related terms) (SW-108ts).

Residential incinerator: Residential incinerator is a predesigned, shop fabricated and assembled unit, shipped as a package, for individual dwellings (See incinerator for more related terms) (EPA-83).

Residential refuse: All those types of solid wastes that normally originate in the private home or apartment house. Also called domestic or household refuse (See refuse for more related terms) (EPA-83).

Residential solid waste: See synonym, residential waste.

Residential tank: A tank located on property used primarily for dwelling purposes (See tank for more related terms) (40 CFR 280.12-91).

Residential use: The use of a pesticide directly:
(1) On humans or pets;
(2) In, on, or around any structure, vehicle, article, surface, or area associated with the household, including but not limited to areas such as nonagricultural outbuildings, non-commercial greenhouses, pleasure boats and recreational vehicles; or
(3) In any preschool or day care facility (40 CFR 152.3-91, See also 40 CFR 157.21-91).

Residential waste (domestic municipal waste, household waste or residential solid waste): The wastes generated by the normal activities of households, including, but not limited to, food wastes, rubbish, ashes, and bulky wastes (*See* waste for more related terms) (40 CFR 243.101-91, *See also* 40 CFR 245.101; 246.101-91).

Residual:
(1) The product or byproduct of a treatment process (EPA-89/12a).
(2) In statistics, the differences between the expected and actual values in a regression analysis (EPA-79/12c).
(3) Amount of a pollutant remaining in the environment after a natural or technological process has taken place, e.g., the sludge remaining after initial wastewater treatment, or particulates remaining in air after it passes through a scrubbing or other process (EPA-94/04).

Residual chlorine: The amount of chlorine left in the treated water that is available to oxidize contaminants if they enter the stream. It is usually in the form of hypochlorous acid of hypochlorite ion or of one of the chloramines. Hypochlorite concentration alone is called free chlorine residual while together with the chloramine concentration their sum is called combined chlorine residual (*See* chlorine for more related terms) (EPA-76/03).

Residual disinfectant concentration (C in CT calculations): The concentration of disinfectant measured in mg/L in a representative sample of water (40 CFR 141.2-91).

Residual moisture:
(1) That moisture remaining in the sample after determining air-dry loss (EPA-83).
(2) The moisture content remaining in the sample after it has been milled down to an analysis sample. Prior to milling, the sample should have been subjected to either a total moisture determination (single stage) or an air drying procedure (EPA-83).
(3) *See* analytical parameters--laboratory for more related terms

Residual noise level: The residual level represents a low-level limit value to which the ambient environmental noise level frequently drops, but below which it seldom goes (DOE-91/04).

Residual oil:
(1) A general term used to indicate a heavy viscous fuel oil (OME-88/12).
(2) Crude oil, fuel oil numbers 1 and 2 that have a nitrogen content greater than 0.05 weight percent, and all fuel oil numbers 4, 5 and 6, as defined by the American Society of Testing and Materials in ASTM D396-78, Standard Specifications for Fuel Oils IBR--*See* 40 CFR 60.17) (40 CFR 60.41b-91, *See also* 40 CFR 60.41c-91).
(3) *See* oil for more related terms.

Residual radioactive material:
(1) Waste (which the Secretary of Energy determines to be radioactive) in the form of tailings resulting from the processing of ores for the extraction of uranium and other valuable constituents of the ores; and
(2) Other waste (which the Secretary of Energy determines to be radioactive) at a processing site which relates to such processing, including any residual stock of unprocessed ores or low-grade materials. This term is used only with respect to materials at sites subject to remediation under Title I of the Uranium Mill Tailings Radiation Control Act of 1978, as amended (10 CFR 40.4-91).
(3) *See* radioactive material for more related terms.

Residual risk: The extent of health risk from air pollutants remaining after application of the Maximum Achievable Control Technology (MACT) (EPA-94/04).

Residual waste: Those materials (solid or liquid) which still require disposal after the completion of a resource recovery activity, e.g., slag and liquid effluents following a pyrolysis operation, plus the discards from front-end separation systems (*See* waste for more related terms) (EPA-83).

Residue:
(1) All the solids that remain after completion of thermal processing, including bottom ash, fly ash, and grate siftings (40 CFR 240.101; 241.101-91).
(2) In metallurgy, dross, skimmings and slag recovered from alloy and metal smelting operations of both the primary and secondary smelters and from foundries (EPA-76/12).
(3) The unusable remainder from any processing system (EPA-83).
(4) The dry solids remaining after the evaporation of a sample of water or sludge (EPA-94/04).
(5) cf. unreclaimable residue.

Residue conveyer: A conveyer, usually a drag- or flight-type, used to remove incinerator residue from a quench trough to a discharge point (*See* conveyer for more related terms) (OME-88/12; SW-108ts).

Resin: There are two types of resins, namely, natural and synthetic. The natural resins are obtained directly from sources such as fossil remains and tree saps. Synthetic resins can be classified by physical properties as thermoplastic or thermosetting. Thermoplastic resins undergo no permanent change upon heating. They can be softened, melted, and molded without change in their physical properties. The thermosetting resins, on the other hand, can be softened, melted, and molded, but with continued heating, they harden or set to a permanent, rigid state and can not be remolded (AP-40). Resin is the main ingredient of paint and ink which binds the various other ingredients together. It also aids adhesion to the surface (EPA-79/12b; 79/12a). Other resin related terms include:
- Acrylic resin
- Alkyd resin
- Bulk resin
- Cold set resin
- Dispersion resin
- Epoxy resin
- Furan resin
- Grade of resin
- Ion exchange resin
- Latex resin
- Phenol formaldehyde resin (*See* synonym, phenolic resin)
- Phenolic resin
- Synthetic resin
- Thermoset resin
- Thermosetting resin (*See* synonym, synthetic resin)

Resin acid: A naturally occurring organic compound in wood (EPA-87/10).

Resin adsorption: *See* definition under adsorption process.

Resistance: For plants and animals, the ability to withstand poor environmental conditions or attacks by chemicals or disease. May be inborn or acquired (EPA-94/04).

Resistance furnace: An electric furnace which is essentially refractory-lined chamber with electrodes, movable or fixed, buried in the charge. the charge itself acts as an electrical resistance that generates heat (*See* furnace for more related terms) (AP-40, p236).

Resistor: A device designed to provide a definite amount of resistance, used in circuits to limit current flow or to provide a voltage drop (EPA-83/03).

Resonance hybrid: The intermediate molecule between two or more valence bond structure.

Resonant frequency: The frequency at which the maximum vibration of a system takes place.

Resorcinol test: A color indicator test used for the determination of the concentration of sucrose in condensate and condenser waters (EPA-75/02d).

Resource: A person, thing, or action needed for living or to improve the quality of life (EPA-89/12).

Resource conservation: The reduction of the amounts of solid waste that are generated, reduction of overall resource consumption, and utilization of recovered resources (RCRA1004-42USC6903-91).

Resource Conservation and Recovery Act (RCRA) of 1976: *See* the term of Act or RCRA.

Resource recovery:
(1) The recovery of material or energy from solid waste (RCRA1004).
(2) A general term used to describe the extraction of materials or energy from wastes (cf. recovery, recycle, reuse or salvage) (EPA-83).
(3) The process of obtaining matter or energy from materials formerly discarded (EPA-94/04).

Resource recovery facility: Any facility at which solid waste is processed for the purpose of extracting, converting to energy, or otherwise separating and preparing solid waste for reuse (RCRA1004, *See also* 40 CFR 51-App/S; 245.101-91).

Resource recovery system: A solid waste management system which provides for collection, separation, recycling, and recovery of solid wastes, including disposal of nonrecoverable waste residues (RCRA1004-42USC6903-91).

Resource recovery unit: A facility that combusts more than 75 percent non-fossil fuel on a quarterly (calendar) heat input basis (40 CFR 60.41a-91).

Respirable dust: The airborne dust in sizes capable of passing through the upper respiratory system to reach the lower lung passages (29 CFR 1910.94a-91).

Respiration: Biological oxidation within a life form; the most likely energy source for animals (the reverse of photosynthesis) (EPA-76/03).

Respiratory bronchiole: Non-cartilagenous airway with lumen open along one side to alveoli; when walls are completely alveolarized it is usually referred to as an alveolar duct. Essentially absent in rats (EPA-90/08).

Respond or response: The remove, removal, remedy, and remedial action, all such terms (including the terms removal and remedial action) include enforcement activities related thereto (SF101, *See also* 40 CFR 300.5-91).

Respondent: Any person proceeded against in the complaint (40 CFR 22.03-91, *See also* 40 CFR 8.33; 32.105; 164.2; 209.3-91).

Response: The efforts to minimize the risks created in an emergency by protecting the people, the environment and property and returning the scene to normal pre-emergency conditions (*See also* 40 CFR 300.5-91) (EPA-85/11).

Response action:
(1) Methods that protect humans health and the environment from asbestos-containing material. Such methods include methods described in chapters 3 and 5 of the Environmental Protection Agency's Guidance for Controlling Asbestos-Containing Materials in Buildings (TSCA202-15USC2642-91).
(2) All activities undertaken to address the problems created by hazardous substances at a National Priorities List site (40 CFR 35.4010-91, *See also* 40 CFR 304.12; 763.83-91; EPA-89/12).
(3) (A) Generic term for actions taken in response to actual or potential health-threatening environmental events such as spills, sudden releases, and asbestos abatement/management problems (EPA-94/04).
    (B) A CERCLA-authorized action involving either a short-term removal action or a long-term removal response. This may include but is not limited to: removing hazardous materials from a site to an EPA-approved hazardous waste facility for treatment, containment or treating the waste on-site, identifying and removing the sources of ground-water contamination and halting further migration of contaminants (EPA-94/04).
    (C) Any of the following actions taken in school buildings in response to AHERA to reduce the risk of exposure to asbestos: removal, encapsulation, enclosure, repair, and operations and maintenance (*See* cleanup) (EPA-94/04)

Response activities: Activities taken to recognize, evaluate, and control an incident (Course 165.5).

Response cost: All costs of removal or remedial action incurred and to be incurred by the United States at a facility pursuant to section 104 of CERCLA, 42USC9604, including, but not limited to, all costs of investigation and information gathering, planning and implementing a response action, administration, enforcement, litigation, interest and indirect costs (*See* cost for more related terms) (40 CFR 304.12-91).

Response factor: The ratio of the known concentration of a VOC compound to the observed meter reading when measured using an instrument calibrated with the reference compound specified in the application regulation (40 CFR 60-App/A(method 21)-91, *See also* 40 CFR 796.1720; 796.1860-91).

Response time: The amount of time required for the measurement system to display 95 percent of a step change in gas concentration on the data recorder (*See* time for more related terms) [40 CFR 60-App/A(method 6C & 7E)-91, *See also* 40 CFR 60-App/A(method 21); 60-App/A(method 25A); 40 CFR 60-App/B-91; LBL-76/07-water).

Response time test: A test that introduces a zero gas into the measurement system at the calibration valve assembly. When the system output has stabilized, switch quickly to the high-level calibration gas. Record the time from the concentration change to the measurement system response, equivalent to 95% of the step change (final value). Repeat the test three times and average the results (EPA-90/04).

Responsible party (RP): A person(s) or company(ies) that the EPA has determined to be responsible for, or to have contributed to, the contamination at a site (EPA-89/12a, *See also* OPA1001; 40 CFR 761.123-91).

Responsiveness summary: A summary of oral and/or written public comments received by EPA during a comment period on key EPA documents, and EPA's response to those comments (EPA-94/04).

Restoration: Measures taken to return a site to pre-violation conditions (EPA-94/04).

Restore: To re-establish a setting or environment in which the natural functions of the floodplain can again operate (40 CFR 6-App/A-91).

Restricted area: Any area access to which is controlled by the licen*See* for purposes of protection of individuals from exposure to radiation and radioactive materials. Restricted area shall not include any areas used as residential quarters, although a separate room or rooms in a residential building may be set apart as a restricted area (10 CFR 20.3-91, *See also* 10 CFR 70.4-91).

Restricted use: A pesticide may be classified (under FIFRA regulations) for restricted use if the it requires special handling because of its toxicity, and, if so, it may be applied only by trained, certified applicators or those under their direct supervision (EPA-94/04).

Restricted use pesticide: A pesticide that is classified for restricted use under the provisions of section 3(d)(l)(C) of the Act (*See* pesticide for more related terms) (40 CFR 171.2-91).

Restricted use pesticide retail dealer: Any person who makes available for use any restricted use pesticide, or who offers to make available for use any such pesticide (40 CFR 171.2-91).

Restriction enzymes: Enzymes that recognize specific regions of a long DNA molecule and cut it at those points (EPA-94/04).

Resurgence: Reappearance of underground water at the surface.

Retan wet finish: The final processing steps performed on a tanned hide including, but not limited to, the following wet processes: retan, bleach, color, and fatliquor (40 CFR 425.02-91).

Retanning: A second tanning process utilizing either the natural tanning materials (chromium or vegetable extracts) or synthetic tanning agents. Retanning imparts specialized properties to the leather (EPA-82/11).

Retarder (active): A device or system for decelerating rolling rail cars and controlling the degree of deceleration on a car by car basis (40 CFR 201.1-91).

Retarder sound: A sound which is heard and identified by the observer as that of a retarder, and that causes a sound level meter indicator at fast meter response 40 CFR 201.1(1) to register an increase of at least ten decibels above the level observed immediately before hearing the sound (40 CFR 201.1-91).

Retention basin: An area designed to retain runoff and prevent erosion and pollution (EPA-89/11).

Retention chamber: A structure within a flow-through test chamber which confines the test organisms, facilitating observation of test organisms and eliminating loss of organisms in outflow water (40 CFR 797.1930-91, *See also* 40 CFR 797.1950-91).

Retention factor: A fraction of aerosol particles removed from a flowing air by a stationary boundary (EPA-88/09a).

Retention index: An index used in the gas chromatography to indicate the retention volume phenomena during analyses.

Retention period: *See* synonym, detention time.

Retention pond: A man-made pond used for temporarily holding the wastewater (cf. detention tank).

Retention time:
(1) In monitoring, the time interval from a step decrease in the input concentration at the instrument inlet to the first corresponding change in the instrument output (LBL-76/07-water).
(2) In incineration, the length of time that solid materials remain in the primary chamber during incineration (EPA-89/03b). It can be expressed as:

$t = 0.19 (L/D)/SN$

where:
t = retention time in minute
L = kiln length in ft
D = kiln diameter in ft

S = slope of kiln ft/ft
N = rotation relocity in rpm (EPA-81/09, p4-104).
(3) *See* time for more related terms.

Retention volume: Retention volume = flow rate x retention time.

Retort:
(1) A system in which substances are distilled, e.g., a vessel in which ore is heated to extract a metal, or coal is heated to produce gas.
(2) A steel vessel in which wood products are impregnated with chemicals that protect the wood from biological deterioration or that impart fire resistance. Also called treating cylinder (EPA-74/04).
(3) Sterilization of food by cooking, usually with steam under pressure (cf. retort incinerator under incinerator, excess-air) (EPA-75/10).

Retort incinerator: *See* definition under excess air incinerator.

Retread tire: A worn automobile, truck, or other motor vehicle tire whose tread has been replaced (40 CFR 253.4-91).

Retrofill: To remove PCB or PCB-contaminated dielectric fluid and to replace it with either PCB, PCB- contaminated, or non-PCB dielectric fluid (40 CFR 761.3-91).

Retrofit:
(1) The addition or removal of an item of equipment, or a required adjustment, connection, or disconnection of an existing item of equipment, for the purpose of reducing emissions (40 CFR 51-App/N-91, *See also* 40 CFR 610.11-91).
(2) Addition of a pollution control device on an existing facility without making major changes to the generating plant (EPA-94/04).

Retrofit device or device:
(1) Any component, equipment, or other device (except a flow measuring instrument or other driving aid, or lubricant or lubricant additive) which is designed to be installed in or on an automobile as an addition to, as a replacement for, or through alteration or modification of, any original component, or other devices; or
(2) Any fuel additive which is to be added to the fuel supply of an automobile by means other than fuel dispenser pumps; and
(3) Which any manufacturer, dealer, or distributor of such device represents will provide higher fuel economy than would have resulted with the automobile as originally equipped, as determined under rules of the Administrator (40 CFR 610.11-1-91).

Retrofitted configuration: The test configuration after adjustment of engine calibrations to the retrofit specifications and after all retrofit hardware has been installed (40 CFR 610.11-91).

Retrospective study: An epidemiological study which compares diseased persons with non-diseased persons and works back in time to determine exposures (Course 165.6).

Return flow (or regenerated water): The part of irrigation water that is not consumed by evapotranspiration and returns to its source or runs off into another body of water (DOI-70/04).

Return on assets (ROA): A measure of potential or realized profit as a percent of the assets (or fixed assets) used to generate the profit (EPA-75/04).

Return on investment (ROI): A measure of potential or realized profit as a percentage of the investment required to generate the profit (EPA-75/04).

Return sludge rate: The ratio between the return sludge to the influent sewage. Return sludge is the sludge that is returned to the aeration tank to mix with the influent sewage.

Returnable beverage container: A beverage container for which a deposit is paid upon purchase and for which a refund of equal value is payable upon return (40 CFR 244.101-91).

Reuse:
(1) The use of a product more than once in its same form for the same purpose; e.g., a soft-drink bottle is reused when it is refined to the bottling company for refilling (cf. recycle) (EPA-89/11).
(2) Using a product or component of municipal solid waste in its original form more than once, e.g., refilling a glass bottle that has been returned or using a coffee can to hold nuts and bolts (EPA-94/04).

Reused or used: A material is used or reused if it is either: (i) Employed as an ingredient (including use as an intermediate) in an industrial process to make a product (for example, distillation bottoms from one process used as feedstock in another process). However, a material will not satisfy this condition if distinct components of the material are recovered as separate end products (as when metals are recovered from metal-containing secondary materials); or (ii) Employed in a particular function or application as an effective substitute for a commercial product (for example, spent pickle liquor used as phosphorous precipitant and sludge conditioner in wastewater treatment) (40 CFR 261.1-91).

Reused water: The process wastewater or treatment facility effluent which is further used in a different manufacturing process (*See* water for more related terms) (EPA-83/03).

Reverberation: The persistence of sound in an enclosed space after the sound source has stopped (EPA-74/11).

Reverberation time: The time that would be required for the mean-square sound pressure level, originally in a steady state, to fall 60 dB after the source is stopped (40 CFR 211.203-91). 60 dB is the threshold of audibility (*See* time for more related terms).

Reverberatory furnace:
(1) A furnace in which the fuel is not in direct contact with the charge (waste or metals) but the heating effect is basically generated by reflection down from a refractory roof (OME-88/12);
(2) A furnace operates by radiating heat from its burner flame, roof, and walls onto the material heated. It can be used to produce aluminum alloy from aluminum scraps (AP-40,

p233; EPA-76/12).
Reverberatory furnace includes the following types of reverberatory furnaces: stationary, rotating, rocking, and tilting (*See* furnace for more related terms) (40 CFR 60.121; 60.131-91).

Reverberatory smelting furnace: Any vessel in which the smelting of copper sulfide ore concentrates or calcines is performed and in which the heat necessary for smelting is provided primarily by combustion of a fossil fuel (*See* furnace for more related terms) (40 CFR 60.161-91).

Reverse: To reverse the direction of flow of gas and air in a regenerative furnace (EPA-83).

Reverse deionization: A technique in which the negative ion exchange unit and positive ion exchange unit are used in sequence to remove all residual ions from a solution.

Reverse incentive: A penalty connected with water use, such as a user charge (based on the amount of water withdrawn from the municipal supply) or an effluent charge (based on the quantity and quality of wastes discharged into a watercourse) to cover damages caused by a user's pollutants (DOI-70/04).

Reverse mutation assay in E. coli: Detects mutation in a gene of tryptophan requiring strain to produce a tryptophan independent strain of this organism (cf. forward mutation) (40 CFR 798.5100-91).

Reverse mutation assay in salmonella typhimurium: Detects mutation in a gene of histidine requiring strain to produce a histidine independent strain of this organism (40 CFR 798.5265-91)

Reverse osmosis (RO): A treatment process used in water systems by adding pressure to force water through a semi-permeable membrane. Reverse osmosis removes most drinking water contaminants. Also used in wastewater treatment. Largescale reverse osmosis plants are being developed (EPA-94/04).

RO is a process in which, if pressure is put on the concentrated side of a liquid system in which liquids with different concentrations of mineral salts are separated by a semi-permeable membrane, molecules of pure water pass out of the concentrated solution to the weak or fresh-water side (contrary to the case of normal osmosis) (DOI-70/04). The RO membrane is permeable to the solvent but impermeable to most dissolved species, both organic and inorganic. The driving force for its separation is an applied pressure gradient.

Conventional forward osmosis transfers solvent through a semi-permeable separator from more dilute to more concentrated solution, driven by the difference in solvent vapor pressure on either side of the separator. RO differs from the conventional filtration in that the flow of the feed is not normal to the membrane surface but tangential to it in an effort to keep it clean of debris and to reduce surface concentration efforts. RO membranes do not become plugged but they may become fouled by film-forming organics or by insoluble salts and scaling.

Reverse well injection: Process in which solutes are injected into an underlying geologic formation through wells (DOE-91/04).

Reversible effect:
(1) An effect which is not permanent, especially adverse effects which diminish when exposure to a toxic chemical is ceased (Course 165.6).
(2) An effect which is not permanent; especially adverse effects which diminish when exposure to a toxic chemical is ceased (EPA-94/04).

Reversible process (or quasi-static process): A process is reversible, if the process progresses in a series of equilibrium states.

Reversible reaction: A chemical reaction capable of proceeding in either direction depending upon the conditions (cf. chemical equilibrium) (EPA-84/08).

Reversing exchanger: A heat-exchanger unit which serves to purify compressed air by the removal of carbon dioxide and water vapor (EPA-77/07).

Revert scrap: *See* synonym, home scrap.

Revolving screen: A screen with a surface that revolves around an axis; the screen surface may be inclined or vertical (*See* screen for more related terms) (EPA-88/08a).

Rewinder: Winder used to salvage unsalable paper rolls and for inspection of offgrade paper; e.g., dished rolls, rolls containing wet streaks, etc., in a salvage operation (EPA-83).

Reynolds analogy: The assumption that the turbulent diffusivities for heat and momentum are equal (NATO-78/10).

Reynolds Number (Re): Re = inertial force/viscous force = (characteristic length)(velocity)(gas density)/viscosity.

The Reynolds number is a dimensionless quantity that characterizes the nature of a fluid flow in a tube or duct. A laminar (smooth) flow (in a tube) is usually encountered at Re number below 2100. Under ordinary conditions of a flow, the flow is turbulent at Re above about 4000. Between 2100 and 4000 a transition region is found, where the type of flow may be either laminar or turbulent (cf. flow regime) (EPA-84/09).
Example:
- Determine Re.

Data:
- Duct diameter = 1.5 ft, gas velocity through the duct = 25 ft/sec., gas viscosity = 1.16 E-05 lb/ft-sec., gas density = 0.075 lb/ft^3.

Solution:
- Re = (1.5 x 25 x 0.075)/1.16 E-05 = 2.42 E+05. Re is greater than 4000, the flow is turbulent.

Reynold stress: The apparent stress on the mean flow caused by turbulent velocity fluctuations which transport momentum. It is calculated as the correlation of this turbulent velocity fluctuations (NATO-78/10).

R_f is the furthest distance traveled by a test material on a thin layer chromatography plate divided by the distance traveled by a solvent front (arbitrarily set at 10.0 cm in soil TLC studies) (40 CFR 796.2700-91).

Rhenium (Re): A transition metal with a.n. 75; a.w. 186.2; d. 21.0 g/cc; m.p. 3180° C and b.p. 5900° C. The element belongs to group VIIB of the periodic table.

Rhodium (Rh): A hard transition metal with a.n. 45; a.w. 102.905; d. 12.4 g/cc; m.p. 1966° C and b.p. 4500° C. The element belongs to group VIII of the periodic table.

Ribbon process: A process whereby molten glass is delivered to a forming unit in a ribbon form (EPA-83).

Ribonucleic acid (RNA): A molecule that carries the genetic message from DNA to a cellular protein-producing mechanisms (EPA-94/04).

Rich combustion: A combustion condition that a combustible mixture contains insufficient oxidant (oxygen or combustion air) for the complete combustion of the mixture (*See* combustion for more related terms).

Richardson number: A dimensionless number expressing the ratio of the production of turbulent energy by buoyancy forces to the production of turbulent energy by shear forces (NATO-78/10).

Ridge and furrow irrigation: A method of irrigation of which water is allowed to flow along surface of fields (EPA-75/02d).

Right-of-way: Includes an easement, lease, permit, or license to occupy, use, or traverse public lands granted for the purpose listed in subchapter V of this chapter (FLPMA103-43USC1702-90).

Rill: A small channel eroded into the soil surface by runoff; can be easily smoothed out obliterated by normal tillage (EPA-94/04).

Rime: A white, opaque granular deposit of ice formed by the rapid freezing of supercooled water drops as they impinge on an exposed object (DOE-91/04).

Ringlemann chart: A series of shaded illustrations used to measure the opacity of air pollution emissions, ranging from light grey through black; used to set and enforce emissions standards (EPA-94/04).

The shades of gray simulate various smoke densities and are assigned numbers ranging from one to five. A clear stack is recorded as 0, and 100% black smoke as 5. Ringelmann No. 1 has 20% density, and 2 through 4 are progressively 20% more dense. Ringelmann charts are used in the setting and enforcement of emission standards (SW-108ts; EPA-74/11).

Rinse: Remove foreign materials from the surface of an object by flow or impingement of a liquid (usually water) on the surface. In the battery industry, rinse may be used interchangeably with wash (EPA-84/08).

Riparian habitat: Areas adjacent to rivers and streams with a high density, diversity, and productivity of plant and animal species relative to nearby uplands (EPA-94/04).

Riparian rights: Entitlement of a land owner to certain uses of water on or bordering his property, including the right to prevent diversion or misuse of upstream waters. Generally a matter of state law (EPA-94/04).

Riprap:
(1) Rough stones of various sizes placed compactly irregularly to prevent erosion (EPA-82/10).
(2) A loose assemblage of stones used in water or soft ground as a foundation (DOE-91/04).

Rise time:
(1) The time required for the spark voltage to increase from 10% to 90% of its maximum value (40 CFR 85.2122(a)(9)(ii)(D)-91).
(2) Rise time means interval between initial response and 95 percent of final response after a step decrease in input concentration (40 CFR 53.23-91).
(3) The time interval between the initial response and a 90% response (unless otherwise specified) after a step increase in the inlet concentration (LBL-76/07-water).
(4) *See* time for more related terms.

Riser: A reservoir of a molten metal connected to the casting to provide an additional metal to the casting. An additional metal is required as the result of shrinkage that occurs before and during solidification (EPA-85/10a).

Riser compounds: Extra strength binders used to reduce the extent of riser erosion. Such materials generally contain lignin, furfuryl alcohol, and phosphoric acid (EPA-85/10a).

Rising current separator: A unit housing a flowing current of water to carry off or wash away organic materials such as food wastes, heavy plastics, and wood from a heavy fraction. The water is pumped upwards causing many materials, which normally would sink, to float and be removed (*See* separator for more related terms) (EPA-83).

Risk:
(1) The probability or a range of probabilities that a specific adverse effect may occur under the conditions of human exposure. It may be expressed in quantitative terms, taking values from zero (certainty that harm will not occur) to one (certainty that it will). In many cases, risk can only be described qualitatively, as high, low, or trivial. The lifetime risk was estimated on the basis of 70-year lifetime and 45-year work exposure (Course 165.6).
(2) A quantitative or qualitative expression of possible loss that considers both the probability that a hazard will cause harm and the consequences of that event (DOE-91/04).
(3) The probability of injury, disease, or death under specific circumstances. In quantitative terms, risk is expressed in values ranging from zero (representing the certainty that harm will not occur) to one (representing the certainty that harm will occur). The following are examples showing the manner in which risk is expressed in IRIS: E-4 = a risk of 1/10,000; E-5 = a risk of 1/100,000; E-6 = a risk of 1/1,000,000. Similarly, 1.3E-3 = a risk of 1.3/1000 = 1/770; 8E-3 = a risk of 1/125; and 1.2E-5 = a risk of 1/83,000 (EPA-92/12).
(4) A measure of the probability that damage to life, health, property, and/or the environment will occur as a result of a given hazard (EPA-94/04).

Other risk related terms include:
- Undue risk
- Unit risk

Risk analysis of incineration emission: An estimate of potential impact on human health and the environment from the emission of incinerator(s). Major factors in making the analysis include carcinogenic potency and threshold toxicity values for organic or heavy metal compounds.

Risk assessment:
(1) The qualitative and quantitative evaluation performed in an effort to define the risk posed to human health and/or the environment by the presence or potential presence and/or use of specific pollutants (EPA-89/12, *See also* NAC-83).
(2) The determination of the kind and degree of hazard posed by an agent, the extent to which a particular group of people has been or may be exposed to the agent, and the present or potential health risk that exists due to the agent (EPA-92/12).
(3) A decision making process that entails considerations of political, social, economic, and engineering information with risk-related information to develop, analyze, and compare regulatory options and to select the appropriate regulatory response to a potential chronic health hazard (EPA-92/12).
(4) Qualitative and quantitative evaluation of the risk posed to human health and/or the environment by the actual or potential presence and/or use of specific pollutants (EPA-94/04).

Risk assessment elements: Including:
(1) Description of potential adverse health effects based on an evaluation of results of epidemiologic, clinical, toxicologic, and environmental research;
(2) Extrapolation from those results to predict the type and estimate the extent of health effects in humans under given conditions of exposure;
(3) Judgments as to the number and characteristics of persons exposed at various intensities and durations;
(4) Summary judgments on the existence and overall magnitude of the public-health problem; and
(5) Characterization of the uncertainties in the process of inferring risk (NAC-83).

Risk assessment steps: Including some or all of the following four steps:
(1) Hazard identification (the determination of whether a particular chemical is or is not causally linked to particular health effects).
(2) Dose-response assessment (the determination of the relation between the magnitude of exposure and the probability of occurrence of the health effects in question).
(3) Exposure assessment (the determination of the extent of human exposure before or after application of regulatory controls).
(4) Risk characterization (the description of the nature and often the magnitude of human risk, including attendant uncertainty) (NAC-83).

Risk based PRGs: Concentration levels set at scoping for individual chemicals that correspond to a specific cancer risk level of risk level of 10^{-6} or an HQ/HI of 1. They are generally selected when ARARs are not available (EPA-91/12).

Risk based targeting: The direction of resources to those areas that have been identified as having the highest potential or actual adverse effects on human health and/or the environment (EPA-94/04).

Risk characterization:
(1) For non-carcinogenic risk, the risk is expressed as a hazard index (HI). For a single compound, HI = E/AL, Where: E = expected exposure that has been obtained through monitoring, sampling or modeling procedures; and AL = Acceptable level.
For a chemical mixture, $HI = E_1/AL_1 + E_2/AL_2 + \ldots$
(2) For carcinogenic risk, the risk is expressed as a "carcinogenic potency factor (CPF) or unit cancer risk (UCR). CPF represents the slope of the dose-response curve. The equation to convert estimated intake directly to incremental cancer risk (P) from a single compound is
P = exposure level (m_g/Kg-day) x CPF (Kg-day/m_g), where m_g = milli-gram (Course 165.6, or Federal Register, Volume 51, 9/24/1986, pp. 34014-34025).
(3) This last step in the risk assessment process characterizes the potential for adverse health effects and evaluates the uncertainty involved (EPA-94/04).

Risk communication: The exchange of information about health or environmental risks among risk assessors and managers, the general public, news media, interest groups, etc (EPA-94/04).

Risk estimate: A description of the probability that organisms exposed to a specific dose of a chemical or other pollutant will develop an adverse response (e.g., cancer (EPA-94/04))

Risk factor:
(1) A correlation of characteristics (e.g., race, sex, age, obesity) or variables (e.g., smoking, occupational exposure level) with increased probability of a toxic effect (Course 165.6).
(2) Characteristic (e.g., race, sex, age, obesity) or variable (e.g., smoking, occupational exposure level) associated with increased probability of a toxic effect (EPA-94/04).

Risk level: The population size on which it is estimated that one additional case of cancer will be reported due to the daily consumption of water and edible aquatic organisms (EPA-85/10).

Risk management: The process of evaluating and selecting alternative regulatory and non-regulatory responses to risk. The selection process necessarily requires the consideration of legal, economic, and behavioral factors (EPA-94/04).

Risk ratio: *See* synonym, relative risk.

Risk regulating agencies: The agencies that have been given primary authority to regulate activities and substances that pose chronic health risks include:
(1) Environmental Protection Agency (EPA).
(2) Food and Drug Administration (FDA), a component of the Department of Health and Human Services.
(3) Occupational Safety and Health Administration (OSHA), a

part of the Department of Labor.
(4) Consumer Product Safety Commission (CPSC) (NAC-83).

Risk retention group: Any corporation or other limited liability association taxable as a corporation, or as an insurance company, formed under the laws of any State:
(1) Whose primary activity consists of assuming and spreading all, or any portion, of the pollution liability of its group members;
(2) Which is organized for the primary purpose of conducting the activity described under subparagraph (1);
(3) Which is chartered or licensed as an insurance company and authorized to engage in the business of insurance under the laws of any State; and
(4) Which does not exclude any person from membership in the group solely to provide for members of such a group a competitive advantage over such a person (SF401-42USC9671-91).

Risk retention group: Any corporation or other limited liability association taxable as a corporation, or as an insurance company, formed under the laws of any State (CERCLA Sec. 401).

Risk specific dose (RSD):
(1) An ambient concentration corresponding to a specified risk. In risk analysis, EPA recommends the health-based levels (known as RSDs) for carcinogens and **Reference Air Concentrations** (RACs) for non-carcinogens. Risk from carcinogens is additive.
(2) The dose associated with a specified risklevel (EPA-94/04).
(3) *See* dose for more related terms.

River basin:
(1) The land area drained by a river and its tributaries (EPA-89/12).
(2) The land area drained by a river and its tributaries (EPA-94/04).

River basin concept: The notion that each river system, from its head-waters to its mouth, is a single unit and should be treated as such. This concept recognizes the inter-relationship of resource elements in a single basin, and assumes that multiple-purpose development can take this interrelationship into account. It extends the principle of ecological balance to the whole of the area and its occupants (EPA-74/11).

River load: The solid matter carried along by a river, including dissolved material, suspended material (mainly mud, silt, and sand), and the larger, heavier material carried along the river bed. The maximum or full load of a river depends on its velocity and volume, and on the size of the particles constituting the load. When the limit of the possible load has been reached, any further addition involves the dropping of an equivalent portion of the original load (*See* load for more related terms) (DOI-70/04).

River profile: A section or curve showing the slope of a river from its source to its mouth (DOI-70/04).

Roadways: The surfaces on which vehicles travel. This term includes public and private highways, roads, streets, parking areas, and driveways (40 CFR 61.141-91).

Roaster: Any facility in which a copper sulfide ore concentrate charge is heated in the presence of air to eliminate a significant portion (5 percent or more) of the sulfur contained in the charge (40 CFR 60.161-91, *See also* 40 CFR 60.171-91).

Roasting: The use of a furnace to heat arsenic plant feed material for the purpose of eliminating a significant portion of the volatile materials contained in the feed (40 CFR 61.181-91).

Rochelle salt: Sodium potassium tartrate, $KNaC_4H_4O_6 \cdot 4H_2O$ (EPA-74/03d).

Rock: To a hydrologist, both hard consolidated formations (such as sandstone, limestone, granite, or lava rocks), and loose unconsolidated sediments (such as gravel, sand, and clay) (DOI-70/04).

Rock crushing and gravel washing facilities: The facilities which process crushed and broken stone, gravel, and riprap. (*See* 40 CFR 436, Subpart B, including the effluent limitations guidelines) (40 CFR 122.27-91).

Rock crystal: Transparent quartz; highly polished blown glassware. handcut or engraved (EPA-83).

Rock wool insulation: The insulation which is composed principally from fibers manufactured from slag or natural rock, with or without binders (40 CFR 248.4-91).

Rocker: An imperfection; a bottle with bottom deformed so it does not stand solidly (rocks) (EPA-83).

Rocket motor test site: Any building, structure, facility, or installation where the static test firing of a beryllium rocket motor and/or the disposal of beryllium propellant is conducted (40 CFR 61.41-91).

Rocking grate: An incinerator stoker with moving and stationary grate bars which are trunnion supported. In operation, the moving bars oscillate on the trunnions, imparting a rocking motion to the bars, thus agitating and conveying the solid fuel and resulting residue through the furnace (*See* grate for more related terms) (EPA-83).

Rocking grate stoker: A stoker with a bed of bars or plates on axles; when the axles are rocking in a coordinated manner, solid waste is lifted and advance along the surface of the grate (*See* stoker for more related terms) (SW-108ts).

Rod proof: A test specimen taken from the melt on an iron rod (EPA-83).

Rodenticide: A chemical or agent used to destroy rats or other rodent pests, or to prevent them from damaging food, crops, etc. (EPA-94/04).

Roentgen (R): A measure of external exposures to ionizing radiation. One roentgen equals that amount of x-ray or gamma radiation required to produce ions carrying a charge of 1 electrostatic unit (esu) in 1 cubic centimeter of dry air under standard conditions. One microroentgen (uR) equals 10^{-6} R (40 CFR

300-App/A-91).

The roentgen (R) is also defined as: 1 R = 2.58×10^{-4} coulombs per kilogram of air. Using the facts that the average energy required to produce one ion pair in air is 34 eV (5.5×10^{-11} ergs, and that a unit charge is 1.6×10^{-19} coulomb, one finds that 1 R = 87.6 ergs per gram of air (*See* radiation unit for more related terms) (LBL-76/07-rad).

Roentgen equivalent man (REM): The unit of dosage equivalent from ionizing radiation to the total body or any internal organ or organ system. A millirem (mrem) is 1/1000 of a rem (40 CFR 141.2-91, *See also* EPA-89/12; DOE-91/04).
(1) **Microrem (μR):** A unit of radiation "dose equivalent" that is equal to one one-millionth of a rem (EPA-88/08a).
(2) **Microrem per hour (μR/hr):** A unit of measure of the rate at which "dose equivalent" is being incurred as a result of exposure to radiation (EPA-88/08a).
(3) **Millirem (mrem):** A unit of radiation "dose equivalent" that is equal to one one-thousandth of a rem (EPA-88/08a).
(4) *See* radiation unit for more related terms.

Roll back model: A simple empirical model which directly relates air quality to emissions. In this model, the contribution to the background concentration by the local sources is considered directly proportional to the total emission. The constant of proportionality is assumed to be independent of the emission changes (NATO-78/10).

Roll bar: Steel protection over the cab of a tractor or any vehicle to prevent injury to the operator (EPA-83).

Roll bonding: The process by which a permanent bond is created between two metals by rolling under high pressure in a bonding mill (co-rolling) (40 CFR 471.02-91).

Roll coater: An apparatus in which a uniform layer of coating is applied by means of one or more rolls across the entire width of a moving substrate (40 CFR 52.741-91).

Roll coating: The process in which the coating is applied by rolls and the coated surface smoothed by means of reverse rolls (EPA-83).

Roll crusher: A reduction crusher consisting of a heavy frame on which two rolls are mounted; the rolls are driven so that they rotate toward one another. Coal is fed in from above and nipped between the moving rolls, crushed, and discharged below (*See* crusher for more related terms) (EPA-82/10).

Roll off container: A large waste container that fits onto a tractor trailer that can be dropped off and picked up hydraulically (EPA-89/11).

Roll on/roll off container: A large container (20 to 40 cubic yards) that can be pulled onto a service vehicle mechanically and carried to a disposal site for emptying (*See* container for more related terms) (SW-108ts).

Roll printer: An apparatus used in the application of words, designs, or pictures to a substrate, usually by means of one or more rolls each with only partial coverage (40 CFR 52.741-91).

Roll printing: The application of words, designs, and pictures to a substrate usually by means of a series of hard rubber or metal rolls each with only partial coverage (40 CFR 52.741-91).

Rollboard: An asbestos-containing product made of paper that is produced in a continuous sheet, is flexible, and is rolled to achieve a desired thickness. Asbestos rollboard consists of two sheets of asbestos paper laminated together. Major applications of this product include: office partitioning; garage paneling; linings for stoves and electric switch boxes and fire-proofing agent for security boxes, safes, and files (40 CFR 763.163-91).

Rolled glass: An optical glass formed by rolling into plates at time of manufacture, as distinguished from transfer glass; flat glass formed by rolling (*See* glass for more related terms) (EPA-83).

Rolled power: A propellant which is formed by forcing a nitrocellulose-nitroglycerin composition between two large steel rolls to form a sheet (EPA-76/03).

Roller coating: A method of applying a coating to a sheet or strip in which the coating is transferred by a roller or series of rollers (40 CFR 52.741-91).

Roller crusher: A machine whose function is to crush material between two opposing steel rollers that rotate slowly on horizontal axes (*See* size reduction machine for more related terms) (EPA-83).

Rolling: The reduction in the thickness or diameter of a workpiece by passing it between rollers (40 CFR 468.02-91, *See also* 40 CFR 467.02; 471.02-91).

Rolling average: The arithmetic mean value over a period of time. The **hourly rolling average (HRA)** is the arithmetic mean of sixty (60) most recent 1-minute average values recorded by the continuous monitoring system. See carbon monoxide (CO) and total hydrocarbon (THC) emission limits and permit formats for example (EPA-90/04).

Rolling terrain: A terrain whose rise within 5 kilometers of the stack is between 10% and 100% of the physical stack height (*See* terrain for more related terms) (EPA-90/04).

Roof coating: An asbestos-containing product intended for use as a coating, cement, adhesive, or sealant on roofs. Major applications of this product include: waterproofing; weather resistance; sealing; repair; and surface rejuvenation (40 CFR 763.163-91).

Roof monitor: That portion of the roof of a potroom where gases not captured at the cell exit from the potroom (40 CFR 60.191-91).

Roofing felt: An asbestos-containing product that is made of paper felt intended for use on building roofs as a covering or underlayer for other roof coverings (40 CFR 763.163-91).

Room and pillar mining: A system of mining in which the distinguishing feature is the mining of 50 percent or more of the coal in the first working. The coal is mined in rooms separated by narrow ribs (pillars); the coal in the pillars can be extracted by subsequent working in which the roof is caved in successive blocks (EPA-82/10).

Root crop: The plants whose edible parts are grown below the surface of the soil (40 CFR 257.3.5-91).

Rosin: A specific kind of a natural resin obtained as a nitrous water-insoluble material from pine oleoresin by removal of the volatile oils, or from tall oils by the removal of the fatty acid components thereof (EPA-79/12). Other rosin related terms include:
- Colophony rosin (*See* synonym, natural rosin)
- Common rosin (*See* synonym, natural rosin)
- Gum rosin (*See* synonym, natural rosin)
- Modified rosin
- Natural rosin
- Pine rosin (*See* synonym, natural rosin)

Rosin size: A coating that renders paper water resistant (EPA-83).

Rotameter: One of liquid flow rate meters. This type of flowmeter is available for a wide range of liquid viscosities including some light-wight slurries. It is calibrated through using a fluid of known density. Reported accuracies are within +/- 5% of hull-scale (*See* flow rate meter for more related terms) (EPA-83/06).

Rotary blast cleaning table: An enclosure where the pieces to be cleaned are positioned on a rotating table and are passed automatically through a series of blast sprays (29 CFR 1910.94a-91).

Rotary kiln incinerator:
(1) An incinerator with a rotating combustion chamber that keeps waste moving, thereby allowing it to vaporize for easier burning (EPA-94/04).
(2) A rotary kiln incinerator is a more versatile incinerator in the sense that it is applicable to the destruction of solid wastes, slurries, and containerized waste as well as liquids. Because of this, the unit is most frequently incorporated into commercial off-site incineration facility design. A rotary kiln incineration system is designed to operate continuously. It must include a device for continuous waste feed to the kiln and continuous ash removal.
(3) *See* incinerator for more related terms.

Rotary kiln incinerator main components: It usually consists of a rotary kiln and an afterburner:
(1) The rotary kiln is a primary combustion chamber which is a cylindrical refractorylined shell that is slightly inclined and rotates, hence the name rotary kiln. The primary function of the kiln is to convert solid wastes to gases, which occurs through a series of volatilization, destructive distillation, and partial combustion reactions. Rotation of the shell provides for transportation of waste through the kiln as well as for enhancement of waste mixing. The residence time of waste solids in the kiln ranges from seconds to hours. This is controlled by the kiln rotation speed (1-5 rpm), the waste feed rate (generally not to exceed 15 to 20 percent of the kiln internal volume) and, in some instances, the inclusion of internal dams. Traditionally, the kiln operates with excess air. However, some manufacturers now have rotary kilns designed to operate with a substoichiometric air in the kiln. These kilns use special kiln seals and air injection schemes. A kiln seal is a sealing ring installed between the rotating kiln and the kiln end plates (EPA-89/03b).

(2) The afterburner is normally referred to the burner at the secondary combustion chamber and is usually required to complete the gas-phase combustion reactions. It is connected directly to the discharge end of the kiln, whereby the gases exiting the kiln turn from a horizontal flow path upwards to the afterburner chamber. Both the afterburner and kiln are usually equipped with an auxiliary fuel firing system to bring the units up to and maintain the desired operating temperatures. The afterburner chamber itself may be horizontally or vertically aligned, and essentially functions much on the same principles as a liquid injection incinerator. In fact, many facilities also fire liquid hazardous waste through separate waste burners in the afterburner chamber (cf. burner, secondary) (Oppelt-87/05).

Rotary kiln stoker: A cylindrical, inclined device that rotates, thus causing the solid waste to move in a slow cascading and forward motion (*See* stoker for more related terms) (SW-108ts).

Rotary lime kiln: A unit with an inclined rotating drum that is used to produce a lime product from limestone by calcination (40 CFR 60.341-91).

Rotary screen: An inclined, meshed cylinder that rotates on its axis and screens material placed in its upper end (*See* screen for more related terms) (SW-108ts).

Rotary sifter: A circular motion applied to a rectangular or circular screen surface (EPA-88/08a).

Rotary spin: A process used to produce wool fiberglass insulation by forcing molten glass through numerous small orifices in the side wall of a spinner to form continuous glass fibers that are then broken into discrete lengths by high velocity air flow (40 CFR 60.681-91).

Rotary vacuum filter: A rotating drum filter which utilizes suction to separate solids from the sludge produced by clarification (*See* filter for more related terms) (EPA-75/02d).

Rotating biological contactor (or rotating biological disc): The process consists of a series of closely spaced flat or lattice-structure plastic disks, usually about 3.0 to 3.5 m in diameter, mounted on a horizontal shaft. Each disk is rotated so that one half of its surface is immersed in the wastewater. During rotation, microbes adhere to the surface of the disk and grow until the entire is coated with a biological slime layer. The disk, then, carries a film of wastewater into the air, which subsequently trickles down the surface and absorbs oxygen. When rotation is complete, this film mixes with the reservoir of wastewater, adding to the oxygen in the reservoir and mixing the treated and partially treated wastewater. When the attached microbial film passes through the reservoir, organics are absorbed and undergo breakdown. As the treated wastewater flows through the reservoir, excess biological growth shears off from the rotating disks and is carried downstream to a settling tank for removal.

Rotating biological disc: *See* synonym, rotating biological contactor.

Rotogravure print station: Any device designed to print or coat inks on one side of a continuous web or substrate using the intaglio printing process with a gravure cylinder (40 CFR 60.581-91).

Rotogravure printing: The application of words, designs, and pictures to a substrate by means of a roll printing technique in which the pattern to be applied is recessed relative to the non-image area (40 CFR 52.741-91).

Rotogravure printing line: A printing line in which each roll printer uses a roll with recessed areas for applying an image to a substrate (40 CFR 52.741-91, *See also* 40 CFR 60.581-91).

Rotogravure printing unit: Any device designed to print one color ink on one side of a continuous web or substrate using a gravure cylinder (40 CFR 60.431-91).

Rotometer: A device, based on the principle of Stoke's law, for measuring rate of fluid flow. It consists of a tapered vertical tube having a circular cross section, and containing a float that is free to move in a vertical path to a height dependent upon the rate of fluid flow upward through the tube (LBL-76/07-air).

Rough fish: Fish not prized for eating, such as gar and suckers. Most are more tolerant of changing environmental conditions than game species (EPA-94/04).

Rougher cell: A flotation cell in which the bulk of the gangue is removed from the ore (EPA-82/05).

Roughing: In mining, upgrading of run-of-mill feed either to produce a low grade preliminary concentrate or to reject valueless tailings at an early stage. Performed by gravity on roughing tables, or in flotation in a rougher circuit (EPA-82/05).

Roughness length: A characteristic length which is a measure of the roughness of a surface. It enters as a parameter in the logarithmic wind velocity profile (NATO-78/10).

Round robin: A formal study between several laboratories, of the characteristics of a material. The study is repeated several times (round) to obtain the precision of a given measurement by a given /. method (EPA-83).

Rounded: A number shortened to the specific number of decimal places in accordance with the Round Off Method specified in ASTM E 29-67 (40 CFR 600.002.85-91).

Route of exposure:
(1) The avenue in which a chemical contaminant enters the body, e.g., inhalation, skin (or eye) absorption, ingestion (oral swallow), and injection (Course 165.5).
(2) The avenue by which a chemical comes into contact with an organism (e.g., inhalation, ingestion, dermal contact, injection. (EPA-94/04))

Routine maintenance area: An area, such as a boiler room or mechanical room, that is not normally frequented by students and in which maintenance employees or contract workers regularly conduct maintenance activities (40 CFR 763.83-91).

Routine method: A method used in a routine measurement. No degree of reliability is implied (*See* method for more related terms) (ACS-87/11).

Routine release: A release that occurs during normal operating procedures or processes (40 CFR 302.3-91).

Routine use: With respect to the disclosure of a record, the use of such record for a purpose which is compatible with the purpose for which it was collected (40 CFR 1516.2-91, *See also* 40 CFR 16.2-91).

Rubber: A high polymer with elastic properties. Other rubber related terms include:
- Butyl rubber
- Nitrile rubber
- Polybutadiene rubber

Rubbish:
(1) A general term for solid waste, excluding food wastes and ashes, taken from residences, commercial establishments, and institutions (40 CFR 243.101-91).
(2) Solid waste, excluding food waste and ashes, from homes, institutions, and work-places (EPA-94/04).

Other rubbish related terms include:
- Combustible rubbish
- Yard rubbish

Rubbish chute: A pipe, duct or trough through which wastes materials are conveyed by gravity from above to a storage area preparatory to burning or compaction (EPA-83).

Rubble: Demolition wastes; broken pieces of masonry, and concrete, asphalt, roofing etc. (cf. waste, construction and demolition) (EPA-83).

Ruben: Developer of the mercury-zinc battery; also refers to the mercury-zinc battery (EPA-84/08).

Rubidium (Rb): A alkali metal with a.n. 37; a.w. 85.47; d. 1.53 g/cc; m.p. 38.9° C and b.p. 688° C. The element belongs to group IA of the periodic table.

Rule out: To eliminate as a possibility (NIOSH-84/10).

Run: The net period of time during which an emission sample is collected. Unless otherwise specified, a run may be either intermittent or continuous within the limits of good engineering practice (40 CFR 60.2-91, *See also* 40 CFR 61.02; 61.61-91).

Run-of-pile triple superphosphate: Any triple superphosphate that has not been processed in a granulator and is composed of particles at least 25 percent by weight of which (when not caked) will pass through a 16 mesh screen (40 CFR 60.231-91).

Runoff:
(1) The portion of precipitation that drains from an area as surface flow (40 CFR 241.101-91).
(2) Any rainwater, leachate, or other liquid that drains over land from any part of a facility (40 CFR 260.10-91, *See also* 40 CFR 419.11-91).
(3) That part of precipitation, snow melt, or irrigation water that

runs off the land into streams or other surface-water. It can carry pollutants from the air and land into receiving waters (EPA-94/04).
Other runoff related terms include:
- Contaminated runoff
- Mining runoff
- Urban runoff

Runoff coefficient: The fraction of total rainfall that will appear at a conveyance as runoff (40 CFR 122.26-91).

Runon: Any rainwater, leachate, or other liquid that drains over land onto any part of a facility (40 CFR 260.10-91).

Runner: A channel through which molten metal flows from one receptacle to another. Runner is often used to refer to the portion of the gate assembly that connects the riser with the casting (EPA-85/10a).

Running changes: Those changes in vehicle or engine configuration, equipment or calibration which are made by an OEM or ICI in the course of motor vehicle or motor vehicle engine production (40 CFR 85.1502-91).

Running loss:
(1) The fuel evaporative emissions resulting from an average trip in an urban area or the simulation of such a trip (40 CFR 86.082.2-91).
(2) Evaporation of motor vehicle fuels from the fuel tank while the vehicle is in use (EPA-94/04).

Rupture of a PCB transformer: A violent or non-violent break in the integrity of a PCB Transformer caused by an overtemperature and/or overpressure condition that results in the release of PCBs (*See* transformer for more related terms) (40 CFR 761.3-91).

Rural area: Outside the limits of any incorporated or unincorporated city, town, village, or any other designated residential or commercial area such as a subdivision, a business or shopping center, or community development (40 CFR 195.2-91).

Rust prevention compounds: Coatings used to protect iron and steel surfaces, against corrosive environment during fabrication, storage, or use (EPA-83/06a).

Ruthenium (Ru): A hard transition metal with a.n. 44; a.w. 101.07; d. 12.2 g/cc; m.p. 2500° C and b.p. 4900° C. The element belongs to group VIII of the periodic table.

Rutile: Titanium dioxide (TiO_2).

Rutherfordium (Rf) (or Kurchatovium (Ku)): A transactinide element (atomic number greater than 103). The element with a.n. 104 is unstable, has very short half live, and belongs to group IVB of the periodic table.

Ss

Saccharide: Water soluble carbohydrates with sweet taste. The compound includes monosaccharide (simple sugar); disaccharide and trisaccharide (*See* sugar for more related terms).

Saccharimeter: An instrument used to measure the sugar content in a solution.

Saccharin ($C_6H_4COSO_2NH$): A sugar substitute for syrups, foods, beverages, etc. (*See* sugar for more related terms).

Sacral vertebrae: The five normally fused vertebrae at the posterior end of the spinal column that form the sacrum (LBL-76/07-bio).

Sacrificial anode: An easily corroded material deliberately installed in a pipe or take to give it up (sacrifice) to corrosion while the rest of the water supply facility remains relatively corrosion free (EPA-94/04).

Safe: Condition of exposure under which there is a practical certainty that no harm will result to exposed individuals (EPA-94/04).

Safe disposal: Discarding pesticides or containers, in a permanent manner so as to comply with these proposed procedures and so as to avoid unreasonable adverse effects on the environment (*See* disposal for more related terms) (40 CFR 165.1-91).

Safe Drinking Water Act (SDWA) of 1974: *See* the term of Act or SDWA.

Safe shutdown earthquake: In the context of a nuclear reactor, as defined by 10 CFR 100, Appendix A: That earthquake which is based upon an evaluation of the maximum earthquake potential considering the regional and local geology and seismology and specific characteristics of local subsurface material. It is that earthquake which produces the maximum vibratory ground motion for which certain structures, systems, and components are designed to remain functional. These structures, systems and components are those necessary to assure:
(1) The integrity of the reactor coolant pressure boundary;
(2) The capability to shut down the reactor and maintain it in a safe shutdown condition; or
(3) The capability to prevent or mitigate the consequences of accidents that could result in potential off-site exposures comparable to the guideline procedures (DOE-91/04).

Safe water: Water that does not contain harmful bacteria, toxic materials, or chemicals and is considered safe for drinking even though it may have taste, and odor, color and certain mineral problems (EPA-94/04).

Safe yield: The annual amount of water that can be taken from a source or supply over a period of years without depleting that source beyond its ability to be replenished naturally in "wet years" (EPA-94/04).

Safe yield (or sustained yield):
(1) The achievement and maintenance in perpetuity of a high level annual or regular periodic output of the various renewable resources of the public lands consistent with multiple use (FLPMA103).
(2) The use of renewable resource at a rate that permits resource regeneration for use continuing undiminished into the future (e.g., timber cut so as to produce the same amount of wood each year; deer hunted without long range damage to the herd) (DOI-70/04).

Safener: A chemical added to a pesticide to keep it from injuring plants (EPA-94/04).

Safety analysis report: A safety document providing a concise but complete description and safety evaluation of a site, design, normal and emergency operation, potential accidents, predicted consequences of such accidents, and the means proposed to prevent such accidents or mitigate their consequences. A safety analysis report is designated as final when it is based on final design information. Otherwise, it is designated as preliminary (DOE-91/04).

Safety class 1: One of three levels assigned to components, systems, or structures that must be designed to provide certain functions to protect operators, the public, or the environment. This class is concerned with function and/or structural integrity for the mitigation of severe events, including design-basis accidents (DOE-91/04).

Safety document: A document prepared specifically to ensure that the safety aspects of part or all of the activities conducted at a reactor are formally and thoroughly analyzed, evaluated, and recorded (for example, technical specifications, safety analysis reports and addenda, and documented reports of special safety reviews and studies) (DOE-91/04).

Safety factor: For a dose response evaluation, a safety factor is a number that reflects the degree of uncertainty that must be considered when data from animal experiments are extrapolated for human application. The following Table shows the safety factor and the facts under consideration (cf. uncertainty factor).
(1) Safety factor 10:
 (A) Valid human, chronic experimental results; and
 (B) No indication of carcinogenicity.
(2) Safety factor 100:
 (A) Valid results from animal tests, but little information about human exposure; and
 (B) No indication of carcinogenicity.
(3) Safety factor 1000:
 (1) No human data, but few animal results; and
 (2) No indication of carcinogenicity (Course 165.6).

Safety glass: A glass so constructed, treated, or combined with other materials, as to reduce, the likelihood of injury to persons by objects from exterior sources or by these safety glasses when they may be cracked or broken (EPA-83). Safety glass types include:
(1) Laminated safety glass
(2) Tempered safety glass
(3) Wire safety glass
(4) See glass for more related terms.

Safety grade (or safety-related): Describes those structures, systems, and components (SSCs) that are necessary to:
(1) Ensure the integrity of a reactor coolant pressure boundary;
(2) Shut down a reactor and maintain it in a safe shutdown condition indefinitely; or
(3) Prevent or mitigate the consequences of a design-basis accident so that the general public and the operating staff are not exposed to radiation in excess of appropriate guideline values. Safety-grade SSCs are designed, built, tested, operated, and maintained in accordance with the highest-level industrial codes and standards, such as those promulgated by the American Society of Mechanical Engineers (ASME) and the Institute of Electrical and Electronics Engineers. SSCs that are not necessary for the above safety-related functions, but for which certain failure modes could adversely affect SSCs performing those functions, are designated safety-grade with respect to such failure modes. Safety-grade SSCs are subject to the quality assurance requirements of ANSI/ASME NQA-1 (Quality Assurance Program Requirements for Nuclear Facilities) (DOE-91/04).

Safety relief valve: A valve which is normally closed and which is designed to open in order to relieve excessive pressures within a vessel or pipe (40 CFR 52.741-91).

Safrole ($C_3H_5C_6H_3O_2CH_2$): A poisonous oil used in medicine, perfumes, insecticides and soaps.

SAFSTOR: Describes a nuclear facility that is placed and maintained in a condition that allows the nuclear facility to be safely stored and subsequently decontaminated. The SAFSTOR option may include the following phases: chemical decontamination, mechanical decontamination and fixing of residual radioactivity, equipment deactivation, preparation for interim care, interim care (surveillance and maintenance), and final dismantlement (DOE-91/04).

Saline water (salt water or sea water): The water which contains salts (See water for more related terms).

Saline estuarine water: Those semi-enclosed coastal waters which have a free connection to the territorial sea, undergo net seaward exchange with ocean waters, and have salinities comparable to those of the ocean. Generally, these waters are near the mouth of estuaries and have cross-sectional annual mean salinities greater than twenty-five (25) parts per thousand (See water for more related terms) (40 CFR 125.58-91).

Salinity:
(1) The degree of salt in water (EPA-89/12).
(2) The relative concentration of salts, usually sodium chloride, in a given water. It is usually expressed in terms of the number of parts per million of chloride.
(3) A measure of the concentration of dissolved mineral substances in water (EPA-82/11f).

Salinometer: An instrument using electrical conductivity or hydrometer for measuring the salinity of a solution.

Salt:
(1) A product resulting from a reaction of an acid and a base, in which the hydrogen in the acid is replaced by a metal or other positive ions. For example, HCl + NaOH ---> NaCl (salt) + H_2O.
(2) Minerals that water picks up as it passes through the air, over and under the ground, or from households and industry (EPA-94/04).

Salt bath descaling: Removing the layer of oxides formed on some metals at elevated temperatures in a salt solution (EPA-83/06a). Other salt bath descaling related terms include:
• Oxidizing salt bath descaling
• Reducing salt bath descaling

Salt bridge: A bridge between two half cells. Potassium chloride is usually used as the bridge salt.

Salt water: See synonym, saline water.

Salt water intrusion (or sea water intrusion): The invasion of fresh surface or ground water by salt water. If the salt water comes from the ocean it may be called sea water intrusion (EPA-94/04, See also EPA-74/11). Because of its greater density, salt water passes below the fresh water. Excessive pumping of deep wells may produce salt containing well water.

Saltcake loss: The loss of cooking chemical from the kraft cycle, primarily at the brownstock washers or screen room (EPA-87/10).

Salvage:
(1) Salvaging means the controlled removal of waste materials for utilization (40 CFR 241.101-91).
(2) The utilization of waste materials (EPA-94/04).
(3) cf. recovery.

Salvage activity: The act of saving or obtaining a secondary material, be it by pickup, sorting, disassembly, or some other activity. Salvage is sometimes used synonymously with extraction, recovery, reclamation, and recycle (EPA-83).

Salvage and reclamation: A refuse disposal process in which the discarded material is separated mechanically or by hand into various categories such as ferrous and nonferrous metals, rags, cardboard, paper and glass, etc. for reuse or recycled as a secondary raw material (EPA-83).

Salvage material: A technical term meaning a quantity of materials, sometimes of mixed composition, no longer useful in its present condition or at its present location, but capable of being recycled, reused, or used in other applications. Salvage also refers to materials recovered after a calamity, such as materials obtained from a ship wrecked at sea or a building destroyed by fire (EPA-83).

Salvage value: The anticipated value of any portion of a facility, including the land at the end of the design period (EPA-80/09).

Samarium (Sm): A rare earth metal with a.n. 62; a.w. 150.35; d. 7.54 g/cc; m.p. 1072° C and b.p. 1900° C. The element belongs to group IIIB of the periodic table.

Sample:
(1) In quality control, a group of samples (chemical) taken from a lot or batch of samples (EPA-84/03).
(2) In monitoring, a representative specimen of air collected for the purpose of determining its pollutant content (LBL-76/07-air).

Other sample related terms include:
- Batch sample
- Batch sample size
- Check sample
- Composite sample
- Composite wastewater sample
- Continuous sample
- Duplicate sample (*See* synonym, replicate sample)
- Environmental sample
- Field sample (*See* synonym, environmental sample)
- Grab sample
- Integrated sample
- Mixed sample (*See* synonym, composite sample)
- Replicate sample
- Representative sample
- Sampling (*See* synonym, continuous sample)
- Split sample
- Standard sample
- Test sample
- Test sample size

Sample average: *See* synonym, sample mean.

Sample division: The process of extracting a smaller sample from a sample so that the representative properties of the larger sample are retained. During this process it is assumed that no change in particle size or other characteristics occurs (*See* analytical parameters--laboratory for more related terms) (EPA-83).

Sample holding time: The storage time allowed between sample collection and sample analysis when the designated preservation and storage techniques are employed.

Sample interface: The portion of a system used for one or more of the following: sample acquisition, sample transport, sample conditioning, or protection of the analyzers from the effects of the stack effluent (*See* gas concentration measurement system for more related terms).

Sample interval: The time period between successive samples for a digital signal or between successive measurements for an analog signal (40 CFR 60-App/A(alt. method 1)-91).

Sample line: A stainless steel or Teflon tube to transport the sample gas to the analyzer. The sample line should be maintained at the temperature between 150 and 175° C to prevent the condensation of samples taken (*See* total hydrocarbon concentration measurement system for more related terms) (EPA-90/04).

Sample loop: A 1/16 in. O.D. (1.6mm) stainless steel tube with an internal volume between 20 and 50 uL. The loop is attached to the sample injection valve of the HPLC and is used to inject standard solutions into the mobile phase of the HPLC when determining the response factor for the recording integrator. The exact volume of the loop must be determined as described in paragraph (b)(3)(ii)(C)(1) of this section when the HPLC method is used (40 CFR 796.1720-91, *See also* 40 CFR 796.1860-91).

Sample mean (or sample average): The average of a population calculated from the sample; it is the most commonly used measure of the center of a distribution. Its value equals the sum of the values of the observations divided by the number of observations (cf. arithmetic mean) (EPA-87/10).

Sample path length: Internal cell or sample length, usually given in centimeters (LBL-76/07-bio).

Sample preparation: The process of that includes drying, size reduction, division, and mixing of a laboratory sample for the purpose of obtaining an unbiased analysis sample (*See* analytical parameters--laboratory for more related terms) (EPA-83).

Sample probe: A stainless steel, or equivalent, three-hole rake type. Sample holes shall be 4 mm in diameter or smaller and located at 16.7, 50, and 83.3 percent of the equivalent stack diameter. Alternatively, a single opening probe may be used so that a gas sample is collected from the centrally located 10 percent area of the stack cross section (*See* total hydrocarbon concentration measurement system for more related terms) (EPA-90/04).

Sample quantitation limit (SQL): The quantity of a substance that can be reasonably quantified given the limits of detection for the methods of analysis and sample characteristic that may affect quantitation (for example, dilution, concentration) (40 CFR 300-App/A-91).

Sample reduction: The process whereby sample particle size is reduced without change in sample weight (*See* analytical parameters--laboratory for more related terms) (EPA-83).

Sample size: The number of units in a sample (EPA-84/03).

Sample system: The system which provides for the transportation of the gaseous emission sample from the sample probe to the inlet of the instrumentation system (40 CFR 87.1-91).

Sampler: A device used with or without flow measurement to obtain any adequate portion of samples such as flue gas, water or waste for analytical purposes. May be designed for taking a single sample (grab), composite sample, continuous sample, or periodic sample (EPA-82/11f).

Sampling:
(1) In statistics, the selection of a finite subset of a population (EPA-83/06).
(2) In monitoring, a process consisting of the withdrawal or isolation of a fractional part of a whole. In air or gas analysis, the separation of a portion of an ambient atmosphere with or without the simultaneous isolation of selected components (LBL-76/07-air).
(3) Sampling (*See* synonym, continuous sample)
Other sampling related terms include:
- Condensation sampling
- Immersion sampling
- Instantaneous sampling
- Intermittent sampling
- Stack sampling

Sampling and analyzing method: *See* synonym, reference method.

Sampling and analytical methods for stack gases in RCRA trial burns: (EPA-89/06, p13 & 22)
I. SAMPLING METHOD
(1) Particulate, Cl⁻(HCl), H_2O, nonvolatile metals (if present in feed): M5 trial with appropriate solutions in one or more of the impingers
(2) Semivolatile and nonvolatile POHCs: Semi-VOST (SW-846 method 0010)
(3) Volatile POHCs: VOST (SW-846 method 0030)
(4) CO_2, O_2: EPA method 3 Orsat analysis of integrated bag sample
(5) CO: Continuous monitor
(6) Volatile metals (if presented in feed): Sampling method: M5 train with appropriate solutions in the second impinger

II. ANALYTICAL METHODS
(1) Sample: M5 train
　(A) Filter, probe rinse
　　- Analytical parameter: Particualte
　　- Analytical method: EPA method 5
　(B) Water impinger and caustic impinger
　　- Analytical parameter: Cl⁻(HCl)
　　- Analytical method: Ion chromatography or EPA 352.2
(2) Multiple metals train
　- Analytical parameter: Metals
　- Analytical method: *See* Table 13 (EPA-89/06)
(3) Semi-VOST
　(A) Filter, probe rinse
　　- Analytical parameter: SV-POHC
　　- Analytical method: *See* Appendix A (EPA-89/06)
　(B) XAD-2
　　- Analytical parameter: SV-POHC
　　- Analytical method: GC/MS

　(C) Condensate
　　- Analytical parameter: SV-POHC
　　- Analytical method: not specified
(4) VOST
　- Analytical parameter: V-POHC
　- Analytical method: GC/MS per SW-846, method 5040
(5) Mylar gas bag
　- Analytical parameter: CO_2, O_2
　- Analytical method: EPA method 3
(6) Tedlar gas bag
　- Analytical parameter: V-POHC
　- Analytical method: Transfer to Tenax trap and GC/MS per SW-846, method 5040

Sampling area: Any area, whether contiguous or not, within a school building which contains friable material that is homogeneous in texture and appearance (40 CFR 763.103-91).

Sampling bag: The bags used for collecting air samples from the field. When the bags are brought back to the laboratory, the collected air is then released for analysis.

Sampling duration: The total time period during which samples of a function are collected (NATO-78/10).

Sampling method: Specified either as continuous, semi-continuous, intermittent, or static (batch) (LBL-76/07-bio).

Sampling station: A location where samples are tapped (taken) for analysis (EPA-82/11f).

Sampling system bias: The difference between the gas concentrations exhibited by the measurement system when a known concentration gas is introduced at the outlet of the sampling probe and when the same gas is introduced directly to the analyzer (40 CFR 60-App/A(method 6C & 7E)-91).

Sampling volume: Range of the amount of sample required to perform the measurement (LBL-76/07-bio).

Sanctions: Actions taken by the federal government for failure to plan or implement a State Improvement Plan (SIP). Such action may be include withholding of highway funds and a ban on construction of new sources of potential pollution (EPA-94/04).

Sand: A course grained soil, the greater portion of which passes through a No. 4 sieve, according to the Unified Soil Classification System (SW-108ts).

Sand bed drying: The process of reducing the water content in a wet substance by transferring that substance to the surface of a sand bed and allowing the processes of drainage through the sand and evaporation to effect the required water separation (EPA-83/06a).

Sand binder: Binder materials are the same as those used in core making. The percentage of binder may vary in core and molds depending on sand strength required, extent of mold distortion from hot metal and the metal surface finish required (EPA-85/10a).

Sand blasting: The process of removing stock including surface film, from a workpiece by the use of abrasive grains pneumatically impinged against the workpiece (EPA-83/06a).

Sand filter (contact bed or contact filter): Devices that remove some suspended solids from sewage. Air and bacteria decompose additional wastes filtering through the sand so that cleaner water drains from the bed (*See* filter for more related terms) (EPA-94/04).

Sand filtration: A process of filtering wastewater through sand. The wastewater is trickled over the bed of sand where air and bacteria decompose the wastes. The clean water flows out through drains in the bottom of the bed. The sludge accumulating at the surface must be removed from the bed periodically (*See* filtration for more related terms) (EPA-83/06a).

Sand holes: Small fractures in the surface of glass, produced by the rough grinding operation, that have not been removed by subsequent fine grinding (EPA-83).

Sanding sealers: Any coatings formulated for and applied to bare wood for sanding and to seal the wood for subsequent application of varnish. To be considered a sanding sealer a coating must be clearly labelled as such (40 CFR 52.741-91).

Sandy loam: A soft, easily worked soil containing 0 to 20% clay, 0 to 50% silt, and 43 to 85% sand, according to the U.S. Department of Agriculture classification code (SW-108ts).

Sanitary landfill:
(1) A facility for the disposal of solid waste which meets the criteria published under section 4004 (RCRA1004, *See also* 40 CFR 165.1; 240.101; 257.2-91).
(2) Land disposal sites for non-hazardous solid wastes at which the waste is spread in layers, compacted to the smallest practical volume, and cover material applied at the end of each operating day (EPA-89/12).
(3) An engineered method of disposing of solid waste on land in a manner that protects the environment, by spreading the waste in thin layers, compacting it to the smallest practical volume, and covering it with soil by the end of each working day (EPA-83).
(4) *See* landfill for more related terms.
Methods of sanitary landfill include:
(1) Area method
(2) Canyon technique
(3) Monofill
(4) Quarry method
(5) Ramp method
(6) Trench technique
(7) Wet or low-lying area technique
(8) *See* landfill (EPA-94/04)

Sanitary landfill compactor: A vehicle equipped with a blade and with rubber tires sheathed in steel or hollow steel cores. Both types of wheels are equipped with load concentrations to provide compaction and a crushing effect (*See* compactor for more related terms) (SW-108ts).

Sanitary landfill liner: An impermeable barrier, manufactured, constructed, or existing in a natural condition, that is utilized to collect leachate (*See* liner for more related terms) (OME-88/12).

Sanitary sewer:
(1) A sewer intended to carry only sanitary or sanitary and industrial waste waters from residences, commercial buildings, industrial plants, and institutions (not storm water) (40 CFR 35.905-91, *See also* 40 CFR 35.2005-91; EPA-82/11e).
(2) Underground pipes that carry off only domestic or industrial waste, not storm water (EPA-94/04).
(3) *See* sewer for more related terms.

Sanitary survey:
(1) An onsite review of the water source, facilities, equipment, operation and maintenance of a public water system for the purpose of evaluating the adequacy of such source, facilities, equipment, operation and maintenance for producing and distributing safe drinking water (40 CFR 141.2, *See also* 40 CFR 142.2-91).
(2) An on-site review of the water sources, facilities, equipment, operation and maintenance of a public water system to evaluate the adequacy of those elements for producing and distributing safe drinking water (EPA-94/04).

Sanitary waste: The wastewater generated by non-industrial processes; e.g., showers, toilets, food preparation operations (*See* waste for more related terms) (EPA-82/11e).

Sanitary wastewater: The wastewater discharging from sanitary conveniences such as toilets, showers, and sinks (*See* wastewater for more related terms) (EPA-85/10).

Sanitary water: The supply of water used for sewage transport and the continuation of such effluents to disposal (*See* water for more related terms) (EPA-83/03).

Sanitary water (also known as gray water): Water discharged from sinks, showers, kitchens, or other nonindustrial operations, but not from commodes (EPA-94/04).

Sanitation:
(1) The control of all the factors in human being's physical environment that exercise or can exercise a deleterious effect on his physical development, health, and survival (SW-108ts).
(2) Control of physical factors in the human environment that could harm development, health, or survival (EPA-94/04).

Sanitized: A version of a document from which information claimed as trade secret or confidential has been omitted or withheld (cf. unsanitized) (40 CFR 350.1-91).

Saponification: The reaction in which caustic material combines with fats or oils to produce soap (EPA-79/12). The hydrolysis of an ester into its corresponding alcohol and soap (cf. soap boiling) (EPA-74/04c).

Saprobe: An organism that eats dead or decaying organic matter for living.

Saprophytes: Organisms living on dead or decaying organic matter that help natural decomposition of organic matter in water (EPA-94/04).

SAROAD site identification form: One of the several forms in the SAROAD system. It is the form which provides a complete description of the site (and its surroundings) of an ambient air quality monitoring station (40 CFR 58.1-91).

Satellite vehicle: A small collection vehicle that transfers its load into a larger vehicle operating in conjunction with it (*See* vehicle for more related terms) (40 CFR 243.101-91).

Saturate: To fill all the voids in a material with fluid; to form the most concentrated solution possible under a given set of physical conditions in the presence of an excess of substance (EPA-83).

Saturated air:
(1) Moist air in which the partial pressure of water vapor equals the vapor pressure of water at the existing temperature. This occurs when dry air and saturated water vapor coexist at the same dry-bulb temperature.
(2) Air which contains the maximum amount of water vapor that it can hold at its temperature and pressure.
(3) *See* air for more related terms.

Saturated calomel electrode: *See* synonym, calomel electrode.

Saturated gas: A mixture of gas and vapor to which no additional vapor can be added, at specified conditions (*See* gas for more related terms) (EPA-89/03b).

Saturated flow: The flow of water through a porous material under saturated conditions (*See* flow for more related terms) (EPA-83).

Saturated hydrocarbon: Those compounds whose carbon atoms are each joined by a single bond (*See* hydrocarbon family for more related terms).

Saturated liquid: The liquid which is present under the saturation curve (*See* liquid for more related terms).

Saturated liquid state: Any state represented by a point on the liquid saturation line (*See* liquid for more related terms) (Wark-p56).

Saturated mixture: A gas-vapor mixture is saturated, when a reduction in temperature would cause part of the vapor to condense.

Saturated phase: Any phase of a pure substance under the saturation state or saturation conditions (Jones-p111).

Saturated pressure: *See* synonym, saturation pressure.

Saturated solution: A solution in which the dissolved solute is in equilibrium with an excess of undissolved solute; or a solution in equilibrium such that at a fixed temperature and pressure, the concentration of the solution is at its maximum value and will not change even in the presence of an excess of solute (40 CFR 796.1840-91, *See also* 40 CFR 796.1860-91).

Saturated temperature: *See* synonym, saturation temperature.

Saturated steam: The steam at the temperature and pressure at which the liquid and vapor phase can exist in equilibrium (*See* steam for more related terms) (EPA-82/11f).

Saturated vapor: The vapor which is present under the saturation curve (*See* vapor for more related terms).

Saturated vapor state: Any state represented by a point on the vapor saturation line (*See* vapor for more related terms) (Wark-p56).

Saturated water: The water at its boiling point (*See* water for more related terms).

Saturated zone:
(1) A subsurface area in which all pores and cracks are filled with water under pressure equal to or greater than that of the atmosphere (EPA-89/12a).
(2) The area below the water table where all open spaces are filled with water (EPA-94/04).

Saturated zone or zone of saturation: That part of the earth's crust in which all voids are filled with water (40 CFR 260.10-91).

Saturation:
(1) In absorption, the maximum amount of pollutant retained by the scrubbing liquid at a given operating condition (EPA-84/09).
(2) In adsorption, the maximum amount of pollutant retained by the adsorbent at a given operating condition (EPA-84/09).
(3) In thermodynamics, the condition for coexistence in stable equilibrium of a vapor and liquid or a vapor and solid phase of the same substance.
(4) The condition of a liquid (water) when it has taken into solution the maximum possible quantity of a given substance at a given temperature and pressure (EPA-94/04).

Saturation curve: The combination of the liquid saturation line and the vapor saturation line.

Saturation conditions: *See* synonym, saturation state.

Saturation pressure (or saturated pressure): The pressure at which two phases of a pure substance can coexist in equilibrium at a given temperature (*See* pressure for more related terms).

Saturation state (or saturation conditions): The conditions under which two or more phases of a pure substance can coexist in equilibrium. At the saturation state, a change of phase may occur without a change of pressure or temperature (Jones-p111; Wark-p56).

Saturation tempreature (or saturated temperature): The temperature at which two phases of a pure substance can coexist in equilibrium at a given pressure (*See* temperature for more related terms).

Saturator: The equipment in which asphalt is applied to felt to make asphalt roofing products. The term saturator includes the saturator, wet looper, and coater (40 CFR 60.471-91).

Saveall: A mechanical device used to recover papermaking fibers and other suspended solids from a wastewater or process stream

(EPA-87/10).

Saw kerf: The wastage of wood immediately adjacent to a saw blade due to the cut-cleaning design of the blade, which enlarges the cut slightly on either side (EPA-74/04).

Sawing: Cutting a workpiece with a band, blade, or circular disc having teeth (40 CFR 471.02-91).

Scalar: A physical quantity which possesses only a magnitude and which therefore can be described by only one numerical value in an arbitrary point of space. An example of a scalar is the concentration (cf. tensor and vector) (NATO-78/10).

Scale: Generally insoluble deposits on equipment and heat transfer surfaces which are created when the solubility of a salt is exceeded. Common scaling agents are calcium carbonate and calcium sulfate (EPA-82/11f).

Scale height: The distance over which the atmospheric pressure decrease with height is by a factor of 1/e.

Scale up: The application of information gathered from a test model to a full-scale prototype facility (EPA-83).

Scalping: Removal of small amounts of oversized material from feed (EPA-88/08a).

Scandium (Sc): A soft metal with a.n. 21; a.w. 44.956; d. 3.0 g/cc; m.p. 1539° C and b.p. 2730° C. The element belongs to group IIIB of the periodic table.

Scarfing: Those steel surface conditioning operations in which flames generated by the combustion of oxygen and fuel are used to remove surface metal imperfections from slabs, billets, or blooms (40 CFR 420.71-91).

Scarification: The process of breaking up the topsoil prior to mining (EPA-82/10).

Scarify: To disturb or break up the natural soil at a borrow area or sanitary landfill (EPA-83).

Scavenger:
(1) One who illegally removes materials at any point in the solid waste management system (EPA-89/11).
(2) One who participates in the uncontrolled removal of materials at any point in the solid waste stream. A term sometimes used in certain sections of the United States to describe an independent solid wastes collector (EPA-83).

Scavenging:
(1) The uncontrolled and unauthorized removal of materials at any point in the solid waste management system (40 CFR 243.101-91, *See also* 40 CFR 241.101-91).
(2) In air, the process in the atmosphere causing the decrease of the concentration of an air pollutant by removal of this pollutant from the atmosphere. Usually applied to the physical process of the capture of air pollutants by falling or floating water droplets (washout, rainout). Chemical processes are usually described by decay or photochemical reactions (NATO-78/10).

scf (standard cubic feet): A volume unit measured under standard pressure and temperature in British unit (cf. normal cubic meter).

scfm (standard cubic feet per minute): A flow rate unit (cubic feet per minute) measured under standard pressure and temperature. Standard conditions in air pollution control applications are usually 60° F and 1 atm. Other standard or reference conditions may be used (cf. acfm) (EPA-84/09).

Schedule and timetable of compliance: A schedule of required measures including an enforceable sequence of actions or operations leading to compliance with an emission limitation, other limitation, prohibition, or standard (CAA302-42USC7602-91).

Schedule of compliance: A schedule of remedial measures, including an enforceable sequence of actions or operations, leading to compliance with an applicable implementation plan, emission standard, emission limitation, or emission prohibition (CAA501, *See also* CWA502; 122.2; 124.2; 144.3; 270.2-91).

Schedule of reinforcement: Specifies the relation between behavioral responses and the delivery of reinforcers, such as food or water (40 CFR 798.6500-91).

Scheduled maintenance: Any periodic procedure, necessary to maintain the integrity or reliability of emissions control performance, which can be anticipated and scheduled in advance. In sulfuric acid plants, it includes among other items the screening or replacement of catalyst, the re-tubing of heat exchangers, and the routine repair and cleaning of gas handling/cleaning equipment (40 CFR 57.103-91, *See also* 40 CFR 86.082.2; 86.084.2; 86.402.78-91).

Schwann cell: One of the large nucleated masses of protoplasm lining the inner surface of the neurilemma, a membrane wrapping the nerve fiber (LBL-76/07-bio).

Science advisory board (SAB): A group of external scientists who advise EPA on science and policy (EPA-94/04).

Scintillation counter: An instrument used for the location of radioactive ore such as uranium. It uses a transparent crystal which gives off a flash of light when struck by a gamma ray, and a photomultiplier tube which produces an electrical impulse when the light from the crystal strikes it (EPA-82/05).

Scope (for environmental impact statement:) Consists of the range of actions, alternatives, and impacts to be considered in an environmental impact statement. The scope of an individual statement may depend on its relationship to other statements (40 CFR 1502.20 and 1508.28). To determine the scope of environmental impact statements, agencies shall consider 3 types of actions, 3 of types of alternatives, and 3 types of impacts. They include:
(1) Actions (other than unconnected single actions) which may be:
 (A) Connected actions, which means that they are closely related and therefore should be discussed in the same impact statement. Actions are connected if they:

(i) Automatically trigger other actions which may require environmental impact statements.
(ii) Cannot or will not proceed unless other actions are taken previously or simultaneously.
(iii) Are interdependent parts of a larger action and depend on the larger action for their justification.
(B) Cumulative actions, which when viewed with other proposed actions have cumulatively significant impacts and should therefore be discussed in the same impact statement.
(C) Similar actions, which when viewed with other reasonably foreSeeable or proposed agency actions, have similarities that provide a basis for evaluating their environmental consequences together, such as common timing or geography. An agency may wish to analyze these actions in the same impact statement. It should do so when the best way to assess adequately the combined impacts of similar actions or reasonable alternatives to such actions is to treat them in a single impact statement.
(2) Alternatives, which include:
(A) No action alternative.
(B) Other reasonable courses of actions.
(C) Mitigation measures (not in the proposed action).
(3) Impacts, which my be:
(A) Direct;
(B) Indirect;
(C) Cumulative (40 CFR 1508.25-91).

Scope: In a document prepared pursuant to the National Environmental Policy Act, the range of actions, alternatives, and impacts to be considered (DOE-91/04).

Scope of work: A document similar in content to the program of requirements but substantially abbreviated. It is usually prepared for small-scale projects (40 CFR 6.901-91).

Scoping: The initial phase of site remediation during which possible site actions and investigative activities are identified (EPA-89/12a).

Scouring: The removal of earth of rock by the action of running water or of a glacier; in wool manufacture, the removal of foreign matter from wool by propelling it through a series of bowls and squeeze rolls by means of reciprocating arms; scouring wastes are the strongest polluting materials in the whole textile industry and the major factor to be considered in dealing with the waste problem at an integrated woolen mill (DOI-70/04).

Scrap: Materials discarded from manufacturing operations that may be suitable for reprocessing (EPA-94/04). Other scrap related terms include:
- Home scrap
- Obsolete scrap
- Old scrap
- Municipal aluminum scrap
- Municipal ferrous scrap
- Prompt industrial scrap
- Revert scrap (*See* synonym, home scrap)

Scrap metal: Bits and pieces of metal parts, e.g., bars, turnings, rods, sheets, wire) or metal pieces that may be combined together with bolts or soldering (e.g., radiators, scrap automobiles, railroad box cars), which when worn or superfluous can be recycled (*See* metal for more related terms) (40 CFR 261.1-91).

Scratch brush wheels: All power-driven rotatable wheels made from wire or bristles, and used for scratch cleaning and brushing purposes (29 CFR 1910.94b-91).

Screen: Screen means a device for separating material according to size by passing undersize material through one or more mesh surfaces (screens) in series and retaining oversize material on the mesh surfaces (screens) (40 CFR 60.381-91).
Other screen related terms include:
- Bar screen
- Revolving screen
- Rotary screen
- Shaking screen
- Side hill screen
- Vibrating screen

Screen chamber: A compartment of the intake of a pump and screen structure in which the screens are located (EPA-76/04).

Screening:
(1) Samples taken often untreated wastewater only to determine the absence or presence of toxic pollutants (EPA-79/12b).
(2) Use of screens to remove coarse floating and suspended solids from sewage (EPA-94/04).

Screening concentration: The media-specific benchmark concentration for a hazardous substance that is used in the HRS for comparison with the concentration of that hazardous substance in a sample from that media. The screening concentration for a specific hazardous substance corresponds to its reference dose of inhalation exposures or for oral exposures, as appropriate, and, if the substance is a human carcinogen with a weight-of-evidence classification of A, B, or C, to that concentration that corresponds to its 10^{-6} individual lifetime excess cancer risk for inhalation exposures or for oral exposures, as appropriate (40 CFR 300-AA-91).

Screening operation: A device for separating material according to size by passing undersize material through one or more mesh surfaces (screens) in series, and retaining oversize material on the mesh surfaces (screens) (40 CFR 60.671-91).

Screw conveyer: A rotating helical shaft that moves materials, such as incinerator siftings, along a trough or tube (*See* conveyer for more related terms) (SW-108ts).

Screw press: A device used to recover spent liquor from cooked wood chips (EPA-87/10).

Scrim designation: The weight and number of yarns of fabric reinforcement per inch of length and width, e.g, a 10 X 10 scrim has 10 yarns per inch in both the machine and cross machine directions (EPA-91/05).

Scrim (or fabric) reinforcement: The fabric reinforcement layer used with some geomembranes for the purpose of increased strength and dimensional stability (EPA-91/05).

Scrubber: An air pollution device that uses a spray of water or reactant or a dry process to trap pollutants in emissions (EPA-94/04; 84/09; 89/11). Types of scrubbers include:
(1) Dry scrubber (EPA-89/02; 89/03b)
  (A) Spray dryer
  (B) Spray dryer absorber (EPA-89/02)
  (C) Dry injection adsorption system (EPA-89/02)
  (D) Spray dryer and dry injection systems combination (EPA-89/02)
(2) Wet scrubber (EPA-84/03b, p1-4)
  (A) Cyclone type scrubber (AP-40)
  (B) Fume scrubber (EPA-74/12a)
  (C) Ionizing wet scrubber (CRWI-89/05)
  (D) Mechanical scrubber (AP-40; Calvert-84)
  (E) Orifice type scrubber (AP-40)
  (F) Packed tower (packed absorber, packed column or packed-bed scrubber) (EPA-81/09, 89/03b; AP-40)
  (G) Plate scrubber (Calvert-84)
  (H) Spray chamber (spray tower) (AP-40)
  (i) Venturi scrubber (EPA-89/03b, AP-40; Hesketh-79)
  (J) Wet filter (AP-40)

Scrubber liquor: An untreated wastewater stream produced by wet scrubbers cleaning gases produced by metal manufacturing operations (*See* liquor for more related terms) (EPA-85/10a).

Scrubbing (or washing): The washing of impurities from any process gas stream (40 CFR 165.1-91).

Scum: A surface deposit sometimes formed on building bricks (cf. skimming) (EPA-83).

Scum chamber: A chamber, a tank or a trough for temporarily holding scum from wastewater treatment.

Scum collector: A device for collecting scum from sedimentation tanks.

Scum baffle: A baffle to prevent scum flowing out with the stream of effluent.

Sea breeze: The breeze due to a local circulation near a shoreline, which blows from the water to the land surface. This local circulation is caused by temperature difference between the water (cold) and the land surface (warm). The sea breeze usually blows during the day time and alternates with the land breeze. Also *See* land breeze for comparison (*See* breeze for more related terms) (NATO-78/10).

Sea level: The level of the surface of the sea between high and low tide, used as a standard in measuring heights and depths (DOI-70/04).

Sea water: *See* synonym, saline water.

Sea water intrusion: *See* synonym, salt water intrusion.

Seacoal: Finely ground bituminous coal used as an ingredient in molding sands to control the thermal expansion of the mold, and to control the composition of the mold cavity gas during pouring (EPA-85/10a).

Sealant: A viscous chemical used to seal the exposed edges of scrim reinforced geomembranes. (Manufacturers and installers should be consulted for the various types of sealant used with specific geomembranes) (EPA-91/05).

Sealant tape: An asbestos-containing product which is initially a semi-liquid mixture of butyl rubber and asbestos, but which solidifies when exposed to air, and which is intended for use as a sealing agent. Major applications of this product include: sealants for buildings and automotive windows, sealants for aerospace equipment components, and sealants for insulated glass (40 CFR 763.163-91).

Sealed cell: A battery cell which can operate in a sealed condition during both charge and discharge (EPA-84/08).

Sealed source: Any byproduct material that is encased in a capsule designed to prevent leakage or escape of the byproduct material (10 CFR 30.4-91, *See also* 40 CFR 52.741-91).

Seam: Mark on glass surface resulting from joint of matching mold parts; - to slightly grind the sharp edges of a piece of glass (EPA-83). Other seam related terms include:
- Extrusion seam
- Factory seam
- Field seam

Seaming board: Smooth wooden planks placed beneath the area to be seamed to provide a uniform resistance to applied roller pressure in the fabrication of seams (EPA-91/05).

Seasoning: Exposing paper to uniform atmospheric conditions to equalize moisture content (EPA-83).

Secator: A separating device that throws mixed materials onto a rotating shaft; heavy and resilient materials bounce off one side of the shaft, while light and inelastic materials land on the other and are cast in the opposite direction (*See* separator for more related terms) (SW-108ts).

Second derivative absorption analyzer: One of continuous emission monitors (*See* continuous emission monitor for various types). This analyzer provides a very specific measurement of compounds with narrow band spectral absorption. Light with a modulated wavelength, is projected through the same gas, or measurement cavity. If the gas to be detected is present, the intensity of the detected light varies at twice the modulation frequency. The resulting signal is processed to determine the second derivatives (d2) value which is directly proportional to the concentration of the gas. Since the technique is sensitive to the curvature of the spectra and not the intensity of the spectra, it provides enhanced sensitivity and permits measurement of a single gas in complex mixtures (EPA-84/03a).

Second order reaction: The rate of a chemical reaction is determined by the concentration of two reactants (*See* chemical reaction for more related terms).

Second side: The final side of a plate glass to be ground and polished (EPA-83).

Secondary air: *See* synonym, secondary combustion air.

Secondary air fan: A fan to provide the principle air used to provide turbulence/mixing so as to assure completing combustion (*See* fan for more related terms) (EPA-83).

Secondary air supply: An air supply that introduces air to the wood heater such that the burn rate is not altered by more than 25 percent when the secondary air supply is adjusted during the test run. The wood heater manufacturer can document this through design drawings that show the secondary air is introduced only into a mixing chamber or secondary chamber outside the firebox (40 CFR 60-App/A(method 28 & 28A)-91).

Secondary alcohol: An alcohol whose OH connected carbon joins with one hydrogen atom. Its molecular structure can be expressed as R_1R_2CHOH, where R_1 and R_2 can be identical or different groups, e.g., propan-2-ol $((CH_3)_2CHOH)$ (*See* alcohol for more related terms).

Secondary amine: An amine whose molecular structure can be expressed as R_1R_2NH, where R_1 and R_2 can be identical or different groups, e.g., dimethylamine $((CH_3)_2NH)$ (*See* amine for more related terms).

Secondary battery (or secondary cell): An electrochemical cell or battery system that can be recharged (*See* battery for more related terms) (EPA-84/08).

Secondary burner (afterburner, catalytic afterburner, fume incinerator, secondary combustion chamber burner, thermal oxidizer or vapor incinerator):
(1) An air pollution control device that incinerates undesirable organic gases (combustible materials) from the primary combustion chamber such as rotary kilns. The combustible materials may be gases, vapors, or entrained particulate matter.
(2) A burner installed at the secondary combustion chamber of an incinerator to maintain a minimum temperature and to complete the combustion of incompletely burned gas (SW-108ts).
(3) *See* burner for more related terms.

Secondary cell: *See* synonym, secondary battery.

Secondary clarifier (secondary settling tank or secondary sedimentation tank): In a waste-treatment plant, a basin or tank that receives liquid from a prior treatment process such as trickling filter or an activated sludge tank; here settleable solids are removed by sedimentation (*See* clarifier for more related terms) (EPA-83/03, *See also* DOI-70/04).

Secondary chamber: *See* synonym, secondary combustion chamber.

Secondary combustion air (or secondary air):
(1) Air for combustion supplied to combustion devices to supplement the primary air.
(2) The air introduced above or below the fuel (waste) bed by a natural, induced or forced draft. It is generally referred to as overfire air if supplied above the fuel bed through the side walls or the bridge wall of the primary chamber (SW-108ts).
(3) *See* combustion air for more related terms.

Secondary combustion chamber (or secondary chamber):
(1) The chamber where a secondary burner is installed and the incompletely burned gas from the primary combustion chamber is re-burned. It operates with an excess air (EPA-89/03b).
(2) Chamber where unburned combustible gases and particulate from the primary chamber are burned to completion (EPA-83).
(3) *See* combustion chamber for more related terms.

Secondary combustion chamber air blower: A forced air blower for providing combustion air to the secondary combustion chamber (*See* blower for more related terms) (EPA-89/03b).

Secondary combustion chamber burner: *See* synonym, secondary burner.

Secondary combustion chamber air port: A port through which combustion air enters the combustion chamber and causes mixing (*See* combustion chamber for more related terms) (EPA-89/03b).

Secondary contact recreation: The activities where a person's water contact would be limited to the extent that bacterial infections of eyes, ears, respiratory, or digestive systems or urogenital areas would normally be avoided (such as wading or fishing) (40 CFR 131.35.d-13-91).

Secondary containment: The process equipment specifically designed to contain material that has breached primary containment before the material is released to the environment and becomes an accidental release. A vent duct and scrubber that are attached to the outlet of a pressure relief device are examples of secondary containment (*See* containment for more related terms) (EPA-87/07a).

Secondary drinking water regulations:
(1) A regulation which applies to public water systems and which specifies the maximum contaminant level which, in the judgment of the Administrator, are requisite to protect the public welfare. Such regulations may apply to any contaminant in drinking water: (a) which may adversely affect the odor or appearance of such water and consequently may cause a substantial number of the persons served by the public water system providing such water to discontinue its use; or (b) which may otherwise adversely affect the public welfare. Such regulations may vary according to geographic and other circumstances (*See* the term of Act or SDWA for more related terms) (SDWA1401-42USC300f-91).
(2) Non-enforceable regulations applying to public water systems and specifying the maximum contamination levels that, in the judgment of EPA, are required to protect the public welfare. These regulations apply to any contaminants that may adversely affect the odor or appearance of such water and consequently may cause people served by the system to discontinue its use (EPA-94/04).

Secondary effluent: The effluent from the secondary clarifiers (*See* effluent for more related terms).

Secondary emission: The emissions which occur as a result of the construction or operation of an existing stationary facility but do not come from the existing stationary facility. Secondary emissions may include, but are not limited to, emissions from ships or trains coming to or from the existing stationary facility (*See* emission for more related terms) (40 CFR 51.301-91, *See also* 40 CFR 51.165; 51.166; 52.21; 52.24; 51.165; 51-App/S; 60.141a; 61.171; 61.181-91).

Secondary emission control system: The combination of equipment used for the capture and collection of secondary emissions, e.g.,:
(1) An open hood system for the capture and collection of primary and secondary emissions from the BOPF, with local hooding ducted to a secondary emission collection device such as a baghouse for the capture and collection of emissions from the hot metal transfer and skimming station; or
(2) An open hood system for the capture and collection of primary and secondary emissions from the furnace, plus a furnace enclosure with local hooding ducted to a secondary emission collection device, such as a baghouse, for additional capture and collection of secondary emissions from the furnace, with local hooding ducted to a secondary emission collection device, such as a baghouse, for the capture and collection of emissions from hot metal transfer and skimming station; or
(3) A furnace enclosure with local hooding ducted to a secondary emission collection device such as a baghouse for the capture and collection of secondary emissions from a BOPF controlled by a closed hood primary emission control system, with local hooding ducted to a secondary emission collection device, such as a baghouse, for the capture and collection of emissions from hot metal transfer and skimming stations (40 CFR 60.141a-91).

Secondary expansion: *See* synonym, permanent expansion.

Secondary hood system: The equipment (including hoods, ducts, fans, and dampers) used to capture and transport secondary inorganic arsenic emissions (40 CFR 61.171-91).

Secondary impact (or indirect impact): The effect of a project arising through induced changes in population, economic growth and land use, and the environmental effects resulting from these changes (EPA-80/08).

Secondary industry category: Any industry category which is not a primary industry category (cf. primary industry category) (40 CFR 122.2-91).

Secondary lead smelter: Any facility producing lead from a leadbearing scrap material by smelting to the metallic form (*See* smelter for more related terms) (40 CFR 60.121-91).

Secondary manufacturing residues: Sawdust, planer shavings, and the like created by converting lumber, plywood, and veneer into manufactured products such as furniture, pallets, flooring, and moldings (EPA-83).

Secondary material:
(1) A material recovered from a waste stream for reprocessing or remanufacturing (OTA-89/10).
(2) A material that is utilized in place of a primary, virgin or or material in manufacturing a product. Materials which might go to waste if not collected and processed for reuse (EPA-83).
(3) Materials that have been manufactured and used at least once and are to be used again (EPA-94/04).
(4) *See* material for more related terms.

Secondary maximum contaminant levels (SMCLs): SMCLs which apply to public water systems and which, in the judgement of the Administrator, are requisite to protect the public welfare. The SMCL means the maximum permissible level of a contaminant in water which is delivered to the free flowing outlet of the ultimate user of public water system. Contaminants added to the water under circumstances controlled by the user, except those resulting from corrosion of piping and plumbing caused by water quality, are excluded from this definition (*See* the term of Act or SDWA for more related terms) (40 CFR 143.2-91). *See* 40 CFR 143.3 for standards.

Secondary metabolites: Products of metabolism that are not associated with propagation of the organism (EPA-88/09a).

Secondary pollutant:
(1) Pollutants produced by the primary pollutants which are directly emitted from a source. Examples of secondary pollutants include ozone and organic nitrates which are produced from primary pollutants such as nitrogen oxides, hydrocarbons by photochemical reactions.
(2) A pollutant formed in the atmosphere by chemical changes taking place between primary pollutants and other substances present in the air.
(3) *See* pollutant for more related terms.

Secondary processor of asbestos: A person who processes for commercial purposes an asbestos mixture (*See* asbestos for more related terms) (40 CFR 763.63-91).

Secondary recycling: The use of a secondary material in an industrial application other than that in which the material originated. An example is the reprocessing of newspapers and old corrugated boxes into container board for packaging or into construction paper (EPA-83).

Secondary sedimentation tank: *See* synonym, secondary clarifier.

Secondary settling: The secondary settling unit for the removal of settleable solids from the prior treatment processes (*See* settling for more related terms) (EPA-83/06a).

Secondary settling tank: *See* synonym, secondary clarifier.

Secondary significance pollutant: The pollutants that are of secondary significance if they are not recommended for regulation, but are specified to be considered on a case-by-case basis for potential deleterious effects on humans and the environment (*See* pollutant for more related terms) (EPA-85/10).

Secondary sludge: The sludge resulting from the secondary clarifiers (*See* sludge for more related terms).

Secondary standard:
(1) A national secondary ambient air quality standard promulgated pursuant to section 109 of the Act (40 CFR 51.100-91).
(2) *See* standard for more related terms.

Secondary standard: For calibration application:
(1) A material having a property that is calibrated against a primary standard (LBL-76/07-air).
(2) Those calibrated against primary or intermediate standards under known conditions of gas type, temperature, and pressure. Accuracies less than 5% are achievable (*See* calibration of air flow for more related terms) (EPA-83/06).
(3) National ambient air quality standards designed to protect welfare, including effects on soils, water, crops, vegetation, manmade materials, animals, wildlife, weather, visibility, and climate, damage to property, transportation hazards, effects on economic values, and on personal comfort and well-being (EPA-94/04).

Secondary treatment:
(2) The second step in most publicly owned waste treatment systems in which bacteria consume the organic parts of the waste. It is accomplished by bringing together waste, bacteria, and oxygen in trickling filters or in the activated sludge process. This treatment removes floating and settleable solids and about 90 percent of the oxygen-demanding substances and suspended solids. Disinfection is the final stage of secondary treatment (*See* primary, tertiary treatment) (EPA-94/04).
(2) *See* synonym, secondary wastewater treatment.

Secondary use: The use of a material in an application other than that in which it originated; however, the material is not changed significantly by processing and retains its identity. Examples are cotton clothing articles that are converted into wiping rags by being washed and cut to size; the use of steel cans tn copper precipitation; and the use of rubber tires as dock bumpers. Materials used in this mode end up as waste after their secondary use is complete (EPA-83).

Secondary treatment: *See* synonym, secondary wastewater treatment.

Secondary wastewater treatment (or secondary treatment):
(1) The second step in most publicly owned waste treatment systems in which bacteria consume the organic parts of the waste. It is accomplished by bringing together waste, bacteria, and oxygen in trickling filters or in the activated sludge process. This treatment removes floating and settleable solids and about 90 percent of the oxygen demanding substances and suspended solids. Disinfection is the final stage of secondary treatment (EPA-89/12).
(2) Wastewater treatment using biological methods (bacterial action) in addition to primary treatment by screening, sedimentation, and flotation. In secondary treatment, bacteria are used to destroy organic wastes as the water trickles over coarse stones. The process removes up to 90 percent of the dissolved pollutants, but leaves many other pollutants untouched. A secondary waste-treatment plant may consist of the following units, in addition to those of the primary treatment plant: trickling filter; aeration or activated sludge; secondary clarifier, secondary settling tank, final settling tank, and final settling basin (DOI-70/04).
(3) *See* treatment or wastewater treatment for more related terms.

Secondary winding: The winding on the load (i.e., output) side of a transformer (EPA-83/03).

Secret: *See* definition under security classification category.

Section 5 notice: Any PMN, consolidated PMN, intermediate PMN, significant new use notice, exemption notice, or exemption application (40 CFR 700.43-91).

Section 13: Refers to section 13 of the Federal Water Pollution Control Act Amendments of 12 (40 CFR 7.25-91).

Section 304(a) criteria: Are developed by EPA under authority of section 304(a) of the Act based on the latest scientific information on the relationship that the effect of a constituent concentration has on particular aquatic species and/or human health. This information is issued periodically to the States as guidance for use in developing criteria (40 CFR 131.3-91).

Section 404 program or State 404 program or 404: An approved State program to regulate the discharge of dredged material and the discharge of fill material under section 404 of the Clean Water Act in State regulated waters (40 CFR 124.2-91).

Section 504: Section 504 of the Rehabilitation Act of 1973 (Pub. L. 93-112, 87 Stat. 394 (29USC794)), as amended by the Rehabilitation Act Amendments of 1974 (Pub. L. 93-516, Stat. 1617); and the Rehabilitation, Comprehensive Services, and Developmental Disabilities Amendments of 1978 (Pub. L. 95-602, 92 Stat. 2955); and the Rehabilitation Act Amendments of 1986 (Pub. L. 99-506, 100 Stat. 1810). As used in this part, section 504 applies only to programs or activities conducted by Executive agencies and not to federally assisted programs (40 CFR 12.103-91).

Section mill: Those steel hot forming operations that produce a variety of finished and semi-finished steel products other than the products of those mills specified below in paragraphs (d), (e), (g) and (h) of this section (40 CFR 420.71-91).

Sectionally supported furnace wall: A furnace or boiler wall which consists of special refractory blocks or shapes that are mounted on and supported at intervals of height by metallic hangers (*See* furnace wall for more related terms) (SW-108ts).

Sector averaging: Over a period of time, all wind direction observations can be placed in a wind direction sector and the number of occurrences within each sector tabulated. For averaging periods greater than 8 hours, it is often assumed that wind direction is uniformly distributed within each sector. It is assumed that there is a uniform distribution of wind direction within each sector. This results in a uniform horizontal distribution of the plume within each sector, rather than assuming a Gaussian distribution along the horizontal axis (EPA-88/09).

Secure chemical landfill: *See* definition under landfill (EPA-94/04).

Secure landfills: Disposal sites for hazardous waste. They are selected and designed to minimize the chance of release of hazardous substances into the environment (EPA-89/12).

Secure maximum contaminant level: Maximum permissible level of a contaminant in water delivered to the free flowing outlet of the ultimate user, or of contamination resulting from corrosion of piping and plumbing caused by water quality (EPA-94/04).

Security classification assignment: The prescription of a specific security classification for a particular area or item of information. The information involved constitutes the sole basis for determining the degree of classification assigned (40 CFR 11.4-e-91).

Security classification category: The specific degree of classification (Top Secret, Secret or Confidential) assigned to classified information to indicate the degree of protection required:
(1) **Top Secret**. Top Secret refers to national security information or material which requires the highest degree of protection. The test for assigning Top Secret classification shall be whether its unauthorized disclosure could reasonably be expected to cause exceptionally grave damage to the national security. Examples of "exceptionally grave damage" include armed hostilities against the United States or its allies; disruption of foreign relations vitally affecting the national security; the compromise of vital national defense plans or complex cryptologic and communications intelligence systems; the revelation of sensitive intelligence operations; and the disclosure of scientific or technological developments vital to national security. This classification shall be used with the utmost restraint.
(2) **Secret**. Secret refers to that national security information or material which requires a substantial degree of protection. The test for assigning Secret classification shall be whether its unauthorized disclosure could reasonably be expected to cause serious damage to the national security. Examples of "serious damage" include disruption of foreign relations significantly affecting the national security; significant impairment of a program or policy directly related to the national security; revelation of significant military plans or intelligence operations; and compromise of scientific or technological developments relating to national security. The classification Secret shall be sparingly used.
(3) **Confidential**. Confidential refers to that national security information or material which requires protection. The test for assigning Confidential classification shall be whether its unauthorized disclosure could reasonably be expected to cause damage to the national security (40 CFR 11.4-f-91).

Sediment:
(1) The unconsolidated inorganic and organic material that is suspended in and being transported by surface water, or has settled out and has deposited into beds (40 CFR 796.2750-91).
(2) Soil, sand, and minerals washed from land into water, usually after rain. They pile up in reservoirs, rivers and harbors, destroying fish and wildlife habitat, and clouding the water so that sunlight cannot reach aquatic plants. Careless farming, mining, and building activities will expose sediment materials, allowing them to wash off the land after rainfall (EPA-94/04).

Sediment yield: The quantity of sediment arriving at a specific location (EPA-94/04).

Sedimentation:
(1) Letting solids settle out of wastewater by gravity during wastewater treatment (40 CFR 141.2-91). In wastewater treatment, sedimentation is a purely physically process where particles suspended in a liquid are made to settle by means of gravitational and inertial forces acting on both the particles suspended in the liquid and the liquid itself. Practically every industry that discharges a process wastewater stream contaminated with suspended and or precipitable pollutants employs some forms of precipitation, flocculation and/or sedimentation. Examples are: removal of heavy metals from iron steel industry wastewater and removal of fluoride from aluminum production waste water.
(2) The setting out of particles in the atmosphere due to their gravitational fall (NATO-78/10).
(3) Letting solids settle out of wastewater by gravity during treatment (EPA-94/04).

Other sedimentation related terms include:
- Final sedimentation
- Plain sedimentation

Sedimendation basin: *See* synonym, settling pond.

Sedimentation tanks:
(1) Wastewater tanks in which floating wastes are skimmed off and settled solids are removed for disposal (EPA-94/04).
(2) *See* synonym, settling pond.

Seed:
(1) Introduction of microorganisms into a culture medium (EPA-87/10a).
(2) An extremely small gaseous inclusion in glass (EPA-83).

Seed protectant: A chemical applied before planting to protect Seeds and Seedlings from disease or insects (EPA-94/04).

Seepage:
(1) The movement of water or gas through soil without forming definite channels (cf. percolation) (SW-108ts).
(2) Percolation of water through the soil from unlined canals, ditches, laterals, watercourses, or water storage facilities (EPA-94/04).

Segregated stormwater sewer system: A drain and collection system designed and operated for the sole purpose of collecting rainfall runoff at a facility, and which is segregated from all other individual drain systems (40 CFR 61.341-91).

Segregated wastewater stream: A wastewater stream generated from part or all of one pesticide process (EPA-85/10).

Seismic: Pertaining to any earth vibration, especially an earthquake (DOE-91/04).

Selective adsorbent: The adsorbents capable of selectively adsorbing (or rejecting) specific components from a multi-component gas or liquid mixture (*See* adsorbent for more related terms).

Selective catalytic reduction (SCR): One of NO_x emission reduction techniques (*See* nitrogen oxide emission control for control structure). SCR is a dry process and has been used extensively in Japan to achieve a 90% reduction in NO_x emissions. This process is based on the preferential reaction of NH_3 with NO_x rather than with SO_2 in the flue gas. The reactions are expressed as:

$$4NH_3 + 4NO + O_2 \longrightarrow 4N_2 + 6H_2O, \text{ and} \quad (a)$$
$$4NH_3 + 2NO + O_2 \longrightarrow 3N_2 + 6H_2O \quad (b)$$

Equation (a) represents the predominant reaction occurring since 90% of NO_x emissions in combustion flue gas are in the form of NO.

This process involves injecting NH_3 into the flue gas and passing this mixture through a catalytic reactor. NO_x emissions are reduced to harmless molecular nitrogen (N_2) and water vapor (H_2O). Ammonia is injected on an NH_3:NO mole ratio of 1:1 attaining a 90% NO_x emission reduction with less than 20 ppm NH_3 leaving the reactor.

The optimum temperature for successful NO_x reduction in the catalytic reactor is between 300 and 400° C. The reactor is usually located between the boiler economizer and air preheater. A bypass around the economizer is used when temperatures begin to fall below 300° C. The reactor can be located before or after the baghouse or electrostatic precipitator used to collect particulate matter.

A number of materials have been used for catalysts. Initially, a platinum metal on an alumina (Al_2O_3) support was used for NO_x control on gas fired boilers. Sulfur oxides, particularly SO_3 and SO_2 poison the alumina. Other catalysts must be used which resist SO_x deterioration when burning fuel containing sulfur. Titanium dioxide (TiO_2) and vanadium (V_2O_5) catalysts are resistant to SO_x attack. Therefore, most catalysts on oil and coal fired boilers contain TiO_2 or V_2O_5. Other active metals including C, CO, Cr, Fe, Mn, Mo, Ni and W have been used as catalysts. The oxides and sulfates of these metals have also been used as catalysts. The exact compositions and constituents of most catalysts are proprietary information.

The biggest problem with the SCR process is the formation of solid ammonium sulfates $(NH_4)_2SO_4$ and liquid ammonium bisulfate (NH_4HSO_4). This problem occurs when using the SCR process on high sulfur oil and coal fired boilers. Ammonia reacts with SO_3 to form corrosive compounds of $(NH_4)_2SO_4$ and NH_4HSO_4 that coat the air preheater. Increased soot blowing in the furnace and water washing of the air preheaters helps remove the materials (EPA-81/12, p7-15).

Selective herbicide: A pesticide intended to kill only certain types of plants, especially broad-leafed weeds, and not harm other plants such as farm crops of lawn grasses. The leading herbicide in the United States is 2,4 D. A related but stronger chemical used mostly for brush control on range, pasture, and forest lands and on utility or highway rights-of-way is 2,4,5-T. Uses of the latter chemical have been somewhat restricted because of laboratory evidence that it or a dioxin contaminant in 2,4,5-T can cause birth defects in test animals (*See* herbicide for more related terms) (EPA-74/11).

Selective medium: *See* synonym, differential medium.

Selective pesticide: A chemical designed to affect only certain types of pests, leaving other plants and animals unharmed (*See* pesticide for more related terms) (EPA-94/04).

Selectivity coefficient: A coefficient related to the ion equilibria in an ion exchange system.

Selenium (Se): The total selenium present in the process wastewater stream exiting the wastewater treatment system (40 CFR 415.361-91, *See also* 40 CFR 415.451; 415.631-91). Seleneium is a metalloid element with a.n. 34; a.w. 78.96; d. 4.79 g/cc; m.p. 217° C and b.p. 685° C. The element belongs to group VIA of the periodic table. Major selenium compounds include:
- Selenic acid (H_2SeO_4): A highly poisonous solid.
- Selenide: Selenium salts, e.g., inorganic salt, silver selenide (Ag_2Se) and organic salt, ethyl selenide (($C_2H_5)_2Se$).

Self-contained breathing apparatus (SCBA): SCBA consists of a face piece and regulator mechanism connected to a cylinder of compressed air or oxygen carried by the wearer (Course 165.5).

Self-purification: Purification of river water by itself. When waste is discharged to a river, bacteria and other micro-organisms in the river digest the waste, thus, reducing the concentration of the waste.

Sellite: Sodium sulfite, used in the finishing operation of TNT (EPA-76/03).

Semi-annual: A 6-month period; the first semiannual period concludes on the last day of the last month during the 180 days following initial startup for new sources; and the first semiannual period concludes on the last day of the last full month during the 180 days after June 6, 1984 for existing sources (40 CFR 61.111-91, *See also* 40 CFR 61.131; 61.241-91).

Semi-closed steaming: A method of steam conditioning in which the condensate formed during open steaming is retained in the retort until sufficient condensate accumulates to cover the coils. The remaining steam required is generated as in closed steaming (EPA-74/04).

Semi-conductor: The solid state electrical devices which perform functions such as information processing and display, power handling, and interconversion between light energy and electrical energy (40 CFR 469.12-91, *See also* EPA-83/03).

Semi-confined aquifer: An aquifer partially confined by soil layers of low permeability through which recharge and discharge can still occur (EPA-94/04).

Semi-chemical paperboard: Paperboard made from a furnish containing not less than 75% virgin wood pulp, the predominant portion of which is produced by a semi-chemical process (EPA-83).

Semi-dry process: A cement manufacturing process in which water, typically 10% to 15%, is added to the ground dry feed material and then fed into the kiln system (*See* process for more related terms) (ETI-92).

Semi-grouser: A crawler track shoe with one or more low cleats (EPA-83).

Semi-metal: *See* synonym, metalloid.

Semi-static test: A test without flow of solution, but with occasional batchwise renewal of test solutions after prolonged periods (e.g., 24 hours) (40 CFR 797.1440-91).

Semi-transparent stain: The stains containing dyes or semi-transparent pigments which are formulated to enhance wood grain and change the color of the surface but not to conceal the surface, including, but not limited to, sap stain, toner, non-grain raising stains, pad stain, or spatter stain (40 CFR 52.741-91).

Semi-volatile: *See* organic compound, semi-volatile.

Semi-volatile organic chemical: A chemical compound with the following properties:
(1) Boiling point (degree C): 100-300
(2) Sampling method: Semi-VOST (volatile organic sampling train)
(3) Capture method: XAD-2 resin
(4) Analytical method: GC/MS (gas chromatographic/mass spectroscopy)
(5) *See* organic chemical for more related terms.

Semi-wet: Those steelmaking air cleaning systems that use water for the sole purpose of conditioning the temperature and humidity of furnace gases such that the gases may be cleaned in dry air pollution control systems (cf. wet) (40 CFR 420.41-91).

Senescence: The aging process. Sometimes used to describe lakes or other bodies of water in advanced stages of eutrophication (EPA-94/04).

Senescent lake: *See* synonym, extinction lake.

Sensible heat:
(1) Heat, the addition or removal of which, results in a change of temperature (cf. latent heat) (EPA-84/09).
(2) The heat which is associated with a change in temperature; specific heat exchange of temperature; in contrast to a heat interchange in which a change of state (latent heat) occurs (*See* heat for more related terms).

Sensitized paper (developing paper or photographic paper): The paper treated with light or chemical sensitive agents (*See* paper for more related terms) (EPA-83).

Sensitivity: The slope of the analytical curve, i.e., functional relationship between emission intensity and concentration (40 CFR 136-App/C-91).

Sensitivity analysis:
(1) An analysis that compares changes in a dependent variable resulting from incremental changes in an independent variable (OTA-89/10).
(2) In air pollution modeling, the system examination of changes in the output variables as a mathematical model due to small variations in the input variables or model parameters (NATO-78/10).
(3) An analysis of a specific theoretical model or hypothesis as a function of the parameters which make up the model; e.g., for a kinetic-based sensitivity analysis, the parameters are residence time, temperature, concentration, and reaction atmosphere (EPA-88/12).
(4) *See* analysis for more related terms.

Sensitization: The process in which a substance other than the catalyst is present to facilitate the start of a catalytic reaction (EPA-83/06a).

Sensor:
(1) A device that measures a physical quantity or the change in a physical quantity such as temperature, pressure, flow rate, pH, or liquid level (40 CFR 60.481-91, *See also* 40 CFR 61.241; 264.1031-91).
(2) A device designed to respond to a physical stimulus (as temperature, illumination, and motion) and transmit a resulting signal for interpretation, or measurement, or for operating a control (LBL-76/07-air).

Separate collection: The collecting recyclable materials which have been separated at the point of generation and keeping those materials separate from other collected solid waste in separate compartments of a single collection vehicle or through the use of separate collection vehicles (40 CFR 246.101-91).

Separate sewer: A sewer that carries wastewater but excludes storm and surface waters (*See* sewer for more related terms) (DOI-70/04).

Separating: Centrifuging, draining, evacuating, filtering, percolating, fitting, pressing, skimming, sorting, and trimming (AP-40, p790).

Separation: The systematic division of solid waste into designated categories (SW-108ts). Types of separation include:
(1) Dense media separation (heavy media separation or sink float)
(2) Gravity separation
(3) Mechanical separation
(4) Metal separation

Separation chamber: *See* synonym, settling chamber.

Separation region: In a flow field, a region distinct from the main flow in which the fluid recirculates behind an obstacle (NATO-78/10).

Separator: A device used to separate materials. For example, a porous material, in a battery system, used to keep plates of opposite polarity separated, yet allowing conduction of ions through the electrolyte (EPA-84/08). Types of separators include:
(1) API separator
(2) Ballistic separator
(3) Centrifugal collector (*See* synonym, mechanical separator)

(4) Centrifugation (*See* synonym, mechanical separator)
(5) Cyclone collector (*See* definition under mechanical separator)
(6) Cyclone separator (*See* definition under mechanical separator)
(7) Eddy current separator
(8) Electrostatic separator
(9) Elutriation
(10) Heavy media separator
(11) Impingement separator
(12) Inertial separator (*See* synonym, mechanical separator)
(13) Magnetic separator
(14) Mechanical separator
(15) Multi-cyclone (*See* definition under mechanical separator)
(16) Multi-stage separator
(17) Oil water separator
(18) Osborne separator
(19) Rising current separator
(20) Secator
(21) Separator (for battery application)

Separator (for battery application): The separator is a porous material, in a battery system, used to keep plates of opposite polarity separated, yet allowing conduction of ions through the electrolyte (EPA-84/08).

Separator tank: A device used for separation of two immiscible liquids (*See* tank for more related terms) (40 CFR 264.1031-91).

Septage: The liquid and solid material pumped from a septic tank, cesspool, or similar domestic sewage treatment system, or a holding tank when the system is cleaned or maintained (40 CFR 122.2-91, *See also* 501.2-91; EPA-80/08).

Septic sludge: The sludge derived from septic tanks (*See* sludge for more related terms).

Septic system: An onsite system designed to treat and dispose of domestic sewage. A typical septic system consists of a tank that receives waste from a residence or business and a system of tile lines or a pit for disposal of the liquid effluent (sludge) that remains after decomposition of the solids by bacteria in the tank and must be pumped out periodically (EPA-94/04).

Septic tank:
(1) A water-tight covered receptacle designed to receive or process, through liquid separation or biological digestion, the sewage discharged from a building sewer. The effluent from such receptacle is distributed for disposal through the soil and settled solids and scum from the tank are pumped out periodically and hauled to a treatment facility (40 CFR 280.12-91).
(2) An underground storage tank for wastes from homes not connected to a sewer line. Waste goes directly from the home to the tank. (*See* septic system (EPA-94/04))
(3) *See* tank for more related terms.

Sequela: Any lesion or affection that follows or is caused by an attack of disease (LBL-76/07-bio).

Sequestering agent: A chemical compound which is added to water systems to prevent the formation of scale by holding the insoluble compounds in suspension (EPA-82/11f).

Serial number: The identification number assigned by the manufacturer to a specific production unit (40 CFR 205.151-91).

Series: In mathematics, the sum of a finite or infinite sequence of terms or functions (NATO-78/10).

Series collection: An operation involving the use of two or more collectors joined in a series (EPA-83/06).

Series resistance: The sum of resistances from the condenser plates to the condensers external connections (40 CFR 85.2122(a)(6)(ii)(B)-91).

Serious acute effect: The human injury or human disease processes that have short latency period for development, result from short term exposure to a chemical substance, or are a combination of these factors and which are likely to result in death or severe or prolonged incapacitation (*See* acute effect for more related terms) (40 CFR 721.3-91).

Serious chronic effect: The human injury or human disease processes that have a long latency period for development, result form long term exposure to a chemical substance, or are a combination of these factors and which are likely to result in death or severe or prolonged incapacitation (*See* chronic effect for more related terms) (40 CFR 721.3-91, *See also* 40 CFR 723.50-91).

Serum: A fluid extracted from an animal for the purpose of inoculation to affect the cure of a disease (EPA-83/09).

Service connector: The pipe that carries tap water from a public water main to a building (EPA-94/04).

Service line: A distribution line that transports gas from a common source of supply to (1) a customer meter or the connection to a customer's piping, whichever is farther downstream, or (2) the connection to a customer's piping if there is no customer meter. A customer meter is the meter that measures the transfer of gas from an operator to a consumer (40 CFR 192.3-91).

Service line sample: A one-liter sample of water collected according to federal regulations that has been standing for at least 6 hours in a service pipeline (EPA-94/04).

Service pipe: The pipeline extending from the water main to the building served or to the consumer's system (EPA-94/04).

Service reservoir: *See* synonym, distribution reservoir.

Service water pump: A pump providing water for auxiliary plant heat exchangers and other uses (EPA-82/11f).

Set of safety relief valves: One or more safety relief valves designed to open in order to relieve excessive pressures in the same vessel or pipe (40 CFR 52.741-91).

Setout/set back collection: The removal of full and the return of empty containers between the on-premise storage point and the curb by a collection crew (*See* waste collection for more related terms)

(SW-108ts).

Setting rate: A comparative term referring to the time required for the glass surface to cool between the limits of the working range. A short time implies a fast setting rate, and a long time implies a slow setting rate (EPA-83).

Settle: The act of considering, ascertaining, adjusting, determining or otherwise resolving a claim (40 CFR 14.2-91).

Settle lime: The precipitation of dissolved solids in wastewater using lime and the subsequent gravity induced deposition of the suspended matter (*See* lime for more related terms) (EPA-82/11e).

Settle mark: *See* synonym, chill mark (EPA-83).

Settleability: An indication of sludge to settle in sedimentation tanks.

Settleable solid:
(1) That matter measured by the volumetric method specified in 40 CFR 434.64 (40 CFR 434.11-91).
(2) The particulate material (both organic or inorganic) which will settle in one hour expressed in milliliters per liter (mL/L) as determined using an Imhoff cone and the method described for Residue Settleable in 40 CFR 136 (40 CFR 440.141-91).
(3) Material heavy enough to sink to the bottom of a wastewater treatment tank (EPA-94/04; *See also* EPA-87/10a; 82/11f).
(4) *See* solid for more related terms.

Settlement: A gradual subsidence of material (cf. differential settlement) (EPA-83).

Settling: The concentration of particulate matter in wastewater by allowing suspended solids to sink to the bottom (EPA-80/10). Other settling related terms include:
- Primary settling
- Secondary settling

Settling chamber (baffle chamber, gravity settler, gravity settling tank or separation chamber):
(1) A chamber following the combustion chamber, in which baffles change the direction of and/or reduce the velocity of the combustion gases in order to promote the settling of fly ash or coarse particulate matter (EPA-83).
(2) Any chamber designed to reduce the velocity of the products of combustion and thus to promote the settling of fly ash from the gas stream (SW-108ts).
(3) A series of screens placed in the way of flue gases to slow the stream of air, thus helping gravity to pull particles into a collection device (EPA-94/04).

Settling pond (sedimentation basin, sedimentation tank, or settling tank):
(1) A container that gravimetrically separates oils, grease, and dirt from petroleum solvent, together with the piping and ductwork used in the installation of this device (40 CFR 60.621-91).
(2) The holding area for wastewater where floating wastes are skimmed off and settled solids are removed for disposal (EPA-89/12).

Settling tank:
(1) A holding area for wastewater, where heavier particles sink to the bottom for removal and disposal (EPA-94/04).
(2) *See also* synonym, settling pond.

Settling velocity (or terminal velocity):
(1) The velocity at which a given dust will fall out of dust-laden gas under the influence of gravity only. Also known as terminal velocity (EPA-83). In air pollution, it is the velocity of a particle calculated by equating the external force acting on the particle (gravity, centrifugal, etc.) with the drag force (EPA-84/09).
(2) The terminal rate of fall of a particle through a fluid as induced by gravity or other external force; the rate at which frictional drag balances the accelerating force (or the external force). Also known as terminal velocity (EPA-83).
(3) Terminal velocity of a small sphere falling under the influence of gravity in a viscous fluid is given by (the equation in this subpart) (40 CFR 796.1520-91).
(4) Terminal velocity is the velocity at which the downward pull of gravity is equal to the viscous force exerted by the particulate suspending fluid or gas (EPA-83).

Seven (7)-day average: The arithmetic mean of pollutant parameter values for samples collected in a period of 7 consecutive days (40 CFR 133.101-91).

Severe: A relative term used to describe the degree to which hazardous material releases can cause adverse effects to human health and the environment (Course 165.5).

Severe accident: An accident with a frequency rate of less than 10/yr that would have more severe consequences than a design-basis accident, in terms of damage to the facility, offsite consequences, or both (DOE-91/04).

Severe property damage: The substantial physical damage to property, damage to the treatment facilities which causes them to become inoperable, or substantial and permanent loss of natural resources which can reasonably be expected to occur in the absence of a bypass. Severe property damage does not mean economic loss caused by delays in production (40 CFR 122.41-91, *See also* 40 CFR 403.17-91).

Sewage:
(1) Human body wastes and the wastes from toilets and other receptacles intended to receive or retain body wastes (cf. wastewater) (40 CFR 140.1-91, *See also* CWA312).
(2) The waste and wastewater produced by residential and commercial sources and discharged into sewers (EPA-94/04).
Other sewage related terms include:
- Domestic sewage
- Raw sewage
- Raw sewage sludge
- Storm sewage

Sewage collection system: For the purpose of section 35.925-13, each, and all, of the common lateral sewers, within a publicly owned treatment system, which are primarily installed to receive waste waters directly from facilities which convey waste water from individual structures or from private property, and which include service connection Y fittings designed for connection with those facilities. The facilities which convey waste water from individual structures, from private property to the public lateral sewer, or its equivalent, are specifically excluded from the definition, with the exception of pumping units, and pressurized lines, for individual structures or groups of structures when such units are cost effective and are owned and maintained by the grantee (40 CFR 35.905-91).

Sewage from vessel: The human body wastes and the wastes from toilets and other receptacles intended to receive or retain body wastes that are discharged from vessels and regulated under section 312 of CWA, except that with respect to commercial vessels on the Great Lakes this term includes graywater. For the purposes of this definition, graywater means galley, bath, and shower water (40 CFR 122.2-91).

Sewage lagoon:
(1) A shallow pond, three to five feet deep, where natural biological processes purify wastewater to a degree comparable to that accomplished in a secondary treatment plant. The organic matter is broken down into simple compounds by bacterial action; these decomposition products are utilized by algae in the course of photosynthesis to produce oxygen, as well as additional algal mass; the oxygen then constitutes the supply needed for aerobic bacterial decomposition. The load of total organic matter that a pond can assimilate depends on many factors. Shallow ponds (three feet) are more effective than deeper ones; those exposed to wind movements are more effective than sheltered ones; other critical factors are temperature and available sunlight, which vary with climate. Loads of 10 to 120 pounds BOD per acre per day have been recorded assimilated in such ponds (cf. lagoon) (DOI-70/04).
(2) *See* lagoon (EPA-94/04).

Sewage pollution: The raw or partially-treated domestic waste (*See* pollution for more related terms) (DOI-70/04).

Sewage sludge:
(1) Any solid, semi-solid, or liquid residue removed during the treatment of municipal wastewater or domestic sewage. Sewage sludge includes, but is not limited to, solids removed during primary, secondary, or advanced waste water treatment, scum, septage, portable toilet pumpings, type III marine sanitation device pumpings (33 CFR 159), and sewage sludge products. Sewage sludge does not include grit or screenings, or ash generated during the incineration of sewage sludge (40 CFR 122.2-91, *See also* 40 CFR 501.2-91).
(2) A semi-liquid substance consisting of settled sewage solids combined with varying amounts of water and dissolved materials (EPA-83).
(3) Sludge produced at a Publicly Owned Treatment Works, the disposal of which is regulated under the Clean Water Act (EPA-94/04).
(4) *See* sludge for more related terms.

Sewage sludge use or disposal practice: The collection, storage, treatment, transportation, processing, monitoring, use, or disposal of sewage sludge (40 CFR 122.2-91).

Sewage treatment residues: Coarse screenings, grit and dewatered or air-dried sludge solids from sewage treatment plants and pumpings of cesspool or septic tank sludges which require disposal with putrescible wastes (EPA-83).

Sewage treatment works (or sewage works): The municipal or domestic waste treatment facilities of any type which are publicly owned or regulated to the extent that feasible compliance schedules are determined by the availability of funding provided by Federal, State, or local governments (40 CFR 220.2-91).

Sewage works: *See* synonym, sewage treatment works.

Sewer: A channel or conduit that carries wastewater and storm-water runoff from the source to a treatment plant or receiving stream. "**Sanitary**" sewers carry household, industrial, and commercial waste. "**Storm**" sewers carry runoff from rain or snow. "**Combined**" sewers handle both (EPA-94/04). Other sewer related terms include:
- Collecting sewer
- Collector sewer
- Combined sewer
- Gravity sewer
- Hydraulics of sewer
- Interceptor sewer
- Lateral sewer
- Municipal separate storm sewer
- Pressure sewer
- Sanitary sewer
- Separate sewer
- Storm drain (*See* synonym, storm sewer)
- Storm sewer
- Trunk main
- Trunk sewer

Sewer appurtenance: The principal appurtenances associated with sanitary sewers are manholes, drop manholes, building connections, and junction chambers (M&EI-72).

Sewer head loss: The pressure head loss that normally occurs in sewerage systems usually consists of those caused by friction, velocity, entrance and exit losses, and losses resulting from the geometric configuration of the sewers and their appurtenances.

Sewer line: A lateral, trunk line, branch line, or other enclosed conduit used to convey waste to a downstream waste management unit (40 CFR 61.341-91).

Sewer lye: Waste sodium hydroxide from reclaiming of scrap soap (EPA-74/04c).

Sewer manhole: *See* synonym, manhole.

Sewer outfall: A sewer that carries wastewater to a point of final discharge (cf. major outfall or outfall) (DOI-70/04).

Sewer system: The system of sewers and related facilities for collection, transportation, and pumping of waste water (DOI-70/04).

Sewerage: The entire system of sewage collection, treatment, and disposal (EPA-94/04, *See also* EPA-74/11).

Sex chromosomes: Chromosomes responsible for the sex determination of organisms.

Sex linked genes: Are present on the sex (X or Y) chromosomes. Sex-linked genes in the context of this guideline refer only to those located on the X-chromosome (40 CFR 798.5275-91).

Shaft horsepower (flywheel or belt horsepower): Actual horsepower produced by an engine, after deducting the drag of accessories (EPA-83).

Shaft power: The only the measured shaft power output of turboprop engine (40 CFR 87.1-91).

Shaft work: In thermodynamics, work can be grouped into flow work and shaft work. The work which is not a flow work is a shaft work. Shaft work can easily be measured by a **dynamometer** (Wark-p124).

Shale: A thinly stratified, consolidated, sedimentary clay (soft rock) with well marked cleavage parallel to the bedding (EPA-83).

Shale oil: Oil derived from shale.

Shakeout: The operation of removing castings from the mold. A mechanical unit is used to separate the mold material from the solidified casting (EPA-85/10a).

Shaking screen: A screen with several screen surfaces in a series, usually slightly inclined, with different apertures and a slow linear motion essentially in place of the screen (*See* screen for more related terms) (EPA-8/88a).

Share: When referring to the awarding agency's portion of real property, equipment or supplies, means the percentage as the awarding agency's portion of the acquiring party's total costs under the grant to which the acquisition costs under the grant to which the acquisition cost of the property was charged. Only costs are to be counted-not the value of third party in-kind contributions (40 CFR 31.3-91).

Sharp:
(1) Any object that can penetrate the skin including, but not limited to, needles, scalpels, and broken capillary tubes (29 CFR 1910).
(2) Hypodermic needles, syringes (with or without the attached needle) pasteur pipettes, scalpel blades, blood vials, needles with attached tubing, and culture dishes used in animal or human patient care or treatment, or in medical, research or industrial laboratories. Also included are other types of broken or unbroken glassware that were in contact with infectious agents, such as used slides and cover slips, and unused hypodermic and suture needles, syringes, and scalpel blades (EPA-94/04).

Sharpeners: Chemicals (such as sodium sulfide and sodium sulfhydrate) used in addition to lime to assist in the unhairing process (EPA-82/11).

Shaving: The waste products generated during the shaving operations. These are essentially small pieces of the tanned hide, which are approximately the size of wood shavings. Shaving is an abrasive, mechanical action used to correct errors in splitting and thus yielding a uniformly thick grain side or split (of hide) (EPA-82/11).

Shear: The variation of a vector quantity in a magnitude and direction along a given line in space (NATO-78/10).

Shear shredder: A size reduction machine that cuts materials between two large blades or between a blade and a stationary edge (*See* size reduction machine for more related terms) (SW-108ts).

Shearing stress: The stress in a fluid tangential to a given surface. This stress is connected to the shear in the fluid (NATO-78/10).

Sheen: An iridescent appearance on the surface of the water (40 CFR 110.1-91).

Sheepsfoot roller: A roller that consists of a steel drum fitted with projecting feet (EPA-83).

Sheet basecoat: A coating applied to metal when the metal is in sheet form to serve as either the exterior or interior of a can for either two-piece or three-piece cans (40 CFR 52.741-91).

Sheet glass: A flat glass (e.g., **window glass**) made by continuous drawing (*See* glass for more related terms) (EPA-83).

Shell deposition: The measured length of shell growth that occurs between the time the shell is ground at test initiation and test termination 96 hours later (40 CFR 797.1800-91).

Shell UOP: One of NO_x emission reduction techniques (*See* nitrogen oxide emission control for control structure). Shell UOP is a dry process that simultaneously removes both NO_x and SO_x emissions. This process can also be designed to remove either compound separately. The process uses a copper oxide (CuO) catalyst supported on alumina. These catalysts are located in two or more parallel passage reactors. Flue gas containing both NO_x and SO_x is introduced into the reactor where the SO_x reacts with copper oxide to form copper sulfate ($CuSO_4$). At the same time ammonia is being injected which reacts with the NO_x. The copper sulfate, and to a less extent the copper oxide, act as catalysts for the NO_x-NH_3 reaction. The following reactions occur in the reactor (SO_2 and NO_x reduction):

$$CuO + 1/2O_2 + SO_2 ---> CuSO_4$$
$$4NO + 4NH_3 + O_2 ---> 4N_2 + 6H_2O$$

When the reactor catalyst is saturated with $CuSO_4$, the flue gas is redirected to a fresh reactor and the spent catalyst is regenerated. Hydrogen is used to regenerate the catalyst by reducing the $CuSO_4$ to copper and producing a concentrated SO_2 gas stream. The SO_2 gas is then used to produce sulfuric acid or elemental sulfur for commercial sale. The copper in the reactor is oxidized to CuO and the process is ready to be put on line again. The reactions that take place in the reactor during catalyst regeneration are:

$$CuSO_4 + 2H_2 ---> Cu + SO_2 + 2H_2O$$
$$Cu + 1/2O ---> CuO$$

The Shell UOP process can be operated as a NO_x emission reduction process by eliminating the regeneration cycle. The process can be operated as a SO_x emission reduction process by eliminating the ammonia injection (EPA-81/12, p7-20).

Shellfish, fish and wildlife: Any biological population or community that might be adversely affected by the applicant's modified discharge (40 CFR 125.58-91).

Shield: A wall that protects workers from harmful radiation released by radioactive materials (EPA-74/11).

Shielded blade: A knife within a housing which protects the blade from being used in an open fashion, i.e., a protected knife (EPA-89/09, *See also* EPA-91/05).

Shiner: A defect in boxboard caused by a contaminant (EPA-83).

Ship canal: A canal deep enough and wide enough to permit the passage of ocean-going vessels (*See* canal for more related terms) (DOI-70/04).

Shipped liquid ammonia: The liquid ammonia commercially shipped for which the Department of Transportation requires 0.2 percent minimum water content (40 CFR 418.51-91).

Shipping losses: The discharges resulting from loading tank cars or tank trucks; discharges resulting from cleaning tank cars or tank trucks; and discharges from air pollution control scrubbers designed to control emissions from loading or cleaning tank cars or tank trucks (40 CFR 418.11-91, *See also* 40 CFR 418.51-91).

Shipping losses: The discharges resulting from loading tank cars or tank trucks; discharges resulting from cleaning tank cars or tank trucks; and discharges from air pollution control scrubbers designed to control emissions from loading or cleaning tank cars or tank trucks (40 CFR 418.21-91).

Shive: A bundle of incompletely separated fibers which may appear in the finished sheet as an imperfection (EPA-87/10).

Shock load:
(1) A quantity of wastewater or pollutant that greatly exceeds the normal discharged into a treatment system, usually occurring over a limited period of time (EPA-75/04).
(2) The arrival at a water treatment plant of raw water containing unusual amounts of algae, colloidal mater, color, suspended solids, turbidity, ore other pollutants (EPA-94/04).

Shop opacity: The arithmetic average of 24 or more opacity observations of emissions from the shop taken in accordance with Method 9 of Appendix A of this part for the applicable time periods (*See* opacity for more related terms) (40 CFR 60.271-91, *See also* 40 CFR 60.271a-91).

Short-circuiting: When some of the water in tanks or basins flows faster than the rest; usually undesirable since it may result in shorter contact, reaction, or settling times in comparison with the calculated or presumed detention times (EPA-94/04).

Short term air quality control: The control of air quality over time periods of the order of one hour to one day (NATO-78/10).

Short term exposure: Multiple or continuous exposures occurring over a week or so (EPA-92/12).

Short term test indicative of carcinogenic potential: Either any limited bioassay that measures tumor or preneoplastic induction, or any test indicative of interaction of a chemical substance with DNA (i.e., positive response in assays for gene mutation, chromosomal aberrations, DNA damage and repair, or cellular transformation) (40 CFR 721.3-91).

Short term test indicative of the potential to cause a developmentally toxic effect: Either any in vivo preliminary development toxicity screen conducted in a mammalian species, or any invitro developmental toxicity screen, including any test system other than the intact pregnant mammal, that has been extensively evaluated and judged reliable for its ability to predict the potential to cause developmentally toxic effects in intact systems across a broad range of chemicals or within a class of chemicals that includes the substance of concern (40 CFR 721.3-91).

Shot blast: A casting cleaning process employing a metal abrasive (grit or shot) propelled by centrifugal or air force (EPA-85/10a).

Shot casting: The production of shot by pouring molten metal in finely divided streams to form spherical particles (40 CFR 471.02-91).

Shotgun: A non-scientific term for the process of breaking up the DNA derived from an organism and then moving each separate and unidentified DNA fragment into a bacterium (EPA-89/12).

Shredded refuse: The solid waste which has been physically reduced to smaller particles (*See* refuse for more related terms) (EPA-83).

Shredder: A size reduction machine which tears or grinds materials to a smaller and more uniform particle size. Shredding process is also called size reduction, grinding, milling, communition, pulverization, hogging, granulating, breaking, maserating, chipping, crushing, cutting, rasping (*See* size reduction machine for more related terms) (EPA-83).

Shredding: Material cut, torn or broken up into small parts (EPA-83/06a).

Shrinkage:
(1) Loss in pulp yield due to the action of the bleaching chemicals.
(2) Presence of cracks or voids in each cover primarily a result of loss of moisture (EPA-83).

Shutdown:
(1) The cessation of operation of an affected facility for any purpose (40 CFR 60.2-91, *See also* 40 CFR 61.161; 61.171; 61.181-91).
(2) For a Department of Energy reactor, that condition in which the reactor has ceased operation and the Department has declared officially that it does not intend to operate it further (*See* DOE Order 5480.6) (DOE-91/04).

Sick building syndrome: A set of symptoms that affect a number of building occupants during the time they spend in the building and diminish or go away during periods when they leave the building. Can not be traced to specific pollutants (contrast with building-related illness) (EPA-88/09b).

Side chain: *See* definition under chain.

Side hill screen: A steeply sloped screen usually used to remove suspensions of stock or other solids from water while retaining the solid on the screen surface (*See* screen for more related terms) (EPA-10/87).

Side reaction: A secondary reaction accompanying to the primary reaction.

Side rite: An iron carbonate ($FeCO_3$).

Side roblasts: Early cells in the red blood cell series that contain granules of free iron as detected by the prussian blue reaction (LBL-76/07-bio).

Side seam spray coat: A coating applied to the seam of a three-piece can (40 CFR 52.741-91).

Sidewall cementing operation: The system used to apply cement to a continuous strip of sidewall component or any other continuous strip component (except combined tread/sidewall component) that is incorporated into the sidewall of a finished tire. A sidewall cementing operation consists of a cement application station and all other equipment, such as the cement supply system and feed and takeaway conveyors, necessary to apply cement to sidewall strips or other continuous strip component (except combined tread/sidewall component) and to allow evaporation of solvent from the cemented rubber (40 CFR 60.541-91).

Siege: Floor of a pot furnace, often called bench (EPA-83).

Siemens: *See* synonym, mho.

Sieve bend: The screens with stationary parallel bars at a right angle to the feed flow; the surface may be straight, with a steep incline, or curved to 300° (EPA-88/08a).

Sifting (or grate sifting): The materials that fall from the solid waste fuel bed through the grate openings (*See* ash for more related terms) (40 CFR 240.101-91).

Sigmoid curve (or S curve): A curve to show the growth of bacteria.

Signal: The volume or product-level change produced by a leak in a tank (EPA-94/04).

Signal spike: An abrupt, momentary increase and decrease in signal amplitude (40 CFR 60-App/A(alt. method 1)-91).

Signal words: The words used on a pesticide label-Danger, Warning, Caution-to indicate level of toxicity (EPA-94/04).

Significance: *See* synonym, statistical significance.

Significant: In reference to a net emissions increase or the potential of a source to emit any of the following pollutions, a rate of emissions that would equal or exceed any of the following rates (40 CFR 51.165-91, *See also* 40 CFR 51-App/S-91; 51.166; 52.21; 52.24-91)
- Carbon monoxide: 100 tons per year (tpy)
- Nitrogen oxides: 40 tpy
- Sulfur dioxide: 40 tpy
- Ozone: 40 tpy of volatile organic compounds
- Lead: 0.6 tpy

Significant adverse environmental effects: The injury to the environment by a chemical substance which reduces or adversely affects the productivity, utility, value, or function of biological, commercial, or agricultural resources, or which may adversely affect a threatened or endangered species. A substance will be considered to have the potential for significant adverse environmental effects if it has one of the following:
(1) An acute aquatic EC50 of 1 mg/L or less.
(2) An acute aquatic EC50 of 20 mg/L or less where the ratio of aquatic vertebrate 24-hour to 48-hour EC50 is greater than or equal to 2.0.
(3) A Maximum Acceptable Toxicant Concentration (MATC) of less than or equal to 100 parts per billion (100 ppb).
(4) An acute aquatic EC50 of 20 mg/L or less coupled with either a measured bioconcentration factor (BCF) equal to or greater than 1,000x or in the absence of bioconcentration data a log P value equal to or greater than 4.3 (40 CFR 721.3-91).

Significant adverse reactions: Reactions that may indicate a substantial impairment of normal activities, or long-lasting or irreversible damage to health or the environment (40 CFR 717.3-i-91).

Significant biological treatment: The use of an aerobic or anaerobic biological treatment process in a treatment works to consistently achieve a 30-day average of a least 65 percent removal of BOD(5) (40 CFR 133.101-91).

Significant deterioration: Pollution resulting from a new source in previously "clean" areas (*See* prevention of significant deterioration) (EPA-94/04).

Significant economic loss: That, under the emergency conditions: for a productive activity, the profitability would be substantially below the expected profitability for that activity; or, for other types of activities, where profits cannot be calculated, the value of public or private fixed assets would be substantially below the expected value for those assets. Only losses caused by the emergency conditions, specific to the impacted site, and specific to the geographic area affected by the emergency conditions are included. The contribution of obvious mismanagement to the loss will not be considered in determining loss. In evaluating the significance of an economic loss for productive activities, the Agency will consider whether the expected reduction in profitability exceeds what would be expected as a result of normal fluctuations over a number of years, and whether the loss would affect the long-term financial viability expected from the productive activity. In evaluating the significance of an economic loss for situations other than productive activities, the Agency will consider reasonable measures of expected loss (40 CFR 166.3-h-91).

Significant environmental effects: Either:
(1) Any irreversible damage to biological, commercial, or agricultural resources of importance to society;
(2) Any reversible damage to biological, commercial, or agricultural resources of importance to society if the damage persists beyond a single generation of the damaged resource or beyond a single year; or
(3) Any known or reasonably anticipated loss of members of an endangered or threatened species (40 CFR 723.50-91).

Significant impairment: For purposes of section 303, visibility impairment which, in the judgment of the Administrator, interferes with the management, protection, preservation, or enjoyment of the visitor's visual experience of the mandatory Class I Federal area. This determination must be made on a case-by-case basis taking into account the geographic extent, intensity, duration, frequency and time of the visibility impairment, and how these factors correlate with (1) times of visitor use of the mandatory Class I Federal area, and (2) the frequency and timing of natural conditions that reduce visibility (40 CFR 51.301-v-91).

Significant loss: Any loss that introduces a bias in final results that is of appreciable importance to concerned parties (*See* analytical parameters--laboratory for more related terms) (EPA-83).

Significant material: Includes, but is not limited to: raw materials; fuels; materials such as solvents, detergents, and plastic pellets; finished materials such as metallic products; raw materials used in food processing or production; hazard substances designated under section 101(14) of CERCLA; any chemical the facility is required to report pursuant to section 313 of title III of SARA; fertilizers; pesticides; and waste products such as ashes, slag and sludge that have the potential to be released with storm water discharges (40 CFR 122.26-91).

Significant municipal facilities: Those publicly owned sewage treatment plants that discharge a million gallons per day or more and are therefore considered by states to have the potential for to substantially effect the quality of receiving waters (EPA-94/04).

Significant new use: The use other than as an intermediate in the production of isodrin or endrin (40 CFR 721.1150-91, *See also* 40 CFR 721.1175; 721.1750-91).

Significant new use notice: Any notice submitted to EPA pursuant to 40 CFR 5(a)(l)(B) of the Act (the Toxic Substances Control Act) in accordance with 40 CFR 721 of this chapter (40 CFR 700.43-91).

Significant new use regulations (SNUR): Under TSCA Section 5, regulations that require a manufacturer to notify EPA in advance for significant new uses of existing chemicals or an appreciable increase in their utilization of existing chemicals (Arbuckle-89).

Significant non-compliance: *See* significant violations (EPA-94/04).

Significant source of groundwater:
(1) An aquifer that:
    (A) Is saturated with water having less than 10,000 milligrams per liter of total dissolved solids;
    (B) Is within 2,500 feet of the land surface;
    (C) Has a transmissivity greater than 200 gallons per day per foot, Provided, That any formation or part of a formation included within the source of ground water has a hydraulic conductivity greater than 2 gallons per day per square foot; and
    (D) Is capable of continuously yielding at least 10,000 gallons per day to a pumped or flowing well for a period of at least a year; or
(2) An aquifer that provides the primary source of water for a community water system as of the effective date of this subpart (40 CFR 191.12-91).
(3) *See* groundwater for more related terms.

Significant violations: Violations by point source dischargers of sufficient magnitude or duration to be a regulatory priority (EPA-94/04).

Significantly: As used in NEPA requires considerations of both context and intensity (for complete, definition, 40 CFR 1508.27-91).

Significantly damaged friable miscellaneous ACM: The damaged friable miscellaneous ACM where the damage is extensive and severe (40 CFR 763.83-91).

Significantly damaged friable surfacing ACM: The damaged friable surfacing ACM in a functional space where the damage is extensive and severe (40 CFR 763.83-91).

Significantly greater effluent reduction than BAT: That the effluent reduction over BAT produced by an innovative technology is significant when compared to the effluent reduction over best practicable control technology currently available (BPT) produced by BAT (40 CFR 125.22-91).

Significant hazard to public health: Any level of contaminant which causes or may cause the aquifer to exceed any maximum, contaminant level set forth in any promulgated National Primary Drinking Water Standard at any point where the water may be used for drinking purposes or which may otherwise adversely affect the health of persons, or which may require a public water system to install additional treatment to prevent such adverse effect (40 CFR 149.101-91).

Significantly lower cost: That an innovative technology must produce a significant cost advantage when compared to the technology used to achieve BAT limitations in terms of annual capital costs and annual operation and maintenance expenses over the useful life of the technology (40 CFR 125.22-91).

Significantly more stringent limitation: BOD5 and SS limitations necessary to meet the percent removal requirements of at least 5 mg/L more stringent than the other-wise applicable concentration-based limitations (e.g., less than 25 mg/L in the case of the secondary treatment limits for BOD5 and SS), or the percent removal limitations in 4 CFR 133.102 and 133.105, if such limits would, by themselves, force significant construction or other significant capital expenditure (40 CFR 133.101-91).

Silica (SiO_2):
(1) Silicon dioxide which occurs in crystalline form as quartz, cristohalite, tridymite. It is a major constituent of fireclay refractories, alone or in chemical combinations (SW-108ts; EPA-83/06a).
(2) The common name for silicon dioxide (SiO_2); one of the necessary ingredients in the production of cement (ETI-92).

Silica glass: An optical glass obtained by cutting or forming the chunk glass into plates or slabs (*See* glass for more related terms) (EPA-83).

Silicate: A chemical compound containing silicon, oxygen, and one or more metals. In soaps and detergents, sodium silicates are added to provide alkalinity and corrosion protection (EPA-74/04c).

Silicomanganese: That alloy as defined by ASTM Designation A48364 (Reapproved 1974) (incorporated by reference-*See* 40 CFR 60.17) (40 CFR 60.261-91).

Silicomanganese zirconium: That alloy containing 60 to 65 percent by weight silicon, 1.5 to 2.5 percent by weight calcium, 5 to 7 percent by weight zirconium, 0.75 to 1.25 percent by weight aluminum, 5 to 7 percent by weight manganese, and 2 to 3 percent by weight barium (40 CFR 60.261-91).

Silicon (Si): A metalloid element with a.n. 14; a.w. 28.086; d. 2.33 g/cc; m.p. 1410° C and b.p. 2680° C. The element belongs to group IVA of the periodic table.

Silicon carbide: A refractory material that has a high melting point, is very dense, and resists abrasion (SW-108ts).

Silicon metal: Any silicon alloy containing more than 96 percent silicon by weight (*See* metal for more related terms) (40 CFR 60.261-91).

Siliconizing: Diffusing silicon into solid metals, usually steels, at an elevated temperature for the purposes of case hardening thereby providing a corrosion and wear-resistant surface (EPA-83/06a).

Sill plate: A horizontal band (typically 2 x 4 or 2 x 6 in.) that rests on top of a block or poured concrete foundation wall and extends around the entire perimeter of the house. The ends of the floor joists which support the floor above the foundation wall rest upon the sill plate. For slab-on-grade, the sill plate is the bottom plate of the wall (EPA-88/08).

Silk screen process: A decorating process in which a design is printed on glass through a silk mesh, woven wire, or similar screen (EPA-83).

Silk screening: A coating method in which an enamel is spread onto a workpiece through a stencil screen (EPA-82/11e).

Silk screen ink: A quick drying, full bodied, volatile ink used in the silk screen printing process (*See* ink for more related terms) (EPA-79/12a).

Silt:
(1) Mineral soil grains intermediate between clay and sand (0.05 to 0.002 mm in diameter) (SW-108ts).
(2) Sedimentary materials composed of fine or intermediate sized mineral particles (EPA-94/04).

Silt basin: A basin to hold silts.

Silver (Ag): A transition metal with a.n. 47; a.w. 107.87; d. 10.5 g/cc; m.p. 960° C and b.p. 2210° C. The element belongs to group IB of the periodic table. Major silver compounds include:
- Silver chloride (AgCl): A toxic powder used in photography, plating, etc.
- Silver nitrate ($AgNO_3$): A toxic crystal used in plating, dyeing, etc.

Silver etch: An application of nitric acid to silver foils for active material support (EPA-84/08).

Silver halide: An inorganic salt of silver in combination with elements from Group 7A of the Periodic Table. Silver halide salts used in photography are silver chloride, silver bromide, and silver iodide. Upon exposure to light, silver halide crystals undergo an internal change making them capable of subsequent reduction to metallic silver by appropriate developing agents (EPA-80/10).

Silver recovery: The removal of silver from used photographic processing solutions and materials so it can be made available for reuse (EPA-80/10).

Silvering: The deposition of thin films of silver on glass, etc. (EPA-83/03).

Silvery iron: Any ferrosilicon, as defined by ASTM Designation A100-69 (Reapproved 1974) (incorporated by reference-*See* 40 CFR 60.17), which contains less than 30 percent silicon (40 CFR 60.261-91).

Silvicultural point source: Any discernible, confined and discrete conveyance related to rock crushing, gravel washing, log sorting, or log storage facilities which are operated in connection with silvicultural activities and from which pollutants are discharged into waters of the United States. The term does not include non-point source silvicultural activities such as nursery operations, site preparation, reforestation and subsequent cultural treatment, thinning, prescribed burning, pest and fire control, harvesting operations, surface drainage, or road construction and maintenance from which there is natural runoff. However, some of these activities (such as stream crossing for roads) may involve point source discharges of dredged or fill material which may require a CWA section 404 permit (*See* 33 CFR 209.120 and part 233) (40 CFR 122.27-1-91).

Silviculture: Management of forest land for timber (EPA-94/04).

Similar composition: Refers to a pesticide product which contains only the same active ingredient(s), or combination of active ingredients, and which is in the same category of toxicity, as a federally registered pesticide product (40 CFR 162.151-91).

Similar in all material respects: That the construction materials, exhaust and inlet air system, and other design features are within the allowed tolerances for components identified in 40 CFR 60.533(k) (40 CFR 60.531-91).

Similar incinerator: Incinerator A is similar to incinerator B if, based on the best engineering judgment, while incinerating identical waste as incinerator B, the stream leaving the combustion chamber of incinerator A contains equal or lower amounts of each, but no additional, potentially hazardous components as the stream leaving the combustion chamber on incinerator B (*See* incinerator for more related terms) (EPA-81/09).

Similar product: A pesticide product which, when compared to a federally registered product, has a similar composition and a similar use pattern (40 CFR 162.151-91).

Similar systems: The engine, fuel metering and emission control system combinations which use the same fuel (e.g., gasoline, diesel, etc.), combustion cycle (i.e., two or four stroke), general type of fuel system (i.e., carburetor or fuel injection), catalyst system (e.g., none, oxidization, three-way plus oxidization, three-way only, etc.), fuel control system (i.e., feedback or non-feedback), secondary air system (i.e., equipped or not equipped) and EGR (i.e., equipped or not equipped) (EGR means exhaust gase recirculation) (40 CFR 86.092.2-91).

Similar use pattern: Refers to a use of a pesticide product which, when compared to a federally registered use of a product with a similar composition, does not require a change in precautionary labeling under 40 CFR 156.10(h), and which is substantially the same as the federally registered use. Registrations involving changed use patterns are not included in this term (40 CFR 162.151-91).

Similar waste: Waste A is similar to waste B if, based on the best engineering judgment, the incineration of waste A in the same facility and under the same operating conditions as those used for waste B would yield a stream leaving the combustion chamber that contains equal or lower amounts of each, (but no additional) potentially hazardous pollutants compared to the amounts yielded by waste B incineration (*See* waste for more related terms) (EPA-81/09).

Simple cycle gas turbine: Any stationary gas turbine which does not recover heat from the gas turbine exhaust gases to preheat the inlet combustion air to the gas turbine, or which does not recover heat from the gas turbine exhaust gases to heat water or generate steam (*See* turbine for more related terms) (40 CFR 60.331-91).

Simple manufacturing operation: All the following unit processes: Desizing, fiber preparation and dyeing (cf. complex manufacturing operation) (40 CFR 410.41-91, *See also* 40 CFR 410.51; 410.61-91).

Simple slaughterhouse: A slaughterhouse which accomplishes very limited byproduct processing, if any, usually no more than two of such operations as rendering, paunch and viscera handling, blood processing, hide processing, or hair processing (40 CFR 432.11-91).

Simulation: A mock accident or release set up to test emergency response methods or for use as a training tool (EPA-11/85).

Single base: A propellant which contains only one explosive ingredient. A propellant consisting essentially of nitrocellulose plus stabilizer and plasticizer, formed by mixing these ingredients with ether and alcohol and extruding the resultant mass through dies and cutters (EPA-76/03).

Single bond: Two atoms which share one pair of electrons (*See* chemical bond for more related terms).

Single coat: One coating application applied to a metal surface (40 CFR 52.741-91).

Single chamber incinerator: A single chamber incinerator is a refractory-lined, cylindrical furnace charged through a door in the upper part of the chamber. Refuse is batch fed periodically (*See* incinerator for more related terms) (EPA-83).

Single failure: An occurrence that results in the loss of a component's capability to perform its intended safety functions. Multiple failures resulting from a single occurrence are considered to be a single failure. Fluid and electric systems are considered to be designed against an assumed single fail re if neither:
(1) A single failure of any active component (assuming that the passive components function properly) nor
(2) A single failure of any passive component (assuming that the active components function properly) results in a loss of the system's capability to perform its safety functions (DOE-91/04).

Single load method: A variation of the daily route method in which areas or routes are laid out that normally provide a full load of solid waste. Each crew usually has at least two such routes for a day's work. The crew quits for the day when the assigned number of routes is completed (*See* waste collection method for more related terms) (SW-108ts).

Single passenger commuter vehicle: A motor-driven vehicle with four or more wheels with capacity for a driver plus one or more passengers which is used by a commuter traveling alone to work or classes and is not customarily required to be used in the course of his employment or studies (40 CFR 52.1161-91, *See also* 40 CFR 52.2297-91).

Single response: All of the concerted activities conducted in response to a single episode, incident or threat causing or contributing to a release or threatened release of hazardous substances or pollutants or contaminants (40 CFR 310.11-91).

Single stand: Those recirculation or direct application cold rolling mills which include only one stand of work rolls (cf. multiple stand) (40 CFR 420.101-91).

Sink: Place in the environment where a compound or material collects (EPA-94/04).

Sink float: *See* synonym, dense media separation.

Sink temperature (or lowest temperature in the surroundings): In thermodynamics, sink temperature means the atmospheric temperature (Jones-p385).

Sinking: Controlling oil spills by using an agent to trap the oil and sink it to the bottom of the body of water where the agent and the oil are biodegraded (EPA-94/04).

Sinter: *See* definition under sintering machine.

Sinter bed: The lead sulfide ore concentrate charge within a sintering machine (40 CFR 60.181-91).

Sintered plate electrode: The electrode formed by sintering metallic powders in a porous structure, which serves as a current collector, and on which the active electrode material is deposited (*See* electrode for more related terms) (EPA-84/08).

Sintering:
(1) A heat treatment that causes adjacent particles of a material to cohere below a temperature that would cause them to melt (SW-108ts).
(2) Forming larger particles, cakes, or masses from small particles by heating alone, or by heating and pressing, so that certain constituents of the particles coalesce, fuse, or otherwise bind together (EPA-75/02). Sintering is a limited form of calcination in which the physical structure, but not the chemical nature, of the solid is changed. For instance, dry powders may be heated to sinter them into a solid mass, usually with some reduction in volume. Additives such as silicates, which also sinter readily, can be added to improve this process (EPA-76/11).
(3) The process of forming a mechanical part from a powdered metal by bonding under pressure and heat but below the melting point of the basis metal (EPA-83/06a).
(4) Fusing materials by heating them to high temperatures (DOE-91/04).

Sintering machine: Any furnace in which calcines are heated in the presence of air to agglomerate the calcines into a hard porous mass called **sinter** (40 CFR 60.171-91, *See also* 40 CFR 60.181-91).

Sintering machine discharge end: Any apparatus which receives sinter as it is discharged from the conveying grate of a sintering machine (40 CFR 60.181-91).

Sintering zone: The thermal zone in the kiln in which clinker is formed (ETI-92).

SIP call: EPA action requiring a state to resubmit all or part of its State Implementation Plan to demonstrate attainment of the required national air quality standards within the statutory deadline. A SIP Revision is a portion of an SIP altered at the request of EPA or on a state's initiative (EPA-94/04).

Siphon: A tubular device which is used to transfer liquid in different levels by air pressure.

Sister chromatid exchanges: Reciprocal interchanges of the two chromatid arms within a single chromosome. These exchanges are visualized during the metaphase portion of the cell cycle and presumably require enzymatic incision, translocation and ligation of at east two DNA helices (40 CFR 798.5900-91, *See also* 40 CFR 798.5915-91).

Site:
(1) The land or water area where any facility or activity is physically located or conducted, including adjacent land used in connection with the facility or activity (40 CFR 122.2-91, *See also* 40 CFR 122.29; 124.2; 124.41; 144.3; 146.3; 190.02; 191.02; 270.2; 300-App/A; 704.3; 710.2; 712.3; 717.3; 721.3; 723.50; 723.175; 763.63-91).
(2) Any location where acutely toxic chemicals are manufactured, processed, stored, handled, used, or disposed; in short, any place where these chemicals may be found. Communities should be aware that chemicals are frequently found at places other than industrial sites (EPA-85/11).
(3) An area or place within the jurisdiction of the EPA and/or a state (EPA-94/04).

Other site related terms include:
- Dump site
- Licensed site

Site area emergency: The events may occur, are in progress, or have occurred that could lead to a significant release of radioactive material and that could require a response by offsite response organizations to protect persons offsite (10 CFR 30.4-91, *See also* 10 CFR 40.4; 70.4-91).

Site assessment program: A means of evaluating hazardous waste sites through preliminary assessments and site inspections to develop a Hazard Ranking System score (EPA-94/04).

Site blank: A blank for evaluating site contamination (*See* blank for more related terms) (ACS-87/11).

Site characterization: The collection and analysis of field data to determine to what extent a site poses a threat to the environment and to begin developing potential remedial alternatives (EPA-89/12a).

Site inspection (SI):
(1) An on-site investigation to determine whether there is a release or potential release and the nature of the associated threats. The purpose is to augment the data collected in the preliminary assessment and to generate, if necessary, sampling and other field data to determine if further action or investigation is appropriate (40 CFR 300.5-91, *See also* EPA-89/12).
(2) The collection of information from a Superfund site to determine the extent and severity of hazards posed by the site. It follows and is more extensive than a preliminary assessment. The purpose is to gather information necessary to score the site, using the Hazard Ranking System, and to determine if it presents an immediate threat requiring prompt removal (EPA-94/04).

Site limited: A chemical substance is manufactured and processed only within a site and is not distributed for commercial purposes as a substance or as part of a mixture or article outside the site. Imported substances are never site limited (40 CFR 710.23-91).

Site limited intermediate: An intermediate manufactured processed, and used only within a site and not distributed in commerce other than as an impurity or for disposal. Imported intermediates cannot be site limited (40 CFR 721.3-91).

Site of construction: The general physical location of any building, highway, or other change or improvement to real property which is undergoing construction, rehabilitation, alteration, conversion, extension, demolition, and repair and any temporary location or facility at which a contractor, subcontractor, or other participating party meets a demand or performs a function relating to the contract or subcontract (40 CFR 8.2-91).

Site safety plan:
(1) The written site specific safety criteria that establishes the requirements for protecting the health and safety of responders during all activities conducted at an incident (Course 165.5).
(2) A crucial element in all removal actions, it includes information on equipment being used, precautions to be taken, and steps to take in the event of an on-site emergency (EPA-94/04).

Siting: The process of choosing a location for a facility (EPA-94/04).

Six (6) 9s: 99.9999 % destruction and removal efficiency (DRE) for PCB waste incineration standards required by the Toxic Substances Control Act (cf. four 9s).

Six center transition state: The transition state where six partial (one electron or three electron) bonds are formed; primarily results in the elimination of a stable molecule from the reactant (EPA-88/12).

Six minute period: Any one of the 10 equal parts of a one-hour period (40 CFR 60.2-91).

Size:
(1) The rated capacity in tons per hour of a crusher, grinding mill, bucket elevator, bagging operation, or enclosed truck or railcar loading station; the total surface area of the top screen of a screening operation; the width of a conveyor belt; and the rated capacity in tons of a storage bin (40 CFR 60.671-91).
(2) Material added to the paper to give it water resistance (EPA-83).

Size classes of discharges: Refers to the following size classes of oil discharges which are provided as guidance to the OSC and serve as the criteria for the actions delineated in subpart D. They are not meant to imply associated degrees of hazard to public health or welfare, nor are they a measure of environmental damage. Any oil discharge that poses a substantial threat to public health or welfare or the environment or results in significant public concern shall be classified as a major discharge regardless of the following quantitative measures:
(1) Minor discharge means a discharge to the inland waters of less than 1,000 gallons of oil or a discharge to the coastal waters of less than 10,000 gallons of oil.
(2) Medium discharge means a discharge of 1,000 to 10,000 gallons of oil to the inland waters or a discharge of 10,000 to 100,000 gallons of oil to the coastal waters.
(3) Major discharge means a discharge of more than 10,000 gallons of oil to the inland waters or more than 100,000 gallons of oil to the coastal waters (40 CFR 300.5-91).

Size classes of releases: Refers to the following size classifications which are provided as guidance to the OSC for meeting pollution reporting requirements in subpart B. The final determination of the appropriate classification of a release will be made by the OSC based on consideration of the particular release (e.g., size, location, impact, etc.):
(1) Minor release means a release of a quantity of hazardous substance(s), pollutant(s), or contaminant(s) that poses minimal threat to public health or welfare or the environment.
(2) Medium release means all releases not meeting the criteria for classification as a minor or major release.
(3) Major release means a release of any quantity of hazardous substance(s), pollutant(s), or contaminant(s) that poses a substantial threat to public health or welfare or the environment or results in significant public concern (under CERCLA) (40 CFR 300.5-91).

Size consistence: The particle size distribution of a product (to be consistent with standard method of sieve analysis (*See* analytical parameters--laboratory for more related terms) (EPA-83).

Size reduction machine: Size reduction machine related terms include:
- Chipper
- Food waste disposer (aee synonym, garbage grinding)
- Garbage grinding
- Hammermill
- Impact mill
- Knife hog
- Pulverization
- RASP (rasper)
- Roller crusher
- Shear shredder
- Shredder
- Trommel
- Wet milling
- Wet pulping (*See* synonym, wet milling)

Siting: The process of choosing a location for a facility (EPA-89/12).

Sizing:
(1) Relating to a property of paper resulting from an alteration of fiber surface characteristics. In terms of internal sizing, it is a measure of the resistance to the penetration of water and various liquids. In terms of surface sizing, it relates to the increase of such properties as water resistance, abrasion resistance, abrasiveness, creasibility, finish, smoothness, surface bonding strength, printability, and the decrease of porosity and surface fuzz.
(2) The addition of materials to a papermaking furnish or the application of materials to the surface of paper and paperboard to provide resistance to liquid penetration and, in the case of surface sizing, to affect one or more of the properties (EPA-87/10).

Skim (non-fat, defatted, or fat free) milk: The milk which fat has been separated as completely as commercially practicable. Typical skim milk contains maximum 0.1% fat, and minimum 8.25% solids (EPA-74/05).

Skimming:
(1) The removal of slag from the molten converter bath (cf. scum) (40 CFR 61.171-91, *See also* EPA-76/12; 74/11).
(2) Using a machine to remove oil or scum from the surface of the water (EPA-94/04).

Skimming station: The facility where slag is mechanically raked from the top of the bath of molten iron (40 CFR 60.141a-91).

Skimming tank: A tank so designed that floating matter will rise and remain on the surface of the wastewater until removed, while the liquid discharges continuously under walls or scum boards (*See* tank for more related terms) (EPA-83/03).

Skin sensitization (or allergic contact dermatitis): An immunologically mediated cutaneous reaction to a substance. In the human, the responses may be characterized by pruritis, erythema, edema, papules, vesicles, bullae, or a combination of these. In other species the reactions may differ and only erythema and edema may be Seen (40 CFR 798.4100-91).

Slab: A layer of concrete, typically about 4 in. thick, which commonly serves as the floor of any part of a house whenever the floor is in direct contact with the underlying soil (EPA-88/08).

Slab below grade: A type of house construction where the bottom floor is a slab which averages between 1 and about 3 ft below grade level on one or more sides (EPA-88/08).

Slab glass: An optical glass obtained by cutting or forming the chunk glass into plates or slabs (*See* glass for more related terms) (EPA-83).

Slab on grade: A type of house construction where the bottom floor of a house is a slab which is no more than about 1 ft below grade level on any side of the house (EPA-88/08).

Slag:
(1) The more or less completely fused and vitrified matter separated during the reduction of a metal from its ore (40 CFR 60.261-91).
(2) A molten inorganic substance, usually high in lime and silica, formed as a waste or by-product during chemical reactions in furnaces. Also applies to the solid material obtained by cooling the liquid. Depending on the rate of cooling, the solid product may be of either a predominantly glassy or a predominantly crystalline nature (cf. fouling) (EPA-83).

Other slag related terms include:
- Air cooled slag
- Granulated slag
- Granulated blast furnace slag
- Oxidizing slag
- Reducing slag

Slag quench: A process of rapidly cooling a molten slag to produce a more easily handled solid material. Usually performed by sudden immersion in a water trough or sump (EPA-85/10a).

Slag tap furnace: A furnace in which the temperature is high enough to maintain ash (slag) in a molten state until it leaves the furnace through a tap at the bottom. The slag falls into the sluicing water where it cools, disintegrates, and is carried away (*See* furnace for more related terms) (EPA-82/11f).

Slagging of refractory: A destructive chemical action that forms slag on refractories subjected to high temperatures. Also a molten or viscous coating produced on refractories by ash particles (SW-108ts).

Slaker: A device used to regenerate white liquor in a green liquor recovery process (EPA-87/10).

Slaking: The process of reacting lime with water to yield a hydrated product, i.e., to convert calcium oxide to calcium hydroxide (EPA-89/02).

Slaughterhouse: A plant that slaughters animals and has as its main product fresh meat as whole, half or quarter carcasses or smaller meat cuts (40 CFR 432.11-91, *See also* 40 CFR 432.21-91).

Sliding damper: A plate normally installed perpendicularly to the flow of gas in a breeching and arranged to slide across it to regulate the flow (*See* damper for more related terms) (SW-108ts).

Slime: A material of extremely fine particle sizes encountered in ore treatment (EPA-82/05).

Slime spots: Semi-transparent defects in the paper caused by bits of slime getting through the machine onto the wire (EPA-83).

Slimicide: A chemical used to prevent slimy growth, as in wood-pulping processes for manufacture of paper and paperboard (EPA-85/10).

Slip gauge: A gauge which has a probe that moves through the gas/liquid interface in a storage or transfer vessel and indicates the level of vinyl chloride in the vessel by the physical state of the material the gauge discharges (40 CFR 61.61-91).

Sliver: Bundles of noncontinuous or short length fibers that have reached that stage of their fabrication into yarn wherein they are parallel and over-lapping and have no twist (EPA-83).

Slope: The deviation of a surface from the horizontal expressed as a percentage, by a ratio, or in degrees (SW-108ts).

Slop (swill): Semi-liquid waste material consisting of putrescible solids and free liquids (EPA-83).

Slop oil: The floating oil and solids that accumulate on the surface of an oil-water separator (*See* oil for more related terms) (40 CFR 60.691; 61.341-91).

Slope factor (or cancer potency factor):
(1) Estimate of the probability of response (for example, cancer) per unit intake of a substance over a lifetime. The slope factor is typically used to estimate upper-bound probability

of an individual developing cancer as a result of exposure to a particular level of a human carcinogen with a weight-of-evidence classification of A, B, or C. [(mg/kg-day)$^{-1}$ for non-radioactive substances and (pC(i))$^{-1}$ for radioactive substances] (40 CFR 300-App/A-91).

(2) A plausible upper-bound estimate of the probability of a response per unit intake of a chemical over a lifetime. The slope factor is used to estimate an upper-bound probability of an individual's developing cancer as a result of a lifetime of exposure to a particular level of a potential carcinogen (EPA-91/12).

(3) The slope of the dose-response curve in the low-dose region. When low-dose linearity cannot be assumed, the slope factor is the slope of the straight line from 0 dose (and 0 excess risk) to the dose at 1% excess risk. An upper bound on this slope is usually used instead of the slope itself. The units of the slope factor are usually expressed as 1/(mg/kg-day) (EPA-92/12).

Slops: *See* synonym, swill.

Slough: Wet or marshy area (SW-108ts).

Sloughings: Trickling filter slimes that have been washed off the filter media. They are generally quite high in BOD5 and will degrade effluent quality unless removed (EPA-74/05).

Slow meter response: That the slow response of the sound level meter shall be used. The slow dynamic response shall comply with the meter dynamic characteristics in paragraph 5.4 of the American National Standard Specification for Sound Level Meters, ANSI S1.4-1971. This publication is available from the American National Standards Institute Inc., 1430 Broadway, New York, New York 10018 (cf. fast meter response) (40 CFR 201.1-91, *See also* 40 CFR 204.2-91).

Slow sand filtration:
(1) A process involving passage of raw water through a bed of sand at low velocity (generally less than 0.4 m/h) resulting in substantial particulate removal by physical and biological mechanisms (*See* filtration for more related terms) (40 CFR 141.2-91, *See also* EPA-89/12).
(2) Passage of raw water through a bed of sand at low velocity, resulting in substantial removal of chemical and biological contaminants (EPA-94/04).

Sludge:
(1) Pumpable materials (semi-solid residue) of naturally occurring or man-made origin possessing. It is a relatively fixed volume and its moisture content ranges from 15 to 90 percent (EPA-89/12a).
(2) Sludge is produced by a treatment plant that processes municipal or industrial waste waters (40 CFR 61.51-91, *See also* RCRA1004; 40 CFR 110.1; 240.101; 241.101; 243.101; 246.101; 257.2; 260.10; 261.1-91).
(3) A semi-solid residue from any of a number of air or water treatment processes; can be a hazardous waste (EPA-94/04).

Other sludge related terms include:
- Activated sludge
- Alum sludge
- Bulky sludge
- Chemical sludge
- De-sludge
- Dewatered sludge
- Digested sludge
- Excess activated sludge
- Excess sludge (*See* synonym, excess activated sludge)
- Surplus sludge (*See* synonym, excess activated sludge)
- Filamentous sludge
- Fresh sludge (*See* synonym, green sludge)
- Green sludge
- Oxidized sludge
- Oxygen activated sludge
- Paint sludge
- Primary sludge
- Raw sludge
- Secondary sludge
- Septic sludge
- Sewage sludge
- Trickling filter sludge
- Undigested sludge
- Waste activated sludge

Sludge age:
(1) The ratio of the weight of volatile solids in the digester to the weight of volatile solids added per day. There is a maximum sludge age beyond which no significant reduction in the concentration of volatile solids will occur (EPA-76/03).
(2) The mean residence time of the activated sludge.

Sludge bank: An accumulation of sewage solid deposits or industrial waste origin on the bed of a waterway (EPA-82/11f).

Sludge bed: *See* synonym, sludge drying bed.

Sludge blanket level: The depth of sludge in a sedimentation tank or a clarifier.

Sludge cake: The material resulting from air drying or dewatering sludge (usually forkable or spadable) (EPA-83/03).

Sludge conditioning: A process employed to prepare sludge for final disposal. Can be thickening, digesting, heat treatment etc. (EPA-84/08).

Sludge density index (SDI): An index for measuring the settleability of activated sludge by density. SDI = 100/(sludge volume index) (cf. sludge volume index).

Sludge dewatering (or sludge drying): The removal of water from sludge by introducing the water sludge slurry into a centrifuge. The sludge is driven outward with the water remaining near the center. The water is withdrawn and the dewatered sludge is usually landfilled (EPA-83/06a).

Sludge digester: Tank in which complex organic substances like sewage sludges are biologically dredged. During these reactions, energy is released and much of the sewage is converted to methane, carbon dioxide, and water (EPA-94/04).

Sludge digestion: A treatment to stabilize raw sludge. The treatment can be either anaerobic process or aerobic process.

Sludge digestion gas: Gas resulting from the sludge digestion process.

Sludge digestion tank: *See* synonym, digester.

Sludge disposal: The final disposal of solid wastes (EPA-83/03).

Sludge dryer: A device used to reduce the moisture content of sludge by heating to temperatures above 65° C (ca. 150° F) directly with combustion gases (40 CFR 61.51-91, *See also* 40 CFR 260.10-91).

Sludge drying: *See* synonym, sludge dewatering.

Sludge drying bed (or sludge bed): A bed on which the humus-like residue from the digester is dried; after being dried, the sludge may be burned or landfilled (DOI-70/04).

Sludge farming: *See* synonym, land treatment.

Sludge filter: A part of the sludge drying bed. The filter can remove a large portion of sludge water.

Sludge loading ratio (SLR): Parameter used in activated sludge treatment systems to estimate imposed waste demand. It is expressed in BOD/day/lb MLVSS (mixed liquor volatile suspended solids) (LBL-76/07-water).

Sludge only facility: Any treatment works treating domestic sewage whose methods of sewage sludge use or disposal are subject to regulations promulgated pursuant to section 405(d) of the CWA, and is required to obtain a permit under 40 CFR 122.1(b)(3) of this Part (40 CFR 122.2-91).

Sludge pond: A basin used for the storage, digestion, or dewatering of sludge (EPA-87/10a).

Sludge requirement: The following statutory provisions and regulations or permits issued thereunder (or more stringent State or local regulations): section 405 of the Clean Water Act; the Solid Waste Disposal Act (SWDA) (including Title II more commonly referred to as the Resource Conservation Recovery Act (RCRA) and State regulations contained in any State sludge management plan prepared pursuant to Sub title D of SWDA); the Clean Air Act; the Toxic Substances Control Act; and the Marine Protection, Research and Sanctuaries Act (40 CFR 403.7-91).

Sludge return rate: The quantity of return activated sludge per unit time.

Sludge thickening: Sludge can be thickened by:
(1) Gravity (sinking to the bottom of treatment tank due to gravity); and
(2) Flotation.

Sludge thickness: The increase in solids concentration of sludge in a sedimentation or digestion tank (EPA-83/03).

Sludge treatment: The sludge management process from the point of the generation of raw sludge to the point of ultimate disposal of treated sludge. The process may include sludge digestion, drying, incineration, etc. Other sludge treatment related terms include:
- Activated sludge process (*See* synonym, activated sludge treatment)
- Activated sludge treatment

Sludge volume index (SVI): An index for measuring the settleability of activated sludge by volume. SVI = 100/(sludge density index) (cf. sludge density index).

Slug: Any non-fibrous glass in a glass fiber product (EPA-83).

Slug flow sampling: A monitoring procedure that follows the same slug of wastewater throughout its transport in the receiving water. Water quality samples are collected at receiving water stations, tributary inflows, and point source discharges only when a dye slug or tracer passes that point (EPA-91/03).

Slugged bottom: An imperfection; very heavy glass on one side and very light on the opposite side of the bottom of a bottle or container. *See also* Heel Tap (EPA-83).

Sluice: An artificial passage for water, fitted with a valve or gate for stopping or regulating the flow; a regulating device for holding water back or letting it flow in or out; a conduit (natural or artificial) to drain or carry off surplus water; a long inclined trough or flume, usually on the ground (EPA-74/11).

Slurry:
(1) A pumpable mixture of solids and liquids (EPA-81/09).
(2) A mixture of liquid and finely divided insoluble solid materials (EPA-89/03b).
(3) A thin suspension of pulp fiber in water (EPA-83).
(4) A watery mixture of insoluble matter resulting from some pollution control techniques (EPA-94/04).

Small capacitor: A capacitor which contains less than 1.36 kg (3 lbs.) of dielectric fluid. The following assumptions may be used if the actual weight of the dielectric fluid is unknown. A capacitor whose total volume is less than 1,639 cubic centimeters (100 cubic inches) may be considered to contain less than 1.36 kgs (3 lbs.) of dielectric fluid and a capacitor whose total volume is more than 3,278 cubic centimeters (200 cubic inches) must be considered to contain more than 1.36 kg (3 lbs.) of dielectric fluid. A capacitor whose volume is between 1,639 and 3,278 cubic centimeters may be considered to contain less then 1.36 kg (3 lbs.) of dielectric fluid if the total weight of the capacitor is less than 4.08 kg (9 lbs.) (*See* capacitor for more related terms) (40 CFR 761.3-91).

Small manufacturer, processor, or importer: A manufacturer or processor who employed no more than 10 full-time employees at any one time in 1981 (40 CFR 763.63-91).

Small processor: An operation that produces up to 2730 kg (6000 lb) per day of any type or combination of finished products (40 CFR 432.51-91, *See also* 40 CFR 704.25; 704.33; 704.104; 704.203-91).

Small quantities for research and development: The quantities of a chemical substance manufactured, imported, or processed or proposed to be manufactured, imported, or processed that:
(1) Are no greater than reasonably necessary for such purposes;

and

(2) After the publication of the revised inventory, are used by, or directly under the supervision of, a technically qualified individual(s) (40 CFR 710.2-91, *See also* 40 CFR 761.3-91).

Small quantity generator: A generator who generates less than 1000 kg of hazardous waste in a calendar month (40 CFR 260.10-91, *See also* RCRA3001; EPA-86/01).

Small quantity generator (SQG-sometimes referred to as "squeegee"): Persons or enterprises that produce 220-2200 pounds per month of hazardous waste; are required to keep more records than conditionally exempt generators. The largest category of hazardous waste generators, SQGs include automotive shops, dry cleaners, photographic developers, and a host of other small businesses (*See* conditionally exempt generators) (EPA94/04).

Small sized plants: The plants which process less than 3,720 kg/day (8,200 lbs/day) of raw materials (40 CFR 428.51-91).

Small source: A source that emits less than 100 tons of regulated pollutants per year, or any class of persons that the Administrator determines, through regulation, generally lack technical ability or knowledge regarding control of air pollution (*See* source for more related terms) (CAA302-42USC7602-91).

Smelt: The molten inorganic cooking chemicals from the recovery boiler (EPA-87/10).

Smelt dissolving tank: A vessel used for dissolving the smelt collected from the recovery furnace (40 CFR 60.281-91).

Smelter: A facility that melts or fuses ore, often with an accompanying chemical change, to separate its metal content. Emissions cause pollution. "Smelting" is the process involved (EPA-94/04). Other smelter related terms include:
- Primary copper smelter
- Primary lead smelter
- Primary zinc smelter
- Secondary lead smelter

Smelter owner and operator: The owner or operator of the smelter, without distinction (40 CFR 57.103-t-91).

Smelting:
(1) Extracting a metal from its ore by firing with a flux at high temperatures. The melt consists of two layers - the slag on top and the separated impure metal below (EPA-83).
(2) Processing techniques for the melting of a copper sulfide ore concentrate or calcine charge leading to the formation of separate layers of molten slag, molten copper, and/or copper matte (40 CFR 60.161-91).

Smelting furnace: Any vessel in which the smelting of copper sulfide ore concentrates or calcines is performed and in which the heat necessary for smelting is provided by an electric current, rapid oxidation of a portion of the sulfur contained in the concentrate as it passes through an oxidizing atmosphere, or the combustion of a fossil fuel (*See* furnace for more related terms) (40 CFR 60.161-91).

Smog:
(1) A term derived from smoke and fog, applied to extensive atmospheric contamination by aerosols, these aerosols arising partly through natural processes and partly from human activities. Now some times used loosely for any contamination of the air (LBL-76/07-air).
(2) Air pollution associated with oxidants (*See* photochemical smog) (EPA-94/04).

Smoke:
(1) The matter in the exhaust emission which obscures the transmission of light (40 CFR 86.082.2-91, *See also* 40 CFR 87.1-91).
(2) Small gasborne particles resulting from incomplete combustion; particles consist predominantly of carbon and other combustible material; present in sufficient quantity to be observable independently of other solids (EPA-89/03b).
(3) Particles suspended in air after incomplete combustion (EPA-94/04).

Smoke alarm: An instrument that continuously measures and records the density of smoke by determining how much light is obscured when a beam is shown through the smoke. An alarm fitted in a flue goes off when the smoke exceeds a preset density (SW-108ts).

Smoke density: The amount of solid matter contained in smoke. It is often measured by systems that relate the grayness of the smoke to an established standard (cf. opacity) (SW-108ts).

Smoke emission: The pollutant generated by combustion in a flare and occurring immediately downstream of the flame. Smoke occurring within the flame, but not downstream of the flame, is not considered a smoke emission (*See* emission for more related terms) (40 CFR 60-App/A(method 22)-91).

Smoke eye: A device consisting of a light source and a photoelectric cell that measures the degree to which smoke in a flue gas obscures light (SW-108ts).

Smoke number (SN): The dimensionless term quantifying smoke emissions (40 CFR 87.1-91).

Smoke stack: *See* synonym, stack.

Smoke stick: A small tube, several inches long, which releases a small stream of inert smoke when a rubber bulb at one end of the tube is compressed. Can be used to visually define bulk air movement in a small area, such as the direction of air flow through small openings in slabs and foundation walls (EPA-88/08).

Smokeless powder: Nitrocellulose-based propellant (EPA-76/03).

Smooth-one-side (S1S) hardboard: The hardboard which is produced by the wet-matting, wet-pressing process (40 CFR 429.11-91).

Smooth-two-sides (S2S) hardboard: The hardboard which is produced by the wet-matting, dry-pressing process (40 CFR 429.11-91). For the subcategories for which numerical limitations are given, the daily maximum limitation is a value that should not be exceeded by any one effluent measurement. The 30-day limitation is a value that should not be exceeded by the average of daily measurements taken during any 30-day period.

Snailing: Streaks or marks on paper caused by air bubbles which disturb fiber alignment in the paper making process (EPA-83).

Snake (snaking): Progressive longitudinal cracking in continuous flat glass operation; variation in width of sheet during the sheet glass drawing process (EPA-83).

Snorkel: A pipe through the furnace roof, or an opening in a furnace roof, used to withdraw the furnace atmosphere (EPA-85/10a).

Snowline: The line of elevation on a mountain or hill slope that marks the lower limit of perpetual snow; below this line, any snow melts during the summer. The altitude of the snowline varies considerably in different regions; in general it occurs progressively lower from the tropics to the polar regions. Its altitude depends largely on the summer temperatures, that determine the rate of melting. Another important factor is the total amount of wintersnow; the snowline will be higher on the southern than the northern side of a mountain. Furthermore, the snowline will be higher on a steep slope, where much of the snow descends as avalanches, than on a gentle slope, where most of it lies where it falls till it melts (DOI-70/04).

Snowout: The capture of gaseous particulate in the air by precipitation in the form of snow (NATO-78/10).

SO_2: *See* synonym, sulfur dioxide.

SO_x: *See* synonym, sulfur oxide.

Soaking: As applied to the firing of ceramic ware, signifies the maintenance of the maximum temperature for a period to effect a desired degree of vitrification, chemical reaction and/or re-crystallization (EPA-83).

Soap: The product resulting from boiling animal fats with sodium hydroxide (cf. neat soap).

Soap boiling: The process of heating a mixture of fats/oils with a caustic solution until the fatty ester is split and the alkaline metal salt formed, glycerine being released in the process. The step where saponification takes place (cf. saponification) (EPA-74/04c).

Societal risk: The aggregated risks to all members of a population located within a specified radius around a potential hazard (DOE-91/04).

Soda: Sodium oxide (Na_2O). Loosely, a carbonate of sodium (EPA-83).

Soda ash: A common name for sodium carbonate (Na_2CO_3) (EPA-77/07).

Soda lime recipe: The glass product composition of the following ranges of weight proportions: 60 to 75 percent silicon dioxide, 10 to 17 percent total R_2O (e.g., Na_2O and K_2O), 8 to 20 percent total RO but not to include any PbO (e.g., CaO, and MgO), 0 to 8 percent total R_2O_3 (e.g., Al_2O_3), and 1 to 5 percent other oxides (40 CFR 60.291-91).

Soda process: The first process for the manufacture of chemical wood pulp. Involves boiling wood in caustic alkali at a high temperature (EPA-87/10).

Sodium (Na): A alkali metal with a.n. 11; a.w. 22.9898; d. 0.97 g/cc; m.p. 97.8° C and b.p. 892° C. The element belongs to group IA of the periodic table. Major sodium compounds include:
- Sodium aluminate ($Na_2Al_2O_4$): White powder used in water purification and soap.
- Sodium carbonate (Na_2CO_3): White powder used in pH control, bleaching and detergents.
- Sodium hydrosulfide ($NaSH \cdot 2H_2O$): Toxic crystals used in paper pulping, etc.
- Sodium hydrosulfite ($Na_2S_2O_4$): Whitish powder used in ore flotation and reducing agents.
- Sodium hydroxide (NaOH): White crystals used in rubber reclaiming, etc.
- Sodium hypochloride (NaOCl): A strongly alkaline solution used in germicides, disinfectants and deodorizers.
- Sodium silicate (Na_2SiO_3): An alkaline solution used in grindstones, abrasive wheels.
- Sodium stearate ($NaC_{18}H_{35}O_2$): A white powder used in toothpastes and waterproofing agents.
- Sodium sulfide (Na_2S): A yellow or red powder used in dehairing hides and wool pulling.
- Sodium thiosulfate ($Na_2S_2O_3 \cdot 5H_2O$): A white powder used in dyeing, pextile printing, etc.
- Sodium tripolyphosphate ($Na_5O_{10}P_3$): A white powder used in water softening.

Sodium absorption ratio (SAR): An indicator of sodium effect on water and crops. SAR = $Na/(0.5(Ca + Mg))$, where Na, Ca and Mg are the concentration of sodium, calcium and magnesium.

Sodium hypochlorite: A water solution of sodium hydroxide and chlorine in which sodium hypochloride is the essential ingredient (EPA-82/11f).

Soft detergent:
(1) Biodegradable detergents (EPA-74/11).
(2) Cleaning agents that break down in nature (EPA-94/04).

Soft lead: The lead produced by the removal of antimony through oxidation. The lead is characterized by low hardness and strength (*See* lead for more related terms) (EPA-83/03a).

Soft water: Any water that does not contain a significant amount of dissolved minerals such as salts of calcium or magnesium (*See* water for more related terms) (EPA-94/04).

Softener: Any device used to remove hardness from water. Hardness in water is due mainly to calcium and magnesium salts. Natural zeolites, ion exchange resins, and precipitation processes are used to remove the calcium and magnesium (EPA-82/11f).

Softwood:
(1) The wood obtained from evergreen or needle bearing trees

such as pines, spruces, and hemlocks (cf. hardwood) (EPA-74/04).
(2) Softwoods are as nonporous woods, the most common fiber source for paper in the northern hemisphere (EPA-83).

Softwood veneer: The veneer which is used in the manufacture of softwood plywood and in some cases the inner plies of hardwood faced plywood (*See* veneer for more related terms) (EPA-74/04).

Soil: All unconsolidated materials normally found on or near the surface of the earth including, but not limited to, silts, clays, sands, gravel, and small rocks (40 CFR 192.11-91, *See also* 40 CFR 761.123; 796.2700; 796.2750-91; EPA-89/12a). Other soil related terms include:
- Intermediate cover soil
- Isotropic soil
- Muck soil
- Night soil
- Plasticity soil
- Tight soil

Soil adsorption field: A sub-surface area containing a trench or bed with clean stones and a system of piping through which treated sewage may Seep into the surrounding soil for further treatment and disposal. Soil and Water Conservation Practices: Control measures consisting of managerial, vegetative, and structural practices to reduce the loss of soil and water (EPA-94/04).

Soil aggregate: The combination or arrangement of soil separates (sand, silt, clay) into secondary units. These units may be arranged in the profile in a distinctive characteristic pattern that can be classified on the basis of size, shape, and degree of distinctness into classes, type, and grades (40 CFR 796.2700-91, *See also* 40 CFR 796.2750-91).

Soil classification: The systematic arrangement of soils into groups or categories. Broad groupings are made on the basis of general characteristics, subdivisions, on the basis of more detailed differences in specific properties. The soil classification system used today in the United States is the 7th Approximation Comprehensive System. The ranking of subdivisions under the system is: order, suborder, great group, family and series (40 CFR 796.2700-91, *See also* 40 CFR 796.2750-91).

Soil cohesion: The mutual attraction exerted on soil particles by molecular forces and moisture films (SW-108ts).

Soil colloid: Clay and humus which are the major components of soil colloids.

Soil conditioner: An organic material like humus or compost that helps soil absorb water, build a bacterial community, and take up mineral nutrients (EPA-94/04).

Soil erodibility: A measure of the soil's susceptibility to raindrop impact, runoff, and other erosional processes (EPA-94/04).

Soil erosion: Detachment and movement of the soil from the land surface by wind or water (EPA-83).

Soil gas: Gaseous elements and compounds in the small spaces between particles of the earth and soil. Such gases can be moved or driven out under pressure (EPA-94/04, *See also* EPA-88/08; 88/08a).

Soil horizon: A layer of soil approximately parallel to the land surface. Adjacent layers differ in physical, chemical, and biological properties or characteristics such as color, structure, texture, consistency, kinds, and numbers of organisms present, and degree of acidity or alkalinity (40 CFR 796.2700-91, *See also* 40 CFR 796.2750-91).

Soil injection: The emplacement of pesticides by ordinary tillage practices within the plow layer of soil (40 CFR 165.1-91).

Soil irrigation: *See* synonym, spray irrigation.

Soil liner: A landfill liner composed of compacted soil used for the containment of leachate (*See* liner for more related terms) (EPA-89/11).

Soil moisture: Water diffused in the upper layers of the soil from which it is taken by plants for transpiration or from which it evaporates into the atmosphere (DOI-70/04).

Soil order: The broadest category of soil classification and is based on general similarities of physical/chemical properties. The formation by similar genetic processes causes these similarities. The soil orders found in the United States are: Alfisol, Aridisol, Entisol, Histosol, Inceptisol, Mollisol, Oxisol, Spodosol, Ultisol, and Vertisol (40 CFR 796.2700-91, *See also* 40 CFR 796.2750-91).

Soil organic matter: The organic fraction of the soil; it includes plant and animal residues at various stages of decomposition, cells and tissues of soil organisms, and substances synthesized by the microbial population (40 CFR 796.2700-91).

Soil percolation: *See* synonym, spray irrigation.

Soil pH: The value obtained by sampling the soil to the depth of cultivation or solid waste placement, whichever is greater, and analyzing by the electrometric method. (Methods of Soil Analysis, Agronomy Monograph No. 9, C.A. Black, ed., American Society of Agronomy, Madison, Wisconsin, pp. 914-926, 1965) (40 CFR 257.3.5-91, *See also* 40 CFR 796.2700-91).

Soil plasticity: The property of a soil that allows it to be deformed or molded in a moist condition without cracking or falling apart (SW-108ts).

Soil pollution: The soil contaminated with heavy metals, inorganic salts, difficult-to-decomposed organic matter or excessive organic matter (*See* pollution for more related terms).

Soil profile: A section through the soil showing the different horizons or layers extending downward from the surface to the parent material (DOI-70/04).

Soil series: The basic unit of soil classification and is a subdivision of a family. A series consists of soils that were developed under comparable climatic and vegetational conditions. The soils comprising a series are essentially alike in all major profile characteristics except for the texture of the "A" horizon (i.e., the surface layer of soil) (40 CFR 796.2700-91, *See also* 40 CFR 2750-91).

Soil sterilant: A chemical that temporarily or permanently prevents the growth of all plants and animals, depending on the chemical (EPA-94/04).

Soil structure: The relation of particles or groups of particles in a soil. It includes crumb structure, block structure, platy structure, and columnar structure (DOI-70/04).

Soil texture: Refers to the classification of soils based on the relative proportions of the various soil separates present. The soil textural classes are: clay, sandy clay, silty clay, clay loam, silty clay loam, sandy clay loam, loam, silt loam, silt, sandy loam loamy sand, and sand (40 CFR 796.2700-91, *See also* 40 CFR 796.2750-91).

Soil swelling: Physical expansion of the soil mass usually caused by an increase in moisture content in an expanding type of clay (EPA-83).

Solar irradiance in water: Related to the sunlight intensity in water and is proportional to the average light flux (in the units of 10^{-3} einsteins cm^{-2}- day^{-1} that is available to cause photoreaction in a wavelength interval centered at (wavelength) over a 24-hour day at a specific latitude and season date (40 CFR 796.3700-91).

Solar radiation: The electromagnetic energy and particles including electrons, protons, etc. emitted from the sun (*See* radiation for more related terms).

Sold or distributed: The aggregate amount of a pesticidal product released for shipment by the establishment in which the pesticidal product was produced (40 CFR 167.3-91).

Solder: Metallic compound used to seal joints between pipes. Until recently, most solder contained 50 percent lead. Use of lead solder containing more than 0.2 percent lead is now prohibited for pipes carrying drinking water (EPA-94/04).

Soldering: The process of joining metals by flowing a thin (capillary thickness) layer of nonferrous filler metal into the space between them. Bonding results from the intimate contact produced by the dissolution of a small amount of base metal in the molten filler metal, without fusion of the base metal. The term soldering is used where the temperature range falls below 425° C (EPA-83/06a).

Sole source aquifer (or principal source): An aquifer that supplies 50 percent or more of the drinking water of an area (*See* aquifer for more related terms) (EPA-94/04, *See also* 40 CFR 146.3; 149.2-91).

Solid: Various types of solids that are commonly determined on water samples. These types of solids are (EPA-74/04):
(1) **Total solids (TS):** The materials left after evaporation and drying a sample at 103-105° C.
(2) **Total suspended solids (TSS):** The materials removed from a sample filtered through a standard glass fiber filter. Then it is dried at 103-105° C.
(3) **Dissolved solids (DS):** The difference between the total and suspended solids.
(4) **Volatile solids (VS):** The materials which are lost when the sample is heated to 550° C.
(5) **Settleable solids:** The materials which settle in an Imhoff cone over a period of time.

Other solid related terms include:
- Settleable solid
- Suspended matter in water (*See* synonym, suspended solid)
- Suspended solid
- Total dissolved solid
- Total solid (*See* synonym, total suspended solid)
- Total suspended residue (*See* synonym, total suspended solid)
- Total suspended solid
- Volatile solid

Solid cone nozzle: One of spray nozzle types (*See* spray nozzles for more types). For this type of nozzle, liquid is force over an insert to break it up into a cone of fine droplets. Cones can be full, hollow, or square with spray angles from 15° to 140°. These nozzles can be made of stainless steel, brass, alloys, and plastic materials (EPA-84/03b, p2-2).

Solid contact clarifier: Solid (sludge) which is in direct contact with influent in a clarifier.

Solid derived fuel: *See* synonym, solid waste derived fuel.

Solid (sludge) rate meter: *See* definition under flow rate meter.

Solid waste: Garbage, refuse, sludges, and other discarded solid materials resulting from industrial and commercial operations and from community activities. It does not include solids or dissolved material in domestic sewage or other significant pollutants in water resources, such as silt, dissolved or suspended solids in industrial wastewater effluents, dissolved materials in irrigation return flows or other common water pollutants (*See* waste for more related terms) (40 CFR 240.101-91).

Solid waste: Non-liquid, non-soluble materials ranging from municipal garbage to industrial wastes that contain complex and sometimes hazardous substances. Solid wastes also include sewage sludge, agricultural refuse, demolition wastes, and mining residues. Technically, solid waste also refers to liquids and gases in containers (EPA-94/04).

Solid waste and medical waste: Shall have the meanings established by the Administrator pursuant to the Solid Waste Disposal Act (CAA129.g-42USC7429-91).

Solid waste boundary: The outermost perimeter of the solid waste (Projected in the Horizontal plane) as it would exist at completion of the disposal activity (40 CFR 257.3.4-91).

Solid waste derived fuel (or solid derived fuel): Any solid, liquid, or gaseous fuel derived from solid fuel for the purpose of creating useful heat and includes, but is not limited to, solvent refined coal, liquified coal, and gasified coal (*See* fuel for more related terms) (40 CFR 60.41a-91, *See also* 40 CFR 247.101-91).

Solid waste disposal:
(1) The disposal of all solid wastes through landfilling, incineration, composting, chemical treatment, and any other method which prepares solid wastes for final disposition (EPA-83).
(2) The final placement of refuse that is not salvaged or recycled (EPA-94/04).
(3) *See* disposal for more related terms.

Solid Waste Disposal Act (SWDA) of 1965: *See* the term of Act or SWDA.

Solid waste incineration unit: A distinct operating unit of any facility which combusts any solid waste material from commercial or industrial establishments or the general public (including single and multiple residences, hotels, and motels). Such term does not include incinerators or other units required to have a permit under section 3005 of the Solid Waste Disposal Act. The term "solid waste incineration unit" does not include:
(1) Materials recovery facilities (including primary or secondary smelters) which combust waste for the primary purpose of recovering metals,
(2) Qualifying small power production facilities, as defined in section 3(17)(C) of the Federal Power Act (16USC769(17)(C)), or qualifying cogeneration facilities, as defined in section 3(18)(B) of the Federal Power Act (16USC796(18)(B)), which burn homogeneous waste (such as units which burn tires or used oil, but not including refuse-derived fuel) for the production of electric energy or in the case of qualifying cogeneration facilities which burn homogeneous waste for the production of electric energy and steam or forms of useful energy (such as heat) which are used for industrial, commercial, heating or cooling purposes, or
(3) Air curtain incinerators provided that such incinerators only burn wood wastes, yard wastes and clean lumber and that such air curtain incinerators comply with opacity limitations to be established by the Administrator by rule (CAA129.g-42USC7429).

Other solid waste incineration unit related terms include:
- Existing solid waste incineration unit
- Modified solid waste incineration unit
- New solid waste incineration unit

Solid waste management:
(1) The systematic administration of activities which provide for the collection, source separation, storage, transportation, transfer, processing, treatment, and disposal of solid waste (RCRA1004-42USC6903-91).
(2) Supervised handling of waste materials from their source through recovery processes to disposal (EPA-94/04).

Solid waste management facility:
(1) Any resource recovery system or component thereof;
(2) Any system, program, or facility for resource conservation; and
(3) Any facility for the collection, source separation, storage, transportation, transfer, processing, treatment or disposal of solid wastes, including hazardous wastes, whether such facility is associated with facilities generating such wastes or otherwise (RCRA1004-42USC6903-91).

Solid waste storage container: A receptacle used for the temporary storage of solid waste while awaiting collection (40 CFR 243.101-91).

Solidification and stabilization (or chemical fixation): Removal of wastewater from a waste or changing it chemically to make it less permeable and susceptible to transport by water (EPA-94/04).
(1) **Solidification:**
 (A) The process of converting a contaminated soil, sludge, or liquid waste into a solid monolithic product that is more easily handled and that reduces the volatilization and leaching of contaminants from the waste (EPA-89/12a).
 (B) In thermodynamics, the change of a pure substance from a liquid phase to its solid phase (*See* latent heat for more related terms).
(2) **Stabilization:** The process of reducing the hazardous potential of a waste by chemically or physically converting the toxic contaminants into their least mobile or reactive form (EPA-89/12a).

Solidification and stabilization process: The solidification and stabilization process usually involves the use of Portland cement or lime and other materials such as ash, cement kiln dust, and blast furnace slag. The wastes are mixed with these materials in a liquid state and when allowed to harden, the hazardous constituents are physically incorporated into the solid matrix. The resulting mass is less permeable to leaching by water but is susceptible to breakdown by acids. Because of the pH of the mixture (normally pH 9 to 11) metals precipitate as relatively insoluble hydroxides, carbonates, or silicates. Soluble silicates are sometimes added to enhance the chemical fixation of heavy metals. There are four major processes in this category as follows:
(1) Cement based process
(2) Lime based process
(3) Thermoplastic process
(4) Organic polymer process

Solifluction: Liquid Seepage (EPA-81/09).

Solubility:
(1) The ability of a solid, liquid, gas or vapor to dissolve in a liquid (e.g., solvent) (Course 165.5). In general, a compound is classified as insoluble to a solvent, if its solubility is < 1 g/L (Course 165.6).
(2) In air pollution control, the capability of a gas to be dissolved in a liquid (EPA-84/03b).

Solute: *See* synonym, absorbate.

Solution: A homogeneous mixture of two or more substances constituting a single (40 CFR 796.1840-91, *See also* 40 CFR 796.1860-91; EPA-83/06a).

Solution heat treatment: The process introducing a workpiece into a quench bath for the purpose of heat treatment following rolling, drawing or extrusion (40 CFR 468.02-91).

Solvent: A liquid substance that is used to dissolve or dilute another substance (40 CFR 52.741-91, *See also* 40 CFR 795.120-91; EPA-89/12; 9/84). Other solvent related terms include:

- Chlorinated solvent
- Exempt solvent
- Makeup solvent
- Organic solvent
- Oxygenated solvent
- Photochemically reactive solvent
- Recovered solvent

Solvent applied in the coating: All organic solvent contained in the adhesive, release, and precoat formulations that is metered into the coating applicator from the formulation area (40 CFR 60.441-91).

Solvent base ink: An ink which uses oils or solvents as the primary vehicle (*See* ink for more related terms) (EPA-79/12a).

Solvent base paint: The paint in which the resin or film former is soluble or dispersed in an organic solvent (*See* paint for more related terms) (EPA-79/12b).

Solvent borne: A coating which contains five percent or less water by weight in its volatile fraction (40 CFR 60.391-91).

Solvent borne ink system: The ink and related coating mixtures whose volatile portion consists essentially of VOC solvent with not more than five weight percent water, as applied to the gravure cylinder (40 CFR 60.431-91).

Solvent cleaning: The process of cleaning soils from surfaces by cold cleaning, open top vapor degreasing, or conveyorized degreasing (40 CFR 52.741-91, *See also* EPA-83/06a).

Solvent degreasing: The removal of oils and grease from a workpiece using organic solvents or solvent vapors (EPA-83/06a).

Solvent extraction: The extraction of selected components from a mixture of two or more components by treating with a substance that preferentially dissolves one or more the components in the mixture (cf. liquid-liquid extraction) (EPA-87/10a).

Solvent extraction operation: An operation or method of separation in which a solid or solution is contacted with a liquid solvent (the two being mutually insoluble) to preferentially dissolve and transfer one or more components into the solvent (40 CFR 264.1031-91).

Solvent feed: The solvent introduced into the spinning solution preparation system or precipitation bath. This feed stream includes the combination of recovered solvent and makeup solvent (40 CFR 60.601-91).

Solvent filter: A discrete solvent filter unit containing a porous medium that traps and removes contaminants from petroleum solvent, together with the piping and ductwork used in the installation of this device (*See* filter for more related terms) (40 CFR 60.621-91).

Solvent inventory variation: The normal changes in the total amount of solvent contained in the affected facility (40 CFR 60.601-91).

Solvent of high photochemical reactivity: Any solvent with an aggregate of more than 20 percent of its total volume composed of the chemical compounds classified below or which exceeds any of the following individual percentage composition limitations in reference to the total volume of solvent:

(1) A combination of hydrocarbons, alcohols, aldehydes, esters, ethers, or ketones having an olefinic or cycloolefinic type of unsaturation: 5 percent;
(2) A combination of aromatic compounds with eight or more carbon atoms to the molecule except ethylbenzene: 8 percent;
(3) A combination of ethylbenzene, ketones having branched hydrocarbon structures, trichloroethylene or toluene: 20 percent. Whenever any organic solvent or any constituent of an organic solvent may be classified from its chemical structure into more than one of the above groups of organic compounds, it shall be considered as a member of the most reactive chemical group, that is, that group having the least allowable percentage of total volume of solvents (40 CFR 52.1145-91).

Solvent recovery: The recovery of volatile solvents that are present in air or a gas stream (EPA-84/09).

Solvent recovery dryer: A class of dry cleaning dryers that employs a condenser to condense and recover solvent vapors evaporated in a closed-loop stream of heated air, together with the piping and ductwork used in the installation of this device (40 CFR 60.621-91).

Solvent recovery system: The equipment associated with capture, transportation, collection, concentration, and purification of organic solvents. It may include enclosures, hoods, ducting, piping, scrubbers, condensers, carbon absorbers, distillation equipment, and associated storage vessels (40 CFR 60.601-91, *See also* 40 CFR 60.431-91).

Solvent spun synthetic fiber: Any synthetic fiber produced by a process that uses an organic solvent in the spinning solution, the precipitation bath, or processing of the sun fiber (40 CFR 60.601-91).

Solvent spun synthetic fiber process: The total of all equipment having a common spinning solution preparation system or a common solvent recovery system, and that is used in the manufacture of solvent-spun synthetic fiber. It includes spinning solution preparation, spinning, fiber processing and solvent recovery, but does not include the polymer production equipment (40 CFR 60.601-91).

Solvent variety: Varieties of solvents include (Markes-67):
(1) **Alcohols:**
  (A) Methyl alcohol (methanol) is made synthetically and is completely miscible with water and most organic liquids;
  (B) Ethyl alcohol (ethanol) is produced by fermentation and synthetically;
  (C) Isopropyl alcohol (isopropanol) is derived mainly from petroleum; and
  (D) Butyl alcohol (normal butanol) is used extensively in lacquer and synthetic resin compositions and also in penetrating oils.

(2) **Esters:**
- (A) Athyl acetate dissolves a large variety of materials, such as nitrocellulose, oils, fats, gums and resins;
- (B) Butyl acetate is the acetic-acid ester of normal butanol and is used extensively for dissolving various cellulose esters, minerals and vegetable oils; and
- (C) Amyl acetate (banana oil) is used mainly in lacquers.

(3) **Hydrocarbons:**
- (A) Aromatic hydrocarbons are derived from coal- tar distillates, the most common of which are benzene, toluene, xylene, and hi-flash naphtha or coal-tar naphtha; and
- (B) Petroleum is hydrocarbons derived from petroleum. Common petroleum solvents includes benzine, mineral spirits and kerosene.

(4) **Chlorinated solvents:**
- (A) Carbon tetrachloride is a colorless non-flammable liquid; and
- (B) Trichlorethylene is similar to carbon tetrachloride but is slower in evaporation rate.

(5) **Ketones:**
- (A) Acetone is an exceptionally active solvent for a wide variety of organic materials, gases, liquids, and solids; and
- (B) Methylethylketone (MEK) is similar to acetone.

Sonic boom: The tremendous booming sound produced as a vehicle, usually a supersonic jet airplane, exceeds the speed of sound, and the shock wave reaches the ground (cf. boom) (EPA-74/11).

Soot:
(1) Agglomerations of particles of carbon formed during the incomplete combustion of carbonaceous material (EPA-83). It is formed primarily under the reducing condition, not under the oxidizing condition (because the oxidizing condition would have enough oxygen to react with carbon for not producing soot (cf. carbon black).
(2) Carbon dust formed by incomplete combustion (EPA-94/04).

Sorbent: A liquid or solid medium in or upon which materials are retained by absorption or adsorption (EPA-83/06).

Sorption: The action of soaking up or attracting substances. A process used in many pollution control systems (EPA-94/04, *See also* EPA-83/06).

Sorted brown kraft: Consists of baled clean sorted brown kraft papers, free from twisted or woven stock, sewed edges and heavy printing (EPA-83).

Sound: Sound and noise are often used interchangeably for anything perceived by means of hearing (cf. noise or car coupling sound).

Sound exposure level: The level in decibels calculated as ten times the common logarithm of time integral of squared A-weighted sound pressure over a given time period or event divided by the square of the standard reference sound pressure of 20 micropascals and a reference duration of one second (40 CFR 201.1-91).

Sound level: The weighted sound pressure level measured by the use of a metering characteristic and weighting B, or C as specified in American National Standard Specification for Sound Level Meters SI.4-11 or subsequent approved revision. The weighting employed must be specified, otherwise A-weighting is understood (40 CFR 204.2-91). Other sound level related terms include:
- A-scale sound level
- Day night sound level
- Equivalent sound level

Sound pressure level: (in stated frequency band) means the level, in decibels, calculated as 20 times the common logarithm of the ratio of a sound pressure to the reference sound pressure of 20 micropascals (40 CFR 201.1-91, *See also* 40 CFR 204.2; 205.2-91).

Sour:
(1) Denotes the presence of sulfur compounds, such as sulfides and mercaptans, that cause bad odors (EPA-74/04b).
(2) Indicates wastewater treatment systems that have a low pH value. The acid condition is favorable to growth of organisms which produce foul smelling by-products, hence is undesirable (EPA-74/03).

Sour water: The wastewater containing sulfur compounds, such as sulfides and mercaptans (*See* water for more related terms) (EPA-79/12c).

Source: Any building, structure, facility, or installation from which there is or may be a discharge of pollutants (CWA306). Other source related terms include:
- Affected source
- Area source
- Continuous source
- Existing source
- Existing OCS (outer continental shelf) source
- Fixed source
- fugitive source
- Industrial source
- Instantaneous source
- Line source
- Major source
- Mobile source
- Modification source
- New source
- New OCS source
- Non-industrial source
- Non-point source
- Point source
- pollution source
- Reconstructed source
- Small source
- Stationary source
- Virtual point
- Volume source

Source assessment sampling system (SASS): In many respects, the SASS is about a five-fold scale-up of the MM5 and collects larger samples, typically 30 dscm over a three-hour sampling period. This sampling train is appropriate whenever a large sample of stack gas (greater than 10 dscm) is required to ensure adequate detection limits (EPA-82/02).

Source configuration: The geographical position and distribution of the area, line and point sources in a region (NATO-78/10).

Source control action: The construction or installation and start-up of those actions necessary to prevent the continued release of hazardous substances or pollutants or contaminants (primarily from a source on top of or within the ground, or in buildings or other structures) into the environment (40 CFR 300.5-91).

Source control maintenance measures: Those measures intended to maintain the effectiveness of source control actions once such actions are operating and functioning properly, such as the maintenance of landfill caps and leachate collection systems (40 CFR 300.5-91).

Source control remedial actions: Those measures that are intended to contain hazardous substances or pollutants or contaminants where they are located or eliminate potential contamination by transporting the hazardous substances or pollutants or contaminants to a new location. Source control remedial actions may be appropriate if a substantial concentration or amount of hazardous substances or pollutants or contaminants remains at or near the area where they are originally located and adequate barriers exist to retard migration of hazardous substances or pollutants or contaminants into the environment. Source control remedial actions may not be appropriate if most hazardous substances or pollutants or contaminants have migrated from the area where originally located or if the lead agency determines that the hazardous substances or pollutants or contaminants are adequately contained.

Source height: The height of a source above the surrounding ground surface (NATO-78/10).

Source material:
(1) Uranium or thorium, or any combination thereof, in any physical or chemical form; or
(2) Ores which contain by weight one-twentieth of one percent (0.05%) or more of:
  (A) Uranium;
  (B) Thorium; or
  (C) Any combination thereof. Source material does not include special nuclear material (10 CFR 20.3-91, See also 10 CFR 30.4; 40.4; 70.4-91).
(3) The meaning contained in the Atomic Energy Act of 1954, 42USC2014 et seq., and the regulations issued thereunder (40 CFR 710.2; 720.3-91).
(4) Source material such as uranium or thorium or ores containing uranium or thorium (DOE-91/04).
(5) See nuclear material for more related terms.

Source reduction:
(1) Any practice which:
  (A) Reduces the amount of any hazardous substance, pollutant, or contaminant entering any waste stream or otherwise released into the environment (including fugitive emissions) prior to recycling, treatment, or disposal; and
  (B) Reduces the hazards to public health and the environment associated with the release of such substances, pollutants, or contaminants. The term includes equipment or technology modifications, process or procedure modifications, reformulation or redesign of products, substitution of raw materials, and improvements in housekeeping, maintenance, training, or inventory control.
  (C) The term source reduction does not include any practice which alters the physical, chemical, or biological characteristics or the volume of a hazardous substance, pollutant, or contaminant through a process or activity which itself is not integral to and necessary for the production of a product or the providing of a service (PPA6603-91, See also EPA-86/10; 89/11).
(2) Reducing the amount of materials entering the waste stream by redesigning products or patterns of production or consumption (e.g., using returnable beverage containers). Synonymous with waste reduction (EPA-94/04).

Source separation:
(1) The setting aside of recyclable materials at their point of generation by the generator (40 CFR 246.101-91, See also EPA-88/12a).
(2) Segregating various wastes at the point of generation (e.g., separation of paper, metal and glass from other wastes to make recycling simpler and more efficient) (EPA-94/04).

Source term: The estimated quantities of radionuclides released to the environment (DOE-91/04).

Source test: A measurement of emissions to determine concentrations and/or mass flow rates (NATO-78/10).

Spade drilling: Drilling with a flat blade drill tip (See drilling for more related terms) (EPA-83/06a).

Spalling: A defect characterized by chipping that occurs without apparent external causes (EPA-83).

Spalling of refractory: The breaking or crushing of a refractory unit due to thermal, mechanical, or structural causes (SW-108ts).

Span: The upper limit of the gas concentration measurement range displayed on the data recorder (40 CFR 60-App/A(method 6C & 7E)-91, See also EPA-90/04).

Span drift:
(1) The percent change in response to an up-scale pollutant concentration over a 24-hour period of continuous unadjusted operation (40 CFR 53.23-91).
(2) The change with time in instrument output over a stated time period of unadjusted continuous operation when the input concentration is a stated value other than zero (expressed as percent of full scale) (LBL-76/07-bio).
(3) See calibration for more related terms.

Span gas: A gas of known concentration which is used routinely to set the output level of an analyzer (40 CFR 86.082.2-91, See also 40 CFR 86.402.78-91).

Span value:
(1) The upper limit of a gas concentration measurement range that is specified for affected source categories in the applicable part of the regulations. The span value is

established in the applicable regulation and is usually 1.5 to 2.5 times the applicable emission limit. If no span value is provided, use a span value equivalent to 1.5 to 2.5 times the expected concentration. For convenience, the span value should correspond to 100 percent of the recorder scale (40 CFR 60-App/A(method 25A)-91, *See also* 40 CFR 60-App/F-91).

(2) The opacity value at which the CEMS is set to produce the maximum data display output as specified in the applicable subpart (40 CFR 60-App/B-91).

Spandex fiber: A manufactured fiber in which the fiber-forming substance is a long chain synthetic polymer comprised of at least 85 percent of a segmented polyurethane (40 CFR 60.601-91).

Spare flue gas desulfurization system module: A separate system of sulfur dioxide emission control equipment capable of treating an amount of flue gas equal to the total amount of flue gas generated by an affected facility when operated at maximum capacity divided by the total number of nonspare flue gas desulfurization modules in the system (40 CFR 60.41a-91).

Sparger: An air diffuser designed to give large amount of bubbles, used singly or in combination with mechanical aeration devices (EPA-76/03).

Sparging:
(1) Removing the volatile constituents of a sample by bubbling an inert gas stream through the sample (EPA-82/02).
(2) Heating a liquid by means of a live steam entering through a perforated or nozzled pipe (used, e.g., to coagulate blood solids in meat processing) (DOI-70/04).

Spark arrester: A screen-like device that keeps sparks, embers, or other ignited materials above a given size within an incinerator (SW-108ts).

Spark ignition powered motor vehicle: A self-propelled over-the-road vehicle that is powered by a spark ignition type of internal combustion engine, including but not limited to engines fueled by gasoline, propane, butane, and methane compounds (*See* vehicle for more related terms) (40 CFR 51.731-91).

Spark plug: A device to suitably deliver high tension electrical ignition voltage to the spark gap in the engine combustion chamber (40 CFR 85.2122(a)(7)(ii) (A)-91).

Special aquatic sites: Those sites identified in Subpart E. They are geographic areas, large or small, possessing special ecological characteristics of productivity, habitat, wildlife protection, or other important and easily disrupted ecological values. These areas are generally recognized as significantly influencing or positively contributing to the general overall environmental health or vitality of the entire ecosystem of a region. (*See* 40 CFR 230.10(a)(3)) (40 CFR 230.3-91).

Special area management plan: A comprehensive plan providing for natural resource protection and reasonable coastal-dependent economic growth containing a detailed and comprehensive statement of policies; standards and criteria to guide public and private uses of lands and waters; and mechanisms for timely implementation in specific geographic areas within the coastal zone (CZMA304-16USC1453-90).

Special local need: An existing or imminent pest problem within a State for which the State lead agency, based upon satisfactory supporting information, has determined that an appropriate federally registered pesticide product is not sufficiently available (40 CFR 162.151-91).

Special news de-ink quality: Baled sorted, fresh, dry newspapers, not sunburned, free from magazines, white blank, pressroom over-issues, and paper other than news, containing not more than the normal percentage of Rotogravure and colored sections (EPA-83).

Special nuclear material:
(1) Plutonium, uranium-233, uranium enriched in the isotope 233 or in the isotope 235, and any other material which the Commission, pursuant to the provisions of section 51 of the act, determines to be special nuclear material, but does not include source material;
(2) Any material artificially enriched by any of the foregoing but does not include source material (10 CFR 20.3-91; *See also* 10 CFR 30.4; 40.4; 70.4-91).
(3) The meaning contained in the Atomic Energy Act of 1954, 42USC2014 et seq., and the regulations issued thereunder (40 CFR 710.2-91, *See also* 40 CFR 720.3-91).
(4) As defined in Section 11 of the Atomic Energy Act of 1954, special nuclear material means:
   (A) Plutonium, uranium enriched in the isotope 233 or in the isotope 235, and any other material which the Nuclear Regulatory Commission determines to be special nuclear material; or
   (B) Any material artificially enriched by any of the foregoing (DOE-91/04).
(5) *See* nuclear material for more related terms.

Special nuclear material of low strategic significance:
(1) Less than amount of special nuclear material of moderate strategic significance, as defined in 40 CFR 70.4(z)(1), but more than 15 grams of uranium-235 (contained in uranium enriched to 20 percent or more in the U^{235} isotope) or 15 grams of uranium-233 or 15 grams of plutonium or the combination of 15 grams when computed by the equation, grams = (grams contained U^{235}) + (grams plutonium) + (grams U^{233}); or
(2) Less than 10,000 grams but more than 1000 grams of uranium-235 (contained in uranium enriched to 10 percent or more but less than 20 percent in the U^{235} isotope); or (3) 10,000 grams or more of uranium-235 (contained in uranium enriched above natural but less than 10 percent in the U^{235} isotope) (*See* nuclear material for more related terms) (10 CFR 70.4-91).

Special nuclear material of moderate strategic significance:
(1) Less than a formula quantity of strategic special nuclear material but more than 1000 grams of uranium-235 (contained in uranium enriched to 20 percent or more in the U^{235} isotope) or more than 500 grams of uranium-233 or plutonium or in a combined quantity of more than 1000 grams when computed by the equation, grams = (grams contained U^{235}) + 2 (grams U^{233} + grams plutonium); or

(2) 10,000 grams or more of uranium-235 (contained in uranium enriched to 10 percent or more but less than 20 percent in the U^{235} isotope) (10 CFR 70.4-91).

(3) *See* nuclear material for more related terms.

Special nuclear material scrap: The various forms of special nuclear material generated during chemical and mechanical processing, other than recycle material and normal process intermediates, which are unsuitable for use in their present form, but all or part of which will be used after further processing (*See* nuclear material for more related terms) (10 CFR 70.4-91).

Special population: Concentrations of people in one area or building for a special purpose or in certain circumstances (e.g., schools, hospitals, nursing homes, orphanages, shopping centers) (EPA-85/11).

Special production area: A demarcated area within which all manufacturing, processing, and use of a new chemical substance takes place, except as provided in paragraph (f) of this section, in accordance with the requirements of paragraph (e) of this section (40 CFR 723.175-91).

Special purpose equipment: The maintenance-of-way equipment which may be located on or operated from rail cars including: Ballast cribbing machines, ballast regulators, conditioners and scarifiers, bolt machines, brush cutters, compactors, concrete mixers, cranes and derricks, earth boring machines, electric welding machines, grinders, grouters, pile drivers, rail heaters, rail layers, sandblasters, snow plows, spike drivers, sprayers and other types of such maintenance-of-way equipment (40 CFR 201.1-91).

Special purpose facility: A building or space, including land incidental to its use, which is wholly or predominantly utilized for the special purpose of an agency and not generally suitable for other uses, as determined by the General Services Administration (40 CFR 6.901-91, *See also* EPA-89/12).

Special review: Formerly known as Rebuttable Presumption Against Registration (RPAR), this is the regulatory process through which existing pesticides suspected of posing unreasonable risks to human health, non-target organisms, or the environment are referred for review by EPA. Such review requires an intensive risk/benefit analysis with opportunity for public comment. If risk is found to outweigh social and economic benefits, regulatory actions ranging from label revisions and use-restriction to cancellation or suspended registration can be initiated (EPA-94/04).

Special source of groundwater: Those Class I groundwaters identified in accordance with the Agency's Groundwater Protection Strategy published in August 1984 that:

(1) Are within the controlled area encompassing a disposal system or are less than five kilometers beyond the controlled area;

(2) Are supplying drinking water for thousands of persons as of the date that the Department chooses a location within that area for detailed characterization as a potential site for a disposal system (e.g., in accordance with section 112(b)(1)(B) of the NWPA); and

(3) Are irreplaceable in that no reasonable alternative source of drinking water is available to that population (40 CFR 191.12-91).

(4) *See* groundwater for more related terms.

Special track work: The track other than normal tie and ballast bolted or welded rail or containing devices such as retarders or switching mechanisms (40 CFR 201.1-91).

Special waste (or unconventional waste):

(1) Non-hazardous solid wastes requiring handling other than that normally used for municipal solid waste (40 CFR 240.101-91).

(2) Waste that requires special or separate handling, such as household hazardous wastes, bulky wastes, tires, and used oil (EPA-89/11).

(3) Items such as household hazardous waste, bulky wastes (refrigerators, pieces of furniture, etc.) tires, and used oil (EPA-94/04).

(4) *See* waste for more related terms.

Specially designated landfill: A landfill at which complete long term protection is provided for the quality of surface and subsurface waters from pesticides, pesticide containers, and pesticide-related wastes deposited therein, and against hazard to public health and the environment. Such sites should be located and engineered to avoid direct hydraulic continuity with surface and subsurface waters, and any leachate or subsurface flow into the disposal area should be contained within the site unless treatment is provided. Monitoring wells should be established and a sampling and analysis program conducted. The location of the disposal site should be permanently recorded in the appropriate local office of legal jurisdiction. Such facility complies with the Agency Guidelines for the Disposal of Solid Wastes as prescribed in 40 CFR 241 (40 CFR 165.1-91).

Specialty hot forming operation (or specialty): Applies to all hot forming operations other than carbon hot forming operations (40 CFR 420.71-91).

Specialty paper: An asbestos-containing product that is made of paper intended for use as filters for beverages or other fluids or as paper fill for cooling towers. Cooling tower fill consists of asbestos paper that is used as a cooling agent for liquids from industrial processes and air conditioning systems (*See* paper for more related terms) (40 CFR 763.163-91).

Specialty steel: Those steel products containing allowing elements which are added to enhance the properties of the steel product when individual alloying elements (e.g., aluminum, chromium, cobalt, columbium, molybdenum, nickel, titanium, tungsten, vanadium, zirconium) exceed 3% or the total of all alloying elements exceed 5% (*See* steel for more related terms) (40 CFR 420.71-91).

Species:

(1) Includes any subspecies of fish or wildlife or plants, and any distinct population segment of any species of vertebrate fish or wildlife which interbreeds when mature (ESA3-16USC1531-90).

(2) A reproductively isolated aggregate of interbreeding organisms (EPA-94/04).

Specific chemical identity: The chemical name, Chemical Abstracts Service (CAS) Registry Number, or any other information that reveals the precise chemical designation of the substance. Where the trade name is reported in lieu of the specific chemical identity, the trade name will be treated as the specific chemical identity for purposes of this part (40 CFR 350.1-91, *See also* 29 CFR 1910.20-91).

Specific conductance:
(1) The property of a solution which allows an electric current to flow when a potential difference is applied (EPA-83/06a). For the measure of a water's capacity to convey an electric current, this property is related to the total concentration of the ionized substances in the water and the temperature of the water. Most inorganic acids, which dissociate readily in aqueous solution, will conduct an electric current well, while organic compounds (such as sucrose and benzene), which do not dissociate in aqueous solution, will conduct a current poorly (DOI-70/04).
(2) Rapid method of estimating the dissolved-solid content of a water supply by testing the capacity of the water to carry an electrical current (EPA-94/04).

Specific energy (or specific head): Sum of the pressure head and the velocity head $V^2/2g$ measured with respect to the channel bottom. The specific-energy concept is especially useful in the analysis of flows in open channels.

Specific gravity:
(1) The ratio of the density of a substance to a the density of a reference substance. Water at 4° C has the maximum density. In general, water at temperature 4° C is used as the reference substance for solids and liquids (Schaum-p9).
- Density of water at 4° C = 62.4 lb/ft^3 in British units
  = 1.0 g/cm^3 in cgs units
  = 1.0 g/cc (cubic centimeter)
  = 1.0 g/ml (milliliter)
(2) The density ratio of two substances--that of the substance of interest to that of a reference substance. The reference substance is normally water. The term is dimensionless (EPA-84/09).
(3) The density of a substance in question to the density of a reference substance (usually, water at 32° F, 1 atm) at specified conditions of temperature and pressure (EPA-89/03b).

Specific head: *See* synonym, specific energy.

Specific heat: The ratio of the heat capacity of a substance to the heat capacity of a reference substance. The reference substance is normally water at 17° C, at which temperature the heat capacity is 1.0 Btu/lb-F. Heat capacity and specific heat are interchangeable (*See* heat for more related terms) (EPA-86/03).

Specific heat at constant pressure (Cp, Btu/(lb-F) or Btu/(lb-R)):
The slope of a constant pressure line on an hT diagram. Cp can be expressed as:
$C_p = (dh/dT)_p$; or $dh = (C_p)dT$
where:
dT = quantity of temperature change (dT)
dh = quantity of enthalpy change due to the change of dT
(Holman-p49; Jones-p124)

Specific heat at constant volume (Cv, Btu/(lb-F) or Btu/(lb-R)):
The slope of a constant pressure line on an hT diagram. Cp can be expressed as:
$C_p = (du/dT)_p$; or $du = (C_p)dT$
where:
dT = quantity of temperature change (dT)
dh = quantity of internal energy change due to the change of dT (Holman-p49; Jones-p124)

Gas	Cp (B/(lb-R))	Cv (B/(lb-R))
Hydrogen	3.41	2.42
Nitrogen	0.248	0.177
Carbon monoxide	0.248	0.177
Carbon dioxide	0.195	0.150

Specific humidity: *See* synonym, humidity ratio (*See also* humidity for more related terms).

Specific property: The value of an extensive property divided by the mass of the system. The value becomes an intensive property and is called a specific property (*See* property for more related terms) (Jones-60, p12).

Specific resistance: *See* synonym, electrical resistivity.

Specific surface: The solid surface area per unit weight or volume of the solid material.

Specific volume: The volume of a system divided by the mass of substances in the system. It is the reciprocal of density or the volume per unit mass. It can be expressed as:
v = (volume)/(mass) = V/m = 1/d, where v = specific volume, V = volume, mass = mass, d = density.

Specific weight: The weight of a substance divided by its volume, or the weight per unit volume. It can be expressed as:
w = (weight)/(volume) = (force)/(volume) = (mg/g$_c$)/V = dg/g$_c$, where: m = mass; g = local acceleration of gravity; V = volume; d = density.
g_c = 32.17 (lbm-ft)/(lbf-sec^2)
= 1 (slug-ft)/(lbf-sec^2)
= 1 (kg mass-m)/(newton-sec^2)
= 1 (g mass-cm)/(dyne-sec^2) (Holman-74, p18)

Specific written consent:
(1) A written authorization containing the following:
  (A) The name and signature of the employee authorizing the release of medical information,
  (B) The date of the written authorization,
  (C) The name of the individual or organization that is authorized to release the medical information,
  (D) The name of the designated representative (individual or organization) that is authorized to receive the released information,
  (E) A general description of the medical information that is authorized to be released,
  (F) A general description of the purpose for the release of the medical information, and
  (G) A date or condition upon which the written authorization will expire (if less than one year).

(2) A written authorization does not operate to authorize the release of medical information not in existence on the date of written authorization, unless the release of future information is expressly authorized, and does not operate for more than one year from the date of written authorization.
(3) A written authorization may be revoked in writing prospectively at any time (29 CFR 1910.20-91).

Specific yield: The amount of water that a unit volume of saturated permeable rock will yield when drained by gravity (EPA-94/04).

Specific yield of water: The amount of water that can be obtained from the pores or cracks of a unit volume of soil or rock (*See* water for more related terms) (DOI-70/04).

Specification: A clear and accurate description of the technical requirements for materials, products or services, identifying the minimum requirements for quality and construction of materials and equipment necessary for an acceptable product. In general, specifications are in the form of written descriptions, drawings, prints, commercial designations, industry standards, and other descriptive references (40 CFR 246.101-91, *See also* 40 CFR 247.101; 248.4; 249.04; 250.4; 252.4; 253.4-91). Other specification related terms include:
- Contract specification
- Guide specification
- Listed specification
- Material specification
- Performance specification
- Prohibit specification
- Withdraw specification

Specified air contaminant: Any air contaminant as to which this Section contains emission standards or other specific limitations (*See* air contaminant for more related terms) (40 CFR 52.741-91).

Specified minimum yield strength (SMYS):
(1) For steel pipe manufactured in accordance with a listed specification, the yield strength specified a minimum in that specification; or
(2) For steel pipe manufactured in accordance with an unknown or unlisted specification, the yield strength determined in accordance with 40 CFR 192.107(b) (40 CFR 192.3-91).

Specified minimum yield strength: The minimum yield strength, expressed in pounds per square inch, prescribed by the specification under which the material is purchased from the manufacturer (40 CFR 195.2-91).

Specified ports and harbors: Those port and harbor areas on inland rivers, and land areas immediately adjacent to those waters, where the USCG acts as predesignated on- scene coordinator. Precise locations are determined by EPA/USCG egional agreements and identified in Federal regional contingency plans (40 CFR 300.5-91).

Specimen: Any material derived from a test system for examination or analysis (40 CFR 160.3-91, *See also* 40 CFR 792.3-91).

Spectral model: The solution of a set of equations describing a given problem by an eigenfunction expansion (NATO-78/10).

Spectrograph: An instrument with one slit that uses photography to obtain a record of a spectral range simultaneously. The radiant power passing through the optical system is integrated over time, and the quantity recorded is a function of radiant energy (LBL-76/07-air). It is a spectroscope equipped with a camera to record the spectrum (cf. spectroscope)

Spectrometer: An instrument with an entrance slit and one or more exit slits, with which measurements are made either by scanning the spectral range, point by point, or by simultaneous measurements at several spectral positions. The quantity measured is a function of radiant power (cf. emission spectrometer) (LBL-76/07-air). It is a spectroscope equipped with a calibrated scale either for measurement of refractive indices of transparent prism materials or equipped with photoelectric photometer to measure radiant intensities at various wavelengths. A spectrometer equipped with the photoelectric photometer (detector) is also known as a spectrophotometer. Other spectrometer related terms include:
- Infrared spectrometer
- Mass spectrometer (MS)
- Optical spectrometer

Spectrometry: Branch of physical science treating the measurement of spectra (cf. mass spectrometry) (LBL-76/07-bio).

Spectrophotometer: A spectrometer with associated equipment, so designed that it furnishes the ratio, or a function of the ratio, of the radiant power of two beams as a function of spectral position. The two beams may be separated in time, space, or both (LBL-76/07-air). *See also* spectrometer.

Spectrophotometric titration: A titration method in which a spectrophotometer is used to measure the radiant energy absorption of a solution at each addition of a titrant until the end point is reached (*See* titration for more related terms).

Spectrophotometry: A method of analyzing a wastewater sample by means of the spectra emitted by its constituents under exposure to light (EPA-83/06a). It is a technique to measure the wavelength range of radiant energy absorbed by a sample under analysis.

Spectroscope: An optical instrument that produces a spectrum for visual observation. The instrument consists of a:
(1) Hollow tube;
(2) A slit at one end of the tube to let light enter into the tube;
(3) A collimator lens at the other end to produce a parallel beam;
(4) A prism or grating to disperse the light; and
(5) A telescope for viewing the spectrum (cf. spectrograph).
(6) cf. mass spectroscope.

Spectroscopy: Studies of producing and analyzing spectra by using various instruments such as spectroscopes, spectrometers, spectrographs, etc. Other spectroscopy related terms include:
- Infrared spectroscopy
- Mass spectroscopy
- Photoelectron spectroscopy

Spectrum: A distribution of radiation intensity as a function of wavelength (light spectrum), mass (mass spectrum), frequency (sound spectrum), etc. (cf. absorption spectrum).

Spent acid solution (or spent pickle liquor): Those solutions of steel pickling acids which have been used in the pickling process and are discharged or removed there-from (40 CFR 420.91-91).

Spent cooking liquor: The cooking liquor after digestion containing lignaceous, as well as chemical, materials (*See* liquor for more related terms) (EPA-87/10).

Spent hypo solution: A solution consisting of a photographic film fixing bath and wash water which contains unreduced silver from film processing (EPA-83/03a).

Spent lubricant: The water or an oil-water mixture which is used in forming operations to reduce friction, heat and wear and ultimately discharged (*See* lubricant for more related terms) (40 CFR 468.02-91).

Spent material: Any material that has been used and as a result of contamination can no longer serve the purpose for which it was produced without processing (*See* material for more related terms) (40 CFR 261.1-91).

Spent nuclear fuel: The fuel that has been withdrawn from a nuclear reactor following irradiation, the constituent elements of which have not been separated by reprocessing (*See* nuclear fuel for more related terms) (40 CFR 191.02-91).

Spent pickle liquor: *See* synonym, spent acid solution.

Sphagnum: A grayish moss growing in dense layers in bogs, that eventually forms peat (DOI-70/04).

Sphalerite: Zinc sulfide (ZnS) (EPA-82/05).

Spider: A wheel-like casting consisting of a rim and radial spokes; assembly of radiating tie rods (EPA-83).

Spike: A sharp point. Other spike related terms include:
- Analysis matrix spike
- Field matrix spike
- Internal standard spike
- Laboratory matrix spike
- Non-target analyte spike
- Target analyte spiking

Spike material: A material of known or established concentration added to samples and analyzed to assess the bias of measurements (EPA-84/03).

Spill: Both intentional and unintentional spills, leaks, and other uncontrolled discharges where the release results in any quantity of PCBs running off or about to run off the external surface of the equipment or other PCB source, as well as the contamination resulting from those releases. This policy applies to spills of 50 ppm or greater PCBs. The concentration of PCBs spilled is determined by the PCB concentration in the material onto which the PCBs were spilled. Where a spill of untested mineral oil occurs, the oil is presumed to contain greater than 50 ppm, but less than 500 ppm, PCBs and is subject to the relevant requirements of this policy (40 CFR 761.123-91; EPA-84/08).

Spill area: The area of soil on which visible traces of the spill can be observed plus a buffer zone of 1 foot beyond the visible traces. Any surface or object (e.g., concrete sidewalk or automobile) within the visible traces area or on which visible traces of the spilled material are observed is included in the spill area. This area represents the minimum area assumed to be contaminated by PCBs in the absence of precleanup sampling data and is thus the minimum area which must be cleaned (40 CFR 761.123-91).

Spill boundaries: The actual area of contamination as determined by postcleanup verification sampling or by precleanup sampling to determine actual spill boundaries. EPA can require additional cleanup when necessary to decontaminate all areas within the spill boundaries to the levels required in this policy (e.g., additional cleanup will be required if postcleanup sampling indicates that the area decontaminated by the responsible party, such as the spill area as defined in this section, did not encompass that actual boundaries of PCB contamination) (40 CFR 761.123-91).

Spill event: A discharge of oil into or upon the navigable waters of the United States or adjoining shorelines in harmful quantities, as defined at 40 CFR 110 (40 CFR 112.2-91).

Spill prevention control and counter measures plan (SPCC): Plan covering the release of hazardous substances as defined in the Clean Water Act (EPA-94/04).

Spindle: The fine threads of achromatic material in the cell nucleus arranged in a spindle-shaped manner during mitosis (LBL-76/07-bio).

Spinning reserve: The sum of the unutilized net generating capability of all units of the electric utility company that are synchronized to the power distribution system and that are capable of immediately accepting additional load. The electric generating capability of equipment under multiple ownership is prorated based on ownership unless the proportional entitlement to electric output is otherwise established by contractual arrangement (40 CFR 60.41a-91, *See also* EPA-82/11f).

Spinning solution: The mixture of polymer, prepolymer, or copolymer and additives dissolved in solvent. The solution is prepared at a viscosity and solvent-to-polymer ratio that is suitable for extrusion into fibers (40 CFR 60.601-91).

Spinning solution preparation system: The equipment used to prepare spinning solutions; the system includes equipment for mixing, filtering, blending, and storage of the spinning solutions (40 CFR 60.601-91).

Spiral classifier: A classifier for separating fine-size solids from coarser solids in a wet pulp consisting of an interrupted-flight screw conveyer, operating in an inclined trough (EPA-75/10c).

Splash loading: A method of loading a tank, railroad tank car, tank truck, or trailer by use of other than a submerged loading pipe (*See* loading for more related terms) (40 CFR 52.741-91).

Split: An imperfection; crack or check going from surface to surface of a glass article (EPA-83).

Split sample: A sample divided into two portions, one of which is sent to a different organization or laboratory and subjected to the same environmental conditions and steps in the measurement process as the one retained in-house (*See* sample for more related terms) (EPA-84/03).

Spoil:
(1) Soil or rock that has been removed from its original location (EPA-83).
(2) Dirt or rock removed from its original location-destroying the composition of the soil in the process-as in strip-mining, dredging, or construction (EPA-94/04).

Spoil material: An overburden that is removed from above the coal seam; usually deposited in previously mined areas (cf. overburden) (EPA-82/10).

Sponge: A highly porous metal powder (EPA-84/08).

Sponsor:
(1) A person who initiates and supports, by provision of financial or other resources, a study;
(2) A person who submits a study to the EPA in support of an application for a research or marketing permit; or
(3) A testing facility, if it both initiates and actually conducts the study (40 CFR 160.3-91, *See also* 40 CFR 790.3; 792.3-91).

Spotter: In truck, the man who directs a driver into a loading or unloading position (SW-108ts).

Sprawl: Unplanned development of open land (EPA-94/04).

Spray: Liquid droplets created by mechanical disintegration, found in absorbers and some particulate control devices (EPA-84/09).

Spray application: A method of applying coatings by atomizing and directing the atomized spray toward the part to be coated (40 CFR 60.311-91, *See also* 40 CFR 60.391; 721.3-91).

Spray booth: A structure housing automatic or manual spray application equipment where prime coat, guide coat, or topcoat is applied to components of automobile or light-duty truck bodies (40 CFR 60.391-91, *See also* 40 CFR 60.451; 60.721-91; EPA-82/11e).

Spray chamber (or spray tower): One of air pollution control devices. A spray chamber is a chamber (or a tower) equipped with water sprays that cool and clean incinerator combustion products passing through the chamber. It removes contaminants by a gas absorption process. The scrubbing liquid is atomized by high pressure spray nozzles into small droplets in the scrubbing chamber. The dirty gases pass through the spray droplets in either countercurrent, concurrent, or cross flow direction. The dirty gas stream velocity decreases as it enters the chamber; the contaminants in the gas stream are wetted and are absorbed by droplets. The wetted particles then settle and are collected at the bottom of the chamber. The outlet of the chamber is sometimes equipped with eliminator plates to help prevent the liquid from being discharged with the clean air stream. Its efficiency is low except for coarse dust.

The spray chamber is a low-energy scrubber used to control large-particle emissions and is extensively used as a gas cooler. A spray tower is a relatively simple scrubber. Its major components consist of a hollow cylindrical steel vessel; and spray nozzles for injecting the scrubbing liquid (*See* scrubber for more related terms) (AP-40).

Spray dryer: A vessel where hot flue gases are contacted with a finely atomized wet alkaline spray. *See also* FGD spray dryer with a baghouse or ESP (*See* scrubber for more related terms) (EPA-81/12, p8-24).

Spray dryer absorber: One of air pollution control devices. This type of scrubber uses the alkaline reagent (pebble lime) as a slurry containing 5 to 20 percent by weight solids. This slurry is atomized by either rotary atomizers or air atomizing nozzles in a large absorber vessel having a residence time of 6 to 20 seconds. Since the primary removal mechanism is absorption into the droplets, drying that is too rapid can reduce pollutant collection efficiency. The absorbers are operated with exist temperatures 90 to 180° F above the saturation temperature (*See* scrubber for more related terms) (EPA-89/02):

Major components of a dry scrubber include:
(1) Lime slaker, if pebble lime is purchased
(2) Sorbent mixing tank
(3) Sorbent feed tank
(4) Atomizer feed tank
(5) Rotary atomizers or air atomizing nozzles
(6) Spray dryer absorber reaction vessel
(7) Solids recycle tank
(8) Particulate control device.

Spray dryer and dry injection system combination: One of air pollution control devices. The acid gas laden flue gas is first treated in an upflow type spray dryer absorber. A series of calcium hydroxide sprays near the bottom of the absorber vessel are used for droplet generation. After the upflow chamber, the partially treated flue gas then passes through a venturi contactor section where it is exposed to a calcium silicate and lime suspension. The purpose of the second reagent material is to remove the dust cake in the downstream fabric filter and to optimize acid gas removal in this dust cake (*See* scrubber for more related terms) (EPA-89/02).

Spray drying tower (or blowing tower): A large vessel in which solids in solution or suspension are dried by falling through a hot gas (EPA-74/04c).

Spray evaporation: A method of wastewater disposal in which water is sprayed into the air to expedite evaporation (EPA-74/04).

Spray finishing operations: Spray finishing operations are employment of methods wherein organic or inorganic materials are utilized in dispersed form for deposit on surfaces to be coated, treated, or cleaned. Such methods of deposit may involve either automatic, manual, or electrostatic deposition but do not include metal spraying or metallizing, dipping, flow coating, roller coating, tumbling, centrifuging, or spray washing and degreasing as conducted in self-contained washing and degreasing machines or systems (29 CFR 1910.94c-91).

Spray in place foam: The rigid cellular polyurethane or

polyisocyanurate foam produced by catalyzed chemical reactions that hardens at the site of the work. The term includes spray applied and injected applications (40 CFR 248.4-91).

Spray in place insulation: The insulation material that is sprayed onto a surface or into cavities and includes cellulose fiber spray-on as well as plastic rigid foam products (40 CFR 248.4-91).

Spray irrigation (soil irrigation or soil percolation): Transport of sludge or wastewater to a distribution system from which it is sprayed over an area of land. The liquid percolates into the soil and/or evaporates. None of the sludge or wastewater runs off the irrigated area (*See* land treatment for more related terms) (EPA-79/12a).

Spray module (or powered spray module): A water cooling device consisting of a pump and spray nozzle or nozzles mounted on floats and moored in the body of water to be cooled. Heat is transferred principally by evaporation from the water drops as they fall through the air (EPA-82/11f).

Spray nozzle (or nozzle): A device used for the controlled introduction of scrubbing liquid at predetermined rates, distribution patterns, pressures, and droplet sizes (EPA-89/03b). Types of spray nozzles include:
(1) Impingement nozzle
(2) Solid cone nozzle
(3) Helical spray nozzle (EPA-84/03b, p2-2).

Spray rinse: A process which utilizes the expulsion of water through a nozzle as a means of rinsing (EPA-83/06a).

Spray room: A spray room is a room in which spray-finishing operations not conducted in a spray booth are performed separately from other areas (29 CFR 1910.94c-91).

Spray tower: *See* synonym, spray chamber.

Spray tower scrubber: A device that sprays alkaline water into a chamber where acid gases present to aid in the neutralizing of the gas (EPA-94/04).

Spray washing: A method of washing film or paper using a spray rather than an immersion tank as a means of conserving water (EPA-80/10).

Spreader stoker incinerator: An incinerator in which solid fuel is introduced to the combustion zone by a mechanism that throws the fuel onto a grate from above. Combustion takes place both in suspension and on the grate (*See* incinerator for more related terms).

Spreader stoker steam generating unit: A steam generating unit in which solid fuel is introduced to the combustion zone by a mechanism that throws the fuel onto a grate from above. Combustion takes place both in suspension and on the grate (40 CFR 60.41b-91).

Spring:
(1) A continuous or intermittent flow of water from the ground (DOI-70/04).
(2) Ground water Seeping out of the earth where the water table intersects the ground surface (EPA-94/04).

Spring melt/thaw: The process by which warm temperatures melt winter snow and ice. Because various forms of acid deposition may have been stored in the frozen water, the melt can result in abnormally large amounts of acidity entering streams and rivers, sometimes causing fish kills (EPA-94/04).

Sprinkling filter: *See* synonym, trickling filter.

Sputtering: A process to deposit a thin layer of a metal on a solid surface in a vacuum. Ions bombard a cathode which emits the metal atoms (EPA-83/03).

Square cut glass: An optical glass cut in small squares, separated and designated by weight (*See* glass for more related terms) (EPA-83).

Square sheet: Papers with equal tensile and tearing strength in machine and cross machine directions (EPA-83).

Square wave thermal pulse: A thermal field reactor characterized by very small (3 milli second) time duration between temperature of inlet and exit gas in the reactor (EPA-88/12).

Squeeze out: *See* synonym, flashing.

Stability:
(1) The state or quality of being stable. In air modeling, atmospheric stability refers to the degree of turbulence in the ambient air. It may vary widely from hour to hour and day to day, and is represented in models by turbulent stability "classes." Under conditions of low turbulence, dispersion of pollutants is limited (cf. atmospheric stability) (EPA-88/09).
(2) A measure of instrument drift. This is a general term and does no specify whether the value is zero drift or span drift (LBL-76/07-bio).
(3) A measure of the length of time. A measuring system, once calibrated, continues to measure the actual parameter value within the calibrated accuracy without the need for adjustment or recalibration. Stability performance is based upon the measurement of standard calibrating solutions with an uncontaminated sensor (LBL-76/07-water).
(4) Property of a bleached pulp to retain its brightness against age, heat, etc (EPA-83).

Stability (of the atmosphere): The atmosphere's tendency to resist or enhance vertical mo on or, alternatively, to suppress or augment existing turbulence. Stability is related to both wind shear and temperature in the vertical, but generally the latter is used as an indicator of the condition (DOE-91/04).

Stabilization: A conversion of the active organic matter in sludge into inert, harmless material (EPA-89/12).

Stabilization lagoon: *See* synonym, lagoon.

Stabilization pond: *See* synonym, lagoon.

Stabilizer:
(1) A substance added to nitrocellulose propellants to prevent decomposition product from catalyzing further decomposition (EPA-76/03).
(2) A chemical bath, usually the last in a processing cycle, that imparts greater life to a processed photographic film or paper through one of several preserving steps (EPA-80/10).

Stable air:
(1) An air mass that remains in the same position rather than moving in its normal horizontal and vertical directions. Stable air does not disperse pollutants and can lead to high build-up of air pollution (*See* air for more related terms) (EPA-74/11).
(2) A motionless mass of air that holds instead of dispersing pollutants (EPA-94/04).

Stabilization: Conversion of the active organic matter in sludge into inert, harmless material (EPA-94/04).

Stabilization ponds: *See* lagoon (EPA-94/04).

Stack (chimney or smoke stack):
(2) Any chimney, flue, conduit, or duct arranged to conduct emissions (or combustion products) to the ambient air (cf. flue) (40 CFR 129.2-91, *See also* 40 CFR 51.100-91; EPA-84/09; 89/03b).
(2) A chimney, smokestack, or vertical pipe that discharges used air (EPA-94/04).
Example:
- Determine the stack cross-sectional area and flue gas discharge velocity.

Data:
- Stack diameter = 1.2 ft, flow rate = 1461.5 acfm (*See* example in actual cubic feet per minute).

Solution:
- Stack cross-sectional area (A) = $3.14(\text{diameter})^2/4$ = $3.14(1.2)^2/4$ = 1.131 sq ft; and discharge velocity = flow rate/A = 1461.5/1.131 = 21.5 ft/sec.

Other stack related terms include:
- Bypass stack
- Dump stack
- Emergency safety stack (*See* synonym, dump stack)
- Evase stack
- Inactive stack
- Multiple stack

Stack effect:
(1) The vertical movement of hot gases in a stack that results because they are hotter (lighter) than the atmosphere (SW-108ts).
(2) The upward movement of house air when the weather is cold, caused by the buoyant force on the warm house air. House air leaks out at the upper levels of the house, so that outdoor air (and soil gas) must leak in at the lower levels to compensate. The continuous exfiltration upstairs and infiltration downstairs maintain the stack effect air movement, so named because it is similar to hot combustion gases rising up a fireplace or furnace or furnace flue stack (EPA-88/08).
(3) Air, as in a chimney, that moves upward because it is warmer than the ambient atmosphere (EPA-94/04).

Stack emission:
(1) Air emissions from a combustion facility stack (EPA-89/11).
(2) The particulate matter captured and released to the atmosphere through a stack, chimney, or flue (40 CFR 60.381-91, *See also* 40 CFR 60.671-91).
(3) *See* emission for more related terms.

Stack gas: *See* synonym, combustion gas or flue gas (EPA-94/04)

Stack gas CO concentration: Each combustion system has a typical operating range for CO. If a stack gas CO concentration goes above this typical range, combustion problems are likely (*See* combustion indicator for more related terms) (EPA-89/03b).

Stack gas oxygen concentration: The stack gas oxygen concentration provides a measure of excess air. Waste incinerators typically operate at 140 to 200% excess air which roughly corresponds to 12 to 14% oxygen concentration in the stack gas (*See* combustion indicator for more related terms) (EPA-89/03b).

Stage II controls: Systems placed on service station gasoline pumps to control and capture gasoline vapors during refuelling (EPA-94/04).

Stack in existence: That the owner or operator had:
(1) Begun, or caused to begin, a continuous program of physical on-site construction of the stack; or
(2) Entered into binding agreements or contractual obligations, which could not be cancelled or modified without substantial loss to the owner or operator, to undertake a program of construction of the stack to be completed within a reasonable time (40 CFR 51.100-91).

Stack sampling: The collecting of representative samples of gaseous and particulate matter that flows through a duct or stack (*See* sampling for more related terms) (SW-108ts).

Stack tip downwash: The downward motion of the plume caused by the formation of a low pressure area in the wake of a stack (EPA-88/09).

Stacked capacitor: A device containing multiple layers of dielectric and conducting materials and designed to store electrical charge (EPA-83/03).

Stage combustion (or off-stoichiometric combustion): One of NO_x emission reduction techniques (*See* nitrogen oxide emission control for control structure). During staged combustion, air and fuel mixtures are combusted in two separate zones. In one zone, the fuel is fired with less than a stoichiometric amount of air. This creates a fuel rich local zone in the regions of the primary flame. The second zone is an air rich zone where the remainder of the combustion air is introduced to complete the combustion of the fuel. The reasons for reducing the NO_x emissions by this technique are:
(1) Lack of available oxygen for NO_x formation.
(2) Temperature is lower, because combustion is incomplete.
(3) The air mixed with the fuel is sub-stoichiometric in the NO_x forming region of the flame, thus creating a low NO_x condition (EPA-81/12, p7-8).

Stage construction: In waste treatment, the building of wastewater treatment plants in steps, so that treatment units serving a small group of homes can be expanded as additional homes are built (DOI-70/04).

Stage aeration: The aeration of activated sludge in more than one stages. Each stage may include both sedimentation tank and sludge return systems (*See* aeration for more related terms).

Stage digestion: The digestion of wastewater in several stages which may all be connected in series (*See* digester for more related terms).

Stage trickling filter: The treatment of wastewater in more than one trickling filters (*See* filter for more related terms).

Stagnant zone: Air mass not flowing with the main stream (EPA-88/09a).

Stagnation:
(1) A meteorological condition which is characterized by the occurrence of a stable air mass with low wind speeds that persists over several days. High concentrations of air pollution may occur caused by the accumulation of pollutants in this air mass (NATO-78/10).
(2) Lack of motion in a mass of air or water that holds pollutants in place (EPA-94/04).

Stain: A solution or suspension of coloring matter in a vehicle designed primarily to be applied to create color effects rather than to form a protective coating. A transparent or semi-opaque coating that colors without completely obscuring the grain of the surface (EPA-79/12b).

Stainless steel: The steels which have good or excellent corrosion resistance. (One of the common grades contains 18% chromium and 8% nickel). There are three broad classes of stainless steels: ferritic; austenitic; and martensitic. These various classes are produced through the use of various alloying elements in differing quantities (*See* steel for more related terms) (EPA-83/06a).

Stall barn: The specialized facilities wherein producing cows and replacement cows are milked and fed in a fixed location (40 CFR 412.11-91).

Stamping: A general term covering almost all press operations. It includes blanking, shearing, hot or cold forming, drawing, bending and coining (EPA-83/06a).

Standard: The norms that impose limits on the amount of pollutants or emissions produced. EPA establishes minimum standards, but states are allowed to be stricter (EPA-94/04).

Standard (or limitation):
(1) In an analytical standard:
 (A) **Primary standard:** A substance or item the value of which can be accepted (within stated limits) and used without question to establish the same value of another substance or item; and
 (B) **Secondary standard:** A substance or item the value of which is based on some primary standards (ACS-87/11).
(2) A specified set of rules or conditions (DOE-91/04).
Other standard related terms include:
- Calibration standard
- Instrument check standard
- Primary standard
- Secondary standard

Standard air: Dry air weighing 0.075 pounds per cubic foot at sea level (29.92 inches barometric pressure) and 70° F (EPA-83).

Standard atmospheric pressure: The average atmospheric pressure at sea level, 45° north latitude at 35° F. It is equivalent to a pressure of 14.696 pound-force per square inch exerted at the base of a column of mercury 29.92 inches high. In the SI system this pressure is equal to 101.325 kPa. In the metric system, the average atmospheric pressure is 760 mm Hg or 1 atm. Weather and altitude are responsible for barometric pressure variations (*See* pressure for more related terms) (EPA-81/12, p2-3; 85/09).

Standard boiling point: The temperature at which the pressure of the saturated vapor of a liquid is the same as the standard pressure. The measured boiling point is dependent on the atmospheric pressure. This dependence can be described quantitatively by the Clausius-Clapeyron equation (provided in this subpart) (40 CFR 796.1220-91).

Standard bushel: A bushel of shelled corn weighing 56 pounds (40 CFR 406.11; 406.21; 406.41-91).

Standard calomel electrode: *See* synonym, calomel electrode.

Standard condition: A temperature of 70° F and a pressure of 14.7 psia (40 CFR 52.741-91, *See also* 40 CFR 60.2; 60.51a-91). It is a specified state. Typical standard conditions include:
(1) In engineering application, standard conditions include:
 (A) For a British system, the standard conditions often used and their corresponding volumes at 1 pound-mole weight are as follows(EPA-84/09):
   - One atm, 32° F. Its corresponding volume can be obtained by using the ideal gas equation: $PV = nR_uT$
   - $V_{60} = nR_uT/P = 1 \times 1545(60 + 460)/(14.6959 \times 144) = 379$ ft^3.
   - $V_{70} = 1 \times 1545(70 + 460)/(14.6959 \times 144) = 387$ ft^3
 (B) For a metric system, at 0° C or 273 K and 1 atm,
   - $V_0 = 1$(kg-mole) $\times 0.08206$(atm-m^3)/(kg-mole-K)$(0 + 273)/1$(atm) $= 22.4$ m^3
 (C) The ideal gas equation can be used to calculate a gas volume at any conditions. For example, assume a constant pressure, the equation can be:
   - Mole basis calculation
     - $V2 = V1(n2)T2/[(n1)T1]$, $n1 = 1$ for standard condition
     - $V2 = V1(n2)T2/T1$
   - Mass basis calculation
     - $PV = nR_uT = 1 \times R_uT$ for standard condition
       $= 28.9RT$, where $R = R_u/M$, $M =$ air molecular weight

$$V2 = V1(m2)T2/(28.9T1)$$

Where: 1 and 2 refer to standard and actual temperature conditions respectively; n = moles, m = mass, V = volume, P = pressure; and T = temperature.

(2) In emission analysis, standard conditions use 29.92 in. w.c. [inch water column, (760 millimeters of mercury)] and 68° F (20° C). A cubic foot measured at this pressure and temperature is known as a standard cubic foot. When a stack test is performed to check the level of emissions from an incinerator, both temperature and pressure are measured during the test in addition to the pollutant of interest. The test results are then converted to standard conditions (grain/dry standard cubic foot) using the temperature and pressure measured. In this way, all test results of all sources including incinerators can be compared on the same basis, i.e., all results are reduced to standard conditions (EPA-83/06).

Example:
- A volume of 20 m³ was drawn from a spirometer at 20° C and 700 mm Hg. What was the standard volume drawn (EPA-83/06, p3-45)?

Solution:
- Using the ideal gas equation, the equation, $V_2 = V_1(P_1T_2/P_2T_1)$, can be applied to this case.

Where: V_2 = volume at condition 2 (standard condition to be determined)
V_1 = volume at condition 1 = 20 m³
P_1 = pressure at condition 1 = 700 mm Hg
P_2 = pressure at condition 2 = 760 mm Hg
T_1 = temperature at condition 1 = 20° C + 273 = 293 K
T_2 = temperature at condition 2 = 25° C + 273 = 298 K
V_2 = (20 x 700 x 298)/(760 x 293) = 18.7 m³

Standard cubic foot (scf): The volume of one cubic foot of gas at standard conditions (40 CFR 52.741-91).

Standard cubic feet per minute: *See* definition under scfm.

Standard curve: A curve which plots concentrations of known analyte standard versus instrument response to the analyte.

Standard day conditions: The standard ambient conditions as described in the United States Standard Atmosphere, 1976, (i.e., Temperature = 15° C, specific humidity = 0.00 kg/H₂O/kg dry air, and pressure = 101325 Pa) (40 CFR 87.1-91).

Standard deviation: A measure of the dispersion about the average (mean) of the elements in a sample. An estimate of the standard deviation of a population (EPA-84/03).

Standard electrode (or reference electrode): An electrode with a known potential in a half cell which is used to measure the electrode potential of other cells (*See* electrode for more related terms).

Standard electrode potential: The measurement of potential relative to a standard hydrogen half cell (*See* electrode for more related terms).

Standard enthalpy of reaction: A change in enthalpy of formation (at a standard state of 298 K and 1 atm) for a given chemical reaction (cf. enthalpy) (EPA-88/12).

Standard equipment: Those features or equipment which are marketed on a vehicle over which the purchaser can exercise no choice (40 CFR 86.082.2-91).

Standard ferromanganese: That alloy as defined by ASTM Designation A9976 (incorporated by reference-*See* 40 CFR 60.17) (40 CFR 60.261-91).

Standard for sewage sludge use or disposal: The regulations promulgated pursuant to section 405(d) of the CWA which govern minimum requirements for sludge quality, management practices, and monitoring and reporting applicable to sewage sludge or the use or disposal of sewage sludge by any person (40 CFR 122.2-91, *See also* 40 CFR 501.2-91).

Standard for the development of test data: A prescription of:
(1) (A) Health and environmental effects; and
 (B) Information relating to toxicity, persistence, and other characteristics which affect health and the environment, for which test data for a chemical substance or mixture are to be developed and any analysis that is to be performed on such data; and
(2) To the extent necessary to assure that data respecting such effects and characteristics are reliable and adequate:
 (A) The manner in which such data are to be developed;
 (B) The specification of any test protocol or methodology to be employed in the development of such data; and
 (C) Such other requirements as are necessary to provide such assurance (TSCA3-15USC2602-91).

Standard industrial classification manual: The Standard Industrial Classification Manual (1987), Superintendent of Documents, U. S. Government Printing Office, Washington, DC 20402 (incorporated by reference as specified in 40 CFR 52.742) (40 CFR 52.741-91).

Standard impinger: A specific instrument employing wet impingement, typically using a liquid volume of 75 ml and a gas flow of 1 cu. ft. per min. (*See* impinger for more related terms).

Standard Industrial Classification (SIC) code: SIC defines industries in accordance with the composition and structure of the economy and covers the entire field of economic activity (EPA-83/06a). SIC code is used by the U.S. Department of Commerce to denote segments of industry (EPA-87/10a).

Standard method: A method of known and demonstrated precision issued by an organization generally recognized as competent to do so (*See* method for more related terms) (ACS-87/11).

Standard of performance: A standard for the control of the discharge of pollutants which reflects the greatest degree of effluent reduction which the Administrator determines to be achievable through application of the best available demonstrated control technology, processes, operating methods,s or other alternatives, including, where practicable, a standard permitting no discharge of pollutants (CWA306, *See also* CAA111; 40 CFR 401.11-91).

Standard operating procedure: A document which describes in detail an operation analysis, or action which is commonly accepted as the preferred method for performing certain routine or repetitive tasks (40 CFR 30.200-91, *See also* 40 CFR 61.61-91; EPA-

86/10a).

Standard or limitation: Any emission standard or limitation established or publicly proposed pursuant to the Act or pursuant to any regulation under the Act (40 CFR 2.301-91, *See also* 40 CFR 2.302-91).

Standard pressure: A pressure of 760 mm of Hg (29.92 in. of Hg) (40 CFR 61.61-91).

Standard raw waste load (SRWL): The raw waste load which characterizes a specific subcategory. This is generally computed by averaging the plant raw waste loads within a subcategory (EPA-76/03).

Standard reference material: *See* definition under reference material.

Standard reference method: A standard method of demonstrated accuracy (*See* method for more related terms) (ACS-87/11).

Standard sample:
(1) The aliquot of finished drinking water that is examined for the presence of coliform bacteria (*See* sample for more related terms) (40 CFR 141.2-91).
(2) The part of finished drinking water that is examined for the presence of coliform bacteria (EPA-94/04).

Standard saybolt universal (SSU): A unit for measuring kinematic viscosity (EPA-81/09).

Standard solution: A solution of known concentration used for volumetric analysis.

Standard temperature: A temperature of 20° C (69° F) (40 CFR 61.61-91).

Standard wipe test: For spills of high-concentration PCBs on solid surfaces, a cleanup to numerical surface standards and sampling by a standard wipe test to verify that the numerical standards have been met. This definition constitutes the minimum requirements for an appropriate wipe testing protocol. A standard-size template (10 centimeters (cm) x 10 cm) will be used to delineate the area of cleanup; the wiping medium will be a gauze pad or glass wool of known size which has been saturated with hexane. It is important that the wipe be performed very quickly after the hexane is exposed to air. EPA strongly recommends that the gauze (or glass wool) be prepared with the hexane in the laboratory and that the wiping medium be stored in sealed glass vials until it is used for the wipe test. Further, EPA requires the collection and testing field blanks and replicates (40 CFR 761.123-91).

Standardized mortality ratio (SMR): The ratio of observed deaths to expected deaths (EPA-92/12).

Standby: That condition in which a reactor facility is neither operable nor declared excess and in which documentary authorization exists to maintain the reactor for possible future operation (*See* DOE Order 5480.6) (DOE-91/04).

STANJAN equilibrium code: A computer program developed at the Stanford University which enables one to quickly calculate the equilibrium high-temperature (or pressure) composition of an organic mixture (EPA-88/12).

Stannous chloride (SnCl): A white crystal used in reducing agents, dyes, etc.

Starch (($C_6H_{10}O_5$)$_n$): An important component of plant cells. Its exact chemical structure is unknown.

Starch coated paper: The paper in which starch is used as an adhesive for pigments (*See* paper for more related terms) (EPA-83).

Stark Einstein law: The second law of photochemistry, states that only one molecule is activated to an excited state per photon or quantum of light absorbed (*See* law for more related terms) (40 CFR 796.3700-91).

Start of a response action: The point in time when there is a guarantee or set-aside of funding either by EPA, other federal agencies, states or Principal Responsible Parties in order to begin response actions at a Superfund site (EPA-94/04).

Start of response action: The point in time when there is a guarantee or set-aide of funding either by EPA, other Federal agencies, States, or PRPs in order to begin response activities at a site. The document which reflects the set-aside of, or formally guarantees, funding during the coming fiscal year, is EPA's annual Superfund Comprehensive Accomplishments Plan (SCAP) (40 CFR 35.4010-91).

Starting material: A substance used to synthesize or purify a technical grade of active ingredient (or the practical equivalent of the technical grade ingredient if the technical grade cannot be isolated) by chemical reaction (40 CFR 158.153-91).

Startup: The setting in operation of a source for any purpose (40 CFR 52.01-91, *See also* 40 CFR 52.741; 60.2; 61.02; 264.1031-91).

Starved air: *See* synonym, substoichiometric combustion air.

Starved air combustion: *See* synonym, substoichiometric combustion air.

Starved air incineration: *See* synonym, substoichiometric combustion air.

Starved air incinerator (controlled air incinerator, fixed hearth incinerator, or pyrolytic incinerator): The term "starved-air" is derived from the principle of combustion most frequently used in this type of incinerator. The combustion air to the chamber into which the waste is fed (primary combustion chamber) is strictly controlled so that the amount of air present is less than that needed for complete combustion, i.e., the chamber is "starved" for air. This type incinerator is often referred to as a controlled air incinerator because the amount and distribution of air to each combustion chamber is controlled to meet the design requirements.

The unit employs a two-stage combustion process, much like rotary kilns. Waste is ram fed into the first stage, or primary chamber, and burned at roughly 50 to 80 percent of stoichiometric air requirements. This starved air condition causes most of the volatile fraction to be destroyed pyrolytically. The resultant smoke and pyrolytic products, consisting primarily of volatile hydrocarbons and carbon monoxide, along with products of combustion, pass to the second stage, or secondary chamber. Here additional air is injected to complete combustion, which can occur either spontaneously or through the addition of supplementary fuels (cf. incinerator, excess-air).

Controlled-air incinerators come in all sizes and shapes. Incinerators are available with design capacities ranging from 50 lb/h (23 kg/h) to 4,000 lb/h (1,800 kg/h). Some are manually controlled, and others are automatically controlled. Some use manual waste loading and ash removal, and others are fully automated (*See* incinerator for more related terms) (EPA-89/03b; Oppelt-87/05).

State: *See* synonym, thermodynamic state.

State 404 program: *See* synonym, section 404 program.

State agency: The State water pollution control agency designated by the Governor having responsibility for enforcing State laws relating to the abatement of pollution (40 CFR 35.905-91).

State authority: The agency established or designated under section 4007 (RCRA1004-42USC6903).

State certifying authority:
(1) For water pollution control facilities, the State pollution control agency as defined in section 502 of the Act;
(2) For air pollution control facilities, the air pollution control agency designated pursuant to 40 CFR 302(b)(1) of the Act; or
(3) For both air and water pollution control facilities, any interstate agency authorized to act in place of the certifying agency of a State (40 CFR 20.2-b-91).

State delayed compliance order: A delayed compliance order issued by a State or by a political subdivision of a State (40 CFR 65.01-91).

State emergency response commission (SERC): Commission appointed by each state governor according to the requirements of SARA Title III. The SERCs designate emergency planning districts, appoint local emergency planning committees, and supervise and coordinate their activities (EPA-94/04).

State/EPA agreement: An agreement between the Regional Administrator and the State which coordinates EPA and State activities, responsibilities and programs including those under the CWA programs (40 CFR 122.2-91, *See also* 40 CFR 144.3; 270.2-91).

State implementation plans (SIP): EPA approved state plans for the establishment, regulation, and enforcement of air pollution standards. Static Water Depth: The vertical distance from the centerline of the pump discharge down to the surface level of the free pool while no water is being drawn rom the pool or water table (EPA-94/04).

State lead agency: *See* synonym, State or State lead agency.

State or local air monitoring stations (SLAMSs): The SLAMS make up the ambient air quality monitoring network which is required by 40 CFR 58.20 to be provided for in the State's implementation plan. This definition places no restrictions on the use of the physical structure or facility housing the SLAMS. Any combination of SLAMS and any other monitors (Special Purpose, NAMS, PSD) may occupy the same facility or structure without affecting the respective definitions of those monitoring station (40 CFR 58.1-91).

State or State lead agency: As used in this subpart means the State agency designated by the State to be responsible for registering pesticides to meet special local needs under Sec. 24(c) of the Act (40 CFR 162.151-j-91).

State primary drinking water regulation: A drinking water regulation of a State which is comparable to a national primary drinking water regulation (40 CFR 142.2-91).

State program:
(1) A program established by a State pursuant to section 1253 of this title to regulate surface coal mining and reclamation operations, on lands within such State in accord with the requirements of this chapter and regulations issued by the Secretary pursuant to this chapter (SMCRA701-30USC1291-90).
(2) *See* Section 404 program (40 CFR 233.2-91).

State program revision: A change in an approved State primacy program (40 CFR 142.2-91).

State regulated waters: Those waters of the United States in which the Corps suspends the issuance of section 404 permits upon approval of a State's section 404 permit program by the Administrator under section 404(h). The program cannot be transferred for those waters which are presently used, or are susceptible to use in their natural condition or by reasonable improvement as a means to transport interstate or foreign commerce shoreward to their ordinary high water mark, including all waters which are subject to the ebb and flow of the tide shoreward to the high tide line, including wetlands adjacent thereto. All other waters of the United States in a State with an approved program shall be under jurisdiction of the State program, and shall be identified in the program description as required by part 233 (40 CFR 232.2-91).

State regulatory authority: The department or agency in each State which has primary responsibility at the State level for administering this chapter (SMCRA701-30USC1291).

State sewage sludge management agency: The agency designated by the Governor as having the lead responsibility for managing or coordinating the approved State program under this Part (40 CFR 501.2-91).

State water pollution control agency: The State agency designated by the Governor having responsibility for enforcing State laws relating to the abatement of pollution (CWA502-33USC1362-91).

Statement: A written approval by EPA, or if appropriate, a State, of the application (40 CFR 21.2-91, *See also* 40 CFR 27.2-91).

Statement of work (SOW): The portion of the cooperative agreement application and/or Superfund State Contract that describes the purpose and scope of activities and tasks to be carried out as a part of the proposed project (40 CFR 35.6015-91).

Static: The test solution is not renewed during the period of the test (40 CFR 797.1400-91).

Static bed: Stationary bed, not a moving bed.

Static calibration: The artificial generation of the response curve of an instrument or method by use of appropriate mechanical, optical, electrical, or chemical means. Often a static calibration checks only a portion of a measurement system. For example, a solution containing a known amount of sulfite compound would simulate an absorbing solution through which has been bubbled a gas containing a known amount of sulfur dioxide. Use of the solution would check out the analytical portion of the pararosaniline method, but would not check out the sampling and flow control parts of the bubbler system (LBL-76/07-air).

Static head: Pressure of a motionless fluid. The pressure is due to the relative positions between two points.

Static loaded radius arc: A portion of a circle whose center is the center of standard tire-rim combination of an automobile and whose radius is the distance from that center to the level surface on which the automobile is standing, measured with the automobile at curb weight, the wheel parallel to the vehicles longitudinal centerline, and the tire inflated to the manufacturer's recommended pressure (40 CFR 86.084.2-91).

Static pressure:
(1) The pressure of a fluid at rest, or in motion, exerted perpendicularly to the direction of a flow.
(2) The pressure exerted by the fluid normal to the streamlines, and it is constant across the flow if the streamlines are parallel (EPA-83/06).
(3) The pressure exerted in all directions by a fluid; measured in a direction normal (perpendicular) to the direction of flow (EPA-89/03b).
(4) *See* pressure for more related terms.

Static system: A test chamber in which the test solution is not renewed during the period of the test (40 CFR 795.120-91, *See also* 40 CFR 797.1050; 797.1300; 797.1930-91).

Static test: A toxicity test with aquatic organisms in which no flow of test solution occurs. Solutions may remain uncharged throughout the duration of the test (40 CFR 797.1350-91, *See also* 40 CFR 797.1440-91).

Static replacement test: A test method in which the test solution is periodically replaced at specific intervals during the test (40 CFR 797.1160-91).

Static water level:
(1) Elevation or level of the water table in a well when the pump is not operating (EPA-94/04).
(2) The level or elevation to which water would rise in a tube connected to an artesian aquifer, or basin, in a conduit under pressure (EPA-94/04).

Station blackout: The complete loss of off-site and on-site AC electric power, except for that generated by inverters fed by batteries (DOE-91/04).

Station wagon: A passenger automobile with an extended roof line to increase cargo or passenger capacity, cargo compartment open to the passenger compartment, a tailgate, and one or more rear seats readily removed or folded to facilitate cargo carrying (40 CFR 600.002.85-91).

Stationary: A condition or process of which the properties are independent of time (NATO-78/10).

Stationary casting: The pouring of molten aluminum into molds and allowing the metal to air cool (40 CFR 467.02-91, *See also* 40 CFR 471.02-91).

Stationary compactor:
(1) A powered machine which is designed to compact solid waste or recyclable materials, and which remains stationary when in operation (40 CFR 243.101-91, *See also* 40 CFR 246.101-91).
(2) A machine that reduces the volume of solid waste by forcing it into a container (SW-108ts).
(3) *See* compactor for more related terms.

Stationary crucible furnace: Stationary crucible furnace is almost identical to a pit furnace except that is not sunk in a pit (*See* furnace for more related terms) (AP-40, p238).

Stationary emission source: A stationary facility that releases combustion gases or vapors to the environment (EPA-83).

Stationary emission source and stationary source: An emission source which is not self-propelled (40 CFR 52.741-91).

Stationary gas turbine: Any simple cycle gas turbine, regenerative cycle gas turbine or any gas turbine portion of a combined cycle steam/electric generating system that is not self propelled. It may, however, be mounted on a vehicle for portability (*See* turbine for more related terms) (40 CFR 60.331-91).

Stationary grate: *See* synonym, fixed grate.

Stationary packer: An adjunct of a refuse container system which compacts refuse at the site of generation into a pull-on detachable container (EPA-83).

Stationary source:
(1) Any building, structure, facility, or installation which emits or may emit any air pollutant (CAA111, *See also* CAA112; CAA302; 40 CFR 51.165; 51.301; 51-App/S; 52.01; 61.02; 65.01-91).
(2) A fixed-site producer of pollution, mainly power plants and other facilities using industrial combustion processes (EPA-94/04).
(3) *See* source for more related terms.

Stationary weight indicator: One of solid or sludge flow rate meters. This method, which includes weigh hoppers/bins and platform scales, determines the dead weight of material loaded into a hopper, bin, or container. After weighing, the contents are then fed as batches into the process. All of these weigh systems give fairly accurate monitoring of weight (within +/- 1%), but one must consider the batch feeding system operations before a true appraisal of the feed rate monitoring can be made (*See* flow rate meter for more related terms) (EPA-89/06).

Stationery: The writing paper suitable for pen and ink, pencil, or typing. Matching envelopes are included in this definition (40 CFR 250.4-91).

Statistic: A constant or coefficient that describes some characteristic of a sample. Statistics are used to estimate parameters of populations (EPA-84/03).

Statistical model: A model which is based on the statistical analysis of a time series of data which may consist of air quality data, metorological data, etc. (e.g., multiple regression models) (NATO-78/10).

Statistical significance (or significance): The statistical significance determined by using appropriate standard techniques of multivariate analysis with results interpreted at the 95 percent confidence level and based on data relating species which are present in sufficient numbers at control areas to permit a valid statistical comparison with the areas being tested (40 CFR 228.2-91, *See also* EPA-79/12c).

Statistical sound level: The level in decibels that is exceeded in a stated percentage (x) of the duration of the measurement period. It is abbreviated as L_x (40 CFR 201.1-91).

Statistical thermodynamics: The study of thermodynamics is coupled with statistical techniques.

Statistically significant increase in a release: An increase in the quantity of the hazardous substance released above the upper bound of the reported normal range of the release (*See* release for more related terms) (40 CFR 302.3-91).

Statistical stability: A condition whereby if a process were to be repeated over time, differences would occur due solely to random processes (EPA-79/12c).

Statistical variance: The sum of the squared deviations about the mean value divided by n-1 (numbers of occurrence minus one). A measure used to identify the dispersion of a set of data (EPA-79/12c).

Statute: *See* definition under law development.

Steady and unsteady flow: A flow is steady, if its properties such as temperature and pressure are invariant with (or independent of) time. Conversely, the flow is unsteady (*See* flow for more related terms) (Perry-73).

Steady flow: A flow in which all thermodynamic properties such as temperature and pressure at each point within an open system remain constant with respect to time (Jones-60, p49).

Steady state:
(1) The time period during which the amounts of test chemical being taken up and depurated by the test oysters are equal, i.e., equilibrium (40 CFR 797.1830-91, *See also* 40 CFR 797.1520-91).
(2) A condition in which the amount of test material being taken up and depurated is equal at a given water concentration. This is condition is also known as the **apparent plateau** (40 CFR 797.1560-91).

Steady state bioconcentration factor: The mean concentration of the test chemical in test organisms during steady-state divided by the mean concentration of the test chemical in the test solution during the same period (40 CFR 797.1830-91).

Steady state model:
(1) In air quality modeling, a model which is based on the assumption of stationary conditions during the period that the model is applied (eg Gaussian plume model) (NATO-78/10).
(2) In water quality modeling, a fate and transport model that uses constant values of input variables to predict constant values of receiving water quality concentrations (EPA-85/09).
(3) A fate and transport model that uses constant values of input variables to predict constant values of receiving water quality concentrations (EPA-91/03).

Steady state permeation: The constant rate of permeation that occurs after breakthrough when all forces affecting permeation have reached equilibrium (NIOSH-84/10).

Steam: A mixture of vapor and liquid. Other steam related terms include:
- Dry saturated steam (*See* synonym, dry steam)
- Dry steam
- Saturated steam
- Super heated steam
- Wet saturated steam
- Steam stripping
- Steam distillation

Steam and water injection: One of NO_x emission reduction techniques (*See* nitrogen oxide emission control for control structure). These techniques are used mainly for NO_x emission reduction in gas turbines and internal combustion engines. Steam or water is injected into the combustion area to lower the peak flame temperature and thus reduce thermal NO_x emissions (EPA-81/12, p7-13).

Steam blister: A defect caused by too high a temperature in a paper machine's dryer section (EPA-83).

Steam electric generating station: An electric generating station utilizing steam turbines for the motive force of its prime movers (*See* electric generating plant for more related terms) (EPA-83).

Steam conditioning: A conditioning method in which unseasoned wood is subjected to an atmosphere of steam at about 120° C (249° F) to reduce its moisture content and improve its permeability preparatory to preservative treatment (EPA-74/04).

Steam distillation: Steam distillation takes advantage of the unique vapor pressure relationship of two immersible liquids. The additive vapor pressures allow distillation at lower temperatures, e.g., toluene boiling point at 1 atm is 111° C, and the mixture of toluene and water has boiling point 84° C. The presence of water lowers the boiling point allowing lower temperature to be used in the distillation.

Steam distillation is usually conducted in a batch operation, i.e., the still is charged and heated to the temperature desired. The distillation is conducted by bubbling steam through the liquid phase. In semi-batch operation, the charge containing a high ratio of volatiles to non-volatiles is fed to the still continuously for a given period while the volatile component is steam-distilled from the mixture (*See* distillation or steam for more related terms).

Steam drum: A vessel in which the saturated steam is separated from the steam-water mixture and into which the feedwater is introduced (EPA-82/11f).

Steam generating unit: Any furnace, boiler, or other device used for combusting fuel for the purpose of producing steam (including fossil-fuel-fired steam generators associated with combined cycle gas turbines; nuclear steam generators are not included) (40 CFR 60.41a-91, *See also* 40 CFR 60.41b; 60.41c; 61.301-91).

Steam generating unit operating day: A 24-hour period between 12:00 midnight and the following midnight during which any fuel is combusted at any time in the steam generating unit. It is not necessary for fuel to be combusted continuously for the entire 24-hour period (40 CFR 60.41b-91, *See also* 40 CFR 60.41c-91).

Steam generator: The equipment which burns fuel and changes water into steam (EPA-83).

Steam stripping: *See* definition under stripping.

Steam stripping operation: A distillation operation in which vaporization of the volatile constituents of a liquid mixture takes place by the introduction of steam directly into the charge (40 CFR 264.1031-91).

Steam turbine: A steam or gas turbine is an enclosed rotary type of prime mover in which heat energy in steam or gas is converted into mechanical energy by the force of a high velocity flow of steam or gases directed against successive rows of radial blades fastened to a central shaft (*See* turbine for more related terms) (EPA-83).

Steaming: Treating a wood material with steam to soften it (EPA-74/04).

Steel:
(1) An iron-base alloy containing carbon, manganese, and often other alloying elements. Steel is defined here to include only those iron-carbon alloys containing less than 1.2 percent carbon by weight (40 CFR 464.31-91).
(2) Refined iron. A typical blast furnace iron and a common steel 1020 have the following impurity composition (EPA-74/06a):

	Iron	Steel
Carbon	3-4.5	0.18-0.23
Silicon	1-3	---
Sulfur	0.04-0.2	<0.05
Phosphorus	0.1-1.0	<0.04
Manganese	0.2-2.0	0.3-0.6

Other steel related terms include:
- Specialty steel
- Stainless steel

Steel basis material: The cold rolled steel, hot rolled steel, and chrome, nickel and tin coated steel which are processed in coil coating (40 CFR 465.02-91).

Steel production cycle: The operations conducted within the BOPF steelmaking facility that are required to produce each batch of steel, including the following operations: scrap charging, preheating (when used), hot metal charging, primary oxygen blowing, sampling (vessel turndown and turnup), additional oxygen blowing (when used), tapping, and deslagging. Hot metal transfer and skimming operations for the next steel production cycle are also included when the hot metal transfer station or skimming station is an affected facility (40 CFR 60.141a-91, *See also* 40 CFR 60.141-91.

STEL: A 15-minute time-weighted average exposure which should not be exceeded at any time during a work day even if the eight-hour time-weighted average is within the TLV. Exposures at the STEL should not be longer than 15 minutes and should not be repeated more than four times per day. There should be at least 60 minutes between successive exposures at the STEL. An averaging period other than 15 minutes may be recommended when this is warranted by observed biological effects (NIOSH-84/10).

Step 1 facilities planning: The preparation of a plan for facilities as described in 40 CFR 35, Subpart E or I (40 CFR 6.501-91).
- **Step 2:** A project to prepare design drawings and specifications as described in 40 CFR 35, Subpart E or I (40 CFR 6.501-91).
- **Step 3:** A project to build a publicly owned treatment works as described in 40 CFR 35, Subpart E or I (40 CFR 6.501-91).
- **Step 2+3:** A project which combines preparation of design drawings and specifications as described in 40 CFR 6.501(b) and building as described in 40 CFR 6.501(c) (40 CFR 6.501-91, *See also* 40 CFR 35.2005-91).

Step aeration: The loading of incoming wastewater into an aeration system is distributed at several inlets (steps) (*See* aeration for more related terms).

Step down transformer: A transformer in which the AC voltages of the secondary windings are lower than those applied to the primary windings (*See* transformer for more related terms) (EPA-83/03).

Step drilling: Using a multiple diameter drill (*See* drilling for more related terms) (EPA-83/06a).

Steppe: An area of grass-covered and generally treeless plains, with a semiarid climate (DOE-91/04).

Steppe climate (semiarid climate): The type of climate in which precipitation is very slight but sufficient for the growth of short, sparse grass (DOE-91/04).

Sterilization:
(1) In pest control, the use of radiation and chemicals to damage body cells needed for reproduction.
(2) The destruction, by chemical or physical, means of a microorganism's ability to reproduce to render something barren (EPA-83).
(3) The removal or destruction of all microorganisms, including pathogenic and other bacteria, vegetative forms and spores (EPA-94/04).

Sterilize: The use of a physical or chemical procedure to destroy all microbial life including highly resistant bacterial endospores (cf. disinfect or pasteurize) (29 CFR 1910, *See also* EPA-83).

Steroid: Any one of a large group of substances chemically related to various alcohols found in plants and animals (EPA-83/09).

Stibnite: An antimony sulfide (Sb_2S_3) which is the most important ore of antimony (EPA-82/05).

Stick water: The water which has been in close contact with fish and has large amounts of organics entrained in it (*See* water for more related terms) (EPA-74/06).

Stickies: Contaminants which stick to paper mill operating equipment (EPA-83).

Stiffness: Stiffness is the ability of paper or paperboard to resist deformation under stress (EPA-83).

Still: A device used to volatilize, separate, and recover petroleum solvent from contaminated solvent, together with the piping and ductwork used in the installation of this device (40 CFR 60.621-91, *See also* EPA-74/04c).

Still bottom: The residue remaining after distillation of a material. The residue can vary from a watery slurry to a thick tar which may turn hard when cooled (EPA-87/10a).

Still rinse: No water flowing in and out as a running rinse and may be a reclaim rinse or dumped periodically to wastewater (EPA-74/03d).

Still water: A pipe, chamber, or compartment with comparatively small inlet or inlets communicating with a main body of water. Its purpose is to dampen waves or surges while permitting the water level within the well to rise and fall with the major fluctuations of the main body of water. It is used with water-measuring devices to improve accuracy of measurement (*See* water for more related terms) (EPA-76/03).

Stillwell: A pipe, chamber, or compartment with small inlets to a main body of water. Its purpose is to dampen waves or surges while permitting the water level within the well to rise and fall with the major fluctuations of the main body of water (EPA-83/09).

Stipend: The supplemental financial assistance, other than tuition and fees, paid directly to the trainee by the recipient organization (40 CFR 45.115-91, *See also* 40 CFR 46.120-91).

Stochastic: Based on the assumption that the actions of a chemical substance result from probabilistic events (Course 165.6).

Stochastic model: A model in which dispersion is described statistically by the random motions of a number of clusters within a turbulence field which is also represented by random moving clusters. The clusters interact through collision and coagulation processes and they are also influenced by decay smaller clusters (NATO-78/10).

Stock on hand: The products which are in the possession, direction or control of a person and are intended for distribution in commerce (40 CFR 763.163-91).

Stock configuration: That no modifications have been made to the original equipment motorcycle that would affect the noise emissiors of the vehicle when measured according to the acceleration test procedure (40 CFR 205.165-91).

Stock solution: The concentrated solution of the test substance which is dissolved and introduced into the dilution water (40 CFR 797.1520-91, *See also* 40 CFR 797.1600-91).

Stoichiometric:
(1) A term applied to a mixture of fuel and air containing precisely the amount of air required for combustion of the fuel. Thus, no oxygen remains in the flue gas following the combustion process. Few combustion processes are designed for operation with a stoichiometric mixture. Excess air is usually provided, the amount varying with the process and type of fuel, e.g., odor control with an afterburner fired by natural gas would employ a minimum of excess air; a garbage incinerator might be designed with introduction of 100% excess air (EPA-84/09).
(2) Characterized by or being a proportion of substances or energy in a specific chemical reaction in which there is no excess of any reactant or product (LBL-76/07-air).

Stoichiometric air: *See* synonym, stoichiometric combustion air.

Stoichiometric coefficient: The number that precedes the chemical symbol in a chemical reaction equation. For example: for the equation, $2H_2 + O_2 \longrightarrow 2H_2O$, the numbers 2 for H_2; 1 for O_2; and 2 for H_2O are the stoichiometric coefficients.

Stoichiometric combustion (or theoretical combustion): A combustion condition that a combustible mixture contains an exact amount of oxidant (oxygen or combustion air) for the complete combustion of the mixture (cf. stoichiometric) (*See* combustion for more related terms).

Stoichiometric combustion air (stoichiometric air, theoretical air or theoretical combustion air):
(1) The quantity of air required to completely burn organic compounds and their associated fuel (if any) with no oxygen

appearing in the products of combustion. The theoretical combustion air is also known as theoretical air, stoichiometric combustion air or stoichiometric air.
(2) Theoretical air is the calculated amount of air required to supply oxygen for complete combustion of a given quantity of a specific combustible material. Also referred to as stoichiometric air (cf. stoichiometry) (EPA-83).
(3) *See* combustion air for more related terms.

Stoichiometry: Material balances involving chemical reactions (EPA-84/09).

Stoke flow: It is used to describe particle settling dynamics in the laminar flow regime, e.g., Reynolds number is less than 2.0 (*See* flow, flow regime or Stokes law for more related terms)) (EPA-84/09).

Stokes law:
(1) The law used to describe particle settling dynamics in the laminar flow regime, i.e., Reynolds number less than 2.0 (cf. flow regime) (EPA-84/09).
(2) The total drag force or resistance of the medium due to fluid motion relative to the particle is the sum of form and friction drag. When particle motion is described by this equation, it is said to be in the Stokes regime (EPA-90/08).
(3) *See* law for more related terms.

Stoker: A movable grate (a mechanical device) designed to feed solid fuel or solid waste to a furnace (cf. grate) (SW-108ts). Other stoker related terms include:
- Chain grate stoker
- Incinerator stoker
- Inertial grate stoker
- Oscillating grate stoker
- Reciprocating grate stoker
- Rocking grate stoker
- Rotary kiln stoker
- Traveling grate stoker

Stone: An imperfection; crystalline contaminations in glass (EPA-83).

Stone feed: Thr limestone feedstock and millscale or other iron oxide additives that become part of the product (40 CFR 60.341-91).

Storing: Piling, stacking, warehousing, and so forth (AP-40, p790).

Storage:
(1) The holding of hazardous waste for a temporary period, at end of which the hazardous waste is treated, disposed of, or stored else where (40 CFR 260.10-93). Storage methods include containers (S01), tanks (S02), waste piles (S03), surface impoundments (S04) and other (S05) (40 CFR 264-App/I-93). (*See also* RCRA1004; 40 CFR 191.02; 243.101; 246.101; 259.10; 260.10; 270.2; 373.4-91).
(2) Methods of keeping raw materials, finished goods, or products while awaiting use, shipment, or consumption (EPA-85/11).
(3) Temporary holding of waste pending treatment or disposal, as in containers, tanks, waste piles, and surface impoundments (EPA-94/04).

Storage and retrieval of aerometric data (SAROAD) system: A computerized system which stores and reports information relating to ambient air quality (40 CFR 58.1-91).

Storage battery: A battery that can store chemical energy with the potential to change to electricity. This conversion of chemical energy to electricity can be reversed thus allowing the battery to be recharged (*See* battery for more related terms) (EPA-84/08).

Storage bin: A facility for storage (including surge bins and hoppers) or metallic minerals prior to further processing or loading (40 CFR 60.381-91, See60.671-91).

Storage coefficient: The volume of water an aquifer releases from or takes into storage per unit surface of the aquifer per unit change in head (Course, 165.7)

Storage container: Any stationary vessel of more than 1,000 gallons (3,785) liters nominal capacity Stationary vessels include portable vessels placed temporarily at the location; e.g., tanks on skids (40 CFR 52.2285-91, *See also* 40 CFR 52.2286-91).

Storage for disposal: The temporary storage of PCBs that have been designated for disposal (40 CFR 761.3-91).

Storage pit: A pit in which solid waste is held prior to processing (SW-108ts).

Storage tank or storage vessel: Any stationary tank, reservoir or container used for the storage of VOL's (40 CFR 52.741-91).

Storage vessel: Any tank, reservoir, or container used for the storage of petroleum liquids, but does not include:
(1) Pressure vessels which are designed to operate in excess of 15 pounds per square inch gauge without emissions to the atmosphere except under emergency conditions;
(2) Subsurface caverns or porous rock reservoirs; or
(3) Underground tanks if the total volume of petroleum liquids added to and taken from a tank annually does not exceed twice the volume of the tank (40 CFR 60.111-91, *See also* 40 CFR 60.111a; 60.111b; 60.691-91).

Storativity: The volume of water that an aquifer releases from or takes into storage per unit surface area of aquifer per unit change in the component of head (fluid pressure plus elevation) normal to the surface (DOE-91/04).

STORET: EPA's computerized water quality data base that includes physical, chemical, and biological data measured in waterbodies throughout the United States (EPA-91/03).

Storm drain: *See* synonym, storm sewer.

Storm sewage:
(1) The liquid flowing in sewers during or following a period of heavy rainfall and resulting therefrom (EPA-76/03).
(2) A system of pipes (separate from sanitary sewers) that carries only water runoff from buildings and land surfaces (EPA-94/04).

(3) *See* sewage for more related terms.

Storm sewer (or storm drain): A sewer intended to carry only storm waters, surface runoff, street wash waters, and drainage (*See* sewer for more related terms) (40 CFR 35.905-91, *See also* 40 CFR 35.2005; 60.691-91; EPA-74/11).

Storm water: The storm water runoff, snow melt runoff, and surface runoff and drainage (*See* water for more related terms) (40 CFR 122.26-91).

Storm water discharge associated with industrial activity: The discharge from any conveyance which is used for collecting and conveying storm water and which is directly related to manufacturing, processing or raw materials storage areas at an industrial plant (for complete definition, *See* 40 CFR 122.26-91).

Storm water lake: A reservoir for storaging storm water runoff collected from plant site; also, auxiliary source of process water (*See* lake for more related terms) (EPA-83/06a).

Storm water or wastewater collection system: The piping, pumps, conduits, and any other equipment necessary to collect and transport the flow of surface water runoff resulting from precipitation, or domestic, commercial or industrial wastewater to and from retention areas or any areas where treatment is designated to occur. The collection of storm water and wastewater does not include treatment except where incidental to conveyance (40 CFR 280.12-91).

Storm water sewer system: A drain and collection system designed and operated for the sole purpose of collecting stormwater and which is segregated from the process wastewater collection system (40 CFR 60.691-91).

Straight chain: *See* definition under chain.

Straight kraft recovery furnace: A furnace used to recover chemicals consisting primarily of sodium and sulfur compounds by burning black liquor which on a quarterly basis contains 7 weight percent or less of the total pulp solids from the neutral sulfite semichemical process or has green liquor sulfidity of 28 percent or less (*See* furnace for more related terms) (40 CFR 60.281-91).

Strain: Elastic deformation due to stress (EPA-83).

Strategic special nuclear material: The uranium-235 (contained in uranium enriched to 20 percent or more in the U^{235} isotope), uranium-233, or plutonium (10 CFR 70.4-91).

Stratification: Separating into layers (EPA-94/04), e.g., (1) The division of a population into sub-populations for sampling purposes. (2) The separation of environmental media (e.g., air or water) into layers as in lakes (Course 165.6)

Stratified fluid: A fluid in which a density variation along the axis of gravity is present, influencing the fluid motions (NATO-78/10).

Stratosphere:
(1) That part of the atmosphere above the tropopause (CAA111-42USC7411-91).

(2) The portion of the atmosphere 10-to-25 miles above the earth's surface (EPA-94/04).

Stratum (plural strata): A single sedimentary bed or layer, regardless of thickness, that consists of generally the same kind of rock material (40 CFR 144.3-91, *See also* 40 CFR 146.3-91).

Stream deposition: The laying down of solid materials carried by a stream, which may take the form of channel deposits, flood deposits, bars, spits, fans, or deltas (DOI-70/04).

Stream flow source zone: The upstream headwaters area which drains into the recharge zone as defined in the December 16, 1975, Notice of Determination (40 CFR 149.101-91).

Stream function: The function in two dimensions, which is constant along each streamline in a fluid (NATO-78/10).

Streamline: *See* synonym, flow line.

Street canyon model: The model which describes the dispersion of air pollutants by the local air circulation in a street between two rows of high buildings. Primarily of importance where the the dispersion of traffic emissions are concerned (NATO-78/10).

Street refuse (or street waste): The wastes materials picked up by manual or mechanical sweeping of streets and sidewalks, litter from public receptacles and dirt removed from catch basins (*See* refuse for more related terms) (EPA-83).

Street waste (or street refuse): Materials picked up by manual or mechanical sweepings of alleys, streets, and sidewalks; wastes from public waste receptacles; and material removed from catch basins (*See* waste for more related terms) (40 CFR 243.101-91).

Strength: A term to indicate relative thickness in sheet glass (EPA-83).

Stress: An internally distributed force; it is the internal mechanical reaction of the material accompanying deformation (Markes-67). In fluid mechanics, the force per unit area on a given surface in a fluid (NATO-78/10).

Stress crack - ASTM D163: An external or internal rupture in a plastic caused by tensile stress less than its short-time mechanical strength (cf. environmental stress crack) (EPA-91/05).

Stress crack - ASTM D3: An external or internal crack in a plastic caused by tensile stresses less than its short-time mechanical strength. Note: The development of such cracks is frequently accelerated by the environmental to which the plastic is exposed. The stresses which cause cracking may be present internally or externally or may be combinations of these stresses (EPA-91/05).

Stress level: The level of tangential or hoop stress, usually expressed as a percentage of specified minimum yield strength (40 CFR 195.2-91).

Stressed water: Those receiving environments in which an applicant can demonstrate to the satisfaction of the Administrator, that the absence of a balanced, indigenous population is caused

Stress relieved: The heat treatment used to relieve the internal stresses induced by forming or heat treating operations. (It consists of heating a part uniformly, followed by cooling slow enough so as not to reintroduce stresses. To obtain low stress levels in steels and cast irons, temperatures as high as 1250-F may be required.) (EPA-83/06a).

solely by human perturbations other than the applicant's modified discharge (*See* water for more related terms) (40 CFR 125.58-91).

Stretching: Any change in length, width or diameter of the cargo tank, or any change to a cargo tank motor vehicle's undercarriage that may affect the cargo tank's structural integrity (40 CFR 180.403-91).

Stria: A cord of low intensity generally of interest only in optical glass (cord) (EPA-83).

Striatum: Corpus striatum (LBL-76/07-bio).

Strict liability: The assessment of liability for damages without requiring a showing of negligence (*See* common law for more related terms) (Sullivan-95/04, p15).

Strike:
(1) For noun: A thin coating of metal (usually less than 0.0001 inch in thickness) to be followed by other coatings (EPA-74/03d).
(2) For noun: A solution used to deposit a strike (EPA-74/03d).
(3) For verb: To plate for a short time, usually at a high initial current density (EPA-74/03d).

Strip: To take off RACM from any part of a facility or facility components (40 CFR 61.141-91).

Strip, sheet, and miscellaneous products: The steel products other than wire products and fasteners (40 CFR 420.121-91).

Strip, sheet and plate: Those acid pickling operations that pickle strip, sheet or plate products (40 CFR 420.91-91).

Strip cropping: Growing crops in a systematic arrangement of strips or bands that serve as barriers to wind and water erosion (EPA-94/04).

Strip mining: A process that uses machines to scrape soil or rock away from mineral deposits just under the earth's surface (EPA-94/04).

Stripper: Includes any vessel in which residual vinyl chloride is removed from polyvinyl chloride resin, except bulk resin, in the slurry form by the use of heat and/or vacuum. In the case of bulk resin, stripper includes any vessel which is used to remove residual vinyl chloride from polyvinyl chloride resin immediately following the polymerization step in the plant process flow (40 CFR 61.61-91).

Stripping: Physical transfer of dissolved molecules from a liquid waste stream to a flowing gas (air stripping) or vapor stream (steam stripping). It is normally carried out as a continuous operation that employs a packed tower for air stripping and a conventional fractional distillation column for steam stripping (cf. ammonia stripping).

(1) For air stripping, liquid waste is pumped to the system near the top of the stripping column and distributed uniformly across the packing. It flows downward by gravity countercurrent to an air flow which is blown into the base of the tower and flows upward. As the air passes through the column, it contacts with the liquid waste, and strips the volatile organics in the waste and carries them to the atmosphere.
(2) For steam stripping, the preheated wastewater enters near the top of the distillation column and then flows down by gravity countercurrent to the steam rising up from the bottom of the column. As the steam passes through the column, it contacts with the liquid waste, and strips the volatile organics in the waste and carries them to a condenser where the mixture of organic vapors and steam is condensed in a water-cooled heat exchanger and collected in an accumulator tank.
(3) For the immiscible organics, the organics and the aqueous liquid (condensed steam) separate due to different gravity. The separated organics might be recovered for re-use as a solvent or a fuel, and the separated aqueous liquid which is saturated with the organic contaminant(s) may be combined with the feed and returned to the top of the stripping column. For the miscible organics, the organics is dissolved in the condensed water and there is no phase separation. The mixture may therefore require further treatment.

Strong acid: An acid which can completely ionize in a solution, e.g., hydrogen chloride (HCl), sulfuric acid (H_2SO_4) or nitric acid (HNO_3) (*See* acid for more related terms).

Strong base: A base which can completely ionize in a solution, e.g., sodium hydroxide (NaOH), potassium hydroxide (KOH) (*See* base for more related terms).

Strong chelating agent: All compounds which, by virtue of their chemical structure and amount present, form soluble metal complexes which are not removed by subsequent metals control techniques such as pH adjustment followed by clarification or filtration (40 CFR 413.02-91).

Strontium (Sr): An alkaline earth metal with a.n. 38; a.w. 87.62; d. 2.6 g/cc; m.p. 768° C and b.p. 1380° C. The element belongs to group IIA of the periodic table.

Structural clay products: Ceramic products used in construction of buildings, walks, roads and pipe lines (EPA-83).

Structural deformation: Distortion in walls of a tank after liquid has been added or removed (EPA-94/04).

Structural glass:
(1) Flat glass, usually colored or opaque, and frequently ground and polished, used for structural purposes.
(2) Glass block, usually hollow, used for structural purposes (EPA-83).
(3) *See* glass for more related terms.

Structural member: Any load-supporting member of a facility, such as beams and load supporting walls; or any nonload-supporting member, such as ceilings and nonload-supporting walls (40 CFR 61.141-91).

Structure:
(1) A microscopic bundle, cluster, fiber, or matrix which may contain asbestos (40 CFR 763-App/A-91).
(2) (*See* building, structure, facility, or installation in 40 CFR 51.165; 51.166; 52.21; 52.24-91).

Strychinine ($C_{21}H_{22}O_2N_2$): A toxic compound used in destroying rodents and predatory animals.

Stuff: Term given to a very fluid mixture of pulp and water - also referred to as slush or stock (EPA-83).

Stuff and burn: A situation in which the charging rate is greater than burning rate of an incinerator (EPA-89/03b).

Study: Any experiment at one or more test sites, in which a test substance is studied in a test system under conditions or in the environment to determine or help predict its effects, metabolism, product performance (efficacy studies only as required by 40 CFR 158.640), environmental and chemical fate, persistence, or other characteristics in humans, other living organisms, or media. The term does not include basic exploratory studies carried out to determine whether a test substance as any potential utility (40 CFR 160.3-91, *See also* 40 CFR 716.3; 720.3; 723.50; 792.3-91).

Stuffing box pressure: The fluid (liquid or gas) pressure inside the casing or housing of a piece of equipment, on the process side of the inboard seal (40 CFR 61.241-91).

Styrene (C_7H_7): A toxic liquid used in plastics, resins and rubbers.

Subagreement: A written agreement between an EPA recipient and another party (other than another public agency) nd ny lower tier agreement for services, supplies, or construction necessary to complete the project. Subagreements include contracts and subcontracts for personal and professional services, agreements with consultants, and purchase orders (40 CFR 30.200-91, *See also* 40 CFR 33.005; 35.936.1; 35.4010-91).

Subbituminous coal: Coal that is classified as subbituminous A, B, or C according to the American Society of Testing and Materials (ASTM) Standard Specification for Classification of Coals by Rank D388-77 (incorporated by reference-*See* 40 CFR 60.17) (*See* coal for more related terms) (40 CFR 60.41a-91).

Subcategorization: The process of segmentation of an industry into groups of plants for which uniform effluent limitations can be established (EPA-85/10a).

Subchronic: Of intermediate duration, usually used to describe studies or levels of exposure between 5 and 90 days (EPA-94/04).

Subchronic delayed neurotoxicity: A prolonged, delayed onset locomotor ataxia resulting from repeated daily administration of the test substance (40 CFR 798.6560-91).

Subchronic dermal toxicity: The adverse effects occurring as a result of the repeated daily exposure of experimental animals to a chemical by dermal application for part (approximately 10 percent) of a life span (40 CFR 798.2250-91).

Subchronic exposure:
(1) Multiple or continuous exposures occurring usually over 3 months (EPA-92/12).
(2) Multiple or continuous exposures occurring over about 10% of an experimental species lifetime, usually over 3 months (EPA-90/08).

Subchronic inhalation toxicity: The adverse effects occurring as a result of the repeated daily exposure of experimental animals to a chemical by inhalation for part (approximately 10 percent) of a life span (40 CFR 798.2450-91).

Subchronic oral toxicity: The adverse effects occurring as a result of the repeated daily exposure of experimental animals to a chemical for a part (approximately 10 percent for rats) of a life span (40 CFR 795.260-91).

Subchronic study: A toxicity study designed to measure effects from subchronic exposure to a chemical (EPA-92/12).

Subclass: A classification of heavy-duty engines or heavy-duty vehicles based on such factors as gross vehicle weight rating, fuel usage, vehicle usage, engine horsepower or additional criteria (for complete definition, *See* 40 CFR 86.1102.87-91).

Subconfiguration: A unique combination within a vehicle configuration of equivalent test weight, road-load horsepower, and any other operational characteristics or parameters which the Administrator determines may significantly affect fuel economy within a vehicle configuration (40 CFR 600.002.85-91).

Subcooled liquid: *See* synonym, compressed liquid.

Subindex: The calculated index value for a single pollutant as described in section 7 (40 CFR 58-App/G-91).

Sublethal: A stimulus below the level that causes death (EPA-91/03; 85/09).

Sublimation: A change of state directly from a solid state to a gas state without appearance of the liquid state (EPA-77/07).

Submerged aquatic vegetation: Vegetation such as sea grasses that cannot withstand excessive drying and therefore live with their leaves at or below the water surface; an important habitat for young fish and other aquatic organisms (EPA-94/04).

Submerged arc furnace: A ferroalloy reduction furnace, the electrodes usually extend to a considerable depth into the charge, hence such furnaces are called submerged-arc furnaces. This name is used for the furnaces whose load is practically entirely of the resistant type (*See* furnace for more related terms) (EPA-75/02).

Submerged loading pipe: Any discharge pipe or nozzle which meets either of the following conditions:
(1) Where the tank is filled from the top, the end of the discharge pipe or nozzle must be totally submerged when the liquid level is 15 cm (6 in.) above the bottom of the tank.

(2) Where the tank is filled from the side, the discharge pipe or nozzle must be totally submerged when the liquid level is 46 cm (18 in.) above the bottom of the tank (40 CFR 52.741-91).

Submerged tube evaporation: The evaporation of feed materials using horizontal steam-heat tubes submerged in solution. Vapors are driven off and condensed while concentrated solution is bled off (EPA-83/06a).

Submission:
(1) A request by a POTW for approval of a Pretreatment Program to the EPA or a Director;
(2) A request by a POTW to the EPA or a Director for authority to revise the discharge limits in categorical Pretreatment Standards to reflect POTW pollutant removals; or
(3) A request to the EPA by an NPDES State for approval of its State pretreatment program (40 CFR 403.3-91).

Subsidence (or land subsidence): The lowering of the natural land surface in response to: Earth movements; lowering of fluid pressure; removal of underlying supporting material by mining or solution of solids, either artificially or from natural causes; compaction due to wetting (hydrocompaction); oxidation of organic matter in soils; or added load on the land (40 CFR 146.3-91, *See also* 40 CFR 147.2902-91).

Subsidence inversion: The inversion caused by the adiabatic warming of downward moving air. It is most common in connection with a high pressure area, in which average downward motions are present (*See* inversion for more related terms) (NATO-78/10).

Subsidy: Direct or indirect payment from government to businesses, citizens, or institutions to encourage a desired activity (OTA-89/10).

Subsoil: The part of the soil beneath the topsoil; usually does not have an appreciable organic matter content (SW-108ts).

Substance: Either a chemical substance or mixture unless otherwise indicated (40 CFR 704.3-91, *See also* 40 CFR 716.3; 717.3-91).

Substantial business relationship: The extent of a business relationship necessary under applicable State law to make a guarantee contract issued incident to that relationship valid and enforceable. A "substantial business relationship" must arise from a pattern of recent or ongoing business transactions, in addition to the guarantee itself, such that a currently existing business relationship between the guarantor and the owner or operator is demonstrated to the satisfaction of the applicable EPA Regional Administrator (40 CFR 264.141-h).

Substantial risk notification: TSCA Section 8(e) places upon chemical manufacturers, etc., the responsibility for reporting any indication of adverse effect. In the words of the Act, any person who obtains information which reasonably supports the conclusion that such substance or mixture presents a substantial risk of injury to health or the environment shall immediately inform the Administrator of such information unless such person has actual knowledge (Arbuckle-89).

Substantiation: The written answers submitted to EPA by a submitter to the specific questions set forth in this regulation in support of a claim in chemical identity is a trade secret (40 CFR 350.1-91).

Substate: Refers to any public regional, local, county, municipal, or intermunicipal agency, or regional or local public (including interstate) solid or hazardous waste management authority, or other public agency below the State level (40 CFR 256.06-91).

Substation: A complete assemblage of a plant at a place where electrical energy is received (from one or more power stations) for conversion (e.g., from AC to DC by means of rectifiers, rotary converters), for stepping-up or down by means of transformers), or for for control (e.g., by means of switch-gear, etc.) (EPA-83/03).

Substitution reaction: A chemical reaction in which replacement of one atom, radical or molecule by another one takes place.

Substoichiometric combustion air (deficient air, starved air, starved combution air, or starved combustion air): The combustion air supplied less than those required at stoichiometric levels. In controlled air incineration, the primary chamber is maintained at less than stoichiometric air conditions (*See* combusiton air for more related terms) (EPA-89/03b).

Substrate:
(1) The surface onto which a coating is applied or into which a coating is impregnated (40 CFR 52.741-91, *See also* 40 CFR 60.741-91).
(2) The reactant portion of any biochemical reaction materials which are transformed into a product.
(3) Any substance used as a nutrient by a microorganism.
(4) The liquor in which activated sludge or other material is kept in suspension (EPA-76/03).

Subtransmission (substation) transformer: At the end of a transmission line, the voltage is reduced to the subtransmission level (at substations) by subtransmission transformers (*See* transformer for more related terms) (EPA-83/03).

Subunit: An atom or group of associated atoms chemically derived from corresponding reactants (40 CFR 704.25-91, *See also* 40 CFR 721.350; 723.250-91).

Sucrose: A disaccharide having the formula $C_{12}H_{22}O_{11}$. The terms sucrose and sugar are generally interchangeable, and the common sugar of commerce is sucrose in varying degrees of purity. Refined cane sugar is essentially 100 percent sucrose (EPA-75/02d).

Suction dredge: A centrifugal pump mounted on a barge to lift the materials (dredge) through a suction pipe (*See* dredge for more related terms) (EPA-82/05).

Suction process: Any process where glass is gathered by vacuum into the mold (EPA-83).

Sudden accidental occurrence: An occurrence which is not continuous or repeated in nature (*See* accidental occurrence for more related terms) (40 CFR 264.141-91, *See also* 40 CFR 265.141-91).

Sufficient evidence: According to the U.S. EPA's Guidelines for Carcinogen Risk Assessment, sufficient evidence is a collection of facts and scientific references which is definitive enough to establish that the adverse effect is caused by the agent in question (EPA-92/12).

Sugar: A sweet, crystallizable substance, colorless or white when pure, occurring in many plant juices and forming an important article of human food. The chief sources of sugar are the sugar cane and the sugar beet, the completely refined products of which are identical and form the granulated sugar of commerce. Chemically, sugar is a disaccharide with the formula $C_{12}H_{22}O_{11}$ formed by union of one molecule of dextrose with one molecule of levulose (EPA-74/01a). Other sugar related terms include:
- Raw sugar
- Raw sugar juice
- Refined sugar
- Saccharide
- Saccharin

Suitable substitute decision: The Administrators decision whether a product which the Administrator has determined to be a low noise-emission product is a suitable substitute for a product or products presently being purchased by the Federal Government (40 CFR 203.1-91).

Sulfate: The final decomposition product of organic sulfur compounds (EPA-76/03).

Sulfate pulp: The pulp produced by chemical methods using alkaline solution of caustic soda and sodium sulfide. The sulfate process produces the strongest fibers. A wide variety of trees can be used, especially those with a high resin content, such as pine and douglas fir. Yields are 40 to 50 percent by weight. Kraft pulps are used principally in paperboard and coarse paper grades; unbleached grades are used in packaging, and bleached grades are used in packaging paperboard and a number of paper grades, including printing grades and tissue. Sulfate pulp is a major pulp product (*See* pulp for more related terms) (EPA-83).

Sulfide:
(1) Ionized sulfur, expressed in mg/L as S (EPA-82/11).
(2) Total sulfide (dissolved and acid soluble) as measured by the procedures listed in 40 CFR 136 (40 CFR 410.01-91).

Sulfidity: Sulfidity is a measure of the amount of sulfur in kraft cooking liquor. It is the percentage ratio of NaS, expressed as NaO, to active alkali (EPA-87/10).

Sulfite cooking liquor: Sulfite cooking liquor shall be defined as bisulfite cooking liquor when the pH of the liquor is between 3.0 and 6.0 and as acid sulfite cooking liquor when the pH is less than 3.0 (*See* liquor for more related terms) (40 CFR 430.101; 430.211-91).

Sulfite pulp: The pulp produced by chemical cooking methods using sulfurous acid and a base in the form of a salt. It is used to process low resin woods such as spruce, fir, or hemlock. Sulfite pulp is used to make printing grades of paper, such as business papers, and for tissue; very little is used in paperboard (*See* pulp for more related terms) (EPA-83).

Sulfur (S): A yellow nonmetallic element with a.n. 16; a.w. 32.064; d. 2.07 g/cc; m.p. 119.0° C and b.p. 444.6° C. The element belongs to group VIA of the periodic table.

Sulfur bacteria: The bacteria that oxidize sulfur compounds, precipitating sulfur or producing noxious sulfur gases such as hydrogen sulfide. In this process they may cause damage to concrete or other structures (*See* bacteria for more related terms) (DOI-70/04).

Sulfur dioxide (SO_2):
(1) A heavy, pungent, colorless gas formed primarily by the combustion of fossil fuels. It can damage the respiratory tract as well as vegetation and materials and is considered a major air pollutant (EPA-74/11, *See also* 40 CFR 58.1-91).
(2) A pungent, colorless, gaseous pollutant formed primarily by the combustion of fossil fuels (EPA-94/04).

Sulfur oxide (SO_x): Two major sulfur oxides, sulfur dioxide (SO_2) and sulfur trioxide (SO_3) are important as air contaminants. The primary source of both is the combination of atmospheric oxygen with the sulfur in certain fuels during their combustion. Both the dioxide and trioxide are capable of producing illness and lung injury even at small concentrations, from 5-10 ppm. Further, each can combine with water in the air to form toxic acid aerosols that can corrode metal surface, fabrics, and the leaves of plants (AP-40, p15).

Sulfur oxide emission control: Sulfur oxide emissions from fossil-fuel-fired combustion sources can be reduced by five techniques:
(1) Fuel desulfurization (EPA-81/12, p8-4)
   (A) Physical coal cleaning
   (B) Chemical coal cleaning
      (a) Microwave desulfurization
      (b) Hydrothermal desulfurization
(2) Combustion of coal and limestone mixtures (EPA-81/12, p8-5)
   (A) Fluidized bed combustion
   (B) Limestone coal pellets as fuel
(3) Coal gasification (EPA-81/12, p8-6)
(4) Coal liquefaction (EPA-81/12, p8-6)
(5) Flue gas desulfurization (FGD) (EPA-81/12, p8-6)
   (A) FGD wet scrubbing process
      (a) FGD nonregenerable (throwaway) process (EPA-81/12, p8-7)
         (i) FGD lime scrubbing (EPA-84/03b, p8-6)
         (ii) FGD limestone scrubbing (EPA-81/12, p8-11)
         (iii) FGD double alkali scrubbing (EPA-84/03b, p8-15)
         (iv) FGD sodium-based once-through scrubbing (EPA-84/03b, p8-20)
      (b) FGD regenerable process (EPA-81/12, p8-17)
         (i) FGD Wellman-Lord (EPA-84/03b, p8-27)
         (ii) FGD magnesium oxide (EPA-81/12, p8-21)
         (iii) FGD citrate (EPA-81/12, p8-22)
   (B) FGD dry scrubbing process (EPA-81/12, p8-25)
      (a) FGD spray dryer with a baghouse or ESP
      (b) FGD dry injection

(c) FGD other dry SO₂ processes

Sulfur production rate: The rate of liquid sulfur accumulation from the sulfur recovery unit (40 CFR 60.641-91).

Sulfur recovery unit: A process device that recovers element sulfur from acid gas (40 CFR 60.641-91).

Sulfuric acid (H_2SO_4): A toxic acid liquid used in fertilizers and explosives.

Sulfuric acid concentrator (SAC): An evaporation process which concentrates weak sulfuric acid (68%) to strong sulfuric acid (92%) (EPA-76/03).

Sulfuric acid pickling: Those operations in which steel products are immersed in sulfuric acid solutions to chemically remove oxides and scale, and those rinsing operations associated with such immersions (40 CFR 420.91-91).

Sulfuric acid plant: Any facility producing sulfuric acid by the contact process by burning elemental sulfur, alkylation acid, hydrogen sulfide, or acid sludge, but not include facilities where conversion to sulfuric acid is utilized primarily as a means of preventing emissions to the atmosphere of sulfur dioxide or other sulfur compounds (40 CFR 51.100-91, *See also* 40 CFR 60.161; 60.171; 60.181-91).

Sulfuric acid production unit: Any facility producing sulfuric acid by the contact process by burning elemental sulfur, alkylation acid, hydrogen sulfide, organic sulfides and mercaptans, or acid sludge, but does not include facilities where conversion to sulfuric acid is utilized primarily as a means of preventing emissions to the atmosphere of sulfur dioxide or other sulfur compounds (40 CFR 60.81-91).

Sulfurous acid (H_2SO_3): A unstable liquid used in manufacture of paper, wine, etc.

Sulphur hydrides: Including H_2S, H_2S_2, H_2S_3 and H_2S_5.

Sump:
(1) Any pit or reservoir that meets the definition of tank and those troughs/trenches connected to it that serves to collect hazardous waste for transport to hazardous waste storage, treatment, or disposal facilities (40 CFR 260.10-91).
(2) A pit or tank that catches liquid runoff for drainage or disposal (EPA-94/04).

Sump pump: A mechanism for removing water or wastewater from a sump or wet well (EPA-89/12).

Sunlight direct aqueous photolysis rate constant: The first-order rate constant in the units of day^{-1} and is a measure of the rate of disappearance of a chemical dissolved in a water body in sunlight (40 CFR 796.3700-91).

Super adiabatic: The condition where the temperature lapse rate is smaller than the dry adiabatic lapse rate resulting in unstable conditions (NATO-78/10).

Super chlorination:
(1) A large quantity of chlorine (up to 20 mg/liter) is used during chlorination.
(2) Chlorination with doses that are deliberately selected to produce free or combined residuals so large as to require dechlorination (EPA-94/04).

Super conductivity: No measurable electrical resistance at temperature approaching to 0 K for certain materials.

Super critical: A state at or above the critical point of water (EPA-82/11f).

Super critical fluid (SCF): SCF is characterized as a form of matter in which the liquid and gaseous states are indistinguishable from one another. It is formed when both temperatures and pressures to which the fluid is subjected exceed the critical point of the state. Under these supercritical states, the character of the fluid becomes very unusual compared to that under ambient conditions. For example, if water is under the supercritical conditions, the density, dielectric constant, hydrogen bonding, and certain other physical properties are so altered that water behaves much as a moderately polar organic liquid. Thus, heptane or benzene could become miscible in all proportions with SCW (supercritical water) which cannot happen with water under sub-critical conditions. Even some types of wood fully dissolve in SCW. On the other hand, the solubility of sodium chloride (NaCl) could be as low as 100 ppm, and that of calcium chloride can be less than 10 ppm. This is the reverse of the solubilities in water that are encountered under sub-critical conditions - under which the two salts' solubilities are about 37 wt % and up to 70 wt %, respectively (Lee-83/07).

Super critical state: The state where the pressure is greater than the critical pressure (Wark-p57).

Super critical water: A type of thermal treatment using moderate temperatures and high pressures to enhance the ability of water to break down large organic molecules into smaller, less toxic ones. Oxygen injected during this process combines with simple organic compounds to form carbon dioxide and water (EPA-94/04).

Super duty fireclay brick: A fireclay brick that has a PCE above Cone 33 on the fire product, shrinks less than 1 % in the ASTM permanent linear change test, Schedule C (2910° F), and does not incur more than 4% loss in the panel spalling test (preheated to 3000° F) (*See* brick for more related terms) (SW-108ts).

Super heated gas: *See* synonym, super heated steam.

Super heated steam (super heated vapor or super heated gas): The steam, vapor or gas at a temperature higher than the saturation temperature at a given pressure (*See* steam for more related terms) (Jones-p143).

Super vapor: *See* synonym, super heated steam.

Super mixed paper: The paper that consists of a baled clean sorted mixture of various qualities of papers containing less than 10% of groundwood stock, coated or uncoated (*See* paper for more related terms) (EPA-83).

Super mixed wastepaper: *See* definition under waste paper.

Super news: Consists of baled sorted fresh newspapers, not sunburned, free from papers other than news, containing not more than the normal percentage of rotogravure and colored sections (EPA-83).

Super news wastepaper: *See* definition under waste paper.

Super saturated solution: *See* synonym, oversaturated solution.

Super saturation: The condition of a solution when it contains more solute (sucrose) than that which would be dissolved under normal pressure and temperature (EPA-75/02d).

Superficial velocity: The velocity through packing (absorber or adsorber) assuming the packing was not present. The actual velocity through the packing is lower since the gas can flow only through the void volume of the packing (EPA-84/09).

Superfund:
(1) Federal authority, established by the Comprehensive Environmental Response, Compensation, and Liability Act (CERCLA) in 1980, to respond directly to releases or threatened releases of hazardous substances that may endanger health or welfare (EPA-92/12).
(2) An act regarding hazardous substance releases into the environment and the cleanup of inactive hazardous waste disposal sites (NRT-87/03).
(3) The program operated under the legislative authority of CERCLA and SARA that funds and carries out EPA solid waste emergency and long-term removal and remedial activities. These activities include establishing the National Priorities List, investigating sites for inclusion on the list, determining their priority, and conducting and/or supervising the cleanup and other remedial actions (EPA-94/04).

Superfund Amendments and Reauthorization Act (SARA) of 1986: *See* the term of Act or SARA.

Superfund innovative technology evaluation (SITE):
(1) A 1986 program established by the EPA's Office of Solid Waste and Emergency Response (OSWER) and Office of Research and Development (ORD) to promote the development and use of innovative treatment technologies during CERCLA response actions (EPA-89/12a).
(2) EPA program to promote development and use of innovative treatment technologies in Superfund site cleanups (EPA-94/04).

Superfund memorandum of agreement (SMOA): A nonbinding, written document executed by an EPA Regional Administrator and the head of a state agency that may establish the nature and extent of EPA and state interaction during the removal, pre-remedial, remedial, and/or enforcement response process. The SMOA is not a site-specific document although attachments may address specific sites. The SMOA generally defines the role and responsibilities of both the lead and the support agencies (40 CFR 300.5-91).

Superfund off-site policy and clean-up standards (OSWER-87):
(1) The new clean-up standards provided for in SARA require that Superfund remedies must be protective of human health and the environment, cost-effective, and utilize permanent solutions, alternative treatment technologies and resource recovery to the maximum extent practicable. The on-site remedies must also meet applicable or relevant and appropriate regulations (ARARs) of other federal statutes including: RCRA, TSCA, SDWA, CAA, and CWA. And, where state standards are more stringent than federal standards, state standards must be met. For wastes remaining on-site, the remedial actions are reviewed every five years.
(2) The new clean-up standards are expected to increase the use of mobile treatment units and stabilization techniques to manage waste on site. However, some concentrated hazardous wastes will likely require off-site treatment and disposal. This could increase the demand for commercial capacity. The off-site disposal provision in SARA restricts disposal of Superfund wastes to those facilities in compliance with RCRA and TSCA and applicable state requirements. Specifically, the unit receiving Superfund wastes must not be releasing any hazardous wastes and releases from other units at the facility must be controlled by a corrective action program. Several Superfund sites have experienced difficulties locating a commercial facility eligible to accept their waste.

Superfund operation: The Superfund system is complex. Sites are identified and enter an inventory because they may require a cleanup. At this point, or at any time, a site may receive a removal action because of emergency conditions that require fast action or because the site could get a lot worse before a remedial cleanup could be implemented. (Most of SARA's requirements for remedial cleanups do not apply to removal actions, even though removal actions can cost several million dollars and resemble a cleanup). In the preremedial process, sites receive a Preliminary Assessment (PA); some then go forward to a Site Inspection (SI), with some of those sites scored by the Hazard Ranking System (HRS). If the score is high enough, the site is placed on the National Priorities List (NPL) and becomes eligible for a remedial cleanup paid for by the government, if necessary, or by responsible parties identified as having contributed to creating the uncontrolled toxic waste site. Under current procedures, only about 10 percent of sites which enter the system are likely to be placed on the NPL. Some States have their own lists of sites which require cleanup; these often contain sites not on the NPL.

NPL sites receive a Remedial Investigation and Feasibility Study (RIFS) to define contamination and environmental problems and to evaluate cleanup alternatives. The public is given an opportunity to comment on the RIFS and EPA's preferred cleanup alternative. Then, EPA issues a Record of Decision (ROD) which says what remedy the government has chosen and the reasons for doing so; the decision may be that no cleanup is necessary. A ROD may only deal with part of a site's cleanup and several RODs may be necessary for a site. The ROD also contains a summary of EPA's responses to public comments. EPA chooses the cleanup goals and technology in the ROD. In actual fact a number of actions involving different technologies are likely to be chosen for any but the simplest sites. The ROD is like a contract in which the government makes a commitment to actions which will render the site safe. If responsible parties agree to clean up the site, they sign a negotiated consent decree with the government; this stipulates the exact details of how the responsible parties will proceed. If the cleanup uses Superfund money, the State must agree to pay 10 percent of the cleanup cost.

In the post-ROD process, the site receives a Remedial Design (RD) study to provide details on how the chosen remedy will be engineered and constructed. The whole process ends with the Remedial Action (RA), the actual implementation of the selected remedial. Many cleanups include long-term monitoring to determine whether the cleanup is effective and if more cleanup is necessary. A ROD may be reopened and amended because of new information discovered or difficulties encountered during the design and remedial action. When a cleanup is deemed complete and effective, the site can be delisted by EPA from the NPL (OTA-89/10a).

Superfund state contract (SSC): A joint, legally binding agreement between EPA and another party(s) to obtain the necessary assurances before a EPA-lead remedial action or any political subdivision-lead activities can begin at a site, and to ensure State or Indian Tribe involvement as required under CERCLA section 121(f) (40 CFR 35.6015-91, *See also* 40 CFR 300.5-91).

Superheated gas: *See* synonym, superheated steam.

Superheated steam (superheated vapor or superheated gas): The steam, vapor or gas at a temperature higher than the saturation temperature corresponding to its pressure (*See* steam for more related terms).

Superheated vapor: *See* synonym, superheated steam.

Supernatant: A substance floating above or on the surface of another substance (EPA-87/10a).

Superphosphoric acid plant: Any facility which concentrates wet-process phosphoric acid to 66 percent or greater P2O5 content by weight for eventual consumption as a fertilizer (40 CFR 60.211-91).

Supplier of water:
(1) Any person who owns or operates a public water system (*See* water for more related terms) (SDWA1401, *See also* 40 CFR 141.2; 142.2; 143.2-91).
(2) Any person who owns or operates a public water system (EPA-94/04).

Supplementary control system (SCS): Any technique for limiting the concentration of a pollutant in the ambient air by varying the emissions of that pollutant according to atmospheric conditions. For the purposes of this part, the term supplementary control system does not include any dispersion technique based solely on the use of a stack the height of which exceeds good engineering practice (as determined under regulations implementing section 123 of the Act) (40 CFR 57.103-91).

Supplier of water: Any person who owns or operates a public water system (SDWA1401).

Supplies: All tangible personal property other than equipment as defined in this part (40 CFR 31.3-91, *See also* 40 CFR 33.005; 35.6015-91).

Supply limited material: A secondary material that is not collected in sufficient amounts or is too highly contaminated for current manufacturing processes (OTA-89/10).

Support facilities: In the context of the new production reactor, all the existing, modified, or new facilities needed to support the production of tritium and/or plutonium, except for the reactor facility. Support facilities include fuel and target fabrication facilities; fuel and target processing facilities; waste management facilities; and facilities providing necessary resources such as power, steam, and water (DOE-91/04).

Support media: The quartz sand or glass beads used to support the plant (40 CFR 797.2800-91, *See also* 40 CFR 797.2850-91).

Supporting studies: Those studies that contain information that is useful for providing insight and support for the conclusions (EPA-92/12).

Suppressed combustion: Those basic oxygen furnace steelmaking wet air cleaning systems which are designed to limit or suppress the combustion of carbon monoxide in furnace gases by restricting the amount of excess air entering the air pollution control system (*See* combustion for more related terms) (40 CFR 420.41-91).

Surface active agent: One which modifies physical, electrical, or chemical characteristics of the surface of solids and also surface tensions of solid or liquid. Used in froth flotation (EPA-82/05).

Surface aeration: The violent agitation of water surface to increase dissolved oxygen in the water (*See* aeration for more related terms).

Surface aerator: An aerator for providing surface aeration (*See* aerator for more related terms).

Surface casing: The first string of well casing to be installed in the well (40 CFR 146.3-91).

Surface charge: Electrical charge (usually negative charge) on the surface of many colloids, bacteria and algae.

Surface coal mining and reclamation operations: The surface mining operations and all activities necessary and incident to the reclamation of such operations after August 3, 1977 (SMCRA701-30USC1291-90).

Surface coal mining operations:
(1) Activities conducted on the surface of lands in connection with a surface coal mine or subject to the requirement of section 1266 of this title surface operations and surface impacts incident to an underground coal mine, the products of which enter commerce or the operations of which directly or indirectly affect interstate commerce. Such activities include excavation for the purpose of obtaining coal including such common methods as contour, strip, auger, mountain-top removal, box cut, open pit, and area mining, the uses of explosives and blasting, and in situ distillation or retorting, leaching or other chemical or physical processing, and the cleaning, concentrating, or other processing or preparation, loading of coal for interstate commerce at or near the mine site: provided, however, that such activities do not include the extraction of coal incidental to the extraction of other

minerals where coal does not exceed 16 2/3 per centum of the tonnage of minerals removed for purposes of commercial use or sale or coal explorations subject to section 1262 of this title; and

(2) The areas upon which such activities occur or where such activities disturb the natural land surface. Such areas shall also include any adjacent land the use of which is incidental to any such activities, all lands affected by the construction of new roads or the improvement or use of existing roads to gain access to the site of such activities and for haulage, and excavations, workings, impoundments, dams, ventilation shafts, entryways, refuse banks, dumps, stockpiles, overburden piles, spoil banks, culm banks, tailings, holes or depressions, repair areas, storage areas, processing areas, shipping areas and other areas upon which are sited structures, facilities, or other property or materials on the surface, resulting from or incident to such activities (SMCRA701-91).

Surface coating: The process of coating a copper workpiece as well as the associated surface finishing and flattening (40 CFR 468.02-91).

Surface coating operation: The system on a metal furniture surface coating line used to apply and dry or cure an organic coating on the surface of the metal furniture part or product. The surface coating operation may be a prime coat or a top coat operation and includes the coating application station(s), flash-off area, and curing oven (40 CFR 60.311-91, See also 40 CFR 60.391; 60.451-91).

Surface collecting agents: Those chemical agents that form a surface film to control the layer thickness of oil (40 CFR 300.5-91).

Surface compaction: Increasing the dry density of surface soil by applying a dynamic load (SW-108ts).

Surface condenser:
(1) A device which removes a substance from a gas stream by reducing the temperature of the stream, without direct contact between the coolant and the stream (40 CFR 52.741-91).
(2) A heat exchanger where the coolant is physically separated from the vapor stream, usually by tubular surfaces and the vapor is condensed (EPA-84/09).
(3) The heat is transferred through a barrier that separates the cooling water and the vapor. The condensate can be recovered separately (EPA-75/02d, AP-40, P199).
(4) See condenser for more related terms.

Surface cracking: Discontinuities that develop in the cover material at a sanitary landfill due to the surface drying or settlement of the solid waste. These discontinuities may result in the exposure of solid waste, entrance or egress of vectors, intrusion of water, and venting of decomposition gases (SW-108ts).

Surface drains: Surface channels which primarily remove surface water (EPA-83).

Surface dump: See synonym, dump.

Surface energy balance: The balance between the incoming and outgoing energy at the earth's surface. The energy consists of radiation (short wave and long wave), turbulent fluxes (sensible heat and latent heat flux) an other fluxes (e.g., ground heat flux) (NATO-78/10).

Surface friction: The drag or shear stress of the earth's surface on the air motion above the surface (NATO-78/10).

Surface impoundment:
(2) Treatment, storage, or disposal of liquid hazardous wastes in ponds (EPA-94/04).
(2) See also synonym, impoundment.

Surface irrigation: The process of wastewater irrigation in which wastewater is applied to and distributed over the surface of the ground (See land treatment for more related terms) (EPA-74/01a).

Surface layer: The lowest part of the atmospheric boundary layer adjacent to the surface. Usually defined as the layer in which the shear stress can be considered to be independent of height (NATO-78/10).

Surface loading rate (SLR): Surface loading rate = (wastewater flow)/(water surface of a treatment tank).

Surface material: Material in a building that is sprayed on surfaces, troweled on surfaces, or otherwise applied to surfaces for acoustical, fireproofing, or other purposes, such as acoustical plaster on ceilings and fireproofing material on structural members (TSCA-AIA1-91).

Surface moisture: The water that is not chemically bound to a metallic mineral or metallic mineral concentrate (40 CFR 60.381-91).

Surface runoff: Precipitation, snow melt, or irrigation in excess of what can infiltrate the soil surface and be stored in small surface depressions; a major transporter of nonpoint source pollutants (EPA-94/04).

Surface sized: Paper treated with a water and ink-resistant material (EPA-83).

Surface tension: A measure of the force opposing the spread of a thin film of liquid (EPA-83/06a).

Surface treatment: A chemical or electrochemical treatment applied to the surface of a metal. Such treatments include pickling, etching, conversion coating, phosphating, and chromating. Surface treatment baths are usually followed by a water rinse. The rinse may consist of single or multiple stage rinsing. For the purposes of this part, a surface treatment operation is defined as a bath followed by a rinse, regardless of the number of stages. Each surface treatment bath, rinse combination is entitled to discharge allowance (See treatment for more related terms) (40 CFR 471.02-91).

Surface uranium mines: Strip mining operations for removal of uraniumbearing ore (EPA-94/04).

Surface water:
(1) All water which is open to the atmosphere and subject to surface runoff (40 CFR 141.2-91, *See also* 40 CFR 131.35-91, EPA-85/10a).
(2) Water on the earth's surface, as distinguished from water in the ground (groundwater) (DOE-91/04).
(3) All water naturally open to the atmosphere (rivers, lakes, reservoirs, ponds, streams, impoundments, seas, estuaries, etc.) and all springs, wells, or other collectors directly influenced by surface water (EPA-94/04).
(4) *See* water for more related terms.

Surface weather observation: The observation of different meteorological parameters at the surface, such as temperature, wind speed and direction. These observations are usually routinely done at a large number of stations at fixed hours each day (NATO-78/10).

Surfacing ACM:
(1) The surfacing material that is ACM (40 CFR 763.83-91).
(2) Asbestos-containing material that is sprayed or troweled on or otherwise applied to surfaces, such as acoustical plaster on ceilings and fireproofing materials on structural members (EPA-94/04).

Surfacing material:
(1) Material in a school building that is sprayed-on, troweled-on, or otherwise applied to surfaces, such as acoustical plaster on ceilings and fireproofing materials on structural members, or other materials on surfaces for acoustical, fireproofing, or other purposes (40 CFR 763.83-91).
(2) Material sprayed or troweled onto structural members (beams, columns, or decking) for fire protection; or on ceilings or walls for fireproofing, acoustical or decorative purposes. Includes textured plaster, and other textured wall and ceiling surfaces (EPA-94/04).

Surfactant:
(1) Surface-active agents used in detergents to cause lathering (EPA-83/06a).
(2) Those methylene blue active substances amendable to measurement by the method described in Methods for Chemical Analysis of Water and Wastes, 1971, Environmental Protection Agency, Analytical Quality Control Laboratory, page 131 (40 CFR 417.91-91, *See also* 40 CFR 417.101; 417.111; 417.121; 417.131; 417.141; 417.151; 417.161; 417.171; 417.181; 417.191-91).
(3) A detergent compound that promotes lathering (EPA-94/04).

Surficial: Of or relating to the surface (EPA-80/08).

Surge: A sudden rise to an excessive value, such as flow, pressure, temperature (EPA-83/06a).

Surge control tank: A large sized pipe or storage reservoir sufficient to contain the surging liquid discharge of the process tank to which it is connected (*See* tank for more related terms) (40 CFR 264.1031-91).

Surge pressure: The pressure produced by a change in velocity of the moving stream that results from shutting down a pump station or pumping unit, closure of a valve, or any other blockage of the moving stream (*See* pressure for more related terms) (40 CFR 195.2-91).

Surge tank: A tank for absorbing and dampening the wavelike motion of a volume of liquid; an in-process storage tank that acts as a flow buffer between process tanks (*See* tank for more related terms) (EPA-87/10a).

Surplus energy: Energy generated that is beyond the immediate needs of the producing system. This energy is frequently obtained from spinning reserve and sold on an interruptible basis (*See* electric energy for more related terms) (EPA-83).

Surplus sludge: *See* synonym, excess activated sludge.

Surrogate: In sampling, a known compound to a sample which is chemically similar to the sample so that an estimate of the accuracy of the analytical measurement and an assessment of the overall efficiency of the analytical procedures can be made (EPA-82/02).

Surroundings: The state or matter outside a system (*See* thermodynamic system for more related terms).

Surveillance system: A series of monitoring devices designed to check on environmental conditions (EPA-94/04).

Survey analysis: The survey analysis is to provide an overall description of the sample in terms of the major organic compounds and major inorganic compounds that are present in the sample. The analysis provides a qualitative description of the overall chemistry of the sample (*See* analysis for more related terms) (EPA-82/02).

Susceptibility: The degree to which an organism is affected by a pesticide at a particular level of exposure (40 CFR 171.2-91).

Suspect material: Building material suspected of containing asbestos, e.g., surfacing material, floor tile, ceiling tile, thermal system insulation, and miscellaneous other materials (EPA-94/04).

Suspended: Those elements which are retained by a 0.45 um membrane filter (40 CFR 136-App/C-91).

Suspended loads: Sediment particles maintained in the water column by turbulence and carried with the flow of water (EPA-94/04).

Suspended matter in air: Particles whose density is closed to atmospheric air, thus remaining in the air for a long time.

Suspended matter in water: *See* synonym, suspended solid.

Suspended metal: The concentration of metals determined in the portion of a sample that is retained by a 0.45 μm filter (*See* metal for more related terms) (Method 3005, SW-846).

Suspended solid (or suspended matter in water): Small particles of solid pollutants that float on the surface of, or are suspended in, sewage or other liquids. They resist removal by conventional means (EPA-94/04).

Suspension: Suspending the use of a pesticide when EPA deems it necessary to prevent an imminent hazard resulting from its continued use. An emergency suspension takes effect immediately; under an ordinary suspension a registrant can request a hearing before the suspension goes into effect. Such a hearing process might take six months (EPA-94/04).

Suspension culture: Cells growing in a liquid nutrient medium (EPA-94/04).

Suspension freezing: *See* synonym, crystallization.

Sustainable development: Development is sustainable. Protection of the environment is to ensure further development in the future. The concept was concluded in a meeting of the World Commission on Environment Development (WCED) under the sponsorship of the United Nations. The meeting was held in Tokyo Japan in 1987.

Sustained yield: *See* synonym, safe yield.

Swaging: A process in which a solid point is formed at the end of a tube, rod, or bar by the repeated blows of one or more pairs of opposing dies (40 CFR 471.02-91).

Swamp: A type of wetland dominated by woody vegetation but without appreciable peat deposits. Swamps may be fresh or salt water and tidal or non-tidal (*See* wetlands) (EPA-94/04).

Sweated pig: An ingot prepared from high iron aluminum alloy (EPA-76/12).

Sweating: Bringing small globules of low-melting constituents to an alloy surface during heat treatment (EPA-83/03a).

Sweet: Term applied to easily workable glass (EPA-83).

Sweet water:
(1) The solution of 8-10 percent crude glycerine and 90-22 percent water that is a by-product of saponification or fat splitting (40 CFR 417.41-91).
(2) A byproduct aqueous glycerine from soap manufacture (EPA-74/04c).
(3) *See* water for more related terms.

Sweetening unit: A process device that separates the H_2S and CO_2 contents from the sour natural gas stream (40 CFR 60.641-91).

Swill (or slops): Semi-liquid waste materials consisting of food scraps and free liquids (SW-108ts).

Swing crew method: A method in which one or more reserve work crews go anywhere help is needed (*See* waste collection method for more related terms) (SW-108ts).

Switcher locomotive: Any locomotive designated as a switcher by the builder or reported to the ICC as a switcher by the operator-owning-railroad and including, but not limited to, all locomotives of the builder/model designations listed in Appendix A to this subpart (40 CFR 201.1-91).

Syngas: The synthetic gas resulting from pyrolysis of organic material produced by incomplete combustion of organic matter. The combustible components of syngas are primarily carbon monoxide and hydrogen; (usually about 300 Btu/SCF, but less than 900 Btu/SCF), also called synthesis or pyrolysis gas (EPA-83).

Synergism:
(1) The characteristic property of a mixture of toxicants that exhibits a greater-than-additive total toxic effect (EPA-91/03).
(2) An interaction of two or more chemicals which results in an effect that is greater than the sum of their effects taken independently (EPA-94/04).

Synergistic effect:
(1) The simultaneous action of separate agents which, together, have greater total effect than the sum of their individual effects (cf. additive effect) (EPA-76/03).
(2) Joint effects of two or more agents, such as drugs, that, when taken together, increase each other's effectiveness (LBL-76/07-bio).

Synoptic scale: In meteorology, a length scale of the order of one thousand kilometers. Applicable to phenomena such as synoptic high or low pressure areas (NATO-78/10).

Syntan: A synthetic tanning material, generally used in combination with vegetable, mineral, or formaldehyde tannages. Syntans are almost exclusively used in retanning rather than tanning operations (EPA-82/11).

Synthesis: The production of a substance by the union of elements or simpler chemical compounds (DOD-78/01).

Synthetic ammonium sulfate manufacturing plant: Any plant which produces ammonium sulfate by direct combination of ammonia and sulfuric acid (40 CFR 60.421-91).

Synthetic detergent: A material which has a cleansing action like soap but are not derived directly from fats and oils. Used in ore flotation (*See* detergent for more related terms) (EPA-82/05).

Synthetic fiber: Any fiber composed partially or entirely of materials made by chemical synthesis, or made partially or entirely from chemically-modified naturally-occurring materials (*See* fiber for more related terms) (40 CFR 60.601-91).

Synthetic natural gas (SNG): A fuel usually in the range of 950 to 1050 Btu per standard cubic foot (EPA-83).

Synthetic organic chemicals (SOCs): Man-made organic chemicals. Some SOCs are volatile, others tend to stay dissolved in water instead of evaporating (EPA-94/04).

Synthetic organic chemicals manufacturing industry: The industry that produces, as intermediates or final products, one or more of the chemicals listed in 40 CFR 60.489 (40 CFR 60.481-91).

Synthetic resin (or thermosetting resin): A combination of chemicals which can be polymerized, e.g., by the application of heat, into a compound which is used to produce the bond or improve the bond (*See* resin for more related terms) (EPA-74/04).

Synthetics: All man-made fibers, including those manufactured from naturally occurring raw materials (regenerated fibers) and chemical synthesis (EPA-82/09).

Syringe: A small medical device for drawing fluids from or injecting fluids into human or animal bodies. Other syringe related terms include:
- Gas tight syringe
- Hypodermic syringe

Syrup: A water solution of sugar, usually sucrose (EPA-74/03).

System:
(1) In equipment design, a collection of interdependent equipment and procedures assembled and integrated to perform a well-defined purpose. It is an assembly of procedures, processes, methods, routines, or techniques united by some form of regulated interaction to form an organized whole (DOE-91/04).
(2) *See* thermodynamic system for more related terms.

System audit: A qualitative on-site evaluation of a measurement system. The objective of the audit is to assess and document all facilities, equipment, systems, recordkeeping, data validation, operating, maintenance, calibration procedures, reporting requirements, and quality control procedures. Since the above items should be defined in the project under evaluation, the Quality Assurance Project Plan provides the basis for the audit (*See* audit for more related terms) (EPA-85/08; 86/10a).

System boundary: The separation wall or layer between a system and its surroundings.

System emergency reserves: An amount of electric generating capacity equivalent to the rated capacity of the single largest electric generating unit in the electric utility company (including steam generating units, internal combustion engines, gas turbines, nuclear units, hydroelectric units, and all other electric generating equipment) which is interconnected with the affected facility that has the malfunctioning flue gas desulfurization system. The electric generating capability of equipment under multiple ownership is prorated based on ownership unless the proportional entitlement to electric output is otherwise established by contractual arrangement (40 CFR 60.41a-91).

System load: The entire electric demand of an electric utility company's service area interconnected with the affected facility that has the malfunctioning flue gas desulfurization system plus firm contractual sales to other electric utility companies. Sales to other electric utility companies (e.g., emergency power) not on a firm contractual basis may also be included in the system load when no available system capacity exists in the electric utility company to which the power is supplied for sale (*See* load for more related terms) (40 CFR 60.41a-91).

System losses: The difference between the system net energy or power input and output resulting from characteristic losses and unaccounted-for between the sources of supply and the metering points of delivery on a system (*See* electric loss for more related terms) (EPA-83).

System state: The conditions of a system.

Systemic effect: Systemic effects are those that require absorption and distribution of the toxicant to a site distant from its entry point, at which point effects are produced. Most chemicals that produce **systemic toxicity** do not cause a similar degree of toxicity in all organs, but usually demonstrate major toxicity to one or two organs. These are referred to as the **target organs of toxicity** for that chemical (EPA-92/12).

Systemic pesticide: A chemical that is taken up from the ground or absorbed through the surface and carried through the system of the organism being protected, making the organism toxic to pests (*See* pesticide for more related terms) (EPA-89/12).

Systemic toxicity: *See* definition under systemic effects.

System with a single service connection: A system that supplies drinking water to consumers via a single service line (EPA-94/04).

Systemic pesticide: A chemical absorbed by an organism that makes the body (EPA-94/04).

Tt

Table classifier: A vibrating, ribbed table designed to separate dense ore or metals from the lighter constituents. Normally the classifier is used with a flow of water (cf. tabling) (EPA-75/02c).

Tablet coating operation: A pharmaceutical coating operation in which tablets are coated (40 CFR 52.741-91).

Tabulating paper: The paper used in tabulating forms for use on automatic data processing equipment (*See* paper for more related terms) (40 CFR 250.4-91).

Tack: Stickiness (EPA-91/05).

Taconite ore: A type of highly abrasive iron ores now extensively mined in the United States (EPA-82/05).

Tag: The stiff paper, metal or other hard material that is tied or otherwise affixed to the packaging of a protector (40 CFR 211.203-91).

Tail water: The runoff of irrigation water from the lower end of an irrigated field (EPA-94/04).

Tailing: Residue of raw materials or waste separated out during the processing of crops or mineral ores (EPA-94/04, *See also* 40 CFR 61.251-91; EPA-88/08a).

Tailing pond: The enclosures or basins constructed for the disposal of mine tailings (the fine rock waste in washings from mills after the grinding and processing of ores), they serve as settling basins and prevent or reduce the contamination of streams and other water bodies by such waste (*See* pond for more related terms) (DOI-70/04).

Tailpipe standards: Emissions limitations applicable to engine exhausts from mobile sources (EPA-94/04).

Take out: A mechanical device for removing a finished article from any glass forming unit (EPA-83).

Talc: A fine, soft powder used as filler in some papers (EPA-83).

Tall oil: The oily mixture of rosin acids, and other materials obtained by acid treatment of the alkaline liquors from the digesting (pulping) of pine wood. Used in drying oils, in cutting oils, emulsifiers, and in flotation agents (*See* oil for more related terms) (EPA-82/05).

Tallow: A product made from beef cattle or sheep fat that has a melting point of 40° C or greater (40 CFR 432.101-91).

Tamper:
(1) To introduce a contaminant into a public water system with the intention of harming persons; or
(2) To otherwise interfere with the operation of a public water system with the intention of harming persons (SDWA1432-42USC300i.1).

Tampering:
(1) The removal or rendering inoperative by any person, other than for purposes of maintenance, repair, or replacement, of any device or element of design incorporated into any product in compliance with regulations under section 6, prior to its sale or delivery to the ultimate purchaser or while it is in use; or the use of a product after such deice or element of design has been removed or rendered inoperative by any person (40 CFR 205.151-91, *See also* 40 CFR 204.2; 204.51; 205.2; 205.51-91).
(2) Adjusting, negating, or removing pollution control equipment on a motor vehicle (EPA-94/04).

Tangible net worth: The tangible assets that remain after deducting liabilities; such assets would not include intangibles such as goodwill and rights to patents or royalties (cf. net worth) (40 CFR 144.61-91, *See also* 40 CFR 264.141; 265.141; 280.92-91).

Tanhouse: *See* synonym, tanyard.

Tank:
(1) A stationary waste management unit that is designed to contain an accumulation of waste and is constructed primarily of nonearthen materials (e.g., wood, concrete, steel, plastic) which provide structural support (40 CFR 61.341-91, *See also* 40 CFR 260.10; 280.12-91).
(2) A melting unit, in which the container for the molten glass is constructed from refractory blocks (EPA-83).

Other tank related terms include:
- Aboveground tank
- Aeration tank
- Breakout tank
- Continuous flow stirred tank
- Dosing tank

- Live tank
- Holding tank
- Imhoff tank
- In process tank
- Inground tank
- Onground tank
- Product tank
- Quench tank
- Residential tank
- Separator tank
- Septic tank
- Skimming tank
- Surge tank
- Surge control tank
- Underground storage tank
- Underground tank
- Wash oil circulation tank

Tank current: The total amperage required to electroplate all the workpieces of a tank load (EPA-74/03d).

Tank fuel volume: The volume of fuel in the fuel tank(s), which is determined by taking the manufacturers nominal fuel tank(s) capacity and multiplying by 0.40, the result being rounded using ASTM E 29-67 to the nearest tenth of a U.S. gallon (40 CFR 86.082.2-91).

Tank load: The total number of workpieces being processed (electroplated) simultaneously in the tank (EPA-74/03d).

Tank system: A hazardous waste storage or treatment tank and its associated ancillary equipment and containment system (40 CFR 260.10-91, See also 40 CFR 280.12-91).

Tank vessel: A vessel that is constructed or adapted to carry, or that carries, oil or hazardous material in bulk as cargo or cargo residue, and that:
(1) Is a vessel of the United States;
(2) Operates on the navigable waters; or
(3) Transfers oil or hazardous material in a place subject to the jurisdiction of the United States (OPA1001-91).

Tank water: The water phase resulting from rendering processes usually occurring in wet rendering (EPA-75/01).

Tankage: The dried animal by-product residues used in feedstuffs (40 CFR 432.101-91).

Tannery waste: The waste from tannery industry. The waste often includes chromium, and high levels of total solids, BOD5, etc. (See waste for more related terms).

Tannic acid ($C_{14}H_{10}O_9$): Yellowish powder used in tanning and textiles.

Tanning: A process for treating animal hides. It includes (1) chemical treatment to prevent them from bacteria attack and (2) fat and grease treatment to make them pliable.

Tannins: Chemicals derived from the leaching of bark, nuts, or other vegetable materials used in the vegetable tanning process (EPA-82/11).

Tantalum (Ta): A transition metal with a.n. 73; a.w. 180.948; d. 16.6 g/cc; m.p. 1996° C and b.p. 5425° C. The element belongs to group VB of the periodic table.

Tantalum foil: A thin sheet of tantalum, usually less than 0.006 inch thickness (EPA-83/03).

Tanyard (or tanhouse): The portion of the tannery in which the bating, pickling, and tanning are performed on the hides or skins (EPA-82/11).

Tap: The pouring of molten steel from an EAF or AOD vessel (40 CFR 60.271a-91, See also 40 CFR 60.271-91).

Tapered aeration: The addition of more air at the entrance of an aeration tank and less air near its outlet (See aeration for more related terms).

Tapping: The removal of slag or product from the electric submerged arc furnace under normal operating conditions such as removal of metal under normal pressure and movement by gravity down the spout into the ladle (40 CFR 60.261-91).

Tapping period: The time duration from initiation of the process of opening the tap hole until plugging of the tap hole is complete (40 CFR 60.261-91, See also 40 CFR 60.271-91).

Tapping station: That general area where molten product or slag is removed from the electric submerged arc furnace (40 CFR 60.261-91).

Tar: See synonym, asphalt.

Tar decanter: Any vessel, tank, or container that functions to separate heavy tar and sludge from flushing liquor by means of gravity, heat, or chemical emulsion breakers. A tar decanter also may be known as a flushing-liquor decanter (40 CFR 61.131-91).

Tar storage tank: Any vessel, tank, reservoir, or other type of container used to collect or store crude tar or tar-entrained naphthalene, except for tar products obtained distillation, such as coal tar pitch, creosotes, or carbolic oil. This definition also includes any vessel, tank, reservoir, or container used to reduce the water content of the tar by means of heat, residence time, chemical emulsion breakers, or centrifugal separation. A tar storage tank also may be known as a tar-dewatering tank (40 CFR 61.131-91).

Tar intercepting sump: Any tank, pit, or enclosure that serves to receive or separate tars and aqueous condensate discharged from the primary cooler. A tar-intercepting sump also may be known as a primary-cooler decanter (40 CFR 61.131-91).

Tare: Waste materials which must be discharged. Also, the empty weight of a container used for weighing or transporting materials (EPA-74/01a).

Tared: Counterweighted before use in the sampling procedure to balance the weight of the filter alone (LBL-76/07-bio).

Target analyte spiking: The spiking with the analyte that is of basic interest in the environmental sample (See spike for more related terms) (EPA-84/03).

Target distance limit: The maximum distance over which targets for the site are evaluated. The target distance limit varies by HRS pathway (40 CFR 300-App/A-91).

Target organ of toxicity: See definition under systemic effect.

Tartaric acid ($(CHOH)_2(COOH)_2$): White crystals used in baking, beverage, textile, processing, etc.

Target risk: A value that is combined with exposure and toxicity information to calculate a risk-based concentration (e.g., PRG). For carcinogenic effects, the target risk is a cancer risk of 10^{-2}. For non-carcinogenic effects, the target risk is a hazard quotient of 1 (EPA-91/12).

Task system (daily route method): A collection crew is assigned a weekly route, divided into daily areas. The crew is then responsible for refuse pickup at all collection points on the assigned daily routes. Weather, refuse quantities and other variables will cause the elapsed time for completion of each daily route to vary. The crew is allowed to go home after completion of the day's route, whether it takes less or more than the established work day to complete. (See also large route collection group task system, single load collection method, method, definite working day method) (EPA-83).

Temporary hardness: The carbonate ($(Ca(HCO_3)_2)$) hardness that can be removed by boiling (See hardness for more related terms).

Taxi/idle (in): Those aircraft operations involving taxi and idle between the time of landing roll out and final shutdown of all propulsion engines (40 CFR 87.1-91).

Taxi/idle (out): Those aircraft operations involving taxi and idle between the time of initial starting of the propulsion engine(s) used for the taxi and turn on to duty runway (40 CFR 87.1-91).

Tear strength: Measurement of resistance of pulp fibers to a tearing force (EPA-83).

Technetium (Tc): A radioactive transition metal with a.n. 43; a.w. 98; d. 11.5 g/cc; m.p. 2140° C and b.p. 4876° C. The element belongs to group VIIB of the periodic table.

Technical assistance grant: As part of the Superfund program, Technical Assistance Grants of up to $50,000 are provided to citizens' groups to obtain assistance in interpreting information related to cleanups at Superfund sites or those proposed for the National Priorities List. Grants are used by such groups to hire technical advisors to help them understand the site-related technical information for the duration of response activities (EPA-94/04).

Technical baseline: A configuration identification document (or documents) formally designated and approved at a specific time. Technical baselines, plus approved changes to those baselines, constitute the current configuration identification. The NPR technical baseline is set forth in the Requirements Document (DOE-91/04).

Technical safety appraisal: A documented, multi-discipline appraisal of selected Department of Energy reactors and nuclear facilities conducted by a team selected by DOE's Deputy Assistant Secretary for Safety, Health, and Quality Assurance. It ensures proper application of particular safety elements of the DOE environment, safety, and health program, nuclear industry lessons learned, and appropriate licensed facility requirements as described in DOE 5482.1B, paragraph 9b (DOE-91/04).

Technical grade of active ingredient: A material containing an active ingredient:
(1) Which contains no inert ingredient, other than one used for purification of the active ingredient; and
(2) Which is produced on a commercial or pilot-plant production scale (whether or not it is ever held for sale) (40 CFR 158.153-91).

Technical support document: The Noncompliance Penalties Technical Support Document which accompanies these regulations. The Technical support Document appears as Appendix A to these regulations (40 CFR 66.3-91).

Technological system of continuous emission reduction:
(1) A technological process for production or operation by any source which is inherently low-polluting or nonpolluting; or
(2) A technological system for continuous reduction of the pollution generated by a source before such pollution is emitted into the ambient air, including precombustion cleaning or treatment of fuels (CAA111-42USC7411-91).

Technology-based limitations: Industry-specific effluent limitations applied to a discharge when it will not cause a violation of water quality standards at low stream flows. Usually applied to discharges into large rivers (EPA-94/04).

Technology based standards: Effluent limitations applicable to direct and indirect sources which are developed on a category-by-category basis using statutory factors, not including water-quality effects (EPA-94/04).

Technology screening: The process of collecting technical information on potentially applicable treatment technologies and determining which technologies to retain as alternatives for consideration in the FS (EPA-89/12a).

Teepee burner: See synonym, conical burner.

Tellurium (Te): A metalloid element with a.n. 52; a.w. 127.60; d. 6.24 g/cc; m.p. 449.5° C and b.p. 989.8° C. The element belongs to group VIA of the periodic table.

Temper: The degree of residual stress in annealed glass measured polarimetrically or by polariscopic comparison with a standard (EPA-83).

Temperature: An indicator of the thermal state of matter. There are two systems of temperature scales: Celsius (C) and Fahrenheit (F) scale. Their relationship is as follows: $F = 1.8° C + 32$. Other temperature related terms include:

- Absolute temperature
- Absolute zero temperature
- Adiabatic flame temperature
- Adiabatic saturation temperature
- Approach temperature
- Auto-ignition temperature
- Base temperature
- Boiling point temperature
- Condensation point temperature
- Critical temperature
- Dew point temperature
- Dry bulb temperature
- Fire point temperature
- Flash point temperature
- Fluid temperature
- Freezing point temperature
- Fusion point temperature
- Ignition temperature
- Liquidus temperature
- Melting point
- Melting point temperature
- Pour point temperature
- Reaction temperature
- Saturated temperature (*See* synonym, saturation temperture)
- Saturation temperature
- Virtual temperature
- Wet bulb temperature

Temperature compensation: An adjustment which corrects for the effect of temperature on the measuring system. In addition it may also adjust the measured value of a parameter to a selected temperature based on a known parameter-temperature relationship. Temperature compensation may be incorporated into the measuring system so that the value reported is the adjusted value. If this approach is not used, then manual or algebraic methods are necessary to obtain the adjusted value (LBL-76/07).

Temperature meter: Types of temperature meters include:
(1) Pyrometer
  (A) Optical pyrometer
  (B) Radiation pyrometer
(2) Thermocouple
(3) Thermometer
  (A) Resistance thermometer
  (B) Vapor bulb thermometer

Temperature profile: The description of the temperature as a function of height (NATO-78/10).

Tempered hardboard: The hardboard that has been specially treated in manufacture to improve its physical properties considerably. Includes, e.g., oil-tempered hardboard. Synonym: superhardboard (EPA-74/04).

Tempered glass: A glass that has been rapidly cooled from near the softening point, under rigorous control, to increase its mechanical and thermal endurance (*See* glass for more related terms) (EPA-83).

Tempered safety glass: A single piece of specially heat treated glass with a stress pattern such that the piece when fractured reduces to numerous granular fragments, with no large jagged edges (*See* safety glass for more related terms) (EPA-83).

Tempering: The process whereby glass is heated near the melting point and then rapidly cooled to increase its mechanical and thermal endurance (40 CFR 426.61-91, *See also* EPA-83/06a).

Tempering air: *See* synonym, cooling air.

Temporary enclosure: A total enclosure that is constructed for the sole purpose of measuring the fugitive VOC emissions from an affected facility (40 CFR 60.741-91, *See also* 40 CFR 60.711-91).

Temporary opening: An opening into an enclosure that is equipped with a means of obstruction, such as a door, window, or port, that is normally closed (40 CFR 60.541-91).

Ten (10) year, 24 hour fall event: A rainfall event with a probable recurrence interval of once in ten years as defined by the National Weather Service in Technical Paper No. 40. Rainfall Frequency Atlas of the United States, May 1961, and subsequent amendments, or equivalent regional or state rainfall probability information developed therefrom (40 CFR 411.31-91).

Ten (10) year, 24 hour precipitation event: The maximum 24-hour precipitation event with a probable reoccurrence interval of once in 10 years. This information is available in Weather Bureau Technical Paper No. 40, May 1961 and NOAA Atlas 2, 1973 for the 11 Western States, and may need to be obtained from the National Climatic Center of Environmental Data Service, National Oceanic and Atmospheric Administration, U.S. Department of Commerce (40 CFR 436.21-91, *See also* 40 CFR 436.31; 436.41; 436.181; 440.132-91).

Ten (10) year, 24 hour rainfall event: The maximum precipitation event with a probable recurrence interval of once in 10 years as defined by the National Weather Service in Technical paper No. 40, Rainfall Frequency Atlas of the United States, May 1961, and subsequent amendments or equivalent regional or State rainfall probability information developed therefrom (40 CFR 129.2-91, *See also* 40 CFR 418.11; 422.41; 422.51; 423.11-91).

Ten (10) year, 24 hour rainfall event and 25 year, 24 hour rainfall event: A rainfall event with a probable recurrence interval of once in ten years or twenty-five years, respectively, as defined by the National Weather Service in Technical Paper Number 40, Rainfall Frequency Atlas of the United States, May 1961, and subsequent amendments, or equivalent regional or state rainfall probability information developed therefrom (40 CFR 412.11-91, *See also* 40 CFR 412.21-91).

Tenax: Polymer materials which are attractive to volatile organic compounds (VOC) and are sensitive to temperature. During the VOST sampling process, the tenax traps VOCs when temperature is high and holds the VOCs when temperature is cold. To retrieve the trapped VOCs, the tenax is reheated and the VOCs is released for analysis. In VOST application, there are two types of sampling traps:
(1) Tenax trap and
(2) Tenax and charcoal trap.
The charcoal in the second trap is a safety feature and is used to trap any leaks of VOCs from the tenax. As a matter of fact,

charcoal can attract VOCs better than tenax, however, it will not release VOCs if reheated.

Tensile strength: Resistance to force parallel to the plane of a specific size sheet of paper (EPA-83).

Tensiometer: A device containing a set of opposing grips used to place a geomembrane seam in tension for evaluating its strength in shear or in peel (EPA-91/05, See also EPA-89/09).

Tensor: A physical quantity which obeys certain laws of transformation in a given space. Scalars can be regarded as tensors of the zeroth order and vectors as tensors of the first order. An example of a tensor of this second order is the stress tensor (NATO-78/10).

Teratogen:
(1) Substance that causes malformation or serious deviation from normal development of embryos and fetuses (EPA-89/12).
(2) A substance causing birth defects in the offspring following exposure of one or both of the parents (EPA-87/10a).

Teratogenesis: The induction of structural of functional development abnormalities by exogenous factors acting during gestation; interference with normal embryonic development (Course 165.6).

Teratogenic:
(1) Inducing birth defects (LBL-76/07-bio).
(2) Affecting the genetic characteristics of an organism so as to cause its offspring to be misshapen or malformed (OME-88/12).

Teratogenic potential: The ability of a chemical to produce effects in the offspring after the pregnant female is exposed. Many chemicals that are teratogenic are also suspected carcinogens (EPA-3/80).

Teratogenicity:
(1) The capacity of a physical or chemical agent to cause non-hereditary congenital malformations (birth defects) in offspring (Course 165.6).
(2) The potential of a substance to cause defects to the developing fetus (ETI-92).
(3) Gross abnormalities, skeletal and visceral malformations, microscopic abnormalities, functional/behavioral deviations (*See* endpoint for more related terms) (EPA-92/12).

Terbium (Tb): A rare earth metal with a.n. 65; a.w. 158.924; d. 8.27 g/cc; m.p. 1356° C and b.p. 2800° C. The element belongs to group IIIB of the periodic table.

Terminal bronchiole: Noncartilagenous airway that conducts airstream to respiratory bronchiole (EPA-90/08).

Terminal velocity: *See* synonym, settling velocity.

Terminology: All terminology used in this part will be consistent with the terms as defined in 40 CFR 1508 (the CEQ Regulations). Any qualifications will be provided in the definitions set forth in each subpart of this regulation (40 CFR 6.101-91).

Terne coating: The coating steel products with terne metal by the hot dip process including the immersion of the steel product in a molten bath of lead and tin metals, and the related operations preceding and subsequent to the immersion phase (40 CFR 420.121-91).

Terpenes: The major chemical components of turpentine. A class of unsaturated organic compounds having the empirical formula, $C_{10}H_{16}$ (EPA-79/12).

Terracing: Dikes built along the contour of sloping farm land that hold runoff and sediment to reduce erosion (EPA-94/04).

Terrain: The natural features of a land tract. Other terrain related terms include:
- Complex terrain
- Flat terrain
- High terrain
- Low terrain
- Rolling terrain

Territorial sea (or territorial water): The belt of the seas measured from the line of ordinary low water along that portion of the coast which is in direct contact with the open sea and the line marking the seaward limit of inland waters, and extending seaward a distance of three miles (CWA502, *See also* 40 CFR 116.3-91, *See also* 40 CFR 230.3-91).

Tertiary alcohol: The alcohol whose OH connected carbon has no hydrogen atom (the carbon is attached to three carbons. Its molecular structure can be expressed as $R_1R_2R_3COH$, where R_1, R_2 and R_3 can be identical or different groups, e.g., 2-methylpropan-2-ol (($CH_3)_3COH$) (*See* alcohol for more related terms).

Tertiary amine: An amine whose molecular structure can be expressed as $R_1R_2R_3N$, where R_1, R_2 and R_3 can be identical or different groups, e.g., trimethylamine (($CH_3)_3NH$) (*See* amine for more related terms).

Tertiary treatment:
(1) Advanced cleaning of wastewater that goes beyond the secondary or biological stage, removing nutrients such as phosphorus, nitrogen, and most BOD and suspended solids (EPA-94/04).
(2) *See* synonym, tertiary wastewater treatment.

Tertiary wastewater treatment (or tertiary treatment): The advanced cleaning of wastewater that goes beyond the secondary or biological stage. It removes nutrients such as phosphorus and nitrogen and most BOD and suspended solids (*See* treatment or wastewater treatment for more related terms) (EPA-89/12).

Tetrachlophenol (C_6HCl_4OH): Brown acid used as a fungicide and for wood preservatives.

Tetraethyl lead ($Pb(C_2H_5)_4$): Poisonous compounds used as a gasoline antiknock agent.

Tetramethrin ($C_{20}H_{25}NO_4$): Yellowish powder used to control houseflies, cattle insects, garden pest, etc.

Tetratogenesis: The introduction of nonhereditary birth defects in a developing fetus by exogenous factors such as physical or chemical agents acting in the womb to interfere with normal embryonic development (EPA-94/04).

Test: Examination, evaluation or observation. Other test related terms include:
- Destructive test
- Nondestructive test

Test analyzer: An analyzer subjected to testing as a candidate method in accordance with Subparts B, C, and/or D of this part, as applicable (40 CFR 53.1-91).

Test chamber: A container in which the test organisms are maintained during the test period (40 CFR 797.1520-91, *See also* 40 CFR 797.1600-91).

Test compressor: A compressor used to demonstrate compliance with the applicable noise emissions standard (*See* compressor for more related terms) (40 CFR 204.51-91).

Test data: Any information which is a quantitative measure of any aspect of the behavior of a device (40 CFR 610.11-91, *See also* 40 CFR 720.3; 723.50; 723.175; 723.250-91).

Test engine: An engine in a test sample (*See* engine for more related terms) (40 CFR 86.1002.84-91).

Test exhaust system: An exhaust system in Selective Enforcement Audit test sample (40 CFR 205.165-91).

Test facility: *See* synonym, testing facility.

Test fuel charge: The collection of test fuel pieces placed in the wood heater at the start of the emission test run (40 CFR 60-App/A(method 28 & 28A)-91).

Test fuel crib: The arrangement of the test fuel charge with the proper spacing requirements between adjacent fuel pieces (40 CFR 60-App/A(method 28 & 28A)-91).

Test fuel loading density: The weight of the as-fired test fuel charge per unit volume of usable firebox (40 CFR 60-App/A(method 28 & 28A)-91).

Test fuel piece: The 2 x 4 or 4 x 4 wood piece cut to the length required for the test fuel charge and used to construct the test fuel crib (40 CFR 60-App/A(method 28 & 28A)-91).

Test hearing protector: A hearing protector that has been selected for testing to verify the value to be put on the label, or which has been designated for testing to determine compliance of the protector with the labeled value (40 CFR 211.203-91).

Test period: The combination of the exposure period and the post-exposure period; or, the entire duration of the test (40 CFR 797.2050-91).

Test product: Any product that must be tested according to regulations published under Part 211 (40 CFR 211.102-91, *See also* 40 CFR 204.2; 205.2-91).

Test request: A request submitted to the manufacturer by the Administrator that will specify the hearing protector category, and test sample size to be tested according to 40 CFR 211.212.1, and other information regarding the audit (40 CFR 211.203-91).

Test rule: Refers to a regulation ordering the development of data on health or environmental effects or chemical fate for a chemical substance or mixture pursuant to TSCA sec. 4(a) (40 CFR 791.3-91).

Test run: An individual emission test which encompasses the time required to consume the mass of the test fuel charge (40 CFR 60-App/A(method 28 & 28A)-91).

Test sample: The collection of vehicles of the same configuration which have been drawn from the population of vehicles of that configuration and which will receive exhaust emission testing (*See* sample for more related terms) (40 CFR 86.602.84-91).

Test sample size: The number of compressors of the same configuration in a test sample (*See* sample for more related terms) (40 CFR 204.51-91, *See also* 40 CFR 205.51-91).

Test sampler: A sampler subjected to testing as part of a candidate method in accordance with Subpart C or D of this part (40 CFR 53.1-91).

Test solution: The test substance and the dilution water in which the test substance is dissolved or suspended (40 CFR 797.1400-91, *See also* 40 CFR 797.1520; 797.1600; 797.2750-91).

Test strips (or test welds): Trial sections of seamed geomembranes used:
(1) To establish machine setting of temperature, pressure and travel rate for a specific geomembrane under a specific set of atmospheric conditions for machine-assisted seaming and
(2) To establish methods and materials for chemical and chemical adhesive seams under a specific set of atmospheric conditions (EPA-91/05, *See also* EPA-89/09).

Test substance: The form of chemical substance or mixture that is specified for use in testing (40 CFR 790.3-91, *See also* 40 CFR 160.3; 792.3; 795.232; 797.1600; 797.2050; 797.2130; 797.2150; 797.2175-91).

Test system: Any animal, plant, microorganism, or chemical or physical matrix, including but not limited to, soil or water, or components thereof, to which the test, control or reference substance is administrated or added for study. Test system also includes appropriate groups or components of the system not treated with the test, control, or reference substances (40 CFR 160.3-91, *See also* 40 CFR 792.3-91).

Test vehicle: A vehicle selected and used to demonstrate compliance with the applicable noise emission standards (*See* vehicle for more related terms) (40 CFR 205.51-91, *See also* 40 CFR 205.151; 86.602.84; 86.1002.84-91).

Test weight: The weight, within an inertia weight class, which is used in the dynamometer testing of a vehicle, and which is based on its loaded vehicle weight in accordance with the provisions of Part 86 (40 CFR 86.082.2-91, *See also* 40 CFR 600.002.85-91).

Test weight and the abbreviation (TW): The vehicle curb weight added to the gross vehicle weight rating (GVWR) and divided by 2 (CAA216-42USC7550-91).

Test weight basis: The basis on which equivalent test weight is determined in accordance with 40 CFR 86.129.94 of subpart B of this part (40 CFR 86.094.2-91).

Test welds: *See* synonym, test strips.

Testing agent: Any person who develops test data on a retrofit device (40 CFR 610.11-91).

Testing exemption: An exemption which may be granted under 40 CFR 203(b)(1) for the purpose of research investigations, studies, demonstrations or training, but not including national security (40 CFR 85.1702-91, *See also* 40 CFR 204.2; 205.2; 211.102-91).

Testing facility (or test facility): The person who actually conducts a study, i.e., actually uses the test substance in a test system. Testing facility encompasses only those operational units that are being or have been used to conduct studies (40 CFR 160.3-91, *See also* 40 CFR 60-App/A(method 28 & 28A); 211.203; 792.3-91).

Teterahedrite: An important ore of copper and silver (EPA-82/05).

Textile: An asbestos-containing product such as: yearn, thread, wick; cord; braided and twisted rope; braided and woven tubing; mat; roving; cloth; slit and woven tape; lap; felt; and other bonded or non-woven fabrics (40 CFR 763.163-91).

Textile fiberglass: The fibrous glass in the form of continuous strands having uniform thickness (40 CFR 60.291-91).

Texture coat: The rough coat that is characterized by discrete, raised spots on the exterior surface of the part. This definition does not include conductive sensitizers or EMI/RFI shielding coatings (40 CFR 60.721-91).

Thalassemia: A hereditary, genetically determined hemolytic anemia with familial and racial incidence; divided into a number of categories based on clinical severity and type(s) of hemoglobin contained in the red blood cells (LBL-76/07-bio).

Thallium (Ti): A soft metallic element with a.n. 81; a.w. 204.37; d. 11.85 g/cc; m.p. 303° C and b.p. 1457° C. The element belongs to group IIIA of the periodic table.

Theoretical air: *See* synonym, stoichiometric combustion air.

Theoretical arsenic emissions factor: The amount of inorganic arsenic, expressed in grams per kilogram of glass produced, as determined based on a material balance (40 CFR 61.161-91).

Theoretical combustion: *See* synonym, stoichiometric combustion.

Theoretical combustion air: *See* synonym, stoichiometric combustion air.

Theoretical flame mode kinetics: A calculation of destruction efficiency of a given compound based on estimation and extrapolation of elementary reaction kinetic data that is available from experiment and theory (EPA-88/12).

Theoretical oxygen demand (TOD): The calculated oxygen demand to completely oxidize organic pollutants in wastewater (*See* oxygen for more related terms).

Therapeutic index:
(1) The ratio of the dose required to produce toxic or lethal effect to dose required to produce non-adverse or therapeutic response (Course 165.6).
(2) The ratio of the dose required to produce toxic or lethal effects to dose required to produce nonadverse or therapeutic response (EPA-94/04).

Thermal analysis: Analysis of heating effects on products.

Thermal bypass: *See* synonym, airflow bypass.

Thermal conductivity: The quantity of heat which flows per unit time across unit area of the subsurface of unit thickness when the temperature of the faces differs by one degree (EPA-74/04).

Thermal conductivity detector: *See* definition under GC/TCD.

Thermal cutting: The process of cutting, slotting or piercing a workpiece using an oxy-acetylene oxygen lance or electric arc cutting tool (EPA-83/06a).

Thermal decomposition curve: The curve describes thermal behavior of a given sample or mixture; usually plotted as mass fraction (initial amount determined by a series of quantitation experiments at temperature where no chemical reaction occurs) on logarithmic scale versus exposure temperature (independent variable) (EPA-88/12).

Thermal discharge: Discharge of unwanted heat to the environment (cf. pollution, thermal).

Thermal dryer: Any facility in which the moisture content of bituminous coal is reduced by contact with a heated gas stream which is exhausted to the atmosphere (40 CFR 60.251-91, *See also* 40 CFR 60.381-91).

Thermal deflection rate: The angular degrees of rotation per degree of temperature change of the thermostatic coil (40 CFR 85.2122(a)(2)(iii)(G)-91).

Thermal efficiency (E): The efficiency of the thermodynamic cycle in producing work from heat. The ratio of usable energy to heat input expressed as a percent (EPA-82/11f; Holman-p398; Jones-p100; Wark-p276). Thermal efficiency can be expressed as:
- $E = W/Q_{in} = 1 - Q_{out}/Q_{in}$
- For a Carnot cycle, **Carnot efficiency (E_c):** $E_c = 1 - T_L/T_H$

Where:
Q_{out} = heat rejected during cycle
Q_{in} = heat added during cycle

W = net work output of cycle
T_L = low temperature of a reservoir
T_H = high temperature of a reservoir

Thermal electric: A term used to identify a type of electric generating station, capacity, or capability, or output in which the source of energy is heat (*See* electric for more related terms) (EPA-83).

Thermal endurance: Ability of glassware to withstand thermal shock (EPA-83).

Thermal fusion: The temporary, thermally induced reorganization in the polymeric make-up of the surface of a polymer geomembrane that, after the application of pressure and the passage of a certain amount of time, results in the two geomembranes being permanently joined together (*See* fusion for more related terms) (EPA-91/05).

Thermal incineration: *See* synonym, thermal incineration.

Thermal infusion: The process of applying a fused zinc, cadmium or other metal coating to a ferrous workpiece by impueing the surface of the workpiece with metal powder or dust in the presence of heat (EPA-83/06a).

Thermal NO$_x$: NO$_x$ formed due to the combustion of air. The nitrogen component in the air can react with oxygen and form NO or NO$_2$ under the combustion temperature. NO and NO$_2$ are often referred to as NO$_x$.

Thermal oxidation: The wet combustion of organic materials through the application of heat in the presence of oxygen (EPA-76/03).

Thermal oxidizer: *See* synonym, secondary burner).

Thermal precipitation: A process consisting of the separation of particulate matter from air and other gases under the influence of a relatively large temperature gradient extending over a short distance. In the "Thermal Precipitator" (a sampling instrument), the air or gas is drawn through a narrow chamber across which extends a heated wire, particulate matter being deposited upon the adjacent collecting surface (*See* precipitation for more related terms) (EPA-83/06).

Thermal pollution:
(1) The addition of large quantities of heat to air, water, or land so that the resulting temperature increase may have harmful effect (*See* pollution for more related terms) (EPA-84/09, *See also* EPA-89/12).
(2) Discharge of heated water from industrial processes that can kill or injure aquatic organisms (EPA-94/04).

Thermal processing: *See* synonym, thermal treatment.

Thermal shock resistance: The ability of a material to withstand sudden heating or cooling or both without cracking or spalling (SW-108ts).

Thermal spring: A stream of warm or hot water issuing from the ground, often after having been heated by buried lava and therefore commonly occurring in volcanic regions when eruptions have ceased (DOI-70/04).

Thermal stability index (or incinerability ranking): Ranking of the relative incinerability of compounds listed in Appendix VIII, 40 CFR 261. Incinerability for a given compound is determined by experimental and theoretical evaluation of thermal stability index (TSI) (temperature for 99% destruction for gas-phase residence time exposure of 2.0 seconds) (EPA-88/12).

Thermal stratification:
(1) The layering of water masses owing to different densities in response to temperature. The condition of a body of water in which the successive horizontal layers have different temperatures (DOD-78/01).
(2) The formation of layers of different temperatures in a lake or reservoir (EPA-94/04).

Thermal system insulation:
(1) The material in a school building applied to pipes, fittings, boilers, breeching, tanks, ducts, or other interior structural components to prevent heat loss or gain, or water condensation, or for other purposes (40 CFR 763.83-91, *See also* TSCA-AIA1; 40 CFR 763.83-91).
(2) Asbestos-containing material applied to pipes, fittings, boilers, breeching, tanks, ducts, or other interior structural components to prevent heat loss or gain or water condensation (EPA-94/04).

Thermal treatment (or thermal processing):
(1) Thermal treatment menas the treatment of hazardous waste in a device which uses elevated temperatures as the primary means to change the chemical, physical, or biological character or composition of the hazardous waste. Examples of thermal treatment processes are incineration, molten salt, pyrolysis, calcination, wet air oxidation, and microwave discharge (cf. incinerator and open burning) (40 CFR 260.10-91).
(2) Thermal processing means the processing of waste material by means of heat (40 CFR 240.101-91, *See also* EPA-89/12a; 1/86).
(3) Use of elevated temperatures to treat hazardous wastes. (*See* incineration; pyrolysis. (EPA-94/04))
(4) *See* treatment for more related terms.

Thermal turbulence: The turbulence produced by buoyancy forces. In the atmospheric boundary layer, usually applied to the turbulence caused by rising air motions due to the heating of air at the earth's surface (NATO-78/10).

Thermionic specific detector: *See* definition under GC/TSD.

Thermochemical transition state theory: The reaction kinetic theory developed largely by S. W. Benson and co-workers where rate coefficients for chemical transformation are estimated by considering the changes in thermodynamic quantities (changes in enthalpy and entropy) as one travels from initial reactants to the transition state; assumes the nature of the transition state determines the extent of reaction (EPA-88/12).

Thermochemistry: A science dealing with heating effect on chemical change.

Thermocline: The middle layer of a thermally stratified lake or reservoir. In this layer there is a rapid decrease in temperature with depth. Also called the metalimnion (EPA-94/04).

Thermocouple:
(1) A thermoelectric device used to measure temperatures (EPA-89/03b).
(2) Two dissimilar electrical conductors so joined as to produce a thermal electromotive force when exposed to temperatures. The electromatic force generated can be calibrated to read in temperature units (EPA-83). Various types of thermocouples are shown below (EPA-89/06):
  - Type J
    - Materials: Iron Constantan
    - Upper limit temperature (F): 1400
    - Thermocouple efficiency (+/- %): 0.75
  - Type E
    - Materials: Chromel Constantan
    - Upper limit temperature (F): 1650
    - Thermocouple efficiency (+/- %): 0.50
  - Type K
    - Materials: Chromel Alumel
    - Upper limit temperature (F): 2300
    - Thermocouple efficiency (+/- %): 0.75
  - Type S
    - Materials: Pt 10% Rhodium Pure Pt
    - Upper limit temperature (F): 2650
    - Thermocouple efficiency (+/- %): 0.25
  - Type B
    - Materials: Pt 13% Rhodium Pure Pt
    - Upper limit temperature (F): 2650
    - Thermocouple efficiency (+/- %): 0.25
  - Type R
    - Materials: Pt 30% Rhodium Pt 6% Rhodium
    - Upper limit temperature (F): 3100
    - Thermocouple efficiency (+/- %): 0.50
(3) *See* temperature meter for more related terms.

Thermodurics: Resist high temperatures (EPA-83).

Thermophilic digestion: That the anaerobic digestion temperature is maintained about 50° C or more (*See* thermophile for more related terms).

Thermodynamic equilibrium (or equilibrium): A thermodynamic system is in equilibrium, if its properties such as pressure, temperature, and density are uniform within the system.

Thermodynamic first law: The energy of an isolated system remains constant. An **isolated system** means one which does not exchange energy with its surroundings. The basic concept underlying the First Law of Thermodynamics is the concept of energy conservation. For a steady-state, steady-flow situation between the entrance and exit of an incineration system. The First Law of Thermodynamics can be expressed as:

heat (Q) - work (W) = enthalpy change + kinetic energy change + potential energy change.

In general, because the velocity of a combustion gas and the difference in height between the feed point and the stack gas exit point of an incinerator are relatively small, the kinetic energy change and potential energy change are generally negligible. Therefore, the first law of thermodynamics can be simplified to:
- $Q - W$ = enthalpy change = $m(h_2 - h_1)$, where: h_2 = Enthalpy value after a chemical reaction takes place; h_1 = enthalpy value before a chemical reaction takes place; and m = mass or mass flow rate.
- Because most incinerators do not involve "Work" per se, the above equation can be further simplified to:
- $Q = m(h_2 - h_1)$ = heat extractable from the combustion gases = $mCp(T_2 - T_1)$ for an ideal gas, where: m = the mass or mass flow rate of the combustion gas; Cp = the specific heat of the gas (in Btu/lb-R) and $(T_2 - T_1)$ = the temperature difference between the firebox temperature and the exhaust gas temperature just upstream of the typical pollution control device.
- *See* thermodynamic law for more related terms.

Thermodynamic process (or process): The path of a system change from one equilibrium state to another over a period of time. Other process related terms include:
- Adiabatic process
- Cycle or cyclic process
- Path function
- point function

Thermodynamic property: *See* synonym, property.

Thermodynamic second law: It is impossible to build a machine which will operate continuously while receiving heat from a single reservoir and producing an equivalent amount of work (*See* thermodynamic law for more related terms).

Thermodynamic state (or state): The conditions specified by the values of its properties.

Thermodynamic system: A three dimensional space bounded by arbitrary geometric surfaces. The bounding surfaces can be real or imaginary and can be at rest or in motion. Other thermodynamic system related terms include:
- Closed system
- Control mass
- Isolated system
- Open system
- Surroundings

Thermodynamic third law: The entrop of a pure substance in a thermodynamic equilibrium approaches zero as the absolute temperature approaches zero (*See* thermodynamic law for more related terms).

Thermodynamic zeroth law: Also known as the zeroth law of thermodynamics. The law means that two bodies are in thermal equilibrium with each other, if each of the two bodies is in thermal equilibrium with a third body (*See* thermodynamic law for more related terms).

Thermodynamics: Thermodynamics is a science which is the study of energy transformation and its relationship to the changes of properties in the system.

Thermometer: A common temperature scale which is based on the freezing and boiling points of water. It includes the gas, liquid (mercury) and solid (bimetallic element) thermometer. Types of thermometers:
(1) Resistance thermometer
(2) Vapor bulb thermometer
(3) *See* temperature meter for more related terms.

Thermometric titration: A titration in which heat is added to a solution until the desired end point is reached (*See* titration for more related terms).

Thermophile: Bacteria or other microorganisms which grow best at temperatures of roughly 45 to 60° C (EPA-83). Other thermophile related terms include:
- Mesophiles
- Psychrophiles
- Thermophilic digestion

Thermoplastic: Having property of softening or fusing when heated and of hardening to a rigid form again when cooled (EPA-75/01a).

Thermoplastic polymer: A polymer that can be heated to a softening point, shaped by pressure, and cooled to retain that shape. The process can be done repeatedly (*See* polymer for more related terms) (EPA-91/05).

Thermoplastic process: Bitumen, paraffin, and polyethylene are used as stabilizing agents in the thermoplastic techniques. The use of thermoplastic solidification systems in radioactive waste disposal has let to the development of waste containment systems that can be adapted to other types of hazardous waste. In processing radioactive waste with bitumen or other thermoplastic material, the waste is dried, heated, and dispersed through a heated plastic matrix. The mixture is subsequently cooled to solidify the mass, and the product is then usually buried in a secondary containment system such as a steel drum (*See* solidification and stabilization for more related terms).

Thermoset polymer: A polymer that can be heated to a softening point, shaped by pressure, and, if desired, removed from the hot mold without cooling. The process cannot be repeated since the polymer can not be re-softened by the application of heat (*See* polymer for more related terms) (EPA-91/05).

Thermoset resin: A resin used as a binding agent in molding sands. Thermoset resins require the addition of heat in order to solidify and set the mold (*See* resin for more related terms) (EPA-85/10a).

Thermosetting: Having the property of becoming permanently hard and rigid when heated or cured (EPA-75/01a).

Thermosetting ink: An ink which polymerizes to a permanently solid and infusible state upon the application of heat (*See* ink for more related terms) (EPA-79/12a).

Thermosetting resin: *See* synonym, synthetic resin.

Thermostatic coil: A spiral-wound coil of thermally-sensitive material which provides rotary force (torque) and/or displacement as a function of applied temperature (40 CFR 85.2122(a)(2)(iii)(D)-91).

Thermostatic switch: An element of thermally-sensitive material which acts to open or close an electrical circuit as a function of temperature (40 CFR 85.2122 (a)(2)(iii)(E)-91).

Thiamine ($C_{12}H_{17}CIN_4OS$): Vitamin B1.

Thiazole (C_3H_3NS): A colorless liquid used as for fungicides, dyes and rubber accelerators manufacturing.

Thickener: A device or system wherein the solid contents of slurries or suspensions are increased by gravity settling and mechanical separation of the phases, or by flotation and mechanical separation of the phases (EPA-83/06a).

Thickening (or sludge dewatering): Thickening or concentration is the process of removing water from sludge after the initial separation of the sludge from wastewater. The basic objective of thickening is to reduce the volume of liquid sludge to be handled in subsequent sludge disposal processes (EPA-83/06a).

Thimble: A refractory shape used for stirring pot-made optical glass (EPA-83).

Thin film evaporation operation: A distillation operation that employs a heating surface consisting of a large diameter tube that may be either straight or tapered, horizontal or vertical. Liquid is spread on the tube wall by a rotating assembly of blades that maintain a close clearance from the wall or actually ride on the film of liquid on the wall (40 CFR 264.1031-91).

Thin layer chromatography: An analysis of liquid compounds using chromatography. The stationary phase is a thin layer of an absorbing solid rather than a column (*See* chromatography for more related terms).

Thinner:
(1) The portion of a paint, varnish, lacquer, or related product that volatilizes during the drying process. The solvents and diluents which act as thinners are used to reduce coating viscosity, and prevent oxidation, polymerization, and drying prior to coating application (EPA-79/12b).
(2) Solvents, diluents, low viscosity oils, and vehicles added to inks to reduce their consistency or tack (EPA-79/12a).

Thiocyanic acid (HSCN): A colorless liquid used as an insecticide.

Third order reaction: The rate of a chemical reaction is determined by the concentration of three reactants (*See* chemical reaction for more related terms).

Thirty day (30 day) average: The arithmetic mean of pollutant parameter values of samples collected in a period of 30 consecutive days (40 CFR 133.101-91).

Thirty day (30 day) limitation: A value that should not be exceeded by the average of daily measurements taken during any 30-day period (40 CFR 429.11-91).

Thirty day rolling average: Any value arithmetically averaged over any consecutive thirty-days (40 CFR 52.741-91).

Thorium (Th): A radioactive metal with a.n. 90; a.w. 232.038; d. 11.7 g/cc; m.p. 1750° C and b.p. 3850° C. The element belongs to group IIIB of the periodic table.

Threat of discharge: *See* definition under discharge.

Threat of release: *See* definition under release.

Threatened species: *See* synonym, endangered species.

Three hour period: Any three consecutive 1-hour periods (each hour commencing on the hour), provided that the number of 3-hour periods during which the vinyl chloride concentration exceeds 10 ppm does not exceed the number of 1-hour periods during which the vinyl chloride concentration exceeds 10 ppm (40 CFR 61.61-91).

Three piece can: A can which is made from a rectangular sheet and two circular ends (40 CFR 52.741-91).

Three process operation facility: The facility including those processes involved with plate stacking, burning or strap casting, and assembly of elements into the battery case (40 CFR 60.371-91).

Three T's of combustion: Three T's mean temperature, time, and turbulence. Combustion principles include that:
(1) High enough temperature to ignite the air and fuel mixture;
(2) Turbulent mixing to bring the air and fuel into contact; and
(3) Sufficient residence time for the reaction to occur (EPA-81/12, p3-2).

Threshold:
(1) The minimum amount needed for a given effect.
(2) In odor studies, the minimum concentration at which a substance can be detected by whatever test method is employed. At the odor threshold, the average person can just barely detect that an odor exists. Higher concentrations can be stated in terms of thresholds:
  (A) Two thresholds means that the odor can be reduced to the threshold level by diluting one part of it with one part of odor free air.
  (B) One hundred thresholds means that one must add 99 parts of odor-free air to bring it down to the threshold concentration.
  (C) In a general sense, a definite odor is about 10 thresholds, a strong odor 100 thresholds, and an overpower odor 1000 thresholds (EPA-84/09).
(3) The dose or exposure below which a significant adverse effect is not expected. Carcinogens are thought to be non-threshold chemicals, to which no exposure can be presumed to be without some risk of adverse effect (EPA-92/12).
(4) The dose or exposure below which a significant adverse effect is not expected. Carcinogenicity is thought to be a non-threshold endpoint, thus, no exposure can be presumed to be without some risk of adverse effect. Non-carcinogenicity is presumed to be a threshold endpoint, thus, some exposures are presumed to be without risk of adverse effects (EPA-90/08).
(5) The lowest dose of a chemical at which a specified measurable effect is observed and below which it is not observed (EPA-94/04).

Threshold dose: The minimum dose of a given substance necessary to produce a measurable physiological or psychological effect (*See* dose for more related terms) (EPA-74/11).

Threshold damage: The minimum pollution necessary to produce a measurable environmental effect (EPA-74/11).

Threshold effects: Results from chemicals that have a safe level (i.e., acute, subacute, or chronic human health effects) (EPA-91/03).

Threshold level: Time-weighted average pollutant concentration values, exposure beyond which is likely to adversely affect human health. (*See* environmental exposure. (EPA-94/04))

Threshold limit median (TLM): The lethal concentration of a pollutant to 50 percent of a tested aquatic species population. TLM is generally expressed in mg/L. Exposure durations may be 24, 48, or 96 hours; most frequently 96-hour are reported. Values vary depending on species tested, the test type (i.e., static or flow through bioassay), and other conditions such as pH or water hardness (EPA-3/80).

Threshold limit value (TLV):
(1) Maximum airborne concentration of a substance to which a worker may be exposed eight hours a day, forty hours a week, for a working lifetime without adverse effect (EPA-83).
(2) The concentration of an airborne substance that an average person can be repeatedly exposed to without adverse effects. TLVs may be expressed in three ways: TLV-TWA-Time weighted average, based on an allowable exposure averaged over a normal 8-hour workday or 40-hour workweek; TLV-STEL-Short-term exposure limit or maximum concentration for a brief specified period of time, depending on a specific chemical (TWA must still be met); and TLV-C-Ceiling Exposure Limit or maximum exposure concentration not to be exceeded under any circumstances (TWA must still be met) (EPA-94/04).
(3) Recommended guidelines for occupational exposure to airborne contaminants published by the American Conference of Governmental Industrial Hygienists (ACGIH). The TLVs represent the average concentration (in mg/cu.m) for an 8-hour workday and a 40-hour work week to which nearly all workers may be repeatedly exposed, day after day, without adverse effect (EPA-92/12).
(4) The recommended concentrations of airborne contaminants to which workers may be exposed according to the American Council of Governmental Industrial Hygienists (DOE-91/04).
(5) TLV is the threshold dose or concentration below which there are no adverse effects. TLV is formulated on the basis of:
  (A) Information from historical experience;
  (B) Experimental human studies; and
  (C) Experimental animal studies. Its value refers to airborne concentrations of substances and represent conditions under which it is believed that nearly all workers may be repeatedly exposed day after day without adverse effects. The exposure levels are stated

as airborne concentrations and durations of exposure, including (EPA-81/09; Course 165.5):

Threshold limit value (TLV) type:
(1) **TLV-time weighted average (TLV-TWA):** A time-weighted average concentration for a normal 8-hour work day and a 40-hour work week (Course 165.5).
(2) **TLV-short term exposure limit (TLV-STEL):** A 15-minute time-weighted average exposure which should not be exceeded at any time during a work day (Course 165.5).
(3) **TLV-ceiling (TLV-C):** A concentration that should not be exceeded, even instantaneously (Course 165.5).

Threshold limit value-ceiling (TLV-C): The concentration that should not be exceeded even instantaneously (NIOSH-84/10).

Threshold limit value-short term exposure limit (TLV-STEL): The concentration to which workers can be exposed continuously for a short period of time without suffering from:
(1) Irritation;
(2) Chronic or irreversible tissue change; or
(3) Narcosis of sufficient degree to increase the likelihood of accidental injury, impair self-rescue or materially reduce work efficiency, and provided that the daily TLVs-TWA also is not exceeded. It is not a separate independent exposure limit, rather it supplements the time-weighted average (TWA) limit where there are recognized acute effects from a substance whose toxic effects are primarily of a chronic nature. STELs are recommended only where toxic effects have been reported from high short-term exposures in either humans or animals (NIOSH-84/10).

Threshold limit value-time weighted average (TLV-TWA): The time-weighted average concentration for a normal 8-hour workday and a 40-hour workweek, to which nearly all workers may be repeatedly exposed, day after day, without adverse effect (NIOSH-84/10).

Threshold odor: *See* odor threshold (EPA-94/04).

Threshold odor number: The number of times a sample needs to be diluted with clean water in order to reach the level that the smell is not detectable. For example, if original sample is 1 liter and the number of times is 3.5, this means that the sample needs to be diluted to 3.5 liters to reach undetectable odor.

Threshold planning quantity (TPQ):
(1) For a substance listed in Appendices A and B, the quantity listed in the column threshold planning quantity for that substance (40 CFR 355.20; 370.2-91).
(2) A quantity designated for each chemical on the list of extremely hazardous substances that triggers notification by facilities to the State Emergency Response Commission that such facilities are subject to emergency planning requirements under SARA Title III (EPA-94/04).

Threshold treatment: In water softening treatment, the level of treatment needed so that precipitation will stop (cf. water softening).

Threshold toxicity: A limit upon which a substance becomes toxic or poisonous to a particular organism (EPA-83/06a).

Throttle: The mechanical linkage which either directly or indirectly controls the fuel flow to the engine (40 CFR 86.082.2-91, *See also* 40 CFR 86.090.2-91).

Throat velocity: The gas velocity through a venturi throat. The typical range is 12,000 to 24,000 ft/min (EPA-84/09).

Thulium (Tm): A rare earth metal with a.n. 69; a.w. 168.934; d. 9.33 g/cc; m.p. 1545° C and b.p. 1727° C. The element belongs to group IIIB of the periodic table.

Tidal marsh: Low, flat marshlands traversed by channels and tidal hollows, subject to tidal inundation; normally, the only vegetation present is salttolerant bushes and grasses (*See* wetlands) (EPA-94/04).

Tidal volume: Air volumes moved in and out of one person's lungs in a respiratory cycle.

Tidal water: The changes of water levels due to periodical tidal actions (*See* water for more related terms).

Tier: One of the three levels of treatability testing (i.e., laboratory screening, bench-scale testing, pilot-scale testing) (EPA-89/12a).

Tiering: Refers to the coverage of general matters in broader environmental impact statements (such as national program or policy statements) with subsequent narrower statements or environmental analyses (such as regional or basinwide program statements or ultimately site-specific statements) incorporating by reference the general discussions and concentrating solely on the issues specific to the statement subsequently prepared. Tiering is appropriate when the sequence of statements or analyses is:
(1) From a program, plan, or policy environmental impact statement to a program plan, or policy statement or analysis of lesser scope or to a site-specific statement or analysis.
(2) From an environmental impact statement on a specific action at an early stage (such as need and site selection) to a supplement (which is preferred) or a subsequent statement or analysis at a later stage (such as environmental mitigation). Tiering in such cases is appropriate when it helps the lead agency to focus on the issues which are ripe for decision and exclude from consideration issues already decided or not yet ripe (40 CFR 1508.28-91).

Tight house: A house with a low air exchange rate. If 0.5 to 0.9 air changes per hour (ach) is typical of modern housing, a tight house would be one with an exchange rate well below 0.5 ach (EPA-88/08).

Tight soil: The soil that is relatively impermeable to water movement (*See* soil for more related terms) (EPA-83).

Tillage: Plowing, Seedbed preparation, and cultivation practices (EPA-94/04).

Tilting furnace: The furnace that is provided with devices for affixing the crucible to the furnace so that the furnace may be tilted with the crucible when the metal is poured (*See* furnace for more related terms) (AP-40, p238).

Timberline: The line of elevation on a mountain or hill slope above which trees do not grow. Its height depends upon local as well as general conditions of climate and soil. It is lower in the temperate than in the tropical zone, lower on the shady than on the sunny side of a mountain, and highest on those slopes which provide the best protection from winds and the longest exposure to the sun (DOI-70/04).

Time: A measure of the sequence of events. Other time related terms include:
- Detention period (*See* synonym, detention time)
- Detention time
- Dwell time
- Fall time
- Ignition delay time
- Lag time
- Residence time
- Response time
- Retention period (*See* synonym, detention time)
- Retention time
- Reverberation time
- Rise time

Time meter: A device such as clocks and timers for measuring the time.

Time period: Any period of time designated by hour, month, season, calendar year, averaging time, or other suitable characteristics, for which ambient air quality is estimated (40 CFR 51.100-91).

Time reference: The time (t_o) when the laser pulse emerges from the laser, used as the reference in all lidar time or range measurements (40 CFR 60-App/A(alt. method 1)-91).

Time response curve: The curve relating cumulative percentage response of a test batch of organisms, exposed to a single dose or single concentration of a chemical, to a period of exposure (40 CFR 797.1350-91, *See also* 40 CFR 797.1440-91).

Time-weighted average (TWA): In air sampling, the average air concentration of contaminants during a given period (EPA-94/04).

Time weighted average example: The average value of a parameter (e.g., concentration of a chemical in air) that varies over time (Course 165.6).
- TWA = $sum[(C_i \times T_i)]/sum(T_i)$, where: i = 1,2,3.....; C = concentration; and T = time of exposure.

Example:
- Calculation of TWA
  Data: Concentration (ppm), Exposure time (hours)

i = 1,	C_i = 15,	T_i = 3
2	21	2.5
3	18	2.5

Solution:
- TWA = [(15 x 3 + 21 x 2.5 + 18 x 2.5)/(3 + 2.5 + 2.5)]
  = 17.8 ppm

Timed delay: A delayed diaphragm displacement controlled to occur within a given time period (40 CFR 85.2122(a)(1)(ii)(B)-91).

Tin (Sn): A soft metallic element with a.n. 50; a.w. 118.69; d. 7.30 g/cc; m.p. 231.9° C and b.p. 2270° C. The element belongs to group IVA of the periodic table.

Tin alloy: The essential constituent of soft solders, type metals, fusible alloys and certain bearing metals, e.g., copper and antimony pewter, bronze and sometimes lead (EPA-83).

Tin can: A can made from tin-plated steel (EPA-83).

Tin free steel (TFS): Cans made from steel coated with chrome rather than tin. This newly developed technique will simplify the recycling of steel cans (EPA-83).

Tint base paint: A noncolored paint shipped to the retailer where colorants are added to meet customer's specifications (*See* paint for more related terms) (EPA-79/12b).

Tipping fee: The price charged for delivering municipal waste to landfill, incinerator, or recycling facility; usually expressed in dollars per ton (*See* fee for more related terms) (OTA-89/10).

Tipping floor: Unloading area for vehicles that are delivering solid waste to an incinerator or other processing facility (SW-108ts and EPA-89/11).

Tire: The following types of tires: passenger car tires, light- and heavy-duty truck tires, high speed industrial tires, bus tires, and special service tires (including military, agricultural, off-the-road, and slow speed industrial) (40 CFR 253.4-91, *See also* 40 CFR 60.541-91).

Tissue: A group of similar cells (Course 165.6).

Tidal volume (VT): Volume of air inhaled/exhaled during normal breathing (EPA-90/08).

Titanium (Ti): A transition metal with a.n. 22; a.w. 47.9; d. 4.51 g/cc; m.p. 1668° C and b.p. 3260° C. The element belongs to group IVB of the periodic table.

Titanium mineral: The main commercial minerals are rutile (TiO_2) and ilmenite ($FeTiO_3$) (EPA-82/05).

Title I design (preliminary design): Continues a design effort using the conceptual design and the project design criteria as a basis for project development. Title I design develops topographical and subsurface data and determines the requirements and criteria that will govern the definitive design. Tasks include preparation of preliminary planning and engineering studies, preliminary drawings and outline specifications, life-cycle cost analysis, preliminary cost estimates, and scheduling for project completion. Preliminary design identifies long-lead-time procurement items and analyzes risks associated with continued project development (DOE-91/04).

Title II design (definitive design): Continues he development of a project based on an approved preliminary design (Title I). Definitive design includes any revisions of the Title I effort; preparation of final working drawings, specifications, bidding documents, and cost estimates; coordination with all parties that night affect the project; development of firm construction and procurement schedules; and assistance in analyzing proposals or bids (DOE-91/04).

Title III inspection (concurrent with construction): Complete architectural and engineering supervision and inspection of construction under the direction of a responsible representative. Includes checking of shop drawings and furnishing of reproducible as-built record drawings and marked-up specifications showing construction as actually accomplished (DOE-91/04).

Title III (superfund):
(1) Title III of the Superfund Amendments and Reauthorization Act of 1986, also titled the Emergency Panning and Community Right-to-Know Act of 1986 (40 CFR 350.1-91, See also 40 CFR 372.3; 372.3-91).
(2) The Emergency Planning and Community Right-to-Know Act of 1986, Specifies requirements for organizing the planning process at the State and local levels for specified extremely hazardous substances; minimum plan content; requirements for fixed facility owners and operators to inform officials about extremely hazardous substances present at the facilities; and mechanisms for making information about extremely hazardous substances available to citizens (NRT-87/03).

Titrand: A substance to be analyzed in a titration process.

Titrant: A substance with known parameters such as component, concentration or volume which is used for analyzing a titrand in a titration process. Other titrand related terms include:
- Acidic titrant

Titration:
(1) A method of measuring acidity of alkalinity.
(2) The determination of a constituent in a known volume of solution by the measured addition of a solution of known strength for completion of the reaction as signaled by observation of an end point (EPA-83/06a).

Other titration related terms include:
- Acid base titration
- Amperometric titration
- Complexometri analysis
- Conductometric titration
- Distillation silver nitrate titration
- Endpoint titration
- Ethylenediamine tetra acetic acid titration
- Potentiometric titration
- Redox titration
- Spectrophotometric titration
- Thermometric titration???
- Turbidimetric titration???

TMI: Three Mile Island Nuclear Plant Unit 2. An accident at TMI led to an action plan by the Nuclear Regulatory Commission (NRC) and the resolution of severe accident and source term issues; much of the safety knowledge gained rom the TMI accident has been incorporated into current NRC regulations (DOE-91/04).

Toe: The bottom of the working face at a sanitary landfill (SW-108ts).

Toilet tissue: A sanitary tissue paper. The principal characteristics are softness, absorbency, cleanliness, and adequate strength (considering easy disposability). It is marketed in rolls of varying sizes or in interleaved packages (40 CFR 250.4-91).

Tolerance:
- The amount of a pesticide residue that legally may be present in or on a raw agricultural commodity under the terms of a tolerance under FFDCA section 408 or a processed food under the terms of a food additive regulation under FFDCA section 409. Tolerances are usually expressed in terms of parts of the pesticide residue per million parts of the food (ppm), by weight (40 CFR 177.3-91).
- The relative capability of an organism to endure or adapt to an unfavorable environmental factor (DOD-78/01).
- Permissible residue levels for pesticides in raw agricultural produce and processed foods. Whenever a pesticide is registered for use on a food or a feed crop, a tolerance (or exemption from the tolerance requirement) must be established. EPA establishes the tolerance levels, which are enforced by the Food and Drug Administration and the Department of Agriculture (EPA-94/04).

Tolerance limit: The numerical value identifying the acceptable range of some variables (EPA-79/12c).

Tolerance limit median: The concentration that kills 50% of the test organisms within a specified time span, usually in 96 hours or less. This system of reporting has been misapplied by some who have erroneously inferred that a TLM value is a safe value, whereas it is merely the level at which half of the test organisms are killed. In many cases, the differences are great between TLM concentrations and concentrations that are low enough to permit reproduction and growth. LC50 has the same numerical value as TLM (EPA-76/03).

Tolerance with regional registration: Any tolerance which is established for pesticide residues resulting from the use of the pesticide pursuant to a regional registration. Such a tolerance is supported by residue data from specific growing regions for a raw agricultural commodity. Individual tolerances with regional registration are designated in separate subsections in 40 CFR 180.101 through 180.999, as appropriate. Additional residue data which are representative of the proposed use area are required to expand the geographical area of usage of a pesticide on a raw agricultural commodity having an established tolerance with regional registration. Persons Seeking geographically broader registration of a crop having a tolerance with regional registration should contact the appropriate EPA product manager concerning additional residue data required to expand the use area (40 CFR 180.1-n-91).

Toluene (C_7H_8): A solvent used in paints and varnishes. It is similar to benzene, but has a higher boiling point and is less toxic (EPA-84/09).

Ton:
(1) Gross ton 2,240 pounds.
(2) Net ton 2,000 pounds (EPA-83).

Tonnage: The amount of waste that a landfill accepts, usually expressed in tons per month. The rate at which a landfill accepts waste is limited by the landfill's permit (EPA-94/04).

Tool steel: Steels used to make cutting tools and dies. Many of these steels have considerable quantities of alloying elements such as chromium, carbon, tungsten, molybdenum and other elements. These form hard carbides which provide good wearing qualities but at the same time decrease machinability. Tool steels in the trade are classified for the most part by their applications, such as hot work die, cold work die, high speed, shock resisting, mold and special purpose steels) (EPA-83/06a).

Tooth: Surface grain of paper (EPA-83).

Top blown furnace: A basic oxygen process furnace (BOPF) in which oxygen is introduced to the bath of molten iron by means of an oxygen lance inserted from the top of the vessel (See furnace for more related terms) (40 CFR 60.141a-91).

Top security: See definition under security classification category.

Top void (block void or void): Air space(s) within masonry walls made of concrete blocks or cinder blocks. Top void specifically refers to the air space in the top course of such walls; that is, the course of block to which the sill plate is attached and on which the walls of the house rest (EPA-88/08).

Topcoat: A coating applied in a multiple coat operation other than prime coat, final repair coat, or prime surfacer coat (40 CFR 52.741-91).

Topcoat operation: All topcoat spray booths, flash-off areas, and bake ovens at a facility which are used to apply, dry, or cure the final coatings (except final off-line repair) on components of automobile or light-duty truck bodies (40 CFR 52.741-91).

Topcoat operation: The topcoat spray booth, flash-off area, and bake oven(s) which are used to apply and dry or cure the final coating(s) on components of automobile and light-duty truck bodies (40 CFR 60.391-91).

Topography: The physical features of a surface area including relative elevations and the position of natural and man-made features (EPA-94/04).

Topographic map: A map indicating surface elevations and slopes (SW-108ts).

Topography: The physical features (configutation) of a surface area including relative elevations and the position of natural and man-made features (EPA-89/12).

Topsoil: The topmost layer of soil; usually refers to soil that contains humus and is capable of supporting good plant growth (SW-108ts).

Torque: Something that causes an object to rotate. It is equal to the product of a force and its perpendicular distance from a rotational center. Other torgue related terms include:
- Full load torque
- Maximum rated torque
- Mechanical torque rate

Torr: A unit of pressure which equals 133.3 pascals or 1 mm Hg at 0 C (40 CFR 796.1950-91).

Tort: Tort is the word used to denote a common law civil wrong for which a court will provide a remedy. A tort arises from the existence of a generalized legal duty to avoid causing harm to others, through acts of omission, as well as of commission (See common law for more related terms) (Sullivan-95/04, p6).

Total: The concentration determined on an unfiltered sample following vigorous digestion (Section 9.3), or the sum of the dissolved plus suspended concentrations. (Section 9.1 plus 9.2) (40 CFR 136-App/C-91).

Total alkalinity: An indicator of the ability to neutralize the acid in water. It represents that alkalinity is measured above 4.5, and bicarbonate ions $(HCO_3)^-$ have reacted with strong acid to form H_2CO_3.

Total annual sales: The total annual revenue (in dollars) generated by the sale of all products of a company. Total annual sales must include the total annual sales revenue of all sites owned or controlled by that company and the total annual sales revenue of that company's subsidiaries and foreign or domestic parent company, if any (40 CFR 704.3-91).

Total carbon: The sum of carbon associated with both inorganic and organic compounds in a sample.

Total chromium: The sum of hexavalent and trivalent chromium as measured by the procedures listed in 40 CFR 136 (See chromium for more related terms) (40 CFR 410.01-91).

Total coliform: The sum of all types of bacteria in a sample, rather than a specific species (See coliform for more related terms).

Total cost bidding: A method of establishing the purchase price of movable equipment. The buyer is guaranteed that maintenance will not exceed a set maximum amount during a fixed period and that the equipment will be repurchased at a set minimum price when the period ends (SW-108ts).

Total combustible: Includes paints, lacquers, coatings, etc. associated with the original ferrous product, as well as combustible materials (paper, plastics, textiles, etc.) which become associated with the ferrous product after it is manufactured (See combustible for more related terms) (SW-108ts) (EPA-83).

Total cyanide:
(1) Total cyanide as determined by the test procedure specified in 40 CFR 136 (Federal Register, Vol. 38 No. 199, October 16, 1973) (EPA-83/09).
(2) The total content of cyanide expressed as the radical CN^- or alkali cyanide whether present as simple or complex ions. The sum of both the combined and free cyanide content of a plating solution. In analytical terminology, total cyanide is the sum of cyanide amenable to oxidation by chlorine and that which is not according to standard analytical methods

(EPA-83/06a).
(3) *See* cyanide for more related terms.

Total dissolved phosphorous: The total phosphorous content o all material that will pass through a filter, which is determined as orthophosphate without prior digestion or hydrolysis. Also called soluble P. or ortho P (EPA-94/04).

Total dissolved solid:
(1) The total dissolved (filterable) solids as determined by use of the method specified in 40 CFR 136 (40 CFR 122.2-91, *See also* 40 CFR 131.35; 144.3; 146.3-91).
(2) All material that passes the standard glass river filter; now called total filtrable reside. Term is used to reflect salinity (EPA-94/04).
(3) *See* solid for more related terms.

Total dynamic head (TDH): The total energy provided by a pump consisting of the difference in elevation between the suction and discharge levels, plus losses due to unrecovered velocity head and friction (EPA-82/11f).

Total enclosure: A structure that is constructed around a source of emissions and operated so that all VOC emissions are collected and exhausted through a stack or duct. With a total enclosure, there will be no fugitive emissions, only stack emissions. The drying oven itself may be part of the total enclosure (40 CFR 60.741-91, *See also* 40 CFR 60.441; 60.711-91).

Total fluoride: The elemental fluorine and all fluoride compounds as measured by reference methods specified in 40 CFR 60.195 or by equivalent or alternative methods (*See* fluoride for more related terms) (40 CFR 60.191-91).

Total hydrocarbon concentration measurement system: The total equipment required for the determination of a gas concentration. For total hydrocarbon concentration (THC) measurement, there are two acceptable types of THC monitoring systems: **heated and unheated systems**.
(1) Heated systems maintain the temperature of the sample gas between 150 to 170° C throughout the system. This requires all system components like probe, calibration valve, filters, sample lines, pump, and the FID analyzer to be kept heated at all times such that no moisture is condensed out of the system.
(2) Unheated systems remove excess moisture from the system and pass it through a gas conditioning system kept at temperatures between 5 to 18° C (40 to 64° F) so that the moisture of the sample gas entering the FID does not exceed 2%.

Either heated or unheated monitors, a THC measurement system consists of the following major subsystems (EPA-90/04):
(1) Sample probe
(2) Sample line
(3) Calibration valve assembly
(4) FID analyzer
(5) Data recorder
(6) *See* measurement for more related terms

Total Kjeldahl nitrogen (TKN) (or Kjeldahl process):
(1) A measure of nitrogen combined in organic and ammonia form in wastewater. Expressed in mg/L as N. It includes ammonia and organic nitrogen but does not include nitrite and nitrate nitrogen.
(2) The sum of free nitrogen and organic nitrogen in a sample (EPA-82/11; 76/03).

Total maximum daily load (TMDL):
(1) The total allowable pollutant load to a receiving water such that any additional loading will produce a violation of water quality standards (EPA-85/09, *See also* 40 CFR 130.2-91).
(2) The sum of the individual waste load allocations and load allocations. A margin of safety is included with the two types of allocations so that any additional loading, regardless of source, would not produce a violation of water quality standards (EPA-91/03).

Total metals:
(1) The sum of the metal content in both soluble and insoluble form (EPA-83/06a).
(2) The concentration of metals determined in a sample following digestion by Methods 3010, 3020, or 3050 specified in SW-846.
(3) (In electroplating), the sum of the concentration or mass of Copper (Cu), Nickel (Ni), Chromium (Cr) (total) and Zinc (Zn) (40 CFR 413.02-91).
(4) *See* metal for more related terms.

Total moisture:
(1) That moisture determined as the loss in weight in an air atmosphere under rigidly controlled conditions of temperature, time and air flow. Total moisture is calculated from the air dry loss and the residual moisture (EPA-83).
(2) The weight loss resulting from drying a sample to constant weight in an oven usually maintained between 103 and 107° C (EPA-83).
(3) *See* analytical parameters--laboratory for more related terms.

Total organic active ingredients: The sum of all organic active ingredients covered by 40 CFR 455.20(a) which are manufactured at a facility subject to this subpart (40 CFR 455.20-91, *See also* 40 CFR 455.21-91).

Total organic active ingredients: The sum of all organic active ingredients covered by 40 CFR 455.20(a) which are manufactured at a facility subject to this subpart (40 CFR 455.21-91).

Total organic carbon (TOC): Total organic carbon (TOC) is a measure of the organic contamination of a water sample. It has an empirical relationship with the biochemical and chemical oxygen demands (*See* carbon for more related terms) (EPA-83; 74/04).

Total organic compounds: Those compounds measured according to the procedures in 40 CFR 60.503 (40 CFR 60.501-91, *See also* 40 CFR 60.561; 60.611; 60.661-91).

Total oxygen demand (TOD): The total oxygen needed for BOD and COD during wastewater treatment (*See* oxygen for more related terms).

Total phenol: The total phenolic compounds as measured by the procedure listed in 40 CFR 136 (distillation followed by colorimetric--4AAP) (*See* phenol for more related terms) (40 CFR 464.02-91).

Total pressure: The pressure representing the sum of static pressure and velocity pressure at the point of measurement (*See* pressure for more related terms) (EPA-83/06).

Total rated capacity: The sum of the rated capacities of all fuel-burning equipment connected to a common stack. The rated capacity shall be the maximum guaranteed by the equipment manufacturer or the maximum normally achieved during use, whichever is greater (40 CFR 52.01-91).

Total recoverable: The concentration determined on an unfiltered sample following treatment with hot, dilute mineral acid (40 CFR 136-App/C-91).

Total recoverable metals: The concentration of metals in an unfiltered sample following treatment with hot diluted mineral acid (*See* metal for more related terms) (Method 3005, SW-846).

Total recycle: The complete reuse of a waste stream, with make-up water added for evaporation losses. There is no blowdown stream from a totally recycled flow and the process water is not periodically or continuously discharged (EPA-83/03a).

Total reduced sulfur (TRS): The sum of the sulfur compounds hydrogen sulfide, methyl mercaptan, dimethyl sulfide, and dimethyl disulfide, that are released during the kraft pulping operation and measured by Reference Method 16 (40 CFR 60.281-91).

Total residual chlorine (TRC) (or total residual oxidants for intake water with bromide):
(1) Free residual plus combined residual (EPA-82/11f).
(2) The value obtained using the amperometric method for total residual chlorine described in 40 CFR 130 (40 CFR 423.11-91; *See also* 420.02-91; EPA-82/11a).
(3) *See* chlorine for more related terms.

Total residual oxidants for intake water with bromide: *See* synonym, total residual chlorine.

Total resource effectiveness (TRE) index value: A measure of the supplemental total resource requirement per unit reduction of TOC associated with an individual air oxidation vent stream, based on vent stream flow rate, emission rate of TOC, net heating value, and corrosion properties (whether or not the vent stream is halogenated), as quantified by the equation given under 40 CFR 60.614(e) (40 CFR 60.611-91).

Total smelter charge: The weight (dry basis) of all copper sulfide ore concentrates processed at a primary copper smelter, plus the weight of all other solid materials introduced into the roasters and smelting furnaces at a primary copper smelter, except calcine, over a one-month period (40 CFR 60.161-91).

Total SO_2 equivalents: The sum of volumetric or mass concentrations of the sulfur compounds obtained by adding the quantity existing as SO_2 to the quantity of $SO2$ that would be obtained if all reduced sulfur compounds were converted to SO_2 (ppmv or kg/DSCM) (40 CFR 60.641-91).

Total solid (TS): *See* synonym, total suspended solid.

Total suspended particles: A method of monitoring particulate matter by total weight (EPA-94/04).

Total suspended particulate: The particulate matter as measured by the method described in Appendix B of Part 50 of this chapter (*See* particulate for more related terms) (40 CFR 51.100-91, *See also* 40 CFR 58.1-91).

Total suspended residue: *See* synonym, total suspended solid.

Total suspended solid (TSS) (total solid or total suspended residue):
(1) The value obtained by the method specified in 40 CFR 136.3 (40 CFR 420.02-91).
(2) The total amount of both suspended and dissolved materials in wastewater. Expressed in mg/L (EPA-82/11).
(3) A measure of the suspended solids in wastewater, effluent, or water bodies, determined by tests for "total suspended nonfilterable solids" (*See* suspended solids (EPA-94/04).
(4) *See* solid for more related terms.

Total test distance: Is defined for each class of motorcycles in 40 CFR 86.427.78 (40 CFR 86.402.78-91).

Total toxic organics (TTO): The total toxic organics, which is the summation of all quantifiable values greater than 0.01 milligrams per liter for the toxic organics listed in 40 CFR 413.02 and 40 CFR 433.11 (40 CFR 433.11-91, *See also* 40 CFR 464.02; 464.11; 464.21; 464.31; 464.41; 465.02; 467.02; 468.02; 469.12; 469.22; 469.31-91).

Total volatile solids: Volatile residue present in wastewater (EPA-83/06a).

Total trihalomethanes (TTHM): The sum of the concentration in milligrams per liter of the trihalomethane compounds (trichloromethane [chloroform], dibromochloromethane, bromodichloromethane and tribromomethane [bromoform]), rounded to two significant figures (40 CFR 141.2-91).

Totally enclosed manner: Any manner that will ensure no exposure of human beings or the environment to any concentration of PCBs (40 CFR 761.3-91).

Totally enclosed treatment facility: A facility for the treatment of hazardous waste which is directly connected to an industrial production process and which is constructed and operated in a manner which prevents the release of any hazardous waste or any constituent thereof into the environment during treatment. An example is a pipe in which waste acid is neutralized (40 CFR 260.10-91).

Touch up coat: The coat applied to correct any imperfections in the finish after color or texture coats have been applied. This definition does not include conductive sensitizers or EMI/RFI shielding coatings (40 CFR 60.721-91).

Toxaphene:
(1) A material consisting of technical grade chlorinated

camphene having the approximate formula of $C_{10}H_{10}C_{18}$ and normally containing 67-69 percent chlorine by weight (40 CFR 129.4-91).
(2) Chemical that causes adverse health effects in domestic water supplies and is toxic to freshwater and marine aquatic life (EPA-94/04).

Toxaphene manufacturer: A manufacturer, excluding any source which is exclusively a toxaphene formulator, who produces, prepares or processes toxaphene or who uses toxaphene as a material in the production, preparation or processing of another synthetic organic substance (40 CFR 129.103-91).

Toxaphene formulator: A person who produces, prepares or processes a formulated product comprising a mixture of toxaphene and inert materials or other diluents into a product intended for application in any use registered under the Federal Insecticide, Fungicide and Rodenticide Act, as amended (7USC135, et seq.) (40 CFR 129.103-91).

Toxic: Harmful to living organisms (cf. poison or toxics) (EPA-89/12).

Toxic air pollutant (or air toxic):
(1) The aggregate emissions of Benzene; 1,3 Butadiene; Polycyclic organic matter (POM); Acetaldehyde and Formaldehyde (CAA211.k-42USC7545-91).
(2) The materials contaminating the environment that cause death, disease, birth defects in organisms that ingest or absorb them. The quantities and length of exposure necessary to cause these effects can vary widely (EPA-89/12).
(3) *See* pollutant for more related terms.

Toxic atmosphere monitor (or colorimetric indicator tube): The monitor consists of a glass tube impregnated with an indicating chemical. The tube is connected to a pump. A known volume of contaminated air is pulled at a predetermined rate through the tube by the pump. The contaminant reacts with the indicator chemical in the tube, producing a change in color whose length is proportional to the contaminant concentration.

The tubes are normally chemical specific. There are different tubes for different gases, e.g., chlorine detector tube for chlorine gas, acrylonitrile tube for acrylonitrile gas, etc. Some manufacturers do produce tubes for groups of gases such as aromatic hydrocarbons and alcohols. Concentration ranges on the tubes may be in the ppm or percent range (*See* air analyzer for more related terms).

Toxic chemical:
(1) A chemical or chemical category listed in 40 CFR 372.65 (toxic substance) (40 CFR 372.3-91, *See also* PPA6603; SF329-91).
(2) Any chemical listed in EPA rules as "Toxic Chemicals Subject to Section 313 of the Emergency Planning and Community Right-to-Know Act of 1986. (EPA-94/04)"

Toxic chemical group: Toxic chemicals can be grouped as follows:
(1) Organic solvents
(2) Toxic metals
(3) Pesticides
(4) Herbicides (Course 165.6).

Toxic chemical release form: Information form required of facilities that manufacture, process, or use (in quantities above a specific amount) chemicals listed under SARA Title III (EPA-94/04).

Toxic chemical use substitution: Replacing toxic chemicals with less harmful chemicals in industrial processes (EPA-94/04).

Toxic cloud: Airborne plume of gases, vapors, fumes, or aerosols containing toxic materials (EPA-94/04).

Toxic concentration low: The lowest concentration of a substance in air to which humans or animals have been exposed for any given period of time that has produce any toxic effect in humans or produce tumorigenic or reproductive effects in animals (*See* dose response for more related terms).

Toxic disease process: Including
(1) Mutagenesis;
(2) Teratogenesis; and
(3) Carcinogenesis (Course 165.6)

Toxic dose low: The lowest dose of a substance introduced by any route, other than inhalation, over any given period of time, and reported to produce any toxic effect in humans or to produce tumorigenic or reproductive effects in animals (*See* dose response for more related terms).

Toxic effect: An adverse change in the structure or function of an experimental animal as a result of exposure to a chemical substance (40 CFR 798.6050; 798.6200-91).

Toxic metals: *See* synonym, heavy metals.

Toxic pollutant:
(1) Those pollutants, or combinations of pollutants, including disease-causing agents, which after discharge and upon exposure, ingestion, inhalation or assimilation into any organism either directly from the environment or indirectly by ingestion through food chains, will, on the basis of information available to the Administrator, cause death, disease, behavioral abnormalities, cancer, genetic mutations, physiological malfunctions (including malfunctions in reproduction) or physical deformations in such organisms or their offspring (CWA502, *See also* 40 CFR 122.2; 125.58; 131.3; 501.2-91).
(2) Under CWA307.a.1, 65 toxic pollutants are listed in CFR 401.15-92.
(3) Materials that cause death, disease, or birth defects in organisms that ingest or absorb them. The quantities and exposures necessary to cause these effects can vary widely (EPA-94/04).
(4) *See* pollutant for more related terms.

Toxic release inventory: Database of toxic releases in the United States compiled from SARA Title III section 313 reports (EPA-94/04).

Toxic response type:
(1) Common local effects: Including irritation; corrosion; and fibrogenesis.

(2) System effect: Inlcuding hepatotoxic (Liver), nephrotoxic (Kidney), neurotoxic (Nerves), hematopoietic (Blood), and reproductive system (Course 165.6).

Toxic substance: A chemical or mixture that may present an unreasonable risk of injury to health or the environment (EPA-94/04).

Toxic Substances Control Act (TSCA) of 1976: *See* the term of Act or TSCA.

Toxic substance or harmful physical agent: Any chemical substance, biological agent (bacteria, virus, fungus, etc.), or physical stress (noise, heat, cold, vibration, repetitive motion, ionizing and non-ionizing radiation, hypo- or hyperbaric pressure, etc.) which:
(1) Is listed in the least printed edition of the National Institute for Occupational Safety and Health (NIOSH) Registry of Toxic Effects of Chemical Substances (RTECS); or (ii) Has yielded positive evidence of an acute or chronic health hazard in testing conducted by, or known to, the employer; or
(2) Is the subject of a material safety data sheet kept by or known to the employer indicating that the material may pose a hazard to human health (29 CFR 1910.20-91).

Toxic units (TUs): A measure of toxicity in an effluent as determined by the acute toxicity units or chronic toxicity units measured (EPA-91/03).

Toxic unit acute (TUA):
(1) The reciprocal of the effluent dilution that causes the acute effect by the end of the acute exposure period (EPA-85/09).
(2) The reciprocal of the effluent concentration that causes 50 percent of the organisms to die by the end of the acute exposure period (i.e., 100 LC50) (EPA-91/03).

Toxic unit chronic (TUC):
(1) The reciprocal of the effluent dilution that causes no unacceptable effect on the test organisms by the end of the chronic exposure period (EPA-85/09).
(2) The reciprocal of the effluent concentration that causes no observable effect on the test organisms by the end of the chronic exposure period (i.e., 100/NOEC) (EPA-91/03).

Toxics use reduction: Refers to the activities grouped under source reduction, where the intent is to reduce, avoid, or eliminate the use of toxics in processes and/or products so as to reduce overall risks to the health of workers, consumers, and the environment without shifting risks between workers. consumers, or parts of the environment (EPA-91/10, p7).

Toxic waste:
(1) A waste which can produce injury upon contact with, or by accumulation in, a susceptible site in or on the body of a living organism (OME-88/12).
(2) A waste that can produce injury if inhaled, swallowed, or absorbed through the skin (EPA-94/04).
(3) *See* waste for more related terms.

Toxicant: A harmful substance or agent that may injure an exposed organism (EPA-94/04).

Toxicant and effect: A toxicant is a chemical which can produce acute (short term effect) and chronic (long term effect) diseases, e.g., sunburn is an acute effect of sun bathing, and a skin cancer is a chronic effect. The toxicants can be classified according to how they affect the respiratory tract.
(1) **Asphyxiants:** Gases that deprive the body tissues of oxygen.
(2) **Irritants:** Chemicals that irritate the air passages.
(3) **Necrosis producers:** Chemicals that result in cell death and edema.
(4) **Fibrosis producers:** Chemicals that produce fibrotic tissue which, if massive, blocks airways and decreases lung capacity.
(5) **Allergens:** Chemicals that induce an allergic response.
(6) **Carcinogens:** Chemicals that are associated with lung cancer (Course 165.5).

Toxicity:
(1) One of the four U.S. EPA hazardous waste characteristics (ETI-92).
(2) The degree of danger posed by a substance to animal or plant life (EPA-89/12, *See also* 40 CFR 131.35; 171.2-91). Example toxicity rate by oral LD50 for rats at concentration mg/Kg is as follows:
   (A) Extremely toxic: 1 mg/Kg or less;
   (B) Highly toxic: 1 to 50 mg/Kg;
   (C) Moderately toxic: 50 to 500 mg/Kg;
   (D) Slightly toxic: 0.5 to 5 g/Kg; (E) Practically nontoxic: 5 to 15 g/Kg (Course 165.6).
(3) A measure of the adverse health effects of exposure to a chemical (EPA-87/07a).
(4) Characterization of the toxicological properties and effects of a chemical, with special emphasis on establishment of doseresponse characteristics (EPA-94/04).

Other toxicity related terms include:
- Acute toxicity
- Chronic toxicity
- Cumulative toxicity
- Dermal toxicity
- Developmental toxicity
- Direct toxicity

Toxicity characteristic leaching procedure (TCLP):
(1) TCLP is designed to determine the mobility of both organic and inorganic contaminants present in liquid, solid, and multiphasic wastes (40 CFR 261-App/II).
(2) A U.S. EPA test developed to evaluate the potential of a component to leach from a substance (ETI-92).

Toxicity curve: The curve produced from toxicity tests when LC50 values are plotted against duration of exposure. (This term is also used in aquatic toxicology, but in a less precise sense, to describe the curve produced when the median period of survival is plotted against test concentrations) (40 CFR 797.1350-91, *See also* 40 CFR 797.1440-91).

Toxicity identification evaluation (TIE): A set of procedures to identify the specific chemicals responsible for effluent toxicity (EPA-91/03).

Toxicity reduction: In municipal waste, eliminating or reducing substances in products that pose risks when the products are discarded as municipal waste (cf. pollution prevention) (OTA-89/10).

Toxicity reduction evaluation (TRE): A site-specific study conducted in a step-wise process designed to identify the causative agents of effluent toxicity, isolate the sources of toxicity, evaluate the effectiveness of toxicity control options, and then confirm the reduction in effluent toxicity (EPA-91/03).

Toxicity test: A procedure to determine the toxicity of a chemical or an effluent using living organisms. A toxicity test measures the degree of effect on exposed test organisms of a specific chemical or effluent (cf. bioassy) (EPA-91/03; 85/09).

Toxicity testing: Biological testing (usually with an invertebrate, fish, or small mammal) to determine the adverse effects of a compound or effluent (EPA-94/04).

Toxicological profile: An examination, summary, and interpretation of a hazardous substance to determine levels of exposure and associated health effects (EPA-94/04).

Toxicology: The science and study of poisons control (EPA-89/12).

Toxics: Those pollutants that have a toxic effect on living organisms. The CWA Section 307(a) priority pollutants are a subset of this group of pollutants (EPA-91/03).

Toxoid: A toxin treated so as to destroy its toxicity, but still capable of inducing formation of antibodies (EPA-83/09).

Trace: A very small but detectable quantity of a chemical substance.

Trace metal: A very small but detectable quantity of metals (*See* metal for more related terms).

Trace method: A method applicable to ppm range (*See* method for more related terms) (ACS-87/11).

Tracer: A traceable substance which is mixed with (or is injected into) a media to show how and where the tracer travels in the media. The purpose is to study the chemical behavior of the media.

Traceable: That a local standard has been compared and certified either directly or via not more than one intermediate standard, to a primary standard such as a National Bureau of Standards Standard Reference Material (NBS SRM), or a USEPA/NBS-approved Certified Reference Material (CRM) (40 CFR 50.1-91, *See also* 40 CFR 58.1-91).

Tracking form: The Federal Medical Tracking Form that must accompany all applicable shipments of regulated medical wastes generated within one of the Covered States (40 CFR 259.10-91).

Tractor: For the purposes of this subpart, any two or three wheeled vehicle used exclusively for agricultural purposes, or for snow plowing, including self-propelled machines used exclusively in growing, harvesting or handling farm produce (40 CFR 205.151-91).

Trade name product: A chemical or mixture of chemicals that is distributed to other persons and that incorporates a toxic chemical component that is not identified by the applicable chemical name or Chemical Abstracts Service Registry number listed in 40 CFR 372.65 (40 CFR 372.3-91).

Trade sales paint: The paint which is sold to the do it yourself market, i.e., the over the counter retail segment of the coatings market. It does not include the paint which is sold to painting contractors or similar professionals (*See* paint for more related terms) (EPA-79/12b).

Traditional pollutant parameters: The pollutant parameters considered, and used, in the development of BPT guidelines. These parameters include BOD, COD, TOC, TSS, and ammonia (*See* pollutant for more related terms) (EPA-79/12c).

Trajectory model: A model which describes the transport and diffusion of a puff of air pollution along a usually horizontal trajectory in the atmosphere. Removal and/or transformation processes can be taken into account (NATO-78/10).

Tramp materials: Contaminants occurring in waste paper which were not originally applied to the paper or paperboard product during production or conversion (EPA-83).

Transactinide element: Chemical elements whose atomic numbers are greater than 103.

Transboundary pollutants: Air pollution that travels from one jurisdiction to another, often crossing state or international boundaries (EPA-94/04).

Transducer: Any device which is used to convert an input signal into an electrical signal.

Transfer: Loading and unloading of chemicals between transport vehicles and storage vessels, and sending chemicals via pipes between storage vessels and process reactors (EPA-85/11).

Transfer and loading system: Any facility used to transfer and load coal for shipment (40 CFR 60.251-91).

Transfer efficiency: The ratio of the amount of coating solids deposited onto a part or product to the total amount of coating solids used (40 CFR 52.741-91, *See also* 40 CFR 60.311; 60.391; 60.451; 60.721-91).

Transfer facility: Any transportation related facility including loading docks, parking areas, storage areas and other similar areas where shipments of hazardous waste are held during the normal course of transportation (40 CFR 260.10-91, *See also* 40 CFR 259.10; 270.2; 761.3-91).

Transfer point: A point in a conveying operation where the nonmetallic mineral is transferred to or from a belt conveyor except where the nonmetallic mineral is being transferred to a stockpile (40 CFR 60.671-91).

Transfer station: A site at which solid wastes are concentrated for transport to a processing facility or land disposal site. A transfer station may be fixed or mobile (40 CFR 243.101-91).

Transferred plasma arc: An arc established between an electrode and the working material as the other electrode. In transferred arc applications, heating occurs by way of conversion, radiation, and electrical resistance (*See* plasma arc for more related terms) (Wittle-93/07).

Transformation: The process of placing new genes into a host cell, thereby inducing the host cell to exhibit functions encoded by the DNA (EPA-89/12).

Transformer: An electrical device with two or more multi-turn coils in a such arrangement that the magnetic field of one can link the other. By using magnetic induction effect, a transformer can transfer electrical energy from one current circuit to another with a possible change in voltage, current, phase or impedance (cf. capacitor). Other transformer related terms include:
- Auto-transformer
- Distribution transformer
- Dry transformer
- Non-PCB transformer
- PCB transformer
- Pole type transformer
- Power transformer
- Primary feeder circuit transformer
- Rupture of a PCB transformer
- Step down transformer
- Subtransmission transformer

Transformer core: A quantity of ferrous material placed in a coil or transformer to provide a better path than air for magnetic flux, thereby increasing the inductance of the coil or increasing the coupling between the windings of a transformer (EPA-83/03).

Transient: Unstable, rapidly changing or a phenomenon before reaching a steady state condition.

Transient water system: A non-community water system that does not serve 25 of the same nonresidents per day for more than six months per year (EPA-94/04).

Transient shipment: A shipment of nuclear material, originating and terminating in foreign countries, on a vessel or aircraft that stops at a United States port (10 CFR 40.4; 70.4-91).

Transit country: Any foreign country, other than a receiving country, through which a hazardous waste is transported (40 CFR 262.51-91).

Transit improvement measures: Those actions or combinations of actions taken by the Metropolitan Transit Division of the Municipality of Metropolitan Seattle (METRO) and the city of Spokane to promote the attractiveness and increased use of public mass transit systems (40 CFR 52.2493-91).

Transition (intermediate) flow: Reynolds number greater than 2.0 but less than 500 for an incompressible flow (*See* flow regime for more related terms).

Translocation: The transference or transport of chemical from the site of uptake to other plant components (40 CFR 797.2850-91).

Transmission class: The basic type of transmission, e.g., manual, automatic, semi-automatic (40 CFR 86.082.2-91, *See also* 40 CFR 600.002.85-91).

Transmission configuration: A unique combination, within a transmission class, of the number of the forward gears and, if applicable, over-drive. The Administrator may further subdivide a transmission configuration (based on such criteria as gear ratios, torque convertor multiplication ratio, stall speed and shift calibration, etc.), if he determines that significant fuel economy or exhaust emission differences exist within that transmission configuration (40 CFR 86.082.2-91, *See also* 40 CFR 600.002.85-91).

Transmission diffraction: An instrument for analyzing the diffraction phenomenon of a substance by transmitting an electronic beam through the very thin film or powder of the substance.

Transmission line: A pipeline, other than a gathering line, that:
(1) Transports gas from a gathering line or storage facility to a distribution center or storage facility;
(2) Operates at a hoop stress of 20 percent of more of SMYS; or
(3) Transports gas within a storage field (40 CFR 192.3-91).

Transmission lines: Pipelines that transport raw water from its source to a water treatment, then to the distribution grid system (EPA-94/04).

Transmissometer: That portion of the CEMS that includes the sample interface and the analyzer (40 CFR 60-App/B-91). It is a device for measuring the visual range of the atmosphere.

Transmissive fault or fracture: A fault or fracture that has sufficient permeability and vertical extent to allow fluids to move between formations (40 CFR 146.61; 148.2-91).

Transmissivity:
(1) The hydraulic conductivity integrated over the saturated thickness of an underground formation. The transmissivity of a series of formations is the sum of the individual transmissivities of each formation comprising the series (40 CFR 191.12-91).
(2) The capacity of an aquifer to transmit water. It is the product of the aquifer thickness and hydraulic conductivity and has a unit of gallon per day per foot (also known as **aquifer transmissivity**) (EPA-87/03).
(3) A measure of a water-bearing unit's capacity to transmit fluid: the product of the thickness and the average hydraulic conductivity of unit. Also, the rate at which water is transmitted through a strip of an aquifer of a unit width under a unit hydraulic gradient at a prevailing temperature and pressure (DOE-91/04).
(4) The ability of an aquifer to transmit water (EPA-94/04).

Transmittance:
(1) The fraction of incident light that is transmitted through an optical medium (40 CFR 60-App/B-91).
(2) The reciprocal of the opacity (cf. opacity).

Transparency: Quality that allows light to pass through without significant deviation of absorption.

Transpiration: The process by which water vapor is lost to the atmosphere from living plants. The term can also be applied to the quantity of water thus dissipated (EPA-94/04).

Transport or transportation:
(1) The movement of a hazardous substance by any mode, including pipeline (as defined in the Pipeline Safety Act), and in the case of a hazardous substance which has been accepted for transportation by a common or contract carrier, the term "transport" or "transportation" shall include any stoppage in transit which is temporary, incidental to the transportation movement, and at the ordinary operating convenience of a common or contract carrier, and any such stoppage shall be considered as a continuity of movement and not as the storage of a hazardous substance (SF101-42USC9601).
(2) The movement of solid waste subsequent to collection (EPA-83).
(3) To carry or convey goods from one place to another using ships, trucks, trains, pipelines, or airplanes (EPA-85/11).

Transport air: Air employed in pneumatic conveying systems to move materials by entrainment in the moving air stream (EPA-83).

Transport mode: Method of transportation: highway (trucks); rail (trains); water (ships/barges); pipelines; air (planes) (EPA-85/11).

Transport phenomena: *See* definition under dispersion.

Transport vehicle: A motor vehicle or rail car used for the transportation of cargo by any mode. Each cargo carrying body (trailer, railroad freight car, etc.) is a separate transport vehicle (*See* vehicle for more related terms) (40 CFR 260.10-91, *See also* 40 CFR 761.3-91).

Transport water: The water used to carry insoluble solids (*See* water for more related terms) (EPA-76/03).

Transportation: The shipment or conveyance or regulated medical waste by air, rail, highway, or water (cf. transport) (40 CFR 259.10-91, *See also* 40 CFR 260.10-91).

Transportation control measure (TCM):
(1) Any measure in an applicable implementation plan which is intended to reduce emissions from transportation sources (40 CFR 51.138-91, *See also* 40 CFR 51.100-91).
(2) Steps taken by a locality to improve air quality by reducing or changing he flow of traffic, e.g., public transit, carpools, HOV lanes, etc (EPA-94/04).

Transportation improvement program (TIP): The staged multiyear program of transportation improvements including an annual (or biennial) element which is required in 23 CFR 450 (40 CFR 51.138-91).

Transportation of gas: The gathering, transmission, or distribution of gas by pipeline or the storage of gas, in or affecting interstate or foreign commerce (40 CFR 192.3-91).

Transporter: A person engaged in the off-site transportation of regulated medical waste by air, rail, highway, or water (40 CFR 259.10-91, *See also* 40 CFR 260.10; 270.2-91).

Transporter of PCB waste: For the purposes of subpart K of this part, any person engaged in the transportation of regulated PCB waste by air, rail, highway, or water for purposes other than consolidation by a generator (40 CFR 761.3-91).

Transuranic radioactive waste: As used in this part, means waste containing more than 100 nanocuries of alpha emitting transuranic isotopes, with half lives greater than twenty years, per gram of waste, except for:
(1) High level radioactive wastes;
(2) Wastes that the Department has determined, with the concurrence of the Administrator, do not need the degree of isolation required by this part; or
(3) Wastes that the Commission has approved for disposal on a case-by-case basis in accordance with 10 CFR 61 (cf. waste, transuranic) (40 CFR 191.02-91).
(4) *See* radioactive waste or waste for more related terms.

Transuranic waste: The radioactive waste that, at the end of institutional control periods, is contaminated with transuranium radionuclides with half-lives greater than 20 years in concentrations greater than 100 nCi/g (*See* radioactive waste or waste for more related terms) (DOE-91/04).

Trapping: The condition when the diffusion of air pollution is confined to a small mixing layer capped by a strong inversion (NATO-78/10).

Trash:
(1) Material considered worthless, unnecessary or offensive that is usually thrown away. Generally defined as dry waste material, but in common usage it is a synonym for garbage, rubbish, or refuse (EPA-89/11).
(2) Material considered worthless or offensive that is thrown away. Generally defined as dry waste material, but in common usage it is a synonym for garbage, rubbish, or refuse (EPA-94/04).

Trash to energy plan: Burning trash to produce energy (EPA-94/04).

Traveling grate: A traveling grate stoker consists of a belt like arrangement of air admitting grate bars similar to a chain grate, but with grate bars mounted on transverse beams usually pulled by chains and sprockets through the furnace (*See* grate for more related terms) (EPA-83).

Traveling grate furnace: A furnace with a moving grate that conveys material through the heating zone. The feed is ignited on the surface as the grate moves past the burners; air is blown in the charge to burn the fuel by downdraft combustion as it moves continuously toward discharge (*See* furnace for more related terms) (EPA-83/03a).

Traveling grate stoker:
(1) A type of boiler where coal is carried and burned (cf. traveling grate furnace) (EPA-84/09).

(2) A type of furnace with a moving grate. A stoker is essentially a moving chain (belt) carried on sprockets and covered with separated, small metal pieces called keys. The entire top surface can act as a grate while moving through the furnace but can flex over the sprocket wheels at the end of the furnace, return under the furnace, and re-enter the furnace over sprocket wheels at the front (SW-108ts).

(3) *See* stoker for more related terms.

Tray aerator: A set of cylindrical trays. While wastewater is induced from the top tray, each tray functions like an aerator housing various packing materials to promote aeration (*See* aerator for more related terms).

Tray column (or plate column): A cylindrical tower housing perforated plates, bubble caps, etc. It may be used for scrubbing polluted air for gaseous (absorption) or particulate control (EPA-84/09).

TRE index value: A measure of the supplemental total resource requirement per unit reduction of TOC associated with an individual distillation vent stream, based on vent stream flow rate, emission rate of TOC net heating value, and corrosion properties (whether or not the vent stream is halogenated), as quantified by the equation given under 40 CFR 60.664(e) (40 CFR 60.661-91).

Tread end cementing operation: The system used to apply cement to one or both ends of the tread or combined tread/sidewall component. A tread end cementing operation consists of a cement application station and all other equipment, such as the cement supply system and feed and takeaway conveyors, necessary to apply cement to tread ends and to allow evaporation of solvent from the cemented tread ends (40 CFR 60.541-91).

Treatability study:
(1) The testing of a remedial alternative in the laboratory or field to obtain data necessary for a detailed evaluation of its feasibility (EPA-89/12a, *See also* 40 CFR 260.10-91).
(2) Tests of potential cleanup technologies conducted in a laboratory (*See* bench-scale tests. (EPA-94/04))

Treatability study sample exemption: A Federal regulation set forth in 40 CFR 261.4(f) that excludes treatability studies conducted offsite from most management and permitting requirements under RCRA (EPA-89/12a).

Treated regulated medical waste:
(1) The regulated medical waste that has been treated to substantially reduce or eliminates its potential for causing disease, but has not yet been destroyed (40 CFR 259.10-91).
(2) Medical waste treated to substantially reduce or eliminate its pathogenicity, but that has not yet been destroyed (EPA-94/04).
(3) *See* medical waste or waste for more related terms.

Treated wastewater: Wastewater that has been subjected to one or more physical, chemical, and biological processes to reduce its pollution of health hazards (EPA-94/04).

Treated water: The raw water or filtered water that has been treated to make it suitable for plant needs such as softening (*See* water for more related terms) (EPA-74/03a).

Treatment:
(1) Any method, technique, or process, including neutralization, designed to change the physical, chemical, or biological character or composition of any hazardous waste so as to neutralize such waste, or so as to recover energy or material resources from the waste, or so as to render such waste non-hazardous, or less hazardous; safer to transport, store, or dispose of; or amenable for recovery, amenable for storage, or reduced in volume (40 CFR 260.10-91).
(2) (A) Any method, technique, or process designed to remove solids and/or pollutants from solid waste, wastestreams, effluents, and air emissions (EPA-94/04).
    (B) Methods used to change the biological character or composition of any regulated medical waste so as to substantially reduce or eliminate its potential for causing disease (EPA-94/04).

Other treatment related terms include:
- Batch treatment
- Biological treatment
- Biological wastewater treatment
- Chemical treatment
- Continuous treatment
- In situ treatment
- Joint treatment
- Physical and chemical
- Physical treatment
- Primary treatment (*See* synonym, primary wastewater treatment)
- Primary wastewater treatment
- Secondary treatment (*See* synonym, secondary wastewater treatment)
- Secondary wastewater treatment
- Surface treatment
- Tertiary treatment (*See* synonym, tertiary wastewater treatment)
- Tertiary wastewater treatment
- Thermal process (*See* synonym, thermal treatment)
- Thermal treatment

Treatment efficiency: The percentage reduction of a specific or group of pollutants by a specific wastewater treatment step or treatment plant (EPA-82/11f).

Treatment facility effluent: Treated process wastewater.

Treatment facility and treatment system: All structures which contain, convey, and as necessary, chemically or physically treat coal rine drainage, coal preparation plant process wastewater, or drainage from coal preparation plant associated areas, which remove pollutants regulated by this Part from such waters. This includes all pipes, channels, ponds, basins, tanks and all other equipment serving such structures (40 CFR 434.11-91).

Treatment plant: A structure built to treat wastewater before discharging it into the environment (EPA-94/04).

Treatment process: A stream stripping unit, thin-film evaporation unit, waste incinerator, or any other process used to comply with

40 CFR 1.348 of this subpart (40 CFR 61.341-91).

Treatment, storage, and disposal (TSD) facility: A site where a hazardous substance is treated, stored, or disposed. TSD facilities are regulated by EPA and states under RCRA (*See* treatment for more related terms) (EPA-94/04).

Treatment technique requirement: A requirement of the national primary drinking water regulations which specifies for a contaminant a specific treatment technique(s) known to the Administrator which leads to a reduction in the level of such contaminant sufficient to comply with the requirements of Part 141 of this chapter (40 CFR 142.2-91).

Treatment technology: Any unit operation or series of unit operations that alters the composition of a hazardous substance or pollutant or contaminant through chemical, biological, or physical means so as to reduce toxicity, mobility, or volume of the contaminated materials being treated. Treatment technologies are an alternative to land disposal of hazardous wastes without treatment (40 CFR 300.5-91). Potential treatment processes and their EPA Codes are as follows. *See* each technology for definitions (40 CFR 264-App/I-Table 2):

(1) **Thermal Treatment--By Alphabetic Order**
T14: Calcination
T16: Cement kiln
T08: Fluidized bed incinerator
T10: Infrared furnace incinerator
T17: Lime kiln
T06: Liquid injection incinerator
T15: Microwave discharge
T11: Molten salt incinerator
T09: Multiple hearth incinerator
T12: Pyrolysis
T07: Rotary kiln incinerator
T13: Wet air oxidation
T18: Other (specify)
- Advanced electrical reactor
- Circulating bed combustor
- Fixed hearth (controlled air, multiple hearth, pyrolytic, or starved air incinerators)
- Flare
- Infrared system
- Oxygen-Enriched Incineration
- Plasma Arc
- Supercritical Fluid Westinghouse/O'Connor Combustor

(2) **Chemical Treatment**
T19: Absorption mound
T20: Absorption field
T21: Chemical fixation (solidification and stabilization)
T22: Chemical oxidation
T23: Chemical precipitation
T24: Chemical reduction
T25: Chlorination
T26: Chlorinolysis
T27: Cyanide destruction
T28: Degradation
T29: Detoxification
T30: Ion exchange
T31: Neutralization
T32: Ozonation
T33: Photolysis
T34: Other (specify)
- Alkaline polyethylene glycol
- Hydrolysis

(3) **Physical Treatment--separation of components**
T35: Centrifugation
T36: Clarification
T37: Coagulation
T38: Decanting
T39: Encapsulation
T40: Filtration
T41: Flocculation
T42: Flotation
T43: Forming
T44: Sedimentation
T45: Thickening
T46: Ultrafiltration
T47: Other (specify)

(4) **Physical Treatment--removal of specific components**
T48: Absorption-molecular sieve
T49: Adsorption (activated carbon and resin adsorption)
T50: Blending
T51: Catalysis
T52: Crystallization (suspension freezing)
T53: Dialysis
T54: Distillation including steam distillation
T55: Electrodialysis
T56: Electrolysis
T57: Evaporation
T58: Magnetic separation
T59: Leaching
T60: Liquid ion exchange
T61: Liquid-liquid extraction
T62: Reverse osmosis
T63: Solvent recovery
T64: Striping
T65: Sand filter
T66: Other (specify)
- Dissolution
- Freeze drying (Lyophilization)
- Zone refining

(5) **Biological Treatment**
T67: Activated sludge
T68: Aerobic lagoon
T69: Aerobic tank
T70: Anaerobic lagoon
T71: Composting
T72: Septic tank
T73: Spray irrigation
T74: Thickening filter
T75: Trickling filter
T76: Waste stabilization pond
T77: Other (specify)
- Enzyme treatment/mutant or adapted microorganisms
- Lagoon (stabilization pond or oxidation pond)
- Land treatment (landfarming)
- Rotational biological contactor (RBC)
- White-rot fungus

Treatment train: A complete treatment process that includes pretreatment, primary treatment, residuals and side-stream treatments, and post-treatment considerations (EPA-89/12a).

Treatment works:
(1) Means any devices and systems used in the storage, treatment, recycling, and reclamation of municipal sewage or industrial wastes of a liquid nature to implement section 201 of this act, or necessary to recycle or reuse water at the most economical cost over the estimated life of the works, including intercepting sewers, outfall sewers, sewage collection systems, pumping, power, and other equipment, and their appurtenances; extensions, improvements, remodeling, additions, and alterations thereof; elements essential to provide a reliable recycled supply such as standby treatment units and clear well facilities; and any works, including site acquisition of the land that will be an integral part of the treatment process (including land use for the storage of treated wastewater in land treatment systems prior to land application) or is used for ultimate disposal of residues resulting for such treatment.
(2) In addition to the definition contained in subparagraph (A) of this paragraph, "treatment works" means any other method or system for preventing, abating, reducing, storing, treating, separating, or disposing of municipal waste, including storm water runoff, or industrial waste, including waste in combined storm water and sanitary sewer systems. Any application for construction grants which includes wholly or in part such methods or systems shall, in accordance with guidelines published by the Administrator pursuant to subparagraph (C) of this paragraph, contain adequate data and analysis demonstrating such proposal to be, over the life of such works, the most efficient alternative to comply with sections 301 or 302 of this act, or the requirements of section 201 of this act.
(3) For the purposes of subparagraph (B) of this paragraph, the Administrator shall, within one hundred and eighty days after the date of enactment of this title, publish and thereafter revise no less than annually, guidelines for the evaluation of methods, including cost effective analysis, described in subparagraph (B) of this paragraph (CWA212-33USC1292).

Treatment works phase or segment: Any substantial Portion of a facility and its interceptors described in a facilities plan under 40 CFR 35.2030, which can be identified as a subagreement or discrete subitem. Multiple subagreements under a project shall not be considered to be segments or phases. Completion of building of a treatment works phase or segment may, but need not in and of itself, result in an operable treatment works (40 CFR 35.2005-91).

Treatment works segment: Any portion of an operable treatment works described in an approved facilities plan under 40 CFR 35.917, which can be identified as a contract or discrete subitem or subcontract for step 1, 2, or 3 work. Completion of construction of a treatment works segment may, but need not, result in an operable treatment works (40 CFR 35.905-91).

Treatment works treating domestic sewage: A POTW or any other sewage sludge or wastewater treatment devices or systems, regardless of ownership (including Federal facilities), used in the storage, treatment, recycling, and reclamation of municipal or domestic sewage, including land dedicated for the disposal of sewage sludge. This definition does not include septic tanks or similar devices. For purposes of this definition, domestic sewage includes waste and waste water from humans or household operations that are discharged to or otherwise enter a treatment works (40 CFR 501.2-91, *See also* 40 CFR 122.2-91).

Treatment zone: A soil area of the unsaturated zone of a land treatment unit within which hazardous constituents are degraded, transformed, or immobilized (40 CFR 260.10-91).

Tremie: Device used to place concrete or grout under water (EPA-94/04).

Trench technique: A method where a trench is excavated specifically for placement of solid wastes and the excavated soil is used as cover material (*See* sanitary landfill for more related terms) (EPA-83).

Trenching or burial operation: The placement of sewage sludge or septic tank pumpings in a trench or other natural or man-made depression and the covering with soil or other suitable material at the end of each operating day such that the wastes do not migrate to the surface (40 CFR 257.3.6-91).

Trespass: Trespass is distinguished from nuisance in that trespass is interference with the possession of property whereas nuisance s interference with the use and enjoyment of property. Trespass is commonly divided into two types:
(1) Trespass to chattels is an injury to or interference with the possession of personal property, with or without the exercise of personal force. This trespass involves destruction of personal property, taking from the possession of another, or a refusal to surrender possession.
(2) Trespass to land is an unlawful, forcible entry on another's realty. An injury to the realty of another or an interference with possession, above or below ground, is a trespass, regardless of the condition of the land and regardless of negligence (*See* common law for more related terms) (Sullivan-95/04, p12).

Trial burn:
(1) A trial burn is the testing that is done to determine whether an incinerator can meet the performance standards and to determine the operating conditions under which this occurs.
(2) A test must be performed for each set of operating conditions for which the applicant desires to be permitted and must include:
 (A) Three replicates or runs must be performed for each test.
 (B) One set of conditions constitutes a test.
 (C) The overall trial burn consists of one test for each set of operating conditions.
(3) Each run of a test must be passed for the incinerator to be permitted to operate at that set of conditions (permit writer training notes for using Guidance Manual (EPA-89/01).
(4) An incinerator test in which emissions are monitored for the presence of specific organic compounds, particulates, and hydrogen chloride (EPA-94/04).

Trial burn plan: A detailed plan which describes the procedure that will be used and the precautions that will be taken during a trial burn (EPA-81/09).

Tribe or tribes: The Colville Confederated Tribes (40 CFR 131.35-91).

Tributary: A stream that contributes its waters to a larger stream by discharging into it (DOI-70/04).

Tributyl phosphate ((C_4H_9)$_3PO_4$): Poisonous liquid used as a heat exchange medium, anti-forming agent, lacquer, plastics, and vinyl resins.

Tributyl tin chloride ((C_4H_9)$_3SnCl$): Colorless liquid used as rodenticide, etc.

Trichlorfon ($C_4H_8Cl_3O_4P$): Colorless crystals used in insecticides for the control of flies and roaches.

Trichloroacetic acid (CCl_3COOH): Colorless crystals used in herbicides, etc.

Trichloroacetonitrile (C_2Cl_3N): Liquids used as in insecticide.

Trichlorobenzene ($C_6H_3Cl_3$, TCB):
(1) 1,2,3-TCB, colorless crystals used as a chemical intermediate.
(2) 1,2,4-TCB, colorless liquid used in transformer oils, lubricants, insecticides, termite contrl.

Trichloroethane ($C_2H_3Cl_3$):
(1) 1,1,1-trichloroethane, poisonous liquid used in solvent, aerosol propellant, pesticide and metal cleaning.
(2) 1,1,2-trichloroethane, colorless liquid used in solvent for fats, waxes and resins.

Trichloroethylene (C_2HCl_3): A stable, low-boiling colorless liquid, toxic by inhalation. It is used as a solvent, metal degreasing agent, and in other industrial applications (EPA-94/04).

Trichlorofluoromethane (CCl_3F): Poisonous liquids used as a solvent, refringent, plastic forms and in fire extinguishers.

Trichloroisocyanuric acid ($C_3Cl_3N_3O_3$): Colorless crystals used as a disinfectant and deodorant.

Trichlorophenol ($C_6H_2Cl_3OH$):
(1) 2,4,5-trichlorophenol, a gray solid used in fungicide, bactericide.
(2) 2,4,6-trichlorophenol, a yellow flake used in fungicide, bactericide, herbicide, defoliant and preservative.

2,4,5-trichlorophenoxyacetic acid ($C_6H_2Cl_3 \bullet OCH_2CO_2H$): A poisonous solid used in herbicide, defoliant and plant hormone.

1,1,2-trichloro-1,2,2-trifluoroethane ($C_2Cl_3F_3$): A colorless liquid used in dry cleaning, fire extinguishers and refrigerants.

Trickle irrigation: Method in which water drips to the soil from perforated tubes or emitters (EPA-94/04).

Trickling filter (bacteria bed, biofilter, biological filter or sprinkling filter):
(1) Wastes are sprayed through the air to absorb oxygen, and then allowed to trickle through a bed of rock or synthetic media coated with a slime of microbial growth. The microbial slime is able to decompose matter in the waste stream.
(2) Wastes are applied at the top of a fixed bed, or tower, of porous media such as stones, slats, or plastic material, through which an upward flow of air passes. The microorganisms adhered to the media decompose the organics dissolved in the liquid-phase.
(3) A coarse treatment system in which wastewater is trickled over a bed of stones or other material covered with bacteria that break down the organic waste and produce clean water (EPA-94/04).
(4) *See* filter for more related terms.

Trickling filter sludge: The sludge from the sedimentation pond of the trickling filter treatment (*See* sludge for more related terms).

Triethanolamine ($C_6H_{15}NO_3$): A viscous liquid used in herbicides, cement additives, cutting oils, etc.

Triethylene glycol ($C_6H_{14}O_4$): A colorless liquid used in air disinfection, solvent, bactericide and fungicide.

Triethyl phosphate (C_2H_5)$_3PO_4$): A poisonous liquid used in insecticides and pesticides.

Trihalomethane (THM):
(1) One of the family of organic compounds, named as derivatives of methane, wherein three of the four hydrogen atoms in methane are each substituted by a halogen atom in the molecular structure (40 CFR 141.2-91).
(2) One of a family of organic compounds named as derivative of methane. THMs are generally by-products of chlorination of drinking water that contains organic material (EPA-94/04).

2,4,6-trinitrotoluene ($CH_3C_5H_2(NO_3)_3$, TNT): A high explosive, exploded by detonators but unaffected by ordinary friction or shock. It is manufactured by reacting toluene (an organic liquid) with nitric acid in the presence of sulfuric acid (EPA-76/03).

Trip blank: *See* synonym, equipment blank.

Triple base: A propellant that contains three explosive ingredients, nitrocellulose, nitroglycerin and nitroguanidine (EPA-76/03).

Triple bond: *See* definition under chemical bond.

Triple cavity process: Any glass-forming process that uses three charges of glass and forms them simultaneously (EPA-83).

Triple point (or triple phase point): The location where solid, liquid and vapor phases are in equilibrium (Jones-p117).

Triple phase point: *See* synonym, triple point.

Triple bond: Two atoms which share three pairs of electrons (*See* chemical bond for more related terms).

Triple rinse: The flushing of containers three times, each time using a volume of the normal diluent equal to approximately ten percent of the containers capacity, and adding the rinse liquid to the spray mixture or disposing of it by a method prescribed for disposing of the pesticide (40 CFR 165.1-91).

Triple superphosphate plant: Any facility manufacturing triple superphosphate by reacting phosphate rock with phosphoric acid. A run-of-pile triple superphosphate plant includes curing and storing (40 CFR 60.231-91).

Tris: Tris(2,3-dibromopropyl) phosphate (also commonly named DBPP, TBPP, and Tris-BP) (40 CFR 704.205-91).

Trisodium phosphate (Na_3PO_4): Water soluble crystals used in water softening.

Tritium: A radioactive isotope of the element hydrogen with two neutrons and one proton. Common symbols for the isotope are H-3 and T (DOE-91/04).

Tritium extraction facility: The tritium-target processing facility that has been already been proposed at the Savannah River Site to support DOE's existing production reactor operations (DOE-91/04).

Tritium target fabrication facility: The land, buildings, equipment, and processes used to make the tritium-target elements from raw material (DOE-91/04).

Tritium target processing facility: The land, building, equipment, and processes used to extract tritium from irradiated tritium-target elements. This facility may also be used to treat remaining materials before transfer to waste treatment, storage, or disposal facilities (DOE-91/04).

Trommel: A perforated rotating horizontal cylinder used to break open trash bags, to screen large pieces of glass and to remove small abrasive items such as stones and dirt. Used in gravel and ore processing, coal preparation and for screening incinerator residue (*See* size reduction machine for more related terms) (EPA-83).

Trophic level: The food supply levels among different living organisms, e.g., the primary producers are the green plants which obtain energy from sun light through photosynthesis. The secondary producers are herbivores that only eating green plants for living. The third producers are carnivores that only eat herbivores for living. There are different trophic levels among carnivores.

Trophic condition: A relative description of a lake's biological productivity based on the availability of plant nutrients. The range of trophic conditions is characterized by the terms of oligotrophic for the least biologically productive, to eutrophic for the most biologically productive (40 CFR 35.1605.6-91).

Tropopause: The layer between the troposphere and stratosphere.

Troposphere: The layer of the atmosphere closest to the earth's surface (EPA-94/04).

Truck: A small vehicle for carrying materials. Other truck related terms include:
- Heavy light duty truck
- Incomplete truck
- Light duty truck
- Light duty truck 1
- Light duty truck 2
- Light duty truck 3
- Light duty truck 4

Truck dumping: The unloading of nonmetallic minerals from movable vehicles designed to transport nonmetallic minerals from one location to another. Movable vehicles include but are not limited to: trucks, front end loaders, skip hoists, and railcars (40 CFR 60.671-91).

Truck loading station: That portion of a metallic mineral processing plant where metallic minerals or metallic mineral concentrates are loaded by a conveying system into trucks (40 CFR 60.381-91).

Truck unloading station: That portion of a metallic mineral processing plant where metallic ore is unloaded from a truck into a hopper, screen, or crusher (40 CFR 60.381-91).

Trucked battery: Batteries moved into or out of the plant by truck when the truck is actually washed in the plant to remove residues left in the truck from the batteries (*See* battery for more related terms) (40 CFR 461.2-91).

True vapor pressure: The equilibrium partial pressure exerted by a volatile organic liquid as determined in accordance with methods described in American Petroleum Institute Bulletin 2517, Evaporation Loss From Floating Roof Tanks, second edition, February 1980 (incorporated by reference as specified in 40 CFR 52.742) (40 CFR 52.741-91, *See also* 40 CFR 60.111; 60.111a-91).

Trunk main: Major water supply pipes from water treatment plants to water towers from where water is supplied to consumers (*See* sewer for more related terms).

Trunk sewer:
(1) A sewer that transports wastewater from collecting sewers to the treatment plant. A trunk sewer does not ordinarily service individual properties, but rather receives tributary branches and serves a larger territory (DOI-70/04).
(2) Large sewer pipes where many branch sewers flow into.
(3) *See* sewer for more related terms.

Trust fund: *See* synonym, fund

Trust fund (CERCLA): A fund set up under the Comprehensive Environmental Response, Compensation and Liability Act (CERCLA) to help pay for cleanup of hazardous waste sites and for legal action to force those responsible for the sites to clean them up (EPA-94/04).

Trustee: An official of a federal natural resources management agency designated in subpart G of the NCP or a designated state official Indian tribe who may pursue claims for damages under section 107(f) of CERCLA (40 CFR 300.5-91).

Tub grinder: A machine which is to grind or chip wood wastes for mulching, composting or size reduction (EPA-89/11).

Tube reducing: An operation which reduces the diameter and wall thickness of tubing with a mandrel and a pair of rolls with tapered grooves (40 CFR 471.02-91).

Tube settler: Device using bundles of tubes to let solids in water settle to the bottom for removal by conventional sludge collection means; sometimes used in sedimentation basins and clarifiers to improve particle removal (EPA-94/04).

Tuberculation: Development or formation of small mounds of corrosion products on the inside of iron pipe. These tubercules roughen the inside of the pipe, increasing its resistance to water flow (EPA-94/04).

Tubular lesions: Lesions of the tubules of the kidney, causing impairment of its reabsorptive capacity (LBL-76/07-bio).

Tumbling or barrel finishing: An operation in which castings, forgings, or parts pressed from metal powder are rotated in a barrel with ceramic or metal slugs or abrasives to remove scale, fins, or burrs. It may be done dry or with an aqueous solution (40 CFR 471.02-91).

Tumbling or burnishing: The process of polishing, deburring, removing sharp corners, and generally smoothing parts for both cosmetic and functional purposes, as well as the process of washing the finished parts and cleaning the abrasion media (40 CFR 468.02-91).

Tumor: An abnormal mass of cells in a body.

Tumor progression: The sequence of changes in which a tumor develops from a microscopic lesion to a malignant stage (EPA-92/12).

Tumorigenicity: The ability of a line of cells to produce (induce) tumors in host tissues (EPA-88/09a).

Tundra: A type of ecosystem dominated by lichens, mosses, grasses, and woody plants. Tundra is found at high latitudes (arctic tundra) and high altitudes (alpine tundra). Arctic tundra is underlain by permafrost and is usually very wet (cf. wetlands) (EPA-94/04).

Tungsten (W) (formerly called wolfram): A hard transition metal with a.n. 74; a.w. 183.85; d. 19.3 g/cc; m.p. 3410 C and b.p. 5930 C. The element belongs to group VIB of the periodic table.

Turbidimeter:
(1) A device that evaluates the amount of suspended solids in a liquid by measuring the intensity difference of a light beam, as the light beam passes through the suspended solids large enough to scatter the light (EPA-82/11f).
(2) A device that measures the density of suspended solids in a liquid (EPA-94/04).

Turbidimetric analysis (or turbidimetry): Studies of using turbidimeters to determine suspended particles.

Turbidimetric titration: The use of titration methods to determine the turbidity. The end point is indicated by the developing turbidity of the titrated solution (*See* titration for more related terms).

Turbidimetry: *See* synonym, turbidimetric analysis.

Turbidity:
(1) Haziness in air caused by the presence of particles and pollutants.
(2) A similar cloudy condition in water due to suspended silt or organic matter (EPA-89/12).
(3) The clarity of water expressed as nephelometric turbidity units (NTU) and measured with a calibrated turbidimeter (40 CFR 131.35-91, *See also* EPA-83/06a).
(2) (A) Haziness in air caused by the presence of particles and pollutants (EPA-94/04).
　　(B) A cloudy condition in water due to suspended silt or organic matter.

Turbine:
(1) Employed in oil/gas production or oil/gas transportation, any stationary gas turbine used to provide power to extract crude oil/natural gas from the earth or to move crude oil/natural gas, or products refined from these substances through pipelines (40 CFR 60.331-91).
(2) A device used to convert the energy of steam or gas into rotational mechanical energy and used as prime mover to drive electric generators (EPA-82/11f).

Other turbine related terms include:
- Combined cycle gas turbine
- Electric utility combined cycle gas turbine
- Emergency gas turbine
- Gas turbine
- Gas turbine model
- Regenerative cycle gas turbine
- Simple cycle gas turbine
- Stationary gas turbine
- Steam turbine

Turbine generator: A rotary-type unit consisting of a turbine and an electric generator (EPA-83).

Turbulence:
(1) A state of fluid flow in which fluid particles (small masses) move in random, irregular paths (instantaneous velocities). The random and irregular flow of the instantaneous velocities can in practice only be described by statistical properties (EPA-88/09).
(2) A fluid motion at high Reynolds numbers characterized by irregular or random flow fluctuations and high fluctuating vorticity resulting in a large dissipation of kinetic energy and a rapid mixing of flow properties and passive contaminants throughout the turbulent flow regime (cf. flow, turbulent) (NATO-78/10).

Other turbulence related terms include:
- Homogeneous turbulence
- Isotropic turbulence
- Mechanical turbulence

Turbulent diffusion:
(1) Diffusion of a property caused by turbulent motions in a fluid as opposed to molecular diffusion.
(2) An aerosol removal mechanism from a turbulent flow (EPA-88/09a; NATO-78/10).

Turbulent flame: A flame produced by a combustion process propagating through a turbulent stream. The stream is an irregular three-dimensional flow in which the transport of heat, mass, and momentum is several orders of magnitude greater than that by molecular conductivity, diffusivity, and viscosity (i.e., greater than that which occurs in a laminar flame). Turbulence causes significant changes in flame speeds, flame stability, and pollutant formation rates. Past scientific investigations in this area have proposed several conceptual models to qualitatively describe turbulent flame characteristics. So far, however, there is no complete fundamental theory which can be used to quantitatively calculate the effects of turbulence on combustion (*See* flame for more related terms).

Turbulent flow (Newton) flow: In a turbulent flow, the fluid is not restricted to parallel paths but moves forward in a haphazard manner. A fully turbulent flow occurs when Reynolds Number is greater than 4000 for a compressible flow (cf. turbulence) (Markes-67). For an incompressible flow, Reynolds number greater than 500 (*See* flow regime for more related terms).

Turndown ratio: The maximum to minimum operating range of a parameter (EPA-81/09).

Turnover: The numbers of workers replaced in a specified period.

Turpentine: A light colored, volatile essential oil from resinous exudates or resinous wood (EPA-79/12).

Tuyeres:
(1) Openings or ports in a grate through which air can be directed to improve combustion (SW-108ts).
(2) Openings in the shell and refractory lining of a furnace through which air is forced (EPA-85/10a).

Twenty five kilotonne party: Any nation listed in Appendix D to this Part (40 CFR 82.3-91).

Twenty five (25) year, 24 hour rainfall event: The maximum 24-hour precipitation event with a probable recurrence interval of once in 25 years as defined by the National Weather Service in technical paper No. 40, Rainfall Frequency Atlas of the United States, May 1961, and subsequent amendments in effect, as of the effective date of this regulation (40 CFR 418.11-91, *See also* 40 CFR 422.41; 422.51-91).

Twenty five (25) hour daily average or 24 hour daily average: The arithmetic or geometric mean (as specified in 40 CFR 60.59a (e), (g), or (h) as applicable) of all hourly emission rates when the affected facility is operating and firing MSW measured over a 24-hour period between 12 midnight and the following midnight (40 CFR 60.51a-91).

Twenty four (24) hour period: The period of time between 12:01 a.m. and 12:00 midnight (40 CFR 60.41a-91, *See also* 40 CFR 60-App/G-91).

Two piece can: A can which is drawn from a shallow cup and requires only one end to be attached (40 CFR 52.741-91, *See also* 40 CFR 60.491-91).

Two stage digestion: *See* definition under anaerobic digestion process.

Two (2) year, 24 hour precipitation event: The maximum 24-hour precipitation event with a probable recurrence interval of once in two years as defined by the National Weather Service and Technical Paper No. 40, Rainfall Frequency Atlas of the U.S., May 1961, or equivalent regional or rainfall probability information developed therefrom (40 CFR 434.11-91).

Tyndall effect: The effect of light scattering as it passes through a solution containing suspended solids.

Type I sedimentation: Particle settling is based on the assumption that the particle size and particle shape remain the same during sedimentation.

Type II sedimentation: Particle settling is based on the assumption that the particle size and particle shape change, because of flocculation during sedimentation.

Type I sound level meter: A sound level meter which meets the Type I requirements of American National Standard Specification S1.4-1971 for sound level meters. This publication is available from the American National Standards Institute, Inc., 1430 Broadway, New York, New York 10018 (40 CFR 204.2-91, *See also* 40 CFR 205.2-91).

Type II water: *See* synonym, deionized water.

Type III foci of transformed cell: Multilayered aggregations of densely staining cells with random orientation and criss-cross arrays at the periphery of the aggregate. They appear as dark stained areas on a light staining background monolayer which is one-cell thick (40 CFR 795.285-91).

Type of pesticidal product: Refers to each individual product as identified by: the product name; EPA Registration Number (or EPA File Symbol, if any, for planned products, or Experimental Permit Number, if the pesticide is produced under an Experimental Use Permit); active ingredients; production type (technical, formulation, repackaging, etc.); and, market for which the product was produced (domestic, foreign, etc.). In cases where a pesticide is not registered, registration is not applied for, or the pesticide is not produced under an Experimental Use Permit, the term shall also include the chemical formulation (40 CFR 167.3-91).

Type of resin: The broad classification of resin referring to the basic manufacturing process for producing that resin, including, but not limited to, the suspension, dispersion, latex, bulk, and solution processes (*See* resin for more related terms) (40 CFR 61.61-91).

Tyuyamunite: A yellow uranium mineral. It is the calcium analogue of carnotite (EPA-82/05).

Uu

U-blade: A dozer blade with an extension on each side. They protrude forward at an obtuse angle to the blade and enable it to handle a larger volume of solid waste than a regular blade (*See* blade for more related terms) (SW-108ts).

U-tube heat exchanger: A heat exchanger device in which one fluid flows through a bundle of U-tubes held by a vessel and the other fluid flows between the vessel shell and U-tubes.

U-tube manometer: A pressure gauge in which a U-shaped tube is partially filled with a liquid with known specific gravity. When both ends of the U-tube are connected to two different pressure sources, the height of different liquid levels shown inside the tube is proportional to the pressure differences between the two sources.

Ultimate analysis (or elemental analysis):
(1) An analysis of elemental chemical composition such as C, O, H, Cl, S, etc. (*See* analysis for more related terms).
(2) The elemental chemical analysis of a solid, liquid, or gaseous fuel. In the cases of coal, coke, or solid waste, the percentages in a dry sample of carbon. Hydrogen, sulfur, nitrogen, ash and chlorine are usually determined. Oxygen is obtained by subtracting the total of the other elements from 100 (*See* analytical parameters--fuels for more related terms) (EPA-83).
(3) For example, anthracite coal has the following ultimate analysis:
- carbon (79.7%)
- hydrogen (2.9%)
- nitrogen (0.9%)
- oxygen (6.1%)
- sulfur (0.8%)
- ash (9.6%)
- heating value (12880 Btu/lb) (Marks, p7-4)

Ultimate biodegradability: The breakdown of an organic compound to CO_2, water, the oxides or mineral salts of other elements and/or to products associated with normal metabolic processes of microorganisms (40 CFR 796.3100-91).

Ultimate consumer (or ultimate purchaser): The first person who purchases an automobile for purposes other than resale or leases an automobile (40 CFR 600.002.85-91, *See also* CAA216; NCA3; 40 CFR 53.1; 85.1902; 205.2-91).

Ultimate oxygen demand (UOD): The sum of carbonaceous BOD and ultimate nitrogenous BOD (*See* oxygen for more related terms).

Ultimate purchaser: *See* synonym, ultimate consumer.

Ultra clean coal (UCC): Coal that is washed, ground into fine particles, then chemically treated to remove sulfur, ash, silicone, and other substances; usually briquetted and coated with a sealant made from coal (EPA-94/04).

Ultra filtration (or micro filtration): A treatment similar to reverse osmosis except that ultra filtration treats solution with larger solute particles so that the solvents can more easily filter through the membrane (*See* filtration for more related terms) (EPA-87/10a).

Ultra trace method: A method applicable below trace levels (*See* method for more related terms) (ACS-87/11).

Ultrasonic agitation: The agitation of a liquid medium through the use of ultrasonic waves (EPA-83/06a).

Ultrasonic cleaning: Immersion cleaning aided by ultrasonic waves which cause micro-agitation (EPA-83/06a).

Ultrasonic machining: Removing materials by means of an ultrasonic-vibrating tool usually working in an abrasive slurry in close contact with a workpiece or having diamond or carbide cutting particles on its end (EPA-83/06a).

Ultrasonic precipitation: A process consisting of the separation of particulate matter from air and other gases following agglomeration induced by an ultrasonic field (*See* precipitation for more related terms) (EPA-83/06).

Ultrasonic testing: A nondestructive test which applies sound, at a frequency above about 20 HJz, to metal, which has been immersed in liquid (usually water) to locate inhomogeneities or structural discontinuities (40 CFR 471.02-91).

Ultraviolet (UV): Pertaining to the region of the electromagnetic spectrum from approximately 10-380 um. The term ultraviolet without further qualification usually refers to the region from 200-380 μm (LBL-76/07-air).

Ultraviolet absorption: One of continuous emission monitors (*See* continuous emission monitor or radiation emission-absorption instrumentation for various types). Aromatic hydrocarbons absorb radiation in the near ultraviolet wavelength region. Aliphatic hydrocarbons typically do not absorb in this region to any great degree. Compounds can absorb over both wide and narrow wavelength bands and continuously absorb over several discrete bands. By determining the absorption maximum, wavelength, and absorption intensity one can identify an organic chromophore (e.g., a carbonyl group) (EPA-84/03a).

Ultraviolet disinfection: The use of ultraviolet waves to disinfect a room.

Ultraviolet radiation: The electromagnetic radiation in the wavelength of 4 to 400 nano meter (nm), 1 nano = 10^{-9} (*See* radiation for more related terms) (EPA-88/09a).

Ultraviolet rays: Radiation from the sun that can be useful or potentially harmful. UV rays from one part of the spectrum (UV-A) enhance plant life and are useful in some medical and dental procedures; UV rays from other parts of the spectrum (UV-B) can cause skin cancer or other tissue damage. The ozone layer in the atmosphere partly shields us from ultraviolet rays reaching the earth's surface (EPA-94/04).

Ultraviolet spectroscopy: A method for the determination of chemical structures by applying the absorption spectroscopy involving electromagnetic wavelengths in the ultraviolet wave range.

Umbo: The narrow end (apex) of the oyster (40 CFR 797.1800; 797.1830-91).

Unacceptable adverse effect: The impact on an aquatic or wetland ecosystem which is likely to result in significant degradation of municipal water supplies (including surface or ground water) or significant loss of or damage to fisheries, shell fishing, or wildlife habitat or recreation areas. In evaluating the unacceptability of such impacts, consideration should be given to the relevant portions of the section 404(b)(1) guidelines (40 CFR 230) (40 CFR 231.2-91).

Unattached radon progeny: Radon decay products which have not yet adhered to other, larger dust particles in the air (or to other surfaces, such as walls). Unattached progeny might result in a higher lung cancer risk than the progeny that are attached to larger particles, because the unattached progeny can selectively deposit in limited areas of the lung (EPA-88/08).

Unattended period: The period of time over which the instrument will meet stated performance specifications without operator intervention (LBL-76/07-bio).

Unauthorized dispersion technique: Refers to any dispersion technique which, under section 123 of the Act and the regulations promulgated pursuant to that section, may not be used to reduce the degree of emission limitation otherwise required in the applicable SIP (40 CFR 57.103-91).

Unavailable energy: The energy difference between the total heat added and the available energy (Holman-p176; Jones-p313; Wark-p269).

Unbleached papers: The paper made of pulp that have not been treated with bleaching agents (*See* paper for more related terms) (40 CFR 250.4-91).

Uncertainty: Sources of uncertainty include (ACS-87/11):
(1) **Accuracy:**
   (A) The difference between a measurement and true value (40 CFR 86.082.2-91).
   (B) The agreement between an experimentally determined value and the accepted reference value. *See* ASTM E180 (EPA-83).
   (C) The degree of agreement of a measurement (or an average of measurements) of a parameter, X, with an accepted reference or true value, T. It is usually expressed as the difference between the two values, X-T, or the difference as a percentage, 100(X-T)/T. It is also sometime expressed as a ratio, X/T. Accuracy is a measure of the bias in a method (EPA-86/10a).
(2) **Bias:**
   (A) A constant or systematic error as opposed to a random error. *See* analytical parameters--laboratory. *See* ASTM E180 (EPA-83).
   (B) An inadequacy in experimental design that leads to results or conclusions not representative of the population under study (Course 165.6).
   (C) The difference between the average value of a set of measurements, X and the accepted reference or true value, T (EPA-84/03).
(3) **Error:** Any deviation of an observed value from the true value (EPA-83).
(4) **Precision:**
   (A) the standard deviation of replicated measurements (40 CFR 86.082.2-91).
   (B) Variation about the mean of repeated measurements of the same pollutant concentration, expressed as one standard deviation about the mean (40 CFR 53.23-91).
   (C) The degree of agreement of repeated measurements of the same property. Also *See* ASTM E180 (EPA-83).

Uncertainty factor:
(1) A factor used in the adjustment of toxicity data to account for unknown variations. Where toxicity is measured on only one test species, other species may exhibit more sensitivity to that effluent. An uncertainty factor would adjust measured toxicity upward and downward to cover the sensitivity range of other, potentially more or less sensitive species (EPA-85/09).
(2) One of several, generally 10-fold factors, used in operationally deriving the Reference Concentration (RfC) from experimental data. UFs are intended to account for:
   (A) The variation in sensitivity among the members of the human population;
   (B) The uncertainty in extrapolating animal data to the case of humans;
   (C) The uncertainty in extrapolating from data obtained in a study that is of less-than-lifetime exposure;
   (D) The uncertainty in using LOAEL data rather than NOAEL data; and
   (E) The inability of any single study to adequately address all possible adverse outcomes in humans (EPA-90/08; 92/12).

Unconfined aquifer (or water table aquifer):
(1) A groundwater reservoir which is continually recharged by water Seeping through the soil from the surface (confined aquifer) (EPA-80/08).
(2) A permeable geological unit having the following properties: a water-filled pore space (saturated), the capability to transmit significant quantities of water under ordinary differences in pressure, and an upper water boundary that is at atmospheric pressure (DOE-91/04).
(3) An aquifer containing water that is not under pressure; the water level in a well is the same as the water table outside the well (EPA-94/04).
(4) *See* aquifer for more related terms

Unconfined groundwater: The groundwater that is bounded by a zone of aeration (*See* groundwater for more related terms).

Unconformity: A lack of vertical continuity between layers of rock representing a gap in a geologic record (DOI-70/04).

Uncontrolled total arsenic emission: The total inorganic arsenic in the glass melting furnace exhaust gas preceding any add-on emission control device (*See* arsenic for more related terms) (40 CFR 61.161-91).

Unconventional waste: *See* synonym, special waste.

Undercoater: Any coatings formulated for and applied to substrates to provide a smooth surface for subsequent coats (40 CFR 52.741-91).

Underfire air:
(1) Any forced or induced air, under control as to quantity and direction, that is supplied from beneath and which passes through the solid wastes fuel bed (40 CFR 240.101-91).
(2) Air for combustion is admitted into a furnace at a point under a grate to promote burning within a fuel bed; or combustion air which enters the fuel bed from orifices in the hearth (EPA-89/03b).
(3) *See* air for more related terms.

Underground area: An underground room, such as a basement, cellar, shaft or vault, providing enough space for physical inspection of the exterior of the tank situated on or above the surface of the floor (40 CFR 280.12-91).

Underground drinking water source:
(1) An aquifer supplying drinking water for human consumption; or
(2) An aquifer in which the ground water contains less than 10,000 mg/L total dissolved solids (40 CFR 257.3.4-91).

Underground injection: The subsurface emplacement of fluids through a bored, drilled or driven well; or through a dug well, where the depth of the dug well is greater than the largest surface dimension (*See* injection well for more related terms) (40 CFR 260.10-91).

Underground injection control (UIC): The program under the Safe Drinking Water Act that regulates the use of wells to pump fluids into the ground (EPA-94/04, *See also* 40 CFR 124.2; 144.3; 146.3; 270.2-91).

Underground release: Any belowground release (40 CFR 280.12-91).

Underground sources of drinking water: Aquifers currently being used as a source of drinking water or those capable of supplying a public water system. They have a total dissolved solids content of 10,000 milligrams per liter or less, and are not "exempted aquifers" (*See* exempted aquifer) (EPA-94/04).

Underground storage tank: A tank located at least partially underground and designed to hold gasoline or other petroleum products or chemicals (*See* tank for more related terms) (EPA-94/04).

Underground tank: A device meeting the definition of tank in 40 CFR 60.10 whose entire surface area is totally below the surface of and covered by the ground (*See* tank for more related terms) (40 CFR 260.10-91).

Underground uranium mine: A man-made underground excavation made for the purpose of removing material containing uranium for the principal purpose of recovering uranium (*See* uranium for more related terms) (40 CFR 61.21-91).

Underground water: All water beneath the surface of the ground, including both groundwater and vadose water, whatever its origin (*See* water for more related terms) (DOI-70/04).

Undertread cementing operation: The system used to apply cement to a continuous strip of tread or combined tread/sidewall component. An undertread cementing operation consists of a cement application station and all other equipment, such as the cement supply system and feed and takeaway conveyors, necessary to apply cement to tread or combined tread/sidewall strips and to allow evaporation or solvent from the cemented tread or combined tread/sidewall (40 CFR 60.541-91).

Undigested sludge: The sludge from a sedimentation prior to treatment (*See* sludge for more related terms).

Undisturbed performance: The predicted behavior of a disposal system, including consideration of the uncertainties in predicted behavior, if the disposal system is not disrupted by human intrusion or the occurrence of unlikely natural events (40 CFR 191.12-91).

Undue risk: A level of risk to the health and safety of the public that exceeds that set forth in the design and operation objectives of the facility (*See* risk for more related terms) (DOE-91/04).

Unexpended consumption allowances: The consumption allowances that have not been used. At any time in any control period, a person's unexpended consumption allowances are the total of the calculated level of consumption allowances he has authorization under this Part to hold at that time for that control period, minus the calculated level of controlled substances that the person has produced and imported in that control period until that time (40 CFR 82.3-91).

Unexpected production allowances: The production allowances that have not been used. At any time in any control period, a person's unexpended production allowances are the total of the calculated level of production allowances he has authorization under this Part to hold at that time for that control period, minus the calculated level of controlled substances that the person has produced in that control period until that time (40 CFR 82.3-91).

Unicellular: An organism that consists of only one cell (DOD-78/01).

Unfit for use tank system: A tank system that has been determined through an integrity assessment or other inspection to be no longer capable of storing or treating hazardous waste without posing a threat of release of hazardous waste to the environment (40 CFR 260.10-91).

Uniform air quality required for the daily reporting of air quality: A modified form of the Pollutant Standards Index (PSI) (40 CFR 58-App/G-91).

Uniform and non-uniform flow: A flow is uniform, if its cross section shape and size are the same throughout the channel or pipe. Conversely, the flow is non-uniform. A temperature or velocity is said to be uniform throughout a region when it has the same value at all parts of the region at a given instant (*See* flow for more related terms) (Perry-73).

Uniform coefficient: A measure of the uniformity of sand grains. It is the ratio of the size where 60 % of the sand by weight passes to the size where 10 % of the sand by wight passes.

Uniform plumbing code: A code of practice frequently adopted by state regulatory authorities as the basis of building codes (EPA-80/08).

Uniformed service: The Army, Navy, Air force, Marine Corps, Coast guard, Public Health Service, or Coast and Geodetic Survey (PHSA2-p).

Uniforming coating: *See* synonym, fog coating.

Uniformity: Even distribution of fibers, color and other properties of a paper (EPA-83).

Unimolecular pathway: *See* synonym, unimolecular reaction.

Unimolecular reaction (or unimolecular pathway): A chemical reaction involving only one molecule, e.g., $N_2O_4 \longrightarrow 2NO_2$. It is a function of temperature, pressure, and concentration and is independent of reaction atmosphere (EPA-88/12).

Unit: *See* dimension and unit.

Unit operation: One treatment technology that is a part of a larger treatment train (EPA-89/12a).

Unit packaging: A package that is labeled with directions to use the entire contents of the package in a single application (40 CFR 157.21-91).

Unit risk: The upper-bound excess lifetime cancer risk estimated to result from continuous exposure to an agent at a concentration of 1 μg/L in water, or 1 μg/cu.m in air (*See* risk for more related terms) (EPA-92/12).

Unit suspended furnace wall: A furnace wall or panel that is hung from a steel structure (*See* furnace wall for more related terms) (SW-108ts).

United States Code (USC): The official publications of the public laws enacted by the Congress such as, Public Law 94-580 for RCRA. It contains the U.S. Constitution and amendments and the codified laws of the United States.

Universal biohazard symbol: The symbol design that conforms to the design shown in 29 CFR 1910.145(f)(8)(ii) (40 CFR 259.10-91).

Universal precaution: A method of infection control in which all human blood and certain human body fluids are treated as if known to be infectious for hepatitis B virus (HBV), human immunodeficiency virus (HIV) and other bloodborne pathogens (29 CFR 1910).

Universe (or population): The totality, finite or infinite, of a set of items, units, elements, measurements, and the like, real or conceptual, that is under consideration (EPA-84/03).

Unleaded gasoline: The gasoline which is produced without the use of any lead additive and which contains not more than 0.05 gram of lead per gallon and not more than 0.005 gram of phosphorus per gallon (40 CFR 80.2-91).

Unloading bulkhead: A steel plate that ejects waste out the rear doors of an enclosed transfer trailer. It is propelled by a telescoping, hydraulically powered cylinder that traverses the length of the trailer (SW-108ts).

Unloading leg: A device which includes a bucket-type elevator which is used to remove grain from a barge or ship (40 CFR 60.301-91).

Unobligated balance: The portion of the funds authorized by the Federal agency that has not been obligated by the grantee and is determined by deducting the cumulative obligations from the cumulative funds authorized (40 CFR 31.3-91).

Unproven emission control systems: The emission control components or systems (and fuel metering systems) that do not qualify as proven emission control systems (40 CFR 86.092.2-91).

Unreasonable adverse effects on the environment: Any unreasonable risk to man or the environment, taking into account the economic, social, and environmental costs and benefits of the use of any pesticide (FIFRA2, *See also* 40 CFR 166.3-91).

Unreasonable degradation of the marine environment:
(1) Significant adverse changes in ecosystem diversity, productivity and stability of the biological community within the area of discharge and surrounding biological communities;
(2) Threat to human health through direct exposure to pollutants

or through consumption of exposed aquatic organisms; or
(3) Loss of esthetic, recreational, scientific or economic values which is unreasonable in relation to the benefit derived from the discharge (40 CFR 125.121-91).

Unreasonable risk: Under the Federal Insecticide, Fungicide, and Rodenticide Act (FIFRA), "unreasonable adverse effects" means any unreasonable risk to man or the environment, taking into account the medical, economic, social, and environmental costs and benefits of any pesticide (EPA-94/04).

Unreclaimable residue: The residual materials of little or no value remaining after incineration (*See* residue for more related terms) (40 CFR 165.1-91).

Unrefined and unprocessed ore: The ore in its natural form prior to any processing, such as grinding, roasting or beneficiating, or refining (*See* ore for more related terms) (10 CFR 40.4-91).

Unregulated hazardous substance:
(1) For which no standard, requirement, criteria, or limitation is in effect under the Toxic Substances Control Act, the Safe Drinking Water Act, the Clean Air Act, or the Clean Water Act; and
(2) For which no water quality criteria are in effect under any provision of the Clean Water Act (SF211).
(3) *See* hazardous substance for more related terms.

Unregulated safety relief valve: A safety relief valve which cannot be actuated by a means other than high pressure in the pipe or vessel which it protects (40 CFR 52.741-91).

Unreviewed safety question: A proposed change, test, or experiment is considered to involve an unreviewed safety question if:
(1) The probability of occurrence or the consequences of n accident or malfunction of equipment important to safety evaluated previously by safety analyses will be significantly increased; or
(2) A possibility for an accident or malfunction of a different type than any evaluated previously by safety analyses will be created that will result in significant safety consequences (DOE-91/04).

Unrestricted area: Any area access to which is not controlled by the licen*See* for purposes of protection of individuals from exposure to radiation and radioactive materials, and any area used for residential quarters (10 CFR 20.3-91).

Unsanitized: A version of a document from which information claimed as trade secret or confidential has not been withheld or omitted (cf. sanitized) (40 CFR 350.1-91).

Unsaturated hydrocarbon: Those compounds which have two or more adjacent carbon atoms joined by a double or triple bond (*See* hydrocarbon for more related terms).

Unsaturated zone (vadose or zone of aeration):
(1) The zone between the land surface and the water table (40 CFR 260.10-91).
(2) A subsurface zone containing water below atmospheric pressure and gases at atmospheric pressure. Also known as the vadose zone (cf. saturatred zone) (EPA-89/12a).
(3) The area above the water table where the soil pores are not fully saturated, although some water may be present (EPA-89/12).
(4) A region in a porous medium in which the pore space is not filled with water (DOE-91/04).
(5) The area above the water table where soil pores are not fully saturated, although some water may be present (EPA-94/04).

Unscheduled DNA synthesis in mammalian cells in culture: The incorporation of tritium labelled thymidine (^{3}H-TdR) into the DNA of cells which are not in the S phase of the cell cycle (40 CFR 798.5550-91).

Unscheduled maintenance: Any adjustment, repair, removal, disassembly, cleaning, or replacement of vehicle components or systems which is performed to correct a part failure or vehicle (if the engine were installed in a vehicle) malfunction (40 CFR 86.082.2-91, *See also* 40 CFR 86.084.2; 86.402.78-91).

Unsolicited proposal: An informal written offer to perform EPA funded work for which EPA did not publish a solicitation (40 CFR 30.200-91).

Untreated regulated medical waste: The regulated medical waste that has not been treated to substantially reduce or eliminate its potential for causing disease (*See* medical waste or waste for more related terms) (40 CFR 259.10-91).

Unusual occurrence report: A written evaluation of an unusual occurrence that is prepared in sufficient detail to enable a reviewer to assess its significance, consequences, or implications and to determine the means of avoiding a recurrence with minimal additional inquiry (DOE-91/04).

Unusual occurrence: Any unusual or unplanned event that adversely affects or potentially affects the performance, reliability, or safety of a facility (DOE-91/04).

Unwarranted failure to comply: The failure of a permittee to prevent the occurrence of any violation of his permit or any requirement of this chapter due to indifference, lack of diligence, or lack of reasonable care, or the failure to abate any violation of such permit or the chapter due to indifference, lack of diligence, or lack of reasonable care (SMCRA701-30USC1291-90).

Upflow coagulation: A technique that after the coagulant has been added to the treatment tank, wastewater enters at the bottom of the tank and leaves at the top to promote mixing with coagulant.

Upflow contact clarifier: In a clarifier, wastewater enters at the bottom of the tank, is forced to flow up through a filter bed and leaves at the top.

Upgrade: The addition or retrofit of some systems such as cathodic protection, lining, or spill and overfill controls to improve the ability of an underground storage tank system to prevent the release of product (40 CFR 280.12-91).

Upper bound: An estimate of the plausible upper limit to the true

value of the quantity. This is usually not a statistical confidence limit (EPA-92/12).

Upper explosive limit: The concentration of fuel (organic) above which combustion will not occur in the presence of a flame or spark (*See* combustion limit for more related terms) (EPA-81/12, p3-6; 84/09).

Upper limit: The emission level for a specific pollutant above which a certificate of conformity may not be issued or may be suspended or revoked (40 CFR 86.1102.87-91).

Upper unsaturated zone: In this zone, most of the pore space is filled with air, but water occurs as soil moisture and in a capillary fringe that extends upward from the water table. Water in this zone is under a negative hydraulic pressure (less than atmospheric pressure) (*See* water table for more related terms) (EPA-87/03).

Uppermost aquifer: The geologic formation nearest the natural ground surface that is an aquifer, as well as lower aquifers that are hydraulically interconnected with this aquifer within the facility's property boundary (*See* aquifer for more related terms) (40 CFR 260.10-91).

Upscale calibration value: The opacity value at which a calibration check of the CEMS is performed by simulating an upscale opacity condition as viewed by the receiver (40 CFR 60-App/B-91).

Upset: An exceptional incident in which there is unintentional and temporary noncompliance with technology based permit effluent limitations because of factors beyond the reasonable control of the permittee. An upset does not include noncompliance to the extent caused by operational error, improperly designed treatment facilities, inadequate treatment facilities, lack of preventive maintenance, or careless or improper operation (40 CFR 122.41.n-91).

Upset condition: For a nuclear facility, anticipated occurrences of moderate frequency that might occur several times during start-up testing or operation of the facility. The equipment, components, and structures must withstand these conditions without damage requiring repair. For ASME-code pressure boundary components, this plant design condition corresponds to the "Level B Service Limits" (*See also* emergency condition and 'faulted condition) (DOE-91/04).

Uptake: The sorption of a test substance into and onto aquatic organisms during exposure (40 CFR 797.1520-91, *See also* 40 CFR 797.1830; 797.1560-91).

Uptake phase: The time during the test when test organisms are being exposed to the test material (40 CFR 797.1560-91, *See also* 40 CFR 797.1520; 797.1830-91).

Uptake rate constant (k_1): The mathematically determined value that is used to define the uptake of test material by exposed test organisms, usually reported in units of liters/gram/hour (40 CFR 797.1560-91).

Upwelling: Flowing upward from a deeper to a shallower depth or to the surface of water, usually as a result of divergence of offshore currents.

Uraninite: Essentially (UO_2). It is a complex uranium mineral containing also rare earths, radium, lead, helium, nitrogen and other elements (EPA-82/05).

Uranium (U):
(1) A radioactive metal with a.n. 92; a.w. 238.03; d. 19.07 g/cc; m.p. 1132° C and b.p. 3818° C. The element belongs to group IIIB of the periodic table.
(2) A radioactive heavy metal element used in nuclear reactors and the production of nuclear weapons. The term refers usually to U-238, the most abundant radium isotope, although a small percentage of naturally-occurring uranium is U-235 (EPA-89/12, *See also* EPA-88/08a).
(3) A heavy (atomic mass = 238.03), silvery-white metal with 14 radioactive isotopes. Uranium-235 is most commonly used, as a fuel or nuclear fission. Another isotope, uranium-238, is transformed into fissionable plutonium-239 following its capture of a neutron in a nuclear reactor (DOE-91/04). Major uranium compounds include:
- Uranium dioxide (UO_2): Very toxic crystals used in packing nuclear fuel rods, ceramics, pigments and photographic chemicals.
- Uranium hexafluoride (UF_6): Very toxic crystals used to separate uranium isotopes in the gaseous diffusion process.
- Uranium hydride (UF_3): Very toxic powder used as a reducing agent.

Other uranium related terms include:
- Depleted uranium
- Underground uranium mine

Uranium byproduct material: The tailings or wastes produced by the extraction or concentration of uranium from any ore processed primarily for its source material content. Ore bodies depleted by uranium solution extraction operations and which remain underground do not constitute byproduct material for the purpose of this subpart (40 CFR 192.31-91, *See also* 40 CFR 61.221; 61.251-91).

Uranium fuel cycle: The operations of milling of uranium ore, chemical conversion of uranium, isotopic enrichment of uranium, fabrication of uranium fuel, generation of electricity by a light-water-cooled nuclear power plant using uranium fuel, and reprocessing of spent uranium fuel, to the extent that these directly support the production of electrical power for public use utilizing nuclear energy, but excludes mining operations, operations at waste disposal sites, transportation of any radioactive material in support of these operations, and the reuse of recovered non-uranium special nuclear and byproduct materials from the cycle (40 CFR 190.02-b-91).

Uranium mill tailings piles: Former uranium ore processing sites that contain leftover radioactive materials (wastes), including radium and unrecovered uranium (EPA-94/04).

Uranium Mill Tailings Radiation Control Act (UMTRCA) Standards: Standards for radionuclides established under sections 102, 104, and 108 of the Uranium Mill Tailings Radiation Control Act, as amended (40 CFR 300-App/A-91).

Uranium mill tailings waste piles: Licensed active mills with tailings piles and evaporation ponds created by acid or alkaline leaching processes (EPA-94/04).

Uranium milling: Any activity that results in the production of byproduct material as defined in this part (10 CFR 40.4-91).

Uranium mineral: More than 150 uranium bearing minerals are known to exist, but only a few are common. The five primary uranium ore minerals are pitchblende, uraninite, davidite, coffinite, and brannerite. These were formed by deep seated hot solutions and are most commonly found in veins or pegmatites. The secondary uranium ore minerals, altered from the primary minerals by weathering or other natural processes, are carnotite, tyuyamunite and metatyuyamunite (both are very similar to carnotite), torbernite and metatorbernite, autunite and meta autunite, and uranophane (EPA-82/05).

Uranium processing facility: The land, buildings, equipment, and processes used to purify uranium recovered from reactor fuel processing. This is the last step of the fuel processing activity that allows recycling of uranium to other uses (DOE-91/04).

Urban bus: A heavy heavy-duty diesel-powered passenger-carrying vehicle with a load capacity of fifteen or more passengers and intended primarily for intra-city operation, i.e., within the confines of a city or greater metropolitan area. Urban bus operation is characterized by short rides and frequent stops. To facilitate this type of operation, more than one set of quick-operating entrance and exit doors would normally be installed. Since fares are usually paid in cash or tokens rather than purchased in advance in the form of tickets, urban buses would normally have equipment installed for collection of fares. Urban buses are also typically characterized by the absence of equipment and facilities for long distance travel, e.g., rest rooms, large luggage compartments, and facilities for stowing carry-on luggage. The useful life for urban buses is the same as the useful life for other heavy heavy-duty diesel engines (40 CFR 86.091.2-91).

Urban runoff: Storm water from city streets and adjacent domestic or commercial properties that carries pollutants of various kinds into the sewer systems and receiving waters (EPA-94/04).

Urban waterfront and port: Any developed area that is densely populated and is being used for, or has been used for, urban residential recreational, commercial, shipping or industrial purposes (CZMA306a-16USC1455a-90).

Urea (CO(NH$_2$)$_2$): A product of protein metabolism.

Urea-formaldehyde foam insulation: A material once used to conserve energy by sealing crawl spaces, attics, etc.; no longer used because emissions were found to be a health hazard (EPA-94/04).

Uremia: Accumulation in the blood of constituents normally eliminated in the urine producing a toxic condition marked by headache, gastrointestinal disorders and especially vomiting, coma and convulsions and commonly associated with severe kidney disorder (LBL-76/07-bio).

Usable: Secondary materials trade term meaning those items recovered from discards that are salable in their existing form as second-hand goods. Examples are steel piping, sinks, door handles, appliances, and clothing (EPA-83).

Usable firebox volume: The volume of the firebox determined using the following definitions:
(1) Height: The vertical distance extending above the loading door, if fuel could reasonably occupy that space, but not more than 2 inches above the top (peak height) of the loading door, to the floor of the firebox (i.e., below a permanent grate) if the grate allows a 1-inch diameter piece of wood to pass through the grate, or, if not, to the top of the grate. Firebox height is not necessarily uniform but must account for variations caused by internal baffles, air channels, or other permanent obstructions.
(2) Length: The longest horizontal fire chamber dimension that is parallel to a wall of the chamber.
(3) Width: The shortest horizontal fire chamber dimension that is parallel to a wall of the chamber (40 CFR 60-AA (method 28-2.11)).

Use attainability analysis: A structured scientific assessment of the factors affecting the attainment of the use which may include physical, chemical, biological, and economic factors as described in 40 CFR 131.10(g) (40 CFR 131.3-91).

Use of asbestos: The presence of asbestos-containing material in school buildings (*See* asbestos for more related terms) (40 CFR 763.103-91).

Use stream: All reasonably anticipated transfer, flow, or disposal of a chemical substance, regardless of physical state or concentration, through all intended operations of industrial, commercial, or consumer use (40 CFR 721.3-91).

Used oil: Any oil which has been:
(1) Refined from crude oil;
(2) Used; and
(3) As a result of such use, contaminated by physical or chemical impurities (RCRA1004-42USC6903-91).
(4) *See* oil for more related terms.

Used oil fuel: Includes any fuel produced from used oil by processing, blending, or other treatment (40 CFR 266.40-91).

Useful life: The estimated period during which a treatment works will be operated (40 CFR 35.905-91).

Useful work: The work difference between the work done by a system and the work done on the atmosphere (Jones-p385).

User charge: A charge levied on users of a treatment works, or that portion of the ad valorem taxes paid by a user, for taxes users proportionate share of the cost of operation and maintenance (including replacement) of such works under sections 2(b)(l)(A) and 201(h)(2) of the Act and this (40 CFR 35.905-91).

User charge for water: A charge for water use based on the amount withdrawn from the public supply (DOI-70/04).

User fee: Fee collected from only those persons who use a particular service, as compared to one collected from the public in general (EPA-94/04).

UST system or tank system: An underground storage tank, connected underground piping, underground ancillary equipment, and containment system, if any (40 CFR 280.12-91).

Utility load: The total electricity demand for a utility district (EPA-94/04).

Utility unit: A unit owned or operated by a utility:
(1) That serves a generator in any State that produces electricity for sale, or
(2) That during 1985, served a generator in any State that produced electricity for sale.
(3) Notwithstanding paragraphs (1) and (2) of this definition, a unit that was in operation during 1985, but did not serve a generator that produced electricity for sale during 1985, and did not commence commercial operation on or after November 15, 1990 is not a utility unit for purposes of the Acid Rain Program.
(4) Notwithstanding paragraphs (1) and (2) of this definition, a unit that cogenerates steam and electricity is not a utility unit for purposes of the Acid Rain Program, unless the unit is constructed for the purpose of supplying, or commences construction after November 15, 1990 and supplies, more than one-third of its potential electrical output capacity and more than 25 MWe output to any power distribution system for sale (40 CFR 72.2-91).

Utilize: Refers to the use of solvent that is delivered to coating mix preparation equipment for the purpose of formulating coatings to be applied on an affected coating operation and any other solvent (e.g., dilution solvent) that is added at any point in the manufacturing process (40 CFR 60.711-91).

UV-VIS absorption spectrum of a solution: A function of the concentration, expressed in mol/L, of all absorbing species present; the path length, of the spectrophotometer cell, expressed in cm; and the molar absorption (extinction) coefficient of each species (40 CFR 796.1050-91).

Vv

Vaccine: A killed or modified live virus or bacteria prepared in suspension for inoculation to prevent or treat certain infectious diseases (EPA-83/09, *See also* EPA-89/12).

Vacuum (or zero pressure): A hypothetical condition denoting the complete absence of matter. Also a condition of zero pressure (EPA-84/09).

Vacuum and blow process: A bottle manufacturing process whereby glass is gathered by vacuum and subsequently blown (EPA-83).

Vacuum box: In seaming a landfill liner, a commonly used type of nondestructive test method which develops a vacuum in a localized region of an FML (flexible membrane liner) seam in order to evaluate the seam's tightness and suitability (EPA-89/09, *See also* EPA-91/05).

Vacuum break (choke pull-off): A vacuum-operated device to open the carburetor choke plate a predetermined amount on cold start (40 CFR 85.2122 (a)(1)(ii)(F)-91).

Vacuum crystallizer: An apparatus for increasing the rate of crystallization through the use of pressure (EPA-77/07).

Vacuum dezincing: A process for removing zinc from a metal by melting or heating the solid metal in a vacuum (EPA-83/03a).

Vacuum evaporization: A method of coating articles by melting and vaporizing the coating material on an electrically heated conductor in a chamber from which air has been exhausted. The process is only used to produce a decorative effect. Gold, silver, copper and aluminum have been used (EPA-83/06a).

Vacuum filter: A filter consisting of a cylindrical drum mounted on a horizontal axis and covered with a filter cloth. The filter revolves with a partial submergence in the liquid, and a vacuum is maintained under the cloth for the larger part of each revolution to extract moisture. The cake is scraped off continuously (*See* filter for more related terms) (EPA-3/76).

Vacuum filtration:
(1) A process used to reduce sludge water contents (EPA-87/10a).
(2) The use of a vacuum filter to separate solids from liquids.
(3) *See* filtration for more related terms.

Vacuum furnace: A furnace in which the charge can be brought to an elevated temperature in a high vacuum. The high vacuum provides an almost completely inert enclosure where the process of reduction and sintering can occur (*See* furnace for more related terms) (EPA-74/03).

Vacuum leakage: The leakage into the vacuum cavity of a vacuum break (40 CFR 85.2122(a)(1)(ii)(E)-91).

Vacuum metalizing: The process of coating a workpiece with metal by flash heating metal vapor in a high-vacuum chamber containing the workpiece. The vapor condenses on all exposed surface (EPA-83/03).

Vacuum pressure: A measure of pressure expressed as a quantity below atmospheric pressure or some other reference pressure (EPA-84/09). vacuum pressure = negative gage pressure (*See* pressure for more related terms).

Vacuum producing system: Any reciprocating, rotary, or centrifugal blower or compressor or any jet ejector or device that creates suction from a pressure below atmospheric and discharges against a greater pressure (40 CFR 52.741-91).

Vacuum purge system: A vacuum system with a controlled air flow to purge the vacuum system of undesirable manifold vapors (40 CFR 85.2122(a)(1)(ii)(H)-91).

Vacuum tube: An electron tube vacated to such a degree that its electrical characteristics are essentially unaffected by the presence of residual gas or vapor (EPA-83/03).

Vacuum water: The water extracted from wood during the vacuum period following steam conditioning (*See* water for more related terms) (EPA-74/04).

Vadose: *See* synonym, unsaturated zone.

Vadose water: The water that lies between the water table and the earth's surface (*See* water for more related terms) (DOI-70/04).

Vadose zone: A subsurface zone containing water below atmospheric pressure and gases at atmospheric pressure. Also known as the unsaturated zone (EPA-89/12a).

Valence: The combining power of an atom or a radical. The power is equal to the number of hydrogen atoms that the atom or the radical can combine with or displace in a chemical compound. The hydrogen atom has a valence of 1.

Valid day: A 24-hour period in which at least 18 valid hours of data are obtained. A valid hour is one in which at least 2 valid data points are obtained (40 CFR 60.101-91).

Valid emission performance warranty claim: The claim in which there is no evidence that the vehicle had not been properly maintained and operated in accordance with manufacturer instructions, the vehicle failed to conform to applicable emission standards as measured by an Office Director-approved type of emission warranty test during its useful life and the owner is subject to sanction as a result of the test failure (40 CFR 85.2102-91).

Valid study: A study that has been conducted in accordance with the Good Laboratory Practice standards of 40 CFR 160 or generally accepted scientific methodology and that EPA has not determined to be invalid (40 CFR 152.83-91).

Validated test: A test determined by the Agency to have been conducted and evaluated in a manner consistent with accepted scientific procedures (40 CFR 154.3-91).

Validation: The testing of a model by comparing its results with experimentally obtained data (NATO-78/10).

Valuable commercial and recreational species: Those species for which catch statistics are compiled on a routine basis by the Federal or State agency responsible for compiling such statistics for the general geographical area impacted, or which are under current study by such Federal or State agencies for potential development for commercial or recreational use (40 CFR 228.2-91).

Value engineering (VE): A specialized cost control technique which uses a systematic and creative approach to identify and to focus on unnecessarily high cost in a project in order to arrive at a cost saving without sacrificing the reliability or efficiency of the project (40 CFR 35.2005-91, *See also* 40 CFR 35.6015; 35.905-91).

Value for pesticide purposes: That characteristic of a substance or mixture of substances which produces an efficacious action on a pest (40 CFR 172.1-91).

Valve not externally regulated: The valves that have no external controls, such as in-line check valves (40 CFR 52.741-91).

Van: A light-duty truck having an integral enclosure, fully enclosing the driver compartment and load carrying device, and having no body sections protruding more than 30 inches head of the leading edge of the windshield (40 CFR 86.082.2-91, *See also* 40 CFR 600.002.85-91).

Van der Waals adsorption: Adsorption due to the van der Waals force.

Van der Waals force: The attractive force between atoms or molecules.

Vanadium (V): A transition metal with a.n. 23; a.w. 50.942; d. 6.1 g/cc; m.p. 1900° C and b.p. 3450° C. The element belongs to group VB of the periodic table. Major vanadium compounds include: vanadium pentoxide (V_2O_5): Yellow crystals used in glass coloring, textile dyeing and nuclear reactors.

Vanadium minerals: Those most exploited for industrial use are patronite (VS4), roscoelite (vanadium mica), vanadinite, carnotite and chlorovanadinite (EPA-82/05).

Vapor: The gaseous phase of substances that are liquid or solid at atmospheric temperature and pressure, e.g., steam (EPA-89/12, *See also* EPA-84/09; 89/03b). Other vapor related terms include:
- Saturated vapor
- Saturated vapor state

Vapor balance system: Any combination of pipes or hoses which creates a closed system between the vapor spaces of an unloading tank and a receiving tank such that vapors displaced from the receiving tank are transferred to the tank being unloaded (40 CFR 52.741-91).

Vapor blasting: A method of roughing plastic surfaces in preparation for plating (EPA-83/06a).

Vapor capture system: Any combination of hoods and ventilation system that captures or contains organic vapors so they may be directed to an abatement or recovery device (EPA-94/04).

Vapor collection system: All piping, seals, hoses, connections, pressure-vacuum vents, and other possible sources between the gasoline delivery vessel and the vapor processing unit and/or the storage tanks and vapor holder (40 CFR 52.741-91, *See also* 40 CFR 60.501; 61.301-91).

Vapor control system: Any system that limits or prevents release to the atmosphere of organic material in the vapors displaced from a tank during the transfer of gasoline (40 CFR 52.741-91).

Vapor dispersion:
(1) The movement of vapor clouds in air due to wind, gravity spreading, and mixing (EPA-89/12, *See also* EPA-3/88).
(2) The movement of vapor clouds in air due to wind, thermal action, gravity spreading, and mixing (EPA-94/04).

Vapor drying: A process in which unseasoned wood is heated in the hot vapors of an organic solvent, usually xylene, to season it prior to preservative treatment (EPA-74/04).

Vapor incinerator: Vapor incinerator is also known as the secondary burner. It is an enclosed combustion device that is used for destroying organic compounds and does not extract energy in the form of steam or process heat (*See* burner for more related terms) (40 CFR 264.1031-91; EPA-89/12).

Vapor mounted seal: A foam-filled primary seal mounted continuously around the circumference of the tank so there is an annular vapor space underneath the seal. The annular vapor space is bounded by the bottom of the primary seal, the tank wall, the liquid surface, and the floating roof (40 CFR 60.111a-91, *See also* 40 CFR 61.341-91).

Vapor plating: Depositing a metal or compound upon a heated surface by reduction or decomposition of a volatile compound at a temperature below the melting points of either the deposit or the basis material (*See* plating for more related terms) (EPA-83/06a).

Vapor plume:
(1) Flue gases that are visible because they contain (condensed) water droplets (EPA-89/12, *See also* EPA-84/09).
(2) Flue gases visible because they contain water droplets (EPA-94/04).

Vapor pressure: The pressure at which a liquid or solid is in equilibrium with its vapor at a given temperature (40 CFR 796.1950-91, *See also* 40 CFR 300-App/A-91; EPA-84/09).

Vapor processing system: All equipment used for recovering or oxidizing total organic compounds vapors displaced from the affected facility (40 CFR 60.501-91).

Vapor recovery system: A vapor gathering system capable of collecting all VOM vapors and gases discharged from the storage tank and a vapor disposal system capable of processing such VOM vapors and gases so as to prevent their emission to the atmosphere (40 CFR 52.741-91, *See also* 40 CFR 60.111-91).

Vapor saturation line: The curve line which separates the vapor region from the liquid-vapor region (Wark-p56).

Vapor tight gasoline tank truck: A gasoline tank truck which has demonstrated within the 2 preceding months that its product delivery tank will sustain a pressure change of not more than 750 pascals (75 mm of water) within 5 minutes after it is pressurized to 4,500 pascals (450 mm of water). This capability is to be demonstrated using the pressure test procedure specified in Reference Method 27 (40 CFR 60.501-91).

Vapor tight marine vessel: A marine vessel with a benzene product tank that has been demonstrated within the preceding 12 months to have no leaks. This demonstration shall be made using method 21 of part 60, appendix A, during the last 20 percent of loading and during a period when the vessel is being loaded at its maximum loading rate. A reading of greater than 10,000 ppm as methane shall constitute a leak. As an alternative, a marine vessel owner or operator may use the vapor-tightness test described in 40 CFR 61.304(f) to demonstrate vapor tightness. A marine vessel operated at negative pressure is assumed to be vapor-tight for the purpose of this standard (40 CFR 61.301-91).

Vapor tight tank truck or vapor-tight railcar: A tank truck or railcar for which it has been demonstrated within the preceding 12 months that its product tank will sustain a pressure change of not more than 750 pascals within 5 minutes after it is pressurized to a minimum of 4,500 pascals. This capability is to be demonstrated using the pressure test procedure specified in method 27 of part 60, appendix A, and a pressure measurement device which has a precision of ± 2.5 mm water and which is capable of measuring above the pressure at which the tank truck or railcar is to be tested for vapor tightness (40 CFR 61.301-91).

Vaporization: The change of a pure substance from a liquid phase to its vapor phase (EPA-89/12). Vaporization is an endothermic process, that is energy must be added to a liquid substance (liquid phase) to convert it to a vapor phase (gas phase). This energy is commonly referred to as the latent heat of vaporization (*See* latent heat for more related terms).

Variable: Something that may change. Other change related terms include:
- Dependent variable
- Deterministic variable
- Dummy variable
- Independent variable
- Probabilistic variable
- Random variable

Variable capacitor: A device whose capacitance can be varied continuously by moving one set of metal plates with respect to another (*See* capacitor for more related terms) (EPA-83/03).

Variable container rate: A charge for solid waste services based on the volume of waste generated measured by the number of containers set out for collection (EPA-89/11).

Variable size crew method: A method in which a variable number of collectors is provided for individual crews, depending on the amount and conditions of work on particular routes (*See* waste collection method for more related terms) (SW-108ts).

Variable wind: A wind for which no specific direction can be defined. The wind speed is usually small (*See* wind for more related terms) (NATO-10/78).

Variability factor: Pollutant-specific peaking factors that relate the numerical limitations for the maximum for any one day and the maximum for monthly average to the long-term average value (EPA-87/10a).

Variance:
(1) The temporary deferral of a final compliance date for an individual source subject to an approved regulation, or a temporary change to an approved regulation as it applies to an individual source (40 CFR 51.100-91, *See also* 40 CFR 122.2; 124.2-91).
(2) Degree of change or difference.
(3) Government permission for a delay or exception in the application of a given law, ordinance, or regulation (EPA-94/04).

Variate: *See* synonym, observed value.

Varnish (or vehicle):
(1) Any agent which facilitates the mixture, dispersion, or solubilization of a test substance with a carrier (40 CFR 160.3; 792.3; 792.226-91).
(2) The volatile and non-volatile liquid portion of a paint or coating which disperses and suspends the pigment whenever the latter is used (EPA-79/12b).
(3) Solution of resin which imparts a thin shiny coat to a paper (EPA-83).

Vector:
(1) A carrier, usually an arthropod, that is capable of

transmitting a pathogen from one organism to another (40 CFR 240.101-91, *See also* 40 CFR 241.101; 243.101-91).
(2) (A) An organism, often an insect or rodent, that carries disease.
(B) An object that is used to transport genes into a host cell (vectors can be plasmids, viruses, or other bacteria). A gene is placed in the vector; the vector then infects the bacterium (EPA-94/04).
(3) A directed line segment that has both magnitude and direction (Markes-67).
(4) cf. disease vector.

Veering wind: A clock-wise rotation of the wind direction (*See* wind for more related terms) (NATO-10/78).

Vegetable parchment: A wet strength paper product used as wrapping for moist materials (EPA-87/10).

Vegetable tan: The process of converting hides into leather using chemicals either derived from vegetable matter or synthesized to produce effects similar to those chemicals (40 CFR 425.02-91).

Vegetative controls: Nonpoint source pollution control practices that involve vegetative cover to reduce erosion and minimize loss of pollutants (EPA-94/04).

Vegetative waste: The plant clippings, prunings, and other discarded material from yards and gardens. Also known as yard rubbish or yard wastes (*See* waste for more related terms) (EPA-83).

Vehicle: A device by which any person or property may be propelled, moved, or drawn upon a highway, excepting a device moved exclusively by human power or used exclusively upon stationary rails or tracks (40 CFR 52.741-91, *See also* 40 CFR 85.2102; 86.602.84; 205.51; 205.151-91). Different vehicle definitions are provided in the varnish term. Other vehicle related terms include:
- Abandoned vehicle
- Acceptance of a vehicle
- Base vehicle
- Baseline vehicle
- Clean fuel vehicle
- Common carrier by motor vehicle
- Contract carrier by motor vehicle
- Controlled vehicle
- Covered fleet vehicle
- Diesel engine vehicle
- Failing vehicle
- Flexible fuel vehicle (or flexible fuel engine)
- Heavy duty vehicle
- Light duty vehicle
- Medium duty vehicle
- Motor vehicle
- Motor vehicle or engine part manufacturer
- New motor vehicle
- Non-conforming vehicle (or non-conforming engine)
- Non-road vehicle
- Packer vehicle (or packer)
- Precontrolled vehicle
- Reconfigured emission data vehicle
- Satellite vehicle
- Spark ignition powered motor vehicle
- Test vehicle
- Transport vehicle

Vehicle configuration: A unique combination of basic engine, engine code, inertia weight class, transmission configuration, and axle ratio (40 CFR 86.082.2-91, *See also* 40 CFR 600.002.85-91).

Vehicle curb weight: The actual or the manufacturers estimated weight of the vehicle in operational status with all standard equipment, and weight of fuel at nominal tank capacity, and the weight of optional equipment computed in accordance with 40 CFR 86.02.24; incomplete light-duty trucks shall have the curb weight specified by the manufacturer (40 CFR 86.082.2-91, *See also* CAA216-42USC7550-91).

Vehicle miles travelled (VMT): A measure of the extent of motor vehicle operation; the total number of vehicle miles travelled within a specific geographic area over a given period of time (EPA-94/04).

Vehicle of infection: Water, food, milk or any substance or article serving as an intermediate means by which the pathogenic agent is transported from a reservoir and introduced into a susceptible host through ingestion, through inoculation or by deposit on the skin or mucous membrane (EPA-83).

Vehicle or engine configuration: The specific subclassification unit of an engine family as determined by engine displacement, fuel system, engine code, transmission and inertia weight class as applicable (40 CFR 85.2113-91).

Vellum: Strong, high quality natural or cream colored parchment like paper (EPA-83).

Velocity: The rate at which a fluid is flowing in a given direction. Air velocity is normally stated in feet per second or feet per minute (EPA-84/09).

Velocity meter: A device for measuring the velocity of a flow (cf. pitot tube).

Velocity pressure:
(1) The kinetic pressure in the direction of flow necessary to cause a fluid at rest to flow at a given velocity. It is usually expressed in inches of water gauge (29 CFR 1910.94b-91).
(2) The pressure caused by and related to the velocity of a fluid flow, a measure of the kinetic energy of the fluid (EPA-83/06).
(3) *See* pressure for more related terms.

Veneer: A thin sheet of wood of uniform thickness produced by peeling, slicing, or sawing logs, bolts, or flitches. Veneers may be categorized as either hardwood or softwood depending on the type of woods used and the intended purpose (EPA-74/04):
(1) Softwood veneer
(2) Hardwood veneer

Vent: An opening through which there is mechanically induced air flow for the purpose of exhausting from a building air carrying particulate matter emissions from one or more affected facilities (40

CFR 60.671-91, *See also* EPA-84/09).

Vent stream: Any gas stream released to the atmosphere directly from an emission source or indirectly either through another piece of process equipment or a material recovery device that constitutes part of the normal recovery operations in a polymer process line where potential emissions are recovered for recycle or resale, and any gas stream directed to an air pollution control device. The emissions released from an air pollution control device are not considered a vent stream unless, as noted above, the control device is part of the normal material recovery operations in a polymer process line where potential emissions are recovered for recycle or resale (40 CFR 60.561-91).

Vented: Discharged through an opening, typically an open-ended pipe or stack, allowing the passage of a stream of liquids, gases, or fumes into the atmosphere. The passage of liquids, gases, or fumes is caused by mechanical means such as compressors or vacuum-producing systems or by process-related means such as evaporation produced by heating and not caused by tank loading and unloading (working losses) or by natural means such as diurnal temperature changes (40 CFR 264.1031-91).

Vented cell: A type of battery cell which has a vent that allows the escape of gas and the addition of water (EPA-84/08).

Ventilated cell composting: Compost is mixed and aerated by being dropped through a vertical series of ventilated cells (*See* composting for more related terms) (SW-108ts).

Ventilation factor: A parameter used in connection with the box model. It is defined as the product of the horizontal wind velocity and the surface of the box perpendicular to the wind direction (NATO-78/10).

Ventilation fever: *See* synonym, humidifier fever.

Ventilation rate: The rate at which outdoor air enters and leaves a building. Expressed in one of two ways: the number of changes of outside air per unit of time (air changes per hour, or ach) or the rate at which a volume of outside air enters per unit of time (cubic feet per minute or cfm) (EPA-88/09b). The ventilation rate depends on the tightness of the house shell, weather conditions, and the operation of appliances (such as fans) influencing air movement. Commonly expressed in terms of air changes per hour, or cubic feet per minute.

Ventilation/suction: The act of admitting fresh air into a space in order to replace stale or contaminated air; achieved by blowing air into the space. Similarly, suction represents the admission of fresh air into an interior space by lowering the pressure outside of the space, thereby drawing the contaminated air outward (EPA-94/04).

Venturi flume: An open channel with a constricted throat which is used in flow measurement.

Venturi meter: *See* synonym, venturi tube.

Venturi scrubber:
(1) One of air pollution control devices. Venturi scrubbers are high-energy scrubbers used for the control of fine particulate emissions. Hydrochloric acid gas, if present, also is controlled by a venturi scrubber. They are designed to maximize turbulence and mixing of water droplets and dirty flue gas to improve pollutant capture efficiency. Liquid is injected into the high velocity gas stream either at the inlet to the converging section or at the venturi throat. The liquid is atomized to form filaments and films which have extremely large surface areas for mass transfer. Gas velocities at the throat are from 15,000 to 20,000 fpm, and pressure drops are from 10 to 30 inches water gage (some systems operate at 80 inches water gage). Venturi scrubbers are the most common types of wet scrubbers and are usually the only technically feasible solution to an air pollution problem. If submicron particulate matter is sticky, flammable, or highly corrosive, for example, precipitators and fabric filters cannot be used and venturi scrubbers become a reasonable choice. Venturi scrubbers are also the only ultra-high efficiency collectors which can simultaneously remove gaseous and particulate matter from a gas stream without any physical modifications (*See* scrubber for more related terms) (EPA-3/89b, AP-40; Hesketh-79).
(2) Air pollution control devices that use water to remove particulate matter from emissions (EPA-94/04).

Venturi scrubber main components: The major components of a venturi scrubber include:
(1) A constriction in the ductwork referred to as a venturi throat. Some venturi scrubbers have an adjustable throat that can be used to vary the size of the opening.
(2) Spray nozzles at the entrance to the venturi throat that supply the scrubbing liquid, usually water
(3) A cyclonic mist eliminator for removing entrained water droplets.
(4) An induced draft fan for moving the flue gas through the scrubber (EPA-3/89b, AP-40; Hesketh-79).

Venturi tube (or venturi meter): A device for measuring a gas or water flow rate in a pipe. It consists of:
(1) A tapered section where the flow cross section is gradually reduced;
(2) A throat section where both the fluid velocity and the pressure drop are at maximum; and
(3) A tapered section where the flow cross section is gradually increased to recover the kinetic energy. By attaching a pressure gauge to the three sections, the pressure drop can be measured and the flow rate through the throat can be calculated.

Venturi scrubber: Venturi scrubbers are high-energy scrubbers used for the control of fine particulate emissions. Hydrochloric acid gas, if present, also is controlled by a venturi scrubber. They are designed to maximize turbulence and mixing of water droplets and dirty flue gas to improve pollutant capture efficiency. Liquid is injected into the high velocity gas stream either at the inlet to the converging section or at the venturi throat. The liquid is atomized to form filaments and films which have extremely large surface areas for mass transfer. Gas velocities at the throat are from 15,000 to 20,000 fpm, and pressure drops are from 10 to 30 inches water gage (some systems operate at 80 inches water gage). Venturi scrubbers are the most common types of wet scrubbers and are usually the only technically feasible solution to an air pollution problem. If submicron particulate matter is sticky, flammable, or highly corrosive, for example, precipitators and fabric filters cannot be used and venturi scrubbers become a reasonable choice. Venturi scrubbers are also the only ultra-high efficiency collectors which can simultaneously remove gaseous and particulate matter from a gas stream without any physical modifications (*See* scrubber for more related terms) (EPA-3/89b, AP-40; Hesketh-79).

The major components of a venturi scrubber include:
(1) A constriction in the ductwork referred to as a venturi throat. Some venturi scrubbers have an adjustable throat that can be used to vary the size of the opening.
(2) Spray nozzles at the entrance to the venturi throat that supply the scrubbing liquid, usually water.
(3) A cyclonic mist eliminator for removing entrained water droplets
(4) An induced draft fan for moving the flue gas through the scrubber.

Verification: A sampling program including samples of untreated and treated wastewater and sludge to determine the levels of classical pollutant and toxic pollutants known to be present, as well as removal efficiencies by various wastewater treatment processes (cf. screening) (EPA-79/12b).

Vermin: Carriers of disease germs, bacteria or viruses, such as rodents, mosquitoes, flies, lice and fleas, which transmit such infectious elements to humans. *See* Vector (EPA-83).

Vertical spindle disc grinder: A grinding machine having a vertical, rotatable, power-driven spindle carrying a horizontal abrasive disc wheel (29 CFR 1910.94b-91).

Vertical wind: The component of the wind velocity along the local vertical direction. This component is usually small compared to the horizontal wind velocity (*See* wind for more related terms) (NATO-10/78).

Very large MWC plant: An MWC plant with an MWC plant capacity greater than 1,000 megagrams per day (1,100 tons per day) of MSW (cf. MWC plant) (40 CFR 60.31a-91).

Very low density polyethylene (VLDPE): A linear polymer of ethylene with other alpha-olefins with a density of 0.900 to 0.910 (EPA-91/05).

Very low sulfur oil: An oil that contains no more than 0.5 weight percent sulfur or that, when combusted without sulfur dioxide emission control, has a sulfur dioxide emission rate equal to or less than 215 ng/J (0.5 lb/million Btu) heat input (*See* oil for more related terms) (40 CFR 60.41b-91).

Vessel: The every description of watercraft or other artificial contrivance used, or capable of being used, as a means of transportation on water other than a public vessel (CWA311). Other vessel related terms include:
- Commercial vessel
- Existing vessel

Viable embryos (fertility): The eggs in which fertilization has occurred and embryonic development has begun. This is determined by candling the egg 11 days after incubation has begun. It is difficult to distinguish between the absence of fertilization and early embryonic death. The distinction can be made by breaking out eggs that appear infertile and examining further. This distinction is especially important when a test compound induces early embryo mortality. Values are expressed as a percentage of eggs set (40 CFR 797.2130-iv-91).

Vibrating screen:
(1) An inclined screen that is vibrated mechanically and screens material placed on it (SW-108ts).
(2) An inclined or horizontal rectangular screening surface with a high-speed vibrating motion that lifts particles off the surface (EPA-88/08a).
(3) *See* screen for more related terms.

Vibration: The periodic motion of friable ACBM which may result in the release of asbestos fibers (40 CFR 763.83-91). (Vibration is the periodic motion of a body).

View port: A sealed glass port for observing the combustion phenomena inside the combustion chamber during incineration (*See* port for more related terms) (EPA-89/03b).

Vinyl asbestos floor tile: An asbestos-containing product composed of vinyl resins and used as floor tile (*See* asbestos for more related terms) (40 CFR 763.163-91).

Vinyl chloride (C_2H_3Cl):
(1) A colorless gas, made by reacting acetylene with hydrogen chloride or by cracking ethylene dichloride. It is used in adhesives and is suspected as a cancer agent for the liver (EPA-74/11).
(2) A chemical compound, used in producing some plastics, that is believed to be oncogenic (EPA-94/04).

Vinyl chloride plant: Includes any plant which produces vinyl chloride by any process (40 CFR 61.61-91).

Vinyl chloride purification: Includes any part of the process of vinyl chloride production which follows vinyl chloride formation (40 CFR 61.61-91).

Vinyl coating: Any topcoat or printing ink applied to vinyl coated fabric or vinyl sheets. Vinyl coating does not include plastisols (40 CFR 52.741-91).

Vinyl coating facility: A facility that includes one or more vinyl coating line(s) (40 CFR 52.741-91).

Vinyl coating line: A coating line in which any protective, decorative or functional coating is applied onto vinyl coated fabric or vinyl sheets (40 CFR 52.741-91).

Violating facility: Any facility that is owned, leased, or supervised by an applicant, recipient, contractor, or subcontractor that EPA lists under 40 CFR 15 as not in compliance with Federal, State, or local requirements under the Clean Air Act or Clean Water Act. A facility includes any building, plant, installation, structure, mine, vessel, or other floating craft (40 CFR 30.200-91).

Virgin aluminum: Aluminum recovered from bauxite (EPA-

76/12).

Virgin material:
(1) A material extracted from nature in its raw form (OTA-89/10).
(2) A raw material used in manufacturing that has been mined or harvested and has not as yet become a product (40 CFR 246.101-91, *See also* RCRA1004; 40 CFR 247.101-91).
(3) Resources extracted from nature in their raw form, such as timber or metal ore (EPA-94/04).
(4) *See* material for more related terms.

Virgin wood pulp: Pulp made from wood, as contrasted to wastepaper sources of fiber (EPA-87/10).

Virtual point source: A description of the emissions from an area source by assuming that this source can be represented by a point source. This approximation will improve with increasing distance from the source (*See* source for more related terms) (NATO-78/10).

Virtual temperature: The temperature of dry air having the same density and pressure as the moist air (*See* temperature for more related terms) (NATO-78/10).

Virtually absent: The subject constituent is present in very low concentrations, and is not objectionable in these barely detectable concentrations (LBL-76/07-water).

Virus:
(1) A virus of fecal origin which is infectious to humans by waterborne transmission (40 CFR 141.2-91).
(2) The smallest form of microorganisms capable of causing disease (EPA-89/12).

Virulence:
(1) The disease-evoking power of a microorganism in a given host (EPA-5/90).
(2) The ease with which a pathogen causes a disease (EPA-88/09a).

Viscose process: The fiber forming process where cellulose and concentrated caustic soda are reacted to form soda or alkali cellulose. This reacts with carbon disulfide to form sodium cellulose xanthate, which is then dissolved in a solution of caustic soda. After ripening, the solution is spun into an acid coagulating bath. This precipitates the cellulose in the form of a regenerated cellulose filament (40 CFR 60.601-91).

Viscosity: A measure of a fluid's resistance to a flow. Viscosities vary greatly, from materials like heavy lubricating oils to water. The viscosity is also a strong function of temperature, increasing with increasing temperature for gases and decreasing with decreasing temperature for liquids. Viscosity is a measure of a fluid's resistance to a flow. Viscosities vary greatly, from materials like heavy lubricating oils to water. The viscosity is also a strong function of temperature, increasing with increasing temperature for gases and decreasing with decreasing temperature for liquids (EPA-84/09).
(1) Definition:
Viscosity is the result of Newton's viscosity law. The law states that in fluid mechanics, the applied shear stress (S_s) is proportional to the rate of deformation or to the velocity gradient normal to the velocity. Mathmatically, it can be expressed as:
$$S_s = \mu(dV/dy)$$
where:
- S_s = shear stress
- V = velocity in differential amount
- y = distance normal to the velocity direction in differential amount
- μ = absolute or dynamic viscosity

(2) Viscosity is the result of two phenomena:
- Intermolecular cohesive forces and
- Momentum transfer between flowing strata caused by molecular agitation perpendicular to the direction of motion. Between adjacent strata of a flowing fluid a shearing stress results that is directly proportional to the velocity gradient. Viscosity is often defined as resistance to flow. *See also* Newton's viscosity law.

(3) Viscosity has the following unit systems
(A) For dynamic viscosity (D_v)
- In the cgs system, the unit of dynamic viscosity is pose (dyne-sec/cm^2)
- 1 centipoise at = 1 poise/10^2
- In the English system, the unit of dynamic viscosity is lb-sec/ft^2
- Dynamic viscosity of water = 1 centipoise at 68.4° F and atmospheric pressure

(B) For kinematic viscosity (K_v)
- Kinematic viscosity = (dynamic viscosity)/density
- In the cgs system, the unit of kinematic viscosity is stoke (cm^2/sec)
- 1 centistoke = 1 stoke/10^2
- In the English system, the unit of dynamic viscosity is ft^2/sec
- 1 centistoke = 1.076x10^{-5} (K_v) ft^2/sec

(4) Viscosity measurement systems
(A) Saybolt Universal viscometer is commonly used for petroleum products and lubricating oils.
(B) Saybolt Furol viscometer is used for heavy oils.

Other viscosity related terms include:
- Kinematic viscosity
- Gas viscosity
- Liquid viscosity

Viscus (pl. viscera): Any internal organ within a body cavity (EPA-74/06).

Visibility: The greatest distance in a given direction at which it is just possible to *See* and identify a prominent dark object in the day time or a light source in the night time (NATO-78/10).

Visibility impairment: Any humanly perceptible change in visibility (visual range, contrast, coloration) from that which would have existed under natural conditions (40 CFR 51.301-91).

Visibility impairment and impairment of visibility: Include reduction in visual range and atmospheric discoloration (CAA169A-42USC7491-91).

Visibility protection area: Any area listed in 40 CFR 81.401-

81.436 (1984) (other identical or similar definitions are provided in 40 CFR 52.28-1) (40 CFR 52.26-1-91).

Visible: Pertaining to radiant energy in the electromagnetic spectral range visible to the normal human eye (approximately 380-780 um) (LBL-76/07-air).

Visible emission: Any emissions, which are visually detectable without the aid of instruments, coming from RACM or asbestos-containing waste material, or from any asbestos milling, manufacturing, or fabricating operation. This does not include condensed, uncombined water vapor (40 CFR 61.141-91).

Visible specific locus mutation: A genetic change that alters factors responsible for coat color and other visible characteristics of certain mouse strains (40 CFR 798.5200-91).

Vitrification (or glassification):
(1) Formation of glassy materials by the application of high temperatures to loose materials.
(2) A process whereby high temperatures affect permanent chemical and physical changes in a ceramic body, most of which is transformed into glass (SW-108ts).

VOC content: The proportion of a coating that is volatile organic compounds (VOC's), expressed as kilograms of VOC's per liter of coating solids (40 CFR 60.311-91, *See also* 40 CFR 60.391; 60.451; 60.461; 60.491-91).

VOC content of the coating applied: The product of Method 24 VOC analyses or formulation data (if the data are demonstrated to be equivalent to Method 24 results) and the total volume of coating fed to the coating applicator. This quantity is intended to include all VOC that actually are emitted from the coating operation in the gaseous phase. Thus, for purposes of the liquid-liquid VOC material balance in 40 CFR 60.713(b)(1), any VOC (including dilution solvent) added to the coatings must be accounted for, and any VOC contained in waste coatings or retained in the final product may be measured and subtracted from the total. (These adjustments are not necessary for the gaseous emission test compliance provisions of 40 CFR 60.713(b)) (40 CFR 60.711-19-91).

VOC emission: The mass of volatile organic compounds (VOCs), expressed as kilograms of VOCs per liter of applied coating solids, emitted from a metal furniture surface coating operation (40 CFR 60.311-91, *See also* 40 CFR 60.451; 60.721-91).

VOC emission control device: The equipment that destroys or recovers VOC (40 CFR 60.541-91).

VOC emission reduction system: A system composed of an enclosure, hood, or other device for containment and capture of VOC emissions and a VOC emission control device (40 CFR 60.541-91).

VOC in the applied coating: The product of Method 24 VOC analyses or formulation data (if those data are demonstrated to be equivalent to Method 24 results) and the total volume of coating fed to the coating applicator (40 CFR 60.741-91).

VOC solvent: An organic liquid or liquid mixture consisting of VOC components (40 CFR 60.431-91).

VOC used: The amount of VOC delivered to the coating mix preparation equipment of the affected facility (including any contained in premixed coatings or other coating ingredients prepared off the plant site) for the formulation of polymeric coatings to be applied to supporting substrates at the coating operation, plus any solvent added after initial formulation is complete (e.g., dilution solvent added at the coating operation). If premixed coatings that require no mixing at the plant site are used, "VOC used" means the amount of VOC delivered to the coating applicator(s) of the affected facility (40 CFR 60.741-91).

Void: *See* synonym, top void.

Void ratio: The volume of the void spaces in soils divided by the volume of the solids (cf. porosity).

Volatile:
(1) A description of any substance that evaporates readily (at a relatively low temperature) (cf. organic compound, volatile) (EPA-88/03).
(2) Any substance that evaporates readily (EPA-94/04).

Volatile acid: Organic acids, generally, those with low molecular weight such as acetic, propionic and butyric acids.

Volatile combustion matter: The relatively light components in a fuel which readily vaporize at a relatively low temperature and which when combined or reacted with oxygen giving out light and heat (EPA-82/11f).

Volatile fraction: The portion of a paint-coating or ink which evaporates from the film during the drying process (EPA-79/12b; 79/12a).

Volatile hazardous air pollutant (VHAP): A substance regulated under this part for which a standard for equipment leaks of the substance has been proposed and promulgated. Benzene is a VHAP. Vinyl chloride is a VHAP (40 CFR 61.241-91).

Volatile liquid:
(1) A liquid which evaporates readily at atmospheric pressure and room temperature (cf. volatile organic compound).
(2) Liquids which easily vaporize or evaporate at room temperature (EPA-94/04).

Volatile matter:
(1) Those products, exclusive of moisture, given off by a material as gas or vapor, determined by definite prescribed methods which may vary according to the nature of the material (*See* analytical parameters--fuels for more related terms) (EDS-24) (EPA-83).
(2) *See* synonym, volatile substance.

Volatile organic compound (VOC) (or volatile organic matter (VOM)):
(1) Any organic compound which participates in atmospheric photochemical reactions; or which is measured by a reference method, an equivalent method, an alternative

method, or which is determined by procedures specified under any subpart (cf. organic compound or volatile liquid) (40 CFR 60.2-91, *See also* 40 CFR 51.165; 51.166; 51-App/S; 52.21; 52.24; 52.741; 60.481; 60.561; 60.711; 60.741-91; EPA-89/12).

(2) A chemical compound with the following properties:
- (A) Boiling point (° C): <100
- (B) Sampling method: VOST (volatile organic sampling train)
- (C) Capture method: tenax, tenax-charcoal
- (D) Analytical method: GC/MS (gas chromatographic/mass spectroscopy)

(3) Any organic compound that participates in atmospheric photochemical reactions except those designated by EPA as having negligible photochemical reactivity (EPA-94/04).

(4) *See* organic chemical for more related terms.

Volatile organic liquid (VOL): Any substance which is liquid at storage conditions and which contains volatile organic compounds (40 CFR 52.741-91, *See also* 40 CFR 60.111b-91).

Volatile organic sampling train: *See* definition under VOST.

Volatile solid:
(1) The sum of the volatile matter and fixed carbon of a sample, as determined by allowing a dried sample to burn to ash in a heated and ventilated furnace (EPA-83).
(2) Those solids in water or other liquids that are lost on ignition of the dry solids at 550° Centigrade (EPA-94/04).
(3) *See* solid for more related terms.

Volatile substance (or volatile matter):
(1) A chemical compound or mixture contained in a solid or liquid which volatilizes or evaporates (due to the application of heat) at or near room temperature (EPA-84/09).
(2) Materials that are readily vaporizable at relatively low temperatures (EPA-85/10a).

Volatile suspended solid (VSS): The quantity of suspended solids lost after the ignition of total suspended solids (EPA-87/10a).

Volatile synthetic organic chemicals: Chemicals that tend to volatilize or evaporate (EPA-94/04).

Volatility: The property of a substance or substances to convert into vapor or gas without chemical change (OME-88/12).

Volatilization: The physical transfer process through which a substance undergoes a change of state from a solid or liquid to a gas (40 CFR 300-App/A-91, *See also* 40 CFR 796.1950-91).

Volt: The voltage which will produce a current of one ampere through a resistance of one ohm (EPA-74/03d).

Voltameter: *See* synonym, coulometer.

Voltametric analyzer: *See* synonym, polarographic analyzer.

Voltage breakdown: The voltage necessary to cause insulation failure (EPA-83/03).

Voltage regulator: Like a transformer, it corrects changes in current to provide continuous, constant current flow (EPA-83/03).

Voltmeter: A device for measuring voltage.

Volume concentration: A concentration expressed in terms of gaseous volume of substance per unit volume of air or other gas, usually expressed in percent or parts per million (EPA-83/06).

Volume of process water used per year: The volume of process water that flows through a contact cooling and heating water process and comes in contact with the plastic product over a period of one year (40 CFR 463.11-91, *See also* 40 CFR 463.21; 463.31-91).

Volume reduction:
(1) The processing of waste materials so as to decrease the amount of space the materials occupy, usually by compacting or shredding (mechanical), incineration (thermal), or composting (biological) (EPA-89/11).
(2) Processing waste materials to decrease the amount of space they occupy, usually by compacting or shredding, incineration, or composting (EPA-94/04).

Volume source: A three dimensional form with the assumption that emissions are being released at a uniform rate from every point in the volume (*See* source for more related terms) (EPA-88/09).

Volumetric analysis:
(1) Addition of a known solution to a unknown solution until the end point has reached (color change or sedimentation) (cf. volumetric method).
(2) The analysis of a gas mixture is based on the volume of each component in the system.
(3) *See* property for more related terms.

Volumetric method: One of solid or sludge flow rate meters.
(1) In measuring settleable solids, a standard method of measuring settleable solids in an aqueous solution (EPA-83/06a).
(2) In measuring solid feed rate, this method includes calibrated augers and pumps, rotary feeders, and belt conveyors. These systems are not generally available precalibrated but must be calibrated by the user for each particular feed material. The accuracy of the method depends upon steady operation at a given speed and assumes appropriate feeders are used to ensure the cavities are always filled to capacity. Most of these methods can provide some kind of tachometer signal to indicate speed, which must be related to feed rate by performing calibration tests. These methods are generally more appropriate as secondary indicators of feed rate (EPA-89/06).
(3) *See* flow rate meter for more related terms.

Volumetric tank test: One of several tests to determine the physical integrity of a storage tank; the volume of fluid in the tank is measured directly or calculated from product-level changes. A marked drop in volume indicates a leak (EPA-94/04).

Voluntary emissions recall: A repair, adjustment, or modification program voluntarily initiated and conducted by a manufacturer to remedy any emission-related defect for which direct notification of vehicle or engine owners has been provided (40 CFR 85.1902-91).

Voluntary exclusion or voluntarily excluded: A status of nonparticipation or limited participation in covered transactions assumed by a person pursuant to the terms of a settlement (40 CFR 32.105-v-91).

Vortex shedding meter: One of liquid flow rate meters. This device is applicable to low-viscosity fluids and gases under turbulent flow conditions. The accuracy is +/- 2% under normal operations (*See* flow rate meter for more related terms) (EPA-89/06).

VOST (volatile organic sampling train): VOST is a sampling method, developed by EPA, to capture volatile organic molecules from combustion stack gases. Sorbent traps capture the organics, and then are taken to ovens to de-sorb the material into an analyzer (SW-846). Generally, VOST uses 2 traps for each pair (in sequence) to capture volatile organics. One trap contains tenax and the other tenax plus charcoal (cf. organic chemical).

Example:
- Determine whether VOST sample size is sufficient to measure 99.99% DRE for POHC, CCl_4 (carbon tetrachloride) (cf. calculation in destruction and removal efficiency) (EPA-89/06, p6).

Data:
- Waste feed rate: 15.2 kg/min (2000 lb/hr).
- POHC feed concentration: 500 ppm (500 ppm = 500kg/10^6kg = 0.5g/kg-feed).
  Stack gas flow rate: 4500 scfm (127.4 m^3/min).
  (1 m^3 = 35.31 ft^3)
- Lower detection limit: 2 ng (10^{-9}g = 10 E-09g) per trap.
  VOST: 3 trap pairs (measued at different time), 500 mL/min (.5 L/min) flow rate per pair, 20 L sample/pair

Solution:
(a) POHC input rate, W_i = waste feed rate x POHC concentration
W_i = 15.2(kg/min) x 0.5(g/kg) = 7.6 g/min.

(b) POHC stack output rate at 99.99% DRE, W_o = W_i(1-DRE/100).
W_o = 7.6(1-0.9999) = 0.00076 g/min.

(c) POHC concentration (Pc) in stack gas at 99.99% DRE = W_o/(stack gas flow rate) = 0.00076/127.4 = 0.0000060 g/m^3. (stack gas flow rate is a given condition).
0.0000060 g/m^3 x (10 E+09 ng/g) x (10 E-03 m^3/L) = 6.0 ng/L.
{1 g = (10 E+09)x(10 E-09)g = (10 E+09)ng}
{1 L = 1000 cc = 10 E+03 cm^3 [m^3/(10 E+06 cm^3)] = 10 E-03 m^3}.

(d) Sample amount collected on one pair of traps = 20(L) x 6.0(ng/L) = 120 ng.

(e) Conclusion: Since the VOST lower detection limit for CCl_4 is 2 ng, the sample is sufficient to detect CCl_4 to determine a DRE of 99.99% or lower. A margin of safety above the detection limit is desirable. This calculation assumes both traps in a pair are combined for analysis. If they are analyzed separately, the distribution of mass on each trap must be considered.

Voticity: A measure of the rotation of a fluid element. It indicates twice the angular velocity at each point in a fluid (NATO-78/10).

Vulcanization: A process which plastic rubber is converted into the elastic rubber or hard rubber state. The process is brought about by linking of macro-molecules at their reactive sites (EPA-74/12a).

Vulnerable zone: An area over which the airborne concentration of a chemical accidentally released could reach the level of concern (EPA-94/04).

Vulnerability analysis: Assessment of elements in the community that are susceptible to damage should a release of hazardous materials occur (EPA-94/04).

Vulnerable zone: An area over which the airborne concentration of a chemical involved in an accidental release could reach the level of concern (EPA-89/12).

Ww

Wake: In a flow field, a region behind an obstacle in which the fluid flow is disturbed due to the presence of this obstacle (NATO-78/10).

Wall fired boilers: Coal-fired furnaces in which burners are installed and fired on opposite walls of the unit (*See* boiler for more related terms) (EPA-94/04).

Wall insulation: A material, primarily designed to resist heat flow, which is installed within or on the walls between conditioned areas of a building and unconditioned areas of a building or the outside, as well as common wall assemblies between separately conditioned units in multiple unit structures (40 CFR 248.4-91).

Warm and cold-water fish: Warm-water fish include black bass, sunfish, catfish, gar and others; whereas cold-water fish include salmon and trout, whitefish, miller's thumb, and blackish. The temperature factor determining distribution is set by adaptation of the eggs to warm or cold water (LBL-76/07-water).

Warm up time: The elapsed time necessary after startup for the instrument to meet stated performance specifications when the instrument has been shut down for at lease 24 hours (LBL-76/07-bio).

Warning device: A sound emitting device used to alert and warn people of the presence of railroad equipment (40 CFR 201.1-91).

Wash coat: A coating containing binders which seals wood surfaces, prevents undesired staining, and controls penetration (40 CFR 52.741-91).

Wash oil circulation tank: Any vessel that functions to hold the wash oil used in light oil recovery operations or the wash oil used in the wash oil final cooler (*See* tank for more related terms) (40 CFR 61.131-91).

Wash oil decanter: Any vessel that functions to separate, by gravity, the condensed water from the wash oil received from a wash-oil final cooler or from a light-oil scrubber (40 CFR 61.131-91).

Wash water: The water used to backwash filter beds, micro-strainers, etc (*See* water for more related terms).

Washboard: An imperfection; ripples, waves, etc., on the surface of glassware (EPA-83).

Washer: A machine which agitates fabric articles in a petroleum solvent bath and spins the articles to remove the solvent, together with the piping and ductwork used in the installation of this device (40 CFR 60.621-91).

Washing: *See* synonym, scrubbing.

Washing cooler: A large vessel where a flowing gas stream is subjected to sprays of water or liquor to remove gas-borne dusts and to cool the gas stream by evaporation (EPA-85/10a).

Washout:
(1) The carrying away of solid waste by waters of the base flood (40 CFR 257.3.1, *See also* 40 CFR 264.18-91).
(2) The capture of gaseous or particulate air pollutants by precipitation (NATO-78/10).

Waste:
(1) Useless, unwanted or discarded material resulting from (agricultural, commercial, community and industrial) activities. Wastes include solids, liquids, and gases (EPA-83).
(2) Any liquid resulting from industrial, commercial, mining or agricultural operations, or from community activities that is discarded or is being accumulated, stored, or physically, chemically, or biologically treated prior to being discarded or recycled (40 CFR 60.111b-91, *See also* 40 CFR 61.341; 191.12; 704.43-91; EPA-89/12).
(3) In theory, the term waste applies to non-product outputs of processes and discarded products, irrespective of the environmental medium affected. In practice, since the passage of the Resource Conservation and Recovery Act (RCRA), most uses of the term "waste" refer exclusively to the hazardous and solid wastes regulated under RCRA, and do not include air emissions or water discharges regulated by the Clean Air Act or the Clean Water Act. The Toxics Release Inventory (TRI) refers to wastes that are hazardous as well as nonhazardous (EPA-91/10, p6).
(4) (A) Unwanted materials left over from a manufacturing process (EPA-94/04).
 (B) Refuse from places of human or animal habitation (EPA-94/04).

Other waste related terms include:
- Agricultural solid waste
- Agricultural waste
- Bulky waste
- Chemical metal cleaning waste

- Classified waste
- Combustible waste
- Commercial solid waste
- Commercial waste (*See* synonym, commercial solid waste)
- Construction and demolition waste
- Consumer waste
- Corrugated container waste
- Dioxin waste
- Domestic municipal waste (*See* synonym, residential waste)
- Explosive waste
- Food waste
- Food processing waste
- Hazardous waste
- High level radioactive waste
- High level waste
- Household hazardous waste
- Household solid waste (*See* synonym, household waste)
- Household waste
- High level radioactive waste
- Incombustible waste
- Incompatible waste
- Industrial refuse (*See* synonym, industrial solid waste)
- Industrial solid waste
- Infectious waste
- Inorganic waste
- waste)
- Institutional waste
- Listed waste
- Low level waste
- Low level radioactive waste
- Medical waste
- Metal cleaning waste
- Mining waste
- Mixed waste
- Mixed municipal refuse (*See* synonym, mixed municipal solid waste)
- Mixed municipal solid waste
- Municipal solid waste (*See* synonym, municipal waste)
- Municipal type solid waste (*See* synonym, municipal waste)
- Municipal waste
- Nitrogenous waste
- Non-combustible rubbish
- Non-combustible waste
- Old scrap (*See* synonym, post consumer waste)
- Organic waste
- Oversized regulated medical waste
- Oversized waste (*See* synonym, bulky waste)
- PCB waste
- Pesticide solid waste (*See* synonym, pesticide waste)
- Pesticide waste
- Pesticide-related waste
- Plant waste (general)
- Post consumer waste
- Pre-consumer waste
- Pre-plating treatment waste
- Problem waste
- Process waste
- Putrescible waste
- Radioactive waste
- Radwaste (*See* synonym, radioactive waste)
- Red bag waste
- Regulated medical waste
- Residential solid waste (*See* synonym, residential waste)
- Residential waste
- Residual waste
- Sanitary waste
- Similar waste
- Solid waste
- Special waste
- Street refuse (*See* synonym, street waste)
- Street waste
- Tannery waste
- Toxic waste
- Transuranic radioactive waste
- Transuranic waste
- Treated regulated medical waste
- Unconventional waste (*See* synonym, special waste)
- Untreated regulated medical waste
- Vegetative waste
- Wood pulp waste
- Yard waste

Waste activated sludge: Excessive activated sludge which must be removed from the treatment system (*See* sludge for more related terms).

Waste category: Either untreated regulated medical waste or treated regulated medical waste (40 CFR 259.10-91).

Waste characterization: Identification of chemical and microbiological constituents of a waste material (EPA-94/04).

Waste collection (or collection):
(1) The act of removing solid waste (or materials which have been separated for the purpose of recycling) from a central storage point (40 CFR 243.101-91, *See also* 40 CFR 246.101-91).
(2) Gathering of MSW (municipal solid waste) for subsequent management (e.g., landfilling, incineration, or recycling) (OTA-89/10).

Other collection related terms include:
- Alley collection
- Carryout collection
- Contract collection
- Curb collection
- Curbside collection (*See* synonym, curb collection)
- Definite working day collection method
- Franchise collection
- Municipal collection
- Private collection
- Setout/setback collection

Waste collection method: Waste collection method related terms include:
- Daily route method
- Definite working day method
- Group task method
- Inter route relief method
- Large route method
- Reservoir route method
- Single load method
- Swing crew method

- Variable size crew method

Waste compatibility: The property of two or more wastes which can remain in contact indefinitely without reaction (cf. compatibility).

Waste disposal:
(1) The orderly process of discarding useless or unwanted materials (*See* disposal for more related terms) (SW-108ts).
(2) A computer and catalog network that redirects waste materials back into the manufacturing or reuse process by matching companies generating specific wastes with companies that use those wastes as manufacturing input: (EPA-89/11)s

Waste exchange: Arrangement in which companies exchange their wastes for the benefit of both parties (EPA-94/04).

Waste feed: The continuous or intermittent flow of wastes into an incinerator (EPA-94/04).

Waste form: The materials comprising the radioactive components of waste and any encapsulating or stabilizing matrix (40 CFR 191.12-91).

Waste generator: Any owner or operator of a source covered by this subpart whose act or process produces asbestos containing waste material (40 CFR 61.141-91).

Waste heat boiler: A boiler used to recover the heat generated during incineration for producing hot water or steam. Addition of a waste heat boiler to an incinerator has several impacts on the incineration system. One impact of adding a boiler to the system is that an induced draft fan must be added to the system in order to move air through the system. An emergency bypass stack is another feature that normally would be added to an incinerator when a waste heat boiler (or air pollution control system) is added to the incinerator system. Since the boiler causes a resistance (blockage to airflow) in the system if the induced draft fan stops, pressure will build up in the incinerator because the hot gases cannot escape quickly enough. The bypass stack is added to allow a route for the hot gases to escape should the fan fail. In other words, it allows the incinerator to go back to a natural draft system. The bypass stack also is used in cases where the boiler must be bypassed for some reason (for example, loss of water flow to the boiler causing heat buildup). The bypass stack usually contains a damper valve in the stack to control direction of the gas flow or a cap on top of the stack to prevent air from being pulled into the system when the fan is operating. When the bypass must be activated, the damper, or cap, is opened. The bypass is usually activated automatically by some type of sensor; for example, if the fan speed falls below a preset level, the bypass opens (*See* boiler for more related terms). (EPA-89/03b).

Waste load allocation (WLA):
(1) The portion of a receiving water's loading capacity that is allocated to one of its existing or future point sources of pollution. WLAs constitute a type of water quality based effluent limitation (cf. load allocation) (40 CFR 130.2-91).
(2) The portion of a receiving water's total maximum daily load that is allocated to one of its existing or future point sources of pollution (EPA-91/03).
(3) The maximum load of pollutants each discharger of waste is allowed to release into a particular waterway. Discharge limits are usually required for each specific water quality criterion being, or expected to be, violated. The portion of a stream's total assimilative capacity assigned to an individual discharge (EPA-94/04).

Waste loading: The total amount of pollutant substance, generally expressed as pounds per day (*See* loading for more related terms) (EPA-75/11).

Waste management unit: A piece of equipment, structure, or transport mechanism used in handling, storage, treatment, or disposal of waste. Examples of a waste management unit include a tank, surface impoundment, container, oil water separator, individual drain system, steam stripping unit, thin film evaporation unit, waste incinerator, and landfill (40 CFR 61.341-91).

Waste minimization:
(1) One of the earliest initiatives in pollution prevention was waste minimization. The initial focus here was on wastes (as defined under RCRA), rather than on a comprehensive evaluation of industrial emissions regulated under all environmental statutes. This term became source of controversy because some considered it to designate approaches to treating waste so as to minimize its volume or toxicity, rather than decreasing the quantity of waste at the source of its generation. The distinction became important because some advocates of decreased waste generation believed that an emphasis on waste minimization would deflect resources away from prevention towards treatment (the EPA Office of Solid Waste Source Reduction and Recycling Action Plan will formally define waste minimization). In the current RCRA biennial report, waste minimization refers to source reduction and recycling activities, and now excludes treatment and energy recovery (EPA-91/10, p6).
(2) In HSWA, the prevention or restriction of waste generation at its source by redesigned products or the patterns of production and consumption. It includes any source reduction or recycling activity undertaken by a generator that results in either:
 (A) The reduction of total volume or quantity of hazardous waste or
 (B) The reduction of toxicity of hazardous waste, or both, so long as such as reduction is consistent with the goal of minimizing present and future threats to human health and the environment (cf. pollution prevention) (EPA-86/10).
(3) Measures or techniques that reduce the amount of wastes generated during industrial production processes; term is also applied to recycling and other efforts to reduce the amount of waste going into the waste stream (EPA-94/04).

Terms related to waste minimization include: extraction; four Rs (reduction, reuse, recycle, and regeneration); material recovery; pollution prevention; prevention; reclamation; recovery; recycle; resource recovery; reuse; salvage; secondary use; separation; sorting; source separation; waste exchange; waste reduction; etc.

Waste minimization policy: In HSWA, this policy involves regulation development for preventing the generation of waste rather than controlling waste after it is generated. It is a national policy that the generation of hazardous waste be reduced as expeditiously as possible (cf. pollution prevention requirements) (OSWER-87)

Waste oil: The used products primarily derived from petroleum, which include, but are not limited to, fuel oils, motor oils, gear oils, cutting oils transmission fluids, hydraulic fluids, and dielectric fluids (40 CFR 761.3-91).

Waste paper: Various recognized grades such as No. 1 news, new kraft corrugated cuttings, old corrugated containers, manila tabulating cards, coated soft white shavings, etc., which are used as a principal ingredient in the manufacture of certain types of paperboard, particularly boxboard made on cylinder machines where the lower grades may go into filler stock, and the higher grades into one or both liners (EPA-87/10, See also 40 CFR 250.4-91). Types of waste paper includes:

(1) **No. 1 mixed wastepaper:** Consists of a baled mixture of various quantities of paper containing less than 25% of ground wood stock, coated or uncoated. Prohibitive materials may not exceed 1%. Total out throws may not exceed 5% (EPA-83).

(2) **No. 2 mixed wastepaper:** Consists of a mixture of various qualities of paper not limited as to type of packing or fiber content. Prohibitive material may not exceed 2%. Total out throws may not exceed 10% (EPA-83).

(3) **Super mixed wastepaper:** Consists of a baled clear sorted mixture of various qualities of papers containing less than 10% of ground wood stock, coated or uncoated. Prohibitive materials may not exceed 1/2 of 1%. Total out throws may not exceed 3% (EPA-83).

(4) **News wastepaper:** Consists of baled newspaper containing less than 5% of other papers. Prohibitive materials may not exceed 1/2 of 1%. Total out throws may not exceed 2% (EPA-83).

(5) **Super news wastepaper:** Consists of baled sorted fresh newspapers, not sunburned, free from papers other than news, containing not more than the normal percentage of rotogravure and colored sections. Prohibitive materials - none permitted. Total out throws may not exceed 2% (EPA-83).

(6) **Corrugated container wastepaper:** Consists of baled corrugated containers having liners of either jute or kraft. Prohibitive materials may not exceed 1%. Total out throws may not exceed S (EPA-83).

(7) **No. 1 sorted colored ledger wastepaper:** Consists of printed or unprinted sheets, shavings and cuttings of colored or white sulphite or sulphate edger bond, writing, and other types of paper which have a similar fiber and filler content. This grade must be.free of treated, coated, padded or heavily printed stock. Prohibitive materials - none permitted. Total out throws may not exceed 2% (EPA-83).

(8) **No. 1 sorted white ledger wastepaper:** Consists of printed or unprinted sheets, shavings, and cuttings of white sulphite or sulphate ledger, bond, writing and other papers which have a similar fiber and filler content. This grade must be free of treated, coated, padded, or heavily printed stock. Prohibitive materials - none permitted (EPA-83).

(9) **New corrugated cutting wastepaper:** Consist of baled corrugated cuttings having two or more liners of either jute or kraft. Non-soluble adhesives, butt soles, slubbed or hogged medium, and treated medium or liners are not acceptable in this grade. Prohibitive materials may not exceed 1. Total out throw may not exceed 5 % (EPA-83).

(10) *See* paper for more related terms.

Waste paper utilizarion rate: The ratio of waste paper consumption to total production of paper and paperboard (OTA-89/10).

Waste processing: An operation such as shredding, compaction, composting, and incineration, in which the physical or chemical properties of wastes are changed (SW-108ts).

Waste reduction:
(1) This term has been used by the Congressional Office of Technology Assessment and INFORM to mean source reduction. On the other hand. many different groups have used the term to refer to waste minimization. Therefore, care must be employed in determining which of these different concepts is implied when the term waste reduction is encountered (EPA-91/10, p7).
(2) Reducing the amount or type of waste generated. Sometimes used synonymously with source reduction (EPA-89/11).
(3) Using source reduction, recycling, or composting to prevent or reduce waste generation (EPA-94/04).

Waste shipment record: The shipping document, required to be originated and signed by the waste generator, used to track and substantiate the disposition of asbestos-containing waste material (40 CFR 61.141-91).

Waste source: Agriculture, residential, commercial, industrial activities that generate wastes (SW-108ts).

Waste stabilization pond: *See* synonym, lagoon.

Waste storage container:
(1) For disposal application, disposable sacks made of wet-strength kraft paper or polyethylene plastic, usually 3 1/2 feet high, with an equivalent capacity of 20 to 35 gallons.
(2) For reuse application, watertight having tight fitting covers and be easy to clean. For rubbish, the containers should be such that the material cannot leak through crevices or be blown from the top. Containers for ashes should be leakproof and fireproof. All containers should be easy to empty and be equipped with suitable handles (EPA-83).
(3) *See* container for more related terms.

Waste stream:
(1) The waste generated by a particular process unit, product tank, or waste management unit. The characteristics of the waste stream (e.g., flow rate, benzene concentration, water content) are determined at the point of waste generation. Examples of a waste stream include process wastewater, product tank drawdown, sludge and slop oil removed from waste management units, and landfill leachate (40 CFR 61.341-91, *See also* EPA-89/11; 87/10a).
(2) The total flow of solid waste from homes, businesses, institutions, and manufacturing plants that are recycled,

burned, or disposed of in landfills, or segments thereof such as the "residential waste stream" or the "recyclable waste stream" (EPA-94/04).

Waste to energy facility: A municipal waste incinerator which converts heat from combustion into energy (i.e., steam or electricity) (OTA-89/10).

Waste treatment lagoon: Impoundment made by excavation or earth fill for biological treatment of wastewater (EPA-94/04).

Waste treatment plant: A facility containing a series of tanks, screens, filters and other processes by which pollutants are removed from water (EPA-94/04).

Waste treatment stream: The continuous movement of waste from generator to treater and disposer (EPA-94/04).

Wastewater: The spent or used water from individual homes, a community, a farm, or an industry that contains dissolved or suspended matter (EPA-94/04, *See also* 40 CFR 268.2-91). Wastewater includes domestic sewage and industrial effluent. Other wastewater related terms include:
- Combined wastewater
- Contact process wastewater
- Contaminated nonprocess wastewater
- Electroplating process wastewater
- Flume wastewater
- Generation of wastewater
- Industrial wastewater
- Industrial effluent (*See* synonym, industrial wastewater)
- Inprocess wastewater
- Non-wastewater
- Non-contact wastewater
- Non-contact process wastewater
- Non-process wastewater (*See* synonym, non-contact process wastewater)
- Oily wastewater
- Process generated wastewater
- Process wastewater
- Raw wastewater
- Sanitary wastewater

Wastewater constituent: Those materials which are carried by or dissolved in a water stream for disposal (EPA-83/06a).

Wastewater discharge factor: The ratio between water discharged from a production process and the mass of product of that production process. Recycle water is not included (EPA-83/03a).

Wastewater infrastructure: The plan or network for the collection, treatment, and disposal of sewage in a community. The level of treatment will depend on the size of the community, the type of discharge, and/or the designated use of the receiving water (EPA-94/04).

Wastewater operations and maintenance: Actions taken after construction to assure that facilities constructed to treat wastewater will be operated, maintained, and managed to reach prescribed effluent levels in an optimum manner (EPA-94/04).

Wastewater system: Any component, piece of equipment, or installation that receives, treats, or processes oily wastewater from petroleum refinery process units (40 CFR 60.691-91).

Wastewater treatment: Types of wastewater treatment include:
(1) Advanced wastewater treatment
(2) Biological treatment
(3) Biological wastewater treatment (*See* synonym, biological treatment)
(4) Complete wastewater treatment system
(5) Central wastewater treatment facility
(6) Conventional wastewater treatment
(7) Intermediate treatment (*See* synonym, intermediate wastewater treatment)
(8) Intermediate wastewater treatment
(9) Primary treatment (*See* synonym, primary wastewater treatment)
(10) Primary wastewater treatment
(11) Secondary treatment (*See* synonym, secondary wastewater treatment)
(12) Secondary wastewater treatment
(13) Tertiary treatment (*See* synonym, tertiary wastewater treatment)
(14) Tertiary wastewater treatment

Wastewater treatment plant: A plant that reduces the harmful and unstable elements in wastewater so they can be disposed of without impairing other essential water uses (DOI-70/04).

Wastewater treatment process: Includes any process which modifies characteristics such as BOD, COD, TSS, and pH, usually for the purpose of meeting effluent guidelines and standards; it does not include any process the purpose of which is to remove vinyl chloride from water to meet requirements of this subpart (40 CFR 61.61-91).

Wastewater treatment system: Any component, piece of equipment, or installation that receives, manages, or treats process wastewater, product tank drawdown, or landfill leachate prior to direct or indirect discharge in accordance with the National Pollutant Discharge Elimination System permit regulations under 40 CFR 122. These systems typically include individual drain systems, oil-water separators, air flotation units, equalization tanks, and biological treatment units (40 CFR 61.341-91).

Wastewater treatment tank: A tank that is designed to receive and treat an influent wastewater through physical, chemical, or biological methods (40 CFR 280.12-91).

Wastewater treatment unit: A device which:
(1) Is part of a wastewater treatment facility that is subject to regulation under either Section 402 or 307(b) of the Clean Water Act;
(2) Receives and treats or stores an influent wastewater that is a hazardous waste as defined in 40 CFR 261.3 of this chapter, or that generates and accumulates a wastewater treatment sludge that is a hazardous waste as defined in 40 CFR 261.3 of this chapter, or treats or stores a wastewater treatment sludge which is a hazardous waste as defined in 40 CFR 261.3 of this Chapter; and
(3) Meets the definition of tank or tank system in 40 CFR

260.10 of this chapter (40 CFR 260.10-91, *See also* 40 CFR 270.2-91).

Water: A transparent, odorless, tasteless liquid, a compound of hydrogen and oxygen, H$_2$O, freezing at 32° F or 0° C and boiling at 212° F or 100° C, which in more or less impure state, constitutes rain, oceans, lakes, rivers, and other such bodies; it contains 11.188 percent hydrogen and 88.812 percent oxygen, by weight. It may exist as a solid, liquid, or gas and, as normally found in the lithosphere, hydrosphere, and atmosphere, may have other solid, gaseous, or liquid materials in solution or suspension. Other water related terms include:

- Acid mine water
- Barometric leg water
- Black water
- Boiler water
- Bound water
- Brackish water
- Break water
- Bromine water
- Category of water
- Cleaning water
- Coastal water
- Condenser water
- Connate water
- Consumptive use of water
- Contact water
- Contact cooling and heating water
- Cooling water
- Cullet water
- Deionized water
- Demineralized water (*See* synonym, deionized water)
- Dilution water
- Distilled water
- Domestic use of water
- Drainage water
- Drinking water (*See* synonym, potable water)
- Extraction water
- Finishing water
- Free water
- Fresh water
- Gland water
- Gravitational water
- Gray water
- Green water
- Gridding of water
- Ground water
- Hard water
- Hardness of water
- Heavy water (D$_2$O)
- Hypertrophic water
- Infiltration water
- Influent water
- Inland water
- Intake water
- Interstate water
- Interstitial water
- Jacket water
- Light water
- Maceration water
- Makeup water
- Mineral water
- Navigable water
- New water
- Non-consumptive use of water
- Non-process water
- Ocean water
- Oligotrophic water
- Perched water
- Perched water body
- Perched water table
- Pink water
- Pore water
- Potable water
- Process water
- Public water
- Pulp screen water
- Raw water
- Receiving water
- Recycled water
- Red water
- Reused water
- Saline water
- Salt water (*See* synonym, saline water)
- Sea water (*See* synonym, saline water)
- Saline estuarine water
- Sanitary water
- Saturated water
- Soft water
- Sour water
- Specific yield of water
- State regulated water
- Stick water
- Still water
- Storm water
- Stressed water
- Supplier of water
- Surface water
- Sweet water
- Tidal water
- Transport water
- Treated water
- Type II water (*See* synonym, deionized water)
- Underground water
- Vacuum water
- Vadose water
- Wash water
- White water
- Yellow water

Water absorption: The weight of water absorbed by a porous ceramic material, under specified conditions, expressed as a percentage of the weight of the dry material (EPA-83).

Water borne preservative: Any one of several formulations of inorganic salts, the most common which are based on copper, chromium, and arsenic (EPA-74/04).

Water balance: An accounting of all water entering and leaving a unit process or operation in either a liquid or vapor form or via raw material, intermediate product, finished product, by-product, waste product, or via process leaks so that the difference in flow between all entering and leaving streams is zero (EPA-84/08).

Water base ink (paint): An ink that uses water as the prime vehicle ingredient (*See* ink for more related terms) (EPA-75/07).

Water base paint: The paint which uses water as the primary vehicle for all other raw materials. It may contain some semi-drying oils, such as soybean oil for desired drying characteristics (*See* paint for more related terms) (EPA-79/12b).

Water based green tire spray: Any mold release agent and lubricant applied to the inside or outside of green tires that contains 12 percent or less, by weight, of VOC as sprayed (40 CFR 60.541-91).

Water budget: An accounting of the inflow to, outflow from, and storage changes in a hydrologic unit (Course 165.7).

Water (bulk shipment): The bulk transportation of hazardous waste which is loaded or carried on board a vessel without containers or labels (40 CFR 260.10-91).

Water consumption: Based on 1954 survey, the average water consumption on a national basis was 147 gallons per capita per day (M&EI-72).

Water cooled furnace wall: A wall having water tubes for extracting or absorbing heat affording cooling (*See* furnace wall for more related terms) (SW-108ts).

Water dumping: The disposal of pesticides in or on lakes, ponds, rivers, sewers, or other water systems as defined in Pub. L. 92-500 (*See* land disposal for more related terms) (40 CFR 165.1-91).

Water finish: A high gloss finish produced by moistening paper as it passes through the calendar stack (EPA-83).

Water glass: Sodium silicate glass that is soluble in water (*See* glass for more related terms) (EPA-83).

Water hammer: A sudden increase in pressure of water due to an instantaneous conversion of momentum to pressure (EPA-83).

Water jet weaving: The internal subdivision of the low water use processing subcategory for facilities primarily engaged in manufacturing woven greige goods through the water jet weaving process (40 CFR 410.31-91).

Water leaf: Unsized paper (EPA-83).

Water level (or water table): Elevation of the top surface of an unconfined aquifer (DOE-91/04).

Water level-drum: Elevation of the surface of the water in a vessel (EPA-83).

Water demineralizing: Removing mineral components from hard water.

Water pollution: The presence in water of enough harmful or objectionable material to damage the water's quality (EPA-94/04).

Water pollution control agency: Any agency which is defined in section 502(1) or section 502(2), 33USC1362(1) or (2), of the CWA (40 CFR 15.4-91).

Water power: Energy obtained from natural or artificial waterfalls, either directly by turning a water wheel or turbine, or indirectly by generating electricity in a dynamo driven by a turbine (DOI-70/04).

Water purveyor: An agency or person that supplies water (usually potable water) (EPA-94/04).

Water quality: The chemical, physical, and biological characteristics of water with respect to its suitability for a particular purpose. The same water may be of good quality for one purpose or use, and bad for another, depending on its characteristics and the requirements for the particular use (LBL-76/07-water).

Water quality assessment: An evaluation of the condition of a waterbody using biological surveys, chemical-specific analyses of pollutants in waterbodies, and toxicity tests (EPA-91/03).

Water Quality Act of 1987: 33 USC1251 et seq., amendment of the Clean Water Act relative to the conduct of research on the effects of water pollutants.

Water quality-based limitations: Effluent limitations applied to dischargers when mere technology-based limitations would cause violations of water quality standards. Usually applied to discharges into small streams (EPA-94/04).

Water quality-based permit: A permit with an effluent limit more stringent than one based on technology performance. Such limits may be necessary to protect the designated use of receiving waters (i.e., recreation, irrigation, industry or water supply) (EPA-94/04).

Water quality criteria:
(1) Specific levels of water quality which, if reached, are expected to render a body of water suitable for its designated use. The criteria are based on specific levels of pollutants that would make the water harmful if used for drinking, swimming, farming, fish production, or industrial processes (EPA-89/12). Water quality criteria are comprised of numeric and narrative criteria. Numeric criteria are scientifically derived ambient concentrations developed by EPA or States for various pollutants of concern to protect human health and aquatic life. Narrative criteria are statements that describe the desired water quality goal (EPA-91/03).
(2) Levels of water quality expected to render a body of water suitable for its designated use. Criteria are based on specific levels of pollutants that would make the water harmful if used for drinking, swimming, farming, fish production, or industrial processes (EPA-94/04).

Water quality limited: Characterizes a stream segment in which it is known that water does not meet applicable water quality standards, and/or is not expected to meet applicable water quality standards even after application of technology-based effluent limitations (EPA-91/03).

Water quality limited segment: Any segment where it is known that water quality does not meet applicable water quality standards, and/or is not expected to meet applicable water quality standards, even after the application of the technology-based effluent limitations required by sections 301(b) and 306 of the Act (40 CFR 130.2-91, *See also* 40 CFR 131.3-91).

Water quality management (WQM) plan: A State or areawide waste treatment management plan developed and updated in accordance with the provisions of sections 205(j), 208 and 303 of the Act and this regulation (40 CFR 130.2-k-91).

Water quality standard (WQS):
(1) Provisions of State or Federal law which consist of a designated use or uses for the water quality criteria for such waters based upon such uses. Water quality standards are to protect the public health or welfare, enhance the quality of water and serve the purposes of the Act (40 CFR 130.2-91, *See also* 40 CFR 121.1; 125.58; 131.3-91).
(2) A law or regulation that consists of the beneficial designated use or uses of a waterbody, the numeric and narrative water quality criteria that are necessary to protect the use or uses of that particular waterbody, and an antidegradation statement (EPA-91/03).
(3) State-adopted and EPA-approved ambient standards for water bodies. The standards prescribe the use of the water body and establish the water quality criteria that must be met to protect designated uses (EPA-94/04).

Water reducible: *See* definition under waterborne or water reducible.

Water rights: The rights acquired under the law to use the water occurring in surface or ground waters for a specified purpose, in a given manner, and usually within the limits of a given period. While these rights may include the use of a body of water for navigation, fishing, and hunting, other recreational purposes, etc., the term is usually applied to the right to divert or store water for some beneficial purpose or use, such as irrigation, generation of hydroelectric power, or domestic or municipal water supply. In some states, a water right by law becomes appurtenant to the particular tract of land to which the water is applied (EPA-74/01a).

Water seal control: A seal pot, p-leg trap, or other type of trap filled with water that has a design capability to create a water barrier between the sewer and the atmosphere (40 CFR 60.691-91, *See also* 40 CFR 61.341-91).

Water softening: Removal of hardness from water. Calcium and magnesium salts of hardness components can be removed by chemical precipitation.

Water solubility:
(1) The maximum concentration of a substance in pure water at a given temperature. For HRS purposes, use the value reported at or near 25° C. [milligrams per liter (mg/L)] (40 CFR 300-App/A-91, *See also* EPA-89/12).
(2) The maximum possible concentration of a chemical compound dissolved in water. If a substance is water soluble it can very readily disperse through the environment (EPA-94/04).

Water spreader: A method of replenishing groundwater. The design and operation of a spreading system are somewhat like those of an irrigation system, except that water is encouraged to percolate rapidly underground instead of being retained within the root zone of irrigated crops (DOI-70/04).

Water storage pond: An impound for liquid wastes designed to accomplish some degree of biochemical treatment (EPA-94/04).

Water supplier: One who owns or operates a public water system (EPA-94/04).

Water supply system: The collection, treatment, storage, and distribution of potable water from source to consumer (EPA-94/04).

Water table:
(1) The upper water level of a body of groundwater (cf. perched water table) (40 CFR 241.101-91).
(2) The level of groundwater (EPA-94/04).

Water table aquifer: *See* synonym, unconfined aquifer.

Water treatment: Treatment of raw water to make water drinkable. Steps of water treatment include screening, coagulation, flocculation, clarification, filtration and disinfection.

Water treatment lagoon: An impound for liquid wastes designed to accomplish some degree of biochemical treatment (EPA-94/04).

Water tube boiler: *See* definition under boiler tube.

Water use: The activities which are conducted in or on the water; but does not mean or include the establishment of any water quality standard or criteria or the regulation of the discharge or runoff of water pollutants except the standards, criteria, or regulations which are incorporated in any program as required by the provisions of section 1456(f) of this title (CZMA304-16USC1453-90).

Water use factor: The total amount of contact water entering a process divided by the amount of products produced by this process. The amount of water involved includes recycle and make-up water (EPA-83/03a).

Water well: An excavation where the intended use is for location, acquisition, development, or artificial recharge of ground water (excluding sandpoint wells) (EPA-94/04).

Water withdrawal or intake: The volume of fresh water removed from a surface or underground water source (stream, lake, or aquifer) by plant facilities or obtained from some source external to the plant (EPA-74/01a).

Waterborne coating: A coating which contains more than 5 weight percent water in its volatile fraction (40 CFR 60.741-91).

Waterborne disease outbreak:
(1) The significant occurrence of acute infectious illness, epidemiologically associated with the ingestion of water from a public water system which is deficient in treatment, as determined by the appropriate local or State agency (40 CFR 141.2-91).

(2) The significant occurrence of acute infection illness associated with drinking water from a public water system that is deficient in treatment, as determined by the appropriate local or state agency (EPA-94/04).

Waterborne ink systems: The ink and related coating mixtures whose volatile portion consists of a mixture of VOC solvent and more than five weight percent water, as applied to the gravure cylinder (40 CFR 60.431-91).

Waterborne or water reducible: A coating which contains more than five weight percent water in its volatile fraction (40 CFR 60.391-91).

Watercourse: A channel in which a flow of water occurs, either continuously or intermittently and if the latter, with some degree of regularity. The flow must be in a definite direction. Watercourses may be either natural or artificial, and the former may occur either on the surface or underground. A different set of legal principles may apply to rights to use water from different classes of watercourses (EPA-74/01a).

Waterfleas (daphnia): Mostly microscopic swimming crustaceans, often forming a major portion of the zooplankton population. The second antennae are very large and are used for swimming (LBL-76/07-water).

Watershed (or catchment basin):
(1) The land area that drains (contributes runoff) into a stream (EPA-89/12).
(2) The land area that drains into a stream; the watershed for a major river may encompass a number of smaller watersheds that ultimately combine at a common delivery point (EPA-94/04).

Waterwall incinerator:
(1) An incinerator whose furnace walls consist of metal tubes through which water passes and absorbs the energy from burning solid waste (OME-88/12).
(2) An incinerator utilizing lined steel tubes filled with circulating water to cool the combustion chamber. Heat from the combustion gases is transferred to the water. The resultant steam is sold or used to generate electricity (EPA-89/11).
(3) *See* incinerator for more related terms.

Watt: An energy rate of one joule per second, or the power of an electric current of one ampere with an intensity of one volt (EPA-74/03d).

Wave length:
(1) The distance, measured along the line of propagation, between two points that are in phase on adjacent waves. The recommended unit of wavelength in the infrared region of the electromagnetic spectrum is the micrometer. The recommended unit in the ultraviolet and visible region of the electromagnetic spectrum is the nanometer or the Angstrom (LBL-76/07-air).
(2) For a specially oscillatory phenomenon, it is the least distance between points moving in the same phase (NATO-78/10).

Wave number:
(1) The number of waves per unit length. The usual unit of wave number is the reciprocal centimeter, cm^{-1}. In terms of this unit, the wave number is the reciprocal of the wavelength, when wavelength is expressed in centimeters (LBL-76/07-air).
(2) In the space Fourier analysis of a function, the wave number is the reciprocal of the wave length (NATO-78/10).

Waxe: The low molecular weight components of some polyethylene compounds which migrate to the surface over time and must be removed by grinding (for HDPE) or be mixed into the melt zone using thermal seaming methods (EPA-91/05).

Waxed paper: The unsized paper processed through a melted wax bath (*See* paper for more related terms) (EPA-83).

Weak acid: An acid which does not ionize completely in a solution, e.g., acetic acid or carbonic acid (cf. strong acid).

Weak nitric acid: The acid which is 30 to 70 percent in strength (cf. nitric acid production unit) (40 CFR 60.71-91).

Weathering:
(1) The mechanical, chemical, and organic decomposition of rock materials under the influence of climatic factors - water, temperature change, and air (DOI-70/04).
(2) The attack on a surface by atmospheric elements (EPA-83).

Web:
(1) A substrate which is printed in continuous roll-fed presses (40 CFR 52.741-91).
(2) A continuous sheet of paper or paperboard (EPA-83).

Web coating: The coating of products, such as fabric, paper, plastic film, metallic foil, metal coil, cord, and yarn, that are flexible enough to be unrolled from a large roll; and coated as a continuous substrate by methods including, but not limited to, knife coating, roll coating, dip coating, impregnation, rotogravure, and extrusion (40 CFR 60.741-91).

Weed: Any plant which grows where not wanted (FIFRA2-7USC136-91).

Week: For reporting analyses of outdoor air on a week rate, results are calculated to a base of 7 consecutive 24-hour days (EPA-83/06).

Weep: A term usually applied to a minute leak in a boiler joint which forms droplets (or tears) of water very slowly (EPA-83).

Weep hole (or weeper): A hole in a retaining structure to drain off accumulated water that might otherwise induce excessive pressure on the structure (DOI-70/04).

Weibull model: A dose-response model of the form: $P(d) = 1 - \exp[-b(d^{**}m)]$; where $P(d)$ is the probability of cancer due to continuous dose rate d, and b and m are constants (EPA-92/12).

Weight: The resultant force of attraction on the mass of a body due to a gravitational field which is 32.1740 ft/sec^2 on the earth (cf. force) (Markes-67).

Weight fraction: A term employed in expressing concentrations of solutions and mixtures. The weight fraction of any component of a mixture or solution is defined as the weight of that component divided by the total weight of the mixture or solution (EPA-84/09).

Weight of evidence: An EPA classification system for characterizing the evidence supporting the designation of a substance as a human carcinogen. EPA weight of evidence groupings include:
- Group A: Human carcinogen - sufficient evidence of carcinogenicity in humans.
- Group B1: Probable human carcinogen - limited evidence of carcinogenicity in humans.
- Group B2: Probable human carcinogen - sufficient evidence of carcinogenicity in animals.
- Group C: Possible human carcinogen - limited evidence of carcinogenicity in animals.
- Group D: Not classifiable as to human carcinogenicity - applicable when there is no animal evidence, or when human or animal evidence is inadequate.
- Group E: Evidence of noncarcinogencity for humans.

Weight of evidence for carcinogenicity: The extent to which the available biomedical data support the hypothesis that a substance causes cancer in humans (EPA-92/12).

Weir:
(1) A control device placed in a channel or tank which facilitates measurement or control of the water flow (EPA-82/11).
(2) A fence or enclosure set in a waterway for taking fish; a dam in a stream to raise the level of the water or divert its flow; a notch in a barrier across or bordering a stream to regulate the flow of water; a device for determining the quantity of water flowing over it from measurements of the depth of the water over the crest and known dimensions of the device (a cipolleti weir is a trapezoidal device of this sort) (EPA-84/09; DOI-70/04).
(3) A device such as a diversion dam that has a crest and some containment of known geometric shape, such as a V, trapezoid, or rectangle and is used to measure flow of liquid. The liquid surface is exposed to the atmosphere. Flow is related to upstream height of water above the crest, to position of crest with respect to downstream water surface, and to geometry of the weir opening (EPA-82/11e).
(4) (A) A wall or plate placed in an open channel to measure the flow of water (EPA-94/04).
    (B) A wall or obstruction used to control flow from settling tanks and clarifiers to assure a uniform flow rate and avoid short-circuiting (See short circuiting) (EPA-94/04).

Weir basin (or weir trough): A wide approach to the upstream side of an irrigation weir constructed so as to minimize the effect of the momentum of the water flowing over the weir (DOI-70/04).

Weir loading: The maximum flow (m³/day) from a tank divided by the length of outlet weir (m).

Welding: The process of joining two or more pieces of material by applying heat, pressure or both, with or without filler material, to produce a localized union through fusion or recrystallization across the interface (EPA-83/03).

Well: A bored, drilled, or driven shaft, or a dug hole whose depth is greater than the largest surface dimension and whose purpose is to reach underground water supplies or oil, or to store or bury fluids below ground (EPA-94/04). Other well related terms include:
- Abandoned well
- Class II well
- Classification of well
- Classification of injection well
- Disposal well
- Dry well
- Existing class II well
- Existing injection well
- Existing well
- Gas well
- Hot well
- Injection well
- New well
- New class II well
- New injection well

Well field: Area containing one or more wells that produce usable amounts of water (or oil) (EPA-94/04).

Well injection:
(1) The subsurface emplacement of fluids through a bored, drilled, or driven well; or through a dug well, where the depth of the dug well is greater than the largest surface dimension (cf. underground injection or injection well) (40 CFR 144.3-91, *See also* 40 CFR 146.3; 147.2902; 165.1; 260.10-91).
(2) The subsurface emplacement of fluids into a well (EPA-94/04) (EPA-94/04).
(3) cf. deep well injection.

Well monitoring: Measurement, by on-site instruments or laboratory methods, of the quality of water in a well (EPA-94/04).

Well plug: A watertight and gastight seal installed in a bore hole or well to prevent movement of fluids (EPA-94/04).

Well stimulation: The several processes used to clean the well bore, enlarge channels, and increase pore space in the interval to be injected thus making it possible for wastewater to move more readily into the formation, and includes:
(1) Surging;
(2) Jetting;
(3) Blasting;
(4) Acidizing;
(5) Hydraulic fracturing (40 CFR 146.3-91).

Well workover: Any reentry of an injection well; including, but not limited to, the pulling of tubular goods, cementing or casing repairs; and excluding any routine maintenance (eg re-seating the packer at the same depth, or repairs to surface equipment) (40 CFR 147.2902-91).

Wellmann Lord process: One of desulfurization processes. The basic reaction process is $Na_2SO_3 + SO_2 + H_2O ---> 2NaHSO_3$.

Wellhead protection area:
(1) The surface and subsurface area surrounding a water well or wellfield, supplying a public water system, through which contaminants are reasonably likely to move toward and reach such water well or wellfield. The extent of a wellhead protection area, within a State, necessary to provide protection from contaminants which may have any adverse effect on the health of persons is to be determined by the State in the program submitted under subsection (a). Not later than one year after the enactment of the Safe Drinking Water Act Amendments of 1986, the Administrator shall issue technical guidance which States may use in making such determinations. Such guidance may reflect such factors as the radius of influence around a well or wellfield, the depth of drawdown of the water table by such well or wellfield at any given point, the time or rate of travel of various contaminants in various hydrologic conditions, distance from the well or wellfield at any given point, the time or rate of travel of various contaminants in various hydrologic conditions, distance from the well or wellfield, or other factors affecting the likelihood of contaminants reaching the well or wellfield, taking into account available engineering pump tests or comparable data, field reconnaissance, topographic information, and the geology of the formation in which the well or wellfield is located (SDWA1428-42USC300h.7).
(2) A protected surface and subsurface zone surrounding a well or wellfield supplying a public water system to keep contaminants from reaching the well water (EPA-94/04).

Westinghouse/O'Connor combustor: The heart of the system is the water-cooled rotary barrel constructed of alternating longitudinal water tubes and flat perforated steel plates welded together to form the perimeter. The combustor is installed on a slight incline and is slowly rotated by a chain and roller drive. The perforations between the water tubes provide controlled distribution of combustion air, while the water cooled walls remove heat and protect the barrel from overheating. Waste is fed directly from the receiving area into the upper end of the tilted combustor. As the waste tumbles down the length of the rotating barrel, it dries and then progressively burns. Ash dropping out of the lower end is about one tenth of the original waste volume; remaining unburned material is more completely combusted in an afterburner grate. The technology can reportedly handle a variety of wastes including municipal wastes, hazardous wastes and possibly even hospital wastes. It can burn liquids, semi-solids, solids, sewage sludges, and residual oils or refinery bottoms (Lee-88/08).

Wet: Those steelmaking air cleaning systems that primarily use water for furnace gas cleaning (cf. semi-wet) (40 CFR 420.41-91).

Wet air oxidation (WAO): Refers to the aqueous phase oxidation of dissolved or suspended organic substances at elevated temperatures and pressures. Water, which represents the aqueous phase, serves to catalyze the oxidation reactions so that they proceed at relatively low temperatures (175° C to 345° C), and at the same time serves to moderate the oxidation rates removing excess heat by evaporation. Water also provides an excellent heat transfer medium which enables the wet air oxidation process to be thermally self-sustaining with relatively low organic feed concentration. The oxygen required by the wet air oxidation reactions is provided by an oxygen-containing gas, usually air, bubbled through the liquid phase in a reactor used to contain the process; thus the commonly used term "wet air oxidation". The process pressure is maintained at a level high enough to prevent excessive evaporation of the liquid phase, generally between 200 and 3,000 psi. Since oxidation takes place in the liquid state, it is not necessary to evaporate the water content of the waste. The process therefore is most useful for treating the wastes which are too dilute to incinerate economically yet too toxic to treat biologically (Lee-83/07).

Wet air pollution control (or wet air scrubber): The technique of air pollution abatement utilizing water as an absorptive media (EPA-87/10a).

Wet air pollution control scrubber: The air pollution control devices used to remove particulates and fumes from air by entraining the pollutants in a water spray (40 CFR 471.02-91).

Wet barking operation: Shall be defined to include hydraulic barking operations and wet drum barking operations which are those drum barking operations that use substantial quantities of water in either water sprays in the barking drums or in a partial submersion of the drums in a tub of water (40 CFR 430.01-91).

Wet bulb temperature:
(1) The temperature measured by a thermometer whose bulb (mecury holder at the bottom of a thermometer) is covered with a cotton wick which is saturated with water (wet). It is indicated by a wet bulb psychrometer (EPA-89/03b).
(2) The lowest temperature which a water wetted body will attain when exposed to an air current. This is the temperature of adiabatic saturation (EPA-83).
(3) *See* temperature for more related terms.

Wet cap: A mechanical device placed on the top of a furnace stack that forms a curtain from a water stream through which the stack gases must pass (EPA-85/10a).

Wet capacitor (or wet slug capacitor): A sintered tantalum capacitor where the anode is placed in a metal can, filled with an electrolyte and then sealed (*See* variable for more related terms).

Wet charge process: A process for the manufacture of lead acid storage batteries in which the plates are formed by electrolysis in sulfuric acid. The plate forming process is usually done with the plates inside the assembled battery case but may be done with the plates in open tanks. In the case of large industrial wet lead acid batteries, problems in formation associated with inhomogeneities in the large plates are alleviated by open tank formation. Wet charge process batteries are shopped with acid electrolyte inside the battery casing (EPA-84/08).

Wet condenser: *See* synonym, jet condenser.

Wet collection device: A variety of methods to wet the contaminant particles in order to remove them from a gas stream (*See* wet scrubber for more details) (AP-40, p99).

Wet collector: *See* synonym, wet scrubber.

Wet cooling tower: A cooling tower in which hot water is sprayed into an air stream. Heat is thus lost through evaporation.

Wet deposition: The material deposited on the surface due to the combined effect of washout, rainout and snowout (NATO-78/10).

Wet desulfurization system: Those systems which remove sulfur compounds from coke oven gases and produce a contaminated process wastewater (40 CFR 420.11-91).

Wet digestion: A solid waste stabilization process in which solid organic wastes are placed in an open digestion pond to decompose anaerobically. The carbonaceous matter is converted into carbon dioxide and methane. The soluble and suspended fraction is converted aerobically by algae in a bio-oxidation pond (EPA-83).

Wet electrostatic precipitator (WEP): Wet ESP is basically same as the dry ESP with the exception of a continuous water flow over the collecting plate (*See* electrostatic precipitator for more related terms).

Wet filter: One of air pollution control devices. A wet filter consists of a spray chamber with filter pads composed of glass fibers, knitted wire mesh, or other fibrous materials. The dust is collected on the filter pads. The sprays are directed against the pads to keep the dust washed off (*See* scrubber for more related terms) (AP-40).

Wet flue gas desulfurization technology: A sulfur dioxide control system that is located downstream of the steam generating unit and removes sulfur oxides from the combustion gases of the steam generating unit by contacting the combustion gas with an alkaline slurry or solution and forming a liquid material. This definition applies to devices where the aqueous liquid material product of this contact is subsequently converted to other forms. Alkaline reagents used in wet flue gas desulfurization technology include, but are not limited to, lime, limestone, and sodium (40 CFR 60.41b-91, *See also* 40 CFR 60.41c-91).

Wet impingement: The process of impingement carried out within a body of liquid, the latter serving to retain the gases or particulate matter (LBL-76/07-air).

Wet lap machine: A machine used to form pulp into thick rough sheets sufficiently dry to permit handling and folding into bundles (laps) convenient for storage or transportation (EPA-87/10).

Wet line kit: A system used in conjunction with an enclosed transfer trailer to power its unloading bulkhead. The bulkhead's hydraulic pump is driven by a power-take-off unit on the semitractor's transmission (SW-108ts).

Wet lot: A confinement facility for raising ducks which is open to the environment with a small portion of shelter area, and with open water runs and swimming areas to which ducks have free access (40 CFR 412.21-91).

Wet milling (or wet pulping): The mechanical size reduction of solid wastes that have been wetted to soften the paper and cardboard constituents (*See* size reduction machine for more related terms) (SW-108ts).

Wet mixture: A water or organic solvent-based suspension, solution, dispersion, or emulsion used in the manufacture of an instant photographic or peel apart film article (40 CFR 723.175-91).

Wet or low-lying area technique: A method of operating in a swampy area where precautions are made to avoid water pollution before proceeding with the area landfill (*See* sanitary landfill for more related terms) (EPA-83).

Wet oxidation: The direct oxidation of organic matter in wastewater liquids in the presence of air under heat and pressure; generally applied to organic oxidation in sludge (EPA-76/03).

Wet oxygen: The oxygen measured at the wet condition (with water vapor) (*See* oxygen for more related terms).

Wet press: The dewatering unit used on a paper-machine between the sheet-forming equipment and the drier section (EPA-87/10).

Wet process: A cement manufacturing process in which water, typically 30% to 40%, is added to the feed material and then fed into the kiln (*See* process for more related terms) (ETI-92).

Wet process phosphoric acid plant: Any facility manufacturing phosphoric acid by reacting phosphate rock and acid (40 CFR 60.201-91).

Wet pulping: *See* synonym, wet milling.

Wet rendering: The cooking with water or live steam added to the material under pressure. This process produces tank water (*See* rendering for more related terms) (EPA-75/01).

Wet saturated steam: Steam at the saturation temperature corresponding to the pressure and containing water particles in suspension (*See* steam for more related terms).

Wet scrubber (absorber or wet collector): One of air pollution control devices. Wet scrubbers are air pollution control devices used to remove particulates and fumes from air by entraining the pollutants in a water spray (40 CFR 467.02-91). The scrubber removes pollutants including particles (e.g., metals and soots) and acid gases (e.g., HCl and SO_2) from combustion gases or industrial exhaust streams. The contaminated gases are forced to enter a tower (usually from the bottom of the tower) containing packing materials, while liquid is introduced over the packing materials (usually from the top of the tower). The pollutants in the gas stream either dissolve or chemically react with the liquid. The principle of pollutant collection includes (*See also* particle collection mechanisms for wet scrubbing systems):

(1) **Particulate matter:** Collection mechanism is primarily through impaction or electrostatic attraction of particles on wetted surfaces or in liquid droplets.
(2) **Gases:** Gaseous collection is through diffusion (mass transfer) and absorption.
(3) The device uses a variety of methods to wet the contaminant particles and then impinge the wetted or unwetted particles on collecting surfaces followed by their removal from the surfaces by a flush with a liquid. It can handle hot gases and sticky particulates and liquids. Scrubbers, which remove

gases by absorption, remove particulate matter mainly by inertial impaction and are effective for particles larger than 0.5 micron meter in size. Smaller particles require a higher pressure drop. It can work for particles less than 0.1 micron at 40 - 80 water column pressure (EPA-84/03b, p1-4).
(4) See scrubber for more related terms.

Wet scrubber category: Wet scrubber categories include:
(1) Scrubbers for particle collection: These types of scrubbers are usually categorized by the gas-side pressure drop of the system. They are:
   (A) Low energy scrubbers having pressure drops less than 12.7 cm (5 in.) of water
   (B) Medium energy scrubbers having pressure drops between 12.7 and 38.1 cm (5 and 15 in.) of water
   (C) High energy scrubbers having pressure drops greater than 38.1 cm (5 in) of water
(2) Scrubbers for collecting either particles or gaseous pollutants, or both: These types of scrubbers are usually categorized by the energy source used for contact. They are:
   (A) For a gas phase contacting wet collector, the energy source is the gas stream
   (B) For a liquid phase contacting wet collector, the energy source is the liquid stream
   (C) For a wet film wet collector, the energy source is the liquid and gas streams
   (D) For the combination of a liquid phase and gas phase wet collector, the energy source is the liquid and gas streams
   (E) For a mechanically aided wet collector, the energy source is a mechanically driven rotor (EPA-84/03b, p1-9).

Wet scrubber components: Several components are used when designing scrubbers to provide gas-liquid contact and separation. Spray nozzles are used to form droplets that, in turn, are used to capture pollutants. Other components are used to enhance gas-liquid contact. These include venturi throats, plates, baffles, packing, orifices, tangential openings, and mechanically driven rotors (EPA-84/03b, p2-1).

Wet scrubber design: Wet scrubbers are uniquely designed to enhance the collection of air pollutants. Variables affecting particulate pollutant collection include particle size, particle velocity, and liquid-droplet size. For gaseous pollutant collection, the pollutant must be soluble in the chosen scrubbing liquid. In addition, the system must be designed to provide good mixing between the gas and liquid phases, and enough time for the gaseous pollutants to dissolve. Another consideration for both particulate and gaseous pollutant collection is the liquid-to-gas (L/G) ratio - the amount of liquid injected into the scrubber per given volume of exhaust flow. Lastly, the system must be designed to remove entrained mists, or droplets, from the cleaned exhaust gas stream before it leaves the stack (EPA-84/03b, p2-1).

Wet scrubber system: Any emission control device that mixes an aqueous stream or slurry with the exhaust gases from a steam generating unit to control emissions of particulate matter or sulfur dioxide (40 CFR 60.41b; 60.41c-91).

Wet shelf life: The period of time that a secondary battery can stand in the charged condition before total degradation (EPA-84/08).

Wet simultaneous NO_x and SO_2 reduction: One of NO_x emission reduction techniques (See nitrogen oxide emission control for control structure). Wet processes for the simultaneous reduction of NO_x and SO_2 emissions have been developed and installed on large oil fired boilers in Japan. These processes use absorbers to reduce both SO_x and NO_x emissions simultaneously (EPA-81/12, p7-24).

Wet slug capacitor: See synonym, wet capacitor.

Wet strength:
(1) The strength of paper after complete saturation with water (EPA-87/10).
(2) Wet strength paper has high resistance to rupture or disintegration when saturated with water, Produced by chemical (resin) treatment of the paper or the fibers with melamine formaldehyde; urea formaldehyde or alkaline or neutral wet strength - nylons or polyamide material (EPA-83).

Wet strength additives: Chemicals such as urea and melanine formaldehydes used in papermaking to impart strength to papers used in wet applications (EPA-87/10).

Wet tantalum capacitor: A polar capacitor whose cathode is a liquid electrolyte (a highly ionized acid or salt solution) (EPA-83/03).

Wet transformer: Having the core and coils immersed in an insulating oil (EPA-83/03).

Wetland:
(1) Those areas that are inundated or saturated by surface or ground water at a frequency and duration sufficient to support, and that under normal circumstances do support, a prevalence of vegetation typically adapted for life in saturated soil conditions. Wetlands generally include swamps, marshes, bogs and similar areas (40 CFR 230.3-91, See also 40 CFR 232.2; 6-App/A; 110.1; 122.2; 257.3.3; 435.41-91).
(2) Land or areas exhibiting hydric soil conditions, saturated or inundated soil during some portion of the year, and plant species tolerant of such conditions (DOE-91/04).
(3) An area that is saturated by surface or ground water with vegetation adapted for life under those soil conditions, as swamps, bogs, fens, marshes, and estuaries (EPA-94/04).

Wetness: The percentage of water in steam; the presence of a water film on heating surface interiors (EPA-83).

Wetting agent: See synonym, wetting compound.

Wetting compound (or wetting agent): A substance that reduces the surface tension of a liquid, thereby causing it to spread more readily on a solid surface (EPA-83).

Whey: A by-product in the manufacture of cheese which remains after separating the cheese curd from the rest of the milk used in the process (EPA-74/05).

White liquor: The liquor made by causticizing green liquor (*See* liquor for more related terms) (EPA-87/10).

White goods: Large, metal household appliances (e.g., stoves, dryers, refrigerators, etc.) (OTA-89/10).

White paper:
(1) Printers term of unprinted paper, even if colored.
(2) Water removed from pulp during the thickening operations (EPA-83).
(3) *See* paper for more related terms.

White water: A general term for all papermill waters which have been separated from the stock or pulp suspension, either on the paper-machine or accessory equipment, such as thickeners, washers, and savealls, and also from pulp grinders (*See* water for more related terms) (EPA-87/10).

Whole effluent toxicity:
(1) The aggregate toxic effect of an effluent measured directly by a toxicity test (40 CFR 122.2-91, *See also* EPA-85/09).
(2) The total toxic effect of an effluent measured directly with a toxicity test (EPA-91/03).

Whole milk: Market milk whose fat content has been standardized to conform to a regulatory definition, typically 3.5% (EPA-74/05).

Whiteness: *See* synonym, brightness (EPA-83).

Wicking: The phenomenon of liquid transmission within the fabric yarns of reinforced geomembranes via capillary action (EPA-91/05).

Wilderness: As used in section 1782 of this title shall have the same meaning as it does in section 1131(c) of title 16 (FLPMA103-43USC1702-90).

Wildlife habitat: The waters and surrounding land areas of the Reservation used by fish, other aquatic life and wildlife at any stage of their life history or activity (40 CFR 131.35-91).

Wildlife refuge: An area designated for the protection of wild animals, within which hunting and fishing are either prohibited or strictly controlled (EPA-94/04).

Wilfley table: A plane rectangle is mounted horizontally and can be slopped about its long axis. Gentle and rapid throwing motion is used on the table longitudinally. Sands usually classified for size range are fed continuously and worked along the table with the aid of feedwater, and across riffles downslope by gravity tilt adjustment, and added washwater. At the discharge end, the sands have separated into bands, the heaviest and smallest uppermost, the lightest and largest lowest (EPA-82/05).

Wind: The motion of the air relative to the earth's surface. Usually applied only to the horizontal component of this motion (NATO-78/10). Other wind related terms include:
- Backing wind
- Cross wind
- Geostrophic wind
- Gradient wind
- Logarithmic wind profile
- Mountain valley
- Power law wind profile
- Variable wind
- Veering wind
- Vertical wind

Wind case: The combination of meteorological conditions, topography and source characteristic for which locally the highest air pollutant concentrations for a given averaging time occur (NATO-78/10).

Wind direction: The direction from which the wind averaged over a certain period of time is blowing (NATO-78/10).

Wind direction sector: The 360 degrees of the compass are divided into 16 equally sized 22.5 degree sectors. The wind direction at any given time is within one of these sectors. The wind sectors are used to define wind direction in order to facilitate compiling meteorological data summaries and to decrease computation time (EPA-88/09).

Wind energy: Energy that is associated with wind and can be used for generating electricity.

Wind field: The description of the three dimensional wind speed and wind direction distribution in a certain region for each time (NATO-78/10).

Wind fluctuation type: A definition of diffusion categories by the width and the appearance of the trace drawn by a continuously recording wind vane. In this way, a direct connection of the diffusion categories with the turbulent wind direction fluctuation is obtained (NATO-78/10).

Wind profile: The description of the wind speed and direction as a function of height (NATO-78/10).

Wind rose: A diagram showing the distribution of wind directions at a certain location for a given period of time (NATO-78/10).

Wind speed class: The combination of several wind speeds in one class. Applied for example in the Gaussian plume model for long term averages to describe the influence of the wind speed on the dispersion during the calculation period (NATO-78/10).

Wind vane: An instrument used to measure the wind direction (NATO-78/10).

Windbox: A chamber below a furnace grate or surrounding a burner, through which air is supplied under pressure to burn the fuel (SW-108ts).

Windbox burner: A plenum chamber around a burner in which an air pressure is maintained to insure proper distribution and discharge of secondary air (*See* burner for more related terms) (AP-40).

Windbox pressure: The static pressure of the air in the windbox of a burner or stoker (EPA-83).

Winding: Winding related terms include:
- Primary winding
- Secondary winding

Window glass: *See* definition under sheet glass.

Windrow: A large, elongated pile of composting materials (EPA-89/11).

Windrow composting: An open air method in which compostable material is placed in windrows, piles or ventilated bins or pits and occasionally turned or mixed. The process may be anaerobic or aerobic (*See* composting for more related terms) (SW-108ts).

Wire: The endless belt or screen on which pulp is formed into paper (EPA-83).

Wire and coil rod: Those acid pickling operations that pickle rod, wire or coiled rod and wire products (40 CFR 420.91-91).

Wire glass: *See* definition under safety glass.

Wire mesh eliminator: A mist eliminator used to collect other mists (*See* mist eliminator for more related terms) (EPA-81/09).

Wire product and fastener: The steel wire, products manufactured from steel wire, and steel fasteners manufactured from steel wire or other steel shapes (40 CFR 420.121-91).

Wire-to-wire efficiency: The efficiency of a pump and motor together (EPA-94/04).

Wire safety glass: A single piece of glass with a layer of meshed wire completely imbedded in the glass, but not necessarily in the center of the sheet. When the glass is broken, the wire mesh holds the pieces together to a considerable extent (*See* safety glass for more related terms) (EPA-83).

With modified processes: Using any technique designed to minimize emissions without the use of add on pollution controls (40 CFR 60.291-91).

Withdraw specification: To remove from designation any area already specified as a disposal site by the U.S. Army Corps of Engineers or by a state which has assumed the section 404 program, or any portion of such area (*See* specification for more related terms) (40 CFR 231.2-91).

Withdrawal: Withholding an area of Federal land from settlement, sale, location, or entry, under some or all of the general land laws, for the purpose of limiting activities under those laws in order to maintain other public values in the area or reserving the area for a particular public purpose or program; or transferring jurisdiction over an area of Federal land, other than "property" governed by the Federal Property and Administrative Services Act, as amended (40USC472) from one department, bureau or agency to another department, bureau or agency (FLPMA103-43USC1702).

Within the impoundment: For all impoundments the term "within the impoundment" for purposes of calculating the volume of process wastewater which may be discharged, shall mean the surface area within the impoundment at the maximum capacity plus the area of the inside and outside slopes of the impoundment dam and the surface area between the outside edge of the impoundment dam and Seepage ditches upon which rain falls and is returned to the impoundment. For the purpose of such calculations, the surface area allowance for external appurtenances to the impoundment shall not be more than 30 percent of the water surface area within the impoundment dam at maximum capacity (40 CFR 421.11-d-91).

Wolfram: *See* synonym, tungsten.

Wood: Wood, wood residue, bark, or any derivative fuel or residue thereof, in any form, including, but not limited to, sawdust, sanderdust, wood chips, scraps, slabs, millings, shavings, and processed pellets made from wood or other forest residues (40 CFR 60.41b-91, *See also* 40 CFR 60.41c-91).

Wood burning stove pollution: Air pollution caused by emissions of particulate matter, carbon monoxide, total suspended particulates, and polycyclic organic matter from wood-burning stoves (EPA-94/04).

Wood extractives: A mixture of chemical compounds, primarily organics, removed from wood (EPA-74/04).

Wood fiber: Elongated, thick walled cells of wood, commonly called fiber (*See* fiber for more related terms) (EPA-83).

Wood fiber furnish subdivision mills: Those mills where cotton fibers are not used in the production of fine papers (cf. cotton fiber furnish subdivision mills) (40 CFR 430.181-91).

Wood flour: Finely ground wood or fine sawdust used chiefly as a filler (EPA-87/10).

Wood furniture coating facility: A facility that includes one or more wood furniture coating line(s) (40 CFR 52.741-91).

Wood furniture coating line: A coating line in which any protective, decorative, or functional coating is applied onto wood furniture (40 CFR 52.741-91).

Wood heater: An enclosed, wood-burning appliance capable of and intended for space heating or domestic water heating, as defined in the applicable regulation (40 CFR 60-App/A(method 28 & 28A)-91).

Wood residue: The bark, sawdust, slabs, chips, shavings, mill trim, and other wood products derived from wood processing and forest management operations (40 CFR 60.41-91).

Wood preparation: A series of operations utilized to prepare wood to a suitable state for further development into pulp, paper, and paperboard. These operations include barking, washing, and chipping (EPA-87/10).

Wood preservative: A chemical or mixture of chemicals with fungistatic and insecticidal properties that is injected into wood to protect it from biological deterioration (EPA-74/04).

Wood pulp: A fibrous raw material derived from wood for use in

most types of paper manufactured by mechanical or chemical means both from hardwood and softwood trees. Classification of wood pulp is as follows:
(1) Mechanical wood pulp includes:
    (A) Groundwood;
    (B) Defibrated/exploded;
    (C) Screenings.
(2) Chemical wood pulp includes:
    (A) Sulfite;
    (B) Neutral Sulfite;
    (C) Sulfate (Kraft);
    (D) Chemical cellulose;
    (E) Soda;
    (F) Semichemical (EPA-83).
(3) *See* pulp for more related terms.

Wood pulp waste: The wood or paper fiber residue resulting from a manufacturing process (*See* waste for more related terms) (SW-108ts).

Wood treatment facility: An industrial facility that treats lumber and other wood products for outdoor use. The process employs chromated copper arsenate, which is regulated as a hazardous material (EPA-94/04).

Woodroom: The area of a pulp mill that handles the barking, washing, chipping or grinding of logs, and processing of purchased chips.

Woodworking: The shaping, sawing, grinding, smoothing, polishing, and making into products of any form or shape of wood (40 CFR 52.741-91).

Woodyard: The area of a mill where roundwood is received and stored prior to transport to the woodroom (EPA-87/10).

Wool:
(1) The dry raw wool as it is received by the wool scouring mill (40 CFR 410.11-91).
(2) Fleecy mass of plair glass fibers (EPA-83).

Wool fiberglass: The fibrous glass of random texture, including fiberglass insulation, and other products listed in SIC 3296 (40 CFR 60.291-91).

Wool fiberglass insulation: A thermal insulation material composed of glass fibers and made from glass produced or melted at the same facility where the manufacturing line is located (40 CFR 60.681-91).

Work (ft-lbf or m-kg units): Work can be defined as:
(1) Work = force x distance or
(2) Work = pressure x volume change

This definition shows that work is an interaction between a system and its surroundings which is caused by a force displacing the boundary between the system and the surroundings. Both heat and work are path functions and thus, to evaluate their magnitude, the entire process must be considered. Typical work units include: ft-lbf or m-kg; where:
- ft = feet
- lbf = pound force
- m = meter
- kg = kilo gram

The conventional signs of work are as follows:
- Work done on the system is negative
- Work done by the system is positive

Work practice controls: Controls that reduce the likelihood of exposure by altering the manner in which a task is performed (29 CFR 1910).

Working charge (or working capacity): A term employed in adsorber calculations. It refers to the net amount of pollutant adsorbed in a cycle. It usually includes the mass transfer zone (MTZ) and heel effects, and is expressed as lb/lb adsorbent or lb/100 lb adsorbent (EPA-84/09).

Working electrode: An electrode in an electrochemical cell which is used for corrosion testing (*See* electrode for more related terms).

Working face: That portion of the compacted solid wastes at a sanitary landfill which will have more refuse placed upon it or is being compacted prior to placement of cover material (EPA-83).

Working level (WL):
(1) Any combination of short-lived radon decay products in one liter of air that will result in the ultimate emission of alpha particles with a total energy of 130 billion electron volts (40 CFR 192.11-91).
(2) A unit of measure for documenting exposure to radon decay products, the so-called "daughters." One working level is equal to approximately 200 picocuries per liter (EPA-94/04).

Working level month (WLM): A unit of measure used to determine cumulative exposure to radon (EPA-94/04).

Working range: The range of surface temperature in which glass is formed into ware in a specific process. The upper end refers to the temperature at which the glass is ready for working, while the lower end refers to the temperature at which it is sufficiently viscous to hold its formed shape (EPA-83).

Workplace: An establishment at one geographic location containing one or more work areas (40 CFR 721.3-91).

Worst case discharge:
(1) In the case of a vessel, a discharge in adverse weather conditions of its entire cargo; and
(2) In the case of an offshore facility or onshore facility, the largest foreSeeable discharge in adverse weather conditions (CWA311-33USC1321-91).

Writing paper:
(1) A paper suitable for pen and ink, pencil, typewriter or printing (40 CFR 250.4-91).
(2) Wide variety of papers suitable for writing or printing (EPA-83).
(3) *See* paper for more related terms.

Xx Yy Zz

X-ray: Electromagnetic radiation with a wavelength of approximately 1 Angstrom (EPA-88/09a).

X-ray diffraction: The scattering (diffraction) of x-rays by a crystal.

Xanthate: A common specific promoter used in flotation of sulfide ores. A salt or ester of xanthic acid which is made of an alcohol, carbon disulfite and an alkalai (EPA-82/05).

Xenobiote: Any biotum displaced from its normal habitat; a chemical foreign to a biological system (EPA-94/04).

Xenobiotic: A term for non-naturally occurring man-made substances found in the environment (i.e., synthetic material solvents, plastics) (EPA-89/12).

Xenon (Xe): A noble gaseous element with a.n. 54; a.w. 131.30; d. 3.06 g/cc; m.p. -111.9° C and b.p. -108.0° C. The element belongs to group VIIIA of the periodic table.

Xerographic/copy paper: Any grade of paper suitable for copying by the xerographic process (a dry method of reproduction) (*See* paper for more related terms).

Xenotrophic viruses: Viruses that are able to infect more than one strain or one species of organisms (i.e., a virus that can infect many strains of mice as well as (or) other rodents) (EPA-88/09a).

Yard rubbish: The prunings, brush, grass, clippings, weeds, leaves and general and garden wastes. *See* Solid Waste (*See* rubbish for more related terms) (EPA-83).

Yard tractor: A small semi-tractor used exclusively for maneuvering transfer trailers into and out of loading position (SW-108ts).

Yard waste:
(1) The plant clippings, prunings, and other discarded materials from yards and gardens. Also known as yard rubbish (SW-108ts).
(2) The part of solid waste composed of grass clippings, leaves, twigs, branches, and garden refuse (EPA-94/04).
(3) *See* waste for more related terms.

Yellow-boy: Iron oxide flocculent (clumps of solids in waste or water); usually observed as orange-yellow deposits in surface streams with excess iron content (*See* floc, flocculation) (EPA-94/04).

Yellow brass: An alloy 70% copper, 30% zinc, also known as 70/30 brass. Used for cartridge cases, condenser tubes, etc. (*See* brass for more related terms) (EPA-83).

Yellow cake:
(1) A term applied to certain uranium concentrates produced by mills. It is the final precipitate formed in the milling process. It is usually considered to be ammonium diuranate, or sodium diuranate, but the composition is variable and depends upon the precipitating conditions.
(2) A common form of triuranium octoxide (U_3O_8) is yellow cake which is the powder obtained by evaporating an ammonia solution of the oxide (EPA-82/05).

Yellow flame: *See* definition under flame combustion.

Yellow water: The effluent coming from the first wash of crude TNT in its purification process (*See* water for more related terms) (EPA-76/03).

Yellowing: Gradual change in paper to yellow due to aging (EPA-83).

Yield:
(1) Useful energy output per unit weight of fuel (EPA-83).
(2) The quantity of water (expressed as a rate of flows or total quantity per year) that can be collected for a given use from surface or groundwater sources (EPA-94/04).

Yield point: The stress at which the material exhibits a permanent set of 0.2 percent (40 CFR 1910.66-App/D-91).

Ytterbium (Yb): A rare earth metal with a.n. 70; a.w. 173.04; d. 6.98 g/cc; m.p. 824° C and b.p. 1427° C. The element belongs to group IIIB of the periodic table.

Yttrium (Y): A soft metal with a.n. 39; a.w. 88.905; d. 4.47 g/cc; m.p. 1509° C and b.p. 2927° C. The element belongs to group IIIB of the periodic table.

Z-list: OSHA's tables of toxic and hazardous air contaminants (EPA-94/04).

Zeolite: Various natural or synthesized silicates used in water softening and as absorbents (EPA-76/03). It is a type of ion exchange resins.

Zeolite catalyst: A catalyst with controlled porosity used as a catalytic cracking catalyst in petroleum refineries and other chemical reactions. The catalyst is hydrated aluminum (or sodium) and calcium silicate such as ($CaO \cdot 2Al_2O_3 \cdot 5SiO_2$ OR $Na_2O \cdot 2Al_2O_3 \cdot 5SiO_2$).

Zeolite process: A process used to reduce hardness by ion exchange. Either a natural mineral or a man-made product, zeolite is a hydrated sodium silicate. The sodium can be used to exchange for calcium or magnesium to soften hard water.

Zeolity filter: A zeolite bed used in the water softening process.

Zero, low-level, and high-level values: The CEMS response values related to the source specific span value. Determination of zero, low-level, and high-level values is defined in the appropriate PS in Appendix B of this part (40 CFR 60-App/F-91).

Zero device miles: The period of time between retrofit installation and the accumulation of 100 miles of automobile operation after installation (40 CFR 610.11-91).

Zero discharge: The prevention of process wastewater from point sources entering navigable waters either directly or indirectly through publicly owned treatment works (EPA-85/10).

Zero drift:
(1) The change in response to zero pollutant concentration, over 12- and 24-hour periods of continuous unadjusted operation (40 CFR 53.23-91, See also 40 CFR 60-App/A(method 25A); 60-App/A(method 6C & 7E); 60-App/B-91).
(2) The change with time in instrument output over a stated time period of unadjusted continuous operation when the input concentration is zero. (expressed as percent of full scale) (LBL-76/07-bio).
(3) See calibration for more related terms.

Zero hour: That point after normal assembly line operations and adjustments are completed and before ten (10) additional operating hours have been accumulated, including emission testing, if performed (40 CFR 86.082.2-91).

Zero gas: For calibration gas application, a high purity air with less than 0.1 ppm by volume of organic material methane or carbon equivalent or less than 0.1% of the span value, whichever is greater (See calibration gas for more related terms) (EPA-90/04).

Zero kilometer: That point after normal assembly line operations and adjustments, after normal dealer setup and preride inspection operations have been completed, and before 100 kilometers of vehicle operation or three hours of engine operation have been accumulated, including emission testing if performed (40 CFR 86.402.78-91).

Zero mile: That point after initial engine starting (not to exceed 100 miles of vehicle operation, or three hours of engine operation) at which normal assembly line operations and adjustments are completed, and including emission testing, if performed (40 CFR 86.082.2-91).

Zero order reaction: A chemical reaction whose rate is independent of reactant concentrations (e.g., a photochemical reaction is dependent on the intensity of light not the concentrations of the reactants).

Zero or alternate discharge: Methods of wastewater discharge from point sources which do not involve discharge to navigable waters either directly or indirectly through publicly owned treatment works. Zero or alternate discharge methods include wastewater reuse, evaporation, and off-site privately owned treatment (EPA-87/10a).

Zero plane displacement: A parameter in the logarithmic wind profile used to indicate the effective height of the surface. This parameter extends the applicability of the logarithmic profile to very rough surroundings, e.g., forest (NATO-78/10).

Zero pressure: See synonym, vacuum.

Zeroth round: Laboratory studies of a method of measuring a given characteristic of a material in preparation of running a round robin series of laboratory tests (EPA-83).

Zinc (Zn):
(1) The total zinc present in the process wastewater stream exiting the wastewater treatment system (40 CFR 415.451-91, See also 40 CFR 415.631; 415.671; 420.02-91).
(2) A hard metallic element with a.n. 30; a.w. 65.38; d. 7.14 g/cc; m.p. 419.5° C and b.p. 906° C. The element belongs to group IIB of the periodic table. Major zinc compounds include:
- Zinc arsenate ($ZnHAsO_4$): Poisonous powder used an insecticide.
- Zinc arsenite ($Zn(AsO_2)_2$): Poisonous powder used an insecticide and wood preservative.
- Zinc oxide (ZnO): White powder used in cosmetics, driers, quick-setting cements.
- Zinc sulfate ($ZnSO_4 \cdot 7H_2O$): Colorless crystals used in wood preserving, and paper bleaching.
- Zinc sulfide (ZnS): Yellow powder used in pigments an television screen.

Zinc casting: The remelting of zinc or zinc alloy to form a cast intermediate or final product by pouring or forcing the molten metal into a mold, except for ingots, pigs, or other cast shapes related to nonferrous (primary) metals manufacturing (40 CFR 421) and nonferrous metals forming (40 CFR 471). Processing operations following the cooling of castings not covered under nonferrous metals forming are covered under the electroplating and metal finishing point source categories (40 CFR 413 and 433) (40 CFR 464.02-91).

Zinc minerals: The main source of zinc is sphalerite (ZnS), but some smithsonite, hemimorphite, zincite, willemite, and franklinite are mined (EPA-82/05).

Zircaloy-4: An alloy of zirconium metal frequently used in nuclear reactors because of its desirable chemical and nuclear properties

(DOE-91/04).

Zircon: A mineral (ZrSiO4), the chief ore of zirconium (EPA-82/05).

Zincon, rutile, ilmenite, monazite: A group of heavy minerals which are usually considered together because of their occurrence as black sand in a natural beach (EPA-82/05).

Zirconium (Zr): A transition metal with a.n. 40; a.w. 91.22; d. 6.49 g/cc; m.p. 1852° C and b.p. 3580° C. The element belongs to group IVB of the periodic table.

Zone control: The control of air flow into individual zones undergrate of a stoker, or plenums of a burner system (EPA-83).

Zone of aeration:
(1) The comparatively dry soil or rock located between the ground surface and the top of the water table (EPA-94/04).
(2) *See* synonym, unsaturated zone.

Zone of capillarity: The area above a water table where some or all of the interstices (pores) are filled with water that is held by capillarity (cf. capillarity water) (SW-108ts).

Zone of engineering control: An area under the control of the owner/operator that, upon detection of a hazardous waste release, can be readily cleaned up prior to the release of hazardous waste or hazardous constituents to ground waster or surface water (40 CFR 260.10-91).

Zone of initial dilution (ZID): The region of initial mixing surrounding or adjacent to the end of the outfall pipe or diffuser ports, provided that the ZID may not be larger than allowed by mixing zone restrictions in applicable water quality standards (40 CFR 125.58-91).

Zone of saturation: *See* definition under saturated zone or zone of saturation (EPA-94/04).

Zone refining: A fractional crystallization technique in which a rod of impure materials is purified by heating it so as to cause a molten zone to pass along its length. Thus, in zone refining, only a part of the material being purified is melted at any one time. Impurities tend to be carried forward in the molten zone. By repeating the process on the rod a number of times, a high degree of purification is possible. Basic equipment consists of a material support or ingot holder to contain the sample; a feed or travel mechanism; and a source of heat.

Zone sedimentation: The zone with settled solids in a sedimentation tank.

Zooglea: Bacteria embedded in a jellylike matrix formed as the result of metabolic activities (LBL-76/07-water).

Zoogleal film: A jelly-like matrix developed by bacteria formed in treatment devices (EPA-75/10).

Zooplankton:
(1) Protozoa and other animal microorganisms living unattached in water. These include small crustacea, such as daphnia and cyclops (LBL-76/07-water).
(2) Tiny aquatic animals eaten by fish (EPA-94/04).

Zooxanthllae: Very small yellow-green algae.

Appendix: Environmental Acronyms

4-AAP: 4-Aminoantipyrine (40 CFR 420.02)
A: Ampere (40 CFR 60)
A: Argon
A/B or AB: Afterburner (40 CFR 60.471)
A/WPR: Air/Water Pollution Report (EPA-94/04)
A&C: Abatement and Control (EPA-94/04)
A&I: Alternative and Innovative (Wastewater Treatment System) (EPA-94/04)
A&R or AR: Air and Radiation (EPA-94/04)
AA: Accountable Area (EPA-94/04)
AA: Adverse Action (EPA-94/04)
AA: Advices of Allowance (EPA-94/04)
AA: Associate Administrator (EPA-94/04)
AA: Attainment Area
AA: Atomic Absorption (EPA-94/04)
AAA: American Automobile Association
AAAS: American Association for the Advancement of Science
AADI: Adjusted Acceptable Daily Intake (EPA-92/12)
AADT: Annual Average Daily Traffic (DOE-91/4)
AAEE: American Academy of Environmental Engineers (EPA-94/04)
AALACS: Ambient Aquatic Life Advisory Concentrations (40 CFR 300)
AAMA: American Academy of Medical Administration
AANWR: Alaskan Arctic National Wildlife Refuge (EPA-94/04)
AAOHN: American Association of Occupational Health Nurses
AAP: Acoustical Assurance Period (40 CFR 205.151)
AAP: Affirmative Action Plan
AAP: Affirmative Action Program
AAP: Asbestos Action Program (EPA-94/04)
AAPCO: American Association of Pesticide Control Officials (EPA-94/04)
AAQSS: Ambient Air Quality Standards (DOE-91/4)
AAR: Applications Analysis Report
AARC: Alliance for Acid Rain Control (EPA-94/04)
AARP: American Association of Retired Persons
AAS: Atomic Absorption Spectrophotometry (40 CFR 266-A9)
AATCC: American Association of Textile Chemists and Colorists
ABA: American Bar Association
ABAG: Association of Bay Area Governments
ABEL: EPA's computer model for analyzing a violator's ability to pay a civil penalty (EPA-94/04)
ABES: Alliance for Balanced Environmental Solutions (EPA-94/04)

ABMA: American Boiler Manufacturers Association
ABS: Alkyl Benzene Sulfonate
Ac: Actinium
AC: Actual Commitment (EPA-94/04)
AC: Advisory Circular (EPA-94/04)
AC: Alternating Current
ACA: American Conservation Association (EPA-94/04)
ACBM: Asbestos Containing Building Material (40 CFR 763.83)
ACE: Alliance for Clean Energy (EPA-94/04)
ACEC: American Consulting Engineers Council
ACEEE: American Council for an Energy Efficient Economy (EPA-94/04)
ACFM: Actual Cubic Feet per Minute (EPA-94/04)
ACGIH: American Conference of Governmental Industrial Hygienists (29 CFR 1910.1200)
AChE: Acetylcholinesterase (EPA-92/12)
ACHP: Advisory Council on Historic Preservation
ACI: American Concrete Institute
ACL: Alternate Concentration Limit (EPA-94/04)
ACL: Analytical Chemistry Laboratory (EPA-94/04)
ACM: Asbestos-Containing Material (40 CFR 763.83)
ACP: Agriculture Control Program (Water Quality Management) (EPA-94/04)
ACP: Air Carcinogen Policy (EPA-94/04)
ACQUIRE: Aquatic Information Retrieval (EPA-94/04)
ACR: Acute-to-Chronic Ratio (EPA-91/3)
ACS: Acid Scrubber
ACS: American Chemical Society (EPA-94/04)
ACSH: American Council on Science and Health
ACT: Action (EPA-94/04)
ACTS: Asbestos Contractor Tracking System (EPA-94/04)
ACUS: Administrative Conference of United States
ACWA: American Clean Water Association (EPA-94/04)
ACWM: Asbestos-Containing Waste Material (EPA-94/04)
ADABA: Acceptable Data Base (EPA-94/04)
ADB: Applications Data Base (EPA-94/04)
ADI: Acceptable Daily Intake (EPA-94/04)
ADP: Adenosine Diphosphate
ADP: AHERA Designated Person (EPA-94/04)
ADP: Automated Data Processing (EPA-94/04)
ADQ: Audits of Data Quality (EPA-94/04)
ADR: Alternate Dispute Resolution (EPA-94/04)
ADSS: Air Data Screening System (EPA-94/04)
ADT: Average Daily Traffic (EPA-94/04)

AEA: Atomic Energy Act (40 CFR 23.9)
AEC: Associate Enforcement Counsels (EPA-94/04)
AEC: Atomic Energy Commission (DOE-91/4)
AECD: Auxiliary emission control device (40 CFR 89.3-94)
AEE: Alliance for Environmental Education (EPA-94/04)
AEERL: Air and Energy Engineering Research Laboratory (EPA-94/04)
AEM: Acoustic Emission Monitoring (EPA-94/04)
AER: Advanced Electrical Reactor
AERE: Association of Environmental and Resource Economists (EPA-94/04)
AEROS: Aerometric and Emissions Reporting System (40 CFR 51.100-l)
AES: Auger Electron Spectrometry (EPA-94/04)
AFA: American Forestry Association (EPA-94/04)
AFCA: Area Fuel Consumption Allocation (EPA-94/04)
AFCEE: Air Force Center for Environmental Excellence (EPA-94/04)
AFRCE: Air Force Regional Civil Engineers
AFS: AIRS Facility Subsystem (EPA-94/04)
AFUG: AIRS Facility Users Group (EPA-94/04)
Ag: Silver
AGC: Associate General Counsels
AGMA: American Gear Manufacturer's Association
AH: Allowance Holders (EPA-94/04)
AHA: American Hospital Association
AHERA: Asbestos Hazard Emergency Response Act of 1986 (40 CFR 22.41)
AHFS: American Hospital Formulary Service (EPA-92/12)
a.i.: Active Ingredient (EPA-92/12)
AI: Artificial Intelligence
AIA: American Institute of Architects
AIA: Asbestos Information Association
AIADA: American International Automobile Dealers Association
AIC: Active to Inert Conversion (EPA-94/04)
AICC: Annual Installed Capital Cost
AICE: American Institute of Civil Engineers
AIChE: American Institute of Chemical Engineers
AICUZ: Air Installation Compatible Use Zones (EPA-94/04)
AID: Agency for International Development (EPA-94/04)
AIDS: Acquired Immune Deficiency Syndrome
AIF: Atomic Industrial Forum
AIG: Assistant Inspector General
AIHA: American Industrial Health Association (EPA-92/12)
AIHC: American Industrial Health Council (EPA-94/04)
AIME: American Institute of Metallurgical, Mining and Petroleum Engineers
AIP: Auto Ignition Point (EPA-94/04)
AIRFA: American Indian Religious Freedom Act (DOE-91/4)
AIRS: Aerometric Information Retrieval System (EPA-94/04)
AISC: American Institute of Steel Construction
AISI: American Iron and Steel Institute
AK: Alaska
Al: Aluminum
AL: Acceptable Level (EPA-94/04)
AL: Administrative Leave
AL: Alabama
AL: Annual Leave
ALA: American League of Anglers
ALA: American Lung Association
ALA: Delta-Aminolevulinic Acid (EPA-94/04)

ALA-O: Delta-Aminolevulinic Acid Dehydrates (EPA-94/04)
ALAPO: Association of Local Air Pollution Control Officers (EPA-94/04)
ALARA: As Low as Reasonably Achievable (EPA-94/04)
ALC: Application Limiting Constituent (EPA-94/04)
ALE: Arid Land Ecology (DOE-91/4)
ALEC: American Legislative Exchange Council
ALJ: Administrative Law Judge (EPA-94/04)
ALMS: Atomic Line Molecular Spectroscopy (EPA-94/04)
ALR: Action Leakage Rate (EPA-94/04)
ALVW: Adjusted Loaded Vehicle Weight (40 CFR 86.094.3-94)
Am: Americium
AMA: American Medical Association (EPA-92/12)
AMBIENS: Atmospheric Mass Balance of Industrially Emitted and Natural Sulfur (EPA-94/04)
AMC: American Mining Congress
AMCA: Air Moment and Control Association
AML: Average Monthly Limit (EPA-91/3)
AMOS: Air Management Oversight System (EPA-94/04)
AMPS: Automatic Mapping and Planning System (EPA-94/04)
AMS: Administrative Management Staff
AMS: American Meteorological Society
AMS: Army Map Service
AMSA: Association of Metropolitan Sewer Agencies (EPA-94/04)
AMSD: Administrative and Management Services Division
a.n.: Atomic Number
ANC: Acid Neutralizing Capacity (EPA-94/04)
ANEC: American Nuclear Energy Council
ANFO: Ammonium Nitrate and Fuel Oil (40 CFR 457.31)
ANL: Argonne National Laboratory
ANOVA: Analysis of Variance (40 CFR 264.97)
ANPR: Advance Notice of Proposed Rulemaking (EPA-94/04)
ANRHRD: Air, Noise, & Radiation Health Research Division/ORD (EPA-94/04)
ANS: American Nuclear Society
ANSI: American National Standards Institute
ANSS: American Nature Study Society (EPA-94/04)
AO: Administrative Officer
AO: Administrative Order (on consent)
AO: Administrator's Office
AO: Area Office
AO: Awards and Obligations
AOAC: Association of Official Analytical Chemists (40 CFR 60.17.b)
AOC: Abnormal Operating Conditions (EPA-94/04)
AOD: Argon-Oxygen Decarbonization (40 CFR 60.271a)
AOML: Atlantic Oceanographic and Meteorological Laboratory (EPA-94/04)
AP: Accounting Point (EPA-94/04)
APA: Administrative Procedures Act (EPA-94/04)
APA: American Pharmaceutical Association (EPA-92/12)
APCA: Air Pollution Control Association (EPA-94/04)
APCD: Air Pollution Control Device
APCD: Air Pollution Control District (EPA-94/04)
APCE: Air Pollution Control Equipment
APCS: Air Pollution Control System
APDS: Automated Procurement Documentation System (EPA-94/04)
APEG: Alkaline Polyethylene Glycol
APHA: American Public Health Association (EPA-94/04)
API: American Paper Institute

API: American Petroleum Industry (40 CFR 86)
APR: Air Purifying Respirator
APRAC: Urban Diffusion Model for Carbon Monoxide from Motor Vehicle Traffic (EPA-94/04)
APT: Associated Pharmacists and Toxicologists
APTI: Air Pollution Training Institute (EPA-94/04)
APWA: American Public Works Association (EPA-94/04)
AQ-7: Non-reactive Pollutant Modelling (EPA-94/04)
AQA: Air Quality Act
AQC: Air Quality Committee
AQCCT: Air-Quality Criteria and Control Techniques (EPA-94/04)
AQCP: Air Quality Control Program (EPA-94/04)
AQCR: Air-Quality Control Region (EPA-94/04)
AQD: Air-Quality Digest (EPA-94/04)
AQDHS: Air-Quality Data Handling System (EPA-94/04)
AQDM: Air-Quality Display Model (EPA-94/04)
AQL: Acceptable Quality Level (40 CFR 86)
AQMA: Air Quality Maintenance Area (40 CFR 51.40)
AQMP: Air-Quality Maintenance Plan (EPA-94/04)
AQMP: Air-Quality Management Plan (EPA-94/04)
AQSM: Air-Quality Simulation Model (EPA-94/04)
AQTAD: Air-Quality Technical Assistance Demonstration (EPA-94/04)
Ar: Argon
AR: Administrative Record (EPA-94/04)
AR: Arkansas
ARA: Assistant Regional Administrator (EPA-94/04)
ARA: Associate Regional Administrator (EPA-94/04)
ARAR: Applicable or Relevant and Appropriate Requirements (RCRA) (40 CFR 300.4)
ARAR: Applicable or Relevant and Appropriate Standards, Limitations, Criteria, and Requirements (EPA-94/04)
ARB: Air Resources Board (EPA-94/04)
ARC: Agency Ranking Committee (EPA-94/04)
ARCC: American Rivers Conservation Council (EPA-94/04)
ARCS: Alternative Remedial Contract Strategy (EPA-94/04)
ARG: American Resources Group (EPA-94/04)
ARIP: Accidental Release Information Program (EPA-94/04)
ARL: Air Resources Laboratory (EPA-94/04)
ARM: Agricultural Runoff Model (EPA-91/3)
ARM: Air Resources Management (EPA-94/04)
ARO: Alternate Regulatory Option (EPA-94/04)
ARPA: Advanced Research Projects Agency
ARPO: Acid Rain Policy Office
ARPS: Atmospheric Research Program Staff
ARRARs: Applicable or Relevant and Appropriate Requirements (EPA-91/12)
ARRP: Acid Rain Research Program (EPA-94/04)
ARRPA: Air Resources Regional Pollution Assessment Model (EPA-94/04)
ARS: Agricultural Research Service (EPA-94/04)
ARTS: Augusta Regional Transportation Study (DOE-91/4)
ARZ: Auto Restricted Zone (EPA-94/04)
As: Arsenic
AS: Absorber
AS: Area Source (EPA-94/04)
ASA: American Standards Association
ASAP: As Soon As Possible
ASC: Area Source Category (EPA-94/04)
ASCE: American Society of Civil Engineers
ASCII: American Standard Code for Information Interchange
ASCP: American Society of Consulting Planners
ASDWA: Association of State Drinking Water Administrators (EPA-94/04)
ASHAA: Asbestos in Schools Hazard Abatement Act (EPA-94/04)
ASHP: American Society of Hospital Pharmacists (EPA-92/12)
ASIFHESE: Alphabet Soup Index for Health and Enviromnental Science and Engineering
ASIWCPA: Association of State and Interstate Water Pollution Control Administrators (EPA-94/04)
ASMDHS: Airshed Model Data Handling System (EPA-94/04)
ASME: American Society of Mechanical Engineers
ASPA: American Society of Public Administration
ASRL: Atmospheric Sciences Research Laboratory (EPA-94/04)
AST: Advanced Secondary (Wastewater) Treatment (EPA-94/04)
ASTHO: Association of State and Territorial Health Officers (EPA-94/04)
ASTM: American Society for Testing and Materials (EPA-94/04)
ASTSWMO: Association of State and Territorial Solid Waste Management Officials (EPA-94/04)
ASV: Anodic Stripping Voltammetry
At: Astatine
AT: Advanced Treatment (EPA-94/04)
AT: Alpha Track Detection (EPA-94/04)
AT: Ash Trap
ATC: Acceptable Tissue Concentration (EPA-91/3)
ATE: Acute Toxicity Endpoint (EPA-91/3)
ATERIS: Air Toxics Exposure and Risk Information System (EPA-94/04)
ATMI: American Textile Manufacturers Institute
ATP: Adenosine Triphosphate
ATS: Action Tracking System (EPA-94/04)
ATS: Administrator's Tracking System
ATSAC: Administrator's Toxic Substances Advisory Committee
ATSDR: Agency for Toxic Substances and Disease Registry (40 CFR 300.4)
ATTF: Air Toxics Task Force (EPA-94/04)
ATTIC: Alternative Treatment Technology Information Center
Au: Gold
AUR: Air Unit Risk (EPA-92/12)
AUSA: Assistant United States Attorney
AUSM: Advanced Utility Simulation Model (EPA-94/04)
AVS: Acid Volatile Sulfides (EPA-91/3)
a.w.: Atomic Weight
AWMA: Air and Waste Management Association
AWQC: Ambient Water Quality Criteria (40 CFR 300-AA)
AWRA: American Water Resources Association (EPA-94/04)
AWS: American Welding Society
AWT: Advanced Waste Treatment
AWT: Advanced Wastewater Treatment (EPA-94/04)
AWWA: American Water Works Association (EPA-94/04)
AWWARF: American Water Works Association Research Foundation (EPA-94/04)
AZ: Arizona

B: Boron
b:a lambda(a): Blood-to-air partition coefficient (EPA-92/12)
Ba: Barium
BAA: Board of Assistance Appeals (EPA-94/04)
BAC: Biotechnology Advisory Committee (EPA-94/04)
BACER: Biological and Climatological Effects Research

BACM: Best Available Control Measures (EPA-94/04)
BACT: Best Available Control Technology (40 CFR 51)
BADCT: Best Available Demonstrated Control Technology
BADT: Best Available Demonstrated Technology (EPA-94/04)
BAF: Bioaccumulation Factor (EPA-91/3)
BaP: Benzo(a)Pyrene (EPA-94/04)
BAP: Benefits Analysis Program (EPA-94/04)
BARF: Best Available Retrofit Facility
BART: Best Available Retrofit Technology (40 CFR 51.301-c)
BASIS: Battelle's Automated Search Information System (EPA-94/04)
BAT: Best Available Technology (40 CFR 141)
BAT: Best Available Technology Economically Achievable (40 CFR 467)
BATEA: Best Available Treatment Economically Achievable (EPA-94/04)
BBS: Bulletin Board System
BBS: OSWER Electronic Bulletin Board System
BCC: Blind Carbon Copy
BCCM: Board for Certified Consulting Meteorologists
BCF: Bioconcentration Factor (40 CFR 797)
BCPCT: Best Conventional Pollutant Control Technology (EPA-94/04)
BCT: Best Control Technology (EPA-94/04)
BCT: Best Conventional Pollutant Control Technology (40 CFR 430)
BCT: Best Conventional Technology (EPA-91/3)
BD: Background Document
BDAT: Best Demonstrated Achievable Technology (EPA-94/04)
BDCT: Best Demonstrated Control Technology (EPA-94/04)
BDT: Best Demonstrated Technology (EPA-94/04)
Be: Beryllium
BEA: Bureau of Economic Advisors
BEA: Bureau of Economic Analysis
BEIR: Biological Effects of Ionizing Radiation (DOE-91/4)
BEJ: Best Engineering Judgement
BEJ: Best Expert Judgment (EPA-94/04)
BEP: Black Employment Program
BF: Bonifide Notice of Intent to Manufacture or Import (IMD/OTS) (EPA-94/04)
BG: Billion Gallons
BGD: Billion Gallons per Day
BHET: Bis-(2-hydroxyethyl)-terephthalate (40 CFR 60.561)
BHNL: Brookhaven National Laboratory
BHP: Biodegradation, Hydrolysis, and Photolysis (EPA-92/12)
BHP: Brake Horsepower (40 CFR 86)
Bi: Bismuth
BI: Brookings Institution
BIA: Bureau of Indian Affairs (40 CFR 147.2902)
BIAC: Business Industry Advisory Committee
BID: Background Information Document (EPA-94/04)
BID: Buoyancy Induced Dispersion (EPA-94/04)
BIOPLUME: Model to Predict the Maximum Extent of Existing Plumes (EPA-94/04)
BIOS: Basic Input Output System (computer)
Bk: Berkelium
BLM: Bureau of Land Management
BLOB: Biologically Liberated Organo-Beasties
BLP: Buoyant Line and Point Source Model
BLS: Bureau of Labor Statistics
BMP: Best Management Practice(s) (40 CFR 122.2)

BMR: Baseline Monitoring Report (EPA-94/04)
BNA: Base, Neutral, and Acid (BNA) Compounds
BO: Budget Obligations (EPA-94/04)
BOA: Basic Ordering Agreement (Contracts) (EPA-94/04)
BOD: Biochemical Oxygen Demand (40 CFR 133; EPA-94/04)
BOD: Biological Oxygen Demand (EPA-94/04)
BOD5: Biochemical Oxygen Demand as Measured in the Standard 5-Day Test (EPA-92/12)
BOF: Basic Oxygen Furnace (EPA-94/04)
BOM: Bureau of Mines or U.S. Bureau of Mines
BOP: Basic Oxygen Process (EPA-94/04)
BOPF: Basic Oxygen Process Furnace (40 CFR 60.141)
BOYSNC: Beginning of Year Significant Non-Compliers (EPA-94/04)
b.p.: Boiling Point
B.P.: Before Present (DOE-91/4)
BP: Boiling Point (EPA-94/04)
BPA: Blanket Purchase Agreement
BPA: Bonneville Power Administration (DOE-91/4)
BPCT: Best Practicable Control Technolgy
BPJ: Best Professional Judgment (EPA-94/04)
BPT: Best Practicable Control Technology Currently Available (40 CFR 430.222)
BPT: Best Practicable Technology (EPA-94/04)
BPT: Best Practicable Treatment (EPA-94/04)
BPWTT: Best Practicable Waste Treatment Technoloy (40 CFR 35.2005-7)
BPWTT: Best Practical Wastewater Treatment Technology (EPA-94/04)
Bq: Becquerel (a radiation unit) (40 CFR 302.4-AB)
BQU: Bacterial Quantity Unit
Br: Bromine
BRS: Bibliographic Retrieval Service (EPA-94/04)
BSCO: Brake Specific Carbon Monoxide (40 CFR 86)
BSHC: Brake Specific Hydrocarbons (40 CFR 86)
BSI: British Standards Institute (EPA-94/04)
BSNO$_x$: Brake Specific Oxides of Nitrogen (40 CFR 86)
BSO: Benzene Soluble Organics (EPA-94/04)
BST: Benzene Study Team (MCA)
BTG: Benzene Task Group (MCA)
Btu: British Thermal Unit (40 CFR 60)
BTX: Benzene-Toluene-Xylene (40 CFR 61.131)
BTZ: Below the Treatment Zone (EPA-94/04)
BU: Bargaining Unit
BUN: Blood Urea Nitrogen (EPA-94/04)
bw: Body Weight (EPA-92/12)
BY: Budget Year

C: Carbon
C: Cyclone
C: Degree Celsius (centigrade) (40 CFR 60-94)
Ca: Calcium
CA: California
CA: Capacity
CA: Carbon Absorber
CA: Citizen Act (EPA-94/04)
CA: Competition Advocate (EPA-94/04)
CA: Cooperative Agreements (EPA-94/04)
CA: Corrective Action (EPA-94/04)
CAA: Clean Air Act (40 CFR 89.3-94)
CAA: Compliance Assurance Agreement (EPA-94/04)

CAAA: Clean Air Act Amendments of 1990 (40 CFR 89.3-94)
CAB: Civil Aeronautics Board
CAD: Computer Assisted Design
CAER: Community Awareness and Emergency Response (EPA-94/04)
CAFE: Corporate Average Fuel Economy (EPA-94/04)
CAFO: Consent Agreement/Final Order (EPA-94/04)
CAG: Carcinogen Assessment Group, U.S. EPA (EPA-92/12)
CAG: Carcinogenic Assessment Group (EPA-94/04)
CAIR: Comprehensive Assessment of Information Rule (40 CFR 704.200)
cal: Calorie (40 CFR 60)
CALINE: California Line Source Model (EPA-94/04)
CAMP: Continuous Air Monitoring Program (EPA-94/04)
CAMU: Corrective Action for Solid Waste Management Unit (40 CFR 264.552)
CAN: Common Account Number (EPA-94/04)
CAO: Corrective Action Order (EPA-94/04)
CAP: Corrective Action Plan (EPA-94/04)
CAP: Cost Allocation Procedure (EPA-94/04)
CAP: Criteria Air Pollutant (EPA-94/04)
CAR: Corrective Action Report (EPA-94/04)
CARB: California Air Resources Board
CAS: Center for Automotive Safety (EPA-94/04)
CAS: Chemical Abstract Service (EPA-94/04)
CASAC: Clean Air Scientific Advisory Committee (EPA-94/04)
CASLP: Conference on Alternative State and Local Practices (EPA-94/04)
CAST: Council on Agricultural Science and Technology
CATS: Corrective Action Tracking System (EPA-94/04)
CAU: Carbon Adsorption Unit (EPA-94/04)
CAU: Command Arithmetic Unit (EPA-94/04)
CB: Carbon Bed
CB: Continuous Bubbler (EPA-94/04)
CBA: Chesapeake Bay Agreement (EPA-94/04)
CBA: Cost Benefit Analysis (EPA-94/04)
CBB: Chesapeake Bay Basin
CBC: Circulating Bed Combustor
CBD: Central Business District (40 CFR 52.2486)
CBD: Commerce Business Daily
CBI: Compliance Biomonitoring Inspection (EPA-94/04)
CBI: Confidential Business Information (EPA-94/04)
CBO: Congressional Budget Office
CBOD: Carbonaceous Biochemical Oxygen Demand (40 CFR 133.101)
CBOD: Carbonaceous Biochemical Oxygen Demand (EPA-94/04)
CBP: Chesapeake Bay Program (EPA-94/04)
CBP: Combustion Byproduct
CBP: County Business Patterns (EPA-94/04)
cc: Cubic Centimeter (40 CFR 60-94)
CC: Carbon Copy
CC: Closed Cup (EPA-92/12)
CC/RTS: Chemical Collection/Request Tracking System (EPA-94/04)
CCA: Competition in Contracting Act (EPA-94/04)
CCA: Council of Chemical Associations
CCAA: Canadian Clean Air Act (EPA-94/04)
CCAP: Center for Clean Air Policy (EPA-94/04)
CCC: Criteria Continuous Concentration (EPA-91/3)
CCDF: Complementary Cumulative Distribution Function (DOE-91/4)

CCE: Carbon Chloroform Extract
CCEA: Conventional Combustion Environmental Assessment (EPA-94/04)
CCERP: Committee to Coordinate Environmental and Related Programs
CCHW: Citizens Clearinghouse for Hazardous Wastes (EPA-94/04)
CCID: Confidential Chemicals Identification System (EPA-94/04)
CCMA: Certified Color Manufacturers Association
CCMS/NATO: Committee on Challenges of a Modern Society/North Atlantic Treaty Organization (EPA-94/04)
CCP: Composite Correction Plan (EPA-94/04)
CCP: Comprehensive Carcinogen Policy (OSHA)
CCS: Counter Current Scrubber
CCTP: Clean Coal Technology Program (EPA-94/04)
CCW: Constituent Concentrations in Waste (40 CFR 268.43)
CCWE: Constituent Concentrations in Waste Extract (40 CFR 268.41)
Cd: Cadmium
CD: Climatological Data (EPA-94/04)
CD: Criterion Document
CDB: Consolidated Data Base (EPA-94/04)
CDBA: Central Data Base Administrator (EPA-94/04)
CDBG: Community Development Block Grant (EPA-94/04)
CDC: Centers for Disease Control (EPA-92/12)
CDC: Communicable Disease Center
CDD: Chlorinated dibenzo-p-dioxin (EPA-94/04)
CDF: Chlorinated dibenzofuran (EPA-94/04)
CDHS: Comprehensive Data Handling System (EPA-94/04)
CDI: Case Development Inspection (EPA-94/04)
CDM: Climatological Dispersion Model (EPA-94/04)
CDM: Comprehensive Data Management (EPA-94/04)
CDMQC: Climatological Dispersion Model with Calibration and Source Contribution (EPA-94/04)
CDNS: Climatological Data National Summary (EPA-94/04)
CDP: Census Designated Places (EPA-94/04)
CDS: Compliance Data System (EPA-94/04)
Ce: Cerium
CE: Categorical Exclusion (EPA-94/04)
CE: Conditionally Exempt Generator (EPA-94/04)
CE: Cost Effectiveness
CEA: Cooperative Enforcement Agreement (EPA-94/04)
CEA: Cost and Economic Assessment (EPA-94/04)
CEA: Council of Economic Advisors
CEAM: Center for Exposure Assessment Modeling (EPA-91/3)
CEARC: Canadian Environmental Assessment Research Council (EPA-94/04)
CEAT: Contractor Evidence Audit Team (EPA-94/04)
CEB: Chemical Element Balance (EPA-94/04)
CEC: Cation Exchange Capacity (40 CFR 257.3.5)
CEC: Clearinghouse on Environmental Carcinogens
CEC: Commission of European Communities
CECATS: CSB Existing Chemicals Assessment Tracking System (EPA-94/04)
CED: Committee for Economic Development
CEDE: Committed Effective Dose Equivalent (DOE-91/4)
CEE: Center for Environmental Education (EPA-94/04)
CEEM: Center for Energy and Environmental Management (EPA-94/04)
CEFIC: *Counseil Europeen Des Federations De L'Industrie Chimique*

CEI: Compliance Evaluation Inspection (CWA) (EPA-94/04)
CELRF: Canadian Environmental Law Research Foundation (EPA-94/04)
CEM: Continuous Emission Monitoring (EPA-94/04)
CEMA: Conveyor Equipment Manufacturer's Association
CEMS: Continuous Emission Monitoring System (40 CFR 60.51a)
CEO: Chief Executive Officer
CEP: Council on Economic Priorities
CEPP: Chemical Emergency Preparedness Plan (EPA-94/04)
CEQ: Council on Environmental Quality (40 CFR 6.101)
CEQA: California Environmental Quality Act
CERCLA: Comprehensive Environmental Response, Compensation, and Liability Act (1980) (EPA-94/04)
CERCLIS: CERCLA Information System (40 CFR 300.5)
CERCLIS: Comprehensive Environmental Response, Compensation, and Liability Information System (EPA-94/04)
CERI: Center for Environmental Research Information (EPA)
CERT: Certificate of Eligibility (EPA-94/04)
CETIS: Complex Effluent Toxicity Information System (EPA-91/3)
CETTP: Complex Effluent Toxicity Testing Program (EPA-91/3)
CEU: Continuing Education Units
cf.: Compare
Cf: Californium
CF: Conservation Foundation (EPA-94/04)
CFA: Consumer Federation of American
CFC: Chlorofluorocarbons (EPA-94/04)
CFC: Combined Federal Campaign
cfh: Cubic feet per hour (40 CFR 86.403.78-94)
cfm: Cubic feet per minute (40 CFR 86.403.78-94)
CFM: Chlorofluoromethanes (EPA-94/04)
CFR : Code of Federal Regulations (EPA-94/04)
CFS: Cubic Feet per Second
CFU: Colony Forming Unit
CFV: Critical Flow Venturi (40 CFR 86)
CFV-CVS: Critical Flow Venturi-Constant Volume Sampler (40 CFR 86)
CGS unit: Centimeter-Gram-Second unit (absolute metric unit)
CHABA: Committee on Hearing and Bio-Acoustics (EPA-94/04)
CHAMP: Community Health Air Monitoring Program (EPA-94/04)
CHC: Chemical of Highest Concern (EPA-91/3)
ChE: Cholinesterase (EPA-92/12)
CHEMNET: A mutual aid network of chemical shippers and contractors (NRT-87/3)
CHEMNET: Chemical Industry Emergency Mutual Aid Network (EPA-94/04)
CHEMTREC: Chemical Transportation Emergency Center (NRT-87/3)
CHESS: Community Health and Environmental Surveillance System (EPA-94/04)
CHIP: Chemical Hazard Information Profiles (EPA-94/04)
CHLOREP: Chlorine Emergency Plan (NRT-87/3)
CHNTRN: Channel Transport Model (EPA-91/3)
CHRIS/HACS: Chemical Hazards Response Information System/Hazard Assessment Computer System (NRT-87/3)
Ci: Curie (40 CFR 190.02)
Ci/L: Curies per Liter (DOE-91/4)
Ci/m^3: Curies per Cubic Meter (DOE-91/4)
Ci/yr: Curies per Year (DOE-91/4)

CI: Color index
CI: Compression-ignition (40 CFR 89.3-94)
CI: Confidence Interval (EPA-94/04)
CIA: Central Intelligence Agency
CIAQ: Council on Indoor Air Quality (EPA-94/04)
CIBL: Convective Internal Boundary Layer (EPA-94/04)
CIBO: Council of Industrial Boiler Owners
CICA: Competition in Contracting Act (EPA-94/04)
CICIS: Chemicals in Commerce Information System (EPA-94/04)
CIDAC: Cancer Information Dissemination and Analysis Center
CIDRS: Cascade Impactor Data Reduction System (EPA-94/04)
CIIT: Chemical Industry Institute of Toxicology (EPA-92/12)
CIMI: Committee on Integrity and Management Improvement (EPA-94/04)
CIRLG: Chemical Industry Regulatory Liaison Group
CIS: Chemical Information System (EPA-94/04)
CIS: Contracts Information System (EPA-94/04)
CJE: Critical Job Element
CJO: Chief Judicial Officer
CKD: Cement Kiln Dust
CKRC: Cement Kiln Recycling Coalition
CL: Chemiluminescence (40 CFR 86)
CL: Chlorine
CLC: Capacity Limiting Constituents (EPA-94/04)
CLEANS: Clinical Laboratory for Evaluation and Assessment of Toxic Substances (EPA-94/04)
CLEVER: Clinical Laboratory for Evaluation and Validation of Epidemiologic Research (EPA-94/04)
CLF: Conservation Law Foundation (EPA-94/04)
CLIPS: Chemical List Index and Processing System (EPA-94/04)
CLP: Contract Laboratory Program (40 CFR 300-AA)
CLS: Chlorine Scrubber
cm: Centimetre(s) (40 CFR 86.403.78-94)
Cm: Curium
CM: Corrective Measure (EPA-94/04)
CMA: Chemical Manufacturers Association (EPA-94/04)
CMAA: Crane Manufacturer's Association of American
CMB: Chemical Mass Balance (EPA-94/04)
CMC: Criteria Maximum Concentration (EPA-91/3)
CME: Comprehensive (groundwater) Monitoring Evaluation
CME: Comprehensive Monitoring Evaluation (EPA-94/04)
CMEL: Comprehensive Monitoring Evaluation Log (EPA-94/04)
CMEP: Critical Mass Energy Project (EPA-94/04)
CMPU: Chemical Manufacturing Process Unit (40 CFR 63 Subpart G Appendix)
CN ratio: Carbon Nitrogen Ratio
CNG: Compressed Natural Gas (EPA-94/04)
CNP ratio: Carbon Nitrogen Phosphorus ratio
CNR: Composite Noise Rating (DOE-91/4)
CNS: Central Nervous System (EPA-92/12)
Co: Cobalt
CO: Carbon Monoxide (40 CFR 88.103-94)
CO: Colorado
CO$_2$: Carbon Dioxide (40 CFR 86.403.78-94)
COA: Corresponding Onshore Area (40 CFR 55.2)
COB: Close of Business
COCO: Contractor-Owned/Contractor-Operated (EPA-94/04)
COD: Chemical Oxygen Demand (EPA-94/04)
COE: U.S. Army Corps of Engineers (DOE-91/4)
COFA: Certification of Fund Availability (EPA)
COG: Council of Governments

COH: Coefficient OfHaze (EPA-94/04)
COHb: Carboxyhemoglobin
COLA: Cost of Living Adjustment
COLIS: Computerized On-Line Information Service
COM: Continuous Opacity Monitor
COMPLEX: Complex Terrain Screening Model
COMPTER: Multiple Source Air Quality Model
COMS: Continuous Opacity Monitoring System
CON: Selected Contractor or "Awardee"
Conc: Concentration (40 CFR 86.403.78-94)
CONG: Congressional Committee
CORMIX 1: Cornell Mixing Zone Expert System (EPA-91/3)
CORPS: Army Corps of Engineers
COS: Conservative Opportunity Society
COWPS: Council on Wage and Price Stability
CP: Construction Permit (DOE-91/4)
CPF: Carcinogenic Potency Factor (EPA-94/04)
CPI: Consumer Price Index
CPK: Creatine Phosphokinase (EPA-92/12)
CPL: Chemistry and Physics Laboratory
CPM: Continuous Particle Monitor
CPO: Certified Project Officer (EPA-94/04)
CPP: Compliance Policy and Planning
CPR: Campaign for Pesticide Reform
CPR: Center for Public Resources
CPR: Coalition for Pesticide Reform
CPS: Compliance Program and Schedule
CPSA: Consumer Product Safety Act (40 CFR 762.3)
CPSC: Consumer Product Safety Commission
CPSDAA: Compliance and Program Staff to the Deputy Assistant Administrator
CPU: Central Processing Unit (computer)
CQA: Construction Quality Assurance (40 CFR 265.19)
Cr: Chromium
CR: Community Relations
CR: Congressional Record
CR: Continuous Radon Monitoring (EPA-94/04)
CRA: Civil Rights Act
CRA: Classification Review Area
CRAC: Chemical Regulations Advisory Committee (CMA)
CRAVE: Carcinogen Risk Assessment Verification Endeavor (EPA-92/12)
CRC: Community Relations Coordinator (40 CFR 300.4)
CRDL: Contract-Required Detection Limit (40 CFR 300-AA)
CRGS: Chemical Regulations and Guidelines System
CRL: Central Regional Laboratory
CRM: Certified Reference Material (40 CFR 50.1-h)
CROP: Consolidated Rules of Practice (EPA-94/04)
CRP: Community Relations Plan (40 CFR 300.4)
CRP: Conservation Reserve Program (EPA-94/04)
CRQL: Contract-Required Quantitation Limit (40 CFR 300-AA)
CRR: Center for Renewable Resources (EPA-94/04)
CRS: Congressional Research Service
CRSTER: Single Source Dispersion Model (EPA-94/04)
CRT: Cathode Ray Tube
CRTK: Community Right-To-Know (40 CFR 372.1)
CRWI: Coalition for Responsible Waste Incineration
Cs: Cesium
CS: Caustic Scrubber
CS: Compliance Staff
CSDT: California State Department of Transportation (DOE-91/4)

CSF: Confidential Statement of Formula (40 CFR 455-table 8)
CSI: Clean Sites, Inc.
CSI: Compliance Sampling Inspection (EPA-94/04)
CSIN: Chemical Substances Information Network (EPA-94/04)
CSMA: Chemical Specialties Manufacturers Association
CSO: Combined Sewer Overflow (40 CFR 35.2024)
CSPA: Council of State Planning Agencies (EPA-94/04)
CSPI: Center for Science in the Public Interest
CSRL: Center for the Study of Responsive Law (EPA-94/04)
CSS: Commodity Stabilization Service
CST: Certification Short Test (40 CFR 86.096.3-94)
CT: Chimney Tray
CT: Closed Throttle (40 CFR 86)
CT: Connecticut
CTAP: Chemical Transport and Analysis Program (EPA-91/3)
CTARC: Chemical Testing and Assessment Research Commission (EPA-94/04)
CTE: Chronic Toxicity Endpoint (EPA-91/3)
CTFA: Cosmetics, Toiletries and Frangances Association
CTG: Control Technique Guideline (40 CFR 52.25)
cu.: Cubic (40 CFR 86.403.78-94)
Cu: Copper
cu-ft: Cubic Feet (40 CFR 60)
cu.m: Cubic Meter (EPA-92/12)
cu.m/day: Cubic meter per day, used in the HEC derivation of an RfC (EPA-92/12)
CV: Chemical Vocabulary (EPA-94/04)
CV: Coefficient of Variation (EPA-91/3)
CVAAS: Cold Vapor Atomic Absorption Spectroscopy (40 CFR 266-A9)
CVS: Constant volume sampler (40 CFR 86.403.78-94)
CW: Congress Watch
CW: Continuous working-level monitoring (EPA-94/04)
CWA: Clean Water Act (aka FWPCA) (EPA-94/04)
CWAP: Clean Water Action Project (EPA-94/04)
cwt: Hundred Weight (40 CFR 406.61)
CWTC: Chemical Waste Transportation Council (EPA-94/04)
CZARA: Coastal Zone Management Act Reauthorization Amendments (EPA-94/04)
CZMA: Coast Zone Management Act (16USC1451-1464)

d.: Density
d: Day (DOE-91/4)
D&D: Decontamination and Decommissioning (DOE-91/4)
DA: Deputy Administrator
DA: Designated Agent
DA: Dilution Air
DAA: Deputy Assistant Administrator
D_{ae}: Aerodynamic equivalent diameter (EPA-8/90b)
D_{ae}: Aerodynamic resistance diameter (EPA-8/90b)
DAFGDS: Dual Alkali Flue Gas Desulfurization System (40 CFR 60-AG)
DAIG: Deputy Assistant Inspector General
DAMDF: Durham Air Monitoring Demonstration Facility
DAPSS: Document and Personnel Security System (IMD) (EPA-94/04)
DAR: Defense Acquisition Regulations
DAR: Direct Assistance Request
DASD: Direct Access Storage Drive
DASE: Dutch Association of Safety Experts (EPA-92/12)
dB: Decibel (40 CFR 201.1)

dBA: Decibel, A weighted (DOE-91/4)
DBA: Design-Basis Accident (DOE-91/4)
DBCP: Dibromochloropropane
DBE: Design-Basis Earthquake (DOE-91/4)
DCA: Document Control Assistant
DCCLC: Dynamic Coupled Column Liquid Chromatographic (40 CFR 796.1720)
dcf: Dry Cubic Feet (40 CFR 60)
DCI: Data Call-In (EPA-94/04)
dcm: Dry Cubic Meter (40 CFR 60)
DCN: Document Control Number
DCO: Delayed Compliance Order (EPA-94/04)
DCO: Document Control Officer (EPA-94/04)
DCP: Direct Current Plasma
DCP: Discrimination Complaints Program
DD: Deputy Director
DDD: Dichloro-diphenyl dichloroethane
DDO: Dispute Decision Official (40 CFR 35.3030)
DDT: Dichloro-Diphenyl Trichloroethane (EPA-94/04)
DE: Delaware
DE: Department of Education
DE: Destruction Efficiency
DEC: Department of Environmental Conservation
DEC: Direct-shell Evacuation Control (40 CFR 60.271a)
DEIS: Draft Environmental Impact Statement (DOE-91/4)
DEMA: Diesel Engine Manufacturers Association
DEPP: Dredge and Fill Permit Program
DERs: Data Evaluation Records (EPA-94/04)
DES: Diethylstilbesterol (EPA-94/04)
DESCON: Computer Program that Estimates Design Condition (EPA-91/3)
DETA: Dyes Environmental and Toxicology Organization
DF: Dilution Factor (EPA-91/3)
DFLOW: Computer Program that Calculates Biologically based design flows (EPA-91/3)
DHEW: U.S. Department of Health, Education, and Welfare (now U.S. Department of Health and Human Services) (EPA-92/12)
DHHS: Department of Health and Human Services
DI: Diagnostic Inspection (EPA-94/04)
DI: Dry Injection
DL: Detection Limit (40 CFR 300-AA)
DLA: Designated Liability Area (40 CFR 57.401)
DM: Demister
DMA: Designated Management Agency (40 CFR 130.2)
DMA: Dimethylaniline (40 CFR 57.302)
DMR: Discharge Monitoring Report (40 CFR 122.2)
DMT: Dimethyl Terephthalate (40 CFR 60.561)
DMT: Dispersion Modeling and Transport (MCA)
DNA: Deoxyribonucleic acid (EPA-94/04)
DNPH: 2,4-dinitrophenylhydrazine (4086.090.3-94)
DNT: Dinitrotoluene
DO: Dissolved Oxygen (EPA-94/04)
DOA: Department of Agriculture
DOC: Department of Commerce (40 CFR 300.4)
DOC/l: Dissolved Organic Carbon/liter (40 CFR 796.3180.a)
DOD: Department of Defense (40 CFR 300.4)
DOE: Department of Energy (40 CFR 300.4)
DOE-RL: Department of Energy, Richland Operations Office (DOE-91/4)
DOE-SR: Department of Energy, Savannah River Operations Office (DOE-91/4)
DOEd: Department of Education
DOI: Department of Interior (40 CFR 300.4)
DOJ: Department of Justice (40 CFR 300.4)
DOL: Department of Labor (40 CFR 300.4)
DOS: Department of State (40 CFR 300.4)
DOS: Disk Operating System (computer)
DOT: Department of Transportation (40 CFR 300.4)
DOTr: Department of Treasury
DOW: Defenders of Wildlife (EPA-94/04)
D_p: Particle diameter (EPA-8/90b)
DPA: Deepwater Ports Act (EPA-94/04)
DPD: Method of measuring chlorine residual in water (EPA-94/04)
DPD: n,n-Diethyl-Paraphenylene Diamine
D.P.H: Doctor of Public Health
D.P.Hy.: Doctor of Public Hygiene
DPLH: Direct Productive Labor Hour (DOE)
DPN: Diphosphopyridine Nucleotide
DQO: Data Quality Objective (EPA-94/04)
DRA: Deputy Regional Administrator
DRC: Deputy Regional Counsel
DRE: Destruction and Removal Efficiency (EPA-94/04)
DRES: Dietary Risk Evaluation System (EPA-94/04)
DRMS: Defense Reutilization and Marketing Service (EPA-94/04)
DRR: Data Review Record (EPA-94/04)
DS: Dichotomous Sampler (EPA-94/04)
DS: Dissolved Solids
DS: Dry Scrubber
DSAP: Data Self Auditing Program (EPA-94/04)
dscf: Dry Standard Cubic Feet (40 CFR 60)
dscm: Dry Standard Cubic Meter (40 CFR 60)
DSS: Decision Support System (EPA-94/04)
DSS: Domestic Sewage Study (EPA-94/04)
DT: Detectors (radon) damaged or lost (EPA-94/04)
DT: Detention Time (EPA-94/04)
DU: Decision Unit (EPA-94/04)
DU: Ducks Unlimited (EPA-94/04)
DUC: Decision Unit Coordinator (EPA-94/04)
DW: Drinking Water (EPA-92/12)
DWEL: Drinking Water Equivalent Level (EPA-94/04)
DWS: Drinking Water Standard (EPA-94/04)
Dy: Dysprosium
DYNHYD4: Hydrodynamic Model (EPA-91/3)
DYNTOX: Dynamic Toxic Model (EPA-91/3)

E: Exponent (e.g., 1.5E-6 = 1.5 x 10 to the power of -6) (EPA-92/12)
E-MAIL: Electronic Mail
E-PERM: Electret-Passive Environmental Radon Monitor
EA: Endangerment Assessment (EPA-94/04)
EA: Enforcement Agreement (EPA-94/04)
EA: Environmental Action (EPA-94/04)
EA: Environmental Assessment (EPA-94/04)
EA: Environmental Audit (EPA-94/04)
EAB: Exclusion Area Boundary (DOE-91/4)
EADS: Environmental Assessment Data Systems
EAF: Electric Arc Furnaces (40 CFR 60)
EAG: Exposure Assessment Group (EPA-94/04)
EAP: Environmental Action Plan (EPA-94/04)
EAR: Environmental Auditing Roundtable (EPA-94/04)
EAS: Economic Analysis Staff

EB: Emissions Balancing (EPA-94/04)
EBCDIC: Extended Binary Coded Decimal Interchange Code
EBR: Experimental Breeder Reactor (DOE-91/4)
EBS: Emergency Broadcasting System (NRT-87/3)
EC: Effect Concentration (EPA-91/3)
EC: Effective Concentration (EPA-94/04)
EC: Emulsifiable Concentrate (EPA-94/04)
EC: Environment Canada (EPA-94/04)
EC: European Communities; or Ecology Subcommittee
EC: European Community (Common Market)
ECA: Economic Community for Africa (EPA-94/04)
ECAO: Environmental Criteria and Assessment Office, Superfund Health Risk (EPA-91/12)
ECAP: Employee Counselling and Assistance Program (EPA-94/04)
ECD: Electron Capture Detector (EPA-94/04)
ECE: Economic Commission for Europe
ECETOC: European Chemical Industry Ecology and Toxicology Centre
ECHH: Electro-Catalytic Hyper-Heaters (EPA-94/04)
ECHO: Each Community Helps Others (EPA)
ECIETC: European Chemical Industry Ecology & Toxicology Center
ECL: Environmental Chemical Laboratory (EPA-94/04)
ECL: Executive Control Language
ECLA: Economic Commission for Latin America
ECR: Enforcement Case Review (EPA-94/04)
ECRA: Economic Cleanup Responsibility Act (EPA-94/04)
ECSL: Enforcement Compliance Schedule Letters
ED: Department of Education
ED: Effective Dose (EPA-94/04)
ED: Electron Diffraction (40 CFR 763-AA)
ED10: 10 Percent Effective Dose (40 CFR 300-AA)
EDA: Economic Development Administration
EDA: Emergency Declaration Area (EPA-94/04)
EDB: Ethylene Dibromide (EPA-94/04)
EDC: Ethylene Dichloride (EPA-94/04)
EDD: Enforcement Decision Document (EPA-94/04)
EDE: Effective Dose Equivalent (DOE-91/4)
EDF: Environmental Defense Fund (EPA-94/04)
EDNA: Environmental Designation for Noise Abatement (DOE-91/4)
EDP: Electrodeposition (40 CFR 60.311)
EDP: Electronic Data Processing
EDRS: Enforcement Document Retrieval System (EPA-94/04)
EDS: Electronic Data System (EPA-94/04)
EDS: Energy Data System (EPA-94/04)
EDT: Edit Data Transmission
EDTA: Ethylene Diamine Triacetic Acid (EPA-94/04)
EDX: Electronic Data Exchange (EPA-94/04)
EDXA: Energy Dispersive X-ray Analysis (40 CFR 763-AA)
EDZ: Emission Density Zoning (EPA-94/04)
EEA: Energy and Environmental Analysis (EPA-94/04)
EEC: European Economic Commission
EECs: Estimated Environmental Concentrations (EPA-94/04)
EED: Estimated Exposure Dose (EPA-92/12)
EEG: Electroencephalogram (EPA-92/12)
EEI: Edison Electrical Institute (DOE-91/4)
EEOC: Equal Employment Opportunity Commission
EER: Excess Emission Report (EPA-94/04)
EERL: Eastern Environmental Radiation Laboratory (EPA-94/04)

EERU: Environmental Emergency Response Unit (EPA-94/04)
EESI: Environment and Energy Study Institute (EPA-94/04)
EESL: Environmental Ecological and Support Laboratory (EPA-94/04)
EETFC: Environmental Effects, Transport, and Fate Committee (EPA-94/04)
EF: Emission Factor (EPA-94/04)
EF: Exposure Frequency (EPA-91/12)
EFO: Equivalent Field Office (EPA-94/04)
EFTC: European Fluorocarbon Technical Committee (EPA-94/04)
e.g.: for Example
EG&G: EG&G Idaho, Inc. (DOE-91/4)
EGR: Exhaust Gas Recirculation (40 CFR 86.403.78-94)
EH: Redox Potential (EPA-94/04)
EHC: Environmental Health Committee (EPA-94/04)
EHS: Extremely Hazardous Substance (EPA-94/04)
EI: Emissions Inventory (EPA-94/04)
EIA: Economic Impact Assessment (EPA-94/04)
EIA: Environmental Impact Assessment (EPA-94/04)
EIL: Environmental Impairment Liability (EPA-94/04)
EIP: Economic Incentive Program (40 CFR 51.490)
EIR: Endangerment Information Report (EPA-94/04)
EIR: Environmental Impact Report (EPA-94/04)
EIS: Environmental Impact Statement (EPA-94/04)
EIS: Environmental Inventory System (EPA-94/04)
EIS/AS: Emissions Inventory System/Area Source (EPA-94/04)
EIS/PS: Emissions Inventory System/Point Source (EPA-94/04)
EKG: Electrocardiogram (EPA-92/12)
EKMA: Empirical Kinetic Modeling Approach (EPA-94/04)
EL: Exposure Level (EPA-94/04)
ELEVEN-AA: 11-Aminoundecanoic Acid (40 CFR 704.25)
ELI: Environmental Law Institute (EPA-94/04)
ELISA: Enzyme-Linked Immunosorbent Assay (EPA-92/12)
ELR: Environmental Law Reporter (EPA-94/04)
ELWK: Equivalent Live Weight Killed (40 CFR 432.11)
EM: Electromagnetic Conductivity (EPA-94/04)
EMAS: Enforcement Management and Accountability System (EPA-94/04)
EMI: Emergency Management Institute (NRT-87/3)
EMI/RFI: Electromagnetic Interference/Radio Frequency Interference (40 CFR 60.721)
EMR: Environmental Management Report (EPA-94/04)
EMS: Enforcement Management System (EPA-94/04)
EMSL: Environmental Monitoring Support Laboratory (EPA-94/04)
EMSL: Environmental Monitoring Systems Laboratory (EPA-94/04)
EMTD: Estimated Maximum Tolerated Dose (EPA-92/12)
EMTS: Environmental Monitoring Testing Site (EPA-94/04)
EMTS: Exposure Monitoring Test Site (EPA-94/04)
ENM: Environmental Noise Model (DOE-91/4)
EO: Ethylene Oxide (EPA-94/04)
EO: Executive Officer
EO: Executive Order
EOB: Executive Office Building
EOC: Emergency Operating Center (EPA-94/04)
EOD: Entrance on Duty
EOE: Equal Opportunity Employer
EOF: Emergency Operations Facility (RTP) (EPA-94/04)
EOJ: End of Job
EOP: Emergency Operations Plan (NRT-87/3)

EOP: End of Pipe (EPA-94/04)
EOT: Emergency Operations Team (EPA-94/04)
EOY: End of Year
EP tox: Extraction procedure toxicity
EP: Earth Protectors (EPA-94/04)
EP: End point (40 CFR 86.403.78-94)
EP: End-use Product (EPA-94/04)
EP: Environmental Profile (EPA-94/04)
EP: Equilibrium Partitioning (EPA-91/3)
EP: Experimental Product (EPA-94/04)
EP: Extraction Procedure (EPA-94/04)
EPA: Environmental Protection Agency (40 CFR 86.403.78-94)
EPAA: Environmental Programs Assistance Act (EPA-94/04)
EPAAR: EPA Acquisition Regulations (EPA-94/04)
EPACASR: EPA Chemical Activities Status Report (EPA-94/04)
EPAYS: EPA Payroll System
EPCA: Energy Policy and Conservation Act (EPA-94/04)
EPCRA: Emergency Planning and Community Right-To-Know Act of 1986 (40 CFR 22.40)
EPCRA: Emergency Preparedness and Community Right to Know Act (EPA-94/04)
EPD: Emergency Planning District (EPA-94/04)
EPI: Environmental Policy Institute (EPA-94/04)
EPIC: Environmental Photographic Interpretation Center (EPA-94/04)
EPNL: Effective Perceived Noise Level (EPA-94/04)
EPO: Estuarian Programs Office (NOAA)
EPRI: Electric Power Research Institute (EPA-94/04)
EPTC: Extraction Procedure Toxicity Characteristic (EPA-94/04)
eq: Equivalent (40 CFR 60)
Er: Erbium
ER: Electrical Resistivity (EPA-94/04)
ER: Extrarespiratory (EPA-92/12)
ERA: Economic Regulatory Agency (EPA-94/04)
ERA: Equal Rights Amendment
ERAB: Energy Research Advisory Board (DOE-91/4)
ERAMS: Environmental Radiation Ambient Monitoring System (EPA-94/04)
ERC: Emergency Response Commission (EPA-94/04)
ERC: Emissions Reduction Credit (EPA-94/04)
ERC: Environmental Research Center (EPA-94/04)
ERCS: Emergency Response Cleanup Services (EPA-94/04)
ERD&DAA: Environmental Research, Development and Demonstration Authorization Act (EPA-94/04)
ERISA: Employee Retirement Income Security Act
ERL: Environmental Research Laboratory (EPA-94/04)
ERNS: Emergency Response Notification System (EPA-94/04)
ERP: Enforcement Response Policy (EPA-94/04)
ERPG: Emergency Response Planning Guideline (DOE-91/4)
ERT: Emergency Response Team (EPA-94/04)
ERT: Environmental Response Team (40 CFR 300.4)
ERTAQ: ERT Air Quality Model (EPA-94/04)
Es: Einsteinium
ES: Enforcement Strategy (EPA-94/04)
ES: Entrainment Separator
ESA: Endangered Species Act (EPA-94/04)
ESA: Environmentally Sensitive Area (EPA-94/04)
ESC: Endangered Species Committee (EPA-94/04)
ESCA: Electron Spectroscopy for Chemical Analysis (EPA-94/04)
ESCAP: Economic and Social Commission for Asia and the Pacific (EPA-94/04)

ESECA: Energy Supply and Environmental Coordination Act (EPA-94/04)
ESH: Environmental Safety and Health (EPA-94/04)
ESP: Electrostatic Precipitator (40 CFR 60.471)
ESU: Electrostatic Unit
et al.: and elsewhere
et seq.: and the following one(s)
ET: Emissions Trading (EPA-94/04)
ET: Extrathoracic Region of the Respiratory Tract (EPA-92/12)
etc. (et cetera): and so forth
ETP: Emissions Trading Policy (EPA-94/04)
ETS: Emergency Temporary Standard
ETS: Environmental Tobacco Smoke (EPA-94/04)
ETV: Electrothermal Vaporization
Eu: Europium
EUP: End-Use Product (EPA-94/04)
EUP: Environmental Use Permit
EUP: Experimental Use Permit (40 CFR 168.22)
EVA: Ethylene Vinyl Acetate
EWCC: Environmental Workforce Coordinating Committee (EPA-94/04)
EX: Executive Level Appointment
EXAMS: Exposure Analysis Modeling System (EPA-94/04)
EXAMS-ll: Exposure Analysis Modeling System (EPA-91/3)
ExEx: Expected Exceedance (EPA-94/04)
EXIMBANK: Export-Import Bank of the U.S.

F: Fahrenheit (40 CFR 86.403.78-94)
F: Fluorine
F/M: Food to Microorganism Ratio (EPA-94/04)
F1: First Filial Generation (in experimental animals) (EPA-92/12)
FAA: Federal Aviation Administration, Department of Transportation (40 CFR 87.2)
FAC: Free Available Chlorine
FACA: Federal Advisory Committee Act (EPA-94/04)
FAME: Framework for Achieving Managerial Excellence
FAN: Fixed Account Number (EPA-94/04)
FAO: Food and Agriculture Organization
FAR: Federal Acquisition Regulation (40 CFR 34.105)
FASB: Financial Accounting Standards Board
FAST: Fugitive Assessment Sampling Train
FATES: FIFRA and TSCA Enforcement System (40 CFR 455-Table 10)
FAV: Final Acute Value (EPA-91/3)
FBC: Fluidized Bed Combustion (EPA-94/04)
FC: Fluorocarbon
FCC: Federal Communications Commission
FCC: Fluid Catalytic Converter (EPA-94/04)
FCCU: Fluid Catalytic Cracking Unit (EPA-94/04)
FCM2: WASP Food Chain Model (EPA-91/3)
FCO: Federal Coordinating Officer (40 CFR 300.4)
FCO: Forms Control Officer (40FR300)
FDA: Food and Drug Administration (40 CFR 160.3)
FDAAL: Food and Drug Administration Action Level (40 CFR 300)
FDF: Fundamentally Different Factors (EPA-94/04)
FDIC: Federal Deposit Insurance Corporation
FDL: Final Determination Letter (EPA-94/04)
FDO: Fee Determination Official (EPA-94/04)
Fe: Iron
FE: Fugitive Emissions (EPA-94/04)

FEA: Federal Energy Administration
FEC: Federal Executive Council
FEDS: Federal Energy Data System (EPA-94/04)
FEFx: Forced Expiratory Flow (EPA-94/04)
FEGLI: Federal Employee Group Life Insurance
FEHB: Federal Employees Health Benefits
FEI: Federal Executive Institute
FEIS: Final Environmental Impact Statement (DOE-91/4)
FEIS: Fugitive Emissions Information System (EPA-94/04)
FEL: Family Emission Limit (40 CFR 86.090.3-94)
FEL: Frank Effect Level (EPA-94/04)
FEMA: Federal Emergency Management Agency (40 CFR 300.5)
FEPCA: Federal Environmental Pesticide Control Act; enacted as amendments to FIFRA (EPA-94/04)
FERC: Federal Energy Regulatory Commission (EPA-94/04)
FERSA: Federal Employee Retirement System Act
FES: Factor Evaluation System (EPA-94/04)
FETRA: Finite Element Transport Model (EPA-91/3)
FEV: Forced Expiratory Volume (EPA-94/04)
FEV$_1$: Forced Expiratory Volume-one second (EPA-94/04)
FEVI: Front End Volatility Index (EPA-94/04)
FEW: Federally Employed Women
FF: Fabric Filter
FF: Federal Facilities (EPA-94/04)
FFA: Flammable Fabrics Act
FFAR: Fuel and Fuel Additive Registration (EPA-94/04)
FFDCA: Federal Food, Drug, and Cosmetic Act (40 CFR 160.3)
FFF: Firm Financial Facility (EPA-94/04)
FFFSG: Fossil-Fuel-Fired Steam Generator (EPA-94/04)
FFIS: Federal Facilities Information System (EPA-94/04)
FFP: Firm Fixed Price (EPA-94/04)
FFPA: Farmland Protection Policy Act
FGD: Flue-Gas Desulfurization (EPA-94/04)
FGETS: Food and Gill Exchange of Toxic Substances (EPA-91/3)
FGR: Flue gas recirculation (EPA-81/12)
FHA: Farmers Home Administration
FHA: Federal Housing Administration
FHLBB: Federal Home Loan Bank Board
FHSA: Federal Hazardous Substances Act
FHWA: Federal Highway Administration (40 CFR 93.101)
FI: Formaldehyde Institute
FIA: Federal Insurance Administration
FIC: Federal Information Center
FICA: Federal Insurance Contributions Act
FID: Flame Ionization Detector (40 CFR 86)
FIFO: First In/First Out
FIFRA: Federal Insecticide, Fungicide, and Rodenticide Act (EPA-94/04)
FIM: Friable Insulation Material (EPA-94/04)
FINDS: Facility Index System (EPA-94/04)
FIP: Federal Implementation Plan (40 CFR 52.741.a)
FIP: Federal Information Plan
FIP: Final Implementation Plan (EPA-94/04)
FIPS: Federal Information Procedures System (EPA-94/04)
FIT: Field Investigation Team (EPA-94/04)
FL: Florida
FL: Full Load (40 CFR 86)
FLETC: Federal Law Enforcement Training Center (EPA-94/04)
FLM: Federal Land Manager (EPA-94/04)
FLOSTAT: U.S. Geological Survey Computer Program that estimates the arithmetic mean flow and 7Q10 of rivers and streams (EPA-91/3)
FLP: Flash Point (EPA-94/04)
FLPMA: Federal Land Policy and Management Act (EPA-94/04)
FLSA: Fair Labor Standards Act
Fm: Fermium
FM: Food Chain Multipliers (EPA-91/3)
FMAP: Financial Management Assistance Project (EPA-94/04)
FMC: Federal Maritime Commission
FMFIA: Federal Managers Financial Integrity Act
FMIA: Federal Meat Inspection Act (21USC60)
FML: Flexible Membrane Liner (40 CFR 258.40-93)
FMO: Financial Management Officer
FMP: Facility Management Plan (EPA-94/04)
FMP: Financial Management Plan (EPA-94/04)
FMS: Financial Management System (EPA-94/04)
FMVCP: Federal Motor Vehicle Control Program (EPA-94/04)
FN: Fog Nozzle
FNSI: Finding of No Significant Impact (40 CFR 1508.13)
FOE: Friends of the Earth (EPA-94/04)
FOI: Freedom of Information (EPA-92/12)
FOIA: Freedom of Information Act (EPA-94/04)
FOISD: Fiber Optic Isolated Spherical Dipole Antenna (EPA-94/04)
FONSI: Finding of No Significant Impact (EPA-94/04)
FORAST: Forest Response to Anthropogenic Stress (EPA-94/04)
FORTRAN: Formula Translation
FP: Fine Particulate (EPA-94/04)
FPA: Federal Pesticide Act (EPA-94/04)
FPAS: Foreign Purchase Acknowledgement Statements (EPA-94/04)
FPC: Federal Power Commission
FPD: Flame Photometric Detector (EPA-94/04)
FPEIS: Fine Particulate Emissions Information System (EPA-94/04)
FPM: Federal Personnel Manual (EPA-94/04)
FPPA: Federal Pollution Prevention Act (EPA-94/04)
FPR: Federal Procurement Regulation (EPA-94/04)
FPRS: Federal Program Resources Statement (EPA-94/04)
FPRS: Formal Planning and Supporting System (EPA-94/04)
Fr: Francium
FR: Federal Register (EPA-94/04)
FR: Final Rulemaking (EPA-94/04)
FRA: Federal Register Act (EPA-94/04)
FRB: Federal Reserve Board
FRC: Federal Records Center
FRC: Functional Reserve Capacity (EPA-92/12)
FRDS: Federal Reporting Data System
FREDS: Flexible Regional Emissions Data System (EPA-94/04)
Freon-113: trichlorotrifluoroethane (DOE-91/4)
FRES: Forest Range Environmental Study (EPA-94/04)
FRM: Federal Reference Methods (EPA-94/04)
FRN: Federal Register Notice (EPA-94/04)
FRN: Final Rulemaking Notice (EPA-94/04)
FRS: Formal Reporting System (EPA-94/04)
FRTIB: Federal Retirement Thrift Investment Board
FS: Feasibility Study (40 CFR 300.4)
FS: Forest Service
FSA: Food Security Act (EPA-94/04)
FSP: Field Sampling Plan
FSS: Facility Status Sheet (EPA-94/04)
FSS: Federal Supply Schedule (EPA-94/04)

ft: Feet (40 CFR 60)
FT: Full Time
ft²: Square Feet (40 CFR 60)
ft³: Cubic Feet (40 CFR 60)
FTA: Federal Transit Administration (40 CFR 93.101)
FTC: Federal Trade Commission
FTE: Full Time Equivalent
FTIR: Fourier Transform Infrared Analyzer
FTP: Federal Test Procedure (40 CFR 89.3-94)
FTS: Federal Telecommunications System
FTS: File Transfer Service (EPA-94/04)
FTS-IR: Fourier transform infrared spectroscopy analyzer
FTT: Full Time Temporary
FTTS: FIFRA/TSCA Tracking System (EPA-94/04)
FTU: Formazin Turbidity Unit
FUA: Fuel Use Act (EPA-94/04)
FURS: Federal Underground Injection Control Reporting System (EPA-94/04)
FVC: Forced Vital Capacity (EPA-8/90b)
FVMP: Federal Visibility Monitoring Program (EPA-94/04)
FWCA: Fish and Wildlife Coordination Act (EPA-94/04)
FWP: Federal Womens Program
FWPCA: Federal Water Pollution and Control Act (40 CFR 220.2)
FWPCA: Federal Water Pollution and Control Administration (EPA-94/04)
FWQC: Federal Water Quality Criteria (EPA-91/12)
FWS: U.S. Fish and Wildlife Service (40 CFR 233.2)
FY: Fiscal Year
FYI: For Your Information

G: Gram (40 CFR 60)
G/DSCM: Grams/(dry standard cubic meter)
G-EQ: gram-equivalent (40 CFR 60)
G/KW-HR: Grams per kilowatt hour (40 CFR 89.3-94)
G/MI: Grams per Mile (EPA-94/04)
G/ML: Grams/milliliter, 1 milliliter = 1 cc = 1 cm^3
G/Ncm: Grams/(normal cubic meter), 1 normal cubic meter = 1 dry standard cubic meter
G&A: General and Administrative Cost
Ga: Gallium
GA: Georgia
GAAP: Generally Accepted Accounting Principles (EPA-94/04)
GAC: Granular Activated Carbon (40 CFR 141.61.b)
GAC: Groundwater Activated Carbon
GACT: Granular Activated Carbon Treatment (EPA-94/04)
gal: Gallon (40 CFR 60)
GAO: General Accounting Office
GATT: General Agreement on Tariffs and Trade
GBL: Government Bill of Lading
GC: Gas chromatograph (40 CFR 86.090.3-94)
GC: Gas Cooler
GC: General Counsel
GC/AFID: Gas Chromatography/Alkali Flame Ionization Detector (EPA-84/03a)
GC/DD: Gas Chromatography/Dual Detector (EPA-84/03a)
GC/ECD: Gas Chromatography/Electrolytic Conductivity Detector (EPA-84/03a)
GC/ECD: Gas Chromatography/Electron Capture Detector (EPA-84/03a)
GC/FID: Gas Chromatography/Flame Ionization Detector (Course 165.5)

GC/HECD: Gas Chromatography/Hall Electrolytic Conductivity Detector (EPA-84/03a)
GC/IR: Gas Chromatography/Infrared Absorption Spectrometer (EPA-84/03a)
GC/MS: Gas Chromatography/Mass Spectrometry (EPA-84/03a)
GC/PID: Gas Chromatography/Photoionization Detector (Course 165.5)
GC/TCD: Gas Chromatography/Thermal Conductivity Detector (EPA-84/03a)
GC/TSD: Gas Chromatography/Thermionic Specific detector (EPA-84/03a)
GCP: Good Combustion Practices
GCWR: Gross Combination Weight Rating (40 CFR 202.10)
Gd: Gadolinium
GD: Guidance Document
GDE: Generic Data Exemption (EPA-94/04)
Ge: Germanium
GEA: Glossary of EPA Acronyms
GEI: Geographic Enforcement Initiative (EPA-94/04)
GEMS: Global Environmental Monitoring System (EPA-94/04)
GEMS: Graphical Exposure Modeling System (EPA-94/04)
GEP: Good Engineering Practice (40FR51)
GF: General Files
GFF: Glass Fiber Filter (EPA-94/04)
GFO: Grant Funding Order (EPA-94/04)
GFP: Government-Furnished Property (EPA-94/04)
GI: Gastrointestinal (EPA-92/12)
GICS: Grant Information and Control System (EPA-94/04)
GIS: Geographic Information Systems (EPA-94/04)
GIS: Global Indexing System (EPA-94/04)
GLBC: Great Lakes Basin Commission
GLC: Gas Liquid Chromatography (EPA-94/04)
GLERL: Great Lakes Environmental Research Laboratory (EPA-94/04)
GLICP: Great Lakes Initiative Contract Program
GLNPO: Great Lakes National Program Office (EPA-94/04)
GLP: Good Laboratory Practices (EPA-94/04)
GLWQA: Great Lakes Water Quality Agreement (EPA-94/04)
GMA: Grocery Manufacturers Association
GMCC: Global Monitoring for Climatic Change (EPA-94/04)
GMT: Greenwich Mean Time
GNP: Gross National Product
GOCM: Goals, Objectives, Commitments, and Measures
GOCO: Government-Owned/Contractor-Operated (EPA-94/04)
GOGO: Government-Owned/ Government-Operated (EPA-94/04)
GOP: General Operating Procedures (EPA-94/04)
GOPO: Government-Owned/Privately Operated (EPA-94/04)
GOTCHA: Generalized Overall Toxics Control and Hazards Act
GPAD: Gallons-per-acre per-day (EPA-94/04)
GPD: Gallons per Day
GPG: Grams-per-Gallon (EPA-94/04)
GPM: Gallons per Minute
GPO: Government Printing Office
GPR: Ground-Penetrating Radar (EPA-94/04)
GPS: Groundwater Protection Strategy (EPA-94/04)
GPT: Glutamic-Pyruvic Transaminase (EPA-92/12)
GR: Grain
GR: Grab Radon Sampling (EPA-94/04)
GR/DSCF: Grains/(dry standard cubic feet), 1 gr/dscf = 2,300 milligrams/dscm
GRAS: Generally Regarded As Safe

GRCDA: Government Refuse Collection and Disposal Association (EPA-94/04)
GRGL: Groundwater Residue Guidance Level (EPA-94/04)
GRI: Gas Research Institute
GS: General Schedule
GSA: General Services Administration
GTN: Global Trend Network (EPA-94/04)
GTR: Government Transportation Request (EPA-94/04)
GVP: Gasoline Vapor Pressure (EPA-94/04)
GVW: Gross Vehicle Weight (40 CFR 51.731)
GVWR: Gross Vehicle Weight Rating (40 CFR 88.103-94)
GW: Grab Working-Level Sampling (EPA-94/04)
GW: Groundwater (EPA-94/04)
GWM: Groundwater Monitoring (EPA-94/04)
GWPS: Groundwater Protection Standard (EPA-94/04)
GWPS: Groundwater Protection Strategy (EPA-94/04)

h: hour (40 CFR 86.403.78-94)
H: Humidifier
H: Hydrogen
H$_2$O: Water (40 CFR 86.403.78-94)
ha: Hectare (40 CFR 401.11)
Ha: Hahnium
HA: Health Advisory (EPA-94/04)
HAC: Heating and Air Conditioning
HAD: Health Assessment Document (EPA-94/04)
HAF: Halogen Acid Furnace (40 CFR 260.10)
HAP: Hazardous Air Pollutant (EPA-94/04)
HAPEMS: Hazardous Air Pollutant Enforcement Management System (EPA-94/04)
HAPPS: Hazardous Air Pollutant Prioritization System (EPA-94/04)
HAS: Health Assessment Summary (EPA-92/12)
HATREMS: Hazardous and Trace Emissions System (40 CFR 51.117)
HAZMAT: Hazardous Materials (EPA-94/04)
HAZOP: Hazard and Operability Study (EPA-94/04)
HB: Health Benefits
HBEP: Hispanic and Black Employment Programs
HBV: Hepatitis B Virus (29 CFR 1910.1030)
HC: Hazardous Constituents (EPA-94/04)
HC: Hydrocarbon(s) (40 CFR 86.403.78-94)
HCA: Hydrogen Chloride Absorber
HCC: House Commerce Committee
HCCPD: Hexachlorocyclo-pentadiene (EPA-94/04)
HCHO: Formaldehyde (40 CFR 88.103-94)
HCP: Hypothermal Coal Process (EPA-94/04)
HCS: Hydrogen Chloride Scrubber
HCT: Hematocrit (EPA-92/12)
HDD: Halogenated Dibenzodioxins (40 CFR 766.3-93)
HDD: Heavy-Duty Diesel (EPA-94/04)
HDE: Heavy-Duty Engine (EPA-94/04)
HDF: Halogenated Dibenzofurans (40 CFR 766.3-93)
HDG: Heavy-Duty Gasoline-Powered Vehicle (EPA-94/04)
HDPE: High Density Polyethylene (EPA-94/04)
HDT: Heavy-Duty Truck (EPA-94/04)
HDT: Highest Dose Tested in a study (EPA-94/04)
HDV: Heavy-Duty Vehicle (40 CFR 88.103-94)
He: Helium
HE: Heat Exchanger
HEAL: Human Exposure Assessment Location (EPA-94/04)

HEAST: Health Effects Assessment Summary Tables (EPA-91/12)
HEC: Human Equivalent Concentration (EPA-92/12)
HECC: House Energy and Commerce Committee (EPA-94/04)
HEEP: Health and Environmental Effects Profile (EPA-92/12)
HEI: Health Effects Institute (EPA-94/04)
HEM: Human Exposure Modeling (EPA-94/04)
HEP: Hispanic Employment Program
HEPA: High-Efficiency Particulate Air (EPA-94/04)
HEPA filter: High-Efficiency Particulate Air filter (40 CFR 763.121)
HERL: Health Effects Research Laboratory
HERS: Hyperion Energy Recovery System (EPA-94/04)
HES: High Energy Scrubber
HEX-BCH: Hexachloronorbornadiene
Hf: Hafnium
HFID: Heated Flame Ionization Detector (40 CFR 86)
HFPO: Hexafluoropropylene oxide (40 CFR 704.104)
Hg: Mercury
Hgb: Hemoglobin (EPA-92/12)
HHC: Human Health Criteria (EPA-91/3)
HHDFLOW: Historic Daily Flow Program (EPA-91/3)
HHE: Human Health and the Environment (EPA-94/04)
HHEM: Human Health Evaluation Manual (EPA-91/12)
HHFA: Housing and Home Finance Agency
HHS: Department of Health and Human Services (40 CFR 26.101)
HHS: Department of Health and Human Services (Formerly HEW) (40 CFR 300.4)
HHV: Higher Heating Value (EPA-94/04)
HI: Hawaii
HI: Hazard Index (EPA-94/04)
HI-VOL: High-Volume Sampler (EPA-94/04)
HIT: Hazard information transmission (NRT-87/3)
HIV: Human Immunodeficiency Virus (29 CFR 1910.1030)
HIWAY: A Line Source Model for Gaseous Pollutants (EPA-94/04)
HLLW: High-Level Liquid Waste (DOE-91/4)
HLRW: High Level Radioactive Waste (40 CFR 191.02)
HLW: High-Level Waste (DOE-91/4)
HMIS: Hazardous Materials Information System (EPA-94/04)
HMS: Hanford Meteorological Station (DOE-91/4)
HMS: Highway Mobile Source (EPA-94/04)
HMTA: Hazardous Materials Transportation Act (EPA-94/04)
HMTR: Hazardous Materials Transportation Regulations (EPA-94/04)
Ho: Holmium
HO: Headquarters Offices
HOC: Halogenated Organic Carbons (EPA-94/04)
HOC: Halogenated Organic Compound (40 CFR 268.2)
HON: Hazardous Organic NESHAP (EPA-94/04)
HOV: High-Occupancy Vehicle (EPA-94/04)
HP: Horse Power (40 CFR 401.11)
HPLC: High-Performance Liquid Chromatography (EPA-94/04)
HPLC: High-Pressure Liquid Chromatography (40 CFR 86.090.3-94)
HPV: High Priority Violator (EPA-94/04)
HQ: Hazard Quotient (EPA-91/12)
HQ: Headquarters
HQCDO: Headquarters Case Development Officer (EPA-94/04)
hr.: Hour(s) (40 CFR 87.2-94)
H.R.: House of Representatives
HRA: Hourly Rolling Average

HRC: Human Resources Council
HRGC: High Resolution Gas Chromatography (40 CFR 766.3-93)
HRMS: High Resolution Mass Spectrometry (40 CFR 766.3-93)
HRS: Hazard Ranking System (40 CFR 300.4)
HRS: Hazardous Ranking System (EPA-94/04)
HRUP: High-Risk Urban Problem (EPA-94/04)
HSA: Historic Sites Act of 1935 (40 CFR 6.301)
HSDB: Hazardous Substance Data Base (EPA-94/04)
HSL: Hazardous Substance List (EPA-94/04)
HSP: Health and Safety Plan
HSPF: Hydrologic Simulation Program (EPA-91/3)
HSWA: Hazardous and Solid Waste Amendments of 1984 (EPA-94/04)
HT: Hypothermally Treated (EPA-94/04)
HTHE: High Temperature Heat Exchanger
HTP: High Temperature and Pressure (EPA-94/04)
HUD: Department of Housing and Urban Development
HVAC: Heating, Ventilating, and Air Conditioning
HVAF: High Velocity Air Filter (40FR60.471)
HVIO: High Volume Industrial Organics (EPA-94/04)
HVL: Highly Volatile Liquid (40 CFR 195.2)
HW: Hazardous Waste (EPA-94/04)
HWDGR: Hazardous Waste Disposal Guidelines and Regulations
HWDMS: Hazardous Waste Data Management System (EPA-94/04)
HWERL: Hazardous Waste Engineering Research Laboratory (See RREL)
HWGTF: Hazardous Waste Groundwater Task Force (EPA-94/04)
HWGTF: Hazardous Waste Groundwater Test Facility (EPA-94/04)
HWLT: Hazardous Waste Land Treatment (EPA-94/04)
HWM: Hazardous Waste Management (40 CFR 270.2)
HWPF: Hazardous Waste Processing Facility (DOE-91/4)
HWR: Heavy-Water Reactor (DOE-91/4)
HWRTF: Hazardous Waste Restrictions Task Force (EPA-94/04)
HWTC: Hazardous Waste Treatment Council (EPA-94/04)
Hz: Hertz (40 CFR 60)

I: Interstate (DOE-91/4)
I: Iodine
I/A: Innovative/Alternative (EPA-94/04)
I/I: Infiltration/inflow (40 CFR 35.905)
I/M: Inspection/Maintenance (40 CFR 51.350)
IA: Interagency Agreement (EPA-94/04)
IA: Iowa
IAAC: Interagency Assessment Advisory Committee (EPA-94/04)
IAEA: International Atomic Energy Agency (DOE-91/4)
IAG: Interagency Agreement (EPA-94/04)
IAP: Incentive Awards Program (EPA-94/04)
IAP: Individual Annoyance Prediction (DOE-91/4)
IAP: Indoor Air Pollution (EPA-94/04)
IARC: International Agency for Research on Cancer (EPA-94/04)
IATA: International Air Transport Association
IATDB: Interim Air Toxics Data Base (EPA-94/04)
IBA: Industrial Biotechnology Association
IBP: Initial Boiling Point (40 CFR 86)
IBR: Incorporation by Reference (40 CFR 60.17)
IBRD: International Bank for Reconstruction and Development
IBT: Industrial Biotest Laboratory (EPA-94/04)
IC: Inhibition Concentration (EPA-91/3)
ICAIR: Interdisciplinary Planning and Information Research (EPA-94/04)
ICAO: International Civil Aviation Organization
ICAP: Inductively Coupled Argon Plasma (40 CFR 266-A9)
ICB: Information Collection Budget (EPA-94/04)
ICBEN: International Commission on the Biological Effects of Noise (EPA-94/04)
ICC: Interstate Commerce Commission
ICE: Industrial Combustion Emissions Model (EPA-94/04)
ICE: Internal Combustion Engine (EPA-94/04)
ICGEC: Interagency Collaborative Group on Environmental Carcinogenesis
ICI: Independent Commercial Importer (40 CFR 89.3-94)
ICIE: International Center for Industry and Environment
ICP: Inductively Coupled Plasma (EPA-94/04)
ICR: Information Collection Request (EPA-94/04)
ICR: Institute of Cancer Research (EPA-92/12)
ICRDB: International Cancer Research Data Bank
ICRE: Ignitability, Corrosivity, Reactivity, Extraction (EPA-94/04)
ICRP: International Commission for Radiological Protection (EPA-92/12)
ICRP: International Commission on Radiological Protection (EPA-94/04)
ICRU: International Commission on Radiological Units (40 CFR 141.2)
ICS: Incident Command System (EPA-94/04)
ICS: Institute for Chemical Studies (EPA-94/04)
ICS: Intermittent Control Strategies (EPA-94/04)
ICS: Intermittent Control System (40 CFR 51.100-nn)
ICTCB: International Congress on Toxic Combustion Byproduct
ICWM: Institute for Chemical Waste Management (EPA-94/04)
ID: Idaho
ID: Inside Diameter (40 CFR 60)
ID50: Infectious Dose 50
ID fan: Induced Draft Fan
IDHW: Idaho Department of Health and Welfare (DOE-91/4)
IDLH: Immediately Dangerous to Life and Health (EPA-94/04)
IDLH: Immediately Dangerous to Life or Health (29 CFR 1910.120)
IDOD: Immediate Dissolved Oxygen Demand
i.e.: That Is
IEB: International Environment Bureau (EPA-94/04)
IEEE: Institute of Electrical and Electronics Engineers
IEMP: Integrated Environmental Management Project (EPA-94/04)
IEMS: Integrated Emergency Management System (NRT-87/3)
IES: Institute for Environmental Studies (EPA-94/04)
IFB: Invitation for Bid (EPA-94/04)
IFCAM: Industrial Fuel Choice Analysis Model (EPA-94/04)
IFIS: Industry File Information System (EPA-94/04)
IFMS: Integrated Financial Management System (EPA-94/04)
IFPP: Industrial Fugitive Process Particulate (EPA-94/04)
IG: Inspector General
IGCC: Integrated Gasification Combined Cycle (EPA-94/04)
IGCI: Industrial Gas Cleaning Institute (EPA-94/04)
IIS: Inflationary Impact Statement (EPA-94/04)
IJC: International Joint Commission (on Great Lakes) (EPA-94/04)
IL: Illinois
ILO: International Labor Organization
i.m.: Intramuscular (EPA-92/12)
IMEP: Indicated Mean Effective Pressure (40 CFR 85.2122)

IMF: International Monetary Fund
IMM: Intersection Midblock Model (EPA-94/04)
IMPACT: Integrated Model of Plumes and Atmosphere in Complex Terrain (EPA-94/04)
IMPROVE: Interagency Monitoring of Protected Visual Environment (EPA-94/04)
in.: Inch(es) (40 CFR 86.403.78-94)
in H$_2$O: Inches of Water (40 CFR 60)
in Hg: Inches of Mercury (40 CFR 60)
In: Indium
IN: Indiana
INEL: Idaho National Engineering Laboratory (DOE-91/4)
INPUFF: A Gaussian Puff Dispersion Model (EPA-94/04)
INT: Intermittent (EPA-94/04)
IO: Immediate Office
IOAA: Immediate Office of the Assistant Administrator
IOAU: Input/Output Arithmetic Unit
IOB: Iron Ore Beneficiation (EPA-94/04)
IOU: Input/Output Unit (EPA-94/04)
i.p.: Intraperitoneal (EPA-92/12)
IP: Inhalable Particles (EPA-94/04)
IPA: Intergovernmental Personnel Act
IPA: Intergovernmental Personnel Agreement
IPCS: International Program on Chemical Safety (EPA-94/04)
IPDWS: Interim Primary Drinking Water Standard (SWDA)
IPM: Inhalable Particulate Matter (EPA-94/04)
IPM: Integrated Pest Management (EPA-94/04)
IPP: Implementation Planning Program (EPA-94/04)
IPP: Independent Power Production (40 CFR 72.2)
IPP: Integrated Plotting Package (EPA-94/04)
IPP: Intermedia Priority Pollutant (document) (EPA-94/04)
Ir: Iridium
IR: Infrared
IRA: Initial Rate of Absorption
IRAC: International Association of Research on Cancer (France)
IRB: Institutional Review Board (40 CFR 26.102)
IRG: Interagency Review Group (EPA-94/04)
IRI: Industrial Risk Insurance
IRIS: Instructional Resources Information System (EPA-94/04)
IRIS: Integrated Risk Information System (EPA-94/04)
IRLG: Interagency Regulatory Liaison Group (Composed of EPA, CPSC, FDA, and OSHA) (EPA-94/04)
IRM: Intermediate Remedial Measures (EPA-94/04)
IRMC: Inter-Regulatory Risk Management Council (EPA-94/04)
IRP: Installation Restoration Program (EPA-94/04)
IRPTC: International Register of Potentially Toxic Chemicals (EPA-94/04)
IRR: Institute of Resource Recovery (EPA-94/04)
IRS: Internal Revenue Service
IRS: International Referral Systems (EPA-94/04)
IS: Interim Status (EPA-94/04)
ISA: Instrument Society of American
ISAC: Industry Sector Advisory Committee
ISAM: Indexed Sequential File Access Method (EPA-94/04)
ISC: Industrial Source Complex (EPA-94/04)
ISCL: Interim Status Compliance Letter (EPA-94/04)
ISCLT: Industrial Source Complex Long Term Model (EPA-94/04)
ISCST: Industrial Source Complex Short Term Model (EPA-94/04)
ISD: Interim Status Document (EPA-94/04)

ISE: Ion-specific electrode (EPA-94/04)
ISMAP: Indirect Source Model for Air Pollution (EPA-94/04)
ISO: International Organization for Standardization
ISO: International Science Organization
ISO: International Standard Organization
ISPF: (IBM) Interactive System Productivity Facility (EPA-94/04)
ISS: Interim Status Standards (EPA-94/04)
ISTEA: Intermodal Surface Transportation Efficiency Act of 1991 (40 CFR 93.101)
ISV: In Situ Vitrification
ITC: Interagency Testing Committee (EPA-94/04)
ITC: International Trade Commission
ITDP: Individual Training and Development Plan
ITII: International Technical Information Institute (EPA-92/12)
ITP: Individual Training Plan
ITP: International Travel Plan
ITSDC: Interagency Toxic Substances Data Committee
IUP: Intended Use Plan (40 CFR 35.3150)
IUPAC: International Union of Pure and Applied Chemistry (29 CFR 1910.1200)
i.v.: Intravenous (EPA-92/12)
IUR: Inventory Update Rule (EPA-94/04)
IW: Infectious Waste
IWC: In-Stream Waste Concentration (EPA-94/04)
IWS: Ionizing Wet Scrubber (EPA-94/04)

J: Joule (40 CFR 60)
JAPCA: Journal of Air Pollution Control Association (EPA-94/04)
JCL: Job Control Language (EPA-94/04)
JEC: Joint Economic Committee (EPA-94/04)
JECFA: Joint Expert Committee of Food Additives (EPA-94/04)
JLC: Justification for Limited Competition (EPA-94/04)
JMPR: Joint Meeting on Pesticide Residues (EPA-94/04)
JNCP: Justification for Non-Competitive Procurement (EPA-94/04)
JOFOC: Justification for Other than Full and Open Competition (EPA-94/04)
JPA: Joint Permitting Agreement (EPA-94/04)
JPL: Jet Propulsion Laboratory
JSD: Jackson Structured Design (EPA-94/04)
JSP: Jackson Structured Programming (EPA-94/04)
JTU: Jackson Turbidity Unit (EPA-94/04)

k: 1000
K: Kelvin (40 CFR 86.403.78-94)
K: Potassium
kg: Kilogram(s) (40 CFR 86.403.78-94)
kkg: 1000 kilogram(s) (40 CFR 401.11)
km: Kilometre(s) (40 CFR 86.403.78-94)
KOV: Knock Out Vessel
kPa: Kilopascals (one thousand newtons per square meter) (40 CFR 52.741-94)
KPEG: potassium polyethylene glycolate
Kr: Krypton
KS: Kansas
Ku: Kurchatovium
kW: Kilowatt (40 CFR 89.3-94)
KWH: Kilowatt Hour (40 CFR 401.11)
KY: Kentucky

L: Liter (40 CFR 401.11)
La: Lanthanum
LA: Load Allocation (40 CFR 130.2)
LA: Louisiana
LAA: Lead Agency Attorney (EPA-94/04)
LADD: Lowest Acceptable Daily Dose (EPA-94/04)
LAER: Lowest Achievable Emission Rate (40 CFR 51)
LAER: Lowest Achievable Emission Rate (EPA-94/04)
LAI: Laboratory Audit Inspection (EPA-94/04)
LAMP: Lake Acidification Mitigation Project (EPA-94/04)
LARC: International Agency for Research on Cancer (29 CFR 1910.1200)
LAS: Linear Alkylate Sulfonate
LASER: Light Amplification by Stimulated Emission of Radiation
lb: Pound(s) (40 CFR 86.403.78-94)
LC LO: Lethal Concentration Low
LC: Lethal Concentration (40 CFR 116)
LC: Liquid Chromatography (EPA-94/04)
LC50: Lethal Concentration 50; concentration lethal to 50% of the animals (EPA-92/12)
LCD: Local Climatological Data (EPA-94/04)
LCL: Lower Control Limit (EPA-94/04)
LCLO: Lethal Concentration Low; the lowest concentration at which death occurred (EPA-92/12)
LCM: Life Cycle Management (EPA-94/04)
LCRS: Leachate Collection and Removal System (EPA-94/04)
LD: Land Disposal (EPA-94/04)
LD: Lethal Dose
LD: Light Duty (EPA-94/04)
LD 0: Lethal Dose Zero
LD5O: Lethal Dose 50; dose lethal to 50% of the animals (EPA-92/12)
LD L0: The lowest dosage of a toxic substance that kills test organisms (EPA-94/04)
LD LO: Lethal Dose Low
LDC: London Dumping Convention (EPA-94/04)
LDCRS: Leachate Detection, Collection, and Removal System (EPA-94/04)
LDD: Light-Duty Diesel (EPA-94/04)
LDH: Lactic Acid Dehydrogenase (EPA-92/12)
LDIP: Laboratory Data Integrity Program (EPA-94/04)
LDR: Land Disposal Restrictions (EPA-94/04)
LDRTF: Land Disposal Restrictions Task Force (EPA-94/04)
LDS: Leak Detection System (EPA-94/04)
LDT: Light-Duty Truck (40 CFR 88.103-94)
LDT: Lowest Dose Tested (EPA-94/04)
LDV: Light-Duty Vehicle (40 CFR 88.103-94)
LEL: Lower Explosive Limit (40 CFR 52.741)
LEL: Lowest Effect Level (EPA-94/04)
LEP: Laboratory Evaluation Program (EPA-94/04)
LEPC: Local Emergency Planning Committee (40 CFR 300.4)
LERC: Local Emergency Response Committee (EPA-94/04)
LEV: Low-Emission Vehicle (40 CFR 88.103-94)
LFL: Lower Flammability Limit (EPA-94/04)
LGR: Local Governments Reimbursement Program (EPA-94/04)
Li: Lithium
LI: Langelier Index (EPA-94/04)
LIDAR: Light Detection and Ranging (40 CFR 60-AA (alt. method 1))
LIFO: Last In/First Out
LIMB: Limestone-Injection Multi-Stage Burner (EPA-94/04)

LLNL: Lawrence Livermore National Laboratory
LLRW: Low Level Radioactive Waste (EPA-94/04)
LLW: Low Level Radioactive Waste (EPA-91/12)
LMFBR: Liquid Metal Fast Breeder Reactor (EPA-94/04)
LMR: Labor Management Relations
LNEP: Low Noise Emission Product
LNG: Liquified Natural Gas
LOAEL: Lowest-Observed-Adverse-Effect-Level (EPA-94/04)
LOC: Letter of Credit (40 CFR 35.3105)
LOC: Level of Concern
LOC: Library of Congress
LOE: Level of Effort
LOEC: Lowest Observed Effect Concentration (EPA-91/3)
LOEL: Lowest Observed Effect Level (EPA-8/90b)
LOIS: Loss of Interim Status (SDWA)
LONGZ: Long Term Terrain Model
LOQ: Level of Quantitation (40 CFR 766.3-93)
LOT: Light off Time (40 CFR 85.2122)
LP: Legislative Proposal
LPC: Limiting Permissible Concentration (40 CFR 227.27)
LPG: Liquefied Petroleum Gas (40 CFR 86.094.3-94)
Lpm: Liter per minute (40 CFR 60)
LRC: Lewis Research Center (NASA)
LRMS: Low Resolution Mass Spectroscopy
LRTAP: Long Range Transportation of Air Pollution
LSI: Legal Support Inspection (CWA)
LSL: Lump Sum Leave
LST: Low Solvent Technology
LTA: Lead Trial Attorney
LTA: Long Term Average (EPA-91/3)
LTD: Land Treatment Demonstration
LTHE: Low Temperature Heat Exchanger
LTO: Landing Takeoff (40 CFR 87.2-94)
LTOP: Lease to Purchase
LTR: Lead Technical Representative
LTU: Land Treatment Unit
Lu: Lutetium
LUIS: Label Use Information System (EPA-94/04)
LUST: Leaking Underground Storage Tank(s) (current usage omits the "L")
LVW: Loaded Vehicle Weight (40 CFR 88.103-94)
Lw: Lawrencium
LWCF: Land and Water Conservation Fund
LWDF: Liquid Waste Derived Fuel
LWK: Live Weight Killed (40 CFR 432.11)
LWOP: Lease With Option to Purchase
LWOP: Leave Without Pay

m: Metre(s) (40 CFR 86.403.78-94)
M: Molar (40 CFR 60)
m^2: Square Meter (40 CFR 60)
m^3: Cubic Meter (40 CFR 60)
M5: Method 5
MA: Massachusetts
MAB or MCA: Monoclonal Antibodies
MAB: Man and Biosphere Program
MAC: Management Advisory Committee
MACT: Maximum Achievable Control Technology (CAA-title III)
MADCAP: Model of Advection, Diffusion, and Chemistry for Air Pollution
MAER: Maximum Allowable Emission Rate

MAG: Management Advisory Group
MAOP: Maximum Allowable Operating Pressure (40 CFR 192.3)
MAPCC: Michigan Air Pollution Control Commission
MAPPER: Maintaining, Preparing, and Producing Executive Reports
MAPS: Multistate Atmospheric Power Production Pollution Study
MAPSIM: Mesoscale Air Pollution Simulation Model (EPA-94/04)
MARC: Mining and Reclamation Council
MARPOL 73/78: International Convention for the Prevention of Oil Pollution from Ships, 1973, as Modified by the Protocol of 1978 (40 CFR 110.1)
MATC: Maximum Acceptable Toxicant Concentration (40 CFR 797.1330)
MATC: Maximum Allowable Toxicant Concentration
MAWP: Maximum Allowable Working Pressure
MBAS: Methylene-Blue-Active Substances (EPA-94/04)
Mbbl: 1000 barrels (one barrel is equivalent to 42 gallons) (40 CFR 419.11)
MBDA: Minority Business Development Agency
MBE: Minority Business Enterprises (40 CFR 35.6015)
MBTA: Massachusetts Bay Transportation Authority (40 CFR 52.1161)
MC: Multiple Cyclones
MCA: Manufacturing Chemists Association
MCEF: Mixed Cellulose Ester Fiber
MCL: Maximum Contaminant Level (40 CFR 142.2)
MCLG: Maximum Continment Level Goal (40 CFR 141.2)
MCP: Methylcyclopentane (40 CFR 799.2155)
MCP: Municipal Compliance Plan (CWA)
Md: Mendelevium
MD: Mail Drop
MD: Maryland
MDA: Methylenedianilline
MDL: Method Detection Limit (40 CFR 300-AA)
ME: Maine
MED: Minimum Effective Dose (EPA-92/12)
MEFS: Midterm Energy Forecasting System
MEFV: Maximum Expiratory Flow Volume (EPA-92/12)
MEG: Multimedia Environmental Goal
MEI: Maximum Exposed Individual
MEK: Methyl Ethyl Ketone
MEM: Modal Emission Model
MENS: Mission Element Needs Statement
MeOH: Methanol (CH3OH) (40 CFR 86.090.3-94)
MEP: Multiple Extraction Procedure (EPA-94/04)
meq: Milliequivalent (40 CFR 60)
MERS: Monticello Ecological Research Station (EPA-91/3)
MESOPAC: Mesoscale Meteorological Reprocesser Program
MESOPLUME: Mesoscale "Bent Plume" Model
MESOPUFF: Mesoscale Puff Model
MESS: Model Evaluation Support System
MEXAMS: Metals Exposure Analysis Modeling System (EPA-91/3)
MF: Membrane Filter (40 CFR 121.21)
MF: Modifying Factor (EPA-92/12)
MFBI: Major Fuel Burning Installation
MFC: Metal Finishing Category
MFL: Million Fibers per Liter (141.23.a)
mg: Milligram = 10^{-3} gram (40 CFR 60)
Mg: Magnesium
Mg: Megagram = 10^6 gram (metric tons or tonnes) (40 CFR 72.741-94)
mg/kg: Milligrams per Kilogram = (approx.) ppm (40 CFR 116)
mg/L: Milligrams per liter = (approx.) ppm (40 CFR 116)
Mgal: One Thousand Gallons (40 CFR 419.11)
MGD: Million Gallons per Day
MH: Man Hours
MHD: Magnetohydrodynamics
MI: Michigan
MIBK: Methyl Isobutyl Ketone
MIC: Methyl Isocaynate
MICE: Management Information Capability for Enforcement
MICH: Michigan fiver Model (EPA-91/3)
MICROMORT: A One-in-a-Million Chance of Death from an Environmental Hazard
min.: Minute(s) (40 CFR 87.2-94)
MINTEQA2: Equilibrium Metals Speciation Model (EPA-91/3)
MIPS: Millions of Instructions per Second
MIS: Management Information System
MIS: Mineral Industry Surveys
MITS: Management Information Tracking System
MJ: Megajoule(s) (1 million joules) (40 CFR 86.090.3-94)
mL: Milliliter = 10^{-3} Liter (40 CFR 60)
mL/L: Milliliters per Liter (40 CFR 434.11)
ML: Meteorology Laboratory
ML: Military Leave
ML: Minimum Level (EPA-91/3)
MLAP: Migrant Legal Action Program
MLSS: Mixed Liquor Suspended Solids
MLVSS: Mixed Liquor Volatile Suspended Solids
mm: Millimeter = 10^{-3} meter (40 CFR 60)
MM5: Modified Method 5 (sampling method)
MMAD: Mass Median Aerodynamic Diameter (EPA-92/12)
mmHg: Millimeters of Mercury; a measure of pressure (EPA-92/12)
MMPA: Marine Mammal Protection Act
MMS: Minerals Management Service (DOI)
MMT: Million Metric Tons
Mn: Manganese
MN: Minnesota
MNT: Mononitrotoluene
Mo: Molybdenium
MO: Missouri
MOA: Memorandum of Agreement
MOBILE: Mobile Source Emission Model
MOD: Miscellaneous Obligation Document
MOD: Modification
MOE: Margin of Exposure (EPA-94/04)
MOI: Memorandum of Intent
mol: Mole (40 CFR 60)
mol. wt.: Molecular weight (40 CFR 60)
MOS: Margin of Safety (EPA-94/04)
MOS: Metal Oxide Semiconductor
MOU: Memorandum of Understanding
MP: Manufacturing-use Product (EPA-94/04)
m.p.: Melting Point
MP: Melting Point (EPA-94/04)
mpc: 1000 Pieces (40 CFR 427.71)
MPC: Max. Permissible Concentration (radiation)
MPG: miles per gallon (40 CFR 600.003.77-94)
mph: Miles per hour (40 CFR 86.403.78-94)
MPN: Maximum Probable Number (EPA-94/04)

MPN: Most Probable Number
MPO: Metropolitan Planning Organization (40 CFR 51.138)
MPP: Merit Promotion Plan
MPRSA: Marine Protection, Research and Sanctuaries Act (40 CFR 2.309)
MPTDS: MPTER Model with Deposition and Settling of Pollutants
MPTER: Multiple Point Source Model with Terrain
MRA: Minimum Retirement Age
mrem: Millirem = 10^{-3} rem (40 CFR 60)
MRF: Materials Recovery Facility (EPA-94/04)
MRID: Master Record Identification number (EPA-94/04)
MRL: Maximum-Residue Limit (Pesticide Tolerance) (EPA-94/04)
MRP: Multi-Roller Press (in sludge drying unit)
MS: Mail Stop
MS: Mass Spectrometry
MS: Mississippi
MSA: Management System Audits
MSA: Metropolitan Statistical Area (40 CFR 60.331)
MSAM: Multi-Keyed Indexed Sequential File Access Method
MSBu: 1000 Standard Bushels (40 CFR 406.11)
MSDS: Material Safety Data Sheet (40 CFR 370.2)
MSEE: Major Source Enforcement Effort
MSHA: Mine Safety and Health Administration (DOL)
MSIS: Model State Information System
MSL: Mean Sea Level
MSPB: Merit System Protection Board
MSW: Medical Solid Waste
MSW: Municipal Solid Waste (40 CFR 240.101)
MSWLF: Municipal Solid Waste Landfill (40 CFR 258.1)
MT: Montana
MTB: Materials Transportation Bureau
MTBE: Methyl Tertiary Butyl Ether
MTD: Maximum Tolerated Dose (EPA-94/04)
MTDDIS: Mesoscale Transport Diffusion and Deposition Model for Industrial Sources
MTF: Multiple-Tube Fermentation (40 CFR 141.21)
MTG: Media Task GroupMTF: Multiple-Tube Fermentation
MTL: Median Threshold Limit (EPA-92/12)
MTP: Maximum Total Trihalomethane Potential (40 CFR 142.2)
MTS: Management Tracking System (OW)
MTSL: Monitoring and Technical Support Laboratory
MTU: Mobile Treatment Unit (40 CFR 261.4(f))
MTZ: Mass Transfer Zone
MUP: Manufacturing-Use Product (EPA-94/04)
MUTA: Mutagenicity (EPA-94/04)
MVA: Multivariate Analysis
MVAC: Motor Vehicle Air Conditioner (40 CFR 82.152)
MVAPCA: Motor Vehicle Air Pollution Control Act
MVEL: Motor Vehicle Emissions Laboratory
MVh: Minute ventilatory volume for human (composite value expressed in cu.m/day) used in the HEC derivation of an RfC (EPA-92/12)
MVho: Minute ventilatory volume for human in an occupational environment, assuming 8 hour/day exposure (composite value expressed in MVI/M) Motor Vehicle Inspection/Maintenance
MVIACSA: Motor Vehicle Information and Cost Saving Act (40 CFR 600.002)
MVICSA: Motor Vehicle Information and Cost Savings Act
MVMA: Motor Vehicle Manufacturers Association
MVRS: Marine Vapor Recovery System

MVTS: Motor Vehicle Tampering Survey
MW: Megawatt(40 CFR 401.11)
MW: Molecular Weight
MWC: Municipal Waste Combustor (40 CFR 60.51a)
MWG: Model Work Group
Mwh: Megawatt Hour(s) (40 CFR 401.11)
MWI: Medical Waste Incinerator
MWL: Municipal Waste Leachate
MWTA: Medical Waste Tracking Act of 1988
MYDP: Multi-Year Development Plans

N: Newton (40 CFR 60)
N: Nitrogen
N: Normal (40 CFR 60)
N/A: Not Applicable
N/A: Not Available
N_2: Nitrogen (40 CFR 86.403.78-94)
n.a.: Not Available (EPA-92/12)
Na: Sodium
NA: National Archives
NA: Nonattainment
NAA: Nonattainment Areas
NAAQS: National Ambient Air Quality Standards Program (CAA) (40 CFR 57.103)
NAAS: National Air Audit System (OAR)
NAC: National Academy of Sciences
NAC: Nitric Acid Concentrator
NACA: National Agricultural Chemicals Association
NAD: Nicotinamide Adenine Dinucleotide
NADB: National Atmospheric Data Bank
NADP: National Atmospheric Deposition Program
NAIS: Neutral Administrative Inspection System)
NALD: Nonattainment Areas Lacking Demonstrations
NAM: National Association of Manufacturers
NAMA: National Air Monitoring Audits
NAMS: National Air Monitoring System (40 CFR 58.1-c)
NANCO: National Association of Noise Control Officials
NAPAP: National Acid Precipitation Assessment Program (CAA402)
NAPBN: National Air Pollution Background Network
NAPBTAC: National Air Pollution Control Technical Advisory Committee
NAPCTAC: National Air Pollution Control Techniques Advisory Committee
NAPIM: National Association of Printing Ink Manufacturers
NAR: National Asbestos Registry
NARA: National Air Resources Act
NARA: National Archives and Records Administration
NARM: Naturally-occurring or Accelerator-produced Radioactive Material
NARS: National Asbestos-Contractor Registry System
NAS: National Academy of Sciences
NAS: National Audubon Society
NASA: National Aeronautics and Space Administration
NATICH: National Air Toxics Information Clearinghouse
NATO: North Atlantic Treaty Organization
NATS: National Air Toxics Strategy
NAWC: National Association of Water Companies
NAWDEX: National Water Data Exchange
Nb: Niobium
NBAR: Non-Binding Allocation of Authority

NBS: National Bureau of Standards (see NIST)
NC: North Carolina
NC fines: Nitrocellulose fines
NCA: National Coal Association
NCA: Noise Control Act of 1972
NCAC: National Clean Air Coalition
NCAF: National Clean Air Fund
NCAMP: National Coalition Against the Misuse of Pesticides
NCAQ: National Commission on Air Quality
NCAR: National Center for Atmospheric Research
NCASI: National Council of the Paper Industry for Air and Stream Improvements
NCC: National Climatic Center
NCC: National Computer Center
NCF: Network Control Facility
NCHS: National Center for Health Statistics (NIH)
nCi/L: Nanocuries per Liter (DOE-91/4)
NCI: National Cancer Institute
NCIC: National Crime Information Center
NCLP: National Contract Laboratory Program
NCM: National Coal Model
NCM: Notice of Commencement of Manufacture (TSCA)
NCN: Nitrocarbonitrate (40 CFR 457.31)
NCO: Negotiated Consent Order
NCP: National Contingency Plan (40 CFR 300.4)
NCP: National Oil and Hazardous Substances Pollution Contingency Plan (EPA-91/12; NRT-87/3)
NCP: Noncompliance Penalties (CAA)
NCP: Nonconformance Penalty (40 CFR 86.1102.87)
NCR: Noncompliance Report (CWA)
NCR: Nonconformance Report
NCRIC: National Chemical Response and Information Center
NCRP: National Council on Radiation Protection and Measurements (DOE-91/4)
NCS: National Compliance Strategy
NCTR: National Center for Toxicological Research
NCV: Nerve Conduction Velocity
NCVECS: National Center for Vehicle Emissions Control and Safety
NCWQ: National Commission on Water Quality
NCWS: Non-Community Water System (EPA-94/04)
Nd: Neodymium
ND: North Dakota
NDD: Negotiation Decision Document
NDDN: National Dry Deposition Network
NDIR: Nondispersive Infrared Analysis (40 CFR 86)
NDS: National Dioxin Study
NDS: National Disposal Site
NDUV: Nondispersion Ultraviolet
NDWAC: National Drinking Water Advisory Council
Ne: Neon
NE: Nebraska
NEA: National Energy Act
NEDA: National Environmental Development Association
NEDS: National Emissions Data System
NEEC: National Environmental Enforcement Council
NEEJ: National Environmental Enforcement Journal
NEIC: National Earthquake Information Center (DOE-91/4)
NEIC: National Enforcement Investigations Center (40 CFR 16.13)
NEMA: National Electrical Manufacturer's Association
NEP: National Energy Plan

NEP: National Estuary Program (40 CFR 35.9000)
NEPA: National Environmental Policy Act of 1969 (40 CFR 1508.21)
NER: National Emissions Report
NEROS: Northeast Regional Oxidant Study
NESCAUM: Northeast States for Coordinated Air Use Management
NESHAP: National Emissions Standards for Hazardous Air Pollutants (CAA) (40 CFR 300)
NETA: National Environmental Training Association (EPA-94/04)
NETTING: Emission Trading Used to Avoid PSD/NSR Permit Review Requirements
NFA: National Fire Academy (NRT-87/3)
NFAN: National Filter Analysis Network
NFFE: National Federation of Federal Employees
NFIP: National Flood Insurance Program (40 CFR 6-AA)
NFPA: National Fire Protection Association
NFRAP: No Further Remedial Action Planned (EPA-94/04)
NFWF: National Fish and Wildlife Foundation
ng: Nanogram = 10^{-9} gram (40 CFR 60)
NGA: Natural Gas Association
NGPA: Natural Gas Policy Act
NGWIC: National Ground Water Information Center
NH: New Hampshire
NHANES: National Health and Nutrition Examination Study
NHMIE: National Hazardous Materials Information Exchange (NRT-87/3)
NHPA: National Historic Preservation Act of 1966
NHTSA: National Highway Traffic Safety Act
NHTSA: National Highway Traffic Safety Administration (DOT)
NHWP: Northeast Hazardous Waste Project
Ni: Nickel
NICS: National Institute for Chemical Studies
NICT: National Incident Coordination Team (EPA-94/04)
NIEHS: National Institute of Environmental Health Sciences
NIEI: National Indoor Environmental Institute
NIH: National Institutes of Health (EPA-92/12)
NIM: National Impact Model
NIMBY: Not In My Backyard
NIOSH: National Institute for Occupational Safety and Health of the U.S. Department of Health and Human Services (29 CFR 1910.20)
NIPDWR: National Interim Primary Drinking Water Regulations (EPA-94/04)
NIS: Noise Information System
NISAC: National Industrial Security Advisory Committee (EPA-94/04)
NIST: National Institute for Standards and Testing (40 CFR 89.3-94)
NIST: National Institute of Standards and Technology (Formerly NBS)
NITEP: National Incinerator Testing and Evaluation Program
Nitrogenous BOD: Nitrogenous Bacteria Oxygen Demand
NJ: New Jersey
NJIT: New Jersey Institute of Technology
NLA: National Lime Association
NLAP: National Laboratory Audit Program
NLETS: National Law Enforcement Teletype Systems
NLGI number: National Lubricating Grease Institute number
NLM: National Library of Medicine (EPA-92/12)
NLT: Not Later Than

nm: Nanometer = 10^{-9} meter
NM: New Mexico
NMC: National Meteorological Center
NMFS: National Marine Fisheries Service (40 CFR 233.2)
NMHC: Non-Methane Hydrocarbons (40 CFR 86.094.3-94)
NMHCE: Non-Methane Hydrocarbon Equivalent (40 CFR 86.094.3-94)
NMOC: Non-Methane Organic Compound
NMOG: Non-Methane Organic Gas (40 CFR 88.103-94)
NMP: National Municipal Policy
NMR: Nuclear Magnetic Resonance
NNC: Notice of Noncompliance
NNPSPP: National Non-Point Source Pollution Program
No: Nobelium
No.: Number (40 CFR 86.403.78-94)
NO: Nitric oxide (40 CFR 89.3-94)
NO$_2$: Nitrogen Dioxide (40 CFR 58.1)
NOA: Nearest Onshore Area (40 CFR 55.2)
NOA: New Obligation Authority
NOA: Notice of Arrival (EPA-94/04)
NOAA: National Oceanic and Atmospheric Administration (DOC) (40FR300.4)
NOAC: Nature of Action Code (EPA-94/04)
NOAEL (NOEL): No Observable Adverse Effect Level (EPA-94/04)
NOC: Notice of Commencement
NOC: Notice of Construction (DOE-91/4)
NOD: Notice of Deficiency (RCRA)
NOD: Same as Nitrogenous BOD
NOEC: No Observed Effect Concentration (40 CFR 797.1600)
NOEL: No Observed Effects Level (40 CFR 798.4350)
NOHSCP: National Oil and Hazardous Substances Contingency Plan (40 CFR 300)
NOI: Notice of Intent
NON: Notice of Noncompliance (TSCA)
NOPES: Non-Occupational Pesticide Exposure Study
NORA: National Oil Recyclers Association
NOS: National Ocean Survey (NOAA)
NOS: Not Otherwise Specified (40 CFR 261-AVIII)
NOV: Notice of Violation
NOV/CD: Notice of Violation/Compliance Demand
NO$_x$: Nitrogen Oxides (CAA302)
Np: Neptunium
NP: Office of New Production Reactors (DOE-91/4)
NPAA: Noise Pollution and Abatement Act
NPCA: National Parks and Conservation Association
NPDES: National Pollutant Discharge Elimination System (40 CFR 110.1)
NPHAP: National Pesticide Hazard Assessment Program (EPA-94/04)
NPIRS: National Pesticide Information Retrieval System
NPL: National Priority(ies) List (40 CFR 300.4)
NPM: National Program Manager
NPN: National Particulate Network
NPR: New Production Reactor (DOE-91/4)
NPRM: Notice of Proposed Rulemaking
NPS: National Park Service
NPS: National Permit Strategy
NPS: National Pesticide Survey (OW)
NPS: Non-Point Source
NPS: Nonpoint Source Model for Urban and Rural Areas (EPA-91/3)
NPUG: National Prime User Group
NRA: National Recreation Area
NRC: National Research Council (EPA-92/12)
NRC: National Response Center (40 CFR 300.4)
NRC: Non-Reusable Containers
NRC: Nuclear Regulatory Commission (40 CFR 61.101)
NRCA: National Resource Council of America
NRDC: Natural Resources Defense Council
NREL: National Renewable Engineering Laboratory (Formerly SERI)
NRHP: National Register of Historic Places (DOE-91/4)
NRMRL: National Risk Management Research Laboratory (Formerly RREL, before June 1995)
NRR: Noise Reduction Rating (40 CFR 211.203)
NRT: National Response Team (40 CFR 300.4)
NRWA: National Rural Water Association
NSC: National Security Council
NSD: No Structure Detected (40 CFR 763-AA)
NSDWR: National Secondary Drinking Water Regulations (EPA-94/04)
NSEC: National System for Emergency Coordination (EPA-94/04)
NSEP: National System for Emergency Preparedness (EPA-94/04)
NSF: National Sanitation Foundation
NSF: National Science Foundation
NSF: National Strike Force (40 CFR 300.4)
NSO: Nonferrous Smelter Orders (CAA) (40 CFR 57.101)
NSPE: National Society of Professional Engineers
NSPS: New Source Performance Standards (CAA) (40 CFR 467.02)
NSR: New Source Review (EPA-94/04)
NSSC: Neutral Sulfite Semi-Chemical
NSTL: National Space Technology Laboratory
NSWMA: National Solid Waste Management Association
NSWS: National Surface Water Survey
NTA: Negotiated Testing Agreement
NTA: Nitrilotriacetic Acid
NTE: Not to Exceed
NTIS: National Technical Information Service (40 CFR 89.3-94)
NTN: National Trends Network
NTNCW: Non-Transient Non-Community Water System (40 CFR 141.2)
NTNCWS: Non-transient Non-community Water System (40 CFR 141)
NTP: National Toxicology Program (29 CFR 1910.1200)
NTS: Nevada Test Site (DOE-91/4)
NTSP: National Transportation Safety Board
NTU: Nephelometric Turbidity Units (40 CFR 131.35)
NURF: National Utility Reference File (CAA402)
NV: Nevada
NVPP: National Vehicle Population Poll
NWA: National Water Alliance
NWF: National Wildlife Federation
NWPA: Nuclear Waste Policy Act of 1982 (40 CFR 191.02)
NWRC: National Weather Records Center
NWS: National Weather Service (NOAA)
NY: New York

O: Oxygen
O/O: Owners or Operators (40 CFR 260-AI-93)
O&G: Oil and Gas

O&G: Oil and Grease (40 CFR 420.02)
O&M: Operation and Maintenance (40 CFR 300.4)
O$_2$: Oxygen (40 CFR 89.3-94)
OAIAP: Office of Atmospheric and Indoor Air Programs (EPA)
OALJ: Office of Administrative Law Judges (EPA)
OAQPS: Office of Air Quality Planning and Standards (EPA)
OARM: Office of Administration and Resources Management (EPA)
OASDI: Old Age and Survivor Insurance
OC: Object Class
OC: Office of Controller (EPA)
OC: Open Cup (EPA-92/12)
OCAPO: Office of Compliance Analysis and Program Operations (EPA)
OCD: Offshore and Coastal Dispersion (EPA-94/04)
OCE: Office of Civil Enforcement (EPA)
OCE: Office of Criminal Enforcement (EPA)
OCEM: Office of Cooperative Environmental Management (EPA)
OCI: Office of Criminal Investigations (40 CFR 16.13)
OCI: Organizational Conflicts of Interest
OCM: Office of Compliance Monitoring (EPA)
OCPSF: Organic Chemicals, Plastics, and Synthetic Fibers (40 CFR 414.11)
OCR: Office of Civil Rights (40 CFR 7.25)
OCR: Optical Character Reader
OCS: Outer Continental Shelf (40 CFR 55.2)
OCSLA: Outer Continental Shelf Lands Act (40 CFR 55.2)
OD: Organizational Development
OD: Outside Diameter (40 CFR 60)
ODBA: Ocean Dumping Ban Act of 1988
ODBU: Office of Disadvantaged Business Utilization (EPA)
OEC: Observed Effect Concentration (40 CFR 797.1600)
OEM: Original Equipment Manufacturer (40 CFR 85.1502)
OEP: Office of External Programs
OER: Office of Exploratory Research (EPA)
OERR: Office of Emergency and Remedial Response (EPA-91/12)
OF: Optional Form (EPA-94/04)
OFA: Office of Federal Activities (EPA)
OFFE: Office of Federal Facilities Enforcement (EPA)
OFR: Office of the Federal Register
OGC: Office of General Counsel (EPA)
OGWDW: Office of Ground Water and Drinking Water (EPA)
OH: Ohio
OHEA: Office of Health and Environmental Assessment (EPA)
OHM/TADS: Oil and Hazardous Materials Technical Assistance Data Systems (EPA-92/12)
ohm: Electrical Resistance Unit (40 CFR 60)
OHR: Office of Health Research (EPA)
OHRM: Office of Human Resources Management (EPA)
OHSS: Occupation Health and Safety Staff
OIA: Office of International Activities (EPA)
OIG: Office of the Inspector General (EPA)
OIRM: Office of Information Resources Management (EPA)
OK: Oklahoma
OLTS: On Line Tracking System (EPA-94/04)
OMB: Office of Management and Budget (40 CFR 31.3)
OMEP: Office of Marine and Esturine Protection (EPA)
OMHCE: Organic Material Hydrocarbon Equivalent (40 CFR 86)
OMMSQA: Office of Modeling, Monitoring Systems, and Quality Assurance (EPA)
OMNMHCE: Organic Material Non-Methane Hydrocarbon Equivalent (40 CFR 86)
OMS: Office of Mobile Sources (EPA)
ONRW: Outstanding National Resource Waters (EPA-91/3)
OP: Office of Prevention (EPA)
OP: Operating Plan
OP Year: Original Production Year (40 CFR 85.1502)
OPA: Office of Policy Analysis (EPA)
OPA: Oil Pollution Act of 1990 (Public Law 101-380)
OPAC: Overall Performance Appraisal Certification
OPAR: Office of Policy Analysis and Review (EPA)
OPF: Official Personnel Folder
OPIDN: Organophosphorus Induced Delayed Neurotoxicity (40 CFR 798.6450)
OPMO: Office of Program Management Operations (EPA)
OPP: Office of Pesticide Programs (EPA)
OPPE: Office of Policy, Planning and Evaluation (EPA)
OPTS: Office of Pesticides and Toxic Substances (40 CFR 179.3)
OR: Office of Enforcement (EPA)
OR: Oregon
ORA: Office of Radiation Programs (EPA)
ORD: Research and Development (EPA)
ORM: Other Regulated Material (EPA-94/04)
ORME: Office of Regulatory Management and Evaluation (EPA)
ORNL: Oak Ridge National Laboratory (DOE-91/4)
ORP: Oxidation-Reduction Potential (EPA-94/04)
ORP recorders: Oxidation Reduction Potential Recorders
ORPM: Office of Research Program Management (EPA)
ORV: Off-road Vehicle
Os: Osmium
OS: Orifice Scrubber
OS/VS: Operating System/Virtual Storage
OSC: On Scene Coordinator (40 CFR 300.4)
OSHA: Occupational Safety and Health Administration (DOL)
OSHAct: Occupational Safety and Health Act
OST: Office of Science and Technology, U.S. EPA (EPA-92/12)
OSTP: Office of Science and Technology Policy (White House)
OSW: Office of Solid Waste (EPA)
OSWER: Office of Solid Waste and Emergency Response (EPA-91/12)
OT: Overtime
OTA: Office of Technology Assessment (U.S. Congress)
OTEC: Ocean Thermal Energy Conversion
OTS: Office of Toxic Substances (EPA)
OTTRS: Office of Technology Transfer and Regulatory Support (EPA)
OUST: Office of Underground Storage Tanks (EPA)
OW: Office of Water (EPA)
OWEC: Office of Wastewater Enforcement and Compliance (EPA)
OWF: On the Weight of the Fiber
OWOW: Office of Wetlands, Oceans and Watersheds (EPA)
OWP: Overall Work Programs (40 CFR 51.138)
OWPE: Office of Waste Programs Enforcement (EPA)
OWRS: Office of Water Regulations and Standards, U.S. EPA (EPA-92/12)
OX: Oxidation (40 CFR 141.61.b)
OY: Operating Year
OYG: Operating Year Guidance
oz: Ounces (40 CFR 60)
OZIPP: Ozone Isopleth Plotting Package
OZIPPM: Modified Ozone Isopleth Plotting Package

p: Probit Dose Extrapolation Model (EPA-92/12)
P: Phosphorus
P.O.: per os (by mouth) (EPA-92/12)
P-TBB: p-tert-butylbenzaldehyde (40 CFR 704.33)
P-TBBA: p-tert-butylbenzoic acid (40 CFR 704.33)
P-TBT: p-tert-butyltoluene (40 CFR 704.33)
P&A: Precision and Accuracy
Pa: Pascal(pressure) (40 CFR 86.403.78-94)
Pa: Protactinium
PA: Pennsylvania
PA: Policy Analyst
PA: Preliminary Assessment (40 CFR 300.4)
PA/SI: Preliminary Assessment/Site Inspection (EPA-91/12)
PAA: Priority Abatement Areas
PAAT: Public Affairs Assist Team (40 CFR 300.5)
PADRE: Particle Analysis and Data Reduction Program
PAG: Protective Action Guide (DOE-91/4)
PAGM: Permit Applications Guidance Manual
PAH: Polycyclic Aromatic Hydrocarbon
PAH: Polynuclear Aromatic Hydrocarbon
PAHO: Pan American Health Organization
PAI: Pesticide Active Ingredient (40 CFR 455-Table 10)
PAI: Performance Audit Inspection (CWA) (EPA-94/04)
PAI: Pure Active Ingredient compound (EPA-94/04)
PAIR: Preliminary Assessment Information Rule
PAL: Point, Area, and Line Source Air Quality Mode
PALDS: PAL Model with Deposition and Settling of Pollutants
PAM: Pesticide Analytical Manual (EPA-94/04)
PAMS: Photochemical Assessment Monitoring Station (40 CFR 58.1)
PAN: Perooxyacetyl Nitrate
PAPR: Powered Air Purifying Respirator
PARS: Precision and Accuracy Reporting System
PASS: Procurement Automated Source System
PAT: Packed Tower Aeration (40 CFR 141.61.b)
PAT: Permit Assistance Team (RCRA) (EPA-94/04)
PATS: Pesticide Action Tracking System (EPA-94/04)
PATS: Pesticides Analytical Transport Solution (EPA-94/04)
Pb: Lead
PBA: Preliminary Benefit Analysis (BEAD) (EPA-94/04)
PBB: Polybromated Biphenyls (40 CFR 704.195)
PBC: Pacific Basin Conference
PBC: Packed Bed Condenser
PBDD: Polybrominated Dibenzo-p-Dioxin (40 CFR 766.3)
PBL: Planetary Boundary Layer
PBLSQ: The Lead Line Source Model
PBPK: Physiologically Based Pharmacokinetic (EPA-92/12)
PBS: Packed Bed Scrubber
PC: Personal Computer
PC: Planned Commitment
PC: Polycarbonate
PC: Position Classification
PC: Pulverized Coal
PC&B: Personnel Compensation and Benefits
PCA: Principle Component Analysis (EPA-94/04)
PCA: Production Compliance Audit (40 CFR 86.1102.87)
PCB: Polychlorinated Biphenyls (40 CFR 129.4)
PCC: Primary Combustion Chamber
PCDD: Polychlorinated Dibenzodioxin (40 CFR 766.3)
PCDF: Polychlorinated Dibendzofuran (40 CFR 766)
PCE: Pollution Control Equipment

PCE: Pyrometric Cone Equivalent
pCi/L: Picocuries per Litre
PCi: Picocurie = 10^{-12} Curie (40 CFR 60)
PCIE: President's Council on Integrity and Efficiency in Government
PCIOS: Processor Common Input/Output System
PCM: Phase Contrast Microscopy (40 CFR 763-AA)
PCN: Policy Criteria Notice (EPA-94/04)
PCNB: Pentachloronitrobenzene
PCO: Pest Control Operator (EPA-94/04)
PCO: Printing Control Officer
PCON: Potential Contractor
PCP: Pentachlorophenyl
PCRC: Pesticides Chemical Review Committee
PCS: Permanent Change of Station
PCS: Permit Compliance System (CWA)
PCS: Primary Coolant System (DOE-91/4)
PCS: Probable Carcinogenic Substances
PCSC: PC Site Coordinator
PCV: Positive Crankcase Ventilation (40 CFR 85.2122)
PCW: Post Consumer Waste (40 CFR 246.101)
Pd: Palladium
PD: Position Description
PD: Position Document (EPA-92/12)
PD: Project Description
PDCI: Product Data Call-In (EPA-94/04)
PDFID: Preconstruction Direct Flame Ionization Detection
PDMS: Pesticide Document Management System (OPP)
PDP: Positive Displacement Pump (40 CFR 86)
PDR: Particulate Data Reduction
PE: Polyethylene
PE: Population Equivalent
PE: Program Element
PEF: Particulate Emission Factor (EPA-91/12)
PEI: Petroleum Equipment Institute
PEL: Permissible Exposure Level (DOE-91/4)
PEL: Permissible Exposure Limit (40 CFR 763.121)
PEL: Personal Exposure Limit
PEM: Partial Equilibrium Multimarket Model
PEM: Personal Exposure Model
PEPE: Prolonged Elevated Pollution Episode
PEQI: Perceived Environmental Quality Indices
PERF: Police Executive Research Forum
PESTAN: Pesticides Analytical Transport Solution
PET: Poly(ethylene terephthalate) (40 CFR 60.561)
PF: Potency Factor
PFCRA: Program Fraud Civil Remedies Act (EPA-94/04)
PFT: Permanent Full Time
PFTE: Permanent Full Time Equivalent
PHC: Principal Hazardous Constituent (EPA-94/04)
PHDD: Polyhalogenated Dibenzo-p-Dioxin (40 CFR 766.3)
PHDF: Polyhalogenated Dibenzofuran (40 CFR 766.3)
PHS: U.S. Public Health Service (EPA-92/12)
PHSA: Public Health Service Act (EPA-94/04)
PI: Preliminary Injunction (EPA-94/04)
PI: Program Information (EPA-94/04)
PIAT: Public Information Assist Team (40 CFR 300.4)
PIC: Products of Incomplete Combustion (EPA-94/04)
PIC: Public Information Center
PID: Photon Ionization Detector
PIGS: Pesticides in Groundwater Strategy (EPA-94/04)

PIMS: Pesticide Incident Monitoring System (EPA-94/04)
PIN: Pesticide Information Network (EPA-94/04)
PIN: Procurement Information Notice (EPA-94/04)
PIP: Public Involvement Program (EPA-94/04)
PIPQUIC: Program Integration Project Queries Used in Interactive Command (EPA-94/04)
PIR: Product of Incomplete Reaction
PIRG: Public Interest Research Group (EPA-94/04)
PIRT: Pretreatment Implementation Review Task Force (EPA-94/04)
PIS: Public Information Specialist
PITS: Project Information Tracking System (EPA-94/04)
PL: Public Law (DOE-91/4)
PLIRRA: Pollution Liability Insurance and Risk Retention Act (EPA-94/04)
PLM: Polarized Light Microscopy (EPA-94/04)
PLO: Public Land Order (DOE-91/4)
PLUVUE: Plume Visibility Model (EPA-94/04)
Pm: Promethium
PM: Particulate Matter (40 CFR 86.094.3-94)
PM: Program Manager
PM10: Particulate Matter (nominally 10 um (micrometers) diameter) and Less (EPA-94/04)
PM15: Particulate Matter (nominally 15 um (micrometers) diameter) (EPA-94/04)
PMA: Pharmaceutical Manufacturers Association
PMEL: Pacific Marine Environmental Laboratory (EPA-94/04)
PMIP: Presidential Management Intern Program
PMIS: Personnel Management Information System (OARM)
PMN: Premanufacture Notification (TSCA) (40 CFR 700.43)
PMNF: Premanufacture Notification Form (EPA-94/04)
PMR: Pollutant Mass Rate (EPA-94/04)
PMRS: Performance Management and Recognition System (EPA-94/04)
PMS: Personnel Management Specialist
PMS: Program Management System (EPA-94/04)
PNA: Polynuclear Aromatic Hydrocarbons (EPA-94/04)
PNA: Probablistic Noise Audibility (DOE-91/4)
PNL: Pacific Northwest Laboratory (DOE-91/4)
Po: Polonium
PO: Project Officer (EPA-94/04)
PO: Purchase Order
POC: Point of Compliance (EPA-94/04)
POC: Program Office Contacts
POE: Point of Exposure (EPA-94/04)
POGO: Privately-Owned/Government-Operated (EPA-94/04)
POHC: Principal Organic Hazardous Constituent (EPA-94/04)
POI: Point of Interception (EPA-94/04)
POLREP: Pollution Report (EPA-94/04)
POM: Particulate Organic Matter (EPA-94/04)
POM: Polycyclic Organic Matter (EPA-94/04)
POR: Program of Requirements (EPA-94/04)
POTW: Publicly Owned Treatment Works (40 CFR 117.1)
POV: Privately Owned Vehicle (EPA-94/04)
PP: Pay Period
PP: Pollution Prevention
PP: Polypropylene (40 CFR 60.561)
PP: Priority Pollutant (CWA)
PP: Program Planning (EPA-94/04)
PPA: Pesticide Producers Association
PPA: Planned Program Accomplishment (EPA-94/04)
PPA: Pollution Prevention Act of 1990
ppb: Parts per Billion (40 CFR 60)
PPIA: Poultry Products Inspection Act (21USC453)
PPIC: Pesticide Programs Information Center (EPA-94/04)
PPIS: Pesticide Product Information System (EPA-94/04)
ppm: parts per million = (approx.) mg/kg (milligrams per kilogram) = (approx.) mg/L (milligrams per liter) (40 CFR 116)
PPM/PPB: Parts per Million/Parts per Billion (EPA-94/04)
PPMAP: Power Planning Modeling Application Procedure (EPA-94/04)
PPPA: Poison Prevention Packaging Act
PPSP: Power Plant Siting Program (EPA-94/04)
PPT: Parts per Trillion (EPA-94/04)
PPT: Permanent Part Time
PPTH: Parts per Thousand (EPA-94/04)
PQL: Practical Quantitation Limits (40 CFR 264-A9)
PQUA: Preliminary Quantitative Usage Analysis (EPA-94/04)
Pr: Praseodymium
PR: Preliminary Review (EPA-94/04)
PR: Procurement Request
PRA: Paperwork Reduction Act (EPA-94/04)
PRA: Planned Regulatory Action (EPA-94/04)
PRA: Probablistic Risk Assessment (DOE-91/4)
PRATS: Pesticides Regulatory Action Tracking System (EPA-94/04)
PRC: Planning Research Corporation (EPA-94/04)
PRG: Preliminary Remediation Goal (EPA-91/12)
PRI: Periodic Reinvestigation (EPA-94/04)
PRM: Prevention Reference Manuals (EPA-94/04)
PRMO: Policy and Resources Management Office (EPA)
PRN: Pesticide Registration Notice (EPA-94/04)
PRP: Potentially Responsible Party (40 CFR 35.4010)
PRP: Potentially Responsible Party (EPA-94/04)
PRZM: Pesticide Root Zone Model (EPA-94/04)
PS: Point Source (EPA-94/04)
PS: Polystyrene (40 CFR 60.561)
PSAM: Point Source Ambient Monitoring (EPA-94/04)
PSC: Program Site Coordinator (EPA-94/04)
PSD: Prevention of Significant Deterioration (40 CFR 52.21)
PSE: Program Subelement
PSES: Pretreatment Standards for Existing Sources (40 CFR 467.02)
psi: Pounds per Square Inch (Pressure)
PSI: Pollutant Standards Index (40 CFR 58-AG-a)
PSI: Pounds per Square Inch (EPA-94/04)
PSI: Pressure per Square Inch (EPA-94/04)
psia: Pounds per Square Inch Absolute (40 CFR 60)
psig: Pounds per Square Inch Gauge (40 CFR 60)
PSIG: Pressure per Square Inch Gauge (EPA-94/04)
PSM: Point Source Monitoring (EPA-94/04)
PSNS: Pretreatment Standards for New Sources (40 CFR 467.02)
PSP: Payroll Savings Plan
PSS: Personnel Staffing Specialist
PSTN: Pesticide Safety Team Network (NRT-87/3)
PSU: Primary Sampling Unit (EPA-94/04)
PSY: Steady-State, Two-Dimensional Plume Model (EPA-91/3)
Pt: Platinum
PT: Packed Tower
PT: Part Time
PTA: Part Throttle Acceleration (40 CFR 86)

PTC: Positive Temperature Coefficient (40 CFR 85)
PTD: Part Throttle Deceleration (40 CFR 86)
PTDIS: Single Stack Meteorological Model in EPA UNAMAP Series (EPA-94/04)
PTE: Permanent Total Enclosure (40 CFR 52.741.a.4)
PTE: Potential to Emit (EPA-94/04)
PTFE: Polytetrafluoroethylene (Teflon) (EPA-94/04)
PTMAX: Single Stack Meteorological Model in EPA UNAMAP series (EPA-94/04)
PTO: Patent and Trademark Office (DOC)
PTPLU: Point Source Gaussian Diffusion Model (EPA-94/04)
Pu: Plutonium
PU: Pulmonary region of the respiratory tract (EPA-92/12)
PUC: Public Utility Commission (EPA-94/04)
PUREX: Plutonium-Uranium Extraction Plant (DOE-91/4)
PV: Project Verification (EPA-94/04)
PVC: Polyvinyl Chloride (EPA-94/04)
PWS: Public Water Supply/System (EPA-94/04)
PWS: Public Water System (SDWA)
PWSS: Public Water Supply System (EPA-94/04)
PWSS: Public Water Supply System (SDWA)
PY: Prior Year

Q: Quench
QA: Quality Assurance (40 CFR 766.3)
QA/QC: Quality Assistance/Quality Control (EPA-94/04)
QAC: Quality Assurance Coordinator (EPA-94/04)
QAMIS: Quality Assurance Management and Information System (EPA-94/04)
QAO: Quality Assurance Officer (EPA-94/04)
QAPP: Quality Assurance Program (or Project) Plan (EPA-94/04)
QAT: Quality Action Team (EPA-94/04)
QBTU: Quadrillion British Thermal Units (EPA-94/04)
QC: Quality Control (40 CFR 766.3)
QC: Quench Column
QCA: Quiet Communities Act (EPA-94/04)
QCI: Quality Control Index (EPA-94/04)
QCP: Quiet Community Program (EPA-94/04)
ql*: Cancer Potency Factor (EPA-91/3)
QNCR: Quarterly Noncompliance Report (EPA-94/04)
QS: Quench Separator
QSAR: Quantitative Structure-Activity Relationships (EPA-91/3)
QSI: Quality Step Increase
QT: Quench Tower
QUA: Qualitative Use Assessment (EPA-94/04)
QUIPE: Quarterly Update for Inspector in Pesticide Enforcement (EPA-94/04)

R: Alkyl (C_nH_{2n+1}) group
R: Degree Rankine (40 CFR 60)
R: Rankine (40 CFR 86.403.78-94)
R&D: Research and Development (EPA-94/04)
Ra: Radium
RA: Reasonable Alternative (EPA-94/04)
RA: Regional Administrator
RA: Regulatory Alternatives (EPA-94/04)
RA: Regulatory Analysis (EPA-94/04)
RA: Remedial Action (40 CFR 300.4)
RA: Resource Allocation (EPA-94/04)
RA: Risk Analysis (EPA-94/04)
RA: Risk Assessment (EPA-94/04)

RAATS: RCRA Administrative Action Tracking System (EPA-94/04)
RAC: Radiation Advisory Committee (EPA-94/04)
RAC: Reference Air Concentration (ug/m^3)
RAC: Reference Ambient Concentration (EPA-91/3)
RAC: Regional Asbestos Coordinator (EPA-94/04)
RAC: Response Action Coordinator (EPA-94/04)
RACM: Reasonably Available Control Measures (EPA-94/04)
RACM: Regulated Asbestos-Containing Material (40 CFR 61.141)
RACT: Reasonably Available Control Technology (40 CFR 51.100-o)
RAD: Radiation Adsorbed Dose (unit of measurement of radiation absorbed by humans) (EPA-94/04)
RADM: Random Walk Advection and Dispersion Model (EPA-94/04)
RADM: Regional Acid Deposition Model (EPA-94/04)
RAGS: Risk Assessment Guidance for Superfund (EPA-91/12)
RAM: Random Access Memory (computer)
RAM: Urban Air Quality Model for Point and Area Source in EPA UNAMAP Series (EPA-94/04)
RAMP: Rural Abandoned Mine Program (EPA-94/04)
RAMS: Regional Air Monitoring System (EPA-94/04)
RAP: Radon Action Program (EPA-94/04)
RAP: Remedial Accomplishment Plan (EPA-94/04)
RAP: Reregistration Assessment Panel (EPA-94/04)
RAP: Response Action Plan (EPA-94/04)
RAPS: Regional Air Pollution Study (EPA-94/04)
RARG: Regulatory Analysis Review Group (EPA-94/04)
RAS: Routine Analytical Service (EPA-94/04)
RAT: Radiological Assistance Team (40 CFR 300.4)
RAT: Relative Accuracy Test (EPA-94/04)
Rb: Rubidium
RB: Red Border
RB: Request for Bid (EPA-94/04)
RBC: Red Blood Cell(s) (EPA-92/12)
RBC: Rotating Biological Contactor (40 CFR 35.2035)
RC: Regional Counsel
RC: Responsibility Center (EPA-94/04)
RCC: Radiation Coordinating Council (EPA-94/04)
RCDO: Regional Case Development Officer (EPA-94/04)
RCO: Regional Compliance Officer (EPA-94/04)
RCP: Regional Contingency Plan (40 CFR 300.4)
RCP: Research Centers Program (EPA-94/04)
RCRA: Resource Conservation and Recovery Act of 1976 (40 CFR 124)
RCRIS: Resource Conservation and Recovery Information System (EPA-94/04)
RD: Remedial Design (40 CFR 300.4)
RD/RA: Remedial Design/ Remedial Action (EPA-94/04)
RD&D: Research, Development and Demonstration (EPA-94/04)
RDD: Regional Deposited Dose (EPA-8/90b)
RDF: Refuse-Derived Fuel (40 CFR 60)
rDNA: Recombinant DNA (EPA-94/04)
RDU: Regional Decision Units (EPA-94/04)
RDV: Reference Dose Values (EPA-94/04)
Re: Rhenium
RE: Reasonable Efforts (EPA-94/04)
RE: Reportable Event (EPA-94/04)
REAG: Reproductive Effects Assessment Group
REAP: Regional Enforcement Activities Plan (EPA-94/04)
redox: Oxidation Reduction

REE: Rare Earth Elements (EPA-94/04)
REEP: Review of Environmental Effects of Pollutants (EPA-94/04)
REF: Reference
REM: Roentgen Equivalent Man (EPA-94/04)
REM: Remedial Engineering Management
REM: Roentgen Equivalent Man (radiation unit)
REM/FIT: Remedial/Field Investigation Team (EPA-94/04)
REMS: RCRA Enforcement Management System (EPA-94/04)
REP: Reasonable Efforts Program (EPA-94/04)
REPS: Regional Emissions Projection System (EPA-94/04)
RESOLVE: Center for Environmental Conflict Resolution (EPA-94/04)
Rf: Rutherfordium
RF: Radio Frequency
RF: Response Factor (40 CFR 796)
RFA: Regulatory Flexibility Act (EPA-94/04)
RFB: Request for Bid (EPA-94/04)
RFBAT: Reasonably Foreseeable Best Available Technology
RfC: Chronic inhalation reference dose (EPA-8/90b)
RfC: Inhalation Reference Concentration (EPA-92/12)
RfC: Reference Concentration (EPA-91/12)
RfC$_s$: Subchronic inhalation reference dose (EPA-8/90b)
RfD: Oral Reference Dose (EPA-92/12)
RfD: Reference Dose (EPA-91/12)
RfD: Reference Dose Values (40 CFR 300)
RFI: Radio Frequency Interference (40 CFR 60)
RFI: Remedial Field Investigation (EPA-94/04)
RFP: Reasonable Further Programs (EPA-94/04)
RFP: Request for Proposal (EPA-94/04)
RFP: Rocky Flats Plant (DOE-91/4)
RFQ: Request for Quote
RgD: Regulatory Dose (EPA-92/12)
RGS: Research Grants Staff
Rh: Rhodium
RH: Reheat
RHRS: Revised Hazard Ranking System (EPA-94/04)

RI: Reconnaissance Inspection (EPA-94/04)
RI: Remedial Investigation (40 CFR 300.4)
RI: Rhode Island
RI/FS: Remedial Information/Feasibility Study (EPA-94/04)
RI/FS: Remedial Investigation/Feasibility Study (EPA-91/12)
RI/FS: Reports of Investigation/Feasibility Studies (DOE-91/4)
RIA: Regulatory Impact Analysis (EPA-94/04)
RIA: Regulatory Impact Assessment (EPA-94/04)
RIC: Radon Information Center (EPA-94/04)
RIC: RTP Information Center
RICC: Retirement Information and Counseling Center (EPA-94/04)
RICO: Racketeer Influenced and Corrupt Organizations Act (EPA-94/04)
RID: Regulatory Integration Division
RIF: Reduction In Force
RIM: Regulatory Interpretation Memorandum (EPA-94/04)
RIN: Regulatory Identifier Number (EPA-94/04)
RIP: RCRA Implementation Plan (EPA-94/04)
RISC: Regulatory Information Service Center (EPA-94/04)
RJE: Remote Job Entry (EPA-94/04)
RJS: Reverse Jet Scrubber
RK: Rotary Kiln

RLL: Rapid and Large Leakage (Rate) (EPA-94/04)
RM: Risk Management (EPA-92/12)
RMCL: Recommended Maximum Contaminant Level (this phrase is being discontinued in favor of MCLG) (EPA-94/04)
RMDHS: Regional Model Data Handling System (EPA-94/04)
RME: Reasonable Maximum Exposure (EPA-91/12)
RMIS: Resources Management Information System (EPA-94/04)
RMO: Records Management Officer
RMP: Revolutions per Minute
Rn: Radon
RNA: Ribonucleic Acid (EPA-94/04)
rO: Rated output (40 CFR 87.2-94)
RO: Regional Office
ROADCHEM: Roadway Version that Includes Chemical Reactions of BI, NO_2, and O_3 (EPA-94/04)
ROADWAY: A Model to Predict Pollutant Concentrations Near a Roadway (EPA-94/04)
ROC: Record of Communication (EPA-94/04)
ROD: Record of Decision (40 CFR 300.4)
RODS: Records of Decision System (EPA-94/04)
ROG: Reactive Organic Gases (EPA-94/04)
ROI: Region of Influence (DOE-91/4)
ROLLBACK: A Proportional Reduction Model (EPA-94/04)
ROM: Read Only Memory (computer)
ROM: Regional Oxidant Model (EPA-94/04)
ROMCOE: Rocky Mountain Center on Environment (EPA-94/04)
ROP: Regional Oversight Policy (EPA-94/04)
ROPA: Record of Procurement Action (EPA-94/04)
RP: Radon Progeny Integrated Sampling (EPA-94/04)
RP: Respirable Particulates (EPA-94/04)
RP: Responsible Party (EPA-94/04)
RP: Responsible Party
RPAR: Rebuttable Presumption Against Registration (FIFRA) (40 CFR 166.3)
rpm: Revolutions per minute (40 CFR 86.403.78-94)
RPM: Reactive Plume Model (EPA-94/04)
RPM: Remedial Project Manager (40 CFR 300.4)
RPO: Regional Planning Officer
RPO: Regional Program Officer
rPR: Rated pressure ratio (40 CFR 87.2-94)
RQ: Reportable Quantities (40 CFR 355-AA)
RRC: Regional Response Center (40 CFR 300.4)
RREL: Risk Reduction Engineering Laboratory (EPA) (Formerly HWERL)
RRT: Regional Response Team (40 CFR 300.4)
RRT: Requisite Remedial Technology (EPA-94/04)
RS: Registration Standard (EPA-94/04)
RSCC: Regional Sample Control Center (EPA-94/04)
RSD: Risk-Specific Dose ($\mu g/m^3$) (EPA-94/04)
RSE: Removal Site Evaluation (EPA-94/04)
RSKERL: Robert S. Kerr Environmental Research Laboratory
RSPA: Research and Special Programs Administration (40 CFR 300.4)
RT: Regional Total
RTCM: Reasonable Transportation Control Measure (EPA-94/04)
RTD: Return to Duty
RTDM: Rough Terrain Diffusion Model (EPA-94/04)
RTECS: Registry of Toxic Effects of Chemical Substances (29 CFR 1910.20)
RTF: Reactivity Task Force (MCA)
RTM: Regional Transport Model (EPA-94/04)

RTP: Research Triangle Park
Ru: Ruthenium
RUP: Restricted Use Pesticide (EPA-94/04)
RV: Residual Volume (EPA-92/12)
RVP: Reid Vapor Pressure (40 CFR 86)
RWC: Receiving Water Concentration (EPA-91/3)
RWC: Residential Wood Combustion (EPA-94/04)

s: Second (40 CFR 60)
S: Scrubber
S: Sulphur
S/cm³: Structures per Cubic Centimeter (40 CFR 763-AA)
S/mm²: Structures per Square Millimeter (40 CFR 763-AA)
S/TCAC: Scientific/Technical Careers Advisory Committee
S&A: Sampling and Analysis (EPA-94/04)
S&A: Surveillance and Analysis (EPA-94/04)
S&E: Salaries and Expenses
S1S: Smooth-One-Side (40 CFR 429.11)
S2S: Smooth-Two-Sides (40 CFR 42.119)
SA: Special Assistant
SA: Sunshine Act
SAB: Science Advisory Board (EPA-94/04)
SAC: Secretarial Advisory Board
SAC: Sulfuric Acid Concentrator
SAC: Support Agency Coordinator (40 CFR 300.4)
SAC: Suspended and Cancelled Pesticides (EPA-94/04)
SACTI: Seasonal/Annual Cooling Tower Impacts (computer code) (DOE-91/4)
SADAA: Science Assistant to the Deputy Administrator
SAE: Society of Automotive Engineers (40 CFR 89.3-94)
SAED: Selected Area Electron Diffraction (40 CFR 763-AA)
SAEWG: Standing Air Emissions Work Group (EPA-94/04)
SAIC: Special-Agents-In-Charge (EPA-94/04)
SAIP: Systems Acquisition and Implementation Program (EPA-94/04)
SAMWG: Standing Air Monitoring Work Group (EPA-94/04)
SANE: Sulfur and Nitrogen Emissions (EPA-94/04)
SANSS: Structure and Nomenclature Search System (EPA-94/04)
SAP: Sampling and Analysis Plan
SAP: Scientific Advisory Panel (EPA-94/04)
SAP: Serum Alkaline Phosphatase (EPA-92/12)
SAR: Sodium Absorption Ratio
SAR: Start Action Request (EPA-94/04)
SAR: Structural Activity Relationship (of a qualitative assessment) (EPA-94/04)
SARA: Superfund Amendments and Reauthorization Act of 1986 (40 CFR 280)
SARAH2: Surface Water Assessment Model for Back Calculating Reductions in Biotic Hazardous Wastes (EPA-91/3)
SAROAD: Storage and Retrieval of Aerometric Data (40 CFR 51.117)
SAS: Special Analytical Service (EPA-94/04)
SAS: Statistical Analysis System (EPA-94/04)
SASS: Source Assessment Sampling System (EPA-94/04)
SATO: Scheduled Airline Traffic Office
SAV: Submerged Aquatic Vegetation (EPA-94/04)
Sb: Antimony
SBA: Small Business Act
SBA: Small Business Administration (40 CFR 21.2)
SBO: Small Business Ombudsman
s.c.: Subcutaneous (EPA-92/12)

Sc: Scandium
SC: Sierra Club (EPA-94/04)
SC: South Carolina
SC: Steering Committee
SCAC: Support Careers Advisory Committee
SCAP: Superfund Comprehensive Accomplishments Plan (40 CFR 35.4010)
SCAP: Superfund Consolidated Accomplishments Plan (EPA-94/04)
SCBA: Self-Contained Breathing Apparatus (EPA-94/04)
SCC: Secondary Combustion Chamber
SCC: Source Classification Code (EPA-94/04)
SCD/SWDC: Soil or Soil and Water Conservation District (EPA-94/04)
SCDHEC: South Carolina Department of Health and Environmental Control (DOE-91/4)
SCE&G: South Carolina Electric and Gas Company (DOE-91/4)
scf: Standard Cubic Feet (40 CFR 60)
SCF: Supercritical Fluid
scfh: Standard Cubic Feet per Hour (40 CFR 60)
SCFM: Standard Cubic Feet per Minute (EPA-94/04)
SCLDF: Sierra Club Legal Defense Fund (EPA-94/04)
scm: Standard Cubic Meter (40 CFR 60)
SCORPIO: Subject Content-Oriented Retriever for Processing Information On-Line
SCR: Selective Catalytic Reduction (EPA-94/04)
SCRAM: State Consolidated RCRA Authorization Manual (EPA-94/04)
SCRC: Superfund Community Relations Coordinator (EPA-94/04)
SCRP: Superfund Community Relations Program
SCS: Soil Conservation Service, U.S. Department of Agriculture
SCS: Supplementary Control Strategy/System (EPA-94/04)
SCS: Supplementary Control System (40 CFR 57.103)
SCSA: Soil Conservation Society of America (EPA-94/04)
SCSP: Storm and Combined Sewer Program (EPA-94/04)
SCW: Supercritical Water Oxidation (EPA-94/04)
SD: South Dakota
SD: Spray Dryer
SD: Standard Deviation
SDBE: Small Disadvantaged Business Enterprise
SDC: Systems Decision Plan (EPA-94/04)
SDI: Sludge Density Index
SDWA: Safe Drinking Water Act (40 CFR 124.2)
SDWC: Secondary Drinking Water Criteria (SDWA)
Se: Selenium
SEA: Selective Enforcement Auditing (40 CFR 89.3-94)
SEA: State Enforcement Agreement (EPA-94/04)
SEA: State/EPA Agreement (EPA-94/04)
SEA: Supplementary Environmental Analysis (like EIS)
SEAM: Surface, Environment, and Mining (EPA-94/04)
SEAS: Strategic Environmental Assessment System (EPA-94/04)
sec.: Seconds (40 CFR 87.2-94)
SEC: Securities and Exchange Commission
SEE: Senior Environmental Employee
SEIA: Socioeconomic Impact Analysis (EPA-94/04)
SEIA: Solar Energy Industries Association
SEM: Scanning Electronic Microscope (40 CFR 763-AA)
SEM: Standard Error of the Means (EPA-94/04)
SEP: Standard Evaluation Procedures (EPA-94/04)
SEP: Structural Integrity Procedure (RCRA)
SEPWC: Senate Environment and Public Works Committee (EPA-

94/04)
SERATRA: Sediment Contaminant Transport Model Simplified Lake/Stream Analysis (EPA-91/3)
SERC: State Emergency Response Commission (40 CFR 300.4)
SERI: Solar Energy Research Institute (See NREL)
SES: Secondary Emissions Standard (EPA-94/04)
SES: Senior Executive Service
SES: Socioeconomic Status
SET: Site Evaluation Team (DOE-91/4)
SETS: Site Enforcement Tracking System (EPA-94/04)
SF: Safety Factor (EPA-92/12)
SF: Slope Factor (EPA-91/12)
SF: Standard Form (EPA-94/04)
SF: Superfund (EPA-94/04)
SFA: Spectral Flame Analyzers (EPA-94/04)
SFDS: Sanitary Facility Data System (EPA-94/04)
SFFAS: Superfund Financial Assessment System (EPA-94/04)
SFIREG: State FIFRA Issues Research and Evaluation Group (EPA-94/04)
SFS: State Funding Study (EPA-94/04)
sft^3: Standard Cubic Feet (DOE-91/4)
SGOT: Serum Glutamic Oxaloacetic Transaminase (EPA-92/12)
SGPT: Serum Glutamic Pyruvic Transaminase (EPA-92/12)
SHORTZ: Short Term Terrain Model (EPA-94/04)
SHPO: State Historic Preservation Officer (DOE-91/4)
SHWL: Seasonal High Water Level (EPA-94/04)
Si: Silicon
SI: International System of Units (EPA-94/04)
SI: Site Inspection (40 CFR 300.4)
SI: Spark Ignition (40 CFR 89.3-94)
SI: Surveillance Index (EPA-94/04)
SI unit: System International (Meter-Kg-Second) unit
SIC: Standard Industrial Classification (EPA-94/04)
SIC: Standard Industrial Code (DOE-91/4)
SICEA: Steel Industry Compliance Extension Act (EPA-94/04)
sigma g: Geometric Standard Deviation (EPA-92/12)
SIMS: Secondary Ion-Mass Spectrometry (EPA-94/04)
SIP: State Implementation Plan (40 CFR 51.350)
SIS: Stay In School
SITE: Superfund Innovative Technology Evaluation (EPA-94/04)
SIWSA: Standard Metropolitan Statistical Area (DOE-91/4)
SL: Sick Leave
SLAEM: Single Layer Analytical Element Model (DOE-91/4)
SLAMS: State/Local Air Monitoring Station (EPA-94/04)
SLAMS: State or Local Air Monitoring Station (40 CFR 58.1-b)
SLANG: Selected Letter and Abbreviated Name Guide
SLRL: Sex Linked Recessive Lethal (40 CFR 798.5275)
SLSM: Simple Line Source Model (EPA-94/04)
Sm: Samarium
sm^3: Standard Cubic Meter (40 CFR 464.02)
SMART: Simple Maintenance of ARTS (EPA-94/04)
SMCL: Secondary Maximum Contaminant Levels (40 CFR 143)
SMCRA: Surface Mining Control and Reclamation Act (EPA-94/04)
SME: Subject Matter Expert (EPA-94/04)
SMO: Sample Management Office (EPA-94/04)
SMOA: Superfund Memorandum of Agreement (40 CFR 300.4)
SMR: Standard Mortality Ratio (EPA-92/12)
SMSA: Standard Metropolitan Statistical Area (EPA-94/04)
SMYS: Specified Minimum Yield Strength (40 CFR 192.3)
Sn: Tin

SN: Smoke Number (40 CFR 87.2-94)
SNA: System Network Architecture (EPA-94/04)
SNAAQS: Secondary National Ambient Air Quality Standards (EPA-94/04)
SNAP: Significant Noncompliance Action Program (EPA-94/04)
SNARL: Suggested No Adverse Response Level (EPA-94/04)
SNC: Significant Noncompliers (EPA-94/04)
SNCR: Selective Non-Catalytic Reduction
SNG: Synthetic Natural Gas
SNL: Sandia National Laboratories (DOE-91/4)
SNUR: Significant New Use Rule (EPA-94/04)
SOC: Synthetic Organic Chemicals (EPA-94/04)
SOCMA: Synthetic Organic Chemical Manufacturers Association
SOCMI: Synthetic Organic Chemical Manufacturing Industry (40 CFR 60.700)
SOEH: Society for Occupational and Environmental Health
SOM: Sensitivity of Method (FDA)
SOP: Standard Operating Procedure (40 CFR 61.61)
SOT: Society of Toxicology
SOTDAT: Source Test Data (EPA-94/04)
SOW: Scope of Work (EPA-94/04)
SOW: Statement of Work (40 CFR 35.6015-45)
SP: Shaft Power (40 CFR 87.2-94)
SPAR: Status of Permit Application Report (EPA-94/04)
SPCC: Spill Prevention, Containment, and Countermeasure (40 CFR 112.3)
SPCCP: Spill Prevention Control and Countermeasure Plan
SPE: Secondary Particulate Emissions (EPA-94/04)
SPECS: Specifications
SPF: Structured Programming Facility (EPA-94/04)
SPI: Society of the Plastics Industry
SPI: Strategic Planning Initiative (EPA-94/04)
SPL: Sound Pressure Level (DOE-91/4)
SPLMD: Soil-Pore Liquid Monitoring Device (EPA-94/04)
SPMS: Special Purpose Monitoring Stations (EPA-94/04)
SPMS: Strategic Planning and Management System (EPA-94/04)
SPOC: Single Point of Contact (EPA-94/04)
SPS: State Permit System (EPA-94/04)
SPSS: Statistical Package for the Social Sciences (EPA-94/04)
SPUR: Software Package for Unique Reports (EPA-94/04)
sq.cm.: Square Centimeters (EPA-92/12)
sq ft: Square Feet
sq m: Square Meter
SQBE: Small Quantity Burner Exemption (EPA-94/04)
SQC: Sediment Quality Criteria (EPA-91/3)
SQG: Small Quantity Generator (40 CFR 261.5)
Sr: Strontium
SR: State Route (DOE-91/4)
SRAB: Source Receptor Analysis Branch (EPA)
SRAP: Superfund Remedial Accomplishment Plan (EPA-94/04)
SRC: Solvent-Refined Coal (EPA-94/04)
SRF: State Water Pollution Control Revolving Fund (40 CFR 35.3105-i)
SRM: Standard Reference Method (EPA-94/04)
SRP: Special Review Procedure (EPA-94/04)
SRR: Second Round Review (EPA-94/04)
SRR: Submission Review Record (EPA-94/04)
SRS: Savannah River Site (DOE-91/4)
SRTS: Service Request Tracking System (EPA-94/04)
SS: Settleable Solids (EPA-94/04)
SS: Spray Saturator

SS: Superfund Surcharge (EPA-94/04)
SS: Suspended Solids (40 CFR 133.101)
SSA: Social Security Administration
SSA: Sole Source Aquifer (40 CFR 149.2)
SSAC: Soil Site Assimilated Capacity (EPA-94/04)
SSAC: Soil Site Assimulated Capacity
SSC: Scientific Support Coordinator (40 CFR 300.4)
SSC: Superfund State Contracts (40 CFR 35.6015-47)
SSD: Standards Support Document (EPA-94/04)
SSE: Safe Shutdown Earthquake (DOE-91/4)
SSEIS: Standard Support and Environmental Impact Statement (EPA-94/04)
SSEIS: Stationary Source Emissions and Inventory System (EPA-94/04)
SSI: Size Selective Inlet (EPA-94/04)
SSMS: Spark Source Mass Spectrometry (EPA-94/04)
SSN: Social Security Number
SSO: Source Selection Official (EPA-94/04)
SST: Supersonic Transport
SSTS: Section Seven Tracking System (EPA-94/04)
SSU: Standard Saybolt Unit (a viscosity unit)
SSURO: Stop Sale, Use and Removal Order (EPA-94/04)
ST: Spray Tower
STALAPCO: State and Local Air-Pollution Control Officials (EPA-94/04)
STAP: State and Territorial Air Pollution (EPA-94/04)
STAPPA: State and Territorial Air Pollution Program Administrators
STAR: Stability Wind Rose (EPA-94/04)
STAR: State Acid Rain Projects (EPA-94/04)
START: Superfund Technical Assistance Response Team
std: Standard (40 CFR 60)
STEL: Short Term Exposure Limit (EPA-94/04)
STEM: Scanning Transmission Electron Microscope (40 CFR 763-AA)
STN: Scientific and Technical Information Network (EPA-94/04)
STORET: Storage and Retrieval of Water Quality Information (EPA-91/3)
STORET: Storage and Retrieval of Water-Related Data (EPA-94/04)
STP: Sewage Treatment Plant (EPA-94/04)
STP: Standard Temperature and Pressure (EPA-94/04)
SUNFED: Special United Nations Fund for Economic Development
SUP: Standard Unit of Processing (EPA-94/04)
SURE: Sulfate Regional Experiment Program (EPA-94/04)
SUS: Saybolt Universal Seconds (29 CFR 1910.1200)
SV: Sampling Visit (EPA-94/04)
SVI: Sludge Volume Index
SW: Slow Wave (EPA-94/04)
SWC: Settlement With Conditions (EPA-94/04)
SWDA: Solid Waste Disposal Act (EPA-94/04)
SWDF: Solid Waste Derived Fuel
SWDF: Solvent Waste Dervide Fuel
SWIE: Southern Waste Information Exchange (EPA-94/04)
SWMU: Solid Waste Management Unit (EPA-94/04)
SWTR: Surface Water Treatment Rule (EPA-94/04)
sy.: Synonym
SYSOP: Systems Operator (EPA-94/04)

t: Metric Ton

T: Temperature, degrees Kelvin (40 CFR 87.2-94)
T-R: Transformer-Rectifier (EPA-94/04)
T&A: Time and Attendance
Ta: Tantalum
TA: Travel Authorization
TACB: Texas Air Control Board
TAG: Technical Assistance Grants (40 CFR 35.4005)
TALMS: Tunable Atomic Line Molecular Spectroscopy (EPA-94/04)
TAMS: Toxic Air Monitoring System (EPA-94/04)
TAMTAC: Toxic Air Monitoring System Advisory Committee (EPA-94/04)
TAO: TSCA Assistance Office
TAP: Technical Assistance Program (EPA-94/04)
TAPDS: Toxic Air Pollutant Data System (EPA-94/04)
TAPP: Time and Attendance, Payroll, and Personnel
TAPPI: Association of the Pulp and Paper Industry (40 CFR 60.17.d)
TARC: Toxics Testing and Assessment Research Committee
TAS: Tolerance Assessment System (EPA-94/04)
Tb: Terbium
TB: Tracheobronchial region of the respiratory tract (EPA-92/12)
TB: Trial Burn
TBD: To Be Determined
TBP: Trial Burn Plan
TBT: Tributyltin (EPA-94/04)
Tc: Technetium
TC: Target Concentration (EPA-94/04)
TC: Technical Center (EPA-94/04)
TC: Total Carbon
TC: Toxic Concentration (EPA-94/04)
TC: Toxicity Characteristics (EPA-94/04)
TCB: Toxic Combustion Byproduct
TCC: Tagliabue closed cup, a standard method of determining flash points (EPA-92/12)
TCDD: Dioxin (Tetrachlorodibenzo-p-dioxin) (EPA-94/04)
TCDF: Tetrachlorodi-benzofurans (EPA-94/04)
TCE: Trichloroethylene (EPA-94/04)
TCLP: Total Concentrate Leachate Procedure (EPA-94/04)
TCLP: Toxicity Characteristic Leachate Procedure (EPA-94/04)
TCM: Transportation Control Measure (40 CFR 51.138)
TCP: Transportation Control Plan (EPA-94/04)
TCP: Trichloroethylene (EPA-94/04)
TCP: Trichloropropane (EPA-94/04)
TCRI: Toxic Chemical Release Inventory (EPA-94/04)
T_D: Dispensed Fuel Temperature (40 CFR 86.098.3-94)
TD: Toxic Dose (EPA-94/04)
TDB: Toxicology Data Base (EPA-92/12)
TDD: Telecommunication Devices for Deaf (40 CFR 12.103)
TDE: Tetrachlorodiphenylethane
TDF: Tire Derived Fuel
TDI: Toluene Diisocyanate
TDS: Total Dissolved Solids (40 CFR 131.35)
TDY: Temporary Duty
Te: Tellurium
TEAM: Total Exposure Assessment Model (EPA-94/04)
TEC: Technical Evaluation Committee (EPA-94/04)
TEG: Tetraethylene Glycol (EPA-94/04)
TEGD: Technical Enforcement Guidance Document (EPA-94/04)
TEL: Tetraethyl Lead (40 CFR 86)
TEM: Texas Episodic Model (EPA-94/04)

TEM: Transmission Electron Microscope (40 CFR 763-AA)
TEP: Technical Evaluation Panel (EPA-94/04)
TEP: Typical End-use Product (EPA-94/04)
TERA: TSCA Environmental Release Application (EPA-94/04)
TES: Technical Enforcement Support (EPA-94/04)
TEXIN: Texas Intersection Air Quality Model (EPA-94/04)
TFS: Tin Free Steel
TFT: Temporary Full Time
TFTE: Temporary Full Time Equivalent
TGAI: Technical Grade of the Active Ingredient (EPA-94/04)
TGIF: Thank God It's Friday!
TGO: Total Gross Output (EPA-94/04)
TGP: Technical Grade Product (EPA-94/04)
Th: Thorium
THC: Total Hydrocarbons (40 CFR 86.094.3-94)
THCE: Total Hydrocarbon Equivalent (40 CFR 86.090.3-94)
THM: Trihalomethane (40 CFR 141.2)
Ti: Titanium
TI: Temporary Intermittent (EPA-94/04)
TI: Therapeutic Index (EPA-94/04)
TIBL: Thermal Internal Boundary Layer (EPA-94/04)
TIC: Technical Information Coordinator (EPA-94/04)
TIC: Tentatively Identified Compounds (EPA-94/04)
TIC: Total Inorganic Carbon
TIE: Toxicity Identification Evaluation (EPA-91/3)
TIM: Technical Information Manager (EPA-94/04)
TIM: Time In Mode (40 CFR 87.2-94)
TIP: Transportation Improvement Program (40 CFR 51.392)
TIS: Tolerance Index System (EPA-94/04)
TISE: Take It Somewhere Else (Solid Waste Syndrome. *See:* NIMBY) (EPA-94/04)
TITC: Toxic Substance Control Act Interagency Testing Committee (EPA-94/04)
TITC: TSCA Interagency Testing Committee
TKN: Total Kjeldahl Nitrogen
Tl: Thallium
TLC: Thin Layer Chromatography (40 CFR 796.2700.a)
TLEV: Transitional Low-Emission Vehicle (40 CFR 88.103-94)
TLM: Tolerance Limit Median
TLV: Threshold Limit Value (EPA-94/04)
TLV-C: TLV-Ceiling (EPA-94/04)
TLV-STEL: TLV-Short Term Exposure Limit (EPA-94/04)
TLV-TWA: TLV-Time Weighted Average (EPA-94/04)
Tm: Thulium
TMDC: Total Maximum Daily Load
TMDL: Total Maximum Daily Load (40 CFR 130.2)
TMI: Three Mile Island
TML: Tetramethyl Lead (40 CFR 86)
TMRC: Theoretical Maximum Residue Contribution (EPA-94/04)
TN: Tennessee
TNCWS: Transient Non-Community Water System (EPA-94/04)
TNT: Trinitrotoluene (40 CFR 457.11)
TO: Task Order (EPA-94/04)
TO: Travel Order
TOA: Trace Organic Analysis (EPA-94/04)
TOC: Total Organic Carbon (40 CFR 268.2)
TOC: Total Organic Carbon/Compound (EPA-94/04)
TOC: Total Organic Compound (40 CFR 60.561)
TOD: Theoretical Oxygen Demand
TOD: Time-Origin-Destination (40 CFR 52.2294)
TOD: Total Oxygen Demand

TODAM: Transport One-Dimensional Degradation and Migration Model (EPA-91/3)
TOT: Time-of-Travel
TOTAL: Total Respiratory Tract (EPA-92/12)
TOX: Tetradichloroxylene (EPA-94/04)
TOX: Total Organic Halogen
TOXI4: A Subset of WASP4 (EPA-91/3)
TOXIC: Toxic Organic Transport and Bioaccumulation Model (EPA-91/3)
TOXIWASP: Chemical Transport and Fate Model (EPA-91/3)
TP: Technical Product (EPA-94/04)
TPA: Terephthalic acid (40 CFR 60.561)
TPC: Testing Priorities Committee (EPA-94/04)
TPD: Tons per Day
TPES: Toxic and Pretreatment Effluent Standards
TPI: Technical Proposal Instructions (EPA-94/04)
TPQ: Threshold Planning Quantity (40 CFR 355.10)
TPSIS: Transportation Planning Support Information System (EPA-94/04)
TPTH: Triphenyltinhydroxide (EPA-94/04)
TPW: Tons per Week
TPY: Tons per Year (EPA-94/04)
TQM: Total Quality Management (EPA-94/04)
TR: Target Risk (EPA-91/12)
TRC: Technical Review Committee (EPA-94/04)
TRC: Total Residual Chlorine (40 CFR 420.02)
TRD: Technical Review Document (EPA-94/04)
TRE: Total Resource Effectiveness (40 CFR 60.611)
TRE: Toxicity Reduction Evaluation (EPA-91/3)
TRI: Toxic Release Inventory (EPA-94/04)
TRIP: Toxic Release Inventory Program (EPA-94/04)
TRIS: Toxic Chemical Release Inventory System (EPA-94/04)
TRLN: Triangle Research Library Network (EPA-94/04)
TRLN: Triangle Research Library Network
TRO: Temporary Restraining Order (EPA-94/04)
TRS: Total Reduced Sulfur (40 CFR 60.281)
TRU: Transuranic, a classification of wastes (DOE-91/4)
TRW: Transuranic Radioactive Waste (40 CFR 191.02)
TS: Total Solids
TSA: Technical Systems Audit (EPA-94/04)
TSCA: Toxic Substances Control Act of 1976 (40 CFR 704.3)
TSCATS: TSCA Test Submissions Database (EPA-94/04)
TSCATS: TSCA Test Submissions Database (OTS)
TSCC: Toxic Substances Coordinating Committee (EPA-94/04)
TSD: Technical Support Document (EPA-94/04)
TSD: Treatment, Storage, and Disposal (40 CFR 260-AI-93)
TSDF: Treatment, Storage, and Disposal Facility (EPA-94/04)
TSDG: Toxic Substances Dialogue Group (EPA-94/04)
TSI: Thermal System Insulation (EPA-94/04)
TSM: Transportation System Management (EPA-94/04)
TSO: Time Sharing Option (EPA-94/04)
TSP: Teleprocessing Services Program
TSP: Thrift Savings Plan
TSP: Total Suspended Particulates (40 CFR 51.100)
TSPC: Toxic Substances Priority Committee
TSS: Terminal Security System
TSS: Total Suspended (non-filterable) Solids (40 CFR 268.2)
TSSC: Toxic Substances Strategy Committee (16 Federal Agencies)
TSSG: Toxic Substances Strategy Group
TSSMS: Time Sharing Services Management System

TTE: Temporary Total Enclosure (40 CFR 52.741.a.4)
TTFA: Target Transformation Factor Analysis (EPA-94/04)
TTHM: Total Trihalomethane (40 CFR 141.30)
TTO: Total Toxic Organics (40 CFR 413.02)
TTY: Teletypewriter (EPA-94/04)
TU: Toxic unit (EPA-91/3)
TU: Turbidity unit (40 CFR 141.13)
TU_a: Acute Toxic Unit (EPA-91/3)
TU_c: Chronic Toxic Unit (EPA-91/3)
TUCC: Triangle University Computer Center
TVA: Tennessee Valley Authority (EPA-94/04)
TVS: Total Volatile Solids
TW: Test Weight (40 CFR 88.103-94)
TWA: Time Weighted Authority
TWA: Time Weighted Average (40 CFR 763.121)
TWMD: Toxics and Waste Management Division
TWS: Transient Water System (EPA-94/04)
TX: Texas
TZ: Treatment Zone (EPA-94/04)

U: Uranium
UAC: User Advisory Committee (EPA-94/04)
UAM: Urban Airshed Model (EPA-94/04)
UAO: Unilateral Administrative Order (EPA-94/04)
UAPSP: Utility Acid Precipitation Study Program (EPA-94/04)
UAQI: Uniform Air Quality Index (EPA-94/04)
UARG: Utility Air Regulatory Group (EPA-94/04)
UCC: Ultra Clean Coal (EPA-94/04)
UCCI: Urea-Formaldehyde Foam Insulation (EPA-94/04)
UCL: Upper Confidence Limit (EPA-92/12)
UCL: Upper Control Limit (EPA-94/04)
UDDS: Urban Dynamometer Driving Schedule (40 CFR 86)
UDKHDEN: Three-Dimensional Model Used for Single or Multiple Port Diffusers (EPA-91/3)
UDMH: Unsymmetrical Dimethyl Hydrazine (EPA-94/04)
UEL: Upper Explosive Limit (EPA-94/04)
UF: Uncertainty Factor (EPA-92/12)
UFL: Upper Flammability Limit (EPA-94/04)
UHWM: Uniform Hazardous Waste Manifest (40 CFR 259.10)
UIC: Underground Injection Control (40 CFR 124.2)
UL: Underwriters Laboratories, 333 Pfingsten Road, North Brook, IL 60062 (40 CFR 60.17.f)
ULEV: Ultra Low-Emission Vehicle (40 CFR 88.103-94)
ULINE: Uniform Linear Density Flume Model (EPA-91/3)
ULP: Unfair Labor Practices
UMERGE: Two-Dimensional Model Used to Analyze Positively Buoyant Discharge (EPA-91/3)
umol: Micromoles (EPA-92/12)
UMTRCA: Uranium Mill Tailings Radiation Control Act (EPA-94/04)
UMTRCA: Uranium Mill Tailings Radiation Control Act of 1978 (40 CFR 23.8; 300-AA)
UMW: United Mine Workers Union
UN: United Nations
UNAMAP: Users' Network for Applied Modeling of Air Pollution (EPA-94/04)
UNEP: United Nations Environment Program (EPA-94/04)
UNEP: United Nations Environment Programme
UNESCO: United Nations Educational, Scientific and Cultural Organization
UNIDO: United Nations Industrial Development Organization

UO_3: Uranium Oxide (DOE-91/4)
UOD: Ultimate Oxygen Demand
UOUTPLM: Cooling Tower Plume Model Adapted for Marine Discharges (EPA-91/3)
UPLUME: Numerical Model That Produces Flux-Average Dilutions (EPA-91/3)
UPS: United Parcel Service (40 CFR 86)
UPWP: Unified Planning Work Program (40 CFR 51.138)
URT: Upper Respiratory Tract (EPA-8/90b)
USAO: United States Attorney's Office
USATHAMA: U.S. Army Toxic and Hazardous Materials Agency (DOD)
USBM: United States Bureau of Mines
USC: Unified Soil Classification (EPA-94/04)
USC: United States Code
USCA: United States Code Annotated
USCG: United States Coast Guard (40 CFR 300.4)
USDA: United States Department of Agriculture (40 CFR 300.4)
USDOI: United States Department of the Interior
USDW: Underground Sources of Drinking Water (40 CFR 144.3)
USEPA: United States Environmental Protection Agency
USFS: United States Forest Service (EPA-94/04)
USFWS: U.S. Fish and Wildlife Service (DOE-91/4)
USGS: United States Geological Survey (DOE-91/4)
USP: U.S. Pharmacopeia (EPA-94/04)
USP: United States Pharmacopeia
USPHS: United States Public Health Service
USPS: United States Postal Service
USS: United States Senate
UST: Underground Storage Tank (40 CFR 280)
USTR: United States Trade Representative
UT: Utah
UTM: Universal Transverse Mercator (EPA-94/04)
UTP: Urban Transportation Planning (EPA-94/04)
UV: Ultraviolet (40 CFR 86.090.3-94)
UV-VIS: Ultraviolet Visible (40 CFR 796.1050)
UZM: Unsaturated Zone Monitoring (EPA-94/04)

V: Vanadium
V: Volt (40 CFR 60)
v/v: Volume per Volume (EPA-92/12; 40 CFR 60)
VA: Veterans Administration
VA: Virginia
VALLEY: Meteorological Model to Calculate Concentrations on Elevated Terrain (EPA-94/04)
VAT: Value Added Tax
VCM: Vinyl Chloride Monomer (EPA-94/04)
VDT: Video Display Terminal
VE: Value Engineering (40 CFR 35.905)
VE: Visual Emissions (EPA-94/04)
VEO: Visible Emission Observation (EPA-94/04)
VF: Volatilization Factor (EPA-91/12)
VHAP: Volatile Hazardous Air Pollutant (40 CFR 61.241)
VHS: Vertical and Horizontal Spread Model (EPA-94/04)
VHT: Vehicle-Hours of Travel (EPA-94/04)
VISTTA: Visibility Impairment from Sulfur Transformation and Transport in the Atmosphere (EPA-94/04)
VKT: Vehicle Kilometers Traveled (EPA-94/04)
VMT: Vehicle Miles Traveled (EPA-94/04)
VOC: Volatile Organic Compounds (40 CFR 60.431-94)
VOL: Volatile Organic Liquid (40 CFR 60.111b)

VOM: Volatile Organic Materials (40 CFR 52.741)
VOS: Vehicle Operating Survey (EPA-94/04)
VOST: Volatile Organic Sampling Train (EPA-94/04)
VP: Vapor Pressure (EPA-94/04)
VQ: Venturi Quench
VS: Venturi Scrubber
VSD: Virtually Safe Dose (EPA-94/04)
VSI: Visual Site Inspection (EPA-94/04)
VSS: Volatile Suspended Solids (EPA-94/04)
V_T: Tidal volume (EPA-90/8b)
VT: Vermont

W: Watt(s) (40 CFR 87.2-94)
W: Wolfram
WA: Washington
WA: Work Assignment (EPA-94/04)
WAC: Washington Administrative Codes (DOE-91/4)
WADTF: Western Atmospheric Deposition Task Force (EPA-94/04)
WAP: Waste Analysis Plan (EPA-94/04)
WASP4: Water Quality Analysis Program (EPA-91/3)
WASTOX: Estuary and Stream Quality Model (EPA-91/3)
WB: Wet Bulb (EPA-94/04)
WB: World Bank
WBC: White Blood Cell(s) (EPA-92/12)
WBE: Women's Business Enterprise (40 CFR 33.005)
WCED: World Commission on Environment and Development (EPA-94/04)
WDF: Waste Derived Fuel
WDOE: Washington Department of Ecology (DOE-91/4)
WDROP: Distribution Register of Organic Pollutants in Water (EPA-94/04)
WENDB: Water Enforcement National Data Base (EPA-94/04)
WERL: Water Engineering Research Laboratory (EPA-94/04)
WF: Weighting Factor (40 CFR 86)
WG: Wage Grade
WG: Work Group
WGI: Within Grade Increase
WHB: Waste Heat Boiler
WHC: Westinghouse Hanford Company (DOE-91/4)
WHO: World Health Organization (EPA-94/04)
WHWT: Water and Hazardous Waste Team (EPA-94/04)
WI: Wisconsin
WIC: Washington Information Center
WICEM: World Industry Conference on Environmental Management (EPA-94/04)
WINCO: Westinghouse Idaho Nuclear Company, Inc. (DOE-91/4)
WIPP: Waste Isolation Pilot Plant (DOE-91/4)
WISE: Women In Science and Engineering
WL: Warning Letter (EPA-94/04)
WL: Working Level (radon measurement) (40 CFR 192.11)
WLA: Wasteload Allocation (40 CFR 130.2)
WLA/TMDL: Wasteload Allocation/Total Maximum Daily Load (EPA-94/04)
WLM: Working Level Months (EPA-94/04)
WMO: World Meteorological Organization (EPA-94/04)
WNP-1: Washington Nuclear Plant No. 1 (DOE-91/4)
WNP-2: Washington Nuclear Plant No. 2 (DOE-91/4)
WOT: Wide Open Throttle (40 CFR 86)
WPCF: Water Pollution Control Federation (EPA-94/04)
WPPSS: Washington Public Power Supply System (DOE-91/4)

WQA: Water Quality Act of 1987
WQAB FLOW: Water Quality Analysis System Flow Data Subroutine (EPA-91/3)
WQC: Water Quality Criteria (EPA-92/12)
WQM: Water Quality Management (40 CFR 130.2)
WQS: Water Quality Standards (40 CFR 130.2)
WRC: Water Resources Council (EPA-94/04)
WRDA: Water Resources Development Act (EPA-94/04)
WRI: World Resources Institute (EPA-94/04)
WS: Wet Scrubber
WS: Work Status (EPA-94/04)
WSF: Water Soluble Fraction (EPA-94/04)
WSRA: Wild and Scenic Rivers Act (EPA-94/04)
WSRC: Westinghouse Savannah River Company (DOE-91/4)
WSTB: Water Sciences and Technology Board (EPA-94/04)
WSTP: Wastewater Sewage Treatment Plant (EPA-94/04)
wt: Weight (40 CFR 86.403.78-94)
WTP: Water Treatment Plant
WV: West Virginia
WWEMA: Waste and Wastewater Equipment Manufacturers Association (EPA-94/04)
WWF: World Wildlife Fund (EPA-94/04)
WWTP: Wastewater Treatment Plant (EPA-94/04)
WWTU: Wastewater Treatment Unit (EPA-94/04)
WY: Wyoming

Xe: Xenon
XRF: X-ray fluorescence

Y: Yttrium
Yb: Ytterbium
yd^2: Square Yard (40 CFR 60)
yr: Year (40 CFR 60)
YTD: Year to Date

ZBB: Zero Base Budgeting
ZEV: Zero-Emission Vehicle (40 CFR 88.103-94)
ZHE: Zero Headspace Extractor (EPA-94/04)
ZID: Zone of Initial Dilution (40 CFR 125.58)
Zn: Zinc
ZOI: Zone of Incorporation (EPA-94/04)
Zr: Zirconium
ZRL: Zero Risk Level (EPA-94/04)

=========================

μCi/L: Microcuries per Liter (DOE-91/4)
μCi/mL: Microcuries per Milliliter (DOE-91/4)
$\mu g/m^3$: Micrograms per Cubic Liter (DOE-91/4)
μg: Microgram = 10^{-6} gram (40 CFR 60)
μL: Microliter = 10^{-6} liter (40 CFR 60)
μm: Micrometer = 10^{-6} meter

References

(10 CFR xxx-91) means 10 Code of Federal Regulations, Parts xxx, 1991 Edition.

(29 CFR xxx-91) means 29 Code of Federal Regulations, Parts xxx, 1991 Edition.

(40 CFR xxx-91) means 40 Code of Federal Regulations, Parts xxx, 1991 Edition.

(29 CFR 1910-89), "Occupational Exposure to Bloodborne Pathogens; Proposed Rule and Notice of Hearing," Federal Register, Tuesday, May 30, 1989.

(ACS-87/11), "Quality Assurance Of Chemical Measurements," An American Chemical Society (ACS) Short Course Prepared in November 1987 and Offered at EPA's Risk Reduction Engineering Laboratory On September 7-8, 1988.

(AP-13), "Atmospheric Emissions from Sulfuric Acid Manufacturing Processes," EPA/AP-13, 1965.

(AP-27), "Atmospheric Emissions from Nitric Acid Manufacturing Processes," EPA/AP-27, 1966.

(AP-40), "Air Pollution Engineering Manual," 2nd Edition, Office of Air Quality Planning and Standards, USEPA, U.S. Government Printing Office, Washington DC, Stock No. 055-003-00059-9, May 1973.

(AP-42), "Compilation of Air Pollution Emission Factors," 1981.

(AP-51), "Control Techniques for Particulate Air Pollutants," US Environmental Protection Agency, 1969.

(API-931), "Manual on Disposal of Refinery Wastes-Volume on Atmospheric Emissions," American Petroleum Institute (API) Publication 931, Chapter 13-Filters and Wet Collectors for the Removal of Particulate Matter, 1974.

(Arbuckle-89), "Environmental Law Handbook," J. G. Arbuckle, et al. Tenth Edition, Government Institutes, Inc. 966 Hungerford Drive #24, Rockville, MD 20850, 1989.

(ASHRAE-77), "American Society of Heating, Refrigerating and Air-Conditioning Engineers Hand Book, 1977 Fundamentals", ASHRAE Inc. 345 E. 47th Street, New York, NY 10017, 1977.

(ATSDR-9/90), "The Public Health Implications of Medical Waste: A Report to Congress," Agency for Toxic Substances and Disease Registry (ATSDR).

(Beachler-87/11), "Bay County, Florida Waste-to-Energy Facility Air Emission Test Results," D. S. Beachler, et.al., Air Pollution Control Association Specialty Conference on Thermal Treatment of Municipal, Industrial and Hospital Wastes, Pittsburgh, PA, November 3-6, 1987.

(Brownell-91/04), "Clean Air Handbook," F. William Brownell & Lee B. Zeugin, Published by Government Institutes, Inc., 966 Hungerford Dr., #24, Rockville, MD 20850, Phone 301 921-2300, April 1991.

(Brunner-84), "Incineration Systems and Selection and Design," C. A. Brunner, Van Nostrand Reinhold Co., 1984.

(Brunner-85), "Hazardous Air Emissions from Incineration," C. A. Brunner, Chapman and Hall, 1985.

(CAAxxx-42USCyyyy-91) means:
- CAAxxx: Clean Air Act including the contents of the 1990 Clean Air Act Amendment, Section xxx;
- 42USCyyyy: Title 42, United States Code (USC), Section yyyy; and
- -91: 1991 Edition, "Environmental Statutes," Published by the Governmental Institutes, Inc., 4 Research Place, Suite 200, Rockville, MD 20850, March 1991.

(Calvert-84), "Handbook of Air Pollution Technology," S. Calvert, and H. M. Englund, John Wiley & Sons, 1984.

(Cheremisinoff-77), "Air Pollution Control and Design Handbook, Part 1 and Part 2" P. N. Cheremisinoff and Richard A. Young, Marcel Dekker, Inc. 1977.

(Coco-86), "Finding the Law," Al Coco, Government Institiute, Inc., 966 Hungerford Drive #24, Rockville, MD 20850, 1986.

(Conn-85), " Pesticide Regulation Handbook," R. L. Conn, et al., Executive Enterprises Publications Co., Inc., 1985.

(Course 165.3), "Incident Mitigation and Treatment Methods," Environmental Response Team, Office of Emergency and Remedial Response, USEPA, U.S. Government Printing Office: 1987-748-121-67038.

(Course 165.4), "Air Surveillance for Hazardous Materials,"

Environmental Response Team, Office of Emergency and Remedial Response, USEPA, DHHS (NIOSH) Publication No. 84-100.

(Course 165.5), "Hazardous Materials Incident Response Operations," Offered by Environmental Response Team, Office of Emergency and Remedial Response, USEPA, U.S. Government Printing Office: 1987-748-121-40722.

(Course 165.6), "Environmental Risk Assessment," Environmental Response Team, Office of Emergency and Remedial Response, USEPA, U.S. Government Printing Office: 1988-548-158-67072.

(Course 165.7), "Introduction to Ground Water Investigation," Environmental Response Team, Office of Emergency and Remedial Response, USEPA, U.S. Government Printing Office: 1978-748-121-67038.

(Course 165.8), "Response Safety, Decision-Making," Environmental Response Team, Office of Emergency and Remedial Response, USEPA.

(Course 165.9), "Sampling for Hazardous Materials," Environmental Response Team, Office of Emergency and Remedial Response, USEPA.

(Course 400), "Introduction to Air Toxics,"Air Pollution Training Institute (APTI), U.S. EPA Environmental Research Center, Research Triangle Park, NC 27711.

(Course 401), "Site Specific Source Monitoring and Evaluation for Air Toxics," Air Pollution Training Institute (APTI), U.S. EPA Environmental Research Center, Research Triangle Park, NC 27711.

(Course 411), "Air Pollutions Dispersion Models-Fundamental Concepts," Air Pollution Training Institute (APTI), U.S. EPA Environmental Research Center, Research Triangle Park, NC 27711.

(Course 413), "Control of Particulate Emissions," Air Pollution Training Institute (APTI), U.S. EPA Environmental Research Center, Research Triangle Park, NC 27711.
- (EPA-80/04), Student Workbook, EPA450-2-80-067, April 1980.
- (EPA-81/10), Student Manual, EPA450-2-80-066, October 1981.

(Course 414), "Quality Assurance for Source Emission Measurement Methods," Air Pollution Training Institute (APTI), U.S. EPA Environmental Research Center, Research Triangle Park, NC 27711.

(Course 415), "Control of Gaseous Emissions," Air Pollution Training Institute (APTI), U.S. EPA Environmental Research Center, Research Triangle Park, NC 27711.
- (EPA-81/05), Student Workbook, EPA450-2-81-006, May 1981.
- (EPA-81/12), Student Manual, EPA450-2-81-005, December 1981.

(Course 416), "Inspection Procedures for Organic Solvent Metal Cleaning (Degreasing) Operations," Air Pollution Training Institute (APTI), U.S. EPA Environmental Research Center, Research Triangle Park, NC 27711.

(Course 420), "Air Pollution Microscopy," Air Pollution Training Institute (APTI), U.S. EPA Environmental Research Center, Research Triangle Park, NC 27711.

(Course 423), "Dispersion of Air Pollution-Theory and Model Application: Air Pollution Training Institute (APTI), U.S. EPA Environmental Research Center, Research Triangle Park, NC 27711.
- Selected Reading Packet, EPA450-2-81-077, October 1981.
- Student Workbook, EPA450-2-81-075, October 1981.
- Instructor's Guide, EPA450-2-81-076, November 1981.

(Course 426), "Statistical Evaluation Methods for Air Pollution Data," Air Pollution Training Institute (APTI), U.S. EPA Environmental Research Center, Research Triangle Park, NC 27711.

(Course 427), "Combustion Evaluation," Air Pollution Training Institute (APTI), U.S. EPA Environmental Research Center, Research Triangle Park, NC 27711.
- Student Manual, EPA450-2-80-063, February 1980.
- Student Workbook, EPA450-2-80-064, February 1980.
- Instructor's Guide, EPA450-2-80-065, February 1980.

(Course 434), "Introduction to Ambient Air Monitoring," Air Pollution Training Institute (APTI), U.S. EPA Environmental Research Center, Research Triangle Park, NC 27711.

(Course 435), "Atmospheric Sampling," Air Pollution Training Institute (APTI), U.S. EPA Environmental Research Center, Research Triangle Park, NC 27711.
- Student Manual (Second Edition), EPA450-2-80-004, June 1983.

(Course 444), "Air Pollution Field Enforcement," Air Pollution Training Institute (APTI), U.S. EPA Environmental Research Center, Research Triangle Park, NC 27711.

(Course 445), "Baseline Source Inspection Techniques," Air Pollution Training Institute (APTI), U.S. EPA Environmental Research Center, Research Triangle Park, NC 27711.

(Course 446), "Inspection Procedures and Safety," Air Pollution Training Institute (APTI), U.S. EPA Environmental Research Center, Research Triangle Park, NC 27711.

(Course 450), "Source Sampling for Particulate Pollutants," Air Pollution Training Institute (APTI), U.S. EPA Environmental Research Center, Research Triangle Park, NC 27711.
- Student Workbook, EPA450-2-79-007, December 1979.

(Course 452), "Principles and Practice of Air Pollution Control," Air Pollution Training Institute (APTI), U.S. EPA Environmental Research Center, Research Triangle Park, NC 27711.

(Course 454), "Effective Permit Writing Workshop," Air Pollution Training Institute (APTI), U.S. EPA Environmental Research

Center, Research Triangle Park, NC 27711.

(Course 464), "Analytical Methods for Air Quality Standards," Air Pollution Training Institute (APTI), U.S. EPA Environmental Research Center, Research Triangle Park, NC 27711.

(Course 468), "Source Sampling and Analysis of Gaseous Pollutants," Air Pollution Training Institute (APTI), U.S. EPA Environmental Research Center, Research Triangle Park, NC 27711.

(Course 470), "Quality Assurance for Air Pollution Measurement Systems," Air Pollution Training Institute (APTI), U.S. EPA Environmental Research Center, Research Triangle Park, NC 27711.

(Course 474), "Continuous Emission Monitoring," Air Pollution Training Institute (APTI), U.S. EPA Environmental Research Center, Research Triangle Park, NC 27711.
- Student Laboratory Workbook, No EPA number, December 1980.
- Regulatory Documents, EPA450-2-83-001, January 1983.

(Course 480), "Control Measures for CO, O_3, and NO_x," Air Pollution Training Institute (APTI), U.S. EPA Environmental Research Center, Research Triangle Park, NC 27711.

(Course 482), "Sources and Control of Volatile Organic Air Pollutants," Air Pollution Training Institute (APTI), U.S. EPA Environmental Research Center, Research Triangle Park, NC 27711.
- Student Workbook, EPA450-2-81-011, March 1981.
- Regulatory Documents, EPA450-2-81-012, March 1981.
- Instructor's Guide, EPA450-2-81-010, May 1981.
- Regulatory Documents Updates and NSPS Regulations, EPA450-2-84-001, April 1984.

(Course 484), "Motor Vehicle Emissions Control - Diagnosis and Repair," Air Pollution Training Institute (APTI), U.S. EPA Environmental Research Center, Research Triangle Park, NC 27711.

(Course 485), "Motor Vehicle Emissions Control - Anti-tampering and Misfueling," Air Pollution Training Institute (APTI), U.S. EPA Environmental Research Center, Research Triangle Park, NC 27711.

(Course 486), "Motor Vehicle Emissions Control - Quality Assurance for I/M Programs," Air Pollution Training Institute (APTI), U.S. EPA Environmental Research Center, Research Triangle Park, NC 27711.

(Course 502), "Hazardous Waste Incineration," Air Pollution Training Institute (APTI), U.S. EPA Environmental Research Center, Research Triangle Park, NC 27711.
- Student Manual (draft), No EPA number, March 1986.
- Student Workbook (draft), No EPA number, March 1986.
- Instructor's Guide (draft), No EPA number, March 1986.

(Course SI:410), "Introduction to Dispersion Modeling: Self-instructional Guidebook," EPA 450/2-82-007, March 1983, Air Pollution Training Institute (APTI), U.S. EPA Environmental Research Center, Research Triangle Park, NC 27711.

(Course SI:412), "Baghouse Plan Review: Student Guidebook," EPA 450/2-82-005, April 1982, Air Pollution Training Institute (APTI), U.S. EPA Environmental Research Center, Research Triangle Park, NC 27711.

(Course SI:412B), "Electrostatic Precipitator Plan Review: Self-instructional Guidebook," EPA 450/2-82-019, July 1983, Air Pollution Training Institute (APTI), U.S. EPA Environmental Research Center, Research Triangle Park, NC 27711.

(Course SI:412C), "Wet Scrubber Plan Review: Self-instructional Guidebook," EPA 450/2-82-020, March 1984, Air Pollution Training Institute (APTI), U.S. EPA Environmental Research Center, Research Triangle Park, NC 27711.

(Course SI:412D), "Control of Gaseous and Particulate Emissions: Self-instructional Problem Workbook," EPA 450/2-84-007, September 1984, Air Pollution Training Institute (APTI), U.S. EPA Environmental Research Center, Research Triangle Park, NC 27711.

(Course SI:428A), "Introduction to Boiler Operation, Self Instructional Guidebook," EPA 450/2-84-010, December 1984.

(Course SI:431), "Air Pollution Control Systems for Selected Industries: Self-instructional Guidebook," EPA 450/2-82-006, June 1983, Air Pollution Training Institute (APTI), U.S. EPA Environmental Research Center, Research Triangle Park, NC 27711.

(Course SI:445), "Introduction to Baseline Source Inspection Techniques: Self-Instructional Guidebook (draft)," no EPA number, December 1985.

(Course SI:476A), "Transmissometer Systems Operation and Maintenance, an Advanced Course," EPA 450/2-84-004, September 1984.

(CRWI-5/89), "Technical Issue Brief," Coalition for Responsible Waste Incineration (CRWI), located at 1330 Connecticut Ave., N.W., Suite 300, Washington DC, 20036, May 1989.

(CWAxxx-33USCyyyy-91) means:
- CWAxxx: Clean Water Act, Section xxx;
- 33USCyyyy: Title 33, United States Code (USC), Section yyyy; and
- -91: 1991 Edition, "Environmental Statutes," Published by the Governmental Governmental Institutes, Inc., 4 Research Place, Suite 200, Rockville, MD 20850, March 1991.

(CZMAxxx-16USCyyyy-90) means:
- CZMAxxx: Coast Zone Management Act (16USC1451-1464), Section xxx;
- 16USCyyyy: Title 16, United States Code (USC), Section yyyy; and
- -90: 1990 Edition, "Environmental Law Deskbook," Published by the Environmental Law Institute, 1616 P. Street, NW, Washington, DC 20036.

(DOD-78/01), "A Glossary of Selected Aquatic Ecological Terms," Aberdeen Proving Ground, Maryland, Department of Defense (DOD), Report No: ARCSL-SP-78002, January 1978.

(DOE-85/06), "Hazardous Chemical Defense Waste Management Program," Prepared by Martin Marietta Energy Systems, Inc. for the Department of Energy (DOE), June 1985.

(DOE-91/04), "Draft Environmental Impact Statement for the Siting, Construction, and Operation of New Production/reactor Capacity, Volume 3: Sections 7-12, appendices A-C," DOE/EIS-0144D, April 1991.

(DOI-70/04), "Glossary of Water Resource Terms," Federal Water Pollution Control Administration, Department of Interior (DOI), NTIS PB-255-156, April 1970.

(EPA-72), "Wet Scrubber Study," EPA R2-72-118a, NTIS PB-213-016, 1972.

(EPA-72a), "Field Operations and Enforcement Manual for Air Pollution Control. Volume III: Inspection Procedures for Specific Industries," APTD-1102, 1972.

(EPA-72/08), "Afterburner Systems Study," EPA R2-72-062, NTIS PB-212-560, August 1972.

(EPA-73), "Package Sorption Device System Study," EPA R2-73-202, 1973.

(EPA-73a), "Field Surveillance and Enforcement Guide for Primary Metallurgical Industries," EPA 450-3-73-002, 1973.

(EPA-73/06), "Field Surveillance and Enforcement Guide: Combustion and Incinerator Sources," APTD-1449, June 1973.

(EPA-73/08), "Development Document for Proposed Effluent Limitations Guidelines and New Source Performance Standards for the Phosphorus Derived Chemicals, Segment of the Phosphate Manufacturing Point Source Category," EPA440-1-73-006, August 1973.

(EPA-73/12), "Recovery of Fatty Materials from Edible Oil Refinery Effluents," EPA660-2-73-015, December 1973.

(EPA-74), "Background Information for New Source Performance Standards: Primary Copper, Zinc, and Lead Smelter," EPA450-2-74-002a, 1974.

(EPA-74/01), "Development Document for Effluent Limitations Guidelines; Building, Construction, and Paper Segment of the Asbestos Manufacturing, Point Source Category," EPA440-1-74-017a, January 1974.

(EPA-74/01), "Development Document for Effluent Limitations Guidelines and Standards of Performance for New Sources, Beet Sugar Processing, Subcategory of the Sugar Processing Point Source Category," EPA440-1-74-002b, January 1974.

(EPA-74/01b), "Development Document for Effluent Limitations Guidelines and New Source Performance Standards for the Cement Manufacturing, Point Source Category," EPA440-1-74-005a, January 1974.

(EPA-74/01c), "Development Document for Effluent Limitations Guidelines and New Source Performance Standards for the Feedlots, Point Source Category," EPA440-1-74-004a, January 1974.

(EPA-74/02), "Development Document for Effluent Limitations Guidelines and New Source Performance Standards for the Red Meat Processing Segment of the Meat Product and Rendering Processing, Point Source Category," EPA440-1-74-012a, Feburary 1974.

(EPA-74/02a), "Development Document for Effluent Limitations Guidelines and New Source Performance Standards for the Smelting and Slag Processing, Segments of the Ferroalloy Manufacturing, Point Source Category," EPA440-1-74-008a, Feburary 1974.

(EPA-74/02b), "Development Document for Effluent Limitations Guidelines and New Source Performance Standards for Tire and Synthetic, Segment of the Rubber Processing, Point Source Category," EPA440-1-74-013a, Feburary 1974.

(EPA-74/03), "Development Document for Effluent Limitations Guidelines and New Source Performance Standards for Apple, Citrus and Potato Processing Segment of the Canned and Preserved Fruits and Vegetables, Point Source Category," EPA440-1-74-027a, March 1974.

(EPA-74/03a), "Development Document for Effluent Limitations Guidelines and New Source Performance Standards for the Basic Fertilizer Chemicals Segment of the Fertilizer Manufacturing, Point Source Category," EPA440-1-74-011a, March 1974.

(EPA-74/03b), "Development Document for Effluent Limitations Guidelines and New Source Performance Standards for the Bauxite Refining Subcategory of the Aluminum Segment of the Nonferrous Metals Manufacturing, Point Source Category," EPA440-1-74-019c, March 1974.

(EPA-74/03c), "Development Document for Effluent Limitations Guidelines and New Source Performance Standards for the Cane Sugar Refining, Segment of the Sugar Processing, Point Source Category," EPA440-1-74-002c, March 1974.

(EPA-74/03d), "Development Document for Effluent Limitations Guidelines and New Source Performance Standards for the Copper, Nickel, Chromium, and Zinc, Segment of the Electroplating, Point Source Category," EPA440-1-74-003a, March 1974.

(EPA-74/03e), "Development Document for Effluent Limitations Guidelines and New Source Performance Standards for the Major Inorganic Products, Segment of the Inorganic Chemicals Manufacturing, Point Source Category," EPA440-1-74-007a, March 1974.

(EPA-74/03f), "Development Document for Effluent Limitations Guidelines and New Source Performance Standards for the Secondary Aluminum Smelting, Subcategory of the Aluminum, Segment of the Nonferrous Metal Manufacturing, Point Source

Category," EPA440-1-74-019e, March 1974.

(EPA-74/03g), "Development Document for Effluent Limitations Guidelines and New Source Performance Standards for the Synthetic Resins, Segment of the Plastics and Synthetic Materials Manufacturing, Point Source Category," EPA440-1-74-010a, March 1974.

(EPA-74/03h), "Development Document for Effluent Limitations Guidelines and New Source Performance Standards for the Grain Processing, Segment of the Grain Mills, Point Source Category," EPA 440-1-74-028a, March 1974.

(EPA-74/04), "Development Document for Effluent Limitations Guidelines and New Source Performance Standards for the Plywood, Hardboard, and Wood Preserving Segment of the Timber Products Processing," EPA440-1-74-023a, April 1974.

(EPA-74/04a), "Development Document for Effluent Limitations Guidelines and New Source Performance Standards for the Major Organic Products, Segment of the Organic Chemical Manufacturing, Point Source Category," EPA440-1-74-009a, April 1974.

(EPA-74/04b), "Development Document for Effluent Limitations Guidelines and New Source Performance Standards for the Petroleum Refining, Point Source Category," EPA440-1-74-014a, April 1974.

(EPA-74/04c), "Development Document for Effluent Limitations Guidelines and New Source Performance Standards for the Soap and Detergent Manufacturing, Point Source Category," EPA440-1-74-018a, April 1974.

(EPA-74/05), "Development Document for Effluent Limitations Guidelines and New Source Performance Standards for the Dairy Product Processing, Point Source Category," EPA440-1-74-021a, May 1974.

(EPA-74/05a), "Development Document for Effluent Limitations Guidelines and New Source Performance Standards for the Unbleached Kraft & Semi-chemical Pulp, Segment of the Pulp, Paper, and Paperboard Mills, Point Source Category," EPA440-1-74-025a, May 1974.

(EPA-74/06), "Development Document for Effluent Limitations Guidelines and New Source Performance Standards for the Catfish, Crab, Shrimp, and Tuna, Segment of the Canned and Preserved Seafood Processing, Point Source Category," EPA440-1-74-020a, June 1974.

(EPA-74/06a), "Development Document for Effluent Limitations Guidelines and New Source Performance Standards for the Steel Making, Segment of the Iron and Steel Manufacturing, Point Source Category," EPA440-1-74-024a, June 1974.

(EPA-74/06b), "Development Document for Effluent Limitations Guidelines and New Source Performance Standards for the Textile Mills, Point Source Category," EPA440-1-74-022a, June 1974.

(EPA-74/07), "Field Surveillance and Enforcement Guide for Petroleum Refineries," EPA450-3-74-042, June 1974.

(EPA-74/08), "Development Document for Effluent Limitations Guidelines and New Source Performance Standards for the Processor, Segment of the Meat Products, Point Source Category," EPA440-1-74-031, August 1974.

(EPA-74/08a), "Development Document for Effluent Limitations Guidelines and New Source Performance Standards for the Wet Storage, Sawmills, Particleboard and Insulation Board, Segment of the Timber Products Processing, Point Source Category," EPA440-1-74-033, August 1974.

(EPA-74/09), "Development Document for Effluent Limitations Guidelines and New Source Performance Standards for the Synthetic Resins, Segment of the Plastics and Synthetic Materials Manufacturing, Point Source Category," EPA440-1-74-036a, September 1974.

(EPA-74/09a), "System Analysis Requirements for Nitrogen Oxide Control of Stationary Sources," EPA650-2-74-091, September 1974.

(EPA-74/11), "Common Environmental Terms," NTIS No. PB-254-630, November, 1974.

(EPA-74/12), "Development Document for Effluent Limitations Guidelines and New Source Performance Standards for the Textile, Fabrication Materials and Sealing Devices, Segment of the Asbestos Manufacturing, Point Source Category," EPA440-1-74-035a, December 1974.

(EPA-74/12a), "Development Document for Effluent Limitations Guidelines and New Source Performance Standards for the Fabricated and Reclaimed Rubber, Segment of the Rubber Processing, Point Source Category," EPA440-1-74-030a, December 1974.

(EPA-74/12b), "Development Document for Effluent Limitations Guidelines and New Source Performance Standards for the Animal Feed, Breakfast Cereal, and Wheat Starch, Segment of the Grain Mills, Point Source Category," EPA440-1-74-039a, December 1974.

(EPA-75/01), "Development Document for Effluent Limitations Guidelines and New Source Performance Standards for the Renderer, Segment of the Meat Products and Rendering Processing, Point Source Category," EPA440-1-74-031d, January 1975.

(EPA-75/01a), "Development Document for Effluent Limitations Guidelines and New Source Performance Standards for the Synthetic Polymers, Segment of the Plastics and Synthetic Materials Manufacturing, Point Source Category," EPA440-1-75-036b, January 1975.

(EPA-75/01b), "Development Document for Effluent Limitations Guidelines and New Source Performance Standards for the Formulated Fertilizer, Segment of the Fertilizer Manufacturing, Point Source Category," EPA440-1-75-042a, January 1975.

(EPA-75/01c), "Inspection Manual for Enforcement of New Source

Performance Standards: Municipal Incinerator," EPA340-1-75-003, January 1975.

(EPA-75/02), "Development Document for Effluent Limitations Guidelines and New Source Performance Standards for the Calcium Carbide, Segment of the Ferroalloy Manufacturing, Point Source Category," EPA440-1-75-038, February 1975.

(EPA-75/02a), "Development Document for Effluent Limitations Guidelines and New Source Performance Standards for the Electric Ferroalloys, Segment of the Ferroalloy Manufacturing, Point Source Category," EPA 440/01-75-038a, February 1975.

(EPA-75/02b), "Development Document for Effluent Limitations Guidelines and New Source Performance Standards for the Primary Copper Smelting Subcategory and the Primary Copper Refining Subcategory of the Copper, Segment of the Nonferrous Metals Manufacturing, Point Source Category," EPA440-1-75-032b, February 1975.

(EPA-75/02c), "Development Document for Effluent Limitations Guidelines and New Source Performance Standards for the Secondary Copper Subcategory of the Copper, Segment of the Nonferrous Metals Manufacturing, Point Source Category," EPA440-1-75-032c, February 1975.

(EPA-75/02d), "Development Document for Effluent Limitations Guidelines and New Source Performance Standards for the Raw Cane Sugar Processing, Segment of the Sugar Processing, Point Source Category," EPA440-1-75-044, February 1975.

(EPA-75/04), "Development Document for Effluent Limitations Guidelines and New Source Performance Standards for the Poultry, Segment of the Meat Product and Rendering Process, Point Source Category," EPA440-1-75-031b, April 1975.

(EPA-75/07), "Development Document for Effluent Limitations Guidelines and New Source Performance Standards for the Oil Base Solvent Wash Subcategories of the Paint Formulating and the Ink Formulating, Point Source Category," EPA440-1-75-050a, July 1975.

(EPA-75/07a), "Development Document for Effluent Limitations Guidelines and New Source Performance Standards for the Paving and Roofing Materials (Tars and Asphalt), Point Source Category," EPA440-1-75-049a, July 1975.

(EPA-75/09), "Development Document for Effluent Limitations Guidelines and New Source Performance Standards for the Fish Meal, Salmon, Bottom Fish, Clam, Oyster, Sardine, Scallop, Herring, and Abalone, Segment of the Canned and Preserved Fish and Seafood Processing Industry, Point Source Category," EPA440-1-75-041a, September 1975.

(EPA-75/09a), "NO_x Combustion Control Methods and Costs for Stationary Sources," EPA600-2-75-046, September 1975.

(EPA-75/09b), "Inspection Manual for Enforcement of New Source Performance Standards: Portland Cement Plants," EPA340-1-75-001, September 1975.

(EPA-75/10), "Development Document for Effluent Limitations Guidelines and New Source Performance Standards for the Fruits, Vegetables and Specialties, Segment of the Canned and Preserved Fruits and Vegetables, Point Source Category," EPA440-1-75-046t, October 1975.

(EPA-75/10a), "Development Document for Effluent Limitations Guidelines and New Source Performance Standards for the Coal Mining, Point Source Category," EPA440-1-75-057, October 1975.

(EPA-75/10b), "Development Document for Effluent Limitations Guidelines and New Source Performance Standards for the Clay, Ceramic, Refractory and Miscellaneous Minerals, Vol. III, Mineral Mining and Processing Industry, Point Source Category," EPA440-1-75-059d, October 1975.

(EPA-75/10c), "Development Document for Effluent Limitations Guidelines and New Source Performance Standards for the Minerals for the Construction Industry, Vol. I, Mineral Mining and Processing Industry, Point Source Category," EPA440-1-75-059b, October 1975.

(EPA-75/11), "Development Document for Effluent Limitations Guidelines and New Source Performance Standards for the Significant Organic Products, Segment of the Organic Chemical Manufacturing, Point Source Category," EPA440-1-75-045, November 1975.

(EPA-76/03), "Development Document for Effluent Limitations Guidelines and New Source Performance Standards for the Explosive, Point Source Category," EPA440-1-76-060j, March 1976.

(EPA-76/03a), "Flare Systems Study," EPA600-2-76-079, March 1976.

(EPA-76/03b), "Inspection Manual for Enforcement of New Source Performance Standards: Asphalt Concrete Plants," EPA340-1-76-003, March 1976.

(EPA-76/03c), "Preliminary Evaluation of Air Pollution Aspects of the Drum-Mix Process," EPA340-1-77-004, March 1976.

(EPA-76/03d), "Workbook for Operators of Small Boilers and Incinerators," EPA450-9-76-001, March 1976.

(EPA-76/04), "Development Document for Best Technology Available for the Location, Design, Construction and Capacity of Cooling Water Intake Structures for Minimizing Adverse Environmental Impact," EPA440-1-76-015a, April 1976.

(EPA-76/06), "Development Document for Effluent Limitations Guidelines and New Source Performance Standards for the Other Non-Fertilizer Phosphate Chemicals, Segment of the Phosphate Manufacturing, Point Source Category," EPA440-1-75-043a, June 1976.

(EPA-76/10), "Environmental Pollution Control Pulp and Paper Industry. Part 1-Air," EPA625-7-76-001, October 1976.

(EPA-76/11), "Physical, Chemical, and Biological Treatment Techniques for Industrial Wastes," Prepared by Arthur D. Little, Inc., NTIS PB-275-054 (Volume I), and PB-275-287 (Volume II), November 1976.

(EPA-76/12), "Supplemental For Pretreatment to the Interim Final Development Document for the Secondary Aluminum, Segment of the Nonferrous Metals Manufacturing," EPA440-1-76-081c, December 1976.

(EPA-76/12a), "Development Document for Effluent Limitations Guidelines (BPCTCA) for the Bleached Kraft, Soda, Deink and Mono-integrated Paper Mills, Segment of the Pulp, Paper, and Paperboard, Point Source Category," EPA440-1-76-047b, December 1976.

(EPA-77), "Electrostatic Precipitator Malfunctions in the Electric Utility Industry," EPA600-2-77-006, 1977.

(EPA-77a), "New Source Performance Standards Inspection Manual for Enforcement of Sulfuric Acid Plants," EPA340-1-77-008, 1977.

(EPA-77b), "A Survey of Sulfate, Nitrate, and Acid Aerosol Emissions and Their Control," EPA600-7-77-041, 1977.

(EPA-77/01), "Municipal Incinerator Enforcement Manual," EPA340-1-76-013, January 1977.

(EPA-77/07), "Supplement for Pretreatment to the Development Document for the Inorganic Chemicals Manufacturing, Point Source Category," EPA440-1-77-087a, July 1977.

(EPA-77/11), "Flue Gas Desulfurization System Manufacturers Survey," EPA450-3-78-043, November 1977.

(EPA-78/01), "Control Techniques for Nitrogen Oxides Emissions From Stationary Sources-2nd Edition," EPA450-1-78-001, January 1978.

(EPA-78/03), "Flue Gas Desulfurization Systems: Design and Operating Considerations, Volume I, Executive Summary," EPA600-7-78-030a, March 1978.

(EPA-78/03), "Operation and Maintenance of Particulate Control Devices on Selected Steel and Ferroalloy Processes," EPA600-2-78-037, March 1978.

(EPA-78/03a), "Flue Gas Desulfurization Systems: Design and Operating Considerations, Volume I, Executive Summary," EPA600-7-78-030a, March 1978.

(EPA-78/03b), "Flue Gas Desulfurization Systems: Design and Operating Considerations, Volume II, Technical Report," EPA600-7-78-030b, March 1978.

(EPA-78/03c), "The Effect of Flue Gas Desulfurization Availability on Electric Utilities, Volume I, Executive Summary," EPA600-7-78-031a, March 1978.

(EPA-78/03d), "The Effect of Flue Gas Desulfurization Availability on Electric Utilities, Volume II, Technical Report," EPA600-7-78-031b, March 1978.

(EPA-78/03e), "Flue Gas Desulfurization System Capabilities for Coal-fired Steam Generators, Volume I, Executive Summary," EPA600-7-78-032a, March 1978.

(EPA-78/03f), "Flue Gas Desulfurization System Capabilities for Coal-fired Steam Generators, Volume II, Technical Report," EPA600-7-78-032b, March 1978.

(EPA-78/06), "A Mathematical Model of Electrostatic Precipitation (Revision 1): Volume I. Modeling and Programming," EPA600-7-78-111a, June 1978.

(EPA-78/06a), "A Mathematical Model of Electrostatic Precipitation (Revision 1): Volume II. User Manual," EPA600-7-78-111b, June 1978.

(EPA-79), "Particulate Control by Fabric Filtration on Coal Fired Industrial Boilers," EPA625-2-79-021, 1979.

(EPA-79a), "A Review of Standards for New Stationary Sources-Nitric Acid Plants," EPA450-3-79-013, 1979.

(EPA-79b), "A Review of Standards of Performance for New Stationary Sources-Sulfuric Acid Plants," EPA450-3-79-003, 1979.

(EPA-79/01), "Guidelines for Particulate Sampling in Gaseous Effluents from Industrial Processes," EPA600-7-79-028, January 1979.

(EPA-79/01a), "Evaluation of Dry Sorbents and Fabric Filtration for FGD," EPA600-7-79-005, January 1979.

(EPA-79/01b), "Design Guidelines for an Optimum Scrubber System," EPA600-7-79-018, January 1979.

(EPA-79/01c), "Enhanced SO_3 Emissions from Staged Combustion," EPA 600-7-79-002, January 1979.

(EPA-79/02), "Summary Report-Sulfur Oxides Control Technology Series: Flue Gas Desulfurization, Wellman-Lord Process," EPA600-8-79-001, February 1979.

(EPA-79/03), "A Review of Standards of Performance for New Sources-Petroleum Refineries," EPA450-3-79-008, March 1979.

(EPA-79/03a), "A Review of Standards of Performance for New Sources-Portland Cement Industry," EPA450-3-79-012, March 1979.

(EPA-79/03b), "A Review of Standards of Performance for New Stationary Sources-Incinerators," EPA450-3-79-009, 1979.

(EPA-79/04), "Guidance for Lowest Achievable Emission Rates from 18 Major Stationary Sources," EPA450-3-79-024, April 1979.

(EPA-79/04a), "Chemical Aspects of Afterburner Systems," EPA600-7-79-096, April 1979.

(EPA-79/05), "Decision Series, Sulfur Emission: Control Technology and Waste Management," EPA600-9-79-019, May 1979.

(EPA-79/06), "A Review of Standards of Performance for New Stationary Sources-Asphalt Concrete Plants," EPA450-3-79-014, June 1979.

(EPA-79/06a), "Continuous Air Pollution Source Monitoring Systems Handbook," EPA625-6-79-005, June 1979.

(EPA-79/08), "Development Document for Existing Source Pretreatment Standards for the Electroplating, Point Source Category," EPA440-1-79-003, August 1979.

(EPA-79/11), "Decision Series, Sulfur Oxides Control in Japan," EPA600-9-79-043, November 1979.

(EPA-79/12), "Development Document for Effluent Limitations Guidelines and Standards for the Gum and Wood Chemicals, Point Source Category," EPA440-1-79-078b, December 1979.

(EPA-79/12a), "Development Document for Effluent Limitations Guidelines and Standards for the Ink Formulating, Point Source Category," EPA440-1-79-090b, December 1979.

(EPA-79/12b), "Development Document for Effluent Limitations Guidelines and Standards for the Painting Formulating, Point Source Category," EPA440-1-79-049b, December 1979.

(EPA-79/12c), "Development Document for Effluent Limitations Guidelines and Standards for the Petroleum Refining, Point Source Category," EPA440-1-79-014b, December 1979.

(EPA-79/12d), "Source Sampling for Particulate Pollutants: Student Workbook," Course 450, EPA450-2-79-007, December 1979.

(EPA-80), "Assessment of Atmospheric Emissions from Petroleum Refining (5 Volumes)," EPA600-2-80-075, 1980.

(EPA-80a), "Industrial Process Profiles for Environmental Use: Chapter 27, Primary Lead Industry," EPA600-2-80-168, 1980.

(EPA-80b), "Industrial Process Profiles for Environmental Use: Chapter 28, Primary Zinc Industry," EPA600-2-80-169, 1980.

(EPA-80c), "Industrial Process Profiles for Environmental Use: Chapter 29, Primary Copper Industry," EPA600-2-80-170, 1980.

(EPA-80/02), "Controlling Nitrogen Oxides," EPA600-8-80-004, February 1980.

(EPA-80/02a), "Survey of Dry SO_2 Control Systems," EPA600-7-80-030, February 1980.

(EPA-80/02b), "Combustion Evaluation: Student Manual," Course 427, EPA450-2-80-063, February 1980.

(EPA-80/02c), "Combustion Evaluation: Student Workbook," Course 427, EPA450-2-80-064, February 1980.

(EPA-80/02d), "Combustion Evaluation: Instructor's Guide," Course 427, EPA450-2-80-064, February 1980.

(EPA-80/03), "Multimedia Environmental Goals for Environmental Assessment, Volume I (Supplement A)," EPA600-7-80-041, March 1980.

(EPA-80/04), "Control of Particulate Emissions: Student Workbook," Course 413, EPA450-2-80-067, April 1980.

(EPA-80/05), "TI-59 Programmable Calculator Programs for In-stack Opacity, Venturi Scrubbers, and Electrostatic Precipitators," EPA600-8-80-024, May 1980.

(EPA-80/05a), "Source Category Survey: Industrial Incinerators," EPA450-3-80-013, May 1980, NTIS PB80-193-303.

(EPA-80/08), "Planning Wastewater Management Facilities for Small Communities," EPA600-8-80-030, August 1980.

(EPA-80/08a), "Research Summary, Controlling Sulfur Oxides," EPA600-8-80-029, August 1980.

(EPA-80/10), "Draft Development Document for Effluent Limitations Guidelines (BATEA) New Source Performance Standards, and Pretreatment Standards for the Photographic Processing Point Source Category, Contract No. 68-01-3273, Prepared by Versar, Inc. for EPA, October 1980.

(EPA-80/11), "Dioxins," EPA600-2-80-197, November 1980.

(EPA-80/12), "Continuous Emission Monitoring: Student Laboratory Workbook," Course 474, No EPA number, December 1980.

(EPA-80/12a), "Hazardous Waste Generation and Commercial Hazardous Waste Management Capacity," SW-894, December 1980.

(EPA-81), "Inspection Manual for Evaluation of Electrostatic Precipitator Performance," EPA 340-1-79-007, 1981.

(EPA-81/03), "Sources and Control of Volatile Organic Air Pollutants: Student Workbook," Course 482, EPA450-2-81-011, March 1981.

(EPA-81/03a), "Sources and Control of Volatile Organic Air Pollutants: Regulatory Documents," Course 482, EPA450-2-81-012, March 1981.

(EPA-81/04), "Control Techniques for Sulfur Oxide Emissions from Stationary Sources," EPA450-3-81-004, April 1981.

(EPA-81/04a), "Emission Volatile Organic Compounds from Drum-Mix Asphalt Plants," EPA600-S2-81-026, April 1981.

(EPA-81/05), "Control of Gaseous Emissions: Student Workbook," Course 415, EPA450-2-81-006, May 1981.

(EPA-81/05a), "Sources and Control of Volatile Organic Air Pollutants: Instructor's Guide," Course 482, EPA450-2-81-010,

May 1981.

(EPA-81/09), "Engineering Handbook for Hazardous Waste Incineration," SW-889, September 1981.

(EPA-81/10), "Control Of Particulate Emissions: Student Manual," Course 413, EPA450-2-80-066, October 1981.

(EPA-81/10a), "Dispersion of Air Pollution-Theory and Model Application: Selected Reading Packet," Course 423, EPA450-2-81-077, October 1981.

(EPA-81/10b), "Dispersion of Air Pollution-Theory and Model Application: Student Workbook," Course 423, EPA450-2-81-075, October 1981.

(EPA-81/11), "Dispersion of Air Pollution-Theory and Model Application: Instructor's Guide, EPA450-2-81-076, November 1981.

(EPA-81/12), "Control of Gaseous Emissions," Student Manual, EPA450-2-81-005, Course 415, December 1981.

(EPA-82/02), "Sampling and Analysis Methods for Hazardous Waste Combustion," EPA600-8-84-002, NTIS PB84-155-845, February 1982.

(EPA-82/04), "Baghouse Plan Review: Student Guidebook," Course SI:412, EPA450-2-82-005, April 1982.

(EPA-82/04a), "Baghouse Plan Review," Course SI:412A, EPA450-2-81-005, April 1982.

(EPA-82/05), "Development Document for Effluent Limitations Guidelines and Standards for the Ore Mining and Dressing," EPA440-1-82-061b, May 1982.

(EPA-82/05a), "Development Document for Effluent Limitations Guidelines and Standards for the Iron and Steel Manufacturing," EPA440-1-82-024, May 1982.

(EPA-82/07), "Development Document for Effluent Limitations Guidelines and Standards for the Electrical and Electronic Components, Point Source Category," EPA440-1-782-075b, July 1982.

(EPA-82/08), "Controlling VOC Emissions from Leaking Process Equipment," EPA450-2-82-015, August 1982.

(EPA-82/09), "Development Document for Effluent Limitations Guidelines and Standards for the Textile Mills, Point Source Category," EPA440-1-82-022, September 1982.

(EPA-82/09a), "Control Techniques for Particulate Emissions from Stationary Sources-Volume 1 and 2," EPA450-3-81-005a & b, September 1982, NTIS PB83-127-498.

(EPA-82/10), "Development Document for Effluent Limitations Guidelines and Standards for the Coal Mining, Point Source Category," EPA440-1-82-057, October 1982.

(EPA-82/10a), "Development Document for Effluent Limitations Guidelines and Standards for the Coil Coating, Point Source Category," EPA440-1-82-071, October 1982.

(EPA-82/11), "Development Document for Effluent Limitations Guidelines and Standards for the Leather Tanning and Finishing," EPA440-1-82-016, November 1982.

(EPA-82/11a), "Development Document for Effluent Limitations Guidelines and Standards and Pretreatment Standards for the Steam Electric," EPA440-1-82-029, November 1982.

(EPA-82/11b), "Development Document for Effluent Limitations Guidelines and Standards for the Aluminum Forming, Point Source Category," EPA440-1-82-073b, November 1982.

(EPA-82/11c), "Development Document for Effluent Limitations Guidelines and Standards for the Metal Molding and Casting (Foundries), Vol. I, Point Source Category," EPA440-1-82-070b, November 1982.

(EPA-82/11d), "Development Document for Effluent Limitations Guidelines and Standards for the Ore Mining and Dressing, Point Source Category," EPA440-1-82-061, November 1982.

(EPA-82/11e), "Development Document for Effluent Limitations Guidelines and Standards for the Porcelain Enameling, Point Source Category," EPA440-1-82-072, November 1982.

(EPA-82/11f), "Development Document for Effluent Limitations Guidelines and Standards and Pretreatment Standards for the Steam Electric, Point Source Category," EPA440-1-82-029, November 1982.

(EPA-83), "Thesaurus on Resource Recovery Terminology," Supported by EPA, Compiled and Published by ASTM, Special Technical Publication 832, Publication Code Number 04-832000-16, 1983.

(EPA-83/01), "Continuous Emission Monitoring: Regulatory Documents," Course 474, EPA450-2-83-001, January 1983.

(EPA-83/02), "Development Document for Effluent Limitations Guidelines and Standards for the Electrical and Electronic Components, Point Source Category," EPA440-1-83-075b, February 1983.

(EPA-83/03), "Development Document for Effluent Limitations Guidelines and Standards for the Electrical and Electronic Components," EPA440-1-83-075, March 1983.

(EPA-83/03a), "Development Document for Effluent Limitations Guidelines and Standards for the Nonferrous Metals, Vol. III, Point Source Category," EPA440-1-83-019b, March 1983.

(EPA-83/03b), "Introduction to Dispersion Modeling: Self-instructional Guidebook," Course SI:410, EPA450-2-82-007, March 1983.

(EPA-83/06), "Atmospheric Sampling," Course 435, EPA450-2-80-004, June 1983.

(EPA-83/06a), "Development Document for Effluent Limitations Guidelines and Standards for the Metal Finishing, Point Source Category," EPA440-1-83-091, June 1983.

(EPA-83/06b), "Air Pollution Control Systems for Selected Industries: Self-instructional Guidebook," Course SI:431, EPA450-2-82-006, June 1983.

(EPA-83/06c), "Atmospheric Sampling: Student Manual," Course 435, EPA450-2-80-004, June 1983.

(EPA-83/06d), "Air Pollution Control Systems for Selected Industries: Self-instructional Guidebook," Course SI:431, EPA450-2-82-006, June 1983.

(EPA-83/07), "Guidance Manual of Hazardous Waste Incinerator Permits," SW-966, NTIS PB84-100-577, July 1983, Volume I of the Hazardous Waste Incineration Guidance Series.

(EPA-83/07a), "Electrostatic Precipitator Plan Review: Self-instructional Guidebook," Course SI:412B, EPA450-2-82-019, July 1983.

(EPA-83/09), "Development Document for Effluent Limitations Guidelines and Standards for the Pharmaceutical Manufacturing, Point Source Category," EPA440-1-83-084, September 1983.

(EPA-83/09a), "Wet Scrubber Inspection and Evaluation Manual," EPA340-1-83-022, September 1983, NTIS PB85-149-375.

(EPA-84/02), "A Profile of Existing Hazardous Waste Incineration Facilities and Manufactures in the United States," NTIS PB84-157-072, February 1984.

(EPA-84/02a), "Fabric Filter Inspection and Evaluation Manual," EPA340-1-84-002, February 1984, NTIS PB86-237-716.

(EPA-84/02b), "A Profile of Existing Hazardous Waste Incineration Facilities and Manufacturers in the United States," NTIS PB84-157-072, February 1984.

(EPA-84/02c), "Trial Burn Protocol Verification at a Hazardous Waste Incinerator," EPA600-2-84-048, February 1984.

(EPA-84/03), "Calculation of Precision, Bias, and Method Detection Limit for Chemical and Physical Measurements," EPA's Office of Monitoring Systems and Quality Assurance, March 1984.

(EPA-84/03a), "Feasibility Study for Adapting Present Combustion Source Continuous Monitoring Systems to Hazardous Waste Incinerators, Vols. I & II," EPA600-8-84-001a, March 1984.

(EPA-84/03b), "Wet Scrubber Plan Review: Self-instructional Guidebook," Course SI:412C, EPA450-2-82-020, March 1984.

(EPA-84/04), "Sources and Control of Volatile Organic Air Pollutants: Regulatory Documents Updates and NSPS Regulations," Course 482, EPA450-2-84-001, April 1984.

(EPA-84/08), "Development Document for Effluent Limitations Guidelines and Standards for the Battery Manufacturing, Vols. I & II, Point Source Category," EPA440-1-84-067, August 1984.

(EPA-84/08a), "Health Assessment Document for Chromium, Final Report," EPA600-8-83-014F, August 1984.

(EPA-84/09), "Control of Gaseous and Particulate Emissions," Course SI:412D, EPA450-2-84-007, September 1984.

(EPA-84/09a), "Transmissometer Systems Operation and Maintenance, an Advanced Course," Course SI:476A, EPA450-2-84-004, September 1984.

(EPA-84/11), "Performance Evaluation of Full-Scale Hazardous Waste Incineration," 5 volumes, EPA600-2-84-181a,b,c,d,e, NTIS PB85-129-500, November 1984.

(EPA-84/12), "Introduction to Boiler Operation, Self-instructional Guidebook," Course SI:428A, EPA450-2-84-010, December 1984.

(EPA-85/08), "Quality Assurance Procedures for the Hazardous Waste Engineering Research Laboratory," Document Control No., QAP-0006-GFS, August 1985.

(EPA-85/09), "Survey of Selected Firms in the Commercial Hazardous Waste Management Industry 1984 Update," Prepared by ICF, Inc. September 1985.

(EPA-85/09a), "Operation and Maintenance Manual for Electrostatic Precipitators," EPA625-1-85-017, September 1985.

(EPA-85/09b), "Air Pollution Source Inspection Safety Procedures: Student Manual," EPA340-1-85-002a, September 1985.

(EPA-85/10), "Development Document for Effluent Limitations Guidelines and Standards for the Pesticide, Point Source Category," EPA440-1-85-079, October 1985.

(EPA-85/10a), "Development Document for Effluent Limitations Guidelines and Standards for the Metal Molding and Casting (Foundries), Point Source Category," EPA440-1-85-070, October 1985.

(EPA-85/11), "Chemical Emergency Preparedness Program Interim Guidance," 9223.0-1A, November 1985.

(EPA-85/12), "Introduction to Baseline Source Inspection Techniques: Self-Instructional Guidebook (draft)," Course SI:445, no EPA number, December 1985.

(EPA-86/01), "RCRA Orientation Manual," EPA530-SW-86-001, January 1986.

(EPA-86/03), "Hazardous Waste Incineration: Steudent Manual (draft)," Course 502, No EPA number, March 1986.

(EPA-86/03a), "Hazardous Waste Incineration: Student Workbook (draft)," Course 502, No EPA number, March 1986.

(EPA-86/03b), "Hazardous Waste Incineration: Instructor's Guide (draft)," Course 502, No EPA number, March 1986.

(EPA-86/05), "EPA Guide for Infectious Waste Management," EPA Office of Solid Waste, EPA530-SW-86-014, NTIS PB86-199-130, May 1986.

(EPA-86/06), "Operation and Maintenance Manual for Fabric Filters," EPA625-1-86-020, June 1986.

(EPA-86/07), "Superfund Treatment Technologies: A Vendor Inventory," EPA540-2-80-004, July 1986.

(EPA-86/09), "Handbook: Control Technologies for Hazardous Air Pollutants," EPA625-6-86-014, September 1986.

(EPA-86/09a), "Permit Writer's Guide to Test Burn Data: Hazardous Waste Incineration," EPA625-6-86-012, September 1986.

(EPA-86/10), "Waste Minimization Issues and Options," EPA 530-SW-86-041.

(EPA-86/10a), " Quality Assurance Procedures for the Hazardous Waste Engineering Research Laboratory," GAP-0006-GFS, October 1986.

(EPA-86/10b), "Technical Resource Document: Treatment Technologies for Dioxin-Containing Wastes," EPA600-2-86-096, October 1986.

(EPA-86/12), "Prevention References Manual - Chemical Specific, Volume f: Carbon Tetrachloride," EPA Contract No. 68-02-3889, Work Assignment 98, December 1986.

(EPA-87/03), "Handbook, Ground Water," EPA625-6-87-016, March 1987.

(EPA-87/06), "Emission Data Base for Municipal Waste Combustors," EPA530-SW-87-021b, June 1987.

(EPA-87/06a), "Municipal Waste Combustion Study: Combustion Control," EPA530 SW-87-021c, June 1987, NTIS PB87-206-090.

(EPA-87/06b), "Municipal Waste Combustion Study: Flue Gas Cleaning Technology," EPA530-SW-87-021d, June 1987, NTIS PB87-206-108.

(EPA-87/06c), "Municipal Waste Combustion Systems, Operation and Maintenance Study," EPA340-1-87-002, June 1987.

(EPA-87/07), "Land Disposal, Remedial Action, Incineration and Treatment of Hazardous Waste," Proceedings of the Thirteenth annual Research Symposium at Cincinnati, Ohio, May 6-8, 1987, EPA600-9-87-015, July 1987.

(EPA-87/07a), "Prevention Release Manual: User's Guide Overview for Controlling Accidental Releases of Air Toxics," EPA600-8-87-028, July 1987.

(EPA-87/08), "The Risk Assessment Guidelines of 1986," EPA600-8-87-045, August 1987.

(EPA-87/08a), "National Dioxin Study," EPA 530-SW-87-025, August 1987.

(EPA-87/08b), "Prevention References Manual: Chemical Specific: Volume 8: Control of Accidental Releases of Hydrogen Fluoride," EPA600-8-87-034h, August 1987.

(EPA-87/09), "The Safe Drinking Water Act," EPA Region 5, September 1987.

(EPA-87/10), "Development Document for Effluent Limitations Guidelines and Standards for the Pulp, Paper, and Paperboard and the Builders' Paper and Board Mills," EPA440-1-87-025, October 1987.

(EPA-87/10a), "Development Document for Effluent Limitations Guidelines and Standards for the Organic Chemicals, Plastics and Synthetic Fibers," EPA440-1-87-009, October 1987.

(EPA-88/08), "Application of Radon Reduction Methods," EPA625-5-88-024, August 1988.

(EPA-88/08a), "Technological Approaches to the Cleanup of Radiologically Contaminated Superfund Sites," EPA540-2-88-002, August 1988.

(EPA-88/09), "Air Pollution Modeling as Applied to Hazardous Waste Incinerator Evaluations, An Introduction for the Permit Writer," Office of Solid Waste, Waste Treatment Branch, September 1988.

(EPA-88/09a), "Final Report on Biosafety in Large-Scale rDNA Processing Facilities," Prepared by Battelle for EPA's Risk Reduction Engineering Laboratory, Contract 68-03-3248, Project Officer, J. O. Burckle, September 1988.

(EPA-88/09b), "The Inside Story - A Guide to Indoor Air Quality," EPA's Office of Air and Radiation, EPA400-1-88-004, September 1988.

(EPA-88/11), "Meeting on Medical Waste, Definition, Tracking, Information Needs," An EPA Meeting Report, November 14-16, 1988, Annapolis, Maryland, November 1988.

(EPA-88/12), "Development of a Thermal Stability Based Index of Hazardous Waste Incinerability," A draft annual report prepared by the University of Dayton Research Institute (UDRI) for EPA's Risk Reduction Engineering Laboratory, December 1988.

(EPA-88/12a), "Hospital Waste Combustion Study: Data Gathering Phase," Office of Air Quality Planning and Standards, EPA450-3-88-017, December 1988.

(EPA-89/01), "Guidance on Setting Permit Conditions and Reporting Trial Burn Results," EPA625-6-89-019, January 1989, Volume II of the Hazardous Waste Incineration Guidance Series.

(EPA-89/02), "Hospital Waste Incinerator Field Inspection and Source Evaluation Manual," EPA340-1-89-001, February 1989.

(EPA-89/03), "Trial Burn Observation Guide," EPA530-SW-89-027, March 1989.

(EPA-89/03a), "Operation and Maintenance of Hospital Medical Waste Incinerators," EPA450-3-89-002, March 1989.

(EPA-89/03b), "Hospital Incinerator Operator Training Course: Volume I, Student Handbook," EPA450-3-89-003, March 1989.

(EPA-89/03c), "Hospital Incinerator Operator Training Course: Volume II, Presentation Slides," EPA450-3-89-003, March 1989.

(EPA-89/03d), "Hospital Incinerator Operator Training Course: Volume III, Instructor Handbook," EPA450-3-89-003, March 1989.

(EPA-89/06), "Hazardous Waste Incineration Measurement Guidance Manual," Volume III of the Hazardous Waste Incineration Guidance Series, EPA625-6-89-021, Prepared by MRI for EPA, June 1989.

(EPA-89/08), "Guidance on Metals and Hydrogen Chloride Controls for Hazardous Waste Incinerators," No EPA number, August 1989, Volume IV of the Hazardous Waste Incineration Guidance Series.

(EPA-89/09), "Technical Guidance Document: The Fabrication of Polyethylene FML (flexible membrane liners) Field Seams," EPA530-SW-89-069, September 1989.

(EPA-89/11), "Decision-Makers Guide to Solid Waste Management," EPA530-SW-89-072, November 1989.

(EPA-89/11a), "Proposed Methods for Stack Emissions Measurement of CO, O_2 THC, HCl, and Metals at Hazardous Waste Incinerators," November 1989, Volume VI of the Hazardous Waste Incineration Guidance Series.

(EPA-89/12), "Glossary of Environmental Terms and Acronym List," EPA Office of Communications and Public Affairs (A-107), 19K-1002, December 1989.

(EPA-89/12a), "Guidance for Conducting Treatability Studies Under CERCLA," EPA540-2-89-058, December 1989.

(EPA-90/04), "Guidance on PIC Controls for Hazardous Waste Incinerators," EPA 530-SW-90-040, April 1990, Volume V of the Hazardous Waste Incineration Guidance Series.

(EPA-90/05), "Medical Waste Management in the United States," EPA 530-SW-90-051a (The first EPA medical waste report to Congress under MWTA), May 1990.

(EPA-90/06), "State-Of-The-Art Assessment of Medical Waste Thermal Treatment," A Draft Report Prepared for EPA's Risk Reduction Engineering Laboratory in Cincinnati, Ohio, June 1990.

(EPA-90/08), "Interim Methods for Development of Inhalation Reference Concentrations," EPA600-8-90-006a, August 1990.

(EPA-90/12), "Methods Manual for Compliance with the BIF Regulations: Burning Hazardous Waste in Boilers and Industrial Furnaces," EPA 530-SW-91-010, December 1990, NTIS PB91-120-006.

(EPA-91/02), "Preparation Aids for the Development of Category III Quality Assurance Project Plans," EPA600-8-91-005, February 1991.

(EPA-91/03), "Technical Support Document For Water Quality-based Toxics Control," EPA505-2-90-001, March 1991.

(EPA-91/05), "Technical Guidance Document: Inspection Techniques for the Fabrication of Geomembrance Field Seams," EPA530-SW-91-051, May 1991.

(EPA-91/12), "Risk Assessment Guidance for Superfund: Volume I-- Human Health Evaluation Manual (Part B, Development of Risk-based Preliminary Remdiation Goals)," EPA540-R-92-003, Publication 9285.7-01B, December 1991.

(EPA-92/12), "Glossary of Risk Assessment-Related Terms." The glossary and an attached list of Acronyms and Abbreviations (dated 06/08/92) were published in the Integrated Risk Information System (IRIS) Program by EPA in Cincinnati, Ohio on December 1, 1992.

(EPA-94/04), "Terms of Environment: Glossary, Abbreviations, and Acronyms," EPA175-B-94-015, April 1994.

(EPS-77/03), "Burning Waste Chlorinated Hydrocarbons in a Cement Kiln," Environmental Protection Service (EPS), Canada, Report No. EPS 4-WP-77-2, March 1977.

(ESAx-16USCyyyy-90) means:
- ESAx: Endangered Species Act, Section x;
- 16USCyyyy: Title 16, United States Code (USC), Section yyyy; and
- -90: 1990 Edition, "Environmental Law Deskbook," Published by the Environmental Law Institute, 1616 P. Street, NW, Washington, DC 20036.

(FIFRAx-7USCyyyy-91) means:
- FIFRAx: Federal Insecticide, Fungicide, and Rodenticide Act, Section x;
- 7USCyyyy: Title 7, United States Code (USC), Section yyyy;
- -91: 1991 Edition, "Environmental Statutes," Published by Government Institutes, Inc., 4 Research Place, Suite 200, Rockville, MD 20850, March 1991.

(FLPMAxxx-43USCyyyy-90) means:
- FLPMAxxx: Federal Land Policy and Management Act, Section xxx;
- 43USCyyyy: Title 43, United States Code (USC), Section yyyy;
- -90: 1990 Edition, "Environmental Law Deskbook," Published by the Environmental Law Institute, 1616 P. Street, NW, Washington, DC 20036.

(Frick-86), "Environmental Glossary," G. William Frick and Thomas F. P. Sullivan, Government Institutes, Inc. Rockville, MD, 1986.

(Gurney-66), "Cooling Towers," J. D. Gurney, 1966.

(Hesketh-79), "Air Pollution Control," H. E. Hesketh, Ann Arbor Science, 1979.

(Holman-69), "Thermodynamic," J. P. Holman, McGraw-Hill Book Company, 1969.

(Huffman-91/10), Personal Communication with George L. Huffman, US EPA, Cincinnati, Ohio, October 1991.

(Isaacs-91), "Concise Science Dictionary," Alan Isaacs; John Daintith and Elizabeth Martin, Oxford University, Second Edition, 1991.

(JMM-88), "Safe Drinking Water Act Highlights," James M. Montgomery (JMM) Consulting Engineers Inc. April 20, 1988.

(Jones-60), "Engineering Thermodynamics," J. B. Jones, and G. A. Hawkins, John Wiley, 1960.

(LBL-76/07), "Instrumentation for Environmental Monitoring," Lawrence Berkeley Laboratory (LBL), University of California, Berkeley, CA 94720, July 1976.

(Lee-83/07), "A Comparison of Innovative Technology for Thermal Destruction of Hazardous Waste," C. C. Lee, Proceedings of the First Annual Hazardous Materials Management Conference, Philadelphia, PA, July 12-14, 1983.

(Lee-85/09), "An Overview of Pilot Scale Research in Hazardous Waste Thermal Destruction," C. C. Lee and G. L. Huffman, Proceedings of International Conference on New Frontiers in Hazardous Waste Management, EPA600-9-85-025, September 1985.

(Lee-88/08), "Update of Innovative Thermal Destruction," C. C. Lee and G. L. Huffman, AIChE 1988 Summer National Meeting, Denver Colorado, August 1988.

(Lee-90/11), "Regulatory Framework for Combustion By-products from Incineration Sources," C.C. Lee and G.L. Huffman, 1990 Pacific Basin Conference on Hazardous Waste, Honolulu, Hawaii, November 12-16, 1990.

(Lee-90/12), "Medical Waste Incineration Handbook," C. C. Lee, Published by Government Institutes, Inc., 966 Hungerford Dr., #24, Rockville, MD 20850. Phone 301 921-2300, December 1990.

(Markes-67), "Standard Handbook for Mechanical Engineers," 17th Edition, McGraw-Hill Book Company, 1967.

(M&EI-72), " Wastewater Engineering, Collection, Treatment, Disposal," Metcalf & Eddy, Inc. (M&EI), McGraw-Hill Book Company, 1972.

(Merck-83), "The Merck Index: An Encyclopedia of Chemicals, Drugs, and Biologicals," Third Edition, Merck & Co., Inc., Rahway, NJ.

(MMPAx-16USCyyyy-90) means:

- MMPAx: Marine Mammal Protection Act, Section x;
- 16USCyyyy: Title 16, United States Code (USC), Section yyyy;
- -90: 1990 Edition, "Environmental Law Deskbook," Published by the Environmental Law Institute, 1616 P. Street, NW, Washington, DC 20036.

(NAC-83), "Risk Assessment in the Federal Government: Managing the Process," Committee on the Institutional Means for Assessment of Risks to Public Health, Commission on Life Sciences, National Research Council, Established by National Academy Council (NAC), National Academy Press, 1983.

(NATO-78/10), "Glossary of Terms on Air Pollution Assessment Methodology and Modeling," North Atlantic Treaty Organization (NATO), NTIS PB-289-376, October 1978.

(NCAx-42USCxxxx-87) means:

- NCAx: Noise Control Act, Section x;
- 42USCyyyy: Title 42, United States Code (USC), Section yyyy;
- -87: 1987 Edition, "Environmental Law," Printed for the Use of the House Committee on Energy and Commerce (HCEC), February 1987.

(NIOSH-84/10), "Personal Protective Equipment for Hazardous Materials Incidents: A Selection Guide," U.S. Department of Health and Human Services, Public Health Service, Centers for Disease Control, National Institute for Occupational Safety and Health (NIOSH), Division of Safety Research, Morgantown, West Virginia 26505, October 1984.

(NJIT-88/05), "Environmental Reporter's Handbook," Edited by David B. Sachsman and supported by the New Jersey Institute of Technology (NJIT), May 1988.

(NRT-87/03), "Hazardous Materials Emergency: Planning Guide," National Release Team (NRT), NRT-1, March 1987.

(OME-87/05), "An Evaluation of Hospital Incinerator Reports", Ontario Ministry of the Environment (OME), May 1987.

(OME-86/10), "Incinerator Design and Operating Criteria - Biomedical Waste Incinerator, Volume II," Ontario Ministry of the Environment (OME), October 1986.

(OME-88/12), "Guidance for Incinerator Design and Operation - General, Volume I," Ontario Ministry of the Environment (OME), December 1988.

(OME-89/04), "Guidance for Incinerator Design and Operation - Cremators, Volume III," Ontario Ministry of the Environment (OME), April 1989.

(OPAxxxx-91) means:

- OPAxxxx: Oil Pollution Act of 1990 (Public Law 101-380, August 18, 1990), Section xxxx;
- -91: 1991 Edition, "Environmental Statutes," Published by Government Institutes, Inc., 4 Research Place, Suite 200, Rockville, MD 20850, March 1991.

(Oppelt-87/05), "Incineration of Hazardous Waste, A Critical Review," E. Timothy Oppelt, Journal of Air Pollution Control Association, Vol. 37, No. 5, pp. 558-586. May 1987.

(OSHAx-29USC651-91) means:

- OSHAx: Occupational Safety and Health Act, Section x;
- 29USC651: Title 29, United States Code (USC), Section yyyy;
- -91: 1991 Edition, "Environmental Statutes," Published by the Government Institutes, Inc., 4 Research Place, Suite 200, Rockville, MD 20850, March 1991.

(OSW-77), "Physical, Chemical, and Biological Treatment Techniques for Industrial Wastes", A. D. Little prepared for EPA's Office of Solid Waste in 1977, NTIS PB-275-287, 1977.

(OSW-86), "RCRA Orientation Manual," Office of Solid Waste, EPA, Washington, DC 20460, 1986.

(OSWER-85), "The New RCRA: A Fact Book," Office of Solid Waste and Emergency Response (OSWER), EPA, Washington DC 20460, EPA 530-SW-85-035, 1985.

(OSWER-87), "The Hazardous Waste System", US EPA, Office of Solid Waste and Emergency Response (OSWER), 1987.

(OTA-83/03), "Technologies and Management Strategies for Hazardous Waste Control," Congress of the United States, Office of Technology Assessment, U.S. Government Printing Office, Washington, DC, OTA-196, March 1983.

(OTA-84/06), "Acid Rain and Transported Air Pollutants: Implications for Public Policy," OTA-O-204, Congress of the United States, Office of Technology Assessment, June 1984.

(OTA-86/08), "Ocean Incineration: Its Role in Managing Hazardous Waste," Congress of the United States, Office of Technology Assessment, OTA-0-313, August 1986.

(OTA-89/10), "Facing American's Trash—What Next for Municipal Solid Waste?", Congress of the United States, Office of Technology Assessment, OTA-O-424, October 1989.

(OTA-89/10a), "Superfund's Problems Can Be Solved...," U.S. Congress, Office of Technology Assessment (OTA), OTA-ITE-433, Washington DC, U.S. Government Printing Office, October 1989.

(Parker-84), "Dictionary of Science and Technical Terms," Sybil P. Parker, Editor in Chief, McGraw-Hill Book Company, Third Edition, 1984.
212-512-2000

(Perry-73), "Chemical Engineers Handbook," R. H. Perry, and C. H. Chilton, 5th edition, McGraw-Hill Chemical Engineering Series, 1973.

(PPAxxxx-91) means:

- PPAxxxx: Pollution Prevention Act of 1990, Section xxxx;
- -91: 1991 Edition of the "Environmental Statutes," Published by Government Institutes, Inc., 4 Research Place, Suite 200, Rockville, MD 20850, March 1991.

(RCRAxxxx-42USCyyyy-91) means:

- RCRAxxxx: Resource Conservation and Recovery Act, Section xxxx;
- 42USCyyyy: Title 42, United States Code (USC), Section yyyy;
- -91: 1991 Edition of the "Environmental Statutes," Published by Government Institutes, Inc., 4 Research Place, Suite 200, Rockville, MD 20850, March 1991.

(Schaum), "Theory and Problems of College Chemistry," Schaum's Outline Series, McGraw-Hill Publisher, 1966.

(Scott-81), "Dictionary of Waste and Water Treatment," John S. Scott and Paul G. Smith, University of Strathclyde, Glasgow, Butterworths Publishing Company, London, UK, 1981.

(SDWAxxxx-42USCyyyy-91) means:

- SDWAxxxx: Safe Drinking Water Act, Section xxxx;
- 42USCyyyy: Title 42, United States Code (USC), Section yyyy;
- -91: 1991 Edition of the "Environmental Statutes," Published by Government Institutes, Inc., 4 Research Place, Suite 200, Rockville, MD 20850, March 1991.

(SFxxx-42USCyyyy-91) means:

- SFxxx: Superfund, Section xxx, Superfund is also known as the Comprehensive Environmental Response, Compensation and Liability Act (CERCLA);
- 42USCyyyy: Title 42, United States Code (USC), Section yyyy;
- -91: 1991 Edition of the "Environmental Statutes," Published by Government Institutes, Inc., 4 Research Place, Suite 200, Rockville, MD 20850, March 1991.

(SMCRAxxx-30USCyyyy-90) means:

- SMCRAxxx: Surface Mining control and Reclamation Act, Section xxx;
- 30USCyyyy: Title 30, United States Code (USC), Section yyyy;
- -90: 1990 Edition, "Environmental Law Deskbook," Published by the Environmental Law Institute, 1616 P. Street, NW, Washington, DC 20036.

(Sonntag-71), "Introduction to Thermodynamics: Classical and Statistical," Richard E. Sonntag and Gordon J. Van Wylen, John Wiley & Sons, Inc. 1971.

(Sullivan-95/04), "Environmental Law Handbook, Thirteenth Edition," Thomas F.P. Sullivan, Editor, Government Institutes, Inc., April 1995.

(SW-108ts), "Solid Waste Management Glossary," Federal Solid

Waste Management Program, NTIS PB-259-501, 1972.

(SW-846), "Test Methods for Evaluating Solid Waste, Volumes 1A - 1C: Laboratory Manual, Physical/Chemical Methods, and Volume II: Field Manual, Physical/Chemical Methods", Third Edition, Office of Solid Waste, US Environmental Protection Agency, Document Control No. 955-001-00000-1, 1986.

(TSCAxxx-15USCyyyy-91) means:

- TSCAxxx: Toxic Substances Control Act, Section xxx;
- 15USCyyyy: Title 15, United States Code (USC), Section yyyy;
- -91: 1991 Edition of the "Environmental Statutes," Published by Government Institutes, Inc., 4 Research Place, Suite 200, Rockville, MD 20850, March 1991.

(TSCA-AIA1-91) means:

- TSCA-AIA1: Toxic Substances Control Act, Section 1, Asbestos Information Act of 1988 (PL100-577, 10/31/88).
- -91: 1991 Edition of the "Environmental Statutes," Published by Government Institutes, Inc., 4 Research Place, Suite 200, Rockville, MD 20850, March 1991.

(Webster-68), "Webster's New World Dictionary of the American Language: Second College Edition," David B. Guralnik, Editor in Chief, The World Publishing Company, 1968.

(Winthrop-89/09), "Summary of Environmental Laws Affecting Incineration of Hazardous Waste," A Report Prepared by Winthrop, Stimson, Putnam & Roberts for the Coalition for Responsible Waste Incineration (CRWI), September 1989.

(Wittle-93/07), A Personal Letter from J. Kenneth Wittle of the Electro-Pyrolysis, Inc. to Dr. C. C. Lee on July 16, 1993. The address of the Electro-Pyrolysis, Inc. is "Suite 1118, 996 Old Eagle School Road, Wayne, PA 19087.

(Wylen-73), "Fundamendals of Classical Thermodynamics," Gordon J. Van Wylen & Richard E. Sonntag, John Wiley & Sons, Inc. 1973.

MAJOR ENVIRONMENTAL LAWS AND ENVIRONMENTAL REGULATIONS	
Environmental Laws	Environmental Regulations (CFR Cites)
• EPA's purpose and functions	• Subchapter A--General; 40 CFR 1-29
• EPA's regulatory authorities	• Subchapter B--Grants and Other Federal Assistance; 40 CFR 30-47
• Clean Air Act (CAA) of 1970	• Subchapter C--Air Programs; 40 CFR 50-99
• Federal Water Pollution Control Act (FWPCA) of 1972	• Subchapter D--Water Programs; 40 CFR 100-140 (oil discharge and prevention related regulations)
• Safe Drinking Water Act (SDWA) of 1974	• Subchapter D--Water Programs; 40 CFR 141-149 (drinking water regulations)
• Federal Insecticide, Fungicide, and Rodenticide Act (FIFRA) of 1947	• Subchapter E--Pesticide Programs; 40 CFR 150-189
• Atomic Energy Act (AEA) of 1954	• Subchapter F--Radiation Protection Programs; 40 CFR 190-192
• Noise Control Act (NCA) 1972	• Subchapter G--Noise Abatement Programs; 40 CFR 201-211
• Marine Protection, Research, and Sanctuaries Act (MPRSA) of 1972	• Subchapter H--Ocean Dumping; 40 CFR 220-238
• Resource Conservation Recovery Act (RCRA) of 1976	• Subchapter I--Solid Wastes; 40 CFR 240-259 (municipal waste, land disposal, and resource recovery regulations)
• Hazardous and Solid Waste Act (HSWA) of 1984	• Subchapter I--Solid Wastes; 40 CFR 260-299 (hazardous waste)
• Comprehensive Environmental Response, Compensation, and Liability Act of 1980 (CERCLA) • Superfund Amendments and Reauthorization Act (SARA) of 1986	• Subchapter J--Superfund, Emergency Planning, and Community Right-To-Know Programs; 40 CFR 300-399
• Clean Water Act (CWA) of 1977	• Subchapter N--Effluent Guidelines and Standards; 40 CFR 400-599
• Motor Vehicle Information and Cost Savings Act (MVICSA) (15USC1901)	• Subchapter P (reserved) • Subchapter Q--Energy Policy; 40 CFR 600-699
• Toxic Substances Control Act (TSCA) of 1976	• Subchapter R--Toxic Substances Control Act; 40 CFR 700-799
• National Environmental Policy Act (NEPA) 1969	• Subchapter V--Council on Environmental Quality; 40 CFR 1500-1517

GOVERNMENT INSTITUTES MINI-CATALOG

PC #	ENVIRONMENTAL TITLES	Pub Date	Price
585	Book of Lists for Regulated Hazardous Substances, 8th Edition	1997	$79
4088	CFR Chemical Lists on CD ROM, 1997 Edition	1997	$125
4089	Chemical Data for Workplace Sampling & Analysis, Single User	1997	$125
512	Clean Water Handbook, 2nd Edition	1996	$89
581	EH&S Auditing Made Easy	1997	$79
587	E H & S CFR Training Requirements, 3rd Edition	1997	$89
4082	EMMI-Envl Monitoring Methods Index for Windows-Network	1997	$537
4082	EMMI-Envl Monitoring Methods Index for Windows-Single User	1997	$179
525	Environmental Audits, 7th Edition	1996	$79
548	Environmental Engineering and Science: An Introduction	1997	$79
578	Environmental Guide to the Internet, 3rd Edition	1997	$59
560	Environmental Law Handbook, 14th Edition	1997	$79
353	Environmental Regulatory Glossary, 6th Edition	1993	$79
625	Environmental Statutes, 1998 Edition	1998	$69
4098	Environmental Statutes Book/Disk Package, 1998 Edition	1997	$208
4994	Environmental Statutes on Disk for Windows-Network	1997	$405
4994	Environmental Statutes on Disk for Windows-Single User	1997	$139
570	Environmentalism at the Crossroads	1995	$39
536	ESAs Made Easy	1996	$59
515	Industrial Environmental Management: A Practical Approach	1996	$79
4078	IRIS Database-Network	1997	$1,485
4078	IRIS Database-Single User	1997	$495
510	ISO 14000: Understanding Environmental Standards	1996	$69
551	ISO 14001: An Executive Repoert	1996	$55
518	Lead Regulation Handbook	1996	$79
478	Principles of EH&S Management	1995	$69
554	Property Rights: Understanding Government Takings	1997	$79
582	Recycling & Waste Mgmt Guide to the Internet	1997	$49
603	Superfund Manual, 6th Edition	1997	$115
566	TSCA Handbook, 3rd Edition	1997	$95
534	Wetland Mitigation: Mitigation Banking and Other Strategies	1997	$75

PC #	SAFETY AND HEALTH TITLES	Pub Date	Price
547	Construction Safety Handbook	1996	$79
553	Cumulative Trauma Disorders	1997	$59
559	Forklift Safety	1997	$65
539	Fundamentals of Occupational Safety & Health	1996	$49
535	Making Sense of OSHA Compliance	1997	$59
563	Managing Change for Safety and Health Professionals	1997	$59
589	Managing Fatigue in Transportation, *ATA Conference*	1997	$75
4086	OSHA Technical Manual, Electronic Edition	1997	$99
598	Project Mgmt for E H & S Professionals	1997	$59
552	Safety & Health in Agriculture, Forestry and Fisheries	1997	$125
613	Safety & Health on the Internet, 2nd Edition	1998	$49
597	Safety Is A People Business	1997	$49
463	Safety Made Easy	1995	$49
590	Your Company Safety and Health Manual	1997	$79

Electronic Product available on CD-ROM or Floppy Disk

PLEASE CALL OUR CUSTOMER SERVICE DEPARTMENT AT (301) 921-2323 FOR A FREE PUBLICATIONS CATALOG.

Government Institutes
4 Research Place, Suite 200 • Rockville, MD 20850-3226
Tel. (301) 921-2323 • FAX (301) 921-0264
E mail: giinfo@govinst.com • Internet: http://www.govinst.com

GOVERNMENT INSTITUTES ORDER FORM

4 Research Place, Suite 200 • Rockville, MD 20850-3226 • Tel (301) 921-2323 • Fax (301) 921-0264
Internet: *http://www.govinst.com* • E-mail: *giinfo@govinst.com*

3 EASY WAYS TO ORDER

1. Phone: **(301) 921-2323**
Have your credit card ready when you call.

2. Fax: **(301) 921-0264**
Fax this completed order form with your company purchase order or credit card information.

3. Mail: **Government Institutes**
4 Research Place, Suite 200
Rockville, MD 20850-3226
USA
Mail this completed order form with a check, company purchase order, or credit card information.

PAYMENT OPTIONS

❑ **Check** (*payable to Government Institutes in US dollars*)

❑ **Purchase Order** (this order form must be attached to your company P.O. <u>Note</u>: All International orders must be pre-paid.)

❑ **Credit Card** ❑ ❑ ❑ AMERICAN EXPRESS

Exp. ___/___

Credit Card No. _____

Signature _____
Government Institutes' Federal I.D.# is 52-0994196

CUSTOMER INFORMATION

Ship To: (Please attach your Purchase Order)

Name: _____
GI Account# (*7 digits on mailing label*): _____
Company/Institution: _____
Address: _____
(please supply street address for UPS shipping)
_____
City: _____ State/Province: _____
Zip/Postal Code: _____ Country: _____
Tel: ( ) _____
Fax: ( ) _____
E mail Address: _____

Bill To: (if different than ship to address)

Name: _____
Title/Position: _____
Company/Institution: _____
Address: _____
(please supply street address for UPS shipping)
_____
City: _____ State/Province: _____
Zip/Postal Code: _____ Country: _____
Tel: ( ) _____
Fax: ( ) _____
E-mail Address: _____

Qty.	Product Code	Title	Price

Subtotal _____
MD Residents add 5% Sales Tax _____
Shipping and Handling (see box below) _____
Total Payment Enclosed _____

❑ **New Edition No Obligation Standing Order Program**
Please enroll me in this program for the products I have ordered. Government Institutes will notify me of new editions by sending me an invoice. I understand that there is no obligation to purchase the product. This invoice is simply my reminder that a new edition has been released.

15 DAY MONEY-BACK GUARANTEE
If you're not completely satisfied with any product, return it undamaged within 15 days for a full and immediate refund on the price of the product.

Within U.S:
1-4 products: $6/product
5 or more: $3/product

Outside U.S:
Add $15 for each item (Airmail)
Add $10 for each item (Surface)

SOURCE CODE: BP01

Government Institutes • 4 Research Place, Suite 200 • Rockville, MD 20850
Internet: *http://www.govinst.com* • E-mail: *giinfo@govinst.com*

REFERENCE BOOK
NOT TO BE TAKEN
FROM THE LIBRARY